CHAPTER 4

Reasoning and Triangle Congruence

CHAPTER 5

Parallel Lines and Quadrilaterals

CHAPTER 6

Similarity

CHAPTER 7

Circles

CHAPTER 8

Coordinate Geometry

CHAPTER 9

Transformation Geometry

COLLEGE GEOMETRY

COLLEGE GEOMETRY

A Problem-Solving Approach with Applications

SECOND EDITION

Gary L. Musser

Oregon State University

Lynn E. Trimpe

Linn-Benton Community College

Vikki R. Maurer

Linn-Benton Community College

Upper Saddle River, New Jersey 07458

Library of Congress Cataloging-in-Publication Data

Musser, Gary L.
College geometry : a problem-solving approach with applications. –2nd ed. / Gary L. Musser, Lynn E. Trimpe, Vikki R. Maurer.
p. cm.
Includes index.
ISBN 0-13-187969-3
1. Geometry. I. Trimpe, Lynn E. II. Maurer, Vikki R. III. Title.

QA455.M87 2008
516–dc22 2006052719

Acquisitions Editor: *Petra Recter*
Vice President and Editorial Director, Mathematics: *Christine Hoag*
Project Manager: *Michael Bell*
Editorial Assistant/Print Supplements Editor: *Joanne Wendelken*
Production Editor: *Raegan Keida Heerema*
Assistant Managing Editor: *Bayani Mendoza de Leon*
Senior Managing Editor: *Linda Mihatov Behrens*
Manufacturing Buyer: *Maura Zaldivar*
Associate Director of Operations: *Alexis Heydt-Long*
Marketing Manager: *Wayne Parkins*
Marketing Assistant: *Jennifer de Leeuwerk*
Art Director: *John Christiana*
Interior Designer: *511 Design*
Cover Designer: *John Christiana*
Art Editor: *Thomas Benfatti*
Creative Director: *Juan R. López*
Manager, Cover Visual Research & Permissions: *Karen Sanatar*
Director, Image Resource Center: *Melinda Patelli*
Manager, Rights and Permissions: *Zina Arabia*
Manager, Visual Research: *Beth Brenzel*
Image Permission Coordinator: *Nancy Seise*
Art Studio: *ICC*
Permission Researcher: *Melinda Alexander*

IMAGE CREDITS
Cover: John Warden/Superstock
Chapter 9 Part Opener: M.C. Escher's "Circle Limit III" © 2006 The M.C. Escher Company-Holland.

Pearson Prentice Hall
Pearson Education, Inc.
Upper Saddle River, NJ 07458

Printed in the United States of America

10 9 8 7 6 5 4 3

ISBN-13: 978-0-13-187969-0
ISBN-10: 0-13-187969-3

Pearson Education, Ltd., *London*
Pearson Education Australia PTY. Limited, *Sydney*
Pearson Education Singapore, Pte., Ltd.
Pearson Education North Asia Ltd., *Hong Kong*
Pearson Education Canada, Ltd., *Toronto*
Pearson Educación de Mexico, S.A. de C.V.
Pearson Education – Japan, *Tokyo*
Pearson Education Malaysia, Pte. Ltd.

To my wonderful wife of 46 years, Irene; my son, Greg;
my granddaughter, Maranda; and my dearly missed mother, Marge, and father, G.L.
—GLM

To my mother and father, Shirley and Howard,
and to my husband, Tim.
—LET

To Janet Narva, my wonderful sister and friend.
—VRM

BRIEF CONTENTS

CONTENTS

PREFACE

PHILOSOPHY

This book has five main goals:

1. To help students develop their problem-solving skills, especially with geometric applications
2. To familiarize students with fundamental properties of two- and three-dimensional geometric shapes and to foster an appreciation for the usefulness of geometry
3. To deepen students' conceptual understandings of geometry by constructing proofs to verify geometric relationships and by using those relationships to solve applied problems
4. To expose students to the axiomatic method of synthetic Euclidean geometry at an appropriate level of sophistication
5. To provide students with a variety of methods, including coordinate geometry and transformation geometry, for solving geometric problems

To accomplish these goals, we have organized the text into three parts:

Part I, Problem Solving, Geometric Shapes, and Measurement (Chapters 1–3)

Part II, Formal Synthetic Euclidean Geometry (Chapters 4–7)

Part III, Alternate Approaches to Plane Geometry (Chapters 8 and 9)

Our rationale for this arrangement of content is based on an understanding of the students who will study the material. First, research on learning geometry suggests that the study of geometry should begin with informal experiences and gradually move toward formal proof. Second, most students reading this book will be using geometry to solve problems encountered in their future vocations as well as in subsequent course work. Because geometric shapes and measurement geometry are particularly relevant to these students, we provide an early development of these ideas in an informal manner. In this way, many ideas that are proved in Part II are introduced informally in Part I and some are justified intuitively in Part I.

Part II provides the core of a standard Euclidean geometry course. We have postulated some results that could have been theorems. This approach allows students to get to the central results more quickly, giving them additional time to use those results other proofs and applied problems.

Part III opens students' eyes to ways of proving results and solving problems in geometry by methods other than traditional deductive approaches. Students grow to appreciate the beauty and efficiency of using these alternate methods.

New in This Second Edition

- Geometry Investigations on the third page of each chapter use hands-on experience to motivate and introduce material presented in the chapter.
- All problem sets have been extensively revised. Most problems appear in matched pairs so that odd-numbered problems (with answers) can be assigned for practice and even-numbered problems can be assigned as homework to be graded.

- Extended Problems have been added to every section. These problems enrich courses and allow a deeper look into geometry.
- Expanded Chapter Reviews now include representative practice problems.
- A new section, Section 6.5, Using Laws of Trigonometry to Solve Geometry Problems, has been added to extend the treatment of trigonometry to oblique triangles.

FEATURES

The following features have been incorporated into the design of this book to enhance student learning.

Pedagogical

- Liberal use of examples throughout
- Two-color format, including the use of a second color in figures to enhance concept development
- Extensive use of figures, many in two colors, to enhance concept development
- Boxes to highlight postulates, theorems, and important definitions
- Boldface type to highlight definitions
- Motivation of many theorems by first considering specific examples
- A distinctive symbol to indicate the end of an example or a proof
- Problem sets that include various combinations of exercises, problems, applied problems, proofs, and extended problems
- Realistic, relevant, and current application problems that motivate students and help them see the utility of geometry
- Extended problems that allow students to dig deeper into a geometric topic, research an area of interest not covered in the text, or explore mathematical connections
- Answers to all odd-numbered exercises, problems, and applied problems
- Answers to odd-numbered proofs, either outlined or else provided in paragraph or statement–reason format
- Problem-solving strategies and clues, given in each chapter together with additional problems that utilize the highlighted strategy
- Writing for Understanding problems, to give students an opportunity to deepen their understanding and to communicate in writing
- Comprehensive Chapter Reviews that engage students in assessing their own understanding of geometric concepts
- A brief table of contents and a second, more complete one listing every section of the text
- Common geometric formulas with figures, a list of symbols, and a table of conversions on the inside back-cover pages
- An Applications by Chapter list on the inside front-cover pages, to help instructors and students seek out desired applications and to see how geometry appears in many facets of our lives
- Topics Sections at the end of the book that provide additional material for customizing courses
- Appendix 1, Getting Started, to which students can refer when starting proofs
- Appendix 2, a quick reference of all the postulates, theorems, and corollaries presented in the text

Motivational

- Chapter openers that set the scene for each chapter by presenting a historical tidbit
- Initial applied problems that motivate the material in each section, with a solution given at the end of the section
- Geometry Around Us features at the end of each section that raise students' awareness of examples of geometry in our world
- People in Geometry vignettes that acquaint students with some well-known geometers

Supplements

Student Activity Manual to Accompany College Geometry—(ISBN 10: 0-13-615798-X, ISBN 13: 978-0-13-615798-4) by Sharon Rodecap (Linn-Benton Community College, Albany, Oregon) and Lyn Riverstone (Oregon State University, Corvallis, Oregon), contains a wealth of hands-on activities correlated with chapters in the text. These activities promote learning of concepts and provide valuable hands-on geometry experience.

Student Solutions Manual to Accompany College Geometry—(ISBN 10: 0-13-187971-5, ISBN 13: 978-0-13-187971-3) by Shirley Buls (St. Cloud State University), includes tips for solving the problems in each section of the text and written solutions to odd-numbered exercises, applications, and proofs in each section. This student resource also contains written solutions to the Chapter Review Problems and Chapter Tests.

The Instructor's Manual—(ISBN 10: 0-13-187970-7, ISBN 13: 978-0-13-187970-6) contains answers to all even-numbered exercises, problems, and applications, as well as outlines of all even-numbered proofs.

Test Item File—(ISBN 10: 0-13-241359-0, ISBN 13: 978-0-13-241359-6) The Test Item File provides over 450 test or quiz questions, keyed to their respective chapters.

ACKNOWLEDGMENTS

We thank the following people for their expert reviews of our manuscripts:

First Edition:
Martin Brown, Jefferson Community College
John Longnecker, University of Northern Iowa
Jill McKenney, Lane Community College
Sue Nolen, Blue Mountain Community College
Bernadette Perham, Ball State University
Arlene Sego, Cuyahoga Community College
Darlene Whitkanack, Northern Illinois University

Second Edition:
Lawrence O'Cannon, Utah State University
Laurie Burton, Western Oregon State University
Shirley Buls, St. Cloud State University
Arlene Blasius, SUNY College at Old Westbury
Catherine Hayes, University of Mobile
Cynthia Piez, University of Idaho
Dorothy Easley, Broward Community College

We thank especially reviewer Shirley Buls for her many helpful suggestions regarding changes to the book—in particular, to Chapters 7 and 8. We thankfully acknowledge the very careful reading of the text by Amanda Blaker (Western Oregon University, Monmouth, Oregon), who served as accuracy checker on this second edition.

We owe special thanks to the following colleagues who used a preliminary version of the text and gave us valuable feedback for the first edition: Ricardo Bell, Candy Drury, Dale Green, Sue Nolen, Wally Reed, Sharon Rodecap, Gayle Smith, Gerry Swenson, and Betty Westfall. Thanks go as well to test bank author Catherine Aune; answer checkers Judy DeSzoeke, Dale Green, Crystal Gilliland, and Roger Maurer, Don Fineran, for permission to use problems from the Oregon Vo-Tech Project; Marv Kirk and Roger Maurer, for their contributions to problem sets; Tommy Bryan, for his superb job of proofreading; and Dave Metz, for preparing the People in Geometry vignettes. We also thank the many students who used our text, encouraged us through their success, and gave us ideas for improvements in the second edition. In addition, we thank users of the first edition for their eagle eyes, employed in catching errors and suggesting improvements over the years. Among these users are Mary Campbell, Hollis Duncan, Debbie Love, Elizabeth Lundy, Erica Rode, and Sharon Rodecap.

Finally, we thank our production team—especially our production editor, Raegan Heerema—for bringing this book to print; our editor, Petra Recter; and Joanne Wendelken, Petra's editorial assistant, for her expert coordination.

—G.L.M.

—L.E.T.

—V.R.M.

CHAPTER 1

PROBLEM SOLVING IN GEOMETRY

George Polya (1887–1985) was a mathematician famous for his lifelong interest in and study of the process of problem solving. Born and educated in Hungary, Polya came to the United States in 1940. He is the author of numerous books and papers on problem solving, the most famous of which is *How to Solve It.* This book has been translated into 17 different languages, and more than 1,000,000 copies have been sold since it was first published in 1945.

Polya defined intelligence as the ability to solve problems and believed that "solving problems is human nature itself." Thus, he felt strongly that a major goal of education should be the development of problem-solving skills. To that end, Polya devised what has become known as the four-step problem-solving process, which will be a focus of this chapter.

Problem-Solving Strategies

1. Draw a Picture
2. Guess and Test
3. Use a Variable
4. Look for a Pattern
5. Make a Table
6. Solve a Simpler Problem

One of the main goals of this book is to help you to become a better problem solver. This chapter introduces six strategies that will be useful in solving geometry problems. In addition, we introduce a new strategy at the beginning of each of the remaining chapters. In this way, your ability to solve problems should grow much as the Problem-Solving Strategies boxes like the one to the left grow throughout the book.

Initial Problem

Chinese checkers is a marble game played by two to six players on a board like the one that follows. This board has a starting pen (triangular region in one point of the star) with four rows of holes that hold 10 marbles for each player up to 6 players-one of the 6 starting pens is shaded. Suppose we desire a more challenging game and we construct a similar board to have six rows (21 marbles) for each starting pen. How many holes, in all, would this expanded board have?

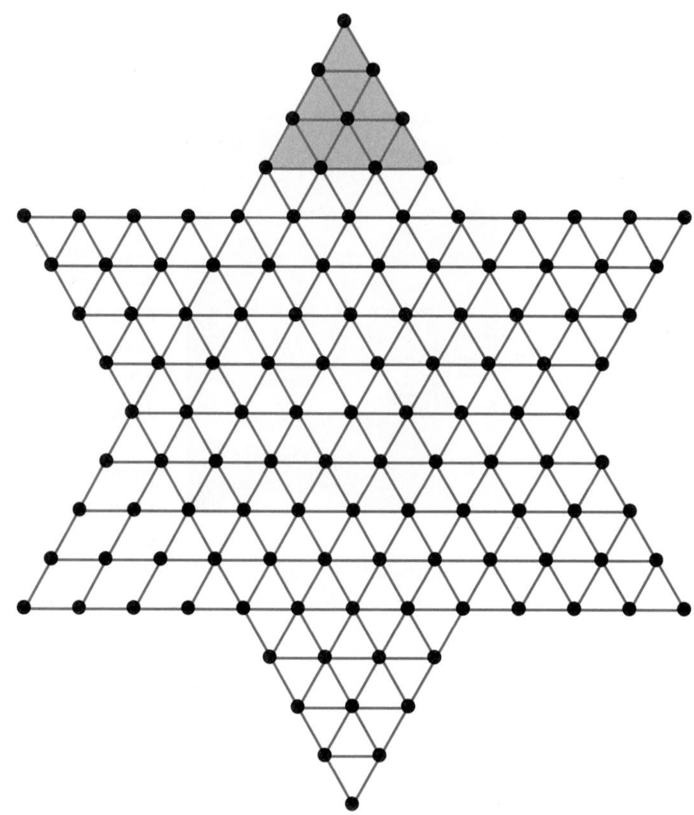

A solution of this Initial Problem is on page 30.

CHINESE TANGRAMS

Tangrams are one of the oldest Chinese puzzles. Their origin is undocumented but dates back to at least 1813. Even Napoleon Bonaparte is said to have carried an ivory set of tangrams. Tangram puzzles are currently available in books and as card games as well as on interactive websites. The following seven pieces make up a tangram puzzle: two large triangles, two small triangles, one medium triangle, one square, and one parallelogram. Trace the pieces and cut them out of stiff paper to create your own set of tangrams.

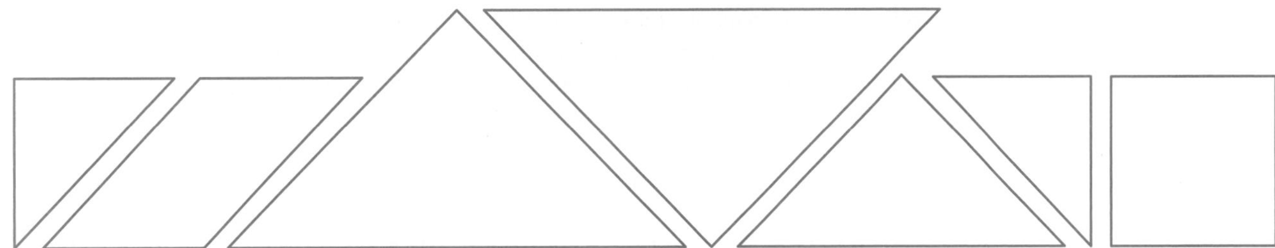

To solve a tangram puzzle, arrange all seven pieces with no gaps or overlap to form a specific shape. For example, to form the following "house" shape, arrange the tangrams as shown.

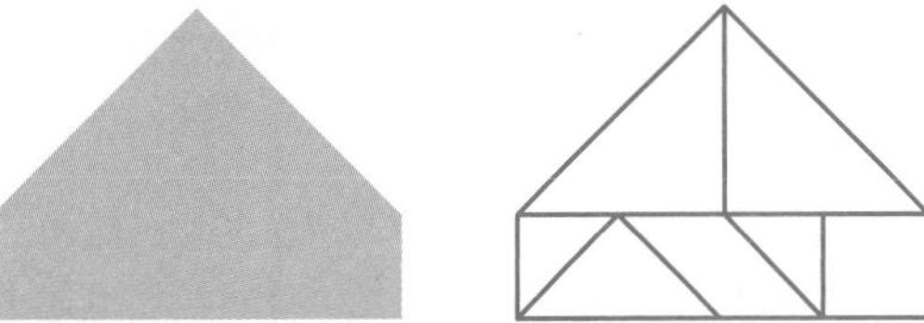

There are at least 1600 different shapes that one can make with tangrams, 17 of which are shown on this page. Arrange all seven tangram pieces to form each of the following shapes.

1. Cat

2. Runner

3. Boat

4. Parallelogram

5. Swan

6. Woman

7. Each letter in the title of this activity is a tangram puzzle. Use all seven tangram pieces to form each letter.

Introduction

Most people consider problem solving to be the main objective of any mathematics course. In a more general sense, problem solving is central to many disciplines, including business and engineering. This chapter provides an introduction to the study of problem solving with an emphasis on problems in geometry.

1.1 PROBLEM-SOLVING STRATEGIES

In our discussion of problem solving, we need to make clear what we mean by a problem. In this book we make a distinction between an "exercise" and a "problem." You can solve an **exercise** by applying a routine procedure. It may be similar to other exercises (or problems) that you have worked or have seen worked. A typical "word problem" in an algebra course is an exercise if you recognize the type of problem and can recall an appropriate procedure to apply.

On the other hand, a **problem** is nonroutine and unfamiliar. To solve a problem, you need to stop and think about how to attack it. You may need to try something completely new and different. You may get stuck several times and have to make new starts. The need for some kind of creative step on your part in solving the problem is what makes it different from an exercise.

George Polya was a mathematician whose name has come to be synonymous with problem solving. In an effort to encourage more students to "experience the tension and enjoy the triumph of discovery" that accompany solving a problem, he presented the following **four-step process** for problem solving.

STEP 1 Understand the Problem.

- Is it clear to you what is to be found?
- Do you understand the terminology used in the problem?
- Is there enough information?
- Is there irrelevant information?
- Are there any restrictions or special conditions to be considered?

STEP 2 Devise a Plan.

- How should the problem be approached?
- Does the problem appear similar to any others you have solved?
- What strategy might you use to solve the problem?

STEP 3 Carry Out the Plan.

- Apply the strategy or course of action chosen in Step 2 until you find a solution or you decide to try another strategy.

STEP 4 Look Back.

- Is your solution correct?
- Do you see another way to solve the problem?
- Can your results be extended to a more general case?

In the four-step process, Step 2 is a critical one. Even if you thoroughly understand a problem (Step 1), you may not be able to progress further. On the other hand, once you have selected a workable strategy, it is usually not difficult to implement it (Step 3). Likewise, once you have obtained a solution, it is usually not difficult to verify whether that solution is correct (Step 4). Therefore, the purpose of this chapter is to focus on Step 2 in Polya's process and to present six general strategies that are frequently useful in solving geometric problems. We will present one new strategy at the beginning of each subsequent chapter of the book.

Strategy 1: *Draw a Picture*

This strategy is a natural choice for solving problems in geometry because many of the problems of geometry are related to figures, shapes, and physical structures. Often drawing one or several pictures can help you to solve the problem or can help you to understand the problem better so that you can formulate a plan for solving it. The following clues may help you to identify situations where the Draw a Picture strategy might be useful.

Clues

The Draw a Picture strategy may be appropriate when

- The problem involves a physical situation.
- The problem involves geometric figures or measurements.
- You want to gain a better understanding of the problem.
- A visual representation of the problem is possible.

As you attempt to solve the following example problems, imagine solving the problems *without* looking at any pictures. Then try solving the problems with a picture to see if the picture helped in the process.

EXAMPLE 1.1 A large cube is formed by arranging 64 smaller cubes of the same size in a stack that measures 4 cubes by 4 cubes by 4 cubes. A cardboard box with no top is constructed so that the large cube fits tightly in the box. How many of the small cubes are *not* touching a side or the bottom of the box?

STEP 1 Understand the Problem.

There are no gaps between the small cubes and the sides of the box. Remember that the box has no top. We must determine how many of the 64 small cubes are in contact with neither a side nor the bottom of the box.

STEP 2 Devise a Plan.

Draw a picture of the cubes inside the box in order to better visualize their arrangement [Figure 1.1(a)]. The 4 by 4 by 4 cube has 64 small cubes in it, arranged in four layers.

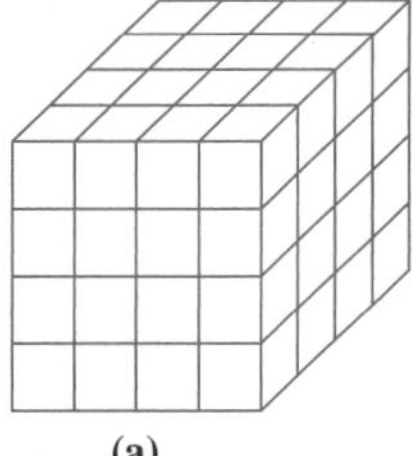
(a)

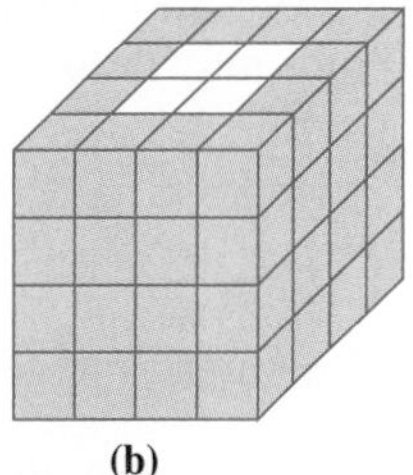
(b)

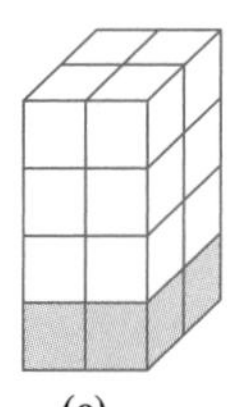
(c)

FIGURE 1.1

STEP 3 Carry Out the Plan.

Now we can start counting the small cubes that touch the sides and bottom of the box. Shading those cubes that are in contact with a side or the bottom of the box may be helpful [Figure 1.1(b)]. Remove the central "core" of cubes [Figure 1.1(c)]. Some of these cubes touch only the bottom of the box. We can see that the top three layers of 4 cubes each, or 12 cubes in all, touch neither the sides nor the bottom of the box.

STEP 4 Look Back.

We also could have solved this problem by subtracting the number of cubes that *do* touch the sides and bottom of the box from 64. What would happen if the large cube measured 5 by 5 by 5? How many small cubes would touch neither the sides nor the bottom of that cube? Can this problem be generalized to an n by n by n cube? If so, how? ■

Additional Problems Where the Strategy "Draw a Picture" Is Useful

1. If the diagonals of a square are drawn in, how many triangles of all sizes are formed?

   To get you started

   a. Do you know what the diagonal of a square is? It is a line segment whose endpoints are two opposite corners of the square. One diagonal is shown in Figure 1.2. (We will give a more general definition of a diagonal later in the book.)

   b. Remember that the triangles may be different sizes. Also, some of the triangles may overlap.

   c. Although one picture may be helpful in formulating a solution to the problem, several pictures might make the solution even clearer.

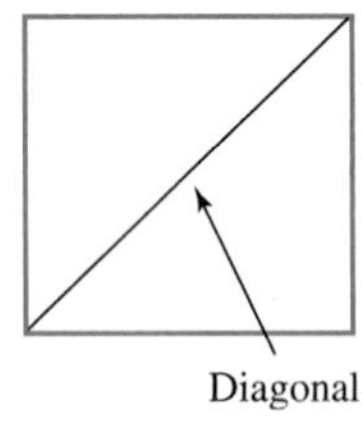

FIGURE 1.2

2. A rectangular milk crate has spaces for 24 bottles in four rows and six columns. Can you put 18 bottles of milk into the crate so that each row and each column of the crate have an even number of bottles in them? (HINT: You might consider the spaces that do *not* have bottles in them, rather than the 18 that do.)

Strategy 2: *Guess and Test*

Also known as "Trial and Error," the Guess and Test strategy is an extremely useful method for solving many problems. Guess and Test is often the very first strategy employed by experienced mathematicians and novice problem solvers alike. Even if it does not lead immediately to a solution, the Guess and Test strategy helps you to get a "feeling" for a problem and may suggest other strategies that could be used to solve the problem.

The Guess and Test method does not necessarily imply random guessing. In fact, often the conditions of the problem will limit the possible guesses. After the first few guesses, you may make some observations that will further limit your guesses. Rather than random Guess and Test, problem solvers more often use a form of systematic Guess and Test. With systematic Guess and Test there is some order to the guesses and a means of recording for which guesses have been tested. The following clues may help you to identify situations where the Guess and Test strategy may be useful.

Clues

The Guess and Test strategy may be appropriate when

- There is a limited number of possible solutions to test.
- You want to gain a better understanding of the problem.

- You have a good idea what the solution is.
- You can systematically try possible solutions.
- Your choices have been narrowed down by first using other strategies.
- There is no obvious strategy to try.

EXAMPLE 1.2 Five friends were sitting on one side of a table. Gary sat next to Bill. Mike sat next to Tom. Howard sat in the third seat from Bill. Gary sat in the third seat from Mike. Who sat on the side of Tom opposite from Mike?

STEP 1 Understand the Problem.

The five seats are arranged in a straight line. We are to assign the five people to the five seats and determine who, in addition to Mike, is next to Tom. The order of the persons may vary and still meet the conditions of the problem. For example, Gary-Bill or Bill-Gary would satisfy the condition "Gary sat next to Bill."

STEP 2 Devise a Plan.

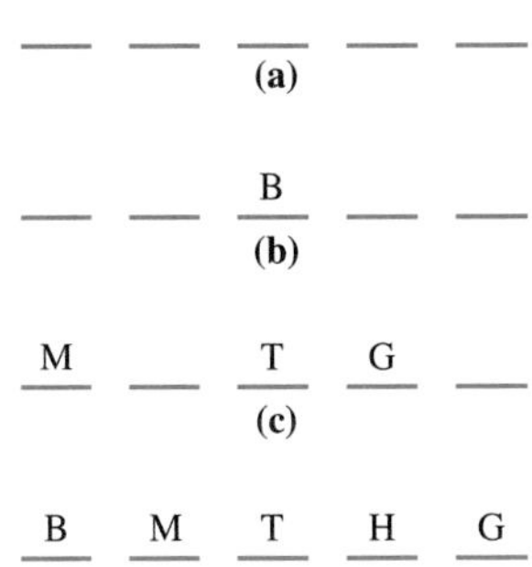

FIGURE 1.3

Draw a picture such as the one in Figure 1.3(a) to represent the seats and use the initial letter of each man's name to represent each of the five men. Test various arrangements of the men using the conditions in the problem to narrow down the guesses.

STEP 3 Carry Out the Plan.

We might start by testing some possibilities for the middle seat. Remember that the problem stated that Howard sat in the third seat away from Bill. Let's try placing Bill in the middle seat [Figure 1.3(b)]. Notice that this makes it impossible for Howard to be placed three seats away from Bill. The same problem arises if Howard is seated in the middle. Likewise, because Gary must be placed in the third seat from Mike, neither Gary nor Mike can be in the middle seat. So Tom must be the man in the middle seat.

Now we will try some arrangements with Tom in the middle seat. In Figure 1.3(c), Gary is in the third seat from Mike as required, but Mike is not next to Tom. In Figure 1.3(d), Mike is now next to Tom and Howard and Bill are placed so that Howard is in the third seat from Bill. But Gary is not seated next to Bill. We can fix that by switching Howard and Bill [Figure 1.3(e)]. Now this arrangement meets all of the stated conditions. To answer the question of this example, Bill must be seated on the other side of Tom.

STEP 4 Look Back.

Is there any other arrangement of the men that works? One other arrangement that satisfies the conditions of the problem is G B T M H. This arrangement still places Bill on the other side of Tom. Are there other solutions? Why or why not? ■

Additional Problems Where the Strategy "Guess and Test" Is Useful

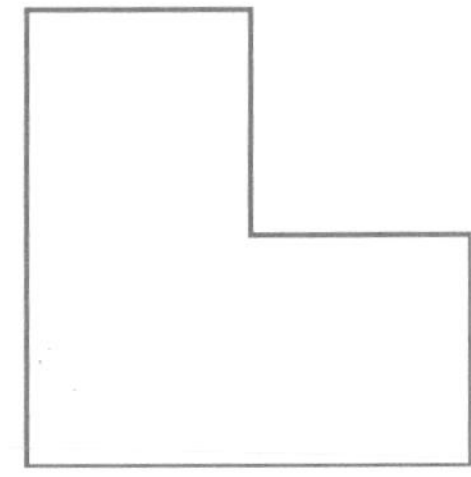
FIGURE 1.4

1. Is it possible to divide the shape in Figure 1.4 into four parts so that the four parts are the same size as each other and have the same shape as the original figure?

   To get you started

   a. The four parts must be exact duplicates of each other, and they must all fit together in the larger shape with no gaps or overlaps. You might verify that the smaller pieces are exact duplicates of each other by cutting them out and fitting them on top of each other.

   b. Since the four parts will be the same shape as the original, you might try starting with the pieces. Draw four copies of the original shape and see if you can put them together to form a larger version of the same shape.

2. Six identical coins are arranged as shown in Figure 1.5. Change the arrangement of coins on the left into the arrangement of coins on the right by moving only two of the coins.

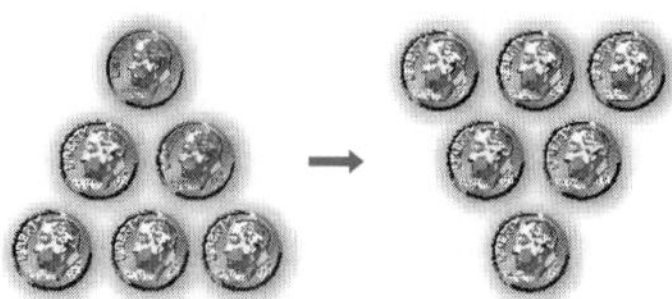

FIGURE 1.5

Strategy 3: *Use a Variable*

The Use a Variable strategy is appropriate when the problem states relationships between unknown quantities, when we can write an equation to model a problem situation, or when we desire a general formula. In applying this strategy, you will be able to use your skills from algebra. The following clues may help you to identify situations where the Use a Variable strategy may be useful.

Clues

The Use a Variable strategy may be appropriate when

- A phrase similar to "for any number" is present or implied.
- A problem suggests an equation.
- You desire a proof or a general solution.
- There are a large number of cases.
- A proof is required in a problem involving numbers.
- An unknown quantity is related to known quantities.
- There are infinitely many numbers involved.
- You are trying to develop a general formula.

EXAMPLE 1.3 The measure of the largest angle of a triangle is three times the measure of the smallest angle. The measure of the third angle of the triangle is 40° more than the measure of the smallest angle. What are the measures of the angles in the triangle? (Use the fact that the sum of the measures of the angles in a triangle is 180°. We will develop this fact later in this book.)

STEP 1 Understand the Problem.

We must use the fact that the measures of the angles in a triangle add up to 180° to help us find the measures of all three angles in the triangle. We also know how the angle measures compare to one another.

STEP 2 Devise a Plan.

We can describe the relationships between the measures of the angles by introducing a variable. If we use x to represent the measure of the smallest angle, we have

x = the measure of the smallest angle

$3x$ = the measure of the largest angle

$x + 40°$ = the measure of the third angle (Figure 1.6)

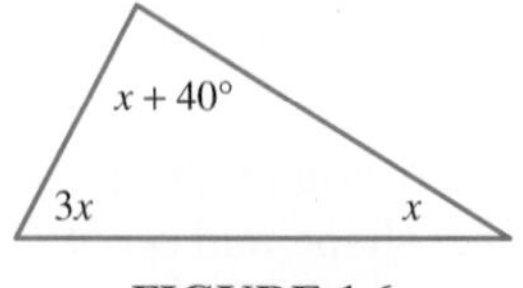

FIGURE 1.6

We must use the relationships we have described to determine the number of degrees in each angle of the triangle. We can use what we know about the angles of a triangle to relate the expressions x, $3x$, and $x + 40°$. Because the sum of the measures of the angles in a triangle is 180°, we can write

$$x + 3x + (x + 40°) = 180°$$

STEP 3 Carry Out the Plan.

Now we can solve the equation to find the measures of the angles.

$$\begin{aligned} 3x + x + (x + 40°) &= 180° \\ 5x + 40° &= 180° \\ 5x &= 140° \\ x &= 28° \end{aligned}$$

This means that $3x = 84°$ and $x + 40° = 68°$. Therefore, the largest angle measures 84°, the smallest measures 28°, and the other angle measures 68°.

STEP 4 Look Back.

Verify that the sum of the measures of the angles is 180°.

Check: $28° + 84° + 68° = 180°$ ✓

Could we have solved the problem another way by using x to represent the measure of one of the other two angles? ■

Additional Problems Where the Strategy "Use a Variable" Is Useful

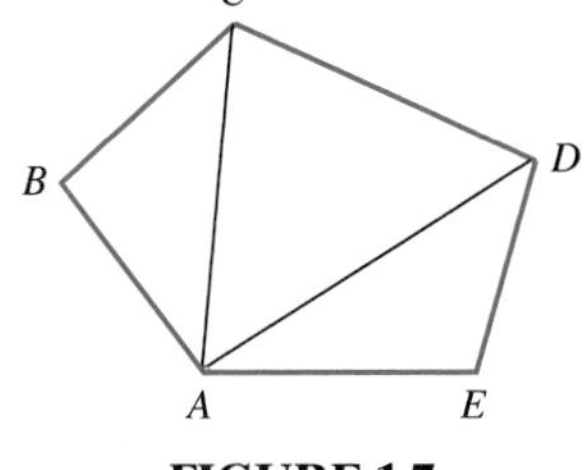

FIGURE 1.7

1. Find a formula for the number of diagonals that can be drawn from one vertex of a polygon with a given number of sides. For example, the pentagon in Figure 1.7 is a polygon with five sides. In that example, two different diagonals can be drawn from vertex A, as shown.

   To get you started

   a. You need to know that a **polygon**, in simple terms, is a geometric figure formed by enclosing a region with straight line segments. (We will give a more formal definition of polygon later in the book.) The line segments are the **sides** of the polygon, and the endpoints of the line segments are the **vertices** (singular is **vertex**) of the polygon.

   b. Draw several other examples of polygons and count the number of diagonals that can be drawn from one vertex. Do you see a relationship between the number of sides in the polygon and the number of diagonals from a vertex? If you let n be the number of sides of the polygon, what expression using n represents the number of diagonals from one vertex?

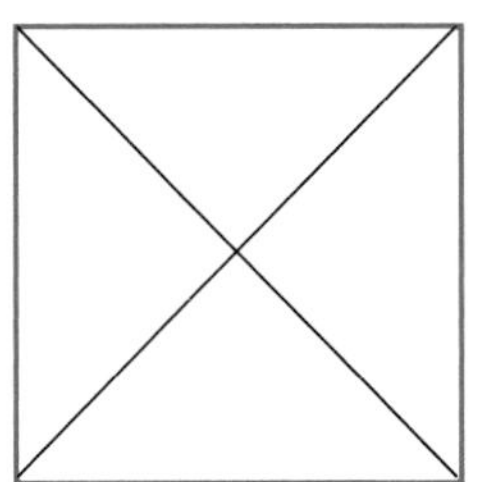

FIGURE 1.8

2. A new football field under construction will have a total length (including end zones) that is 40 feet more than twice its width. We must calculate the area of the football field in order to determine the amount of grass seed necessary to seed the field. Find the area of the field if its perimeter will be 1040 feet. (NOTE: Recall that the area of a rectangle is $A = lw$ and the perimeter is $P = 2l + 2w$.)

O	O	O	O	O	O
X	X	O	O	O	O
X	O	X	O	O	O
O	X	X	O	O	O

FIGURE 1.9

Solutions of Additional Problems for Strategies 1–3

Draw a picture

1. There are four large triangles and four smaller ones (Figure 1.8). Thus, there are eight triangles in all.
2. Many solutions are possible; one of which is shown at left. X's mark the six positions in the crate that are *empty* (Figure 1.9).

Guess and test

1.

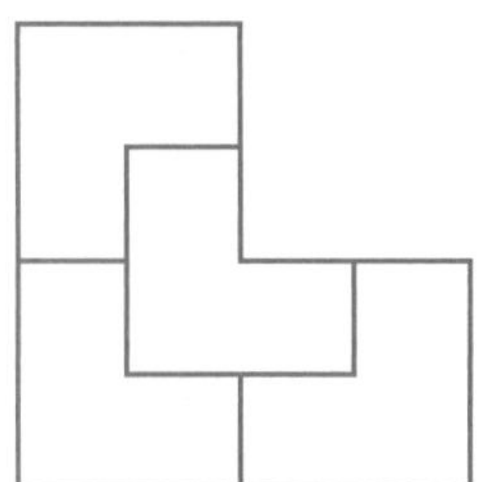

2. One solution follows.

Use a variable

1. The number of diagonals from one vertex is $n - 3$.

2. $w = 160$ ft, $l = 360$ ft, $A = 57{,}600$ ft^2.

GEOMETRY AROUND US

Of the polygons that appear in nature, such as squares and triangles, one of the most frequently seen is the regular hexagon, a polygon with six sides all the same length and six angles all the same measure. The most familiar examples are probably the snowflake and cells of a honeycomb. It is interesting to note that although it is said of snowflakes that "no two are alike," every such crystal can be outlined with a regular hexagon. Also, although the cells of a honeycomb are hexagonal, they are cylindrical tubes when the bees first make them. The pressure of the tubes against one another results in their hexagonal shape.

PROBLEM SET 1.1

EXERCISES/PROBLEMS

1. How many equilateral triangles of all sizes are there in the 3 by 3 by 3 equilateral triangle shown next? (An **equilateral triangle** is a triangle with three sides of the same length. We will discuss equilateral triangles later in the book.)

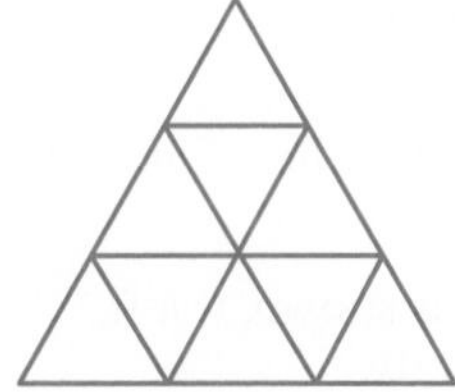

2. How many squares of all sizes are there in the 3 by 3 square shown next?

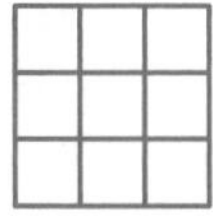

3. A **tetromino** is a shape made up of four squares where two squares are joined along an entire side.

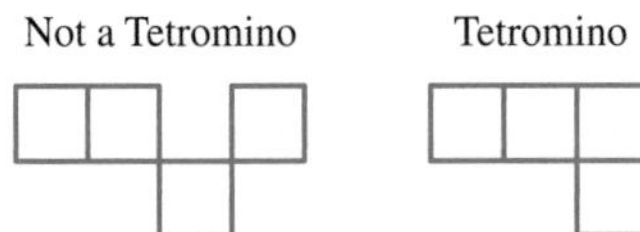

Two tetromino shapes are the same if they can be matched by turning or flipping. For example, the following two tetrominos are the same as the preceding tetromino.

How many different tetromino shapes are possible?

4. A **pentomino** is made up of five squares where two squares are joined along an entire side. Two pentominos are the same if they can be matched by turning or flipping (see problem 3). How many different pentomino shapes are possible?

5. Find five different pentomino shapes that fit together to form a 5 by 5 square. Sketch your solution.

6. Find three different pentomino shapes that fit together to form a 3 by 5 rectangle. Sketch your solution.

7. A **hexomino** is made up of six squares where two squares are joined along an entire side. Each of the following hexomino shapes can be folded to form a cube. After folding, what number will be directly opposite the 1?

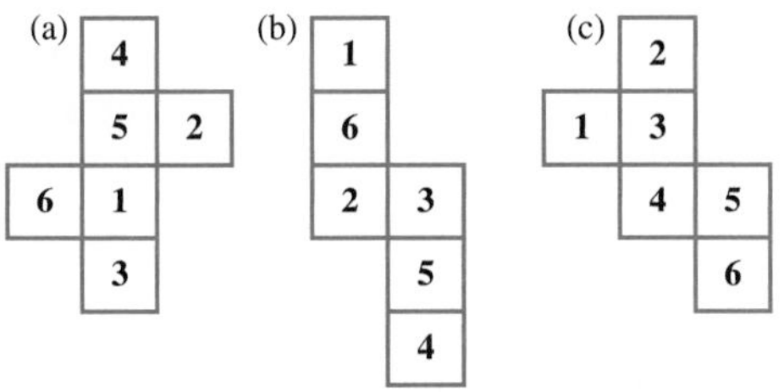

8. Each of the following hexomino shapes can be folded to form a cube. After folding, what number will be directly opposite the 1?

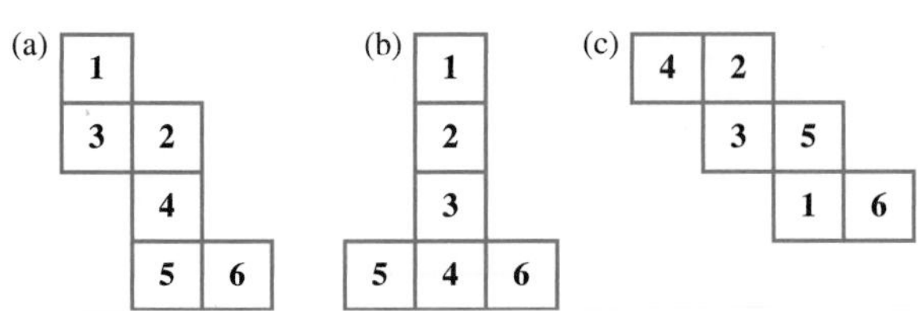

9. Color the following 4 by 4 grid of squares so that there are four red squares and there are three squares each of the colors blue, green, white, and black. No color may appear more than once in a row, in a column, or in a diagonal.

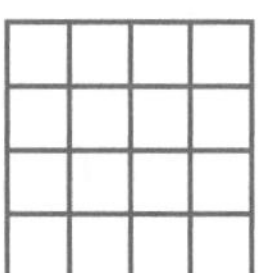

10. **a.** Determine if exactly six squares of the 3 by 3 grid shown can be shaded so that no three shaded squares line up horizontally, vertically, or diagonally.

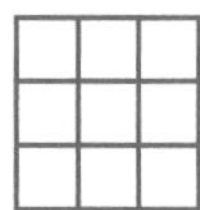

b. Determine if exactly 12 squares of a 4 by 4 grid can be shaded so that no four shaded squares line up horizontally, vertically, or diagonally.

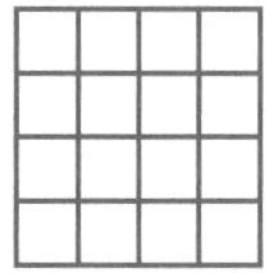

11. Identical white cubes will be selected from a bag one at a time, and at least one side will then be painted black. How many unique cubes can be created?

12. How can 18 cubes be stacked so that the minimum number of sides is exposed? How can 18 cubes be stacked so that the maximum number of sides is exposed?

13. Determine if a circular pizza can be cut into 11 pieces with four straight cuts. (HINT: The pieces need not be the same size or shape.)

14. Determine how to cut a triangular region into 22 regions of varying sizes using six straight cuts.

15. A man has a 10-meter by 10-meter rectangular garden that he wants to fence. How many fence posts will he need if each corner has a post and posts are spaced 1 meter apart along the sides?

16. An organization has access to 10 rectangular tables that will each seat four people on a side and one person on each end. For an awards banquet the tables are to be lined up end to end to form one long table. How many people can be seated at this long table? How many people could be seated at a long table made from 25 of the rectangular tables?

17. Toothpicks are arranged to form the following figure.

a. Remove exactly six toothpicks and leave two squares. (There should be no "leftover" toothpicks.)

b. Remove exactly five toothpicks and leave three squares. (Again, there should be no "leftover" toothpicks.)

18. Use the toothpick arrangement in problem 17 to do the following.

a. Remove exactly seven toothpicks to leave two squares.

b. Remove two toothpicks to leave six squares.

19. Can you trace the following figure without lifting your pencil and without retracing any lines? If so, describe how. If not, why not?

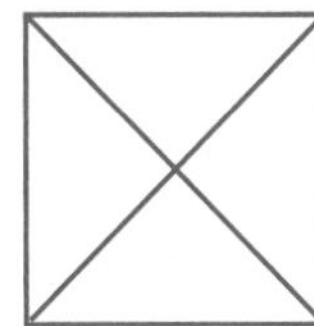

20. Can you trace the following figure without lifting your pencil and without retracing any lines? If so, explain how. If not, why not?

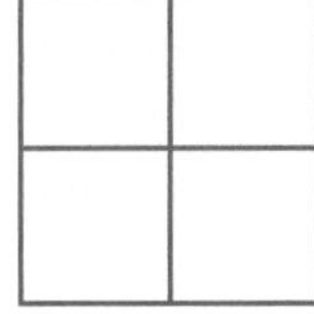

21. Trace the following figure without lifting your pencil and without retracing any lines.

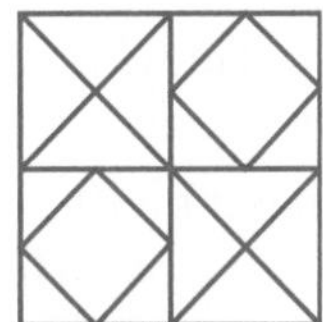

22. Trace the following figure without lifting your pencil and without retracing any lines.

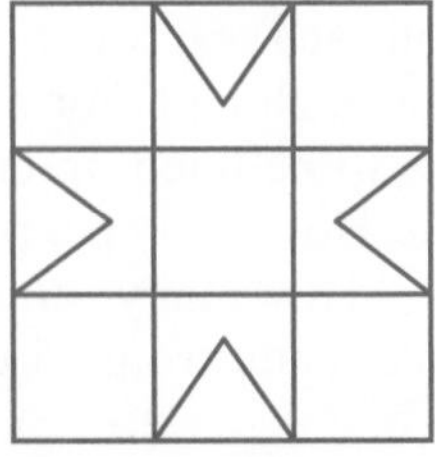

23. Laura bends a piece of wire 72 inches long into the shape of a rectangle that is three times as long as it is wide. What are the dimensions of the rectangle?

24. Greg has 1002 meters of fencing. He will use all the fencing to enclose a rectangular region that is four times as long as it is wide. One of the longer sides is bordered by a river, so that side will not be fenced. What will be the dimensions of the region?

25. Joseph has a certain length of chain link fence with which to make an enclosure for his dog. He plans to use all of the fence material to form a square or a triangle with three sides the same length. He noted that the side of the triangle would be 11 feet longer than the side of the square. How many feet of chain link fence does he have?

26. Tables at a wedding reception have six sides or eight sides and will be trimmed around the edges with lace. Each side on a given table is the same length, but a side on a six-sided table is 8 inches longer than a side on an eight-sided table. One package of lace trims the eight-sided table with no excess. When the same package of lace trims the six-sided table, there are 2 inches of lace left over. What is the length of lace in one package?

27. A child has 10 blocks with heights of 1 cm, 2 cm, 3 cm, 4 cm, 5 cm, 6 cm, 7 cm, 8 cm, 9 cm, and 10 cm. The child wants to use only these 10 blocks to build two separate towers that are exactly the same height. Can this be done? If so, how? If not, why not?

28. A mail order company must ship fragile boxed items to a customer and would like to pack them in stacks that are the same height to avoid movement during shipping. The shipment contains 10 boxes with heights 1, 1, 2, 2, 3, 4, 5, 6, 7, and 9 inches. If the tallest available shipping container is 18 inches, how many stacks must be used and how can the boxes be stacked?

29. Make four copies of the following figure and rearrange them to form a larger version of the original figure. Sketch your solution.

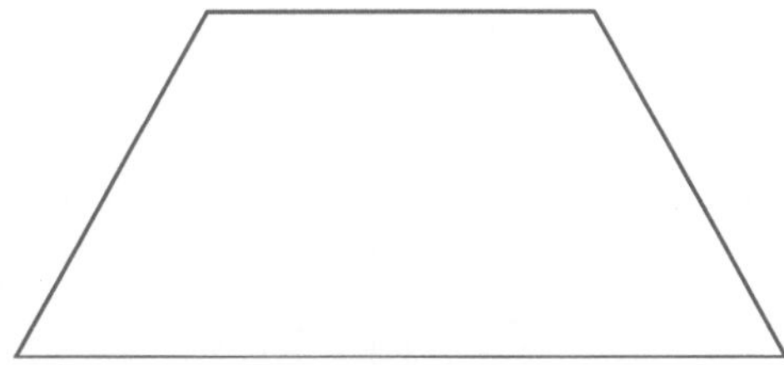

30. Make three copies of the figure from the previous problem and rearrange them to form a triangle. Sketch your solution.

31. Cut the figure into two pieces using one straight cut and rearrange the pieces to form a square. Sketch your solution.

32. Cut the figure into two pieces using one straight cut and rearrange the pieces to form a triangle. Sketch your solution.

33. Henry Dudeney created the following figure in 1902. Trace the figure and cut it into four pieces along the indicated lines. Rearrange the pieces to form a square. You must use all four pieces and have no gaps or overlaps, and you cannot turn the pieces over. Sketch your solution.

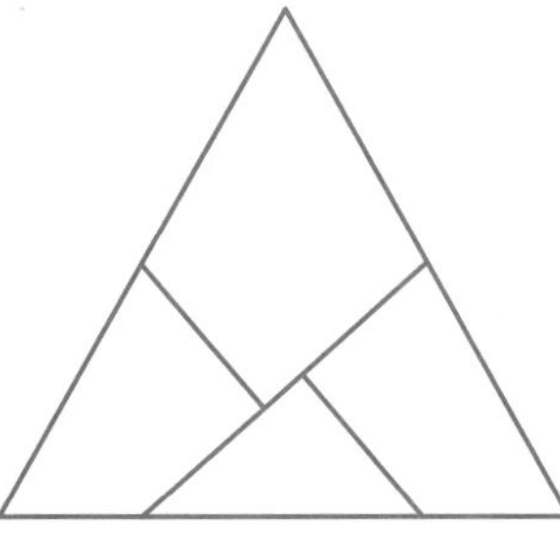

34. Trace the following figure and cut it into five pieces along the indicated lines. Rearrange the pieces to form a square. You must use all five pieces and have no gaps or overlaps, and you cannot turn the pieces over.

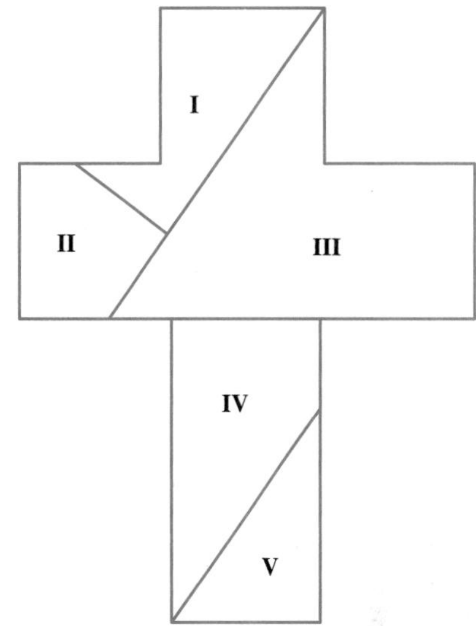

35. Use all seven tangrams from the Chinese tangram activity at the beginning of the chapter to construct a square, a triangle, and a rectangle. Sketch your solution.

36. Use all seven tangrams from the Chinese tangram activity at the beginning of the chapter to construct three different six-sided shapes. Sketch your solution.

37. a. Arrange nine toothpicks to form exactly five equilateral triangles not necessarily of the same size. Toothpicks may not be bent or broken.

b. Arrange six toothpicks to form exactly four equilateral triangles of the same size. Toothpicks may not be bent or broken.

38. Six coins are arranged in a triangular shape as shown. Rearrange the coins to form a circular shape by moving only two coins. To "move" a coin, you must slide it along the surface, without disturbing any other coins, and place it in a new position so that it is touching another coin.

APPLICATIONS

39. A delivery driver must deliver packages to homes on each street in the following map. Lines represent streets, and letters identify intersections. List the intersections in the order the driver will pass through them so that she travels each street exactly once. Can she begin her route and end at the same intersection? Explain.

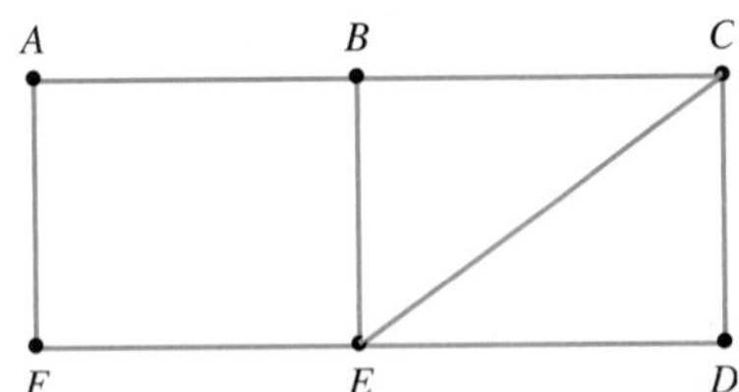

40. A standard tennis court is shown next. How many rectangles of all sizes are there on a tennis court? (NOTE: Treat the line under the net as one of the lines of the court.)

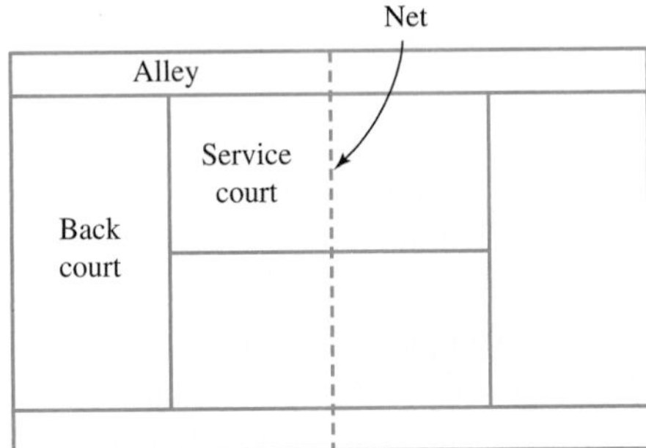

41. Puzzler Henry Dudeney posed a famous problem, presenting the following situation. Suppose that houses are located at points *A*, *B*, and *C*. We want to connect each house to water, electricity, and gas located at points *W*, *G*, and *E* without any of the pipes or wires crossing each other. (NOTE: This problem is related to a branch of mathematics called graph theory.)

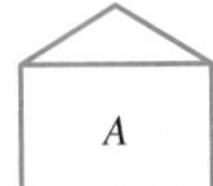

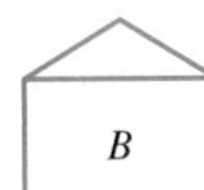

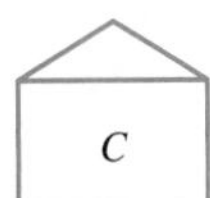

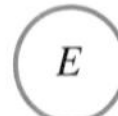

a. Try making all of the connections. Can it be done? If so, how?

b. Suppose that the owner of one of the houses, say *B*, is willing to let the pipe for one of his neighbors' connections pass through his house. Then can all of the connections be made? If so, how?

42. Workers are using mechanical joint pipe to lay a water main for a street. The pipe comes in 18-foot and 20-foot sections, and a designer determines that the water main would require 14 fewer sections of 20-foot pipe than if 18-foot sections were used. Find the total length of the water main.

43. A man living in the Northwest wants to mail a Christmas tree to a friend living in the South. He plans to bundle the tree tightly and pack it in a long box with a square end. The mailing service told him that the length plus the girth of his package cannot exceed 130 inches. If the Christmas tree is 5′ 10″ tall, what is the maximum width and height for his package? (NOTE: The **girth** is the distance around the package as shown.)

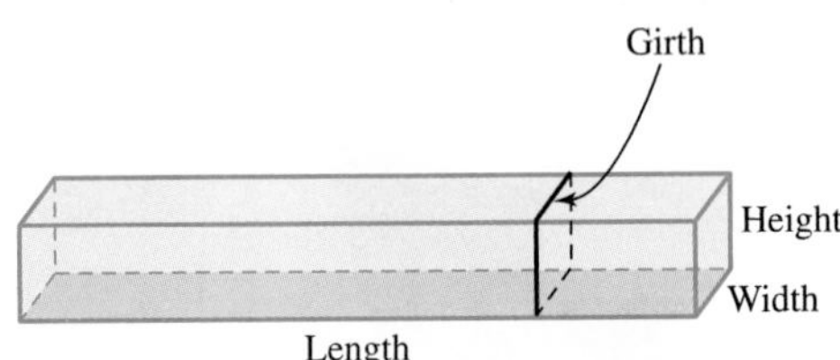

44. Many products are shrink wrapped and shipped in corrugated trays. The following figure shows one way a popular product is shipped. There are 24 bottles arranged in four rows and six columns.

a. If 10 bottles are removed, how can you arrange the remaining 14 bottles so that each row and each column contain an odd number of bottles?

b. If 14 bottles are removed, how can you arrange the remaining 10 bottles along the edges of the tray so that there is the same number of bottles along each edge?

45. A problem that challenged mathematicians for many years concerns the coloring of maps. That is, what is the minimum number of colors necessary to color any map? In 1976, mathematicians finally proved, with the aid of a computer, that no map requires more than four colors. Determine the smallest number of colors necessary to color each map shown. (NOTE: Two "countries" that share only one point can be colored the same color, but if they have more than one point in common, they must be colored differently.)

a.

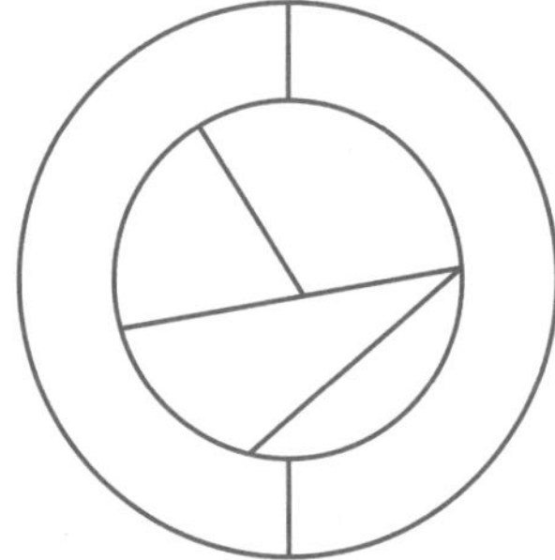

b.

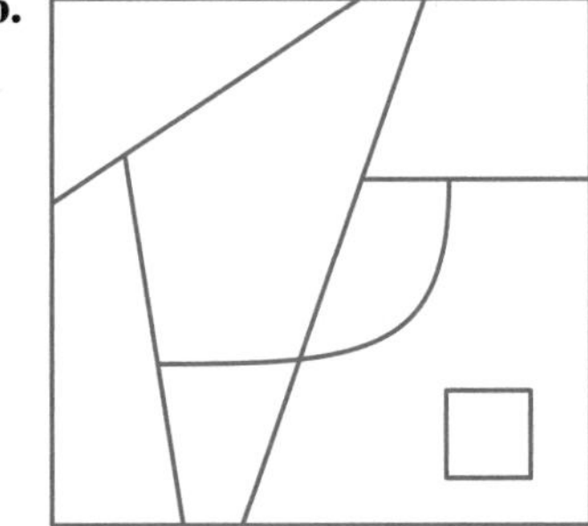

c.

EXTENDED PROBLEMS

46. Sam Loyd created a famous dissection puzzle by cutting a square into five pieces as shown in the following figure.

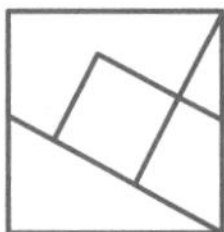

Research Loyd's puzzle and explain his rationale for the cuts he made. Cut your own set of pieces out of cardboard. Show how to arrange them to form a triangle and a rectangle. Then create a new dissection puzzle of your own by cutting a square into six pieces. Form two geometric shapes and challenge a friend to solve your puzzles.

47. Research the four-color map problem discussed in problem 45. To what kinds of maps does the solution apply? Write an essay that summarizes your findings. Locate an outline map of the countries in Europe and show how to color the map using at most four colors.

48. On April 14, 1970, Apollo 13 suffered an explosion in space. The planned moon landing was lost and the crew had to work quickly to solve a series of critical problems with life-threatening consequences and a surprising geometric twist. Research the Apollo 13 mission and focus on the lithium hydroxide filter problem. Write a summary of the problem, the solution, and the items used. Then create your own solution to the problem using the same items. To simulate the filter and receptacle, use a toothpaste box and bath tissue roll. Summarize the problem-solving strategies you used.

1.2 MORE PROBLEM-SOLVING STRATEGIES

In this section we will continue to examine strategies useful in solving problems in geometry. You will often use the three strategies presented in this section together in solving a given problem. In fact, at times it may be difficult to solve a problem using only one of these strategies and not employing any others. The same is true for many of the strategies we will discuss in this book. It is probably more common to use a combination of strategies in solving a problem than to use one particular strategy. For example, frequently the first step in solving a problem is to "play with" the problem using the Guess and Test strategy or the Draw a Picture strategy. After we spend some time on a problem in this way, we might decide to use a variable to write an equation or begin to look for a pattern.

Strategy 4: *Look For a Pattern*

Much of mathematics consists of observing and describing patterns. We can observe patterns in sequences of numbers, relationships between numbers, and in geometric figures. We often use the Look for a Pattern method in conjunction with other problem-solving strategies, such as Make a Table or Solve a Simpler Problem, which we will discuss soon. The following clues may help you to identify situations where the Look for a Pattern strategy may be useful.

Clues

The Look for a Pattern strategy may be appropriate when

- The problem gives a list of data.
- The problem involves a sequence of numbers or figures.
- Listing special cases helps you to deal with a complex problem.
- The problem asks you to make a prediction or a generalization.
- You can express or view information in an organized manner, such as in a table.

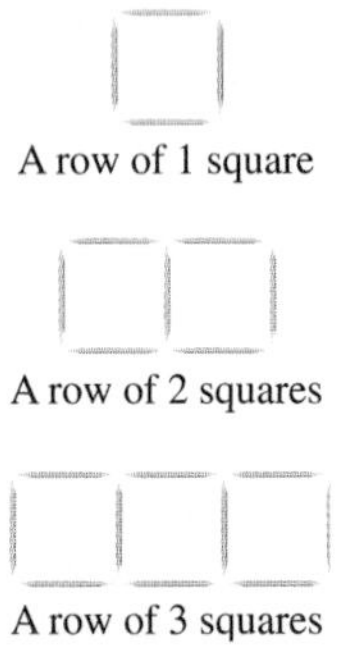

A row of 1 square

A row of 2 squares

A row of 3 squares

FIGURE 1.10

EXAMPLE 1.4 We can arrange toothpicks to form rows of squares as shown in Figure 1.10. How many toothpicks are required to form a row of 100 squares in this manner?

STEP 1 Understand the Problem.

We cannot break the toothpicks. We must arrange the toothpicks in a straight line of squares, not in a stacked array of squares. For example, the arrangement in Figure 1.11 would not be acceptable. We must determine the number of toothpicks to form a row of 100 squares.

FIGURE 1.11

STEP 2 Devise a Plan.

We might try a few more rows of squares, say four squares and then five squares (Figure 1.12). Then we could examine the results to see if there appears to be a relationship between the number of squares in a row and the number of toothpicks used.

STEP 3 Carry Out the Plan.

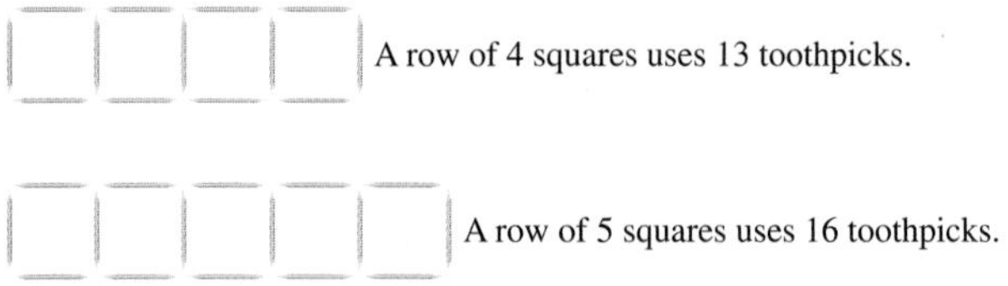

A row of 4 squares uses 13 toothpicks.

A row of 5 squares uses 16 toothpicks.

FIGURE 1.12

So far, from Figures 1.10 and 1.12, we have the following:

1 square	requires	4 toothpicks
2 squares	require	7 toothpicks
3 squares	require	10 toothpicks
4 squares	require	13 toothpicks
5 squares	require	16 toothpicks

It appears that for each additional square, three more toothpicks are required. Therefore, we can say that the row with 100 squares will require three more toothpicks than the row with 99 squares. But how many toothpicks are required for the row with 99 squares? We would prefer not to count so many toothpicks. Looking at the pictured rows of squares we might also observe a visual pattern. A row of two squares consists of two sets of three toothpicks plus a single toothpick necessary to complete the last square as shown in Figure 1.13.

FIGURE 1.13

Similarly, a row of three squares consists of three sets of three toothpicks plus a single toothpick and a row of four squares consists of four sets of three toothpicks plus a single toothpick, as shown in Figure 1.14.

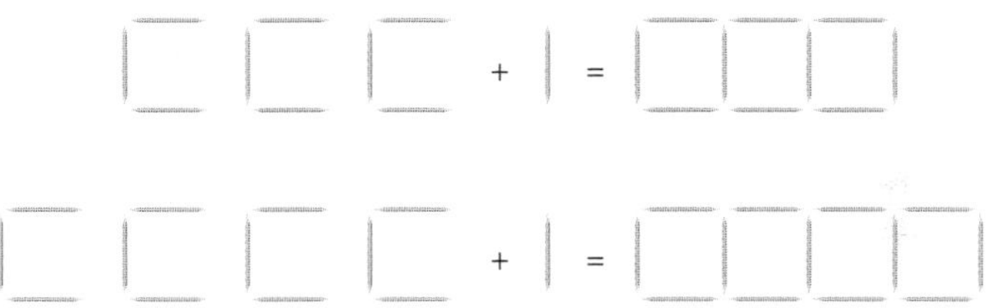

FIGURE 1.14

So a row of five squares would require five sets of three toothpicks plus one single toothpick, or a total of $5(3) + 1 = 16$ toothpicks.

This means that a row of 100 squares would require 100 sets of 3 toothpicks plus one more toothpick to complete the last square. So the total number of toothpicks necessary for a row of 100 squares would be $100(3) + 1 = 301$. You may observe other visual patterns for this problem, but they should all lead to the same final answer, namely 301 toothpicks.

STEP 4 Look Back.

How many toothpicks would be required for other rows of squares? We could use a variable to generalize the pattern we have observed. Let n be the number of squares in a row. Then we can calculate the number of toothpicks required as in Step 3. The row would require n sets of three toothpicks plus one more toothpick, so the total number of toothpicks required would be $n(3) + 1 = 3n + 1$. We can also use this formula to determine the number of toothpicks required to form a row of any size. For example, a row of 200 squares would require $3(200) + 1$, or 601 toothpicks. ■

In the preceding toothpick problem, we observed five different examples of rows of toothpicks, noticed a pattern, and then concluded that this pattern would hold for a row of any length. When we look at specific examples and draw a general conclusion from the pattern observed, we are using what is called **inductive reasoning**.

We often use inductive reasoning to draw conclusions. For example, scientists reason inductively when they perform experiments, observe the results, and form a hypothesis to explain those results. We may observe the sequence of numbers 5, 7, 11, 13, 17, 19, 23, . . . , and notice a pattern. Namely, the differences between successive terms alternate between 2 and 4, that is, $7 - 5 = 2$, $11 - 7 = 4$, $13 - 11 = 2$, and so on. Reasoning inductively, we may predict that the next number in this sequence will be 2 more than 23, or 25.

In the situations involving inductive reasoning described so far, we observed specific examples and drew a general conclusion. This transition from the specific to the general is the essence of inductive reasoning and distinguishes it from deductive reasoning, which is used extensively in geometry. In **deductive reasoning** some general statements are accepted as true and then specific conclusions are drawn from those statements. We use both inductive and deductive reasoning in geometry. Deductive reasoning will be a major focus of our study in Chapter 4.

We must consider one pitfall of inductive reasoning. Based on our observations of a few examples, we conclude that the pattern we notice will continue and that it is the *only* correct one. This assumption may not be warranted. For example, in the case of the numerical sequence 5, 7, 11, 13, 17, 19, 23, . . . , it may seem reasonable to assume that the next term will be 25. However, it is just as reasonable to predict that the next term will be 29 because all numbers listed are prime numbers (positive integers that have only two factors) and the next prime number after 23 is 29. Thus, in this example, more than one valid conclusion is possible. Due to this difficulty, we often use a combination of inductive and deductive reasoning to justify a conclusion.

Additional Problems Where the Strategy "Look for a Pattern" Is Useful

1. Examine the sequence of rectangles with subdivisions and shading as shown in Figure 1.15. Draw the figure that would come next in the sequence.

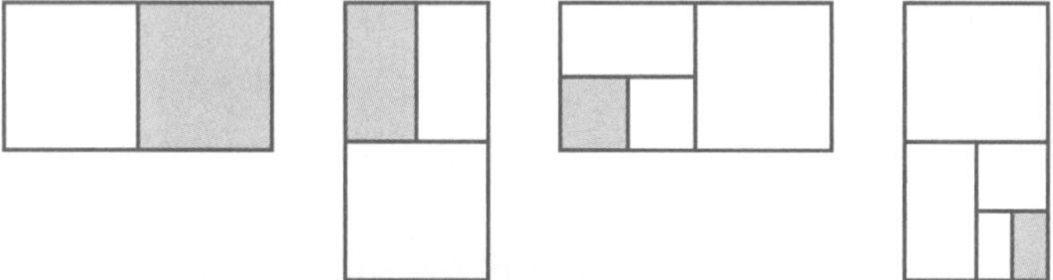

FIGURE 1.15

To get you started

a. Compare consecutive figures and consider the size and position of the shaded region in each of them.
b. What do you observe about the orientation of the large rectangle?
c. Try focusing on only one attribute at a time and follow it through the sequence of figures from left to right.

2. The rectangular numbers are the positive integers that represent the total number of dots in the arrays shown in Figure 1.16.

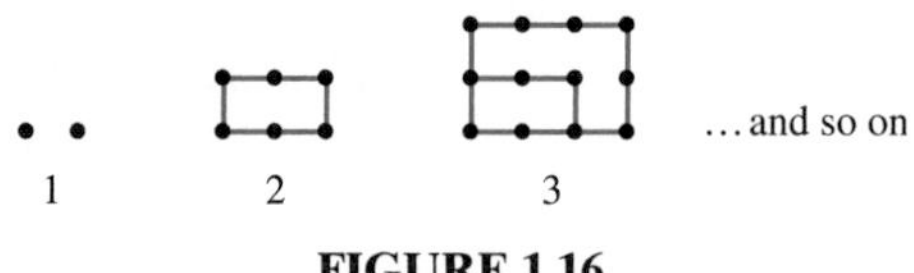

FIGURE 1.16

Notice that there are six dots in the second array, so the second rectangular number is 6. There are 12 dots in the third array so the third rectangular number is 12. Find the number of dots in the nth rectangular number.

Strategy 5: *Make a Table*

It is often helpful to make a list or a table when searching for a pattern. For instance, in Example 1.4 we calculated and recorded the number of toothpicks for rows of different sizes. We could have neatly summarized these data in a table such as Table 1.1, shown next.

Number of Squares in a Row	Number of Toothpicks Required
1	4
2	7
3	10
4	13
5	16

TABLE 1.1

Arranging data in a table such as Table 1.1 may make patterns easier to recognize. For example, we might observe that adding 2 to each number in the right column yields a column of numbers that are multiples of 3. You will often see tables combined with the Look for a Pattern strategy.

A table can also provide a convenient means for systematically recording guesses when applying the Guess and Test strategy. A table is a good way to organize data as they are generated, summarize information given in a problem, or list possible values of variables. The following clues may help you to identify situations where the Make a Table strategy might be useful.

Clues

The Make a Table strategy may be appropriate when

- You can easily organize and present information.
- You will generate data.
- You want to list the results obtained by using Guess and Test.
- The problem asks "in how many ways" something can be done.
- You are trying to learn about a collection of numbers or figures generated by a rule or formula.

EXAMPLE 1.5 A man wishes to enclose a rectangular area of 100 square feet for his dog (Figure 1.17). He is considering rectangles with *whole number* dimensions. What dimensions would require the least amount of fencing?

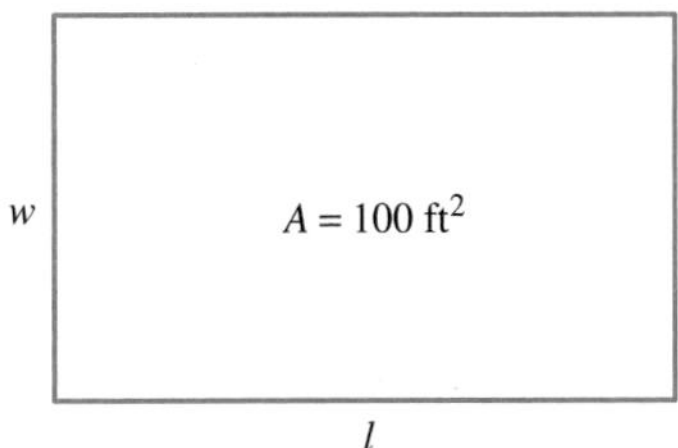

FIGURE 1.17

STEP 1. Understand the Problem.

The rectangle must have an area of exactly 100 square feet, and its dimensions must be whole numbers. In order for the amount of fencing to be as small as possible, the perimeter of the rectangle must be as small as possible. Recall that the area of a rectangle is $A = l \times w$ and that the perimeter of a rectangle is $P = 2l + 2w$, or $P = 2(l + w)$. We must determine the length and width that give the smallest perimeter.

STEP 2. Devise a Plan.

What lengths and widths give an area of 100 square feet? Make a table summarizing the possible dimensions and showing the perimeter in each case.

STEP 3. Carry Out the Plan.

Possible lengths and widths are summarized in Table 1.2.

Width	Length	Area $= l \times w$	Perimeter $= 2(l + w)$
1	100	100	202
2	50	100	104
4	25	100	58
5	20	100	50
10	10	100	40

TABLE 1.2

Because 1, 2, 4, 5, 10, 20, 25, 50, and 100 are the only whole number factors of 100, the table shown contains all possible dimensions for the rectangle. Therefore, the minimum perimeter is 40 feet, and it occurs when the rectangle measures 10 feet by 10 feet. Notice that the rectangle with the minimum perimeter in this case is a square. A square is a special type of rectangle.

STEP 4. Look Back.

Will a square always give the minimum perimeter? Try rectangles with different areas. Would it make a difference if only three sides of the enclosure were fenced and the fourth side was against a garage? ■

Additional Problems Where the Strategy, "Make a Table" Is Useful

1. A square piece of plywood measures 5 inches by 5 inches [Figure 1.18(a)]. In how many ways can it be cut into smaller pieces such that each of the smaller pieces is a square with whole-number dimensions? (NOTE: Two solutions are considered to be the same if they result in the same number and size of pieces.)

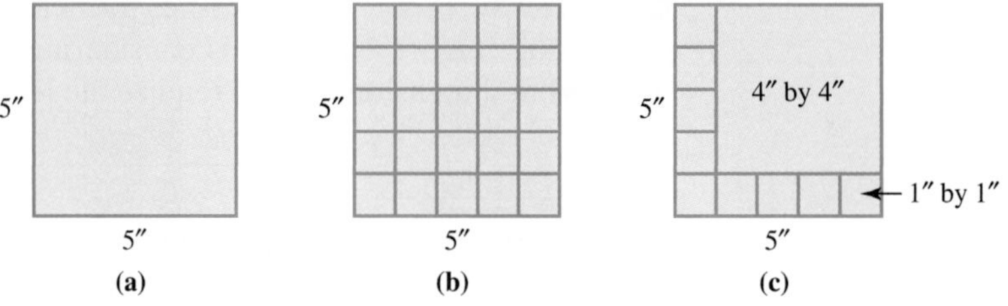

FIGURE 1.18

To get you started

a. Here you might want to use both the Draw a Picture and Make a Table strategies.

b. There are many solutions. One possibility is that the square could be cut into 25 squares 1 inch on a side [Figure 1.18(b)].

c. The squares that result need not have the same dimensions. For example, some of the pieces might be one by one squares, and one piece might be a 4 by 4 square [Figure 1.18(c)].

d. What dimensions are possible for the smaller squares? You might list those possibilities as headings in a table.

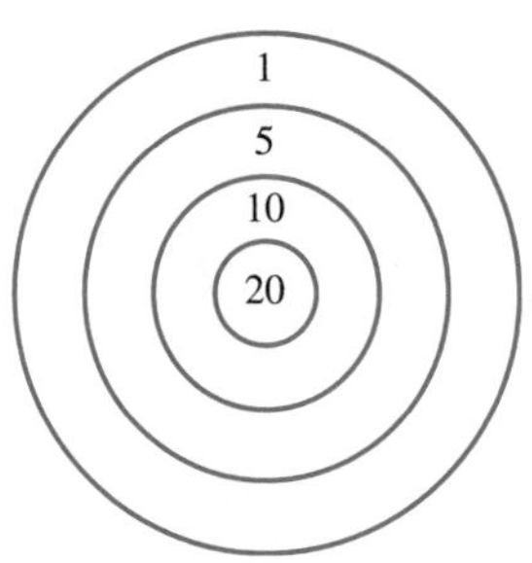

FIGURE 1.19

2. A dart board consisting of four circles is shown in Figure 1.19. Three colors—red, yellow, and black—are available to repaint the board. If the bull's-eye of the dartboard is to be red, and no two bordering circles can be the same color, how many different color schemes are possible?

George Polya was a proponent of the next method. He is quoted as saying "If you can't solve a problem, then there is an easier problem you can solve - find it."

Strategy 6: *Solve a Simpler Problem*

If the numbers in a problem are very large, very small, or difficult to work with, it is often helpful to solve a simpler version of the same problem. For example, large numbers you may replace with smaller, more manageable numbers, or numbers involving many decimal places you may replace with whole numbers. Once a technique is found for solving the simpler problem, you can often use the same technique to solve the original problem.

For instance, in Example 1.4 we wanted to determine the number of toothpicks necessary to form 100 squares. Rather than solve the problem with 100 squares, we studied several simpler problems. We examined rows consisting of 3, 4, and 5 squares and we observed a pattern that could be extended to the case of 100 squares.

We can also use the Solve a Simpler Problem strategy when a figure is complex. We can look at simpler versions of the same figure and see if a pattern emerges that will lead us to a solution of the original problem. The following clues may help you to identify situations where the Solve a Simpler Problem strategy might be useful.

Clues

The Solve a Simpler Problem strategy may be appropriate when

- The problem involves complicated computations.
- The problem involves very large or very small numbers.
- A direct solution is too complex.
- You want to gain a better understanding of the problem.
- The problem involves a large array or diagram.

EXAMPLE 1.6 There are 20 people at a party. If each person shakes hands with every other person at the party, how many handshakes will there be?

STEP 1. Understand the Problem.

Each person must shake hands with every other person exactly once. If person A shakes hands with person B, then person B does not shake hands with person A again. We must count the number of handshakes that occur and avoid duplication.

STEP 2. Devise a Plan.

We might represent the people as points on a circle and draw a line segment to represent each handshake. For example, the line segment joining A and B represents persons A and B shaking hands (Figure 1.20).

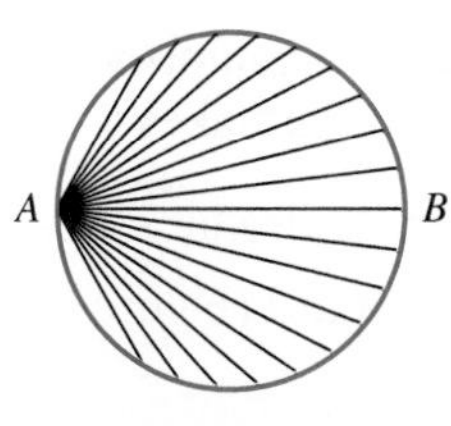

FIGURE 1.20

STEP 3. Carry Out the Plan.

Figure 1.20 shows all the handshakes for A. If we draw the line segments for the remaining handshakes, it will be very difficult to count all of them. However, we can more easily solve simpler problems involving parties with fewer people. Figure 1.21 shows the results for parties with two, three, and four people.

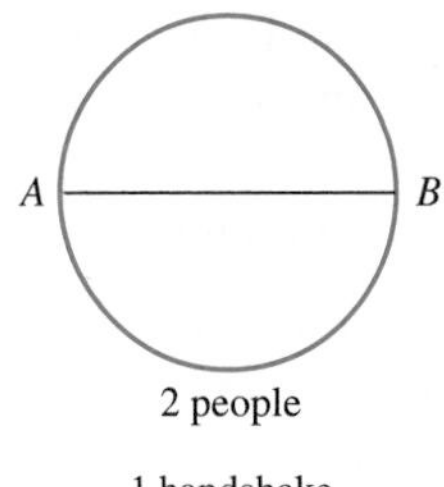

2 people

1 handshake

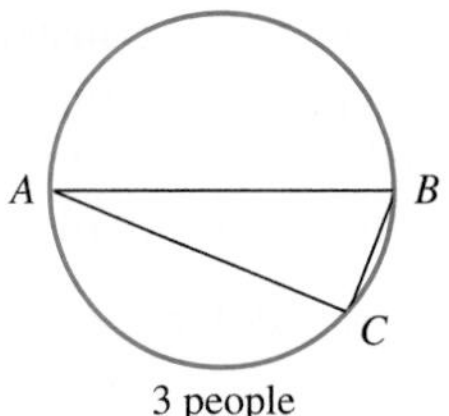

3 people

2 handshakes for A and 1 additional one for B.

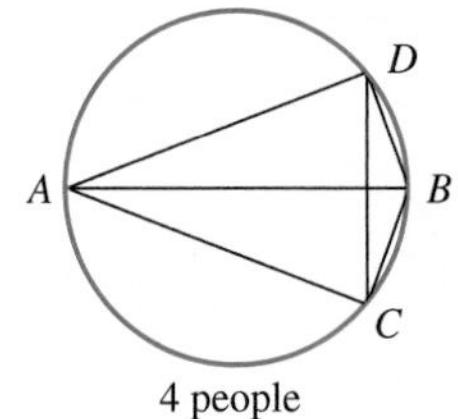

4 people

3 handshakes for A, 2 additional ones for B, and 1 additional one for C.

FIGURE 1.21

We summarize the results in Table 1.3 and look for a pattern.

Number of People	Number of Handshakes
2	1
3	$2 + 1 = 3$
4	$3 + 2 + 1 = 6$
5	$4 + 3 + 2 + 1 = 10$

TABLE 1.3

Based on the pattern in the table, we might calculate that the number of handshakes for six people would be $5 + 4 + 3 + 2 + 1 = 15$. Therefore, for any number of people, say n, the number of handshakes would be

$$(n - 1) + (n - 2) + \cdots + 2 + 1$$

So, for 20 people, the number of handshakes would be

$$19 + 18 + 17 + \cdots + 3 + 2 + 1$$

We can use a pattern to find this sum. Notice that if we pair terms as shown next, the sum of each pair of terms is 20, with 10 being an unpaired number in the middle.

$$19 + 18 + 17 + 16 + \cdots + 10 + \cdots + 4 + 3 + 2 + 1$$

$$19 + 1 = 20,\ 18 + 2 = 20,\ \ldots,\ 11 + 9 = 20,\ 10$$

There would be a total of nine pairs of terms, where each pair adds up to 20, plus the "leftover" middle term of 10. So we could say that the total number of handshakes would be $9(20) + 10 = 180 + 10 = 190$.

STEP 4. Look Back.

Is there another way we can solve this problem? Also, if n represents the number of people at the party, can we find a general formula for the number of handshakes? ■

Additional Problems Where the Strategy "Solve a Simpler Problem" Is Useful

1. How many squares of all sizes are there on an 8 by 8 checkerboard (Figure 1.22)?

   To get you started

   a. You might start by drawing a smaller 2 by 2 or 3 by 3 checkerboard and counting the number of squares of all sizes formed. Several checkerboards of each size may be easier to use than a single checkerboard.

FIGURE 1.22

b. Listing the number of squares of various sizes in a table may help you to better keep track of the number of squares. For example, in the case of the 3 by 3 checkerboard you would want to list the number of different 1 by 1, 2 by 2, and 3 by 3 squares on the board.

c. Can you see a pattern in the number of squares of each size for each checkerboard?

2. A schoolroom has 13 different desks and 13 different chairs. Suppose you want to arrange the desks and chairs so that each desk has a chair with it. How many such arrangements are possible? (Assume that desk number 1 paired with chair number 1 is a different arrangement from desk number 1 paired with chair number 2, etc.)

Solutions of Additional Problems for Strategies 4–6

Look for a pattern

1. 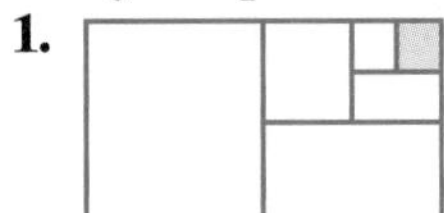

2. $n(n + 1)$

Make a table

1. There are 10 possibilities, as shown in Table 1.4.

4 by 4	3 by 3	2 by 2	1 by 1
1			9
	1	3	4
	1	2	8
	1	1	12
	1		16
		4	9
		3	13
		2	17
		1	21
			25

TABLE 1.4

2. There are eight possibilities, as shown in Table 1.5.

20	10	5	1
Red	Yellow	Red	Black
Red	Yellow	Red	Yellow
Red	Yellow	Black	Yellow
Red	Yellow	Black	Red
Red	Black	Red	Black
Red	Black	Red	Yellow
Red	Black	Yellow	Black
Red	Black	Yellow	Red

TABLE 1.5

Solve a simpler problem

1. $1 + 4 + 9 + 16 + 25 + 36 + 49 + 64 = 204$
2. $13(12)(11)(10)\cdots(1) = 6{,}227{,}020{,}800$

GEOMETRY AROUND US As the shell of a chambered nautilus grows, it creates a new wall behind its body every few months, forming a spiral array of chambers. We can approximate a nautilus spiral formed in this way, using a sequence of squares whose side lengths follow what is known as the Fibonacci sequence: 1, 1, 2, 3, 5, 8, 13, ... (The Fibonacci is discussed in Problem Set 1.2.)

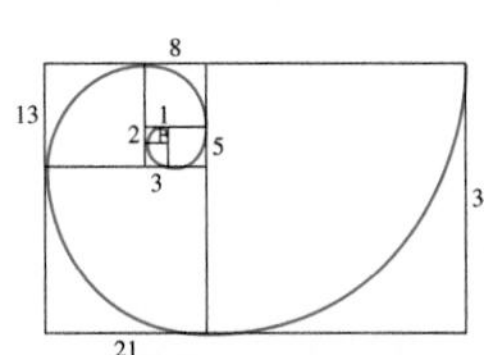

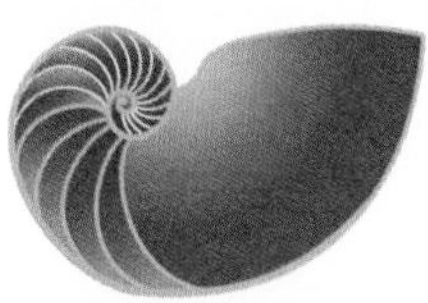

PROBLEM SET 1.2

EXERCISES/PROBLEMS

1. The x's in half of each figure can be counted in two ways.

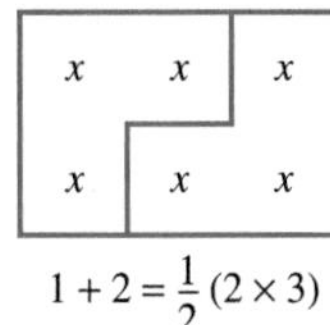

$1 + 2 = \frac{1}{2}(2 \times 3)$

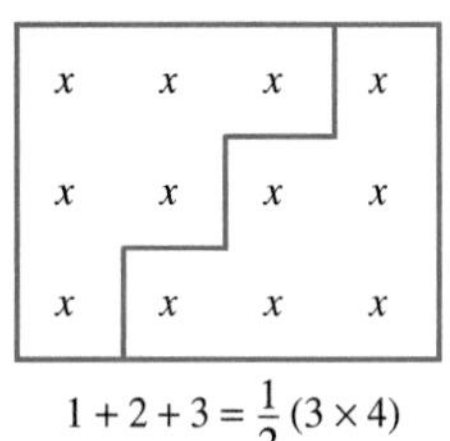

$1 + 2 + 3 = \frac{1}{2}(3 \times 4)$

   a. Draw a similar figure to find $1 + 2 + 3 + 4$.
   b. Express the sum $1 + 2 + 3 + 4$ in a similar way. Check to see that the sum is correct.
   c. Look for a pattern in the sums. Use it to find the sum of the whole numbers from 1 to 50 and from 1 to 75.
   d. Generalize your findings in part (c) by writing a formula for
   $1 + 2 + 3 + \cdots + (n - 1) + n$

2. a. Find the number of toothpicks required to form each of the following patterns and the number required to form the next five patterns.

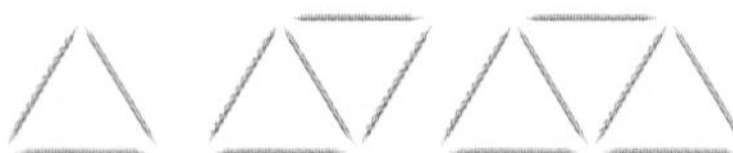

   b. If 41 toothpicks are used in one of the patterns, then how many triangles make up the pattern?
   c. Find the number of toothpicks in a pattern with n triangles.

3. Find the number of toothpicks required to form each of the following patterns and then predict the number of toothpicks needed to form the nth pattern.

4. Find the number of toothpicks required to form each of the following patterns and then predict the number of toothpicks needed to form the nth pattern.

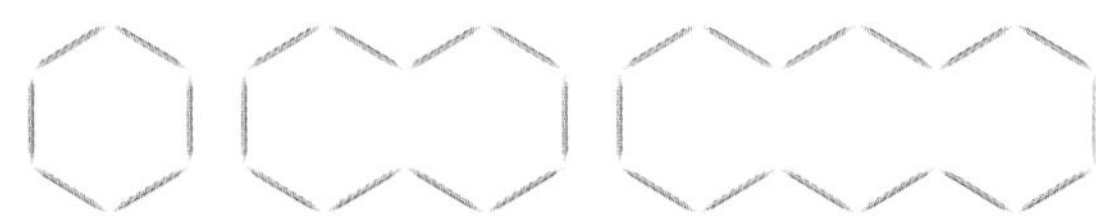

5. Find the number of toothpicks required to form each of the following patterns and then predict the number of toothpicks needed in the nth pattern.

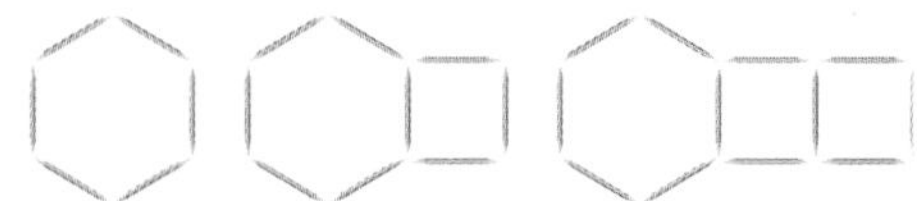

6. Find the number of toothpicks required to form each of the following patterns and then predict the number of toothpicks needed in the nth pattern.

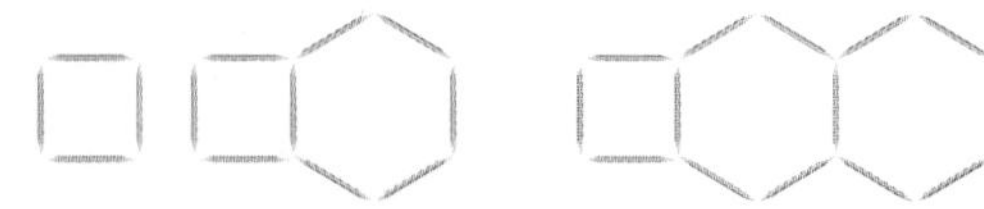

7. The **triangular numbers** are the whole numbers that represent the following shapes.

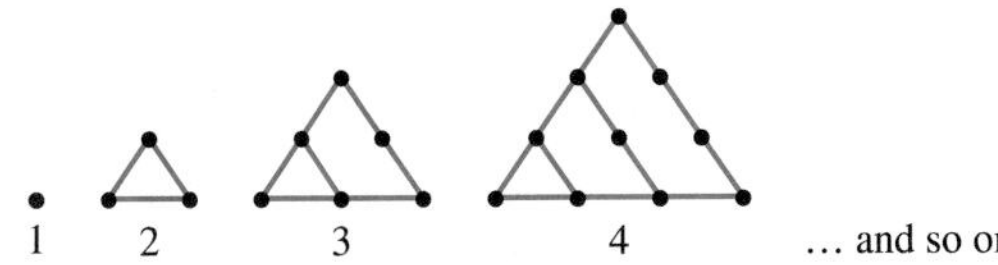

Find the number of dots in the nth triangular number.

8. The **pentagonal numbers** are the whole numbers that represent the following shapes

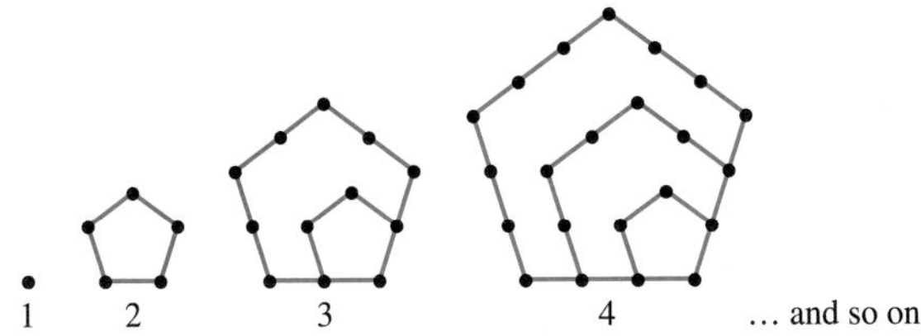

Find the number of dots in the nth pentagonal number.

9. The **hexagonal numbers** are the whole numbers that represent the following shapes.

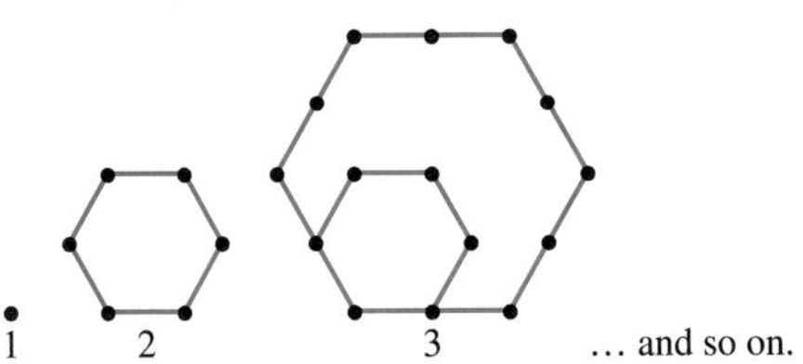

Find the number of dots in the nth hexagonal number.

10. The **octagonal numbers** are the whole numbers that represent the following shapes.

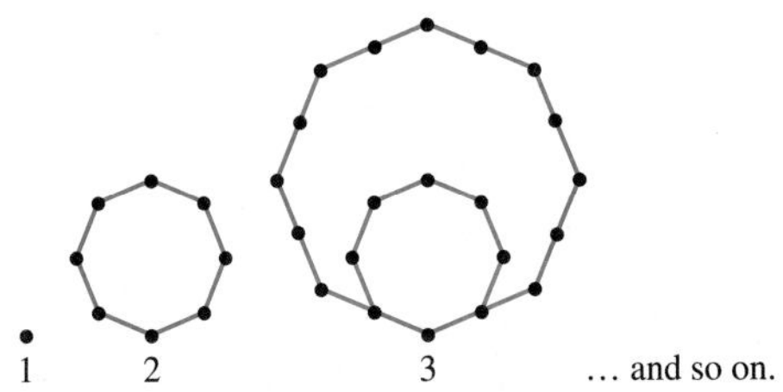

Find the number of dots in the nth octagonal number.

11. If A belongs with B, then X belongs with Y. Which of (i), (ii) or (iii) is the best choice for Y?

a.

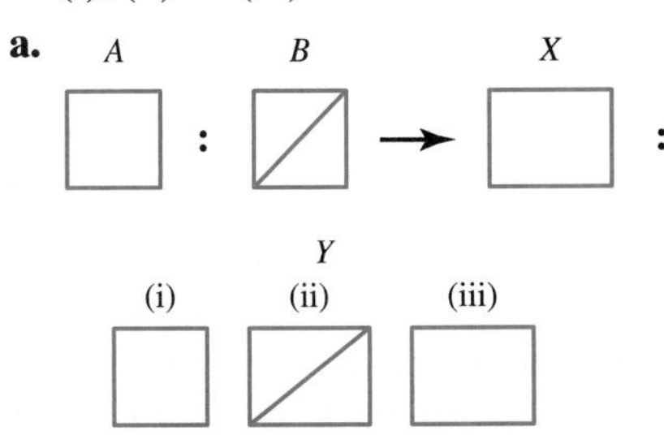

b.

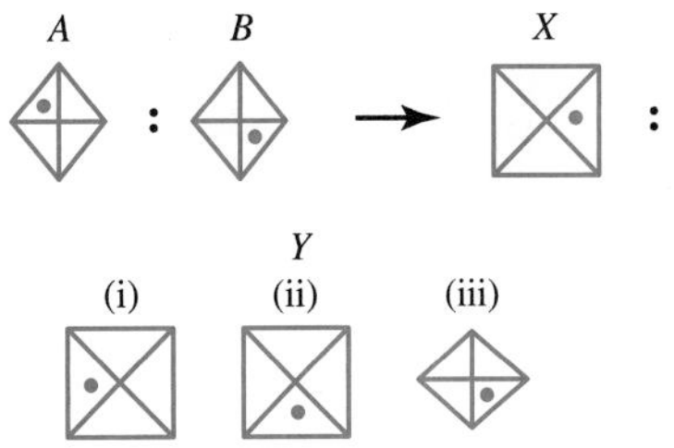

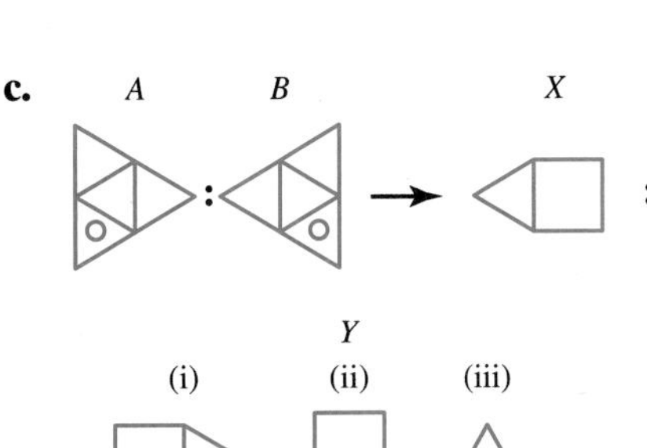

c.

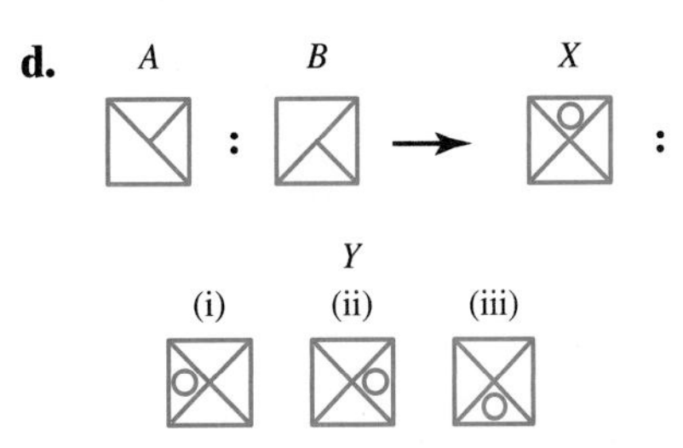

d.

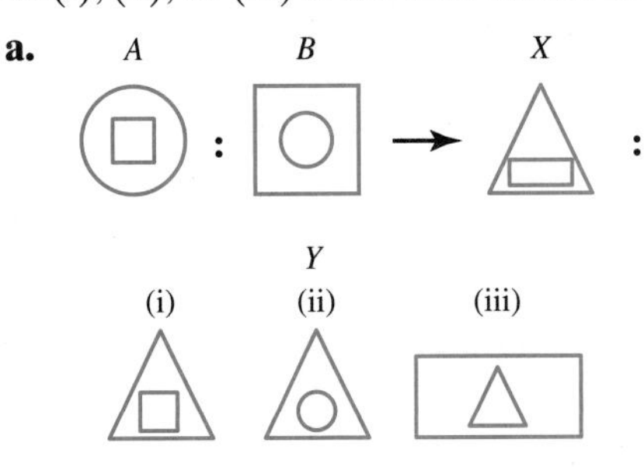

12. If A belongs with B, then X belongs with Y. Which of (i), (ii), or (iii) is the best choice for Y?

a.

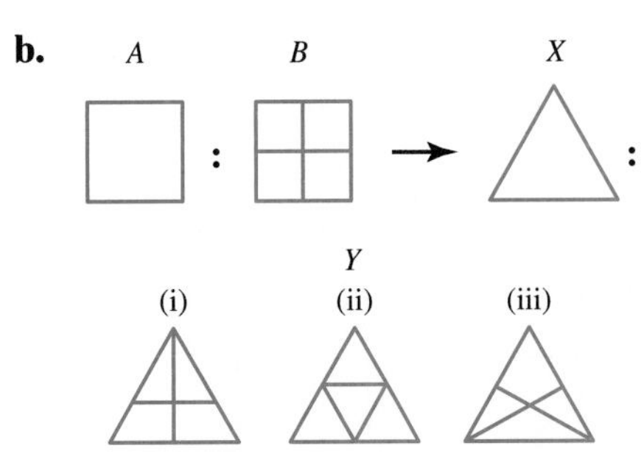

b. A : B → X : (i) (ii) (iii)

c.

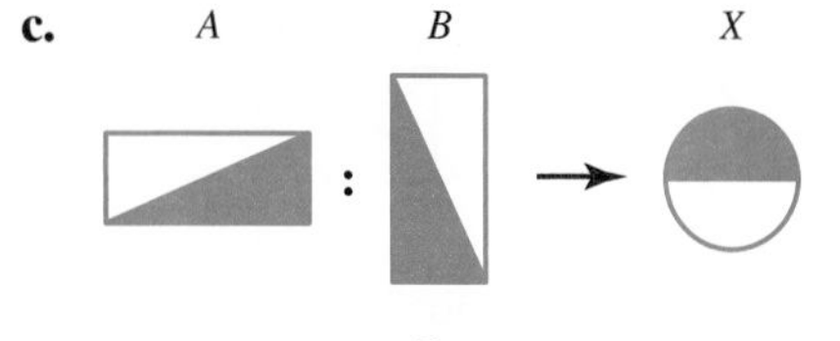

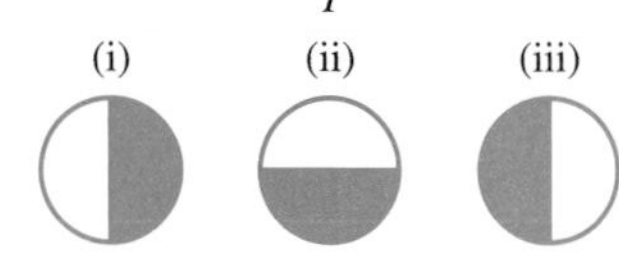

d.

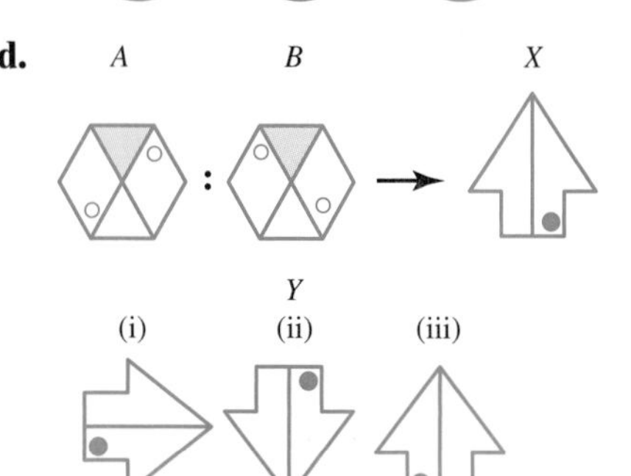

13. Use inductive reasoning to draw the next figure in each of the following sequences of figures.

a.  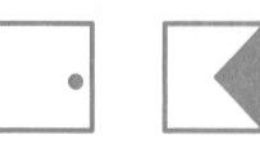 

b. 

c. 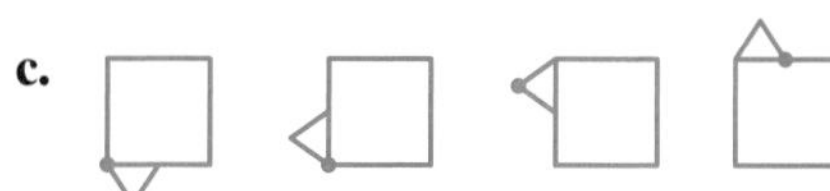

d. 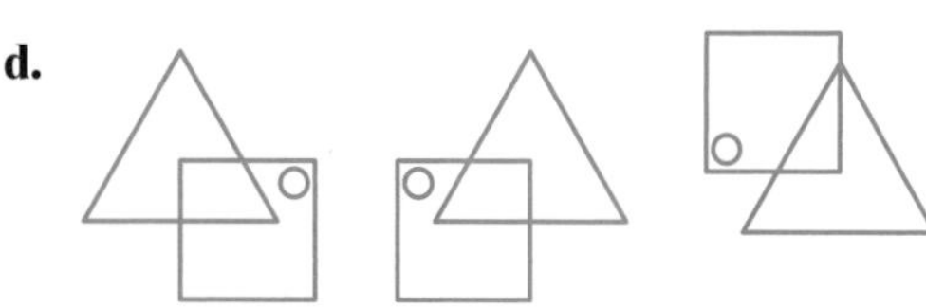

e. 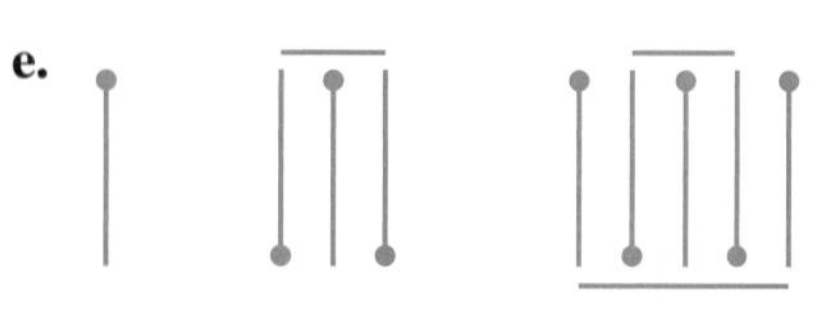

14. Use inductive reasoning to draw the next figure in each of the following sequences of figures.

a. 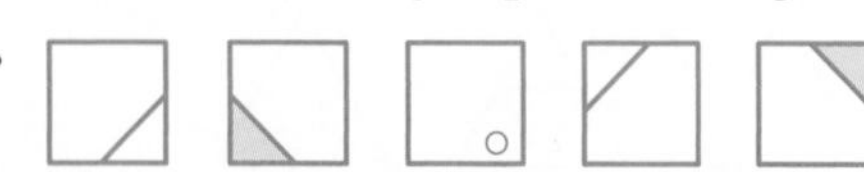

b. 

c. 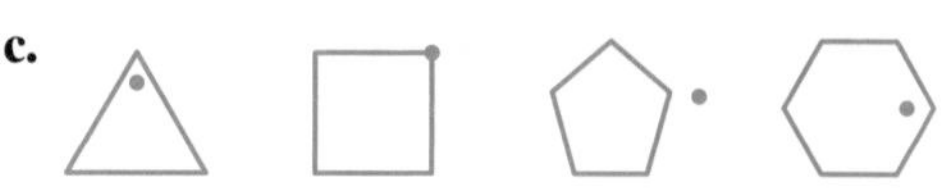

d. 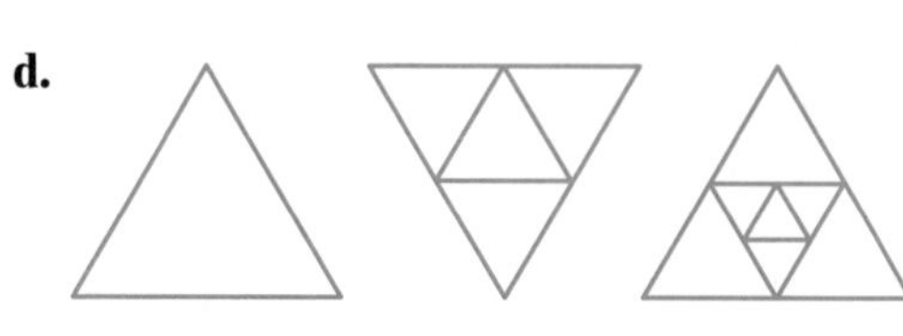

e. 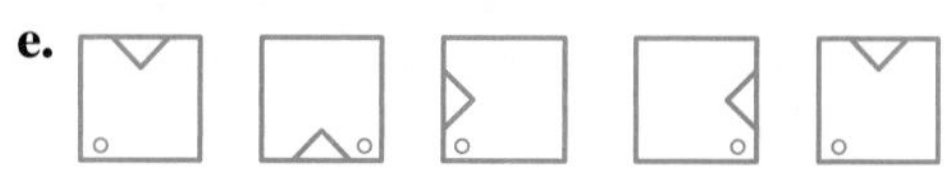

15. Each of the following problems gives the first five terms of a sequence. Use inductive reasoning to determine the 6th, 7th, and 12th terms of each sequence. Then write a formula for the nth term.

a. $1, 4, 9, 16, 25, \ldots$

b. $-2, -4, -6, -8, -10, \ldots$

c. $1, 3, 9, 27, 81, \ldots$

d. $3, 5, 7, 9, 11, \ldots$

16. Each of the following problems gives the first five terms of a sequence. Use inductive reasoning to determine the 6th, 7th, and 12th terms of each sequence. Then write a formula for the nth term.

a. $4, 7, 10, 13, 16, \ldots$

b. $0, 3, 8, 15, 24, \ldots$

c. $1, 2, 4, 8, 16, \ldots$

d. $\frac{-1}{2}, \frac{-2}{3}, \frac{-3}{4}, \frac{-4}{5}, \frac{-5}{6}, \ldots$

17. There are five couples at a party. If each person shakes hands with every other person except the person they came to the party with, how many handshakes will there be?

18. Six cities A, B, C, D, E, and F are located as shown next. If bikepaths were constructed so that each pair of cities would be connected by a direct path, how many paths would be required?

19. Julie walks to school at point B from her house at point A, a distance of six blocks. For variety, she likes to try different routes each day. How many different paths of exactly six blocks can she take? One path is shown.

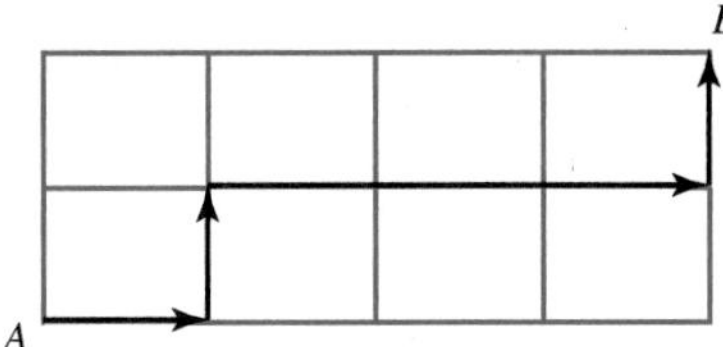

20. During your lunch hour, you need to walk from work to the bank, a distance of six blocks. One block is closed due to sidewalk construction, yielding the following arrangement of available sidewalks. How many different six-block paths can you take?

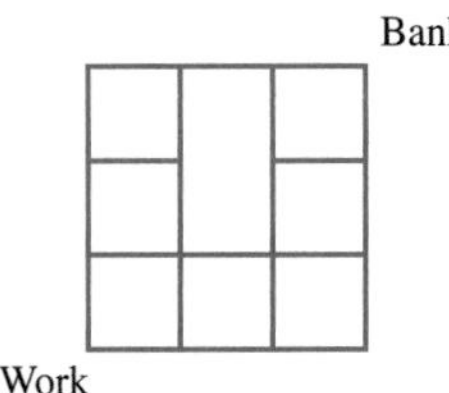

21. A well-known sequence of numbers is shown next.

$$1, 1, 2, 3, 5, 8, 13, 21, 34, \ldots$$

This sequence is called the **Fibonacci sequence** and is named for a 10th-century Italian mathematician whose pen name was Fibonacci and who posed a problem involving this sequence. This sequence arises in the shapes of nature, art, and architecture, as well as in music and probability. Write the next five terms of this sequence.

22. Determine the missing numbers in each of the following Fibonacci-type sequences.

a. $2, 5, 7, 12, \_, \_, \_, \ldots$

b. $1, \_, 7, \_, 20, \_, \ldots$

c. $3, \_, \_, 19, \_, \ldots$

d. $10, \_, \_, \_, \_, 95, \ldots$

23. How many triangles of all sizes are contained in this figure?

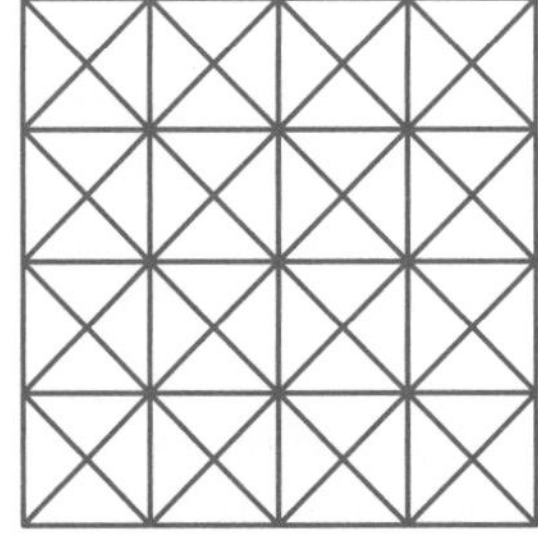

24. How many triangles of all sizes and shapes are in the picture?

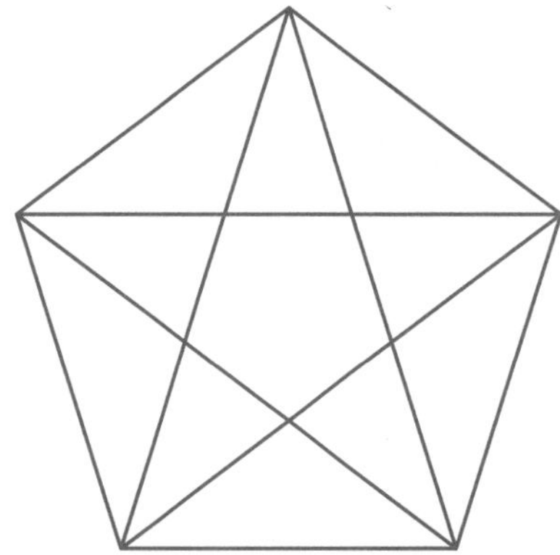

25. How many squares will be in the 10th set of squares in the following sequence? In the nth set?

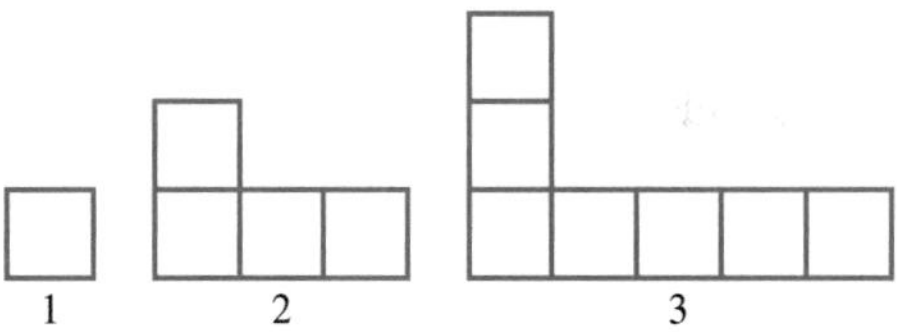

26. How many squares will be in the 10th set of squares in the following sequence? In the nth set?

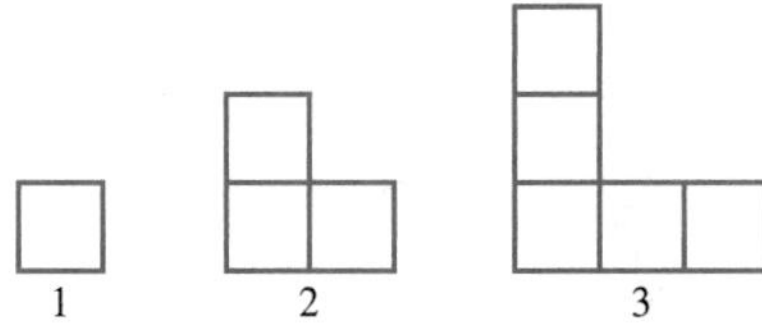

27. How many squares will be in the 10th set of squares in the following sequence? In the nth set?

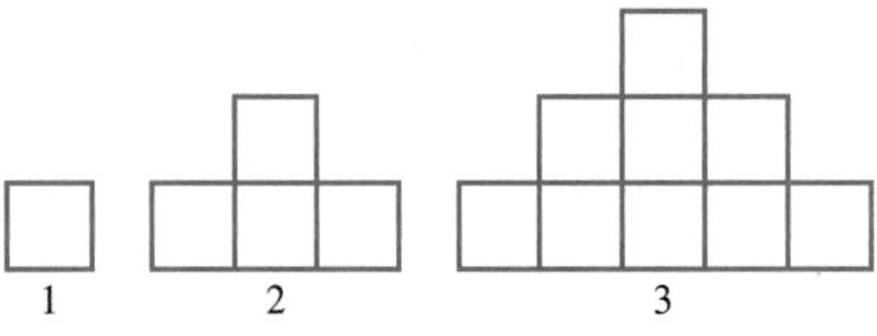

28. How many squares will be in the 10th set of squares in the following sequence? In the nth set?

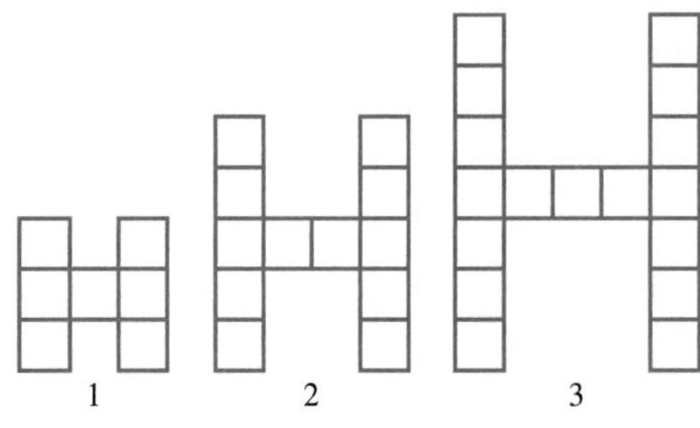

29. How many cubes are in the 10th set of cubes in the following sequence? In the nth set?

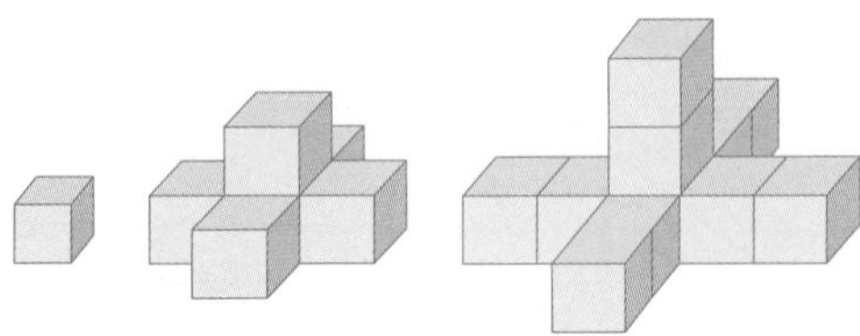

30. How many cubes are in the 10th set of cubes in the following sequence? In the nth set?

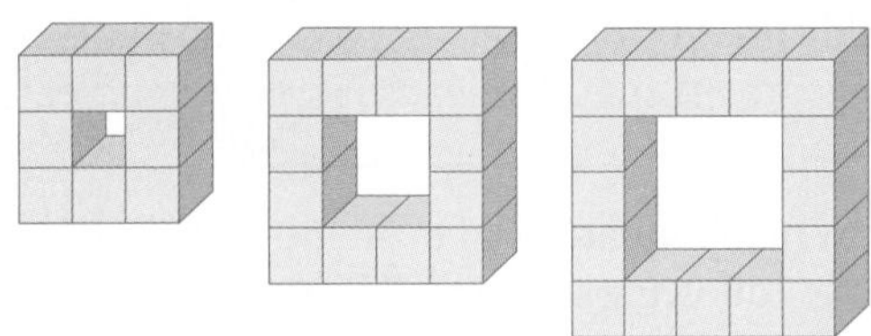

31. How many cubes are in the 10th set of cubes in the following sequence?

32. How many cubes are in the 10th set of cubes in the following sequence?

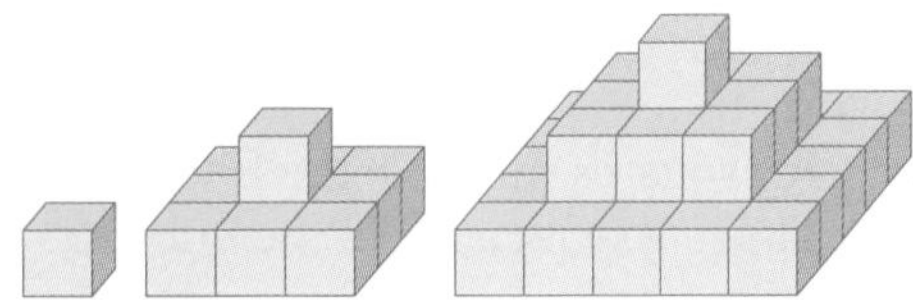

33. a. If the following pattern is continued, how many letters will be in the "M" column?

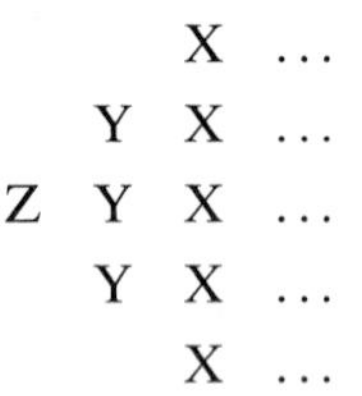

b. Which column will contain 45 letters?

34. A 3 by 3 by 3 cube was painted on all faces and then cut apart into twenty-seven 1 by 1 by 1 cubes.

a. How many 1 by 1 by 1 cubes had no faces painted? One face painted? Two faces painted? Three faces painted? Four or more faces painted?

b. Answer the same questions for a 4 by 4 by 4 cube and for a 5 by 5 by 5 cube.

c. Answer the same questions for an n by n by n cube, where n is any whole number greater than 1.

35. What fraction of the large square region is shaded? Assume that the pattern of shading continues forever.

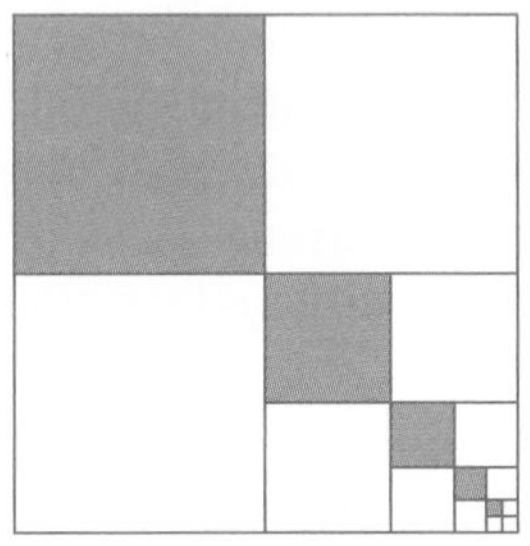

36. We cut two corner squares from an 8 by 8 checkerboard to obtain the following figure. Can this new board be exactly covered by 31 dominoes which measure 2 by 1? (NOTE: You may not cut any of the dominoes.)

37. If the pattern in the following figures continues indefinitely, the resulting figure is called the **Sierpinski triangle** or the **Sierpinski gasket**.

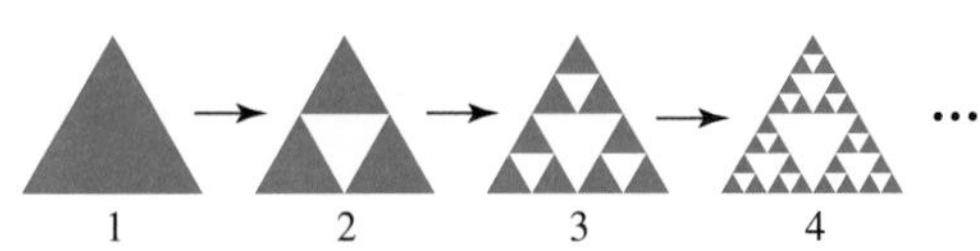

a. How many black triangles are there in each figure? How many white triangles?

b. How many black triangles do you expect would appear in the fifth figure? How many white triangles?

c. How many black triangles do you expect would appear in the nth figure?

38. If the terms of the Fibonacci sequence (see problem 21) are squared and then added as shown next, an interesting pattern emerges.

$$1^2 + 1^2 = 2$$
$$1^2 + 1^2 + 2^2 = 6$$
$$1^2 + 1^2 + 2^2 + 3^2 = 15$$
$$1^2 + 1^2 + 2^2 + 3^2 + 5^2 = 40$$

Use this pattern to predict the following sum without performing the addition.

$$1^2 + 1^2 + 2^2 + 3^2 + \cdots + 144^2$$

Check your answer using a calculator.

APPLICATIONS

39. A certain type of gutter comes in 6-ft, 8-ft, and 10-ft sections. How many different lengths of gutter can be formed using three sections of gutter?

40. Alice purchased 30 decorative stone square blocks to create an enclosed rectangular garden area. Each square block measures 1 foot by 1 foot, and one possible arrangement is shown. Including the arrangement shown, in how many distinct ways can Alice arrange the blocks to enclose a garden, and for which arrangement is the garden area greatest?

41. Workers will construct a decorative concrete floor for an outdoor picnic area by first pouring two side-by-side square slabs with side lengths of 1 foot, thus forming a rectangle. They will then pour a new square slab next to the long side of the rectangle. This process will continue as shown next.

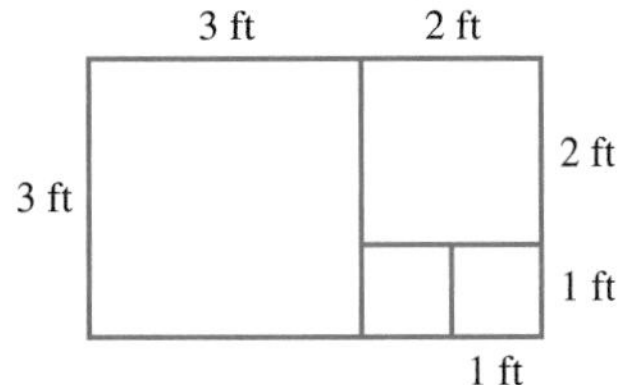

a. Sketch the floor as it will look after the next square slab is poured.

b. The dimensions of the rectangular floor formed by the first two squares are 1 foot by 2 feet. Give the dimensions of the rectangular floors after each of the next five squares are added.

c. What do you notice about the sequence of side lengths of the square slabs poured at each step?

d. How many squares will have been poured when the area of the floor is 4895 square feet?

EXTENDED PROBLEMS

42. The Fibonacci sequence, introduced in problem 21, contains many patterns. Revisit problem 21. Explore the sums of odd-numbered terms in the sequence and predict the sum of the first n odd-numbered terms. Do the same for the even-numbered terms. Add any 10 consecutive Fibonacci numbers and compare the sum to various Fibonacci numbers multiplied by 11. For consecutive pairs of Fibonacci numbers, divide the larger number by the smaller number and look for a pattern in the quotients. Summarize your discoveries and explain the relationships you found.

43. Workers will construct a rectangular brick retaining wall using bricks that measure 1 unit by 2 units. The height of the finished wall will be 2 units. Acceptable walls will have no bricks that hang over the edge of another as shown next.

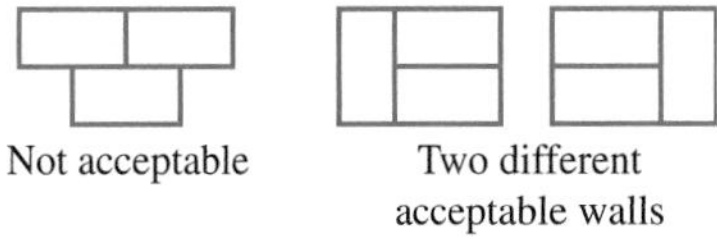

In how many ways may a wall be constructed using just one brick if the wall must be 2 units tall? Two bricks? Three bricks? Four bricks? Five bricks? Sketch all the patterns for the wall in each case and count them. List the numbers in the sequence formed and explain how the sequence is generated. Will a similar sequence result if the bricks used measure 1 unit by 3 units and the wall height must be 3 units? Explain.

44. Create your own sequence of geometric figures that follows a pattern. Have another student do the same. Each of you should sketch at least three figures of the sequence you create and write a description of the steps you used to create the sequence. Then trade sketches and try to discover the pattern in the sequence. Write an explanation of the steps you used to discover the pattern and compare your explanation to the description your partner wrote about that sequence. Were your explanations the same? What kind of reasoning did each of you use?

Solution of Initial Problem

We can look at simpler versions of the problem and see if we can observe a pattern. Suppose that a board had only two or three rows of holes in each starting pen as shown next.

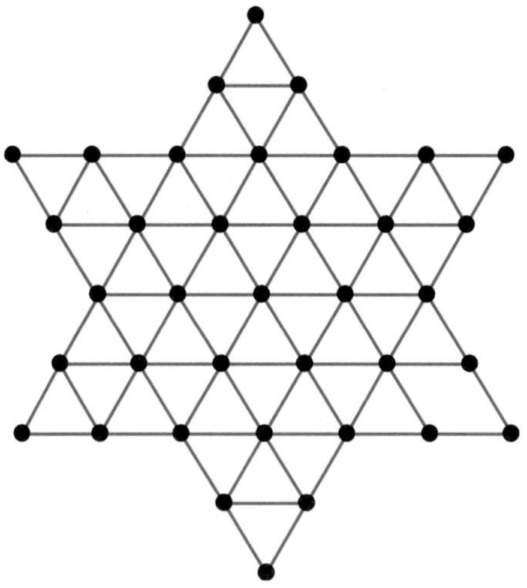

The total number of holes in all six pens is $6(1 + 2) = 6(3) = 18$. The number of holes in the inner hexagonal region is

$$3 + 4 + 5 + 4 + 3 = 2(3) + 2(4) + 5$$

We can summarize and extend this information in Table 1.6.

Number of Rows in a Starting Pen	Number of Holes in Starting Pens	Number of Holes in the Central Hexagon	Total Number of Holes
2	$6(1 + 2) = 18$	$2(3) + 2(4) + 5 = 19$	$18 + 19 = 37$
3	$6(1 + 2 + 3) = 36$	$2(4) + 2(5) + 2(6) + 7 = 37$	$36 + 37 = 73$
4	$6(1 + 2 + 3 + 4) = 60$	$2(5) + 2(6) + 2(7) + 2(8) + 9 = 61$	$60 + 61 = 121$
n	$6(1 + 2 + \cdots + n)$	$2(n + 1) + 2(n + 2) + \cdots + 2(2n) + 2n + 1$	The sum of the numbers in columns 2 and 3

TABLE 1.6

Substituting 6 for n in the last row in Table 1.6, we see that a game with six rows of holes (or 21 holes) in a starting pen has the following total number of holes:

$$T = 6(1 + 2 + 3 + 4 + 5 + 6) + 2(7) + 2(8) + 2(9) + 2(10) + 2(11) + 2(12) + 13 = 253 \text{ holes}$$

Writing for Understanding

1. Find a problem that you want to spend time solving, making sure that it *is* a problem for you, not an exercise. Then begin to solve the problem, writing down all of your steps and writing down your insights into the problem as they occur to you. Also,

write down the things you say to yourself as you solve the problem. Focus on your approach to the problem, the strategies you use, and how you are feeling about your success with the problem. The goal here is for you to learn about your own problem-solving process, not necessarily to find the solution to the problem at this time.

2. Select six problems from this chapter's problem set, one for each of the strategies. Identify the clue that led you to select the strategy you used and explain how that clue was useful.
3. Select three problems from this chapter's problem set. For each of the three problems, use the Look Back step to construct two new problems that are extensions or minor modifications of the original problem.
4. Revisit a real-world problem you have solved. Explain how you could apply Polya's four-step process to solve that problem.

PEOPLE IN GEOMETRY

Euclid of Alexandria (circa 300 B.C.E.) has been called the "father of geometry." He authored many works, but he is most famous for *The Elements.* Although many of the results of plane geometry were already known, Euclid's unique contribution was the use of definitions, postulates, and axioms with statements to be proved, called propositions. Most of the theorems and much of the development contained in a typical high school geometry course today are taken from *The Elements.*

Euclid founded the first school of mathematics in Alexandria. As one story goes, a student who had learned the first theorem asked Euclid, "But what shall I get by learning this?" Euclid called his slave and said, "Give him three pence, since he must make gain out of what he learns." As another story goes, a king asked Euclid if there was not an easier way to learn geometry than by studying *The Elements.* Euclid replied by saying, "There is no royal road to geometry."

CHAPTER REVIEW

For each section of this chapter, you will find a list of vocabulary and notation, questions to assess your understanding of key concepts, and review problems similar to the problems you worked in your homework. Review each item in the *Vocabulary/Notation* list mentally, and, if necessary, refer back to the indicated page and write a definition. Then answer the *Concept Check Questions,* looking back at the section if you need help. Work the *Review Problems* as practice before you move on to the *Chapter Test.* Answers to the *Review Problems* and the *Chapter Test* can be found at the back of the book.

SECTION 1.1 Problem-Solving Strategies

Vocabulary/Notation

Polya 1
Exercise 4
Problem 4
Four-step process 4
Polygon 9
Side(s) 9
Vertex (vertices) 9

Concept Check Questions

1. What is the difference between an exercise and a problem? 4
2. What are the steps in Polya's four-step problem-solving process? 4
3. Which of Polya's problem-solving steps is most difficult and why? 4
4. Under what conditions might the Draw a Picture strategy be appropriate? 5
5. Under what conditions might the Guess and Test strategy be appropriate? 6–7
6. Under what conditions might the Use a Variable strategy be appropriate? 8

Review Problems

1. Shown next are five different views of the same block in different positions. Each one is a different front view and a top view of the block. Sketch the arrangement of the figures on the flattened block to match the arrangement on the block when folded.

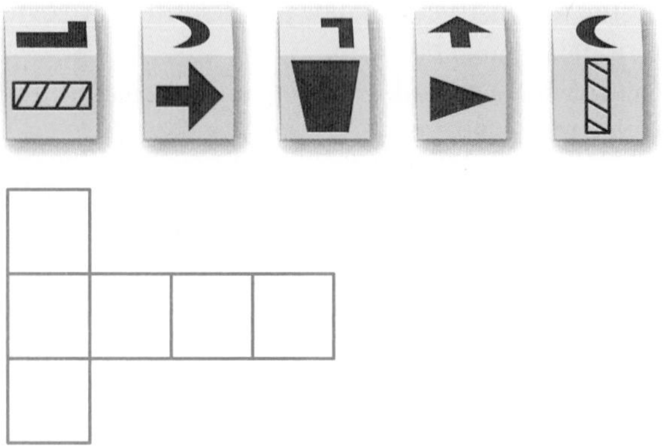

2. In how many different ways can the following 4 by 4 grid be cut along the lines to obtain two identical pieces? (NOTE: Two cuts are the same if they look alike when the figure is rotated or flipped.)

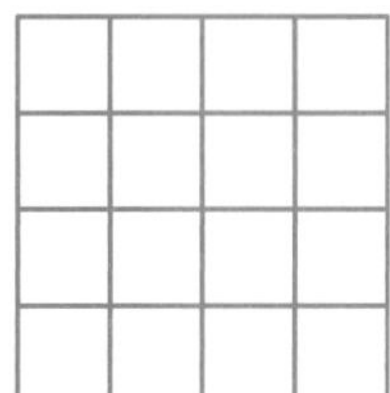

3. A triangle with equal side lengths is cut from a square as shown so that the distance around the triangle is one-half the distance around the remaining figure. If the square has a side of length 10 inches, what are the side lengths of the triangle?

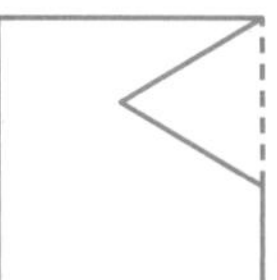

4. Connect the dots in the following figure using only four straight lines and without picking up your pencil.

• • •

• • •

• • •

5. Make four copies of the following figure, cut them out, and arrange them to form a larger version of the original figure. Pieces may be flipped over.

SECTION 1.2 More Problem-Solving Strategies

Vocabulary/Notation

Inductive reasoning 17

Deductive reasoning 18

Concept Check Questions

1. Under what conditions might the Look for a Pattern strategy be appropriate? 16
2. What is the difference between inductive and deductive reasoning? 18
3. Under what conditions might the Make a Table strategy be appropriate? 19
4. Under what conditions might the Solve a Simpler Problem strategy be appropriate? 21

Review Problems

1. Use inductive reasoning to predict the next three terms in the sequence
   a. 1, 3, 3, 9, 27, ___, ___, ___
   b. 3, 14, 25, 36, ___, ___, ___
   c.

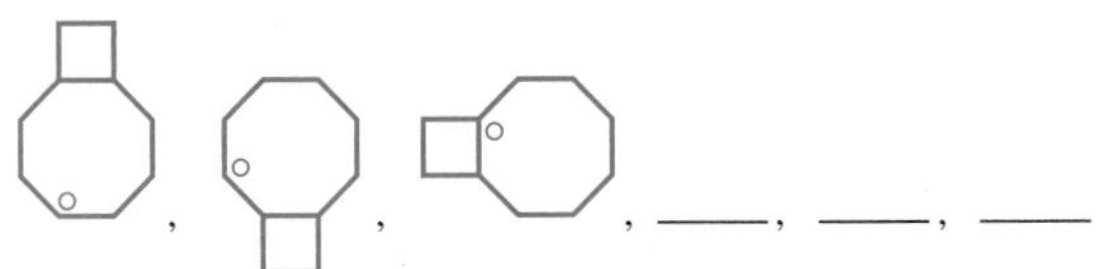

2. Greg's grandmother just returned from a vacation in Mexico and brought him four different hand-painted tiles. Greg plans to hang the tiles side by side on his wall. In how many ways can hang the tiles?
3. How many paths of length 3 units are there along the edges of the cube from A to B where the path always travels toward B?

4. Consider the following patterns made up of toothpicks. Find the number of toothpicks required to make each pattern and predict the number of toothpicks that would be required to make the nth pattern, composed of n hexagons.

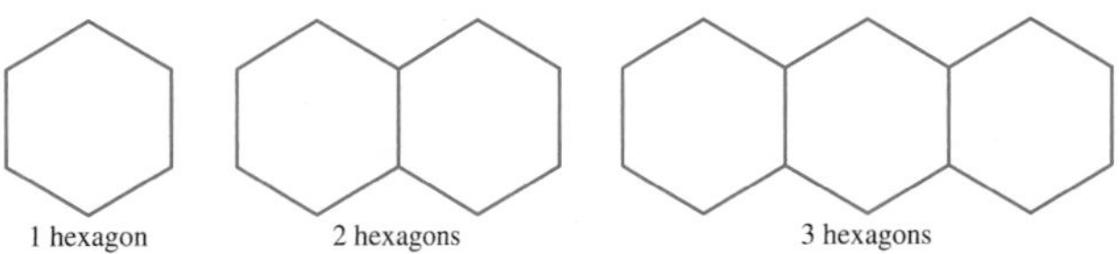

5. What is the maximum number of points in which three distinct lines can intersect? Four lines? Five lines? Nine lines? n lines?

CHAPTER 1 TEST

1. List the six strategies for problem solving you have learned in this chapter.
2. Given the following two "clues," which problem-solving strategy should you attempt?
   a. A phrase similar to "for any number" is present or implied.
   b. You are trying to develop a general formula.
3. Using the following figure, move just four toothpicks to make three squares.

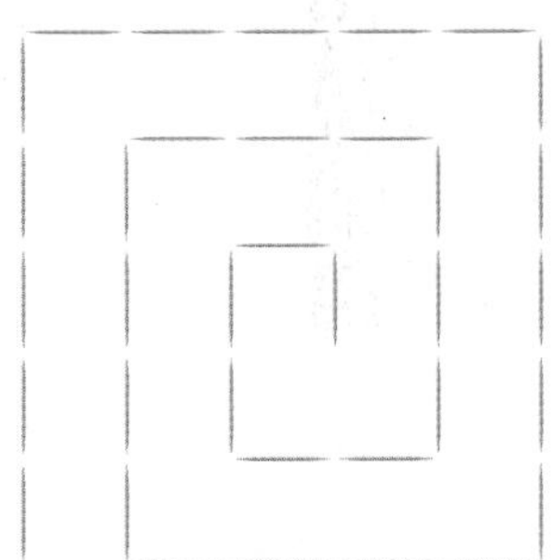

4. Twelve toothpicks form six equilateral triangles. Using the following figure, move four toothpicks to leave three equilateral triangles.

5. Two angles of a triangle have the same measure. The measure of the third angle is five less than three times the measure of each of the other angles. Find the measures of the angles in the triangle. (Use the fact that the sum of the measures of the angles in a triangle is 180°.)
6. In the following figure, the big square represents a farm and the shaded square represents a house and yard. The farmer wishes to retire and remain in the house. She also wants to divide the rest of the farm into pieces having the same size and shape for her five sons. How can she do this?

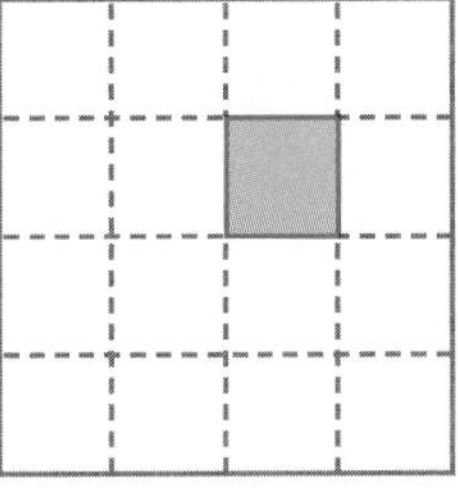

7. Find the maximum number of pieces a round cake can be cut into using

a. three straight cuts

b. four straight cuts

c. five straight cuts

Assume all cuts are perpendicular to the bottom of the cake.

8. How many squares will be in the 10th set of squares in the following sequence? In the nth set?

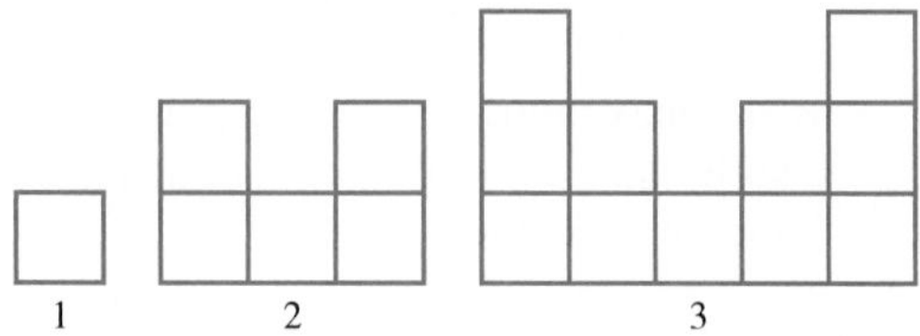

9. Use inductive reasoning to write the 10th term in each of the following sequences. Then write a formula for the nth term.

a. $4, 13, 22, 31, 40, \ldots$

b. $3, 5, 9, 17, 33, \ldots$

10. A large cube is made up of 1 by 1 by 1 cubes. Find the total number of cubes of all sizes in a large cube having dimensions:

a. 2 by 2 by 2

b. 3 by 3 by 3

c. 4 by 4 by 4

d. n by n by n

11. Trace the following figure and cut it into five pieces along the indicated lines. Rearrange the pieces to form a square. You must use all five pieces and have no gaps or overlaps, and you cannot turn the pieces over. Sketch your solution.

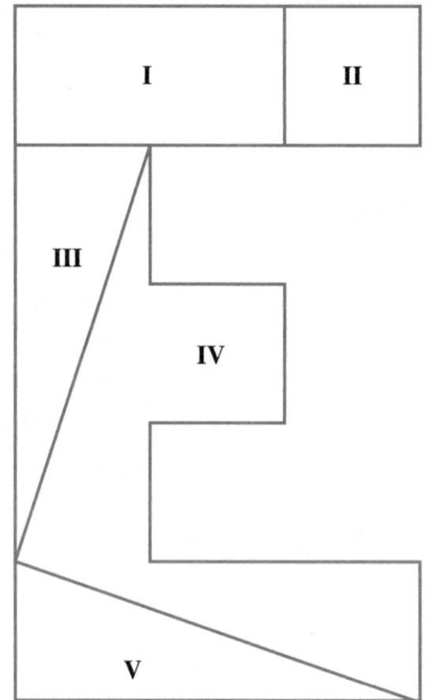

CHAPTER 2

GEOMETRIC SHAPES AND MEASUREMENT

Geometers used informal reasoning until approximately 300 B.C.E. when Euclid wrote his famous set of books called *The Elements.* In *The Elements*, Euclid made two very important contributions to the way we study mathematics. First, he organized existing results in a manner that linked them together. In this way, we can verify or prove later results using earlier results. Second, Euclid introduced us to an axiomatic system that was built upon common notions, definitions, postulates, axioms, and theorems. Later mathematicians replaced some of Euclid's definitions with undefined terms to form what is now known as an axiomatic system.

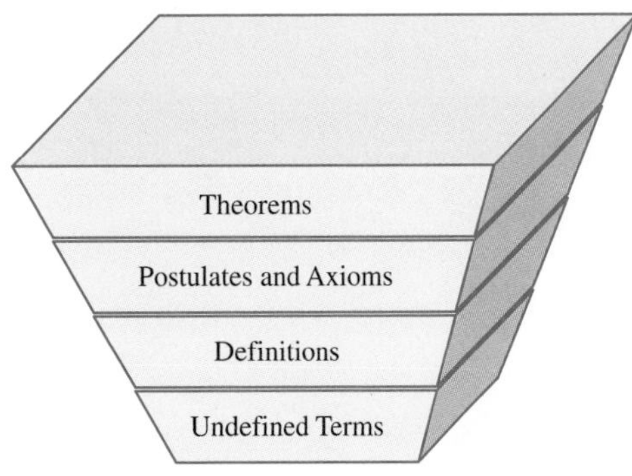

Elements of an Axiomatic System

Problem-Solving Strategies

1. Draw a Picture
2. Guess and Test
3. Use a Variable
4. Look for a Pattern
5. Make a Table
6. Solve a Simpler Problem
7. *Look for a Formula*

Strategy 7: *Look for a Formula*

Initial Problem

Find the number of squares of all sizes in the 10 by 10 grid of squares shown. (NOTE: To find the total number of squares, you must count the number of different 1 by 1 squares, 2 by 2 squares, etc.)

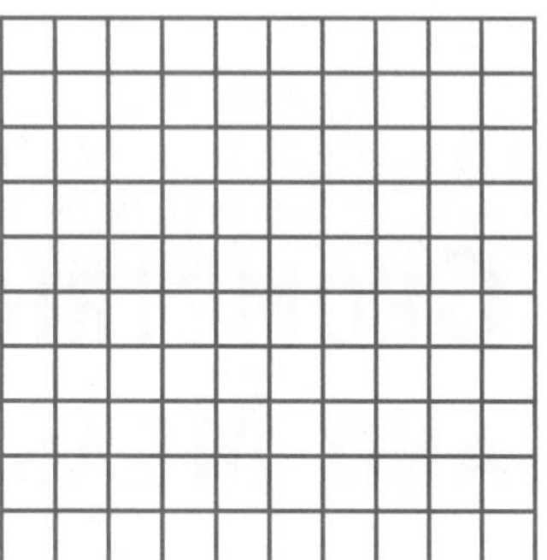

Clues

The Look for a Formula strategy may be appropriate when

- A problem suggests a pattern that can be generalized.
- Ideas such as distance, area, volume, or other measurable attributes are involved.
- Applications in science, business, and so on are involved.
- A problem requires counting a number of possibilities.

A solution of this Initial Problem is on page 93.

ESTIMATION AND MEASUREMENT

OBJECT: To test your estimation skills by "hitting" a "golf ball" into the hole for the three holes shown in the fewest number of "strokes."

RULES: To play this game, make a copy of this page. Begin at the first tee and, without using a ruler or protractor, estimate the distance (in centimeters and/or millimeters) to the first hole and the angle (in degrees) measured counterclockwise from the horizontal. Record your estimate on the scorecard. Then plot where your stroke lands using a ruler and a protractor. Take your next stroke from the new position and continue until you land in hole 1. Play all three holes in order. Assess yourself a penalty stroke if a ball lands in the rough. Be sure to record all estimates, strokes, and penalties in the scorecard. Your score is the sum of the strokes for all three holes, including the penalty strokes. (You may trace and cut out a protractor from Figure 2.18 in Section 2.1.)

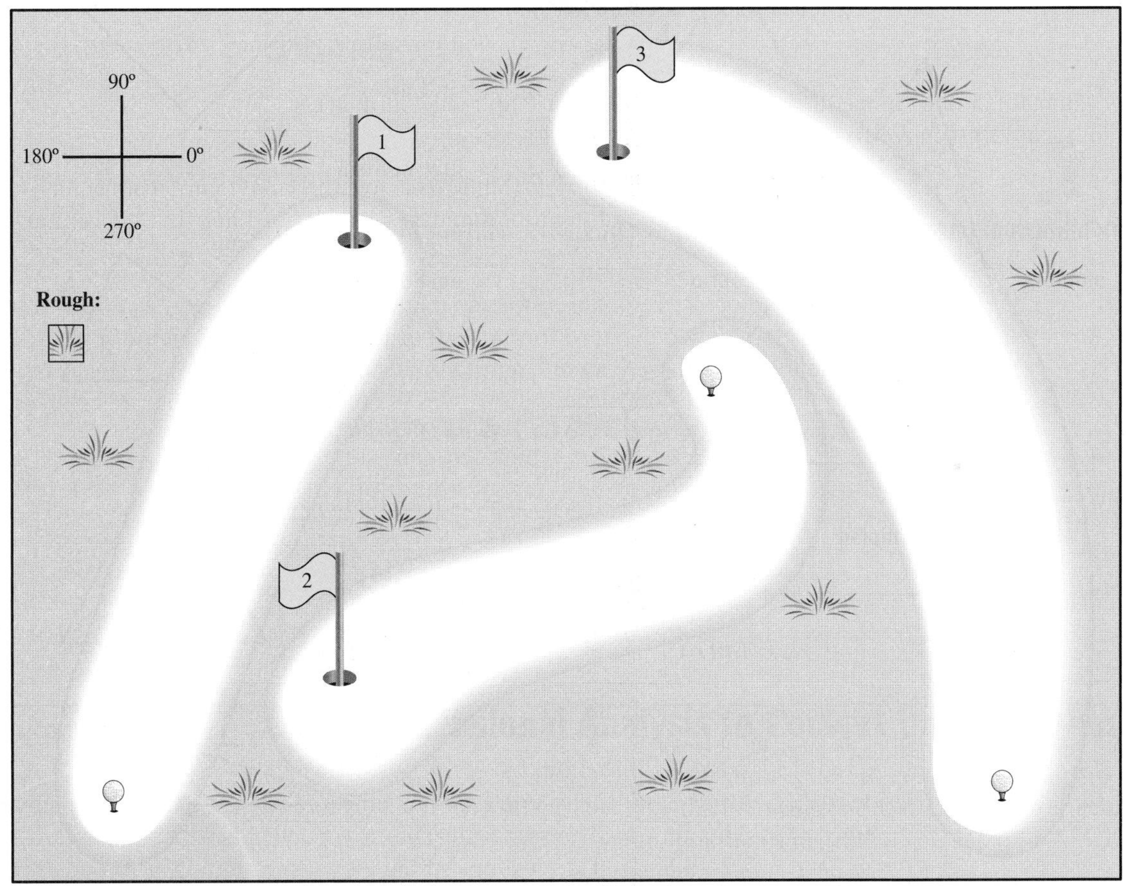

Scorecard

Hole 1			**Hole 2**			**Hole 3**		
Distance	Angle	Stroke/ Penalty	Distance	Angle	Stroke/ Penalty	Distance	Angle	Stroke/ Penalty
Total Hole 1:			**Total Hole 2:**			**Total Hole 3:**		

Introduction

Geometry was initially used to solve problems involving measurements on the earth ("geo" means earth and "metry" means measure). Although the earth is essentially spherical in shape, when we view the earth in small regions it appears to be flat (Figure 2.1).

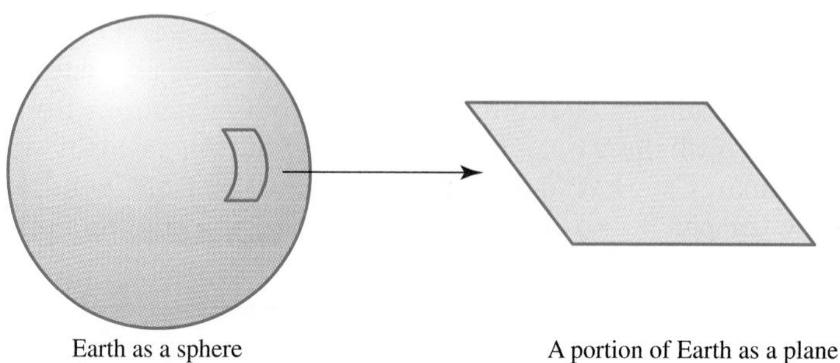

FIGURE 2.1

Thus, in many practical situations involving measurement, a small portion of the surface of the earth is assumed to be flat and considered to be part of a plane. This is one of the reasons plane geometry has been a main focus of school mathematics for the past several centuries.

In Chapter 1, we solved problems related to geometry by reasoning informally. In this chapter, we begin to formalize the study of geometry in the spirit of Euclid. A modern **axiomatic system** begins with a small set of undefined terms and builds through the addition of definitions and postulates to the point where many rich mathematical theorems can be proved (Figure 2.2).

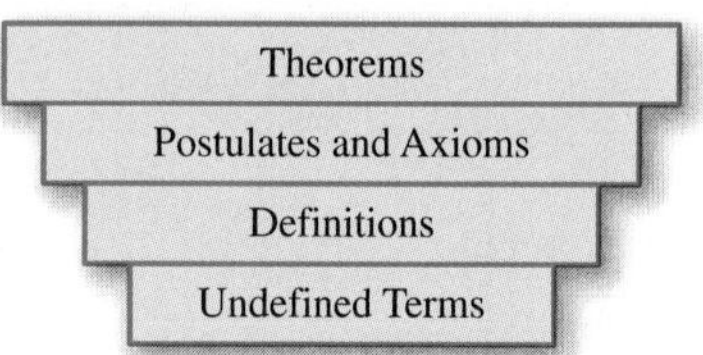

FIGURE 2.2

In Section 2.1 we introduce some undefined terms and postulates, and many definitions of plane geometry. Then we study common two-dimensional geometric shapes in Section 2.2. In Section 2.3 we look at angle measure in polygons and at how we can arrange polygons to cover a plane. In Section 2.4 we examine three-dimensional shapes. Finally, in Section 2.5 we introduce dimensional analysis, a useful technique for solving problems involving units of measure.

2.1 UNDEFINED TERMS, DEFINITIONS, POSTULATES, SEGMENTS, AND ANGLES

Applied Problem

A **compass card**, or **compass rose**, is marked with the 32 equally spaced points of the compass and is used in navigation. An example with some of the directions labeled is shown. If a ship sailing WNW changes its course to NE, it has turned through an angle of how many degrees?

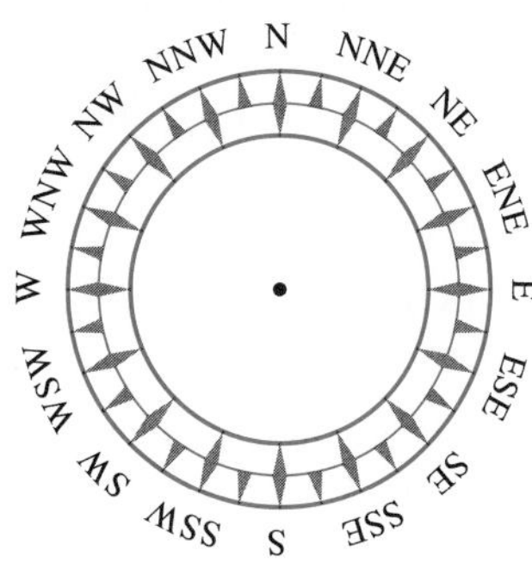

Elements of an Axiomatic System

Definitions, as they appear in dictionaries, are circular. That is, if you continue to look up key words in a definition, you will likely revisit one of the words you have found already. By contrast, in an axiomatic system, we agree to leave some terms undefined and then build definitions of other terms and postulates from those undefined terms.

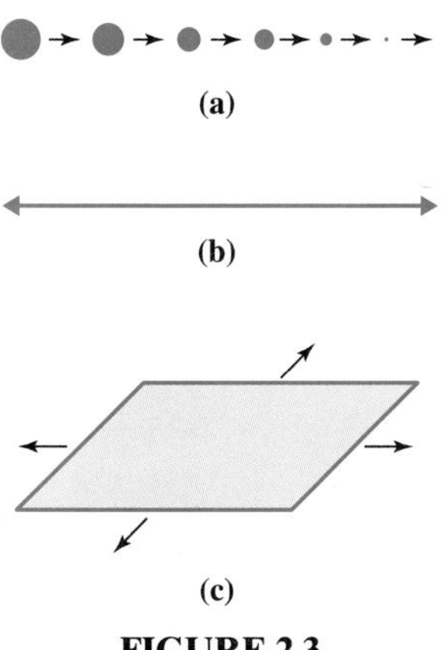

FIGURE 2.3

Undefined terms can be *described* but cannot be given precise definitions using simpler known terms. Three undefined terms in geometry are point, line, and plane. We think of a **point** as a circular dot that is shrunk until it has no size [Figure 2.3(a)]. By a **line**, we can think of a "wire of points" stretched as tightly as possible (hence "straight"), of infinite length (denoted by arrowheads), and having no thickness [Figure 2.3(b)]. Finally, we can think of a **plane** as a "sheet of paper of points" with no thickness, stretched tightly, and extending infinitely in all directions [Figure 2.3(c)]. These descriptions provide a way for us to visualize points, lines, and planes, but they are *not* definitions. We *define* **space** to be the set of all points. Any collection of points is called a **geometric figure**. In particular, lines and planes are composed of points, and hence are geometric figures.

Postulates are statements that we *assume* to be true. Postulates state relationships among defined and undefined terms. The purpose of stating postulates is to establish some first principles upon which a geometry is based. This technique of developing a system, such as geometry, by beginning with a few first principles was perhaps the most significant achievement attributed to Euclid. After introducing postulates and some definitions, we may *deduce* new results. Results that we deduce from undefined terms, definitions, or postulates, and/or results that follow from them are commonly called **theorems**.

Relationships Among Points, Lines, and Planes

Now we begin to examine some of the postulates of geometry that will be assumed in this book.

POSTULATE 2.1

Every line contains at least two distinct points.

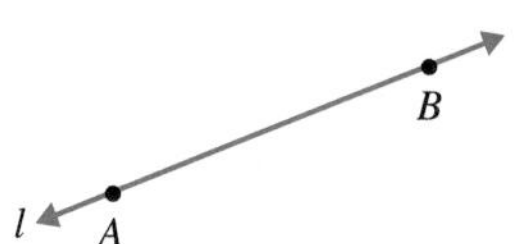

FIGURE 2.4

When a line contains a point, we say that the point is "on" the line. Lines are named by any two of their points together with a double arrow over the letters. For example, $\overleftrightarrow{AB}$ (or $\overleftrightarrow{BA}$) is shown in Figure 2.4. A line also may be named by a single lowercase letter. For example, $\overleftrightarrow{AB}$ in Figure 2.4 can be called line l. Although the preceding postulate states that any line contains at least two points, in fact we will see that a line contains infinitely many points. This result follows from a postulate to be stated soon.

As you may remember from your experience with graphing lines in algebra, two points determine a unique line. The next postulate formalizes this idea.

POSTULATE 2.2

Two points are contained in one and only one line.

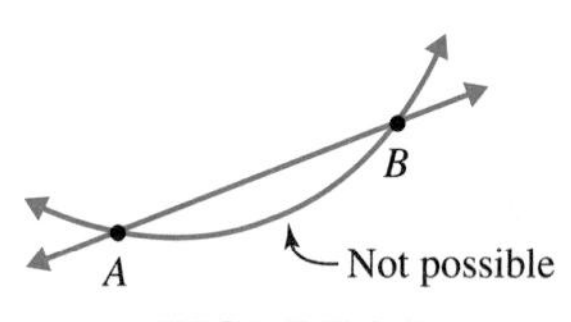

FIGURE 2.5

Postulate 2.2 suggests that all lines are "straight" because otherwise more than one line could be drawn through two points (Figure 2.5). Also, because we can imagine two points arbitrarily far apart, this postulate suggests that lines are infinite in length.

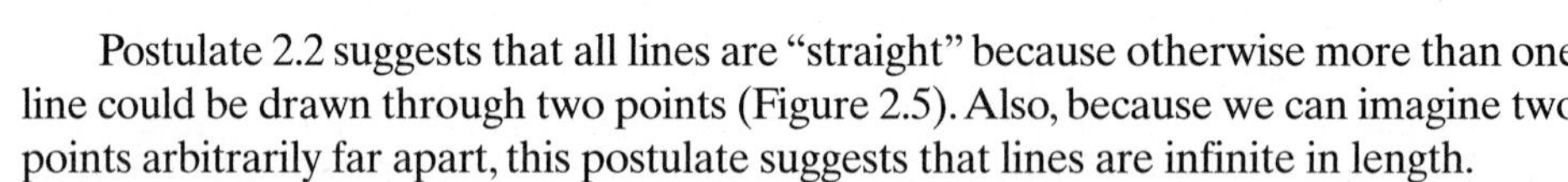

The next postulate, as illustrated in Figure 2.6, suggests a connection between the "straightness" of lines and the "flatness" of planes. That is, just as lines are "straight," planes are "flat."

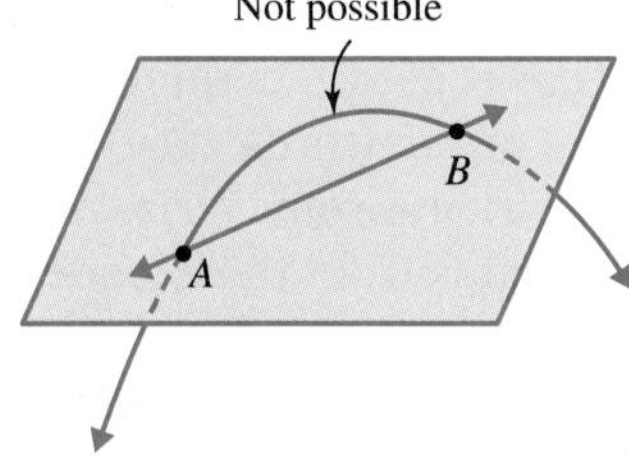

FIGURE 2.6

POSTULATE 2.3

If two points are in a plane, then the line containing these points is also in the plane.

Points are said to be **collinear** if there is a line containing all of the points and **noncollinear** if there is no line that contains all of the points. In Figure 2.7, A, B, and C are collinear. Also A, B, and D are noncollinear because they do not lie on one line. Just as two points determine a unique line, three noncollinear points determine a unique plane. The next postulate states this fact.

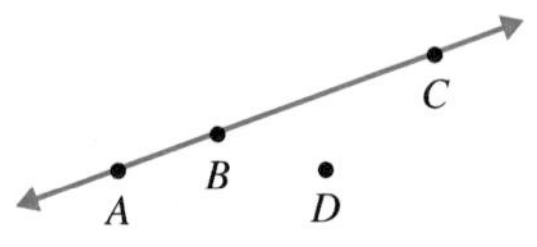

FIGURE 2.7

POSTULATE 2.4

Three noncollinear points are contained in one and only one plane, and every plane contains at least three noncollinear points.

Postulate 2.4 guarantees that a plane consists of more than a single line.

Points are called **coplanar** if there is a plane containing all of the points, and are said to be **noncoplanar** if there is no plane that contains all of the points. In Figure 2.8, for example, D is above the plane containing A, B, and C, so points A, B, and C are coplanar whereas A, B, C, and D are noncoplanar.

The next postulate makes a similar statement about space.

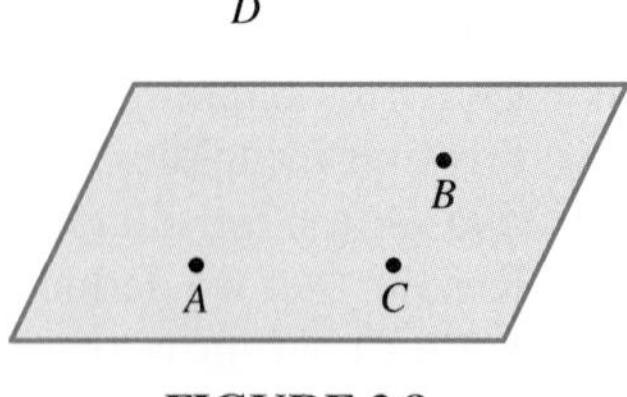

FIGURE 2.8

POSTULATE 2.5

In space, there exist at least four points that are not all coplanar.

Postulate 2.5 assures us that space is not just a single plane; that is, space is three-dimensional.

Line Segments and Their Measure

The next postulate involves the real number line and is the one that will allow us to draw many conclusions about plane geometry as well as allow us to find lengths of line segments.

> **POSTULATE 2.6 The Ruler Postulate**
>
> Every line can be made into an exact copy of the real number line using a 1-1 correspondence.

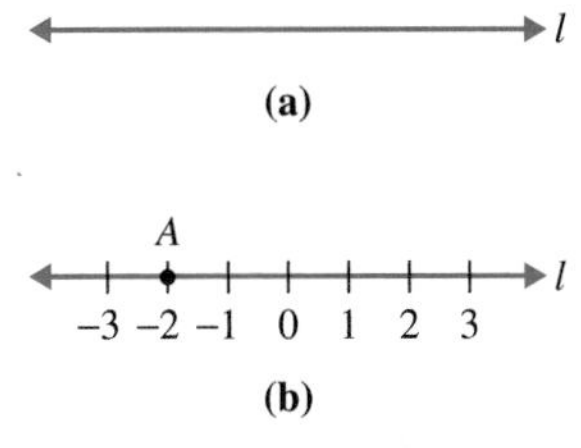

FIGURE 2.9

Figure 2.9(a) shows a line l, and Figure 2.9(b) shows l labeled as a real number line using the Ruler Postulate. The real number associated with a point on a line is called the **coordinate** of that point. For example, in Figure 2.9(b) the coordinate of point A is -2. Only integers are labeled in Figure 2.9(b) because it is impossible to label all the real numbers. Because the set of real number is infinite, Postulate 2.6 implies that every line has infinitely many points. Also, because the real number line is infinite in length, this postulate implies that *all* lines are infinite in length.

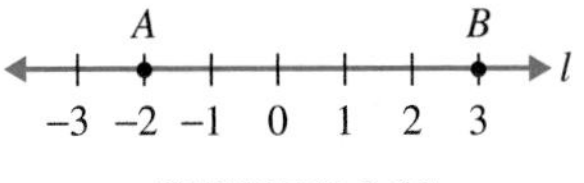

FIGURE 2.10

We measure distances between points using the coordinates of the points. We define the **distance from *A* to *B*** as the nonnegative difference between their coordinates. In Figure 2.10, the distance from A to B is $3 - (-2) = 5$.

FIGURE 2.11

The Ruler Postulate allows us to *define* a line segment as well as to find its length. If A and B are two points on a line l, then the **line segment** determined by A and B, written $\overline{AB}$ or $\overline{BA}$, is points A and B together with all points on the line whose coordinates are between those of A and B (Figure 2.11). Points A and B are called the **endpoints** of the segment $\overline{AB}$. The **length of the line segment** $\overline{AB}$, written as $\boldsymbol{AB}$, is the distance from A to B. In Figure 2.10, $AB = 5$.

Two segments are said to be **congruent** if they have the same length. When the segments $\overline{AB}$ and $\overline{CD}$ are congruent, we write $\boldsymbol{\overline{AB} \cong \overline{CD}}$, which is read "$\overline{AB}$ is **congruent** to $\overline{CD}$," or we write $AB = CD$ to mean that the distance from A to B equals the distance from C to D. We indicate congruent line segments using marks as illustrated in Figure 2.12(a). If more than one pair of segments is congruent, we use double or triple marks [Figure 2.12(b)].

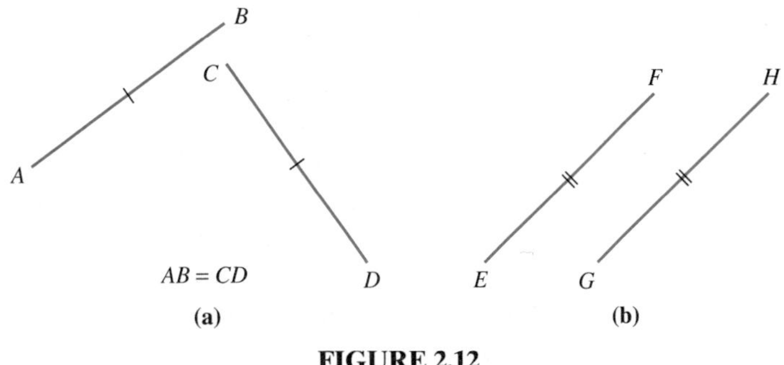

FIGURE 2.12

FIGURE 2.13

A portion of a line that has one endpoint and extends indefinitely in one direction is called a **ray** (Figure 2.13). More precisely, if points A and B are on a line l, then ray $\boldsymbol{\overrightarrow{AB}}$ consists of the points A and B and all points X on l whose coordinates satisfy one of the following: Either (1) the coordinate of X is between the coordinates A and B, or (2) the coordinate of B is between the coordinates of A and X.

Angles and Their Measure

In addition to measuring lengths, it is important to be able to measure regions, such as plots of land or walls of a building. Such regions are often bounded by shapes composed of line segments (Figure 2.14).

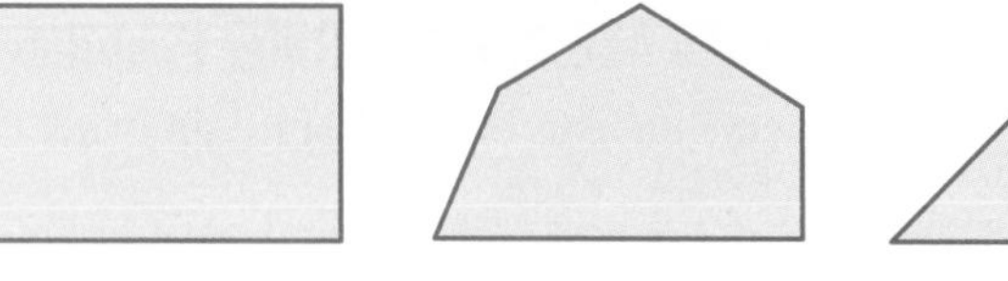

FIGURE 2.14

The sides of these shapes meet to form corners of various sizes. Two line segments or rays meeting at a common endpoint form an **angle**. The common endpoint is called the **vertex** (plural **vertices**) of the angle and the segments or rays are called the **sides** of the angle. Angles are usually named using the angle symbol "$\angle$" and three points: a point on one side, then the vertex, then a point on the other side. Figure 2.15(a) shows $\angle ABC$ (or $\angle CBA$) and $\angle DEF$ (or $\angle FED$).

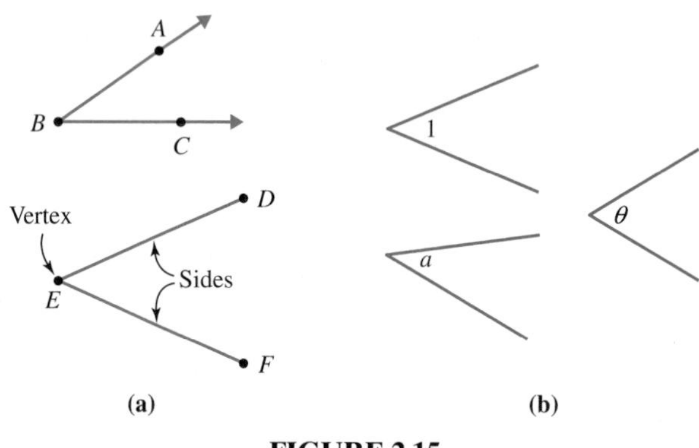

FIGURE 2.15

Points B and E are the vertices of $\angle ABC$ and $\angle DEF$, respectively. Rays $\overrightarrow{BA}$ and $\overrightarrow{BC}$ are the sides of $\angle ABC$, and $\overline{ED}$ and $\overline{EF}$ are the sides of $\angle DEF$. When the meaning is clear, we may name an angle using the letter at its vertex. For example, $\angle ABC$ may be called $\angle B$. We may also name angles by numbers, lowercase letters, or Greek letters, as illustrated in Figure 2.15(b).

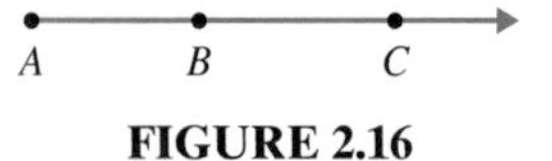

FIGURE 2.16

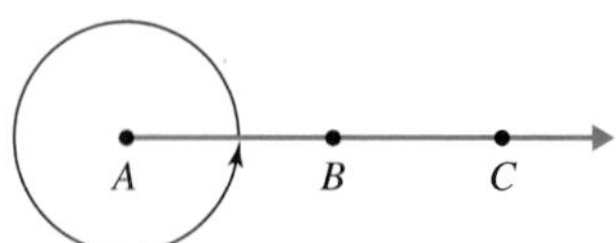

FIGURE 2.17

Angles are usually measured in terms of degrees or radians. In this book, we will usually use degree measure. If two rays coincide, such as $\overrightarrow{AB}$ and $\overrightarrow{AC}$ in Figure 2.16, the angle they form has measure 0°. If $\overrightarrow{AC}$ is rotated counterclockwise one complete revolution as shown in Figure 2.17 until it coincides with $\overrightarrow{AB}$ again, we say the measure of this angle is 360°.

The measure of an angle that is 1/360 of a complete revolution is one **degree**, written 1°. Degrees are represented by the small, raised circle "°". Angles can be measured using a **protractor**, such as the one shown in Figure 2.18, together with the following postulate.

> **POSTULATE 2.7 The Protractor Postulate**
>
> If we place one ray of an angle at 0° on a protractor and we place the vertex at the midpoint of the bottom edge, then there is a 1-1 correspondence between all other rays that can serve as the second side of the angle and the real numbers between 0° and 180° inclusive, as indicated by a protractor.

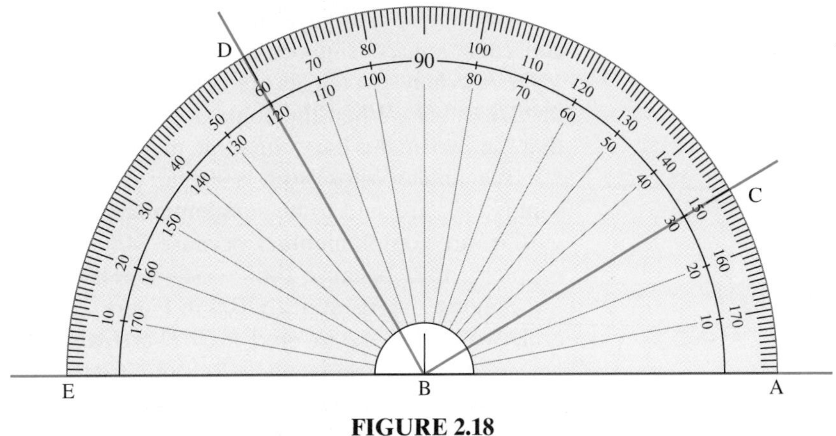

FIGURE 2.18

If we place the vertex of an angle at the center of the bottom of the protractor and we place one ray (or segment) of the angle at 0° at either end of the protractor, then the **measure of the angle** is given by the number that falls on the other ray (or segment). We use the inner numbers on the protractor for angles measured counterclockwise from the right, and we use the outer numbers for angles measured clockwise from the left. For example, in Figure 2.18 the measure of $\angle ABC$ is 30°, written as $m(\angle ABC) = 30°$, or $\angle ABC = 30°$ when the context is clear. The measure of $\angle ABD$ is 120°, and the measure of $\angle EBC$ is 150°. Because the measure of $\angle ABC$ is less than 90°, it is an example of what is called an acute angle. On the other hand, $\angle ABD$, whose measure is greater than 90° but less than 180°, is an example of an obtuse angle. In fact, there are five special classes of angles; they are defined next and illustrated in Figure 2.19.

DEFINITION

Type of Angle	Measure
Acute	Between 0° and 90°
Right	90°
Obtuse	Between 90° and 180°
Straight	180°
Reflex	Between 180° and 360°

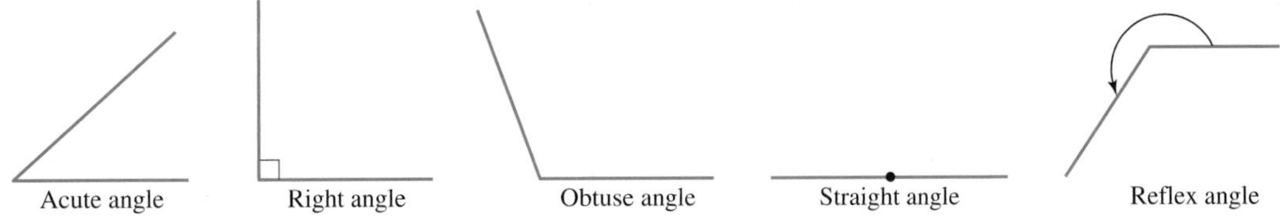

FIGURE 2.19

(NOTE: Unless an arrow is included as shown in the reflex angle, we will assume an angle is not a reflex angle.)

Notice that in Figure 2.19 a right angle is indicated by the "square corner" placed in the right angle. Two lines that form a right angle are said to be **perpendicular**. If lines l and m are perpendicular, we write $\boldsymbol{l \perp m}$.

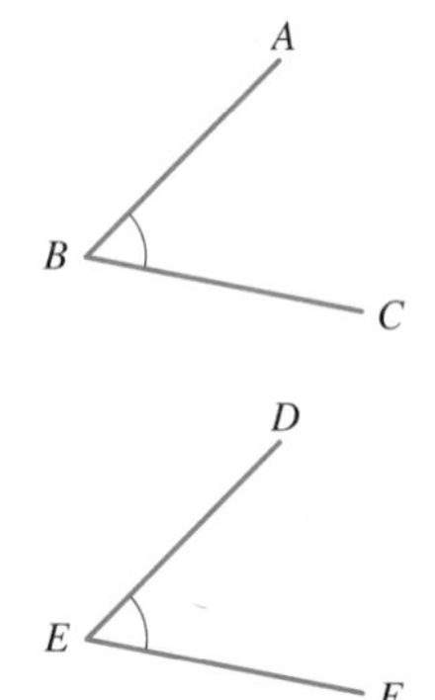

FIGURE 2.20

Two angles that have the same measure are said to be **congruent**. We designate congruent angles using one or more small arcs. In Figure 2.20, $\angle ABC$ is congruent to $\angle DEF$. When two angles are congruent, as in this case, we write $\angle \mathbf{ABC} \cong \angle \mathbf{DEF}$ or $\angle \mathbf{B} \cong \angle \mathbf{E}$. When the context is clear, we also may write $\angle B = \angle E$ to indicate that the two angles have the same measure.

Two angles whose sum is 90° are said to be **complementary**. Two angles whose sum is 180° are said to be **supplementary**. For example, in Figure 2.21, $\angle CBD$ and $\angle BDC$ are complementary because $34° + 56° = 90°$, and $\angle ABC$ and $\angle CBD$ are supplementary because $146° + 34° = 180°$.

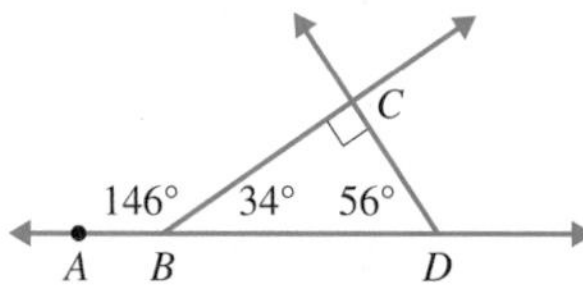

FIGURE 2.21

Angles $\angle ABC$ and $\angle CBD$ in Figure 2.21 are a pair of adjacent angles. In general, two angles $\angle ABC$ and $\angle CBD$ are said to be **adjacent** if they have a common vertex, B, and a common side, $\overline{BC}$, and if $\overline{BC}$ is between $\overline{BA}$ and $\overline{BD}$ (that is, the sum of the measures of $\angle ABC$ and $\angle CBD$ must equal the measure of $\angle ABD$). In Figure 2.22, $\angle POQ$ and $\angle QOR$ are adjacent angles. The angles $\angle POR$ and $\angle QOR$ are *not* adjacent angles because the common side $\overline{OR}$ is *not* between $\overline{OP}$ and $\overline{OQ}$.

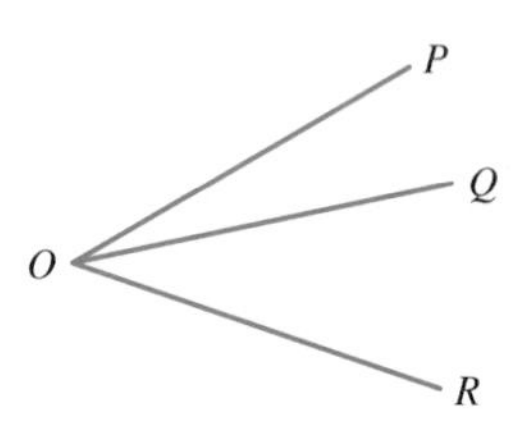

FIGURE 2.22

EXAMPLE 2.1 In Figure 2.23, $\overline{CG} \perp \overline{GE}$, $\angle AGB = 77°$, $\angle CGD = 54°$, and $\angle FGC = 157°$. Use this information to do the following.

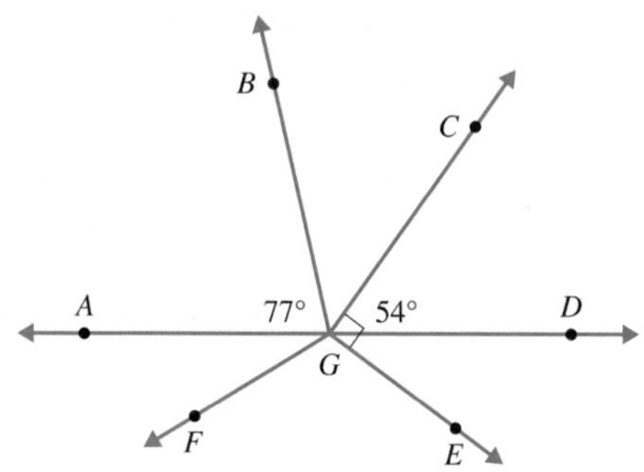

FIGURE 2.23

a. Determine the measures of $\angle DGE$, $\angle BGC$, $\angle AGF$, and $\angle FGE$.

b. Name one pair of complementary angles.

c. Name four pairs of supplementary angles.

d. Name four pairs of adjacent angles.

SOLUTION

a. **i.** Find $\angle DGE$: Because $\overline{CG} \perp \overline{GE}$, we know $\angle CGE = 90°$. In addition, $\angle CGD + \angle DGE = \angle CGE$. So we know that $54° + \angle DGE = 90°$, or $\angle DGE = 90° - 54° = 36°$.

ii. Find $\angle BGC$: Because $\angle AGD$ is a straight angle, we know that

$$\begin{aligned} \angle AGB + \angle BGC + \angle CGD &= 180°, \text{so} \\ 77° + \angle BGC + 54° &= 180° \\ \angle BGC &= 180° - 131° \\ \angle BGC &= 49° \end{aligned}$$

iii. Find $\angle AGF$: Because $\angle FGC$ was given as 157°, we know that

$$\begin{aligned} \angle AGF + \angle AGB + \angle BGC &= 157°, \text{so} \\ \angle AGF + 77° + 49° &= 157° \\ \angle AGF &= 157° - 126° \\ \angle AGF &= 31° \end{aligned}$$

iv. Find $\angle FGE$: Because the sum of the measures of all six nonoverlapping angles is 360°, we can find the measure of the remaining angle by subtracting from 360°.

$$\angle FGE = 360° - 77° - 49° - 54° - 36° - 31° = 113°$$

b. Complementary angles are angles the sum of whose measures is 90°. Because $\angle CGE = 90°$, $\angle CGD$ and $\angle DGE$ are complementary.

c. Supplementary angles are angles the sum of whose measures is 180°. Because $\angle AGD = 180°$, we can say that the following pairs of angles are supplementary: $\angle AGB$ and $\angle BGD$, $\angle AGC$ and $\angle CGD$, $\angle AGF$ and $\angle FGD$, and $\angle AGE$ and $\angle EGD$.

d. Adjacent angles share a common side and vertex, but do not overlap. Four examples of adjacent angles in the figure are $\angle AGB$ and $\angle BGC$, $\angle BGC$ and $\angle CGD$, $\angle AGE$ and $\angle EGC$, and $\angle EGF$ and $\angle FGC$. There are other pairs that also could have been listed. (NOTE: All of the pairs of angles in part (c) are also adjacent angles.) ■

In surveying, navigation, and other applications of angle measure, we must often measure angles more precisely than to the nearest degree. In those cases, we may measure angles to the nearest tenth or hundredth of a degree, or in minutes and seconds, where 60 minutes = 1 degree, and 60 seconds = 1 minute. We denote minutes by a single tick mark (for example, $45' = 45$ minutes), and seconds by a double tick mark (for example, $45'' = 45$ seconds). So we might write an angle measure as $32°15'40''$. We could also write this same angle measure as 32° plus a decimal fraction of a degree as the next example shows.

EXAMPLE 2.2 Perform the following conversions.

a. Express the angle measure $32°15'40''$ in degrees plus a decimal fraction of a degree. Round to the nearest thousandth of a degree.

b. Express the angle measure 75.3° in degrees and minutes.

SOLUTION

a. We first consider how to convert $15'$ to a decimal part of a degree. Because we know that 60 minutes = 1 degree, we know that each minute is $\frac{1}{60}$ of a degree. Thus, $15' = \frac{15}{60}$ of a degree.

As you convert $40''$ to a decimal part of a degree, remember that 60 seconds = 1 minute, so 1 second $= \frac{1}{60}$ of a minute $= \frac{1}{60} \cdot \frac{1}{60} = \frac{1}{3600}$ of a degree. Thus, $40''$ is equivalent to $\frac{40}{3600}$ of a degree.

Now we can perform the necessary conversion.

$$32°15'40'' = 32 + \frac{15}{60} + \frac{40}{3600} \approx 32 + 0.25 + 0.011 = 32.261°$$

b. Because 1 degree = 60 minutes, we know that $0.3° = 0.3(60') = 18'$. Thus, an angle measure of 75.3° is equivalent to $75°18'$. ■

Solution of Applied Problem

Because there are 32 equally spaced directions on the compass card shown, the points are $360°/32 = 11.25°$ apart. Therefore, the angle from WNW to N is $6(11.25°) = 67.5°$, and the angle from N to NE is $4(11.25) = 45°$. Thus, the ship has turned through an angle of $67.5° + 45° = 112.5°$.

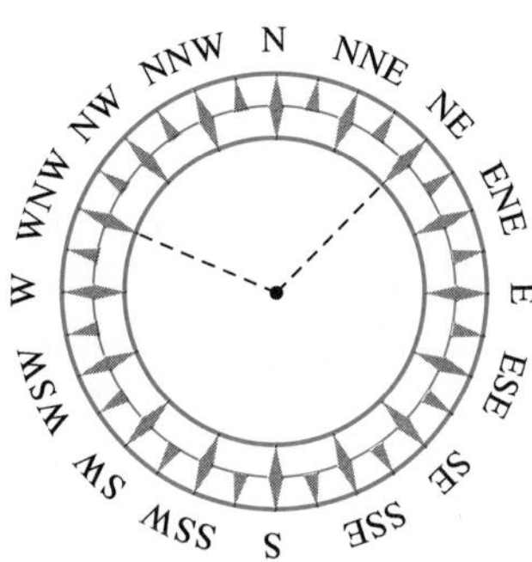

GEOMETRY AROUND US Cameras and artists' canvases are often set upon some form of tripod. One explanation for the use of a tripod is that three points determine one and only one plane. This means that a three-legged stand will be more stable than one with four or more legs even if the floor is not perfectly level. In fact, when dairy barns had dirt floors, milking stools were three-legged for this reason.

PROBLEM SET 2.1

EXERCISES/PROBLEMS

Exercises 1–6 refer to the following figure.

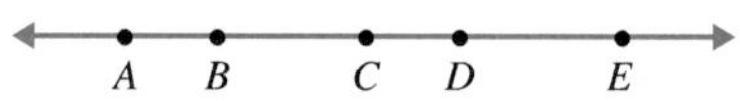

1. How many different line segments appear in the figure? Name them.
2. How many different line segments have B as one endpoint? Name them.
3. How many different rays appear in the figure? Name them.
4. How many different rays have C as their initial point? Name them.
5. Name the line $\overleftrightarrow{AE}$ in nine other ways.
6. Name the ray $\overrightarrow{EA}$ in three other ways.
7. Points P, Q, R, and S are located on line l. Their corresponding coordinates are $P = -3.78$, $Q = -1.35$, $R = 0.56$, and $S = 2.87$. Find each of the following distances.

   a. PR **b.** RQ **c.** PS **d.** QS

   P Q R S
   −4 −3 −2 −1 0 1 2 3 4 l

8. Points A, B, C, and D are located on line l. Their corresponding coordinates are $A = -5\sqrt{2}$, $B = -\sqrt{2}$, $C = \sqrt{2}$, and $D = 4\sqrt{2}$. Find each of the following distances.

   a. AC **b.** BD **c.** BC **d.** AD

   A B C D
   −8 −6 −4 −2 0 2 4 6 8 l

9. **a.** Three different angles are pictured in the following figure. Name each of them in two different ways.

b. Name two adjacent angles in the figure.

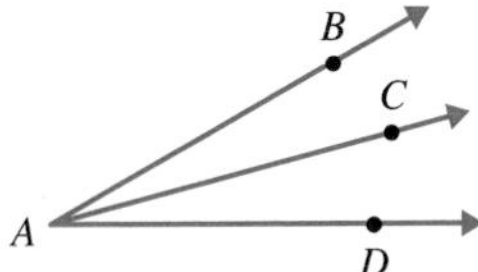

10. **a.** Three different angles in the following figure have Q as a vertex. Name each of them in two different ways.

b. Name two pairs of adjacent angles in the figure.

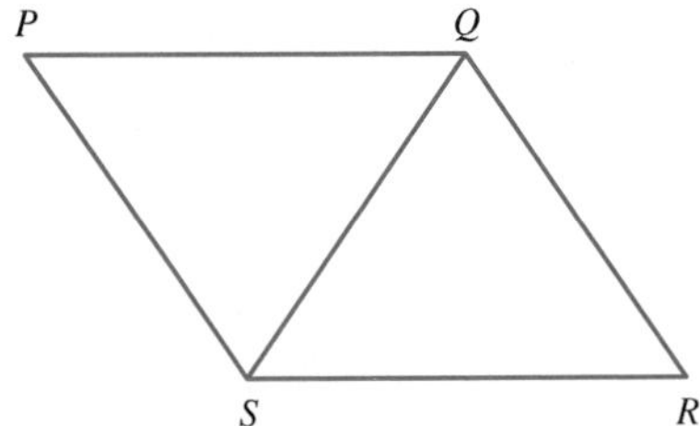

11. **a.** How many different angles less than 180° are shown in the following figure?

b. How many of them are obtuse?

c. How many of them are acute?

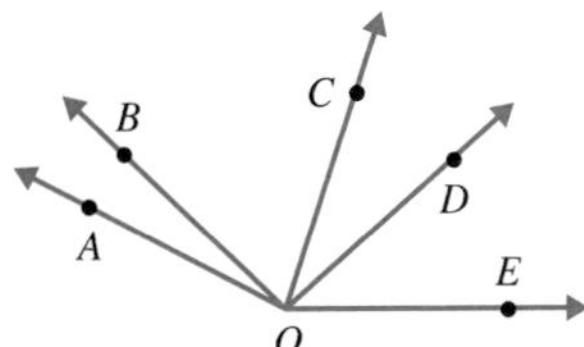

12. **a.** How many different angles are shown in the following figure, where T, S, and W are collinear?

b. How many of them are obtuse?

c. How many pairs of angles are supplementary? Name them.

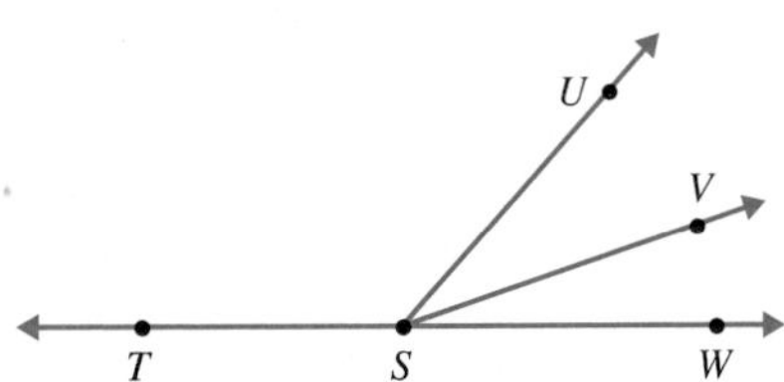

13. Use your protractor to measure the following angles. Classify each of them as acute, right, or obtuse. If necessary, extend the sides of an angle using a straightedge so that you can measure the angle more accurately.

a.

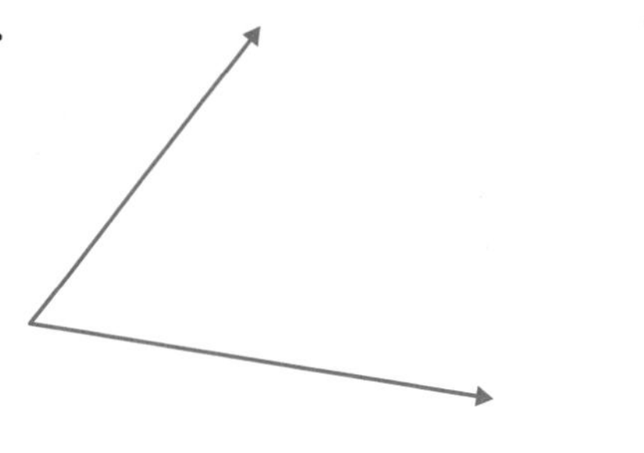

b.

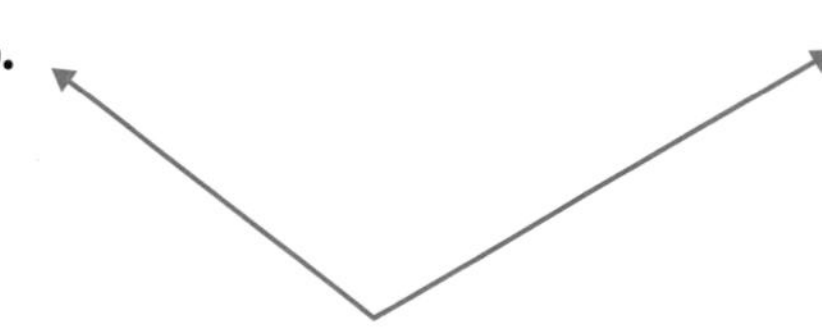

c.

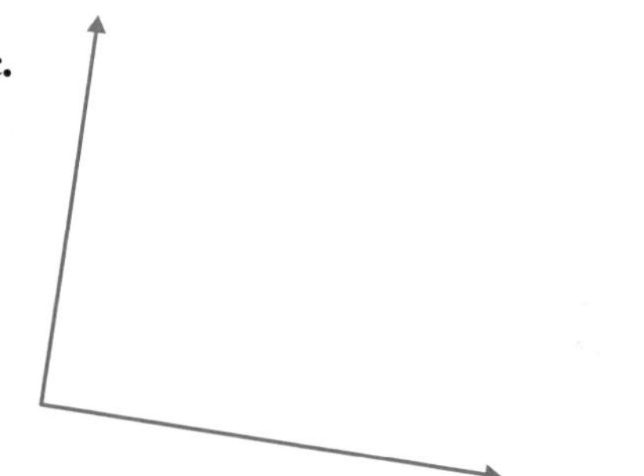

14. Use your protractor to measure the following angles. Classify each of them as acute, right, or obtuse. If necessary, extend the sides of an angle.

a.

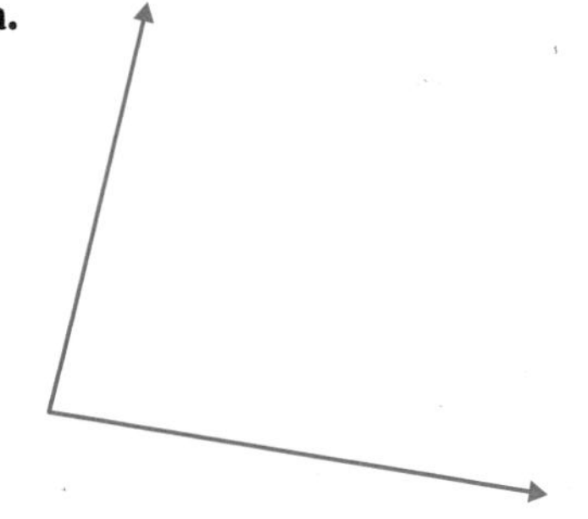

b.

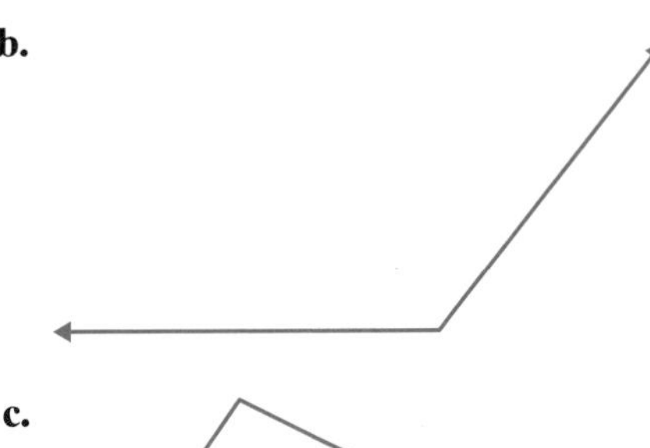

c.

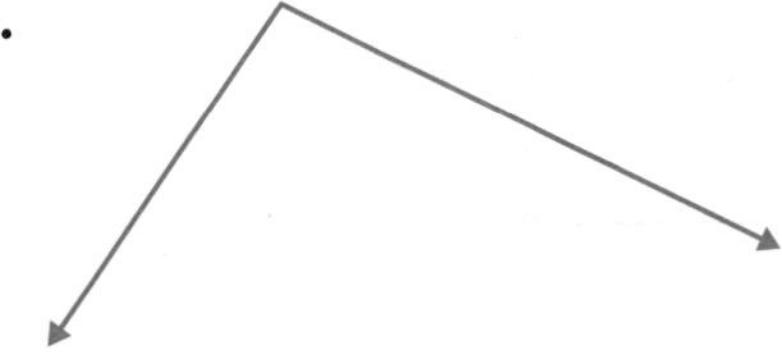

15. In the figure, $\angle BFC = 55°$, $\angle AFD = 150°$, $\angle BFE = 120°$ and $\angle AFE = 180°$. Determine the measures of $\angle AFB$ and $\angle CFD$ without using a protractor.

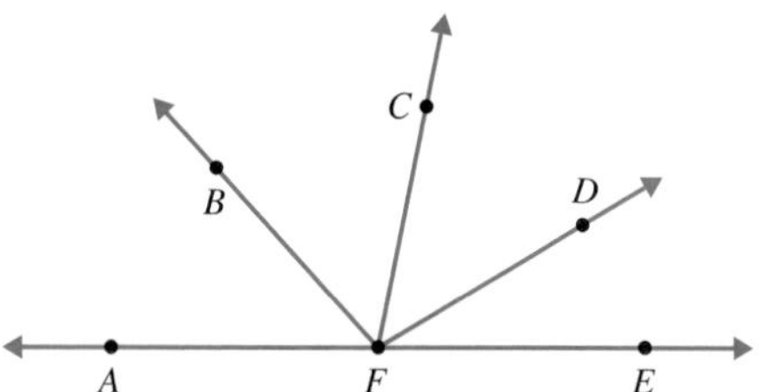

16. In the following figure $\overline{AO}$ is perpendicular to $\overline{CO}$. If $\angle AOD = 150°$ and $\angle BOD = 72°$, determine the measures of $\angle AOB$ and $\angle BOC$ without using a protractor.

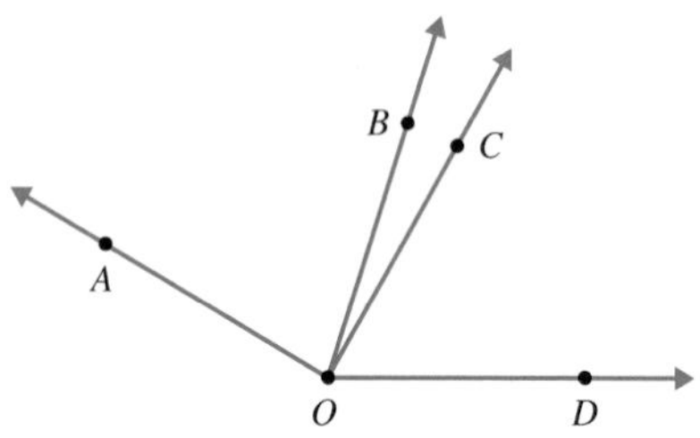

17. In the following figure, the measure of $\angle 1$ is 9° less than half the measure of $\angle 2$. Find the measures of $\angle 1$ and $\angle 2$.

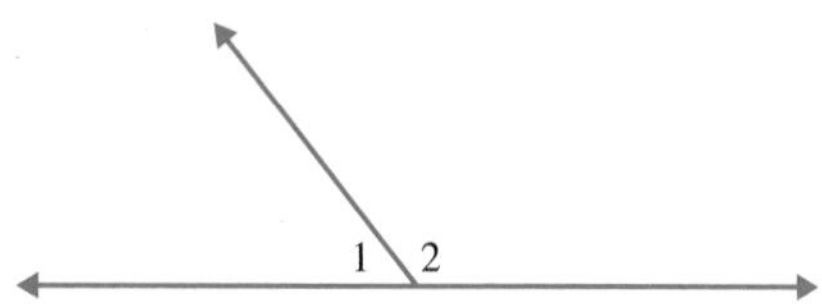

18. In the following figure, the measure of $\angle 1$ is 14° less than twice the measure of $\angle 2$. Find the measures of $\angle 1$ and $\angle 2$. Round your answers to the nearest hundredth of a degree.

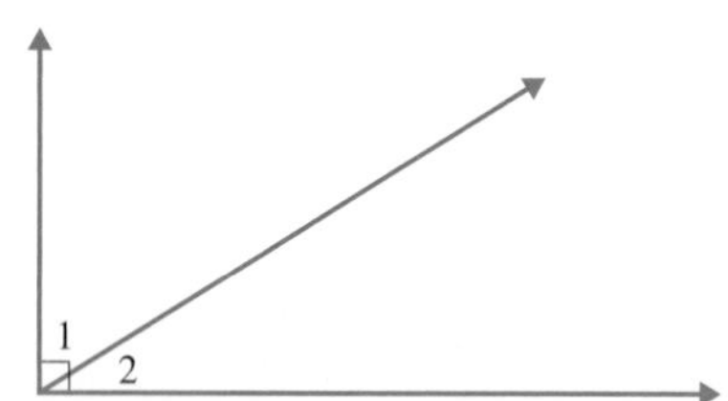

19. If $\angle A$ and $\angle B$ are complementary but $\angle A$ is four times as large as $\angle B$, find the measure of $\angle A$.

20. The measure of $\angle X$ is 9° more than twice the measure of $\angle Y$. If $\angle X$ and $\angle Y$ are supplementary angles, find the measure of $\angle X$.

21. Two lines drawn in a plane separate the plane into three different regions if the lines are parallel.

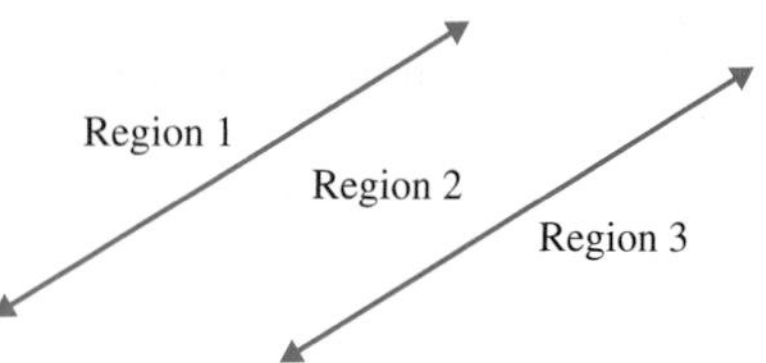

If the lines are intersecting, then they will divide the plane into four regions.

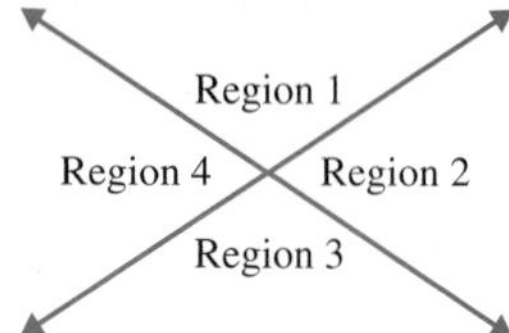

Thus, the greatest number of regions into which two lines can divide a plane is four. Determine the greatest number of regions into which a plane can be divided by 3 lines, 4 lines, 5 lines, and 10 lines. Generalize to n lines.

22. Suppose three points are located in a plane. There is one line that can be drawn through them if the points are collinear.

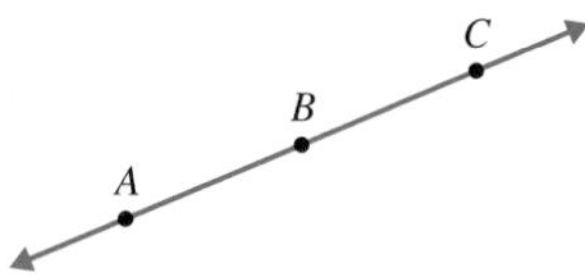

If the three points are noncollinear, then there are three lines that can be drawn through pairs of points.

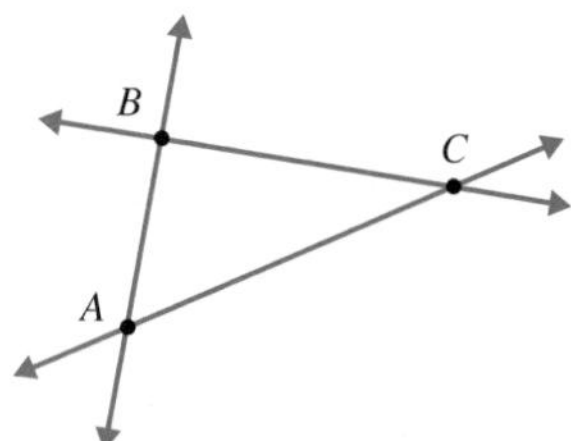

Therefore, for three points, three is the greatest number of lines that can be drawn through pairs of points. Determine the greatest number of lines that can be drawn for four points, five points, and six points. Generalize to n points.

APPLICATIONS

23. Convert each of the following angle measures to degrees and decimal fractions of a degree. Round to the nearest thousandth of a degree if necessary.

a. $19° 3'$ **b.** $12° 6' 36''$

c. $247° 56'$ **d.** $3° 31' 58''$

24. Convert each of the following angle measures to degrees and decimal fractions of a degree. Round to the nearest thousandth of a degree if necessary.

a. $189° 51'$ **b.** $220° 36' 54''$

c. $95° 34'$ **d.** $353° 25' 49''$

25. Convert each of the following angle measures in degrees to degrees and minutes. Round to the nearest minute if necessary.

a. $31.6°$ **b.** $95.75°$ **c.** $241.32°$ **d.** $25.48°$

26. Convert each of the following angle measures in degrees to degrees and minutes. Round to the nearest minute if necessary.

a. $175.95°$ **b.** $11.4°$ **c.** $87.19°$ **d.** $270.66°$

27. Convert each of the following angle measures in degrees to degrees, minutes, and seconds. Round to the nearest second if necessary.

a. $16.51°$ **b.** $0.33°$ **c.** $91.993°$ **d.** $58.029°$

28. Convert each of the following angle measures in degrees to degrees, minutes, and seconds. Round to the nearest second if necessary.

a. $15.73°$ **b.** $2.16°$ **c.** $112.222°$ **d.** $244.326°$

29. In surveying, directions are usually specified in terms of bearings. The **bearing** of a line segment is the acute angle that the line segment makes with a north-south line. The angle and the quadrant are specified. For example, the bearing of $\overline{AB}$ is N 20° E. We also could say that the bearing of $\overline{BA}$ is S 20° W. (In this case, imagine placing B at the origin rather than A.)

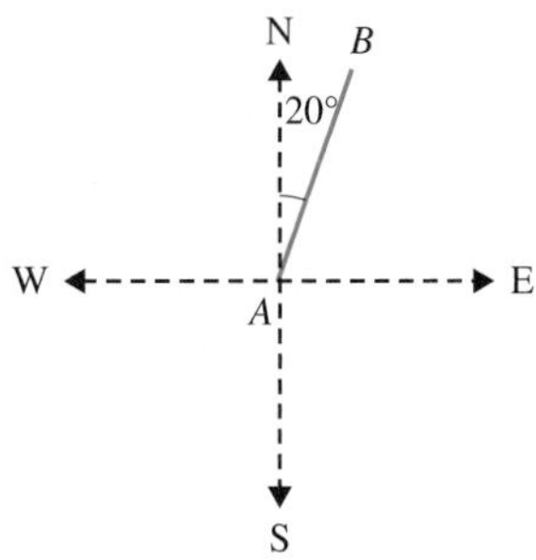

Sketch $\overline{AB}$ with the given bearing. Also give the bearing of $\overline{BA}$.

a. N 17° W **b.** S 48° E

c. N 78° E **d.** S 65° W

30. An air traffic control tower is located at the origin of the following figure. The locations of four planes A, B, C, and D are indicated by dots. For each plane, give the bearing from the control tower. Also give the bearing of the control tower from each plane.

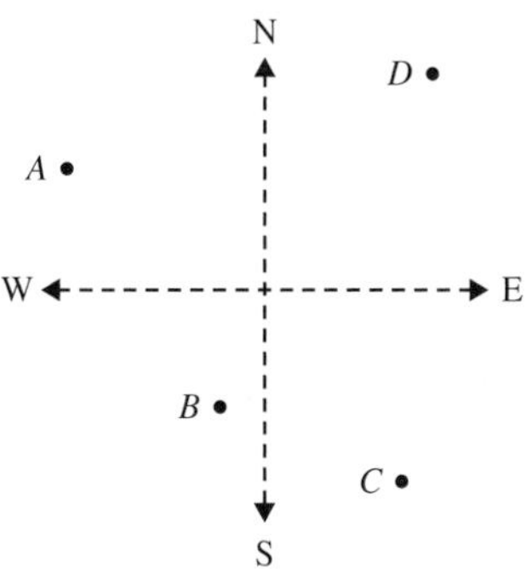

31. An airplane flies due north a certain number of miles. The plane then flies due east three times as far. Find the approximate bearing of the plane from its starting point. Draw a diagram and measure the angle with a protractor.

32. A child leaves home and arrives at school after walking six blocks due south and eight blocks due west. Find the approximate bearing of the child's home from the school. Draw a diagram and measure the angle with a protractor.

33. The **azimuth** of a line segment is the angle that the line segment makes with a north-south line. It is measured from 0° to 360° in a clockwise direction from the north. In the following example, the azimuth of $\overline{AB}$ is 250°.

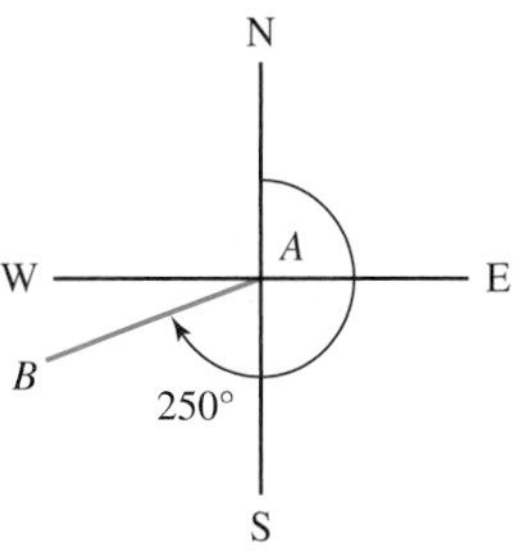

Sketch $\overline{AB}$ with the following azimuths.

a. $115°$ **b.** $225°$ **c.** $72°$ **d.** $329°25'$

34. For any location on the earth, the apparent position of the sun at any time can be given by its azimuth (see problem 33) and its altitude. The altitude is the angle of the sun measured up from the horizon. Because the sun is at the horizon at sunrise or sunset, we can ignore altitude at these times of day and use just azimuth to locate the sun. Use a circle as shown to represent the horizon and locate the apparent positions of the sun at sunrise and sunset for the following cities on July 1, 2006.

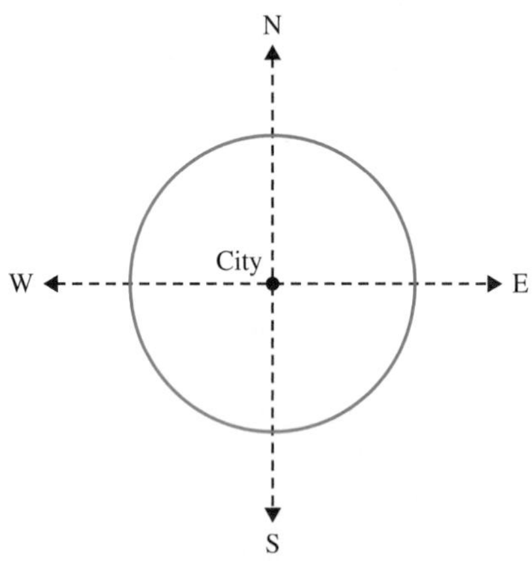

a. Oslo, Norway: sunrise at 3:03 AM, azimuth 36.7°; sunset at 9:38 PM, azimuth 323.1°

b. Cuiaba, Brazil: sunrise at 6:07 AM, azimuth 66.8°; sunset at 5:48 PM, azimuth 293.2°

c. Ushuaia, Tierra del Fuego: sunrise at 9:00 AM, azimuth 48.1°; sunset at 4:14 PM, azimuth 311.7°

35. Convert the following bearings to azimuths.

a. N 40° E **b.** N 40° W

c. N 82°17′ W **d.** S 24°35′ E

36. Convert the following azimuths to bearings.

a. 163° **b.** 241°

c. 123°48′ **d.** 329°25′

37. Determine the measure of $\angle ABC$ in the following figure. The bearings of $\overline{AB}$ and $\overline{BC}$ are indicated.

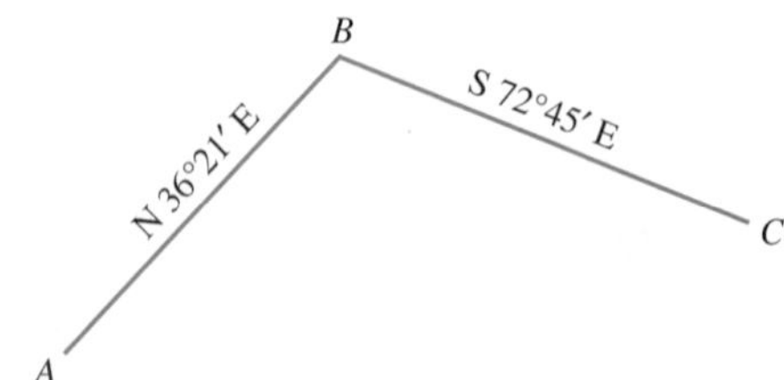

38. Given the bearings of $\overline{AB}$ and $\overline{BC}$ as shown next, determine the measure $\angle ABC$. (NOTE: Be careful with the order of the points here. The bearing of $\overline{AB}$ is not the same as the bearing of $\overline{BA}$. See problem 29.)

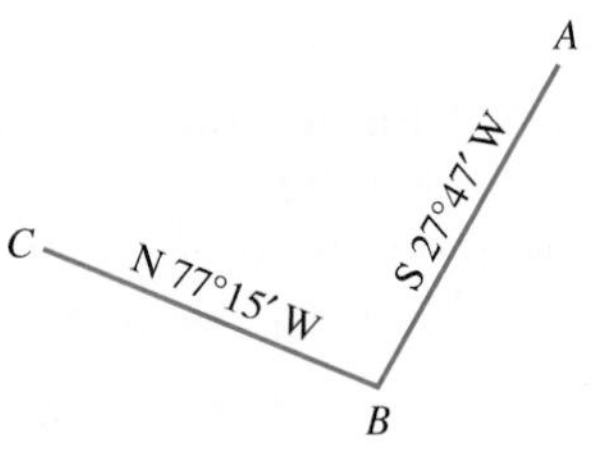

39. A standard clock face is divided into 12 hours or 60 minutes.

a. After one hour, through what angle has the hour hand rotated? The minute hand?

b. After one minute, through what angle has the minute hand rotated? The hour hand?

c. Determine the measure of the obtuse angle formed by the hour and minute hands at 9:10 and 5:56.

d. During what hour and minute after midnight will the hour and minute hands first form a right angle?

40. Bézier curves are smooth curves specified by a small set of values and are important in computer graphics. Pierre Bézier, who worked in the automotive industry in France, introduced them in the 1960s. You can create an approximation to a Bézier curve by using an angle and several lines. Begin by using a straightedge to draw any $\angle ABC$. Mark off 10 equal lengths along $\overline{BA}$ labeling the points 1 to 10 from B toward A. Using the same length, mark off 10 equal lengths along $\overline{BC}$ labeling the points 10 to 1 from B toward C. Use a straightedge to connect each pair of points labeled with the same number. The "ghost" of a curve will appear inside the angle; this is the approximation to a Bézier curve. Use the method just described to create approximations to Bézier curves by starting with a 45° angle, a 90° angle, and a 120° angle.

41. You can separate white light into a full spectrum of colors using a triangular prism. Upon entering the prism, each color that makes up white light is bent or refracted by an angle θ_r. The angle of refraction depends on the angle of incidence θ_i, which is the angle at which light enters the prism measured from a perpendicular line drawn from the side of the prism. (See the following figure.) The angle of emergence θ_e is measured from a perpendicular line drawn at the point where the light ray exits the prism. The angles of refraction and emergence for three spectrum colors are given in the following table, assuming an angle of incidence of 70° and a prism made of diamond. Use a straightedge and a protractor to construct a large diagram similar to the one given. Your diagram will show the prism with a 45° angle, the path of white light as it enters the prism, three refracted rays, and three emergent rays. Add color to your final diagram.

Color	θ_r	θ_e
Red	23.0°	64.5°
Yellow	22.9°	65.5°
Violet	22.5°	69.6°

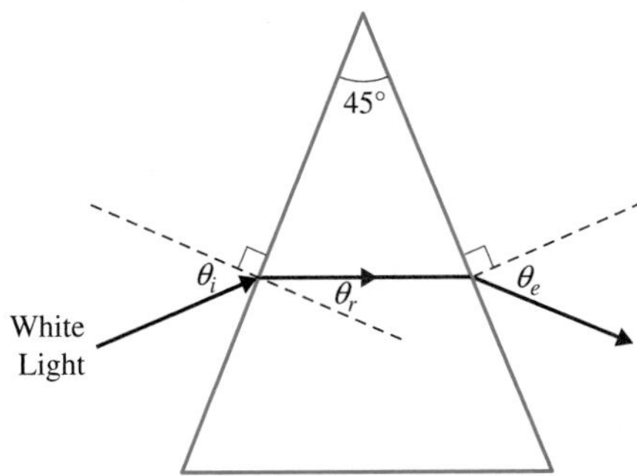

42. We have seen that angles can be measured in degrees or in degrees, minutes, and seconds. Another unit of angle measure that is especially useful in calculus is the **radian**. To define one radian, consider an angle situated so that its vertex is at the center of a circle. Such an angle is called a **central angle**. One radian is the measure of the central angle, θ, that cuts off a portion of the distance around the circle that is the same length as the radius of the circle. The figure shows a central angle θ with a measure of 1 radian.

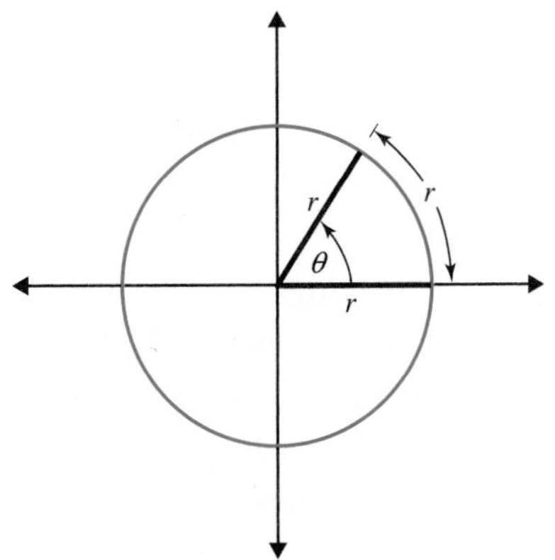

The radian measure of one full revolution is 2π, so $2\pi = 360°$. It follows that an angle with a measure of π radians corresponds to 180°, one-half of a revolution.

a. Use a compass to draw a circle such as the one shown centered on a set of axes. Use a protractor to measure and mark the following angles: 30°, 45°, 60°, 90°, 135°, 180°, 225°, 270°, 315°, and 360°. For each angle marked in degrees, identify the radian measure and express it as a fraction in lowest terms.

b. Two angles are complementary. One of them has a measure of $\frac{\pi}{7}$ radians. Find the measure of the other angle in radians.

c. One angle measure is twice another. The angles are supplementary. Find the measures of the two angles expressed in radians.

EXTENDED PROBLEMS

43. Sailors have used celestial navigation for centuries to find their way across the oceans. For an observer on earth, the position of an object clockwise from true north and above the horizon can be measured by its azimuth and its altitude (see problems 33 and 34). Research celestial navigation and write a summary explaining how fixed stars are used. Why is Polaris important and what is its azimuth? Locate Polaris in the night sky and estimate its altitude. Locate a star pattern such as the Big Dipper and estimate the altitude and azimuth of each star in the pattern. Use the fact that the width of your thumbnail held at arm's length is approximately 1°, and the width of your palm is approximately 10°. Create a diagram using your altitude and azimuth estimates that shows the horizon, Polaris, and each star in the pattern.

44. **Orienteering** is the sport of land navigation through a series of control points with maps and a compass. Research the sport of orienteering and write a summary of your findings. In which country did it originate and for what purpose? When did orienteering become a sport? What type of map is typically used? Print a sample orienteering map from the Defense Mapping Agency Hydrographic Topographic Center. Select a starting point, at least six control points, and a finish point. Give the bearing and distance from each control point to the next.

45. The **semaphore flag signaling system** is a communication system based on the positions of a pair of handheld flags. Research the semaphore flag signaling system and print the flag positions that represent the letters of the alphabet. For each letter, the signaler holds his arms in a certain position forming an angle. Suppose the left arm represents the initial side of the angle and the right arm represents the terminal side of the angle. If the angle is measured counterclockwise, for which letters does the angle measure 45°, 90°, 135°, 180°, 225°, 270°, and 315°?

2.2 POLYGONS AND CIRCLES

Applied Problem

A box-end wrench has a six-point opening at one end. What is the name of the polygonal opening, and why would this shape be used?

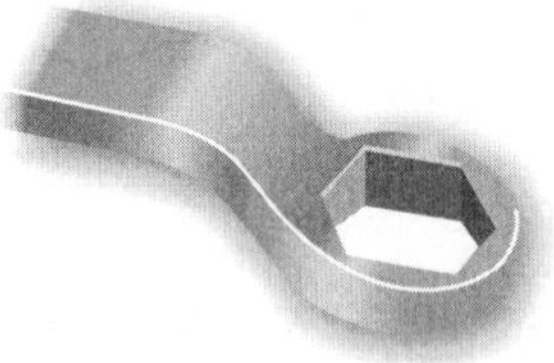

Triangles

Many geometric figures made up of line segments can be described by the lengths of their segments or by the measures of the angles formed. The simplest of these figures is the triangle. Given three *noncollinear* points A, B, and C, the **triangle** determined by these three points, written $\triangle ABC$, is formed by the **sides** $\overline{AB}$, $\overline{AC}$, and $\overline{BC}$ (Figure 2.24). A triangle has three angles and three sides.

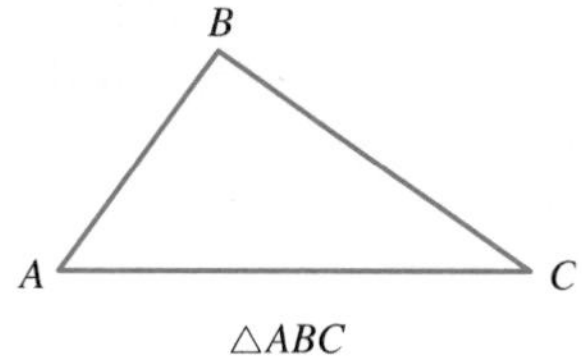

$\triangle ABC$

FIGURE 2.24

We can classify triangles by both the lengths of their sides and the measures of their angles. We have defined many special triangles next. Figure 2.25 shows an example of each of them.

DEFINITION

Types of Triangles	Definition
Equilateral	All three sides are congruent.
Isosceles	At least two sides are congruent.
Scalene	No two sides are congruent.
Acute	All three angles are less than 90°.
Right	One angle is a right angle.
Obtuse	One angle is an obtuse angle.
Equiangular	All three angles are congruent.

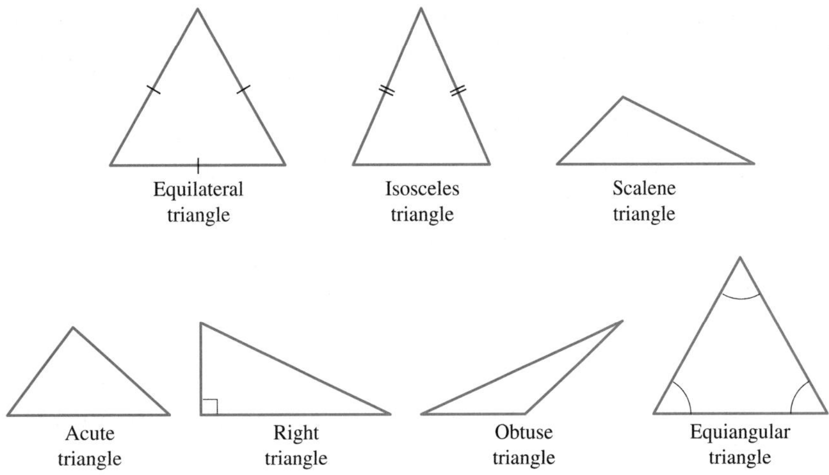

FIGURE 2.25

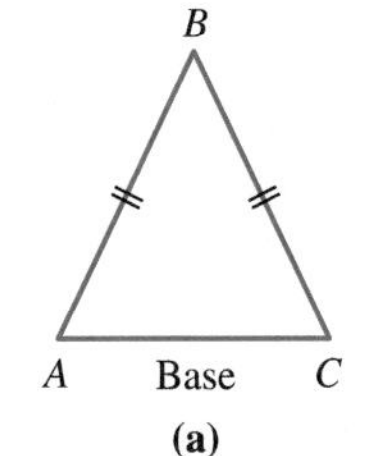

(a)

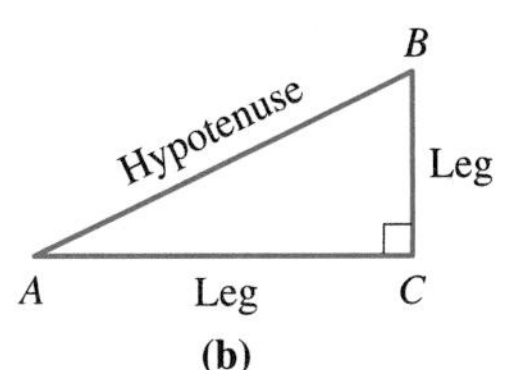

(b)

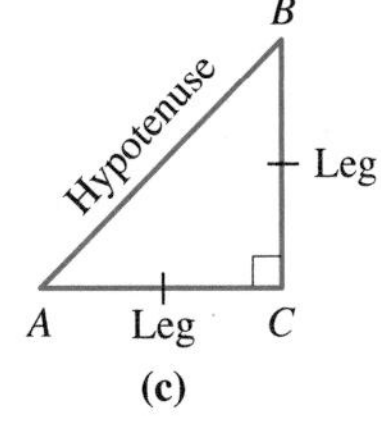

(c)

FIGURE 2.26

In an isosceles triangle, the angles opposite the congruent sides are called **base angles**, and the side between these two angles is usually called the **base** of the triangle. In Figure 2.26(a), the base angles are $\angle A$ and $\angle C$ and the base is $\overline{AC}$. The angle between the two congruent sides is often called the **vertex angle**. In Figure 2.26(a), $\angle B$ is the vertex angle. Notice that because an isosceles triangle has *at least* two sides congruent, any equilateral triangle also can be classified as an isosceles triangle. In a right triangle, the side opposite the right angle is called the **hypotenuse** and the other two, shorter sides are called the **legs** of the triangle. In Figure 2.26(b), $\overline{AC}$ and $\overline{BC}$ are legs and $\overline{AB}$ is the hypotenuse of right triangle $\triangle ABC$.

In general, we use more than one of the terms just defined to describe a triangle. For example, a triangle with two congruent sides and a right angle is called an isosceles right triangle. Figure 2.26(c) shows right isosceles triangle $\triangle ABC$. A triangle with one right angle but with no two sides congruent is called a right scalene triangle [Figure 2.26(b)]. We will show in Chapter 4 that every equilateral triangle is also equiangular, and vice versa.

Simple Closed Curves

Triangles are part of a more general class of geometric figures called simple closed curves. A **simple closed curve** is a figure that lies in a plane and can be traced so that the starting and ending points are the same and no part of the curve is crossed or retraced (Figure 2.27).

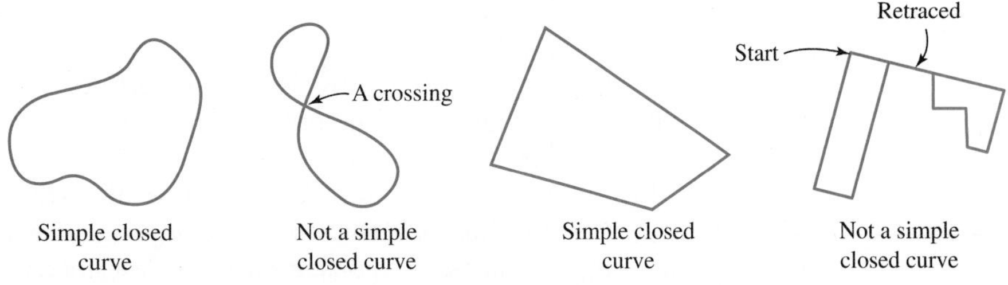

FIGURE 2.27

A **circle** is a simple closed curve that consists of the set of all points equidistant from a given point, called the **center** of the circle. A circle is named by its center. Circle O is pictured in Figure 2.28(a).

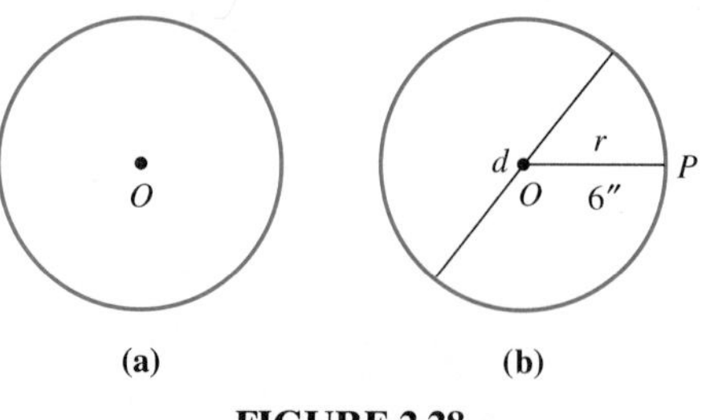

FIGURE 2.28

A line segment joining the center with a point on the **circle** is called a **radius** (plural is **radii**) of the circle; its length is called **the radius** of the circle. A line segment containing the center and with endpoints on the circle is called a **diameter** of the circle; its length is called **the diameter** of the circle. Notice that we use the words radius and diameter to refer to line segments as well as to the lengths of those line segments. In Figure 2.28(b), r (or $\overline{OP}$) is *a* radius and d is *a* diameter when viewed as segments. We also say that *the* radius of circle O is 6 inches and *the* diameter is 12 inches.

Polygons

As you can see in Figure 2.27, a simple closed curve need not be "curved." For example, a triangle is a simple closed curve composed of three line segments. In general, a **polygon** is a simple closed curve composed of line segments. Figure 2.29 shows some examples of polygons.

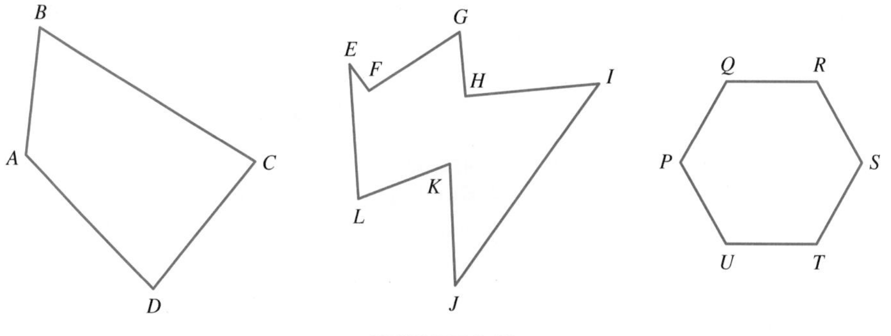

FIGURE 2.29

Polygons are named by listing the vertices in order around the figure. For instance, we might name the polygon in Figure 2.29(a) as $ABCD$, $BCDA$, or $DABC$. Two vertices that are endpoints of a side of a polygon are called **adjacent vertices**. Thus, in Figure 2.29(a), A and B are adjacent vertices, as are B and C. Two sides of a polygon that share a vertex are called **adjacent sides**. In Figure 2.29(a), $\overline{AD}$ and $\overline{DC}$ are adjacent sides. Polygons are usually described by the number of sides they contain. We have listed some names of common polygons next.

DEFINITION

Type of Polygon	Number of Sides
Triangle	3
Quadrilateral	4
Pentagon	5
Hexagon	6
Octagon	8
Decagon	10
***n*-gon**	n

By these definitions, in Figure 2.29, $ABCD$ is a quadrilateral, $EFGHIJKL$ is an octagon, and $PQRSTU$ is a hexagon. Notice that in a polygon, the number of vertices is the same as the number of sides.

Quadrilaterals

Many quadrilaterals are named by the lengths of their sides, the measures of their angles, or some other attribute such as parallelism. We say that two lines are **parallel** if they are in the same plane and do not intersect. Two line segments are parallel if the lines containing the segments are parallel. Parallel lines are identified using one or more arrowheads in the middle of the lines. In Figure 2.30, line l is parallel to line m, and we write $\boldsymbol{l \parallel m}$.

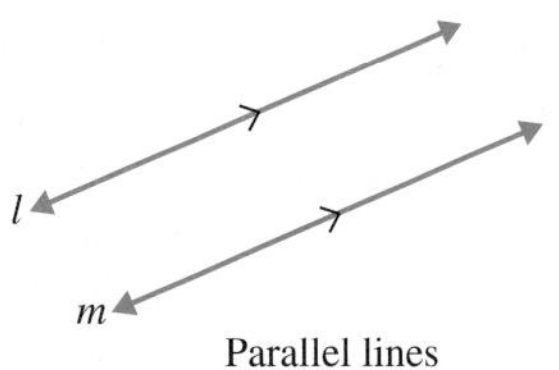

FIGURE 2.30

We describe several types of quadrilaterals next and Figure 2.31 displays an example of each type.

DEFINITION

Type of Quadrilateral	Definition
Square	All sides are congruent and all angles are right angles.
Rectangle	All angles are right angles.
Rhombus	All sides are congruent.
Parallelogram	Opposite sides are parallel.
Trapezoid	Exactly one pair of sides is parallel.
Isosceles trapezoid	A trapezoid in which nonparallel sides are congruent.
Kite	There are two pairs of adjacent sides where the sides in each pair are congruent, but the pairs have no side in common.

The parallel sides in a trapezoid are called **bases** of the trapezoid. The sides that are not parallel are called the **legs** of the trapezoid. Thus, in each trapezoid in Figure 2.31, the bases of the trapezoid are indicated by the parallel line marks. The other two sides are the legs.

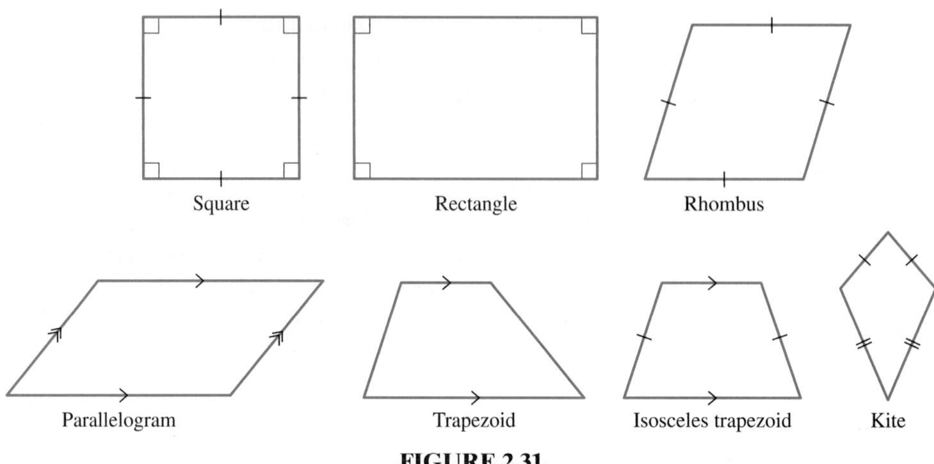

FIGURE 2.31

You may have noticed that there is some overlap in the definitions. For example, every square is also a rectangle because a square has four right angles. We will explore several other such relationships in the problem set and in Chapter 5.

Symmetry

We have seen that we can describe polygons in terms of the number of sides they have or in terms of attributes of their sides and angles. Also, we can use symmetries of various types to describe polygons.

There are two common types of symmetry a geometric figure may possess: reflection symmetry and rotation symmetry. A figure has **reflection symmetry** if there is a line along which the figure may be folded so that one half of the figure matches exactly with the other half. The fold line is called the **line of symmetry** or the **axis of symmetry**.

The isosceles triangle shown in Figure 2.32(a) has reflection symmetry because if we fold it along the axis of symmetry, the two portions of the triangle will match exactly. A figure can have more than one axis of symmetry, as the fold lines for the square in Figure 2.32(b) and the equilateral triangle in Figure 2.32(c) demonstrate. The triangle in Figure 2.32(d) has no line of symmetry. Hence, it has no reflection symmetry.

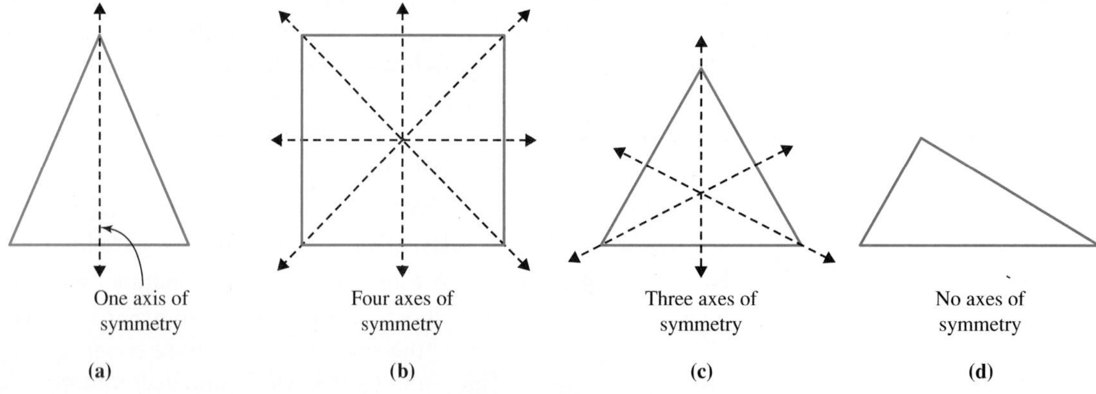

FIGURE 2.32

Because an isosceles triangle has reflection symmetry, we can establish a relationship between its base angles. When the triangle in Figure 2.32(a) is folded along the axis of symmetry, the base angles will match. This means that these two angles have the same measure. We usually state this result as "base angles of an isosceles triangle are congruent."

An equilateral triangle is a special type of isosceles triangle, so it also has base angles congruent. But an equilateral triangle has three axes of symmetry, one through each vertex [Figure 2.32(c)]. Thus, any two angles of the equilateral triangle will have the same measure. This means that an equilateral triangle is also equiangular. We discuss these angle relationships more formally in Chapter 4.

A geometric figure has **rotation symmetry** if the figure can be rotated about a point less than a full turn, so that the image is identical to the original figure. The point is called the **center of rotation symmetry**. The equilateral triangle in Figure 2.33 has rotation symmetry because it can be rotated through an angle of 120° (one third of a full turn) and the image will match exactly with the original triangle. The equilateral triangle also has a rotation symmetry of 240°, so an equilateral triangle has two rotation symmetries.

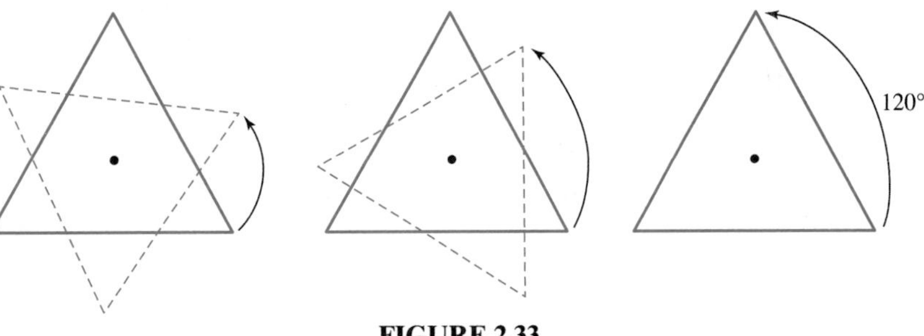

FIGURE 2.33

FIGURE 2.34

The trapezoid in Figure 2.34 does not have rotation symmetry because it must be rotated through one full turn of 360° before the image will match the original figure.

We can test figures for both reflection and rotation symmetries. To test for reflection symmetry, try folding the figure in half. To test for rotation symmetry, trace a copy of the figure and try to superimpose the tracing on the original figure by rotating it about a point.

EXAMPLE 2.3 Quilters construct quilts by starting with a single square block and then stitching many identical blocks together to form a finished quilt with an eye-pleasing pattern. Traditional quilt blocks often incorporate geometric shapes and exhibit reflection and/or rotation symmetry. Three different quilt block patterns are shown in Figure 2.35. Describe as completely as possible the geometric shapes that make up each pattern and the symmetries of the block.

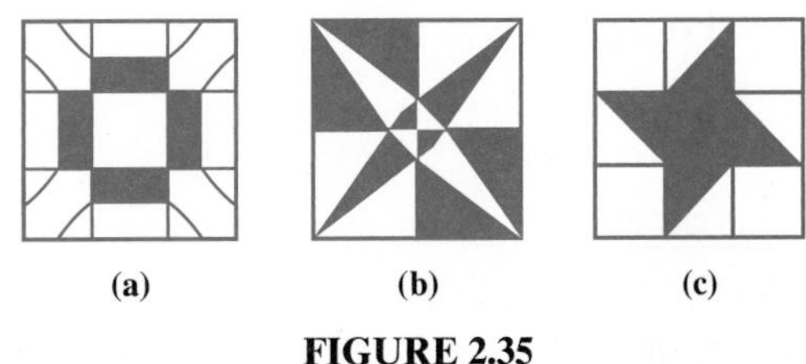

FIGURE 2.35

SOLUTION

a. The quilt block is made up of rectangles, right isosceles triangles, and hexagons. It has both reflection and rotation symmetry. There are four lines of symmetry and three rotations between 0° and 360° around the center of the square through which the block may be turned to match the original: 90°, 180°, and 270°. We say this block has three rotation symmetries.

b. The large triangles in the quilt block are right scalene triangles, the smallest triangles are right isosceles triangles, and the long, narrow triangles are acute isosceles triangles. The figure has both reflection and rotation symmetry. There are two axes of reflection symmetry, namely the two diagonals, and one rotation symmetry of 180°.

c. The quilt block is made up of squares and right isosceles triangles. It has no reflection symmetry, but it does have three rotation symmetries of 90°, 180°, and 270°. ■

Solution of Applied Problem

The opening in the wrench has six sides, so it is a hexagon. The wrench has six rotational symmetries and fits a hexagonal bolt in any of six positions, each 60° apart.

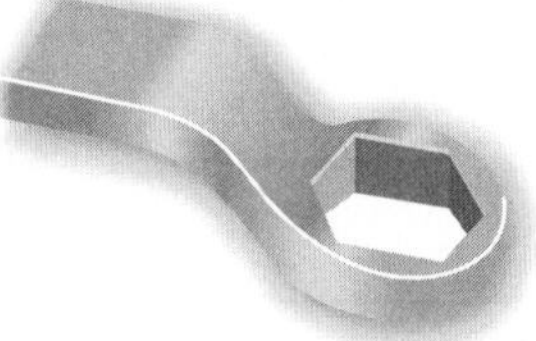

GEOMETRY AROUND US An excellent example of a pentagon is the Pentagon in Arlington, VA, home to the U.S. Department of Defense. This building has the largest ground area of any office building in the world. Each of the outer sides of the Pentagon measures 921 feet, and it covers an area of 1,263,240 square feet. Each day 26,000 people work in the building and travel along its 17 miles of hallways.

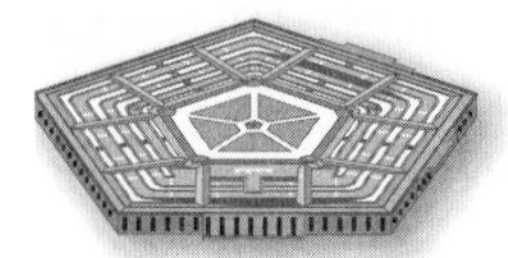

PROBLEM SET 2.2

EXERCISES/PROBLEMS

1. Several triangles are shown with congruent sides, congruent angles, and right angles as indicated.
   - **a.** Which triangles are isosceles?
   - **b.** Which triangles are equilateral?
   - **c.** Which triangles are scalene?
   - **d.** Which triangles are right triangles?
   - **e.** Which triangles are obtuse?

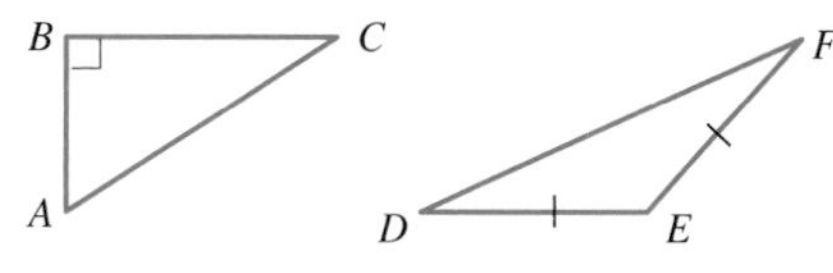

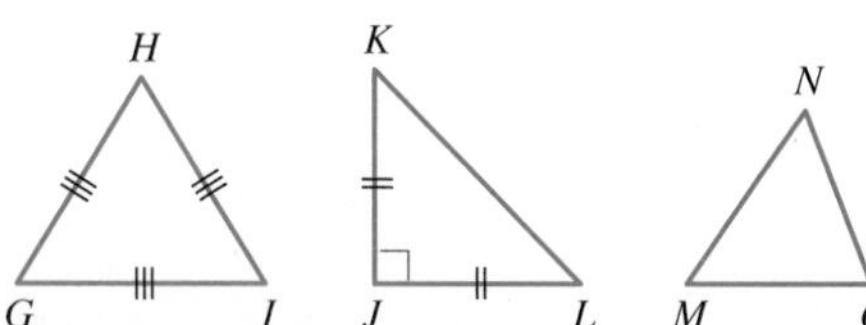

2. Several triangles are shown with congruent sides, congruent angles, and right angles as indicated.

a. Which triangles are scalene?

b. Which triangles are isosceles?

c. Which triangles are right triangles?

d. Which triangles are equiangular?

e. Which triangles are obtuse?

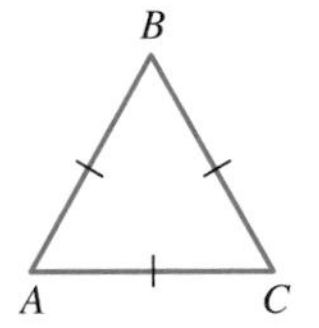

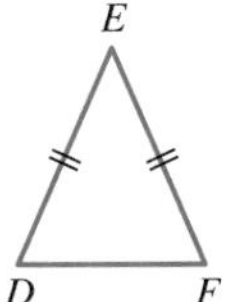

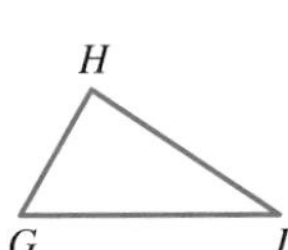

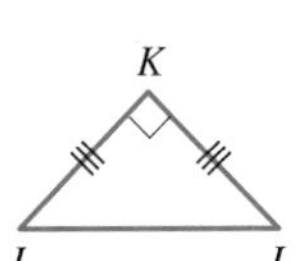

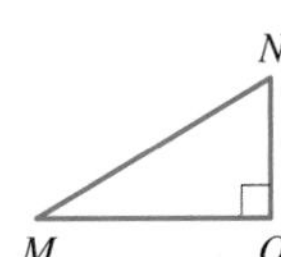

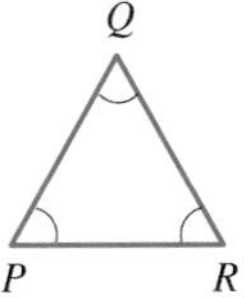

3. Draw examples, if possible, of each of the following types of triangles. If a triangle is not possible, explain why. Be sure to mark congruent sides, congruent angles, and right angles.

a. Acute scalene triangle

b. Obtuse scalene triangle

c. Right equilateral triangle

d. Scalene equilateral triangle

4. Draw examples, if possible, of each of the following types of triangles. If a triangle is not possible, explain why. Be sure to mark congruent sides, congruent angles, and right angles.

a. Right scalene triangle

b. Right isosceles triangle

c. Obtuse isosceles triangle

d. Acute right triangle

5. The following scalene triangle is formed from 14 toothpicks. We can describe this triangle as a 3-5-6 triangle.

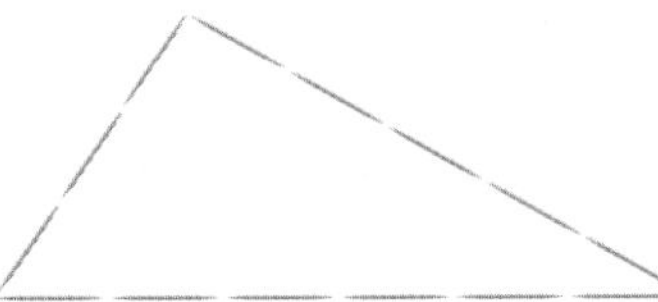

Use toothpicks to answer each of the following questions.

a. How many different isosceles triangles can you form using 20 toothpicks? (NOTE: You may have no gaps or overlapping toothpicks, and you may not break toothpicks.)

b. How many different scalene triangles can you form using 20 toothpicks?

c. How many different equilateral triangles can you form using 20 toothpicks?

6. a. How many different isosceles triangles can you form using 24 toothpicks? (See the directions for the preceding problem.)

b. How many different scalene triangles can you form using 24 toothpicks?

c. How many different equilateral triangles can you form using 24 toothpicks?

7. Consider the square lattice shown next.

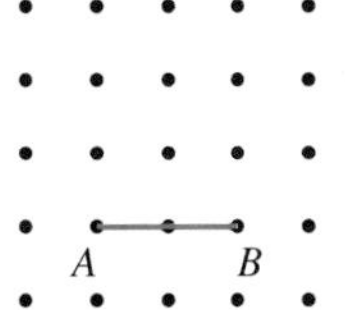

a. How many different triangles can you draw that have $\overline{AB}$ as one side?

b. How many of these are isosceles?

c. How many are right triangles?

d. How many are acute?

e. How many are obtuse? (HINT: Use your answers to parts (a), (c), and (d).)

8. Consider the square lattice shown next.

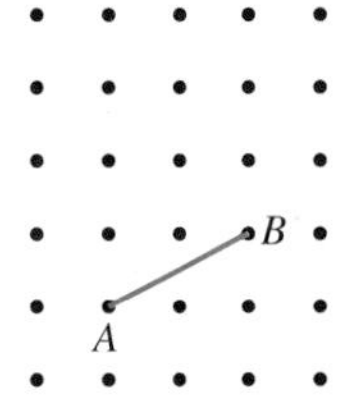

a. How many different triangles can you draw that have $\overline{AB}$ as one side?

b. How many of these are isosceles?

c. How many are right triangles?

d. How many are acute?

e. How many are obtuse? (HINT: Use your answers to parts (a), (c), and (d).)

9. Modify the following figure so that it forms a simple closed curve.

10. Modify the figure from problem 9 so that it forms a nonsimple closed curve.

11. Given the following simple closed curves, draw an example, if possible, where they intersect in exactly the number of points given.

a. One point
b. Two point
c. Three points
d. Four points

12. Given the following simple closed curves, draw an example, if possible, where they intersect in exactly the number of points given.

a. One point
b. Two point
c. Three points
d. Six points

13. Which of the following shapes are polygons? If one is not, explain why.

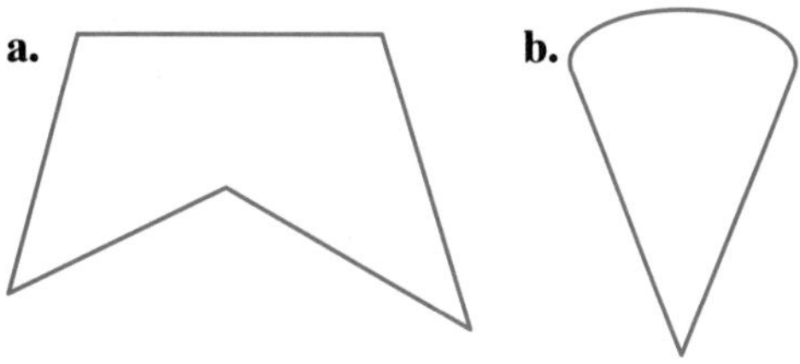

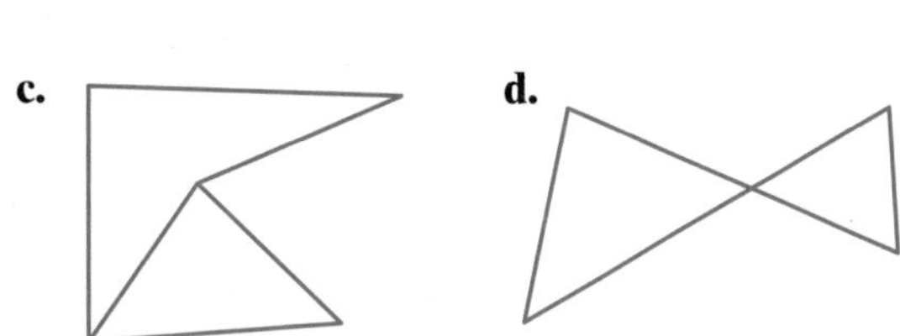

14. Which of the following shapes are polygons? If one is not, explain why.

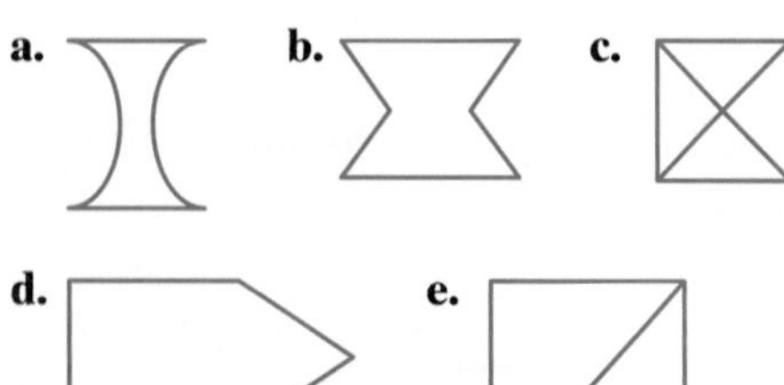

15. Draw two copies of the hexagon shown.

a. Divide one hexagon into three identical parts so that each part is a rhombus.
b. Divide the second hexagon into six identical kites.

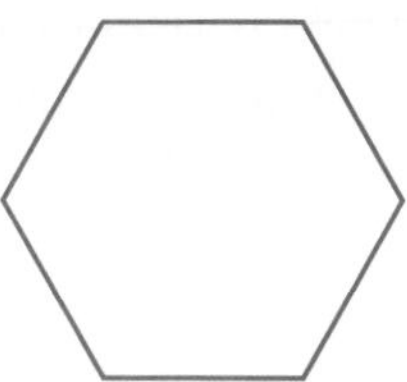

16. Draw two copies of the hexagon shown.

a. Divide one hexagon into four identical trapezoids.
b. Divide the second hexagon into eight identical polygons.

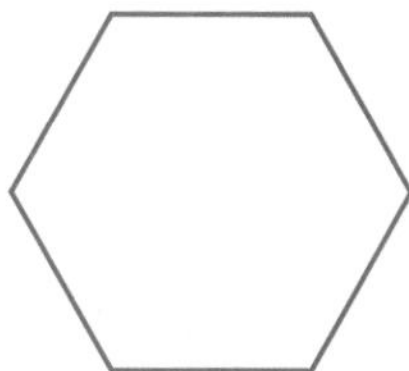

17. Find one example of each of the specified shapes in the following figure. Assume that angles that appear to be right angles are right angles, segments that appear to be parallel are parallel, and segments that appear to be congruent are congruent.

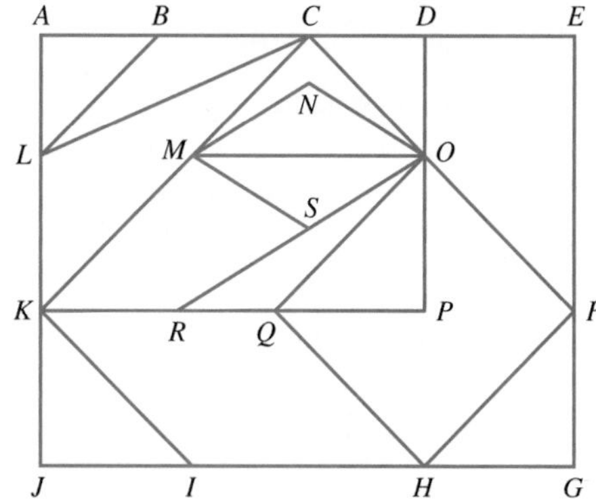

a. A square
b. A rectangle that is not a square
c. A parallelogram that is not a rectangle
d. An isosceles right triangle
e. An isosceles triangle with no right angles
f. A rhombus that is not a square
g. A kite that is not a rhombus
h. A scalene triangle with no right angles
i. A right scalene triangle
j. A trapezoid that is not isosceles
k. An isosceles trapezoid

18. Find one example of each of the specified shapes in the following figure. Assume that angles that appear to be right angles are right angles, segments that appear to be parallel are parallel, and segments that appear to be congruent are congruent.

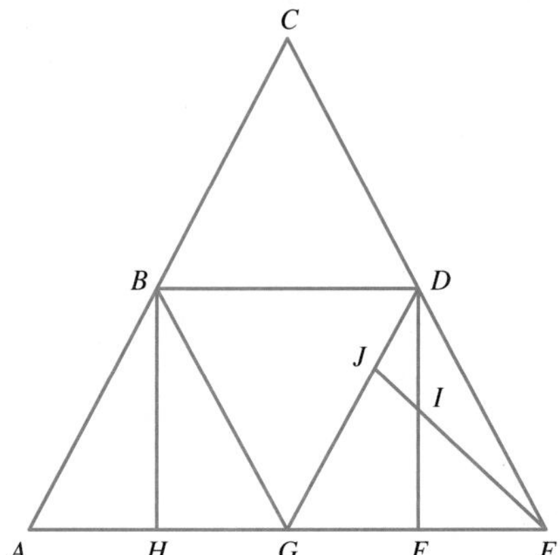

a. A rectangle
b. A kite
c. An isosceles triangle with no right angle
d. An equilateral triangle
e. A right isosceles triangle
f. A right scalene triangle
g. A parallelogram that is not a rectangle
h. An isosceles trapezoid
i. An obtuse scalene triangle
j. A trapezoid that is not isosceles
k. A pentagon

19. Given the square lattice shown, draw quadrilaterals having $\overline{AB}$ as a side.

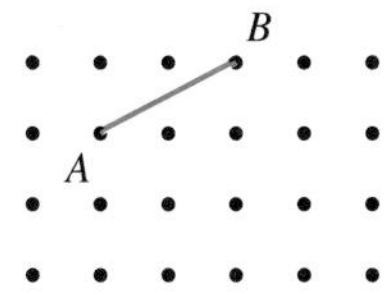

a. How many different parallelograms are possible?
b. How many rectangles are possible?
c. How many rhombuses?
d. How many squares?

20. Given the triangular lattice shown, draw quadrilaterals having $\overline{AB}$ as a side.

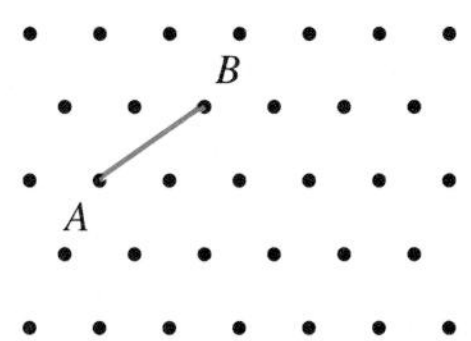

a. How many different parallelograms are possible?
b. How many rectangles are possible?
c. How many rhombuses?
d. How many trapezoids?

21. Consider the equally spaced points on the following circle.

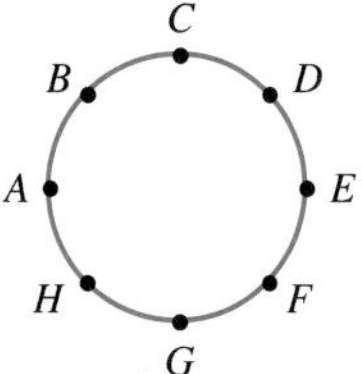

a. How many different kites can you draw using four of the points on the circle as vertices?
b. How many of the kites from (a) are rhombuses?
c. How many of the possible rhombuses from (b) are not squares?

22. Consider the equally spaced points on the following circle.

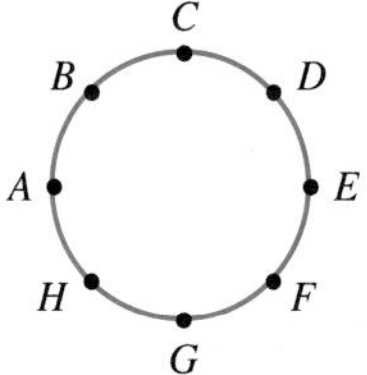

a. How many different rectangles can you draw using four of the points on the circle as vertices?
b. How many different trapezoids can you draw using four of the points on the circle as vertices?
c. How many of the trapezoids from (b) are not isosceles?

23. Fold a rectangular piece of paper on the dotted line as shown in each of the following figures. Then make cuts in the paper as indicated. Sketch what you think the shape will be when you unfold the paper. Then unfold your paper to check your picture.

a.

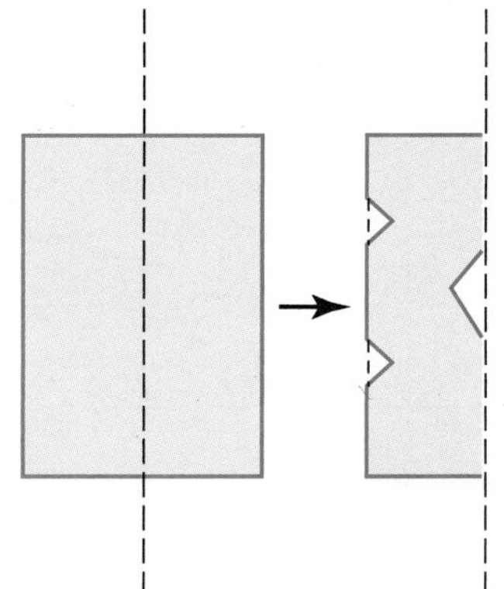

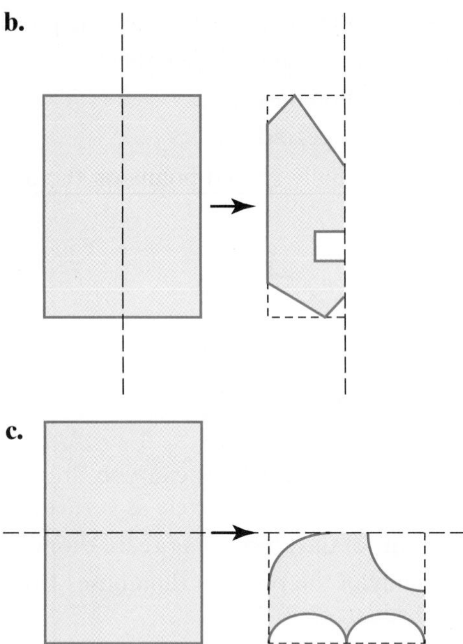

24. Fold a rectangular piece of paper on the dotted line as shown in each of the following figures. Then make cuts in the paper as indicated. Sketch what you think the shape will be when you unfold the paper. Then unfold your paper to check your picture.

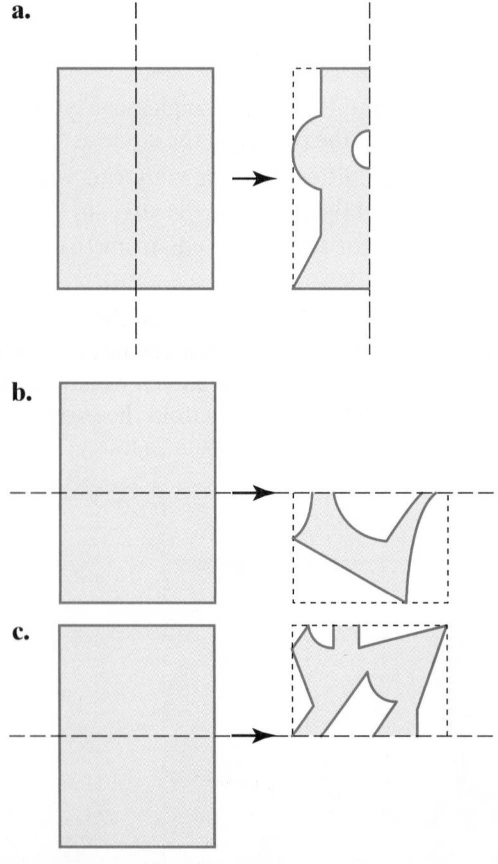

25. Each shape shown next was obtained by folding a rectangular piece of paper in half vertically (lengthwise) and then making appropriate cuts in the paper. For each figure, draw the folded paper and show the cuts that must be made to make the figure. Try folding and cutting a piece of paper to check your answer.

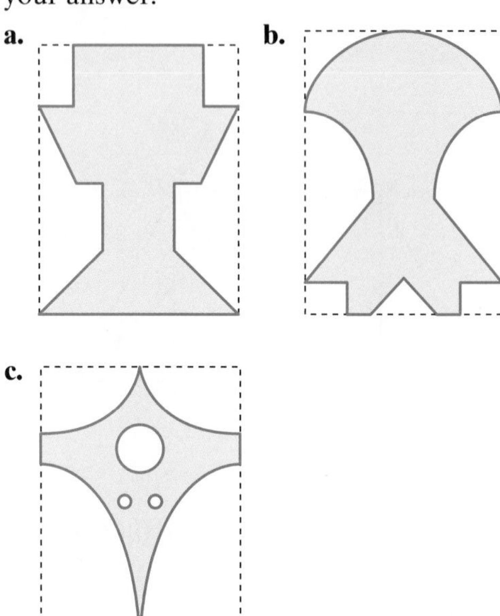

26. Each shape shown next was obtained by folding a rectangular piece of paper in half horizontally and then making appropriate cuts in the paper. For each figure, draw the folded paper and show the cuts that must be made to make the figure. Try folding and cutting a piece of paper to check your answer.

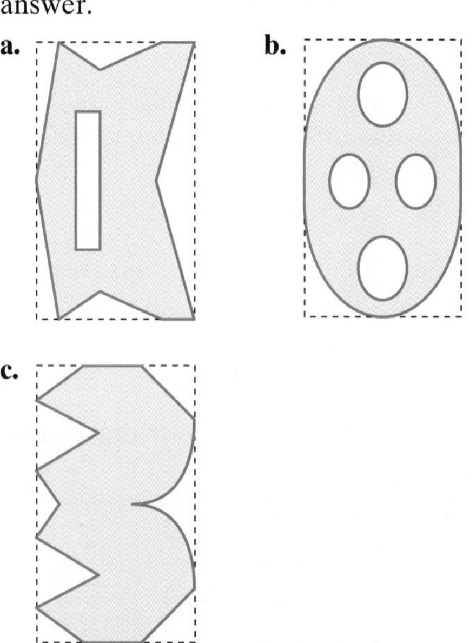

27. a. Does the rectangle shown have reflection symmetry? If so, how many axes of symmetry does it have?

b. Does the rectangle have rotation symmetry? If so, how many different rotation symmetries does it have?

28. a. Does the rhombus shown have reflection symmetry? If so, how many axes of symmetry does it have?

b. Does the rhombus have rotation symmetry? If so, how many different rotation symmetries does it have?

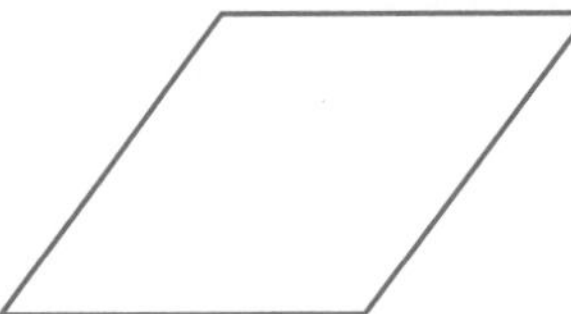

29. a. Does the isosceles trapezoid shown have reflection symmetry? If so, how many axes of symmetry does it have?

b. Does the isosceles trapezoid have rotation symmetry? If so, how many different rotation symmetries does it have?

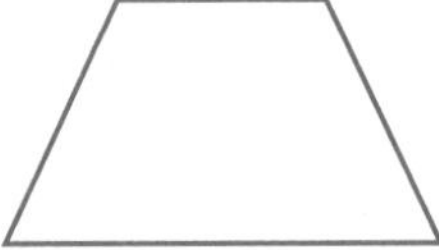

30. a. Does the right isosceles triangle shown have reflection symmetry? If so, how many axes of symmetry does it have?

b. Does the right isosceles triangle have rotation symmetry? If so, how many different rotation symmetries does it have?

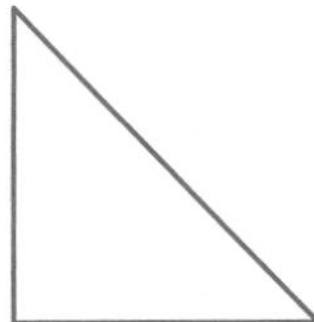

31. a. Draw the lines of symmetry in the regular n-gons in (i)–(iii). How many does each have?

b. How many lines of symmetry does a regular n-gon have?

(i)

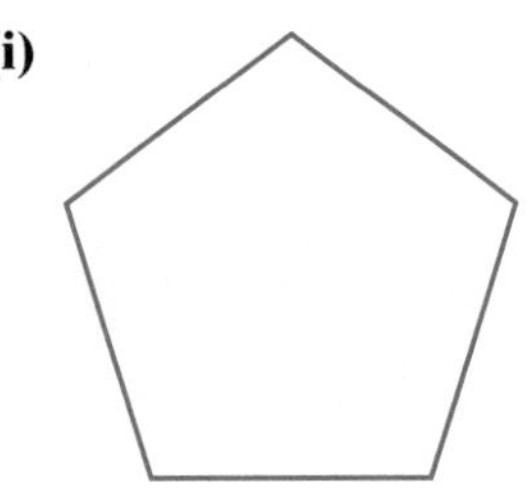

(ii)

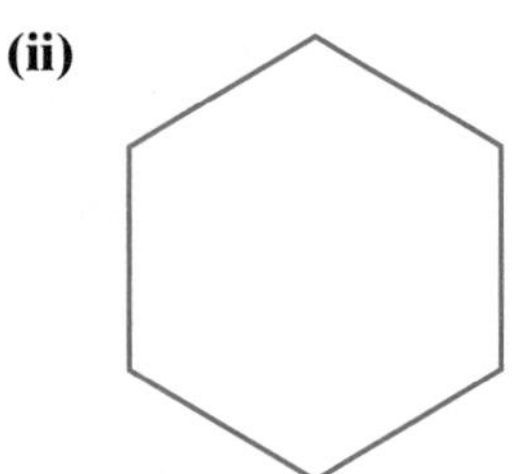

(iii)

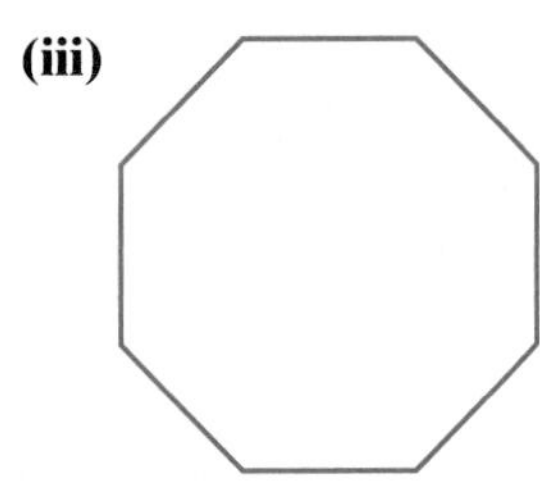

32. a. Trace the regular n-gons in problem 31 to find all of the rotation symmetries for a regular pentagon, a regular hexagon, and a regular octagon.

b. How many rotation symmetries does a regular n-gon have?

APPLICATION

33. Bingo is played on a 5 by 5 grid of squares in which certain squares must be covered in order to win, usually five squares in a row vertically, horizontally, or diagonally.

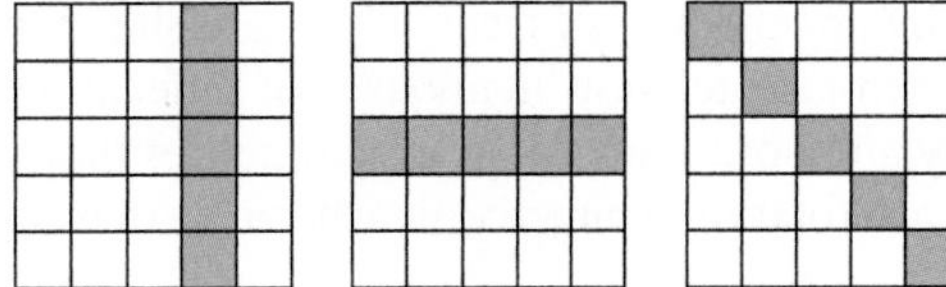

To make the games more interesting, other patterns on the grid are often chosen to be winners. Several examples are shown. For each one, tell whether the pattern has reflection symmetry, rotation symmetry, or both.

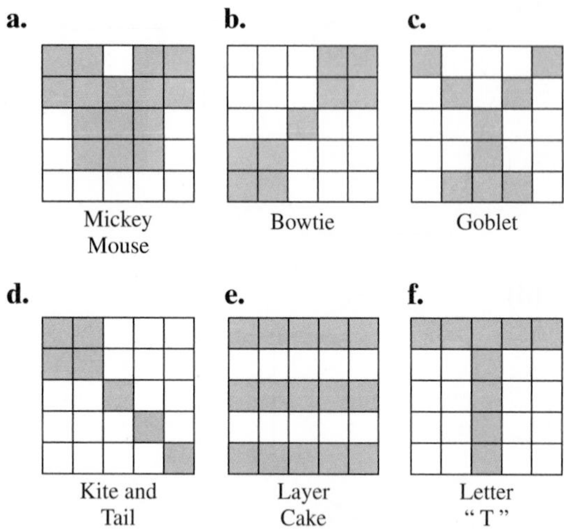

34. The smooth, flat surfaces on a cut diamond are called **facets**. Facets are cut to optimize the reflective properties of the diamond. Tell whether each diamond has reflection symmetry, rotation symmetry, or both.

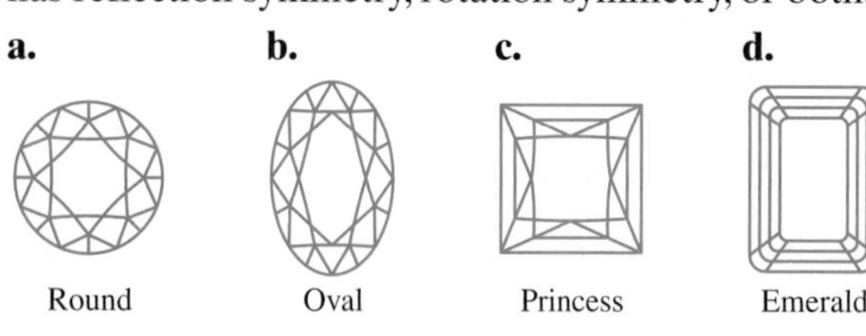

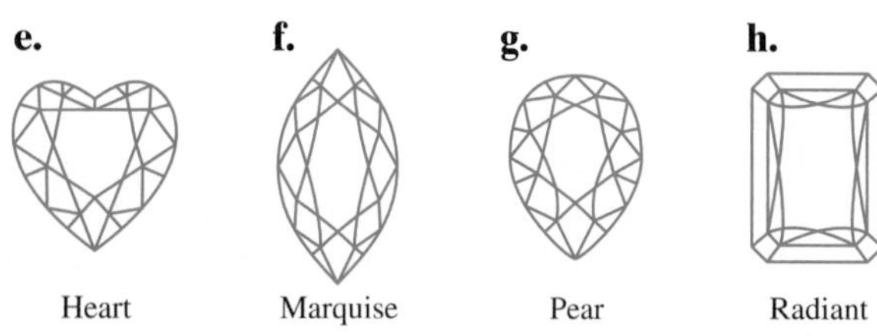

35. A standard die has six faces, which are shown next. For each face, tell how many axes of reflection symmetry and how many rotation symmetries there are. For each rotation symmetry, give the angle of rotation.

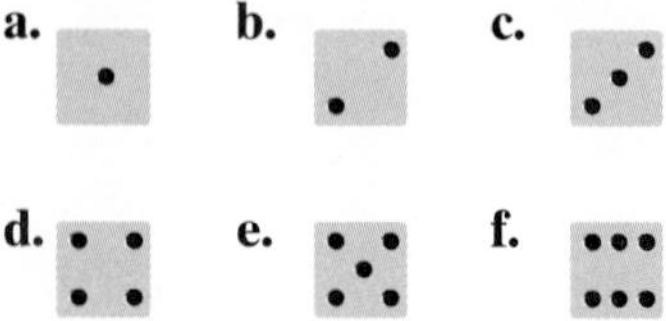

36. For each of the following quilt blocks, describe as completely as possible the geometric shapes that make up each pattern and the symmetries of the block.

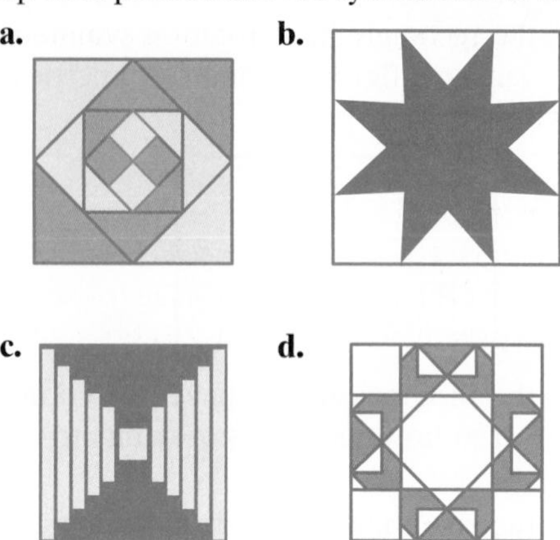

c. d.

37. In 1845, pioneers in Utah constructed wagon wheels so that front wheels and rear wheels had a different number of evenly spaced spokes. Spokes are rods that extend from the center of the wheel to the circular traction surface of the wheel.

a. If a rear wheel had exactly seven axes of reflection symmetry, then how many spokes were used and how many rotation symmetries did the wheel have?

b. If the measure of the angle between two adjacent front-wheel spokes was 30°, then how many spokes were used and how many axes of reflection symmetry did the wheel have?

EXTENDED PROBLEMS

38. Another name for a simple closed curve is a **Jordan curve**. The Jordan curve theorem states that every simple closed curve divides the plane into two regions, an inside region and an outside region, and the curve is the boundary between the regions. In order to pass from one region to the other, it is necessary to cross through the curve. Revisit problem 41 from Section 1.1 and explain how the Jordan curve theorem can be used to show that the problem has no solution.

39. While driving, we rely on traffic signs to help us arrive at our destinations safely. Each type of traffic sign communicates a message through its shape as well as its color. The Federal Highway Administration publishes the Manual on Uniform Traffic Control Devices, MUTCD, which defines traffic sign standards. Research traffic signs to determine the meaning and most common uses of traffic signs in the following shapes: circle, rectangle, square, octagon, pentagon, rhombus, trapezoid, equilateral triangle, and isosceles triangle. Download or sketch an example of a sign of each shape.

40. The U.S. State Department recognizes 192 countries in the world. Each country is represented by a flag, many of which display symmetry. Research flags of the world and determine which country has a flag with exactly four axes of reflection symmetry and exactly three rotation symmetries. List five countries whose flags have exactly two axes of reflection symmetry and one rotation symmetry. List five countries whose flags have exactly one axis of reflection symmetry and zero rotation symmetries. List five countries whose flags have no symmetry. You can find a collection of flags of the world by using search keyword "World Factbook" on the Internet. Include a sketch or print a copy of each flag included in your list.

2.3 ANGLE MEASURE IN POLYGONS AND TESSELLATIONS

Applied Problem

A decorative floor tiling pattern is shown. It uses two types of polygons to cover the floor without gaps or overlaps. Explain why this arrangement of tiles is possible.

Tessellations

A **polygonal region** is a polygon together with the portion of the plane that is enclosed by the polygon (Figure 2.36).

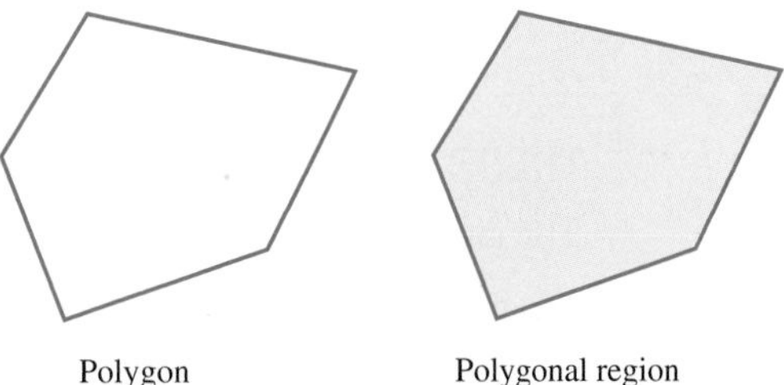

FIGURE 2.36

We can arrange polygonal regions to cover a plane just as we can arrange ceramic tiles or carpet squares to cover a countertop or a floor. Such an arrangement is called a **tessellation** or a **tiling** if (1) we can cover the entire plane without gaps, and (2) no two polygons overlap; that is, polygons share only common sides. Figure 2.37 shows some partial tessellations of the plane.

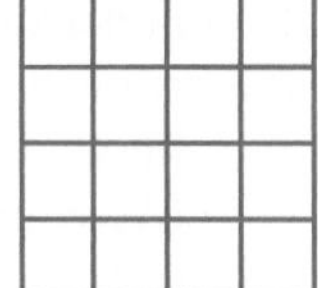
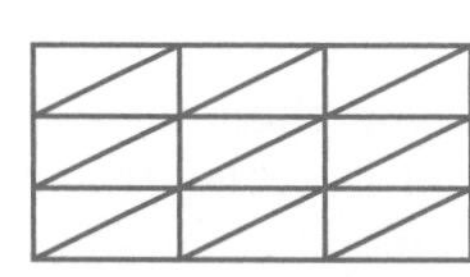

FIGURE 2.37

Angle Measure in a Triangle

We may use a tessellation composed of triangles to develop a useful relationship among the angle measures of a triangle. Figure 2.38(a) shows a triangle that we can use to tessellate the plane. That is, we can arrange identical copies of the triangle to cover the plane [Figure 2.38(b)].

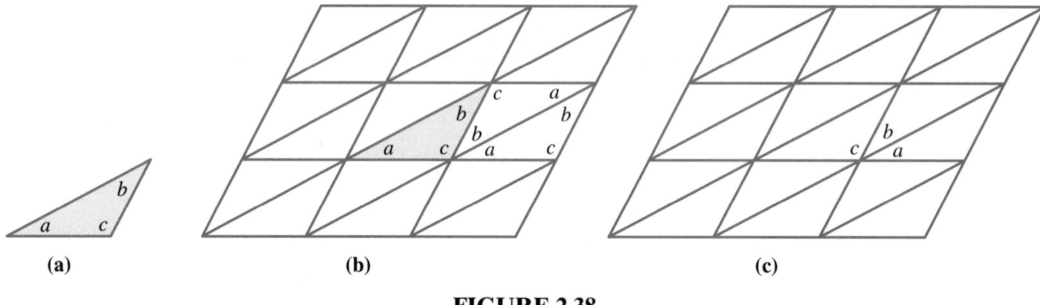

FIGURE 2.38

Notice that the sum of the angle measures in each of the triangles is $a + b + c$. However, because of the arrangement of triangles, we can see that the angles labeled c, b, and a in Figure 2.38(c) form a straight angle. Hence, $c + b + a = 180°$. However, c, b, and a are the angle measures in any one of the triangles. Thus, the sum of the angle measures in each of the triangles is 180°. This relationship holds true for any triangle, and we state it in the next theorem. We will prove this theorem formally in Chapter 5.

THEOREM 2.1 Angle Measure in a Triangle

The sum of the measures of the angles in a triangle is 180°.

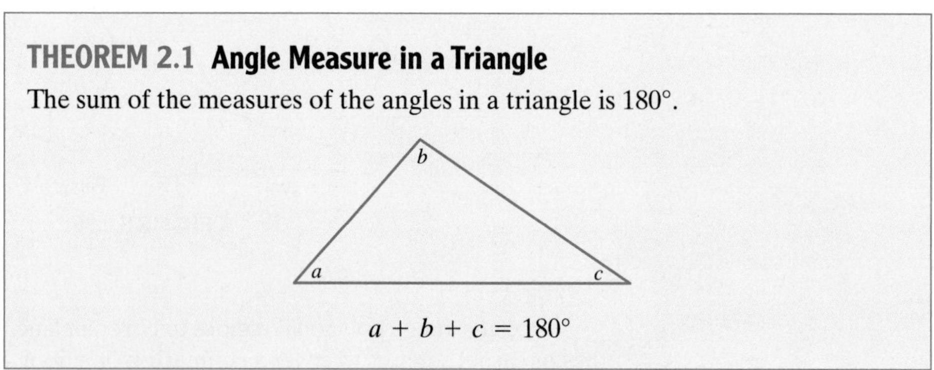

$$a + b + c = 180°$$

We can use this theorem to find missing angle measures in many geometric figures.

EXAMPLE 2.4 Find the missing angle measures in Figure 2.39. Congruent angles and right angles are indicated.

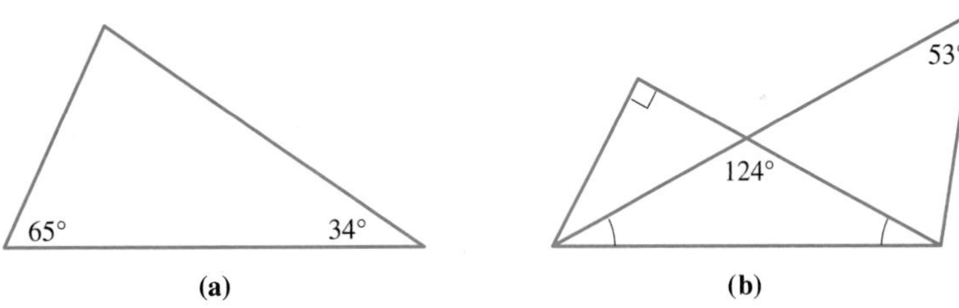

FIGURE 2.39

SOLUTION

a. Let $x =$ missing angle measure [Figure 2.40(a)]. We know that

$$x + 65° + 34° = 180°$$

Therefore, $x + 99° = 180°$, or $x = 81°$.

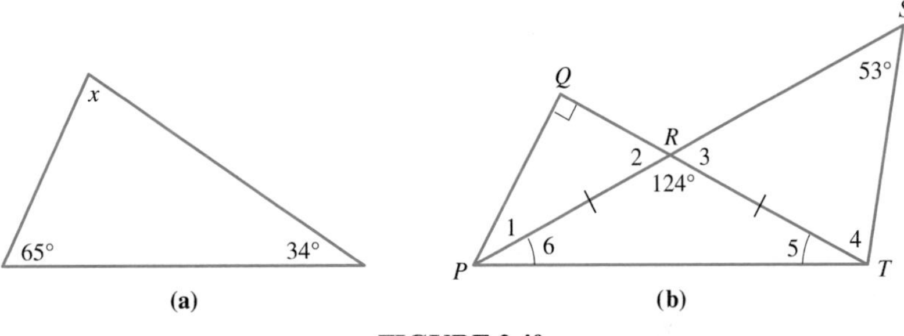

FIGURE 2.40

b. First number the angles with missing measures and label the vertices [Figure 2.40(b)]. Consider $\triangle PRT$. Because the sum of the angle measures is 180° and $\angle 5 = \angle 6$, we can say

$$\angle 5 + \angle 6 + 124° = 180°$$
$$\angle 5 + \angle 6 = 56°$$
$$2(\angle 6) = 56°$$
$$\angle 6 = 28°$$

Hence, $\angle 5$ and $\angle 6$ both have measure 28°.

We also might notice that $\angle QRT$ is a straight angle. Therefore, $\angle 2 + 124° = 180°$, so $\angle 2 = 56°$. Likewise, $\angle 3 = 56°$. Now we know the measures of two of the three angles in $\triangle PQR$. Thus, we can say

$$\angle 1 + \angle 2 + 90° = 180°$$
$$\angle 1 + 56° + 90° = 180°$$
$$\angle 1 = 34°$$

In a similar way we can use $\triangle RST$ to find the measure of $\angle 4 = 71°$ because we know the measures of two angles in the triangle are 53° and 56°. Therefore, we have found the following:

$$\angle 1 = 34°, \angle 2 = 56°, \angle 3 = 56°, \angle 4 = 71°, \angle 5 = 28°, \angle 6 = 28°$$ ■

Vertex Angles in a Polygon

The angles in a polygon are called its **vertex angles**. For example, in Figure 2.41(a), the vertex angles in the pentagon are $\angle V$, $\angle W$, $\angle X$, $\angle Y$, and $\angle Z$. Line segments joining nonadjacent vertices in a polygon are called **diagonals**. In Figure 2.41(b), $\overline{WZ}$ and $\overline{WY}$ are two of the diagonals of $VWXYZ$. We can find the sum of the measures of the vertex angles of a polygon by subdividing the polygon into triangles using diagonals.

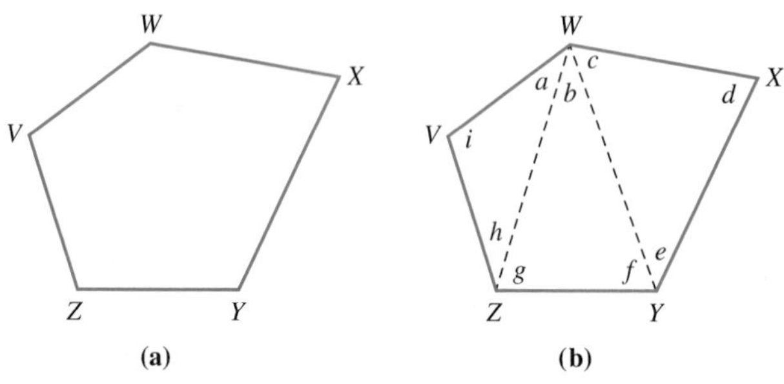

FIGURE 2.41

EXAMPLE 2.5 Find the sum of the measures of the vertex angles of a pentagon.

SOLUTION

Draw a pentagon and then divide it into triangles by drawing in all of the diagonals from any *one* vertex [Figure 2.41(b)]. Because the sum of the angle measures in a triangle is 180°, in Figure 2.41(b) we have $a + h + i = 180°$, $b + g + f = 180°$, and $c + d + e = 180°$. Thus, the sum of the vertex angles in pentagon $VWXYZ$ is $3(180°) = 540°$. ■

We can generalize the technique used in Example 2.5 to find the sum of the measures of the vertex angles in any polygon. In the pentagon, three triangles were formed by drawing diagonals from one vertex. In a polygon with n sides, $n - 2$ triangles will be formed. This leads to the next theorem.

> **THEOREM 2.2 Angle Measure in a Polygon**
> The sum of the measures of the vertex angles in a polygon with n sides is $(n - 2)180°$.

Regular Polygons

Regular polygons are polygons in which all of the sides are congruent and all of the angles are congruent. Squares and equilateral triangles are examples of regular polygons. Three other regular polygons are shown in Figure 2.42.

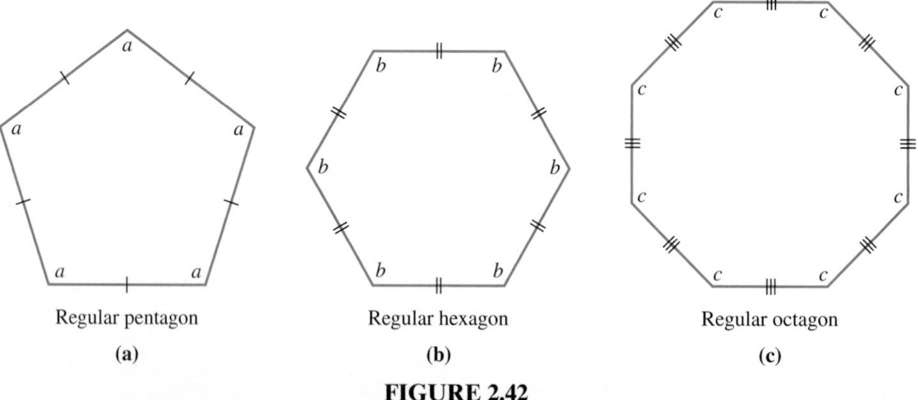

Regular pentagon (a) Regular hexagon (b) Regular octagon (c)

FIGURE 2.42

EXAMPLE 2.6 Find the measure of a vertex angle of a regular pentagon [Figure 2.42(c)].

SOLUTION

The sum of the angle measures in a regular pentagon is 540°, as shown in Example 2.5. Because the vertex angles in a regular polygon are congruent and there are five such angles, the measure of *each* vertex angle is $540°/5 = 108°$. ■

We can generalize the technique illustrated in Examples 2.5 and 2.6 to any regular n-gon. The next theorem summarizes this more general result. It will be verified in the problem set.

THEOREM 2.3 Vertex Angle Measure in a Regular Polygon

The measure of a vertex angle in a regular n-gon is $\dfrac{(n-2)180°}{n}$.

Using Theorem 2.3, we can calculate the measure of a vertex angle in any regular polygon. Table 2.1 contains a list of several regular n-gons together with the measures of their vertex angles.

n	Measure of a Vertex Angle in a Regular n-gon
3	$\dfrac{(3-2)180°}{3} = 60°$
4	$\dfrac{(4-2)180°}{4} = 90°$
5	$\dfrac{(5-2)180°}{5} = 108°$
6	$\dfrac{(6-2)180°}{6} = 120°$
8	$\dfrac{(8-2)180°}{8} = 135°$
10	$\dfrac{(10-2)180°}{10} = 144°$

TABLE 2.1

Notice that as the number of sides in the regular polygon increases, the measure of one of the polygon's vertex angles also increases.

We can use theorem 2.3 to determine the number of sides in a regular polygon if we know the measure of a vertex angle. The next example demonstrates this.

EXAMPLE 2.7 A vertex angle of a regular polygon measures 157.5°. How many sides does the polygon have?

SOLUTION

We know that the measure of the vertex angle will be $\dfrac{(n-2)180°}{n}$, where n is the number of sides of the polygon. In this case, $\dfrac{(n-2)180°}{n} = 157.5°$. We can find n by solving this equation.

$$\frac{(n - 2)180^\circ}{n} = 157.5^\circ$$
$$(n - 2)180^\circ = 157.5^\circ n$$
$$180^\circ n - 360^\circ = 157.5^\circ n$$
$$-360^\circ = -22.5^\circ n$$
$$\frac{-360^\circ}{-22.5^\circ} = n$$
$$n = 16$$

Therefore, the polygon has 16 sides. ■

A **regular tessellation** is a tessellation composed of regular polygons that are all the same size and shape. Figure 2.43 shows three regular tessellations.

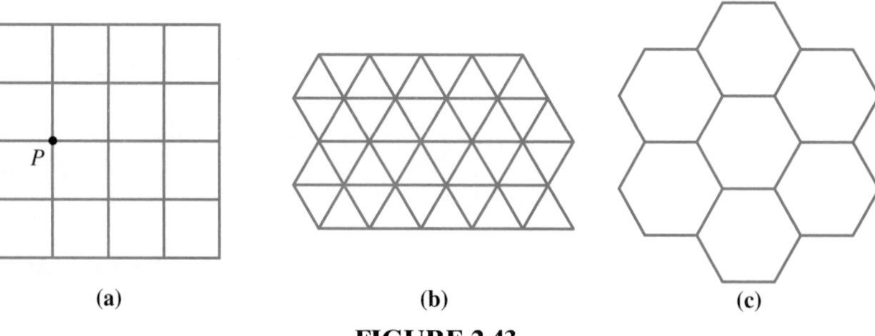

(a) (b) (c)

FIGURE 2.43

To form a regular tessellation, the polygonal regions must fit together exactly around each vertex. In other words, the sum of the vertex angle measures must equal 360°. For example, in the case of the tessellation of squares in Figure 2.43(a), four 90° angles fit together around vertex P and add up to 360°. This tessellation is possible because 4(90°) = 360°; that is, 90° is a factor of 360°. If the number of degrees in the vertex angle of a regular polygon is a factor of 360°, then a regular tessellation can be formed using that regular polygon. Using Table 2.1, we can see that 60°, 90°, and 120° are measures of vertex angles of regular polygons that are also factors of 360°. Also, when $n > 6$, the vertex angle measure is between 120° and 180°; hence, it will not divide evenly into 360°. Thus, the only possible regular tessellations are composed of equilateral triangles, squares, and regular hexagons, which are the regular tessellations shown in Figure 2.43.

Solution of Applied Problem

The tessellation pictured is composed of squares and regular octagons. Each vertex angle of a square measures 90° and each vertex angle of a regular octagon measures 135°. When two octagons and one square are placed together at a common vertex, the sum of the measures of their vertex angles is $2(135^\circ) + 90^\circ = 360^\circ$. Thus, the region surrounding the common vertex is covered completely, without any overlap. This tiling composed of regular polygons of different types is an example of what is called a semiregular tessellation.

GEOMETRY AROUND US In 1985, chemists at Rice University discovered a new form of carbon with some unique geometric properties. Carbon-60 is also known as Buckminsterfullerene or Buckyball because its shape is similar to the geodesic domes of Buckminster Fuller. Buckyball molecules are composed of regular pentagons and hexagons and resemble soccer balls or symmetrical spheres. Their unique properties prompted physicists and chemists to envision many exciting potential applications, including lubricants, drugs, fuels, batteries, and high strength materials. As yet no such products have been developed; the research continues.

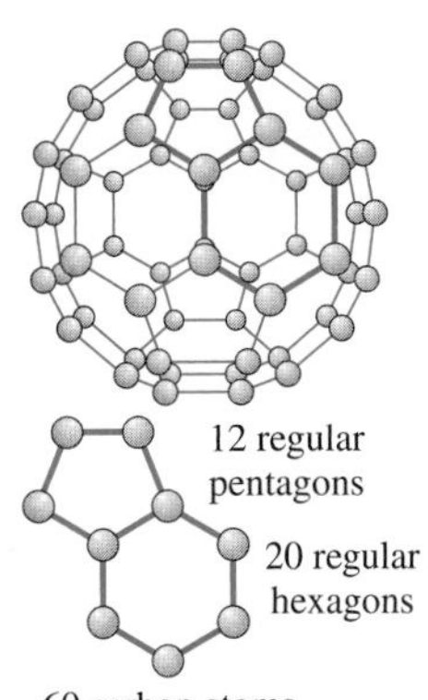

PROBLEM SET 2.3

1. Make six copies of the following triangle and show how they can be used to create a tessellation.

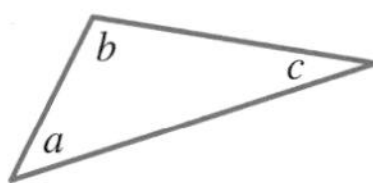

2. Make six copies of the following triangle and show how they can be used to create a tessellation.

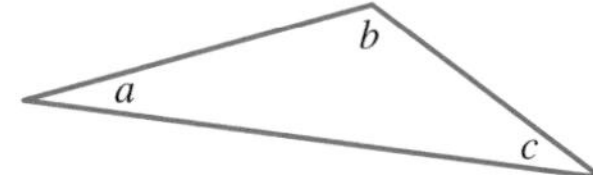

3. On a square lattice, draw a tessellation using each of the following polygons. It may be helpful to cut out a copy of each polygon.

a. b. c.

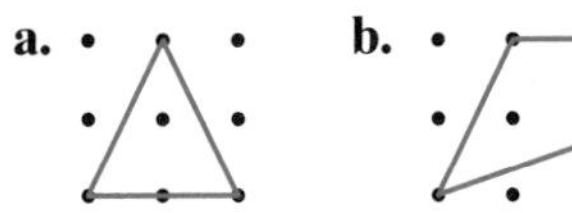
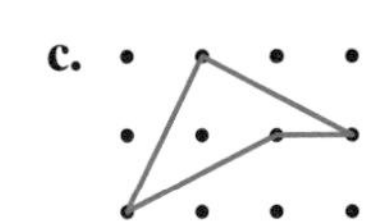

4. On a square lattice, draw a tessellation using each of the following polygons. It may be helpful to cut out a copy of each polygon.

a. b. c.

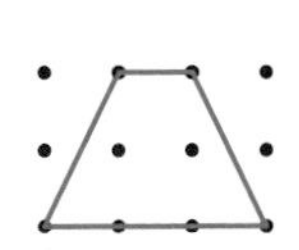
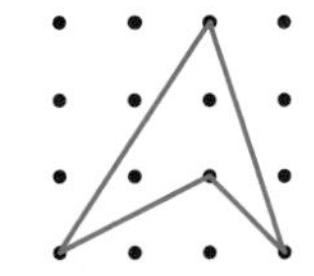

5. Use your protractor to measure each vertex angle in each polygon shown. If necessary, extend the sides of the angles in order to measure them. Then find the sum of the measures of the vertex angles in each polygon. What *should* the angle sum be in each case?

a.

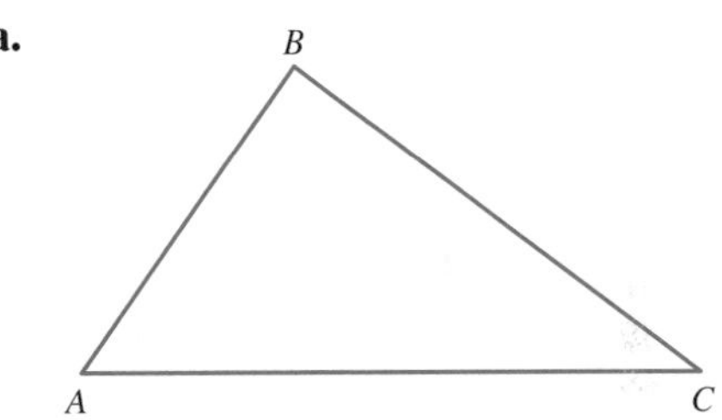

b.

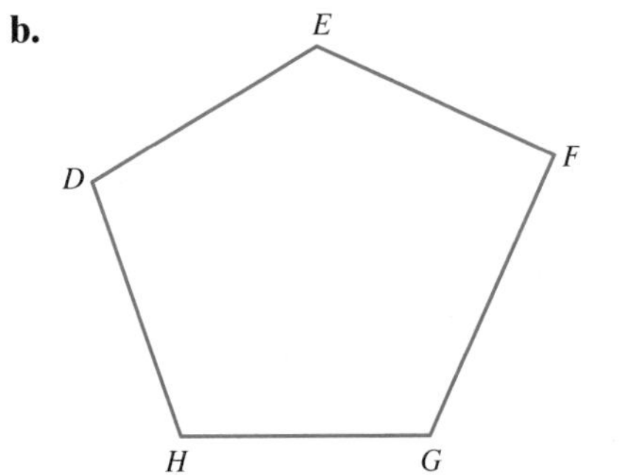

6. Use your protractor to measure each vertex angle in each polygon shown. If necessary, extend the sides of the angles in order to measure them. Then find the sum of the measures of the vertex angles in each polygon. What *should* the angle sum be in each case?

a.

Y X W Z

b.

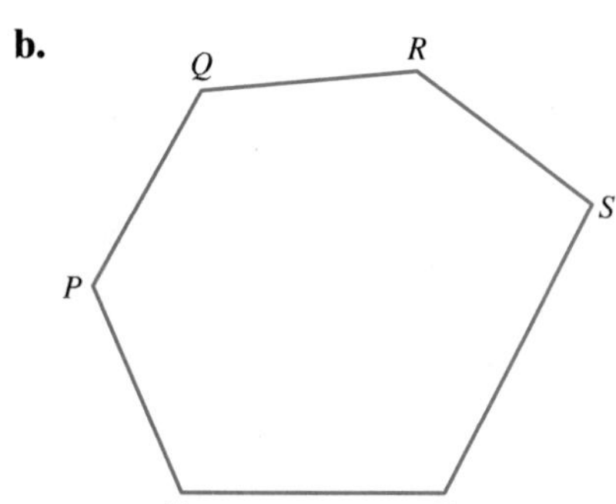

7. The measures of two angles in a triangle are given. Find the missing measure of the angle.

a. 90°, 60° **b.** 120°, 40°

c. 85°, 33° **d.** 79°, 67°

8. The measures of $\angle A$, $\angle B$, and $\angle C$ are given. Can you make a triangle $\triangle ABC$ with the given angle measures? Explain.

a. $\angle A = 36°, \angle B = 78°, \angle C = 66°$

b. $\angle A = 124°, \angle B = 56°, \angle C = 20°$

c. $\angle A = 90°, \angle B = 74°, \angle C = 18°$

9. Calculate the measure of each lettered angle. Congruent angles and right angles are indicated.

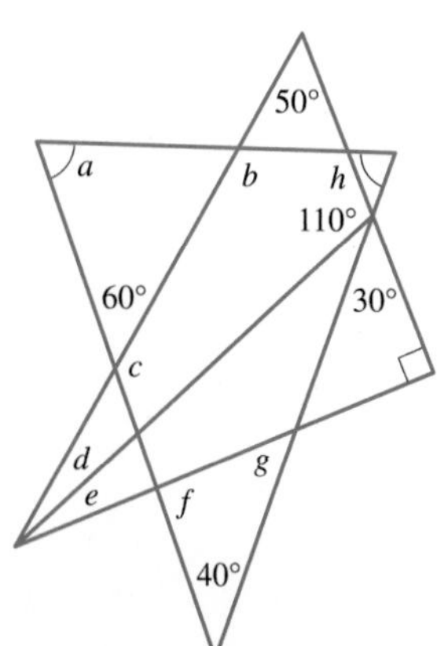

10. Calculate the measure of each lettered angle. Congruent angles and right angles are indicated.

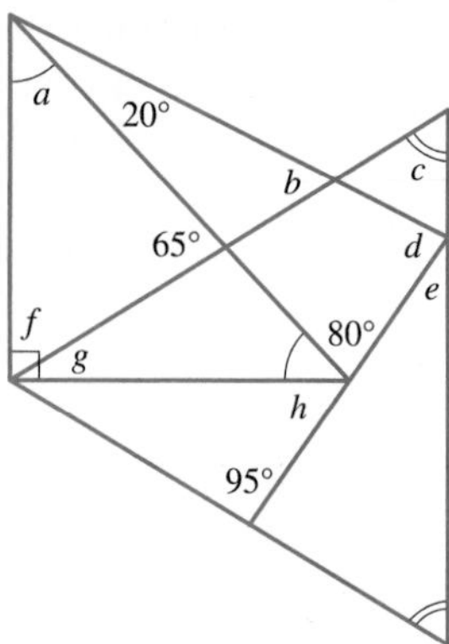

11. Find the missing angle measures in the quadrilaterals shown.

a. 100° 72°

b. 55° 55°

c. 60°

d. $x - 10°$ x $x + 5°$ $x - 15°$

12. Find the missing angle measures in the quadrilaterals shown.

a. 108°

b. x $x + 15°$

c. **d.**

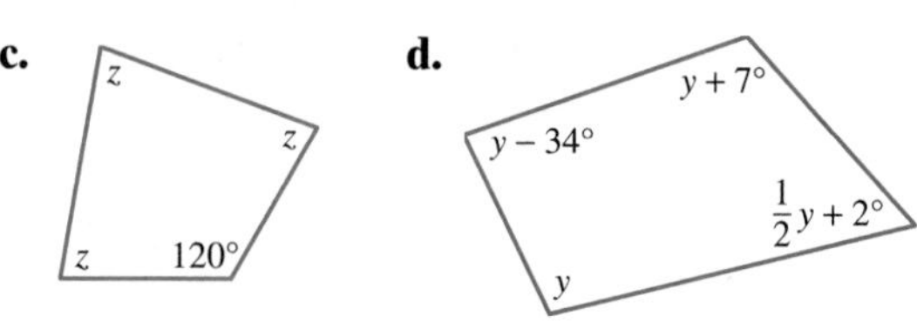

13. Suppose that the sum of the measures of the vertex angles of a polygon is 1620°. How many sides does the polygon have?

14. Suppose that the sum of the measures of the vertex angles of a polygon is 3420°. How many sides does the polygon have?

15. What is the sum of the measures of the vertex angles of a 29-sided polygon?

16. What is the sum of the measures of the vertex angles of a 35-sided polygon?

17. For the following regular n-gons, give the measure of a vertex angle.

a. 12-gon **b.** 16-gon **c.** 10-gon
d. 20-gon **e.** 18-gon **f.** 36-gon

18. For the following regular n-gons, give the measure of a vertex angle.

a. 15-gon **b.** 24-gon **c.** 40-gon
d. 32-gon **e.** 25-gon **f.** 30-gon

19. Measures of a vertex angle of regular polygons are given. Determine how many sides each polygon has.

a. 172.5°
b. 175°
c. 171°
d. 176.4°

20. Measures of a vertex angle of regular polygons are given. Determine how many sides each polygon has.

a. 150° **b.** 156° **c.** 174° **d.** 178°

21. The following circle has 12 evenly spaced dots. Name and sketch examples of all possible regular polygons that can be created by connecting dots to form congruent sides.

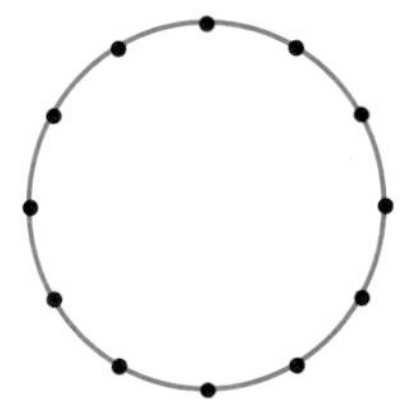

22. Suppose a circle has 24 evenly spaced dots. Name and sketch examples of all possible regular polygons that can be created by connecting dots to form congruent sides.

23. A regular n-gon must have n congruent angles and n congruent sides. Show that having n congruent angles is not sufficient for a polygon to be regular by sketching examples of

a. An equiangular quadrilateral that is not regular
b. An equiangular pentagon that is not regular

24. Show that having n congruent sides is not sufficient for a polygon to be regular by sketching examples of

a. An equilateral quadrilateral that is not regular
b. An equilateral hexagon that is not regular

25. Explain why a regular tessellation cannot be composed of regular heptagons.

26. Explain why a regular tessellation cannot be composed of regular octagons.

27. Verify that the measure of a vertex angle in a regular polygon is $(n - 2)180°/n$ (Theorem 2.3), by completing the given table as follows: (i) Sketch regular polygons with four, five, six, and more sides, and (ii) divide each polygon into triangles by drawing all possible diagonals from one vertex. Use inductive reasoning to develop the formula given in Theorem 2.3.

Number of Sides and Angles in a Polygon	Number of Triangles into Which the Polygon Can Be Divided	Sum of Angle Measures in the Polygon	Measure of One Angle in the Polygon
3	1	180°	$\frac{180°}{3} = 60°$
4	2	$2(180°) = 360°$	$\frac{360°}{4} = 90°$
5			
6			
8			
10			
n			

28. Complete the following table. Let V represent the number of vertices, D the number of diagonals from each vertex, and T the total number of diagonals in the polygon.

Polygon	V	D	T
Triangle			
Quadrilateral			
Pentagon	5	2	5
Hexagon			
Octagon			
n-gon			

APPLICATIONS

29. From a plane at point A in the following picture, the angle of depression (the acute angle between the horizontal and the line of sight) to a control tower at point C is 20°35′50″, as shown. Find the measure of $\angle ACB$ in degrees, minutes, and seconds.

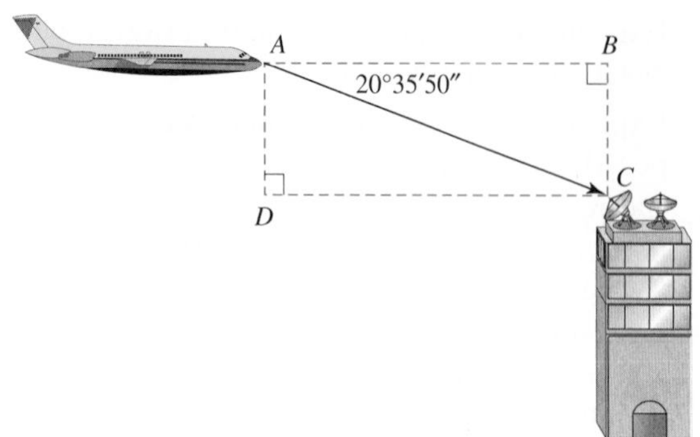

30. A forester determines that the **angle of elevation** (the acute angle between the horizontal and the line of sight) to the top of a fir tree is 42°15′, as shown next. Find the measure of $\angle CBA$ in degrees, minutes, and seconds.

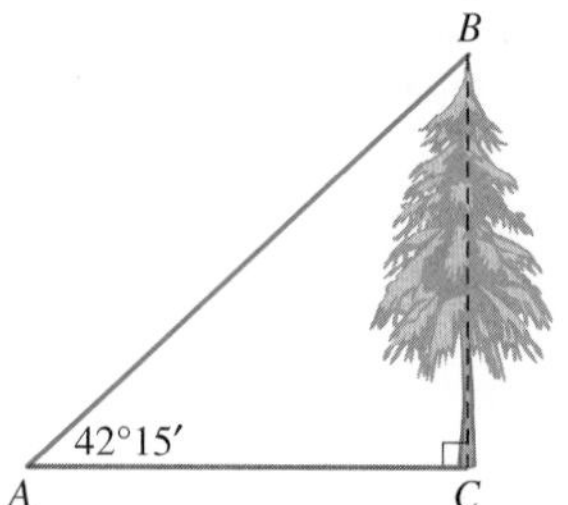

31. In surveying, a **traverse** is the measurement of a polygonal region, made in a series of legs, or line segments. The end of one leg is the start of the next, so the end of the final leg should connect with the beginning of the first leg. When this happens, the traverse is said to "close." The traditional way to check the accuracy of a closed traverse is to measure the interior angles turned at each point and compare the sum of their measures to $(n - 2)180°$.
   a. A survey party traversed a boundary of a proposed timber sale that contained five sides. The interior angles formed were 141°32′, 78°17′, 63°25′, 97°10′, and 159°36′. Determine the error, if any, in this traverse.
   b. A closed traverse with six sides has interior angles of 109°28′, 92°15′, 136°46′, 112°53′, 145°35′, and 121°8′. Determine the error, if any, in this traverse.

32. The interior angles of a certain traverse are given next. Determine whether each traverse closes (that is, forms a closed polygon). (HINT: You might begin by drawing each figure by using a protractor.)
   a. $\angle A = 97.8°$, $\angle B = 61.3°$, $\angle C = 115.5°$, and $\angle D = 82.6°$.
   b. $\angle A = 116°15'$, $\angle B = 89°45'$, $\angle C = 103°30'$ $\angle D = 128°45'$, and $\angle E = 101°45'$.

33. A contractor would like to tile a floor using square tiles and tiles that are equilateral triangles. Both tiles have the same side length. The homeowner would like the same configuration of tiles to surround each vertex. Show two different ways the contractor can lay the floor.

34. Copy and cut out several of each of the following shapes and arrange them to form an aesthetically pleasing quilt block in the given square. You do not need to use all the pieces. Select a vertex from your completed quilt block and use angle measurements to explain why the region surrounding the vertex is completely covered. Also explain why the quilt block will tessellate the plane.

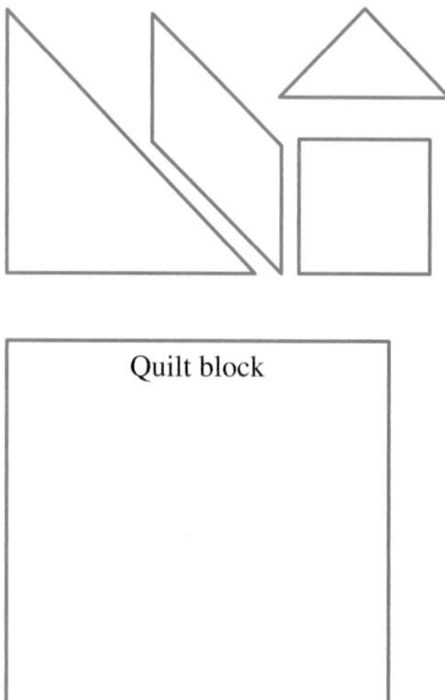

EXTENDED PROBLEMS

35. Draw a large copy of $\triangle ABC$ on scratch paper and cut it out.

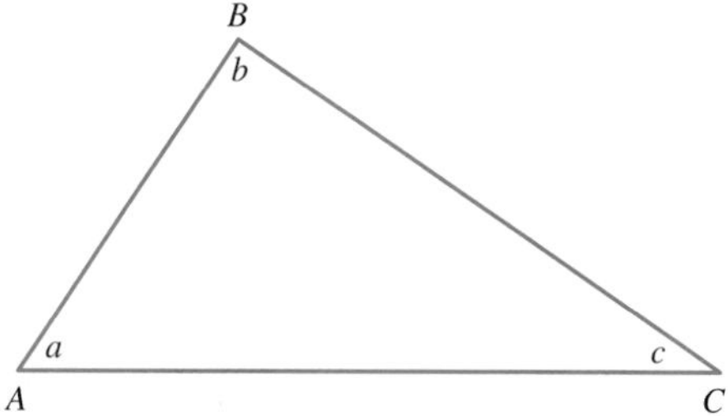

Fold down vertex B so that it lies on $\overline{AC}$ and so that the fold line is parallel to $\overline{AC}$.

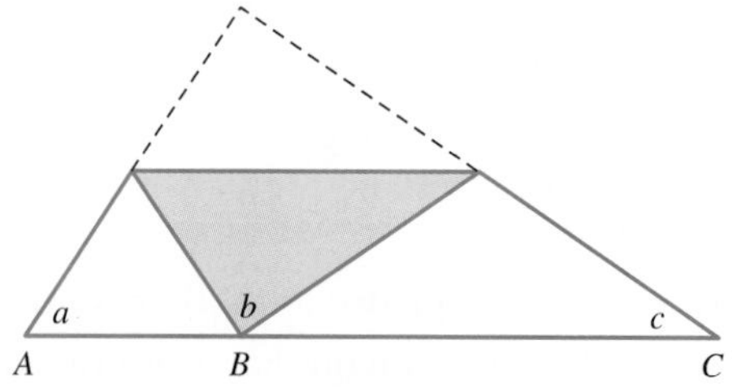

Now fold vertices A and C into point B.

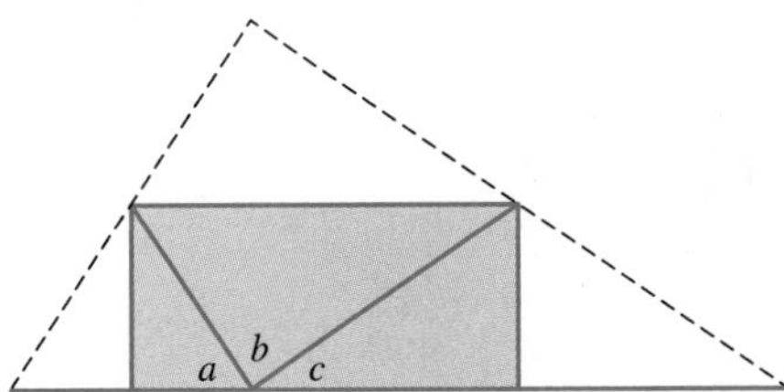

a. What does the resulting figure tell you about the measures of $\angle A$, $\angle B$, and $\angle C$? Explain.
b. What kind of polygon is the folded shape and what is the length of its base?
c. Try this same procedure with two other types of triangles. Are the results the same?

36. Regular pentagons will not tessellate the plane, but there are many irregular pentagons that will. We currently know 14 types of irregular pentagons will tessellate the plane. Students, a physicist, and a homemaker discovered these tessellations independently. Research irregular pentagon tessellations and write a report. For five of the irregular pentagon tessellations, list the angle and side length requirements for the irregular pentagons and provide sketches of the tessellations they will form.

37. Any regular hexagon will form a tessellation, but there are exactly three types of convex, irregular hexagons that tessellate the plane. Research irregular, convex hexagon tessellations. For each hexagon, list the angle and side length requirements and provide sketches of the tessellations it will generate.

2.4 THREE-DIMENSIONAL SHAPES

Applied Problem

A restaurant packages take-out food in cardboard containers like the one shown. The box shapes are convenient in that when assembled they stack inside one another. They also can be unfolded and laid flat. If you ignore the flaps to hold the cover shut, what is the shape of the box when unfolded?

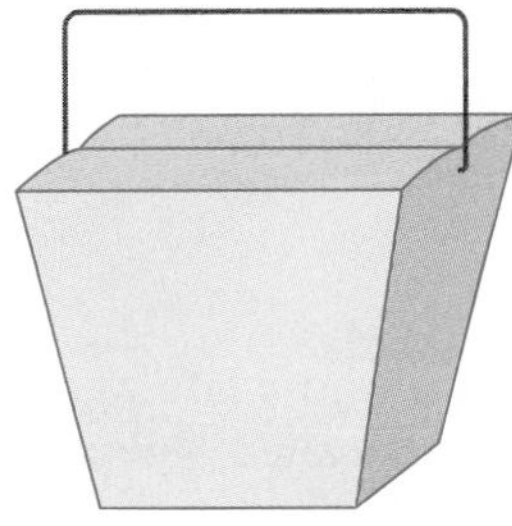

Polyhedra

Thus far we have been studying figures and regions on the plane; that is, figures that are zero dimensional (points), one dimensional (such as lines), and two dimensional (such as triangles). Now we begin a study of three-dimensional shapes.

A **polyhedron** (plural **polyhedra**) is a three-dimensional shape composed of polygonal regions, any two of which have at most a common side. In addition, it is an enclosed, connected finite portion of space without holes. Figure 2.44 shows three examples of polyhedra.

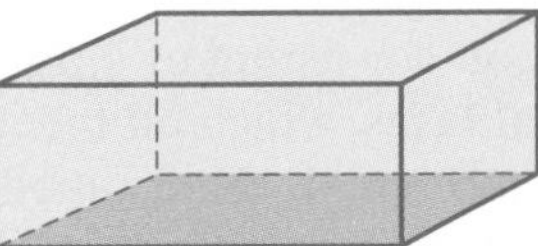
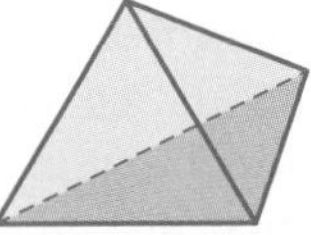
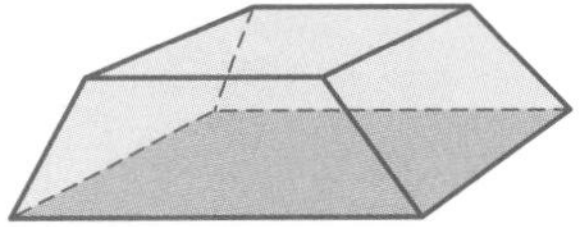

FIGURE 2.44

The polygonal regions of a polyhedron are called **faces**, the common line segments are called **edges**, and a point where edges meet is called a **vertex** (Figure 2.45).

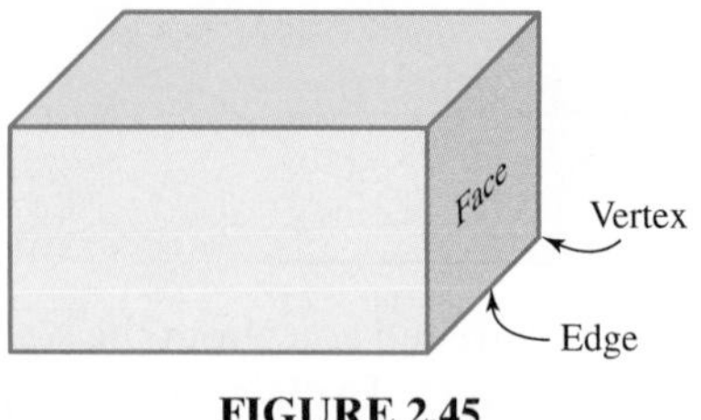

FIGURE 2.45

Prisms

Two common types of polyhedra we will study are prisms and pyramids. A **prism** is a polyhedron with two opposite faces, called **bases**, that are identical polygonal regions in parallel planes; that is, planes that do not have any common points. The other faces are called **lateral faces**. The **height** of a prism is the perpendicular distance between the bases. Prisms are named by the shape of their bases. Some examples are shown in Figure 2.46.

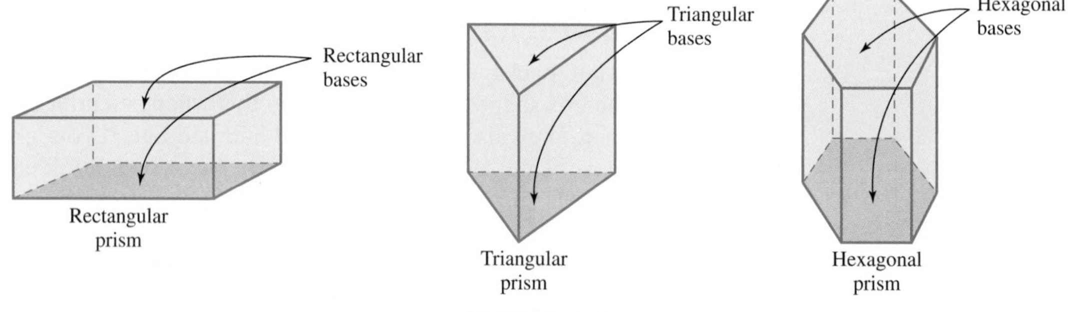

FIGURE 2.46

Prisms whose lateral faces are rectangular regions are called **right prisms**; otherwise they are called **oblique prisms** (Figure 2.47).

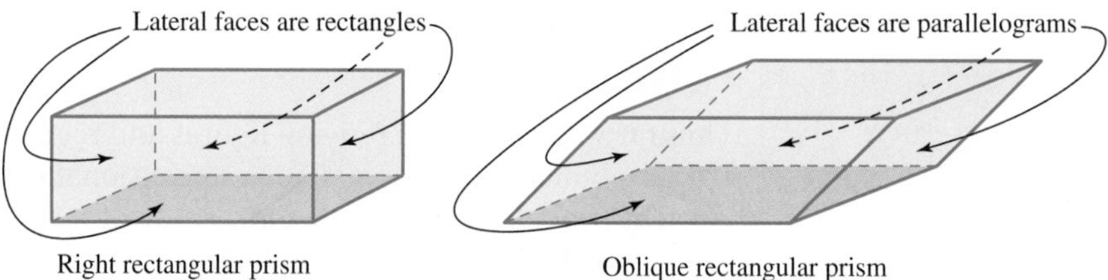

FIGURE 2.47

Pyramids

A **pyramid** is a polyhedron consisting of a polygonal region for its **base** and triangular regions as **lateral faces**, which all meet at a point called the **apex** of the pyramid. The name of a pyramid is determined by the shape of its base. The perpendicular distance from the apex to the base of a pyramid is called the **height** of the pyramid (Figure 2.48).

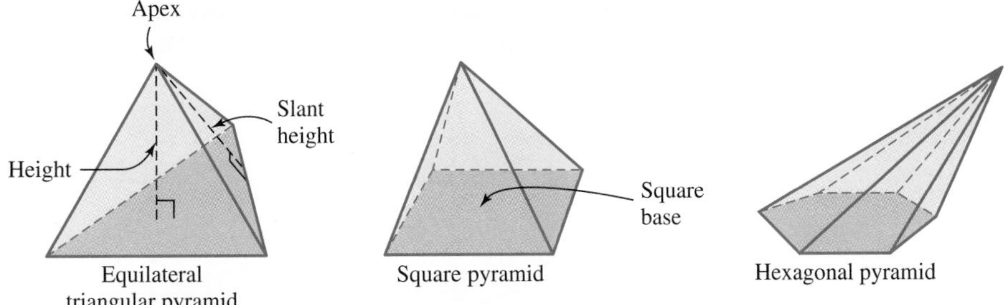

FIGURE 2.48

When the base of a pyramid is a regular polygon and its lateral faces are *isosceles* triangles, the pyramid is called a **right regular pyramid**, but if its lateral faces are not isosceles triangles, the pyramid is called an **oblique regular pyramid**. The **slant height** of a right regular pyramid is the height of any of its faces. The triangular and square pyramids in Figure 2.48 are right regular pyramids.

EXAMPLE 2.8 Name each of the pyramids or prisms shown in Figure 2.49. Give as complete a description as possible.

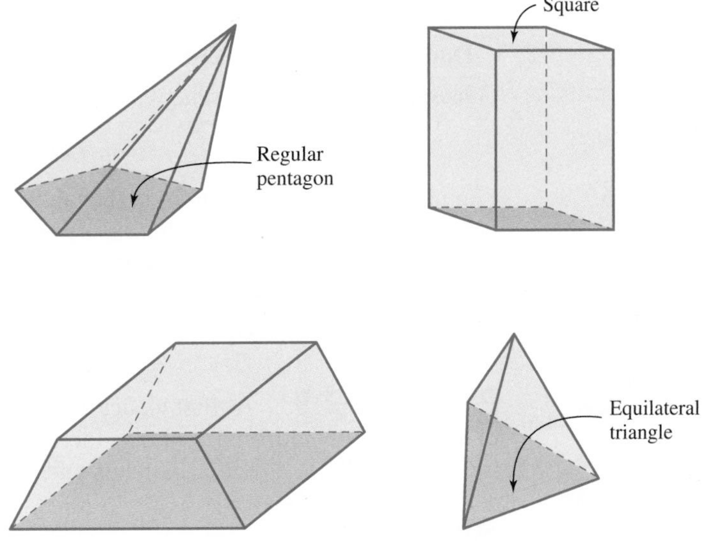

FIGURE 2.49

SOLUTION

a. Oblique regular pentagonal pyramid

b. Right square prism

c. Right trapezoidal prism
(NOTE: The base of the prism is not the "bottom" of the prism, that is, the prism is "resting on" one of its lateral faces.)

d. Right regular triangular pyramid, or right equilateral triangular pyramid ■

Regular Polyhedra

Polyhedra that have regular polygons as faces form a group of especially interesting figures. The Greeks studied the regular polyhedra and imbued them with special significance. A **regular polyhedron** is a polyhedron in which all faces are regular polygons of exactly the same size and shape. The Greeks discovered early on that only five such regular polyhedra exist. Those five are shown in Figure 2.50.

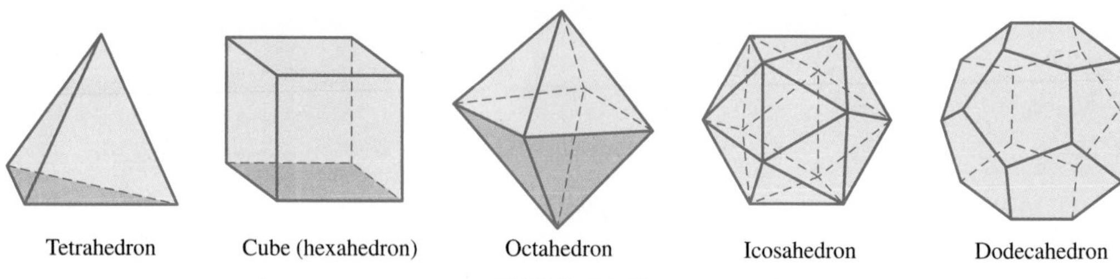

FIGURE 2.50

Table 2.2 summarizes information about the regular polyhedra.

Polyhedron	Shape of Face	Number of Faces, F	Number of Vetrices, V	Number of Edges, E
Tetrahedron	Triangle	4	4	6
Hexahedron	Square	6	8	12
Octahedron	Triangle	8	6	12
Dodecahedron	Pentagon	12	20	30
Icosahedron	Triangle	20	12	30

TABLE 2.2

Examine the last three columns in the table and see if you notice a pattern in the numbers. The mathematician Leonhard Euler noticed a pattern in the relationship among F, V, and E. He observed that in every case $F + V = E + 2$. This relationship has come to be known as **Euler's formula**, and it holds for *all* polyhedra, not just for regular polyhedra. Verify for yourself that Euler's formula holds for the shapes in Figure 2.50.

FIGURE 2.51

EXAMPLE 2.9 At first glance, it might appear that a soccer ball is an example of a regular polyhedron because its faces are all regular polygons (Figure 2.51). Why is a soccer ball not a regular polyhedron?

SOLUTION

The faces of a soccer ball are regular pentagons (the black faces) and regular hexagons (the white faces). Thus, it is not a regular polyhedron because its faces are not identical. ■

If the faces of a soccer ball were all identical regular pentagons, the ball would take on the shape of a dodecahedron and would not be sufficiently "round." If the faces were all identical regular hexagons, the ball would be a flat surface, because as we saw in Section 2.3, regular hexagons tessellate the plane.

The arrangement of faces on the soccer ball does make it an example of a semiregular polyhedron. A **semiregular polyhedron** has several different regular polygons as faces, but the same arrangement of polygons appears at each vertex.

Cylinder

There are common three-dimensional curved shapes analogous to the prism and pyramid. A **circular cylinder** is the shape formed by two identical circular regions in parallel planes together with the surface formed by line segments joining corresponding points of the two circles (Figure 2.52).

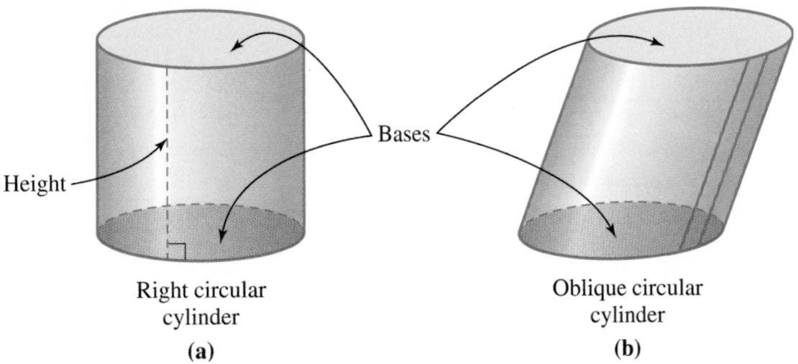

FIGURE 2.52

The two circular regions are called the **bases** of a cylinder. When the line segments joining corresponding points of the two bases are perpendicular to the planes containing the bases, the cylinder is called a **right circular cylinder** [Figure 2.52(a)]; otherwise it is an **oblique circular cylinder** [Figure 2.52(b)]. The **height** of a cylinder is the perpendicular distance between the bases.

Cones

A **circular cone** is the shape formed by a circular region, called the **base**, together with the surface formed when the **apex**, a point not in the same plane as the base, is joined by line segments to every point on the circle (Figure 2.53).

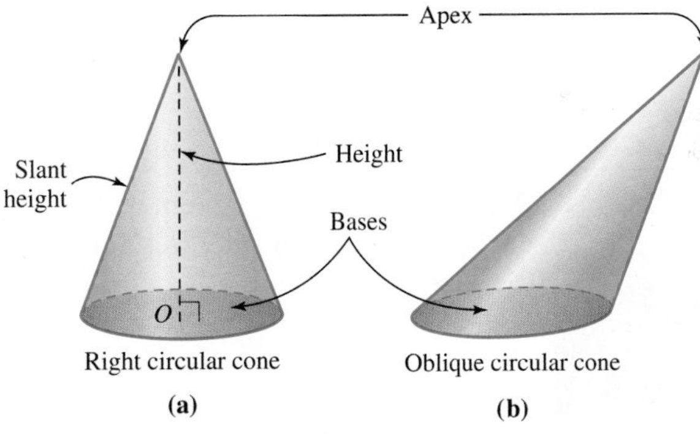

FIGURE 2.53

A **right circular cone** is a circular cone with its apex on the line that is perpendicular to the base at its center [Figure 2.53(a)]; otherwise circular cones are **oblique** [Figure 2.53(b)]. In the right circular cone, the perpendicular distance from the apex to the center of the base is called the **height** of the cone. For a right circular cone, the distance from the apex to a point on the edge of the base is called the **slant height** of the cone.

Spheres

A **sphere** is the set of all points in space that are a fixed distance from a given point (the **center**) of the sphere. A line segment whose endpoints are the center and a point on the sphere is called **a radius** of the sphere; the length of any radius is called **the radius** of the sphere. A line segment that contains the center of the sphere and whose endpoints are on the sphere is called **a diameter** of the sphere; the length of any diameter is called **the diameter** of the sphere (Figure 2.54).

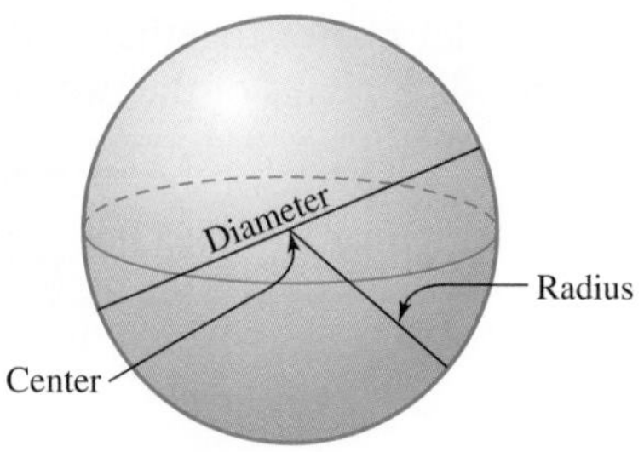

FIGURE 2.54

Solution of Applied Problem

The shape of the container when it is folded up is called a frustum. The top and bottom are rectangles and the lateral faces are all isosceles trapezoids. So the shape of the box when unfolded (ignoring flaps for securing the sides and top) is as shown, where the rectangle for the top has been decomposed into two rectangles.

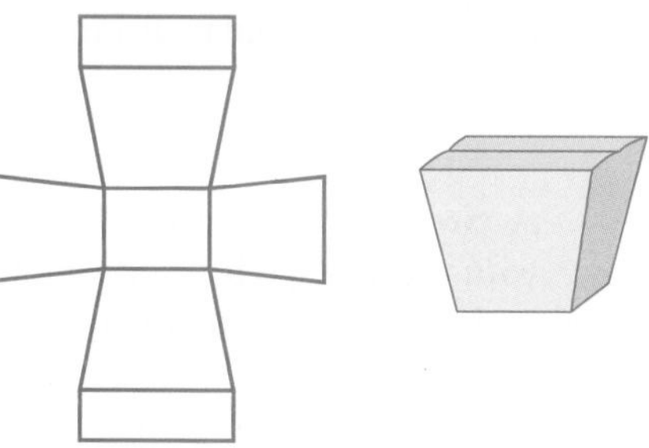

GEOMETRY AROUND US

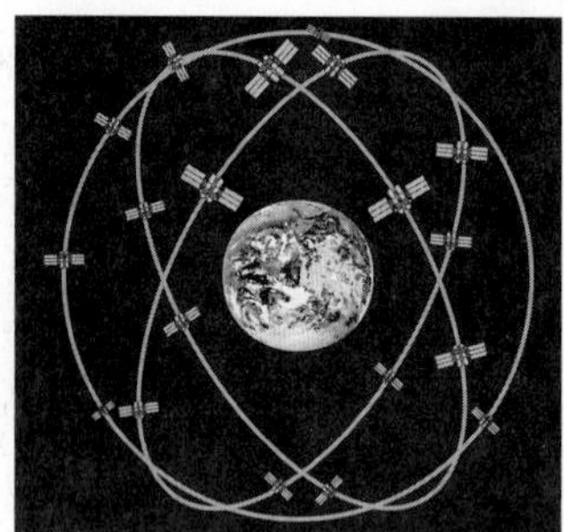

A global positioning system (GPS) receiver works by determining its distance from each of at least four satellites. It uses a process called trilateration, which is a method of determining the relative positions of objects. The GPS receiver uses the fact that it is located a certain distance from a satellite. We can think of the receiver as being located on the surface of an imaginary sphere that has its center at the satellite. Two spheres intersect in a circle. Three spheres intersect in two points. Four spheres intersect in one point. Thus, by using four satellites and four spheres, the GPS receiver can determine its location.

PROBLEM SET 2.4

1. Which of the following shapes are polyhedra? If one is not, explain why.

a. 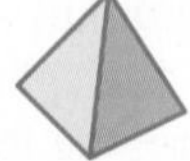 b. 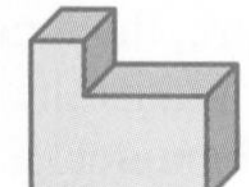 c. 

2. Which of the following shapes are polyhedra? If one is not, explain why.

a.  b. 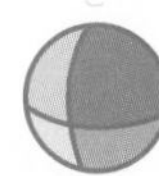 c. 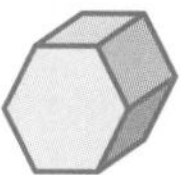

3. Pictured is a stack of cubes. Also given are the top view, the front view, and the right side view. (Assume that the only hidden cubes are ones that support a pictured cube.)

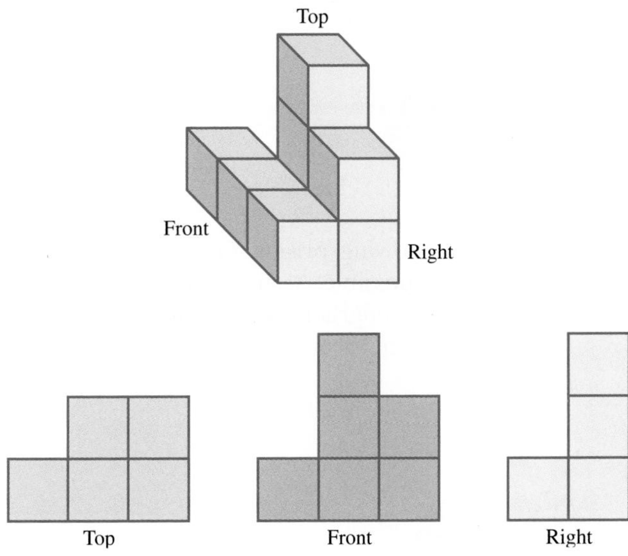

Draw the three views of each of the following stacks of cubes.

a.

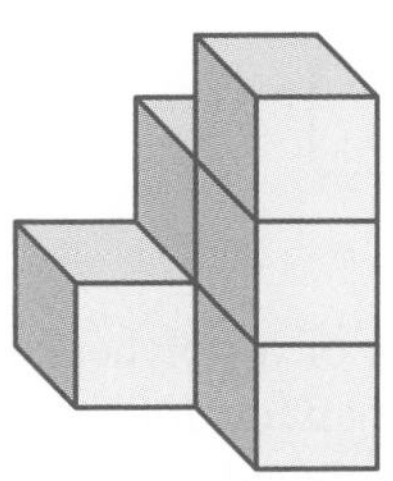

b.

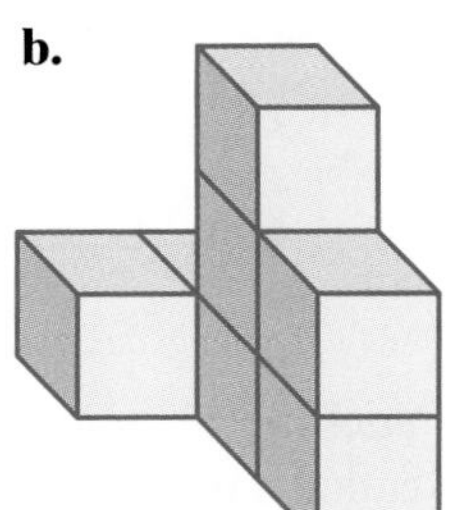

c.

d.

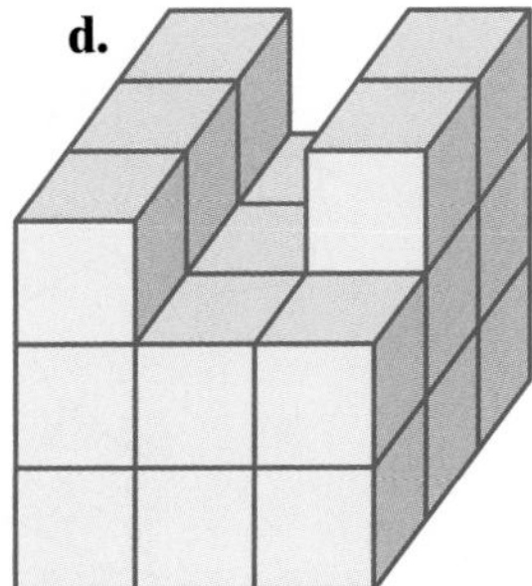

4. In the following figure, the drawing on the left shows a shape. The drawing on the right tells you how many cubes are in each stack forming the shape. It is called a base design.

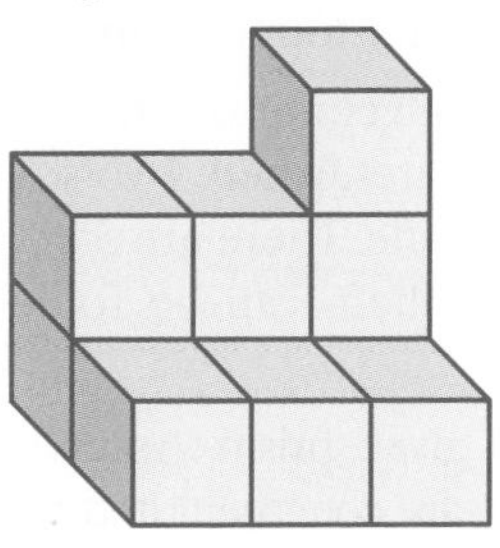

2	2	3
1	1	1

a. Which is the correct base design for the following shape?

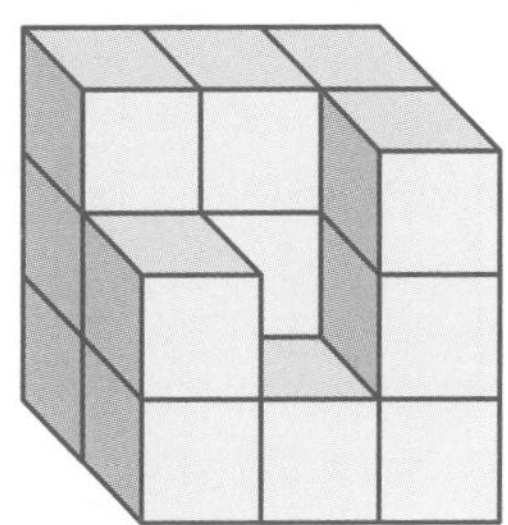

(i)

3	2	3
2	1	2

(ii)

3	3	3
2	1	3

(iii)

3	3	3
3	2	1

b. Make a base design for the next shape.

c. Draw a cube picture for this base design.

3	3	2	1
1	1	2	

5. Recall that a pentomino is a geometric shape composed of five connected squares, where two sides are joined along an entire side. (See problem 4 in Section 1.1). Which of the 12 different pentominos can be folded to form an open-topped cube?

6. Recall that a **hexomino** is a geometric shape composed of six connected squares, where two squares are joined along an entire side. There are 35 different hexominos. Which of them can be folded to form a cube?

7. Do the following for the given prism. Assume segments that look congruent are congruent and assume angles that look like right angles are right angles.

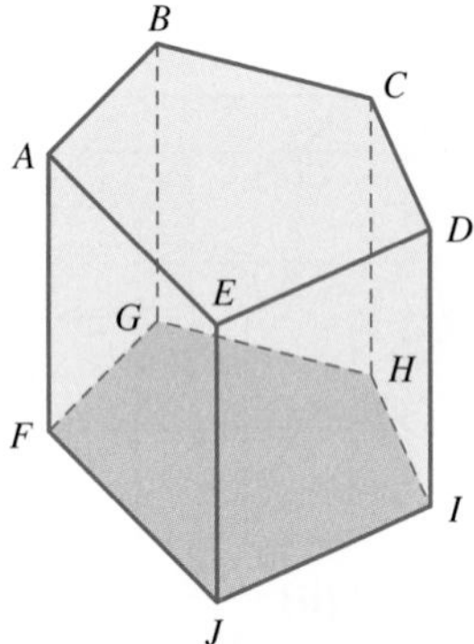

   a. Name the bases of the prism.
   b. Name the lateral faces of the prism.
   c. Name the faces that are hidden from view.
   d. Name the prism by type. Give as complete a description as possible.

8. Do the following for the given prism. Assume segments that look congruent are congruent and assume angles that look like right angles are right angles.

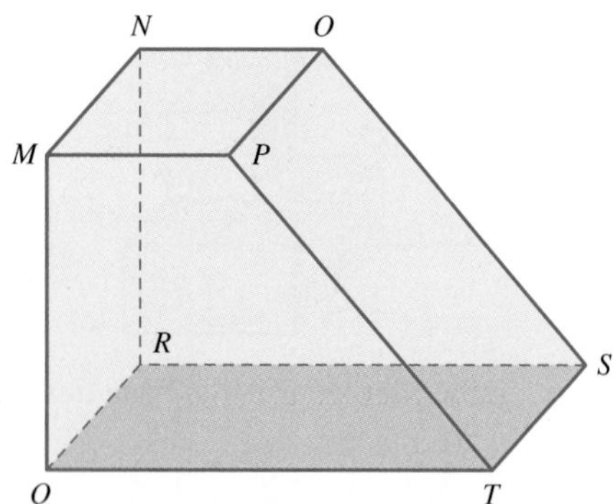

   a. Name the bases of the prism.
   b. Name the lateral faces of the prism.
   c. Name the faces that are hidden from view.
   d. Name the prism by type. Give as complete a description as possible.

9. Name the following prisms. Give as complete a description as possible. Assume segments that look congruent are congruent and assume angles that look like right angles are right angles.

   a. 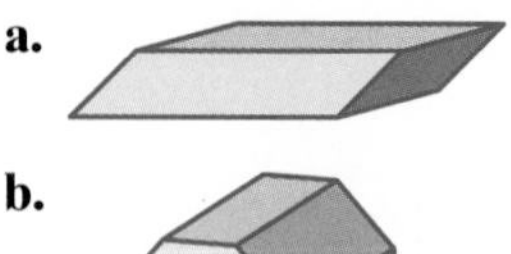

   b. 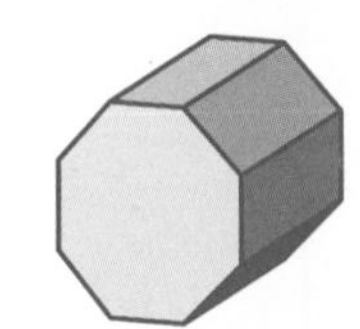

10. Name the following prisms. Give as complete a description as possible. Assume segments that look congruent are congruent and assume angles that look like right angles are right angles.

   a. 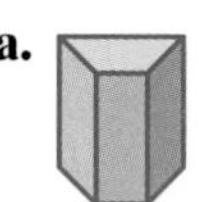

   b. 

11. A cube when unfolded might look like the following.

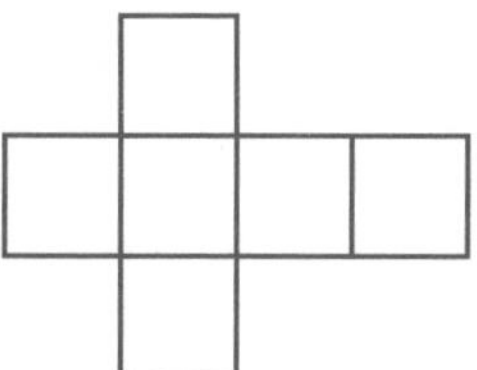

Such an arrangement of faces of a polyhedron is called a **net**. When the following net is folded, what is the name of the resulting polyhedron? Give as complete a description as possible.

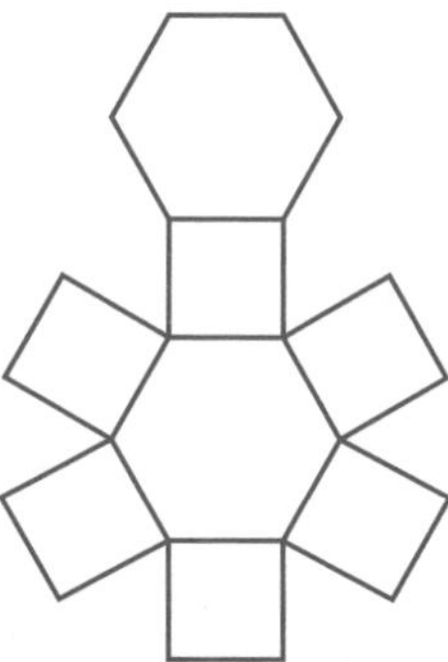

12. Draw a net for the right triangular prism shown.

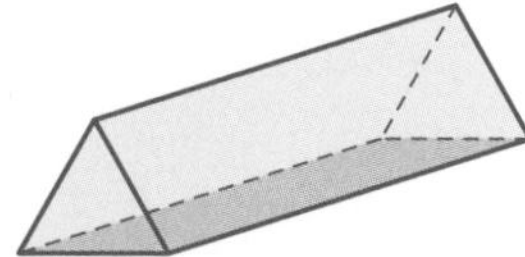

13. Name the following pyramids. Give as complete a description as possible.

a.

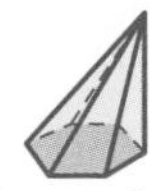

The base is a regular hexagon.

b.

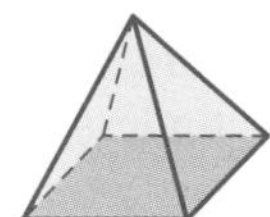

The base is a square.

14. Name the following pyramids. Give as complete a description as possible.

a.

The base is a regular pentagon.

b.

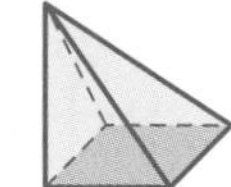

The base is a square.

15. Which of the following nets can be folded to form a regular tetrahedron?

a.

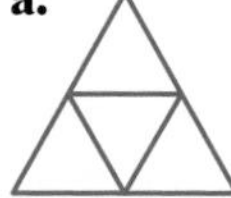

b.

c.

16. Which of the following nets can be folded to form a square pyramid?

a.

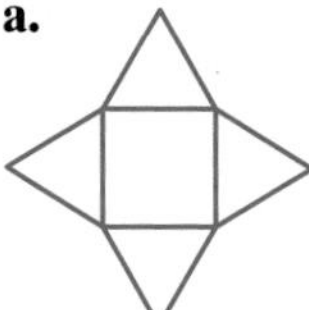

b. **c.**

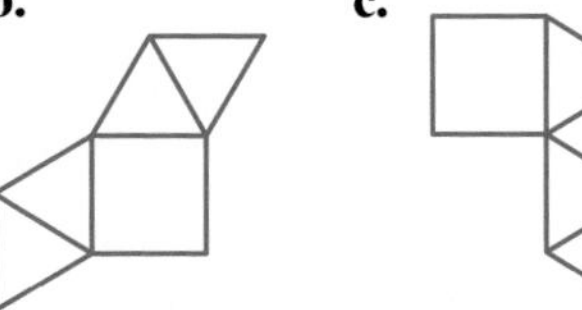

17. When the following net is folded, what is the name of the resulting polyhedron?

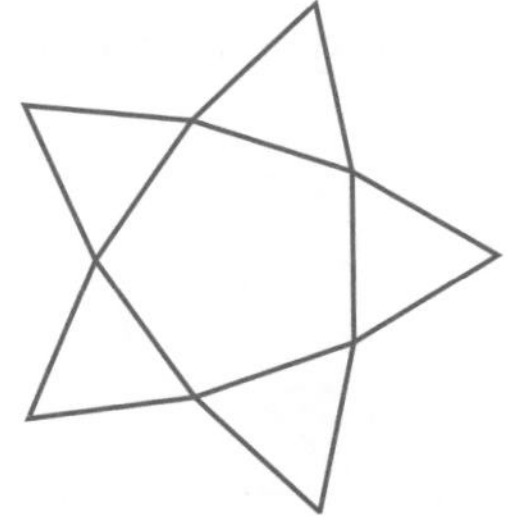

18. Draw a net for the rectangular pyramid shown.

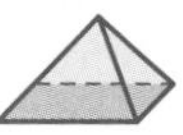

19. Draw a net for a regular octahedron.

20. Draw a net for a regular hexahedron.

21. Which of the three cubes represents a different view of the cube on the left? All faces have different letters.

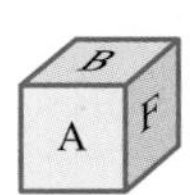

(i)

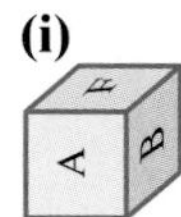

(ii)

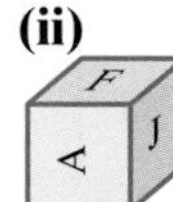

(iii)

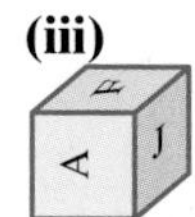

22. All the faces of the cube on the left have different figures on them. Which of the three other cubes represents a different view of that cube?

(i)

(ii)

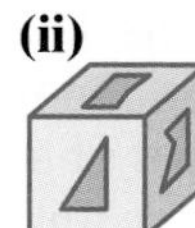

(iii)

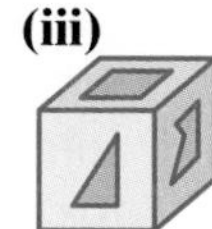

23. Given are several prisms. Use these to complete the following table to see if Euler's formula holds for prisms.

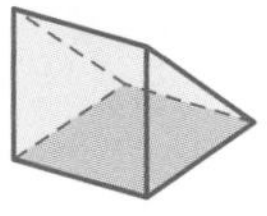

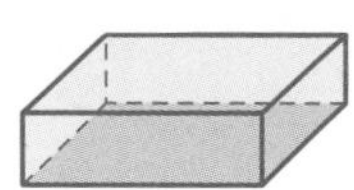

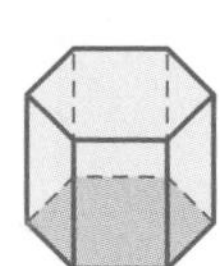

Base	***F***	***V***	***F* + *V***	***E***
Triangle				
Quadrilateral				
Hexagon				
n-gon				

24. Given are several pyramids. Use these to complete the following table to see if Euler's formula holds for pyramids.

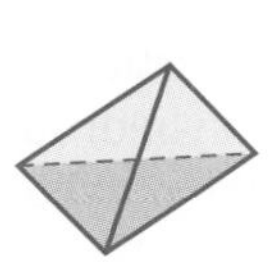

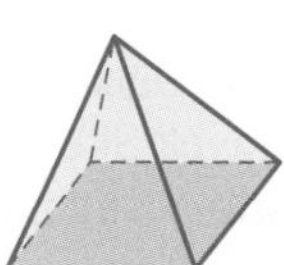

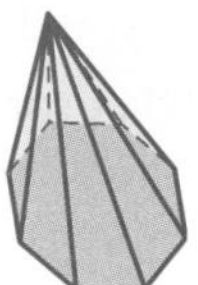

Base	***F***	***V***	***F* + *V***	***E***
Triangle				
Quadrilateral				
Octagon				
n-gon				

25. Given are three arbitrary three-dimensional shapes. Use these to complete the following table to see if Euler's formula holds for these shapes.

a.  **b.**  **c.** 

Figure	F	V	$F + V$	E
a.				
b.				
c.				

26. An **antiprism** is a polyhedron that resembles a prism except that its lateral faces are triangles. Three antiprisms are shown. Notice how the bases are rotated images of each other. Use these antiprisms to complete the following table to see if Euler's formula holds for these shapes.

a. 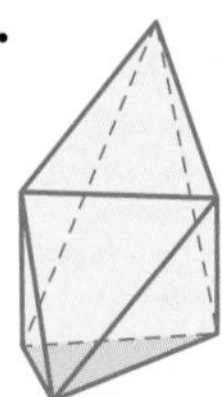 **b.** 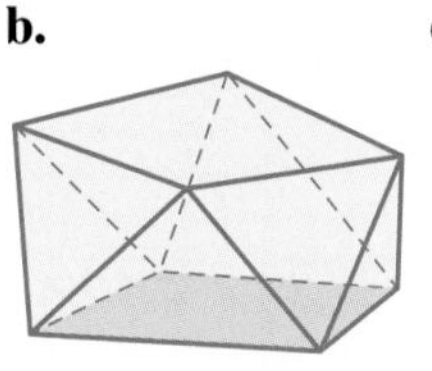 **c.** 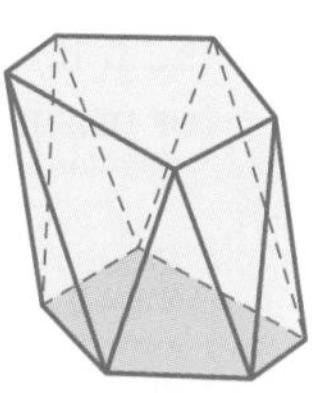

Base	F	V	$F + V$	E
Triangle				
Quadrilateral				
Pentagon				
n-gon				

27. A polyhedron has 18 edges and eight faces. How many vertices does it have? Sketch a polyhedron that satisfies these conditions and name the polyhedron you drew.

28. A polyhedron has 10 edges and six vertices. How many faces does it have? Sketch a polyhedron that satisfies these conditions and name the polyhedron you drew.

29. Identify these figures by name. Give as complete a description as possible.

a. 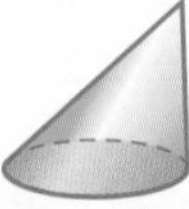

Base is a circular region.

b. 

Base is a circular region.

30. Identify these figures by name. Give as complete a description as possible.

a.

All cross-sections are circles.

b. 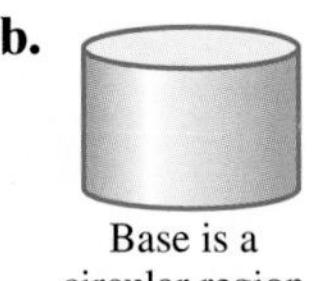

Base is a circular region.

31. Describe the cross sections formed when a plane cuts the following circular cylinders as shown.

a. **b.** 

32. Describe the cross sections formed when a plane cuts the following right circular cones as shown.

a. 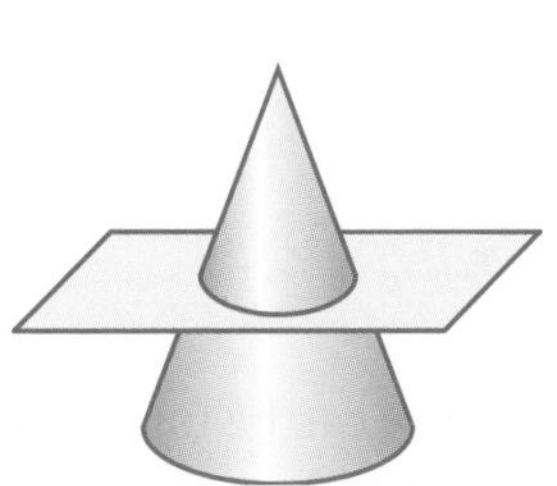 **b.** 

33. If the cube illustrated is cut by a plane midway between opposite faces and the front portion is placed against a mirror, the entire cube appears to be formed. The cutting plane is called a **plane of symmetry** and the figure is said to have **reflection symmetry**.

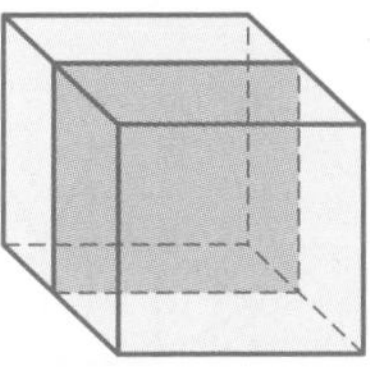

a. How many planes of symmetry of this type are there for a cube?

b. A plane passing through pairs of opposite edges as shown next is also a plane of symmetry. How many planes of symmetry of this type are there in a cube?

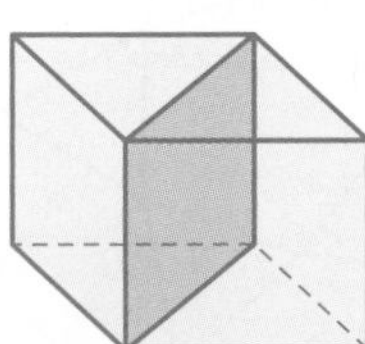

c. How many planes of symmetry of all types are there for a cube?

34. The line connecting centers of opposite faces of a cube is an axis of rotational symmetry because the cube can be turned about the axis and appears to be in the same position as it was initially. In fact, as it rotates around the axis, the cube can be placed in three different positions that match the original position. This axis of symmetry is said to have order 4. How many axes of symmetry of order 4 are there in a cube?

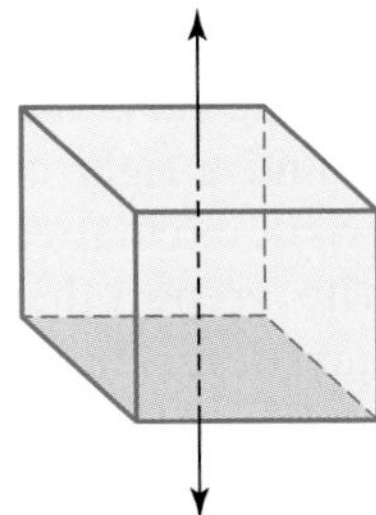

35. How many planes of symmetry do the following figures have? (Models may help.)

a. Regular tetrahedron

b. Right square pyramid

c. Right regular pentagonal prism

d. Right circular cylinder

36. Find the axes of symmetry for the following figures. Indicate the order of each type. (Models may help.)

a. Regular tetrahedron

b. Right regular pentagonal pyramid

37. Imagine a regular tetrahedron cut with a plane through the midpoints of three edges as shown. Describe the shape of each intersection.

a.

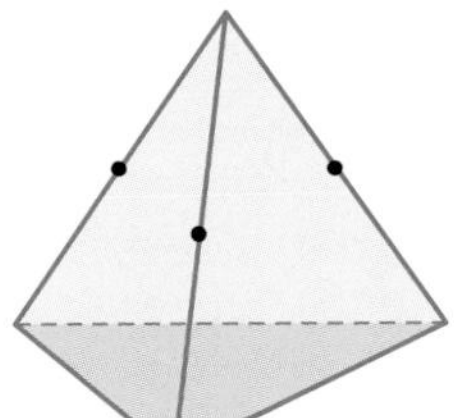

b.

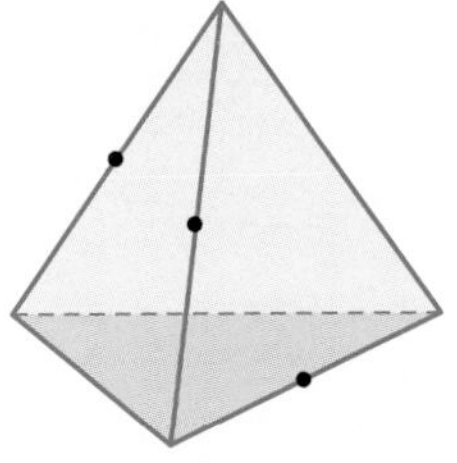

38. Imagine a cube cut with a plane through the midpoints of three edges as shown. Describe the shape of each intersection.

a.

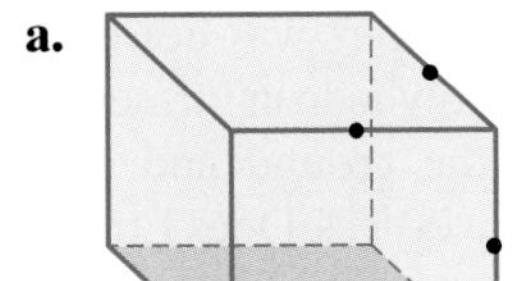

b.

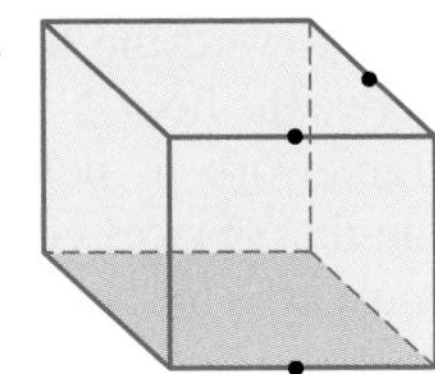

c.

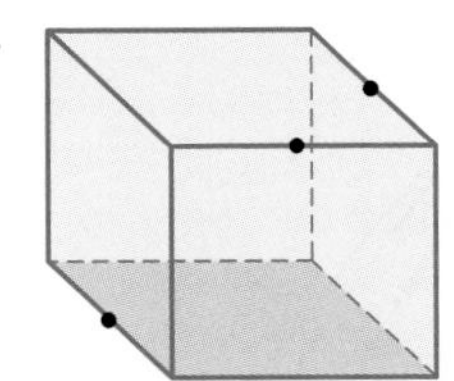

d.

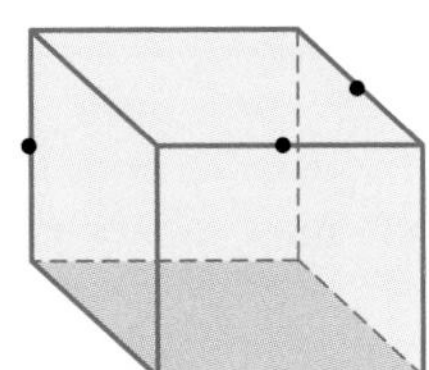

APPLICATIONS

39. Give the name of the three-dimensional shape that corresponds to each of the following common objects. Give as complete a description as possible.

a.

b.

c.

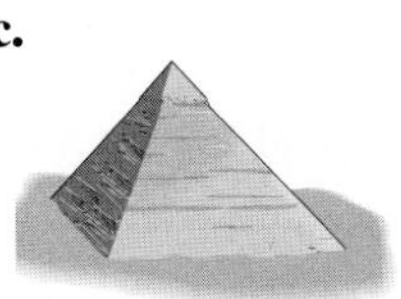

d.

40. Give the name of the three-dimensional shape that corresponds to each of the following common objects. Give as complete a description as possible.

a.

b.

c.

d.

41. A tipi is a portable structure made of poles and covered in hides. Native American tipis varied in size and the number of poles used. A 12-pole tipi framework is shown in the following figure. The framework of a tipi

lean toward the west so the apex was not centered above the floor. Name the polyhedron formed by the framework of poles and the ground and give the number of faces, edges, and vertices. Does Euler's formula hold for Native American tipis?

EXTENDED PROBLEMS

42. Search your home to find two examples each of prisms, pyramids, cones, and cylinders. Make a sketch for each object you find. Name each object, describing it as completely as possible using the vocabulary from this section. For each polyhedron example you find, specify the shape of the base(s) and the lateral faces; list the number of edges, vertices, and faces; and verify that Euler's formula holds.

43. Most minerals occur naturally as crystals. A crystal is a three-dimensional structure whose shape depends on how its atoms combine. When minerals crystallize in an environment without interference, they develop into shapes we recognize as polyhedra. Research the following shape classifications of crystals: cubic, tetragonal, orthorhombic, hexagonal, trigonal, and monoclinic. For each, describe the shapes of the faces, verify that Euler's formula holds, and list three examples of crystals of that type.

44. A conic section is the curve formed by the intersection of a plane and a right circular cone, or of a plane and two right circular cones. Research conic sections and write a report that includes answers to the following questions. How many different conic sections are possible? How must the two cones be oriented? How must the plane intersect the cone(s) to produce each conic section? To whom is credit given for discovering the conic sections? What are some applications of conic sections? Be sure to print or sketch examples of each conic section.

2.5 DIMENSIONAL ANALYSIS

Applied Problem

David is planning a summer motorcycle trip around the United States. He drew his route on a map and estimated its length to be about 265 cm. The scale on his map is 1 cm = 39 km. His motorcycle averages about 50 miles per gallon. If he figures gasoline will cost about $3.00 per gallon, how much will the gasoline for his trip cost him?

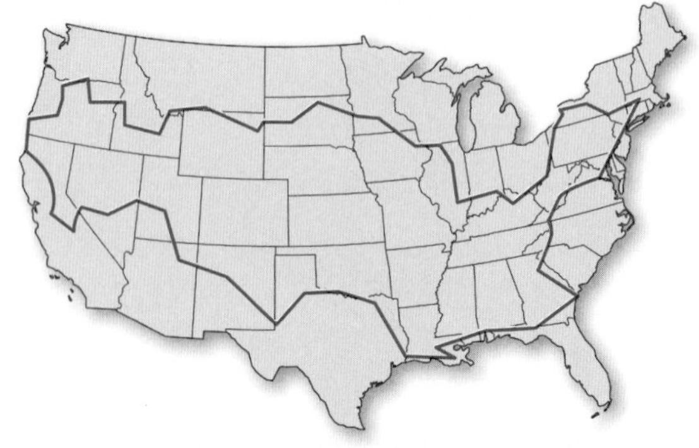

Converting Using Dimensional Analysis

Various units of measure and systems of measure are available for measuring the length of a line segment, size of an angle, or area of a polygonal region. Conversions within these systems or between two systems are frequently necessary. **Dimensional analysis** is a problem-solving strategy that we can use to make conversions. This strategy may be appropriate when units of measure are involved, the problem involves physical quantities, or conversions are required.

In the technique of dimensional analysis, we use fractions equivalent to 1 that include units of measure. These fractions are called **unit ratios** or **unit fractions**. For example, because 1 foot = 12 inches, we can write unit ratios of $\frac{1 \text{ foot}}{12 \text{ inches}}$ or $\frac{12 \text{ inches}}{1 \text{ foot}}$. In each case, the numerator equals the denominator, so each fraction is equivalent to 1. Other examples of unit ratios are $\frac{60 \text{ min}}{1 \text{ hr}}$, $\frac{1 \text{ hr}}{60 \text{ min}}$, $\frac{4 \text{ qt}}{1 \text{ gal}}$, and $\frac{1 \text{ kg}}{1000 \text{ g}}$.

To perform a conversion, we treat a unit ratio as a fraction and multiply it by a measurement or another unit fraction. We must carefully choose the unit ratio in order to obtain the desired units.

EXAMPLE 2.10

Convert a length of 9 feet to inches.

SOLUTION

First we write the given measurement as a fraction: $9 \text{ ft} = \frac{9 \text{ ft}}{1}$. If we multiply this fraction by any other fraction equivalent to 1, we will obtain a new fraction equivalent to the original one. Two unit ratios that we might use are $\frac{12 \text{ in.}}{1 \text{ ft}}$ and $\frac{1 \text{ ft}}{12 \text{ in.}}$ because they both relate feet to inches. However, using the first ratio will allow us to "cancel" units as follows. (NOTE: We abbreviate inch using a period "in." to distinguish it from "in" as in "don't fall in.")

$$9 \text{ ft} = \frac{9 \cancel{\text{ft}}}{1} \cdot \frac{12 \text{ in.}}{1 \cancel{\text{ft}}} = \frac{9(12)}{1(1)} \text{ in.} = 108 \text{ in.}$$

This means that a length of 9 ft is the same as 108 in. ■

Notice that because we wished to convert from feet to inches and the original fraction had ft in the numerator, we chose a unit ratio, $\frac{12 \text{ in.}}{1 \text{ ft}}$, having 1 ft in the denominator. In this way the units of feet canceled and we were left with the units we desired, namely inches.

The method of dimensional analysis is especially helpful if several conversions must be made. In that case, we may chain unit ratios together in a multiplication process. The next example demonstrates such a conversion within the metric system.

EXAMPLE 2.11

A rectangular garden plot measures 345 cm on one side. What is this length in kilometers?

SOLUTION

Because 100 cm = 1 m and 1000 m = 1 km, we can write

$$345 \text{ cm} = \frac{345 \cancel{\text{cm}}}{1} \cdot \frac{1 \cancel{\text{m}}}{100 \cancel{\text{cm}}} \cdot \frac{1 \text{ km}}{1000 \cancel{\text{m}}} = \frac{345(1)(1)}{1(100)(1000)} \text{km} = 0.00345 \text{ km}$$ ■

Notice again that we chose to use $\frac{1 \text{ m}}{100 \text{ cm}}$ rather than the unit ratio $\frac{100 \text{ cm}}{1 \text{ m}}$ in order for the appropriate units to "cancel." Next we chose $\frac{1 \text{ km}}{1000 \text{ m}}$ to cancel the meters.

Dimensional analysis works equally well when we are required to convert between the English and metric systems. Although converting from feet to inches or cm to km without using dimensional analysis is fairly straightforward, it may be more difficult to decide whether to multiply or divide when converting between systems or when working with unfamiliar units. At those times, dimensional analysis is an excellent choice of strategy.

To convert between the English and metric units for length, only one relationship between the two systems is really necessary: 1 inch equals exactly 2.54 centimeters. All other conversions can be performed using this relationship.

EXAMPLE 2.12 The radius of the earth is approximately 6380 km. What is the radius of the earth in miles?

SOLUTION

We will use the relationship 1 in. = 2.54 cm and will multiply several unit ratios together. First we convert km to cm.

$$6380 \text{ km} = \frac{6380 \cancel{\text{km}}}{1} \cdot \frac{1000 \cancel{\text{m}}}{1 \cancel{\text{km}}} \cdot \frac{100 \cancel{\text{cm}}}{1 \cancel{\text{m}}} = 638{,}000{,}000 \text{ cm}$$

Next we convert from metric to English units.

$$\frac{638{,}000{,}000 \text{ cm}}{1} \cdot \frac{1 \text{ in.}}{2.54 \cancel{\text{cm}}} \approx 251{,}000{,}000 \text{ in.}$$

Finally, we convert from inches to miles.

$$\frac{251{,}000{,}000 \cancel{\text{in.}}}{1} \cdot \frac{1 \cancel{\text{ft}}}{12 \cancel{\text{in.}}} \cdot \frac{1 \text{ mi}}{5280 \cancel{\text{ft}}} \approx 3964 \text{ mi}$$

The previous three steps can be compressed into one step as follows:

$$\begin{aligned} 6380 \text{ km} &= \frac{6380 \cancel{\text{km}}}{1} \cdot \frac{1000 \cancel{\text{m}}}{1 \cancel{\text{km}}} \cdot \frac{100 \cancel{\text{cm}}}{1 \cancel{\text{m}}} \cdot \frac{1 \cancel{\text{in.}}}{2.54 \cancel{\text{cm}}} \cdot \frac{1 \cancel{\text{ft}}}{12 \cancel{\text{in.}}} \cdot \frac{1 \text{ mi}}{5280 \cancel{\text{ft}}} \\ &= \frac{6380(1000)(100)(1)(1)(1)}{1(1)(1)(2.54)(12)(5280)} \text{ mi} \approx 3964 \text{ mi} \end{aligned}$$

■

So the radius of the earth is about 3964 mi. It is usually best to perform the conversion in one step, doing all the multiplications at once to avoid rounding errors.

We can use the same steps as in Example 2.12 to derive a new conversion ratio that allows us to convert directly from mi to km or km to mi.

EXAMPLE 2.13 Find the number of miles in a kilometer; that is, 1 km is the same as how many mi?

SOLUTION

$$\begin{aligned} 1 \text{ km} &= \frac{1 \cancel{\text{km}}}{1} \cdot \frac{1000 \cancel{\text{m}}}{1 \cancel{\text{km}}} \cdot \frac{100 \cancel{\text{cm}}}{1 \cancel{\text{m}}} \cdot \frac{1 \cancel{\text{in.}}}{2.54 \cancel{\text{cm}}} \cdot \frac{1 \cancel{\text{ft}}}{12 \cancel{\text{in.}}} \cdot \frac{1 \text{ mi}}{5280 \cancel{\text{ft}}} \\ &= \frac{1(1000)(100)(1)(1)(1)}{1(1)(1)(2.54)(12)(5280)} \text{ mi} \approx 0.6214 \text{ mi} \end{aligned}$$

So, for future reference, we have these *approximate* conversion ratios:

$$\frac{1 \text{ km}}{0.6214 \text{ mi}} \quad \text{and} \quad \frac{0.6214 \text{ mi}}{1 \text{ km}}$$

■

Converting Rates Using Dimensional Analysis

Different units of measure are often mixed to form some kind of **rate**, such as 35 mi/gal, 88 ft/sec, or 1.49 dollars/liter. Dimensional analysis can be helpful in converting from one rate to another equivalent rate, as shown next.

EXAMPLE 2.14 A well produces water at a rate of 30 gallons per minute. Express this rate in liters per day.

SOLUTION

We must convert the rate gal/min to liters/day. This means we must multiply by unit fractions to convert gallons to liters and minutes to days. We can perform either of these conversions first or do them simultaneously. We need a conversion ratio that relates gallons and liters. We will use the facts that 1 liter = 1.057 quarts and that 1 gallon = 4 quarts.

We can think of the rate 30 gal/min as the ratio $\frac{30 \text{ gal}}{1 \text{ min}}$. That is, 30 gal of water are produced for every 1 min of time that passes. To convert from gallons per minute to liters per minute, we can use the following sequence of fractions:

$$\frac{30 \text{ gal}}{1 \text{ min}} = \frac{30 \cancel{\text{gal}}}{1 \text{ min}} \cdot \frac{4 \cancel{\text{qt}}}{1 \cancel{\text{gal}}} \cdot \frac{1 \text{ L}}{1.057 \cancel{\text{qt}}} \approx \frac{113.5 \text{ L}}{1 \text{ min}}$$

If these fractions are multiplied, the result is in L/min. Now we must convert from minutes to days.

$$\begin{aligned} \frac{30 \text{ gal}}{1 \text{ min}} &\approx \frac{113.5 \text{ L}}{1 \cancel{\text{min}}} \cdot \frac{60 \cancel{\text{min}}}{1 \cancel{\text{hr}}} \cdot \frac{24 \cancel{\text{hr}}}{1 \text{ day}} \\ &= \frac{113.5(60)(24)}{1(1)(1)} \text{ L/day} \\ &\approx 163{,}000 \text{ L/day} \end{aligned}$$

■

We could have found the solution in Example 2.14 using one set of fractions as follows:

$$\begin{aligned} \frac{30 \text{ gal}}{1 \text{ min}} &= \frac{30 \cancel{\text{gal}}}{1 \cancel{\text{min}}} \cdot \frac{4 \cancel{\text{qt}}}{1 \cancel{\text{gal}}} \cdot \frac{1 \text{ L}}{1.057 \cancel{\text{qt}}} \cdot \frac{60 \cancel{\text{min}}}{1 \cancel{\text{hr}}} \cdot \frac{24 \cancel{\text{hr}}}{1 \text{ day}} \\ &\approx 163{,}000 \text{ L/day} \end{aligned}$$

Notice that we chose the unit ratio $\frac{60 \text{ min}}{1 \text{ hr}}$ because the original rate, $\frac{30 \text{ gal}}{1 \text{ min}}$, had minutes in the denominator.

The next and last two examples demonstrate how we can use the technique of dimensional analysis in more general problem-solving situations.

Problem Solving Using Dimensional Analysis

EXAMPLE 2.15 The sun is 93,000,000 miles from the earth and light travels at a speed of approximately 186,000 miles/second. How long, in minutes, does it take for light from the sun to reach the earth?

SOLUTION

We can express the rate 186,000 miles/second as $\frac{186{,}000 \text{ mi}}{1 \text{ sec}}$ or as $\frac{1 \text{ sec}}{186{,}000 \text{ mi}}$. However, because we are interested in the unit time, we use the latter fraction since time (sec) is in the numerator.

$$\frac{(93{,}000{,}000)\,\cancel{\text{mi}}}{1} \cdot \frac{1\,\cancel{\text{sec}}}{(186{,}000)\,\cancel{\text{mi}}} \cdot \frac{1 \text{ min}}{60\,\cancel{\text{sec}}} = \frac{(93{,}000{,}000)(1)(1)}{(186{,}000)(60)} \text{ min} = 8\tfrac{1}{3} \text{ min}$$

So light from the sun hits the earth in about $8\frac{1}{3}$ minutes. ■

EXAMPLE 2.16 A do-it-yourselfer is building a backyard barbecue. The project requires 525 bricks and they must be moved to the building site in a wheelbarrow that will carry a maximum of 150 lb of bricks. If the bricks weigh $2\frac{1}{4}$ tons per thousand, how many trips with the wheelbarrow will be necessary?

SOLUTION

We start with the 525 bricks and multiply by ratios of the appropriate form to convert to wheelbarrow loads.

$$\begin{aligned} 525 \text{ bricks} &= \frac{525\,\cancel{\text{bricks}}}{1} \cdot \frac{2.25\,\cancel{\text{ton}}}{1000\,\cancel{\text{bricks}}} \cdot \frac{2000\,\cancel{\text{lb}}}{1\,\cancel{\text{ton}}} \cdot \frac{1 \text{ wheelbarrow load}}{150\,\cancel{\text{lb}}} \\ &= \frac{525(2.25)(2000)(1)}{1(1000)(1)(150)} \text{ wheelbarrow loads} \\ &= 15.75 \text{ wheelbarrow loads} \end{aligned}$$

So it will take 16 trips with the wheelbarrow to move all of the bricks. ■

Solution of Applied Problem

We can use dimensional analysis to solve this problem if we chain the unit fractions required. First, we determine the actual distance to be traveled according to the scale and his measurements.

$$\text{Distance traveled in km} = \frac{265\,\cancel{\text{cm}}}{1} \cdot \frac{39 \text{ km}}{1\,\cancel{\text{cm}}}$$

To convert from km to mi, we could use the fact that 1 in. = 2.54 cm. However, we can obtain an approximate solution with fewer calculations if we use 1 km $\approx$ 0.6214 mi from Example 2.13.

$$\text{Distance traveled in mi} \approx \frac{265\,\cancel{\text{cm}}}{1} \cdot \frac{39\,\cancel{\text{km}}}{1\,\cancel{\text{cm}}} \cdot \frac{0.6214 \text{ mi}}{1\,\cancel{\text{km}}}$$

We can find the total cost by multiplying by two more unit ratios to find the number of gallons of gas needed and the cost for that amount of gas.

$$\frac{265\,\cancel{\text{cm}}}{1} \cdot \frac{39\,\cancel{\text{km}}}{1\,\cancel{\text{cm}}} \cdot \frac{0.6214\,\cancel{\text{mi}}}{1\,\cancel{\text{km}}} \cdot \frac{1\,\cancel{\text{gal}}}{50\,\cancel{\text{mi}}} \cdot \frac{\$3.00}{1\,\cancel{\text{gal}}} \approx \$385.33$$

GEOMETRY AROUND US

In 1866, the United States adopted the metric system as a legal system. However, as of 2006, our customary system was still in use. The United States remains almost an island in a metric world. Two other countries that are not exclusively metric are Liberia and Myanmar. While the United States has made progress, and product packaging now displays our customary units as well as their metric equivalents, Americans still think in terms of feet, miles, gallons, and degrees Fahrenheit. In 1988, President Reagan signed a bill that required all government agencies to be metric by 1992. Although many U.S. industries, including the automobile and pharmaceutical industries, use metric tools and measures, the change to everyday use of the metric system by the public continues to be gradual.

PROBLEM SET 2.5

A summary of units of measurement is given at the end of the book near the inside back cover.

EXERCISES/PROBLEMS

1. Give the appropriate unit ratios to convert each of the following measures.

a. inches to feet **b.** miles to feet

c. pounds to ounces **d.** quarts to gallons

2. Give the appropriate unit ratios to convert each of the following measures.

a. meters to centimeters

b. millimeters to centimeters

c. kilograms to grams

d. milliliters to liters

3. Give the appropriate unit ratios to convert each of the following measures.

a. 17 hours to minutes

b. 360 seconds to minutes

c. 720 inches to yards

d. 1440 work-hours to work-days

4. Give the appropriate unit ratios to convert each of the following measures.

a. 45 degrees to minutes

b. 0.4 meters to millimeters

c. 60 hours to days

d. 15¢/foot to cents/inch

5. Use the following units of measurement and create the appropriate unit ratios to convert from ounces to pecks.

1 gallon = 4 quarts

1 cup = 8 ounces

1 quart = 4 cups

1 peck = 2 gallons

6. Use the following units of measurement and create the appropriate unit ratios to convert from slugs to glugs, which are rarely used units of mass.

1 hyl ≈ 9.8 kilograms

1 slug ≈ 32.17 pounds

1 glug = 0.1 hyls

1 kilogram ≈ 2.2 pounds

7. Using dimensional analysis, perform the following conversions.

a. 4.5 lb to oz

b. 3744 min to deg

c. 25 mi to in.

d. 10,500 mL to kL

8. Using dimensional analysis, perform the following conversions.

a. 58 mg to g

b. 0.7 m to km

c. 3 tons to oz

d. 2400 sec to days

9. Using the fact that 1 in. = 2.54 cm, perform the following conversions.

a. 6-in. snowfall to cm

b. 100-yd football field to m

c. 420-ft home run to m

d. 1-km racetrack to mi

10. Using the fact that 1 in. = 2.54 cm, perform the following conversions.

a. 26.2-mi marathon to m

b. 1-m ruler to in.

c. 195-mm tire width to in.

d. 8-ft ceiling to cm

11. Using dimensional analysis, perform the following conversions.

a. 45 mi/hr to ft/sec

b. $200/day to dollars/hour

c. 0.3 in./year to ft/century

d. 16 km/L to m/mL

12. Using dimensional analysis, perform the following conversions.

a. 40 mi/gal to mi/qt

b. 64 in./sec to mi/hr

c. 50 kg/m to g/cm

d. $1.99/L to cents/mL

13. Using the fact that 1 in. = 2.54 cm and 1 kg $\approx$ 2.205 lb, perform the following conversions.

a. 100 ft/sec to m/sec

b. 0.4 kg/m to lb/yd

14. Using the fact that 1 in. = 2.54 cm, perform the following conversions.

a. 48¢/ft to cents/cm

b. 60 km/hr to ft/min

APPLICATIONS

15. The speed limit on some U.S. highways is 65 mph. If metric speed limit signs are posted, what will they read? Round to the nearest whole number.

16. The maximum default speed limit on the Autobahn in Germany is 100 km/hr. What is this speed in mi/hr? Round to the nearest whole number.

17. Mary Decker has run a mile on an indoor track in 4 min 17.6 sec. What was her speed in km/hr?

18. Bamboo can grow as much as 90 cm per day. Express this rate in ft/hr. Round to the nearest hundredth.

19. From age two until puberty humans grow an average of 0.12 cm per week. Express this rate in inches per year. Round to the nearest tenth.

20. A seamstress has agreed to sew four bridesmaids' dresses for an upcoming wedding. Each dress requires $1\frac{5}{8}$ yds of lace that sells for 89¢ per inch. Use dimensional analysis to find the total cost for the lace.

21. In 2005, the United States used approximately 20 million barrels of crude oil per day. One barrel of crude oil contains 42 gallons. How many liters of crude oil did the United States use per day in 2005? Round to the nearest liter.

22. The gas tank of a 2006 Audi A3 automobile has a capacity of 55 liters. If regular gasoline is selling for $2.75 per gallon, how much will it cost to fill the tank? Round your answer to the nearest cent.

23. Human hair typically grows about one-half inch in four weeks. At what rate, in mph, does hair grow?

24. A mason laying brick for a wall 8′9″ tall allows $2\frac{5}{8}$″ for each "course," or row, of bricks, including $\frac{1}{2}$″ for the mortar joint. Use dimensional analysis to determine how many courses of brick must be laid.

25. We see lightning before we hear thunder because light travels faster than sound. If sound travels at about 1000 km/hr through air at sea level at 15°C and light travels at 186,282 mi/sec, how many times faster is the speed of light than the speed of sound? Use dimensional analysis.

26. A DC-9, which burns approximately 7000 lb of fuel per hour, makes a daily round trip between St. Louis and San Francisco. The trip from St. Louis to San Francisco takes 4 hr 15 min and the return trip takes 3 hr 45 min because of prevailing winds. How many tons of fuel are burned by the DC-9 in a 30-day month?

27. Light travels 186,282 miles per second. Use dimensional analysis to answer each of the following questions.

a. Based on a 365-day year, how far, in miles, will light travel in one year? This unit of distance is called a **light year**.

b. If a star in Andromeda is 76 light years away from earth, how many miles will light from the star travel on its way back to earth?

c. The planet Jupiter is about 480,000,000 miles from the sun. How long, in hours, does it take for light to travel from the sun to Jupiter?

28. Angles are measured in degrees or radians, where 180° = π radians. (Refer to problem 42 in Section 2.1) Use unit fractions and the technique of dimensional analysis to convert the following angle measures from degrees to radians or from radians to degrees. Round to the nearest hundredth if necessary.

a. 1 radian

b. $\frac{\pi}{4}$ radian

c. 3 radians

d. 20°

e. 300°

f. 155°

29. The **smoot** is a nonstandard unit of length named for Oliver R. Smoot, who was 5 feet 7 inches tall. The Harvard Bridge was found to be 364.4 smoots long.

a. Give a unit ratio that can be used to convert inches to smoots.

b. Give a unit ratio that can be used to convert smoots to meters.

c. How long was the Harvard Bridge in meters? Round to the nearest meter.

d. An American football field is 120 yards long. How long is the field in smoots? Round to the nearest tenth.

EXTENDED PROBLEMS

30. In personal computers, the hard disk drive is the primary storage location for data. In 1956, IBM invented the first computer hard disk storage system that could hold 5 megabytes. Today, it is common to see 60 to 180 gigabyte hard disk drives. Research the history of computer hard disk drive capacity and create a timeline. Include definitions and conversion facts for the following units: bit, kilobit, megabit, byte, kilobyte, megabyte, and gigabyte. If a music CD can store 700 megabytes, investigate how many music CDs can be stored on a typical hard disk drive.

31. The change to the metric system in the United States is a slow process. Today, most product labels list an English measure and, in parentheses, give the metric equivalent. Research the metric system and focus on answering the following questions. What is the difference between "soft metric" and "hard metric"? What would have to happen in order for the United States to become a hard metric country? Next consider five products you find at home such as milk, soda, cereal, canned food, and frozen vegetables. List the English measure and the metric measure from the packages. If these products were packaged in a hard metric country, what would their measures likely be and by what percentage would the weight or volume of each product increase or decrease?

32. The Mars Climate Orbiter was one of a series of missions in Mars exploration. Early in the morning on September 23, 1999, the spacecraft fired its main engine to go into orbit around Mars. All systems looked normal. The orbiter passed behind Mars, but at the point when it was expected to emerge from behind the planet, there was no signal. The orbiter had been lost, and, as it turned out, the cause was directly related to units of measure. Research the loss of the Mars Climate Orbiter and write an essay that explains what went wrong.

Solution of Initial Problem

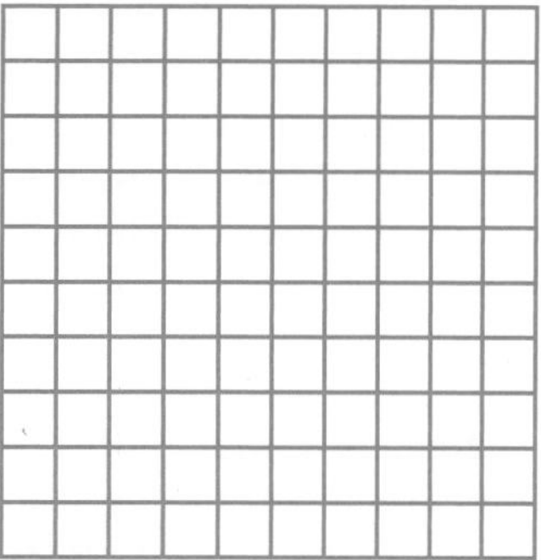

One way to keep count of the various overlapping squares is to count the upper right-hand corners of all of the squares of a particular size. Let's begin a table:

Length of a Side	10	9	8	7
Number of Squares	1	4	9	16

Notice that the number of squares in the second row formed at each stage is itself a square. As we can see, the fifth such number will be $5^2 = 25$, the sixth $6^2 = 36$, and, in general, the formula for the nth such number is n^2. In this particular case, the number in the bottom row that will correspond with 1 in the top row is 10^2. Thus, the total number of squares will be the sum of the squares $1^2 + 2^2 + 3^2 + 4^2 + 5^2 + 6^2 + 7^2 + 8^2 + 9^2 + 10^2 = 385$.

Additional Problems Where the Strategy "Look for a Formula" Is Useful

1. Jack's beanstalk increases in height by $\frac{1}{2}$ the first day, $\frac{1}{3}$ the second day, $\frac{1}{4}$ the third day, and so on. What is the fewest number of days it would take to become at least 100 times as tall as its original height?
2. How many different (nonzero) angles are formed in a fan of rays like the one pictured next, but one having 100 rays?

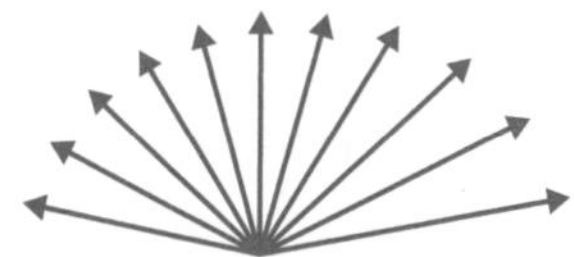

Answers for the additional problems can be found in the Answers section near the end of the book.

Writing for Understanding

1. Describe as many applications of angle measure in other fields as you can. For example, angles are important in sports such as golf, where the flight of the ball is related to the angle at which it is hit as well as to the angle of the clubface.
2. Describe the types of symmetry you see in the room around you. Be sure to consider both types of symmetry and two-dimensional as well as three-dimensional shapes.
3. **Fermi problems** are problems to be solved by estimation without seeking exact numbers from references. Describe how to use estimation and your current knowledge, including dimensional analysis, to solve the following Fermi problem:
   Suppose all the newspapers sold in the United States in one year were stacked in one pile. Estimate the height of that stack in miles. Explain your reasoning at each step.
4. "Geometry Around Us" at the end of Section 2.5 discusses the metric system, which is formally referred to as the Système International (SI). Write a report contrasting this system with our customary system, which uses feet, pounds, and so on. Make a case for the retention of our customary system or for converting to the metric system. Be sure to support your arguments.

PEOPLE IN GEOMETRY

Leonhard Euler (1707–1783) was one of the most prolific of all mathematicians. He published 530 books and papers during his lifetime and left much unpublished work at the time of his death. From 1771 on, he was totally blind, yet his mathematical discoveries continued. He would work mentally, then dictate to assistants, sometimes writing the formulas for them on a large chalkboard. His writings are a model of clear exposition, and his style has influenced much of our modern mathematical notation. For instance, the modern use of the symbol π is due to Euler. In geometry, he is best known for the Euler line of a triangle and the formula $V - E + F = 2$, which relates the number of vertices, edges, and faces of any simple closed polyhedron.

CHAPTER REVIEW

For each section of this chapter, you will find a list of vocabulary and notation, questions to assess your understanding of key concepts, and review problems similar to the problems you worked in your homework. Review each item in the *Vocabulary/Notation* list mentally, and, if necessary, refer back to the indicated page and write a definition. Then answer the *Concept Check Questions,* looking back at the section if you need help. Work the *Review Problems* as practice before you move on to the *Chapter Test.* Answers to the *Review Problems* and *Chapter Test* can be found at the back of the book.

SECTION 2.1 Undefined Terms, Definitions, Postulates, Segments, and Angles

Vocabulary/Notation

Axiomatic system 38
Undefined terms 39
Point 39
Line ($\overleftrightarrow{AB}$) 39
Plane 39
Space 39
Geometric figure 39
Postulate 39
Theorem 39
Collinear 40
Noncollinear 40
Coplanar 40
Noncoplanar 40
Ruler Postulate 41
Coordinate 41
Distance from A to B 41
Line segment ($\overline{AB}$) 41
Endpoint 41
Length of a line segment 41
Congruent segments ($\cong$) 41
Ray ($\overrightarrow{AB}$) 41
Angle ($\angle$) 42
Vertex (vertices) 42
Sides (of an angle) 42
Degree 42
Protractor 42
Protractor Postulate 42
Measure of an angle 43
Acute angle 43
Right angle 43
Obtuse angle 43
Straight angle 43
Reflex angle 43
Pependicular lines ($\perp$) 43
Congruent angles ($\cong$) 44
Complementary angles 44
Supplementary angles 44
Adjacent angles 44

Concept Check Questions

1. What is the relationship between undefined terms, postulates, and theorems in an axiomatic system? 39
2. How is the term "point" used in geometry? 39
3. How is the term "line" used in geometry? 39
4. How is the term "plane" used in geometry? 39
5. How is the term "space" used in geometry? 39
6. How does the Ruler Postulate allow us to find lengths of line segments? 41
7. How can we find the distance between two points on a line? 41
8. How does the Protractor Postulate allow us to find the measure of an angle? 42
9. How is a protractor used to measure an angle? 43
10. What does it mean for two angles to be congruent? 44

Review Exercises

1. Draw and label a line, a line segment, and a ray.
2. Use a protractor to draw a 150° angle, a 90° angle, a 245° angle, a 180° angle, and a 38° angle. Label each angle with its degree measure and identify each angle by type as acute, obtuse, right, reflex, or straight.
3. Points A, B, and C are located on line l. Give the coordinate for each point and calculate the distances AB, CB, and AC.

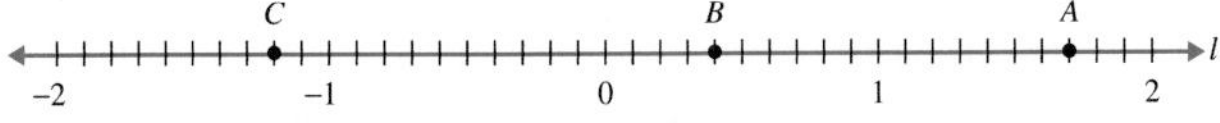

4. Two angles are both complementary and congruent. Find the measures of the angles.
5. In the following figure, points A, F, and D are collinear; points B, F, and E are collinear; and $\overline{BE} \perp \overline{CF}$.

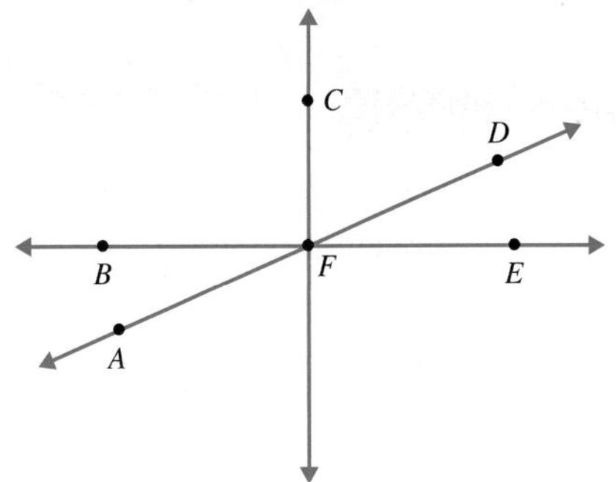

a. Name the acute angle adjacent to $\angle EFD$.
b. Name a pair of adjacent, complementary angles.
c. If the measure of $\angle CFD$ is 62°, then find the measure of $\angle DFE$.
d. Name an angle congruent to $\angle BFC$.
e. Name two obtuse angles that are supplementary to $\angle BFA$.
f. If the measure of reflex $\angle EFA$ is 214°, then find the measure of $\angle AFB$.
g. Name two straight angles.

SECTION 2.2 Polygons and Circles

Vocabulary/Notation

Triangle ($\triangle ABC$) 52
Sides of a triangle 52
Equilateral triangle 52
Isosceles triangle 52
Scalene triangle 52
Acute triangle 52
Right triangle 52
Obtuse triangle 52
Equiangular triangle 52
Base angles of an isosceles triangle 53
Base of an isosceles triangle 53
Vertex angle of an isosceles triangle 53
Hypotenuse 53
Leg 53
Simple closed curve 53
Circle 54
Center (of a circle) 54
Radius (of a circle) 54
Diameter (of a circle) 54
Polygon 54
Adjacent vertices 54
Adjacent sides 54
Quadrilateral 55
Pentagon 55
Hexagon 55
Octagon 55
Decagon 55
n-gon 55
Parallel lines (segments) ($\|$) 55
Square 55
Rectangle 55
Rhombus 55
Parallelogram 55
Trapezoid 55
Isosceles trapezoid 55
Kite 55
Bases of a trapezoid 56
Legs of a trapezoid 56
Reflection symmetry 56
Line of symmetry 56
Axis of symmetry 56
Rotation symmetry 57
Center of rotation symmetry 57

Concept Check Questions

1. How are sides and angle measures used to classify a triangle? 52
2. What makes a figure a simple closed curve? 53
3. In what ways can a figure fail to be a simple closed curve? 53
4. Why is a circle not classified as a polygon? 54
5. How are polygons classified by type? 55
6. Which quadrilaterals have four right angles? 55
7. Can a trapezoid be a parallelogram? 55
8. What two common types of symmetries can a geometric figure possess? 56–57
9. How are rotation symmetries counted? 57

Review Exercises

1. Give the number of rotation symmetries and the number of axes of reflection symmetry in each of the following figures.

a.
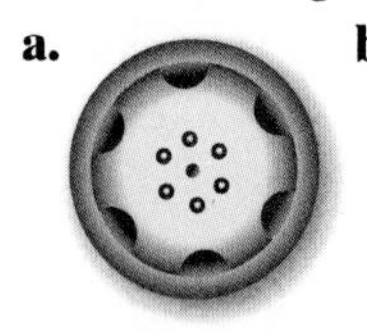
Hubcap

b.

Shell

c.

Flag of the country Georgia

2. For each of the following polygons, specify the number of axes of reflection symmetry.
   a. An isosceles trapezoid
   b. An equilateral triangle
   c. A scalene triangle
   d. A rectangle
3. Describe the rotation symmetries in the specified polygon by giving the angle(s) of rotation.
   a. A square
   b. A rhombus that is not a square
   c. An isosceles triangle
4. Remember that not every quadrilateral is a parallelogram.
   a. Name two types of quadrilaterals that are not parallelograms.
   b. Give the minimum requirements for a quadrilateral to be a parallelogram.
   c. What additional condition(s) must be satisfied for a parallelogram to be a rectangle?
   d. What additional condition(s) must be satisfied for a parallelogram to be a rhombus?

e. What additional condition(s) must be satisfied for a parallelogram to be a square?

5. If possible, sketch an example of the triangle described. If it is not possible, then explain why. Be sure to mark right angles and congruent segments.

   a. An obtuse isosceles triangle

   b. An acute right triangle

   c. A right scalene triangle

   d. A scalene isosceles triangle

6. A square has four axes of reflection symmetry. Two of them divide the square into two congruent triangles. Describe the triangles as completely as possible.

SECTION 2.3 Angle Measure in Polygons and Tessellations

Vocabulary/Notation

Polygonal region 65
Tessellation (tiling) 65
Vertex angle in a polygon 68
Diagonal of a polygon 68
Regular polygon 68
Regular tessellation 70

Concept Check Questions

1. What is the difference between a polygon and a polygonal region? 65
2. What are the two requirements for an arrangement of polygonal regions to be called a tessellation? 65
3. How can we use a tessellation composed of triangles to develop a relationship between the measures of the angles in a triangle? 66
4. Why does any triangle tessellate the plane? 66
5. How many triangles are formed when all possible diagonals are drawn from one vertex of a polygon? 68
6. What two conditions must a polygon satisfy to be a regular polygon? 68
7. What regular polygons will form regular tessellations? 70
8. Why will a regular heptagon (7-gon) not form a regular tessellation? 70

Review Exercises

1. a. Sketch an example of a nonregular quadrilateral with four congruent vertex angles.

   b. Sketch an example of a nonregular quadrilateral with four congruent sides.

2. a. If the sum of the measures of the vertex angles in a polygon is 2520°, then how many sides does the polygon have?

   b. If the polygon from part (a) is regular, what is the measure of a vertex angle?

3. In the following figure, each polygon is regular. Find the angle measures x and y.

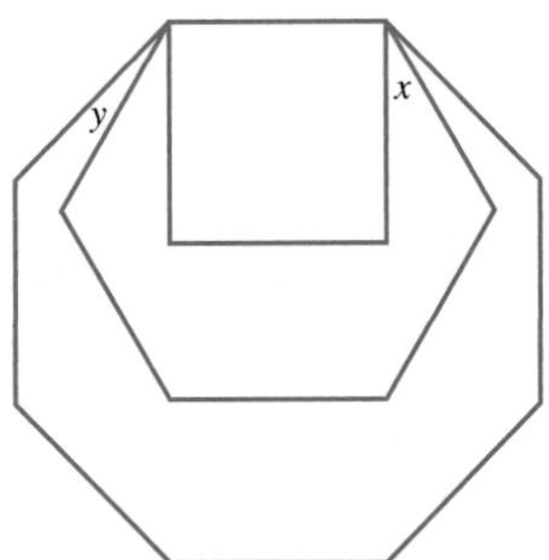

4. If the measures of the angles in a triangle are x, $x + 3$, and $2x$, then find the value of x.

5. Make six copies of $\triangle ABC$, cut them out, and arrange them on your paper to show how they can form a tessellation of the plane.

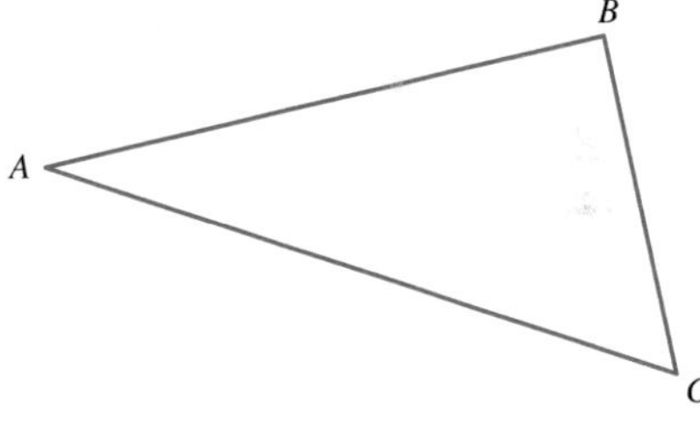

6. In the five-pointed star shown, what is the sum of the angle measures at A, B, C, D, and E? The pentagon inside the star is regular.

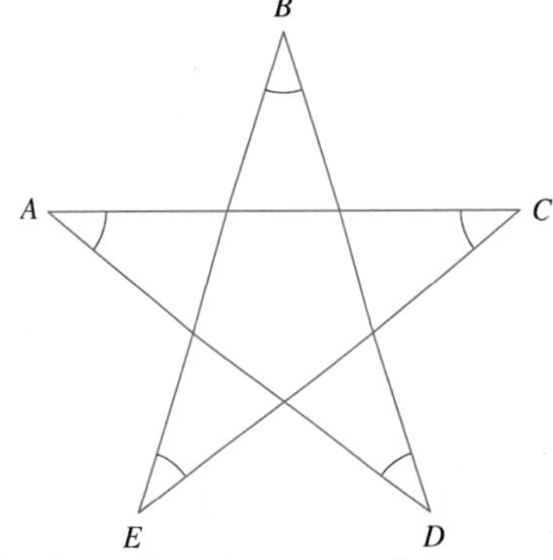

7. Use the fact that the sum of the measures of the vertex angles in a quadrilateral is 360° to explain why any quadrilateral, not just a square or a rectangle, can be used to tessellate the plane.

SECTION 2.4 Three-Dimensional Shapes

Vocabulary/Notation

Polyhedron (polyhedra) 75
Face of a polyhedron 76
Edge of a polyhedron 76
Vertex of a polyhedron 76
Prism 76
Base of a prism 76
Lateral face of a prism 76
Height of a prism 76
Right prism 76
Oblique prism 76
Pyramid 76
Base of a pyramid 76
Lateral face of a pyramid 76
Apex of a pyramid 76
Height of a pyramid 76
Right regular pyramid 77
Oblique regular pyramid 77
Slant height of right regular pyramid 77
Regular polyhedron 77
Tetrahedron 78
Hexahedron 78
Octahedron 78
Dodecahedron 78
Isosahedron 78
Euler's formula 78
Semiregular polyhedron 78
Circular cylinder 78
Base of a cylinder 79
Right circular cylinder 79
Oblique circular cylinder 79
Height of a cylinder 79
Circular cone 79
Base of a cone 79
Apex of a cone 79
Right circular cone 79
Oblique circular cone 79
Height of a cone 79
Slant height of a cone 79
Sphere 79
Center of a sphere 79
Radius of a sphere 79
Diameter of a sphere 79

Concept Check Questions

1. What is the difference between a right prism and an oblique prism? 76
2. How do you determine which faces of a right regular prism are the bases? 76
3. What is the difference between the height of a right regular pyramid and the slant height of a right regular pyramid? 76–77
4. If the lateral faces of a polyhedron are identical isosceles triangles, then what type of polyhedron is it? 77
5. What are the requirements for a polyhdron to be regular? 77
6. How are the numbers of edges, vertices, and faces of a polyhedron related? 78
7. What are the features of a semiregular polyhedron? 78
8. What is the difference between a right circular cylinder and an oblique circular cylinder? 79

Review Exercises

1. Which of the following are prisms? Pyramids? Neither?

a. 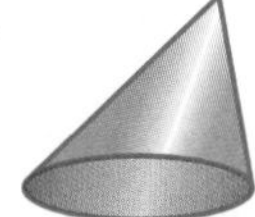

b. 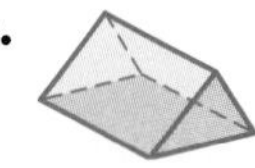

c. 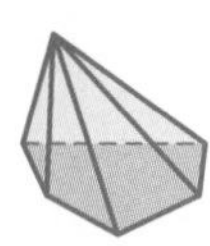

d. 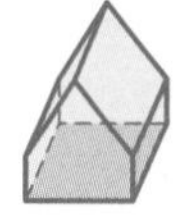

2. Name each of the following shapes as precisely as possible.

a. 

A tennis ball

b. 

A juice box

c. 

A candle

d. 

A standard die

3. Suppose that after studying a certain prism, you note the bases are squares, two of the lateral faces are rectangles, and two of the lateral faces are parallelograms. What is the name of the prism?
4. Given an oblique equilateral triangular prism, what is the name of each base, and what is the name of each lateral face?
5. Given a right regular hexagonal pyramid, what is the name of the base, and what is the name of each lateral face?
6. If a polyhedron has 31 edges and twice as many vertices as faces, then how many vertices does it have? How many faces does it have?

SECTION 2.5 Dimensional Analysis

Vocabulary/Notation

Dimensional analysis 87 Unit ratio (unit fraction) 87 Rate 89

Concept Check Questions

1. How can two different unit ratios be created using the fact that 1 m = 100 cm? 87
2. When converting from mi to ft, what difference does it make whether you use $\frac{5280\text{ ft}}{1\text{ mi}}$ or $\frac{1\text{ mi}}{5280\text{ ft}}$, given that they are both unit ratios? 87
3. Why is 1 in. = 2.54 cm such an important relationship? 88
4. What is an example of a rate? 89

Review Exercises

1. Use dimensional analysis to convert 39.48 wingnuts to marlocks, assuming that there are 2.8 wingnuts in a blog and 0.5 blogs in a marlock.
2. Use dimensional analysis to convert 0.68 gallons to tablespoons.
3. The distance from Boise, Idaho, to Portland, Oregon, is about 430 miles. Use dimensional analysis to convert this distance to kilometers. Round to the nearest kilometer.
4. It is generally advised that humans drink eight 8-ounce glasses of water each day. Use dimensional analysis to convert this to liters per year. Round to the nearest liter.
5. In 2005, the European Union called for a 100 km/hr speed limit in Europe. Use dimensional analysis to convert this rate to m/sec and round to the nearest tenth.

CHAPTER 2 TEST

TRUE-FALSE

Mark as true any statement that is always true. Mark as false any statement that is never true or that is not necessarily true. Be able to justify your answers.

1. Any three points lie in one and only one plane.
2. If two adjacent angles are supplementary, then their noncommon sides form a straight line.
3. An isosceles triangle has reflection symmetry but not rotation symmetry.
4. The vertex angle of a hexagon measures 120°.
5. The radius of a circle is twice its diameter.
6. An equiangular triangle must be an acute triangle.
7. A circle is the set of all points in a plane that are the same distance from a fixed point.
8. A parallelogram with four congruent sides is a rhombus.
9. A triangle is an example of a simple closed curve.
10. A right circular cylinder is a polyhedron.

EXERCISES/PROBLEMS

11. All squares are kites, but not all kites are squares. What additional conditions must be satisfied for a kite to be a square?
12. Find the measures of the lettered angles. Congruent angles are marked.

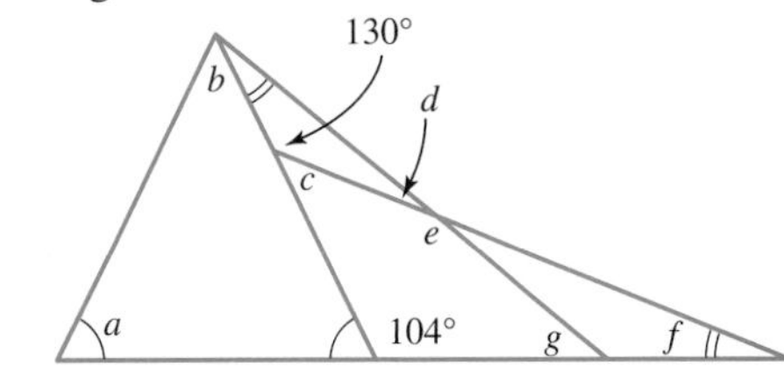

13. Find the measure of a vertex angle for a regular 24-gon.

14. Find the specified shapes in the following figure. Assume that angles which appear to be right angles are right angles, segments which appear to be parallel are parallel, and segments which appear to be congruent are congruent.

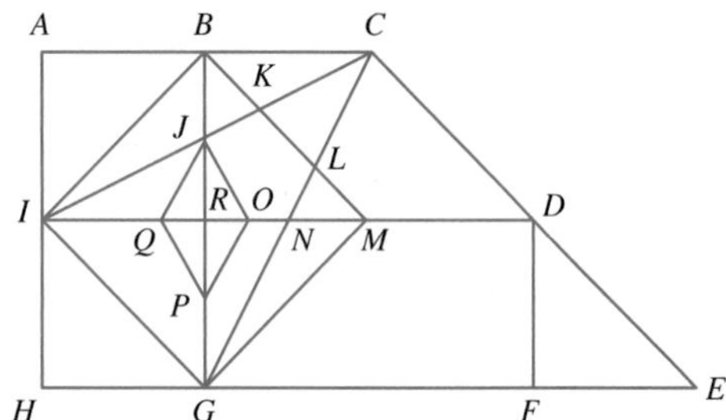

a. Three squares

b. A rectangle that is not a square

c. A parallelogram that is not a rectangle

d. Seven congruent right isosceles triangles

e. An isosceles triangle not congruent to those in (d)

f. A rhombus that is not a square

g. A kite that is not a rhombus

h. A scalene triangle with no right angles

i. A right scalene triangle

j. A trapezoid that is not isosceles

k. An isosceles trapezoid

15. In the following figure, find one example of the following angles.

a. supplementary **b.** complementary
c. right **d.** adjacent
e. acute **f.** obtuse

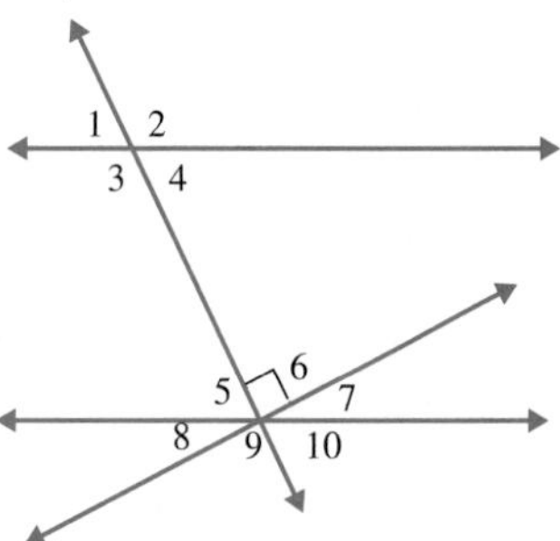

16. Determine the number of rotation and reflection symmetries for each of the following figures.

a.

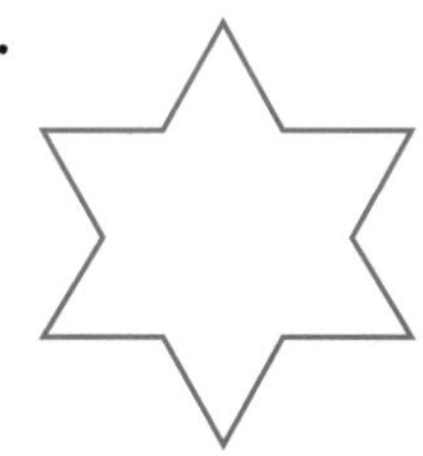

b.

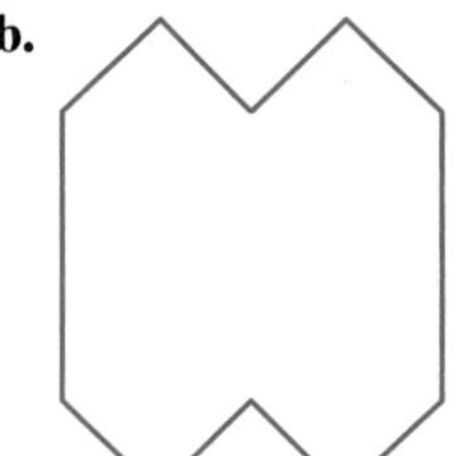

c.

17. Form a tessellation of the plane using the quadrilateral $ABCD$.

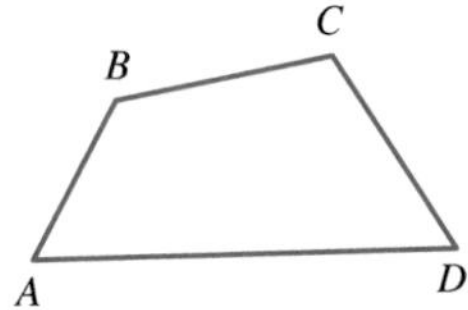

18. Given the triangular lattice shown, draw triangles having $\overline{AB}$ as a side.

a. How many different triangles are possible?

b. How many are isosceles triangles?

c. How many are obtuse triangles?

d. How many are right triangles?

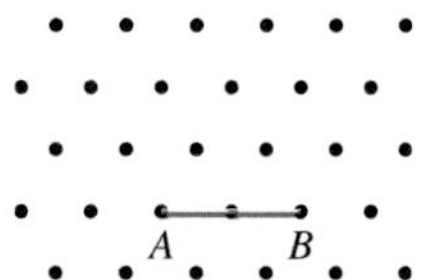

19. One millisecond is 1/1000 of a second. Use dimensional analysis to convert $3\frac{1}{4}$ hours to milliseconds.

20. Use your protractor and a ruler to draw an example of each of the following.

a. A quadrilateral with three angles that each measure 100° and with no two sides congruent.

b. A hexagon that is equiangular but not regular.

APPLICATIONS

21. Use dimensional analysis to convert the rate 6 ounces per cubic foot to grams per liter.

22. A traverse with five sides has interior angles with the following measures: 121°45′, 118°25′, 87°32′, 98°34′, and 113°54′. Does the traverse close? If not, determine the error in the traverse.

23. An audio CD makes about 500 revolutions per minute near the center while it reads the innermost music track and about 180 revolutions per minute near the outer edge. If you listen to a song from the innermost track for 45 seconds then switch and listen to a 2 minute and 20 second song from the outer edge, about how many revolutions did the CD make? Round to the nearest whole revolution.

24. The bed of a gravel truck has the shape shown next where the lateral faces are rectangles. Describe the shape as completely as possible.

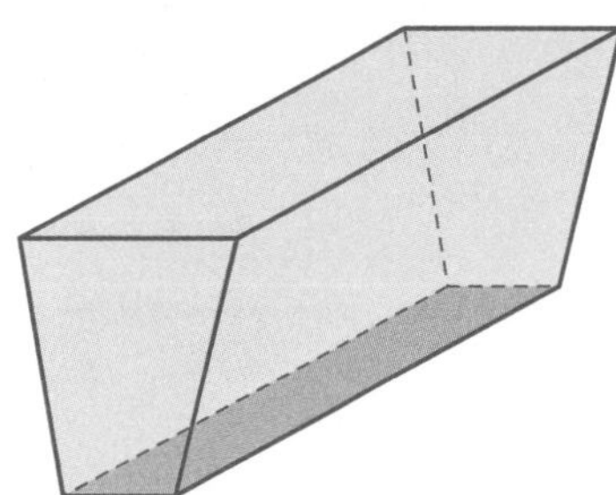

CHAPTER 3

PERIMETER, AREA, AND VOLUME

Archimedes (287–212 B.C.E.) is considered to have been the greatest mathematician in antiquity. One of his discoveries was that, by using appropriately dimensioned cross sections, a solid cone and solid sphere placed two units from the fulcrum of a lever would balance a solid cylinder placed one unit from the fulcrum. Because the volume of the cylinder and the cone were known, he then found the formula for the volume of the sphere by using a law of physics.

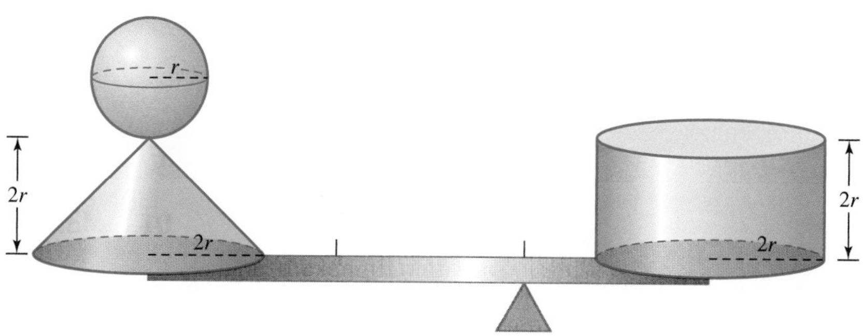

Problem-Solving Strategies

1. Draw a Picture
2. Guess and Test
3. Use a Variable
4. Look for a Pattern
5. Make a Table
6. Solve a Simpler Problem
7. Look for a Formula
8. *Use a Model*

Strategy 8: *Use a Model*

The Use a Model strategy is useful in problems involving figures. A model may be as simple as a paper, wooden, or plastic shape, or as complicated as a carefully constructed replica that an architect or engineer might use.

Initial Problem

Describe a solid shape that will fill each of the holes in this template as well as pass through each hole.

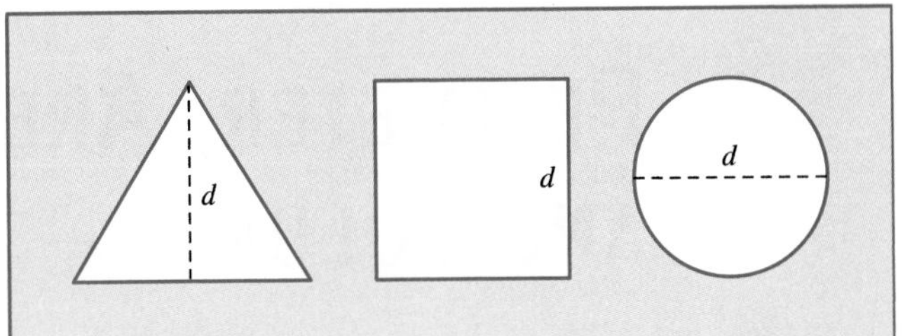

Clues

The Use a Model strategy may be appropriate when

- We can use physical objects to represent the ideas involved.
- A drawing is either too complex or inadequate to provide insight into the problem.
- A problem involves three-dimensional objects.

A solution of this Initial Problem is on page 174.

PΨTHAGORΣAN PUZZLΣ

A scalene right triangle with legs of lengths a and b and hypotenuse of length c has squares drawn on its three sides as shown.

Make a copy of the preceding figure, which consists of the triangle and three squares, and shade the squares as shown. Locate the center of the shaded square with side length b by lightly drawing in its diagonals with a straightedge. Label the intersection of the two diagonals O, and then erase the diagonals. Using a protractor and a straightedge, divide this square into four pieces as follows:

1. Draw a line segment through point O that is perpendicular to the hypotenuse (labeled c) of the right triangle, where the line segment drawn has its endpoints on opposite sides of the square. This line segment separates the square into two equal pieces.
2. Next, divide the square into four equal pieces by drawing a line segment through O perpendicular to the line segment you drew in Step 1, where this new line segment also has its endpoints on opposite sides of the square.

Cut out the two shaded squares. Cut the larger one into four pieces along the line segments you drew. Arrange these four pieces, together with the small shaded square, to fit exactly into the unshaded square on the hypotenuse of the triangle.

What conclusion can you draw concerning the areas of the two shaded squares compared to the area of the unshaded square? Write an equation that describes the relationship you discovered.

Introduction

Knowledge of geometry is required in problems that involve perimeter, area, and volume. Perimeter involves measurement in one dimension. Area and volume, which we find by using linear measure, are important ideas in two and three dimensions, respectively. Section 3.1 and 3.2 develop the notions of perimeter and area as applied to polygons and circles. In Section 3.3, we use the concept of area to develop an important relationship between the sides of a right triangle. Then, in Sections 3.4 and 3.5, we apply ideas involving linear and area measure in three dimensions to find the surface area and volume of common three-dimensional shapes.

3.1 PERIMETER, CIRCUMFERENCE, AND AREA OF RECTANGLES AND TRIANGLES

Applied Problem

A family plans to invest in new carpet for a living room. The floor plan and dimensions shown include a brick hearth and a hardwood entry, which will not be carpeted. If the carpet costs \$22.50/yd^2 installed and 5 percent extra area is allowed for cutting, how much will it cost to carpet the room?

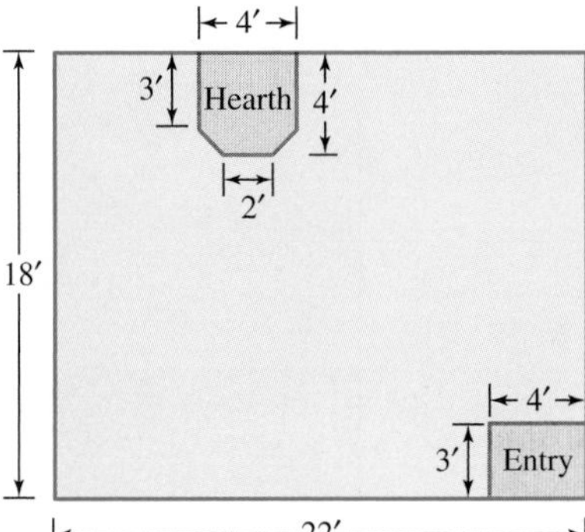

Perimeter of a Polygon

The **perimeter** of a polygon is the sum of the lengths of its sides. "Peri" means around and "meter" means measure; hence, perimeter literally means the measure around. We can develop perimeter formulas for some common quadrilaterals. A square and a rhombus both have four congruent sides. If one side is of length s, then the perimeter of each of them can be represented by $4s$ (Figure 3.1).

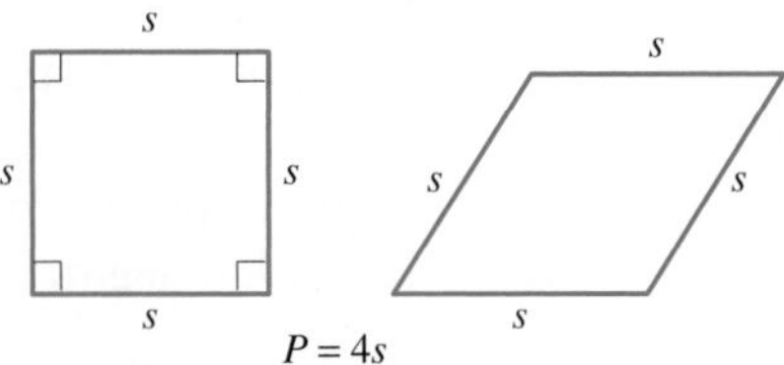

FIGURE 3.1

In rectangles and parallelograms, pairs of opposite sides are congruent. We will verify this formally in Chapter 5. Thus, if the lengths of its sides are a and b, then the

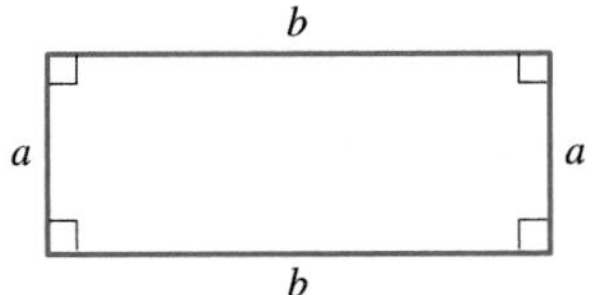

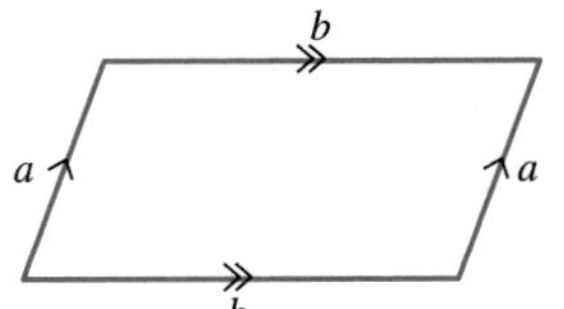

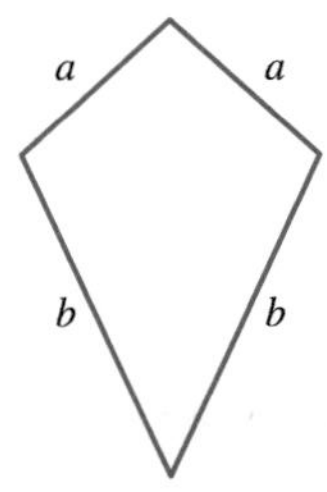

$P = 2a + 2b$

FIGURE 3.2

perimeter of a rectangle or a parallelogram is $2a + 2b$. We can use a similar formula to find the perimeter of a kite (Figure 3.2). The following theorem summarizes these results. (Recall that a theorem is a result that is proved or derived from other known facts or results.)

THEOREM 3.1 Perimeters of Common Quadrilaterals

Figure	Perimeter
Square with sides of length s	$4s$
Rectangle with sides of lengths a and b	$2a + 2b$
Parallelogram with sides of lengths a and b	$2a + 2b$
Rhombus with sides of length s	$4s$
Kite with sides of lengths a and b	$2a + 2b$

EXAMPLE 3.1 Find the perimeters of the following (Figure 3.3):

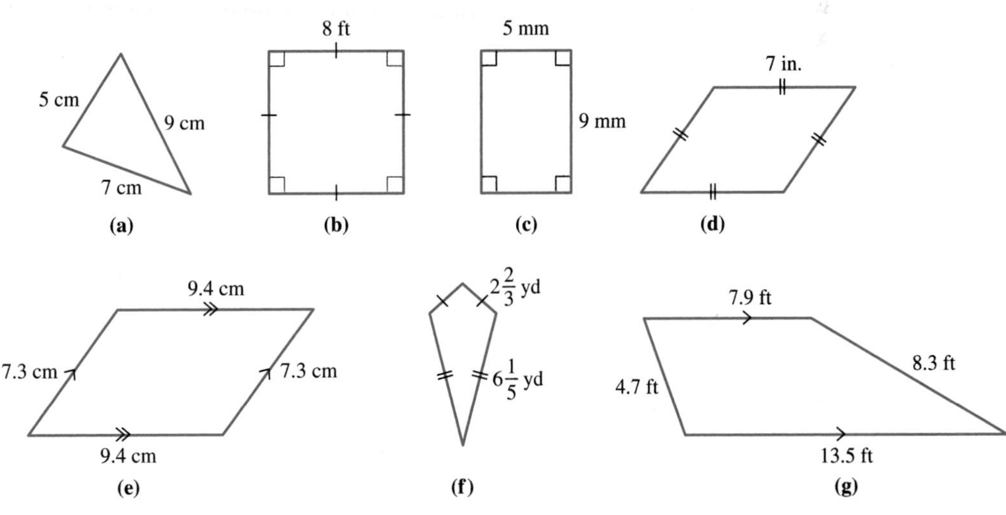

FIGURE 3.3

SOLUTION

a. $P = 5\text{ cm} + 7\text{ cm} + 9\text{ cm} = 21\text{ cm}$

b. $P = 4(8\text{ ft}) = 32\text{ ft}$

c. $P = 2(9\text{ mm}) + 2(5\text{ mm}) = 28\text{ mm}$

d. $P = 4\ (7\text{ in.}) = 28\text{ in.}$

e. $P = 2(7.3\text{ cm}) + 2(9.4\text{ cm}) = 33.4\text{ cm}$

f. $P = 2(2\frac{2}{3}\text{ yd}) + 2(6\frac{1}{5}\text{ yd}) = 5\frac{1}{3}\text{ yd} + 12\frac{2}{5}\text{ yd} = 17\frac{11}{15}\text{ yd}$

g. $P = 13.5\text{ ft} + 7.9\text{ ft} + 4.7\text{ ft} + 8.3\text{ ft} = 34.4\text{ ft}$ ■

Circumference of a Circle

The perimeter of a circle, that is, the distance around the circle, is called its **circumference**. Since ancient times, geometers have known that the ratio of the circumference to the diameter in any circle is a constant. In about 225 B.C.E., Archimedes determined that this ratio was between $3\frac{10}{71}$ and $3\frac{1}{7}$. That is, if C represents the circumference and d represents the diameter of any circle, then $\frac{C}{d}$ can be *approximated* by $3\frac{1}{7}$, or $\frac{22}{7}$. In decimal form, $\frac{C}{d}$ is *approximately* 3.14.

The ratio $\frac{C}{d}$ has come to be represented by the Greek letter π (**pi**), which has been determined to be an **irrational number**, that is, a nonterminating, nonrepeating decimal. An approximation of π more precise than 3.14 is 3.14159265. By 2002, the value of π had been calculated to more than one trillion places with the aid of a computer.

The definition of π as $\frac{C}{d}$ allows us to calculate the circumference of a circle. That is, because $\frac{C}{d} = \pi$, we can write $C = \pi d = \pi(2r) = 2\pi r$.

POSTULATE 3.1 Circumference of a Circle

The circumference of a circle is the product of π and the diameter of the circle.

$$C = \pi d = 2\pi r$$

EXAMPLE 3.2 Find the perimeter of each geometric figure in Figure 3.4. The figure on top of the square in Figure 3.4(b) is a semicircle.

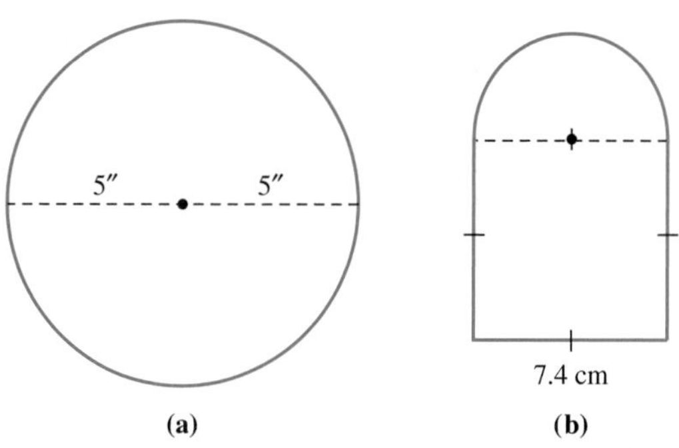

FIGURE 3.4

SOLUTION

a. $C = 2\pi r = 2\pi(5 \text{ inches}) = 10\pi$ inches, or approximately 31.42 inches. (NOTE: The exact answer is 10π, whereas 31.42 is a decimal approximation.)

b. The figure consists of a square topped by a semicircle, or half of a circle. The distance around this figure is the sum of the lengths of three sides of the square plus half of the circumference of a circle, or

$$P = 3s + \tfrac{1}{2}\pi d = 3(7.4 \text{ cm}) + \tfrac{1}{2}\pi(7.4 \text{ cm}) \approx 33.82 \text{ cm}$$

Area

We can use tessellations to determine the area of the region enclosed by a simple closed curve. For convenience, we use a square tessellation, where one of the small squares is taken to be our unit square [Figure 3.5(a)]. By counting the full squares

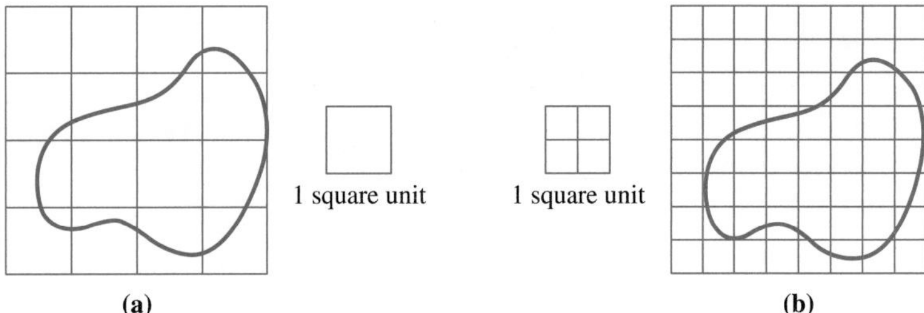

FIGURE 3.5

and adjusting for the partial squares, we can estimate the area of the region enclosed by the simple closed curve in Figure 3.5(a) to be about 6 square units. If smaller squares are used in the tessellation, such as four per unit square as in Figure 3.5(b), then a better estimate of the area may be possible.

This discussion leads to our first area postulate.

POSTULATE 3.2 Area of a Region Enclosed by a Simple Closed Curve

a. For every simple closed curve and unit square, there is a positive real number that gives the number of unit squares (and parts of unit squares) that exactly tessellate the region enclosed by the simple closed curve.

b. The area of a region enclosed by a simple closed curve is the sum of the areas of the smaller regions into which the region can be subdivided.

The real number described in Postulate 3.2(a) is commonly called the **area** of the region enclosed by the simple closed curve. For example, if a region has area 2.3 square units, this means that the region can be covered exactly by an equivalent of 2.3 unit squares.

Area of a Rectangle

Postulate 3.2 leads to some convenient formulas when we apply it to certain polygonal regions. For example, suppose a 3 cm by 5 cm rectangle is covered by a square tessellation where the squares are 1 cm by 1 cm (Figure 3.6).

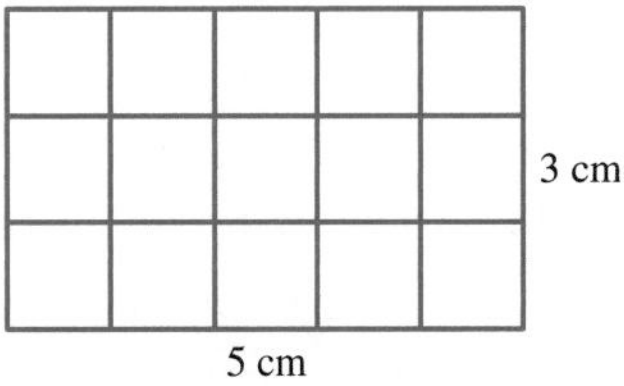

FIGURE 3.6

Three rows of five squares, or exactly 15 square centimeters, cover the rectangular region. We say that the area of the rectangular region is 15 square centimeters, which we can also write as 15 cm^2. When the length and width of a rectangle are whole numbers, as in Figure 3.6, the area of the rectangular region (or area of the rectangle, for short) is the product of its length and its width. Moreover, this relationship holds when the length and width of the rectangle are not necessarily whole numbers, but may be any real numbers.

DEFINITION

Area of a Rectangle

The area of a rectangle is the product of its length and its width.

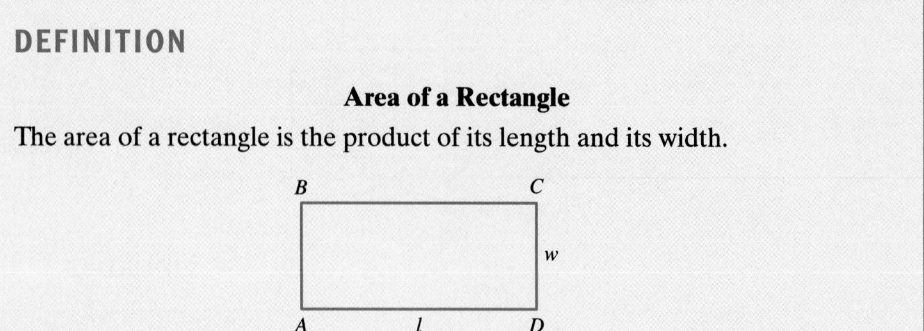

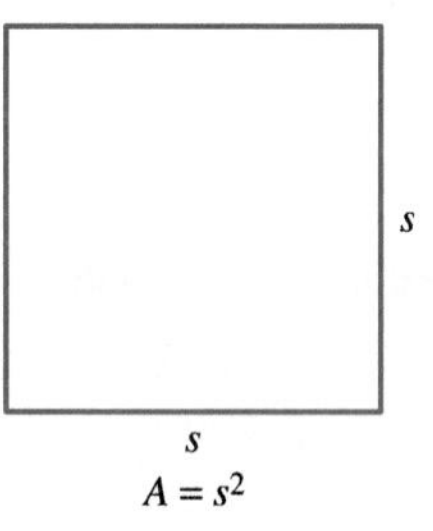

FIGURE 3.7

A square is a special rectangle where the length and width are the same. So, in the case of a square with side of length s, its area is $s \cdot s$, or s^2 (Figure 3.7).

Area of a Triangle

Before finding the area of an arbitrary triangle, we will find the area of a right triangle. First, however, we will assume the following postulate. It essentially states that two triangles that are exact duplicates of each other have the same area.

POSTULATE 3.3

Any two triangles whose corresponding angles and sides are congruent have the same area.

Using Postulates 3.2 and 3.3, we can find the area of right triangle $\triangle ABC$ in Figure 3.8 by making a copy, $\triangle A'B'C'$, of $\triangle ABC$ and arranging the two triangles to form a quadrilateral. Because the sum of the angle measures in a triangle is 180°, we can show that all four angles of the quadrilateral are right angles. Hence, $AC'BC$ is a rectangle (Figure 3.8).

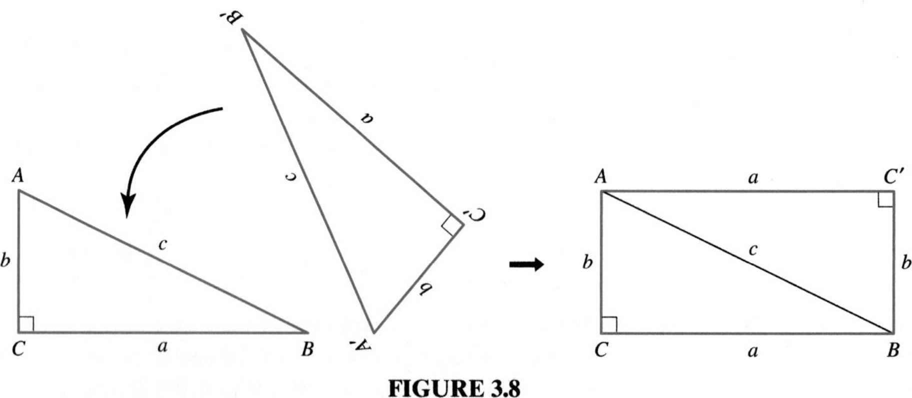

FIGURE 3.8

According to Postulates 3.2 and 3.3, because two identical right triangles make up rectangle $AC'BC$, each of their areas must be half the area of $AC'BC$. This leads to the following theorem.

THEOREM 3.2 Area of a Right Triangle

The area of a right triangle is half the product of the lengths of its legs.

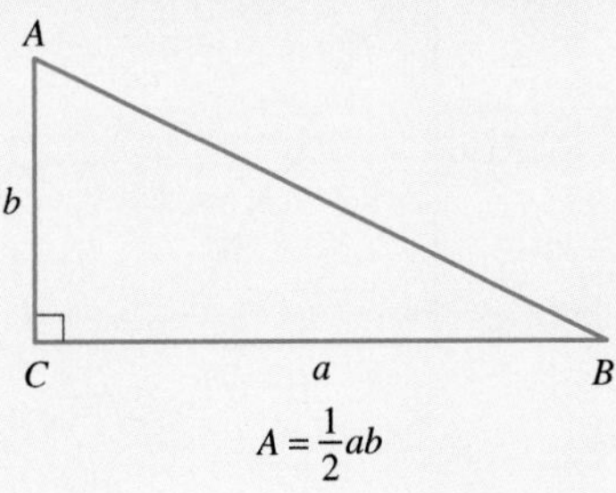

EXAMPLE 3.3 Determine the areas of the following triangles (Figure 3.9):

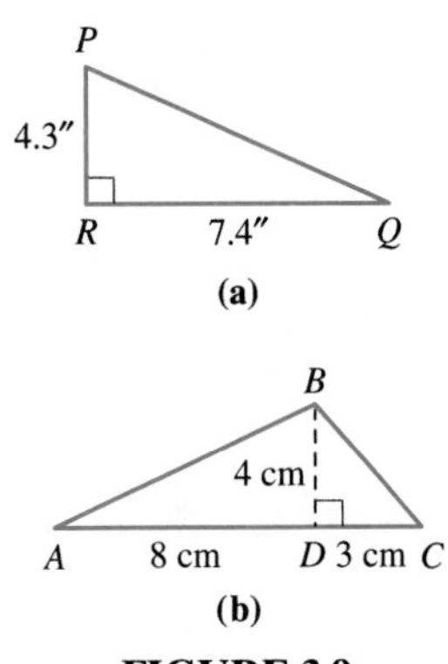

FIGURE 3.9

SOLUTION

a. Because $\triangle PQR$ is a right triangle, $A = \frac{1}{2}(7.4'')(4.3'') = 15.91 \text{ in}^2$.

b. Both $\triangle ABC$ and $\triangle BCD$ are right triangles. Thus, the area of $\triangle ABD$ is $\frac{1}{2}(8 \text{ cm})(4 \text{ cm})$, or 16 cm^2, and the area of $\triangle BCD$ is $\frac{1}{2}(3 \text{ cm})(4 \text{ cm})$, or 6 cm^2. Then, by Postulate 3.2(b), the area of $\triangle ABC$ is $16 \text{ cm}^2 + 6 \text{ cm}^2$, or 22 cm^2. ■

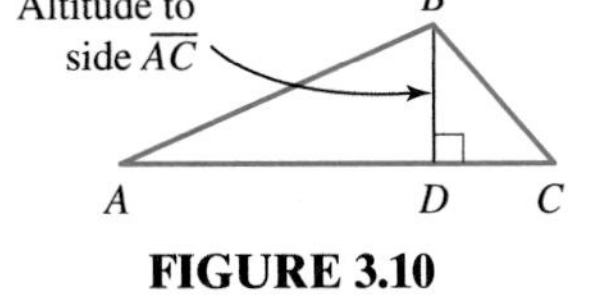

FIGURE 3.10

A line segment from a vertex of a triangle that is perpendicular to the opposite side of the triangle is called **an altitude** of the triangle (Figure 3.10). The *length* of an altitude is also referred to as **the altitude** or **height** to a particular side. The side to which the altitude is drawn is called the **base** of the triangle associated with that altitude. So in Figure 3.9, $\overline{BD}$ is the altitude to base $\overline{AC}$. In Example 3.3(b), we could have written the sum of the two areas as $\frac{1}{2}(8)(4) + \frac{1}{2}(3)(4)$, or $\frac{1}{2}(8 + 3)(4)$. This suggests the next theorem, whose verification is left for the problem set.

THEOREM 3.3 Area of a Triangle

The area of a triangle is half the product of the length of one side and the height to that side.

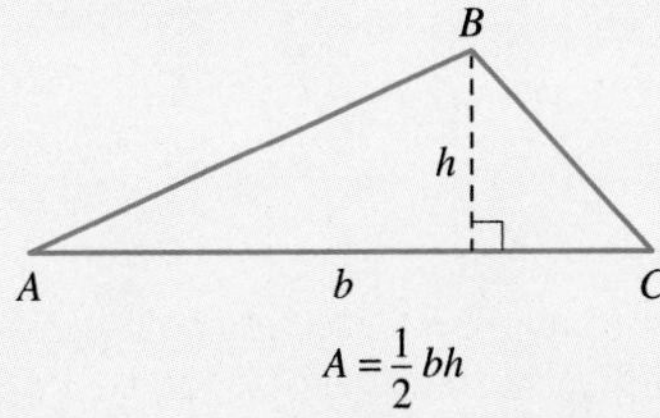

EXAMPLE 3.4 Calculate the areas of the triangles in Figure 3.11.

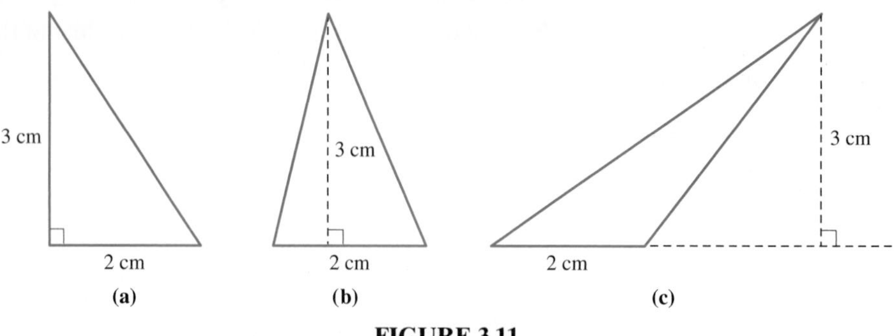

FIGURE 3.11

SOLUTION

Although the shapes of the three triangles are very different, each of the triangles has a base of 2 cm with a height of 3 cm. Thus, all three triangles will have exactly the same area, namely

$$A = \tfrac{1}{2}bh = \tfrac{1}{2}(2\text{ cm})(3\text{ cm}) = 3\text{ cm}^2$$

■

We should emphasize that any side can serve as "the base" of a triangle when the area is to be determined. The base need not be the "bottom" side of a triangle.

EXAMPLE 3.5 Calculate the area of $\triangle XYZ$ in Figure 3.12.

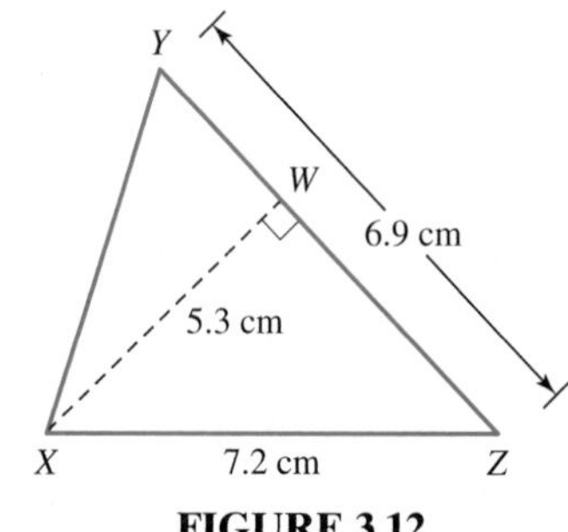

FIGURE 3.12

SOLUTION

We will use $\overline{YZ}$ as the base, where $\overline{XW}$ is the altitude to that base. Thus, $A = \tfrac{1}{2}bh = \tfrac{1}{2}(6.9\text{ cm})(5.3\text{ cm}) \approx 18.3\text{ cm}^2$. ■

Any triangle has three different altitudes, or heights, one to each side (Figure 3.13). The area of a triangle does not depend on which side and corresponding height are chosen. They all yield the same area.

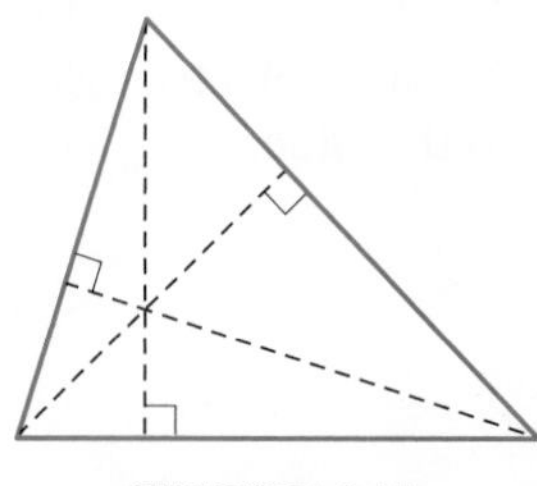

FIGURE 3.13

Converting Units of Area Using Dimensional Analysis

In Section 2.5, we used dimensional analysis to convert from one unit of measure to another. We can use this technique to convert from one unit of area measure to another as shown in the next example.

EXAMPLE 3.6 Convert 10 yd^2 to ft^2.

SOLUTION

First notice that we are converting square units to square units. It makes no sense to convert *square* yards to *linear* feet, for example.

One square yard, or 1 yd^2, can be represented as a square that measures 1 yd by 1 yd [Figure 3.14(a)]. Since 1 yd = 3 ft, we can also view the 1 yd by 1 yd square as a 3 ft by 3 ft square [Figure 3.14(b)].

Thus,

$$1 \text{ yd}^2 = (1 \text{ yd})(1 \text{ yd}) = (3 \text{ ft})(3 \text{ ft}) = 9 \text{ ft}^2$$

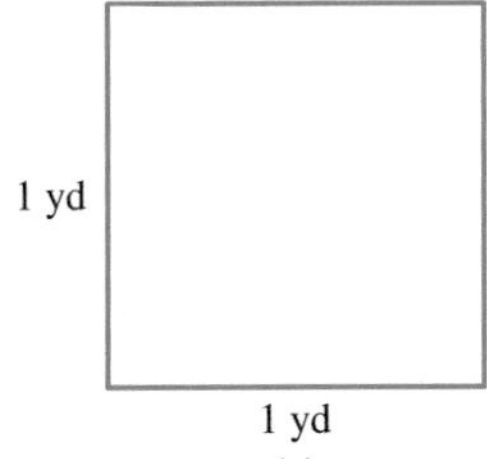

(a)

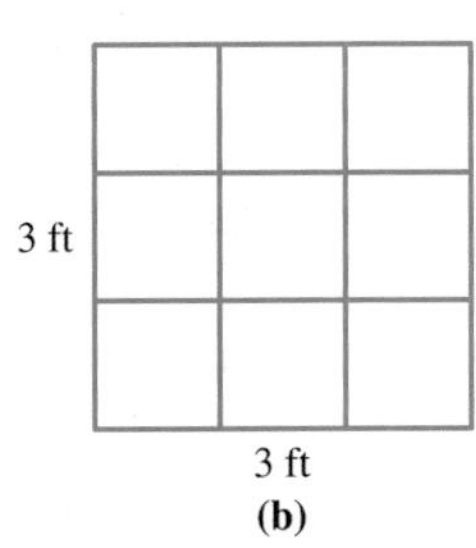

(b)

FIGURE 3.14

Notice that in Figure 3.14(b), there are 9 square feet in the square yard. So, to convert 10 yd^2 to ft^2, we can write

$$10 \text{ yd}^2 = \frac{10 \cancel{\text{yd}^2}}{1} \cdot \frac{3 \text{ ft} \cdot 3 \text{ ft}}{1 \cancel{\text{yd}} \cdot 1 \cancel{\text{yd}}} = \frac{10(3)(3)}{1(1)(1)} \text{ft}^2 = 90 \text{ ft}^2$$

In Example 3.6, because 1 yd^2 had two dimensions, we used the conversion factor 1 yd = 3 ft *twice* to make the conversion.

We can also use dimensional analysis to convert between English and metric units of area as shown in the next example.

EXAMPLE 3.7 An Olympic-sized swimming pool of uniform depth measures 50 m by 21 m. The bottom of the pool is to be treated with a sealer, and one gallon of the sealer covers 300 ft^2 of surface. How many gallons of sealer will be required for the job? (Recall: 1 in. = 2.54 cm)

SOLUTION

First we must determine the area of the bottom of the pool in square feet. We can calculate the area and perform the conversion at the same time as follows.

$$\begin{aligned} A &= lw \\ &= \frac{50 \cancel{\text{m}} \cdot 21 \cancel{\text{m}}}{1} \cdot \frac{100 \cancel{\text{cm}} \cdot 100 \cancel{\text{cm}}}{1 \cancel{\text{m}} \cdot 1 \cancel{\text{m}}} \cdot \frac{1 \cancel{\text{in.}} \cdot 1 \cancel{\text{in.}}}{2.54 \cancel{\text{cm}} \cdot 2.54 \cancel{\text{cm}}} \cdot \frac{1 \text{ ft} \cdot 1 \text{ ft}}{12 \cancel{\text{in.}} \cdot 12 \cancel{\text{in.}}} \\ &= \frac{50(21)(100)(100)(1)(1)(1)(1)}{1(1)(1)(2.54)(2.54)(12)(12)} \text{ft}^2 \approx 11{,}302 \text{ ft}^2 \end{aligned}$$

Now we can use dimensional analysis again to determine the number of gallons of sealer needed.

$$\frac{11{,}302 \cancel{\text{ft}^2}}{1} \cdot \frac{1 \text{ gal}}{300 \cancel{\text{ft}^2}} = \frac{11{,}302\,(1)}{1(300)} \text{gal} \approx 38 \text{ gal}$$

Therefore, approximately 38 gallons of sealer would be needed.

Solution of Applied Problem

The total area of the room is $18'(22') = 396 \text{ ft}^2$. The areas that will not be carpeted consist of (i) the entry: $3'(4') = 12 \text{ ft}^2$, and (ii) the hearth, which has been divided here into two rectangles and two triangles:

$$3'(4') + 2'(1') + 2(\tfrac{1}{2})(1')(1') = 15 \text{ ft}^2$$

Thus, the area to be carpeted is $396 \text{ ft}^2 - 12 \text{ ft}^2 - 15 \text{ ft}^2 = 369 \text{ ft}^2$. Taking into account the additional 5 percent, $369 \text{ ft}^2 + (0.05)(369 \text{ ft}^2) = 387.45 \text{ ft}^2$ must be ordered. The total cost at \$22.50 per square yard is

$$\frac{387.45 \cancel{\text{ft}^2}}{1} \cdot \frac{1 \cancel{\text{yd}}}{3 \cancel{\text{ft}}} \cdot \frac{1 \cancel{\text{yd}}}{3 \cancel{\text{ft}}} \cdot \frac{\$22.50}{1 \cancel{\text{yd}^2}} = \frac{(387.45)(1)(\$22.50)}{1(9)(1)} \approx \$968.63.$$

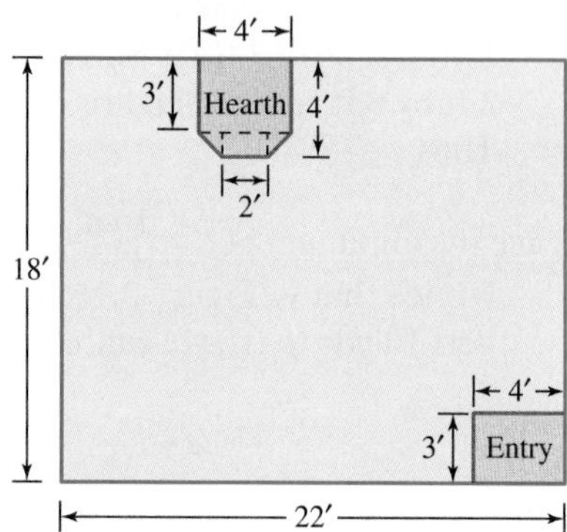

GEOMETRY AROUND US

In 1997, a U.S. flag created by Ski Demski was listed in the Guinness Book of World Records® as the "Largest Flag in the World." This huge flag measured 255 ft × 505 ft (larger than two football fields), and it weighed 3000 lb. It has frequently been displayed at special events such as Super Bowls, World Series games, and NASCAR races. Unfurling the flag requires the coordinated work of 600 workers.

PROBLEM SET 3.1

EXERCISES/PROBLEMS

1. Determine the perimeter of each figure shown.

a.

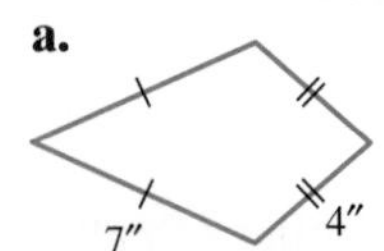

b.

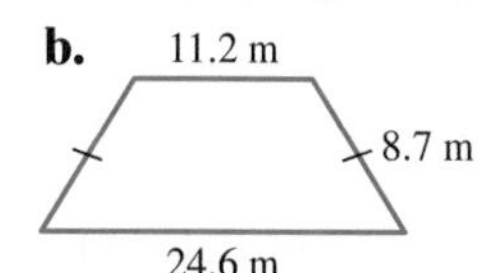

2. Determine the perimeter of each figure shown.

a.

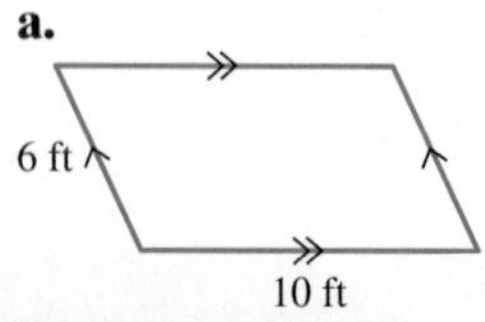

b.

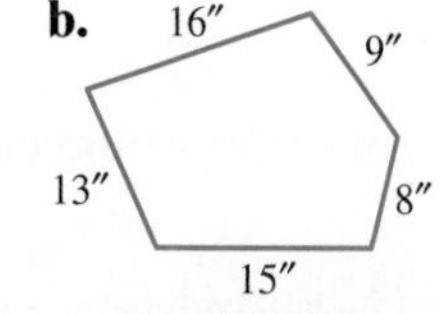

3. Determine the perimeter of each regular polygon shown.

a.

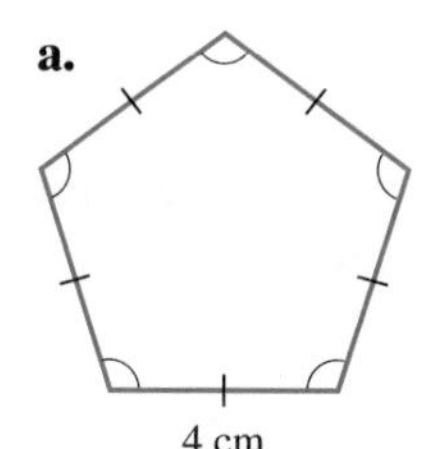

b.

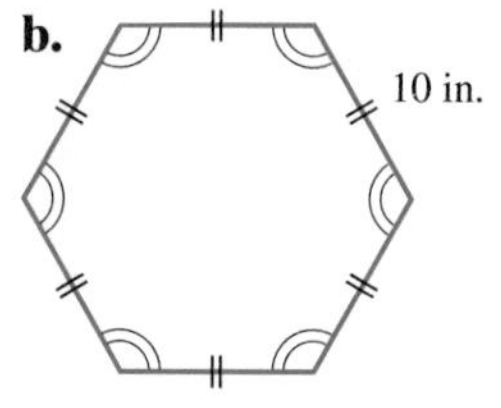

c. Write a formula for the perimeter of a regular polygon with n sides, each of length s.

4. Determine the perimeter of each figure shown.

a.

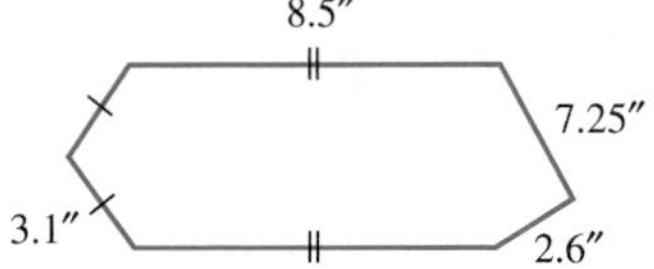

b.

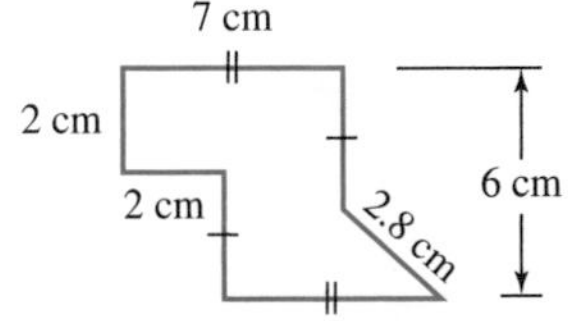

5. Recall that a tetromino is a geometric shape composed of four connected squares, where two squares are joined along an entire side. (See problem 3 in Section 1.1.)

a. Which tetrominos have the maximum perimeter?

b. Which tetrominos have the minimum perimeter?

6. A pentomino is a geometric shape composed of five connected squares, where two squares are joined along an entire side. (See problem 4 in Section 1.1.)

a. Which pentominos have the maximum perimeter?

b. Which pentominos have the minimum perimeter?

c. Generalize your results from problems 5, 6(a), and 6(b) for a shape formed of eight squares.

7. In Section 1.2, toothpicks were arranged to form rows of squares and equilateral triangles. If one toothpick represents one unit, determine the perimeter if we arrange n polygons of the following types in a row.

(a) n equilateral triangles:

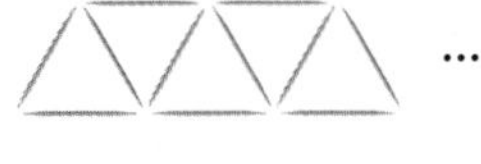

(b) n squares:

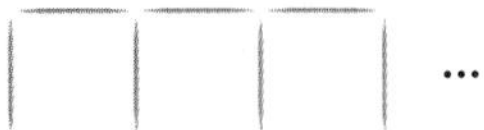

(c) n pentagons:

(d) n hexagons:

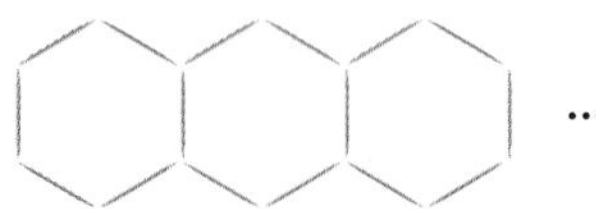

(e) n octagons:

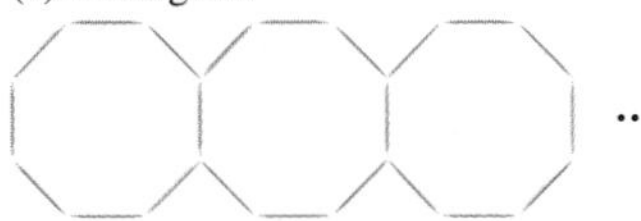

(f) n m-sided polygons

8. Suppose we form a large square by arranging toothpicks as shown.

a. Find the perimeter and area of this square.

b. Suppose now we rearrange the toothpicks in the corners to form a small square one tooth-pick per side in every corner. Find the perimeter and area for this new figure.

c. Suppose we rearrange the toothpicks again to form squares with two toothpicks on a side in each corner. Find the perimeter and area of the resulting figure.

d. If the squares in the corners have three toothpicks per side, find the perimeter and area of the figure.

e. What pattern do you observe? Suppose the original square had m toothpicks on a side and the small squares in each corner had n toothpicks on a side. Write a formula for the perimeter and a formula for the area of the figure.

9. Calculate the circumference of each circle shown. Round to the nearest hundredth.

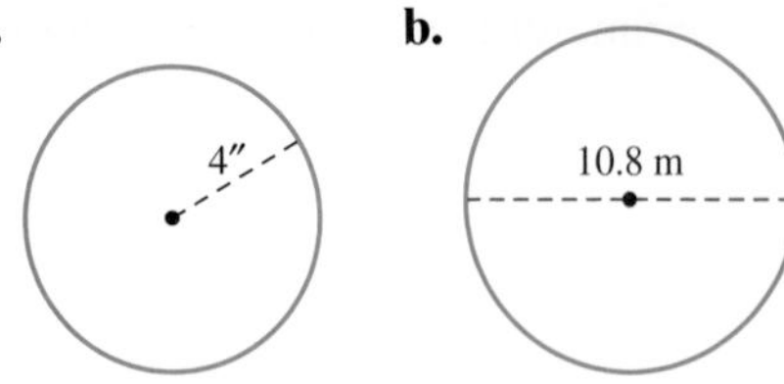

10. For each circle, the circumference is given. Find the radius and diameter for each circle. Round to the nearest hundredth.

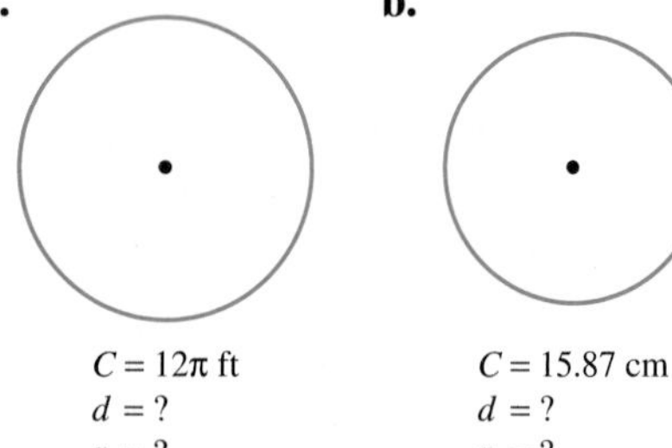

11. A wire bent into the shape of a square encloses an area of 25 cm^2. Then the same wire is cut and bent into two identical circles. What is the radius of one of the circles? Round to the nearest hundredth.

12. A square and a circle both made out of wire enclose the same area, 18.45 cm^2. When the wires are flattened out, which is longer and by how much? Round to the nearest hundredth.

13. By counting squares and parts of squares, find the area of each figure shown on the following square lattices. Express your answer in square units, where 1 square unit is the region shown.

□ = 1 square unit

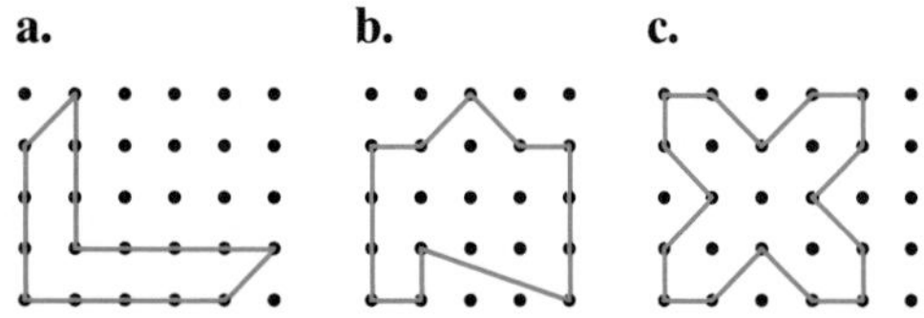

14. By counting squares and parts of squares, find the area of each figure shown on the following square lattices.

□ = 1 square unit

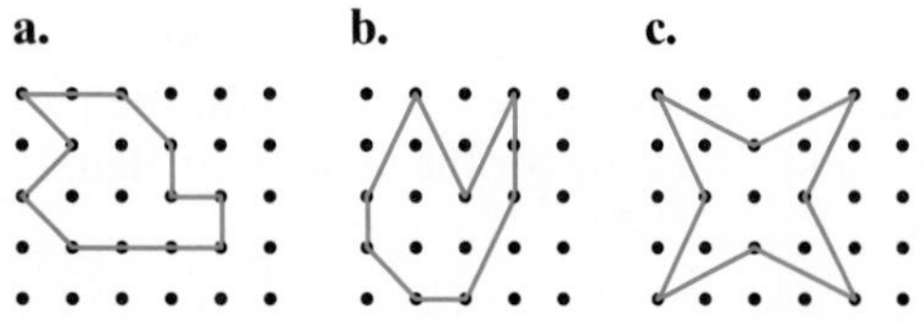

15. Find the area and perimeter of each rectangle.

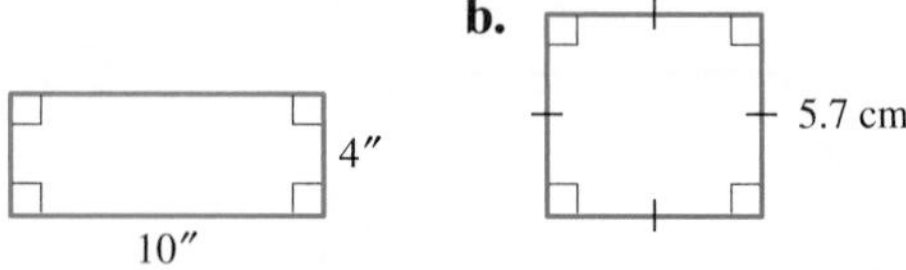

16. Find the area and perimeter of each rectangle.

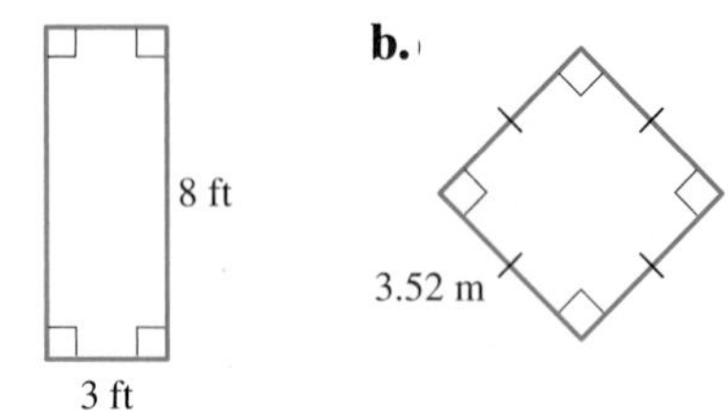

17. The length of a rectangle is 6 in. less than twice its width. If the perimeter of the rectangle is 72 in., find the dimensions of the rectangle.

18. The width of a rectangle is 18 in. less than its length. If the area of the rectangle is 1440 in^2, find the dimensions of the rectangle.

19. What happens to the perimeter of a rectangle if we multiply both its length and width by 2? by 3? by 10? by n?

20. What happens to the area of a rectangle if we multiply both its length and width by 2? by 3? by 10? by n?

21. Suppose two rectangles both have a perimeter of 40 m. Must they also have the same area? Try several examples to test your conjecture.

22. Suppose two rectangles both have areas of 144 cm^2. Must they also have the same perimeter? Try several examples to test your conjecture.

23. On the square lattice shown, sketch examples of four different triangles each of which has an area of 2 square units.

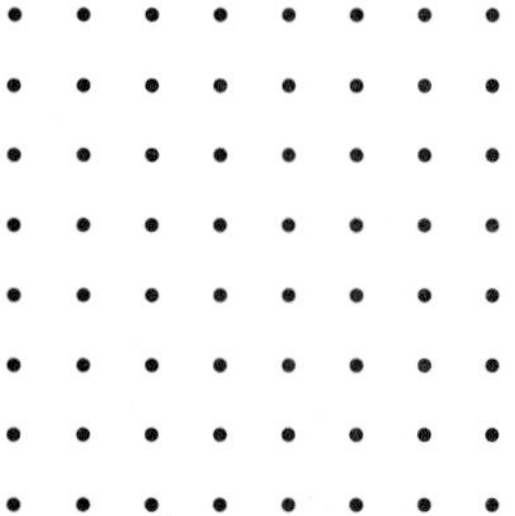

24. On a square lattice like the one given in problem 23, sketch examples of four different triangles, each of which has an area of 3 square units.

25. Determine the perimeter and area of each of the following right triangles.

a.

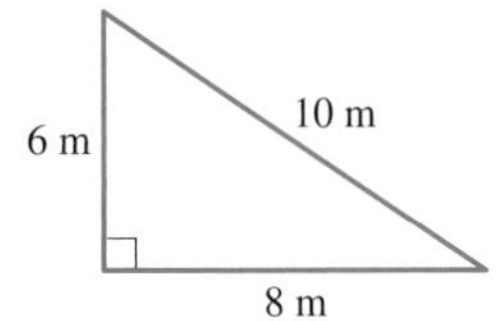

b.

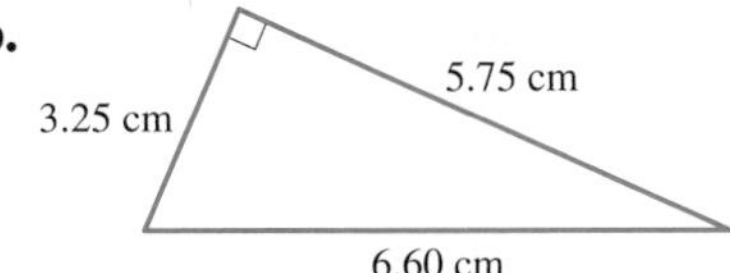

c.

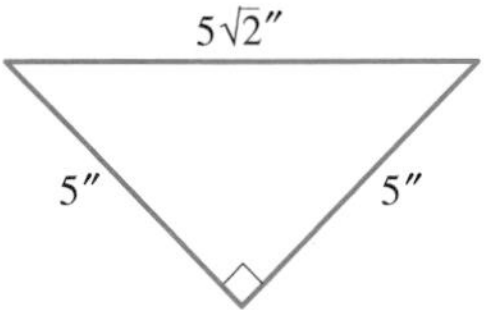

26. Determine the perimeter and area of each of the following triangles. Round to the nearest hundredth.

a.

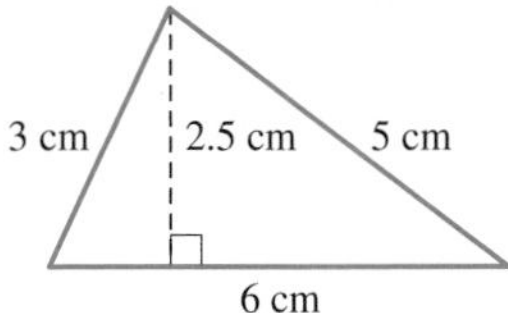

b.

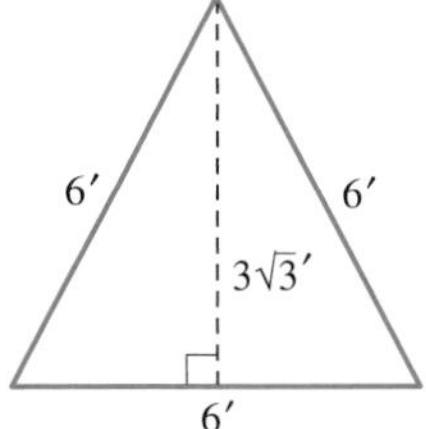

c.

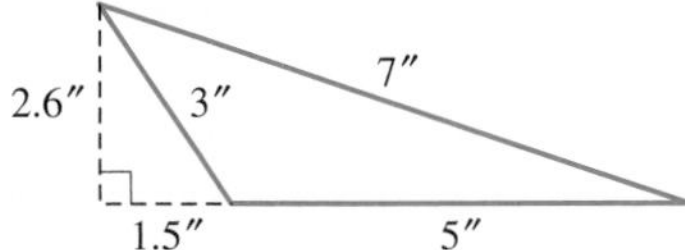

27. Side $\overline{ED}$ in $\triangle AED$ is twice the length of $\overline{CD}$ in rectangle $ABCD$. How do the areas of rectangle $ABCD$ and $\triangle AED$ compare? Explain.

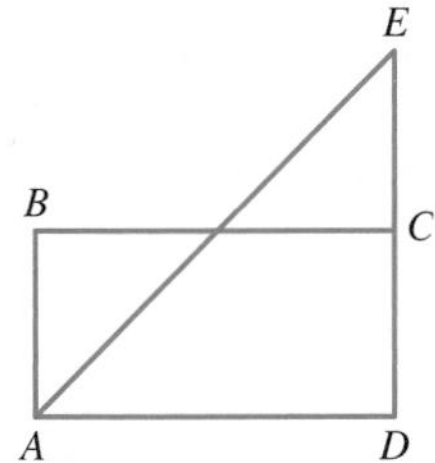

28. The area of the rectangle $ABCD$ is 40 in^2. What is the area of the unshaded region? Explain.

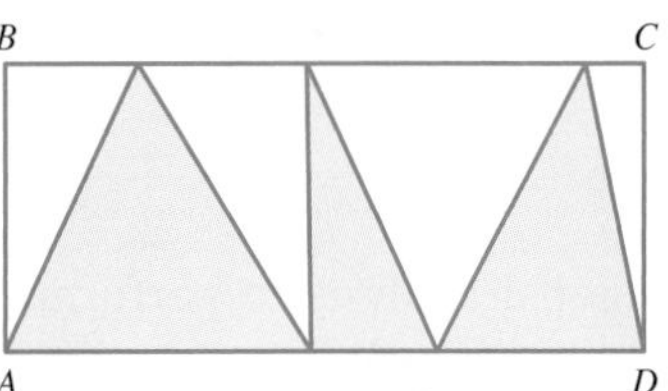

29. For the following triangle, measure the length of each side in mm and then measure the height to each of those sides. Record your results in a table like the following one. Use each pair to calculate the area of the triangle. Did you get approximately the same area in each case?

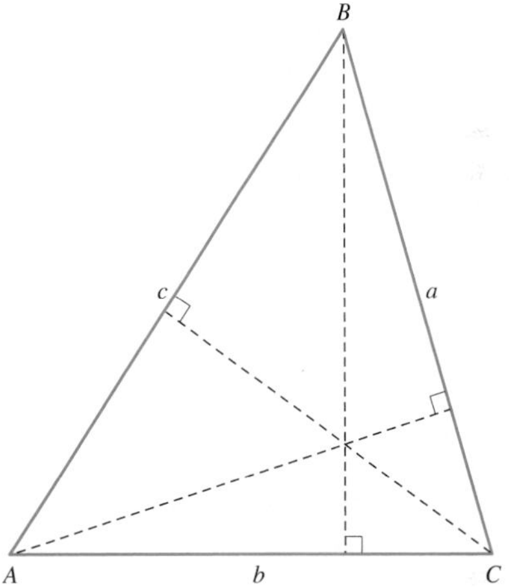

side	height	area
$a =$		
$b =$		
$c =$		

30. For the following triangle, measure the length of each side in mm and then measure the height to each of those sides. Record your results in a table like the following one. Use each pair to calculate the area of the triangle. Did you get approximately the same area in each case?

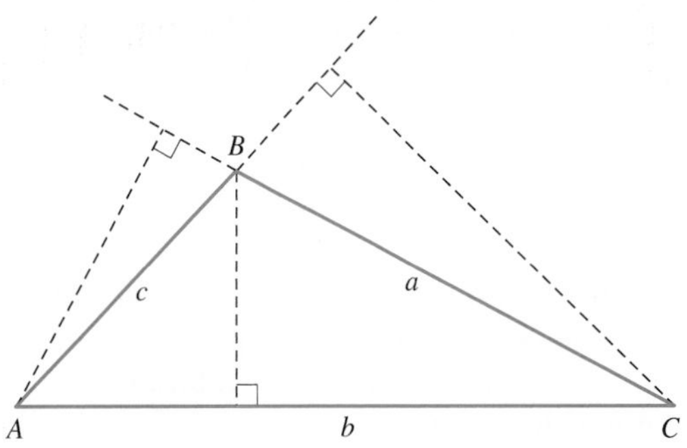

side	height	area
$a =$		
$b =$		
$c =$		

31. Use dimensional analysis to perform each of the following conversions. Where answers are approximate, round to the nearest hundredth.

a. 50 in^2 to ft^2

b. 2000 m^2 to cm^2

c. 3.5 yd^2 to m^2

32. Use dimensional analysis to perform each of the following conversions. Where answers are approximate, round to the nearest hundredth.

a. 0.25 mi^2 to ft^2

b. 400 m^2 to km^2

c. 248 cm^2 to yd^2

33. Find the area of the rectangle shown in (a) square feet and (b) square meters. Round to the nearest hundredth.

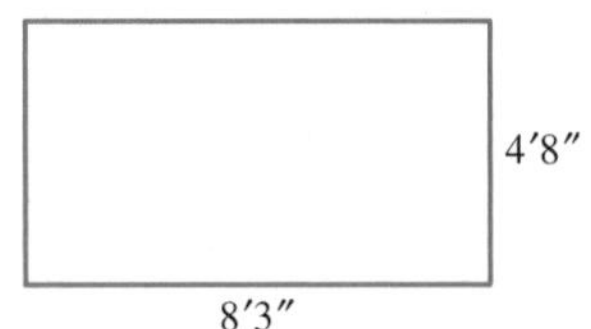

34. Find the area of the triangle shown in (a) square meters and (b) square inches. Round to the nearest hundred thousandth.

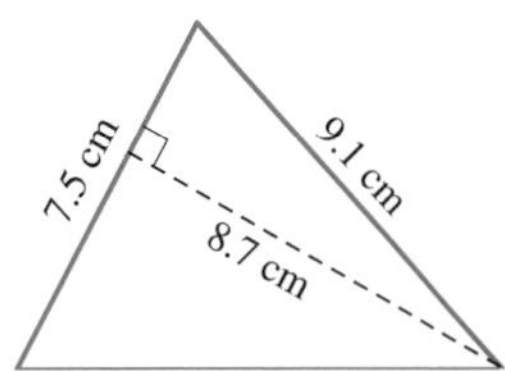

35. Imagine that a cable is wrapped tightly around the Earth at the equator. Then suppose that the cable is cut, an additional 30 ft are added to the cable, and that this longer cable is lifted so that it is equidistant from the equator around the Earth. Because the cable is now a little longer than the circumference of the Earth, there will be a gap between the cable and the surface of the Earth. How large is the gap? Would you be able to crawl under the cable? Walk under it? (NOTE: The equatorial diameter of the Earth is about 7926 mi.) Round to the nearest hundredth.

36. The two coins shown have radii of 3 cm and 1 cm.

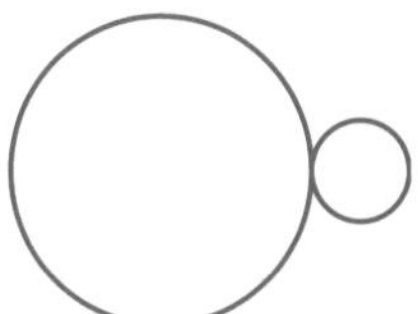

The smaller coin rolls along the circumference of the fixed larger coin until the smaller coin returns to its original position. How many revolutions has the small coin made?

37. Heron of Alexandria, an early Greek mathematician, provided another method for finding the area of a triangle. We can use Heron's formula to find the area of a triangle if we know the lengths of the three sides. **Heron's formula** gives the area of a triangle as $A = \sqrt{s(s-a)(s-b)(s-c)}$, where a, b, and c are the lengths of the three sides and s is half the perimeter, namely $s = \dfrac{a+b+c}{2}$. The number s is called the **semiperimeter** of the triangle. Use this formula to find the areas of triangles with the following side lengths. For answers that are not exact, round to the nearest hundredth.

a. 5 cm, 12 cm, 13 cm

b. 4 m, 5 m, 6 m

c. 4 ft, 5 ft, 8 ft

38. The area of a polygon drawn on a square lattice or geoboard can be found using **Pick's theorem**, which states that if I represents the number of interior lattice points and B represents the number of border lattice points, then the area A of the polygon is given by $A = I + \dfrac{B}{2} - 1$.

a. Use Pick's theorem to determine the area of each figure in problem 13.

b. Use Pick's theorem to approximate the area of Oklahoma as shown next if the distance between two lattice points is 40 miles.

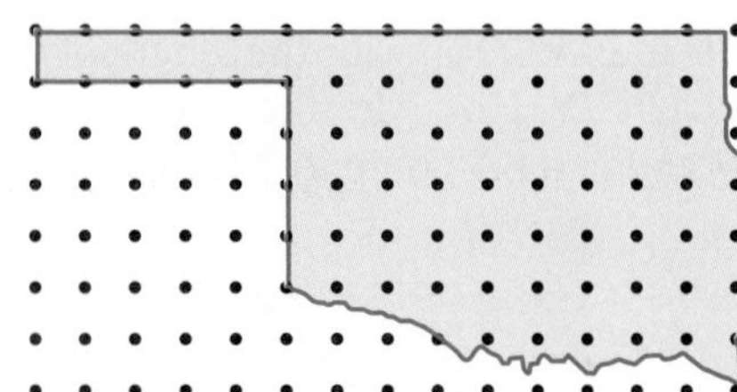

39. Use the method in Example 3.3(b) to verify Theorem 3.3 by answering the following questions about $\triangle ABC$.

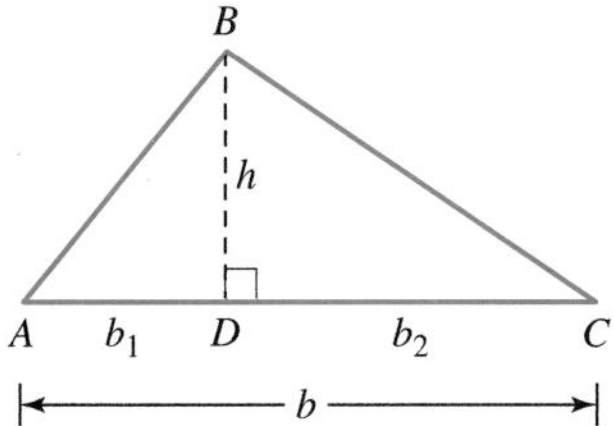

a. In $\triangle ABC$, b is the length of $\overline{AC}$ and h is the height to that side. If $AD = b_1$ and $DC = b_2$, what is the area of right triangle $\triangle ABD$? (Use Theorem 3.2.)

b. What is the area of right triangle $\triangle CBD$?

c. Express the area of $\triangle ABC$ as the sum of the areas of $\triangle ABD$ and $\triangle CBD$.

d. Simplify your answer in (c) by factoring out common factors. What does the sum $b_1 + b_2$ represent? Substitute that expression for $b_1 + b_2$ in your answer.

e. Using your answer in (d), what is the area of $\triangle ABC$? Does this result agree with Theorem 3.3?

40. a. The first two stages of a fractal called the **Sierpinski carpet** are shown, in which a square is divided into nine squares, and the middle square is removed. Construct the next stage in the process by subdividing each square in Stage 1 into nine smaller squares and removing each middle square.

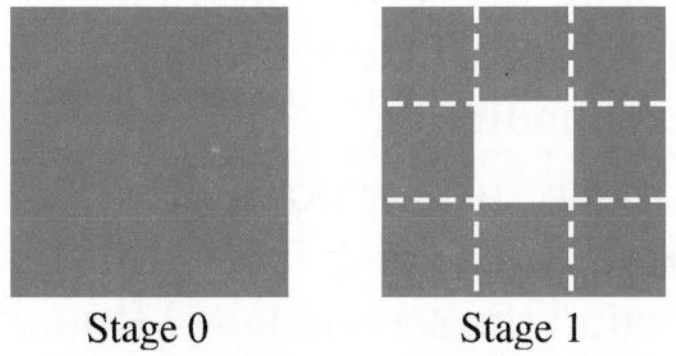

b. Fill in the following table for Stages 0 to 3, assuming that in Stage 0 the square has side lengths of 243 units. Look for a pattern in the number of shaded squares and the area of one shaded square. Using your observations, fill in the table for Stages 4 and 5.

Stage	Number of Shaded Squares	Area of One Shaded Square	Total Shaded Area
0			
1			
2			
3			
4			
5			

APPLICATIONS

41. You may have heard that a basketball hoop is large enough for two basketballs to go through the ring side by side. The inside diameter of a basketball rim is 18″. The circumference of a basketball must be no less than 29.5″ and no greater than 30″. Is the hoop wide enough to accommodate two basketballs at the same time?

42. One of the world's largest synchrotrons, or atom smashers, is at the Fermi National Accelerator Laboratory in Illinois. The accelerator is in a circular underground tunnel 2 km in diameter. Find the distance that atomic particles travel in one pass through the tunnel. Round to the nearest meter.

43. The floor plan of a room in a house is shown. The room will be carpeted, and quarter-round molding will be laid around the edges of the room.

a. Determine the length of quarter round that will be required. Allow for the 7′ patio doors and the two 3′ wide doors in the figure. Round to the nearest foot.

b. Determine the number of square yards of carpet that will be needed. Round to the nearest square yard.

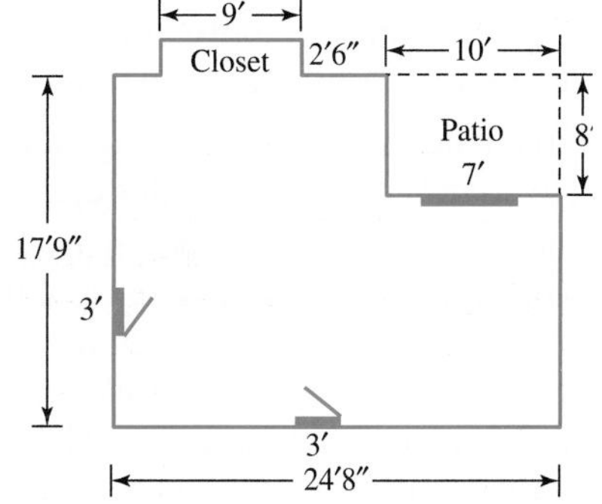

44. According to the directions on a 20-lb bag of fertilizer/weed killer, one bag will cover 5000 ft^2 of lawn. If a house, driveway, and lot have dimensions as shown in the following diagram, will one bag be enough? Assume that all of the *unshaded* region is to be fertilized.

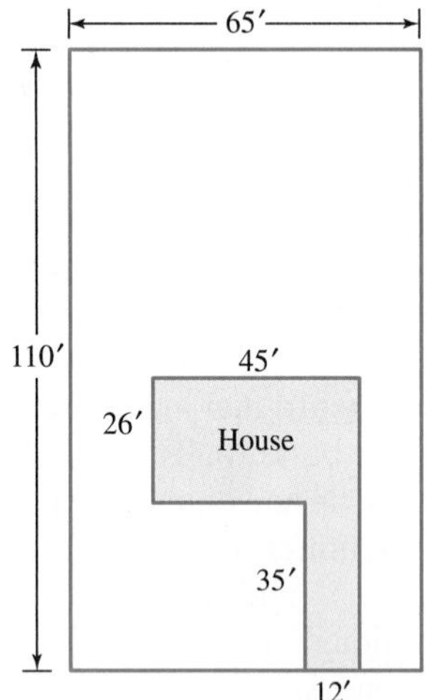

45. A photographer plans to use a 10-ft strip of molding to make square frames for two photographs. The side of one frame will be three times as long as a side of the smaller frame. If the entire 10-ft strip of molding will be used and 10 percent of the strip is allowed for cuts and waste, how large can the two picture frames be?

46. A mason is laying tile on three walls surrounding a shower. Two of the walls to be tiled are 4 feet wide, and one wall is 3 feet wide. If tile will extend 7 feet up each wall, approximately how many 4-inch by 4-inch tiles will be needed for the job?

47. At the end of 2006, there were about 6.5 billion people on Earth. If they all lined up and joined hands, how many times would the line of people wrap around the equator? Assume that each person takes up about 2 yards of space with arms extended and that the radius of the Earth is about 3960 miles.

48. The state of Wyoming is approximately a rectangle with a length of 360 miles and a width of 280 miles.

a. Find the land area of Wyoming if water takes up approximately 700 square miles of space.

b. By the end of the year 2006, there were about 6.5 billion people on Earth. If each of the world's 6.5 billion people stood on a 20-foot by 20-foot square, then how many square miles of land would be needed to accommodate them all? How many states the size of Wyoming would they cover?

49. The speedometer in a car uses the circumference and revolutions of a tire to measure the speed at which the car is traveling. Car manufacturers set the speedometer on a car based on the tire diameter. If the speedometer on your car correctly reads 65 mph for a tire diameter of 29.78 in., then what will the true speed of the car be if you change the tire diameter to 30.78 in. and the speedometer reads 65 mph? Round to the nearest tenth.

50. A coil of cable consists of 42 loops. The inside diameter of the coiled cable is 16 inches, and the outside diameter is 30 inches. The total length of the cable can be calculated using the average circumference as follows:

$$\text{average circumference} = \pi(\text{average diameter})$$
$$\text{length} = (\text{average circumference}) \times (\text{number of loops})$$

Determine the total length of the cable, in feet.

51. A car tire has a diameter of 22 in. When the car is travelling at 30 mph, how many revolutions per minute is the tire turning? Round to the nearest tenth. (HINT: Dimensional analysis may be helpful here.)

52. A track is designed as shown, with semicircular ends and inner dimensions as given. Suppose that each lane is 3 feet wide.

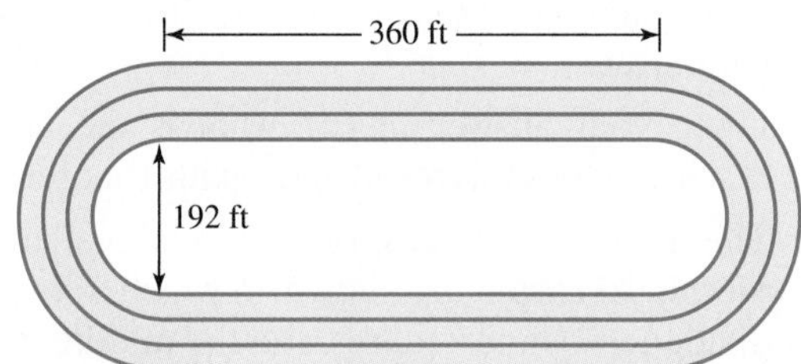

a. Calculate the distance (to the nearest hundredth of a foot) covered in one lap around the innermost lane. Then express your answer in miles, rounded to the nearest hundredth.

b. How much further does a runner travel in one lap around the second lane than in a lap around the first lane (innermost lane)? Round to the nearest hundredth.

c. How much further does a runner travel in one lap around the track in the third lane than in a lap around the second lane? Round to the nearest hundredth.

d. If a race were to be run and starting points in each lane staggered so that all runners would travel exactly the same distance, how much of a head start should runners in the outer lanes be given?

EXTENDED PROBLEMS

53. a. Trace the following square and calculate its area. Cut along the solid lines, and rearrange the pieces into a rectangle that is not a square and calculate the area of the rectangle. What happened?

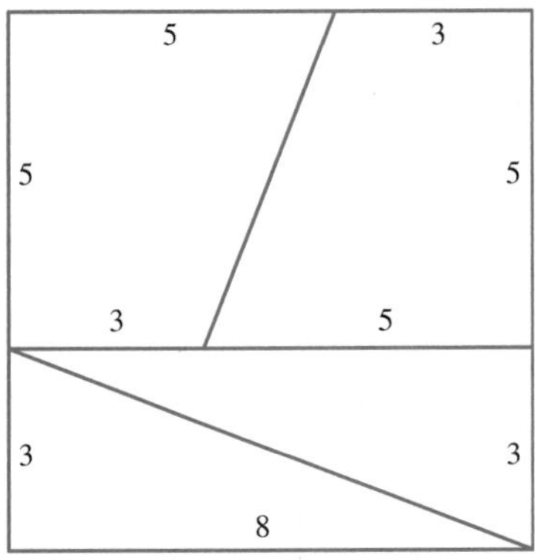

b. Draw another square with a side length of 13 cm. Subdivide the square as shown in part (a) but use the lengths 5 cm, 8 cm, and 13 cm rather than 3, 5, and 8. Cut out the pieces and rearrange to form a rectangle and then compare areas.

c. Draw another square using lengths 2, 3, and 5 respectively. Cut out, rearrange to form a rectangle, and compare areas.

d. Using your results from parts (a) through (c), explain what is happening.

54. A **cyclic quadrilateral** is one in which all four vertices can be positioned on a circle. For example, quadrilateral *ABCD* shown next is a cyclic quadrilateral.

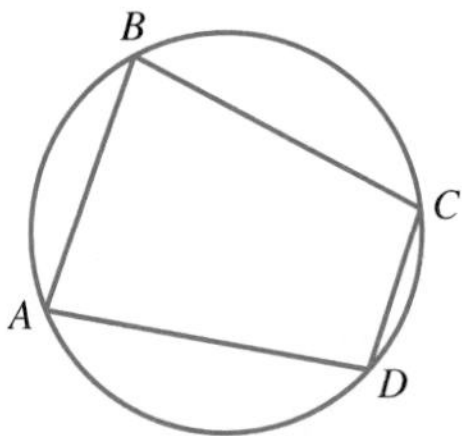

We can use **Brahmagupta's theorem** to find the areas of cyclic quadrilaterals. The theorem states that if a, b, c, and d are the lengths of the sides of a cyclic quadrilateral, and the semiperimeter of the quadrilateral is $s = \frac{a + b + c + d}{2}$, then the area of the quadrilateral is

$$A = \sqrt{(s - a)(s - b)(s - c)(s - d)}$$

a. A square is a cyclic quadrilateral. Use Brahmagupta's theorem to find the area of a square with side lengths of 9 in.

b. A rectangle is a cyclic quadrilateral. Use Brahmagupta's theorem to find the area of a rectangle with sides of length 4 cm and 10 cm.

c. Simplify the area formula given in Brahmagupta's theorem for the area of a square with side length x.

d. Simplify the area formula given in Brahmagupta's theorem for the area of a rectangle with side lengths x and y.

e. Research the properties of cyclic quadrilaterals and list five properties they share. Draw a cyclic quadrilateral that is not a square or a rectangle and show how to use Brahmagupta's theorem to find the area. Verify the area using Heron's formula from problem 37.

55. As far back as Archimedes in 225 B.C.E., mathematicians have been approximating the value of π, and they continue to do so today using computers. Research the history of the calculation of π and create a timeline to show who made advances in the approximation of π and how many digits of accuracy each mathematician achieved. Describe some of the methods used to obtain the approximations. What is the current record for the number of digits of π calculated? List 10 ways in which π has been used, both historically and currently. Discuss why researchers today continue to seek more digits of π.

3.2 MORE AREA FORMULAS

Applied Problem

A single 4″ diameter water pipe is to be replaced by smaller 2″ pipes. How many 2″ pipes will be required to carry at least as much water as the 4″ pipe?

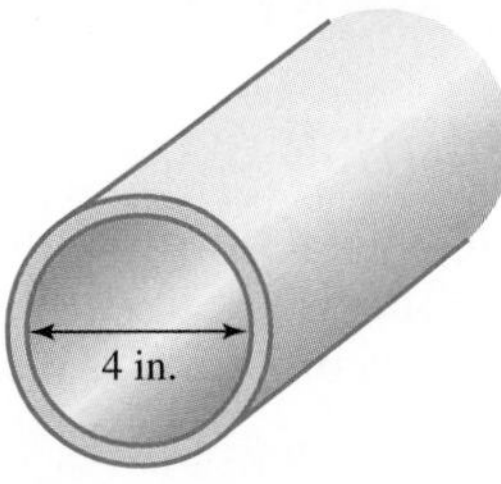

Area of a Parallelogram

Now that we can use Theorem 3.3 to find the area of any triangle, we can use Postulate 3.2 to find the areas of other common polygons.

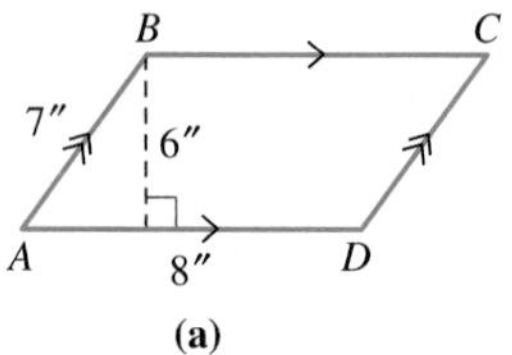

(a)

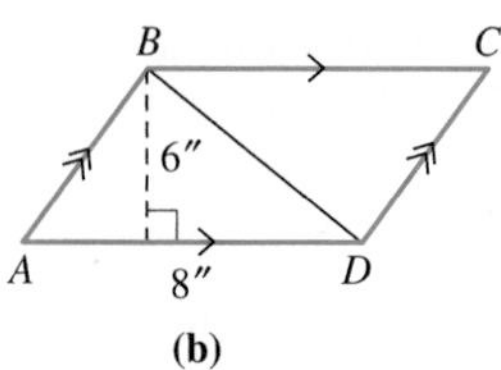

(b)

FIGURE 3.15

EXAMPLE 3.8 Find the area of the parallelogram in Figure 3.15(a).

SOLUTION

If we draw diagonal $\overline{BD}$ [Figure 3.15(b)], we form two triangles having the same size and shape. (We prove this in Chapter 5.) Thus, the area of $ABCD$ is twice the area of $\triangle ABD$. Because the height of $\triangle ABD$ is 6 in. and the base has measure 8 in., its area is $\frac{1}{2}(6 \text{ in.})(8 \text{ in.}) = 24 \text{ in}^2$. Because $\triangle ABD$ and $\triangle CDB$ are identical triangles, each with area 24 in^2, by Postulate 3.2 the area of the parallelogram is $2(24) = 48 \text{ in}^2$. ■

Notice that we could have found the area in Example 3.8 by multiplying 6 in. (the height of one of the triangles) by 8 in. (the length of the base of one of the triangles).

In a parallelogram, the length of a perpendicular segment having one end point at a vertex and the other end point on the opposite parallel side (or the side extended) is called the **height** of the parallelogram. As is the case with triangles, a height is associated with a particular side, called the corresponding **base** of the parallelogram.

Example 3.8 leads to the following general result. We will leave the verification for the problem set.

THEOREM 3.4 Area of a Parallelogram

The area of a parallelogram is equal to the product of the length of one of its sides and the height to that side.

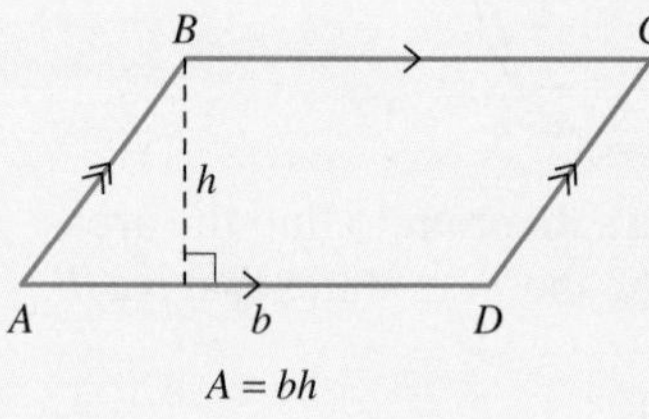

In Chapter 5, we will prove that a rhombus is a special type of parallelogram. Thus, the area of a rhombus can be found using the formula in Theorem 3.4.

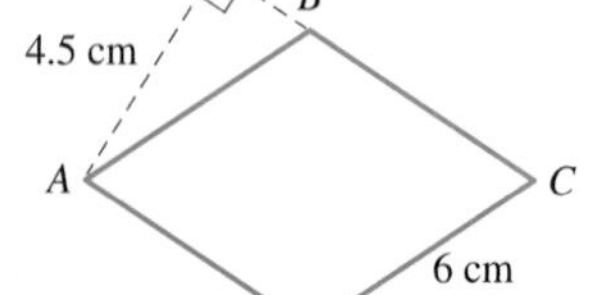

FIGURE 3.16

EXAMPLE 3.9 Find the area of the rhombus $ABCD$ in Figure 3.16.

SOLUTION

Any side of the rhombus may be used as the base. We will use $\overline{BC}$ since the height to that side is given. A rhombus has four congruent sides, so $BC = CD = 6$ cm. Thus, the area of $ABCD$ is given by

$$A = bh = (6 \text{ cm})(4.5 \text{ cm}) = 27 \text{ cm}^2$$ ■

Area of a Trapezoid

The same technique used to derive the formula for the area of a parallelogram can be used to find a formula for the area of a trapezoid.

EXAMPLE 3.10 Find the area of the trapezoid in Figure 3.17(a).

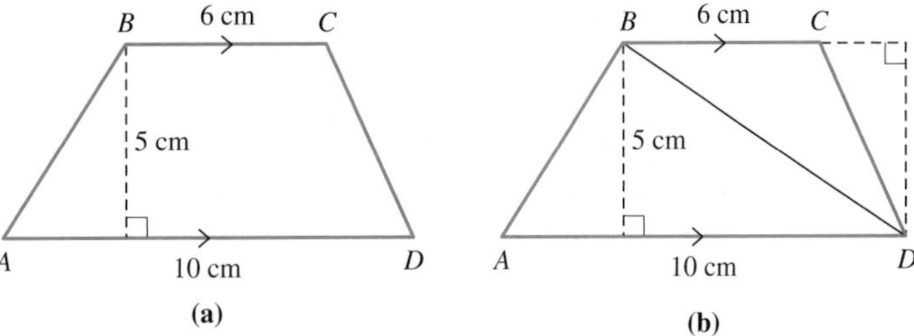

FIGURE 3.17

SOLUTION

If we draw diagonal $\overline{BD}$ [Figure 3.17(b)], we form $\triangle ABD$ and $\triangle BCD$. Thus, the area of $ABCD$ is the sum of the areas of the two triangles. Because the height of both triangles is 5 cm and the bases are 10 cm and 6 cm, the area of the trapezoid is

$$\tfrac{1}{2}(10 \text{ cm})(5 \text{ cm}) + \tfrac{1}{2}(6 \text{ cm})(5 \text{ cm}) = 25 \text{ cm}^2 + 15 \text{ cm}^2 = 40 \text{ cm}^2$$ ■

Notice that $\frac{1}{2}(10)(5) + \frac{1}{2}(5)(6)$ can be rewritten as $\frac{1}{2}(5)(10 + 6)$. This leads to the following general result, whose verification we will leave for the problem set. The **height** of a trapezoid is the perpendicular distance between the two bases.

THEOREM 3.5 Area of a Trapezoid

The area of a trapezoid is equal to half the product of the height and the sum of the lengths of its two bases.

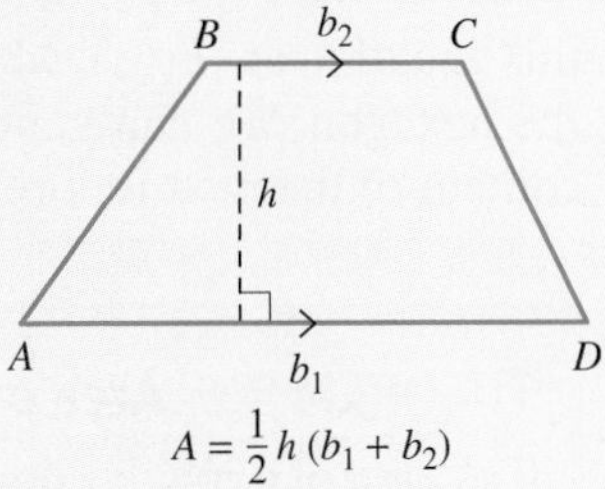

$$A = \frac{1}{2}h(b_1 + b_2)$$

EXAMPLE 3.11 Find the area of the trapezoid in Figure 3.18(a) in two different ways.

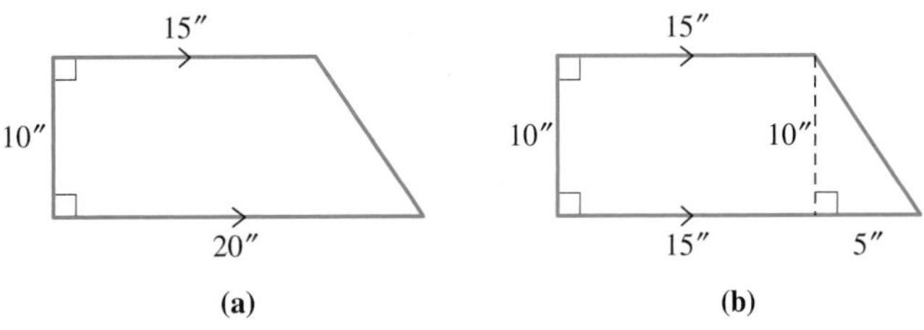

FIGURE 3.18

SOLUTION

i. Using Theorem 3.4, we have

$$A = \tfrac{1}{2}h(b_1 + b_2) = \tfrac{1}{2}(10 \text{ in.})(15 \text{ in.} + 20 \text{ in.}) = 175 \text{ in}^2$$

ii. Subdividing the trapezoid into a rectangle and a right triangle as in Figure 3.18(b), we have

$$A = lw + \tfrac{1}{2}bh = 15 \text{ in.}(10 \text{ in.}) + \tfrac{1}{2}(5 \text{ in.})(10 \text{ in.}) = 150 \text{ in}^2 + 25 \text{ in}^2 = 175 \text{ in}^2 \quad ■$$

Area of a Regular Polygon

We can show that there is a point in the plane that is equidistant from all vertices of a regular polygon and from all the sides of the polygon. This point is called the **center** of the polygon. In Figure 3.19(a), point O is the center of the regular hexagon.

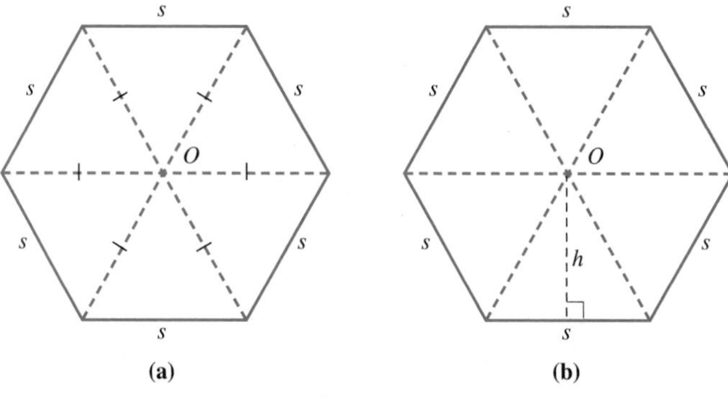

FIGURE 3.19

We can conveniently find the area of any regular polygon by subdividing the polygon into triangles as suggested by the regular hexagon in Figure 3.19(a). The area of each of the six triangles is $\frac{1}{2}sh$, where h is the perpendicular distance from the center of the polygon to any side [Figure 3.19(b)]. (The line segment with length h is called an **apothem** of the polygon.) Thus, the area of the hexagon is $6(\frac{1}{2}sh)$. We can rewrite this equation as $\frac{1}{2}h(6s)$, which is equivalent to $\frac{1}{2}hP$, where $P = 6s$ is the perimeter of the hexagon. We can generalize this technique to any *regular* polygon. This is the content of the next theorem.

> **THEOREM 3.6 Area of a Regular Polygon**
>
> The area of a regular polygon is equal to half the product of the perimeter of the polygon and the perpendicular distance from its center to one of its sides.
>
> 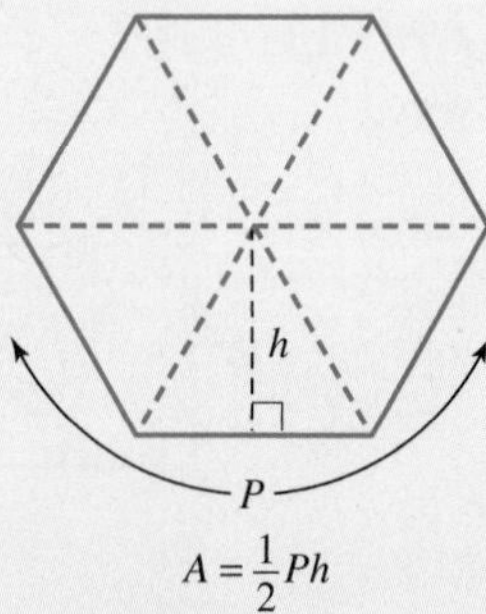
>

EXAMPLE 3.12 A bandstand in a city park is to be constructed in the shape of a regular octagon (Figure 3.20). To estimate the lumber needed, we must calculate the area of the platform. Using the dimensions given in the figure, where point O is the center, determine the area of the octagon.

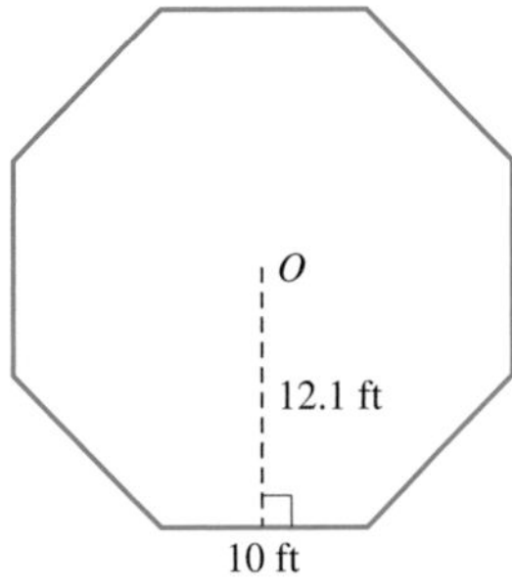

FIGURE 3.20

SOLUTION

Using Theorem 3.6, we have

$$A = \frac{1}{2}Ph = \frac{1}{2}(8 \cdot 10 \text{ ft})(12.1 \text{ ft}) = 484 \text{ ft}^2$$ ■

Area of a Circle

Now imagine a series of regular polygons, as shown in Figure 3.21, where h remains constant but the number of sides increases to approach a circle.

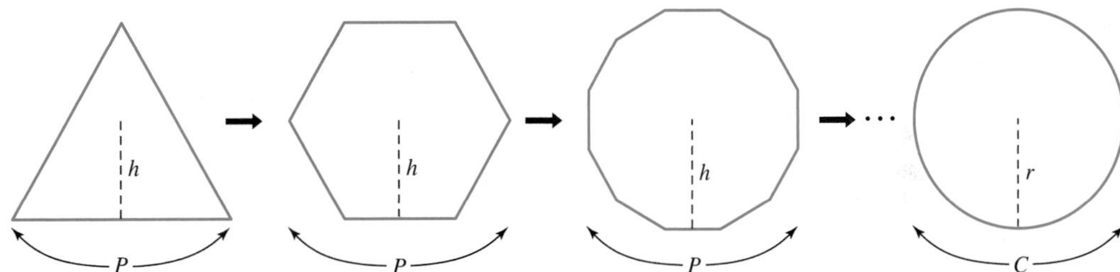

FIGURE 3.21

The h in Theorem 3.6 becomes the radius, r, of the circle and the perimeter P becomes the circumference C of the circle. Thus, by substituting r for h and C for P in the area formula in Theorem 3.6, we find that the area of a circle of radius r is equal to $\frac{1}{2}rC = \frac{1}{2}r(2\pi r) = \pi r^2$.

THEOREM 3.7 Area of a Circle

The area of a circle is equal to the product of π and the square of the radius.

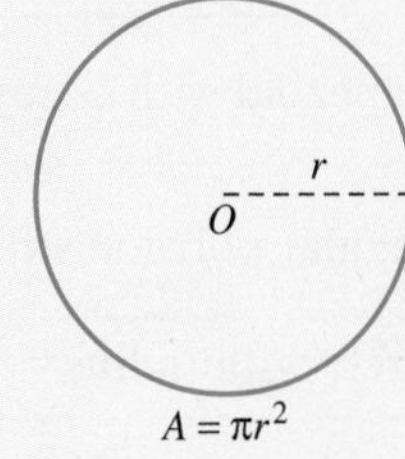

$A = \pi r^2$

EXAMPLE 3.13 Figure 3.22 is composed of one large and four small semicircles. Find the area enclosed by these five semicircles.

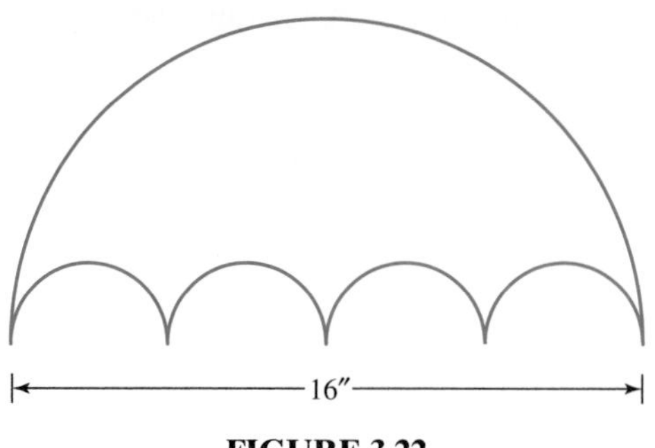

FIGURE 3.22

SOLUTION

The diameter D of the larger semicircle is 16 inches, so its radius R is 8 inches. The diameter d of each small semicircle is $\frac{16}{4} = 4$ inches, so the radius r is 2 inches. The area of the large semicircle is

$$A_L = \tfrac{1}{2}\pi R^2 = \tfrac{1}{2}\pi(8'')^2 = 32\pi \text{ in}^2$$

The area cut out by the four small semicircles is

$$A_s = 4 \cdot \tfrac{1}{2}\pi r^2 = 2\pi r^2 = 2\pi(2'')^2 = 8\pi \text{ in}^2$$

Thus, the total area is given by

$$A_{\text{Total}} = A_L - A_S = 32\pi \text{ in}^2 - 8\pi \text{ in}^2 = 24\pi \text{ in}^2$$

The exact area of the shaded region is 24π in^2. For many practical applications, a decimal approximation is more appropriate. In this case,

$$A_{\text{Total}} = 24\pi \text{ in}^2 \approx 75.4 \text{ in}^2 \quad \text{(to one decimal place)}$$ ■

The following table summarizes the area formulas we have discussed thus far.

Summary of Area Formulas for Two-Dimensional Shapes	
Geometric Shape	**Area**
Rectangle with length l and width w	$A = lw$
Right triangle with legs of length a and b	$A = \frac{1}{2}ab$
Triangle with side length b and height h to that side	$A = \frac{1}{2}bh$
Parallelogram with side length b and height h	$A = bh$
Trapezoid with bases of length b_1 and b_2 and height h	$A = \frac{1}{2}h(b_1 + b_2)$
Regular polygon with perimeter P and apothem h	$A = \frac{1}{2}Ph$
Circle with radius r	$A = \pi r^2$

TABLE 3.1

Solution of Applied Problem

To replace a 4″ pipe with 2″ pipes, the total cross-sectional area of the 2″ pipes must be at least as great as the cross-sectional area of the 4″ pipe. A 4″ pipe has cross-sectional area of $\pi(r^2) = \pi(2\text{ in.})^2 = 4\pi\text{ in}^2$. A 2″ pipe has cross-sectional area $\pi(1\text{ in.})^2 = \pi\text{ in}^2$. Hence, four 2″ pipes will have the same cross-sectional area as one 4″ pipe.

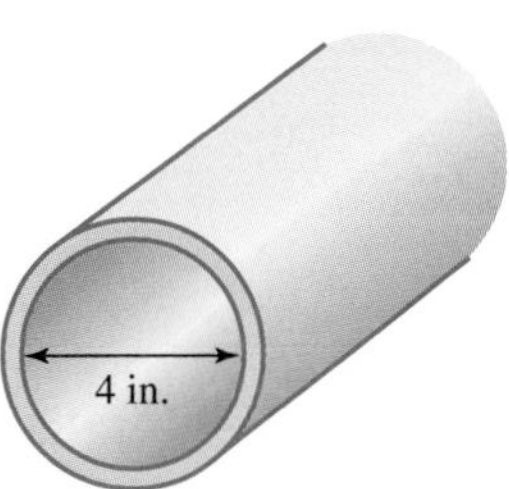

GEOMETRY AROUND US The pictograph to the left compares the number of cars that two different countries produce in a year. This graph is deceptive, however. Using the numbers given, we see that the production of cars in the United States is roughly twice that of Germany. However, the car symbol for the United States appears to be about *four* times as large. The reason is that both the length and the height of the car were doubled, which results in the area being quadrupled.

To present an accurate two-dimensional comparison using the car symbols where one represents n times as many cars as the other, either of the following is correct: (a) One of the dimensions of the first car should be displayed as n times the corresponding dimension of the second car and the other dimensions of the two cars should be exactly the same, or (b) both dimensions of the first car should be $\sqrt{n}$ times those of the second car.

CARS PRODUCED PER YEAR

United States
11,900,000

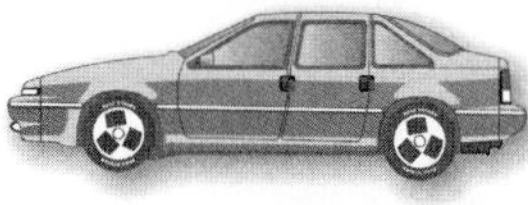

Germany
5,600,000

PROBLEM SET 3.2

EXERCISES/PROBLEMS

1. Find the area of each figure.

a.

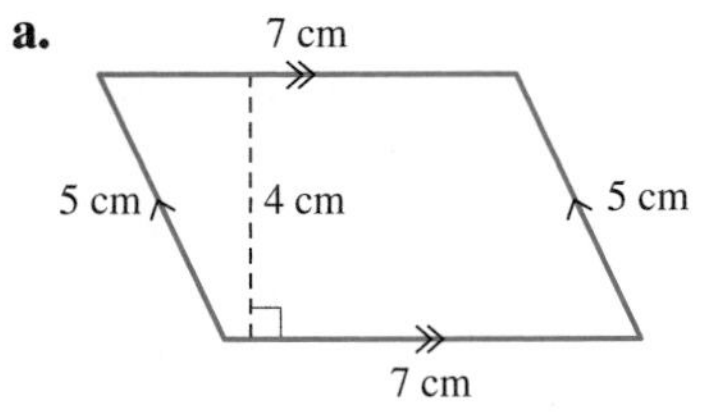

b.

2. Find the area of each figure. Round to the nearest hundredth.

a.

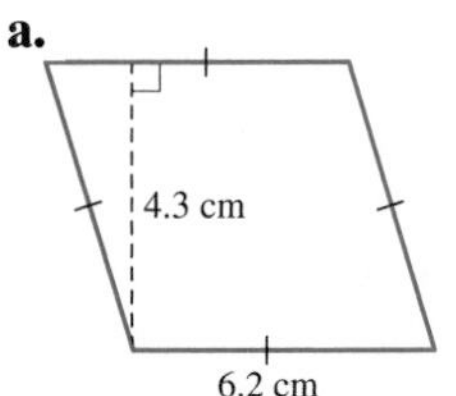

b.

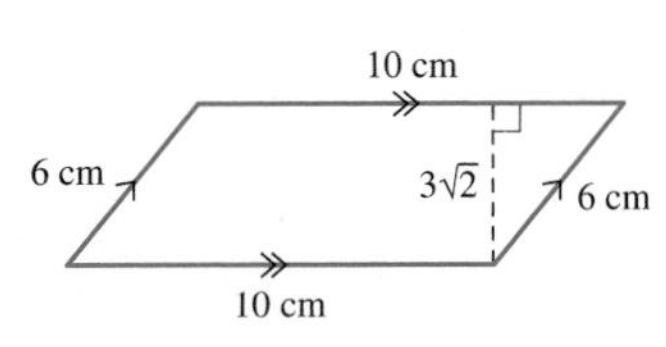

3. Complete the following table, which gives information about a parallelogram with base b, height h, and area A. Round to the nearest hundredth.

	b	h	A
a.		6	40
b.	$3\sqrt{2}$	$2\sqrt{3}$	

4. Complete the following table, which gives information about a parallelogram with base b, height h, and area A. Round to the nearest hundredth.

	b	h	A
a.	13.6		74.9
b.	$3\sqrt{5}$	$\sqrt{5}$	

5. For the parallelogram shown, measure the length of each side in millimeters. Then measure the height to each side. Record your results in a table like the one shown. Use each pair of measurements to calculate the area of the parallelogram. Were your results equal or nearly equal?

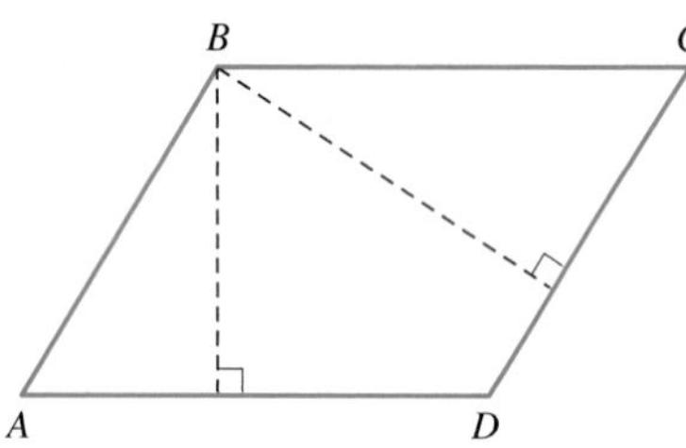

Side	Height	Area

6. Apply the directions for problem 5 to the following parallelogram.

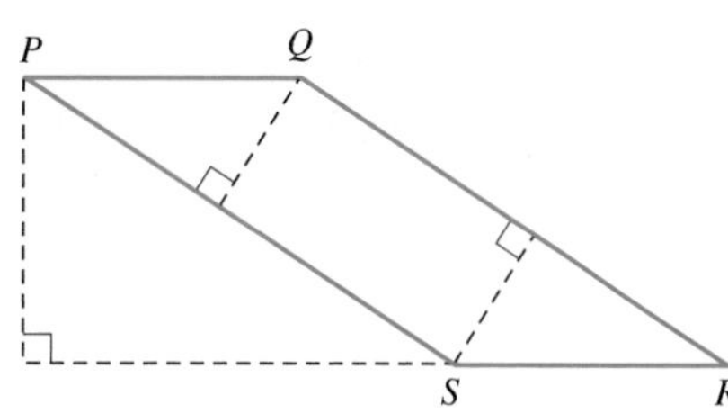

Side	Height	Area

7. Find the area of the trapezoid.

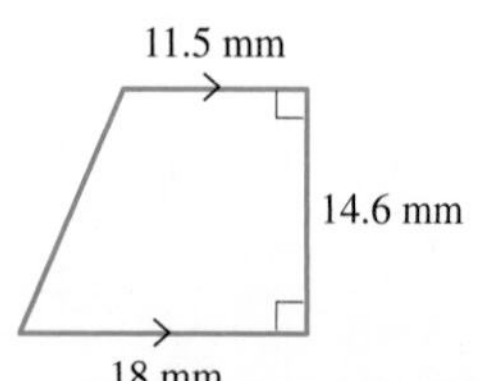

8. Find the area of the trapezoid.

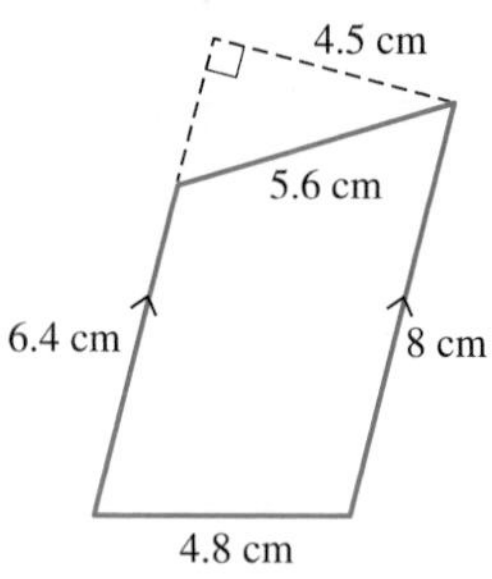

9. Find the area of the shaded region.

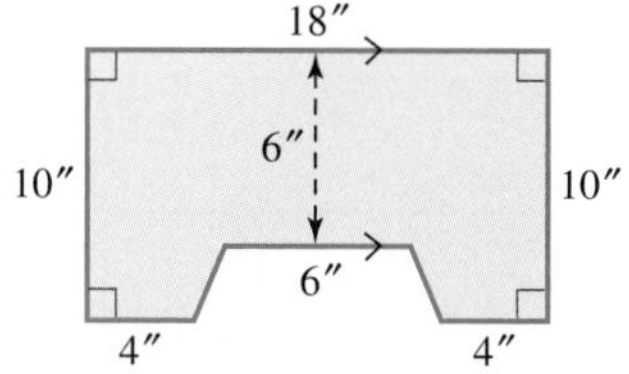

10. Find the area of the shaded region.

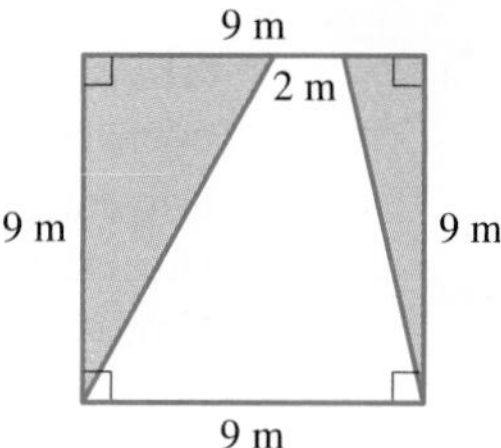

11. The area of a trapezoid is 69 in^2. Its bases have lengths 9 in. and 14 in. Find the height of the trapezoid.

12. A trapezoid has a height of 5 cm and one base that measures 8 cm. If the area of the trapezoid is 28 cm^2, find the length of the other base.

13. Find the area of the shaded region. The hexagon is regular. Round to the nearest hundredth.

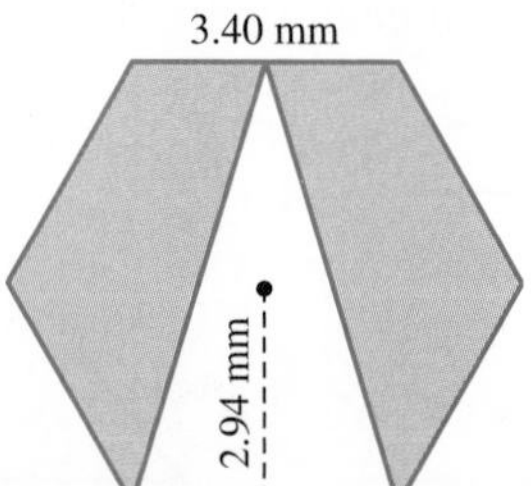

14. Find the area of the shaded region. The two regular octagons have the same center, $OP = 5.5$ cm, and $PQ = 1$ cm.

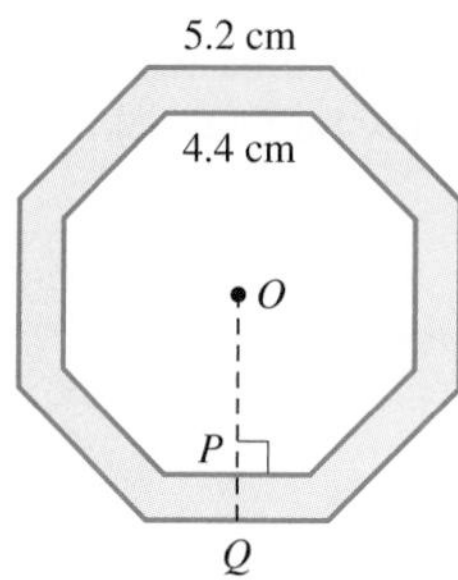

15. Find the area of each figure. Round to the nearest hundredth.

a.

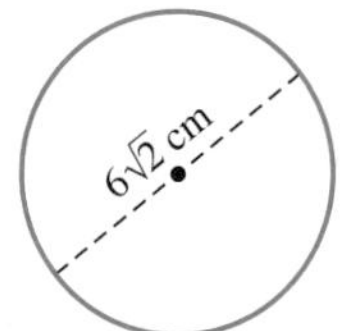

b.

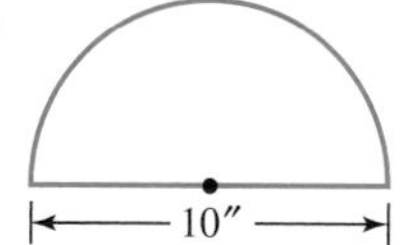

16. Find the area of the shaded machined piece shown, which is made by cutting a small semicircle out of a larger semicircle. The radius of the larger semicircle is twice the radius of the smaller. Round to the nearest square centimeter.

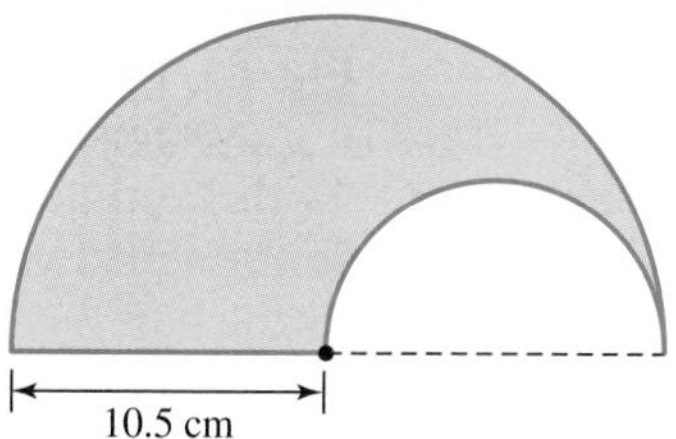

17. Find the area of each of the figures shown. $\triangle ABC$ is equilateral and all curves are semicircles. Round to the nearest hundredth. (HINT: See problem 37 in Section 3.1.)

a.

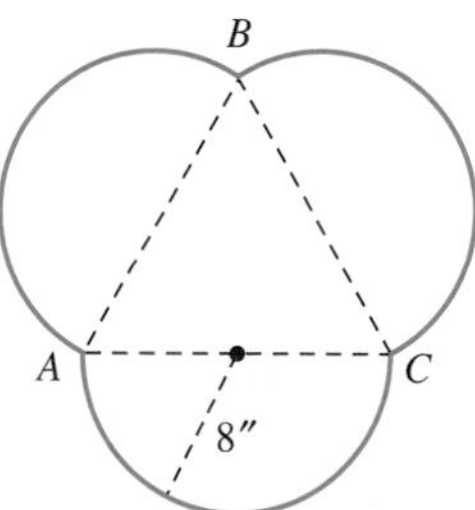

b.

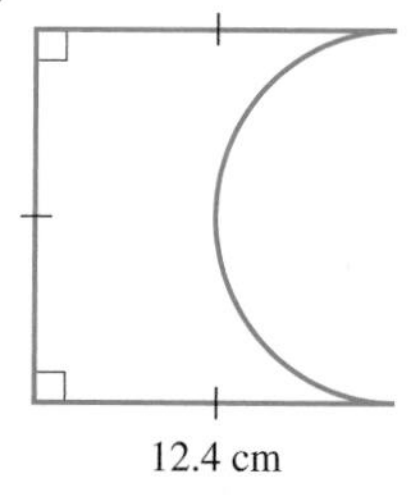

18. Determine the area of the shaded region in each of the following figures. Round to the nearest hundredth.

a.

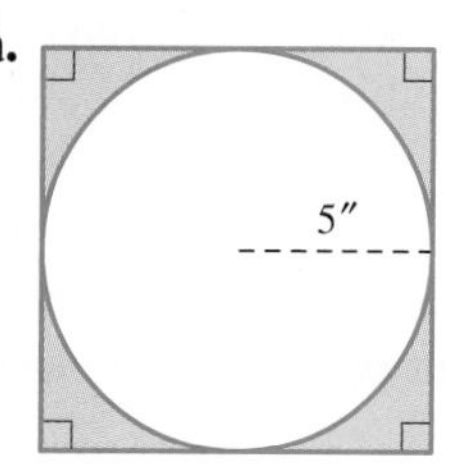

b.

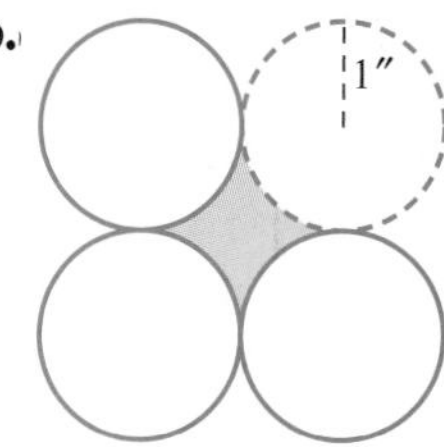

19. In the following figure, each petal inside the square is formed by the intersection of two semicircles. Find the area of the shaded region. Give the *exact* answer.

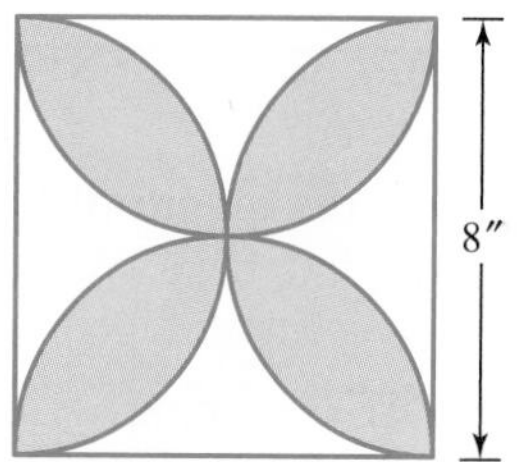

20. The circles in the figure have radii of 6″, 4″, 4″, and 2″. Which is larger, the shaded area inside the large circle or the shaded area outside the large circle?

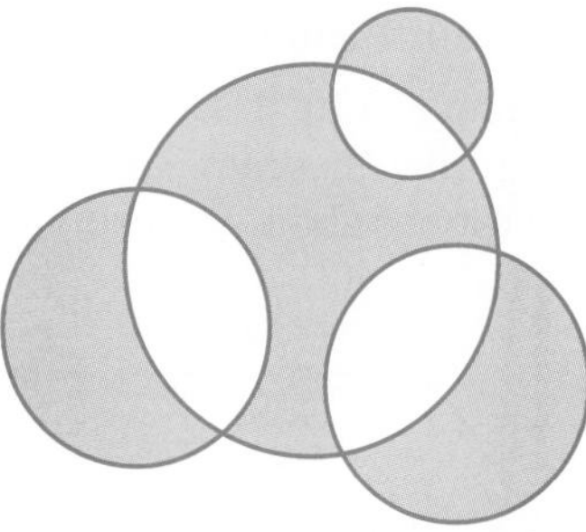

21. Which has the larger area, a square with a perimeter of 20 inches or a circle with a circumference of 20 inches?

22. **Concentric circles** are circles that have the same center. Two concentric circles have radii of 10 cm and 14 cm.

a. Which is larger, the area inside the smaller circle or the area between the two circles?

b. What would the radius of the larger circle have to be in order for the area of the inner circle to equal the area between the two circles?

23. If the radius of a circle is doubled, what is the area of the resulting circle? If the radius is multiplied by 5, what is the new area? What happens if the radius is multiplied by n?

24. The base and height of a parallelogram are both multiplied by 3. What is the area of the resulting parallelogram? If both the base and the height are multiplied by 7, what is the new area? What happens if the base is multiplied by m and the height is multiplied by n?

25. Review Example 3.8. Use the method of that example to verify Theorem 3.4 by answering the following questions about parallelogram $ABCD$.

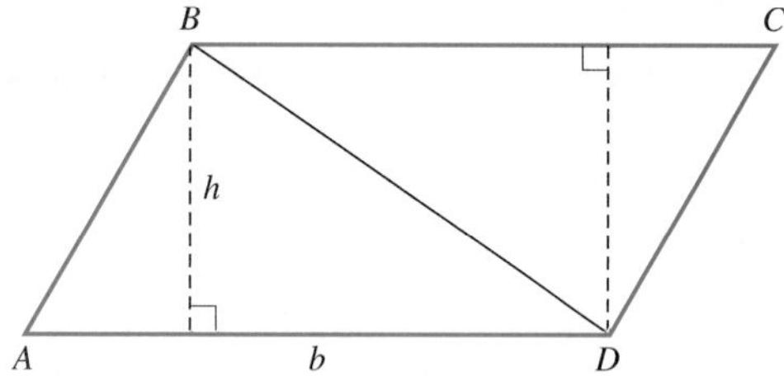

a. In $ABCD$, b is the length of base $\overline{AD}$ and h is the height to that base. What is the area of $\triangle ABD$?

b. How does $\triangle ABD$ compare with $\triangle CDB$? What is the area of $\triangle ABD$? of $\triangle CDB$?

c. Express the area of parallelogram $ABCD$ in terms of the areas of $\triangle ABD$ and $\triangle CDB$.

d. Considering your answer in (c), how can you find the area of a parallelogram? Does your result agree with Theorem 3.4?

26. Review Example 3.10. Use the method of that example to verify Theorem 3.5 by answering the following questions about trapezoid $ABCD$.

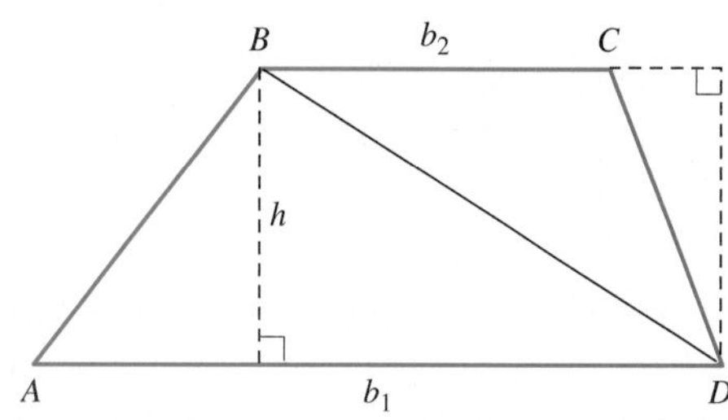

a. In $ABCD$, b_1 and b_2 are the lengths of its bases, and h is its height. What is the area of $\triangle ABD$?

b. What is the height of $\triangle BCD$? What is the area of $\triangle BCD$?

c. Express the area of trapezoid $ABCD$ in terms of the areas of $\triangle ABD$ and $\triangle BCD$.

d. Simplify your answer in (c) by factoring out common factors.

e. Considering your answer in (d), how can you find the area of a trapezoid? Does your answer agree with Theorem 3.5?

27. Determine the approximate area of the figure shown by measuring appropriate dimensions in millimeters. Round your answer to the nearest square millimeter. (HINT: Try subdividing the shape into smaller regions. Heron's formula from problem 37 in Section 3.1 may be useful.)

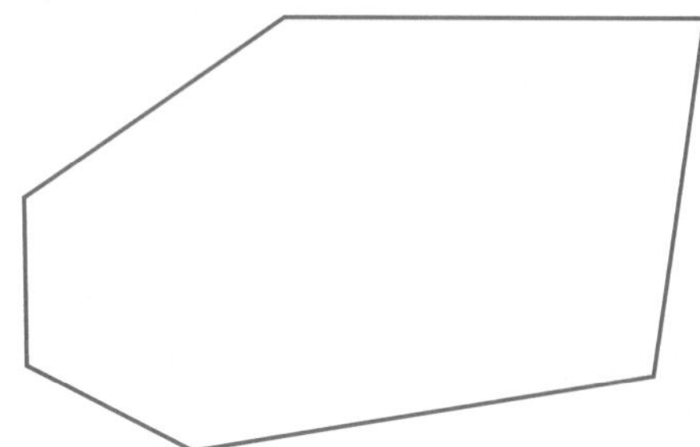

28. Determine the approximate area of the figure shown by measuring appropriate dimensions in millimeters. Round to the nearest square millimeter.

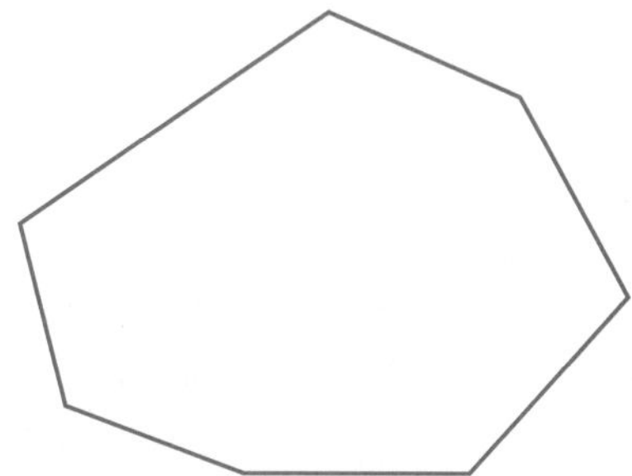

APPLICATIONS

29. A take-and-bake pizza store offers three sizes of circular pizzas: small (10″ diameter), medium (12″ diameter), and large (16″ diameter). A special combination pizza comes "with everything" and costs \$10.99 for a small, \$12.75 for a medium, and \$15.25 for a large. Which pizza is the best buy? Justify your response.

30. The dimensions and costs for three different pine gazebos are given. The floors are regular octagons. Which gazebo costs the least per square inch? Justify your response.

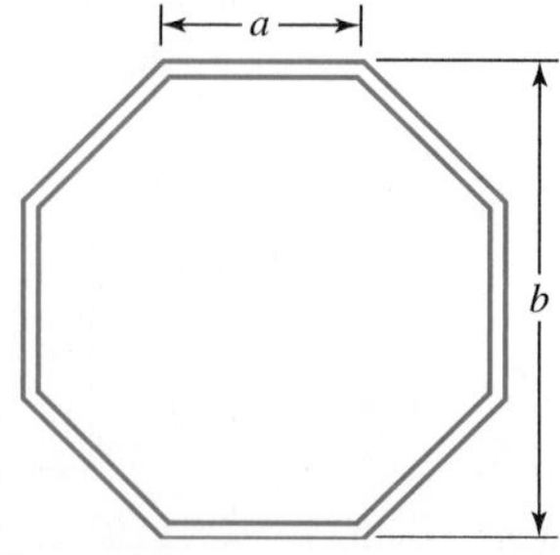

	Gazebo I	Gazebo II	Gazebo III
a	$36\frac{7}{16}''$	$57\frac{3}{8}''$	$77\frac{7}{16}''$
b	$88''$	$138\frac{5}{8}''$	$186\frac{7}{8}''$
Total Cost	\$4195	\$5195	\$9719

31. An official National Hockey League rink measures 200 feet by 85 feet and has rounded corners formed by quarter circles with radii 28 feet, as shown. Find the area of the rink. Round to the nearest square foot.

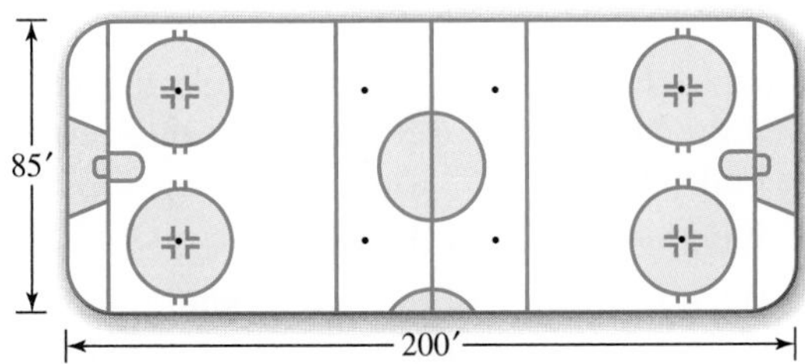

32. A circular table has a diameter of 48″. The table can be enlarged by inserting two rectangular leaves, each 18″ wide and 48″ long. What is the area of the expanded table? Round to the nearest tenth.

33. A soybean farmer uses an irrigation system in which eight sections of 165-ft pipe are connected end to end, and the resulting length of sprinkler pipe travels in a circle around a well as shown in the next figure.

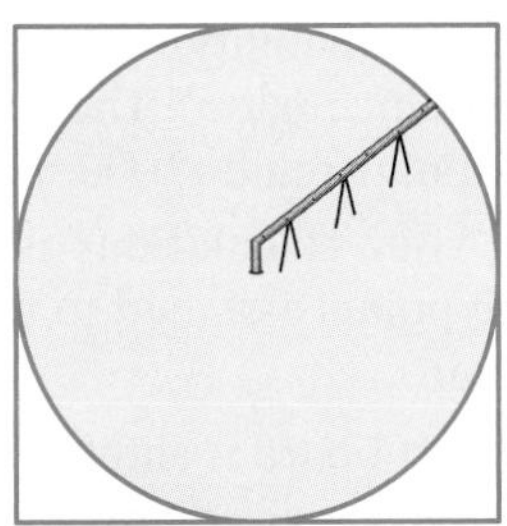

a. How many acres of the square field are irrigated? (One acre is 43,560 ft^2.)

b. The farmer usually plants wheat in the corners of the field, which are not irrigated. How many acres of wheat are planted in this field?

34. In the past, a goalie in the National Hockey League could play the puck from anywhere behind the goal line. Beginning in the 2005–2006 season, the goalie could play the puck only from within a trapezoidal area behind the goal (see figure). What is the area of the trapezoid?

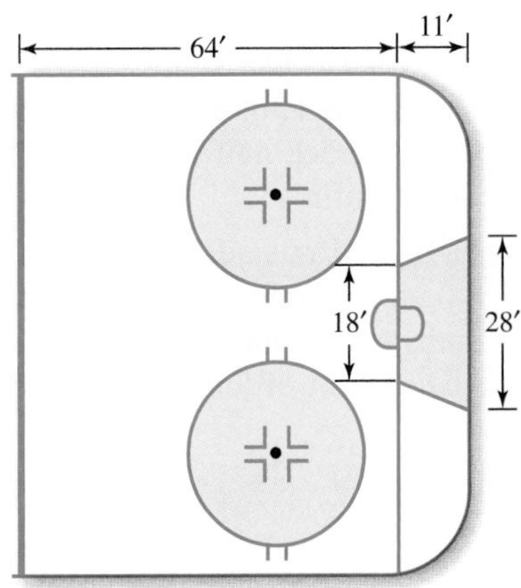

35. A rectangular metal plate 8.4″ by 10.2″ has four holes drilled in it. Two holes have a diameter of $\frac{3}{4}''$ and two holes have a diameter of $\frac{1}{2}''$. If the plate material weighs 7.5 lb/ft^2, find the weight of the plate. Round to the nearest tenth.

36. Two 1-inch pipes empty into one larger pipe. What is the minimum diameter of the larger pipe that can accommodate the two smaller pipes?

37. A stretch of old highway shown next will be widened to include 6-foot bike lanes on each shoulder. Each existing lane is 12 feet wide. How many square feet of surface will be paved when the entire newly widened stretch of highway is resurfaced? (NOTE: Not drawn to scale.)

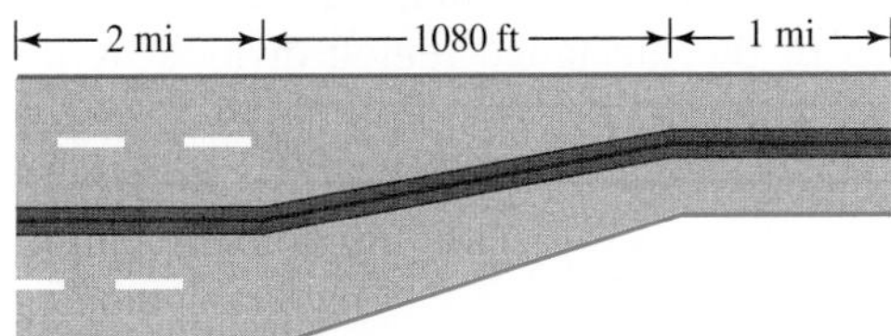

38. An indoor running track has the dimensions shown. The ends of the track are semicircles. If the track is to be resurfaced, how many square feet of material will be required? Round to the nearest square foot.

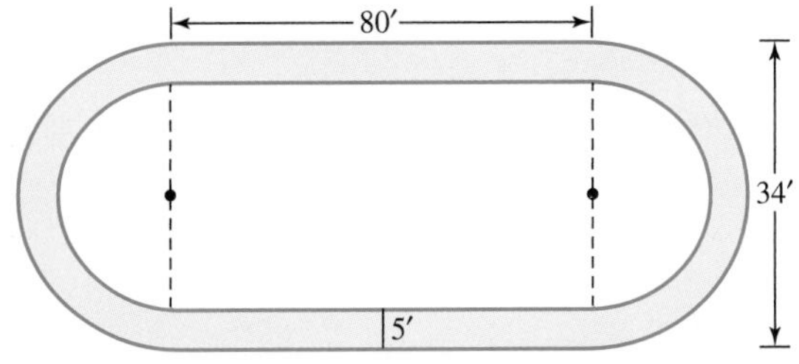

39. A goat is tied to one corner of a square pen 20 ft on a side.

a. If the goat's tether is 10 ft long, what percentage of the grass in the pen can the goat reach?

b. What should the length of the tether be so that the goat can reach half of the grass? Round to the nearest tenth.

40. The flag of Brunei was adopted on September 29, 1959. (See figure). Excluding the design in the middle, how many square inches of each color material would be needed in order to construct an official Brunei flag?

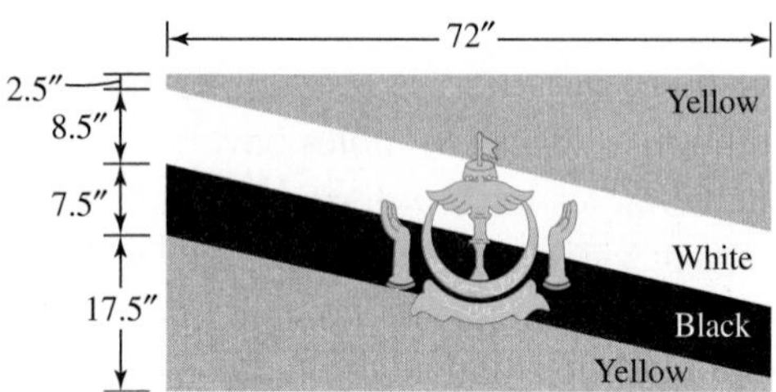

41. A paraglider wants to land in the unshaded region in the square field illustrated because the shaded regions (four quarter circles) are briar patches. If the paraglider hits the field randomly due to unexpected wind currents, what is the probability that she misses the briar patches? Use the fact that the probability of landing in the unshaded region is the area of the unshaded region divided by the total area of the field.

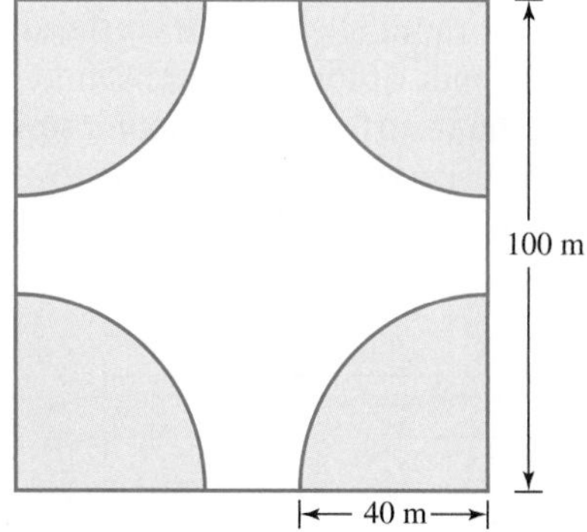

42. An archery target consists of five concentric circles as shown.

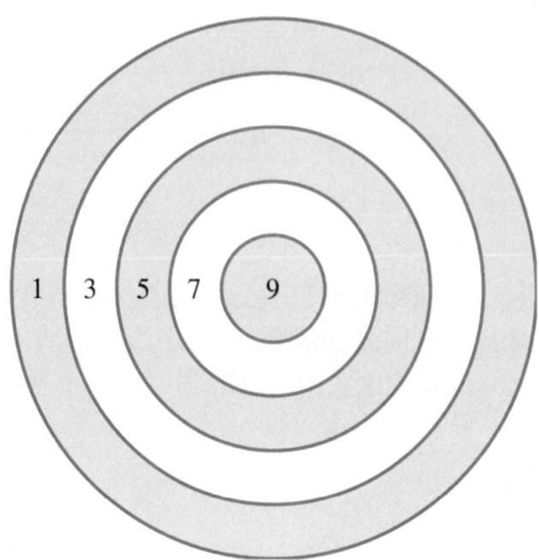

Points for each region on the target are indicated. The radius of the bull's-eye is 4.5″, and each ring surrounding the bull's-eye is 4.625″ wide.

a. What is the area of the bull's-eye?

b. What is the area of the seven-point region?

c. What is the area of the five-point region?

d. What is the area of entire target?

e. The probability of hitting the bull's-eye on a random shot that hits the target is

$$P(\text{bull's-eye}) = \frac{\text{area of bull's-eye}}{\text{area of entire target}}$$

Calculate the probability of hitting the bull's-eye. Express your answer as a percent. This gives the percentage of the time in which a shot that hits the target at random would be expected to hit the bull's-eye.

f. What is the probability that a random shot hits the seven-point region? The five-point region? The three-point region? The one-point region?

g. Based on your calculations in part (f), determine if the points assigned to the bull's-eye and rings are fair.

43. A clock face dart board is shown. Suppose the dart board is 16 inches in diameter and has other dimensions as follows:

no-score ring: 1 inch wide
double-score ring: 0.5 inch wide
triple-score ring: 0.5 inch wide
25-point ring: 0.25 inch wide
50-point ring (bull's-eye): 1 inch in diameter

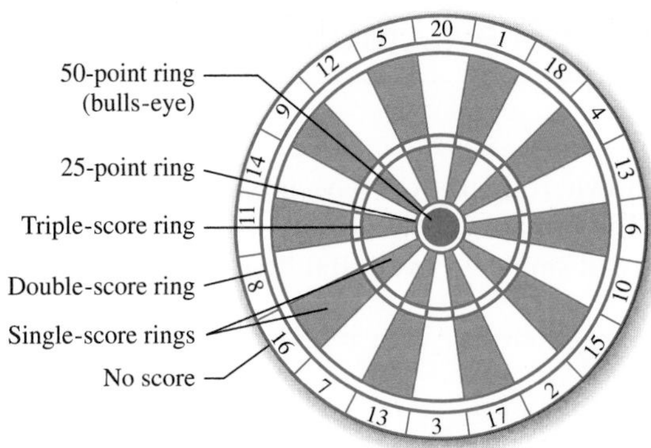

a. What is the area of the entire dart board?

b. What is the area of the no-score ring?

c. What is the area of the double-score ring?

d. What is the probability that a dart hitting the target at random lands in the no-score ring?

e. What is the probability that a dart hitting the target at random lands in the double-score ring?

EXTENDED PROBLEMS

44. A computer hard disk is divided into concentric circles called **tracks**. Tracks are further partitioned into **sectors**. Data are stored on the hard disk in sectors and tracks.

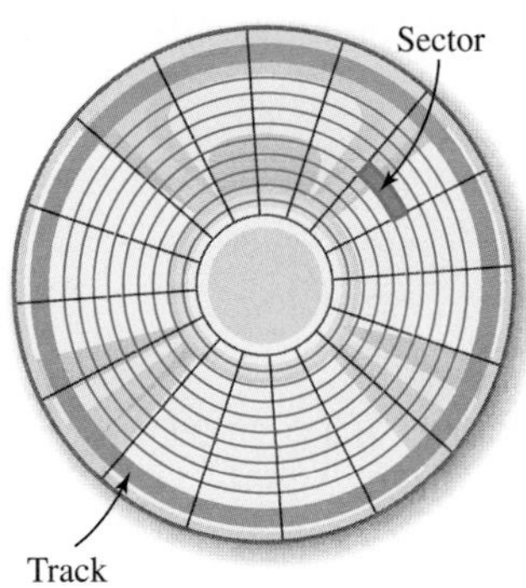

Research computer hard disks and answer the following questions. What is the maximum number of tracks there can be on a 3.5-inch hard disk? What is the area of the outermost track? How does the area of a sector near the center of the hard disk compare to the area of a sector near the outer edge and does this difference affect the amount of data stored in a sector?

45. An alternative approach to developing the formula for the area of a circle is to cut a circle into an even number of congruent pie-shaped pieces and arrange them as shown. Now imagine the number of pieces increasing without end. Explain how you can derive the formula for the area of a circle using this method.

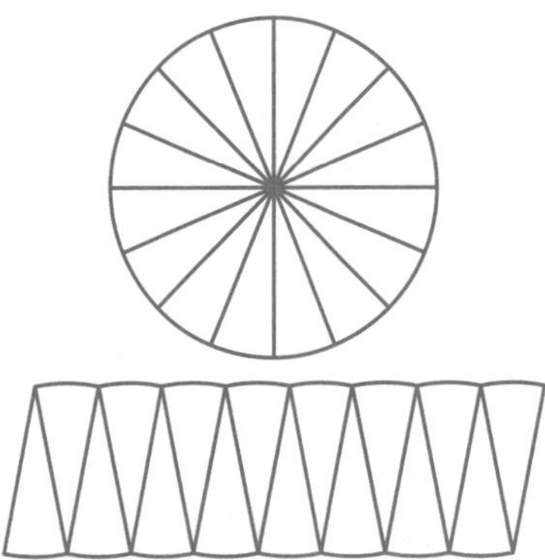

46. A circle is a special case of another simple closed curve called an ellipse. One ellipse is given next.

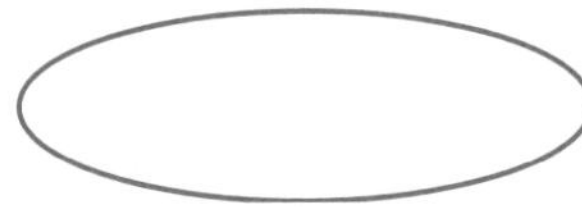

Remember that a circle was defined in Chapter 2 to be the set of all points equidistant from a given point called its center. An ellipse is defined in terms of distances from two points.

a. Research the ellipse and describe the set of points that make up an ellipse. What is the area formula for an ellipse and how is it similar to the area formula of a circle? How can you construct an ellipse using two thumbtacks and a piece of string? Use this method to create an ellipse and label the center, foci, major axis, and minor axis.

b. For many years planets were thought to orbit the sun in circular paths, but Johannes Kepler proposed that planetary paths are actually ellipses, not circles. Investigate Johannes Kepler's model of the solar system and write a report. Where is the sun in Kepler's model? Describe Kepler's first two laws of motion and draw or download figures to support your explanation.

3.3 THE PYTHAGOREAN THEOREM AND RIGHT TRIANGLES

Applied Problem

A building has the dimensions shown. If the rafters have a 2-ft overhang, determine the length of a rafter and the number of squares of shingles required to roof the building. (NOTE: A square of shingles covers 100 ft^2.)

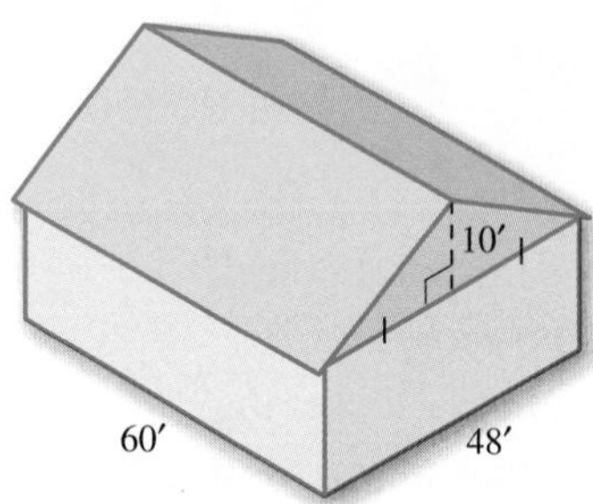

The Pythagorean Theorem

The Pythagorean theorem, which relates the lengths of the legs and the hypotenuse of a right triangle, is perhaps the most famous of all theorems in geometry. It is named for the Greek mathematician, Pythagoras (circa 500 B.C.E.), who presented a proof of the relationship. However, there is no question that the results of the theorem were known and used for hundreds of years before his time.

THEOREM 3.8 Pythagorean Theorem

The sum of the squares of the lengths of the legs of a right triangle is equal to the square of the length of the hypotenuse.

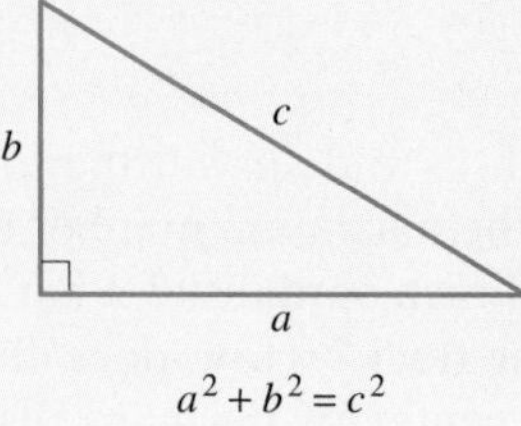

$$a^2 + b^2 = c^2$$

We can verify theorem 3.8 using areas of triangles and squares as shown next.

EXAMPLE 3.14 Show that the Pythagorean theorem holds for right triangle $\triangle ABC$ [Figure 3.23(a)].

SOLUTION

Arrange four copies of $\triangle ABC$ to form a large square as shown in Figure 3.23(b).

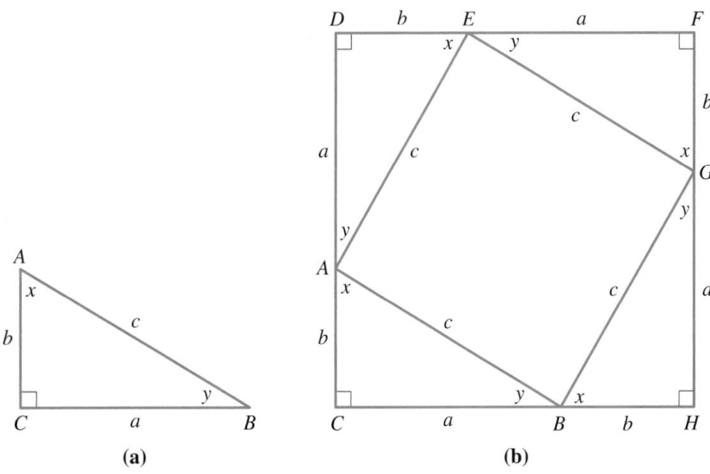

FIGURE 3.23

Because $\triangle ABC$ has a right angle at C, $x + y = 90°$. Because $\angle CBH$ is a straight angle and $x + y = 90°$, we can conclude that $\angle ABG$ is a right angle. Using similar reasoning, we find that $\angle BGE$, $\angle AEG$, and $\angle EAB$ are also right angles. Thus, $AEGB$ is a square.

Now we will compute (a) the area of square $CDFH$, and (b) the area of square $AEGB$ together with the areas of the four triangles:

$$\text{area of } CDFH = (a + b)^2 = a^2 + 2ab + b^2$$

$$\text{area of } AEGB + 4(\text{area of } \triangle ABC) = c^2 + 4\left(\tfrac{1}{2}ab\right) = c^2 + 2ab$$

The area of the larger square must be the same as the area of the smaller square plus the areas of the four right triangles. Therefore, $a^2 + 2ab + b^2 = c^2 + 2ab$. Subtracting $2ab$ from both sides, we have $a^2 + b^2 = c^2$. ■

Many geometric figures as well as real-life designs and constructions include right angles, providing opportunities to apply the Pythagorean theorem.

EXAMPLE 3.15 Find the missing side lengths in each triangle in Figure 3.24.

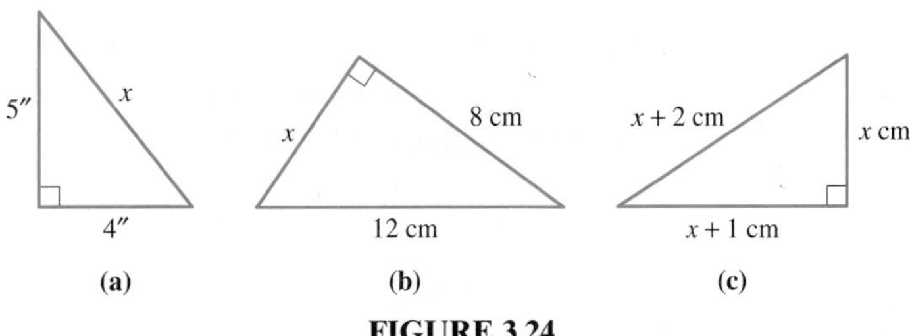

FIGURE 3.24

SOLUTION

a. In the triangle in Figure 3.24(a), the length of the hypotenuse is unknown. However, by the Pythagorean theorem, we have

$$\begin{aligned} 4^2 + 5^2 &= x^2 \\ 16 + 25 &= x^2 \\ 41 &= x^2 \\ x &= \sqrt{41} \text{ in.} \qquad \text{(exact answer), or} \\ x &\approx 6.4 \text{ in.} \end{aligned}$$

(NOTE: Because the length of a side cannot be negative, we chose the positive square root of 41.)

b. In the triangle in Figure 3.24(b), the length of one of the legs is unknown. So, we have

$$\begin{aligned} x^2 + 8^2 &= 12^2 \\ x^2 + 64 &= 144 \\ x^2 &= 80 \\ x &= \sqrt{8}\text{ cm} = 4\sqrt{5}\text{ cm} \qquad \text{(exact answer)} \\ x &\approx 8.9 \text{ cm} \qquad \text{(to the nearest tenth)} \end{aligned}$$

c. In the triangle in Figure 3.24(c), none of the lengths of the sides is known but a relationship between them is given. To find the lengths of all three sides, we will use the Pythagorean theorem.

$$\begin{aligned} x^2 + (x + 1)^2 &= (x + 2)^2 \\ x^2 + x^2 + 2x + 1 &= x^2 + 4x + 4 \\ x^2 - 2x - 3 &= 0 \\ (x - 3)(x + 1) &= 0 \\ x = 3 \text{ cm} \quad \text{or} \quad x &= -1 \text{ cm} \end{aligned}$$

Because the length of a side cannot be negative, x must be 3 cm. Hence, $x + 1 = 4$ cm and $x + 2 = 5$ cm. ■

We can use the Pythagorean theorem to calculate areas of geometric figures, as illustrated by the next example.

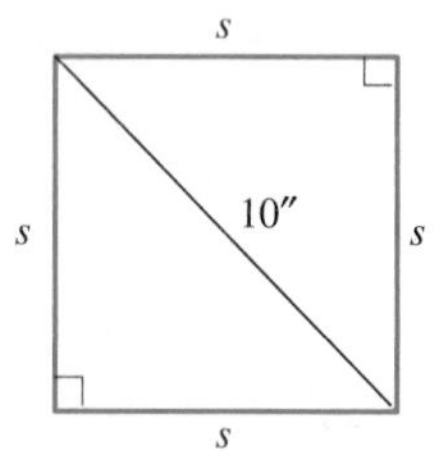

FIGURE 3.25

EXAMPLE 3.16

Find the area of a square that has a diagonal 10 inches long (Figure 3.25).

SOLUTION

The diagonal and the sides of the square form two right triangles (Figure 3.25). To find the length of a side of the square, we will use the Pythagorean theorem.

$$\begin{aligned} s^2 + s^2 &= 10^2 \\ 2s^2 &= 100 \\ s^2 &= 50 \end{aligned}$$

Hence, the area of the square is $A = s^2 = 50 \text{ in}^2$. ■

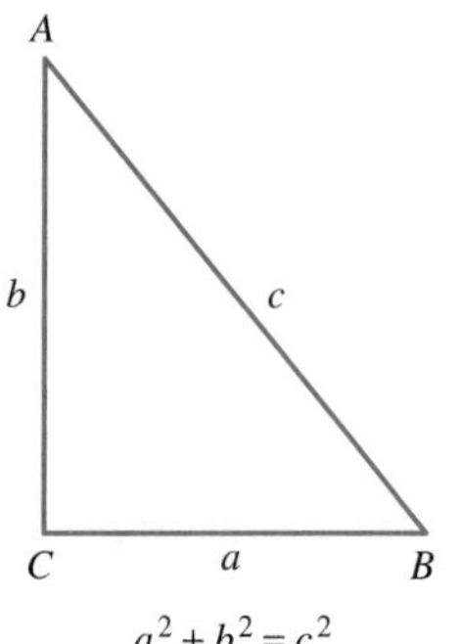

FIGURE 3.26

Test For Right Triangles

The Pythagorean theorem states that in a right triangle with legs of lengths a and b and hypotenuse of length c, we have $a^2 + b^2 = c^2$. On the other hand, sometimes we need to determine whether a triangle is in fact a right triangle. There is a test based on the Pythagorean theorem that indicates whether or not a triangle contains a right angle. This test depends on the relationship among the three sides of a triangle. If $\triangle ABC$ has sides of length a, b, and c as in Figure 3.26 and if $a^2 + b^2 = c^2$, then $\triangle ABC$ has a right angle at C. If $a^2 + b^2 \neq c^2$, then $\triangle ABC$ is *not* a right triangle. We will verify this test in Chapter 4.

EXAMPLE 3.17 Lengths of sides of two triangles are given. Determine if either triangle is a right triangle.

a. 11, 15, 19 **b.** 7, 16.8, 18.2

SOLUTION

a. Let $a = 11$, $b = 15$, and $c = 19$. Then $a^2 + b^2 = 11^2 + 15^2 = 346$ and $c^2 = 19^2 = 361$. Because $a^2 + b^2 \neq c^2$, the triangle is not a right triangle.

b. $a^2 + b^2 = 7^2 + (16.8)^2 = 331.24$ and $c^2 = (18.2)^2 = 331.24$. Because $a^2 + b^2 = c^2$, a triangle with sides of lengths 7, 16.8, and 18.2 is a right triangle. ■

Special Right Triangles

Next, we will discuss two special right triangles that appear frequently in geometric applications. First consider an equilateral triangle. Recall that an equilateral triangle is also equiangular [Figure 3.27(a)].

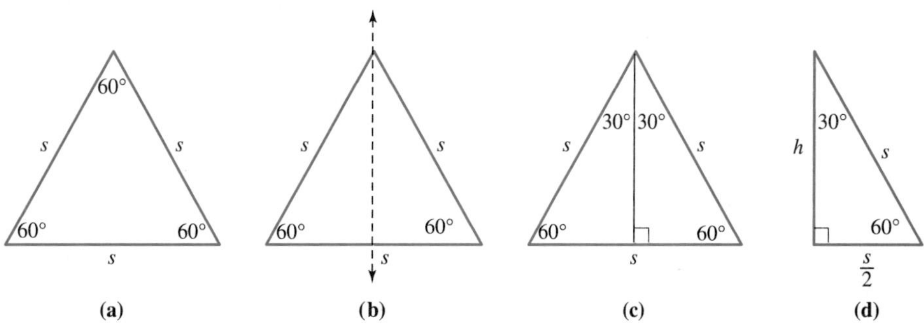

FIGURE 3.27

Also, in Chapter 2 we saw that an equilateral triangle has three lines of symmetry. If we fold the triangle on a line of symmetry as shown in Figure 3.27(b), we form two identical triangles. Because the angles on either side of the line of symmetry must be congruent, the angles in each triangle measure 30°, 60°, and 90° [Figure 3.27(c)]. Thus, each triangle that is formed is called a **30°-60° right triangle** or a **30°-60°-90° triangle**. Because the base of the equilateral triangle is bisected, that is, cut in half, by the fold line as shown in Figure 3.27(c), each 30°-60° right triangle has a base of $\frac{s}{2}$. Using the Pythagorean theorem, we can determine the height, h, of any 30°-60° right triangle in terms of s as follows [Figure 3.27(d)].

$$h^2 + \left(\frac{s}{2}\right)^2 = s^2 \quad \text{so} \quad h^2 + \frac{s^2}{4} = s^2 \quad \text{or} \quad h^2 = \frac{3s^2}{4}$$

Therefore, $h = \sqrt{\frac{3s^2}{4}} = \frac{s\sqrt{3}}{2}$. The next theorem summarizes this discussion.

THEOREM 3.9 30°-60° Right Triangle

In a 30°-60° right triangle, the length of the leg opposite the 30° angle is half the length of the hypotenuse. The length of the leg opposite the 60° angle is $\frac{\sqrt{3}}{2}$ times the length of the hypotenuse.

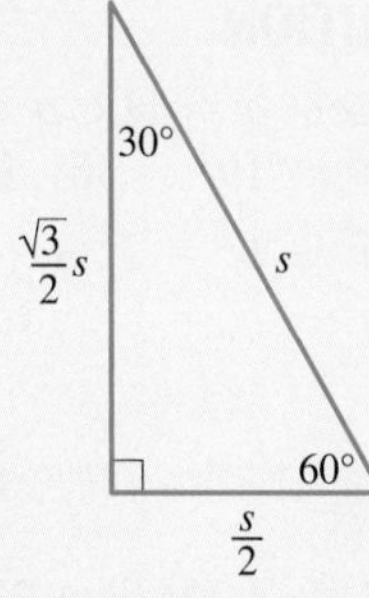

The fact that the relationships stated in Theorem 3.9 are true for any 30°-60° right triangle means that if we know the length of *any* one side of such a triangle, we can find the lengths of the other two.

EXAMPLE 3.18 Find the missing side lengths for each of the 30°-60° right triangles in Figure 3.28.

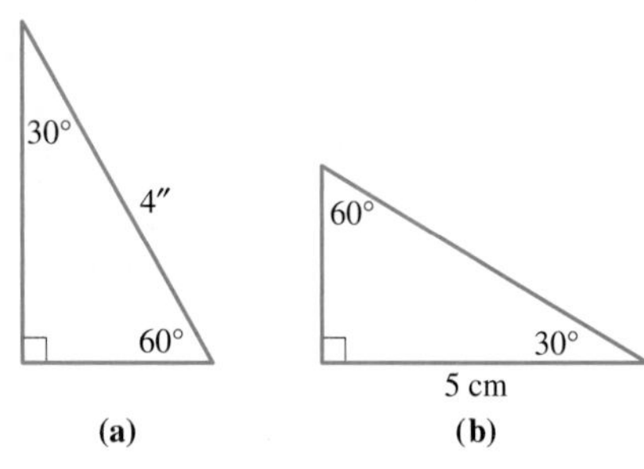

FIGURE 3.28

SOLUTION

a. Using the notation in Theorem 3.9, we find that $s = 4''$, so $\frac{s}{2} = 2''$ and $\frac{\sqrt{3}}{2}s = \frac{\sqrt{3}}{2}(4'') = 2\sqrt{3}$ in. Therefore, the leg opposite the 30° angle is 2″, and the leg opposite the 60° angle is $2\sqrt{3}$ in.

b. Here, $\frac{\sqrt{3}}{2}s = 5$ cm, so $s = 5\left(\frac{2}{\sqrt{3}}\right) = \frac{10}{\sqrt{3}} = \frac{10\sqrt{3}}{3}$ cm, and $\frac{s}{2} = \frac{5\sqrt{3}}{3}$ cm.

Thus, the leg opposite the 30° angle has length $\frac{5\sqrt{3}}{3}$ cm and the hypotenuse has length $\frac{10\sqrt{3}}{3}$ cm. ■

A similar relationship exists between the lengths of the sides of the right triangle formed by folding a square along a line of symmetry containing one of its diagonals [Figure 3.29(a)].

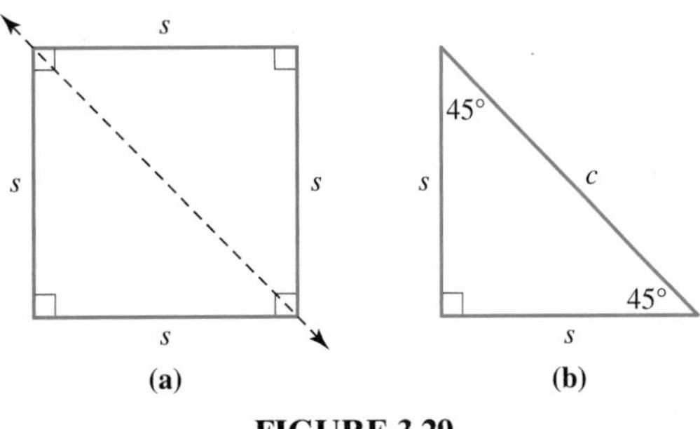

FIGURE 3.29

Because the angles formed on either side of the fold line must be congruent and because they also are complementary, each of these acute angles must measure 45°. Thus, each of the triangles formed is called a **45°-45° right triangle** or **45°-45°-90° triangle** [Figure 3.29(b)]. Notice that each triangle formed is also a right isosceles triangle.

We can apply the Pythagorean theorem to determine the length of the hypotenuse, c, in a 45°-45° right triangle in terms of the length of one of the sides, s, as follows [Figure 3.29(b)]:

$$s^2 + s^2 = c^2 \quad \text{so} \quad c^2 = 2s^2 \quad \text{or} \quad c = s\sqrt{2}$$

This result is summarized in our next theorem.

THEOREM 3.10 45°-45° Right Triangle

In a 45°-45° right triangle, the length of the hypotenuse is $\sqrt{2}$ times the length of a leg.

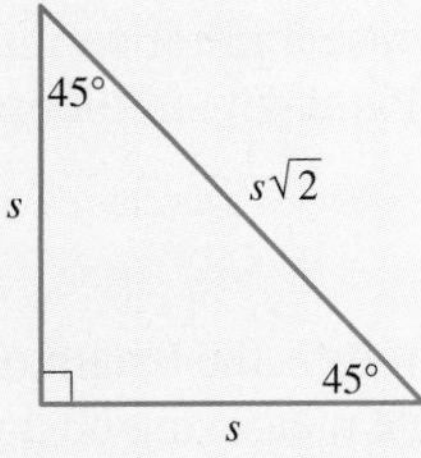

As was true for the 30°-60° right triangle, this theorem allows us to find the missing lengths of two sides of a 45°-45° right triangle if we know the length of any one side.

EXAMPLE 3.19 Find the area, to the nearest square inch, of parallelogram $ABCD$ in Figure 3.30(a).

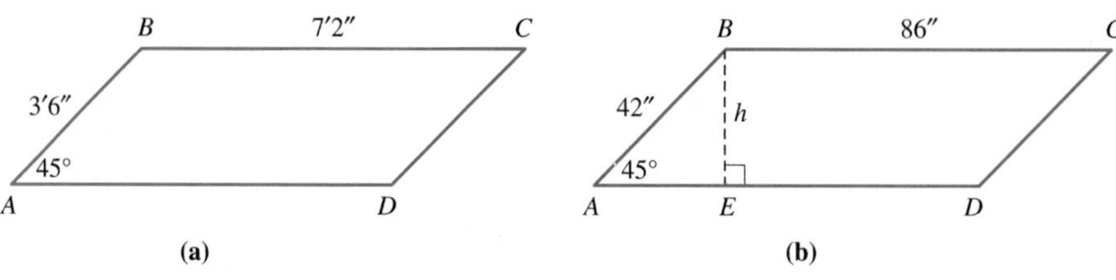

FIGURE 3.30

SOLUTION

Convert the dimensions to inches and draw in the height to the longer side [Figure 3.30(b)]. $\triangle ABE$ is a 45°-45° right triangle so we have

$$h\sqrt{2} = 42 \qquad \text{or} \qquad h = \frac{42}{\sqrt{2}} \text{ in.}$$

Therefore, the area of $ABCD$ is $A = bh = (86)\left(\frac{42}{\sqrt{2}}\right) \approx 2554 \text{ in}^2$. ■

We have shown that if a triangle is a 30°-60° right triangle or a 45°-45° right triangle and if we know the length of one side of the triangle, then we can determine the lengths of the other two sides. We will generalize this technique in Chapter 6 so that given the measure of one angle of *any* size and the length of one side of a right triangle, we can find the lengths of the other two sides.

Solution of Applied Problem

The following figure is an end view of the roof portion of the house.

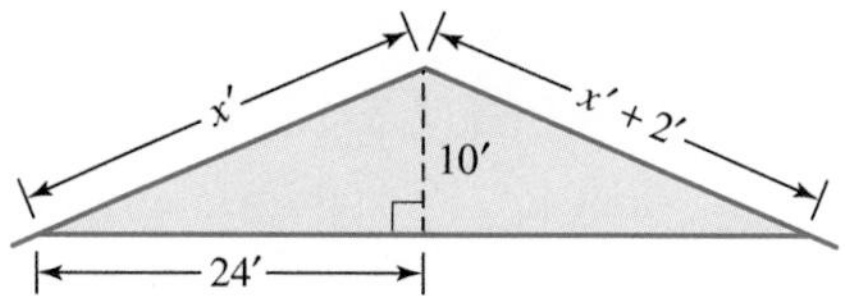

Because a rafter forms a right triangle with the end of the house, we can apply the Pythagorean theorem to find the length of a rafter, $x + 2$. We have

$$24^2 + 10^2 = x^2$$
$$676 = x^2 \qquad \text{or} \qquad x = 26'$$

Therefore, the length of the rafter is $x + 2 = 28$ ft. Each half of the roof is a rectangle measuring 60′ by 28′. Hence, the total area of the roof is given by

$$A = 2(28')(60') = 3360 \text{ ft}^2$$

The number of 100 ft^2 in 3360 ft^2 is 33.60. Thus, the roof requires 33.6 squares of shingles.

GEOMETRY AROUND US Carpenters, bricklayers, and other builders frequently make use of the Test for Right Triangles to lay out a square corner. They may mark off a distance of 3 feet on one wall and a distance of 4 feet on the adjacent wall as shown. If the corner at point A is to be a right angle, then the distance from B to C must be exactly 5 feet since $3^2 + 4^2 = 5^2$. This 3-4-5 triangle is probably the most commonly used right triangle, but any multiple such as 6-8-10 could be used. Other common right triangle ratios, such as 5-12-13 or 8-15-17, can also be used to "square a corner."

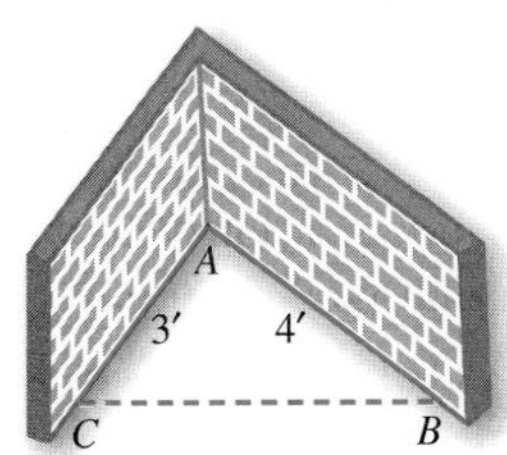

PROBLEM SET 3.3

EXERCISES/PROBLEMS

1. Find the missing side length in each right triangle.

a.
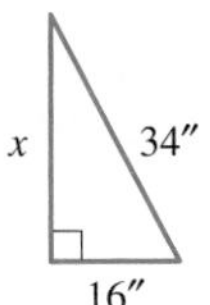

b.
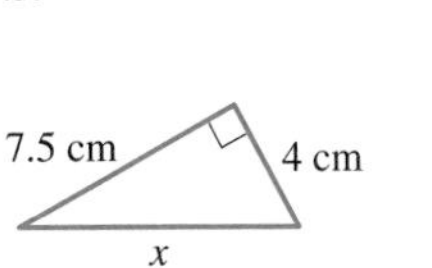

c.
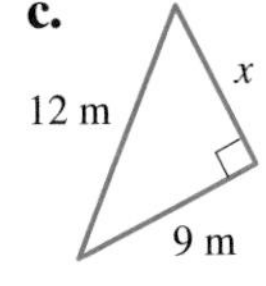

2. Find the missing side lengths in each right triangle.

a.
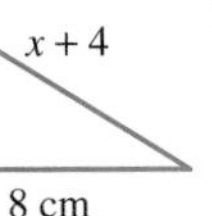

b.

c.
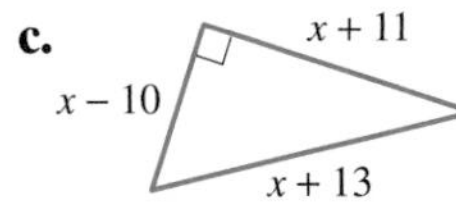

3. Find the missing lengths in the following figures. Round to the nearest tenth.

a.
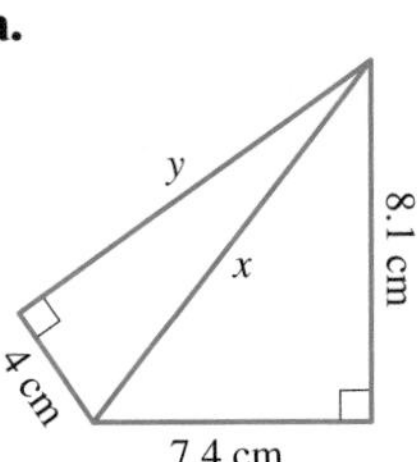

b.
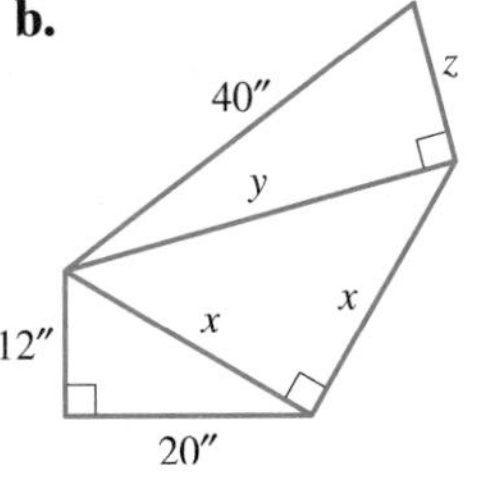

4. Find the areas of the polygons with dimensions as shown.

a.
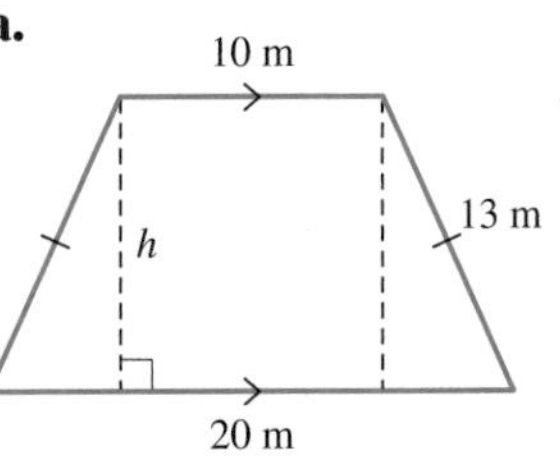

b.
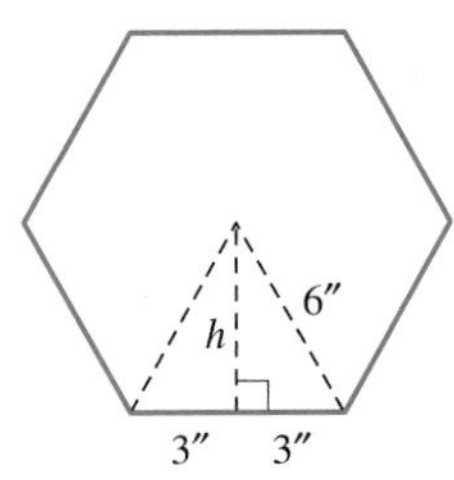

5. Complete the following table, where information is given about a right triangle with legs a and b, hypotenuse c, perimeter P, and area A. Round to the nearest hundredth.

	a	b	c	P	A
a.	2.30			21.02	10.60
b.		6.1			30.5
c.	$\sqrt{7}$	$\sqrt{7}$			
d.		$\frac{x}{2}$	x		$\frac{\sqrt{3}}{2}$

6. Complete the following table, where information is given about a right triangle with legs a and b, hypotenuse c, perimeter P, and area A. Round to the nearest hundredth.

	a	b	c	P	A
a.		4.81	7.65		
b.	10.55				78.23
c.	$\sqrt{5}$	$\sqrt{15}$			
d.	x	x		60	

7. The area of a rectangle is 24 in^2. A diagonal of the rectangle measures exactly $\sqrt{73}$ in. Find the length and width of the rectangle.

8. In the figure shown, the areas of the two large circles are both 25π cm^2. The area of $\triangle OPQ$ is 6 cm^2. Find the area of circle P.

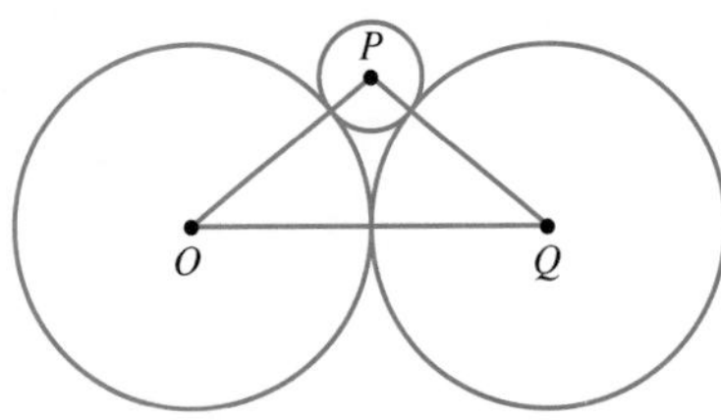

9. Which of the following numbers could be the lengths of sides of a right triangle?

a. 7, 24, 25 **b.** 12, 24, 26

c. 28, 21, 35

10. Which of the following numbers could be the lengths of sides of a right triangle?

a. 11, 60, 61 **b.** 8, 9, 15

c. 10, 22, 26

11. A triangle has sides of lengths a, b, and c, where $a < b < c$. Consider the relationship between a^2, b^2, and c^2 for each of the following questions.

a. When will the triangle be a right triangle?

b. When will the triangle be an acute triangle?

c. When will the triangle be an obtuse triangle?

12. Given are the lengths of sides of a triangle. Indicate whether each triangle is a right triangle. If not, specify whether it is acute or obtuse. (HINT: See the previous problem.)

a. 70, 54, 90 **b.** 63, 16, 65

c. 24, 48, 52 **d.** 27, 36, 45

e. 48, 46, 50 **f.** 9, 40, 46

13. Find the missing lengths in each triangle.

a.

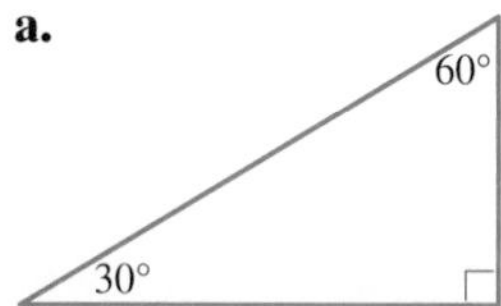

b. 45° 45° 12″

14. Find the missing lengths in each triangle.

a.

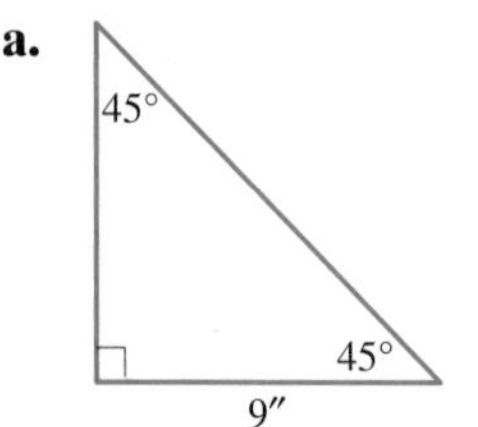

b. 20 cm 30° 60°

15. Determine the height of trapezoid $ABCD$. What is the area of the trapezoid to the nearest tenth?

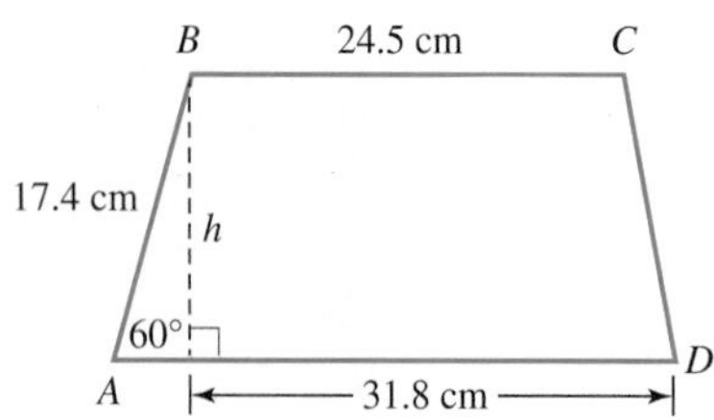

16. Determine the lengths of the legs of isosceles trapezoid $PQRS$. Calculate the perimeter and area of the trapezoid.

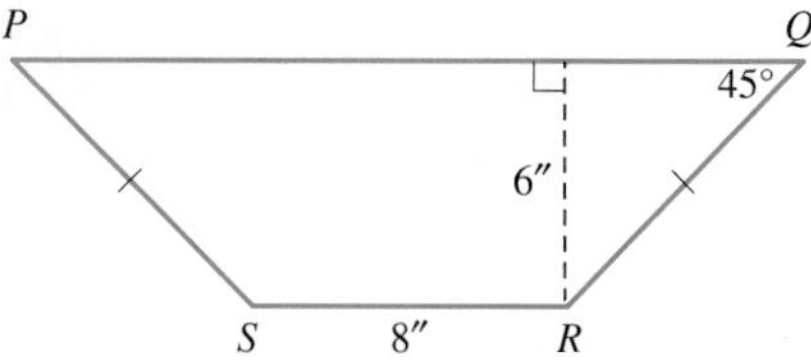

17. Find the area of parallelogram $ABCD$ given $AB = 12$ cm, $AD = 20$ cm, and $\angle A = 30°$.

18. Find the area of isosceles trapezoid $ABCD$ if the bases measure 10″ and 16″ and two base angles each measure 60°. Give an exact answer.

19. Find the area of a rhombus with a side of length 9.4 m and a vertex angle that measures 45°. Round your answer to the nearest hundredth.

20. Find the area of trapezoid $PQRS$ if $\angle P = 60°$, $\angle S = 45°$, $PS = 16$ in., and the height is 5 in. Round to the nearest hundredth.

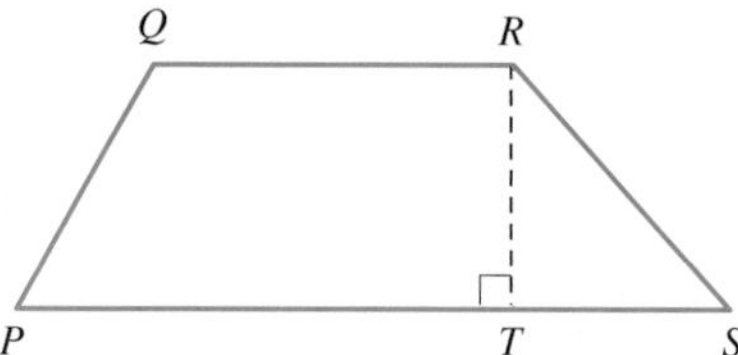

21. The distance from the middle of a regular hexagon to the midpoint of a side is $10\sqrt{3}$ cm. Find the area of the hexagon. Round to the nearest tenth.

22. The distance from the center of a regular hexagon to a vertex is 15 cm. Find the area of the regular hexagon. Round to the nearest tenth.

23. Let h represent the length of the altitude of an equilateral triangle. Find a formula for the area of an equilateral triangle in terms of its altitude.

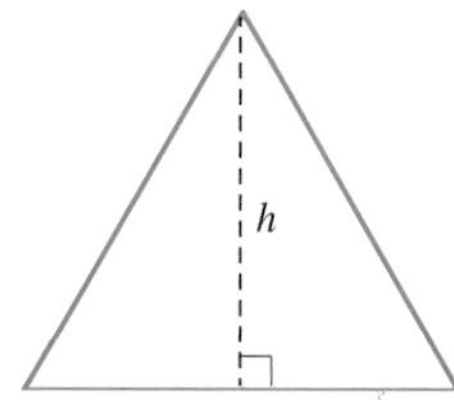

24. The altitude of an equilateral triangle cuts the base in half. Use this fact to find a formula for the area of the following equilateral triangle in terms of s.

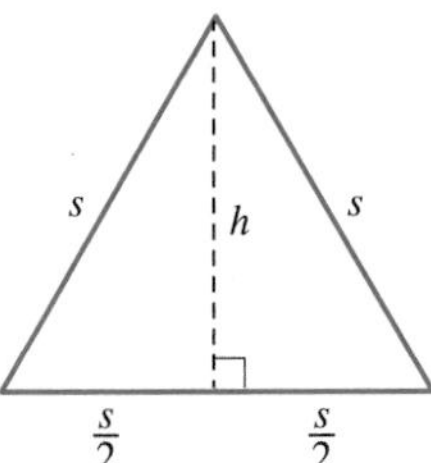

25. Find a formula for the distance d between opposite sides of a regular octagon in terms of the length of one side x.

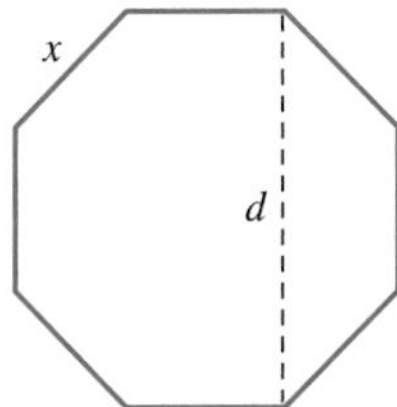

26. Find a formula for the distance v between two opposite vertices of a regular octagon in terms of the length of one side x. (HINT: Use your result from problem 25.)

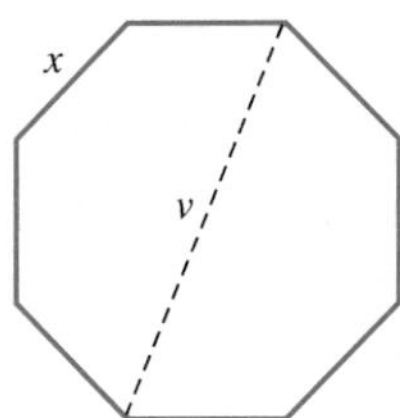

27. The vertices of an equilateral triangle with side length $10\sqrt{3}$ cm lie on a circle as shown. Find the side length of the regular hexagon whose vertices lie on the same circle.

28. An equilateral triangle with side length 8 cm can be partitioned into a smaller equilateral triangle and an isosceles trapezoid as shown. Find the height of the trapezoid so that the area of the trapezoid is the same as the area of the smaller triangle. Round to the nearest hundredth.

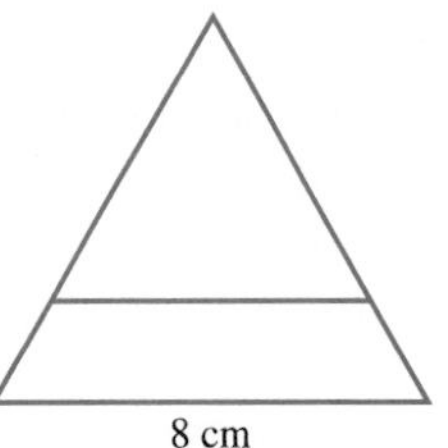

29. The proof of the Pythagorean theorem that we presented in this section depended upon a figure containing squares and right triangles. Another famous proof of the Pythagorean theorem also uses squares and right triangles. It is attributed to a 12th-century Hindu mathematician, Bhaskara, and it is said that he simply wrote the word "Behold" above the figure, believing that the proof was evident from the drawing. Use algebra and the areas of the right triangles and squares in the figure to verify the Pythagorean theorem.

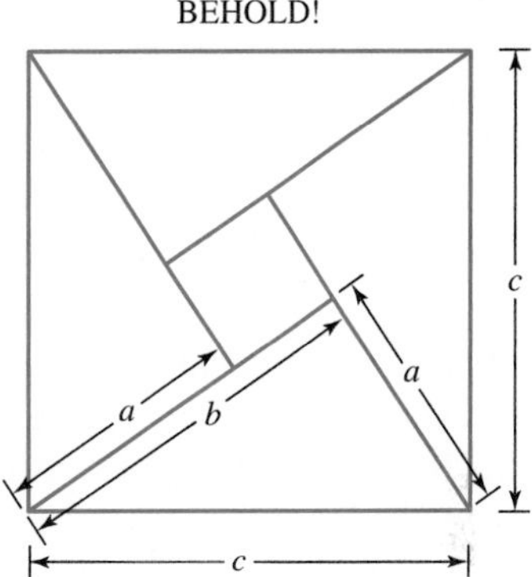

30. Another famous proof of the Pythagorean theorem is credited to former President James Garfield. Two identical right triangles are positioned as shown next and two vertices of the right triangles are connected to form a trapezoid. Use algebra, the area of a trapezoid, and the area of the right triangles to verify the Pythagorean theorem.

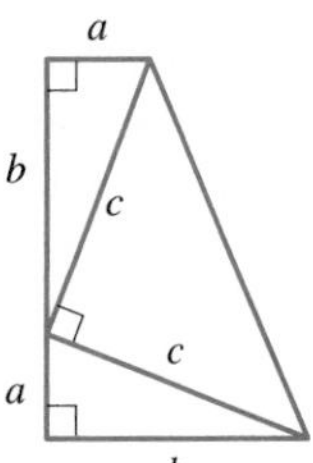

APPLICATIONS

31. A square ruler designed for quilters is shown next. The dashed line along the diagonal is marked as 6.5 inches long. What is the side length of the square? Round to the nearest tenth.

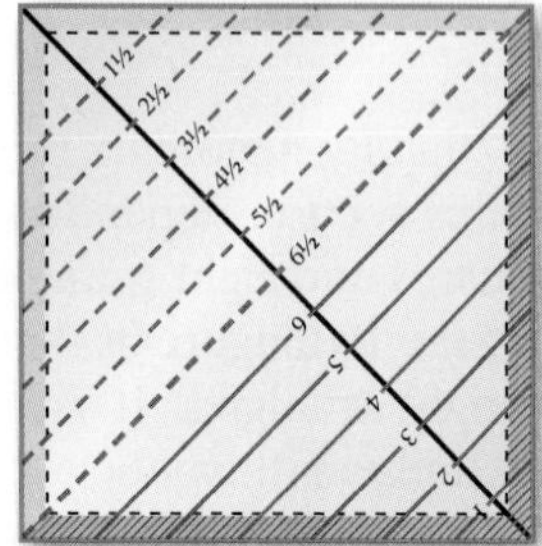

32. A baseball diamond is a square that measures 90 feet on a side. How far must the catcher throw the ball from home plate to second base to pick off a runner?

33. Painters will use a 16-foot ladder to paint a house. If the painters must place the foot of the ladder at least 4 feet away from the house to avoid flowers and shrubs, what is the highest point on the house that the ladder will reach? Round to the nearest tenth.

34. Two boats leave a dock at the same time. One heads due east at 20 mph, and one heads due south at 15 mph. How far apart are the two boats after 45 minutes have elapsed?

35. The longest escalator in the Western Hemisphere is near Washington, D.C., at Wheaton Station. It moves passengers 115 feet below street level and has an incline of 30°. If the escalator ride takes three minutes, then find the rate (in miles per hour) at which passengers move on the escalator. Round to the nearest hundredth.

36. A conveyor belt is to be set up to move bundles of shingles up to a roof. If the edge of the roof is 10 feet above the ground and the conveyor can make no more than a 45° angle with the ground, how long a conveyor belt will be needed? How far away from the house will the bottom of the conveyor be placed? Round to the nearest foot.

37. As an airplane takes off from a runway, it makes an angle of 30° with the ground. When it has traveled 1100 feet at this take-off angle, what is its altitude?

38. A rectangular field measures 325 m by 240 m. How much shorter would it be to walk diagonally across the field than to walk around two sides of the field? Round to the nearest meter.

39. A faucet is located at one corner of a yard that measures 50 ft by 65 ft. How long a garden hose will be necessary to be able to water plants in any part of the yard? Round to the nearest foot.

40. In the figure shown, the center of one pulley is 11′8″ above and 18′3″ to the right of the center of the other pulley. Find the distance between their centers to the nearest inch.

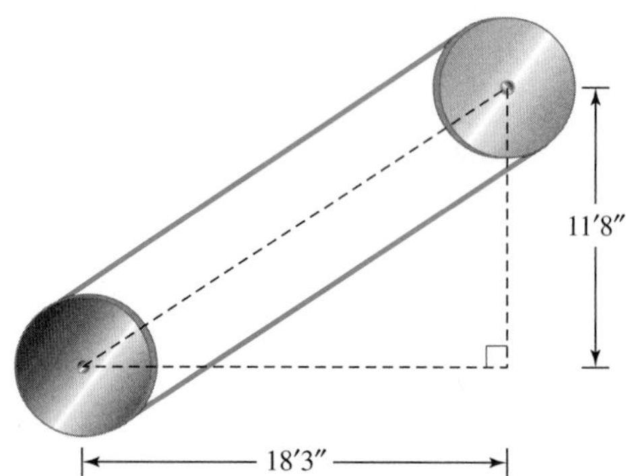

41. To determine a distance across the lake shown, workers took the indicated measurements. To the nearest foot, what is the distance PR?

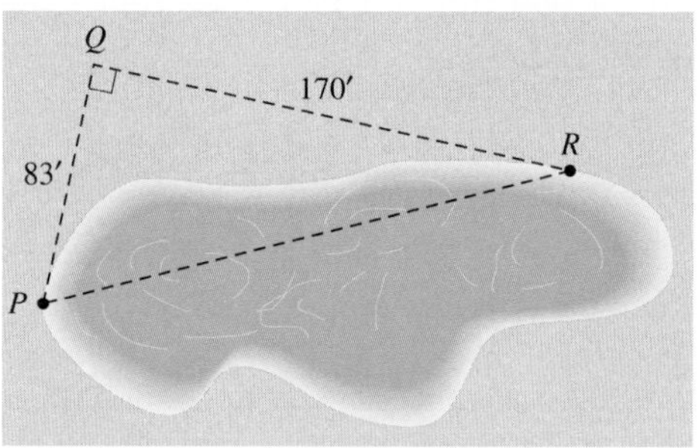

42. A 19″ TV screen measures 19″ diagonally across the rectangular face of the screen. If the screen is about 4″ wider than it is high, find the dimensions of the screen. Round to the nearest tenth.

43. Square plugs are often used to check the diameter of a hole. What must the length of the side of the plug be to test a hole with diameter 3.16 cm? Round down to the nearest tenth.

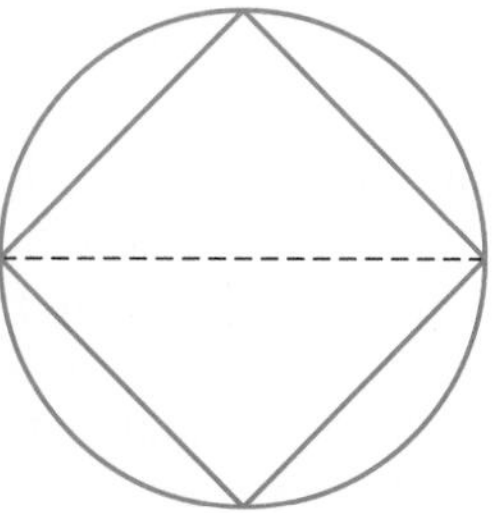

44. Workers will mill a log 30″ in diameter into a single piece of lumber with a square end.

a. What are the dimensions of the largest piece that can be cut from the log? (NOTE: We will show in Chapter 7 that the diagonals of the square must pass through the center of the circle.)

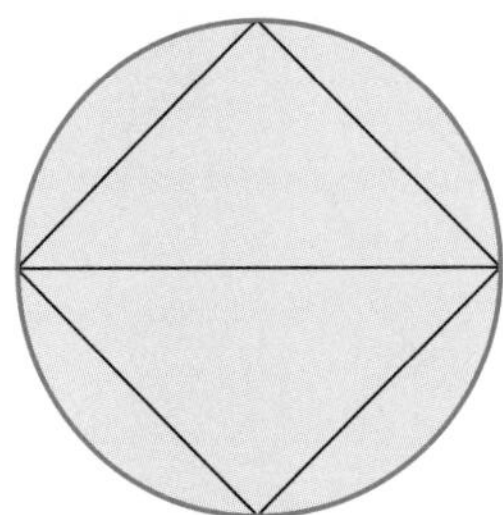

b. What percent of the log is used in the piece of lumber?

45. A 40-foot antenna is to be held in place by four guy wires that will be attached to the antenna halfway up and anchored to the flat roof at points 15 feet from the base of the antenna. If you allow 20 feet of wire for fastening the wires to the antenna and to the roof, how many feet of wire will be required? Round to the nearest foot.

46. The roof of the house shown is known to have a pitch of "4 in 12." That is, the roof rises 4 feet for each 12 feet of span. Two dimensions on the end of the house have been measured. Use them to calculate the gable height x of the roof and the length of the rafter y. (Assume an 18-inch overhang.)

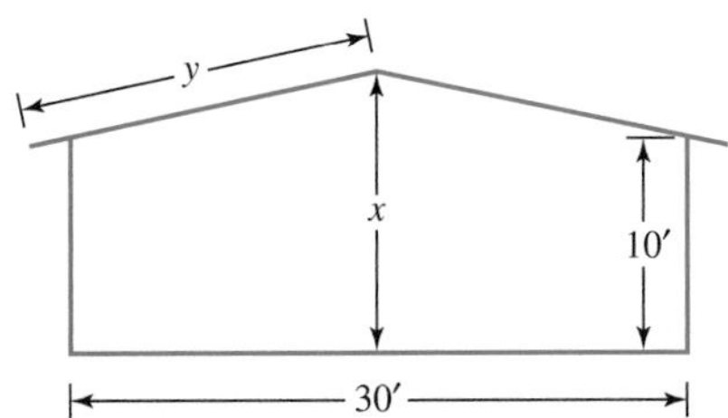

47. A rectangular packing box measures 24″ by 18″ by 15″ as shown. Will an umbrella 32″ long fit into this box? Explain.

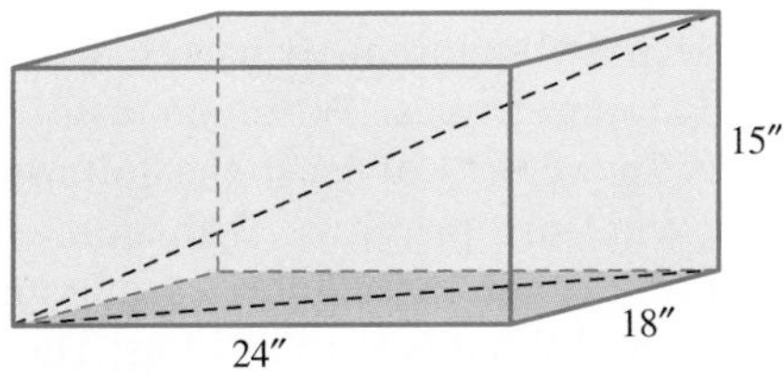

48. Sign makers can apply real gold to raised letters using gold leaf paper. If it costs $33 for 70 cm^2 of gold leaf paper, then how much does the following sheet of gold leaf paper cost? Round to the nearest penny. (HINT: Heron's formula may be useful here. See problem 37 in Section 3.1.)

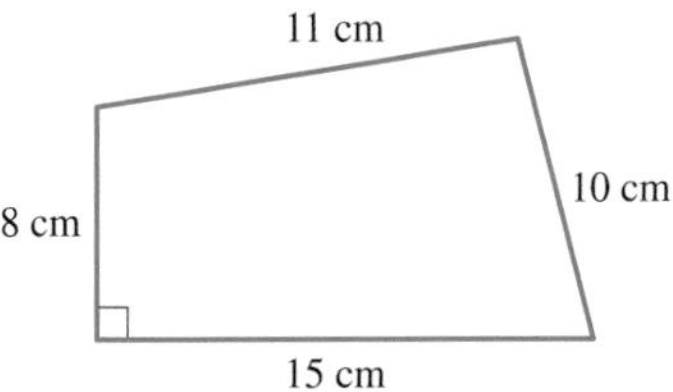

49. A circular tube, or conduit, is needed to enclose four wires, each with a diameter of 1/4 inch. Determine the smallest diameter conduit that will contain the cables. Round to the nearest hundredth of an inch.

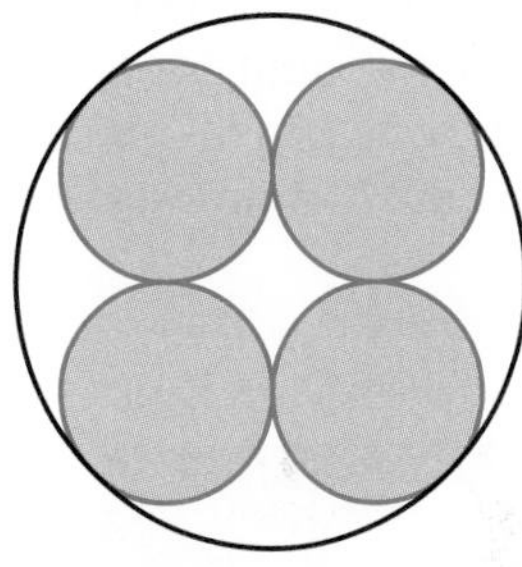

50. Three pipes are stacked lengthwise on a flat-bed truck as shown. If each pipe has a diameter of 60 inches, how high is the stack, to the nearest inch?

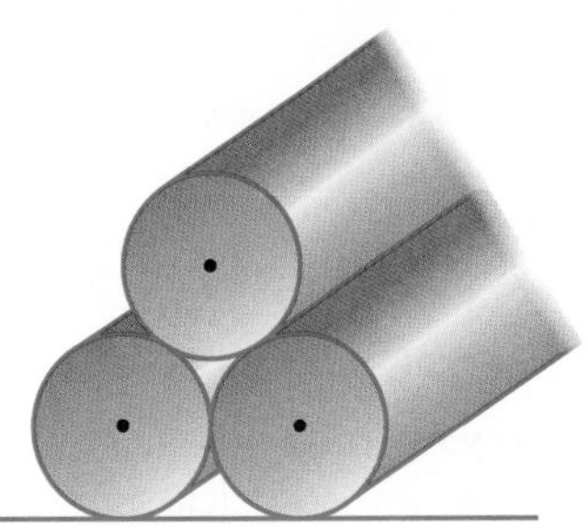

51. **Origami** is the art of paper folding. To make the origami whale shown, begin with a square paper with side length 6 inches. Fold and crease one diagonal; then unfold the paper and lay it flat. The next fold takes one side of the square to the diagonal (see shaded fold). Crease the paper and then unfold and lay the square flat. The second crease divides the original square into a right triangle and a right trapezoid. Find the area of the trapezoid. Round to the nearest hundredth.

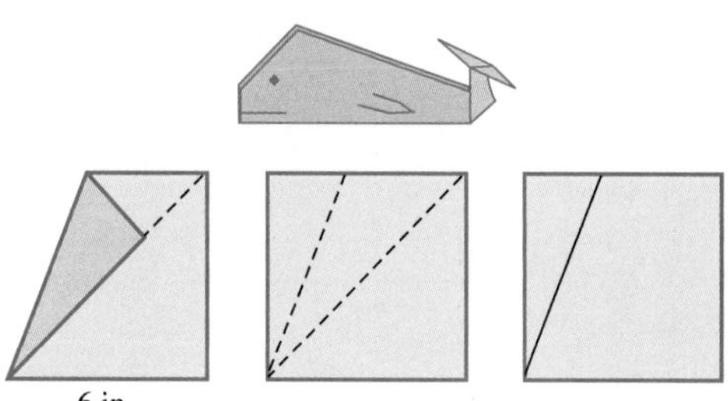

52. A pattern for a square block of a quilt is shown. The quilt will be made of 80 blocks 9″ on a side like the one shown, and four colors will be used. Fabric to be used for the quilt is 36″ wide. How many inches of fabric of each color must be purchased to construct the quilt? Ignore waste.

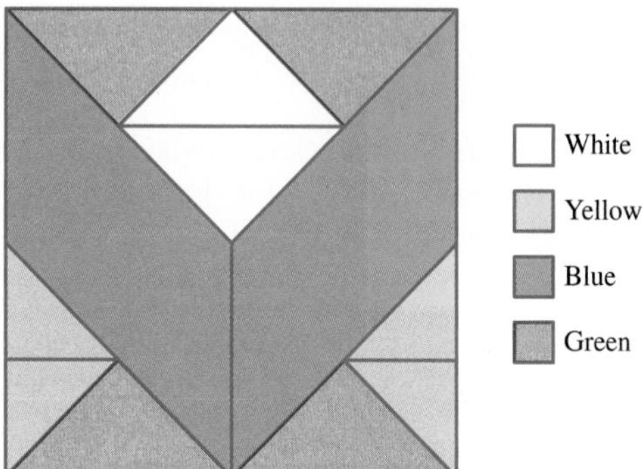

EXTENDED PROBLEMS

53. **a.** The following root spiral is created by first constructing an isosceles right triangle. Each new right triangle is created using the hypotenuse of the previous right triangle as one of its legs. The other leg of each new right triangle is equal in length to a leg of the original isosceles right triangle. Notice that only the first right triangle in the figure is an isosceles right triangle.

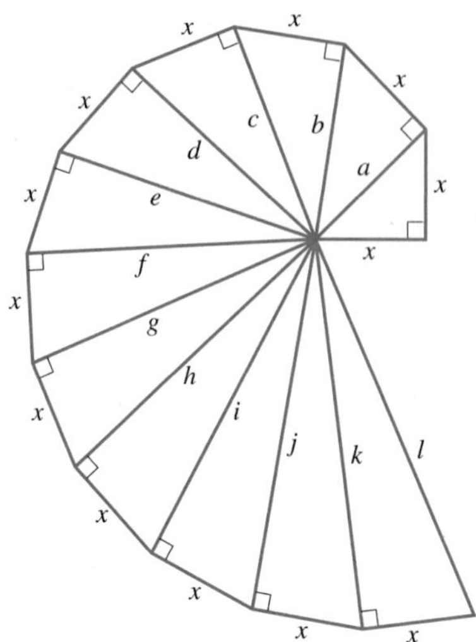

Research the root spiral and write a summary of its history including the name of the person credited with its discovery and at least two interesting facts about the spiral.

b. Find the lengths a through l if $x = 1$ inch and use a protractor and a ruler to construct the first 12 triangles in the spiral. Label each length.

54. One of the most famous theorems in mathematics is the Pythagorean theorem, for which hundreds of proofs have been written. Find two geometric proofs of the Pythagorean theorem different from those presented in this text and create a visual display along with explanations of each proof. Use the book *The Pythagorean Proposition* from your library or search for proofs on the Internet.

55. A **Pythagorean triple** is a set of three positive integers such that $a^2 + b^2 = c^2$. For example, (3, 4, 5) is a Pythagorean triple. We can generate other Pythagorean triples by the following process:

(i) Choose any two positive integers, u and v, such that $u > v$.

(ii) Let $a = 2uv$, $b = u^2 - v^2$, and $c = u^2 + v^2$.

a. Generate a, b, and c for several pairs of values for u and v. Then verify that (a, b, c) forms a Pythagorean triple in each case.

b. Using algebra, show why the expressions given for a, b, and c will always satisfy the equation $a^2 + b^2 = c^2$.

c. Research Pythagorean triples. List at least 30 Pythagorean triples. Write an essay that includes answers to at least the following questions. What are primitive Pythagorean triples? Are there other Pythagorean triple generating algorithms? What is Fermat's Last Theorem?

3.4 SURFACE AREA

Applied Problem

A hopper designed to store wood chips and empty them into trucks has the shape shown. How many square feet of metal were used in its construction, including material for the top?

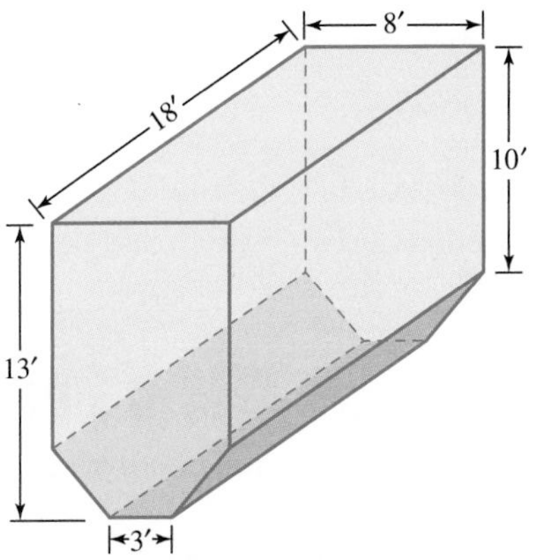

In Sections 3.1 and 3.2, we discussed areas of regions in the plane. Now we will extend that idea to shapes in three dimensions.

Surface Area of a Right Prism

The **surface area** of a polyhedron is the sum of the areas of its lateral faces and bases. SA is an abbreviation for "surface area."

EXAMPLE 3.20 Find the surface area of the right rectangular prism in Figure 3.31.

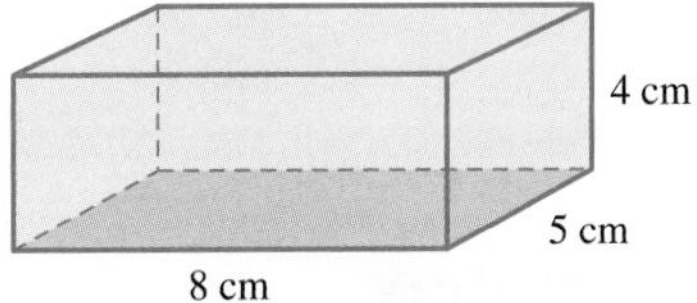

FIGURE 3.31

SOLUTION

The area of each base is $8 \text{ cm} \cdot 5 \text{ cm} = 40 \text{ cm}^2$. Two of the lateral faces each have an area of $4 \text{ cm} \cdot 5 \text{ cm} = 20 \text{ cm}^2$ and two have area $8 \text{ cm} \cdot 4 \text{ cm} = 32 \text{ cm}^2$, so the surface area is $2 \cdot 40 + 2 \cdot 20 + 2 \cdot 32 = 2(40 + 20 + 32) = 184 \text{ cm}^2$. ■

Notice that we could express the surface area of the right rectangular prism in Example 3.20 as follows:

$$\begin{aligned} SA &= 2(8 \cdot 5 + 4 \cdot 5 + 4 \cdot 8) \\ &= 2(8 \cdot 5) + 2(4 \cdot 5 + 4 \cdot 8) \\ &= 2(8 \cdot 5) + (2 \cdot 5 + 2 \cdot 8)4 \\ &= 2(\text{area of the base}) + (\text{perimeter of the base})(\text{height}) \end{aligned}$$

This result holds for any right prism; that is, the base need not be a rectangle.

THEOREM 3.11 Surface Area of a Right Prism

The surface area of a right prism is the sum of twice the area of its base plus the product of the perimeter of the base and the height of the prism.

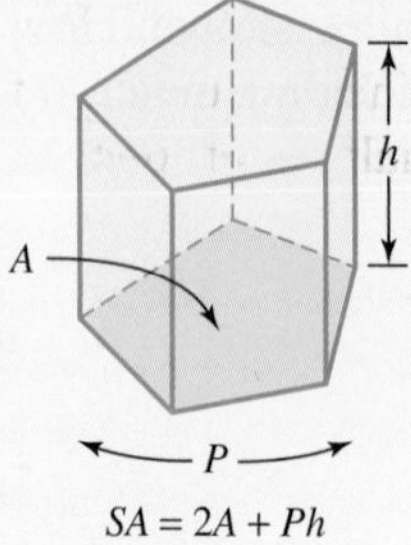

$SA = 2A + Ph$

Surface Area of a Right Pyramid

The surface area of a pyramid is the sum of the areas of its base and lateral faces.

EXAMPLE 3.21 Find the surface area of a right regular triangular pyramid if the sides of the base are 6 cm and the slant height is 8 cm [Figure 3.32(a)].

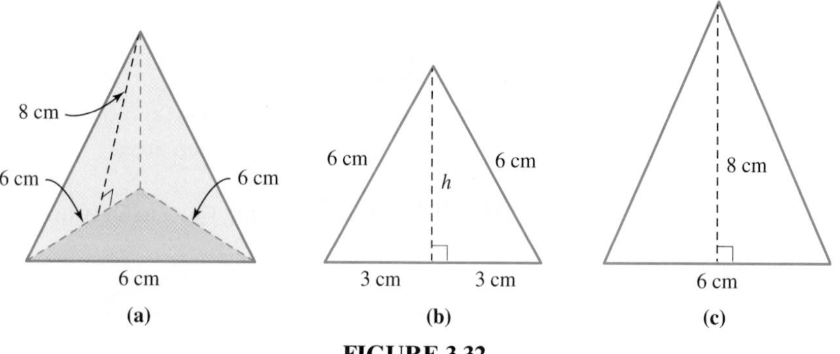

FIGURE 3.32

SOLUTION

First, the pyramid is a right *regular triangular* pyramid. Therefore, its base is an equilateral triangle [Figure 3.32(b)]. Using the Pythagorean theorem, we have $h^2 + 3^2 = 6^2$, or $h^2 = 36 - 9 = 27$. Thus, $h = \sqrt{27} = 3\sqrt{3}$ cm and the area of the base is $\frac{1}{2}bh = \frac{1}{2}(6)(3\sqrt{3}) = 9\sqrt{3}\,\text{cm}^2$. (NOTE: We could also find the height of the base by observing that the base is divided into two 30°-60° right triangles.) The three lateral faces are identical isosceles triangles [Figure 3.32(c)]. The area of one face is $\frac{1}{2}(6\,\text{cm})(8\,\text{cm}) = 24\,\text{cm}^2$. Thus, the area of all three lateral faces is $3 \cdot 24\,\text{cm}^2 = 72\,\text{cm}^2$. The surface area, the sum of the areas of all faces, is $9\sqrt{3}\,\text{cm}^2 + 72\,\text{cm}^2$, or approximately 87.6 cm^2 to one decimal place. ■

We can express the surface area of the pyramid in Example 3.21 as follows:

$$\begin{aligned} SA &= 9\sqrt{3} + 3[\tfrac{1}{2}(6 \cdot 8)] \\ &= 9\sqrt{3} + \tfrac{1}{2}(3 \cdot 6)8 \\ &= \text{area of the base} + \tfrac{1}{2}(\text{perimeter of the base})(\text{slant height}) \end{aligned}$$

This result holds for any right regular pyramid, regardless of the shape of the base. We find the surface area of a pyramid with a base that is not regular by adding the areas of the base and the lateral faces.

THEOREM 3.12 Surface Area of a Right Regular Pyramid

The surface area of a right regular pyramid is the sum of the area of its base plus half the product of the perimeter of the base and the slant height.

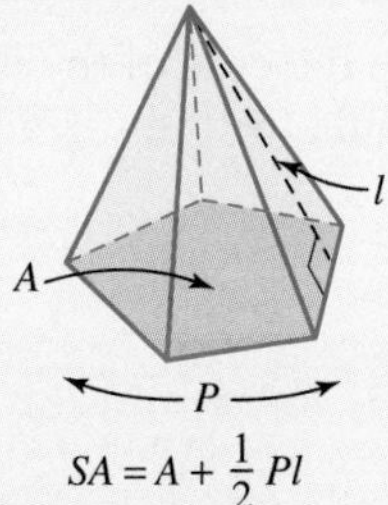

Surface Area of Right Cylinder

We can find the surface area of a cylinder by adapting the formula for the surface area of a prism as suggested by Figure 3.33. Imagine increasing the number of sides in the bases of the prism so they approach a circle.

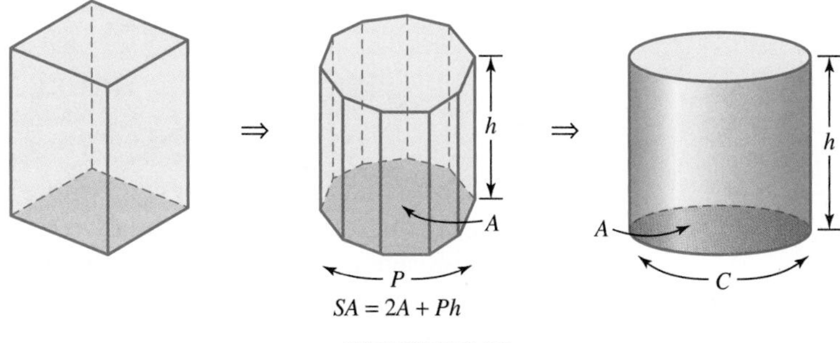

FIGURE 3.33

In the case of a right circular cylinder whose base has radius r, A represents the area of the base, namely πr^2, and P represents the perimeter (or C the circumference), namely $2\pi r$. This leads to the following theorem.

THEOREM 3.13 Surface Area of a Right Circular Cylinder

The surface area of a right circular cylinder is twice the area of its base plus the circumference of its base times its height.

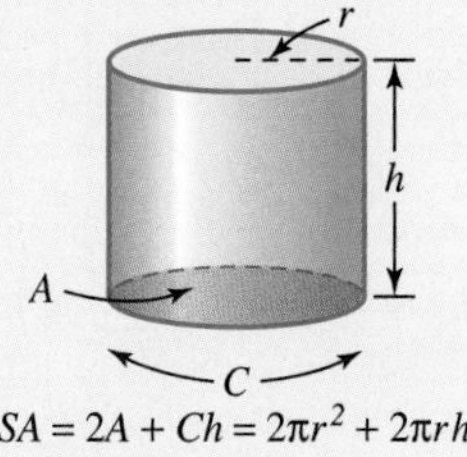

$SA = 2A + Ch = 2\pi r^2 + 2\pi rh$

EXAMPLE 3.22 Find the surface area of the right circular cylinder whose height is 7 cm and whose base has diameter 12 cm.

SOLUTION

$$SA = 2\pi(6^2) + 2\pi(6)(7) = 72\pi + 84\pi = 156\,\pi \text{ cm}^2$$ ■

Surface Area of a Right Circular Cone

As in the case of the cylinder, we may find the surface area of a cone by adapting the formula for the surface area of a pyramid as suggested in Figure 3.34.

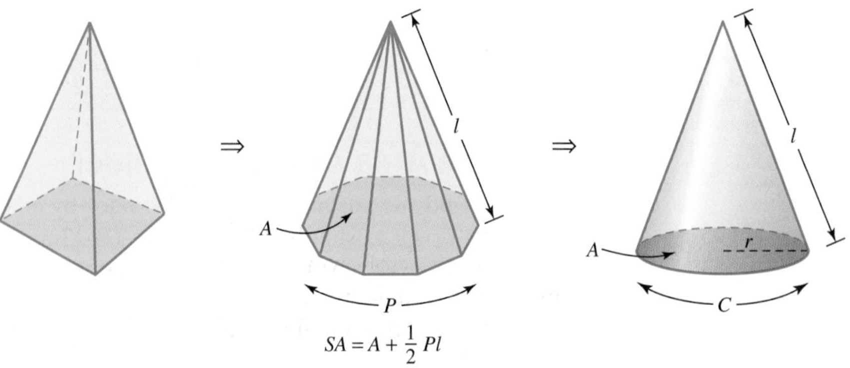

FIGURE 3.34

In the case of a right circular cone where r is the radius of the base, A represents the area of the base, namely πr^2, and C represents the circumference of the base, $2\pi r$. Also, l represents the slant height. By substituting for A and P in the formula $A + \frac{1}{2}Pl$, we can arrive at a formula for surface area of a right circular cone.

$$SA = A + \tfrac{1}{2}Cl = \pi r^2 + \tfrac{1}{2}(2\pi rl) = \pi(r + l)$$

This discussion leads to our next result.

> **THEOREM 3.14 Surface Area of a Right Circular Cone**
>
> The surface area of a right circular cone is the sum of the area of its base plus half the circumference of its base times its slant height.
>
> 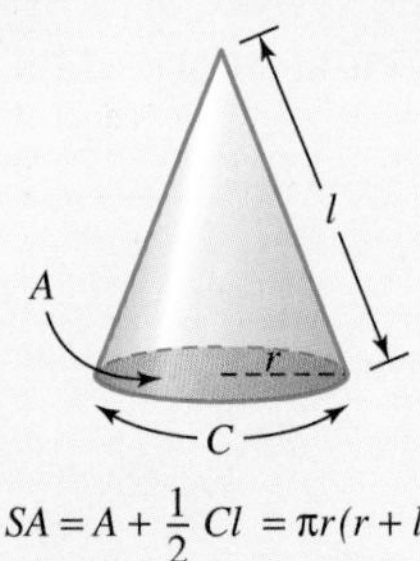
>

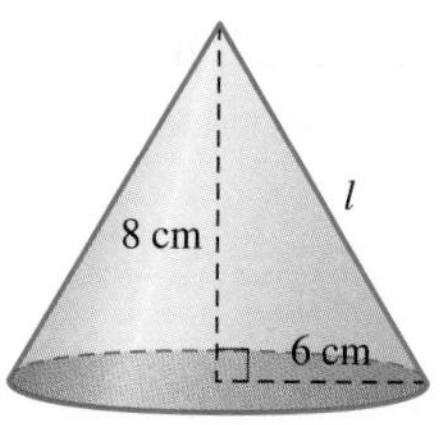

FIGURE 3.35

EXAMPLE 3.23 Find the surface area of the right circular cone whose height is 8 cm and whose base has radius 6 cm (Figure 3.35).

SOLUTION

First we must find the slant height, l.

$$l^2 = 6^2 + 8^2 = 36 + 64 = 100 \quad \text{or} \quad l = 10 \text{ cm}$$

(NOTE: We also could have observed that $6 = 2 \cdot 3$ and $8 = 2 \cdot 4$. Thus, $l = 2 \cdot 5 = 10$ because it is the hypotenuse of a right triangle whose sides are multiples of the sides of a 3-4-5 right triangle.)

Therefore,

$$SA = \pi(6^2) + \tfrac{1}{2}(2\pi)(6)(10) = 36\pi + 60\pi = 96\pi \text{ cm}^2$$ ■

Using the result that $SA = \pi r(r + l)$ in Theorem 3.14 we also find

$$SA = \pi \cdot 6(6 + 10) = \pi \cdot 6 \cdot 16 = 96\pi \text{ cm}^2$$

Surface Area of a Sphere

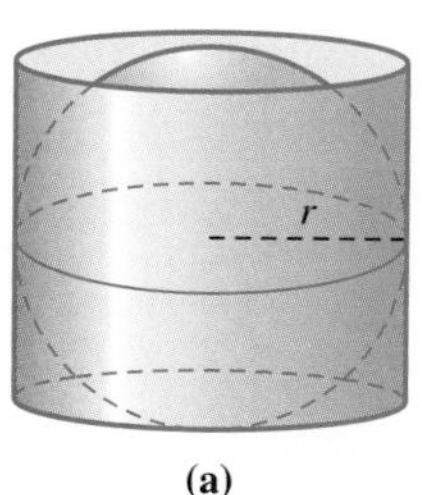

(a)

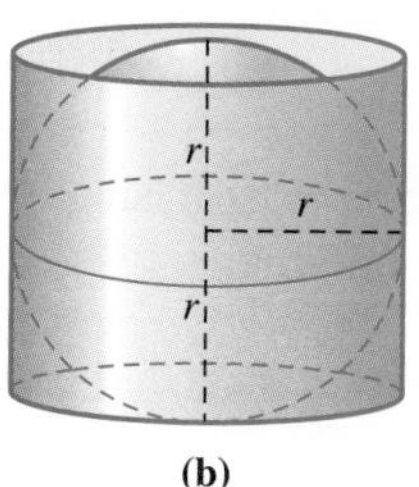

(b)

FIGURE 3.36

The derivation of the formula for the surface area of a sphere is quite complex. However, the following observation due to Archimedes simplifies the task of remembering the formula.

Consider the smallest right circular cylinder containing a sphere of radius r [Figure 3.36(a)]. Because the sphere has radius r, the base of the cylinder has radius r and the height of the cylinder is $2r$ [Figure 3.36(b)]. Therefore, the surface area of the *cylinder* is as follows:

$$SA = 2A + Ch = 2\pi r^2 + 2\pi r(2r) = 6\pi r^2$$

Archimedes's observation was that *both* the surface area and the volume of a sphere are two-thirds the respective surface area and volume of the smallest right circular cylinder containing the sphere. Thus, the surface area of the sphere of radius r can be found using the previous equation as follows:

$$SA_{\text{Sphere}} = \tfrac{2}{3} SA_{\text{Cylinder}} = \tfrac{2}{3}(6\pi r^2) = 4\pi r^2$$

THEOREM 3.15 Surface Area of a Sphere

The surface area of a sphere is 4π times the square of its radius.

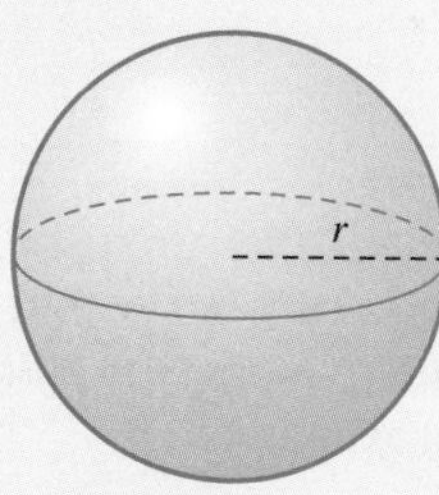

EXAMPLE 3.24 Although estimates vary, water covers about 70 percent of the earth's surface. About how many square miles of the earth's surface are composed of land?

SOLUTION

In Section 2.5, we found that the radius of the earth is approximately 3964 mi. Thus, the total surface area of the earth is

$$SA = 4\pi r^2 \approx 4\pi(3964)^2 \approx 197{,}459{,}101 \text{ mi}^2$$

If 70 percent of that area is water, then land covers about

$$0.30(197{,}459{,}101 \text{ mi}^2) \approx 59{,}240{,}000 \text{ mi}^2$$

Solution of Applied Problem

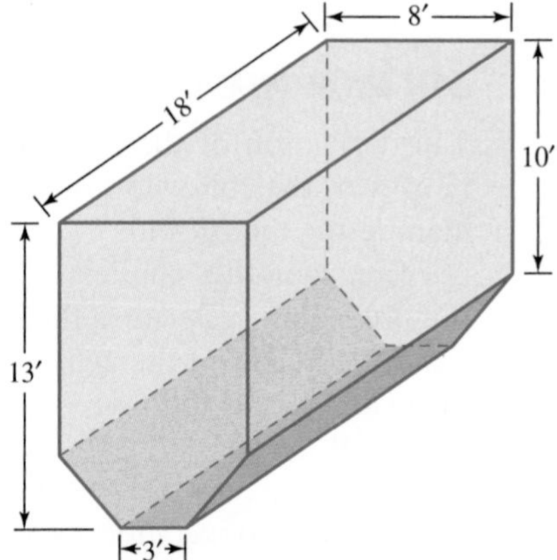

We need to find the surface area of this right hexagonal prism. Using the Pythagorean theorem to solve for x in the following figure, we find that

$$3^2 + (2.5)^2 = x^2 \quad \text{or} \quad x \approx 3.91 \text{ ft}$$

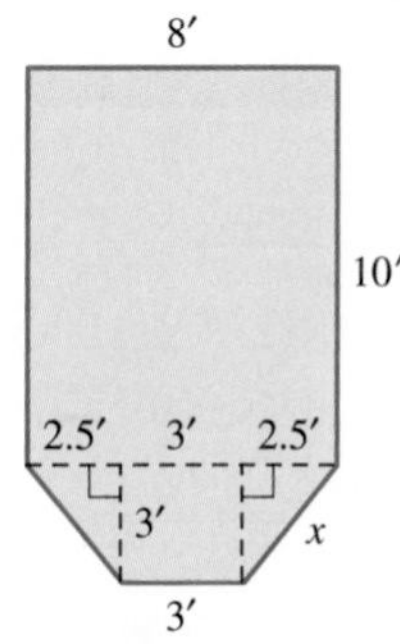

Thus, the area of the six rectangular faces is given by $A = Ph$, or $A = (8 + 10 + 3.91 + 3 + 3.91 + 10)18$, which equals 698.76 ft^2. The area of the two bases is $2[8(10) + \frac{1}{2}(3)(3 + 8)] = 193 \text{ ft}^2$. Thus, the surface area is approximately

$$193 \text{ ft}^2 + 698.76 \text{ ft}^2 = 891.76 \text{ ft}^2$$

GEOMETRY AROUND US Porous materials with large internal surface areas are important in many applications including gas storage and the production of fuel cells. In 2004, U.S. researchers created block-shaped crystals (see figure) that have surface areas of up to 4500 m^2 per gram. By comparison, a human lung has a surface area of about 0.25 m^2 per gram and the surface area of a tennis court is 196 m^2. One obstacle faced by the automobile industry in the transition from fossil fuels to hydrogen is finding an efficient way to store the gas. The fact that the new crystals can bind large volumes of gas in a small space offers a possible solution to the hydrogen storage problem.

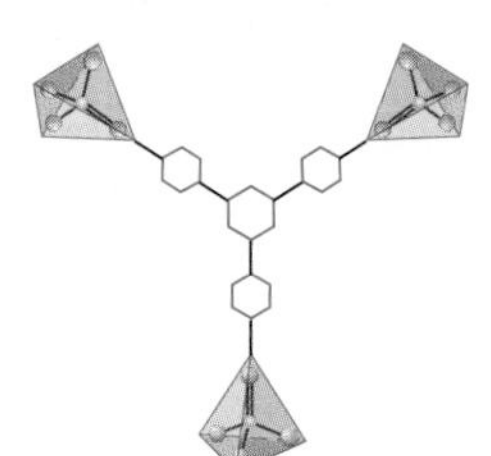

PROBLEM SET 3.4

EXERCISES/PROBLEMS

1. Find the surface area of each of the following right prisms.

a.

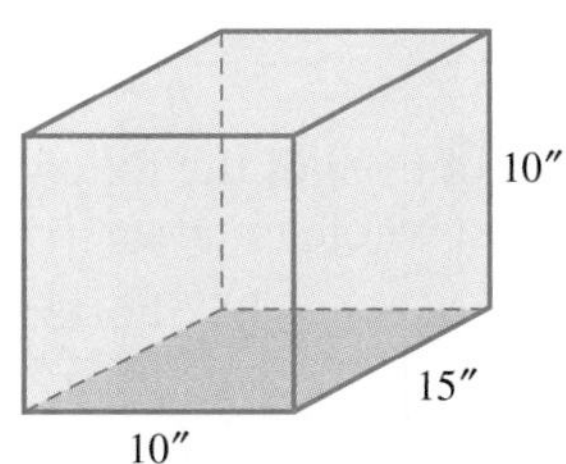

b.

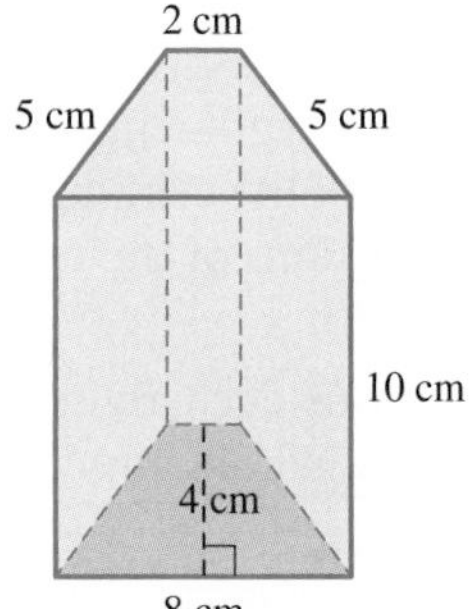

2. Find the surface area of each of the following right prisms. Round to the nearest whole number.

a.

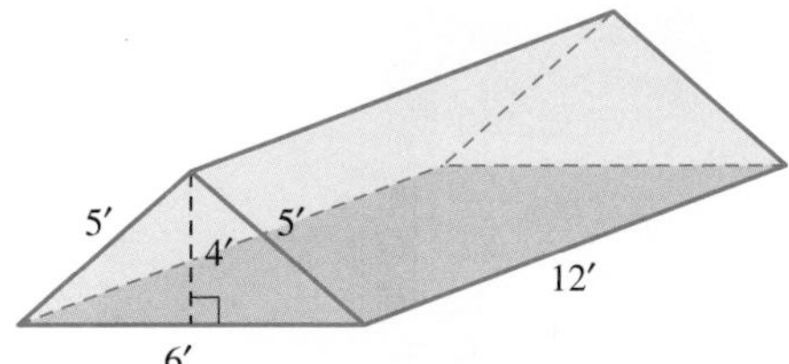

b.

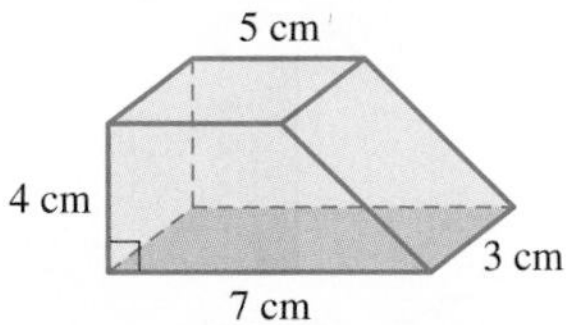

3. Find the surface area of each right prism with the given features. Round to the nearest whole number.

a. The bases are equilateral triangles with sides of length 8 ft, and the height of the prism is 10 ft.

b. The bases are trapezoids with parallel sides of lengths 7 cm and 9 cm perpendicular to one side of length 6 cm. The height of the prism is 12 cm.

4. Find the surface area of each right prism with the given features. Round to the nearest whole number.

a. The base is a right triangle with legs of length 5 cm and 12 cm, and the height of the prism is 20 cm.

b. The bases are isosceles trapezoids with bases of 4 inches and 12 inches and with nonparallel side lengths of $4\sqrt{2}$ inches. The height of the prism is 16 inches.

5. The top of a rectangular box has an area of 96 square inches. Its side has area 72 square inches and its end has area 48 square inches. What are the dimensions of the box?

6. The top of a rectangular box has an area of 63 cm^2. Its side has an area of 42 cm^2 and its end has an area of 54 cm^2. What are the dimensions of the box?

7. Find the surface area of each of the following right square pyramids.

a.

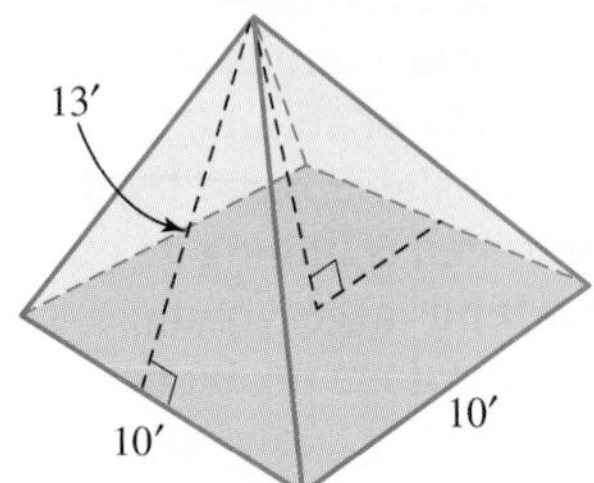

b.

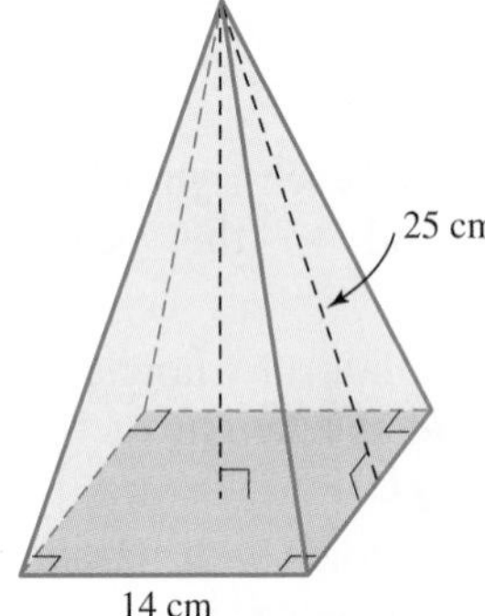

8. Find the surface area of each of the following right square pyramids. Round to the nearest whole number.

a.

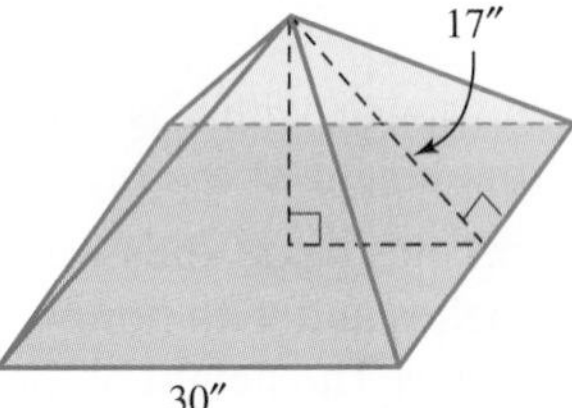

b.

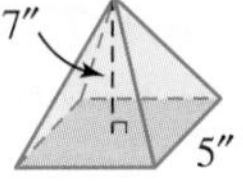

9. Find the surface area of a pyramid if its height is 4 cm and its base is a rectangle with side lengths 4 cm and 6 cm. Round to the nearest whole number.

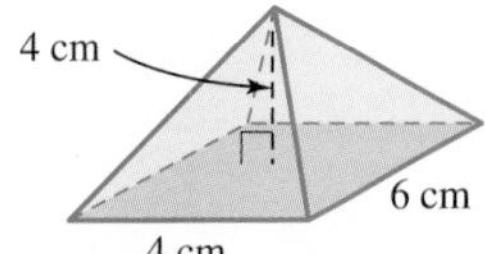

10. A right regular pyramid has a height of 10″ and a base that is a regular octagon with sides of length 3″. Find the surface area of the pyramid, rounding to the nearest whole number.

11. Find the surface area of a right pyramid if its height is 14″ and the base is a regular hexagon with sides of length 12″. Round to the nearest whole number.

12. Find the surface area of a pyramid if its base is a 10″ by 18″ rectangle. Assume the apex is centered over the base.

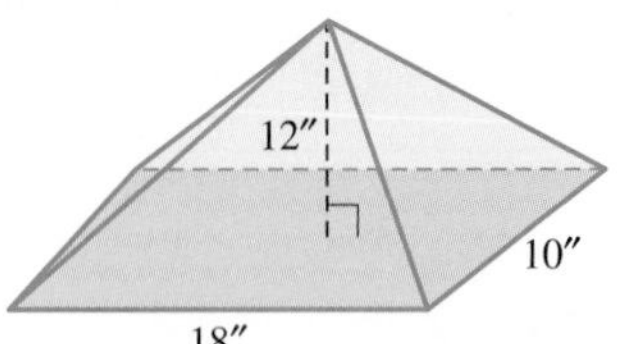

13. Find the surface area of each of the following cans rounding to the nearest whole number.

a. Coffee can
$r = 7.6$ cm
$h = 16.3$ cm

b. Soup can
$r = 3.3$ cm
$h = 10$ cm

14. Find the surface area of each of the following cans rounding to the nearest whole number.

a. Juice can
$r = 5.3$ cm
$h = 17.7$ cm

b. Shortening can
$r = 6.5$ cm
$h = 14.7$ cm

15. A right circular cylinder has a total surface area of 112π. If the height of the cylinder is 10, find the diameter of the base.

16. The surface area of a right circular cylinder is 1200 cm^2. If the height of the cylinder is twice its diameter, find the dimensions of the cylinder. Round to the nearest hundredth.

17. Find the surface area of each right circular cone, rounding to the nearest whole number.

a.

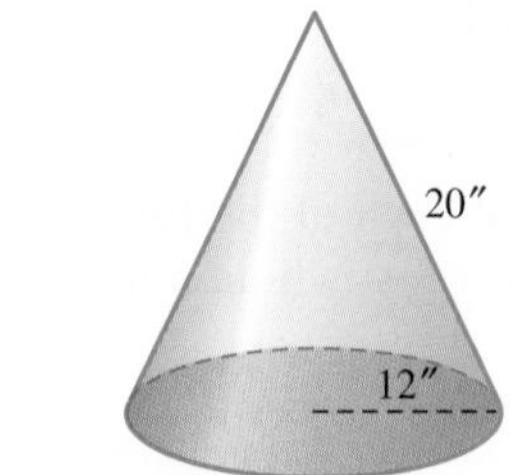

b.

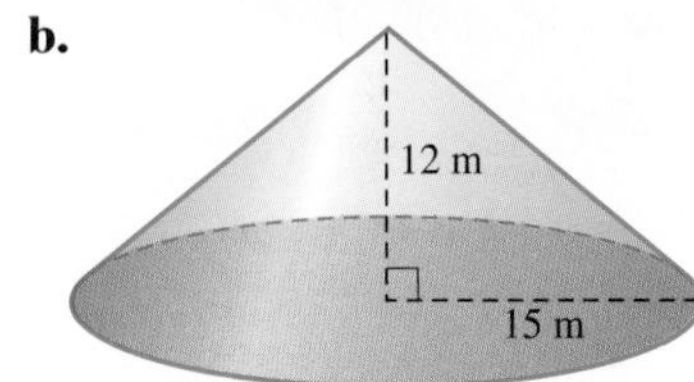

18. Find the surface area of each right circular cone, rounding to the nearest whole number.

a.

b.

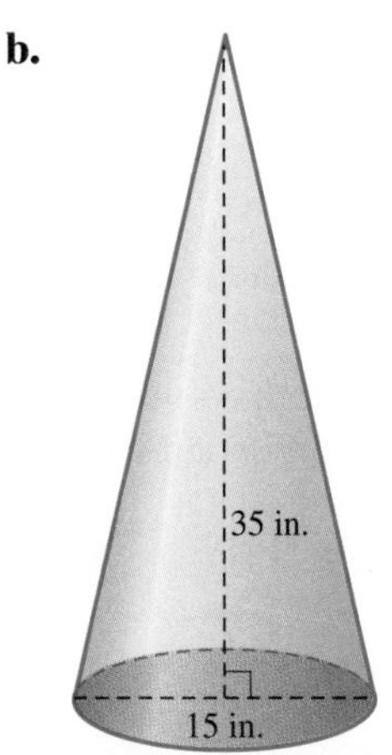

19. Find the surface area of spheres having the following measurements. Round to the nearest whole number.

a. $r = 6$ in. **b.** $d = 24$ m **c.** $C = 7\pi$ cm

20. Find the surface area of spheres having the following measurements. Round to the nearest whole number.

a. $r = 2.3$ in.
b. $d = 6.7$ km
c. $C = 23\pi$ mm

21. Find the diameter, to the nearest mm, of a sphere with surface area 215.8 cm^2.

22. The radius of one sphere is 4 cm and the radius of a second sphere is 12 cm. How do the surface areas of the two spheres compare?

23. A spherical balloon has a diameter of 8 feet. By how much should the radius of the balloon be increased if the surface area of the balloon is to be doubled? Round to the nearest hundredth.

24. A rubber ball has a circumference of 6π meters. By how much should the circumference be increased if the surface area of the ball is to be tripled? Round to the nearest tenth.

25. Find the surface area of a regular tetrahedron with edge length 5 cm. Round to the nearest whole number.

26. Find the surface area of a regular octahedron with edge length $4\sqrt{3}$ cm. Round to the nearest whole number.

27. Find the total surface area. Round to the nearest whole unit.

a.

b.

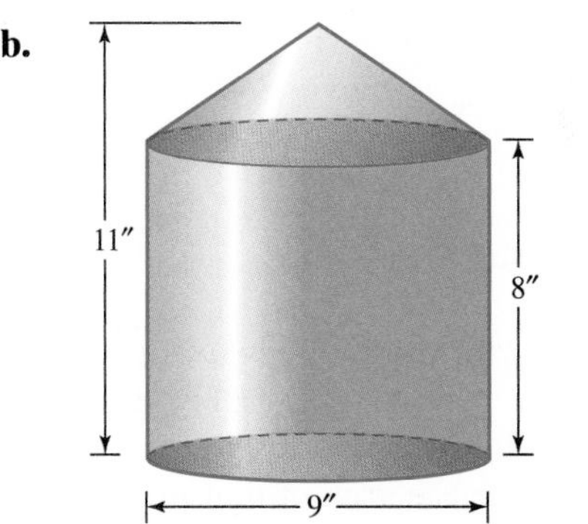

28. Find the total surface area of the following solid shapes. Round to the nearest hundredth.

a.

b.

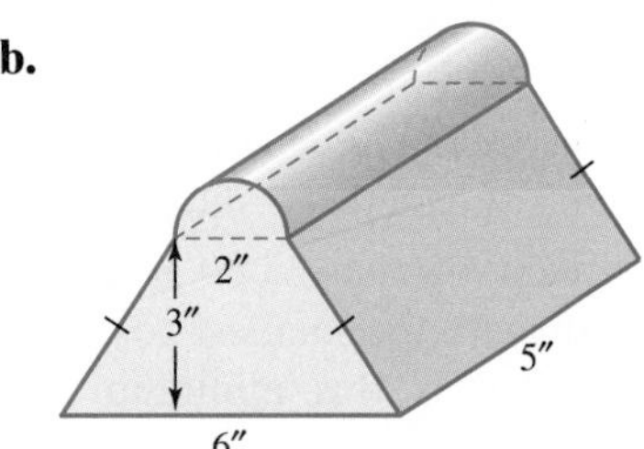

29. Find the total surface area of the hollowed-out hemisphere shown.

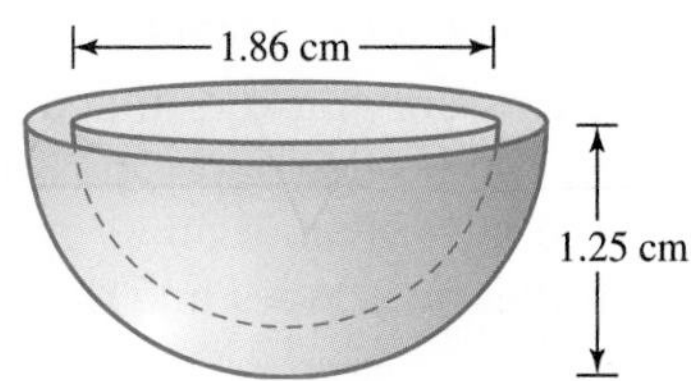

30. Find the total surface area of the hollowed-out cylinder shown.

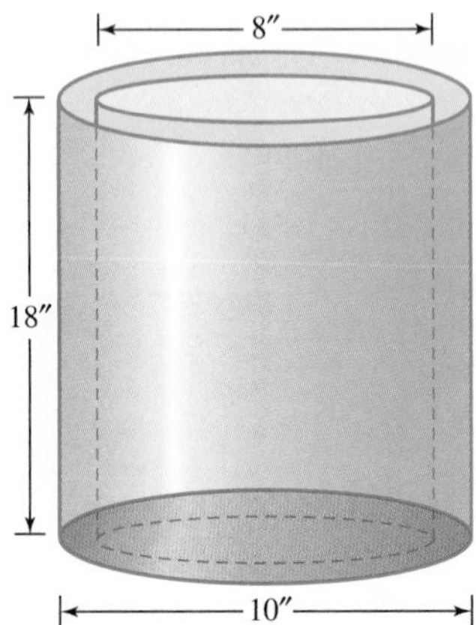

31. Suppose you have 36 unit cubes like the one shown below.

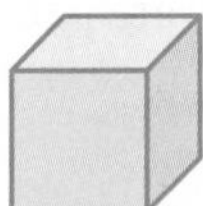

You could arrange those cubes to form right rectangular prisms of various sizes. For example, you could arrange the cubes in nine rows of four cubes as shown. The surface area of that prism would be 98 square units.

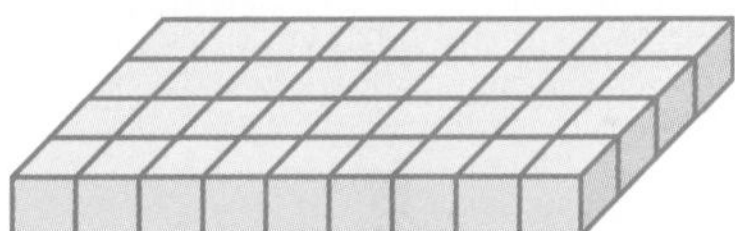

a. Sketch an arrangement of the cubes in the shape of a right rectangular prism that has a total surface area of exactly 96 square units.

b. Sketch an arrangement of the cubes in the shape of a right rectangular prism that has a total surface area of exactly 80 square units.

c. What arrangement of the cubes gives a right rectangular prism with the smallest possible total surface area? What is this minimum surface area?

d. What arrangement of the cubes gives a right rectangular prism with the largest possible total surface area? What is this maximum surface area?

32. Thirty unit cubes are stacked in square layers to form a tower. The bottom layer measures 4 cubes by 4 cubes, the next layer 3 cubes by 3 cubes, the next layer 2 cubes by 2 cubes, and the top layer is a single cube.

a. Determine the total surface area of the tower of cubes.

b. Suppose the number of cubes and the height of the tower were increased so that the bottom layer of cubes measured 8 cubes by 8 cubes. What would be the total surface area of this tower?

c. What would be the total surface area of the tower if the bottom layer measured 20 cubes by 20 cubes?

APPLICATIONS

33. A box without a top is to be made from sheet metal. If the box will measure 5′6″ by 2′10″ by 1′8″ high, how many square feet of sheet metal will be needed? Round to the nearest tenth.

34. When determining heating and cooling requirements for a greenhouse, it is important to calculate the surface area. Determine the surface area of the greenhouse with the dimensions shown. Round to the nearest whole number.

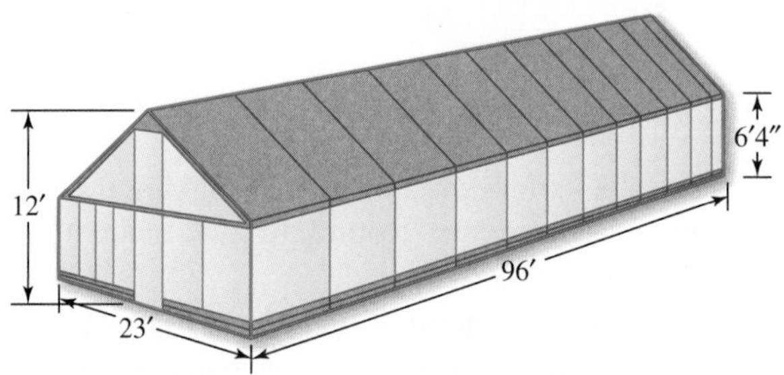

35. Calculate the surface area of the following greenhouse. Round to the nearest tenth.

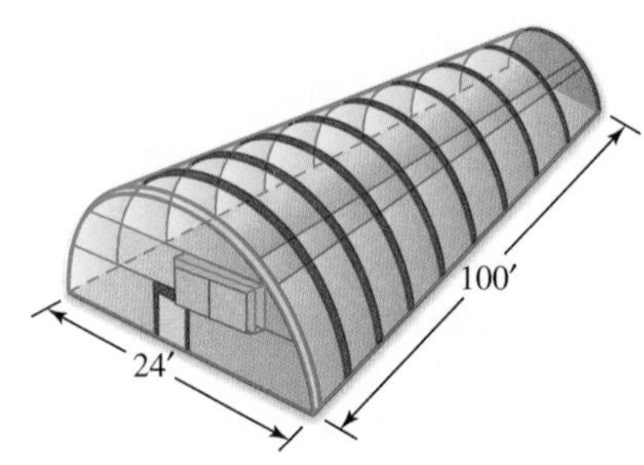

36. The inside of a cylindrical tank used to hold resin must be coated with a corrosion preventative. The tank measures 7 ft in diameter and 5 ft in height. One gallon of corrosion preventative will cover 10 ft^2. How many gallons will be needed to coat the tank? Assume that the sides, top, and bottom of the tank will be treated.

37. A room measures 4 m by 7 m and the ceiling is 3 m high. A liter of paint covers 20 m^2. How many liters of paint will it take to paint all but the floor of the room?

38. A golf ball has a diameter of 1.68 inches. A box in, the shape of a square prism to hold three golf balls, will be made out of cardboard. What is the least amount of cardboard that will be needed, assuming the centers of the balls will be collinear when in the box? Ignore flaps to be folded in on the ends and round your answer to the nearest tenth of a square inch.

39. A machined piece is a right regular hexagonal prism with a hole 1/2 inch in diameter drilled through it. Find the total surface area, rounding to the nearest tenth of a square inch.

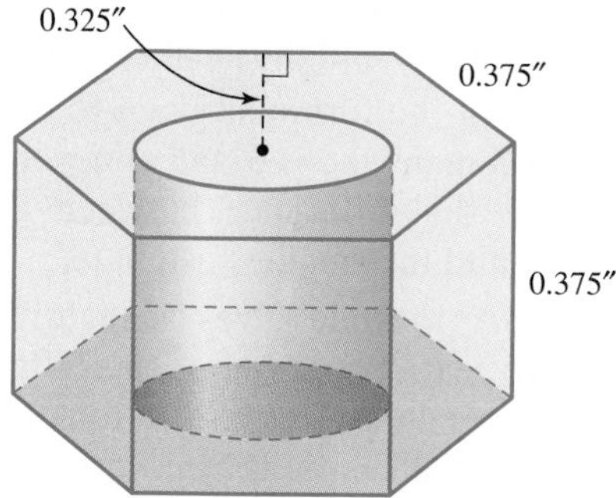

40. A gutter is to be formed from a length of tin bent so that the cross section is a semicircle 4 inches in diameter. If a total length of 210 feet is needed and an extra 3/4 inch at each edge is required to finish the edge, calculate the number of square feet of tin needed. Round to the nearest square foot.

41. The cylindrical storage tank with hemispherical ends as shown must be painted. If the paint costs $21.95 per gallon and one gallon of paint covers 200 square feet, what will the paint cost? Round to the nearest cent.

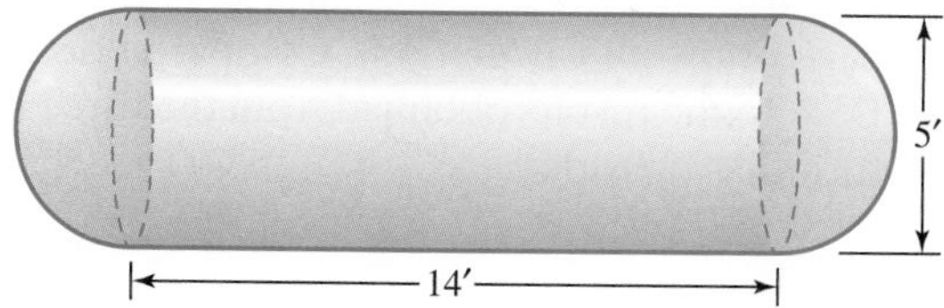

42. A water tower in the shape of a right circular cylinder has a stairway to the top that winds around the outside of the tower. The tower is 50 ft in diameter and 40 ft tall, and the stairs wrap around the cylinder exactly twice. How long, to the nearest foot, is the stairway?

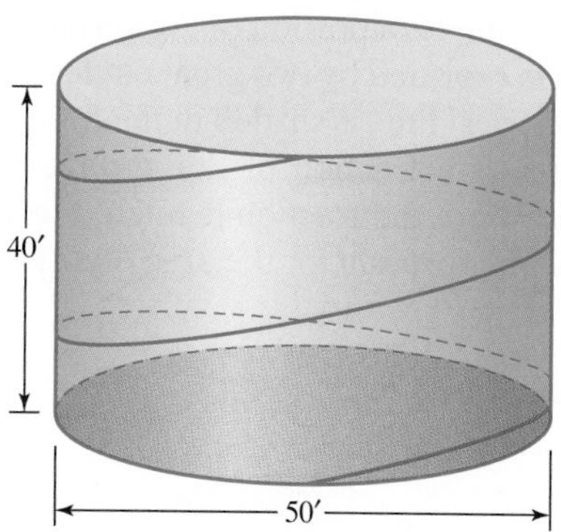

43. A barber pole consists of a cylinder of radius 10 cm on which one red, one white, and one blue helix, each of equal width, are painted. The cylinder is 1 m high. If each stripe makes a constant angle of 60° with the vertical axis of the cylinder, how much surface area is covered by the red stripe?

44. A room measures 4 m wide by 4 m high by 8 m long. At point P in the center of the front and 0.5 m from the floor, is a DVD player. At a point Q in the center of the back wall and 0.5 m down from the ceiling, is a stereo speaker. The speaker is to be connected to the DVD player. The speaker wire must be in contact with a wall, the ceiling, or the floor at all times. What is the shortest length of wire that can be used? (HINT: Draw a two-dimensional picture.)

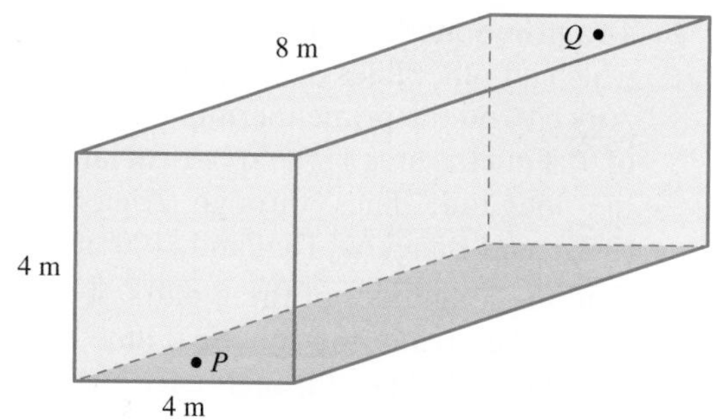

EXTENDED PROBLEMS

45. Create paper models of a regular tetrahedron with side length 3 inches and of a right square pyramid with equilateral faces such that the faces of the tetrahedron and the lateral faces of the pyramid are identical triangles. Find the total surface area of the polyhedrons. If the tetrahedron is attached to a face of the pyramid, what is the fewest number of faces possible for the resulting polyhedron and what is the surface area?

46. Cross sections of a sphere that have as their center the center of the sphere are called **great circles**. A **spherical lune** is a portion of the surface of a sphere of radius r formed by two great circles. An example of a spherical lune is shaded in the following figure. The area is a fraction of the surface area of the sphere, where the fraction is related to the angle θ shown. The formula for the area of a spherical lune is $S = 4\pi r^2\left(\frac{\theta}{360°}\right)$ if the angle θ is measured in degrees.

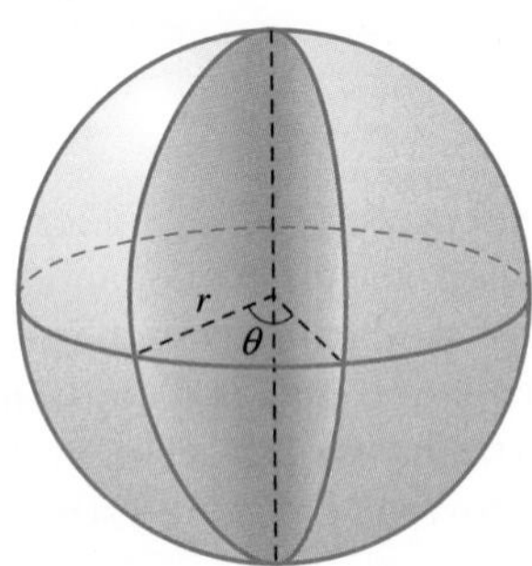

a. On a globe, an example of a spherical lune is the area between two lines of longitude. (NOTE: Each "line" of longitude is actually a great circle.) If there are 24 lines of longitude labeled on a globe, then how many degrees separate each adjacent pair of lines? If the earth has a diameter of approximately 3500 km, find the area of the spherical lune between two of these adjacent lines of longitude.

b. The **prime meridian** is the longitude line from which all other lines of longitude are measured. Any city on the prime meridian has a longitude of 0°. Find the area of the spherical lune between the longitude lines through Algiers, Algeria (3°E), and Bangkok, Thailand (100°30′E).

c. Find the longitudes of three pairs of cities. Calculate the area of the spherical lune created by the longitude lines through each pair of cities.

47. The human body's largest organ is the skin. Doctors often need an estimate of the human body surface area (BSA) to calculate drug dosages and fluid requirements. Accurate BSA estimates are especially critical for cancer patients. There are several formulas used to calculate the BSA. Each formula gives BSA in m^2.

Mosteller Formula:

$$\text{BSA} = \sqrt{\frac{(\text{height in cm})(\text{weight in kg})}{3600}}$$

DuBois and DuBois Formula:

$$\text{BSA} = 0.20247(\text{height in m})^{0.725}(\text{weight in kg})^{0.425}$$

Haycock Formula:

$$\text{BSA} = 0.024265(\text{height in cm})^{0.3964}(\text{weight in kg})^{0.5378}$$

a. Use each of the formulas given to estimate the BSA for a woman who is 5′9″ tall and weighs 147 lb. Calculate the difference between the largest estimate and the smallest estimate.

b. Use each of the formulas given to estimate the BSA for a man who is 6′ tall and weighs 192 lb. Calculate the difference between the largest estimate and the smallest estimate.

c. Each BSA formula gives a slightly different result. To learn more about the different BSA formulas, read the Standardization of Body Surface Area Calculations report by Thanh Vu, B.Sc. (pharm), Department of Pharmacy, Cross Cancer Institute, 1999, on the Internet. Which formula does Vu recommend and why? How were body surface areas measured in order to create the different formulas? How well do the formulas predict BSA of people who are obese? Do these formulas apply to children as well as adults? Summarize your findings in an essay.

d. Approximate your own body surface area by using surface areas of prisms, pyramids, cylinders, cones, and spheres. For example, each finger can be approximated by a right circular cylinder with one base missing. Have someone help you take careful measurements. Draw a sketch of your body indicating which shape you will use to estimate the surface area of each part. Show all calculations. Compare your final BSA to the results obtained from the three BSA formulas provided.

3.5 VOLUME

Applied Problem

At a steel mill a thick slab of steel 1 m by 1 m by 15 cm will be rolled into a long strip 1 m wide and 0.5 cm thick. How long will the rolled strip be?

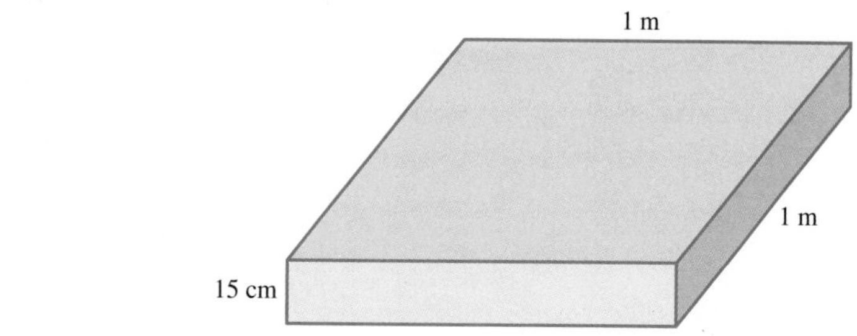

Volume

Volume is the three-dimensional analog of area. To find the volume of a shape, we must designate a unit cube. Then we must determine how many unit cubes completely fill a given three-dimensional shape (Figure 3.37).

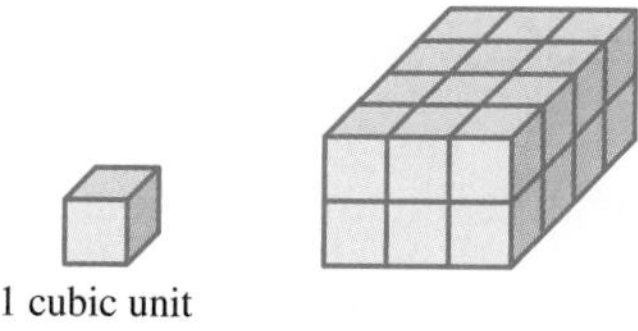

FIGURE 3.37

The following postulate for volume is analogous to the Area Postulate.

POSTULATE 3.4 Volume Postulate

a. For every polyhedron and unit cube, there is a real number that gives the number of unit cubes (and parts of unit cubes) that exactly fill the region enclosed by the polyhedron.

b. The volume of the region enclosed by a polyhedron is the sum of the volumes of the smaller regions into which the region can be subdivided.

The real number described in Postulate 3.4 is called the **volume** of the region enclosed by the polyhedron. For the sake of simplicity, we usually say "the volume of the polyhedron" to mean the volume of the region enclosed by the polyhedron. The next example shows how to find the volume of a right rectangular prism.

EXAMPLE 3.25 Using a unit cube of 1 cm by 1 cm by 1 cm, find the volume of a right rectangular prism 3 cm by 4 cm by 2 cm by counting the number of unit cubes that will fill the prism exactly (Figure 3.38).

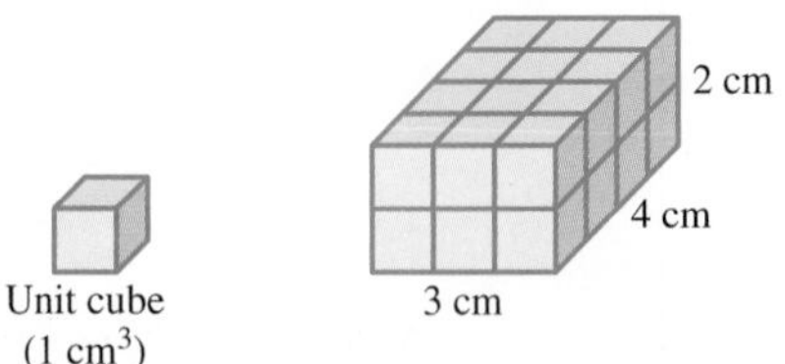

FIGURE 3.38

SOLUTION

As we can see in Figure 3.38, the prism would hold two layers of cubes that have dimensions 3 cm by 4 cm. Because there are $(3)(4) = 12$ cubes in each layer, there are $(3)(4)(2) = 24$ unit cubes in all. Thus, the volume is 24 cm^3. We may view this result as the product of the area of the base of the prism $3 \text{ cm} \cdot 4 \text{ cm}$, and its height, 2 cm. ■

Volume of a Prism

As in the case of finding the area of a rectangle, we can find the volume of a prism having real number dimensions using the technique of Example 3.25.

DEFINITION

Volume of a Prism

The volume of a prism is the product of the area of its base and its height.

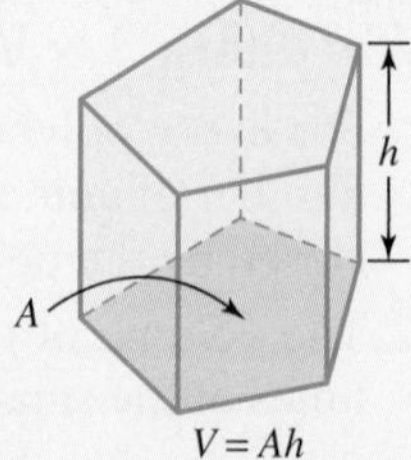

The following two results are immediate consequences of this definition of volume.

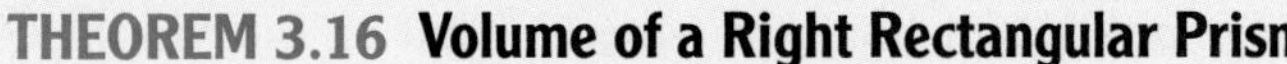

THEOREM 3.16 Volume of a Right Rectangular Prism

The volume of a right rectangular prism is the product of its length, width, and height.

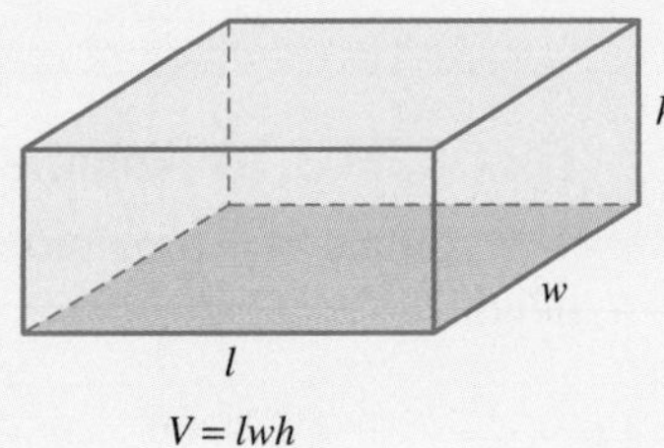

$V = lwh$

COROLLARY 3.17 Volume of a Cube

The volume of a cube is the cube of the length of its side.

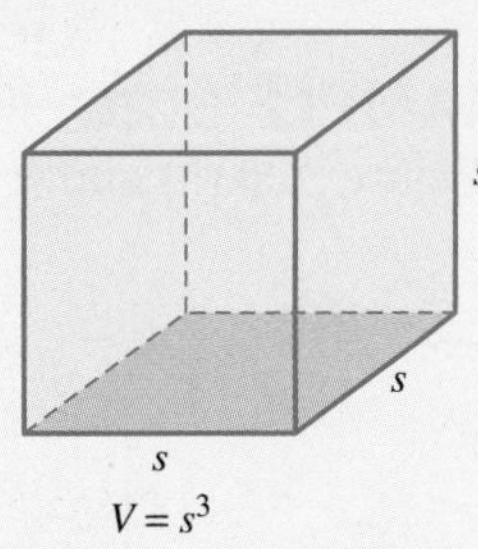

$V = s^3$

(NOTE: A **corollary** is a theorem that is a direct result or a special case of a preceding theorem. In this case, a cube is a special type of right rectangular prism.)

Volume of a Pyramid

We may infer the volume of a pyramid from Figure 3.39, where three pyramids identical to the one on the left are shown to be filling a right prism completely.

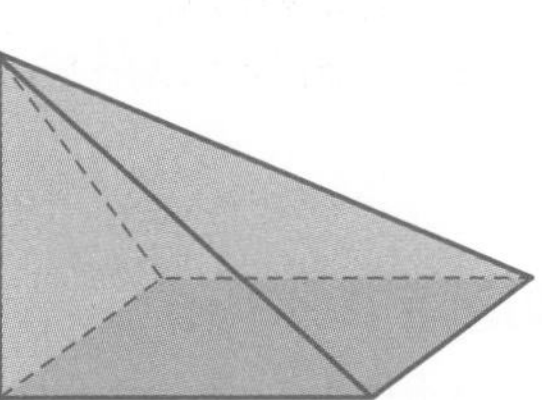

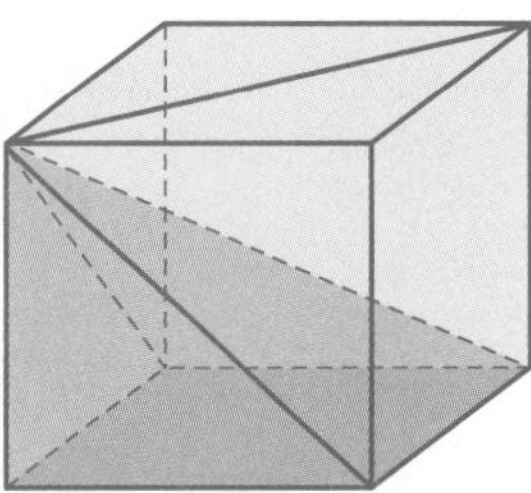

FIGURE 3.39

Even though Figure 3.39 pictures a square pyramid, the ratio of three pyramids to one prism holds in general. That is, the volume of a pyramid is one-third the volume of the corresponding prism determined by the base and height of the pyramid.

THEOREM 3.18 Volume of a Pyramid

The volume of a pyramid is one-third the product of the area of its base and its height.

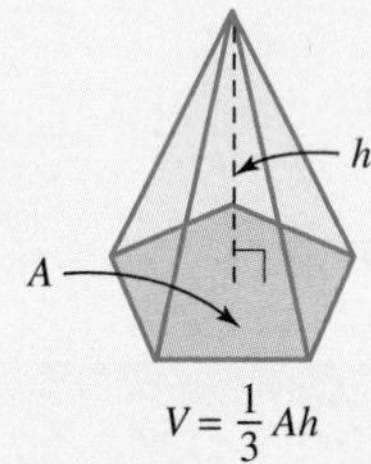

EXAMPLE 3.26 Find the volumes of the polyhedra with dimensions shown in Figure 3.40 (a) and (b).

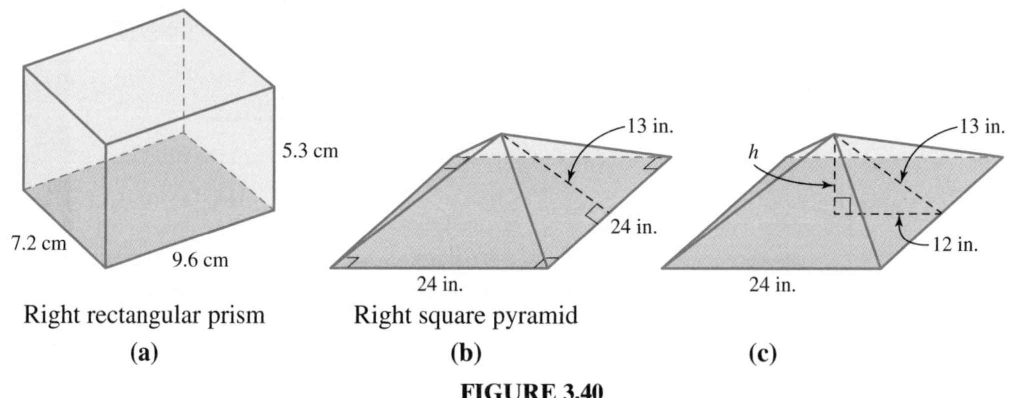

Right rectangular prism (a) Right square pyramid (b) (c)

FIGURE 3.40

SOLUTION

a. $V = lwh = (7.2\text{ cm})(9.6\text{ cm})(5.3\text{ cm}) = 366.336\text{ cm}^3$

b. The slant height is 13 in. The distance from the center of the base to an edge is $\frac{1}{2}(24\text{ in.}) = 12\text{ in.}$ [Figure 3.40(c)]. Thus, the height, h, satisfies the equation $h^2 + 12^2 = 13^2$, or $h = 5$ in. The volume, then, is given by

$$V = \frac{1}{3}Ah = \frac{1}{3}(24\text{ in.})^2(5\text{ in.}) = 960\text{ in}^3$$

Volumes of Cylinders and Cones

We obtain the formulas for the volumes of cylinders and cones in a manner similar to that used to derive the formulas for their surface areas (Figure 3.41).

In the case of the circular cylinder with a base of radius r, A represents the area of the base, or πr^2. The height of the cylinder is represented by h. Figure 3.41 suggests the following theorems.

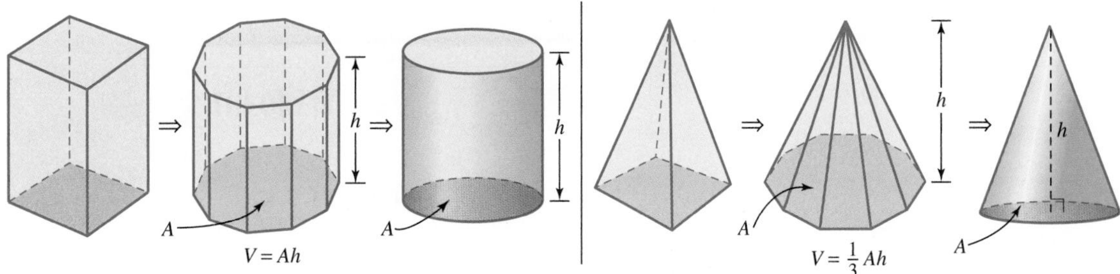

FIGURE 3.41

THEOREM 3.19 Volume of a Right Circular Cylinder

The volume of a right circular cylinder is the product of the area of its base and its height.

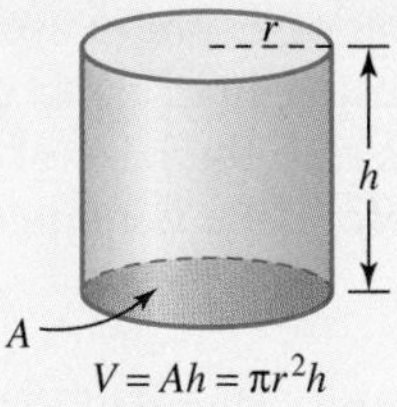

THEOREM 3.20 Volume of a Right Circular Cone

The volume of a circular cone is one-third the product of the area of its base and its height.

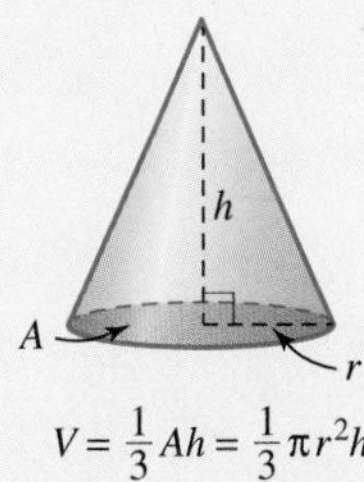

EXAMPLE 3.27 Find the volumes of the right circular cylinder and right circular cone with dimensions given in Figure 3.42.

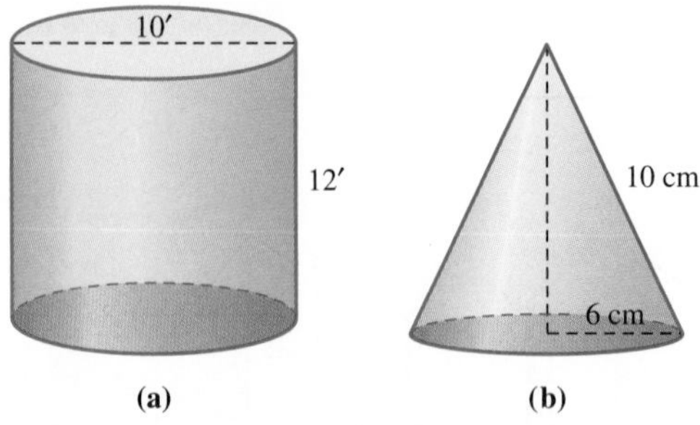

FIGURE 3.42

SOLUTION

a. $V = \pi r^2 h = \pi(5^2)(12) = 300\pi \text{ ft}^3$

b. $V = \frac{1}{3}\pi r^2 h = \frac{1}{3}\pi(6^2)\sqrt{10^2 - 6^2} = \frac{1}{3}\pi(6^2)(8) = 96\pi \text{ cm}^3$ ■

Although we gave formulas for the volumes of right prisms, pyramids, cylinders, and cones, the same formulas hold for similar oblique shapes.

Volume of a Sphere

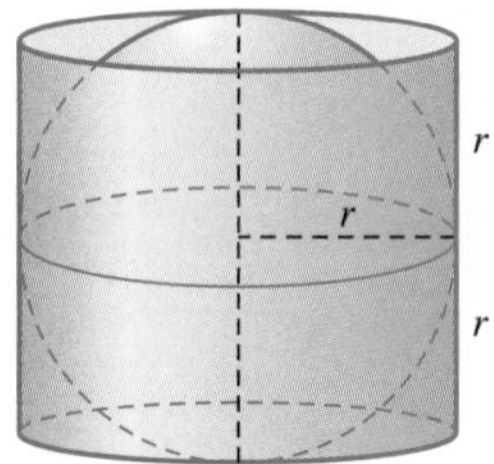

FIGURE 3.43

We stated in Section 3.4 that Archimedes observed that both the surface area and volume of a sphere are two-thirds of the respective surface area and volume of the smallest right circular cylinder containing the sphere (Figure 3.43). The sphere has radius r, so the base of the cylinder has radius r and the height of the cylinder is $2r$. Therefore, the volume of the cylinder is

$$V = Ah = \pi r^2(2r) = 2\pi r^3$$

Thus, the volume of the sphere of radius r can be found as

$$V_{\text{Sphere}} = \tfrac{2}{3}V_{\text{Cylinder}} = \tfrac{2}{3}(2\pi r^3) = \tfrac{4}{3}\pi r^3$$

We summarize this result in the following theorem.

THEOREM 3.21 Volume of a Sphere

The volume of a sphere is $\frac{4}{3}\pi$ times the cube of its radius.

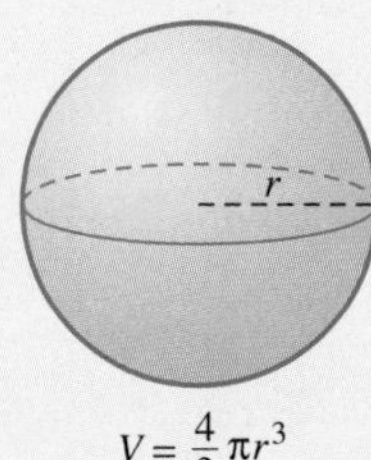

$V = \frac{4}{3}\pi r^3$

EXAMPLE 3.28 Find the radius of the sphere for which the surface area and volume are the same number.

SOLUTION

Let r be the radius of the sphere. Then $4\pi r^2 = \frac{4}{3}\pi r^3$. Next solve for r.

$$\begin{aligned} 4\pi r^2 &= \tfrac{4}{3}\pi r^3 \\ 12\pi r^2 &= 4\pi r^3 \\ 4\pi r^3 - 12\pi r^2 &= 0 \\ 4\pi r^2(r - 3) &= 0 \end{aligned}$$

Therefore, $r = 0$ or $r = 3$. Because $r = 0$ produces a sphere consisting of a single point, the answer is $r = 3$.

As a check, a sphere of radius 3 has surface area $4\pi(3^2) = 36\pi$ and volume $\frac{4}{3}\pi(3^3) = 36\pi$. Of course, the units are square units for the surface area and cubic units for the volume. ■

The following table summarizes the surface area and volume formulas we have studied. Connections formed by observing the similarities and differences in the various formulas will help you remember them more easily.

Summary of Surface Area and Volume Formulas for Three-Dimensional Shapes

Geometric Shape	Surface Area	Volume
Right prism	$SA = 2A + Ph$	$V = Ah$
Right circular cylinder	$SA = 2A + Ch$ $= 2\pi r(r + h)$	$V = Ah$ $= \pi r^2 h$
Right regular pyramid	$SA = A + \frac{1}{2}Pl$	$V = \frac{1}{3}Ah$
Right circular cone	$SA = A + \frac{1}{2}Cl$ $= \pi r(r + l)$	$V = \frac{1}{3}Ah$ $= \frac{1}{3}\pi r^2 h$
Sphere	$SA = 4\pi r^2$	$V = \frac{4}{3}\pi r^3$

TABLE 3.2

Using Dimensional Analysis to Convert Units of Volume

In Section 3.1, we used dimensional analysis to convert units of area measure. The next example shows how we can use dimensional analysis to convert between units of volume measure.

EXAMPLE 3.29 Calculate the volume of dirt that must be removed from a 200-foot-long trench with its cross-section dimensions shown in Figure 3.44. Find the volume in cubic inches first, and then convert to cubic yards.

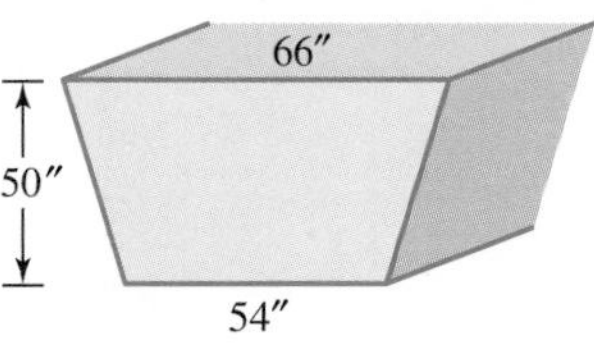

FIGURE 3.44

SOLUTION

The trench in Figure 3.44 is a right prism with a base in the shape of a trapezoid and a height equal to the length of the trench, namely 200 ft or 2400 in. Thus, we have

$$\begin{aligned} V &= Ah \\ &= \tfrac{1}{2}[(50)(66 + 54)](2400) \\ &= 7{,}200{,}000 \text{ in}^3 \end{aligned}$$

To convert from cubic inches to cubic feet, we use the fact that $1 \text{ ft}^3 = (1 \text{ ft})(1 \text{ ft})(1 \text{ ft}) = (12 \text{ in.})(12 \text{ in.})(12 \text{ in.}) = 1728 \text{ in}^3$. Hence,

$$V = \frac{7{,}200{,}000 \cancel{\text{in}^3}}{1} \cdot \frac{1 \cancel{\text{ft}^3}}{1{,}728 \cancel{\text{in}^3}} \cdot \frac{1 \text{ yd}^3}{27 \cancel{\text{ft}^3}} \approx 154 \text{ yd}^3$$

So approximately 154 yd^3 of dirt will be removed from the trench. ■

In the previous example, we calculated the volume first and then performed the conversion. However, we could have converted to yards first and then found the volume in cubic yards.

Solution of Applied Problem

The volume of the initial slab in cubic meters is

$$V = lwh = (1 \text{ m})(1 \text{ m})(0.15 \text{ m}) = 0.15 \text{ m}^3$$

Because the volume of the new, longer strip must also be 0.15 m^3, we can find the length, l, of the new strip.

$$\begin{aligned} 0.15 \text{ m}^3 &= l(1 \text{ m})(0.005 \text{ m}) \\ l &= \frac{0.15 \text{ m}^3}{0.005 \text{ m}^2} = 30 \text{ m} \end{aligned}$$

Therefore, the new strip will be 30 m long.

GEOMETRY AROUND US The sphere is one of the most commonly occurring shapes in nature. Frog eggs, tomatoes, oranges, the Earth, and the moon are a few examples. The regular occurrence of the spherical shape is not coincidental. One explanation for the frequency of the sphere's appearance is that for a given surface area, the sphere encloses the greatest volume. In other words, a sphere requires the least amount of natural material to surround a given volume.

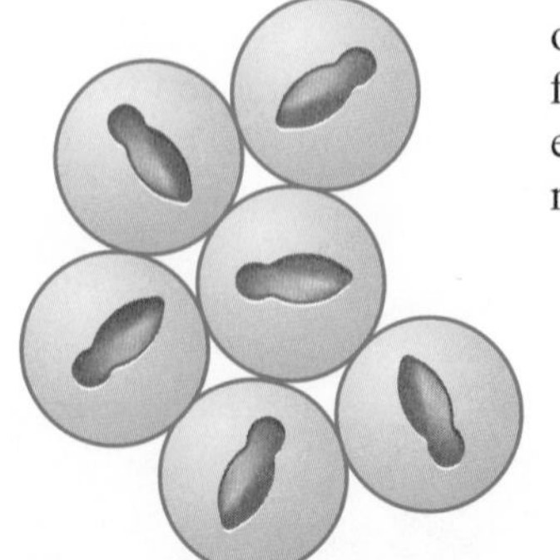

PROBLEM SET 3.5

EXERCISES/PROBLEMS

1. Shown next are base designs for stacks of unit cubes (see problem set in Section 2.4). For each one, determine the volume and total surface area (including the bottom) of the stack described.

a.

3	4	2
1	1	3

b.

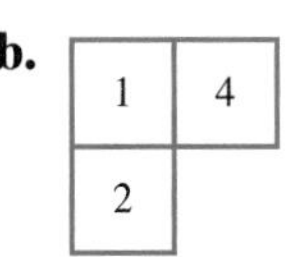

c.

3	3	3
1	2	3
1	2	3

2. Shown next are base designs for stacks of unit cubes (see problem set in Section 2.4). For each one, determine the volume and total surface area (including the bottom) of the stack described.

a.

1	2	4	3

b.

2	3
1	5

c.

1	4	5	6
2	3		

3. Find the volume of each of the following right prisms.

a.

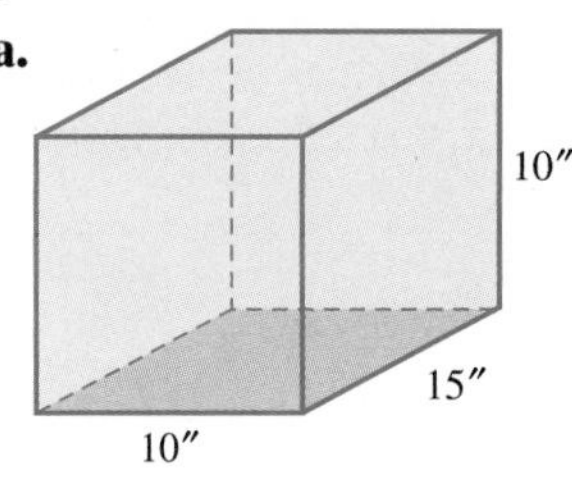

b.

2 cm
5 cm
5 cm
10 cm
4 cm
8 cm

4. Find the volume of each of the following right prisms.

a.

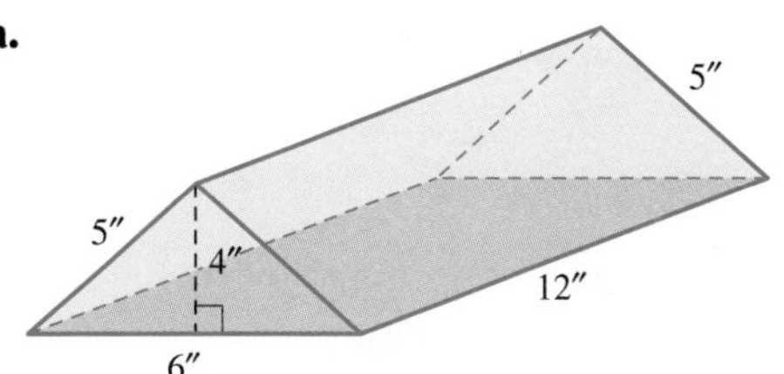

b.

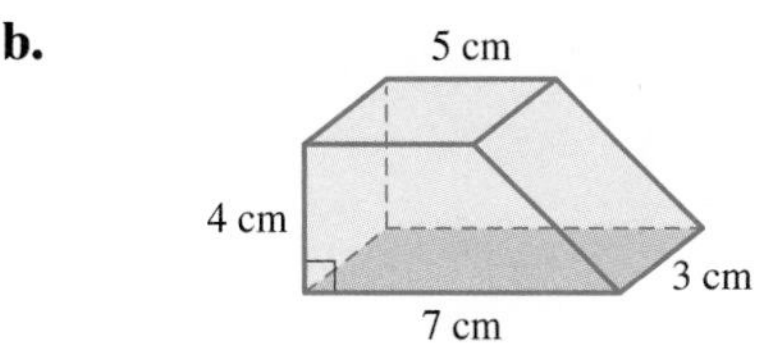

5. Find the volume of each right prism with the given features. Round to the nearest whole number.
   a. The bases are equilateral triangles with sides of length 8 ft, and the height of the prism is 10 ft.
   b. The bases are trapezoids with bases of lengths 7 cm and 9 cm perpendicular to one side of length 6 cm, and the height of the prism is 12 cm.

6. Find the volume of each right prism with the given features.
   a. The base is a right triangle with legs of length 5 in. and 12 in., and the height of the prism is 20 in.
   b. The bases are rhombi with sides of length 12 m and a height of 6 m. The height of the prism is 3.5 m.

7. Find the volume of each of the following pyramids.

a.

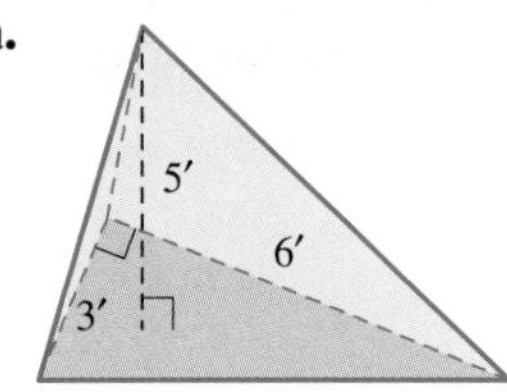

b.

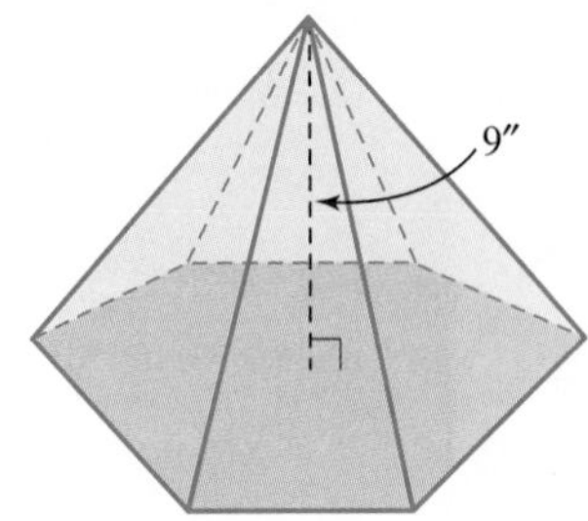

Area of the base is 38 in^2

8. Find the volume of each of the following pyramids.

a.

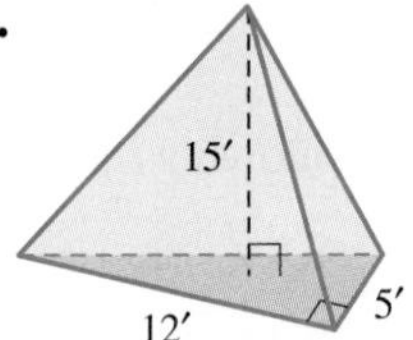

b.

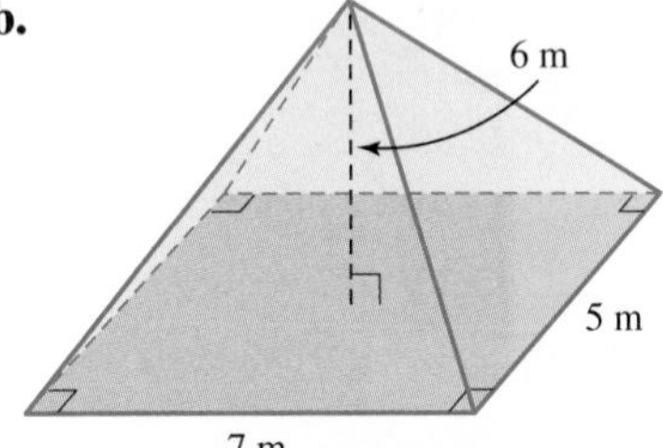

9. Find the volume of a pyramid if the base is a rectangle with side lengths 23.5 cm and 18.7 cm. The height of the pyramid is 9.9 cm.

10. Find the volume of a right pyramid if the base is a regular pentagon such that its side length is 16 ft and the perpendicular distance from its center to one of its sides is 11 ft. The height of the pyramid is 19.2 ft.

11. Find the volume of each of the following cans. Round to the nearest whole number.

a. Coffee can

$r = 7.6$ cm

$h = 16.3$ cm

b. Soup can

$d = 6.6$ cm

$h = 10$ cm

12. Find the volume of each of the following cans. Round to the nearest whole number.

a. Juice can

$r = 5.3$ cm

$h = 17.7$ cm

b. Shortening can

$d = 13$ cm

$h = 14.7$ cm

13. Find the volume of each right circular cone. Round to the nearest whole number.

a.

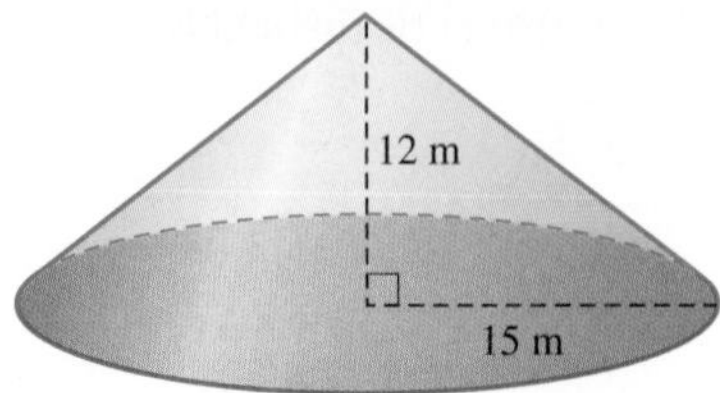

b.

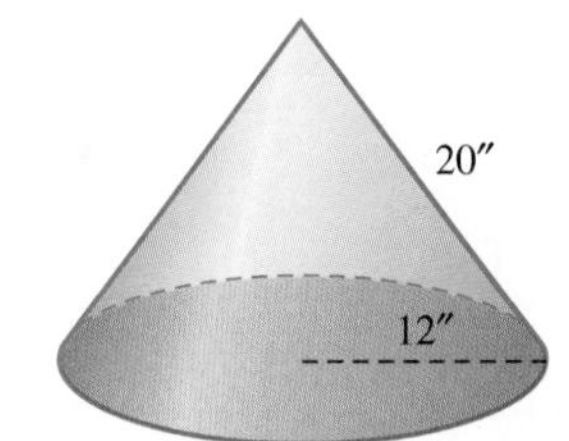

14. Find the volume of each right circular cone. Round to the nearest whole number.

a.

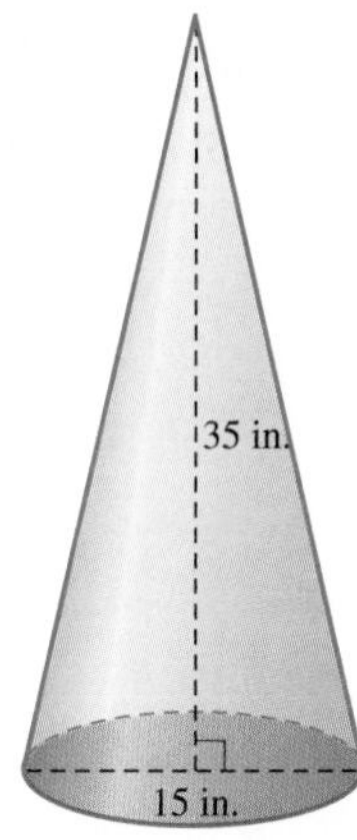

b.

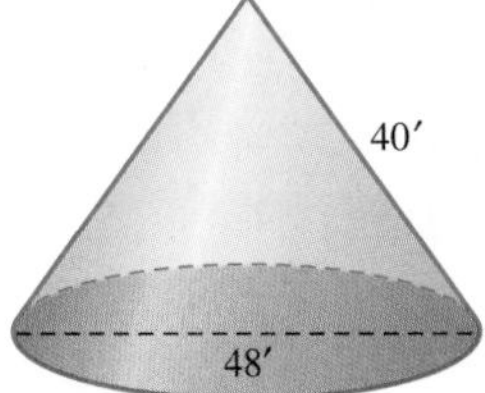

15. Find the volume of the following spheres. Round to the nearest whole number.

a. $r = 6$ in.

b. $d = 24$ m

c. $C = 18\pi$ cm

16. Find the volume of the following spheres. Round to the nearest whole number.

a. $r = 2.3$ in.

b. $d = 6.7$ km

c. $C = 108\pi$ m

17. Use dimensional analysis to perform each of the following conversions. For approximate answers, round to the nearest hundredth.

a. 400 in^3 to ft^3

b. 1.2 m^3 to cm^3

c. 0.4 ft^3 to mm^3

18. Use dimensional analysis to perform each of the following conversions. For approximate answers, round to the nearest hundredth.

a. 115 yd^3 to ft^3

b. 42 mm^3 to cm^3

c. 2.3 m^3 to yd^3

19. The following sketch is a pattern (or net) for a three-dimensional figure. Use the dimensions given to calculate the volume and the total surface area of the three-dimensional figure formed by the pattern. Round to the nearest whole number.

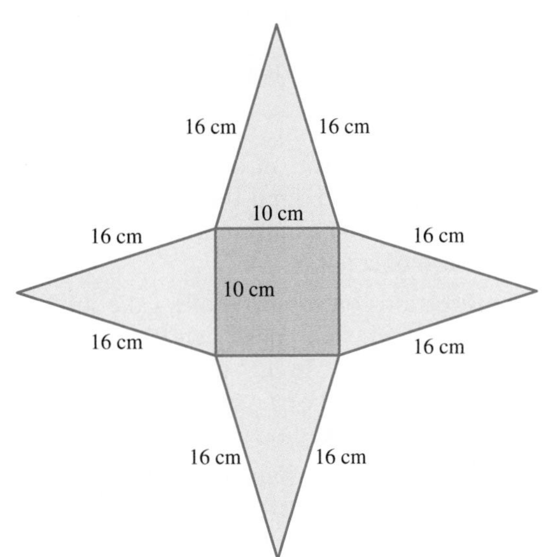

20. The following sketch is a pattern (or net) for a three-dimensional figure. Use the dimensions given to calculate the volume and the total surface area of the three-dimensional figure formed by the pattern. Round to the nearest whole number.

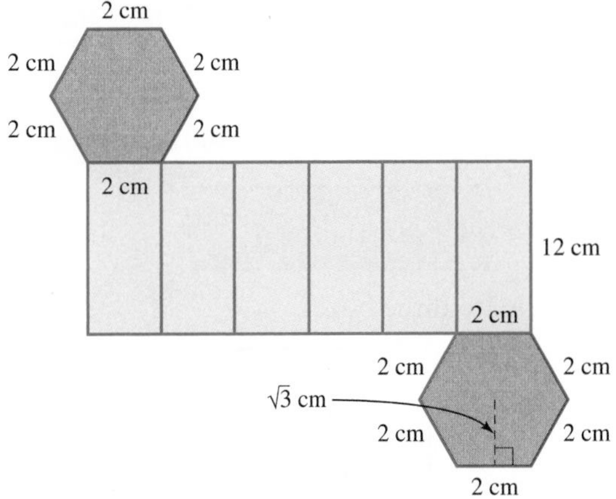

21. Suppose all the dimensions of a square prism are doubled. How would the volume change?

22. How does the volume of a sphere change if its radius is doubled?

23. a. If the side of one square is three times as long as the side of a second square, how do their areas compare?

b. If the side of a cube is three times as long as the side of a second cube, how do their volumes compare?

c. If all the dimensions of a rectangular box are doubled, what happens to its volume?

24. a. How does the volume of a right circular cylinder change if its radius is doubled?

b. How does the volume of a right circular cylinder change if its height is doubled?

c. How does the volume of a right circular cylinder change if its radius and height are both doubled?

25. A right rectangular prism has a volume of 324 cubic units. One edge has a measure twice that of a second edge and nine times that of the third edge. What are the dimensions of the prism?

26. A right circular cone has a volume of 140 in^3. The height of the cone is the same length as the diameter of the base. Find the radius and height. Round to the nearest tenth of an inch.

27. A rectangular piece of $8\frac{1}{2}''$ by $11''$ paper can be rolled into a cylinder in two different directions. If there is no overlapping, which cylinder has the greater volume, the one with the long side of the rectangle as its height, or the one with the short side of the rectangle as its height?

28. Given are three boxes.

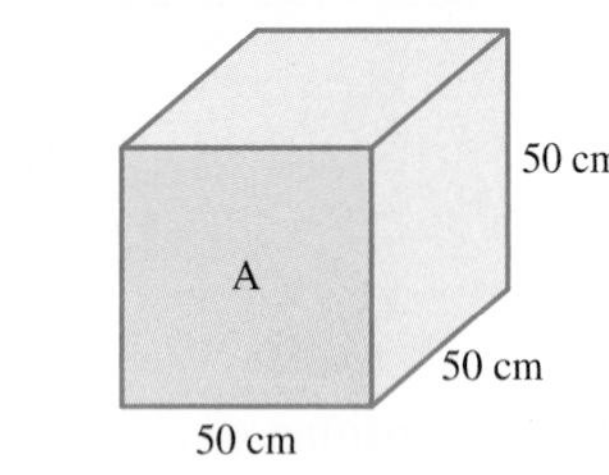

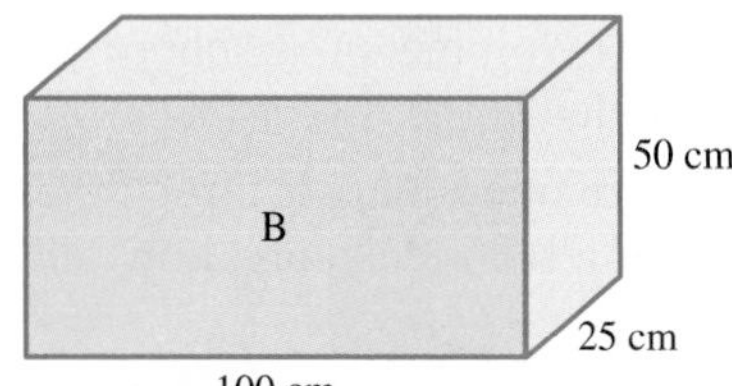

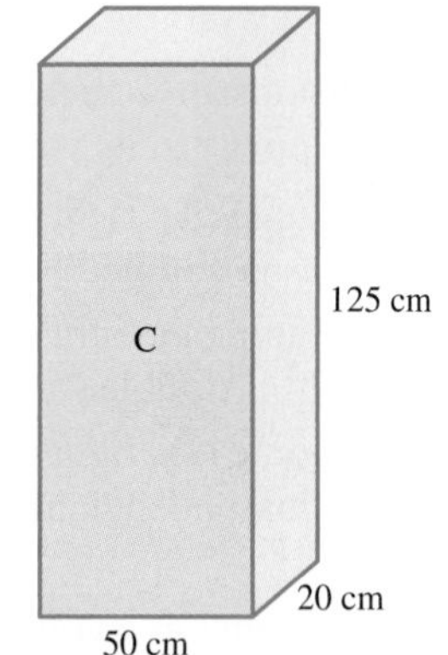

a. Find the volume of each box.

b. Find the surface area of each box.

c. Do boxes with the same volume always have the same surface area?

d. Which box used the least amount of cardboard?

APPLICATIONS

29. The Great Pyramid at Giza is the largest of the original Seven Wonders of the World and was completed in about 2580 B.C.E. A side of its square base originally measured 230.5 m and the original height was 146.5 m. Today, a side length of the base measures 227.5 m and the height is 137 m. How much volume has been lost over time? Round to the nearest whole number.

30. On October 28, 1965, the Gateway Arch in St. Louis was completed as a monument to the spirit of western pioneers. The arch is made up of equilateral triangular sections (see figure) with double walls of steel three feet apart. Space between the walls is filled with reinforced concrete. Sizes of the sections vary, but a section at the bottom of the Arch has 54-foot sides and is 12 feet tall. About how much concrete was used between the double walls in one section at the base of the monument? Round to the nearest whole number.

31. A 4-inch-thick concrete slab is being poured for a circular patio 10 feet in diameter. Concrete costs \$50 per cubic yard. Find the cost of the concrete. (Assume that you cannot purchase a fraction of a cubic yard of concrete.)

32. The directions on the back of a cake mix box call for two round cake pans, each 9″ in diameter, or one rectangular 9″ by 13″ pan. Which method will yield the taller cake? (Assume the two 9″ round cakes are *not* stacked on top of one another.)

33. A waffle cone has a diameter of 9 cm and a height of 15 cm. How much frozen yogurt, to the nearest cubic centimeter, will the cone hold if filled so that the yogurt is level with the top of the cone?

34. Human blood transports oxygen to each cell in the body, removes waste, and helps regulate temperature. Before surgery, it is important to estimate blood volume (BV) to determine the maximum allowable blood loss. We can estimate BV in liters using **Allen's formulas**.

$BV_{\text{women}} = 0.3561(\text{height in m})^3 + 0.03308(\text{weight in kg}) + 0.1833$

$BV_{\text{men}} = 0.3669(\text{height in m})^3 + 0.03219(\text{weight in kg}) + 0.6041$

a. Estimate the BV for a woman who is 5′9″ tall and weighs 147 lb. Round to the nearest tenth.

b. Estimate the BV for a man who is 6′ tall and weighs 192 lb. Round to the nearest tenth.

35. a. If one inch of rain fell over one acre, how many cubic inches of water would that be? How many cubic feet?

b. If one cubic foot of water weighs approximately 62 pounds, what is the weight, in tons, of one inch of rain over one acre of ground?

c. The weight of one gallon of water is about 8.3 pounds. A rainfall of one inch over one acre of ground means about how many gallons of water?

36. A father and his son working together can cut 48 ft^3 of firewood per hour.

a. If they work an eight-hour day and are able to sell all the wood they cut at $100 per cord, how much money can they earn? A **cord** is defined as 4 ft by 4 ft by 8 ft.

b. If they split the money evenly, at what hourly rate should the father pay his son?

c. If the delivery truck can hold 100 ft^3, how many trips would it take to deliver all the wood cut in a day?

d. If they sell their wood for $85 per truckload, what price are they getting per cord?

37. We can determine the volume of an object with an irregular shape by measuring the volume of water it displaces.

a. A rock placed in an aquarium measuring $2\frac{1}{2}$ ft long by 1 ft wide causes the water level to rise 1/4 in. What is the volume of the rock?

b. With the rock in place, the water level in the aquarium is one-half inch from the top. The owner wants to add to the aquarium 200 solid marbles, each with a diameter of 1.5 cm. Will the addition of these marbles cause the water in the aquarium to overflow?

38. The **density** of a substance is the ratio of its mass to its volume.

$$\text{density} = \frac{\text{mass}}{\text{volume}}$$

Density is usually expressed in terms of grams per cubic centimeter (g/cm^3). For example, the density of copper is 8.94 g/cm^3.

a. Express the density of copper in kg/dm^3.

b. A chunk of oak firewood has a mass of about 2.85 kg and has a volume of 4100 cm^3. Determine the density of oak in g/cm^3, rounding to the nearest thousandth.

c. A piece of iron weighs 45 oz and has a volume of 10 in^3. Determine the density of iron in g/cm^3, rounding to the nearest tenth.

39. A sculpture made of iron has the shape of a right square prism topped by a sphere. The metal in each part of the sculpture is 2 mm thick. If the outside dimensions are as shown and the density of iron is 7.87 g/cm^3, calculate the approximate mass of the sculpture in kilograms.

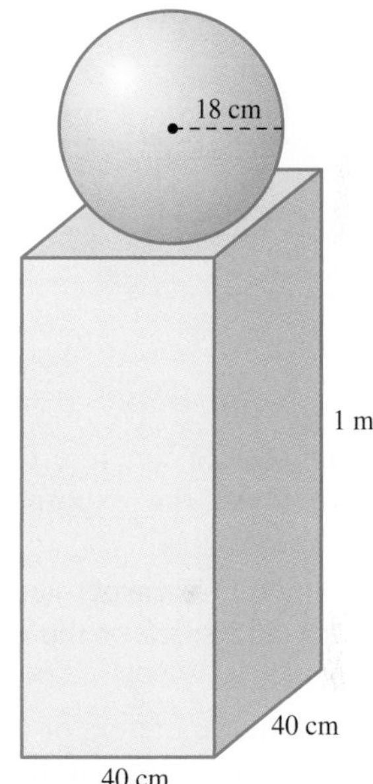

40. While rummaging in his great aunt's attic, Bernard found a small figurine that he believed to be made out of silver. To test his guess, he looked up the density of silver in his chemistry book and found it to be 10.5 g/cm^3. He found that the figure had a mass of 149 g. To determine its volume, he dropped it into a cylindrical glass of water. If the diameter of the glass was 6 cm and the figurine was pure silver, by how much did the water level in the glass rise?

41. A hollow ball has a circumference of 22 cm and is made of rubber which is 0.6 cm thick. Find the volume of rubber in the ball, rounding to the nearest cm^3.

42. A standard tennis ball can is a cylinder that holds three tennis balls.

a. Which is greater, the circumference of the can or its height?

b. If the radius of a tennis ball is 3.5 cm, what percentage of the can is occupied by air?

43. A tank full of a liquid is in the shape of a sphere 6 ft in diameter. If 200 gal of liquid are pumped out of the tank, how many gallons remain in the tank? (Recall that 1 $ft^3 \approx 7.48$ gal.)

44. A cylindrical steel pipe has an inside diameter of 2 in. and an outside diameter of 2.5 in. If the pipe is 15 ft long, how many cubic inches of steel does the pipe contain? If steel weighs 490 lb/ft^3, what is the weight of the pipe? What volume of water will the pipe hold when full? Round to the nearest whole unit.

45. The first three steps of a 10-step staircase are shown.

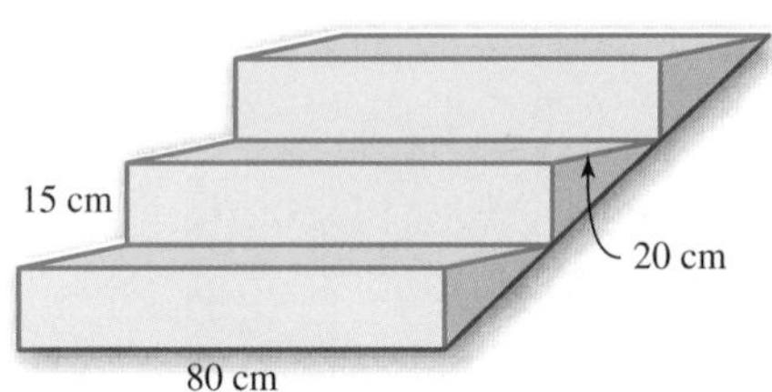

a. Find the amount of concrete, to the nearest cm^3, needed to make the exposed portion of the staircase.

b. Find the amount of carpet needed to cover the fronts, tops, and sides of the concrete steps, to the nearest cm^2.

46. a. How many square meters of tile are needed to tile the sides and bottom of the swimming pool illustrated?

b. How much water does the pool hold?

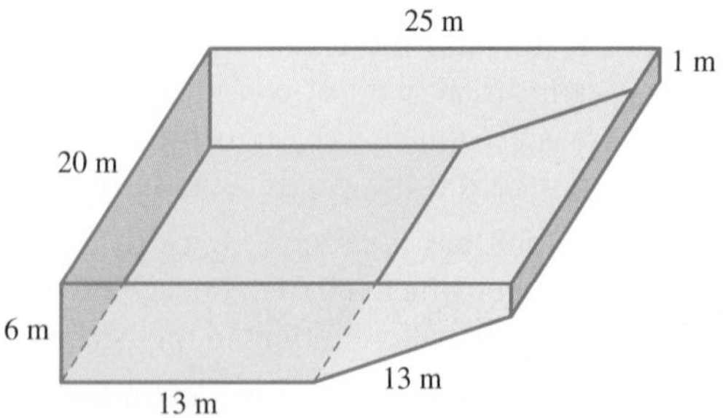

47. Find the volume, in cubic inches, of the simplified I-beam whose cross section is shown and whose length is 25′.

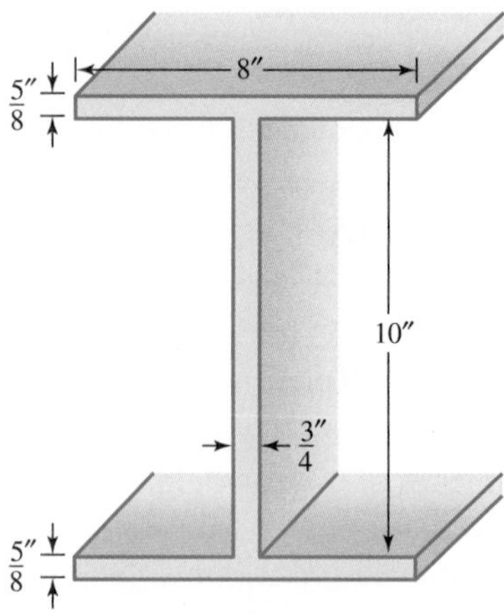

48. At a water treatment plant, a cylindrical digester 25 feet in diameter and 20 feet deep is to be emptied into two rectangular drying beds. If the beds are 60 feet long by 20 feet wide and will be filled to a depth of 15 inches, by how many inches will the level in the digester be lowered?

49. Plywood glue is stored in a tank that has the shape shown. How many cubic feet of glue will this tank hold? Round to the nearest whole number.

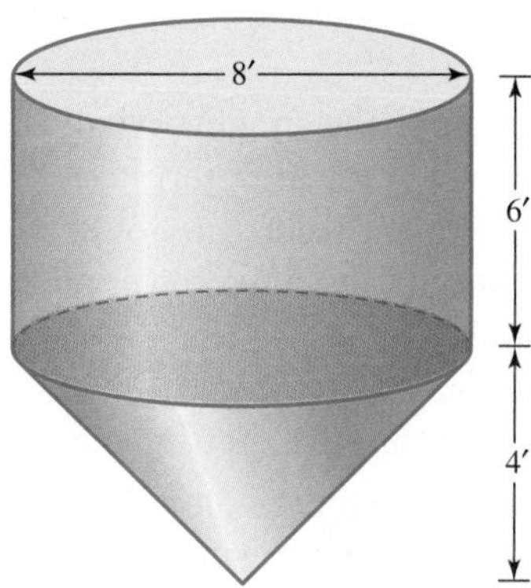

50. Lumber is measured in board feet. A **board foot** is the volume of a square piece of wood measuring 1 foot long, 1 foot wide, and 1 inch thick. A surfaced "two by four" actually measures 1.5 inches by 3.5 inches, a "two by six" measures 1.5 by 5.5, and so on. Plywood is sold in *exact* dimensions and is differentiated by thickness. Find the number of board feet in the following pieces of lumber.

a. 6-foot-long two by four

b. 10-foot two by eight

c. 4-foot by 8-foot sheet of $\frac{3}{4}$-inch plywood

d. 4-foot by 6-foot sheet of $\frac{5}{8}$-inch plywood

51. A farmer's irrigation system sprays water in a circle with a radius of 1320 ft at the rate of 1000 gal/min. At this rate, how long, to the nearest hour, must the sprinkler run to distribute the equivalent of 3/4 in. of rainfall over the area?

52. A bucket has the shape of a truncated circular cone, or a **frustum**. The formula for the volume of a frustum is

$$V = \frac{1}{3}h(B_L + B_S + \sqrt{B_L B_S})$$

where h is the height of the frustum, B_L is the area of the larger base, and B_S is the area of the smaller base. (This formula will be verified in Chapter 6.)

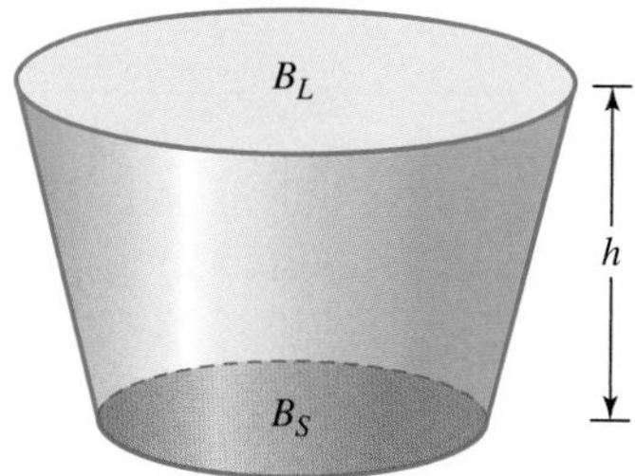

If the diameter of the top of the bucket is 12 in., the diameter of the bottom is 9 in., and the bucket is 10 in. tall, calculate the volume of the bucket in gallons. Use the fact that 1 gal $\approx$ 231 in^3.

53. Two designs for an oil storage tank are being considered: spherical and cylindrical. The two tanks would have the same capacity and would each have an inside diameter of 60 feet.

a. What would be the height of the cylindrical tank?

b. If one cubic foot holds 7.5 gallons of oil, what is the capacity of each tank in gallons?

c. Which of the two designs has the smallest surface area and would thus require less material in its construction?

EXTENDED PROBLEMS

54. In 2005, the world produced and consumed approximately 80 million barrels of crude oil per day. Research the peak oil theory. What volume of oil, in barrels, is estimated to be in reserve worldwide? List the top five oil-producing countries and what percent of known reserves of oil they have. Also, list the five top oil-consuming countries and what percent of the 80 million barrels each uses. What does Hubbert's Peak Oil Theory predict for the future? Then, based on our current rate of usage, determine how long the Arctic Wildlife Oil Reserves would serve the United States, assuming that it was our only source of oil.

55. Suppose you have 10 separate unit cubes. The total volume is 10 cubic units and the surface area is 60 square units. If you arrange the cubes as shown, the volume is still 10 cubic units, but the surface area is only 36 square units. For each of the following problems, assume that all cubes are stacked to form a single shape sharing complete faces (no loose cubes allowed).

a. How can you arrange 10 cubes to get a surface area of 34 square units? Sketch your answer.

b. What is the greatest possible surface area you can obtain with 10 cubes? Sketch your answer.

c. How can you arrange the 10 cubes to get the smallest possible surface area? What is this area?

d. Answer the questions in parts (a), (b), and (c) for 27 and 64 cubes.

e. What arrangement has the greatest surface area for a given number of cubes?

f. What arrangement seems to have the least surface area for a given number of cubes?

g. Biologists have found that an animal's surface area and volume are important factors in its regulation of body temperature. Research animal heat loss and write a summary. Select two animals from hot regions and two animals from cold regions, and discuss how their surface areas and volumes are suited to their climates. Include pictures or sketches.

56. Research the solar system to find the mass and diameter of each planet. Assume each planet is a perfect sphere and calculate the density of each planet using the formula: density $= \dfrac{\text{mass}}{\text{volume}}$. By how much would the density of each planet have to change if each planet increased in size or decreased in size so that it had the same volume as the Earth?

Solution of Initial Problem

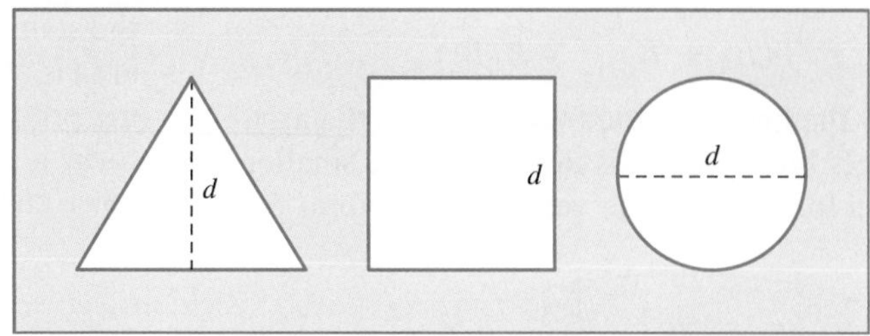

A cylinder whose base has diameter d and whose height is d will pass through the square and circle exactly [Figure 3.45(a)]. We will modify a model of this cylinder to get a shape with a triangular cross section.

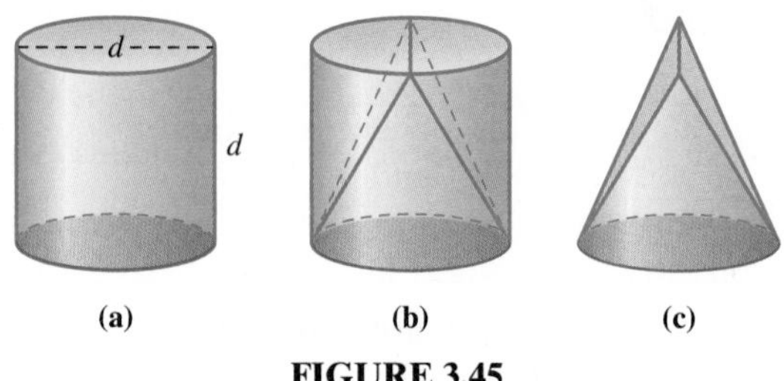

FIGURE 3.45

If we slice the cylinder along the heavy lines in Figure 3.45(b), the resulting model will pass through the triangular hole exactly. Figure 3.45(c) shows the resulting model that will pass through all three holes exactly.

Additional Problems Where the Strategy "Use a Model" Is Useful

1. A store is promoting a sale on cola, and a display is being set up at the end of an aisle. Workers have arranged 144 six-packs in a 12 by 12 square on the floor to form the first layer of a square pyramid. What is the greatest number of six-packs that can be stacked on top of this to form a pyramid?
2. If we slice a cube with a plane parallel to one of the faces, the cross section that is revealed will be a square. Describe how to slice a cube with a plane to obtain each of the following cross sections. (NOTE: Only *one* cut is allowed in each case.)
   a. A rectangle
   b. An isosceles triangle
   c. A rhombus other than a square

Writing for Understanding

1. Describe how you would solve the following Fermi problem (see Chapter 2, "Writing for Understanding," for a description of a Fermi problem): Suppose two shoes for each resident of your state were lined up toe to heel. About how long in meters would this line of shoes be? What would be the dimensions of the smallest cubical box that could contain all of the shoes? Explain your reasoning.

2. Discuss how a change in the dimensions of a figure affects the area, perimeter, volume, and surface area where appropriate. For example, what happens when one dimension is doubled? Include a rationale and specific examples in your answer.
3. The Pythagorean theorem is one of the most famous theorems in mathematics. Why do you think that it is so well-known and what makes it so useful? Include examples in your discussion.
4. The book *The Pythagorean Proposition* contains more than 100 different proofs of the Pythagorean theorem. Find this book in your library. Select three of your favorite proofs and explain why you prefer them.

PEOPLE IN GEOMETRY

Pythagoras (circa 580–500 B.C.E.) was a Greek mathematician who founded a mystical school in southern Italy. The members of the school, known as the Pythagoreans, were a brotherhood with secret religious rites. Their motto was "all is number." They studied the properties of numbers, believing that whole numbers were the basis of the structure of the universe. The Pythagoreans were the first to discover irrational numbers, numbers that cannot be expressed as the quotient of two whole numbers. This discovery was deeply disturbing for the Pythagoreans, for it invalidated some of their mathematical proofs and challenged the assumption that whole numbers were the ultimate reality. Of course, Pythagoras is most famous for the theorem named after him. Although he likely did not discover the Pythagorean theorem, he may have been the first to prove it.

CHAPTER REVIEW

For each section of this chapter, you will find a list of vocabulary and notation, questions to assess your understanding of key concepts, and review problems similar to the problems you worked in your homework. Review each item in the *Vocabulary/Notation* list mentally, and, if necessary, refer back to the indicated page and write a definition. Then answer the *Concept Check Questions*, looking back at the section if you need help. Work the *Review Problems* as practice before you move on to the *Chapter Test.* Answers to the *Review Problems* and *Chapter Test* can be found at the back of the book.

SECTION 3.1 Perimeter, Area, and Volume

Vocabulary/Notation

Perimeter 106
Circumference 107
Pi (π) 108
Irrational number 108
Area 109
Altitude of a triangle 111
Height of a triangle 111
Base of a triangle 111

Concept Check Questions

1. If every square is a parallelogram, why are their perimeter formulas different? 106–107
2. How is the number π related to circles? 107–108
3. What does it mean to say that 3.14 is approximately equal to π? 108
4. How can you use a tessellation to determine the area of a simple closed curve? 108–109
5. What does it mean to say that a region has an area of x square units? 109

6. How can you derive the area formula for a triangle from the area formula for a rectangle? 110
7. How are the base and altitude of a triangle related? 111
8. How many bases does a triangle have? 112
9. How many altitudes does a triangle have? 112
10. What is the difference between converting units of length and units of area? 113

Review Problems

1. Find the perimeter of each of the following.
   a. A rhombus with a side length of 3.2 mm.
   b. A parallelogram with side lengths 9.5 cm and 2.1 cm.
   c. A kite with side lengths of 12 ft and 20 ft.
   d. A circle with diameter 6 cm. Round to the nearest tenth.
   e. An isosceles trapezoid whose bases have lengths 4.24 ft and 7 ft and whose legs have length 2.5 ft.
2. A professional football field measures 360 feet by 160 feet.
   a. Find the perimeter in yards. Round to the nearest tenth.
   b. Find the area in acres. Round to the nearest hundredth.
3. There are many variations of the children's game of hopscotch. One diagram of the playing surface is given next. Spaces 1 and 2 are squares. Spaces 3, 4, 5, and 6 are identical isosceles right triangles with leg lengths of $30\sqrt{2}$ inches. Space 7 is a rectangle, and space 8 is a semicircle.

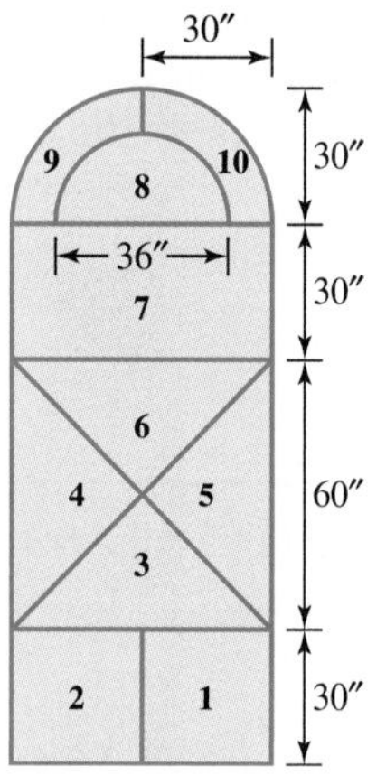

   a. If the hopscotch playing surface will be laid out in tape, then how much tape will be needed? Round to the nearest whole number.
   b. Determine the area of space 3 in square feet.
4. A rectangular picture frame measures 8 inches by 10 inches. A square picture frame has the same perimeter. What is the area inside the square frame?
5. If the circumference of a circle is 10 inches, then find the radius.
6. What happens to the circumference of a circle if the radius is doubled?
7. Find the perimeter and area of each of the following triangles.
   a.

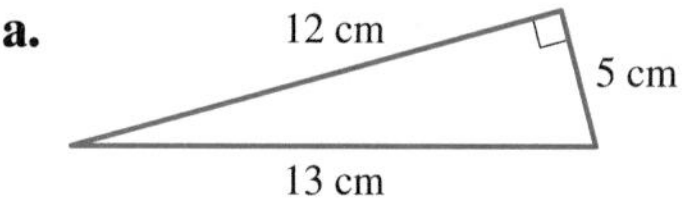

   b. All measurements are in millimeters. Round to the nearest tenth.

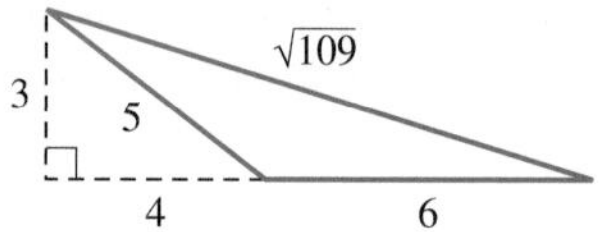

SECTION 3.2 More Area Formulas

Vocabulary/Notation

Height of a parallelogram 122
Base of a parallelogram 122
Height of a trapezoid 123
Center of a regular polygon 124
Apothem 124

Concept Check Questions

1. How is the area formula for a parallelogram different from that of a rectangle? 122
2. How is the area formula for a triangle used to create the area formula for any trapezoid? 123
3. The area formula for a trapezoid requires the lengths of the two bases. Does it matter how the bases are labeled? 123
4. How can you subdivide a regular n-gon to form n congruent triangles? 124
5. Why does the area formula for a regular polygon use the perpendicular distance from its center to one of its sides? 124
6. How can you generalize the area formula for a regular polygon to create the area formula for a circle? 125

Review Problems

1. Find the area of each figure.

a.

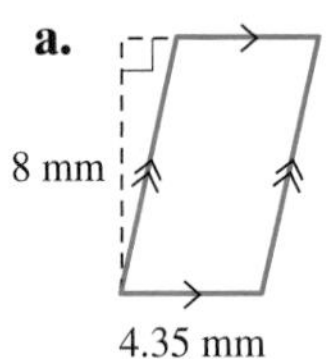

b.

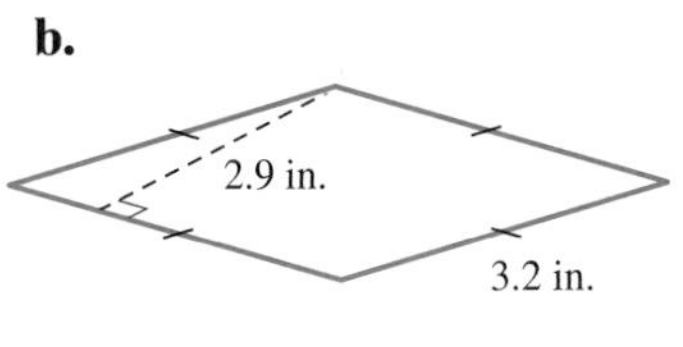

2. Find the area of each figure.

a.

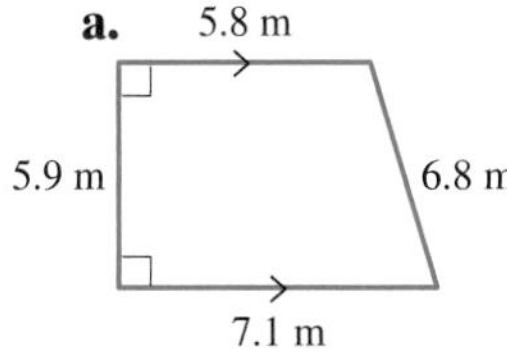

b.

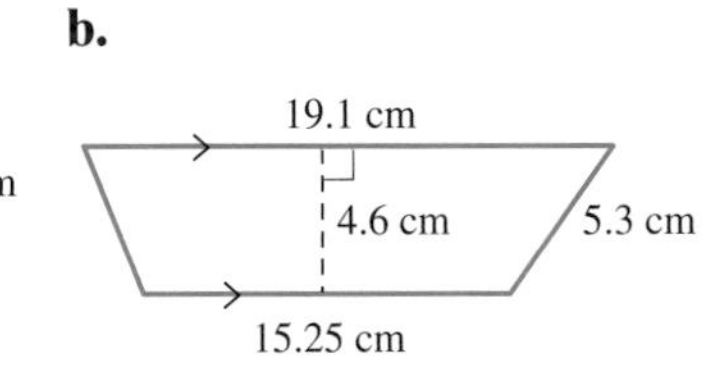

3. Find the area of a semicircle with diameter 5 cm. Round to the nearest tenth.
4. Find the area of a circle with circumference 19.25 in. Round to the nearest hundredth.
5. A stop sign is a regular octagon. The dimensions for a stop sign can be as small as 24 in. by 24 in. with a side length of 9.94 in. or as large as 48 in. by 48 in. with a side length of 19.88 in. as shown in the following figure. What is the difference in area between the largest and smallest stop signs? Round to the nearest whole number.

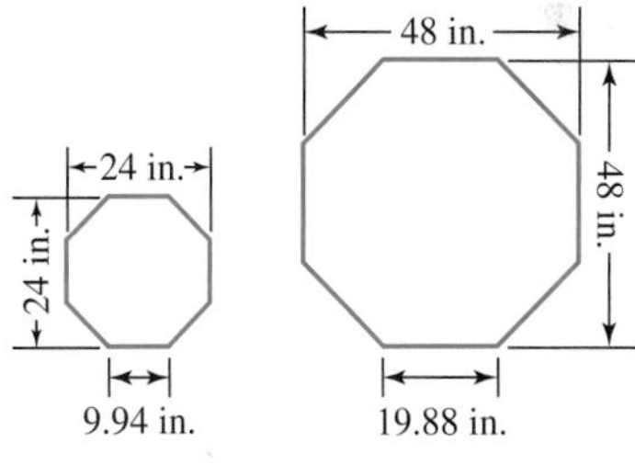

6. A regular pentagon with side length 6.11 mm is inscribed in a circle with radius $3\sqrt{3}$ mm. Find the area of the shaded region if $OP = 4.2$ mm. Round to the nearest thousandth.

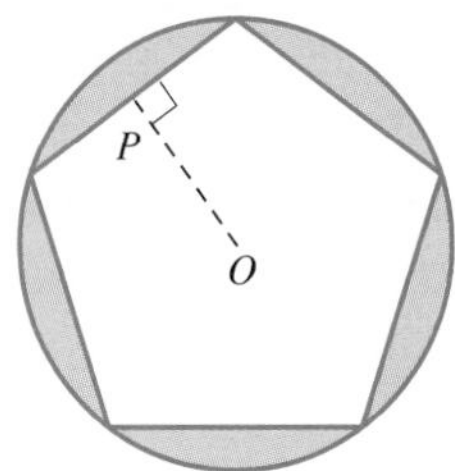

SECTION 3.3 The Pythagorean Theorem and Right Triangles

Vocabulary/Notation

30°-60° right triangles 137
45°-45° right triangles 139

Concept Check Questions

1. How are the lengths of the legs and the hypotenuse of a right triangle related? 134
2. How can you use a test based on the Pythagorean theorem to determine whether a triangle is a right triangle? 137

3. What special right triangles are formed by a line of symmetry for an equilateral triangle? 137
4. How are the lengths of the legs and the hypotenuse of a 30°-60° right triangle related? 138
5. What special right triangles are formed by a line of symmetry for a square? 139
6. How are the lengths of the legs and the hypotenuse of a 45°-45° right triangle related? 139

Review Problems

1. Verify that a triangle with leg lengths 12 cm and 35 cm and hypotenuse length 37 cm is a right triangle.
2. Is a triangle with side lengths of 3.1, 4.6, and 7.3 cm a right triangle? If not, then what kind of triangle is it?
3. Find the missing side length in each of the following right triangles. Round to the nearest hundredth.
   a. A leg has length 10.22 in. and the hypotenuse has length 19.5 in.
   b. The two legs have lengths $\sqrt{5}$ cm and $\sqrt{11}$ cm
4. A right triangle has legs of lengths of $x - 9$ and $3x$ and hypotenuse of length $3x + 1$. What is the value of x?
5. a. A right triangle has a 5-in. hypotenuse and a 2.5-in. leg. What are the measures of the two acute angles in the triangle and the exact length of the other leg?
   b. Both legs of a right triangle measure 7 cm. What are the measures of the two acute angles in the triangle and the exact length of the hypotenuse?
6. The distance from the center of a regular hexagon to a vertex is 6 meters. Find the area of the hexagon. Round to the nearest tenth.
7. In the triangle shown, $\angle P = 45°$, $\angle Q = 105°$, and $QR = 12$ inches. Find the lengths PQ and PR. Round to the nearest tenth.

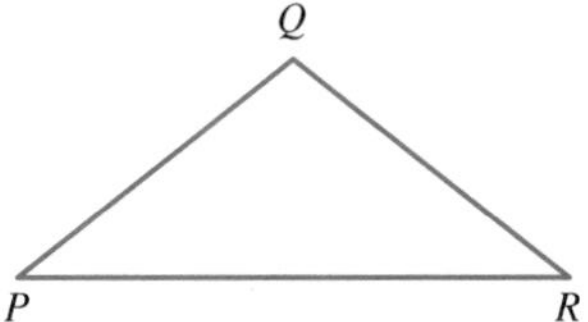

SECTION 3.4 Surface Area

Vocabulary/Notation

Surface area 147

Concept Check Questions

1. What does it mean to find the surface area of a polyhedron? 147
2. What does each variable in the surface area formula for a right prism, $2A + Ph$, represent? 148
3. How can you find the surface area of a pyramid with a base that is not regular? 149
4. How is the formula for the surface area of a right prism modified to create the surface area formula for a right circular cylinder? 149
5. How is the formula for the surface area of a right pyramid modified to create the surface area formula for a right circular cone? 150

Review Problems

1. Find the surface area of a right rectangular prism that measures 3.5 m by 2.8 m by 9.1 m.
2. Find the surface area of a right regular hexagonal prism. The base has a side length of 10 in. and the height of the prism is 14 in. Round to the nearest tenth.
3. A spherical balloon is inflated to a diameter of 10 cm. Air will be let out of the balloon. What must the diameter of the smaller balloon be if its surface area will be one-third as large as the original? Round to the nearest hundredth.
4. Find the surface area of a right square pyramid with base area 1056.25 m^2 and height 45 m. Round to the nearest hundredth.
5. An open-topped settlement tank for treating raw sewage is shown. A corrosion-resistant liner will be

installed in the tank. How many square meters of liner will be needed? Round to the nearest whole number.

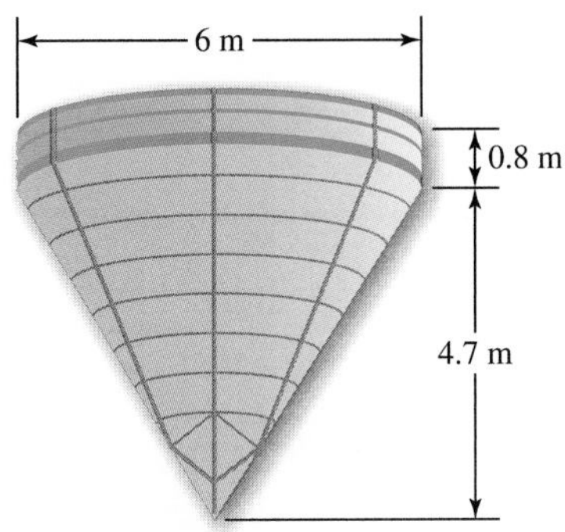

6. A 2″ by 2″ by 2″ block of wood has a 1″ diameter hole drilled through the center. Find the total surface area. Round to the nearest tenth.

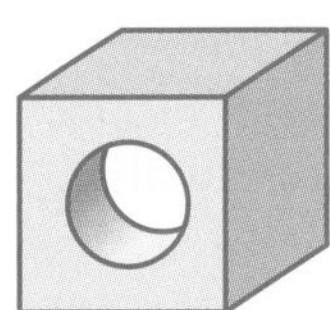

7. Use dimensional analysis to perform each of the following conversions. Where answers are approximate, round to four decimal places.
   a. 38,500 mm^2 to m^2
   b. 0.00074 mi^2 to in^2
   c. 10 m^2 to yd^2

SECTION 3.5 Volume

Vocabulary/Notation

Volume 159

Concept Check Questions

1. How is the formula for the volume of a right rectangular prism related to the formula for the volume of a cube? 161
2. How can you modify the formula for the volume of a prism to create the formula for the volume of a pyramid? 161–162
3. How can you derive the formula for the volume of a right circular cylinder from the formula for the volume of a prism? 163
4. How can you derive the formula for the volume of a right circular cone from the formula for the volume of a pyramid? 163
5. How is the formula for the volume of a right circular cylinder related to the formula for the volume of a sphere? 164
6. How is converting units of volume different from converting units of length? 166

Review Problems

1. Find the volume of each of the following prisms. Round to the nearest whole number.
   a. The base is a right triangle with legs of length 2.1 cm and 6.4 cm. The height is 15 cm.
   b. The base is a regular hexagon with side length 9 in. The height is 6 in.
2. The diameter of a sphere is 5 cm. Find the height of a right circular cylinder that has the same volume as the sphere if the diameter of the cylinder is the same as its height. Round to the nearest hundredth.
3. Use dimensional analysis to perform each of the following conversions. Where answers are approximate, round to the nearest thousandth.
   a. 0.015 km^3 to m^3
   b. 650π in^3 to yd^3
   c. 1800 mm^3 to in^3
4. A 20-m length of gutter has a cross section that is a trapezoid as shown. If the downspouts are clogged with leaves, then find the volume of water, in cubic meters, the gutter will hold before it overflows. Round to the nearest hundredth.

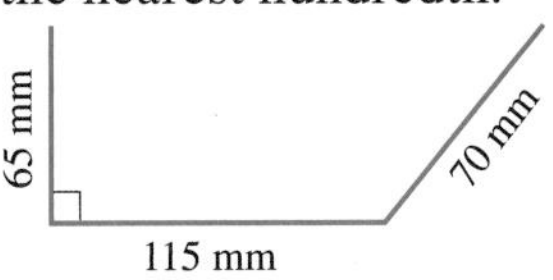

5. A 2″ by 2″ by 2″ block of wood has a 1″ diameter hole drilled through the center. Find the total volume. Round to the nearest tenth.

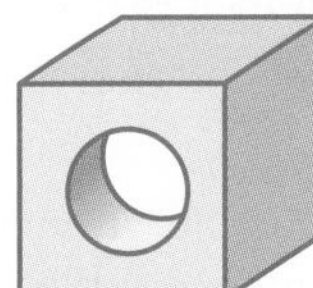

6. The Pyramid Arena in Memphis, Tennessee, has a square base that measures approximately 544 feet on a side. The pyramid is 321 feet high.
   a. Calculate its volume.
   b. Calculate its lateral surface area and round to the nearest whole number.

7. A soft-drink cup is in the shape of a right circular cone with capacity 250 cm^3. The radius of the circular base is 5 cm. How deep is the cup? Round to the nearest tenth.

CHAPTER 3 TEST

TRUE-FALSE

Mark as true any statement that is always true. Mark as false any statement that is never true or that is not necessarily true. Be able to justify your answers.

1. The formula for the area of a rectangle is $A = 2l + 2w$.
2. If two sides of a triangle are known, the third side can always be found using the Pythagorean theorem.
3. To find the area of a trapezoid, you need the height of the trapezoid and the lengths of each of the parallel sides.
4. If two rectangles have equal perimeters, then their areas are equal.
5. If the formula for the volume of a three-dimensional shape is $\frac{1}{3}Ah$, then the shape is a cone.
6. If a sphere and the base of a cylinder have the same radius and the height of the cylinder is the same as the diameter of its base, then their surface areas are the same.
7. Every altitude of a triangle has the same length.
8. There are infinitely many great circles of a sphere.
9. A triangle with sides measuring 6.75 cm, 17.55 cm, and 16.2 cm is a right triangle.
10. If the circumference of a circle is doubled, then the area of the circle is also doubled.

EXERCISES/PROBLEMS

11. Express the formula for the area of a circle in terms of its
    a. Diameter
    b. Circumference
12. An isosceles right triangle has an area of 32 cm^2. Find the lengths of its legs.
13. a. Find the volume of the right square pyramid shown.
    b. What happens to the volume if the length of a side of the square is doubled?

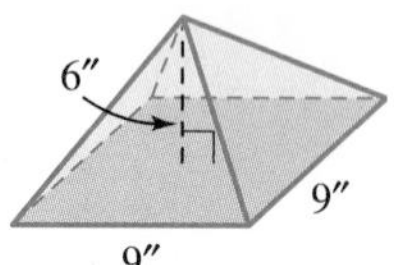

    c. What happens to the volume if the length of a side of the square is tripled?
    d. Find the surface area of the pyramid.
14. Develop a formula for the area of a regular hexagon with side length x by partitioning it into two trapezoids.
15. Complete the following table for a sphere:

	Diameter	Radius	*SA*	Circumference	*V*
a.				56π	
b.			100π		
c.					972π
d.	6				

16. Find the area and perimeter of the following figure where the curved piece is a semicircle. Round to two decimal places.

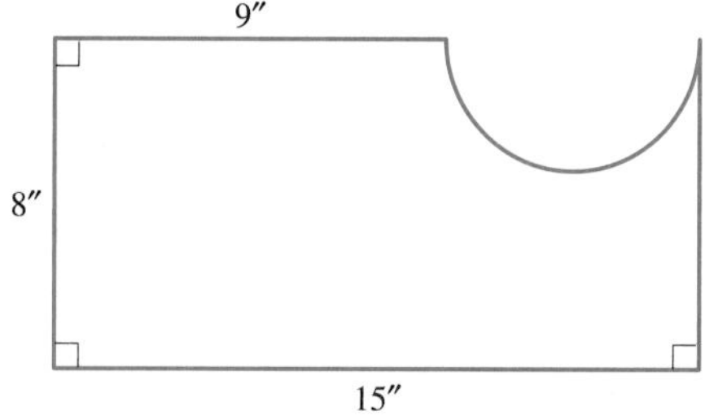

17. A new car owner needs to convert the following measurements:

    Weight: 3000 lb to kg
    Length: 4000 mm to ft
    Tire pressure: 27 lb/in^2 to g/cm^2

Use dimensional analysis to perform each of these conversions given 1 in. = 2.54 cm and 1 kg ≈ 2.2 lb.

18. Find the area of the figure shown on the square lattice.

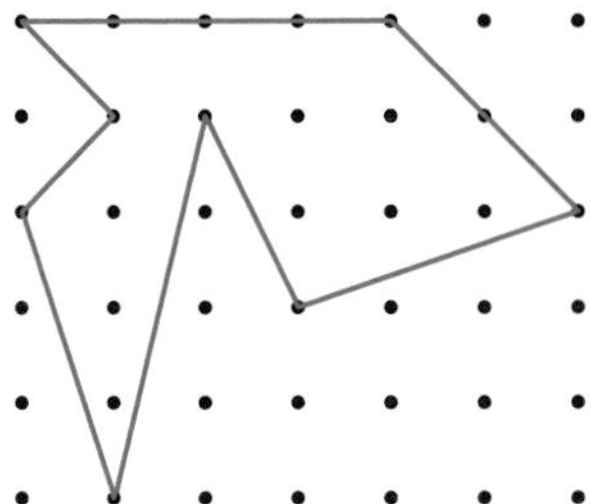

19. Find the surface area and volume of the solid figure shown, rounding to the nearest whole number.

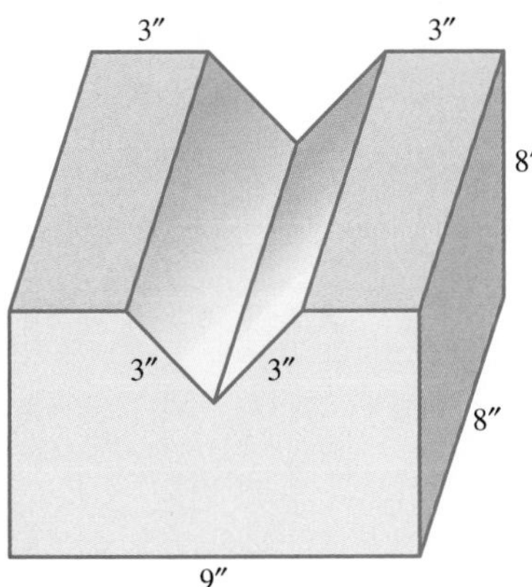

APPLICATIONS

20. A piece of rubber tubing is in the shape of a hollowed-out cylinder. If the outside radius is 24 inches, the inside radius is 20 inches, and the length of the tube is 10 feet, find the total surface area of the tube.

21. Alice and Bob took a 20-mile bicycle ride together. Alice rode a touring bicycle and Bob rode his mountain bike. If the outside diameter of the tires on Alice's bike is 27 inches and on Bob's bike is 26 inches, how many more revolutions did Bob's wheels make on the trip than Alice's wheels?

22. An Oregon wheat silo is in the shape of a right circular cylinder with a hemisphere on the top. If the radius of the cylinder is 10 feet and the cylinder is 70 feet tall, what volume of wheat can be stored inside?

23. The Great Wall of China is about 1500 miles long. The cross section of the wall is a trapezoid 25 feet high, 25 feet wide at the bottom, and 15 feet wide at the top. How many cubic yards of material make up the wall?

24. A boy flying a kite has 100 feet of string out when the kite lodges in a tree. He reels in 20 feet of the string as he walks 25 feet closer to the tree. How high in the tree is his kite stuck? Round your answer to the nearest foot.

CHAPTER 4

REASONING AND TRIANGLE CONGRUENCE

EUCLID—THE FATHER OF GEOMETRY

Euclid of Alexandria (circa 300 B.C.E.) has been called the "father of geometry." Euclid's major contribution to mathematics came when he organized much of the geometry of his day into one book called *The Elements. The Elements* consisted of 13 "books," 5 on plane geometry, 3 on solid geometry, and 5 on geometric explanations of the mathematics now studied in algebra. Because much of the geometry he organized continues to be the heart of high school geometry courses, it is often titled Euclidean geometry. His most famous original contribution to the book was a proof of the Pythagorean theorem. The diagram he used to prove it, sometimes called the "windmill," is displayed below.

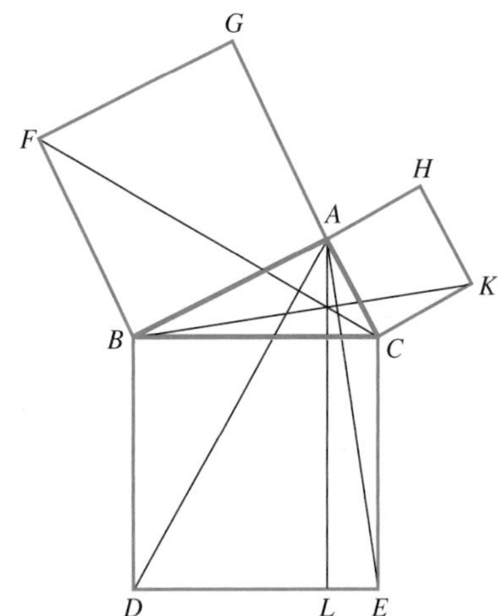

Problem-Solving Strategies

1. Draw a Picture
2. Guess and Test
3. Use a Variable
4. Look for a Pattern
5. Make a Table
6. Solve a Simpler Problem
7. Look for a Formula
8. Use a Model
9. ***Identify Subgoals***

Strategy 9: *Identify Subgoals*

Often we can solve a problem more easily by breaking it down into smaller problems. Then we solve the original problem by solving the various parts, called **subgoals**, and assembling those solutions in a meaningful sequence.

Initial Problem

The following figure shows a regular octagon adjacent to a regular pentagon. Find the measure of $\angle ABC$.

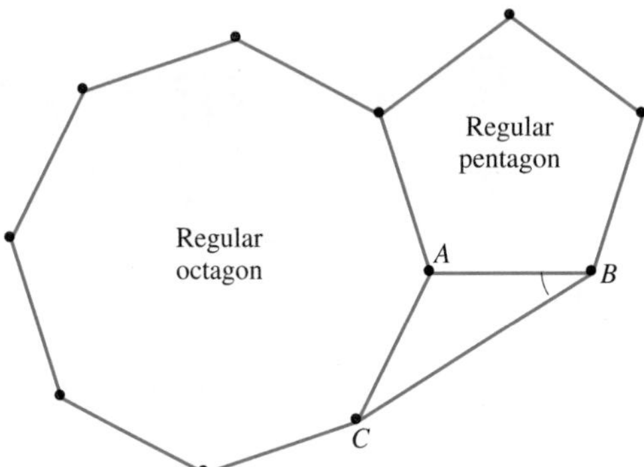

Clues

The Identify Subgoals strategy may be appropriate when

- You can break down a problem into a series of simpler problems.
- The statement of the problem is very long and complex.
- You can say, "If I only knew . . . , then I could solve the problem."
- There is a simple, intermediate step that would be useful.
- You wish the problem contained other information.

A solution of this Initial Problem is on page 228.

Hexagon Construction

Using a compass, construct a circle and label the center as point O—call the radius of the circle 1 unit. Keep the compass open to the same setting throughout this construction. Set the point of the compass anywhere on the circle and label this point A. Using your pencil point, swing an arc that intersects the circle—label this point B. (See the figure).

Set the point of the compass on point B and swing another arc that intersects the circle in a point other than A. Label this point C. Continue making arcs around the circle until you have labeled point F.

Then set the point on F and swing an arc toward A. Did your pencil point arc coincide with point A? If not, repeat the procedure on another circle. Once points A and F coincide, use a straightedge to connect point A to B, B to C, $\ldots$, and F to A. Then connect each point to the center O to create six triangles.

Based on the kind of triangles that you formed and their vertex angle measures, explain why your compass pencil point had to have ended precisely at A after forming six arcs. Also, explain why hexagon $ABCDEF$ is regular.

What is the perimeter of the hexagon? Is the circumference of the circle larger or smaller than the perimeter of the hexagon? Why?

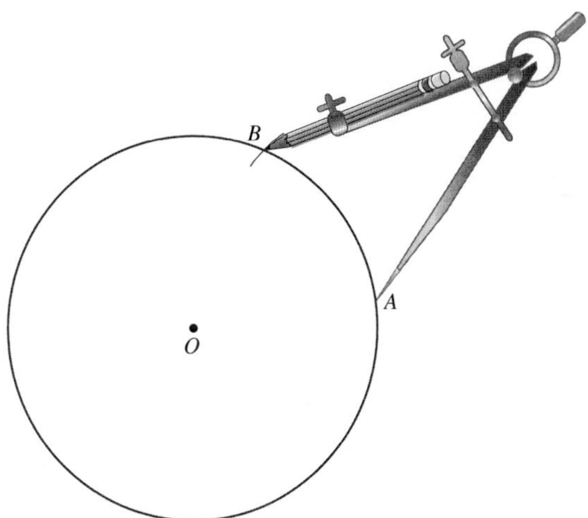

Using a protractor, draw lines that are perpendicular to each of the six radii at points A, B, C, D, E, and F. Using these six lines, form a hexagon surrounding the circle.

Use what you know about 30°–60° right triangles to find the perimeter of this hexagon surrounding the circle.

Using the perimeters of the inside and the outside hexagons, estimate the circumference of the circle. How does your estimate compare with the actual circumference of 2π? Describe a way to get a closer estimate.

Introduction

In Chapters 1–3, we often derived results from known results by reasoning informally. In this chapter we begin to reason more formally by constructing proofs. We apply these proofs to the study of triangle congruence. We then use the triangle congruence results to draw conclusions about other shapes in later chapters. In Section 4.1, we discuss reasoning and proof. In Sections 4.2 and 4.3, we study congruent triangles and relationships that can be derived from them. Section 4.4 focuses on geometric constructions using a compass and straightedge. Note that we will use some of the results that we proved informally in Chapters 2 and 3 even though we will not prove them formally until Chapters 5 or 6. This approach has the virtue of allowing us to explore a richer collection of applications, yet it does not introduce any circular reasoning.

4.1 REASONING AND PROOF IN GEOMETRY

Applied Problem

A surveyor lays out a traverse with four vertices. As measured, the interior angles of the quadrilateral would be as shown. How can we use deductive reasoning to determine whether the traverse "closes"?

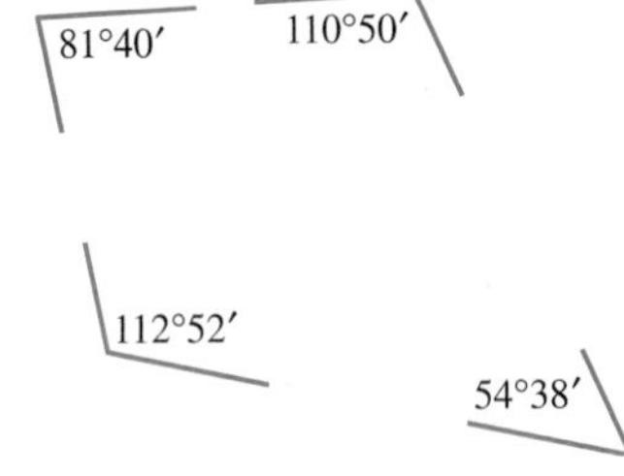

Conditional Statements and Deductive Reasoning

In the first three chapters, we solved many problems using direct reasoning. **Direct reasoning**, or **deductive reasoning**, is used to draw a conclusion from a series of statements where statements are often represented by single letters such as p and q. **Conditional statements**, statements of the form "if p, then q," play a central role in direct reasoning. The conditional "if p, then q" is written symbolically as "$p \Rightarrow q$." The statement "p" is called the **hypothesis** of the conditional and "q" is called its **conclusion**. We also read "$p \Rightarrow q$" as "p implies q" or "p only if q."

EXAMPLE 4.1 What can we deduce from the following two statements?

1. If it is raining, then the street is wet.
2. It is raining.

SOLUTION

In conditional statement (1), p, the hypothesis, is the statement "it is raining," and q, the conclusion, is the statement "the street is wet." Because (2) states that it is raining—that is, that p holds—we can conclude from conditional (1) that q also must hold, or that the street is wet. ■

Example 4.1 illustrates one of the most common forms of direct reasoning. If we let p represent "it is raining" and q represent "the street is wet," then we may represent Example 4.1 and its solution as follows:

If p, then q.		$p \Rightarrow q$
p	or symbolically as	p
Therefore q.		$\therefore q$

The **three-dot triangle**, namely $\therefore$, is a symbolic abbreviation for the word "therefore." The preceding form of argument is called the **Law of Detachment**. Notice how the q is "detached" from the conditional $p \Rightarrow q$ once we know p.

Besides displaying conditional statements in verbal and symbolic form, we also can represent them visually using what are called **Venn diagrams** or **Euler circles** (Figure 4.1).

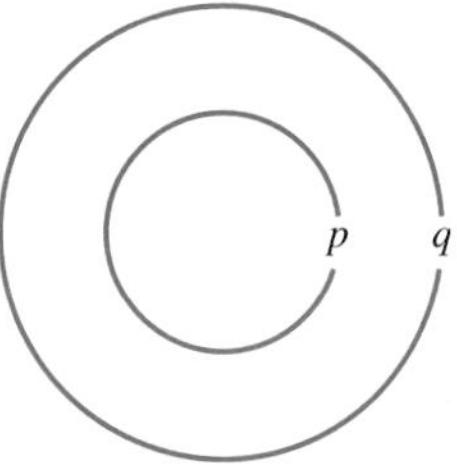

FIGURE 4.1

In Figure 4.1, because the interior of circle p is contained within the interior of circle q, we can say this diagram represents "if p, then q."

A second common form of direct reasoning, called the Law of Syllogism, plays a central role in mathematics. The **Law of Syllogism** relates several if . . . , then statements and takes the following form:

If p, then q.		$p \Rightarrow q$
If q, then r.	or symbolically	$q \Rightarrow r$
Therefore if p, then r.		$\therefore p \Rightarrow r$

EXAMPLE 4.2 What can we deduce from the following two statements?

1. If it is raining, then the street is wet.
2. If the street is wet, then the street is slippery.

SOLUTION

Here, p is the statement "it is raining," q is the statement "the street is wet," and r is the statement "the street is slippery." Both (1) and (2) are conditionals *and* the conclusion of (1) is the same as the hypothesis of (2). Therefore, by the Law of Syllogism, we can conclude that "if it is raining, then the street is slippery." ■

The Laws of Detachment and Syllogism are commonly used together. Consider the statements presented in the next example.

EXAMPLE 4.3 What can we deduce from the following statements?

1. If the circumference of a circle is 8π inches, then its diameter is 8 inches.
2. If the diameter of a circle is 8 inches, then its radius is 4 inches.
3. If the radius of a circle is 4 inches, then its area is 16π square inches.
4. The circumference of circle O is 8π inches.

SOLUTION

From statements (1) and (2), we know by the Law of Syllogism that if the circumference of a circle is 8π, then its radius is 4 inches. This result with statement (3) leads us to conclude that if the circumference of a circle is 8π, then its area must be 16π square inches. This statement with statement (4) and the Law of Detachment leads us to finally conclude that the area of circle O is 16π square inches. ■

We can present the verbal argument in Example 4.3 symbolically as follows:

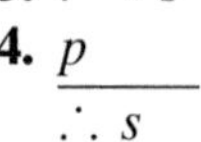

1. $p \Rightarrow q$
2. $q \Rightarrow r$
3. $r \Rightarrow s$
4. $\underline{p}$
$\therefore s$

Using (1) and (2), we can deduce that $p \Rightarrow r$ using the Law of Syllogism. Using the Law of Syllogism with $p \Rightarrow r$ and $r \Rightarrow s$ from (3), we can conclude that $p \Rightarrow s$. Using the Law of Detachment with $p \Rightarrow s$ and p from (4), we can finally conclude s. We can represent this argument using Euler circles (Figure 4.2). Notice that the nested circles represent statements (1)–(3). Then, because circle p is within circle s, we can state "if p, then s."

FIGURE 4.2

When presented with a sequence of statements such as that given in Example 4.3, we do not always deliberate over each statement separately. Sometimes the conclusion will seem fairly automatic. However, it is important to recognize why such conclusions can be drawn and to know how to begin to analyze more complex arguments. In addition, some conclusions that may seem obvious may not really be valid. Symbolic presentations such as the previous one are often easier to analyze than their verbal counterparts.

The next example shows again how we can use direct reasoning to draw conclusions.

EXAMPLE 4.4 What can we deduce from the following statements?

1. If quadrilateral $ABCD$ is a square, then the opposite sides of $ABCD$ are parallel.
2. If the opposite sides of a quadrilateral are parallel, then the quadrilateral is a parallelogram.
3. $ABCD$ is a square.

SOLUTION

We can apply the Law of Syllogism to (1) and (2) to deduce that "if $ABCD$ is a square, then it is a parallelogram." Then, using the Law of Detachment with this latter statement and (3), we can conclude that $ABCD$ is a parallelogram.

Alternatively, using the Law of Detachment with (1) and (3), we can conclude that the opposite sides of $ABCD$ are parallel. Combining this latter statement with (2) and using the Law of Detachment again, we can conclude that $ABCD$ is a parallelogram. Thus, using either method, we can conclude that $ABCD$ is a parallelogram. ■

Other Forms of the Conditional

Many conditional statements do not appear explicitly in the recognizable "if . . . , then . . ." form, but we can restate them in that form and treat them as conditionals.

EXAMPLE 4.5 Rewrite each of the following statements in "if . . . , then . . ." form.

a. The sum of the squares of the lengths of the legs of a right triangle is equal to the square of the length of the hypotenuse.

b. All squares are rectangles.

SOLUTION

a. In the statement given, it is assumed that the triangle being discussed is a right triangle. This assumption is the hypothesis. Thus, we could say the following:

> If a triangle is a right triangle, then the sum of the squares of the lengths of its legs is equal to the square of the length of its hypotenuse.

b. If a polygon is a square, then it is a rectangle. ■

There are three commonly used variants of the conditional statement "If p, then q."

Converse of $p \Rightarrow q$:	$q \Rightarrow p$
Inverse of $p \Rightarrow q$:	not $p \Rightarrow$ not q
Contrapositive of $p \Rightarrow q$:	not $q \Rightarrow$ not p

EXAMPLE 4.6 Write the converse, inverse, and contrapositive of the statement "If $ABCD$ is a square, then it has four right angles."

SOLUTION

Converse: If $ABCD$ has four right angles, then it is a square.

Inverse: If $ABCD$ is not a square, then it does not have four right angles.

Contrapositive: If $ABCD$ does not have four right angles, then it is not a square. ■

Because the converse of a conditional will arise often, it is important to understand how a conditional relates to its converse. Notice that in Example 4.6, the original statement was true. Its converse was false because there are quadrilaterals with four right angles that are *not* squares.

On the other hand, there are instances when a conditional and its converse both hold. In the case when $p \Rightarrow q$ and $q \Rightarrow p$ both hold, we write $p \Leftrightarrow q$. This is read "p if and only if q." This form is called a **biconditional**. The following are examples of biconditionals.

1. The three sides of a triangle are congruent if and only if the three angles of the triangle are congruent.
2. The opposite sides of a quadrilateral are congruent if and only if the opposite angles of the quadrilateral are congruent.
3. Two triangles have the same shape if and only if their angles are congruent, respectively.

Remember that, for example, statement (1) means that *both* of the following statements are true:

> If the three sides of a triangle are congruent, then the three angles of the triangle are congruent, and
>
> If the three angles of a triangle are congruent, then the three sides of the triangle are congruent.

Proof

At the heart of Euclid's axiomatic system of mathematics is the concept of proof. Put simply, a **proof** is a convincing mathematical argument. This means that any person who understands the terminology, accepts the definitions and premises of the mathematics involved, and thinks in a logically correct fashion could not deny the validity of the conclusions drawn.

The Laws of Detachment and Syllogism will form the basis of our proofs or direct reasoning. When we write out proofs, we will use two different forms, paragraph and statement-reason forms. Our first example of a proof concerns angles formed by intersecting lines.

Two intersecting lines form many angles with the same vertex. Two angles that are opposite each other, such as $\angle 1$ and $\angle 3$ or $\angle 2$ and $\angle 4$ in Figure 4.3, are called **vertical angles**. Each pair of vertical angles appears to be congruent.

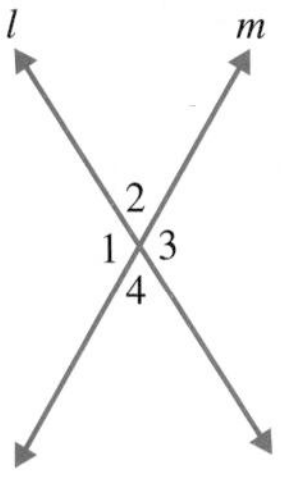

FIGURE 4.3

Next, we will display a **paragraph proof** and a **statement-reason proof** of this fact. In each case, it is helpful to state what is given and what is to be proved. The "given" is the hypothesis of the conditional to be proved and the "prove" is the conclusion. We will prove the following result in two ways.

THEOREM 4.1

If x and y are the measures of a pair of vertical angles, then $x = y$.

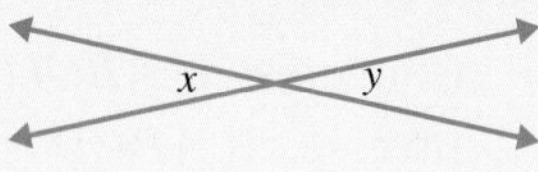

PARAGRAPH PROOF

Given Intersecting lines l and m with two vertical angles having measures x and y [Figure 4.4(a)]

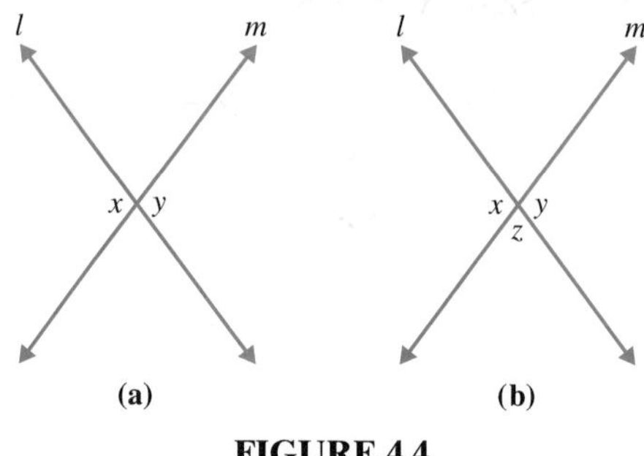

FIGURE 4.4

Prove $x = y$

Proof In Figure 4.4(a), lines l and m intersect to form a pair of vertical angles having measures x and y. Let one of the other two angles have measure z [Figure 4.4(b)]. Then $x + z = 180°$ and $y + z = 180°$ because each of these pairs of angles forms a straight angle. So by substitution $x + z = y + z$. Subtracting z from both sides of this equation yields $x = y$, which is what we wanted to prove.

STATEMENT-REASON PROOF

Given Intersecting lines l and m with two vertical angles having measures x and y [Figure 4.4(a)]

Prove $x = y$

We first label another angle as having measure z as shown in Figure 4.4(b).

PROOF

Statement	Reason
1. $x + z = 180°$	**1.** x and z are adjacent angles whose nonadjacent sides form a straight angle.
2. $y + z = 180°$	**2.** y and z are adjacent angles whose nonadjacent sides form a straight angle.
3. $x + z = y + z$	**3.** Substitution from steps 1 and 2.
4. $x = y$	**4.** z is subtracted from both sides of the equation in step 3. ■

Both the paragraph and statement-reason forms of proof are acceptable. The advantage of paragraph proofs is that they are conversational. That is, you write out the proof in much the same way that you would make a convincing verbal argument. However, reasons are sometimes omitted in paragraph proofs, so when you read a proof, you must be able to justify parts of the proof for yourself. Statement-reason proofs have the advantage of being easier to read. In addition, precise reasons are provided for each step.

The next example illustrates another result proved using both forms of proof.

EXAMPLE 4.7 Prove the following conditional statement in two ways: If two angles are congruent and supplementary, then they are right angles.

SOLUTION

Given $\angle A$ and $\angle B$ where $\angle A \cong \angle B$ and $\angle A + \angle B = 180°$

Prove $\angle A = \angle B = 90°$

PARAGRAPH PROOF

Let $\angle A = \angle B$ and $\angle A + \angle B = 180°$. By substitution, we have that $180° = \angle A + \angle B = \angle A + \angle A$. Thus, $2(\angle A) = 180°$, or $\angle A = 90°$. Because $\angle A = \angle B$, $\angle B = 90°$ also.

STATEMENT-REASON PROOF

Statement	Reason
1. $\angle A = \angle B$	**1.** Given
2. $\angle A + \angle A = \angle A + \angle B$	**2.** The same number may be added to both sides of an equation.
3. $\angle A + \angle B = 180°$	**3.** Given
4. $\angle A + \angle A = 180°$, or $2(\angle A) = 180°$	**4.** Substitution from steps 2 and 3, and simplification.
5. $\angle A = 90°$	**5.** Both sides of an equation may be divided by the same number.
6. $\angle B = 90°$	**6.** Substitution from steps 1 and 5. ■

Solution of Applied Problem

The surveyor knows that the traverse closes if and only if the sum of the measures of the angles is 360°. The sum of the angles is

$$81°40' + 110°50' + 54°38' + 112°52' = 360°$$

So the traverse does close.

GEOMETRY AROUND US

If you watch and listen carefully, you will observe many examples of the use (and misuse) of conditionals and deductive reasoning in advertising. For example, a popular advertising jingle went something like, "If you're out of Crunchy's, you're out of chips." Here the advertiser is attempting to get the viewer to think of Crunchy's and chips as equivalent.

A similar example is "Lo-Lo Yogurt is the breakfast that real women eat." This statement is equivalent to "If you are a real woman, then you eat Lo-Lo Yogurt." Here again, the advertiser wants the viewer to accept the converse statement "If you eat Lo-Lo Yogurt, then you are a real woman." However, the converse may or may not be true. Analyzing advertising for correct logic can help to make you a more discriminating consumer.

PROBLEM SET 4.1

EXERCISES/PROBLEMS

Draw a conclusion for each of Exercises 1–10. If impossible, write "no conclusion possible."

1. If three points are not collinear, then they lie in one and only one plane.

   The three vertices of a triangle are not collinear.
2. If a quadrilateral is an isosceles trapezoid, then it has two congruent sides.

   $PQRS$ has two congruent sides.
3. If a cube has a volume of 27 in^3, then it has an edge of length 3 in.

   If a cube has an edge of length 3 in., then it has a surface area of 54 in^2.

   $ABCDEFGH$ is a cube with a volume of 27 in^3.
4. If a quadrilateral has four right angles, then it is a rectangle.

   If a quadrilateral is a rectangle, then its opposite sides are congruent.

   $ABCD$ is a rectangle.
5. If lines l and m are in the same plane and are not parallel, then l and m intersect.

   Lines l and m are in the same plane and are not parallel.
6. If a point is equidistant from the endpoints of a segment, then the point lies on the perpendicular bisector of the segment.

   Point P is equidistant from points A and B.
7. If two lines are perpendicular, then the lines form right angles.

   If two lines form right angles, then the lines are perpendicular.
8. If $XYZW$ is a square, then the diagonals bisect each other.

   If the diagonals of a quadrilateral bisect each other, then the quadrilateral is a parallelogram.

   $XYZW$ is a square.
9. If a triangle is scalene, then the triangle has no two sides congruent.

   $\triangle ABC$ is a scalene triangle.
10. If a quadrilateral has four congruent sides, then it is a rhombus.

    If a quadrilateral is a rhombus, then its diagonals are perpendicular.

    A quadrilateral has four congruent sides.

If possible, write a statement that can be deduced from the statements in Exercises 11–16. If impossible, write "no deduction possible."

11. If two lines are parallel, then the lines do not intersect.

    Two lines do not intersect.
12. If the sum of two angles is 180 °, then the angles are supplementary.

    The sum of $\angle A$ and $\angle B$ is 90°.
13. If each angle in a triangle is less than 90°, then the triangle is acute.

    $\triangle QRS$ is equiangular.

14. If $\triangle PQR$ is equilateral, then all three angles of the triangle are less than 90°.

A triangle has a 60° angle.

15. If $ABCD$ is a square, then it is a quadrilateral.

If $ABCD$ is a quadrilateral with four congruent sides, then it is a rhombus.

$ABCD$ is a square.

16. If $XYZW$ is a square, then it has four right angles.

If a quadrilateral has four right angles, then it is a rectangle.

If a quadrilateral is a rectangle, then its opposite sides are parallel.

If a quadrilateral has opposite sides parallel, then it is a parallelogram.

$XYZW$ is a square.

For Exercises 17–32, decide if the statement is true or false, form the converse, and decide if the converse is true or false. If both the statement and its converse are true, write a biconditional equivalent to the two statements.

17. If a square has an area of 25 cm^2, then the length of its side is 5 cm.

18. If an angle measures 117°, then it is an obtuse angle.

19. If two angles are adjacent, then they share a common side.

20. If a triangle is equilateral, then it is isosceles.

21. If the side of a square measures 10 in., then the diagonal of the square is $10\sqrt{2}$ in.

22. If a polygon is a regular polygon, then all its sides are congruent.

23. If $\triangle ABC$ as shown has a right angle at C, then $a^2 + b^2 = c^2$.

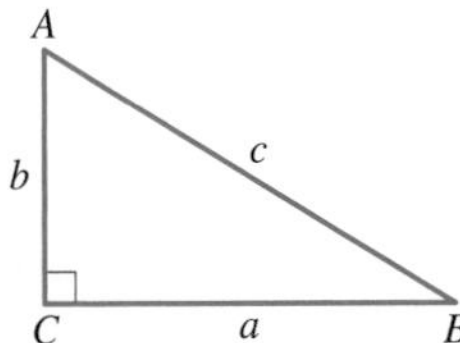

24. If $a = b$, then the kite shown is a rhombus.

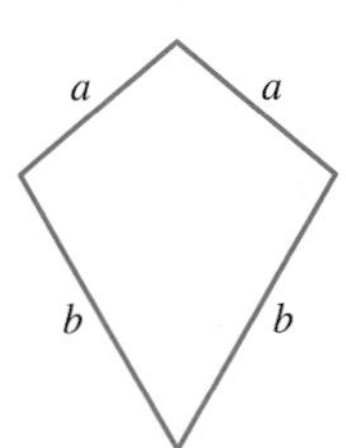

25. If a triangle has side lengths 3, 4, and 5, then it is a right triangle.

26. If two angles have a common vertex, then they are adjacent.

27. If a triangle is equilateral, then the triangle has rotation symmetry.

28. If two angles of a triangle are complementary, then the triangle is a right triangle.

29. If a triangle has three equal sides, then the triangle is equiangular.

30. If a triangle is equilateral, then the triangle has a line of symmetry.

31. If a pyramid has a base with n sides, then the pyramid has $n + 1$ vertices.

32. If a prism has bases with n sides each, then the prism has a total of $3n$ edges.

State a conclusion that can be drawn from each of Exercises 33–38.

33. $r \Rightarrow s$
$\underline{r\quad\quad}$
$\therefore$

34. not $p \Rightarrow t$
$\underline{\text{not } p\quad\quad}$
$\therefore$

35. $r \Rightarrow s$
$s \Rightarrow p$
$\underline{p \Rightarrow t}$
$\therefore$

36. $p \Leftrightarrow q$
$\underline{q \Leftrightarrow r}$
$\therefore$

37. $p \Rightarrow$ not q
not $r \Rightarrow t$
$\underline{\text{not } q \Rightarrow \text{not } r}$
$\therefore$

38. $p \Rightarrow q$
$t \Rightarrow q$
$q \Rightarrow s$
$\underline{p\quad\quad}$
$\therefore$

For each statement given in problems 39–44, write an equivalent conditional statement in "if . . . , then . . ." form.

39. All parallelograms are quadrilaterals.

40. The height of an equilateral triangle with a side of length s is $\frac{s}{2}\sqrt{3}$.

41. A prism with a base having n sides has $2n$ vertices.

42. The only polyhedron with four vertices is a triangular pyramid.

43. Every regular polyhedron has faces that are regular polygons.

44. A cylinder is not a polyhedron.

PROOF

In problem 45 and 46, give reasons that justify each step.

45. Given $\triangle PQR$ with a right angle at R, prove that $\angle P$ and $\angle Q$ are complementary.

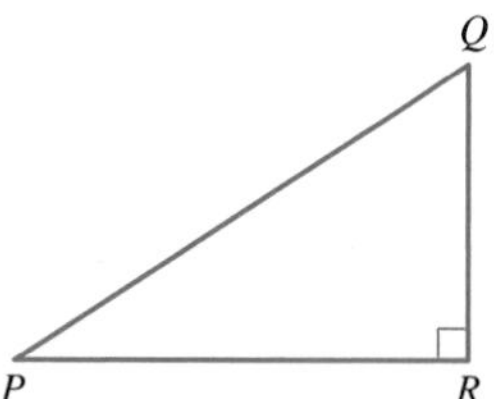

PROOF

Statement	Reason
1. $\triangle PQR$ is a right triangle.	1. ________
2. $\angle P + \angle Q + \angle R = 180°$	2. ________
3. $\angle R = 90°$	3. ________
4. $\angle P + \angle Q + 90° = 180°$	4. ________
5. $\angle P + \angle Q = 90°$	5. ________
6. $\angle P$ and $\angle Q$ are complementary.	6. ________

46. Given $\angle ABD$ and $\angle DBC$ are adjacent angles and A, B, and C are collinear, prove that $\angle ABD$ and $\angle DBC$ are supplementary.

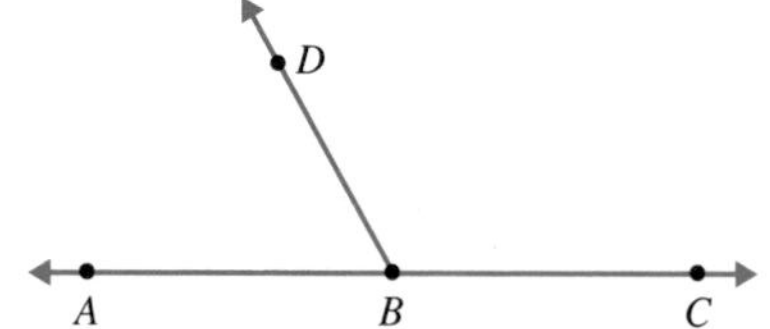

PROOF

Statement	Reason
1. $\angle ABD$ and $\angle DBC$ are adjacent angles.	1. ________
2. $\angle ABD + \angle DBC = \angle ABC$	2. ________
3. A, B, and C are collinear.	3. ________
4. $\angle ABC = 180°$	4. ________
5. $\angle ABD + \angle DBC = 180°$	5. ________
6. $\angle ABD$ and $\angle DBC$ are supplementary.	6. ________

APPLICATIONS

47. A farmer has 1300 cubic feet of compost and wants to add an inch of compost to her rectangular garden that measures 100 feet by 150 feet. How can she use deductive reasoning to determine whether she has enough compost?

48. The bed of a pickup measures 8′3″ by 5′0″. You want to haul 9′ lengths of pipe so that they lay flat in the bed of the pickup. How can you use deductive reasoning to determine whether you can lay the pipes in the bed of the pickup?

49. Advertising slogans often use conditional statements, which may or may not include the words "if" and "then." Sometimes the conditional is implied. Restate each of the following in "If. . . , then. . ." form. Decide if the statement is true or false. Write the converse, and decide if the converse is true or false. Justify your responses.

a. "If you believe in peanut butter, you gotta believe in Peter Pan." (Peter Pan Peanut Butter)

b. "You've got questions. We've got answers." (Radio Shack)

c. "It takes a licking and keeps on ticking." (Timex)

50. Advertising slogans often use conditional statements, which may or may not include the words "if" and "then." Sometimes the conditional is implied. Restate each of the following in "If. . . , then. . ." form. Decide if the statement is true or false. Write the converse, and decide if the converse is true or false. Justify your responses.

a. "Choosy mothers choose Jif." (Jif Peanut Butter)

b. "If it says Libby's, Libby's, Libby's on the label, label, label, you will like it, like it, like it on your table, table, table." (Libby's Food)

c. "If you don't get it, you don't get it." (The Washington Post)

EXTENDED PROBLEMS

51. Four famous paradoxes are credited to philosopher and mathematician Zeno of Elea from the fifth century B.C.E. Define the term paradox and research the first of Zeno's four paradoxes. State the paradox in "if. . . , then. . ." form and explain the logic behind the paradox. Include a discussion of how this paradox has been resolved.

52. Collect three examples of conditional statements, or statements that can be written as conditional statements, from newspaper or magazine advertisements. Discuss whether each conditional statement is true or false. Form the converse and discuss whether it is true or false. In each case, what is the advertiser trying to imply?

53. When programming on a graphing calculator, the If-Then statement directs the flow of the program. If a condition holds, then the program carries out a certain command; otherwise, it skips to the next line. The following is a program for a TI-92 calculator that determines whether a triangle is right, obtuse, or acute after the lengths of the sides have been input.

```
:Prgm
:Request "Enter a",a
:expr(a)→a
:Request "Enter b",b
:expr(b)→b
:Request "Enter c",c
:expr(c)→c
:If a² + b² = c² Then
:Disp "Right Triangle"
:End If
:If a² + b² < c² Then
:Disp "Obtuse Triangle"
:End If
:If a² + b² > c² Then
:Disp "Acute Triangle"
:End If
:EndPrgm
```

For each of the following inputs, what will the program display?

a. a = 11, b = 20, c = 23

b. a = 14, b = 6, c = 15

c. a = 60, b = 11, c = 61

4.2 TRIANGLE CONGRUENCE CONDITIONS

Applied Problem

A saw blade is made by cutting six right triangles out of a regular hexagon as shown. If a segment of length AB is cut at a right angle at each tooth, show that all the triangles that are cut out are congruent.

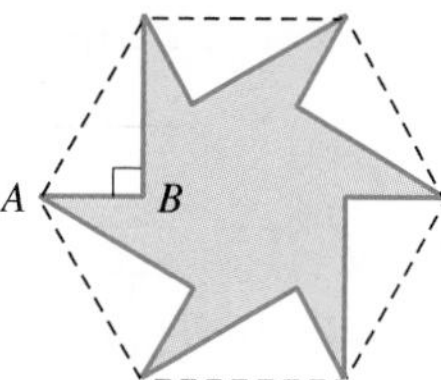

Congruent Triangles

Recall that two line segments $\overline{AB}$ and $\overline{CD}$ are congruent if they have the same length, and two angles $\angle EFG$ and $\angle HIJ$ are congruent if they have the same angle measure. Also, remember that line segments of the same length and angles of the same measure are identified using special marks as illustrated in Figure 4.5.

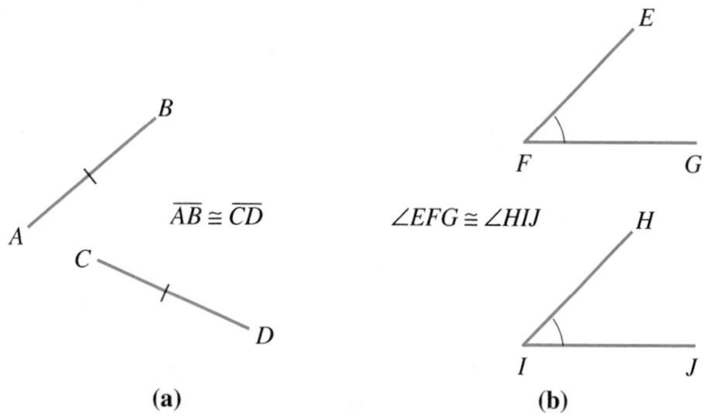

FIGURE 4.5

Because the lengths of sides and measures of angles determine the size and shape of a triangle, congruent triangles can be defined using their various parts as follows.

DEFINITION

Congruent Triangles

$\triangle ABC$ is **congruent** to $\triangle DEF$ (written $\triangle ABC \cong \triangle DEF$) under the correspondence $A \leftrightarrow D, B \leftrightarrow E, C \leftrightarrow F$ if and only if

1. all three pairs of corresponding angles are congruent, and
2. all three pairs of corresponding sides are congruent.

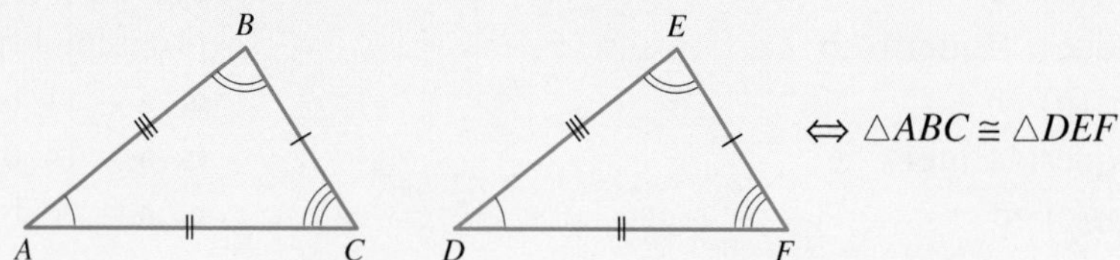

$\triangle ABC \cong \triangle DEF$ if and only if $\angle A \cong \angle D$, $\angle B \cong \angle E$, and $\angle C \cong \angle F$, and $\overline{AB} \cong \overline{DE}$, $\overline{BC} \cong \overline{EF}$, and $\overline{AC} \cong \overline{DF}$.

The following relationships follow from the definition:

1. **Reflexive Property**: $\triangle ABC \cong \triangle ABC$ for all triangles $\triangle ABC$.
2. **Symmetric Property**: If $\triangle ABC \cong \triangle DEF$ then $\triangle DEF \cong \triangle ABC$.
3. **Transitive Property**: If $\triangle ABC \cong \triangle DEF$ and $\triangle DEF \cong \triangle GHI$, then $\triangle ABC \cong \triangle GHI$.

EXAMPLE 4.8 Decide if the following pairs of triangles are congruent under the implied correspondences in Figure 4.6.

a. $\triangle ABC$ and $\triangle QRP$ **b.** $\triangle GHI$ and $\triangle JKL$

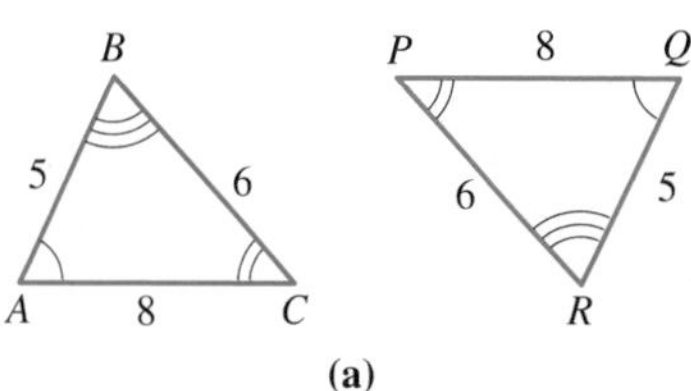

(a)

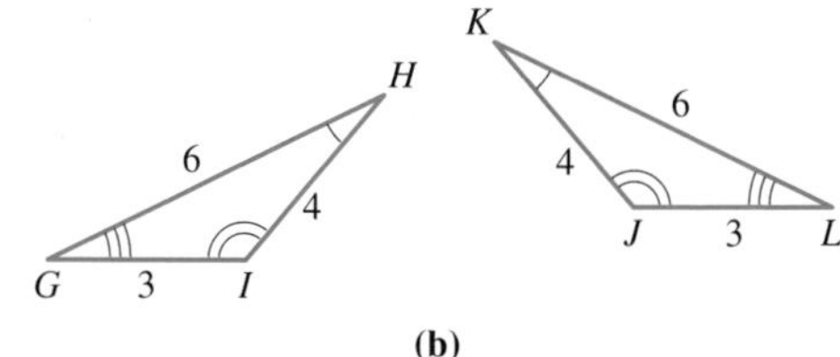

(b)

FIGURE 4.6

SOLUTION

a. $\triangle ABC \cong \triangle QRP$ because $\angle A \cong \angle Q$, $\angle B \cong \angle R$, $\angle C \cong \angle P$, $\overline{AB} \cong \overline{QR}$, $\overline{BC} \cong \overline{RP}$, and $\overline{AC} \cong \overline{QP}$.

b. $\triangle GHI$ is not congruent to $\triangle JKL$ because, for example, corresponding sides $\overline{GH}$ and $\overline{JK}$ are not congruent. However, we do have $\triangle GHI \cong \triangle LKJ$ because all pairs of corresponding angles and sides are congruent under the correspondence $G \leftrightarrow L, H \leftrightarrow K, I \leftrightarrow J$. ■

Every triangle $\triangle ABC$ is congruent to itself by the reflexive property under the correspondence $A \leftrightarrow A$, $B \leftrightarrow B$, $C \leftrightarrow C$, but it is interesting to observe that a

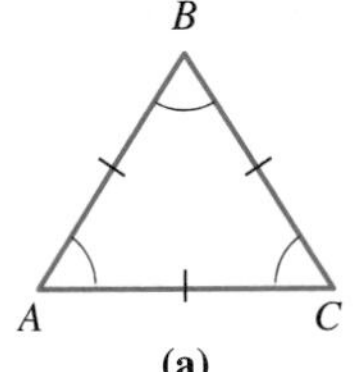

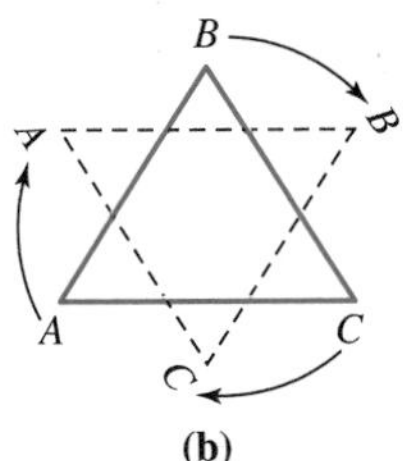

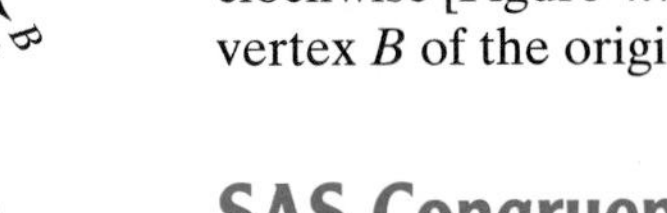

FIGURE 4.7

triangle may be congruent to itself under different correspondences. For example, consider the equilateral triangle in Figure 4.7(a). Here $\triangle ABC \cong \triangle BCA$ because six pairs of the corresponding angles and sides are congruent under the correspondence $A \leftrightarrow B$, $B \leftrightarrow C$, $C \leftrightarrow A$. We could show in a similar fashion that $\triangle ABC \cong \triangle CBA$.

The different correspondences between a triangle and itself can be represented physically. When we say $\triangle ABC \cong \triangle ABC$, we can imagine a tracing of $\triangle ABC$ placed over $\triangle ABC$ with vertex A paired with A and so on. The two images will coincide. When we say $\triangle ABC \cong \triangle BCA$, we can imagine a tracing of $\triangle ABC$ rotated 120° clockwise [Figure 4.7(b)] and placed over $\triangle ABC$. In this way, vertex A is paired with vertex B of the original triangle and so on. Again, the two triangles will coincide.

SAS Congruence

Now consider the partial triangle in Figure 4.8(a). To complete this triangle, we need only draw $\overline{BC}$ [Figure 4.8(b)]. That is, if two sides and the included angle between them are designated, the size and shape of the triangle are completely determined. Thus, to decide whether two triangles are congruent, it is sufficient to find only three appropriate pairs of parts (here, two sides and the included angle) that are congruent. This is the essence of the next postulate, called the Side-Angle-Side Congruence Postulate and abbreviated as SAS.

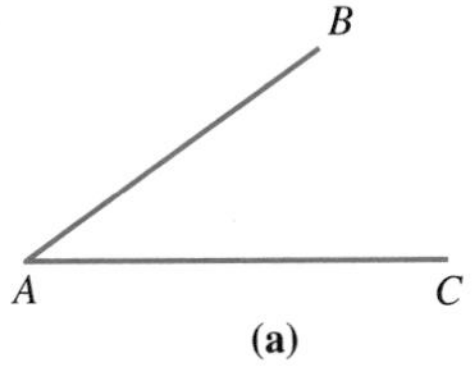

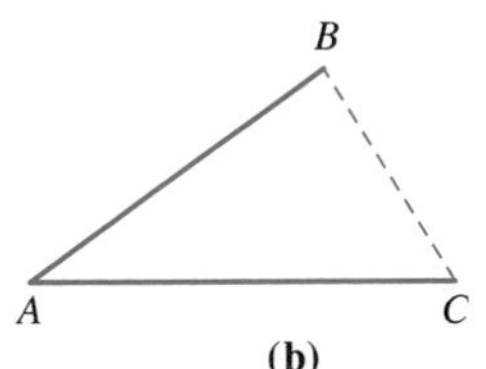

FIGURE 4.8

POSTULATE 4.1 SAS Congruence Postulate

If two sides and the included angle of one triangle are congruent respectively to two sides and the included angle of another triangle, then the two triangles are congruent.

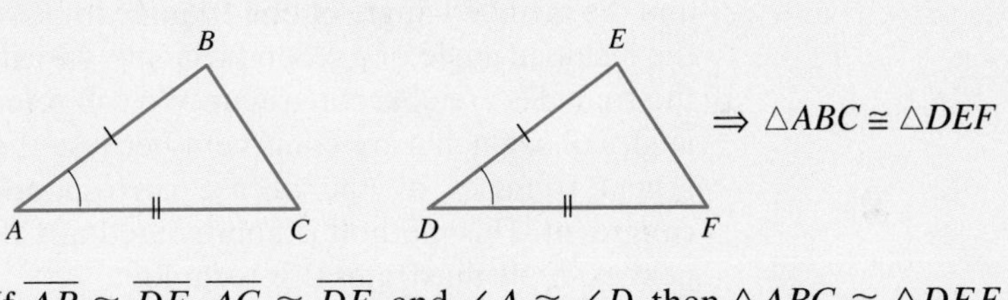

If $\overline{AB} \cong \overline{DE}$, $\overline{AC} \cong \overline{DF}$, and $\angle A \cong \angle D$, then $\triangle ABC \cong \triangle DEF$.

The next theorem, the Leg-Leg Congruence Theorem and abbreviated by LL, is an immediate consequence of Postulate 4.1 where the included angle is a right angle.

THEOREM 4.2 LL Congruence Theorem

If two legs of one right triangle are congruent respectively to two legs of another right triangle, then the two triangles are congruent.

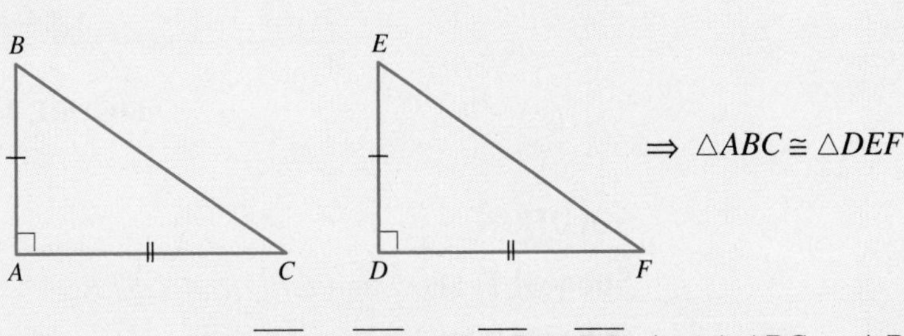

If $\angle A = \angle D = 90°$, $\overline{AB} \cong \overline{DE}$, and $\overline{AC} \cong \overline{DF}$, then $\triangle ABC \cong \triangle DEF$.

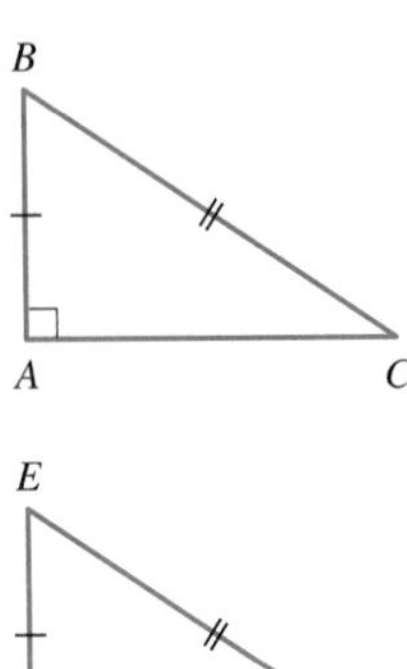

FIGURE 4.9

A similar congruence result holds if the hypotenuse and one leg of one right triangle are congruent, respectively, to the hypotenuse and the corresponding leg of a second right triangle (Figure 4.9). We can use the Pythagorean theorem to show that the other two corresponding legs are congruent. Therefore, the two triangles are congruent by SAS. We leave the proof of this next theorem, called the Hypotenuse-Leg Congruence Theorem (abbreviated HL), for the problem set.

THEOREM 4.3 HL Congruence Theorem

If the hypotenuse and a leg of one right triangle are congruent respectively to the hypotenuse and a leg of another right triangle, then the two triangles are congruent.

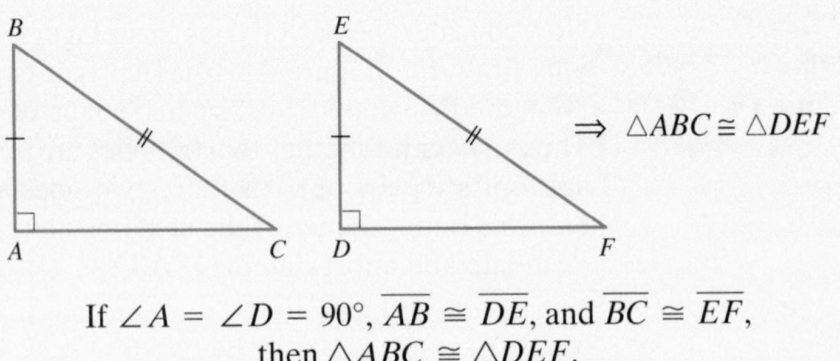

If $\angle A = \angle D = 90°$, $\overline{AB} \cong \overline{DE}$, and $\overline{BC} \cong \overline{EF}$, then $\triangle ABC \cong \triangle DEF$.

We can use the SAS Postulate to verify many other relationships that appear in geometric figures. The advantage of using this postulate is that once we find two sides and the included angle of one triangle to be congruent respectively to two sides and the included angle of a second triangle, we can conclude that the other three parts of the triangles are also congruent. We will refer to this technique of proving that two angles or segments are congruent because they are corresponding parts of two congruent triangles by the phrase **corresponding parts of congruent triangles are congruent**. This method is abbreviated as **C.P.** (short for ***c***orresponding ***p***arts). The next example illustrates this technique.

EXAMPLE 4.9 Congruent parts are indicated in Figure 4.10. Prove that $\angle P \cong \angle S$.

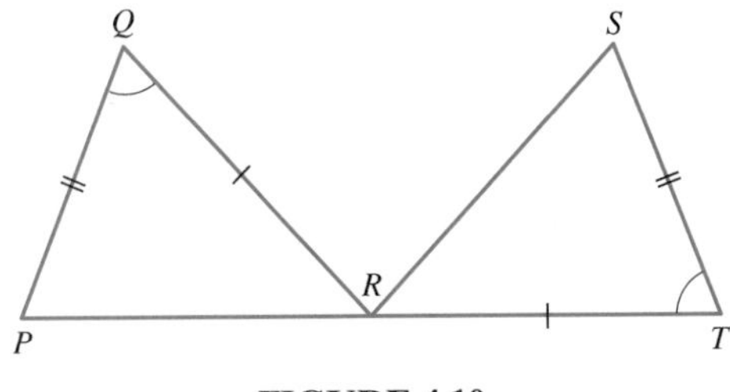

FIGURE 4.10

SOLUTION

Subgoal 1 Show that $\triangle PQR \cong \triangle STR$.

Subgoal 2 Show that $\angle P \cong \angle S$ by corresponding parts.
We will write this proof in paragraph form.

Proof of Subgoal 1 From Figure 4.10, $\angle Q \cong \angle T$, $\overline{PQ} \cong \overline{ST}$, and $\overline{QR} \cong \overline{TR}$. Thus, we have two sides and the included angle of one triangle congruent to two sides and the included angle of the other triangle. Therefore, $\triangle PQR \cong \triangle STR$ by the SAS Congruence Postulate.

Proof of Subgoal 2 Because the two triangles are congruent, all six parts of one triangle are congruent to the six corresponding parts of the other triangle. In particular, $\angle P \cong \angle S$ by C.P. ■

ASA Congruence

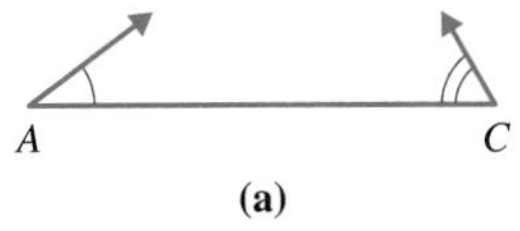

(a)

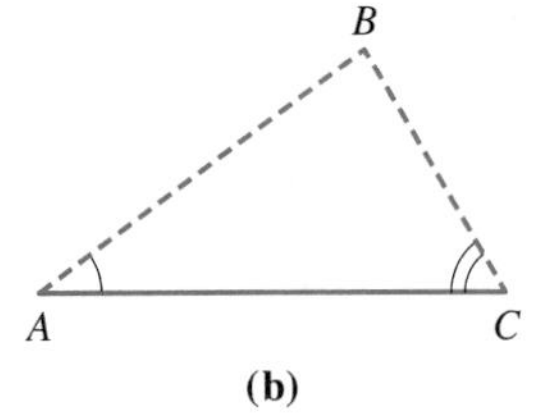

(b)

FIGURE 4.11

Now consider the partial triangle given in Figure 4.11(a), where two angles and their included side are specified. Notice how the third vertex B is established by extending the two sides [Figure 4.11(b)]. That is, only one triangle is possible for the given two angles and their included side. This figure suggests the following result, the Angle-Side-Angle Congruence Postulate (ASA).

POSTULATE 4.2 ASA Congruence Postulate

If two angles and the included side of one triangle are congruent respectively to two angles and the included side of another triangle, then the two triangles are congruent.

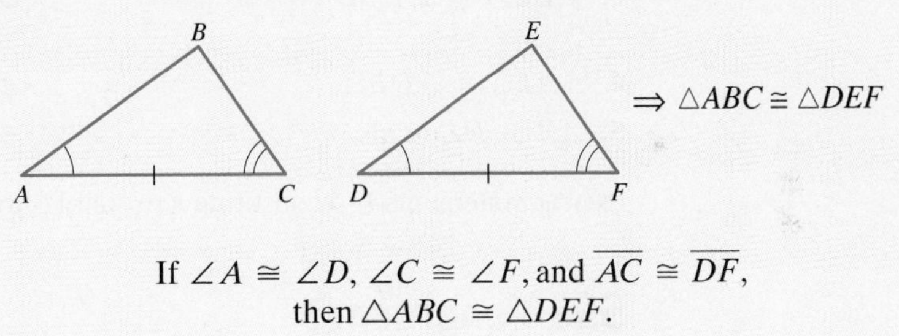

If $\angle A \cong \angle D$, $\angle C \cong \angle F$, and $\overline{AC} \cong \overline{DF}$, then $\triangle ABC \cong \triangle DEF$.

Although we have assumed ASA Congruence as a postulate, we can prove it as a theorem using the SAS Congruence Postulate. However, the proof is somewhat complex.

EXAMPLE 4.10 In Figure 4.12(a), $\angle A \cong \angle D$ and $\overline{AC} \cong \overline{DC}$. Prove that $\overline{AB} \cong \overline{DE}$.

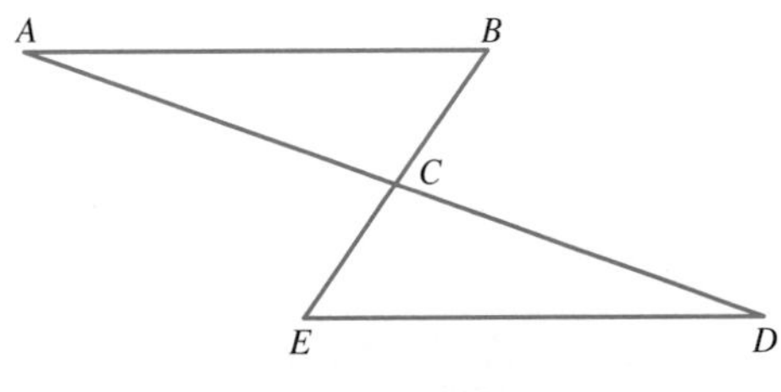

FIGURE 4.12(a)

SOLUTION

First, we will label the figure using the information given [Fig. 4.12(b)], and then establish subgoals.

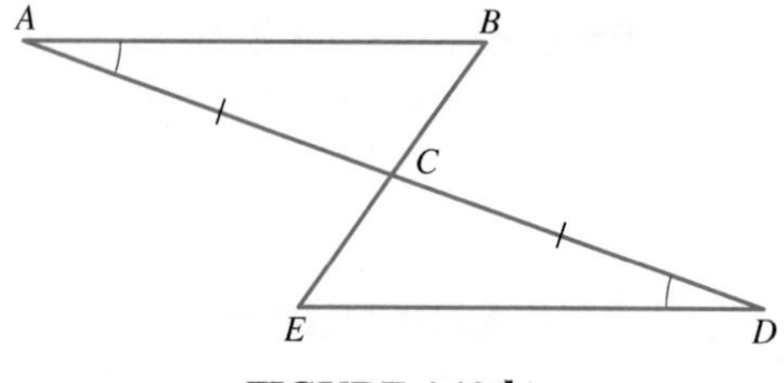

FIGURE 4.12(b)

Subgoal 1 Show that $\triangle ABC \cong \triangle DEC$.

Subgoal 2 Show that $\overline{AB} \cong \overline{DE}$ by corresponding parts.
We will write this proof in statement-reason form.

PROOF

Statement	Reason
1. $\angle A \cong \angle D$	**1.** Given
2. $\overline{AC} \cong \overline{DC}$	**2.** Given
3. $\angle BCA \cong \angle ECD$	**3.** Vertical angles formed by intersecting lines are congruent.
4. $\triangle ABC \cong \triangle DEC$	**4.** ASA Congruence Postulate
5. $\overline{AB} \cong \overline{DE}$	**5.** C.P. ■

(NOTE: Statements 1–4 constitute a proof of Subgoal 1. Statement 5 proves Subgoal 2.)

SSS Congruence

Next consider the three line segments given in Figure 4.13(a). (Notice that the sum of the lengths of any two segments exceeds the length of the third segment.)

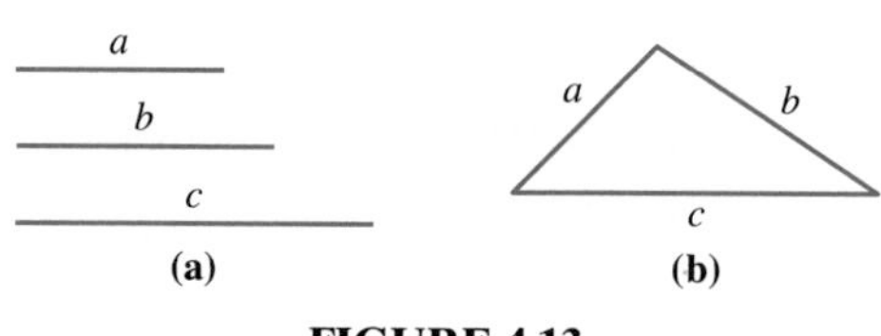

FIGURE 4.13

If we try to draw a triangle using these three segments for sides, the size and shape of the triangle are determined [Figure 4.13(b)]. In other words, only one triangle can be constructed. This fact suggests the Side–Side–Side Congruence Postulate (SSS). It is possible to prove SSS, but due to the complexity of the proof, we state it as a postulate next.

POSTULATE 4.3 SSS Congruence Postulate

If three sides of one triangle are congruent respectively to three sides of another triangle, then the two triangles are congruent.

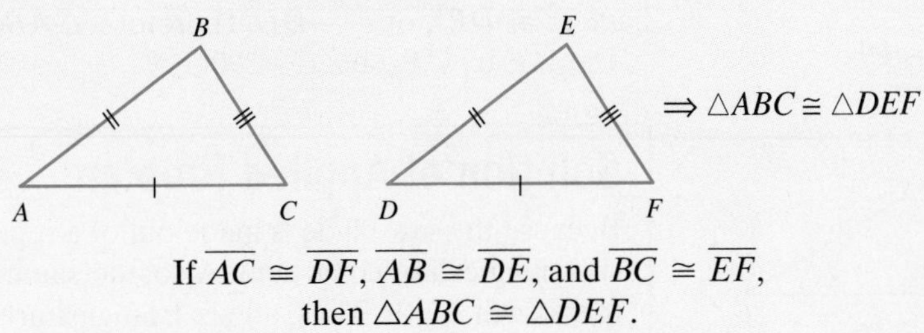

If $\overline{AC} \cong \overline{DF}$, $\overline{AB} \cong \overline{DE}$, and $\overline{BC} \cong \overline{EF}$, then $\triangle ABC \cong \triangle DEF$.

The following example shows how we can use Postulate 4.3 to learn more about geometric figures.

EXAMPLE 4.11 For the kite in Figure 4.14(a), prove that $\angle A \cong \angle C$.

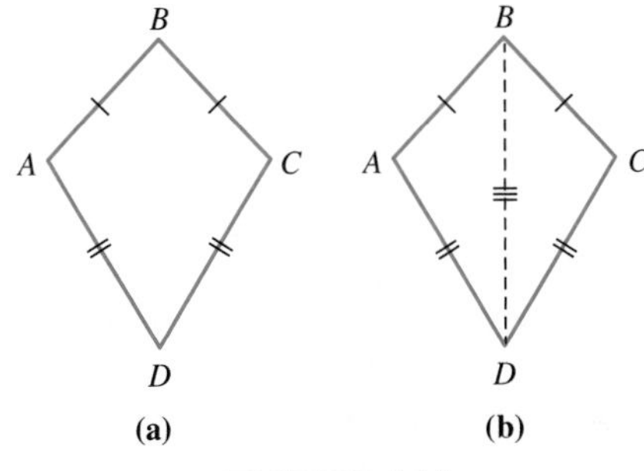

FIGURE 4.14

SOLUTION

First, draw in diagonal $\overline{BD}$ [Figure 4.14(b)]. Then $\triangle ABD \cong \triangle CBD$ by SSS because $\overline{AB} \cong \overline{CB}$, $\overline{AD} \cong \overline{CD}$, and $\overline{BD} \cong \overline{BD}$. Therefore, $\angle A \cong \angle C$ by C. P. ■

Now that we have the SSS Congruence Postulate to use, we can prove the converse of the Pythagorean theorem, which we introduced informally in Chapter 3 as the Test for Right Triangles. It is fascinating to observe that we can use the Pythagorean theorem to prove its converse.

THEOREM 4.4 Converse of the Pythagorean Theorem

If the sum of the squares of the lengths of two sides of a triangle equals the square of the third side, then the triangle is a right triangle.

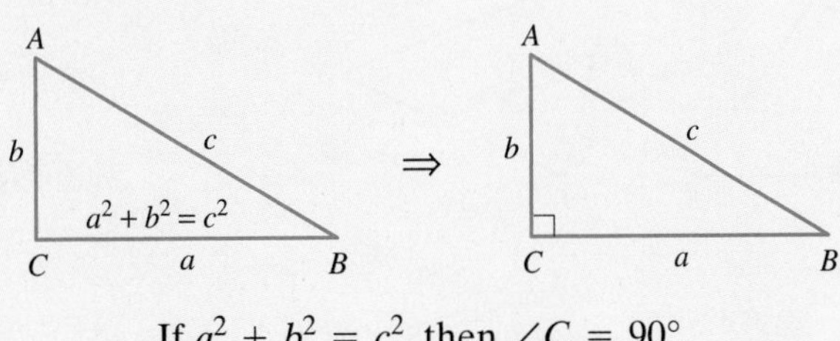

If $a^2 + b^2 = c^2$, then $\angle C = 90°$.

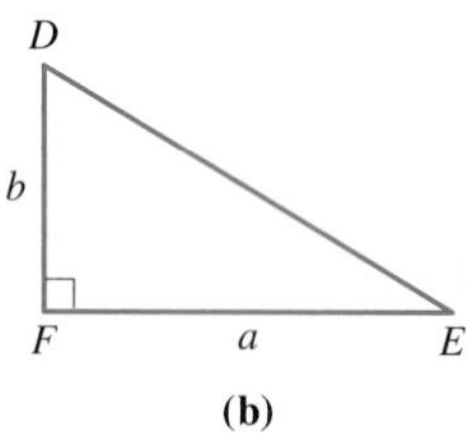

FIGURE 4.15

Given $\triangle ABC$ with $a^2 + b^2 = c^2$ [Figure 4.15(a)]

Prove $\angle C = 90°$

Proof Construct a right triangle $\triangle DEF$ with legs of length a and b [Figure 4.15(b)]. By the Pythagorean theorem, $a^2 + b^2 = DE^2$. Because $a^2 + b^2 = c^2$ is given, we have $c^2 = DE^2$, or $c = DE$. Therefore, $\triangle ABC \cong \triangle DEF$ by SSS. As a consequence, $\angle C \cong \angle F$ by C.P., so $\angle C = 90°$. ■

Solution of Applied Problem

Because the saw blade is made out of a regular hexagon, the hypotenuses of the six triangles are congruent. Also, the same length AB is cut at a right angle to form each tooth. Thus, all six triangles are congruent by the HL Congruence Theorem.

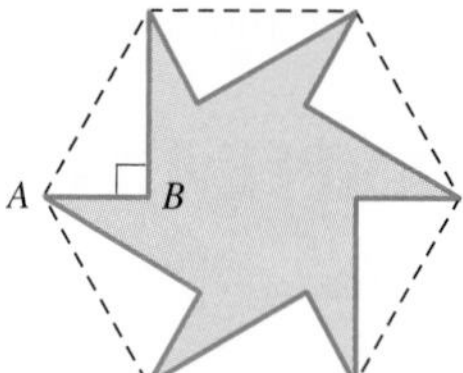

GEOMETRY AROUND US The triangle is frequently seen in construction—in trusses, bridges, scaffolding, braces, and so on. The reason for its frequent use is its rigid shape. Any other polygon will shift and flex and change its shape. Only a triangle retains its fixed shape. Other polygons can be made rigid by adding diagonal sections to form triangles as in railroad bridges such as the one illustrated.

PROBLEM SET 4.2

EXERCISES/PROBLEMS

1. Given that $\triangle PQR \cong \triangle XYZ$, list the six pairs of congruent parts for the two triangles.

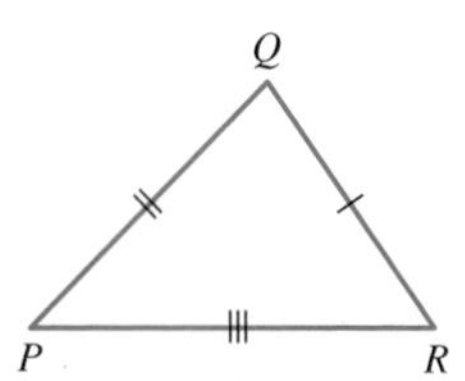

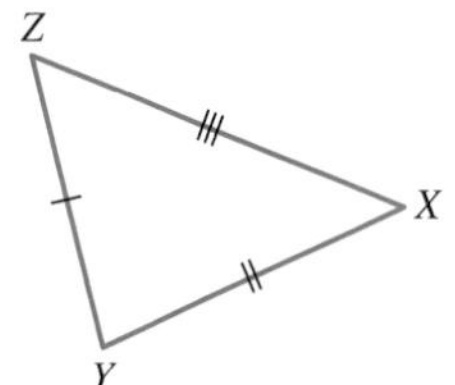

2. Suppose $\triangle DEF \cong \triangle JKL$. List the six pairs of congruent parts for the two triangles.
3. For the pair of triangles shown, write an appropriate congruence statement. Be careful about the order in which you list the vertices.

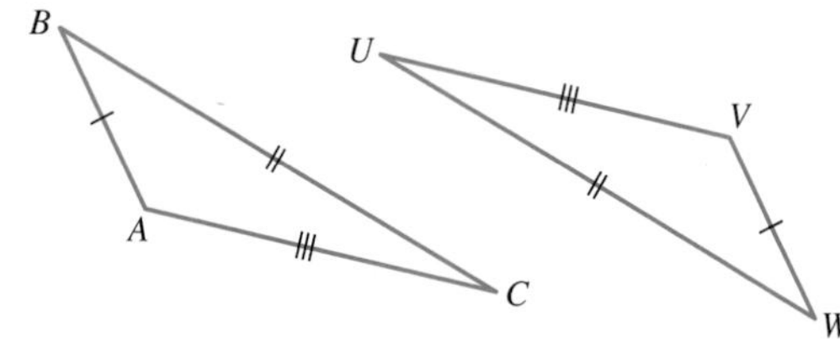

4. For the pair of triangles shown, write an appropriate congruence statement. Be careful about the order in which you list the vertices.

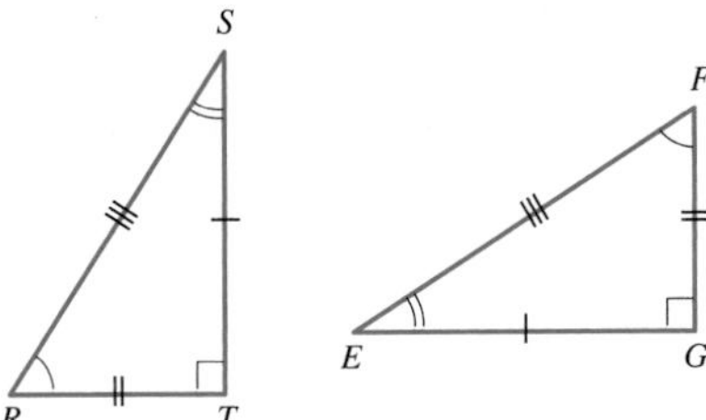

For each figure in Exercises 5–18, decide whether a pair of triangles is necessarily congruent. If so, write an appropriate congruence statement and specify which congruence principle applies. Be careful not *to make any assumptions about triangles that just "appear" to be congruent.*

5.

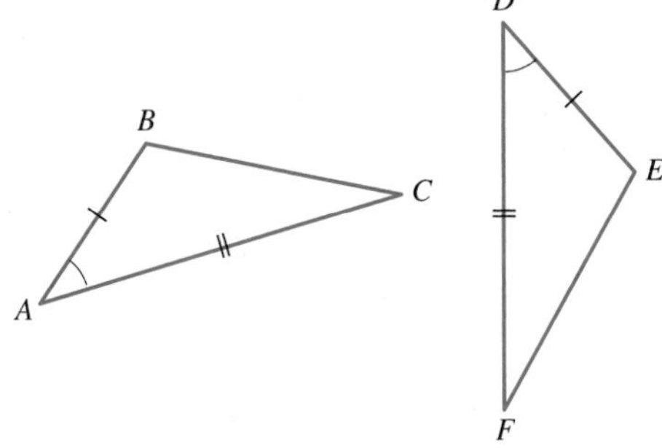

6.

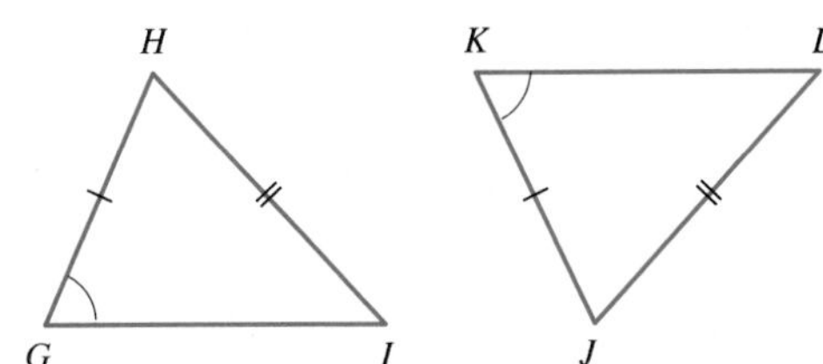

7.

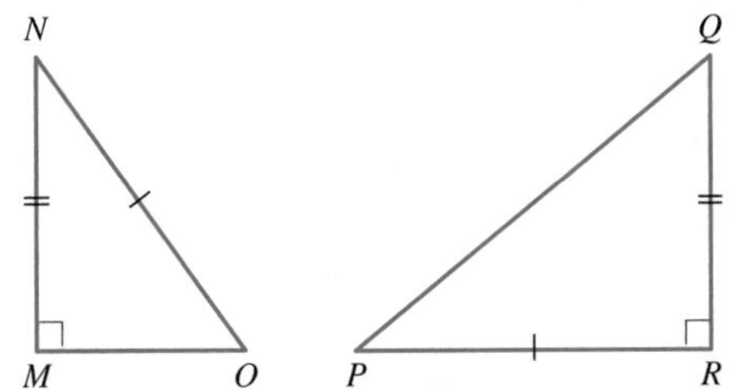

8.

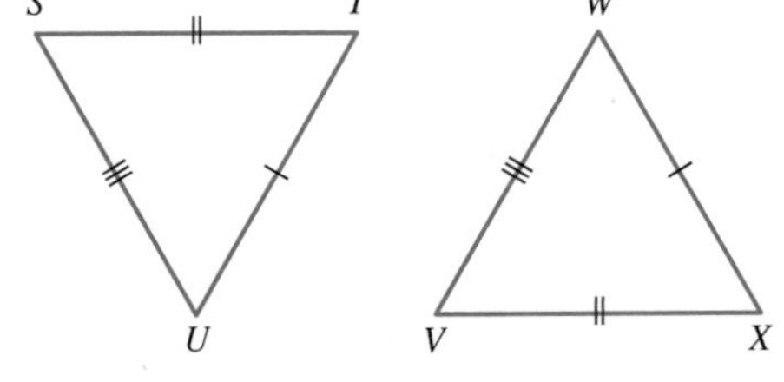

9.

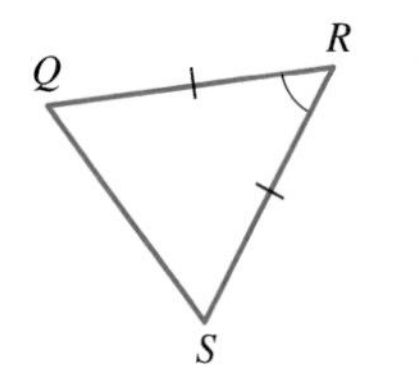

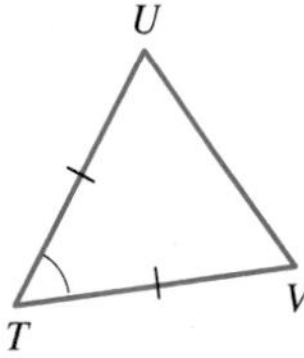

10.

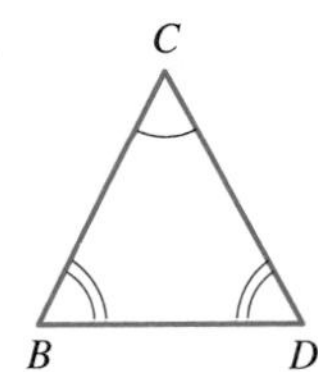

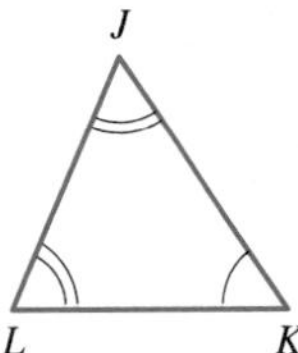

11.

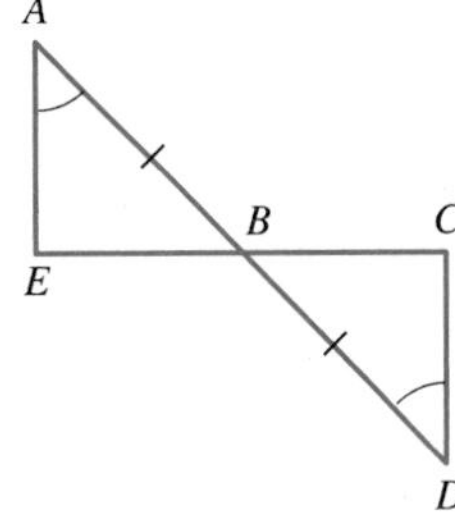

12.

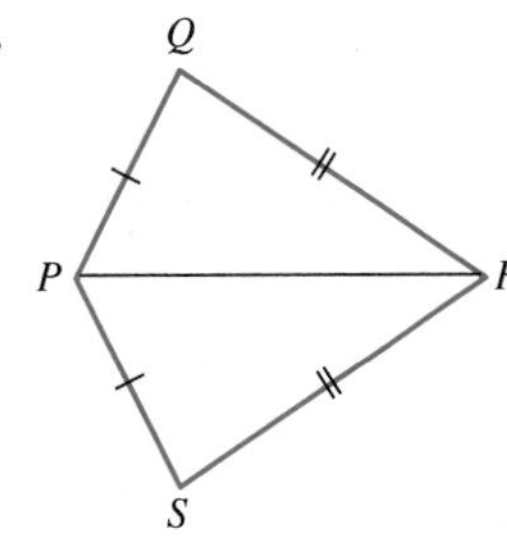

13.

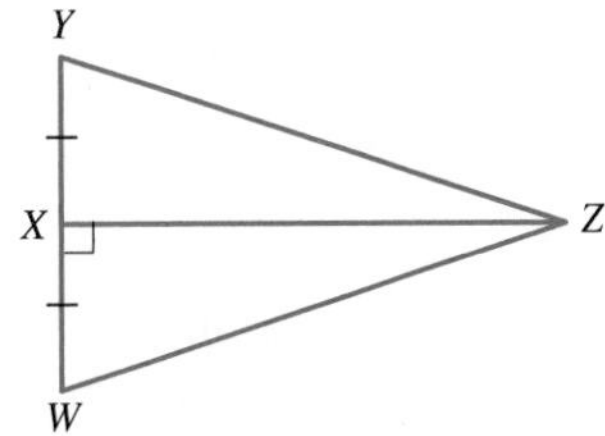

14.

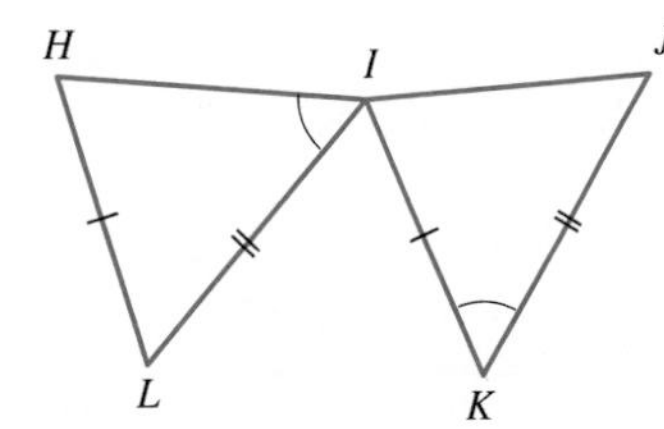

15.

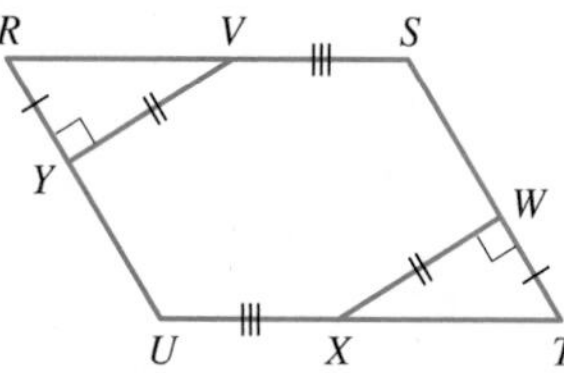

16.

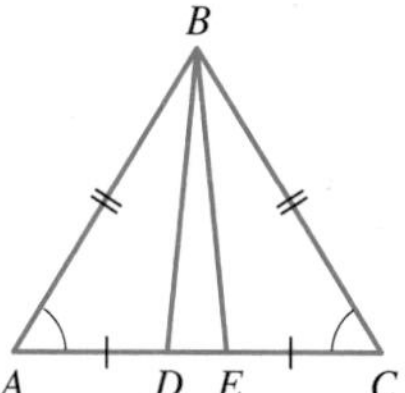

17.

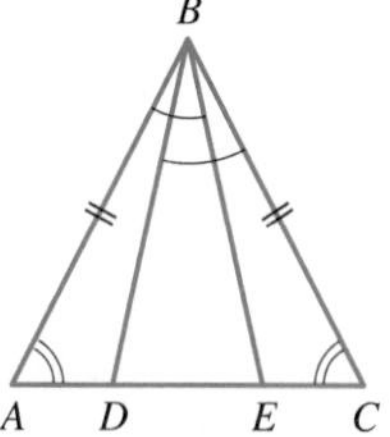

18.

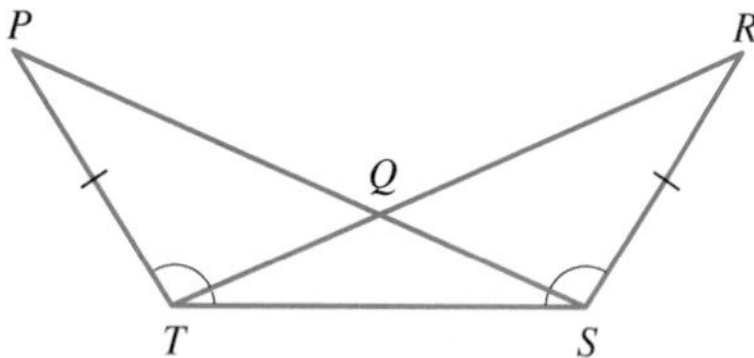

19. Two pairs of triangles are congruent in this figure. Give both of them, and state which congruence principles apply.

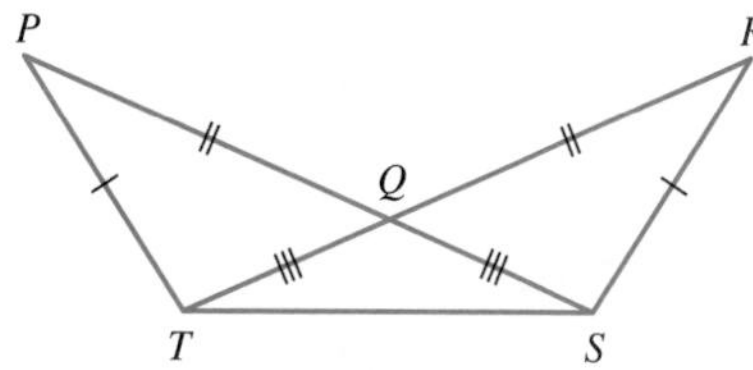

20. Two pairs of triangles are congruent in this figure. Give both of them, and state which congruence principles apply.

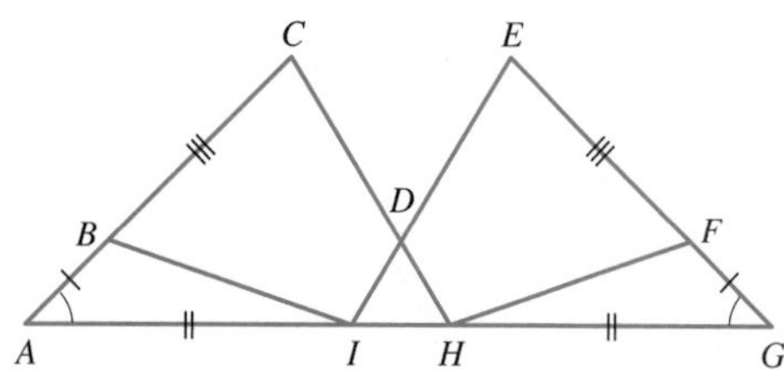

Refer to the following triangles for Exercises 21–26.

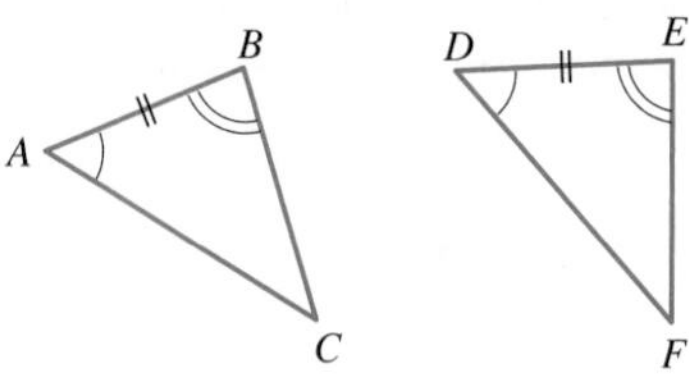

21. $\overline{AB} \cong$ _____

22. $\angle A \cong \angle$ _____

23. $\angle E \cong \angle$ _____

24. $\triangle FED \cong \triangle$ _____. Which congruence principle guarantees this?

25. $\overline{CB} \cong$ _____. Why?

26. $\angle F \cong \angle$ _____. Why?

Refer to the following regular hexagon for Exercises 27–32.

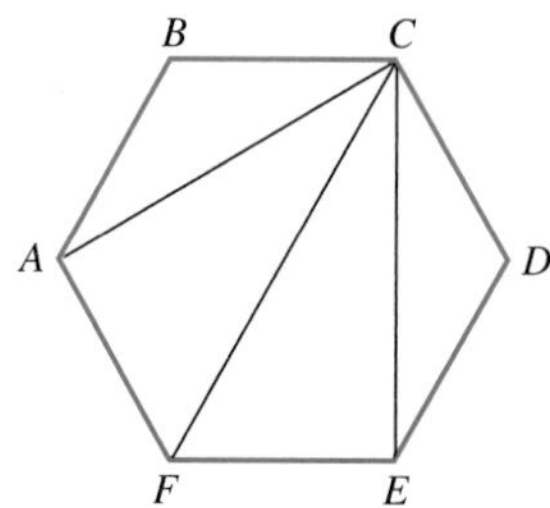

27. $\angle B \cong \angle$ _____

28. $\triangle BAC \cong \triangle$ _____. Which congruence principle guarantees this?

29. $\overline{AC} \cong$ _________. Why?

30. $\angle BAC \cong \angle$ _____

31. $\angle FEC \cong \angle$ _____

32. $\triangle FEC \cong \triangle$ _____. Which congruence principle guarantees this?

PROOFS

33. In the following figure, $\overline{UW} \cong \overline{YW}$ and $\overline{VW} \cong \overline{XW}$. Prove that $\angle U \cong \angle Y$. (HINT: First show that $\triangle UVW \cong \triangle YXW$.)

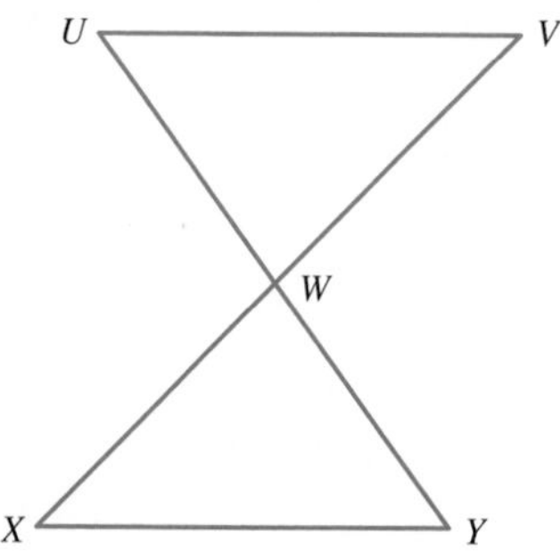

34. In the figure shown, $\angle BAC \cong \angle DAC$ and $\angle ACB \cong \angle ACD$. Prove that $\overline{AB} \cong \overline{AD}$. (HINT: First show that $\triangle ABC \cong \triangle ADC$.)

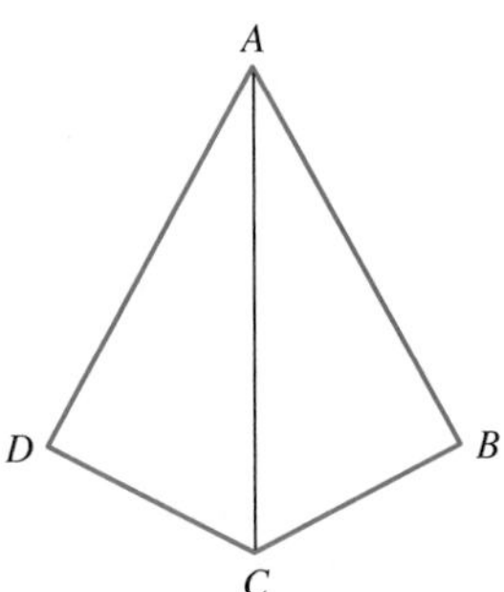

35. $ABCD$ is a rhombus with diagonal $\overline{BD}$. Prove that the diagonal divides the rhombus into two congruent triangles.

36. In the triangles shown, congruent parts are marked. Prove that $\angle Q \cong \angle T$.

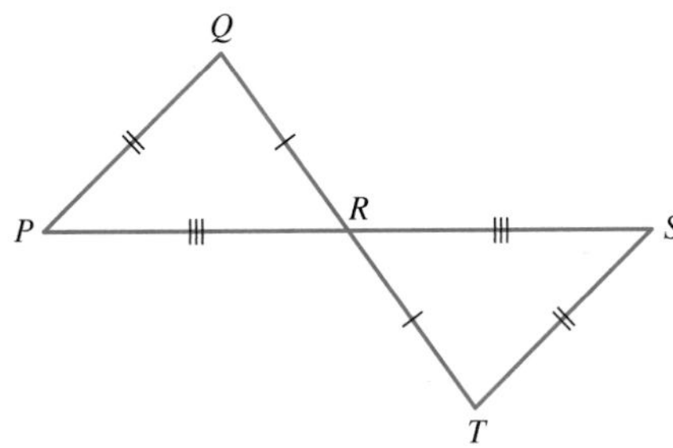

37. In $ABCD$, line segments $\overline{AC}$ and $\overline{BD}$ divide each other in half at E. Prove that $\angle DBC \cong \angle BDA$.

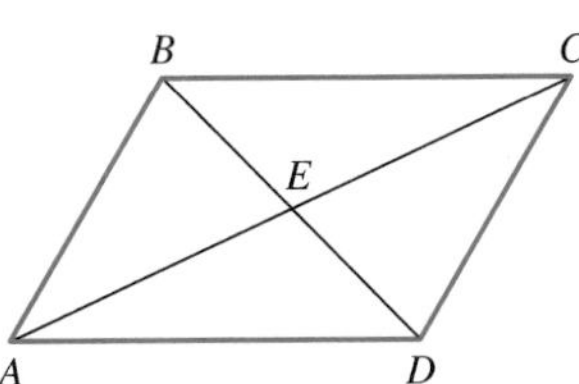

38. In the following figure, $\angle EDH = \angle GHD = 90°$, and $\overline{DE} \cong \overline{HG}$. Prove that $\angle GDH \cong \angle EHD$.

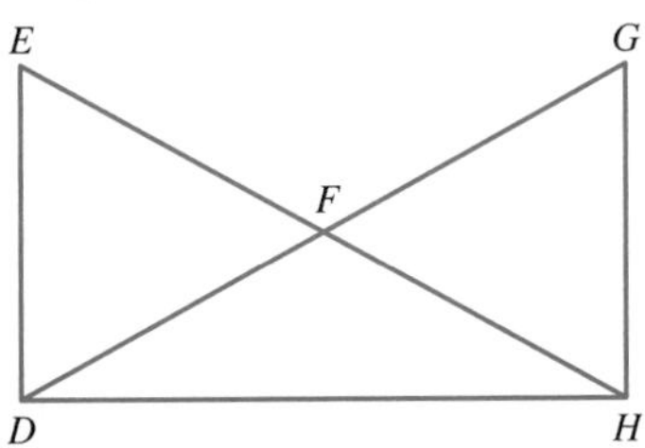

39. In $ABCD$, line segments $\overline{AC}$ and $\overline{BD}$ divide each other in half at E. Prove $\triangle DBC \cong \triangle BDA$.

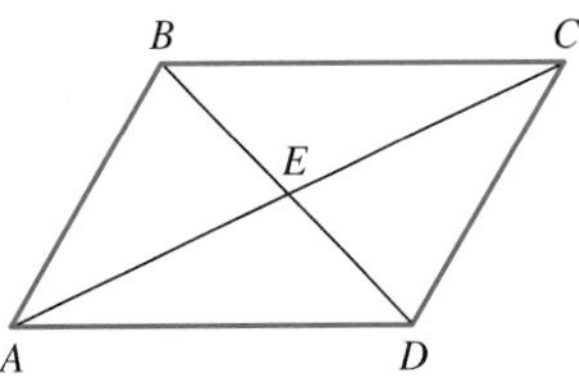

40. In $ABCD$, $\overline{BC} \cong \overline{AD}$ and $\overline{AB} \cong \overline{CD}$. Prove $\angle BCA \cong \angle DAC$.

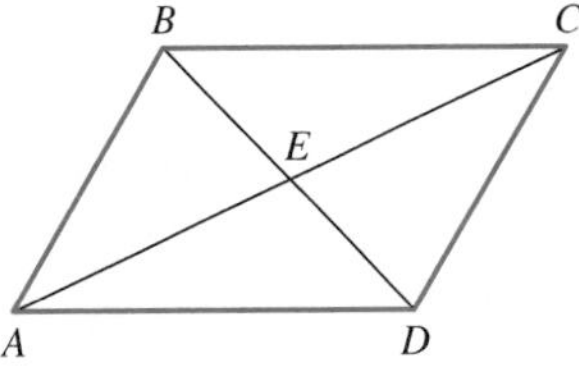

41. In the following figure, congruent segments are marked. Prove $\angle ACF \cong \angle AEB$.

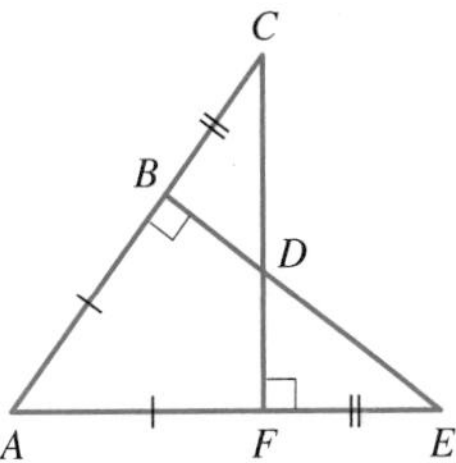

42. Given $\triangle ABC$, where $\overline{BD}$ is an altitude and $AD = CD$. Prove $\triangle ADB \cong \triangle CDB$.

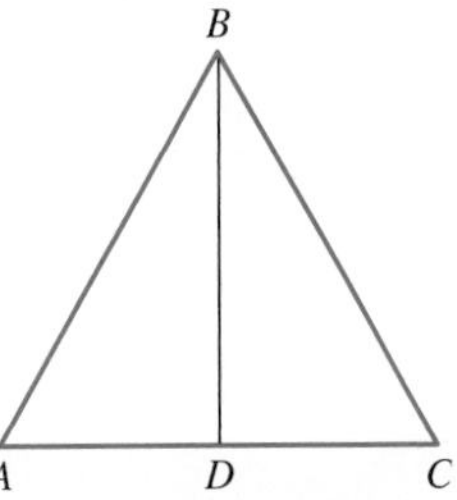

43. Prove Theorem 4.3: If the hypotenuse and a leg of one right triangle are congruent respectively to the hypotenuse and a leg of another right triangle, then the two triangles are congruent.

APPLICATIONS

44. The rules for a sailboat race stipulate that the triangular sail, labeled in the following figure, must be the same size and shape on each boat. List three measurements the judge can take on each sail so she can verify the sails are the same size without taking them down to measure them. Which triangle congruence property applies?

45. A **truss** is a rigid framework of wood or metal used in construction. A simple version used in home construction is the **king post truss**, labeled in the following figure. A builder must make 20 of these trusses and begins by cutting 20 beams of length AC and 20 of length DB. He then attaches each $\overline{DB}$ at the midpoint B of $\overline{AC}$ and perpendicular to $\overline{AC}$. Explain why the builder can cut 40 beams of length DC to complete the trusses. Which triangle congruence property applies?

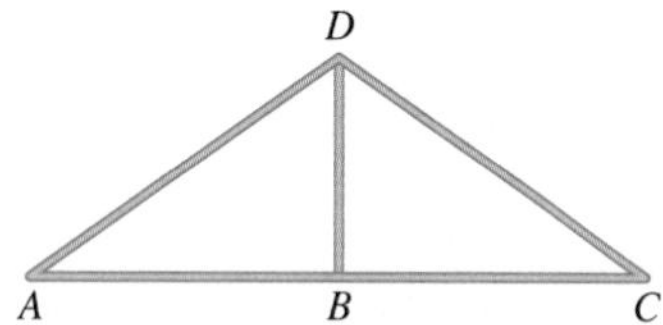

46. Young trees are staked to prevent toppling during strong winds while the root system is developing. A tree stake kit comes with equal-length ropes that are attached to the tree at the same height as shown in the following figure. If the ends of the ropes are anchored in opposite directions and at equal distances from the base of the tree, explain why the tree will stay perpendicular to the ground. Which triangle congruence property applies?

EXTENDED PROBLEMS

47. When six pairs of corresponding parts of two triangles are congruent, the triangles are congruent. If possible, draw an example of two *noncongruent* triangles that satisfy the following conditions. If not possible, then explain why.
    a. Three pairs of corresponding parts are congruent.
    b. Four pairs of parts are congruent.
    c. Five pairs of parts are congruent.

48. One of the most famous proofs of the Pythagorean theorem is by Euclid. He labeled it Proposition 47. The proof hinges on the ability to show two pairs of triangles are congruent. Research Proposition 47. Study Euclid's proof and draw a diagram that illustrates his proof. Write a paragraph outlining the steps in the proof.

49. In computer animation, different methods are used to create smooth, three-dimensional images from less detailed surfaces. One such method, created by Charles Loop and based on triangles, is called the **Loop subdivision method**. Research the Loop subdivision method. Suppose that a surface is modeled on a computer and is covered with a mesh of triangles. Describe the steps of the Loop subdivision method that would lead to a smooth-looking surface. How does the Loop method compare to other surface-smoothing methods? At what point in the process are congruent triangles formed?

4.3 PROBLEM SOLVING USING TRIANGLE CONGRUENCE

Applied Problem

It is recommended that gardeners plant tomatoes with three feet between plants. A truck farmer planting many tomato plants might use a square arrangement of plants in rows, as shown next.

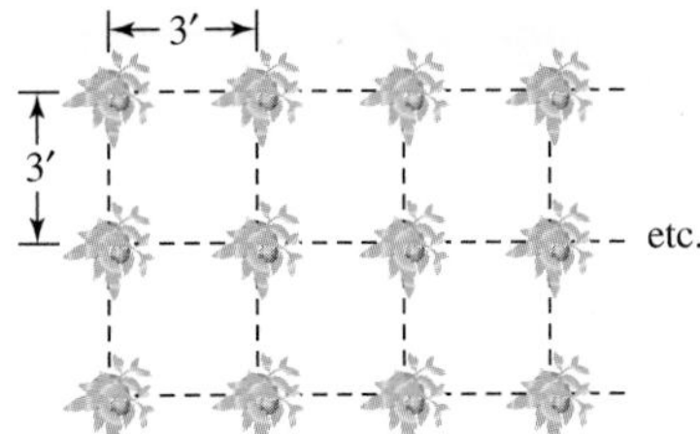

On the other hand, the farmer may make maximal use of the land by planting the tomatoes in a triangular grid, as shown next.

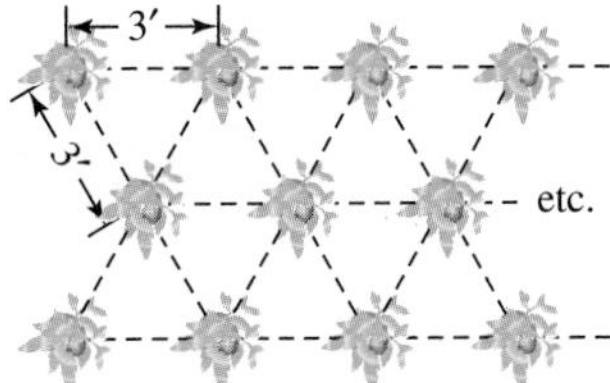

How much space between rows is saved by using the triangular planting arrangement?

In this section, we use triangle congruence to prove properties of geometric figures. In particular, we will derive some relationships that will be useful in later chapters.

Isosceles Triangles

In Chapter 2, we used the reflection symmetry of an isosceles triangle to establish a relationship between its base angles. The next theorem is a formal verification of that relationship.

THEOREM 4.5

In an isosceles triangle, the angles opposite the congruent sides are congruent.

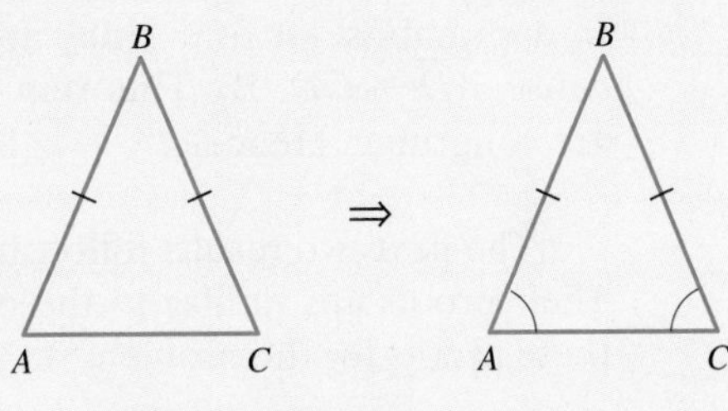

If $\overline{AB} \cong \overline{CB}$, then $\angle C \cong \angle A$.

We can summarize Theorem 4.5 as follows: Base angles of an isosceles triangle are congruent.

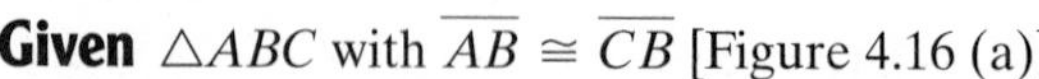

Given $\triangle ABC$ with $\overline{AB} \cong \overline{CB}$ [Figure 4.16 (a)]

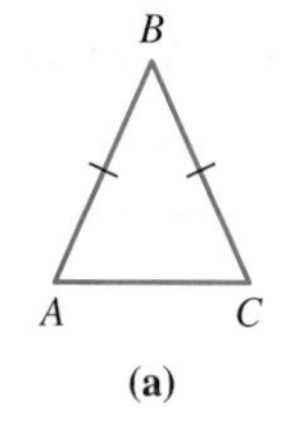

(a)

Prove $\angle C \cong \angle A$

Subgoal 1 Prove that $\triangle ABC$ is congruent to itself using the correspondence $\triangle ABC \cong \triangle CBA$.

Subgoal 2 Prove that the base angles are congruent by C.P.

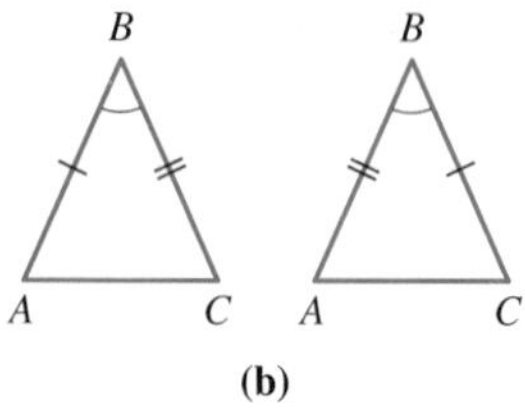

(b)

FIGURE 4.16

PROOF

Statement	Reason
1. $\overline{AB} \cong \overline{CB}$	**1.** Given
2. $\angle B \cong \angle B$	**2.** Reflexive property
3. $\overline{CB} \cong \overline{AB}$	**3.** Symmetric property
4. $\triangle ABC \cong \triangle CBA$	**4.** SAS Postulate [Figure 4.16(b)]
5. $\angle A \cong \angle C$	**5.** C.P. ■

Because every equilateral triangle is also an isosceles triangle, Theorem 4.5 applies to equilateral triangles also. In fact, by viewing an equilateral triangle in two different ways, the following corollary shows that all the angles in an equilateral triangle are congruent. We leave the proof of this corollary for the problem set.

COROLLARY 4.6

Every equilateral triangle is equiangular.

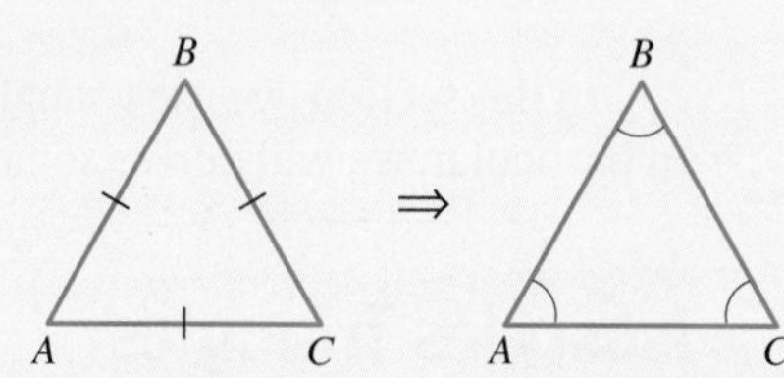

If $\overline{AB} \cong \overline{BC} \cong \overline{AC}$, then $\angle A \cong \angle B \cong \angle C$.

EXAMPLE 4.12 The rhombus $ABCD$ in Figure 4.17 has diagonal $\overline{AC}$, which forms $\angle 1$ and $\angle 2$. Show $\angle 1 \cong \angle 2$.

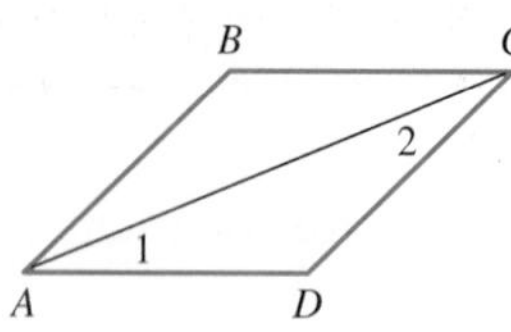

FIGURE 4.17

SOLUTION

In a rhombus, all four sides are congruent. Therefore, $\triangle ADC$ is isosceles because $\overline{AD} \cong \overline{CD}$. By Theorem 4.5, we find that the angles opposite these sides are congruent. Hence, $\angle 1 \cong \angle 2$. ■

The next two results follow from Postulate 4.2, the ASA Congruence Postulate. Their proofs are similar to the ones for Theorem 4.5 and Corollary 4.6. We leave these proofs for the problem set.

THEOREM 4.7

If two angles of a triangle are congruent, then the sides opposite those angles are congruent.

If $\angle C \cong \angle A$, then $\overline{AB} \cong \overline{CB}$.

We can summarize this theorem as follows: If two angles of a triangle are congruent, then the triangle is isosceles.

By extension, if all three angles of a triangle are congruent, then the triangle is equilateral. This result is stated in the next corollary.

COROLLARY 4.8

Every equiangular triangle is equilateral.

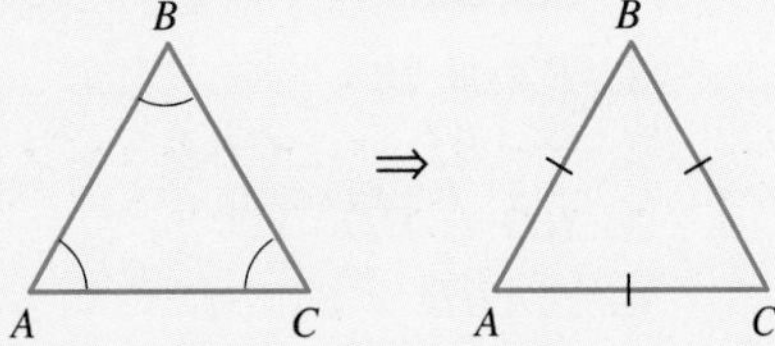

If $\angle A \cong \angle B \cong \angle C$, then $\overline{AB} \cong \overline{BC} \cong \overline{AC}$.

Perpendicular Bisector of a Segment

Before considering the next theorem, we need to state some definitions. The word **bisect** means to divide in half. Thus, an **angle bisector** is a line, ray, or line segment that divides an angle into two congruent angles. The **perpendicular bisector of a line segment** is a line, ray, or line segment that passes through the midpoint of the line segment and is perpendicular to the line segment.

The next two theorems display a relationship between isosceles triangles and perpendicular bisectors.

THEOREM 4.9

In an isosceles triangle, the ray that bisects the vertex angle bisects the base and is perpendicular to it.

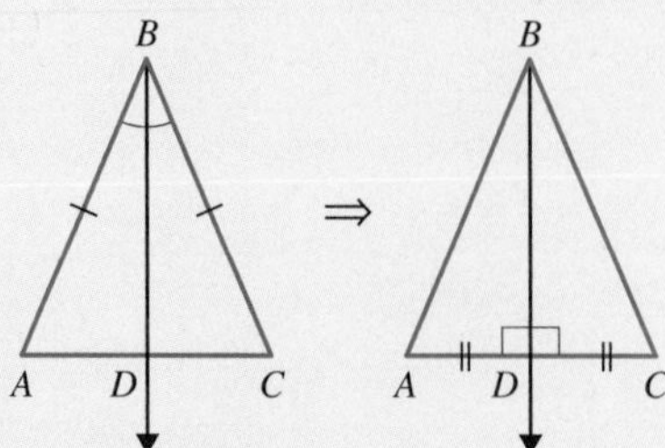

If $\overline{BA} \cong \overline{BC}$ and $\angle ABD \cong \angle CBD$, then $\overrightarrow{BD}$ is the perpendicular bisector of $\overline{AC}$.

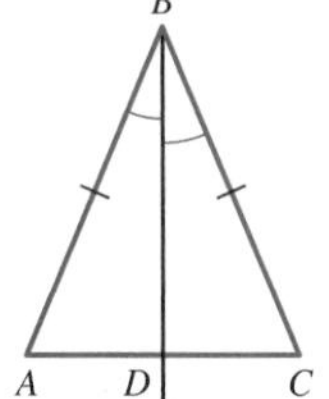

FIGURE 4.18

Given $\triangle ABC$ with $\overline{BA} \cong \overline{BC}$ and $\overrightarrow{BD}$ the angle bisector of $\angle B$ (Figure 4.18)

Prove $\overline{AD} \cong \overline{CD}$ and $\overline{BD} \perp \overline{AC}$

Subgoal Prove that $\triangle ABD \cong \triangle CBD$. Then use C.P.

PROOF

Statement	Reason
1. $\overline{BA} \cong \overline{BC}$	1. Given
2. $\angle ABD \cong \angle CBD$	2. $\overrightarrow{BD}$ is the angle bisector.
3. $\overline{BD} \cong \overline{BD}$	3. Reflexive property
4. $\triangle ABD \cong \triangle CBD$	4. SAS Congruence
5. $\overline{AD} \cong \overline{CD}$	5. C.P.
6. $\angle ADB \cong \angle CDB$	6. C.P.
7. $\angle ADB$ and $\angle CDB$ are right angles, therefore, $\overline{BD}$ is perpendicular to $\overline{AC}$.	7. Two congruent angles that are supplementary are right angles.

■

A **median** of a triangle is a line segment whose endpoints are a vertex of the triangle and the midpoint of the side opposite that vertex. In Figure 4.19, $\overline{AD}$ is a median of $\triangle ABC$. Therefore, we can summarize Theorem 4.9 as follows: In an isosceles triangle, the bisector of the vertex angle is also the altitude to the base, the perpendicular bisector of the base, and the median to the base. Because every equilateral triangle is isosceles, this theorem also holds for any equilateral triangle. In the equilateral triangle, any angle may be considered the "vertex angle" and any side the "base."

FIGURE 4.19

EXAMPLE 4.13

a. In Figure 4.20(a), B is on the perpendicular bisector of $\overline{AC}$. Find AB and BC.

b. In Figure 4.20(b), H is the midpoint of $\overline{EG}$. Find $\angle EHF$.

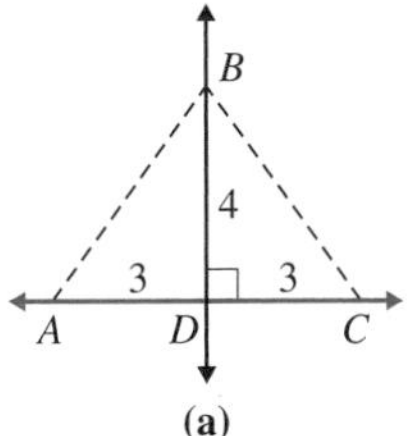

(a)

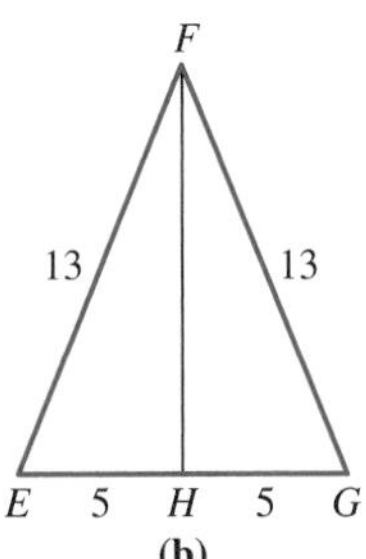

(b)

FIGURE 4.20

SOLUTION

a. By the Pythagorean theorem, $AB = \sqrt{3^2 + 4^2} = 5$. In a similar fashion, $BC = 5$. Thus, $AB = BC$; that is, B is equidistant from A and C.

b. First, $\triangle EFH \cong \triangle GFH$ by SSS Congruence. Thus, $\angle EHF \cong \angle GHF$ by corresponding parts. Because these angles are supplementary, $\angle EHF = \angle GHF = 90°$. This means that $\overline{FH}$ is the perpendicular bisector of $\overline{EG}$. ■

We can generalize this example in the following theorem.

> **THEOREM 4.10 Perpendicular Bisector Theorem**
>
> A point is on the perpendicular bisector of a line segment if and only if it is equidistant from the endpoints of the segment.
>
> 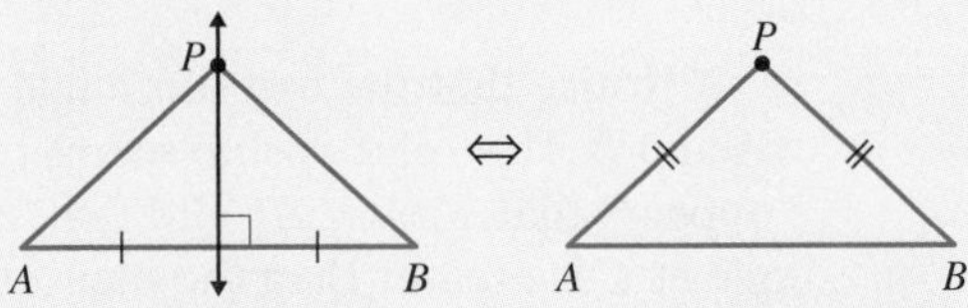
>
>
> 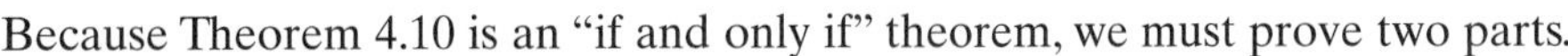
>
> P is on the perpendicular bisector of $\overline{AB}$ if and only if $AP = BP$.

Because Theorem 4.10 is an "if and only if" theorem, we must prove two parts.

Subgoal 1 Prove that if a point is on the perpendicular bisector of a segment, then it is equidistant from the endpoints of the segment.

Subgoal 2 Prove that if a point is equidistant from the endpoints of a segment, then it is on the perpendicular bisector of the segment.

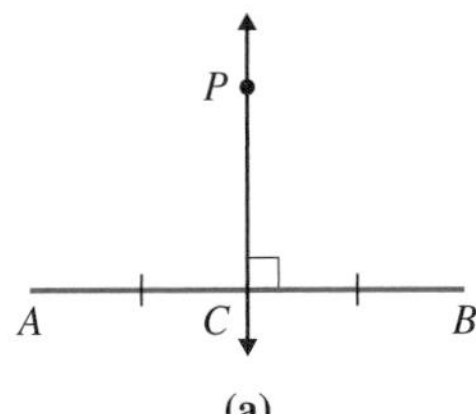

(a)

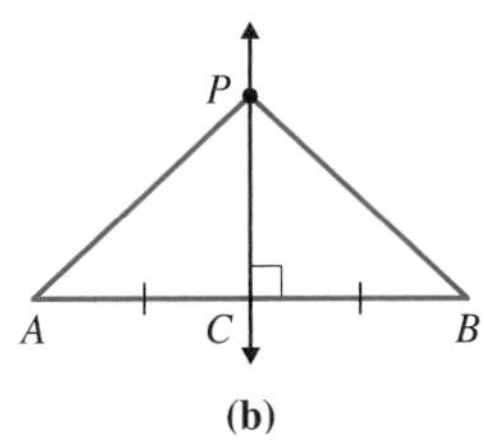

(b)

FIGURE 4.21

Proof of Subgoal 1 Let P be a point on the perpendicular bisector of $\overline{AB}$ [Figure 4.21(a)]. Then, in Figure 4.21(b), we can see that $\triangle ACP \cong \triangle BCP$ by LL Congruence. Therefore, $AP = BP$ by corresponding parts, which means P is equidistant from A and B.

Proof of Subgoal 2 Let P be equidistant from the endpoints of $\overline{AB}$ [Figure 4.22(a)]. Let C be the midpoint of $\overline{AB}$, and draw $\overline{PC}$ [Figure 4.22(b)]. Then, $\triangle ACP \cong \triangle BCP$ by SSS Congruence. By corresponding parts, $\angle PCA \cong \angle PCB$. Also, $\angle PCA$ and $\angle PCB$ are supplementary, so $\angle PCA = \angle PCB = 90°$. Thus, $\overline{PC}$ is the perpendicular bisector of $\overline{AB}$, so P lies on the perpendicular bisector of $\overline{AB}$. ■

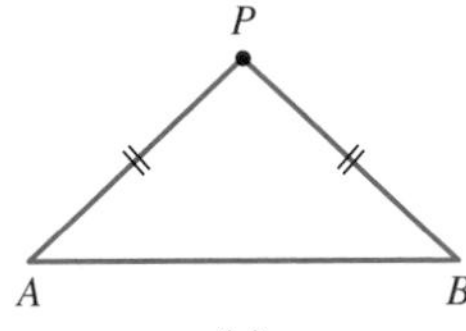

(a)

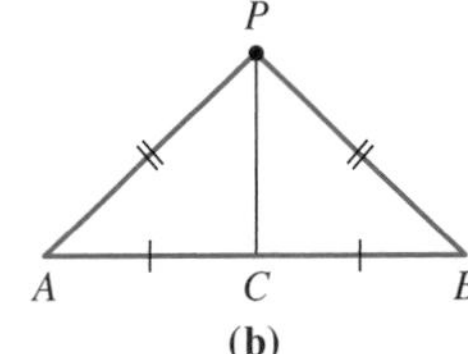

(b)

FIGURE 4.22

We will apply Theorem 4.10 in Section 7.4.

We proved Theorem 4.10 in two parts due to the biconditional "if and only if." Because our goal was to prove this theorem, each of these two parts was considered as a subgoal to reach our goal. We will also organize the next two theorems into subgoals whose proofs will be combined to form a proof of the theorem.

Exterior Angle of a Triangle

Although we showed informally in Chapter 2 that the angle sum in a triangle is 180°, we will not be able to prove this fact formally until Chapter 5. However, we can prove a theorem that shows a relationship between an exterior angle of a triangle and the two nonadjacent interior angles. An **exterior angle** is formed by one side of a triangle and the extension of an adjacent side as illustrated by $\angle 3$ in Figure 4.23.

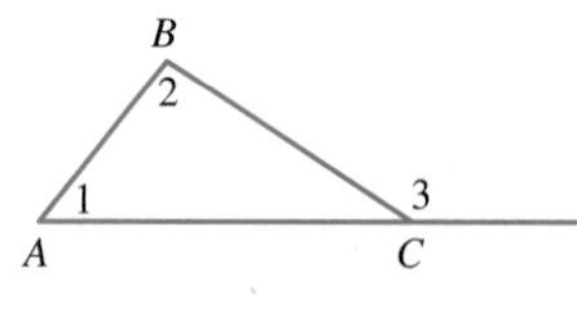

FIGURE 4.23

Notice that the exterior angle is larger than either of the two interior angles at A and B. This relationship seems true for any triangle. For example, in Figure 4.23 it appears that $\angle 3 > \angle 1$ and that $\angle 3 > \angle 2$. That is, the measure of an exterior angle at C is greater than the measure of either nonadjacent, interior angle. This relationship is stated in the next theorem.

THEOREM 4.11 Exterior Angle Theorem

The measure of an exterior angle of a triangle is greater than the measure of either of the nonadjacent interior angles.

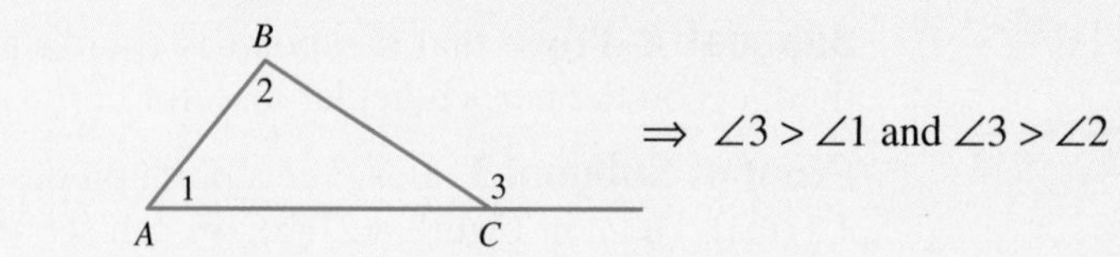

Given $\triangle ABC$ with an exterior angle $\angle BCD$ at C [Figure 4.24(a)]

Prove $\angle BCD > \angle A$ and $\angle BCD > \angle B$

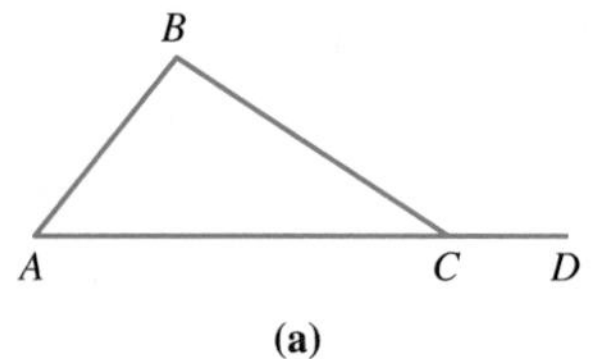

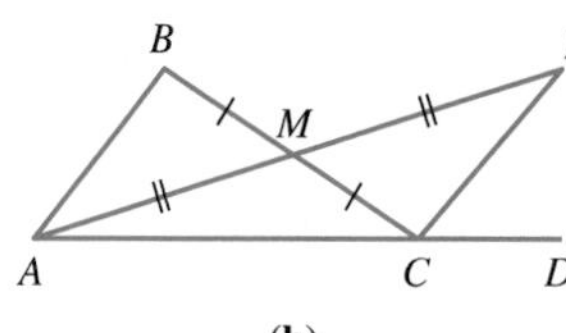

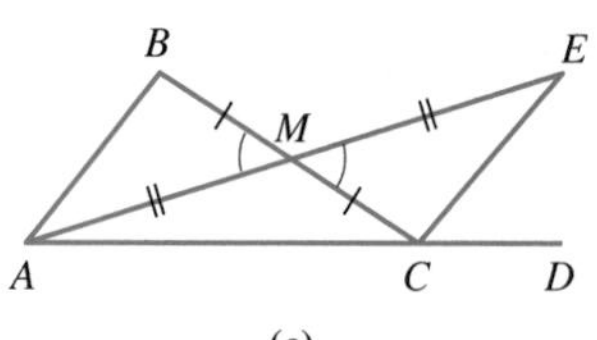

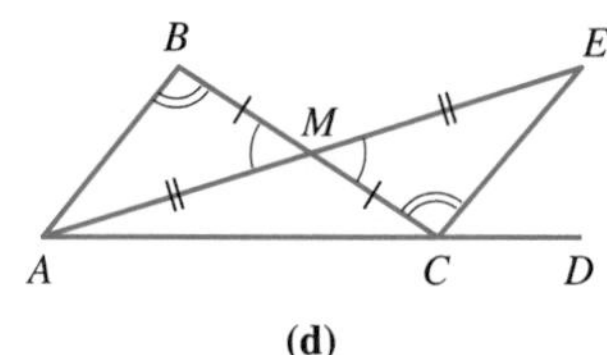

FIGURE 4.24

Plan Because this proof requires several steps, we will decompose it into subgoals that we can prove more easily. Then the proof of the theorem will follow from the proofs of the subgoals. Locate the midpoint M of $\overline{BC}$. Then draw $\overline{AE}$ through M so that $AM = EM$ [Figure 4.24(b)].

Subgoal 1 Prove that $\triangle AMB \cong \triangle EMC$.

Subgoal 2 Using the result of Subgoal 1, prove that $\angle BCD > \angle B$.

Proof of Subgoal 1 Because $\angle AMB$ and $\angle EMC$ are vertical angles, they are congruent [Figure 4.24(c)]. Also $BM = CM$ and $AM = EM$ because of our choice of point M. Therefore, $\triangle AMB \cong \triangle EMC$ by SAS Congruence.

Proof of Subgoal 2 By corresponding parts, $\angle B \cong \angle BCE$ [Figure 4.24(d)]. Also, in Figure 4.24(d), $\angle BCD = \angle BCE + \angle ECD$, so $\angle BCD > \angle BCE$. Therefore, by substitution, $\angle BCD > \angle B$. ■

In a similar fashion, we can show that $\angle BCD > \angle A$. We leave this proof for the problem set.

Solution of Applied Problem

In the square arrangement there are 3 ft between rows. Consider the situation where the plants are 3 ft apart in triangular regions such as $\triangle ABC$.

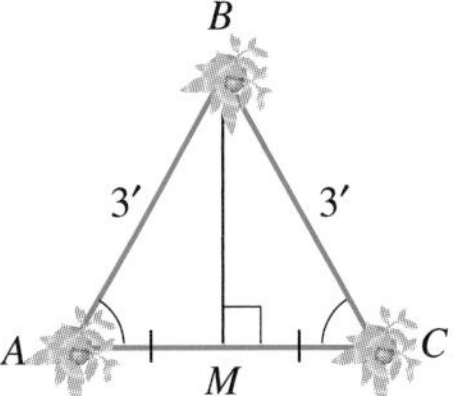

In equilateral $\triangle ABC$, $AB = BC = AC = 3$ ft. If M is the midpoint of $\overline{AC}$, then $\overline{BM}$ is the perpendicular bisector of $\overline{AC}$. Therefore, $AM = MC = \frac{3}{2}$. We can determine BM, which represents the distance between rows, using the Pythagorean theorem.

$$AB^2 = AM^2 + BM^2$$
$$3^2 = \left(\tfrac{3}{2}\right)^2 + BM^2$$
$$9 = \tfrac{9}{4} + BM^2$$
$$\tfrac{27}{4} = BM^2$$

Therefore, $BM = \sqrt{\frac{27}{4}} \approx 2.6$ ft to the nearest tenth of a foot. So, by using the triangular arrangement, we find that $3 - 2.6 = 0.4$ ft, or about 5 in., is saved between rows.

GEOMETRY AROUND US The geodesic domes designed by Buckminster Fuller are portions of special polyhedra composed of many congruent triangular faces of several types. The triangles provide rigidity to the structure. Also, the nearly spherical shape allows a minimum amount of material to contain a given volume inside the structure.

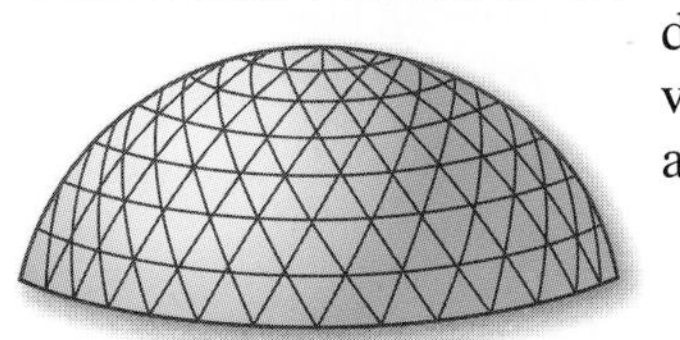

PROBLEM SET 4.3

EXERCISES/PROBLEMS

1. In $\triangle XYZ$, $\angle X = 70°$ and $\overline{YP}$ is the perpendicular bisector of $\overline{XZ}$. Find $\angle Z$. Justify your answer.

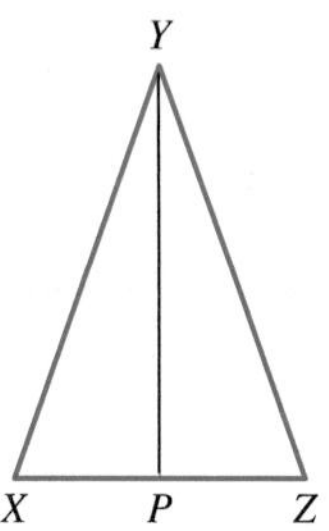

2. In $\triangle PQR$, $\angle P \cong \angle R$, $\overline{QT}$ bisects $\angle PQR$, and $\angle R = 61°$. Find $\angle PQT$. Justify your answer.

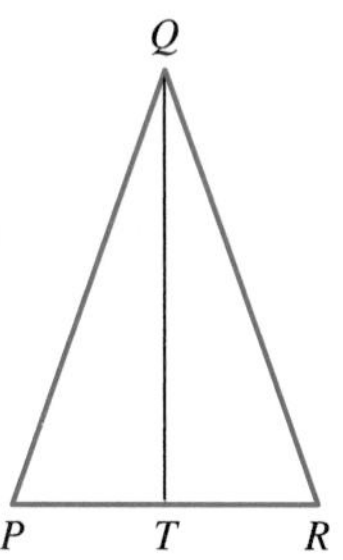

3. In the figure shown, $\overline{AB} \cong \overline{AC}$, $\overline{EC} \cong \overline{ED}$, and $\overline{BC} \cong \overline{DC}$. If $\angle BEC = 86°$, find the measures of $\angle BCA$ and $\angle ABE$.

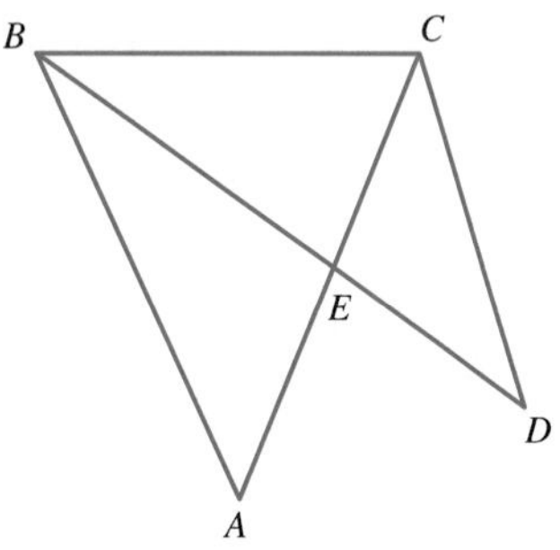

4. In the figure shown, $\triangle ADE$ and $\triangle ABE$ are isosceles with $\overline{AD} \cong \overline{DE}$ and $\overline{AB} \cong \overline{AE}$. If $\angle ADE = 50°$ and $\angle BAC = 45°$, find the measures of $\angle DEC$ and $\angle ACE$.

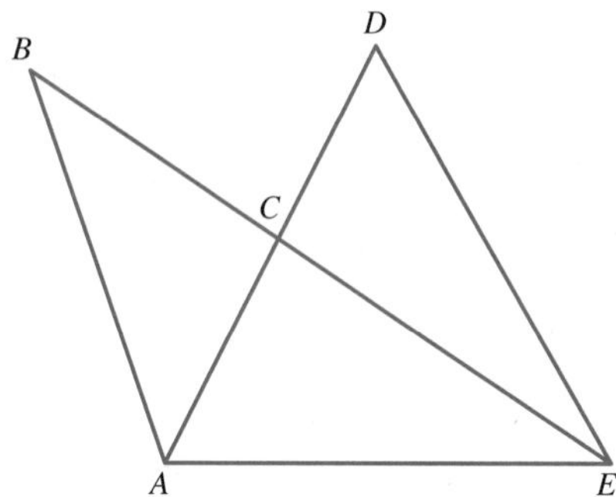

5. In the following figure, in addition to the congruent segments indicated, $\overline{BD} \cong \overline{DG}$. If A, G, and E are collinear, F, G, and C are collinear, and $\angle ADG = 28°$, find the measures of the following angles.

a. $\angle GBD$ **b.** $\angle ACG$ **c.** $\angle AGB$

d. $\angle FGE$ **e.** $\angle CGD$ **f.** $\angle BGC$

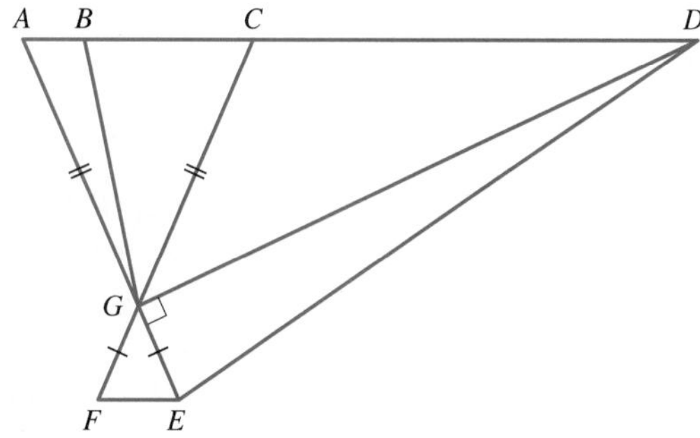

6. In the following figure, in addition to the congruent segments indicated, $\overline{AB} \cong \overline{FB}$. If $\angle ABF = 20°$, find the measures of the following angles.

a. $\angle FAB$ **b.** $\angle FDE$ **c.** $\angle FCB$

d. $\angle BFD$ **e.** $\angle CFB$ **f.** $\angle AFC$

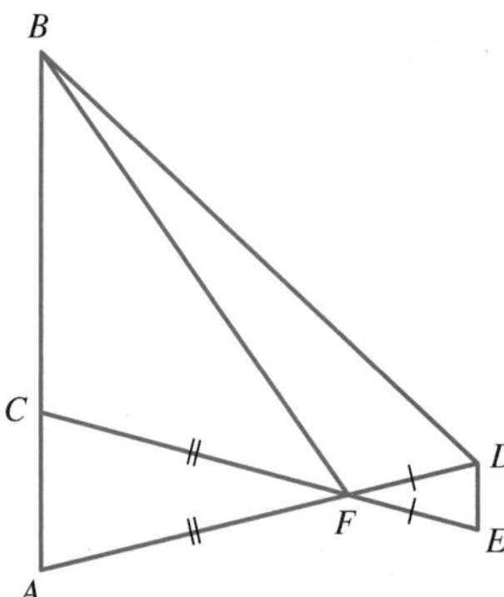

7. In isosceles $\triangle ABC$, $\angle ABD \cong \angle CBD$. Complete the following statements.

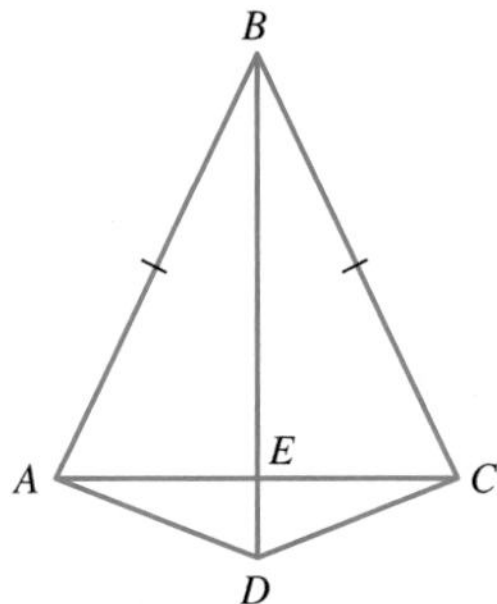

a. $\angle AEB = \angle CEB =$ _____

b. $\overline{AE} \cong \overline{CE}$ by _____

c. $\overline{AD} \cong \overline{CD}$ by _____

8. In the figure shown, $\overline{UX}$ is the perpendicular bisector of $\overline{SW}$. If UX is 4 in., $SX = 4$ in., and $TU = 3$ in., complete the following statements.

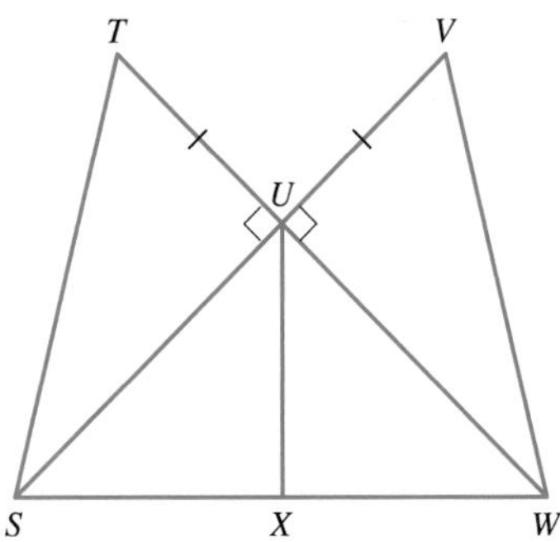

a. $\overline{SU} \cong \overline{WU}$ by _____

b. $\triangle STU \cong \triangle$ _____ by _____

c. $VW =$ _____ in.

PROOFS

9. In the following figure, congruent angles are marked. Prove $\overline{WX} \cong \overline{YV}$.

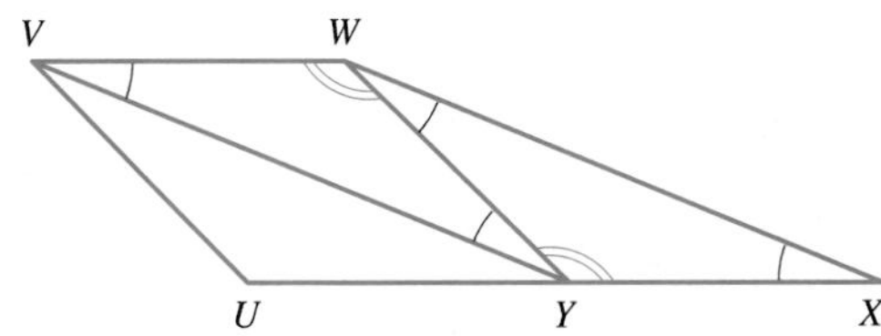

10. In the following figure $\overline{NO} \cong \overline{PO}$, $\overline{MS} \cong \overline{RQ}$, and $\angle OSR \cong \angle ORS$. Prove that $\triangle MNR \cong \triangle QPS$.

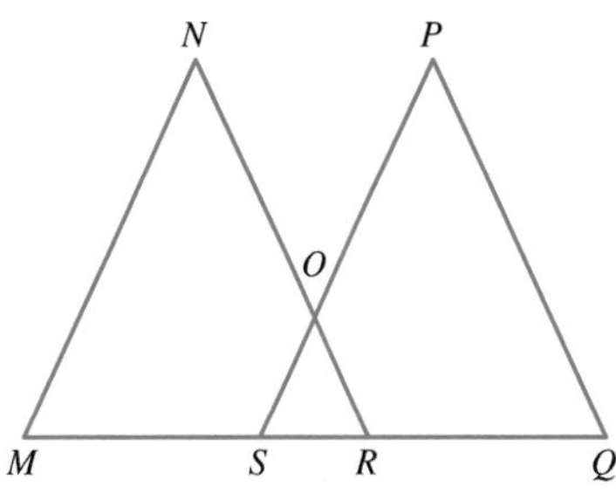

11. In isosceles $\triangle PQR$, $\overline{PY}$ and $\overline{RX}$ are altitudes, and $\overline{PX} \cong \overline{RY}$. Prove that $\triangle PXR \cong \triangle RYP$.

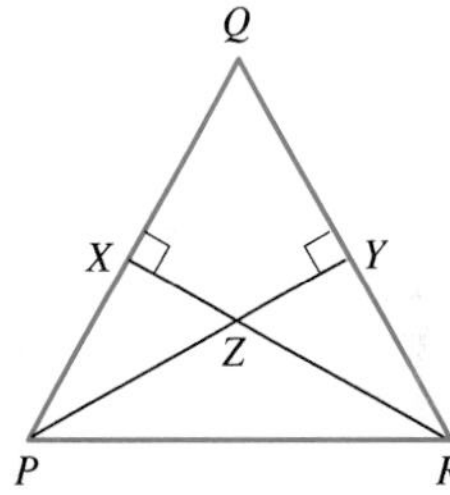

12. In $\triangle ABC$, $\overline{BD}$ is an altitude of $\triangle ABC$ and also the bisector of $\angle B$. Prove that $\triangle ABC$ is isosceles.

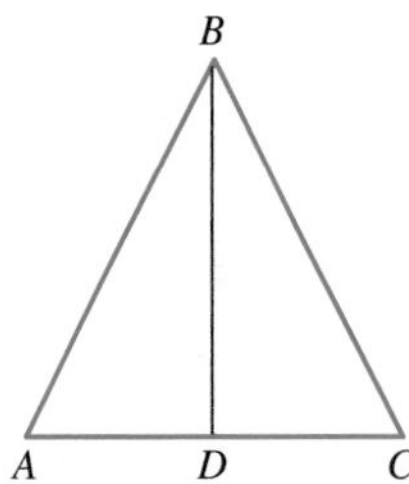

13. In $\triangle ABC$, $\overline{BD}$ is an altitude of $\triangle ABC$ and also the bisector of $\angle B$. Prove that $\overline{BD}$ is the perpendicular bisector of $\overline{AC}$.

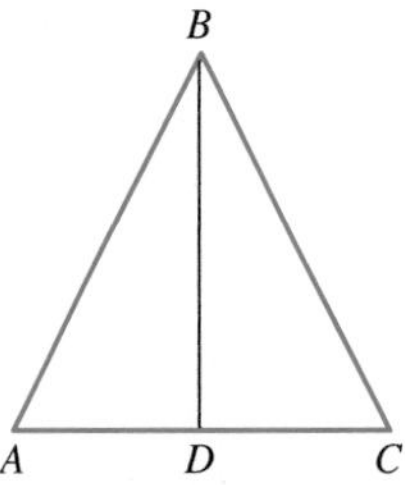

14. Prove or disprove: For any triangle $\triangle ABC$, if $\triangle ABC \cong \triangle CBA$, then $\triangle ABC$ is isosceles.

15. Prove that in any isosceles triangle, the median drawn from the vertex angle is the perpendicular bisector of the opposite side.

16. Prove that in any isosceles triangle, the median drawn from the vertex angle is the bisector of that angle.
17. Prove Corollary 4.6: Every equilateral triangle is equiangular.
18. Prove Theorem 4.7: If two angles of a triangle are congruent, then the sides opposite those angles are congruent.
19. Prove Corollary 4.8: Every equiangular triangle is equilateral.
20. Prove that $\angle BCD > \angle A$ in Theorem 4.11.

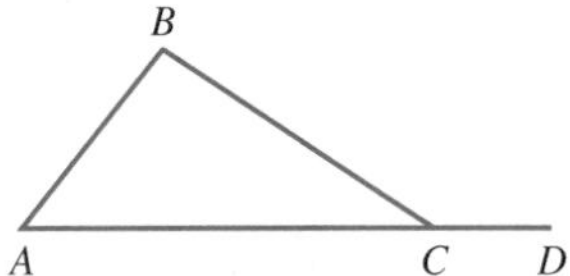

APPLICATIONS

21. A womans want to align three identical pictures vertically along the middle of a wall as shown.

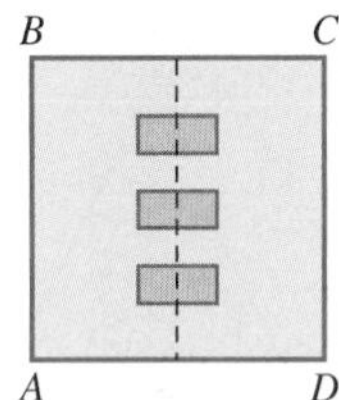

Once she has determined approximate locations for the pictures, how can she check the positions by using only a tape measure and a pencil?

22. A construction worker positions the end of one board at the center of another board and must make sure the boards are perpendicular to each other before nailing them together. Explain how the worker can verify the boards are perpendicular using only a tape measure.

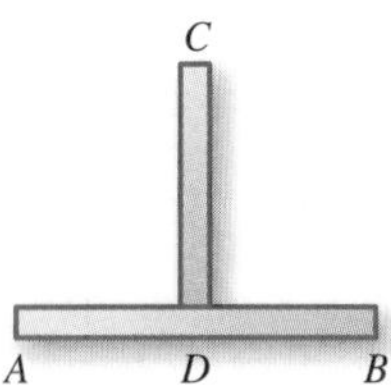

EXTENDED PROBLEMS

23. Use a ruler to draw an isosceles triangle with side lengths 7 cm, 7 cm, and 6 cm. Label the three vertices as "1 or 2", "3 or 4", and "5 or 6" respectively. Place a dot randomly in the interior of the triangle and roll a single die. Measure the distance from the dot to the vertex labeled with the number on the die and place a new dot at the midpoint of the segment connecting the dot to the vertex, but do not draw the segment.
    a. Predict whether a pattern or shape will emerge in the array of dots if the process is repeated many times. To test your prediction, use this method to plot 25 additional points in the triangle, always using the previous dot to locate the next dot. What do you observe?
    b. The process described in this problem and applied many times in part (a) is called the "Chaos Game." Research the Chaos Game on the Internet. Who is credited with its creation? Search for a Chaos Game applet and observe what happens if the game is played 1000 times. What geometric figure would result if the game was played indefinitely? How would the result change if a different polygon was used initially or the distance used to place a new point was changed?
24. In bridges, trusses are required to support heavy weights and span great distances.
    a. Research the Warren truss and write a summary of your findings that includes answers to the following questions. Who is credited with the design and when was the patent obtained? What kinds of triangles are used in a Warren truss? What distance is a Warren truss designed to span? Under what conditions is a vertical beam added to certain triangles in the truss? Where is the longest Warren truss bridge located?
    b. Draw an example of a Warren truss with vertical beams. Label a triangle in the truss using the vocabulary from this section. Identify congruent beams, congruent angles, and right angles.
25. In cities, emergency services and businesses are often located so that they serve nonoverlapping areas in the community. One way to determine the service areas is to use a **Voronoi diagram**. A Voronoi diagram uses the perpendicular bisector of the segment connecting two locations to establish a service area around a location.
    a. Create the Voronoi diagram for the case in which there are two pizza delivery companies in a town. Draw a segment between the two

locations and use a protractor and ruler to construct the perpendicular bisector of the segment. The perpendicular bisector creates two service areas, one on each side of the line. Shade the service region for store B. If a customer lived on the perpendicular bisector of the segment, what could you say about his or her access to the pizza delivery service companies?

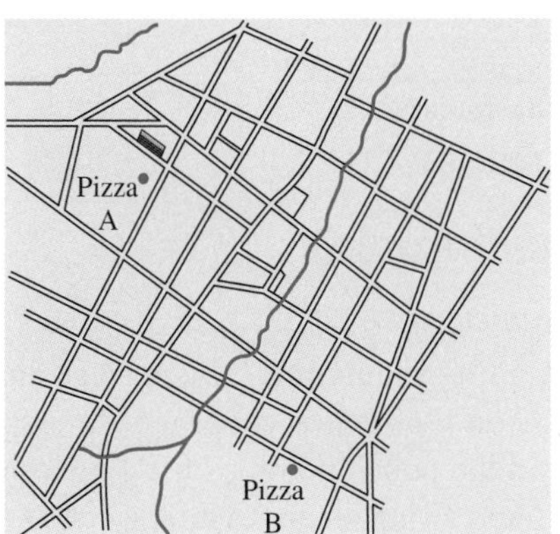

b. Research Voronoi diagrams and describe how the method can be extended to three or more locations. Download a street map and place six points on the map to represent locations for fire stations. Create the Voronoi diagram to show the service area for each fire station. Color the final diagram.

4.4 THE BASIC GEOMETRIC CONSTRUCTIONS

Applied Problem

A drafting technician wishes to calculate the shortest distance from point P to the roadway using the drawing shown. If the scale is 1 cm : 20 m, how can she calculate the distance?

P

The ancient Greeks used a compass and straightedge to construct geometric figures. A **compass** is a device that is used to draw circles or arcs of circles (Figure 4.25). A **straightedge** is a device that is used to draw straight line segments. Neither of these two devices has marks for measuring. (A ruler is commonly used as a straightedge, but all markings are ignored.) In this section, we introduce some basic constructions made using a compass and straightedge. We will use triangle congruence conditions to prove that those constructions are correct. Then the basic constructions will be combined to make more complex constructions.

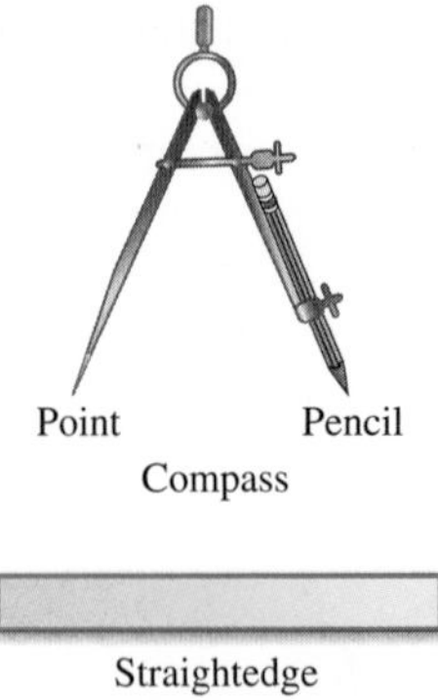

Compass

Straightedge

FIGURE 4.25

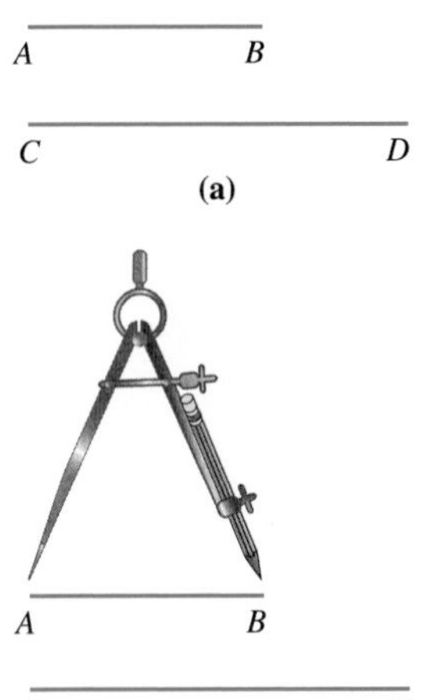
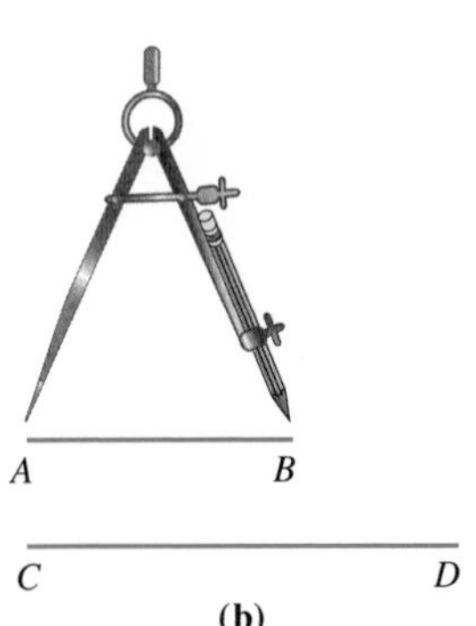

FIGURE 4.26

Copying Line Segments and Angles

Construction 1. *To copy a line segment*

Procedure To copy $\overline{AB}$, first draw a line segment $\overline{CD}$, using a straightedge [Figure 4.26(a)]. To make a segment $\overline{CE}$ on $\overline{CD}$ that is congruent to $\overline{AB}$, spread your compass so that the point is on A and the pencil point is on B [Figure 4.26(b)]. Pick up your compass, put the point on C and swing an arc to intersect $\overline{CD}$ at some point E [Figure 4.26(c)]. Then $\overline{CE}$ will be congruent to $\overline{AB}$.

Justification Because the compass is rigid, the lengths AB and CE will be the same. Therefore, $\overline{AB} \cong \overline{CE}$. ■

Construction 2. *To copy an angle*

Procedure Consider $\angle A$ and $\overline{DE}$ [Figure 4.27(a)].

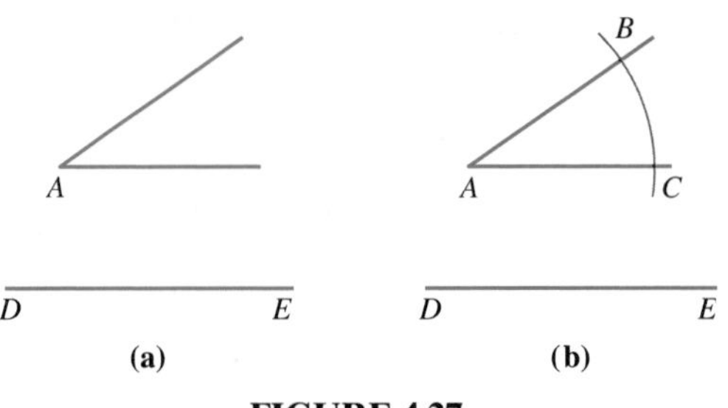

FIGURE 4.27

To copy $\angle A$ so that point A corresponds to D and one side of $\angle A$ corresponds to $\overline{DE}$, first swing an arc crossing the sides of $\angle A$ forming points B and C [Figure 4.27(b)].

Pick up your compass, put the point on D and, using the radius AC, swing an arc intersecting $\overline{DE}$ forming point F [Figure 4.27(c)].

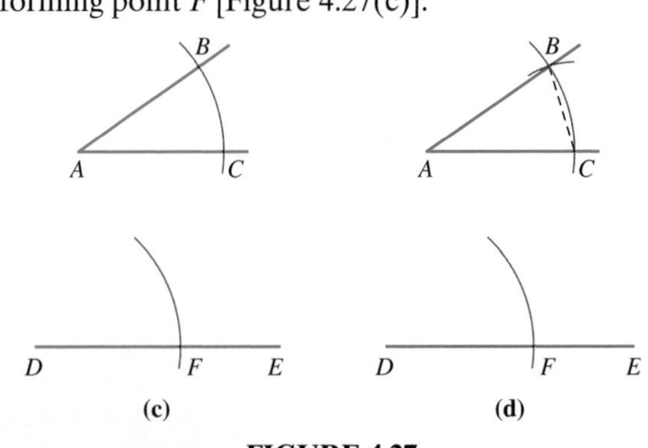
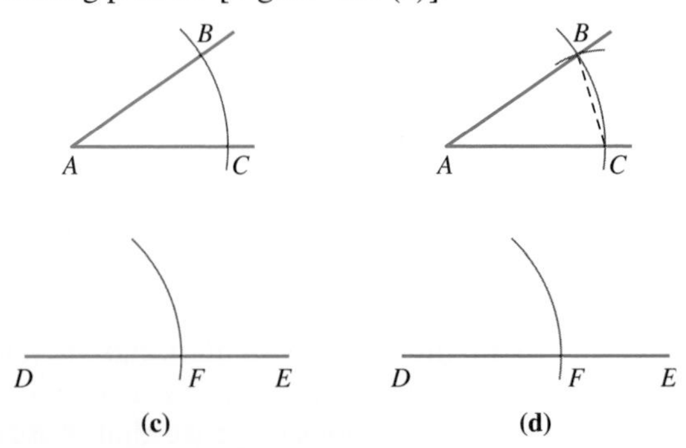

FIGURE 4.27

Use your compass to copy the distance BC [Figure 4.27(d)]. Set the opening of the compass as BC. Pick up your compass, put the point on F, and swing an arc intersecting your arc through F forming G [Figure 4.27(e)].

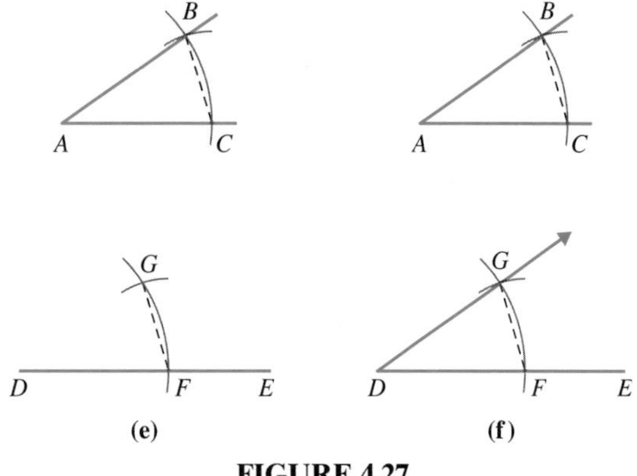

FIGURE 4.27

Draw $\overrightarrow{DG}$. Then $\angle D$ will be congruent to $\angle A$ [Figure 4.27(f)].

Justification Using the larger arcs in Figure 4.27(f), we can conclude that $AB = AC = DG = DF$. Also, in Figure 4.27(f), because $\overline{GF}$ is copied from $\overline{BC}$, we can conclude that $GF = BC$. Thus, $\triangle ABC \cong \triangle DGF$ by SSS and $\angle A \cong \angle D$ by C.P. ■

Bisecting Angles

Construction 3. *To bisect an angle*

Procedure Consider $\angle A$ [Figure 4.28(a)]. To bisect $\angle A$, swing an arc crossing the sides of $\angle A$ forming B and C [Figure 4.28(b)].

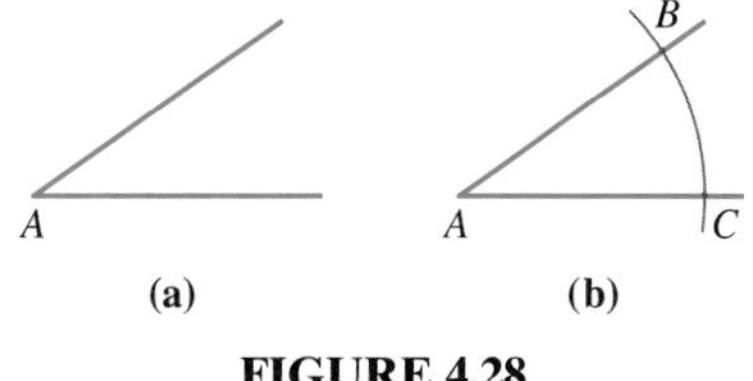

FIGURE 4.28

Pick up your compass, put the point on B, and swing an arc [Figure 4.28(c)]. Then put the point on C and, using the same opening, swing an arc intersecting your first arc forming D [Figure 4.28(d)].

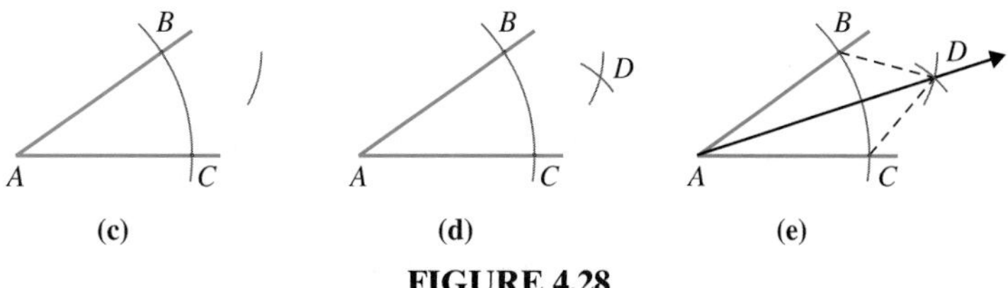

FIGURE 4.28

Draw $\overrightarrow{AD}$ [Figure 4.28(e)]. Then $\overrightarrow{AD}$ is the bisector of $\angle A$.

Justification By the construction in Figures 4.28(a)–(d), $AB = AC$ and $BD = CD$. Also, $AD = AD$ [Figure 4.28(e)]. Thus, $\triangle ABD \cong \triangle ACD$ by SSS. Therefore, $\angle BAD \cong \angle CAD$ by C.P., so $\overrightarrow{AD}$ bisects $\angle BAC$. ■

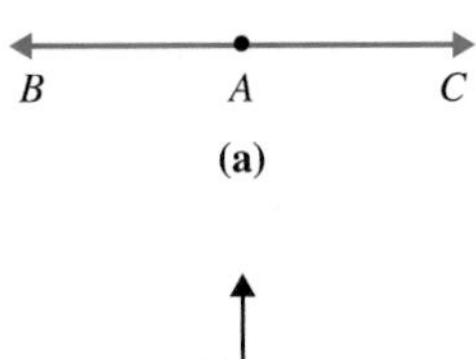

(a)

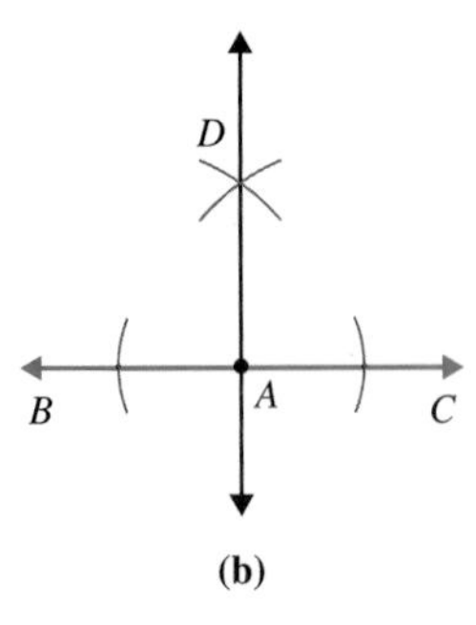

(b)

FIGURE 4.29

Constructing Perpendiculars

Construction 4. *To construct a perpendicular to a point on a line*

Procedure View $\angle BAC$ as a straight angle [Figure 4.29(a)]. Then we can construct a line perpendicular to $\overleftrightarrow{BC}$ at point A by bisecting $\angle BAC$. Follow the same procedure as in Construction 3 [Figure 4.29(b)].

Justification This is a special case of Construction 3 where the angle is a straight angle. Thus, $\angle BAD \cong \angle CAD$ by Construction 3. Since these angles are supplementary, they must both be 90°. ■

Construction 5. *To construct a perpendicular to a line from a point not on the line*

Procedure Given point A and line l [Figure 4.30(a)], swing an arc with center A that intersects l forming B and C [Figure 4.30(b)].

FIGURE 4.30

Now bisect $\angle BAC$ using Construction 3 [Figure 4.30(c)]. Then $\overrightarrow{AD}$ is perpendicular to l at E [Figure 4.30(d)].

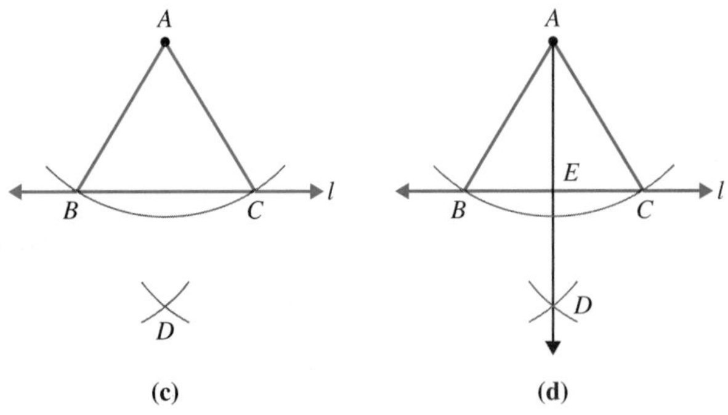

FIGURE 4.30

Justification According to Construction 3, $\overrightarrow{AD}$ bisects $\angle BAC$, so $\angle BAE \cong \angle CAE$ in Figure 4.30(d). Also, $AB = AC$ and $AE = AE$. So, by SAS, $\triangle BEA \cong \triangle CEA$, and $\angle BEA \cong \angle CEA$ by C.P. Because $\angle BEA$ and $\angle CEA$ are congruent and supplementary, they are right angles. Hence, $\overline{AD}$ is perpendicular to $\overline{BC}$. ■

Construction 6. *To construct the perpendicular bisector of a segment*

Procedure To bisect $\overline{BC}$ [Figure 4.31(a)], first place the compass point at B and make arcs above and below $\overline{BC}$ [Figure 4.31(b)], then place the point of the compass at C and make arcs above and below using the same radius, forming points A and D [Figure 4.31(c)].

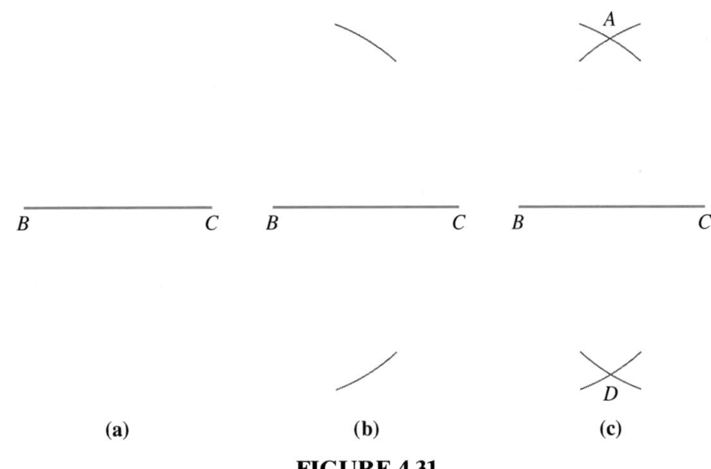

FIGURE 4.31

Draw $\overleftrightarrow{AD}$ [Figure 4.31(d)]. Then $\overleftrightarrow{AD}$ is perpendicular to $\overline{BC}$ at E.

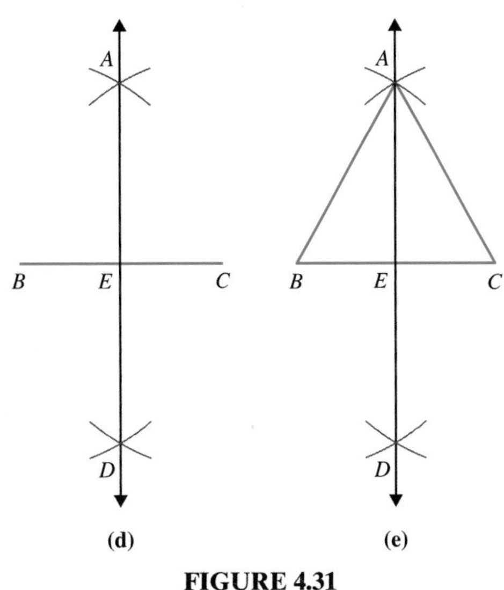

FIGURE 4.31

Justification Using an argument similar to that for Construction 5, we can conclude that $\triangle ABE \cong \triangle ACE$ [Figure 4.31(e)] and that $\overline{AD}$ is perpendicular to $\overline{BC}$. By C.P., we can conclude that $BE = CE$. Thus, AD is the perpendicular bisector of $\overline{BC}$. ■

We also can use Theorem 4.10 to justify this construction. Because we used the same opening on the compass for all arcs, we know that $AB = AC$. Thus, A is equidistant from B and C, so A lies on the perpendicular bisector of $\overline{BC}$. Likewise, D must lie on the perpendicular bisector of $\overline{BC}$ because $DB = DC$. Therefore, $\overleftrightarrow{AD}$ is the perpendicular bisector of $\overline{BC}$.

EXAMPLE 4.14 Using only a compass and straightedge, construct a 30°–60° right triangle with hypotenuse of length c as given in Figure 4.32(a).

c

FIGURE 4.32(a)

SOLUTION

There are several ways to do this construction. We will first construct a right angle using Construction 4 [Figure 4.32(b)].

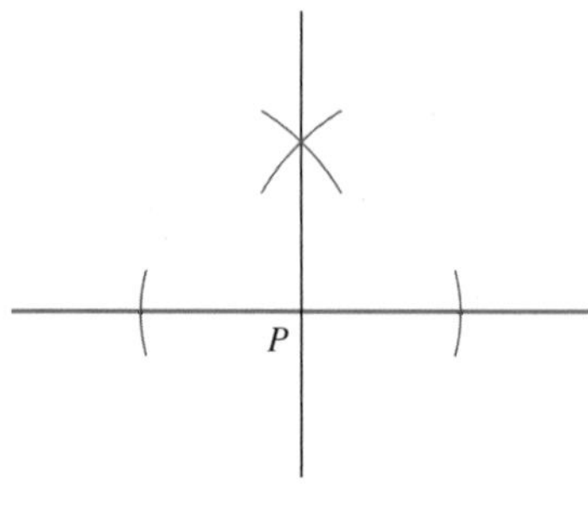

FIGURE 4.32(b)

Now, we can use the fact that in a 30°–60° right triangle the length of the side opposite the 30° angle is half the length of the hypotenuse. So one leg in this triangle must have length $\frac{c}{2}$. We bisect the given segment of length c using Construction 6 [Figure 4.32(c)] and copy half of the segment, making it leg $\overline{PQ}$ in the desired triangle [Figure 4.32(d)].

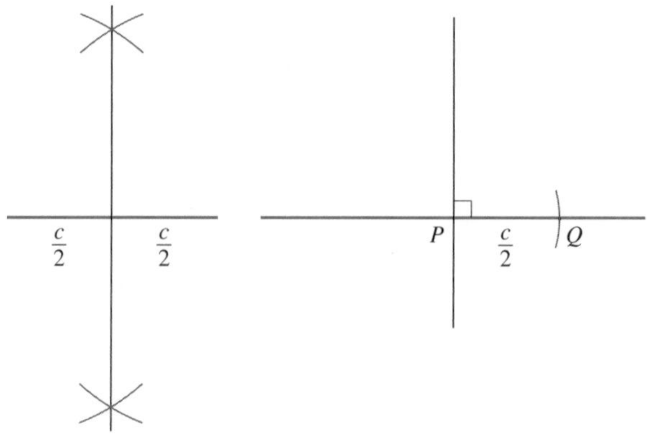

FIGURE 4.32(c) **FIGURE 4.32(d)**

Now the hypotenuse of the triangle must have length c. Placing the point of the compass at Q, we make an arc with radius c. The point R where this arc intersects the perpendicular through P is the third vertex of the triangle [Figure 4.32(e)].

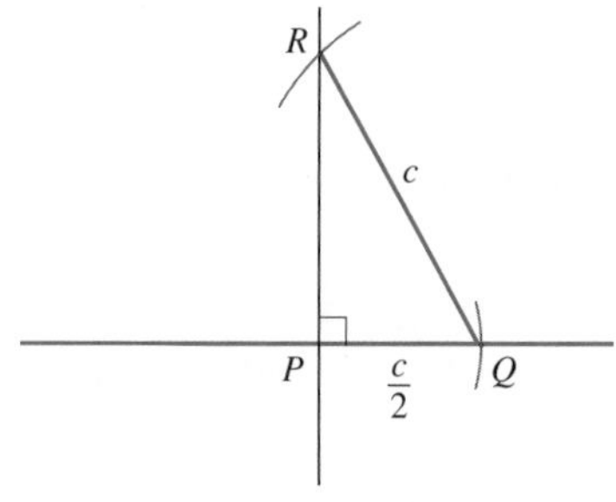

FIGURE 4.32(e)

■

EXAMPLE 4.15 Construct $\triangle ABC$ given two angles, $\angle A$ and $\angle B$, and their included side of length c as shown. Then construct the altitude from vertex C.

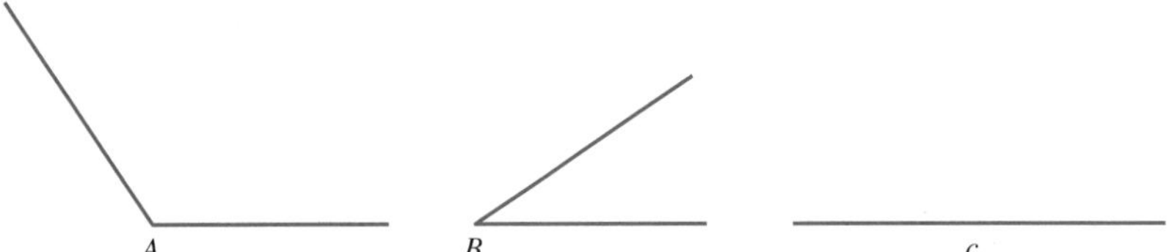

SOLUTION

We start by copying $\angle A$ [Figure 4.33(a)].

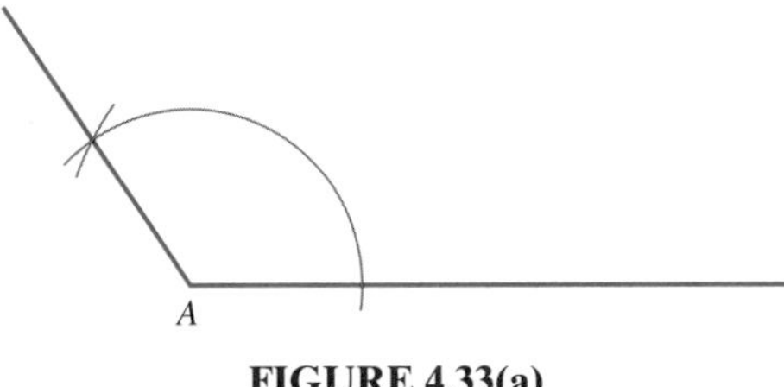

FIGURE 4.33(a)

Then mark off length c on one leg of A [Figure 4.33(b)]. The point of intersection will be vertex B.

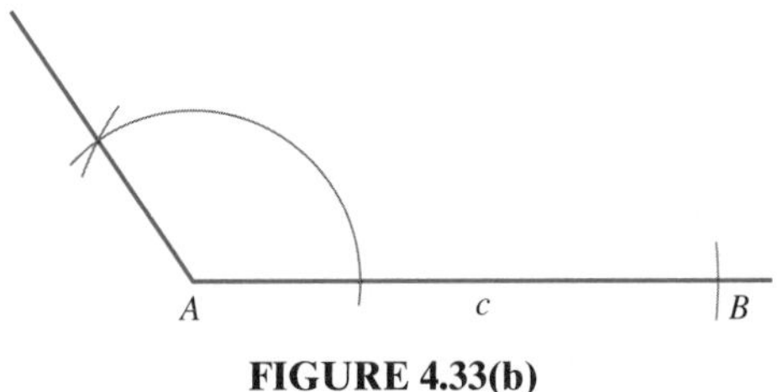

FIGURE 4.33(b)

Now copy $\angle B$. Extend the side of the angle at B so that it intersects the side from A forming point C [Figure 4.33(c)]. This produces $\triangle ABC$.

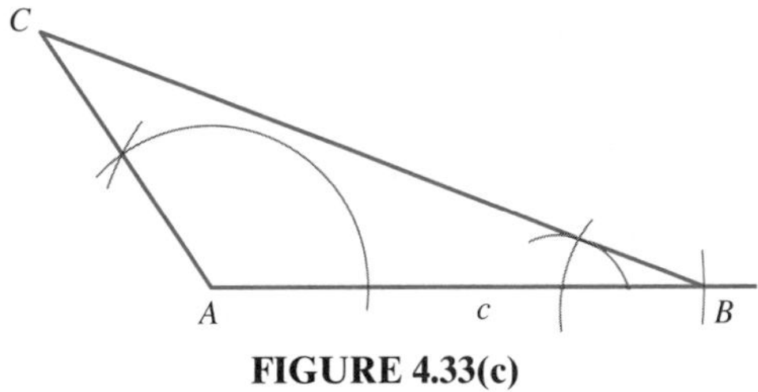

FIGURE 4.33(c)

The altitude from C is the line segment drawn from C that is perpendicular to $\overleftrightarrow{AB}$. We extend $\overline{AB}$ through A and construct a perpendicular from C to $\overleftrightarrow{AB}$ forming point D [Figure 4.33(d)]. Then $\overline{CD}$ is an altitude of $\triangle ABC$.

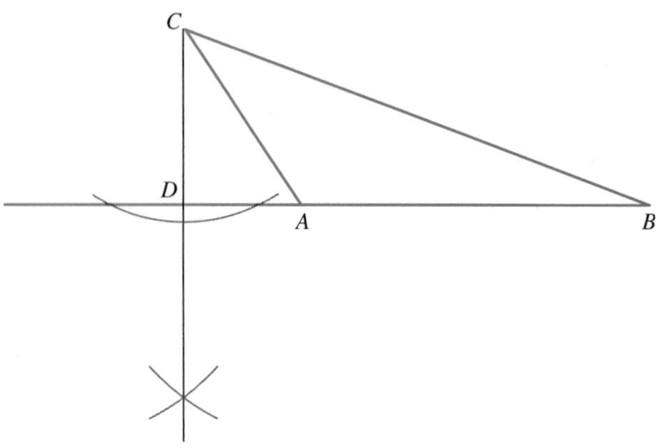

FIGURE 4.33(d)

Solution of Applied Problem

The shortest distance from P to the roadway is the length of the perpendicular segment from P to the roadway. She can construct a perpendicular from P to the line determined by the road and measure the length on the drawing.

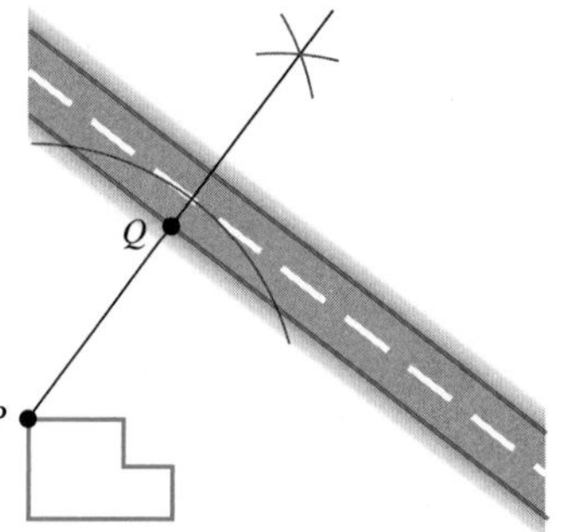

If $PQ = 2.3$ cm, for example, then we find the actual distance d as follows:

$$\frac{1\text{ cm}}{20\text{ m}} = \frac{2.3\text{ cm}}{d}, \qquad \text{or} \qquad d = 20(2.3) = 46\text{ m}$$

GEOMETRY AROUND US The Penrose Triangle, a familiar example of an optical illusion, is composed of three triangles with congruent corresponding angles. The figure appears to be three-dimensional, but is actually impossible to construct.

PROBLEM SET 4.4

EXERCISES/PROBLEMS

1. Draw a line segment, $\overline{AB}$. Use your compass and straightedge to construct another line segment whose length is $4(AB)$.

2. Draw a line segment, $\overline{AB}$. Use your compass and straightedge to construct another line segment whose length is $3(AB)$.

3. Draw an acute angle, $\angle A$. Use your compass and straightedge to construct another angle whose measure is twice that of $\angle A$.

4. Draw an obtuse angle, $\angle A$. Use your compass and straightedge to construct another angle whose measure is twice that of $\angle A$.

5. Draw an angle, $\angle A$. Use your compass and straightedge to bisect $\angle A$.

6. Draw a segment, $\overline{AC}$, and label point B between points A and C. Use your compass and straightedge to bisect $\angle ABC$.

7. Draw a line segment, $\overline{AB}$. Use a compass and straightedge to find the midpoint of $\overline{AB}$.

8. Draw a line segment, $\overline{PQ}$. Use a compass and straightedge to divide it into four congruent segments.

9. Draw a line, $\overleftrightarrow{AB}$, and a point, P, that is not on $\overleftrightarrow{AB}$. Use your compass and straightedge to construct a line through P perpendicular to $\overleftrightarrow{AB}$.

10. Draw a line, $\overleftrightarrow{AB}$. Choose another point, C, on $\overleftrightarrow{AB}$ Use your compass and straightedge to construct a line that is perpendicular to $\overleftrightarrow{AB}$ and that passes through C.

11. Construct an equilateral triangle given a side of length a.

a

12. Construct a triangle given three sides of lengths a, b, and c.

a

b

c

13. Construct a triangle given two sides of lengths a and b, and their included angle, $\angle C$.

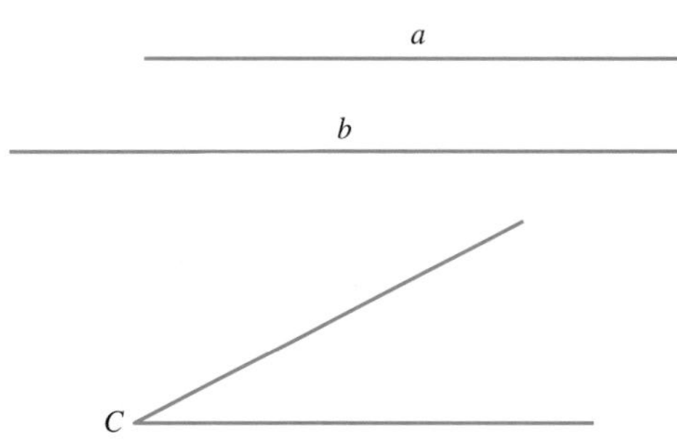

14. Construct a triangle given two angles, $\angle A$ and $\angle B$, and their included side of length c.

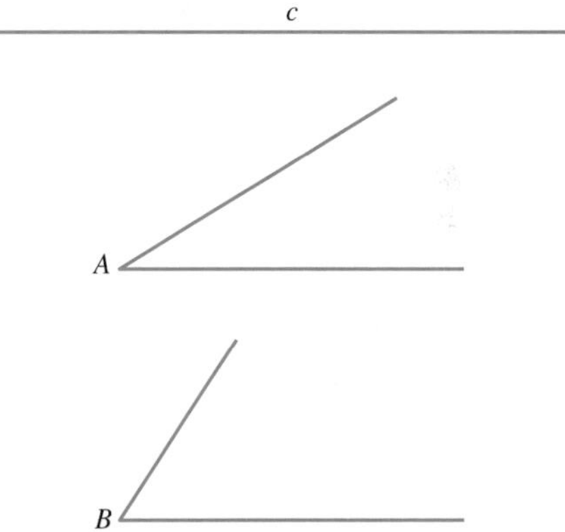

15. Use a compass and straightedge to construct a 60° angle.

16. Use a compass and straightedge to construct a 30° angle.

17. Use a compass and straightedge to construct a 45° angle.

18. Use a compass and straightedge to construct a 15° angle.

19. Use a compass and straightedge to construct a 120° angle.

20. Use a compass and straightedge to construct a 75° angle.

21. Construct a right triangle with a hypotenuse of length c and a leg of length a.

a

c

22. Construct a right isosceles triangle with legs of length a.

a

23. Trace $\triangle ABC$ and construct the altitude from B.

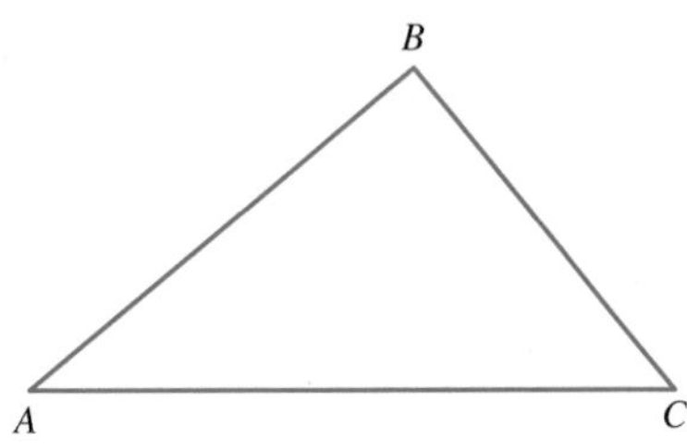

24. Trace $\triangle ABC$ and construct the median from B.

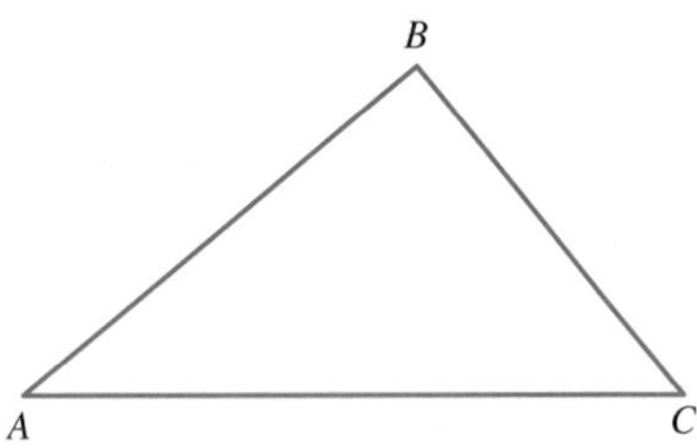

25. Trace $\triangle ABC$ and construct the perpendicular bisector to $\overline{AC}$ where $\triangle ABC$ is isosceles. Also, construct the bisector of $\angle B$ in $\triangle ABC$. What do you observe?

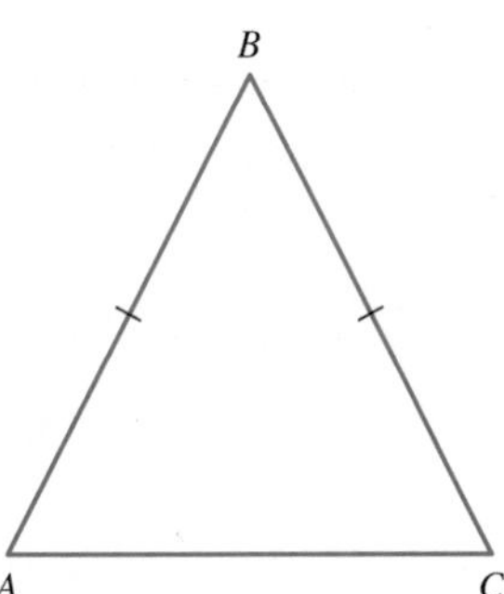

26. Trace square $ABCD$ and construct the bisectors of $\angle B$ and $\angle D$. What do you observe?

27. Trace $\triangle ABC$ and construct the perpendicular bisector of each side of $\triangle ABC$. What do you observe?

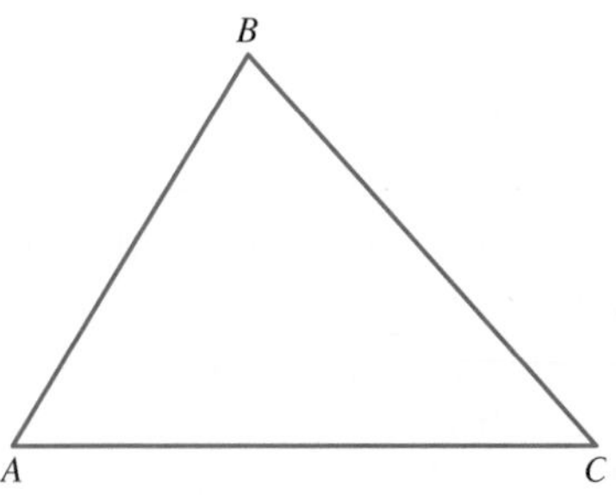

Repeat this construction with two other triangles of your choice. What generalization can you make?

28. Trace $\triangle ABC$ and construct the bisector of each angle of $\triangle ABC$. What do you observe?

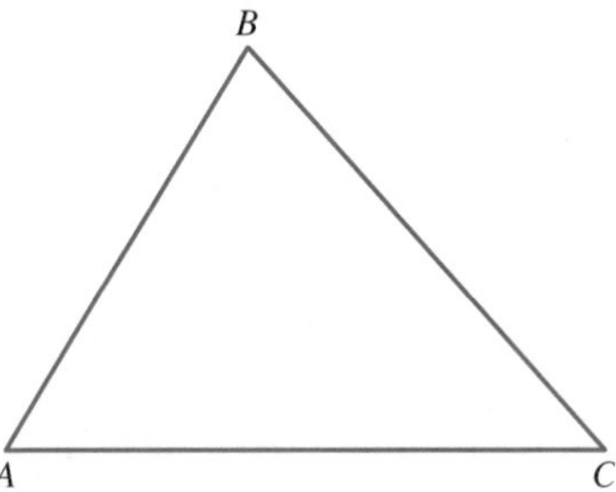

Repeat this construction with two other triangles of your choice. What generalization can you make?

29. Trace $\triangle ABC$ and construct the altitude to each side of $\triangle ABC$. What do you observe?

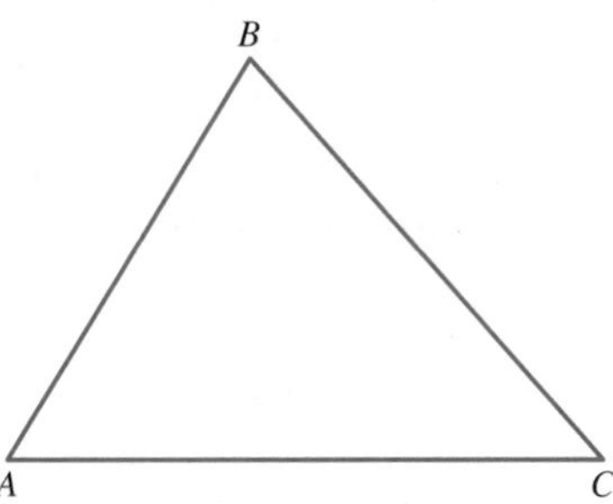

Repeat this construction with two other triangles of your choice. What generalization can you make?

30. Trace $\triangle ABC$ and construct the median to each side of $\triangle ABC$. What do you observe?

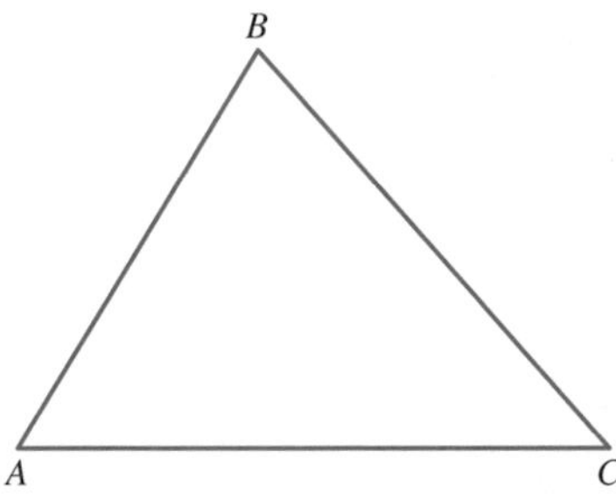

Repeat this construction with two other triangles of your choice. What generalization can you make?

PROOFS

31. Justify your construction in Exercise 11.

32. Justify your construction in Exercise 12.

33. Justify your construction in Exercise 13.

34. Justify your construction in Exercise 14.

35. Justify your construction in Exercise 23.

36. Justify your construction in Exercise 24.

APPLICATIONS

37. Draftpersons sometimes employ the following method to bisect a line segment $\overline{AB}$.

i. Angles of 45° are drawn at A and at B.

ii. A perpendicular is drawn from the point of intersection C.

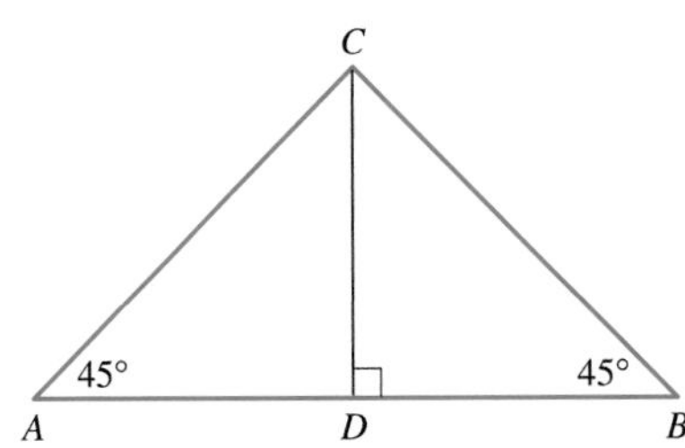

iii. Point D divides $\overline{AB}$ into congruent segments. Explain why this method works.

38. Two neighbors at points A and B will share the cost of extending electric service from a road to their houses. A line will be laid that is equidistant from the two houses. How can they decide where the line should run?

•B

A•

39. A well will be dug to serve three residences located at points A, B, and C. So that the cost to each homeowner is about the same, the well should not be closer to one house than another.

•B

A•

•C

a. Where can the well be dug if it must be equidistant from A and B?

b. Where can the well be dug if it must be equidistant from A and C?

c. Where should the well be dug if it must be equidistant from A, B, and C? How can this location be determined?

EXTENDED PROBLEMS

40. There are three classic compass and straightedge constructions that have challenged and frustrated mathematicians for centuries. Mathematicians have attempted and failed to perform these constructions until they were finally proved to be impossible. Research these three impossible constructions and determine when interest in the constructions arose. What motivated mathematicians to first attempt these constructions? Describe the goal of each construction and explain the conditions imposed that make the constructions impossible. When were they each proven impossible, and what was the basic argument behind each proof?

41. The **quadratrix** is a curve discovered by mathematician Hippias in 420 B.C.E. Research the quadratrix.

a. Why was the curve created? Describe how the curve is generated. Is it possible to construct the curve with just a compass and an unmarked straightedge?

b. Use a compass and straightedge, the angle bisector construction, and the segment midpoint construction to create a curve that approximates a quadratrix.

c. How can you use this curve to divide any acute angle into any number of equal angles? What impossible compass and straightedge construction is possible using a quadratrix?

42. A formal Euclidean construction placed a restriction on the compass that we do not use today. In the time of Euclid, it was important to think of the compass as collapsible. Research the idea behind a collapsible compass. Describe what it is and how it would change the way we use a compass today. Research, describe, and demonstrate how to copy a segment using a collapsible compass.

Solution of Initial Problem

The following figure shows a regular octagon adjacent to a regular pentagon. Find the measure of $\angle ABC$.

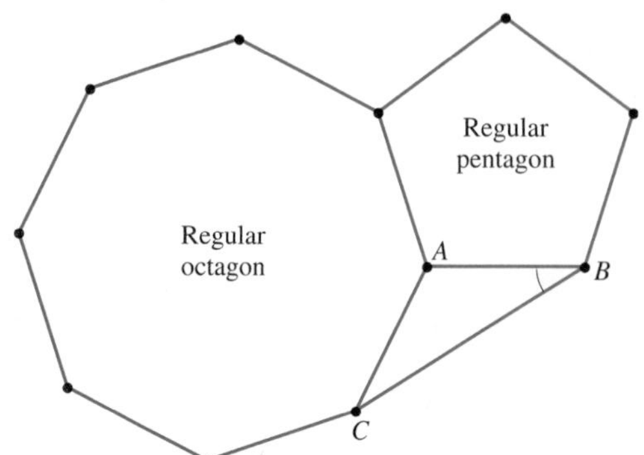

Because both the octagon and the pentagon are regular and they have a side in common, all their sides must be the same length. Therefore, $\triangle ABC$ is an isosceles triangle and $\angle ACB = \angle ABC$. If we can find the measure of $\angle BAC$, then we also can find the measure of $\angle ABC$; namely, $\angle ABC = \frac{1}{2}(180° - \angle BAC)$. However, $\angle BAC$ is 360° minus the sum of the measures of an angle of the octagon and an angle of the pentagon. Hence, if we solve the following two subgoals, we can solve the problem.

Subgoal 1 Find the measures of an angle of the octagon and an angle of the pentagon.

Subgoal 2 Find the measure of $\angle BAC$.

Solution

The measure of an angle of a regular octagon is $\frac{(8 - 2)(180°)}{8} = 135°$, and the measure of an angle of a regular pentagon is $\frac{(5 - 2)(180°)}{5} = 108°$. Therefore, $\angle BAC = 360° - (135° + 108°) = 117°$. So, $\angle ABC = \frac{1}{2}(180° - 117°) = 31.5°$ ■

Additional Problems Where the Strategy "Identify Subgoals" Is Useful

1. In the figure shown, $AB = BC = CD$, $AE = 4''$, and the area of $\triangle CDE =$ 6 square inches. Find EC.

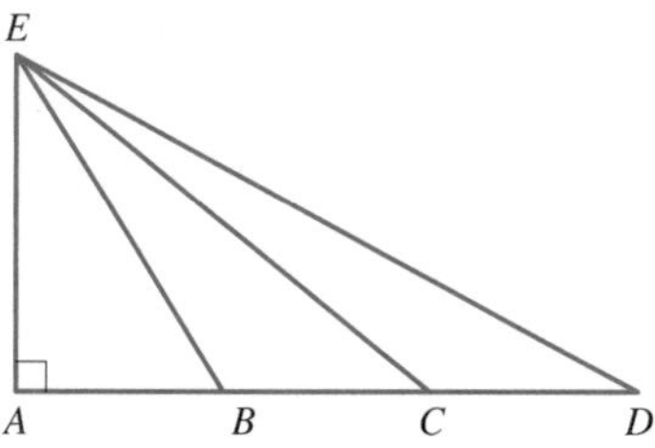

2. Asphalt roofing felt is sold in 36-inch rolls as shown. According to the label, one roll covers 324 square feet. Use the dimensions shown to determine the approximate thickness of a sheet of felt.

Writing for Understanding

1. Describe as many ways as you can to show that two triangles are congruent. Select appropriate proofs or problems, and explain how you decide which congruence condition to apply in a given situation.
2. During the course of one day, pay close attention to conditionals and uses of deductive reasoning that you hear. Write at least five examples. Determine whether the arguments presented were valid.
3. Determine if the following are triangle congruence conditions. If they are, explain why; if they are not, explain why not. You may use examples and counterexamples.
   a. AAA **b.** SSA **c.** AAS
4. The "Geometry Around Us" section at the end of Section 4.2 discusses the construction of bridges. Do some research on the construction of at least three different types of bridges, focusing on the role geometry plays. Summarize your findings.

PEOPLE IN GEOMETRY

Blaise Pascal (1623–1662) was a precocious child, discovering for himself that the sum of the measures of the angles of a triangle is 180° by experimenting with folded paper triangles. At the age of 18, he invented the first mechanical calculating machine, which worked with a series of toothed wheels. He designed it to assist his father, who was a tax official in France. Pascal went on to make important discoveries in the study of fluid pressure, probability, and conic sections (circles, ellipses, parabolas, and hyperbolas). One of his most remarkable results in geometry is the Mystic Hexagram Theorem. It states that if one inscribes a hexagon in a conic section so that no two opposite sides of the hexagon are parallel, and if one extends the pairs of opposite sides until they intersect, then the three points of intersection will be collinear.

CHAPTER REVIEW

For each section of this chapter, you will find a list of vocabulary and notation, questions to assess your understanding of key concepts, and review problems similar to the problems you worked in your homework. Review each item in the *Vocabulary/Notation* list mentally, and, if necessary, refer back to the indicated page and write a definition. Then answer the *Concept Check Questions*, looking back at the section if you need help. Work the *Review Problems* as practice before you move on to the *Chapter Test*. Answers to the *Review Problems* and *Chapter Test* can be found at the back of the book.

SECTION 4.1 Reasoning and Proof in Geometry

Vocabulary/Notation

Direct (deductive) reasoning 186
Conditional statement 186
Hypothesis 186
Conclusion 186
$\Rightarrow$ 186
$\therefore$ 187
Law of Detachment 187
Venn diagram (Euler circle) 187
Law of Syllogism 187
Converse 189
Inverse 189
Contrapositive 189
Biconditional 189
$\Leftrightarrow$ 189
Proof 189
Vertical angles 190
Paragraph proof 190
Statement-reason proof 190

Concept Check Questions

1. How do we reach a conclusion through deductive reasoning? 186
2. How can we read the statement $p \Rightarrow q$? 186
3. Given the statement $p \Rightarrow q$, what must hold in order to conclude q? 186
4. How must statements be related to apply the Law of Syllogism? 187
5. How does a conditional statement relate to its converse? 189
6. What must be true in order to state $p \Leftrightarrow q$? 189
7. How are the measures of vertical angles related? 190
8. What is the difference between a paragraph proof and a statement-reason proof? 190–191

Review Problems

1. Consider the following statement: If a triangle is equilateral, then it is isosceles.
   a. Identify the hypothesis.
   b. Identify the conclusion.
   c. Suppose a triangle is equilateral. What conclusion (if any) can we draw?
   d. Suppose a triangle is isosceles. What conclusion (if any) can we draw?
2. What can we deduce from the following statements?

   If an equilateral triangle has perimeter 18 cm, then it has a base of 6 cm and a height of $3\sqrt{3}$ cm.

   If a triangle has a base of 6 cm and a height of $3\sqrt{3}$ cm, then it has an area of $9\sqrt{3}$ cm^2. *ABC* is an equilateral triangle with perimeter 18 cm.
3. State conclusions that can be drawn in each case.

   a. $t \Rightarrow v$
   $v \Rightarrow w$
   $\underline{t}$
   $\therefore$

   b. $s \Rightarrow q$
   $r \Rightarrow p$
   $q \Rightarrow t$
   $\underline{s}$
   $\therefore$

   c. $r \Leftrightarrow s$
   $\underline{r \Leftrightarrow t}$
   $\therefore$
4. A polygon with congruent angles and congruent sides is regular.
   a. Rewrite the statement in "if . . . , then . . ." form.
   b. Form the converse, and decide if the converse is true or false.
5. What two statements must be true in order for the following statement to hold?

   Two line segments have the same length if and only if the two line segments are congruent.
6. Given $\angle ABC$ and $\angle DEF$ are congruent and supplementary, prove that $\angle ABC$ and $\angle DEF$ are both right angles.

PROOF

Statement	Reason
1. $\angle ABC \cong \angle DEF$	1. ________
2. $\angle ABC = \angle DEF$	2. ________
3. $\angle ABC$ and $\angle DEF$ are supplementary.	3. ________
4. $\angle ABC + \angle DEF = 180°$	4. ________

5. $\angle ABC + \angle ABC = 180°$ **5.** __________
6. $2(\angle ABC) = 180°$ **6.** __________
7. $\angle ABC = 90°$ **7.** __________
8. $\angle DEF = 90°$ **8.** __________
9. $\angle ABC$ and $\angle DEF$ are right angles. **9.** __________

SECTION 4.2 Triangle Congruence Conditions

Vocabulary/Notation

Congruent triangles ($\triangle ABC \cong \triangle DEF$) 196
Reflexive Property 196
Symmetric Property 196
Transitive Property 196
SAS Congruence 197
LL Congruence 197
HL Congruence 198
Corresponding parts (C.P.) 198
ASA Congruence 199
SSS Congruence 201
Converse of the Pythagorean Theorem 201

Concept Check Questions

1. What does it mean to say that two triangles are congruent? 196
2. According to the reflexive property, $\triangle DEF$ is congruent to what triangle? 196
3. If $\triangle PQR \cong \triangle STU$, what does the symmetric property guarantee? 196
4. When considering pairs of congruent parts of triangles, what is the minimum number of pairs that can be used to establish congruence? 197
5. Why is the LL Congruence Theorem a special case of the SAS Congruence Postulate? 197
6. What must we establish in order to conclude a triangle is a right triangle? 201

Review Problems

1. Given that $\triangle PQR \cong \triangle STU$, complete each of the following statements.
   a. $\angle R \cong$ _____
   b. $\overline{PR} \cong$ _____
   c. $\overline{UT} \cong$ _____
2. For each pair of triangles given, write an appropriate triangle congruence statement and state which congruence principle applies.

a.

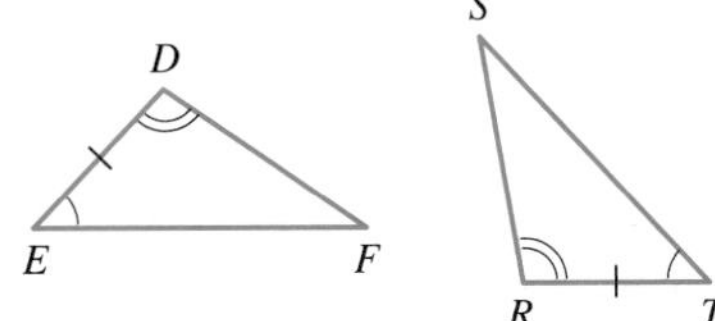

b.

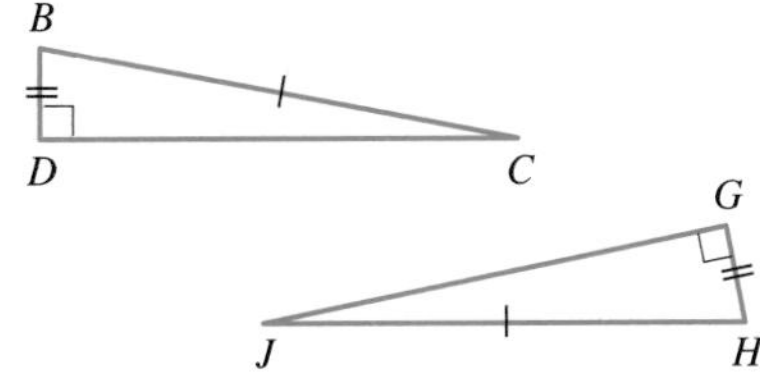

3. Given the following conditions, determine whether $\triangle ABC$ is necessarily congruent to $\triangle DEF$. If the triangles are congruent, specify the congruence principle that applies. If the triangles are not necessarily congruent, sketch a figure to show why not.
   a. $\overline{BC} \cong \overline{EF}, \overline{AB} \cong \overline{DE}, \angle B \cong \angle E$
   b. $\overline{AC} \cong \overline{DF}, \angle A \cong \angle D, \overline{BC} \cong \overline{EF}$
   c. $\overline{BC} \cong \overline{EF}, \angle C = \angle F = 90°, \overline{AC} \cong \overline{DF}$
4. Two pairs of triangles are congruent in the following figure. Give both pairs and state which congruence principles apply.

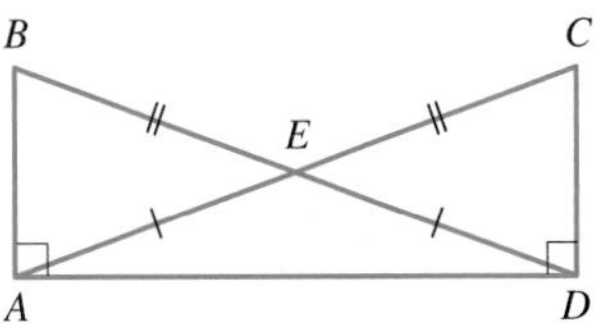

5. In the following figure, congruent angles and segments are marked. Prove $\angle C \cong \angle A$.

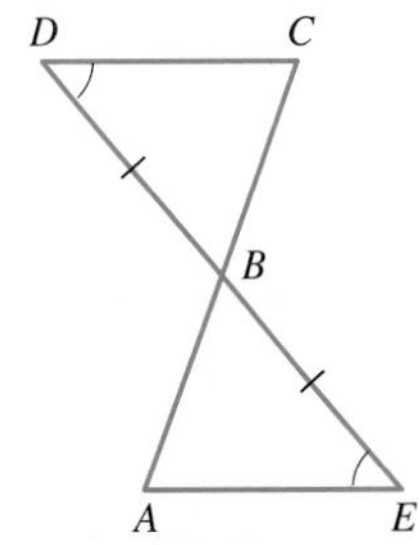

SECTION 4.3 Problem Solving Using Triangle Congruence

Vocabulary/Notation

Bisect 209
Angle bisector 209
Perpendicular bisector 209
Median 210
Exterior angle 212

Concept Check Questions

1. Which angles are congruent in an isosceles triangle? 207
2. If two angles in a triangle are congruent, then what kind of triangle must it be? 209
3. If three angles in a triangle are congruent, then what kind of triangle must it be? 209
4. The endpoints of the median of a triangle correspond to what two points? 210
5. In any isosceles triangle, how are the vertex angle bisector, the median from the vertex angle, and the altitude from the vertex angle related? 210
6. If a point is on the perpendicular bisector of a line segment, how is the point related to the endpoints of the line segment? 211
7. How is an exterior angle of a triangle related to the two opposite interior angles? 212

Review Problems

1. In $\triangle ABC$, $\overline{AB} \cong \overline{CB}$, and $\angle C = 50°$. Find the measures of $\angle A$ and $\angle B$.

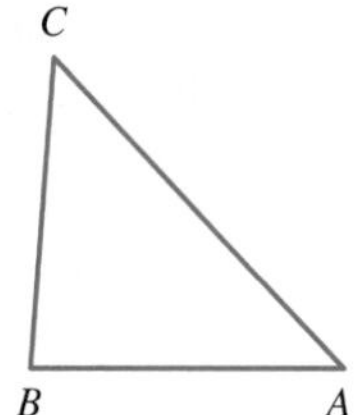

2. In $\triangle PQR$, $\overline{PR} \cong \overline{QR}$, $\angle PRS \cong \angle QRS$, and $PQ = 15$ cm. Find the length of $\overline{SQ}$ and the measure of $\angle RSQ$.

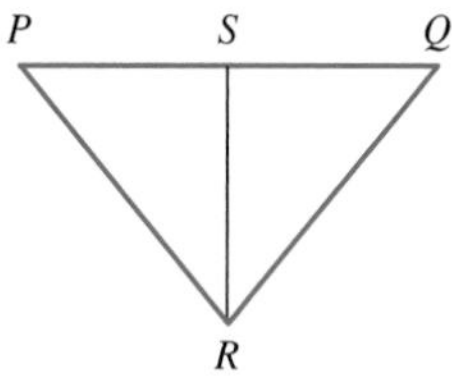

3. In $\triangle DEF$, $\overline{DE} \cong \overline{FE}$ and $\overline{DG} \cong \overline{FG}$. What must be true about $\overline{EG}$? Explain.

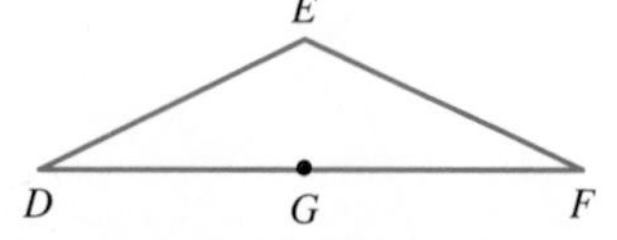

4. In the following figure, congruent segments and congruent angles are marked. If $\angle BAD = 24°$, then find the measures of the following angles.

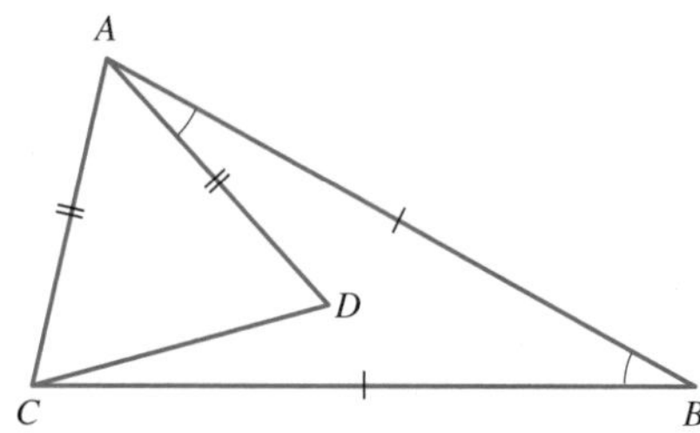

a. $\angle DAC$ b. $\angle ADC$ c. $\angle ACD$
d. $\angle ABC$ e. $\angle DCB$

5. Given that $\overline{XZ}$ bisects $\angle WZY$, $\overline{WX} \perp \overline{WZ}$, and $\overline{YX} \perp \overline{YZ}$, prove that $\triangle XWZ \cong \triangle XYZ$.

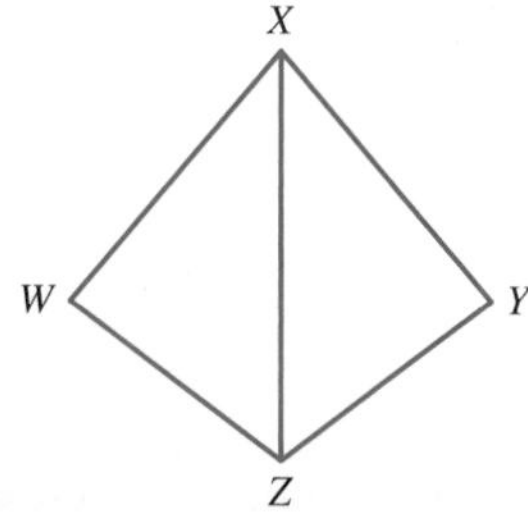

SECTION 4.4 The Basic Geometric Constructions

Vocabulary/Notation

Compass 217 Straightedge 217

Concept Check Questions

1. What kinds of figures can we construct with just a compass? 217
2. What kinds of figures can we construct with just a straightedge? 217
3. How can we use a compass and straightedge to copy a segment? 218
4. How can we use a compass and straightedge to copy an angle? 218–219
5. How can we use the properties of isosceles triangles to verify the construction of a perpendicular bisector of a segment? 221

Review Problems

1. Use a compass and straightedge to construct a segment three times as long as the given segment.

A *B*

2. Use a compass and straightedge to copy and bisect the following angle.

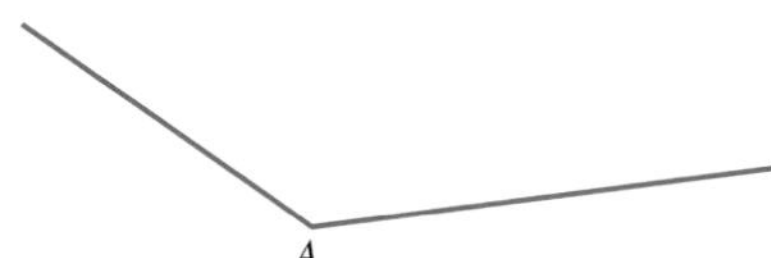

3. Use a compass and straightedge to construct a 105° angle.
4. Use a compass and straightedge to construct a right isosceles triangle given a hypotenuse of length *a*.

a

5. Use a compass and straightedge to construct a 30°–60° right triangle.
6. Copy the following triangle and use a compass and straightedge to construct the altitude from *B* and the median from *A*.

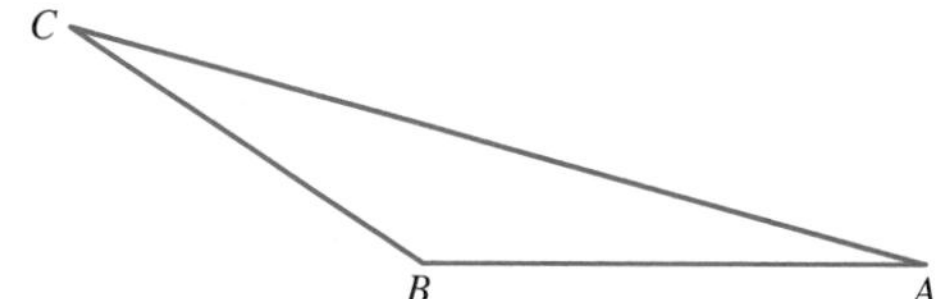

CHAPTER 4 TEST

TRUE-FALSE

Mark as true any statement that is always true. Mark as false any statement that is never true or that is not necessarily true. Be able to justify your answers.

1. Every median of an equilateral triangle is also an altitude of the triangle.
2. If two angles of a triangle are congruent, then two sides of the triangle are congruent.
3. If a conditional statement and its inverse are both true, then they can be combined to form a true biconditional statement.
4. If four pairs of corresponding parts of two triangles are congruent, then the two triangles are congruent.
5. If one right triangle has a hypotenuse of 34 inches and one leg of 16 inches, and another right triangle has legs of 16 inches and 30 inches, then the triangles are congruent.
6. The "if" part of a conditional statement is called the conclusion.

7. If three angles of one triangle are congruent, respectively, to three angles of a second triangle, then the two triangles are congruent.
8. If P is a point on the perpendicular bisector of $\overline{MN}$ and $PM = 5$ cm, then $PN = 5$ cm.
9. Any two intersecting lines form two pairs of congruent vertical angles.
10. If two sides and one angle of one triangle are congruent, respectively, to two sides and one angle of a second triangle, then the two triangles are congruent.

EXERCISES/PROBLEMS

11. a. Write the statement "Every integer is a real number" as a conditional in "if . . . , then . . ." form.
    b. Write the converse of the conditional statement that you wrote in part (a).
    c. Can the statements you wrote for parts (a) and (b) be combined to form a biconditional that is true?
12. In the figure shown next, BD is the perpendicular bisector of EC, and BE is the perpendicular bisector of AD. If $BC = 50$ cm, $EF = 13.02$ cm, and $AB = 43$ cm, find FD. Round your answer to the nearest tenth.

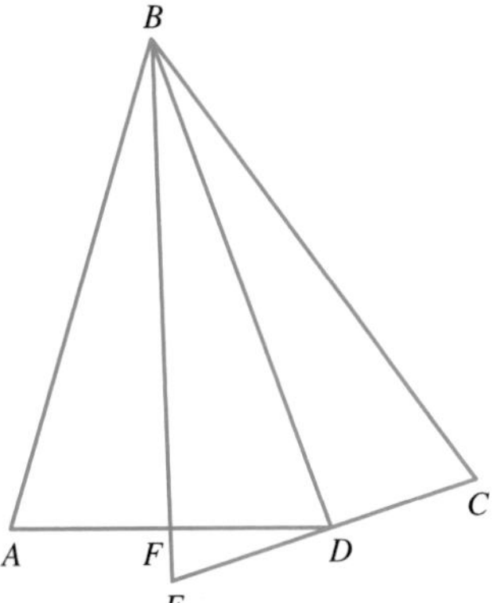

13. What conclusion, if any, can you deduce from the following statements?

    If Jan walks to work, then she carries her work shoes.

    If Jan carries her work shoes, then she does not carry a lunch.

    If Jan does not carry a lunch, then she eats out for lunch.

    If Jan eats out for lunch, then she overeats.

    If Jan overeats, then she gains weight.

    Jan walks to work.

For problems 14–16, decide whether a pair of triangles in the figure is congruent. If so, write an appropriate congruence statement and specify which congruence principle applies. Be careful not to make any assumptions about triangles that just "appear" to be congruent.

14.

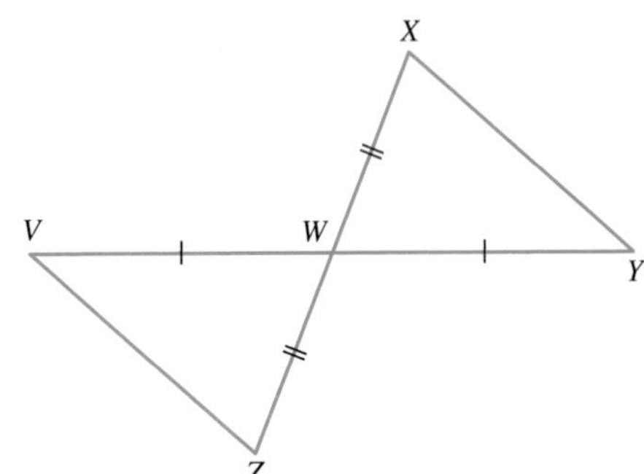

15.

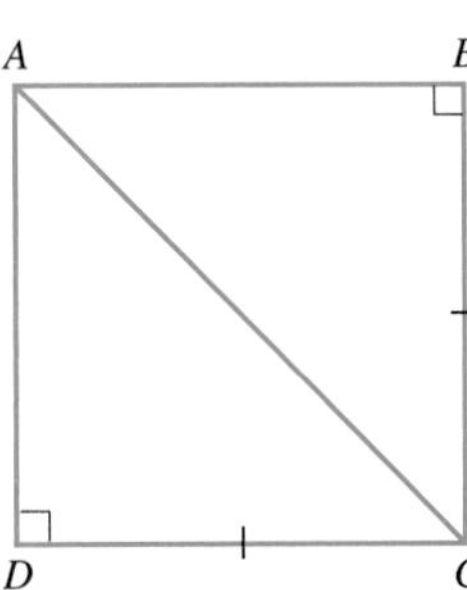

16.

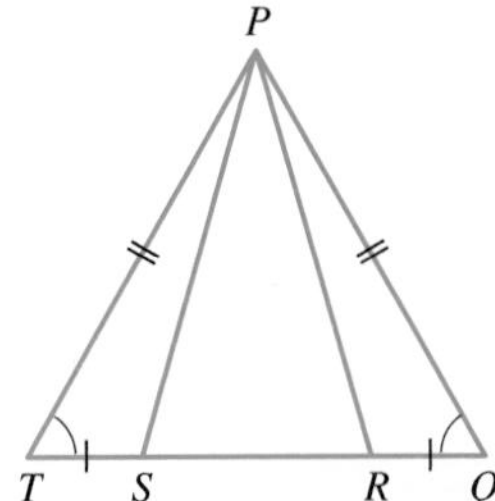

17. Use your compass and straightedge to construct an angle of 22.5°.
18. Use your compass and straightedge to construct the three medians of $\triangle PQR$.

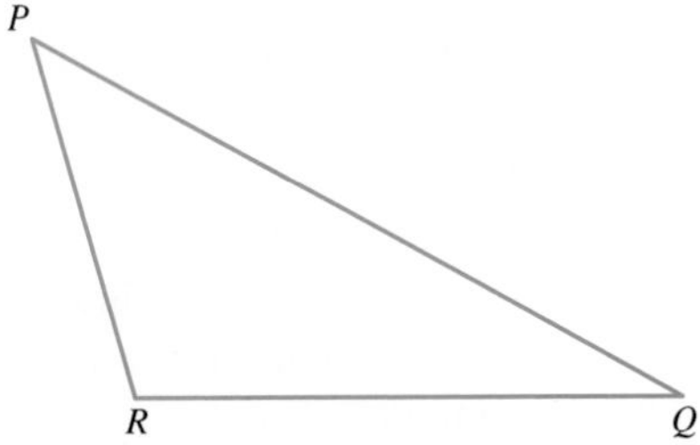

19. Use your compass and straightedge to construct a triangle with sides of length $2a$, $3a$, and $4a$, where length a is given.

a

20. By copying line segments and angles using a compass and straightedge, construct a copy of quadrilateral $WXYZ$ as shown.

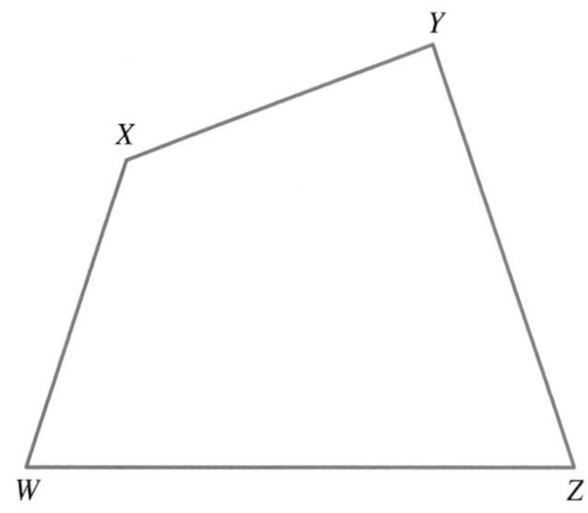

PROOFS

21. In the next figure, congruent segments are marked. Prove that $\triangle PST \cong \triangle RTS$.

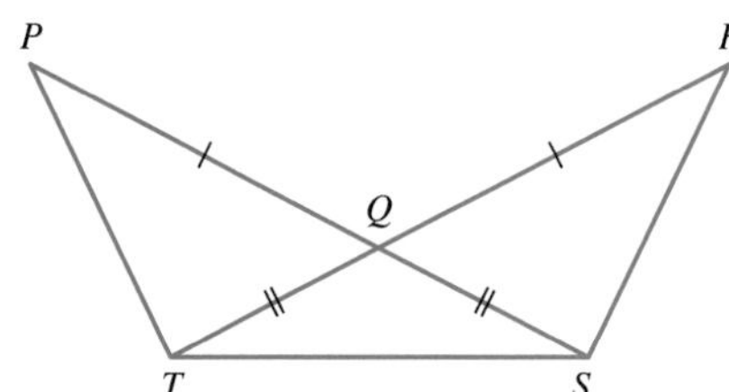

22. In isosceles trapezoid $ABCD$, $\overline{QA}$ bisects $\angle BAD$ and $\overline{PD}$ bisects $\angle CDA$. If $\angle BAD \cong \angle CDA$ and $\angle B \cong \angle C$, prove each of the following:

a. $\triangle ARD$ is isosceles.

b. $\triangle PCD \cong \triangle QBA$

c. $\triangle PQR$ is isosceles.

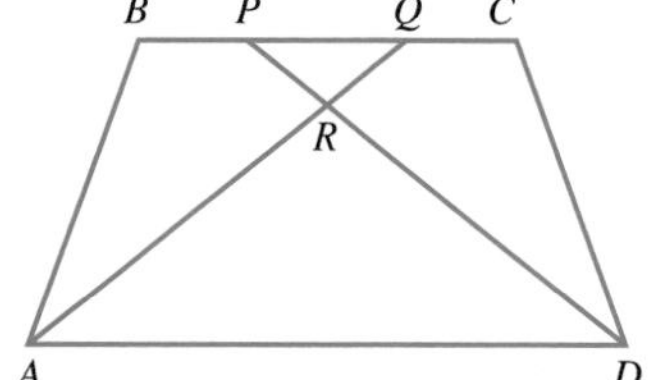

APPLICATIONS

23. The Gateway Arch in St. Louis, Missouri, has cross sections that are equilateral triangles pointing in at the base and pointing down at the top. The triangles at the base measure 54 feet on a side, and those at the top measure 17 feet on a side. Find the difference in the cross-sectional areas at the base and at the top of the arch. Round to the nearest whole number.

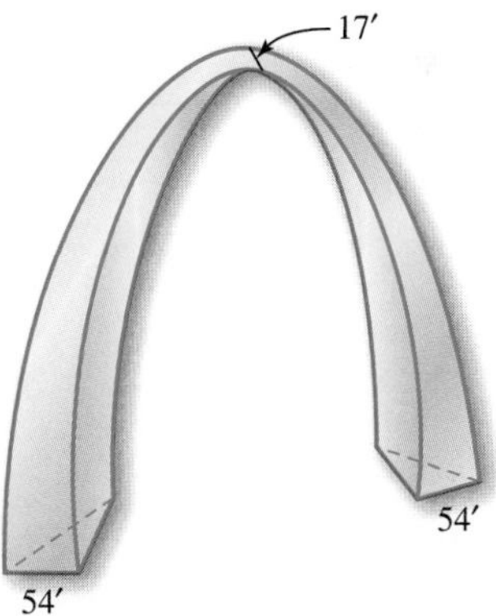

24. A man starts jogging toward a friend's house. When he gets halfway there, he cannot decide whether to continue or return home. What path from that point should he follow so that at any point in his run he will be the same distance from his home and his friend's house?

CHAPTER 5

PARALLEL LINES AND QUADRILATERALS

EUCLID'S PARALLEL POSTULATE

The statement of one of Euclid's postulates was unusually long and complex. It stated the following:

> If a straight line meets two straight lines, so as to make the two interior angles on the same side of it taken together less than two right angles, these straight lines, being continually produced, shall at length meet on the side on which are the angles which are less than two right angles.

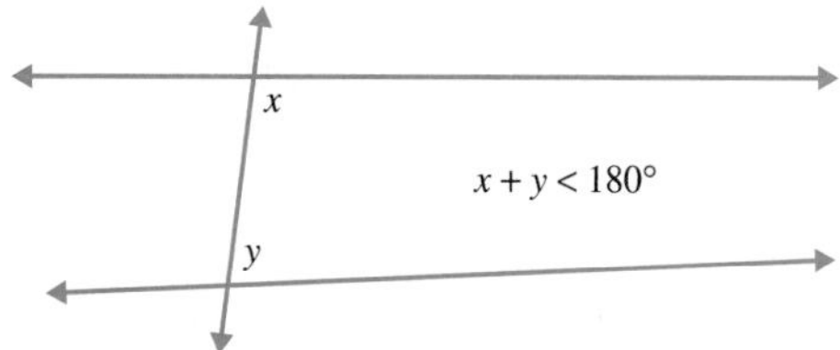

Many mathematicians who used *The Elements* were certain that this statement was provable from previously known theorems and postulates. However, it was shown that one could accept or reject this postulate.

Another way of thinking about parallel lines is to imagine three lines in a plane where two of the lines are perpendicular to the third line. It would seem that the two lines would be parallel. However, there are other geometries where this is not true. One of these geometries is spherical geometry. If you consider the Earth to be a sphere and great circles to be "lines," any two great circles from the North Pole form right angles with the equator. However, these two "lines" are not parallel. In fact, all "lines" perpendicular to the equator intersect at the North Pole. We study other interesting types of geometries in Topic 3 near the end of the book.

Problem-Solving Strategies

1. Draw a Picture
2. Guess and Test
3. Use a Variable
4. Look for a Pattern
5. Make a Table
6. Solve a Simpler Problem
7. Look for a Formula
8. Use a Model
9. Identify Subgoals
10. *Use Indirect Reasoning*

Strategy 10: *Use Indirect Reasoning*

Consider the following situation. A room has two doors, say door X and door Y, as its only entrances, and we want to decide if a person enters the room through door X. We could sit inside and watch door X until someone walked through—this would be a direct verification. We could alternatively watch door Y. If we observed a person in the room who did not come through Y, we could conclude that the person used door X—this is an indirect verification. Often in geometry there are proofs that are difficult to construct using direct reasoning. In those situations, indirect reasoning is especially useful.

Initial Problem

We can arrange the numbers 1 through 9 in a 3 by 3 square array so that the sum of the numbers in each row, column, and diagonal is 15. Show that 1 cannot be in a corner. (HINT: Assume that 1 *can* be in a corner and show that this assumption leads to an impossible situation.)

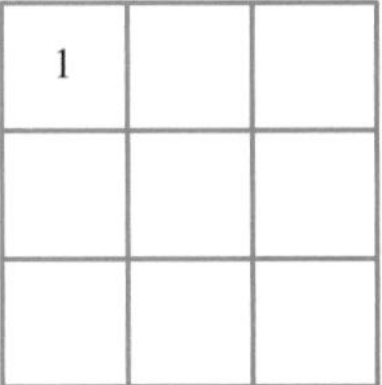

Clues

The Use Indirect Reasoning strategy may be appropriate when

- Direct reasoning seems too complex or does not lead to a solution.
- Assuming the negation of what you are trying to prove narrows the scope of the problem.

A solution of this Initial Problem is on page 287.

One of the classic Greek construction problems is that of trisecting an angle using only a compass and an unmarked straightedge. In 1873, Pierre Wantzel proved that such a construction is impossible. Some special angles, such as a right angle and a straight angle, can be trisected using only a compass and straightedge, but the methods used cannot be generalized to trisect any angle. However, there is an origami technique that we can use to trisect any acute angle. The following trisection method is credited to Hisashi Abe.

Begin with any rectangular piece of paper; a square piece is usually used in origami. Fold and crease the paper through the lower left vertex of the square. This creates an acute angle that will be trisected as shown. (Your angle may look quite different from the one shown here.) Then make two horizontal folds as shown on the right below so the creases are evenly spaced.

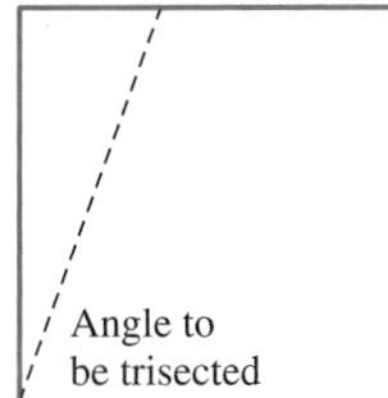

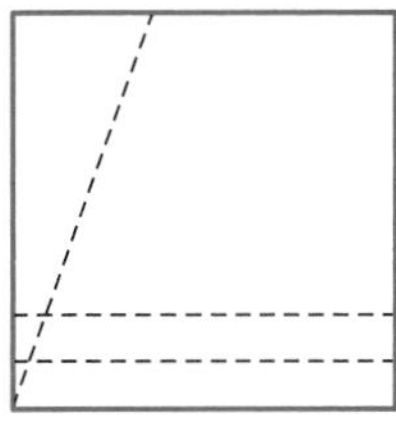

The left-hand endpoints of the horizontal creases and the lower left vertex of the square are marked in the following figure. Mark your paper similarly on the front and on the back, so the dots are clearly visible. Fold the paper so the top and bottom dots align with the original fold and a horizontal crease, respectively, as shown. Mark the position of the middle dot on the paper and then unfold the paper.

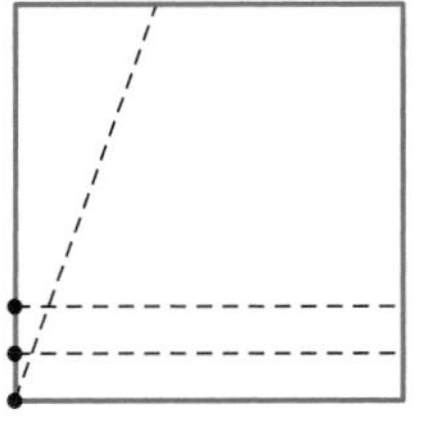

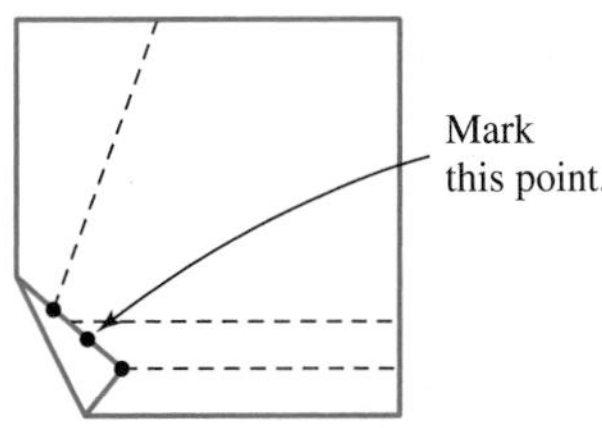

Fold and crease the paper from the lower left vertex of the square through the marked point. This crease, together with the original crease, creates an acute angle that is one-third the measure of the original angle. Next, fold the bottom of the square onto your new fold to bisect the remaining angle. When you unfold the paper, you will see that the angle has been trisected.

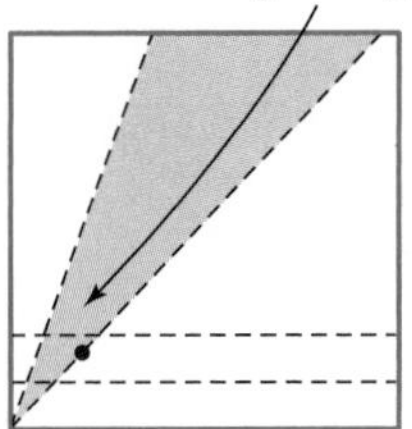

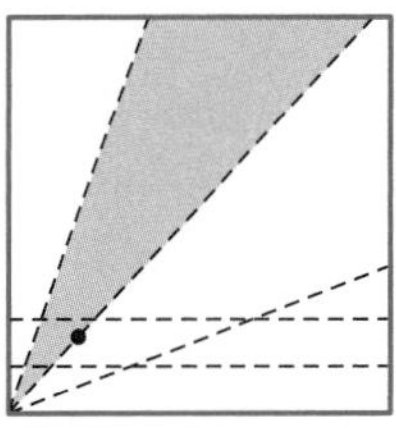

Use this paper-folding method to trisect three different acute angles. For each angle, check the angle measures with a protractor to verify the trisection.

Introduction

We present many important concepts, relationships, and results in this chapter. First, we discuss indirect reasoning. This important technique can be used to construct proofs where direct reasoning is cumbersome, too difficult, or even impossible. Next, we introduce the Parallel Postulate. This postulate is the most famous and was the most controversial in Euclidean geometry. Then we prove many results that follow from the Parallel Postulate and are related to parallelograms, rhombuses, rectangles, squares, and trapezoids.

5.1 INDIRECT REASONING AND THE PARALLEL POSTULATE

Applied Problem

In about 275 B.C.E., Eratosthenes developed a method for approximating the size of the earth. He observed that at noon on June 21 in Syene, Egypt, the sun was directly overhead. He determined that at the same time 500 miles north in Alexandria, the inclination of the sun's rays to the vertical was approximately 7.2°. Assuming that the sun's rays hit the earth in parallel lines, how did Eratosthenes calculate the circumference of the earth?

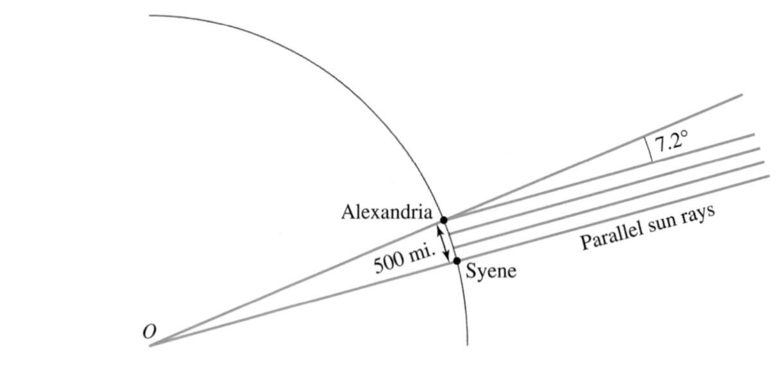

Indirect Reasoning

In Chapter 4, we used direct reasoning to derive and prove many useful relationships in geometry. There are situations in mathematics where direct reasoning is difficult to apply. For this reason, we will begin to use another form of reasoning, indirect reasoning, in this chapter. The first example illustrates the use of both direct and indirect reasoning to prove the same result.

EXAMPLE 5.1 In $\triangle ABC$ and $\triangle DEF$, two angles of one triangle are congruent to two angles of the other triangle (Figure 5.1).

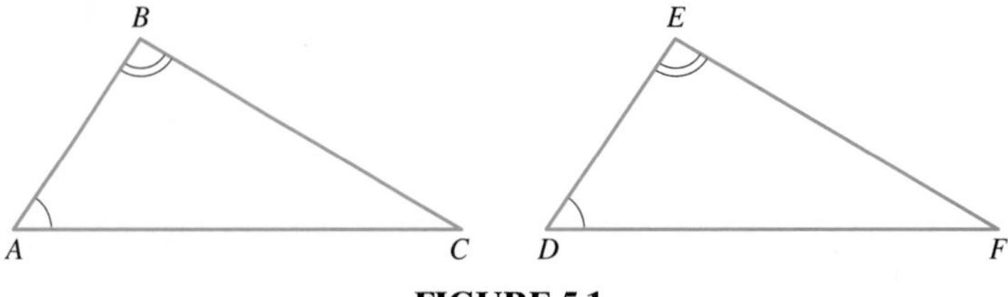

FIGURE 5.1

Show that the third angles are congruent. (Recall that in Chapter 2 we used an informal argument to show that the angle sum in a triangle is 180°.)

Given $\angle A = \angle D$ and $\angle B = \angle E$

Prove $\angle C = \angle F$

SOLUTION

USING DIRECT REASONING

Proof First we have $\angle A + \angle B + \angle C = 180°$ and $\angle D + \angle E + \angle F = 180°$. So we have

$$\text{(i) } \angle A + \angle B + \angle C = \angle D + \angle E + \angle F$$

Because $\angle A = \angle D$ and $\angle B = \angle E$, we have

$$\text{(ii) } \angle A + \angle B = \angle D + \angle E$$

Subtracting (ii) from (i) we have $\angle C = \angle F$.

USING INDIRECT REASONING

Plan Assume that the given holds; that is, $\angle A = \angle D$ and $\angle B = \angle E$. We know that one, and only one, of the following can be true: (1) $\angle C < \angle F$, (2) $\angle C > \angle F$, or (3) $\angle C = \angle F$. We will assume that (1) is true and show that a contradiction results. Then we will do the same with (2). If neither of these statements is true, then (3) must be true, which is what we wanted to prove.

Proof *Assume* that the given holds, namely $\angle A = \angle D$ and $\angle B = \angle E$, and also assume that $\angle C < \angle F$. If $\angle C < \angle F$ and $\angle A = \angle D$, then $\angle A + \angle C < \angle D + \angle F$. Because $\angle B = \angle E$, we also have

$$\text{(iii) } \angle A + \angle B + \angle C < \angle D + \angle E + \angle F$$

But $\angle A + \angle B + \angle C = 180°$ and $\angle D + \angle E + \angle F = 180°$. Thus, by substitution in (iii), we have $180° < 180°$, a contradiction. If we assume that $\angle C > \angle F$, a similar contradiction arises. Thus, neither $\angle C < \angle F$ nor $\angle C > \angle F$ is possible. Hence, we must conclude that $\angle C = \angle F$. ■

Notice that **indirect reasoning** (or an **indirect proof**) assumes the hypothesis of the result that we are trying to prove and the *negation* (or opposite) of the conclusion. Then we show that this situation leads to a contradiction. Because this situation is not possible, we conclude that our assumption must be incorrect and the conclusion *as given in the original statement* must follow from the hypothesis.

We will use the technique of indirect reasoning often in this section. To use the technique to prove a theorem, we always assume that the hypothesis *and* the negation of the conclusion are true. Then, from these assumptions, we reach a statement that contradicts an earlier result or a known fact.

Parallel Lines

Recall that two lines in a plane are parallel if they do not intersect. It is important to note that the two parallel lines are said to lie in the same plane. Lines that do not lie in the same plane and do not intersect are called **skew lines**. For example, in

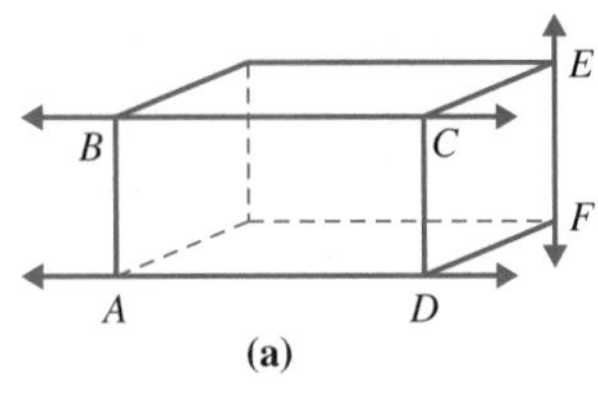

(a)

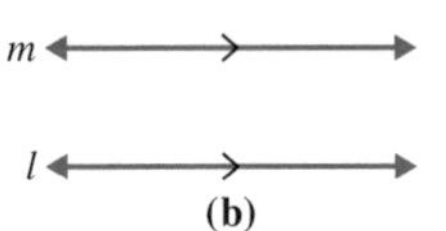

(b)

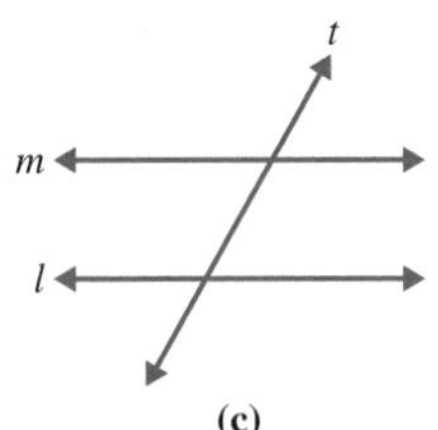

(c)

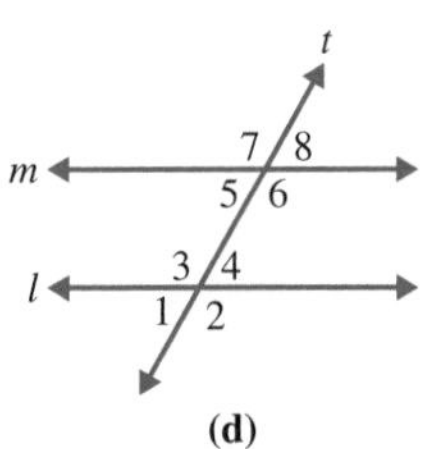

(d)

FIGURE 5.2

Figure 5.2(a), $\overleftrightarrow{AD}$ and $\overleftrightarrow{BC}$ are parallel lines and $\overleftrightarrow{BC}$ and $\overleftrightarrow{EF}$ are skew lines. Throughout the rest of this book, we will be studying geometry of the *plane*. Therefore, we will not be considering skew lines. Recall that parallel lines are indicated by placing an arrowhead (or arrowheads) on each line [Figure 5.2(b)].

If two lines are intersected by a third line, the third line is called a **transversal**. In Figure 5.2(c), a pair of lines l and m and a transversal t are pictured. There are several pairs of angles formed that are of special interest. In Figure 5.2(d), the pair $\angle 3$ and $\angle 6$ and the pair $\angle 4$ and $\angle 5$ are called **alternate interior angles**. The pair $\angle 1$ and $\angle 8$ and the pair $\angle 2$ and $\angle 7$ are called **alternate exterior angles**. The pair $\angle 3$ and $\angle 5$ and the pair $\angle 4$ and $\angle 6$ are called **interior angles on the same side of the transversal**. The pair $\angle 1$ and $\angle 7$ and the pair $\angle 2$ and $\angle 8$ are called **exterior angles on the same side of the transversal**. In addition, the following pairs are called **corresponding angles**: $\angle 1$ and $\angle 5$, $\angle 3$ and $\angle 7$, $\angle 2$ and $\angle 6$, and $\angle 4$ and $\angle 8$.

In Figure 5.2(d), many pairs of angles appear to be congruent. In particular, it appears that $\angle 4 \cong \angle 5$. In addition, lines l and m appear to be parallel. In fact, if a pair of lines and a transversal form congruent alternate interior angles, we can conclude that the lines are parallel. We prove this result next using indirect reasoning.

THEOREM 5.1

If two lines are cut by a transversal to form a pair of congruent alternate interior angles, then the lines are parallel.

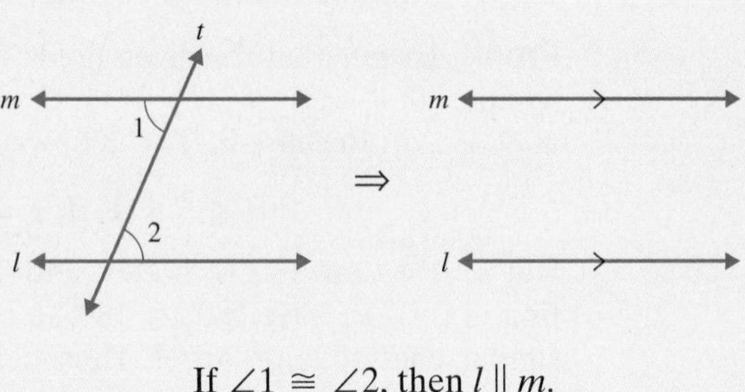

If $\angle 1 \cong \angle 2$, then $l \parallel m$.

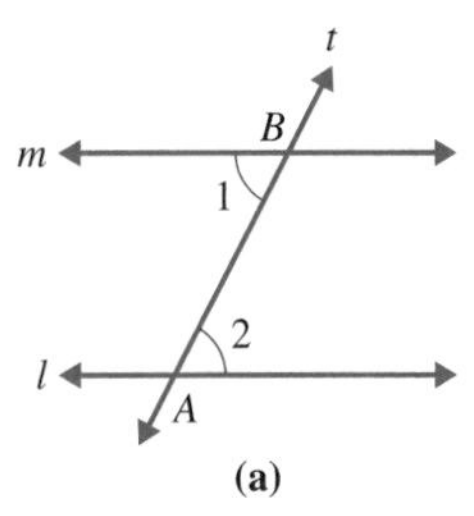

(a)

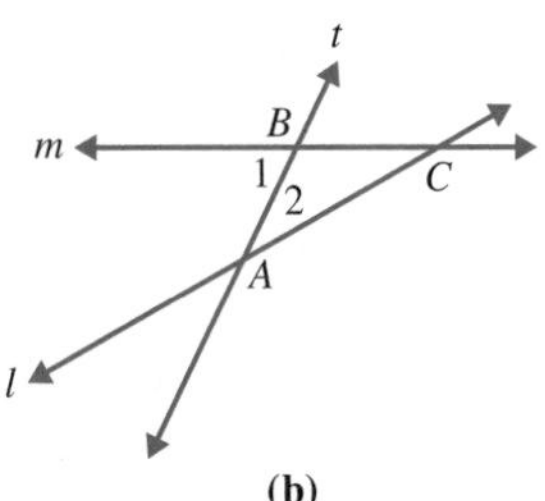

(b)

FIGURE 5.3

Given Lines l and m with transversal t, $\angle 1 \cong \angle 2$, and A and B points of intersection of t with l and m, respectively [Figure 5.3(a)].

Prove $l \parallel m$

Plan We will use indirect reasoning. We will assume that l and m are not parallel and try to arrive at a contradiction.

Proof If we assume l and m are not parallel, then they intersect in a point C [Figure 5.3(b)]. From the given, we have $\angle 1 \cong \angle 2$ for $\triangle ABC$. However, by Theorem 4.11, $\angle 1 > \angle 2$ because $\angle 1$ is an exterior angle and $\angle 2$ is a nonadjacent interior angle. This is a contradiction because we cannot have $\angle 1 = \angle 2$ and $\angle 1 > \angle 2$. By indirect reasoning, we can conclude that l and m do *not* intersect. So they must be parallel. ■

The corollaries listed next follow from Theorem 5.1. We leave their proofs for the problem set.

COROLLARY 5.2

If two lines are both perpendicular to a transversal, then the lines are parallel.

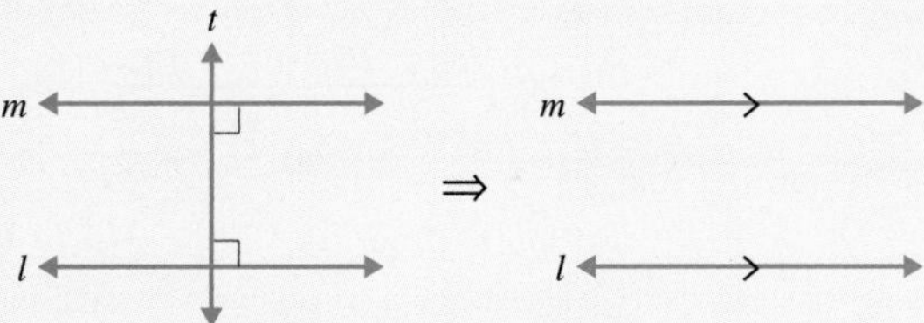

If $l \perp t$ and $m \perp t$, then $l \parallel m$.

COROLLARY 5.3

If two lines cut by a transversal form a pair of congruent corresponding angles with the transversal, then the lines are parallel.

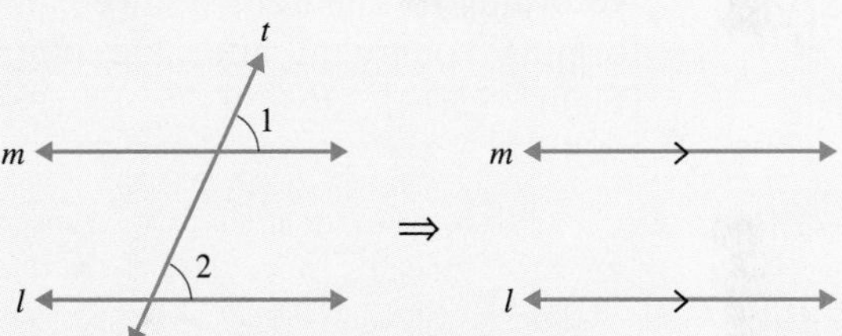

If $\angle 1 \cong \angle 2$, then $l \parallel m$.

COROLLARY 5.4

If two lines cut by a transversal form a pair of supplementary interior angles on the same side of the transversal, then the lines are parallel.

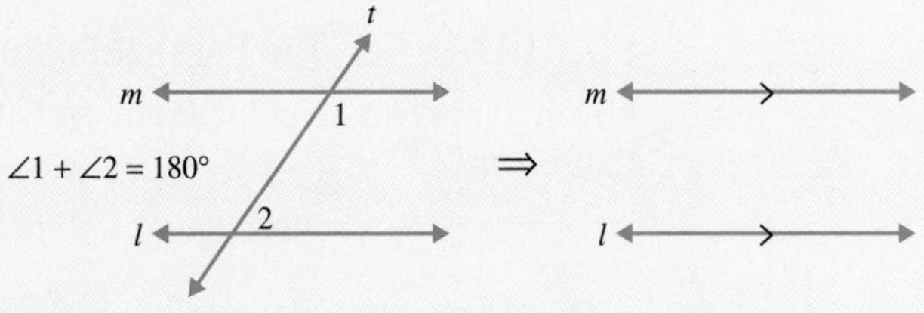

If $\angle 1 + \angle 2 = 180°$, then $l \parallel m$.

We can use any one of Theorem 5.1 or Corollaries 5.2–5.4 to show that two lines are parallel. We will make frequent use of them.

EXAMPLE 5.2 Prove that $ABCD$ is a parallelogram [Figure 5.4(a)].

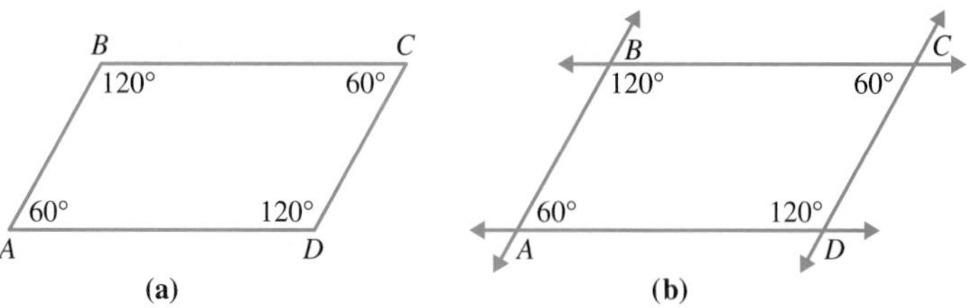

FIGURE 5.4

SOLUTION

Think of $\overline{AB}$ as a transversal for sides $\overline{AD}$ and $\overline{BC}$ [Figure 5.4(b)]. Because $60° + 120° = 180°$, we have supplementary interior angles on the same side of the transversal, and $\overline{AD} \parallel \overline{BC}$ by Corollary 5.4. By a similar argument, $\overline{AB} \parallel \overline{CD}$. Because both pairs of opposite sides are parallel, $ABCD$ is a parallelogram. ■

The Parallel Postulate

The preceding theorems and corollaries began with a pair of lines and a transversal. Suppose that we are given one line l and a point P not on that line [Figure 5.5(a)]. According to Theorem 5.1, we can construct a line m through P parallel to l by forming congruent alternate interior angles with any transversal through P [Figure 5.5(b)].

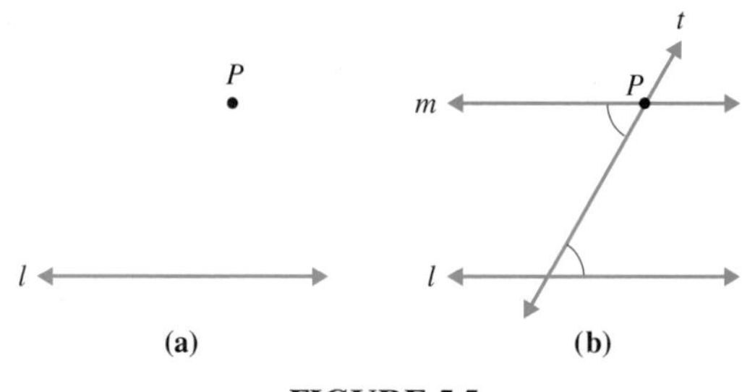

FIGURE 5.5

From this figure, it would appear that there is *only one* such line. For hundreds of years, mathematicians tried to prove this result from Euclid's postulates that preceded it. It was finally shown that this result could not be proved in this way, so it was added as a postulate.

> **POSTULATE 5.1 The Parallel Postulate**
>
> Given a line l and a point P not on l, there is only one line m containing P such that $l \parallel m$.

The Parallel Postulate has several equivalent forms. However, the version stated here is popular because mathematicians have developed other geometries by replacing the phrase "only one line" with "no line" and "infinitely many lines." These geometries are called **elliptic** and **hyperbolic** geometries, respectively. We discuss them in Topic 3 near the end of the book.

Now that we have the Parallel Postulate, we can prove the converse of Theorem 5.1.

> **THEOREM 5.5**
>
> If a pair of parallel lines is cut by a transversal, then the alternate interior angles formed are congruent.
>
> 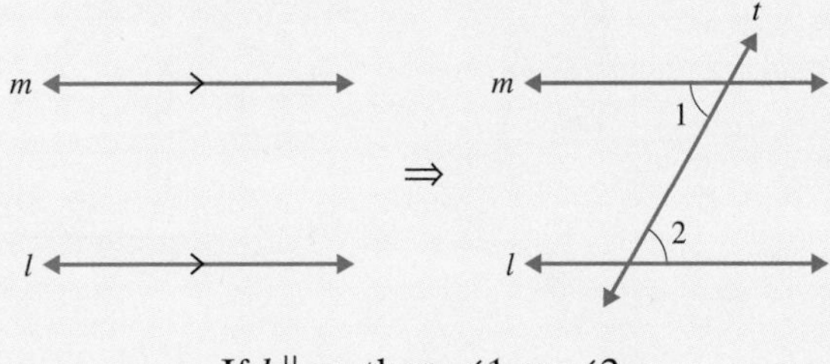
>
>
> If $l \parallel m$, then $\angle 1 \cong \angle 2$.

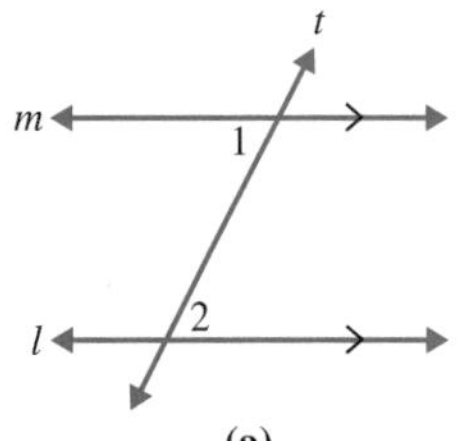

(a)

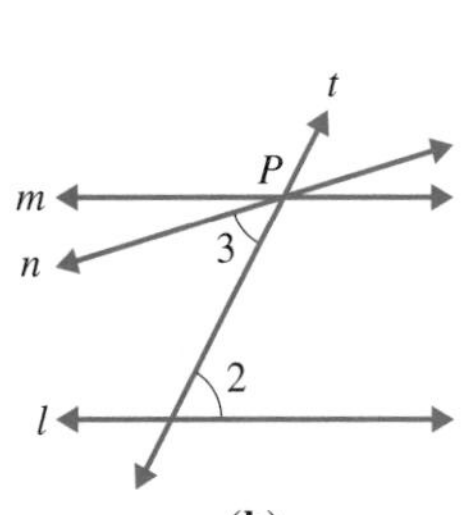

(b)

FIGURE 5.6

Given Parallel lines l and m, and transversal t [Figure 5.6(a)]

Prove $\angle 1 = \angle 2$

Plan Use indirect reasoning. We will assume that $\angle 1 \neq \angle 2$ and try to arrive at a contradiction.

Proof Assume that $\angle 1 \neq \angle 2$. Then $\angle 1 > \angle 2$ or $\angle 1 < \angle 2$. Suppose that $\angle 1 > \angle 2$. Then there is a line n through P that forms $\angle 3$, where $\angle 3 = \angle 2$. [Figure 5.6(b)]. By Theorem 5.1, $n \parallel l$. However, because $m \parallel l$ is given, we now have two lines (m and n) containing P, each parallel to l. This contradicts the Parallel Postulate, which states that there can be only one such line. Thus, by indirect reasoning, we have $\angle 1 = \angle 2$. (NOTE: In this proof, we assumed that $\angle 1 > \angle 2$ and then reached a contradiction. We would have reached the same contradiction if we had assumed that $\angle 1 < \angle 2$.) ■

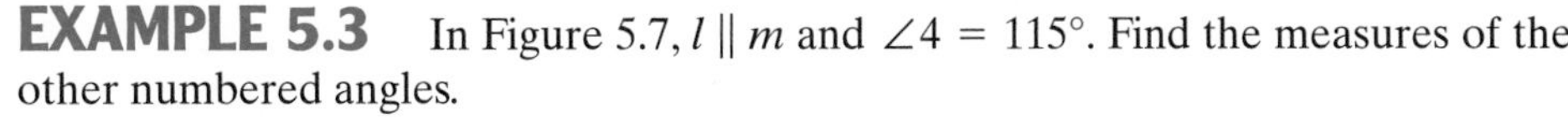

EXAMPLE 5.3 In Figure 5.7, $l \parallel m$ and $\angle 4 = 115°$. Find the measures of the other numbered angles.

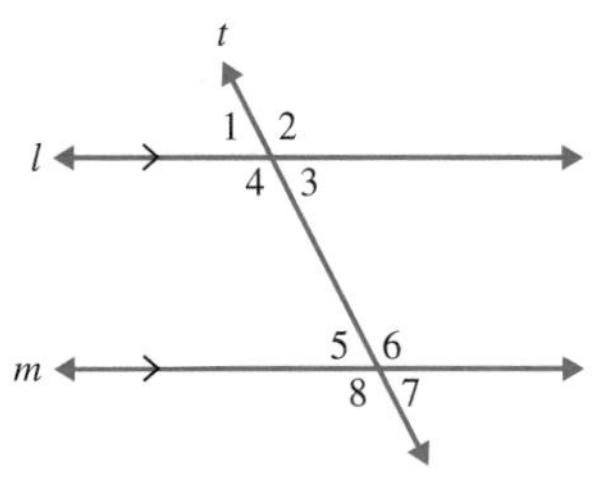

FIGURE 5.7

SOLUTION

Because $\angle 1$ and $\angle 4$ are supplementary,

$$\angle 1 = 180° - \angle 4 = 180° - 115° = 65°$$

In the same way, $\angle 3 = 65°$. Notice that $\angle 2$ and $\angle 4$ are vertical angles, so $\angle 2 = \angle 4 = 115°$. By Theorem 5.5, $\angle 4 \cong \angle 6$ because they are alternate interior angles and $l \parallel m$, so $\angle 6 = 115°$. Then $\angle 5$ and $\angle 7$ are supplements of $\angle 6$, so each of them measures 65°. At last, $\angle 6$ and $\angle 8$ are vertical angles, so $\angle 8 = \angle 6 = 115°$. We can summarize these results as follows:

$$\angle 1 = \angle 3 = \angle 5 = \angle 7 = 65° \text{ and } \angle 2 = \angle 4 = \angle 6 = \angle 8 = 115°$$ ■

Notice in Example 5.3 that each pair of corresponding angles was congruent. For instance, $\angle 1 \cong \angle 5$ and $\angle 3 \cong \angle 7$. Also, the pairs of interior angles on the same side of the transversal, such as $\angle 4$ and $\angle 5$, were supplementary. The corollaries stated next summarize results like these, which are the converses of Corollaries 5.2–5.4. We leave their proofs for the problem set.

COROLLARY 5.6

If two lines are parallel and a line is perpendicular to one of the two lines, then it is perpendicular to the other line.

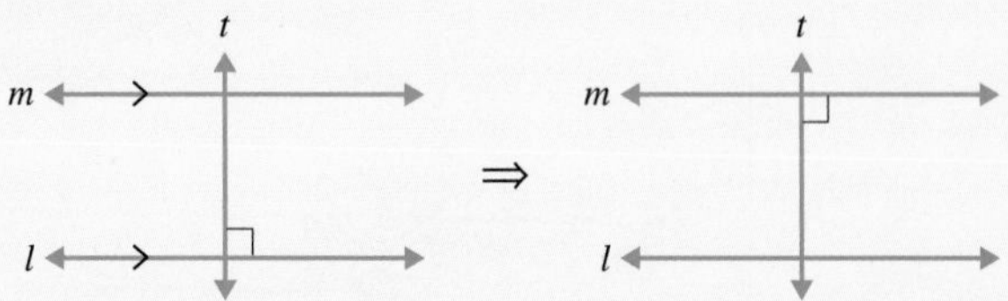

If $l \parallel m$ and $t \perp l$, then $t \perp m$.

COROLLARY 5.7

If two parallel lines are cut by a transversal, then each pair of corresponding angles formed is congruent.

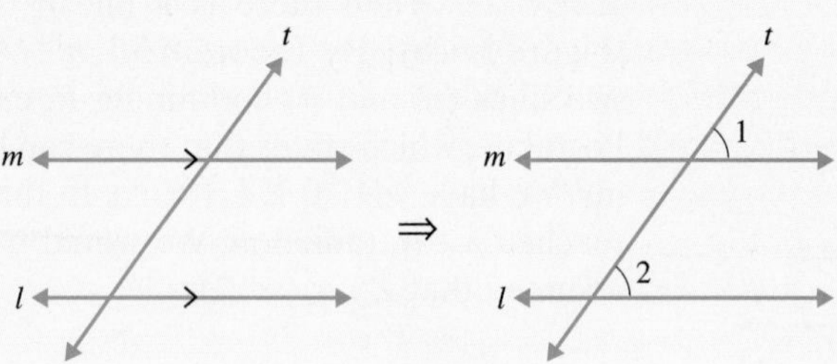

If $l \parallel m$, then $\angle 1 \cong \angle 2$.

COROLLARY 5.8

If two parallel lines are cut by a transversal, then both pairs of interior angles on the same side of the transversal are supplementary.

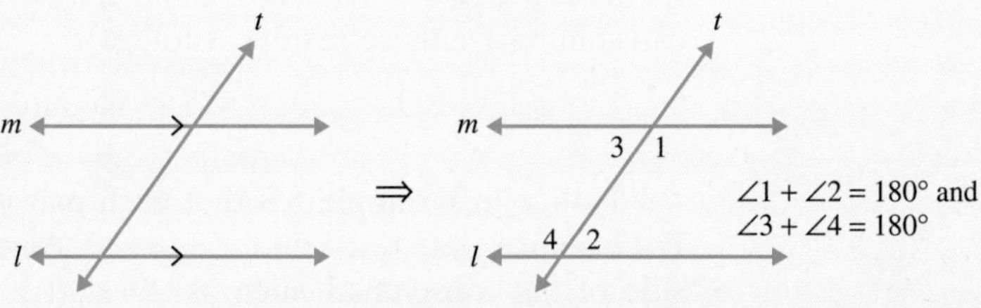

If $l \parallel m$, then $\angle 1 + \angle 2 = 180°$ and $\angle 3 + \angle 4 = 180°$.

Solution of Applied Problem

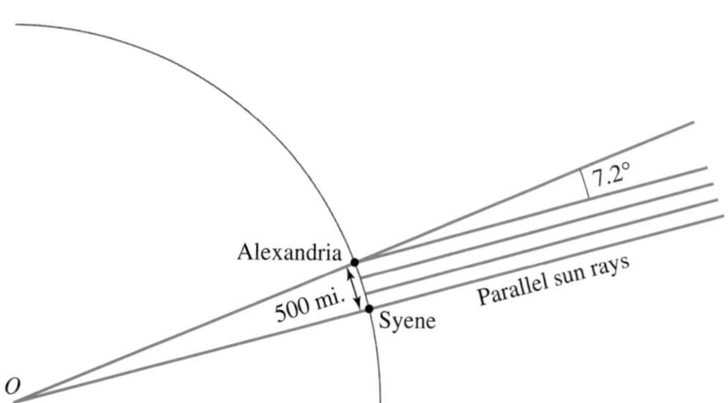

Because we assume that the sun's rays are parallel, the angle formed at the center of the earth and the 7.2° angle are congruent corresponding angles. Thus, $\angle O = 7.2°$, so the distance from Syene to Alexandria is $\frac{7.2}{360} = \frac{1}{50}$ of the circumference of the earth. This means that the circumference of the earth is about 50(500 mi), or about 25,000 mi. A more recent calculation showed the circumference to be about 24,820 mi.

GEOMETRY AROUND US A simple periscope uses two parallel mirrors placed at each end of a tube. One mirror receives light from an object and reflects it to the other mirror. The second mirror then reflects the light to the viewer's eye. Periscopes recently have been made of glass fibers that can be bent and twisted around corners. The glass fibers reflect the image down the length of the tube, so mirrors are unnecessary.

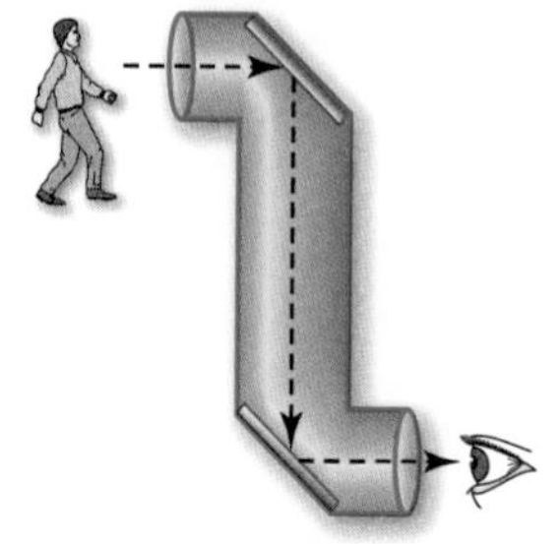

PROBLEM SET 5.1

EXERCISES/PROBLEMS

Use the following figure for Exercises 1–10.

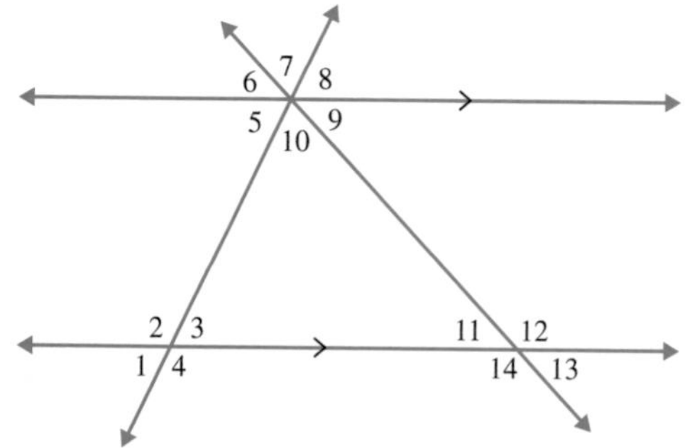

1. Identify two pairs of alternate interior angles.
2. Identify two pairs of interior angles on the same side of the transversal.
3. Identify two pairs of corresponding angles.
4. Identify two pairs of exterior angles on the same side of the transversal.
5. Find the measure of $\angle 6 + \angle 12$.
6. Find the measures of $\angle 2$ and $\angle 4$ if $\angle 5 = 52°$.
7. Find the measures of $\angle 3$ and $\angle 8$ if $\angle 5 = 65°$.
8. Find the measures of $\angle 13$, $\angle 12$, and $\angle 9$ if $\angle 6 = 40°$.
9. Find the measures of all the numbered angles given that $\angle 5 = 60°$ and $\angle 6 = 45°$.
10. Find the measures of all the numbered angles given that $\angle 14 = 140°$ and $\angle 1 = 50°$.

Exercises 11–18 refer to the following figure.

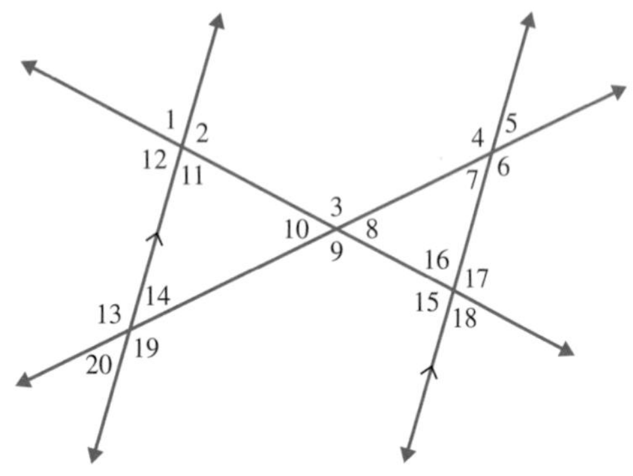

11. ∠16 and ∠ _______ are a pair of alternate interior angles.

12. ∠5 and ∠ _______ are a pair of alternate exterior angles.

13. ∠2 and ∠ _______ are a pair of corresponding angles.

14. ∠14 and ∠ _______ are a pair of interior angles on the same side of the transversal.

15. Find the measures of ∠13 and ∠5 if ∠6 = 140°.

16. Find the measures of ∠1 and ∠2 if ∠16 = 72°.

17. Find the measures of all the numbered angles if ∠1 = 80° and ∠4 = 125°.

18. Find the measures of all the numbered angles if ∠10 = 58° and ∠18 = 75°.

Use the next figure for Exercises 19–22, where congruent angles are marked. Decide if each statement is true or false. If true, explain why. Be careful not to make any assumptions about angles that just "appear" to be congruent.

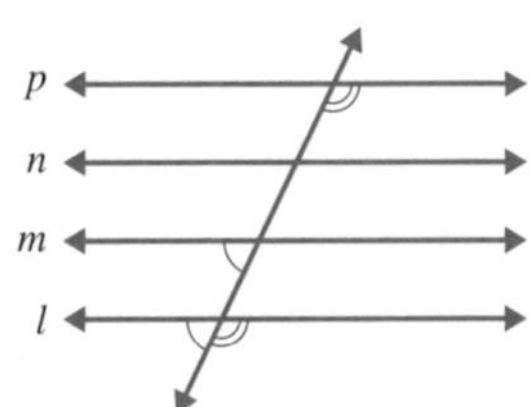

19. $n \parallel p$

20. $n \parallel l$

21. $m \parallel l$

22. $p \parallel l$

Use the following figure to find the measures of the angles specified in Exercises 23–28. Parallel segments are marked.

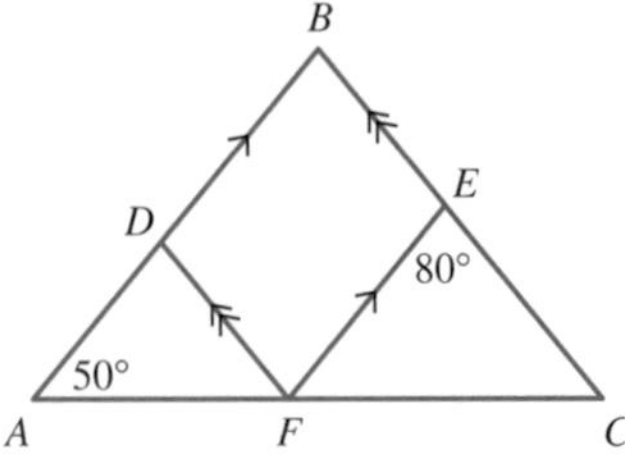

23. ∠B = _______. Why?

24. ∠BDF = _______. Why?

25. ∠ADF = _______. Why?

26. ∠AFD = _______. Why?

27. ∠DFE = _______. Why?

28. ∠C = _______. Why?

29. Find the measures of the numbered angles.

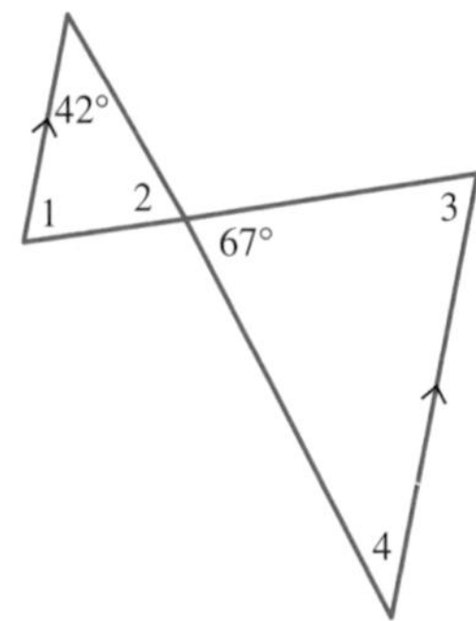

30. Find the measures of the numbered angles.

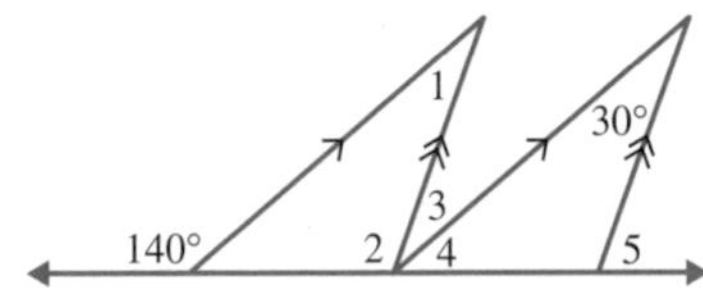

31. Find the measures of the numbered angles.

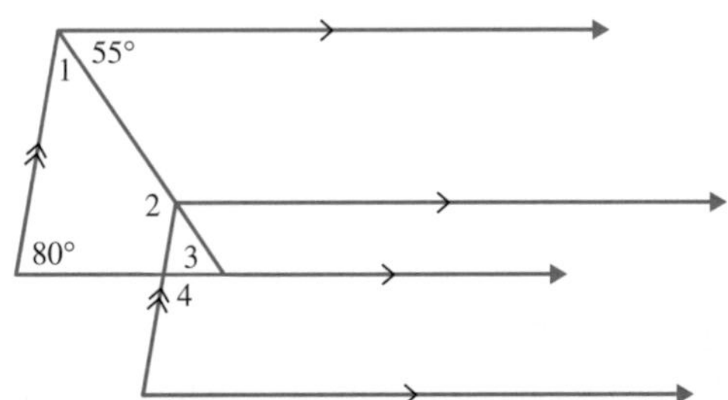

32. Find the measures of $\angle 1$, $\angle 2$, $\angle 3$, and $\angle 4$ if some angles formed are related as shown.

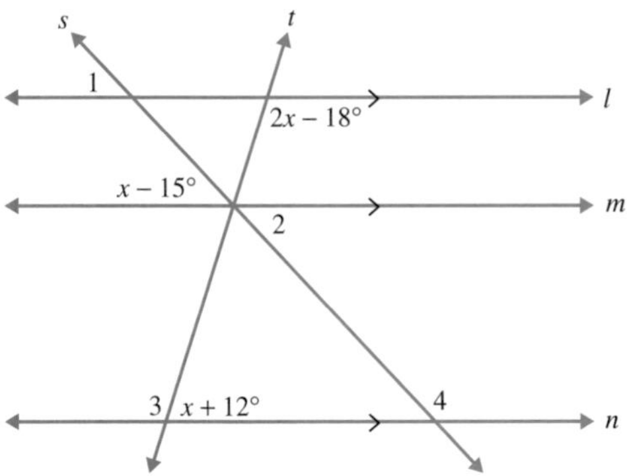

PROOFS

33. Prove Corollary 5.2: If two lines are both perpendicular to a transversal, then the lines are parallel.

34. Prove Corollary 5.3: If two lines cut by a transversal form a pair of congruent corresponding angles with the transversal, then the lines are parallel.

35. Prove Corollary 5.4: If two lines cut by a transversal form a pair of supplementary interior angles on the same side of the transversal, then the lines are parallel.

36. Prove: If two lines cut by a transversal form a pair of supplementary exterior angles on the same side of the transversal, then the lines are parallel.

37. Prove Corollary 5.6: If two lines are parallel and a line is perpendicular to one of the two lines, then it is perpendicular to the other line.

38. Prove Corollary 5.7: If two parallel lines are cut by a transversal, then each pair of corresponding angles formed is congruent.

39. Prove Corollary 5.8: If two parallel lines are cut by a transversal, then both pairs of interior angles on the same side of the transversal are supplementary.

40. Prove: If two parallel lines are cut by a transversal, then they form a pair of congruent alternate exterior angles with the transversal.

APPLICATIONS

41. When a ray of light strikes a flat mirror, $\overline{AB}$, at point C, the angle at which the ray hits the mirror, called the **angle of incidence**, is congruent to the angle at which the ray is reflected away from the mirror, called the **angle of reflection**. Those angles are measured with respect to $\overleftrightarrow{CD}$, which is perpendicular to $\overline{AB}$, as shown in the following figure.

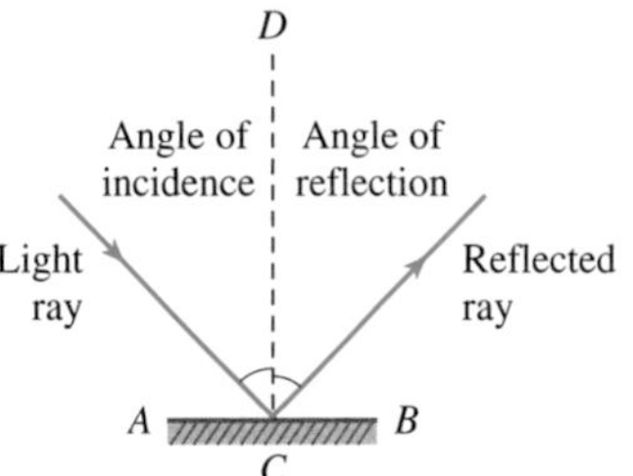

a. If the angle of incidence of a ray of light is 25.6°, find the angle between the reflected ray of light and the mirror.

b. Two mirrors $\overline{AC}$ and $\overline{DE}$ are situated so that they are parallel, as shown in the figure. A ray of light strikes mirror $\overline{AC}$ so that $\angle ABF$ is 70°. Find the measure of $\angle CEG$.

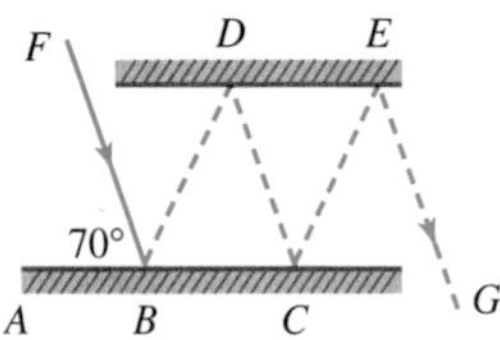

42. A periscope is a device that allows the user to view objects not in the user's direct line of sight. Some periscopes work by using two flat mirrors, as shown in the following diagram. If the ray of light that enters the periscope is parallel to the ray that exits the periscope, find the angle of incidence and angle of reflection of the light ray for each mirror. (HINT: See Problem 41.) Explain your reasoning.

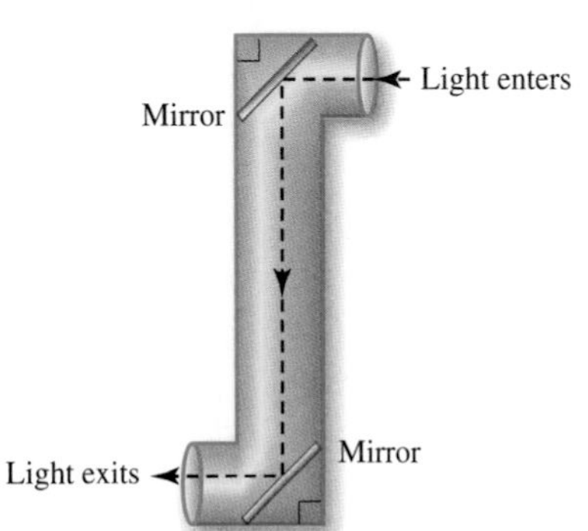

EXTENDED PROBLEMS

43. Three mirrors placed at right angles to each other create a **retroreflector**. Research the retroreflector and explain how it creates parallel rays of light. Because of their unique reflective properties, retroreflectors are used in common objects. Name four uses for retroreflectors. For what reason were retroreflectors placed on the surface of the moon?

44. Euclid strived to derive as many results as he could from a minimum number of postulates. Mathematicians have determined that the Parallel Postulate cannot be derived from the first four postulates. Thus, it has been made a postulate. There are many equivalent forms of the Parallel Postulate other than those stated in this section. Research the Parallel Postulate and make a list of eight equivalent forms.

45. The geometry we have been exploring in this text is commonly referred to as Euclidean geometry. However, there exist other types of geometry, with different assumptions and different, sometimes surprising, results. Research another geometry, such as spherical, elliptical, hyperbolic, or taxicab geometry. Write a summary of your findings that includes information about who developed the geometry you selected and when. Describe which postulates in the alternative geometry differ from those of Euclidean geometry, and list two results that arise from these differences.

5.2 IMPORTANT THEOREMS BASED ON THE PARALLEL POSTULATE

Applied Problem

Two roads intersect as shown next. A rest stop will be constructed between them at a point that is equidistant from the two roads. Where can the rest stop be located, and how can the planners lay it out?

Angle Sum in a Triangle

In this section we apply the Parallel Postulate and the postulates and theorems related to parallel lines from Section 5.1. We will use them to prove some important theorems about triangles. For example, in Chapter 2 we determined informally that the sum of the angle measures in a triangle is 180°. Now that the Parallel Postulate has been assumed, we can give a deductive proof of that result.

THEOREM 5.9 Angle Sum in a Triangle Theorem

The sum of the angle measures in a triangle is 180°.

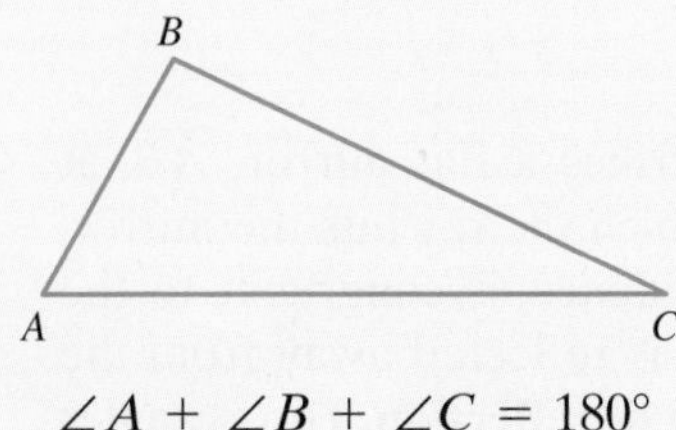

$$\angle A + \angle B + \angle C = 180°$$

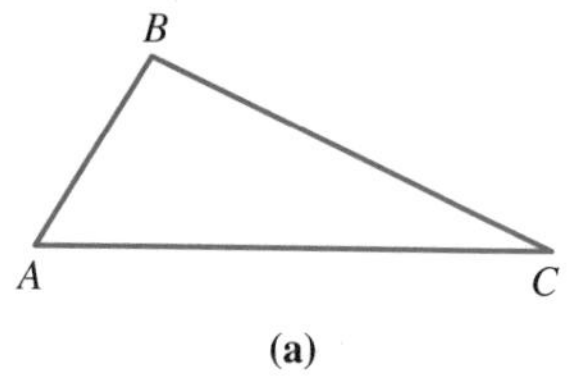

(a)

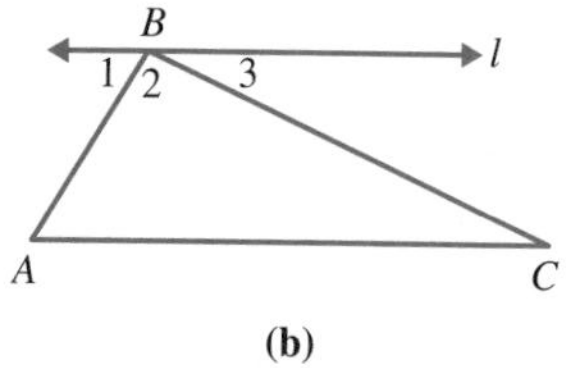

(b)

FIGURE 5.8

Given $\triangle ABC$ [Figure 5.8(a)]

Prove $\angle A + \angle B + \angle C = 180°$

Plan Draw a line l through B parallel to $\overline{AC}$ [Figure 5.8(b)] and use properties of parallel lines.

Proof

Statements	Reasons
1. $\angle 1 = \angle A$	1. Alternate interior angles (Theorem 5.5)
2. $\angle 3 = \angle C$	2. Alternate interior angles (Theorem 5.5)
3. $\angle 1 + \angle 2 + \angle 3 = 180°$	3. A straight angle has measure 180°.
4. $\angle A + \angle 2 + \angle C = 180°$	4. Substitution from Steps 1, 2, and 3
5. $\angle A + \angle B + \angle C = 180°$	5. Substitution from Step 4 ■

Exterior Angle of a Triangle

In Chapter 4, Theorem 4.11 states that an exterior angle of a triangle is greater than any nonadjacent interior angle. The next example will suggest a refinement of that theorem.

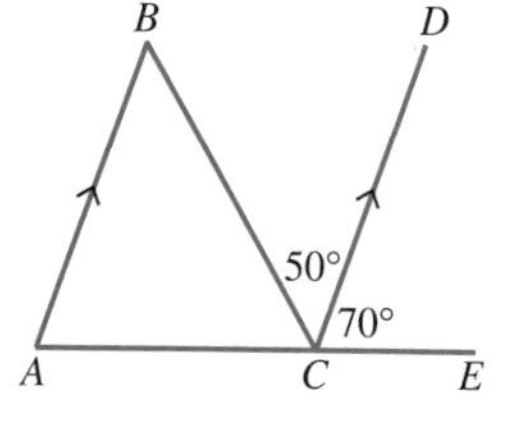

FIGURE 5.9

EXAMPLE 5.4 In Figure 5.9, $\overline{AB}$ is parallel to $\overline{CD}$. Show $\angle A + \angle B = 120°$.

SOLUTION

$\overline{BC}$ is a transversal for segments $\overline{AB}$ and $\overline{CD}$. Thus, $\angle B$ and $\angle BCD$ are alternate interior angles; hence, $\angle B = 50°$. Consider $\overline{AE}$ as a transversal for $\overline{AB}$ and $\overline{CD}$. Then $\angle A$ and $\angle DCE$ are corresponding angles. Hence, $\angle A = 70°$. Thus, $\angle A + \angle B = 70° + 50° = 120°$. ■

Notice that we have just shown that the measure of $\angle BCE$ is the same as the sum of the measures of $\angle A$ and $\angle B$. Using the Angle Sum in a Triangle Theorem, we now prove the following refinement of Theorem 4.11.

COROLLARY 5.10 The Exterior Angle Theorem

An exterior angle of a triangle is equal to the sum of the measures of the two nonadjacent interior angles.

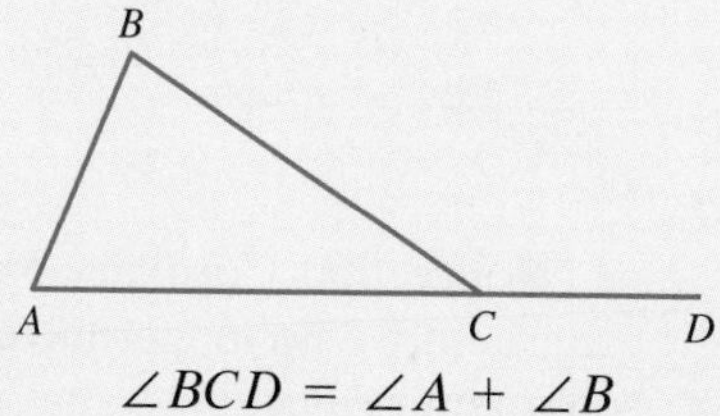

$\angle BCD = \angle A + \angle B$

Given $\triangle ABC$ (Figure 5.10)

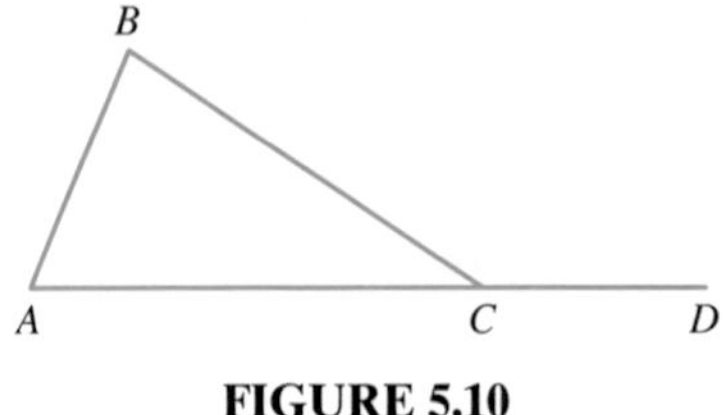

FIGURE 5.10

Prove $\angle BCD = \angle A + \angle B$

Proof

Statements	Reasons
1. $\angle BCA + \angle BCD = 180°$	**1.** A straight angle has measure 180°.
2. $\angle A + \angle B + \angle BCA = 180°$	**2.** Angle Sum in a Triangle Theorem
3. $\angle BCA + \angle BCD = \angle A + \angle B + \angle BCA$	**3.** Substitution from Steps 1 and 2
4. $\angle BCD = \angle A + \angle B$	**4.** Subtracting $\angle BCA$ from both sides in Step 3 ■

AAS Congruence

The next two corollaries also follow from the Angle Sum in a Triangle Theorem. We will leave the proofs of the Angle-Angle-Side Congruence Theorem (AAS) and the Hypotenuse-Angle Congruence Theorem (HA) for the problem set.

COROLLARY 5.11 AAS Congruence Theorem

If two angles and a side of one triangle are congruent respectively to two angles and a side of another triangle, then the two triangles are congruent.

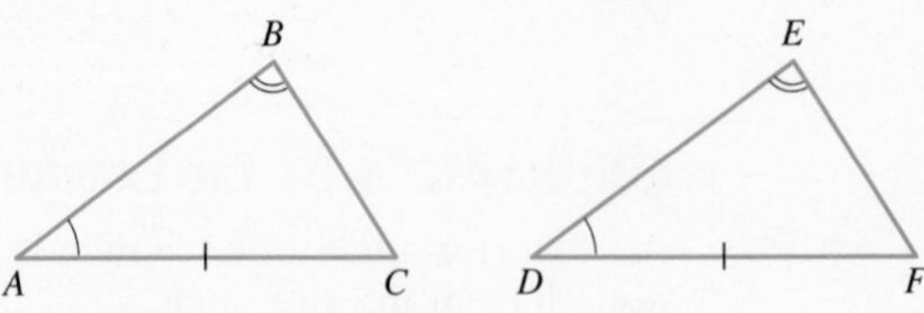

If $\angle A \cong \angle D$, $\angle B \cong \angle E$, and $\overline{AC} \cong \overline{DF}$, then $\triangle ABC \cong \triangle DEF$.

Notice that Corollary 5.11 states that in order to show that two triangles are congruent we no longer have to insist on having two angles and their *included* side congruent respectively. That is, two angles and *any* side will suffice. Next is a special case of this corollary.

COROLLARY 5.12 HA Congruence Theorem

If the hypotenuse and an acute angle of one right triangle are congruent respectively to the hypotenuse and an acute angle of another right triangle, then the two triangles are congruent.

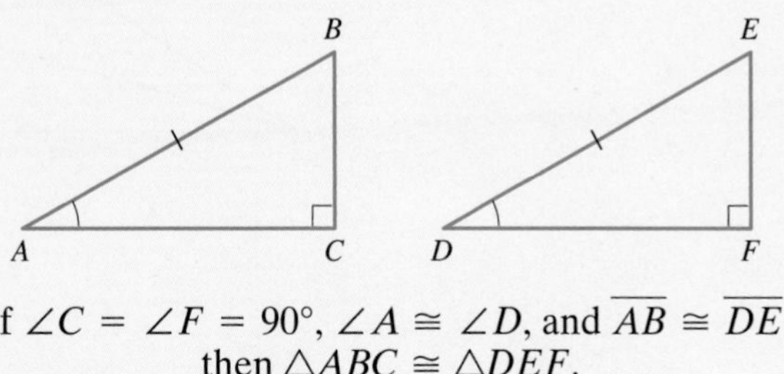

If $\angle C = \angle F = 90°$, $\angle A \cong \angle D$, and $\overline{AB} \cong \overline{DE}$, then $\triangle ABC \cong \triangle DEF$.

Bisector of an Angle

Earlier we showed that any point on the perpendicular bisector of a segment was equidistant from the endpoints of the segment. The next example and theorem show an analogous property of angle bisectors.

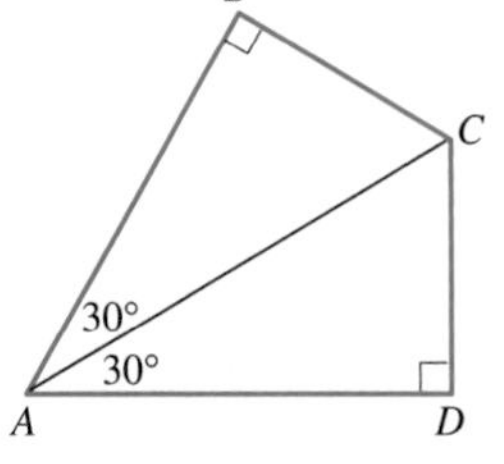

FIGURE 5.11

EXAMPLE 5.5 In Figure 5.11, $AC = 10$ inches. Find BC and DC.

SOLUTION

$\triangle ABC$ is a 30°–60° right triangle with hypotenuse of length 10″. Because $\overline{BC}$ is the leg opposite the 30° angle, $BC = \frac{1}{2}(10'') = 5''$. Similarly, since $\triangle ACD$ is a 30°–60° right triangle, $DC = 5''$. ■

Notice in Example 5.5 that $\overline{AC}$ is the bisector of $\angle A$ and point C is equidistant from B and D. The next theorem states this result more generally.

THEOREM 5.13 Angle Bisector Theorem

A point is on the bisector of an angle if and only if it is equidistant from the sides of the angle.

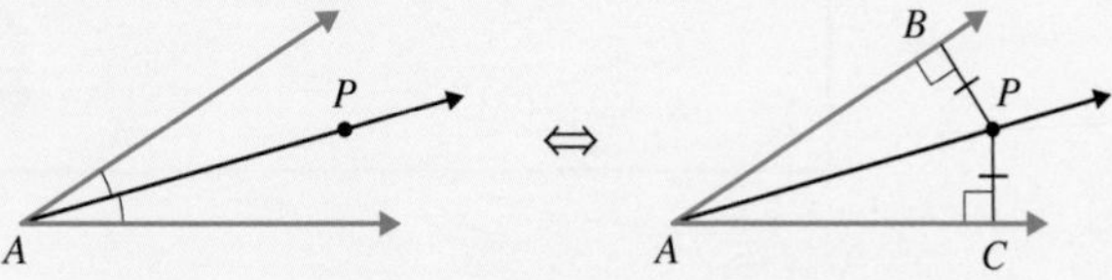

P is on the bisector of $\angle A$ if and only if $PB = PC$.

Because Theorem 5.13 is an "if and only if" theorem, we must prove two parts.

Subgoal 1 Prove: If a point is on the bisector of an angle, then it is equidistant from the sides of the angle.

Subgoal 2 Prove: If a point is equidistant from the sides of an angle, then it is on the bisector of the angle.

We will prove Subgoal 1 and leave the proof of Subgoal 2 for the problem set.

Given Point P on the bisector of $\angle A$ [Figure 5.12(a)]

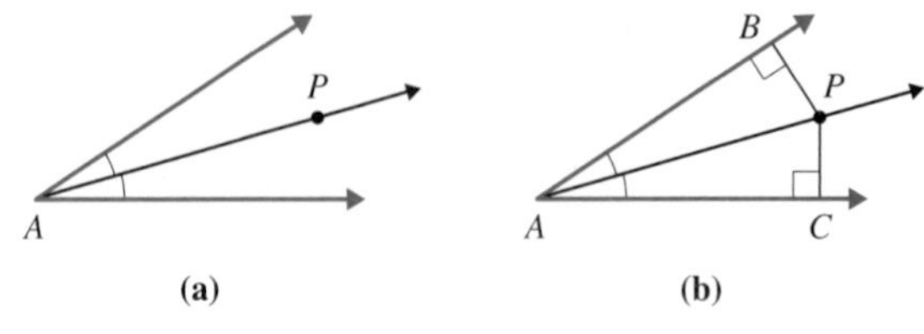

FIGURE 5.12

Prove P is equidistant from the sides of $\angle A$.

Proof The distances from P to the sides of $\angle A$ are the lengths of the segments $\overline{PB}$ and $\overline{PC}$, which are perpendicular to the respective sides [Figure 5.12(b)]. Because $\overrightarrow{AP}$ bisects $\angle A$, we have $\angle BAP \cong \angle CAP$. Also, $\angle B = \angle C = 90°$ and $\overline{PA} \cong \overline{PA}$ [Figure 5.12(b)]. Therefore, we have $\triangle APB \cong \triangle APC$ by AAS Congruence (or by HA Congruence) and $PB = PC$ by corresponding parts. ■

Solution of Applied Problem

Because the rest stop will be equidistant from the two roads, it must be placed at a point that is equidistant from the lines representing the two roads. Thus, by Theorem 5.13, it should be somewhere on the bisector of the angle formed by the two roads. The planners can construct the angle bisector on the drawing to examine the possible locations of the rest stop.

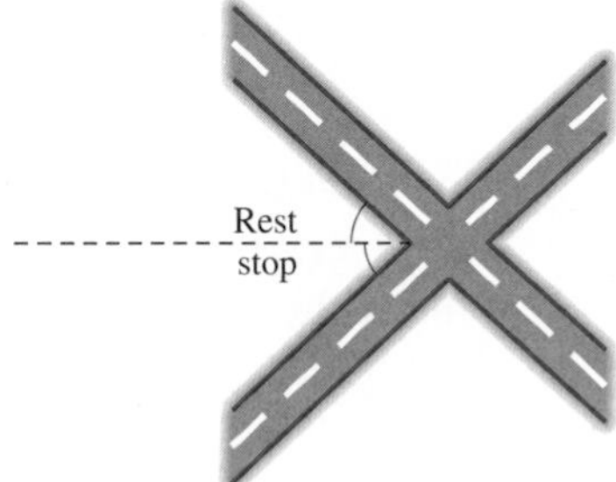

GEOMETRY AROUND US Flashlights and headlights use reflectors that are designed to take advantage of a unique property of parabolas. The reflector behind the bulb is made in the shape of a paraboloid. The bulb is located at a special point called the focus of the paraboloid so that the light rays from the bulb will be reflected in parallel lines, yielding one strong beam of light.

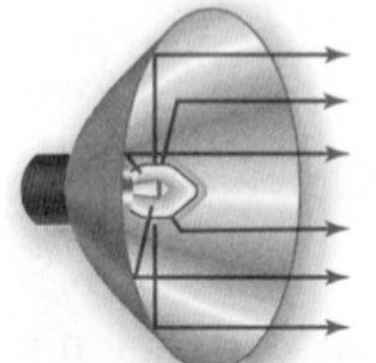

PROBLEM SET 5.2

EXERCISES/PROBLEMS

Use the following figure for Exercises 1–6.

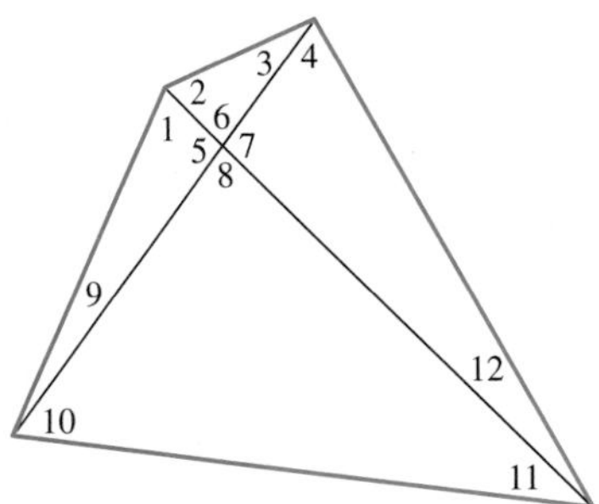

Determine whether each statement is true or false. If true, explain why.

1. $\angle 5 = \angle 2 + \angle 3$
2. $\angle 7 = \angle 1 + \angle 9$
3. $\angle 2 + \angle 3 = \angle 10 + \angle 11$
4. $\angle 1 + \angle 9 = \angle 4 + \angle 12$
5. $\angle 2 + \angle 3 + \angle 4 + \angle 12 = 180°$
6. $\angle 6 + \angle 10 + \angle 11 = 180°$

Use the following figure for Exercises 7–12.

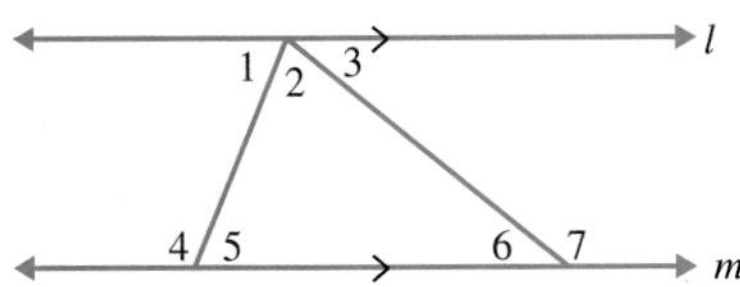

If $l \parallel m$, determine whether each statement is true or false. If true, explain why.

7. $\angle 4 = \angle 2 + \angle 6$
8. $\angle 3 = \angle 2 + \angle 5$
9. $\angle 1 + \angle 2 + \angle 6 = 180°$
10. $\angle 4 + \angle 1 = 180°$
11. $\angle 2 + \angle 6 = 180°$
12. $\angle 7 = \angle 1 + \angle 2$

Use the following figure for Exercises 13–20. Congruent segments and parallel lines are indicated.

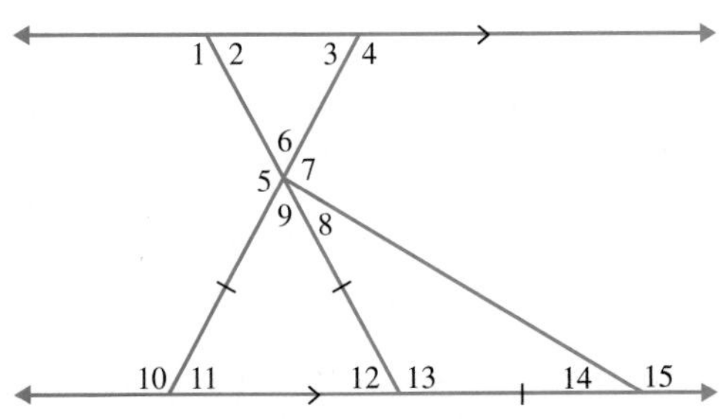

Determine whether each statement is true or false. If true, explain why.

13. $\angle 12 = \angle 8$
14. $\angle 5 = \angle 11 + \angle 14$
15. $\angle 15 = \angle 8 + \angle 9 + \angle 11$
16. $\angle 7 + \angle 8 = \angle 11 + \angle 12$
17. $\angle 12 = 2(\angle 8)$
18. $\angle 2 = \angle 3$
19. $\angle 13 + \angle 11 = 180°$
20. $\angle 4 + \angle 7 = 180°$
21. Find the measure of each numbered angle in the following figure, where $l \parallel m$ and congruent segments are marked.

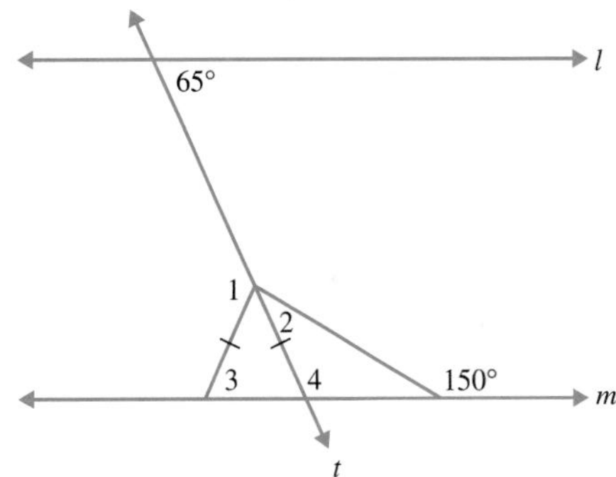

22. Find the measure of each numbered angle in the next figure, where $l \parallel m$.

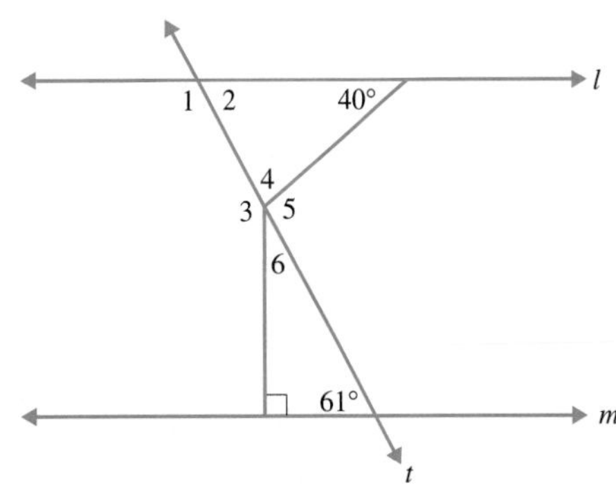

23. In the figure shown, $\angle ABF = \angle CBF$, $AB = 11''$, $AD = 3''$, $EC = 4''$, and $BF = 10''$. Find the area of $\triangle ABC$.

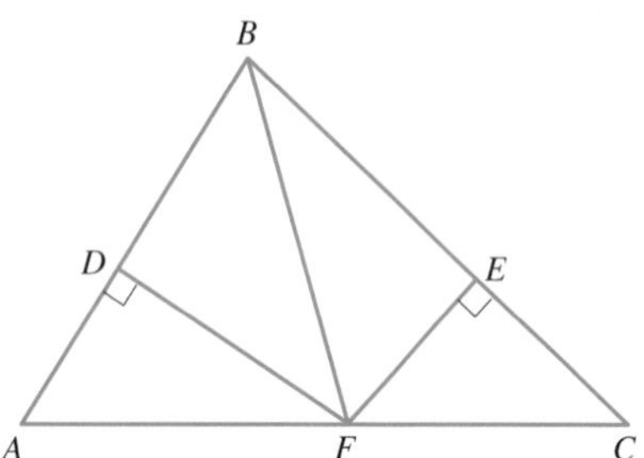

24. In the following figure, $ABCD$ is a trapezoid, $BC = 14$ cm, $AE = 11.9$ cm, $BG = 13$ cm, and $CD = 12$ cm. If $\angle EBG = \angle FBG$, find the area of trapezoid $ABCD$.

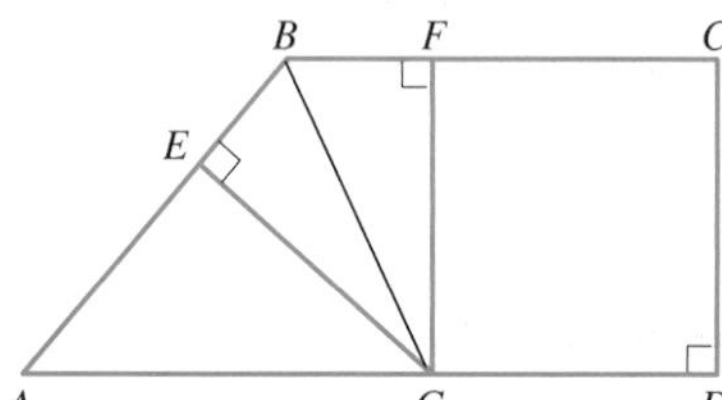

PROOFS

25. In the figure shown, $\angle ACD \cong \angle ACB$ and $\angle ADC \cong \angle ABC$. Prove that $CD \cong CB$.

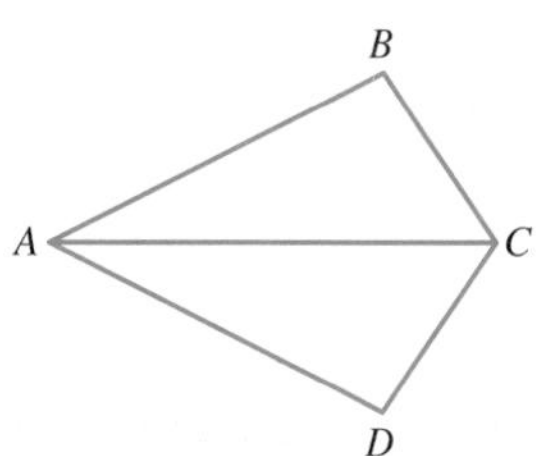

26. In isosceles $\triangle ABC$ with $\overline{AC} \cong \overline{AB}$, and $\angle BDC \cong \angle CEB$, prove that $\overline{BD} \cong \overline{CE}$.

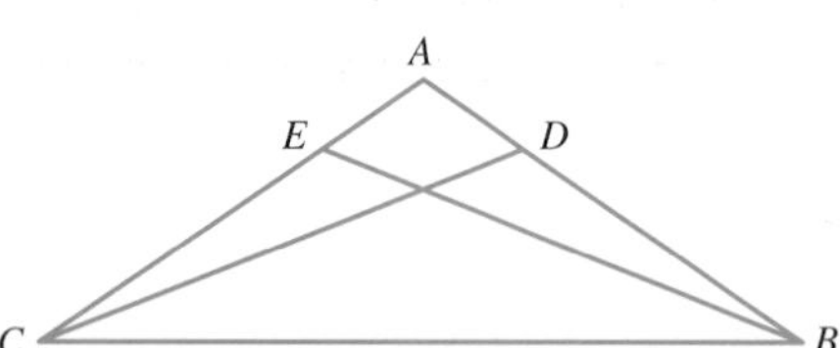

27. Prove Corollary 5.11: If two angles and a side of one triangle are congruent respectively to two angles and a side of another triangle, then the two triangles are congruent.

28. Prove Corollary 5.12: If the hypotenuse and an acute angle of one right triangle are congruent respectively to the hypotenuse and an acute angle of another right triangle, then the two triangles are congruent.

29. Prove Subgoal 2 of Theorem 5.13: If a point is equidistant from the sides of an angle, then it is on the bisector of the angle.

30. The Geometry Investigation at the beginning of this chapter illustrated how we can use origami paper folding to trisect an angle. Review the sequence of folds in the process, and use what you have learned about congruent triangles, parallel lines, and so on, to justify the technique.

APPLICATION

31. A walking trail/bike path has a turn as shown. A marker is to be erected that will be the same distance from all three portions of the path. How can the position of the marker be determined?

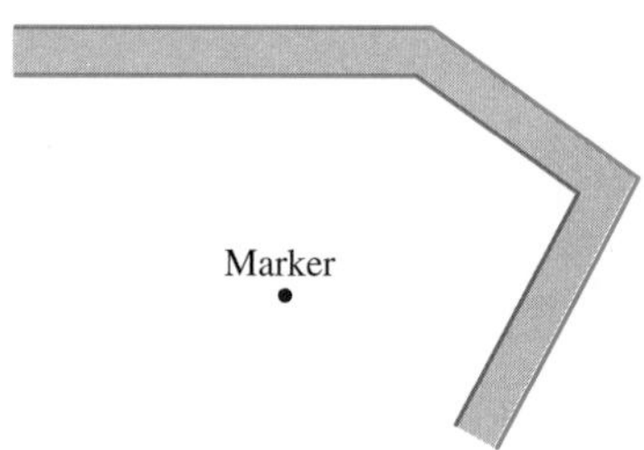

EXTENDED PROBLEMS

32. The Japanese art of origami is intricately linked with geometry. Many of the basic folds depend on parallel lines or on angle bisectors. Research the basic origami folds. Highlight the use of parallel lines or angle bisectors as you explain how to complete the valley fold, kite base, and diamond base. Locate a set of instructions for using origami to make a regular heptagon, and then make one. Explain which folds in the sequence require angle bisectors or parallel lines.

33. In 1899, Frank Morley, a Harvard College mathematics professor, discovered a surprising relationship between the angle trisectors of any triangle and the triangle formed by their intersections. Research and state Morley's theorem. Illustrate the theorem for the case of an equilateral triangle by drawing an equilateral triangle, using a protractor to trisect each angle, and then drawing and measuring the sides of the triangle formed by the intersections of the angle trisectors. Repeat for an isosceles triangle and a scalene triangle. Locate an applet on the Internet that demonstrates Morley's theorem.

Manipulate the original triangle and observe the resulting triangle. Has Morley's theorem been generalized to any other polygons?

34. The angle bisector of an angle in a triangle is a segment that extends from a vertex to the opposite side and bisects the vertex angle. If a triangle is isosceles, then the angle bisectors of the congruent angles are congruent. Is the converse of this statement true? If two angle bisectors of a triangle are congruent, is the triangle isosceles? In 1840, a professor from Berlin named C.L. Lehmus asked Jacob Steiner to prove the converse. Research the Steiner-Lehmus theorem, which has many proofs. Select one and summarize the proof.

5.3 PARALLELOGRAMS AND RHOMBUSES

Applied Problem

A pool table measures 9 ft by $4\frac{1}{2}$ ft. A player hits a ball from corner A at an angle of 60° as shown. How many times will the ball hit a side of the table before it goes into a pocket?

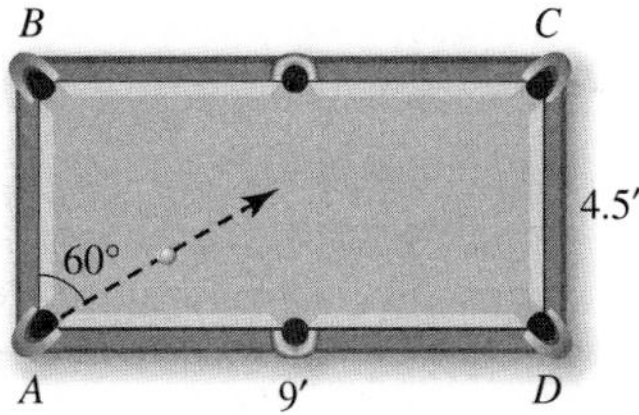

Parallelograms

In Chapter 4 and Sections 5.1 and 5.2, we derived many results involving triangles and parallel lines. In this section, we focus on the study of parallelograms, that is, quadrilaterals with opposite sides parallel.

In Chapter 3, we observed that the diagonal of a parallelogram divides it into two congruent triangles. Here we will examine that idea more formally. Consider the parallelogram $ABCD$ with diagonal $\overline{AC}$ in Figure 5.13(a). Because $\overline{BC} \parallel \overline{AD}$, we have $\angle BCA \cong \angle DAC$ since they are alternate interior angles [Figure 5.13(b)]. Because $\overline{AB} \parallel \overline{CD}$, we also have $\angle BAC \cong \angle DCA$ [Figure 5.13(c)].

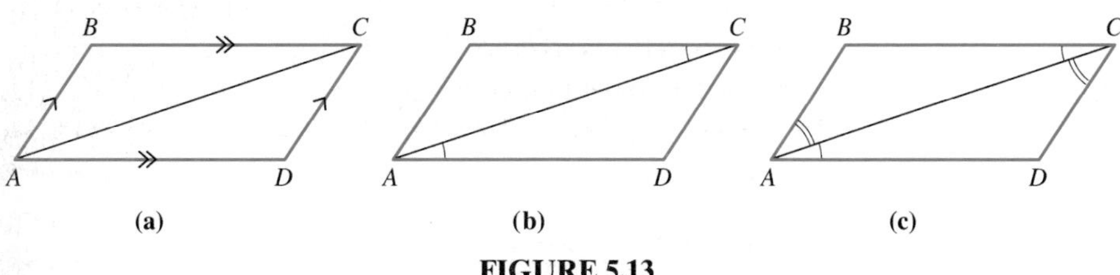

FIGURE 5.13

Notice that $\overline{AC}$ is common to $\triangle ABC$ and $\triangle CDA$. Therefore, we can conclude that $\triangle ABC \cong \triangle CDA$ by ASA. We state this result next.

THEOREM 5.14

A diagonal of a parallelogram forms two congruent triangles.

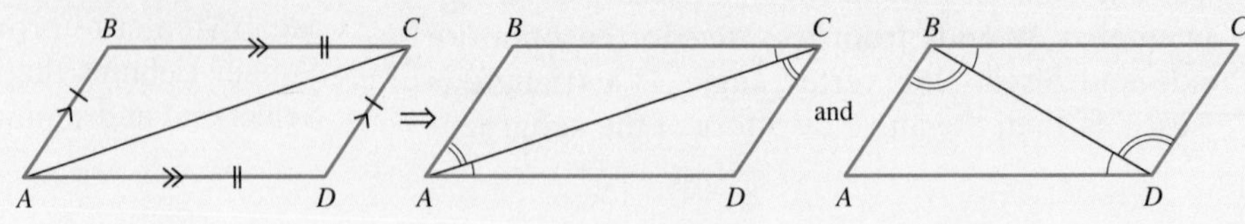

If $ABCD$ is a parallelogram, then $\triangle ABC \cong \triangle CDA$ and $\triangle ABD \cong \triangle CDB$.

Notice that by corresponding parts of $\triangle ABC$ and $\triangle CDA$, $\angle B \cong \angle D$, $\overline{BC} \cong \overline{AD}$, and $\overline{AB} \cong \overline{CD}$. In a similar manner, using $\triangle ABD$ and $\triangle CDB$, we can conclude that $\angle A \cong \angle C$. This observation leads us to the next corollary.

COROLLARY 5.15

In a parallelogram, the opposite sides are congruent and the opposite angles are congruent.

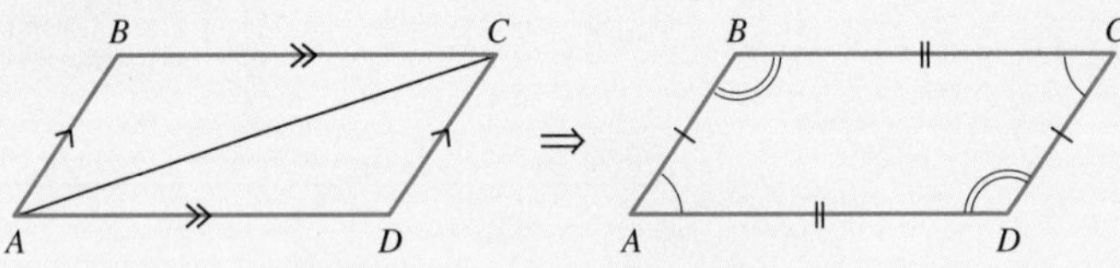

If $ABCD$ is a parallelogram, then $\overline{AB} \cong \overline{CD}$, $\overline{AD} \cong \overline{BC}$, $\angle A \cong \angle C$, and $\angle B \cong \angle D$.

(a)

(b)

FIGURE 5.14

Another important result follows from Corollary 5.15. Let $l \parallel m$ [Figure 5.14(a)]. Let $\overline{AD}$ and $\overline{BC}$ be any two segments that are perpendicular to l (and hence m) [Figure 5.14(b)]. Therefore, $\overline{AD} \parallel \overline{BC}$ because corresponding angles are congruent. This means that $ABCD$ is a parallelogram. As a consequence, $AD = BC$ and the two distances between l and m are equal. We summarize this conclusion next.

COROLLARY 5.16

Parallel lines are everywhere equidistant.

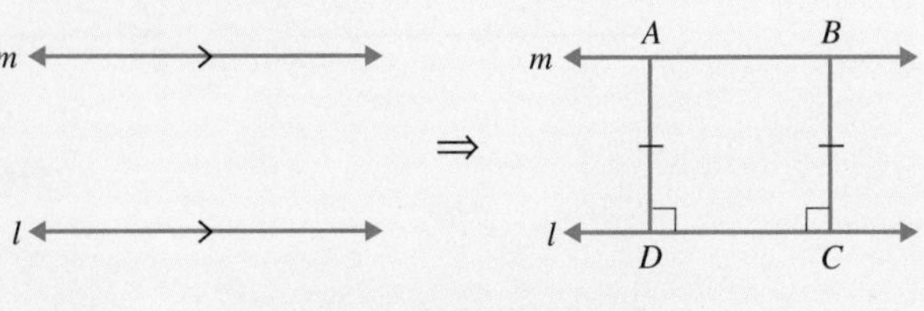

If $l \parallel m$, then $AD = BC$.

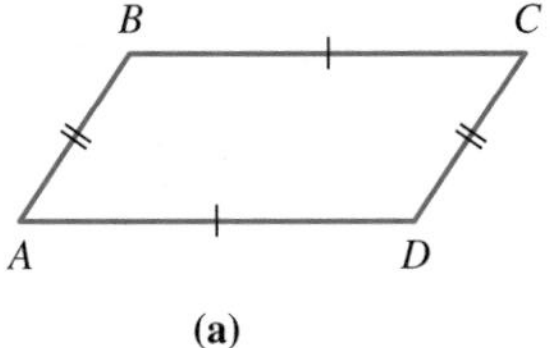

(a)

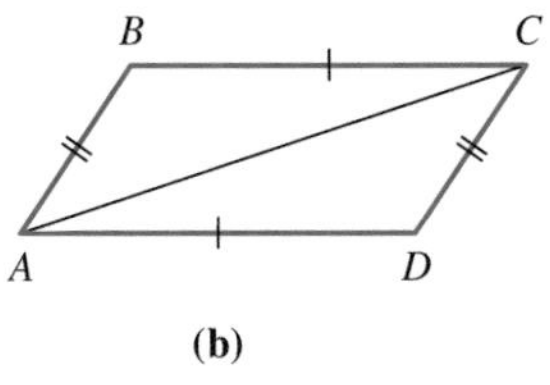

(b)

FIGURE 5.15

Now let's consider the converse of Corollary 5.15. That is, if the opposite sides of a quadrilateral are congruent, will it be a parallelogram? Consider quadrilateral $ABCD$ in which the opposite sides are congruent [Figure 5.15(a)]. If we draw $\overline{AC}$ [Figure 5.15(b)], then two congruent triangles are formed, namely $\triangle ABC \cong \triangle CDA$, by SSS. By corresponding parts, $\angle BCA \cong \angle DAC$, so $\overline{BC} \parallel \overline{AD}$. In a similar fashion, by drawing $\overline{BD}$ in Figure 5.15(a), we could show that $\overline{AB} \parallel \overline{CD}$. Thus, $ABCD$ is a parallelogram. We state this result next.

THEOREM 5.17

If both pairs of opposite sides of a quadrilateral are congruent, then the quadrilateral is a parallelogram.

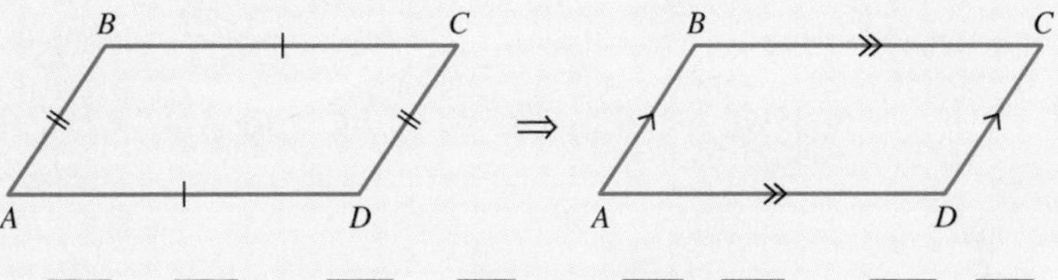

If $\overline{AB} \cong \overline{CD}$ and $\overline{AD} \cong \overline{BC}$, then $\overline{AB} \parallel \overline{CD}$ and $\overline{AD} \parallel \overline{BC}$.

It also turns out to be the case that if the opposite angles of a quadrilateral are congruent, then it is a parallelogram. We state this result next and leave its proof for the problem set.

THEOREM 5.18

If both pairs of opposite angles of a quadrilateral are congruent, then the quadrilateral is a parallelogram.

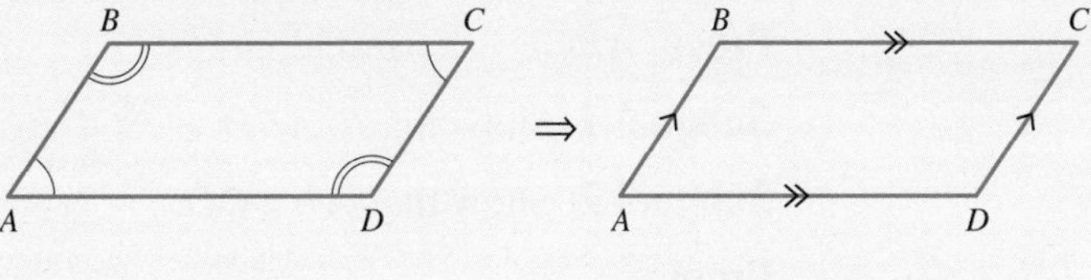

If $\angle A \cong \angle C$ and $\angle B \cong \angle D$, then $\overline{AB} \parallel \overline{CD}$ and $\overline{AD} \parallel \overline{BC}$.

What can we say about a quadrilateral in which two sides are both congruent and parallel? Consider the pair of parallel line segments $\overline{AD}$ and $\overline{BC}$ in Figure 5.16(a), where $\overline{AD} \cong \overline{BC}$.

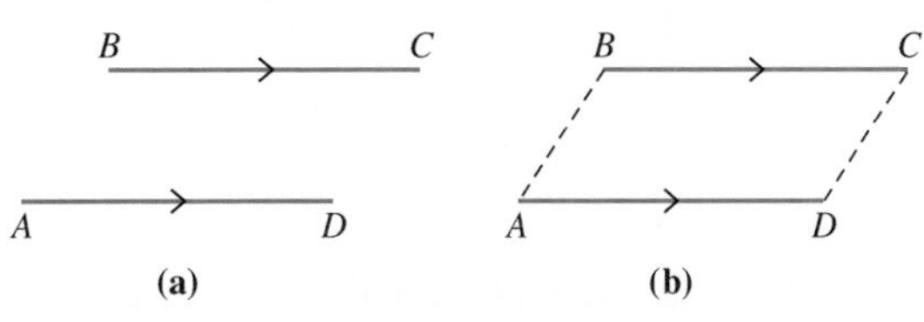

FIGURE 5.16

If we draw $\overline{AB}$ and $\overline{CD}$ [Figure 5.16(b)], it appears that $ABCD$ is a parallelogram. We prove this fact next.

THEOREM 5.19

If a quadrilateral has two sides that are parallel and congruent, then it is a parallelogram.

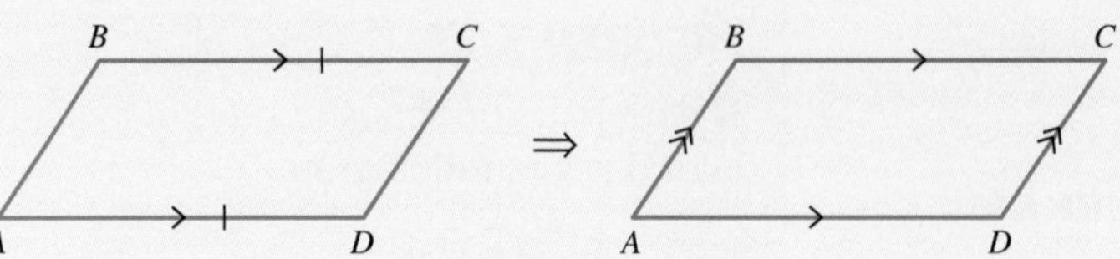

If $\overline{AD} \parallel \overline{BC}$ and $\overline{AD} \cong \overline{BC}$, then $\overline{AB} \parallel \overline{CD}$ and $\overline{AD} \parallel \overline{BC}$.

Given $\overline{AD} \parallel \overline{BC}$ and $\overline{AD} \cong \overline{BC}$ [Figure 5.17(a)]

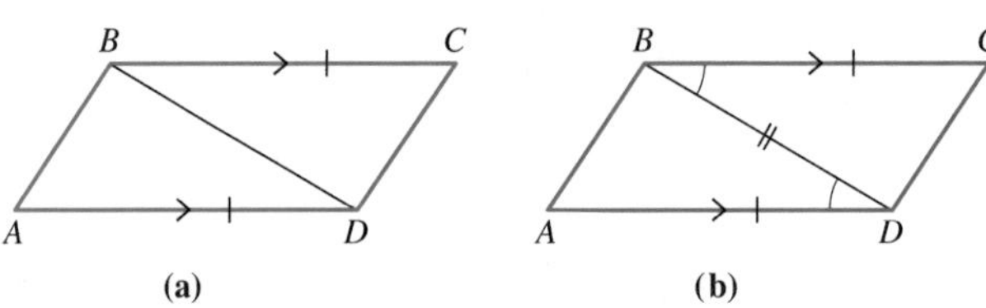

FIGURE 5.17

Prove $ABCD$ is a parallelogram.

Subgoal 1 Show that $\triangle ABD \cong \triangle CDB$.

Subgoal 2 Show that $\overline{AB} \parallel \overline{CD}$.

Proof

Statements	Reasons
1. $\overline{AD} \parallel \overline{BC}$ and $\overline{AD} \cong \overline{BC}$	**1.** Given
2. $\angle ADB \cong \angle CBD$	**2.** Alternate interior angles [Figure 5.17(b)]
3. $\overline{BD} \cong \overline{BD}$	**3.** Reflexive property [Figure 5.17(b)].
4. $\triangle ABD \cong \triangle CDB$	**4.** SAS (Subgoal 1)
5. $\angle ABD \cong \angle CDB$	**5.** C.P.
6. $\overline{AB} \parallel \overline{CD}$	**6.** Congruent alternate interior angles (Subgoal 2)
7. $ABCD$ is a parallelogram.	**7.** Definition of a parallelogram (Opposite sides are parallel.) ■

EXAMPLE 5.6 Given parallelogram $ABCD$ with midpoints $E, F, G,$ and H on sides $\overline{AB}$, $\overline{BC}$, $\overline{CD}$, and $\overline{AD}$, respectively, prove that $EFGH$ is a parallelogram [Figure 5.18(a)].

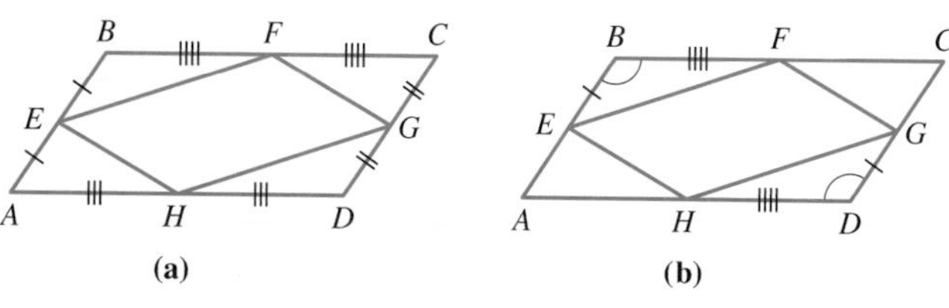

FIGURE 5.18

Plan Show that the opposite sides of $EFGH$ are congruent using congruent triangles.

SOLUTION

First, $\angle B = \angle D$ because opposite angles of a parallelogram are congruent. Next, $AB = CD$ and $AD = BC$ because opposite sides of a parallelogram are congruent. Also, $EB = \frac{1}{2}AB$ and $DG = \frac{1}{2}CD$. Hence, $EB = DG$. In a similar fashion, $BF = DH$ [Figure 5.18(b)]. Therefore, $\triangle BEF \cong \triangle DGH$ by SAS. By corresponding parts, $EF = GH$. In a similar manner, we can show that $FG = EH$. Thus, because the opposite sides of $EFGH$ are congruent, $EFGH$ is a parallelogram. ■

In Chapter 6, we will consider the figure formed by connecting the midpoints of the sides of any quadrilateral.

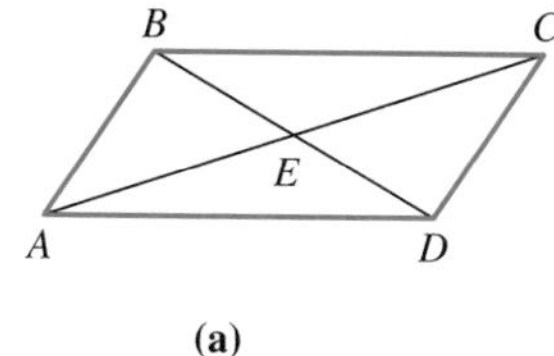

(a)

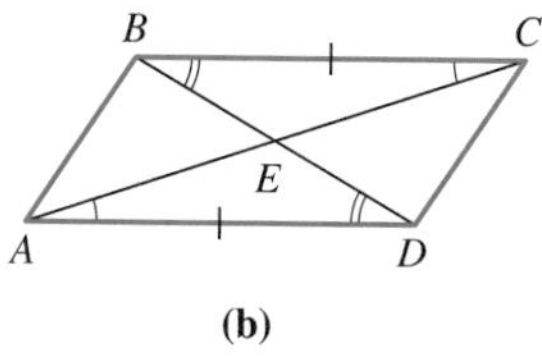

(b)

FIGURE 5.19

Diagonals of a Parallelogram

Next we will look at the diagonals of a parallelogram. Consider parallelogram $ABCD$ with diagonals $\overline{AC}$ and $\overline{BD}$ [Figure 5.19(a)]. Because $\overline{AD} \parallel \overline{BC}$ and $\overline{CA}$ is a transversal, $\angle EAD = \angle ECB$. In a similar fashion, $\angle EBC = \angle EDA$ [Figure 5.19(b)]. Because $ABCD$ is a parallelogram, $\overline{BC} \cong \overline{AD}$. Thus, $\triangle ADE \cong \triangle CBE$ by ASA. By corresponding parts, $\overline{BE} \cong \overline{DE}$ and $\overline{AE} \cong \overline{CE}$, which means that E is the midpoint of both $\overline{BD}$ and $\overline{AC}$. We summarize this result next.

THEOREM 5.20

The diagonals of a parallelogram bisect each other.

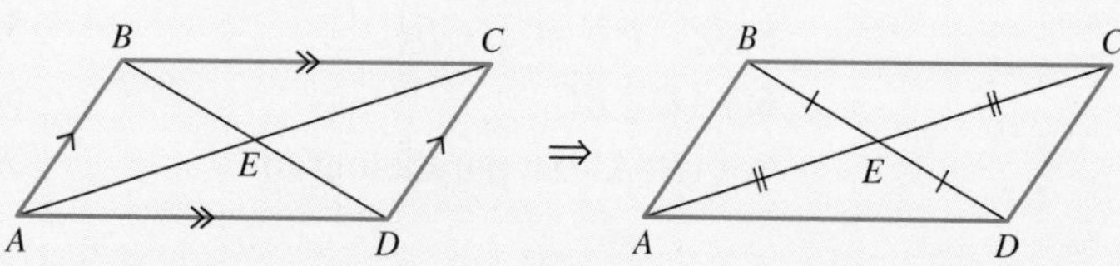

If $ABCD$ is a parallelogram, then $AE = CE$ and $BE = DE$.

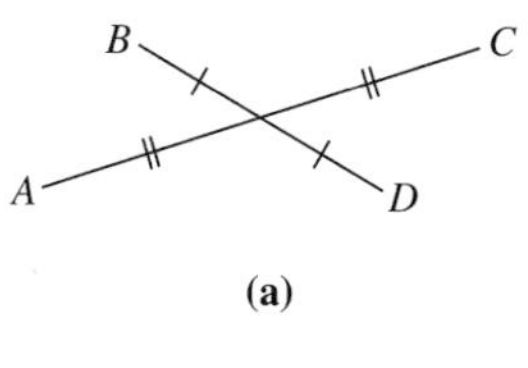

(a)

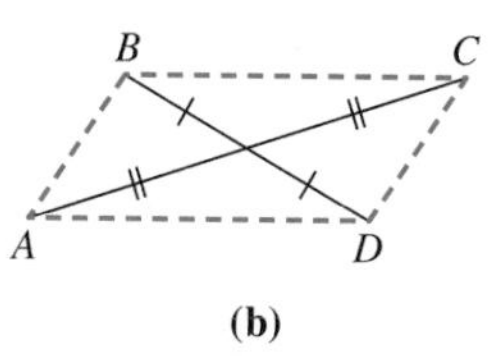

(b)

FIGURE 5.20

Now consider the two line segments that intersect at their midpoints in Figure 5.20(a). If we draw segments $\overline{AB}$, $\overline{BC}$, $\overline{CD}$, and $\overline{AD}$ [Figure 5.20(b)], the resulting quadrilateral appears to be a parallelogram. This result, which is the converse of the previous theorem, is proved next.

THEOREM 5.21

If the diagonals of a quadrilateral bisect each other, then it is a parallelogram.

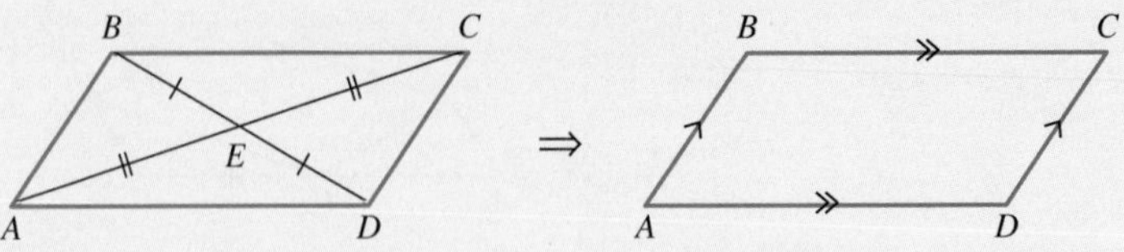

If $AE = CE$ and $BE = DE$, then $\overline{AB} \parallel \overline{CD}$ and $\overline{AD} \parallel \overline{BC}$.

Given $ABCD$ with $AE = CE$ and $BE = DE$ [Figure 5.21(a)]

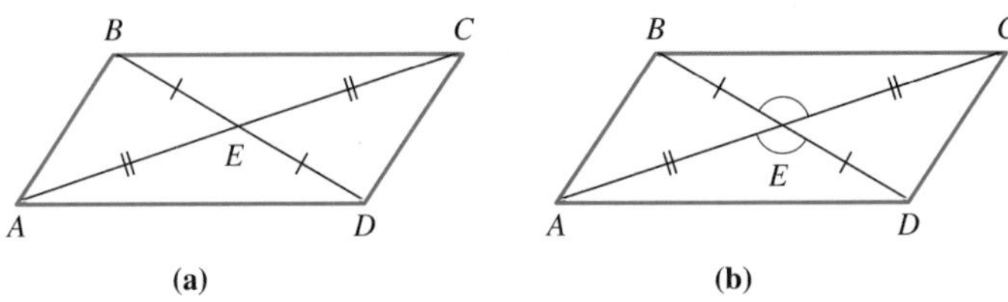

FIGURE 5.21

Prove $ABCD$ is a parallelogram.

Plan Use congruent triangles to show that two sides of $ABCD$ are parallel and congruent.

Proof

Statements	Reasons
1. $AE = CE$ and $BE = DE$	1. Given
2. $\angle AED = \angle CEB$	2. Vertical angles [Figure 5.21(b)]
3. $\triangle AED \cong \triangle CEB$	3. SAS
4. $AD = BC$	4. C.P.
5. $\angle ADE = \angle CBE$	5. C.P.
6. $\overline{AD} \parallel \overline{BC}$	6. Congruent alternate interior angles
7. $ABCD$ is a parallelogram.	7. Two opposite sides of the quadrilateral are parallel and congruent. ■

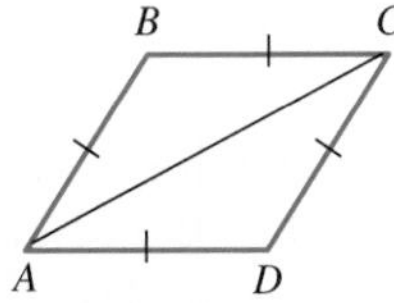

FIGURE 5.22

Rhombuses

Recall that a rhombus was defined to be a quadrilateral with all sides congruent. Consider rhombus $ABCD$ (Figure 5.22), where the diagonal $\overline{AC}$ is shown. By SSS, $\triangle ABC \cong \triangle CDA$. Then the corresponding parts of these congruent triangles are congruent. In particular, we have $\angle BAC \cong \angle DCA$. Taking these as alternate interior angles with $\overline{AC}$ as a transversal, we can conclude that $\overline{AB} \parallel \overline{CD}$. Because $\overline{AB}$ and $\overline{CD}$ are both congruent and parallel, $ABCD$ is a parallelogram. Thus, we have the following theorem.

THEOREM 5.22

Every rhombus is a parallelogram.

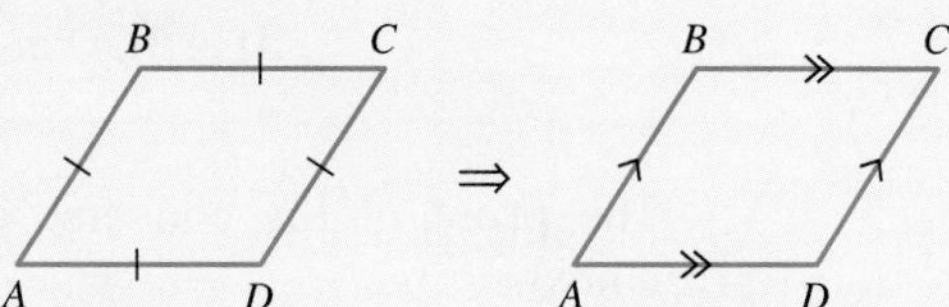

If $ABCD$ is a rhombus, then $\overline{AB} \parallel \overline{CD}$ and $\overline{AD} \parallel \overline{BC}$.

Now that we have shown that every rhombus is a parallelogram, we can conclude that all the theorems about parallelograms also apply to rhombuses. That is, we can say that

1. The diagonals of a rhombus bisect each other.
2. The opposite angles of a rhombus are congruent.
3. A diagonal of a rhombus forms two congruent triangles.

Consider rhombus $ABCD$ with diagonals $\overline{AC}$ and $\overline{BD}$ [Figure 5.23(a)]. Because $ABCD$ is a rhombus (hence a parallelogram), the diagonals bisect each other at E [Figure 5.23(b)].

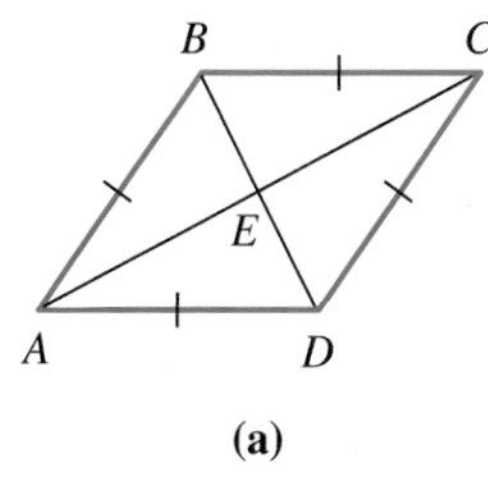

(a)

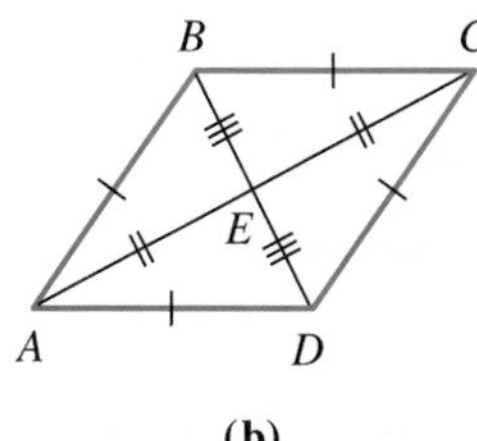

(b)

FIGURE 5.23

By SSS, the two diagonals divide the rhombus into four congruent triangles. In particular, we have $\triangle AEB \cong \triangle AED$. Therefore, $\angle AEB = \angle AED$ by corresponding parts. Because they are also supplementary adjacent angles, they must be right angles. So $\overline{AC}$ is perpendicular to $\overline{BD}$. We summarize this result next.

THEOREM 5.23

The diagonals of a rhombus are perpendicular to each other.

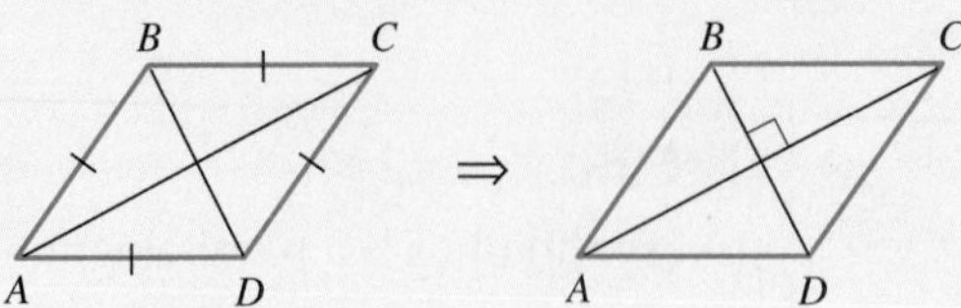

If $ABCD$ is a rhombus, then $\overline{AC} \perp \overline{BD}$.

The proof of the converse of this theorem, restricted to parallelograms, is given next.

THEOREM 5.24

If the diagonals of a parallelogram are perpendicular to each other, then the parallelogram is a rhombus.

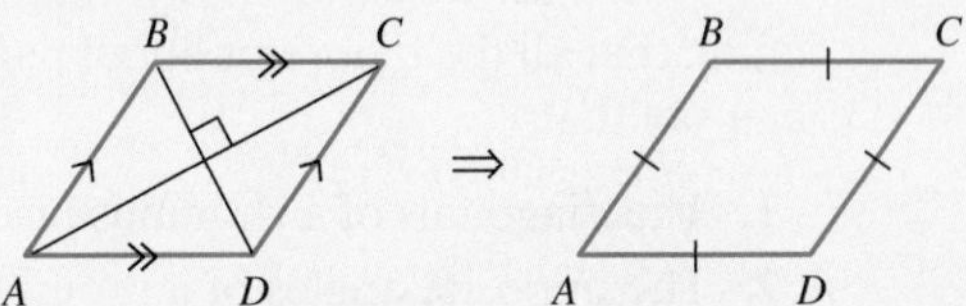

If $ABCD$ is a parallelogram with $\overline{AC} \perp \overline{BD}$, then $AB = BC = CD = AD$.

Given Parallelogram $ABCD$ with $\overline{AC}$ perpendicular to $\overline{BD}$ [Figure 5.24(a)]

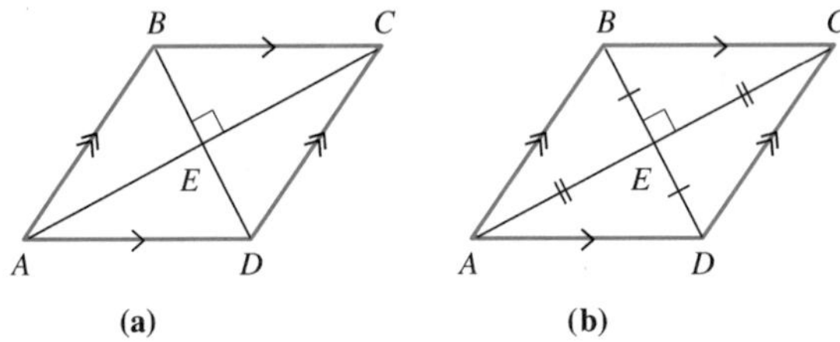

FIGURE 5.24

Prove $ABCD$ is a rhombus.

Proof

Statements	Reasons
1. $\overline{AC}$ is perpendicular to $\overline{BD}$.	**1.** Given
2. $AE = CE$ and $BE = DE$	**2.** The diagonals of a parallelogram bisect each other [Figure 5.24(b)].
3. $\triangle AED \cong \triangle AEB \cong \triangle CED \cong \triangle CEB$	**3.** LL Congruence (or SAS)
4. $\overline{AB} \cong \overline{BC} \cong \overline{CD} \cong \overline{AD}$	**4.** C.P.
5. $ABCD$ is a rhombus.	**5.** Definition of a rhombus (All four sides are congruent.) ■

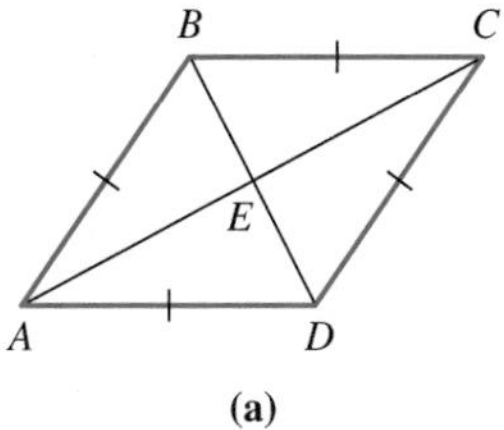

(a)

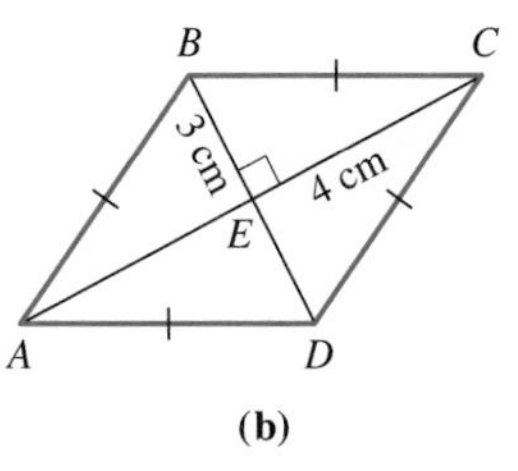

(b)

FIGURE 5.25

EXAMPLE 5.7 The diagonals in rhombus $ABCD$ have lengths 6 cm and 8 cm. Find the length of the sides of the rhombus.

SOLUTION

Let E be the point where the diagonals intersect [Figure 5.25(a)]. Then, because the diagonals of a rhombus bisect each other, $CE = \frac{1}{2}(CA) = 4$ cm and $BE = 3$ cm [Figure 5.25(b)]. Because the diagonals of a rhombus are perpendicular, we can use the Pythagorean theorem to find the length of a side. Thus, we have the following:

$$BC^2 = 3^2 + 4^2 = 9 + 16 = 25$$

so $BC = 5$ cm. Because the four sides of a rhombus are congruent, their lengths are 5 cm. ■

Using the proof given for Theorem 5.24, we can also conclude that the diagonals bisect the opposite angles of a rhombus. This is the final result of the section. We leave its proof for the problem set.

> **THEOREM 5.25**
>
> A parallelogram is a rhombus if and only if its diagonals bisect the opposite angles.
>
> 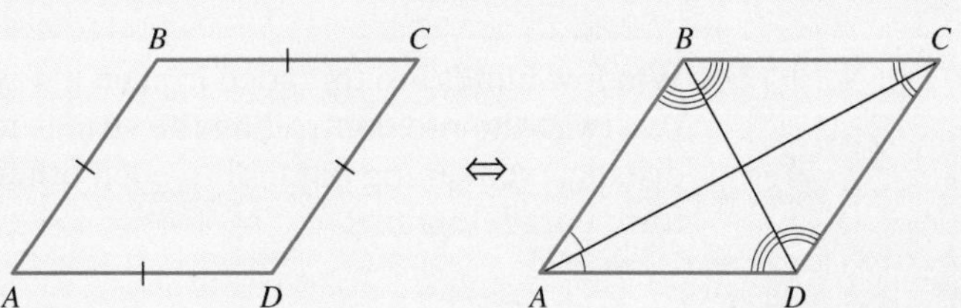
>
>
> Parallelogram $ABCD$ is a rhombus if and only if $\overline{AC}$ bisects $\angle A$ and $\angle C$, and $\overline{BD}$ bisects $\angle B$ and $\angle D$.

EXAMPLE 5.8 In rhombus $ABCD$, prove $\angle BAC \cong \angle BCA$ and $\angle DBA \cong \angle BDA$ (Figure 5.26).

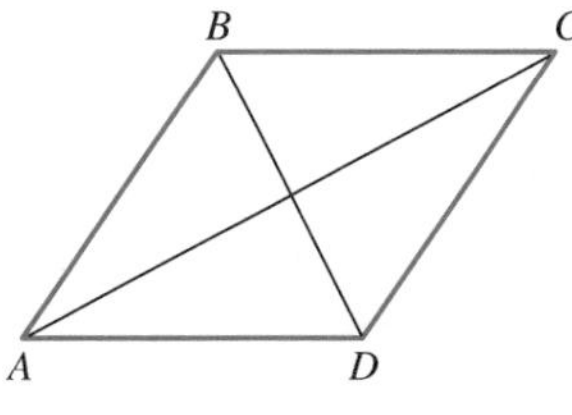

FIGURE 5.26

SOLUTION

Because $ABCD$ is a rhombus, it is a parallelogram. Thus, opposite angles are congruent, and $\angle A = \angle C$. So, by Theorem 5.25, $\frac{1}{2}\angle A = \frac{1}{2}\angle C$ or $\angle BAC = \angle BCA$. In a similar manner, $\angle DBA = \angle BDA$. ■

Solution of Applied Problem

The ball travels so that the angle of incidence always equals the angle of reflection, as illustrated on the sides $\overline{BC}$ and $\overline{CD}$. That is, the angle at which the ball hits the side is the same as the angle at which the ball bounces off the side.

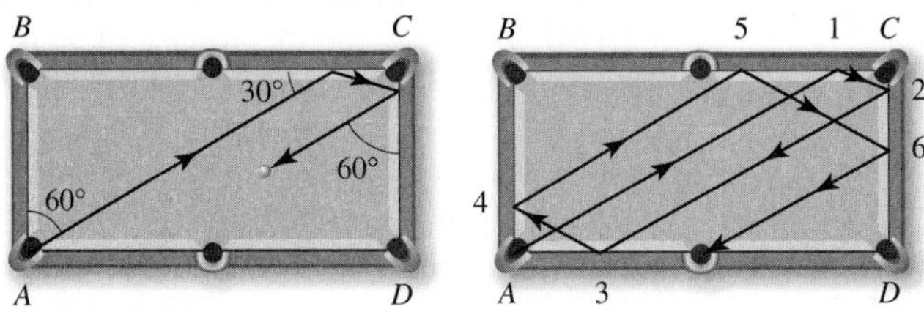

The ball will continue to hit a side and rebound at the same angle it hits, always traveling along parallel lines and forming parallelograms as shown. The ball will hit the sides 6 times before going into the pocket in the middle of side $\overline{AD}$. You can verify this result by making a scale drawing and sketching the path of the ball.

GEOMETRY AROUND US Window blinds are threaded in such a way that the blinds always move in parallel. If you twist the rod controlling the strings to adjust the blinds, they all move the same distance, remaining parallel to each other. At any given time the blinds and strings form parallelograms.

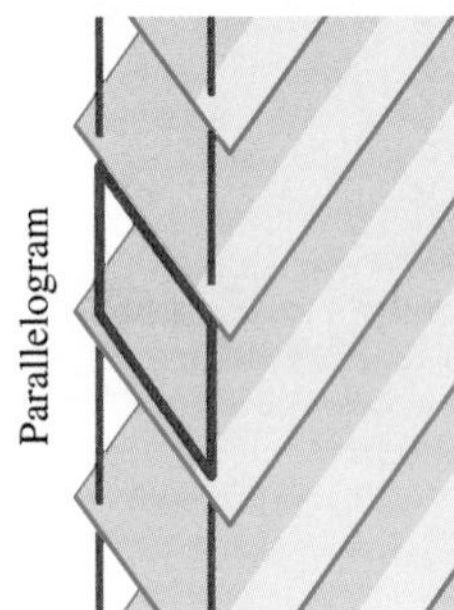

PROBLEM SET 5.3

EXERCISES/PROBLEMS

Use the following figure for Exercises 1–12.

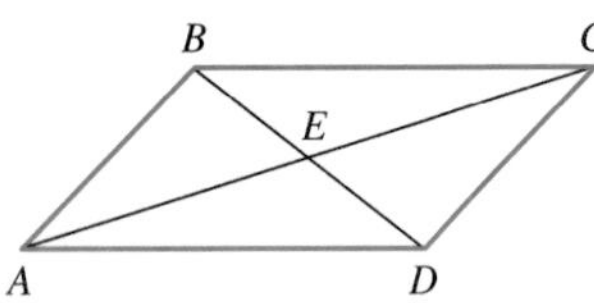

Determine whether the statement is true or false for any parallelogram ABCD that is not a rectangle. If true, explain why.

1. $AC = BD$
2. $AB = CD$
3. $AD = BC$
4. $AE = CE$
5. $AE = BE$
6. $\angle A \cong \angle C$
7. $\angle A \cong \angle D$
8. $\angle A + \angle D = 180°$
9. $\triangle AED \cong \triangle BEC$
10. $\triangle ABE \cong \triangle CDE$
11. $\triangle ADC \cong \triangle CBA$
12. $\triangle BCA \cong \triangle CBD$

13. Find the measures of the angles in the following parallelogram.

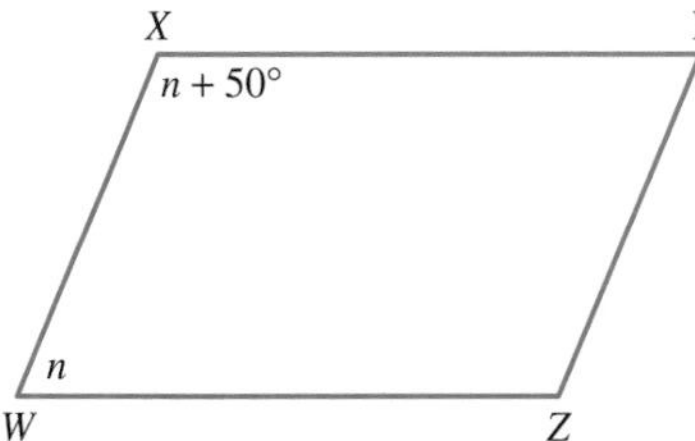

14. Use the following figure to find the measures of the angles $\angle PSR$ and $\angle SRQ$ and side length PS in parallelogram $PQRS$.

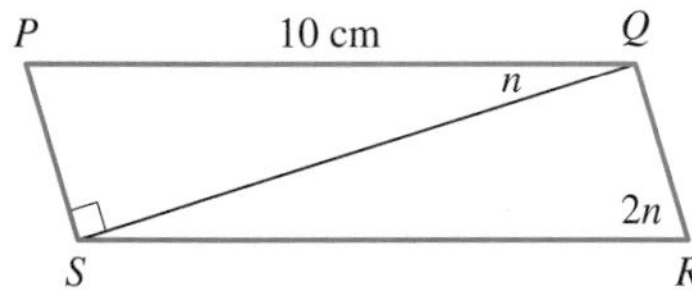

15. Find the measures of the numbered angles in parallelogram $ABCD$.

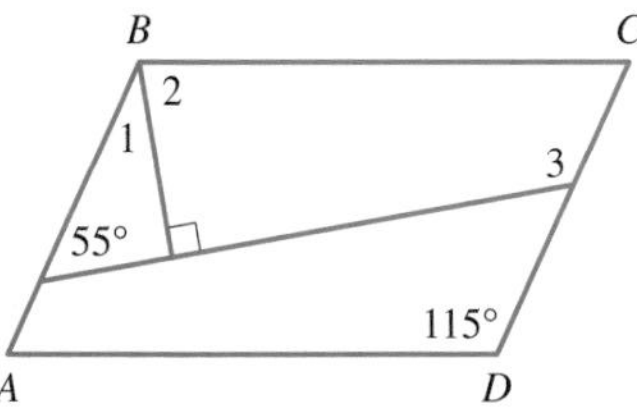

16. Find the measures of $\angle 1$ and $\angle 2$ in parallelogram $PQRS$.

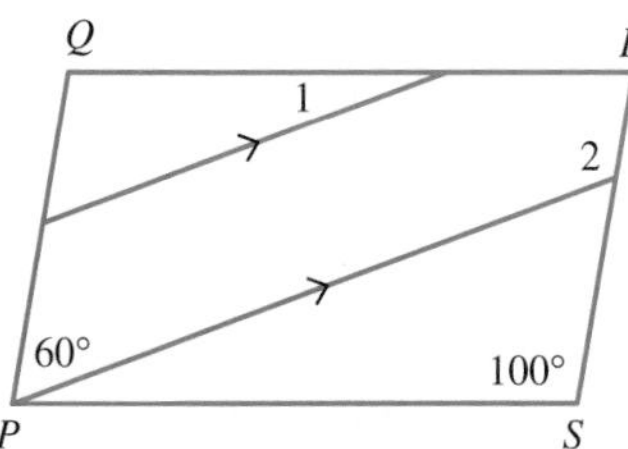

17. The measures of consecutive angles in a parallelogram are in the ratio 2:3. Find the measures of the angles of the parallelogram.

18. The measure of one vertex angle of a parallelogram is 18° more than twice the measure of the next vertex angle. Find the measures of the angles of the parallelogram.

Use the following figure for Exercises 19–26.

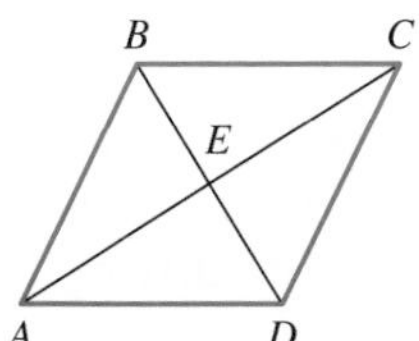

Determine whether the statement is true or false for rhombus ABCD. If true, explain why.

19. $AE = CE$

20. $AB = BC$

21. $\triangle ABC \cong \triangle ADC$

22. $\triangle BEA \cong \triangle DEA$

23. $\angle EBC + \angle ECB = 90°$

24. $\angle DAC \cong \angle BCA$

25. $\angle ABC \cong \angle CDA$

26. $AE = ED$

27. Find the measures of the numbered angles in the rhombus $PQRS$.

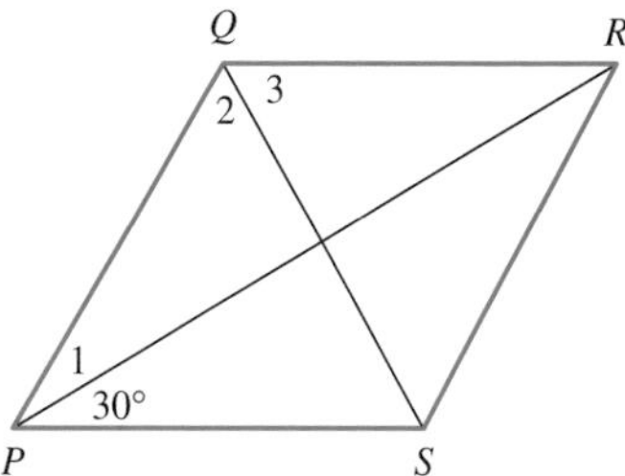

28. Find the measures of the numbered angles in rhombus $ABCD$ if $BE = 7''$ and $CD = 14''$.

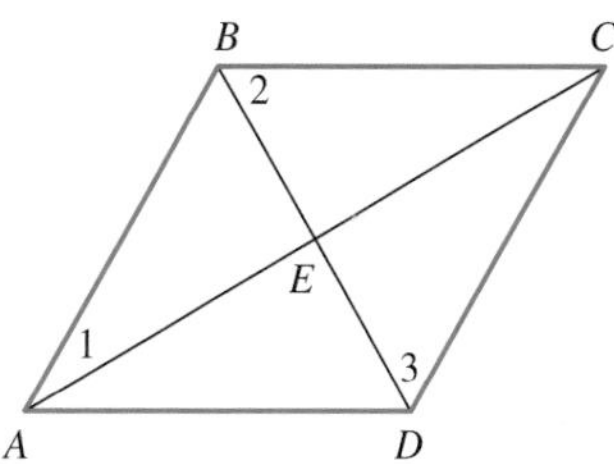

29. Find the length of a side of rhombus $ABCD$ if $AE = 6''$.

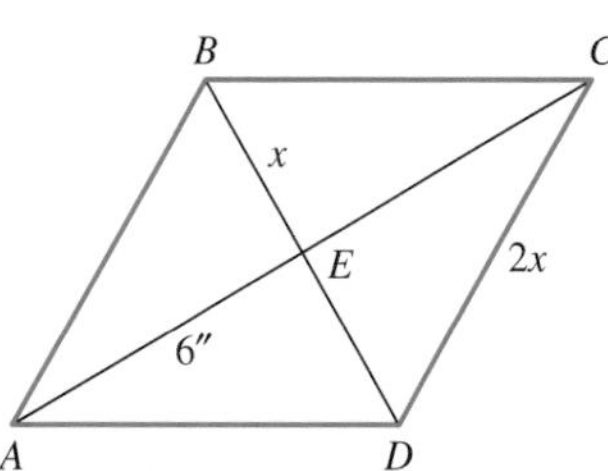

30. In rhombus $ABCD$ shown next, $BC = BD$ and $AC = 16$ cm. Find the measures of $\overline{AB}$, $\overline{DE}$, $\angle ABC$, and $\angle DCA$.

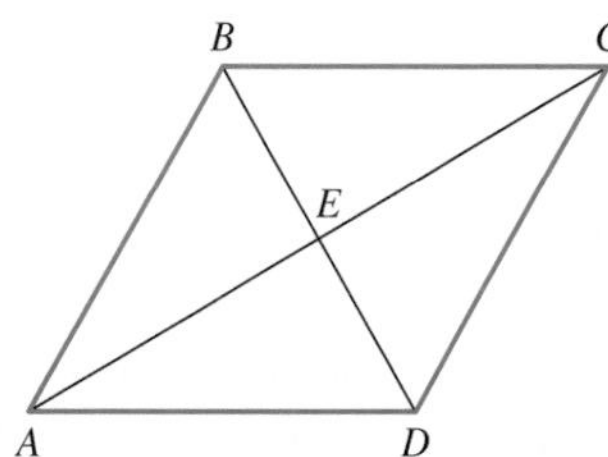

31. Find the measures of the numbered angles in the next figure, where $\overline{CD} \parallel \overline{BE} \parallel \overline{AF}$, $\overline{BC} \parallel \overline{DE}$ and $\overline{AB} \parallel \overline{EF}$. Congruent segments are marked.

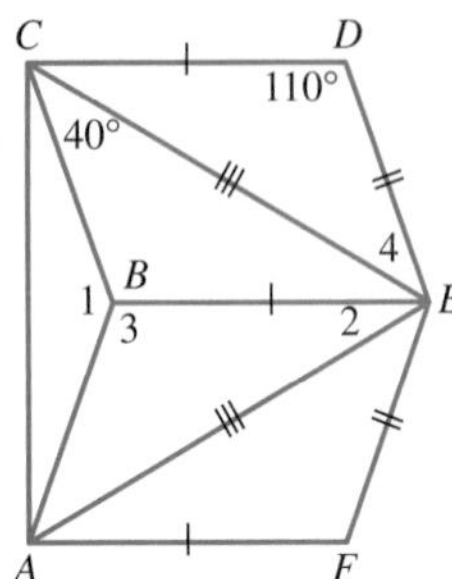

32. In the figure shown, $ABDF$ and $BCEF$ are parallelograms. If $\angle CBD = 22°$, $\angle E = 56°$, $\angle BAF = 60°$, and $\overline{AD} \parallel \overline{FE}$, what are the measures of the numbered angles?

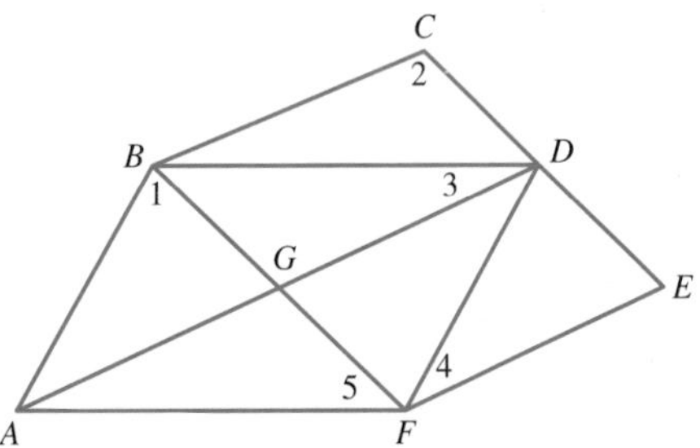

33. In parallelogram $PQRS$, find the height h. Then calculate the area of $PQRS$.

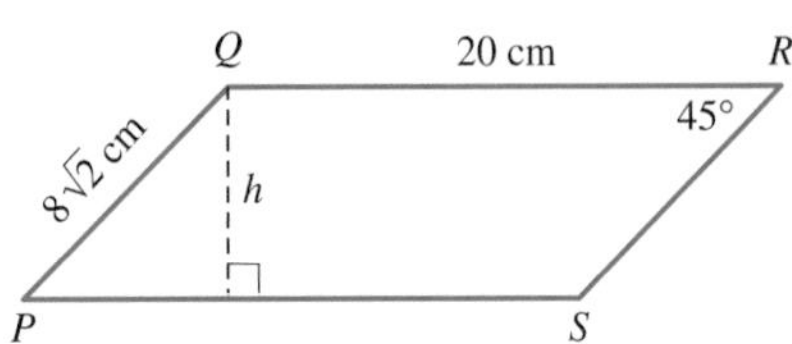

34. In rhombus $WXYZ$, find the height h. Then calculate the area of $WXYZ$.

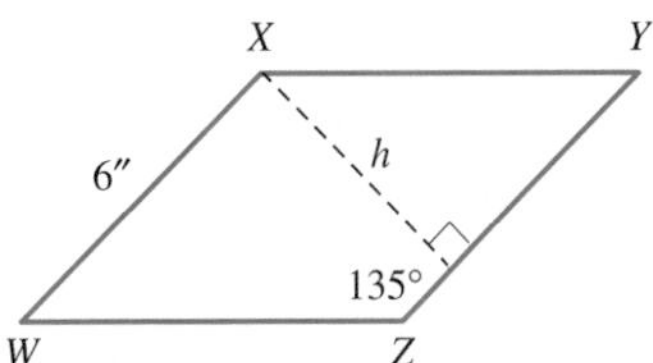

35. The side of a rhombus measures 10 cm and one diagonal measures 12 cm. Find the length of the other diagonal.

36. The diagonals of a rhombus measure 10 inches and 24 inches. Find the length of a side of the rhombus.

PROOFS

37. Prove Theorem 5.18: If both pairs of opposite angles of a quadrilateral are congruent, then the quadrilateral is a parallelogram.

38. Prove that the diagonals of a rhombus bisect each other.

39. Prove: If $ABCD$ is a rhombus, then $\angle CAD$ and $\angle BDA$ are complementary.

40. Prove: In parallelogram $ABCD$, if $\angle CAD$ and $\angle BDA$ are complementary, then $ABCD$ is a rhombus.

41. Prove: The bisectors of any two consecutive angles of a parallelogram are perpendicular.

42. Prove Theorem 5.25: A parallelogram is a rhombus if and only if its diagonals bisect the opposite angles.

43. In parallelogram $ABCD$, $\overline{AE}$ bisects $\angle A$ and $\overline{CF}$ bisects $\angle C$. Prove that $AECF$ is a parallelogram.

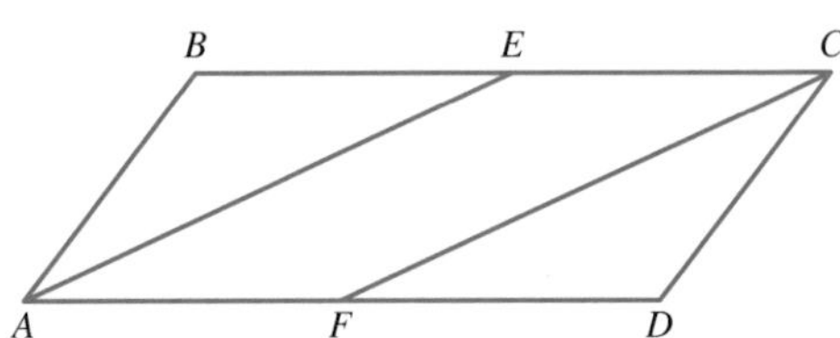

APPLICATIONS

44. A **linkage** is a system of rods and joints used to transmit motion. A four-bar linkage in which opposite links are of equal length is called a parallel-motion linkage. Using the following diagram, explain why the connecting link will always move in such a way that it is parallel to the fixed link.

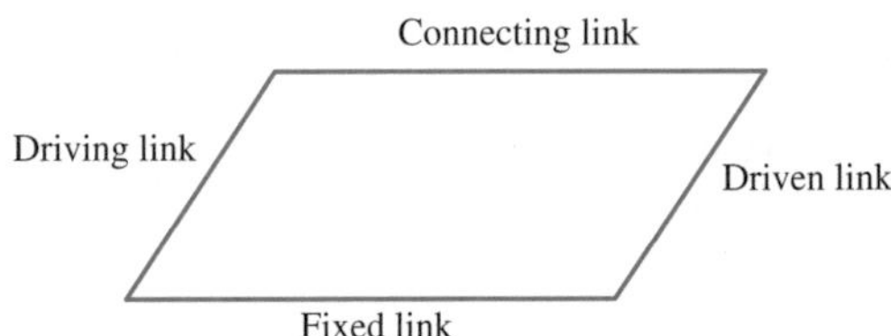

An example of this kind of linkage is the rod on the wheels of a locomotive. In this case, the two wheels take the place of the driving link and the driven link.

45. An ironing board and a scissor-bed truck, use the same type of linkage. For example by positioning the legs appropriately, we can adjust an ironing board to different heights.

In the picture, suppose that the legs of the ironing board are connected at their midpoints and are able to pivot there. Why will the ironing board surface always be parallel to the floor, regardless of the height of the board from the floor? The same principle keeps the bed of a scissor-bed truck parallel to the ground.

46. Suppose a player hits a ball from point A at an angle of 45° on a pool table that measures 9′ by 4.5′ as shown.

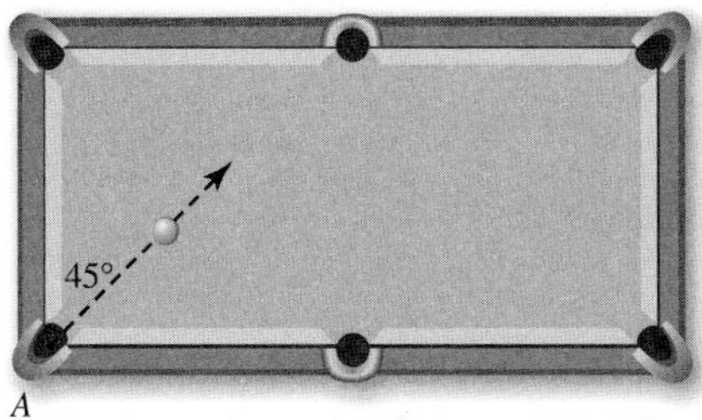

How many times will the ball hit a side of the table before going into a pocket? If a player hits the ball at the same angle but the table measures 9′ by 6′, how many times will it hit a side? (HINT: Try making a scale drawing.)

47. Suppose a player hits a ball from A at an angle of 20° on a pool table measuring 9′ by 4.5′. How many times will the ball hit a side of the table before going into a pocket?

48. In physics, a **force** is a push or a pull on an object. We can represent a force by a ray called a vector or a **directed line segment**. The length of the vector denotes the magnitude of the force. We can specify the direction of the force by giving the angle measure, where the angle is measured counterclockwise from the positive x-axis. For example, a force of 10 pounds applied to an object at a 30° angle can be represented by the following vector v, which has length 10 units.

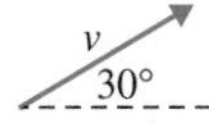

When two forces v and w act upon an object, the two vectors are usually drawn as shown.

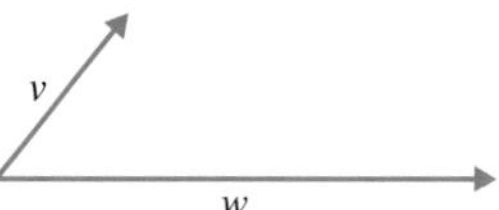

We can represent the two force vectors by a single vector, called the **resultant vector**, which is the diagonal of a parallelogram, as shown in the following figure. This method of determining the resultant of two vectors is called the **parallelogram law**.

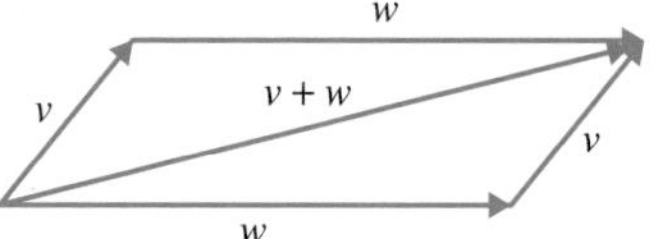

a. Explain why the figure is a parallelogram.

b. When would the resultant force be the diagonal of a rhombus?

EXTENDED PROBLEMS

49. a. An angle trisection tool developed in 1875 and credited to C. A. Laisant uses two hinged rhombuses and their diagonals to trisect an angle. Research Laisant's angle trisection tool and study the following figure, in which $ABFD$ and $ACGE$ are rhombuses. Explain how the tool works, and discuss its limitations. Be sure to explain why we must use rhombuses and not general parallelograms.

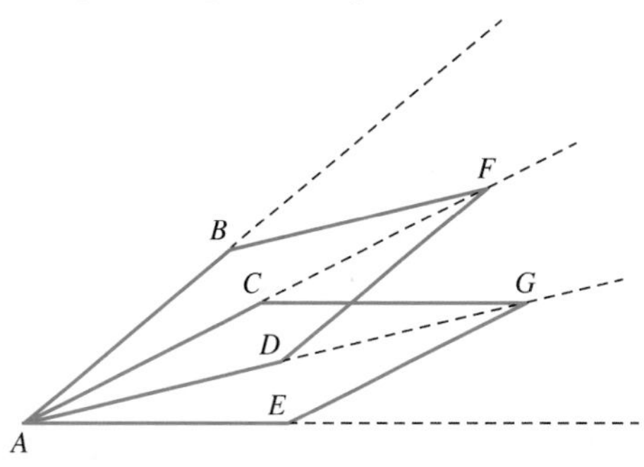

b. Research other angle trisection tools. Archimedes created one other such tool. Make a sketch of the tool and explain how we may use it to trisect an angle.

50. A **pantograph** is a tool that uses properties of parallelograms to copy figures. Investigate the pantograph.

a. When was the pantograph invented and who is credited with the design? Use parallelogram properties to explain how the tool works. Draw a picture of a pantograph and label all congruent and parallel parts.

b. Build a simple pantograph using strips of cardboard or straws together with brads, thumbtacks, or screws.

c. Explore and discuss how this tool has been adapted for use in steel milling, sculpting, and engraving.

51. Revisit the applied problem at the beginning of this section in which a rectangular pool table measuring 9 ft by $4\frac{1}{2}$ ft was shown. Explore other shapes for pool tables and other angles at which a player might hit the ball from a corner. What would happen if the pool table was a square and a player hit the ball from a vertex at an angle of 60°, 45°, or 30° measured from a side? What would happen if the pool table was a rhombus with a 60° vertex angle and a player hit the ball from the vertex at an angle of 60°, 45°, or 30° measured from a side? Create a pool table that is a parallelogram but not a rectangle or a rhombus and explore what happens when players take shots at various angles. Make scale drawings of each table and mark off each shot.

5.4 RECTANGLES, SQUARES, AND TRAPEZOIDS

Applied Problem

Harry wants to construct a frame for his favorite photograph. He cuts two pieces of oak 9″ long and two pieces 12″ long, miters the corners, and attaches the pieces. How can he verify that the resulting frame is "square," that is, that it has 90° corners?

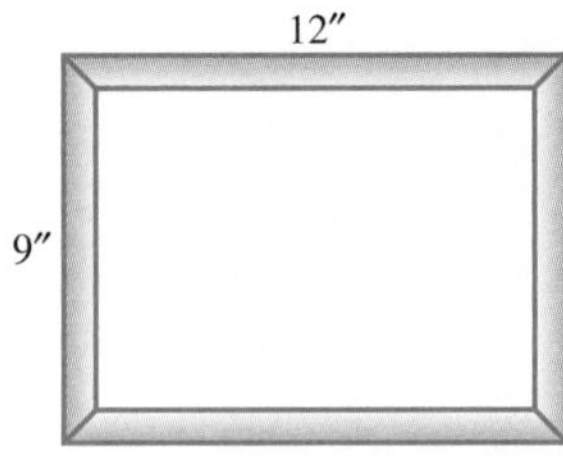

Rectangles

Recall that a rectangle was defined to be a quadrilateral with four right angles (Figure 5.27). By viewing $\angle A$ and $\angle B$ as supplementary angles formed by the transversal $\overline{AB}$, we can see that $\overline{AD} \parallel \overline{BC}$. In a similar manner, $\overline{AB} \parallel \overline{CD}$. Thus, we can say that $ABCD$ is also a parallelogram.

FIGURE 5.27

THEOREM 5.26

Every rectangle is a parallelogram.

If $ABCD$ is a rectangle, then $\overline{AB} \parallel \overline{CD}$ and $\overline{AD} \parallel \overline{BC}$.

Because every rectangle is a parallelogram, we know the following results for parallelograms also hold for rectangles.

1. The opposite sides are parallel.
2. The opposite sides are congruent.
3. The diagonals bisect each other.

A rectangle is a quadrilateral with four right angles. However, if we know that a quadrilateral is a parallelogram, then one right angle is sufficient to show that the parallelogram is also a rectangle. We leave the proof of this fact for the problem set.

THEOREM 5.27

A parallelogram with one right angle is a rectangle.

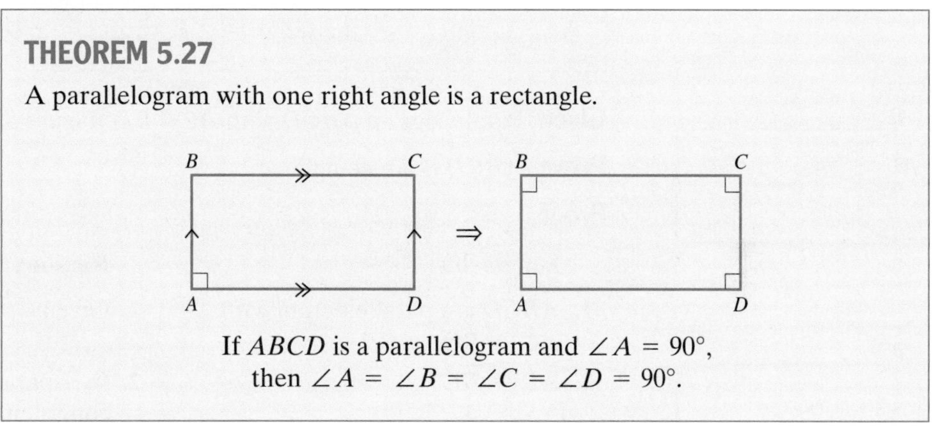

If $ABCD$ is a parallelogram and $\angle A = 90°$, then $\angle A = \angle B = \angle C = \angle D = 90°$.

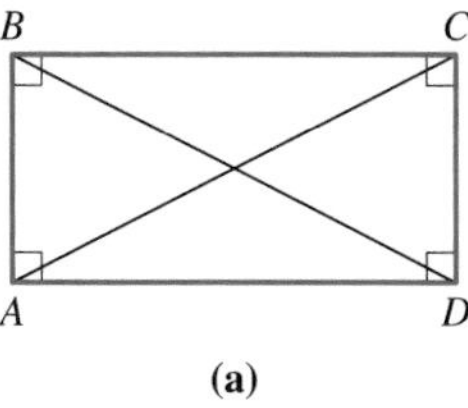

(a)

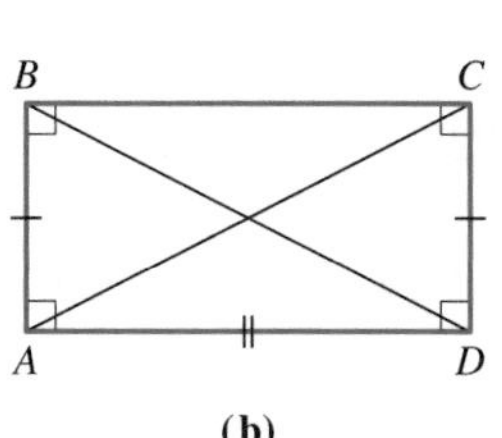

(b)

FIGURE 5.28

This theorem gives an alternative definition of a rectangle. That is, we could have defined a rectangle to be a parallelogram with one right angle.

Now let's consider rectangle $ABCD$ having diagonals $\overline{AC}$ and $\overline{BD}$ [Figure 5.28(a)]. Because opposite sides are congruent, we know that $\overline{AB} \cong \overline{CD}$. Because $\angle BAD = 90° = \angle CDA$, we can conclude that $\triangle BAD \cong \triangle CDA$ by SAS (or LL) [Figure 5.28(b)]. Thus, $\overline{AC} \cong \overline{BD}$ by corresponding parts. We summarize this discussion in the next theorem.

THEOREM 5.28

The diagonals of a rectangle are congruent.

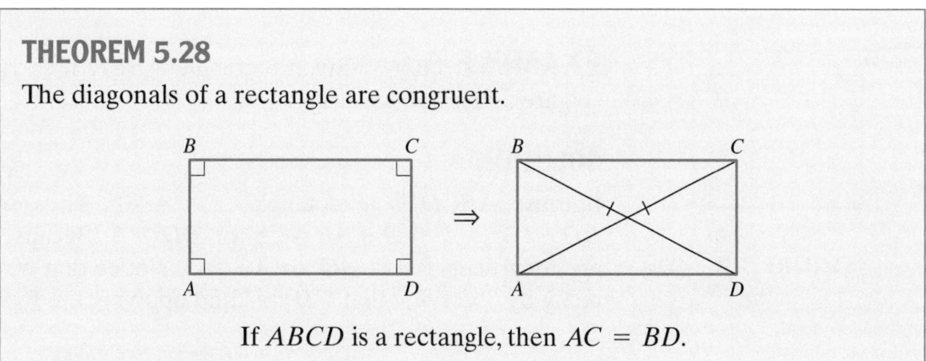

If $ABCD$ is a rectangle, then $AC = BD$.

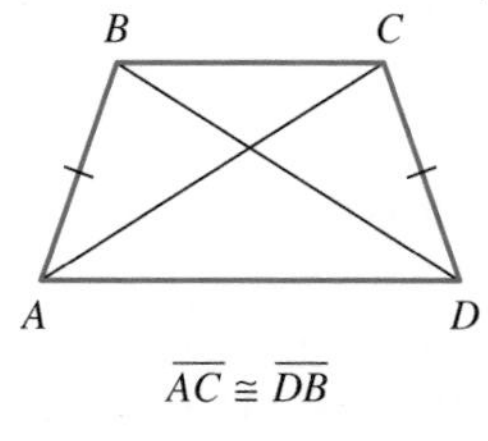

$\overline{AC} \cong \overline{DB}$

FIGURE 5.29

Does the converse of this theorem hold? That is, suppose that a quadrilateral has congruent diagonals. Is it necessarily a rectangle? Figure 5.29 shows an isosceles trapezoid that has congruent diagonals. Because a trapezoid is *not* a rectangle, a quadrilateral with congruent diagonals is not necessarily a rectangle. However, the following theorem *is* true.

THEOREM 5.29

If a parallelogram has congruent diagonals, then it is a rectangle.

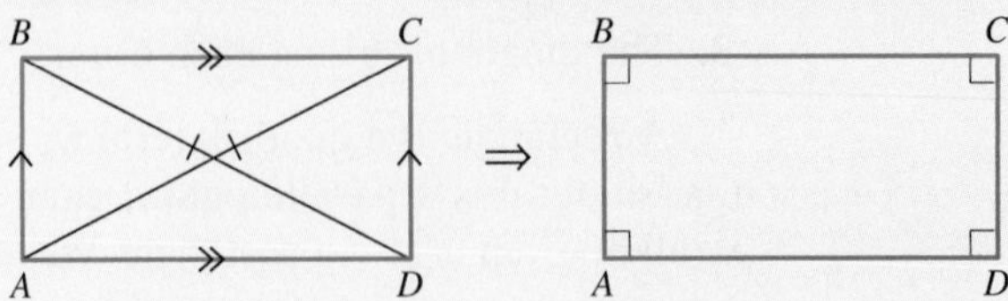

If $ABCD$ is a parallelogram with $AC = BD$, then $\angle A = \angle B = \angle C = \angle D = 90°$.

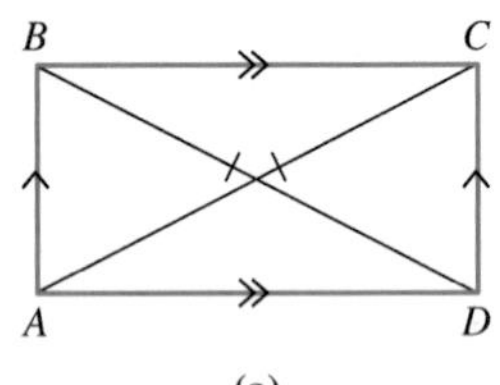

(a)

(b)

FIGURE 5.30

Given Parallelogram $ABCD$ with $AC = BD$ [Figure 5.30(a)]

Prove $ABCD$ is a rectangle.

Proof

Statements	Reasons
1. $ABCD$ is a parallelogram with $AC = BD$.	1. Given
2. $\overline{AB} \cong \overline{DC}$	2. In a parallelogram, opposite sides are congruent.
3. $\overline{AD} \cong \overline{AD}$	3. Reflexive property
4. $\triangle BAD \cong \triangle CDA$	4. SSS [Figure 5.30(b)]
5. $\angle BAD \cong \angle CDA$	5. C.P.
6. $\angle BAD$ and $\angle CDA$ are supplementary.	6. Interior angles on the same side of the transversal
7. $\angle BAD$ and $\angle CDA$ are right angles.	7. Steps 5 and 6
8. $ABCD$ is a rectangle.	8. A parallelogram with one right angle is a rectangle. ■

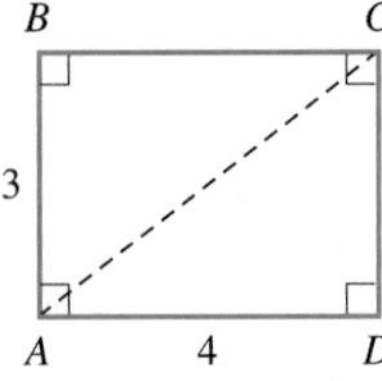

FIGURE 5.31

EXAMPLE 5.9 Given rectangle $ABCD$ with $AB = 3$ and $AD = 4$, find AC (Figure 5.31).

SOLUTION

Because $ABCD$ is a rectangle, $\angle A = 90°$. Thus, by the Pythagorean theorem, $BD^2 = 3^2 + 4^2 = 9 + 16 = 25$, or $BD = 5$. Because the diagonals of a rectangle are congruent, $BD = AC$, or $AC = 5$. Notice that we could also have found AC directly by observing that $CD = 3$ and applying the Pythagorean theorem to $\triangle ADC$. ■

Squares

In Chapter 2, we defined a square to be a quadrilateral with all sides congruent and four right angles. Because all of its sides are congruent, a square is a rhombus, and because all four of its angles are right angles, a square is also a rectangle. Because both a rhombus and a rectangle are parallelograms, a square is also a parallelogram. Therefore, all the theorems that hold for rectangles, parallelograms, and rhombuses also apply to squares.

We choose definitions to best describe shapes or relationships. However, as in the case of the rectangle, we could have used different definitions for the square. Next we will state several theorems that could have served as definitions of a square. We will leave their proofs for the problem set.

1. A square is a rhombus with one right angle.
2. A square is a rectangle with two adjacent sides congruent.
3. A square is a rhombus with congruent diagonals.
4. A square is a rectangle with perpendicular diagonals.

Isosceles Trapezoids

Let's finally turn our attention to trapezoids, Recall that a trapezoid is a quadrilateral with exactly one pair of parallel sides [Figure 5.32(a)]. An isosceles trapezoid has congruent legs [Figure 5.32(b)]. In Figure 5.32(b), $\angle A$ appears to be congruent to $\angle D$. That figure suggests the next theorem.

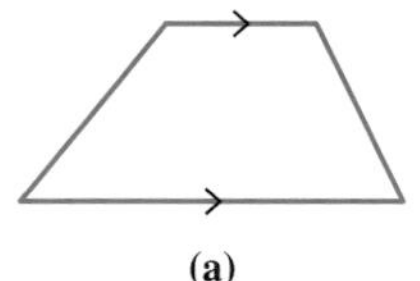

(a)

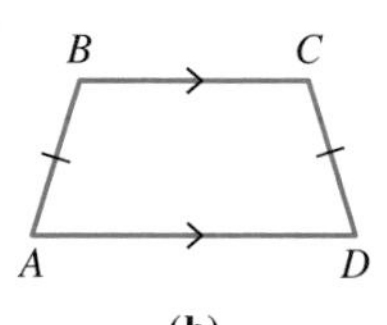

(b)

FIGURE 5.32

THEOREM 5.30

The base angles of an isosceles trapezoid are congruent.

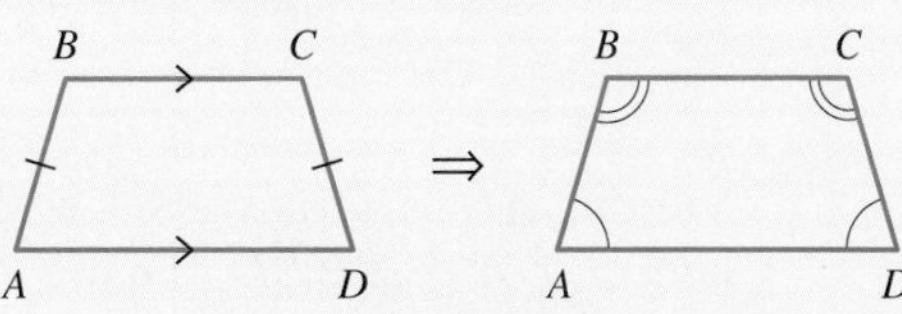

In $ABCD$, if $\overline{AD} \parallel \overline{BC}$ and $AB = CD$, then $\angle A \cong \angle D$ and $\angle B \cong \angle C$.

Given Isosceles trapezoid $ABCD$ with $\overline{AD} \parallel \overline{BC}$ and with $AB = CD$ [Figure 5.33(a)].

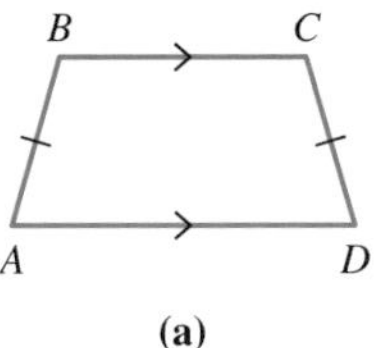

(a)

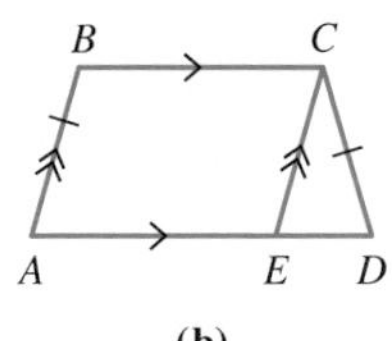

(b)

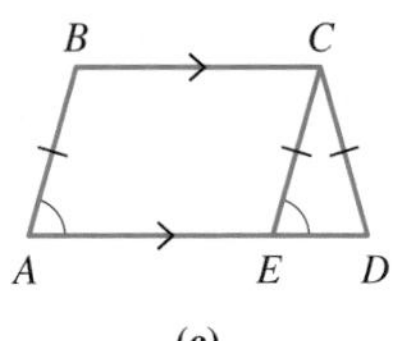

(c)

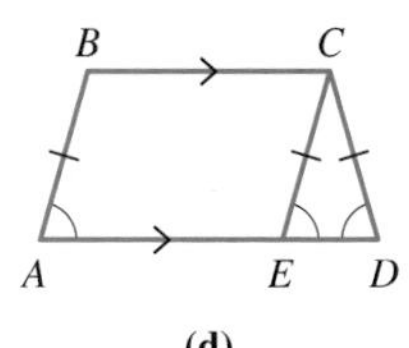

(d)

FIGURE 5.33

Prove $\angle A \cong \angle D$ and $\angle B \cong \angle C$

Plan Draw a segment through C parallel to $\overline{AB}$ to form a parallelogram and a triangle.

Proof Let $\overline{CE}$ be parallel to $\overline{AB}$ [Figure 5.33(b)]. Then $ABCE$ is a parallelogram, so $AB = CE$. Therefore, $\angle A \cong \angle CED$ because they are corresponding angles [Figure 5.33(c)]. However remember that $AB = CD$, so we must have $CE = CD$. ■

Thus, $\triangle CDE$ is an isosceles triangle, so $\angle CED \cong \angle D$ [Figure 5.33(d)]. Therefore, $\angle A \cong \angle D$. Because $\overline{AD} \parallel \overline{BC}$, it follows that $\angle B$ and $\angle C$ are supplementary to $\angle A$ and $\angle D$, respectively. Therefore, they are equal.

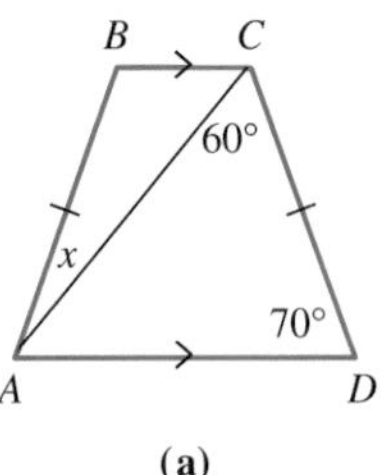

(a)

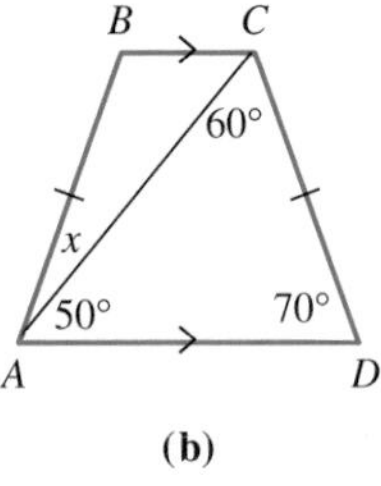

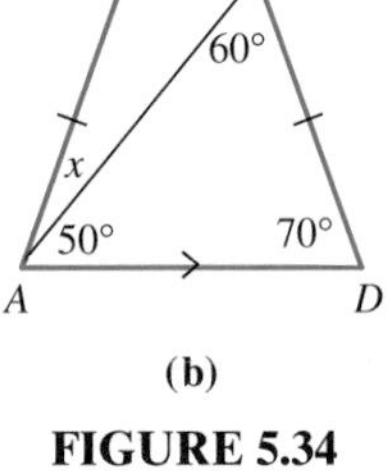

(b)

FIGURE 5.34

EXAMPLE 5.10 Trapezoid $ABCD$ is isosceles. Find the angle measure x [Figure 5.34(a)].

SOLUTION

In $\triangle ACD$, the sum of the angles is 180°. Hence, $\angle CAD + 60° + 70° = 180°$, and $\angle CAD = 50°$ [Figure 5.34(b)]. Because $ABCD$ is isosceles, $\angle A = \angle D$. This means that $x + \angle CAD = 70°$, or $x + 50° = 70°$. Thus, $x = 20°$. ■

Next, let's consider the converse of Theorem 5.30. Suppose that $ABCD$ is a trapezoid with base angles congruent [Figure 5.35(a)]. Is it an isosceles trapezoid? Let $\overline{CE}$ be parallel to $\overline{AB}$ [Figure 5.35(b)]. Then $ABCE$ is a parallelogram, so $\overline{AB} \cong \overline{CE}$.

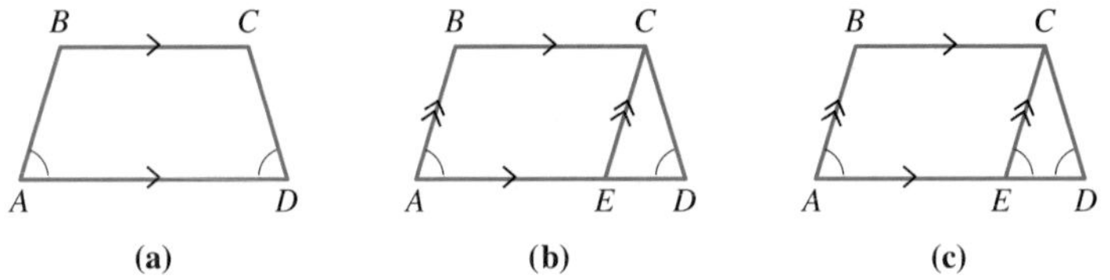

(a)

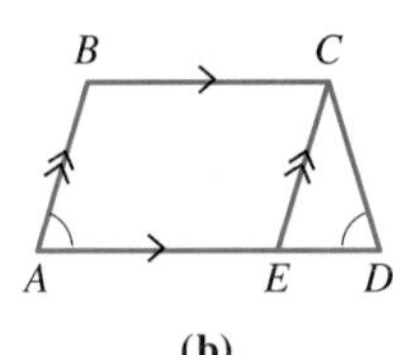

(b)

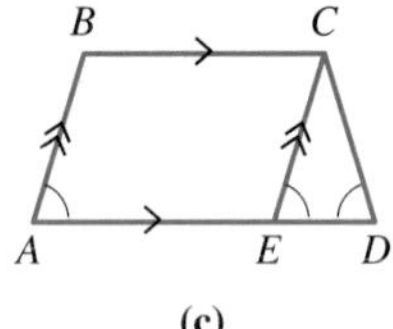

(c)

FIGURE 5.35

Viewing $\overline{AD}$ as a transversal, we see that $\angle A$ and $\angle CED$ are corresponding angles, hence, $\angle A \cong \angle CED$ [Figure 5.35(c)]. Because $\angle A \cong \angle D$ is given, we have $\angle CED \cong \angle D$. This means that $\overline{CE} \cong \overline{CD}$ in $\triangle CED$. From $\overline{AB} \cong \overline{CE}$ and $\overline{CE} \cong \overline{CD}$, we can conclude that $\overline{AB} \cong \overline{CD}$. This proves the following theorem.

(a)

THEOREM 5.31

If the base angles of a trapezoid are congruent, then it is isosceles.

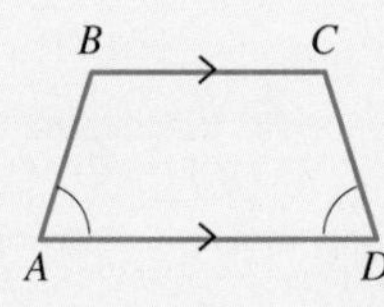

⇒

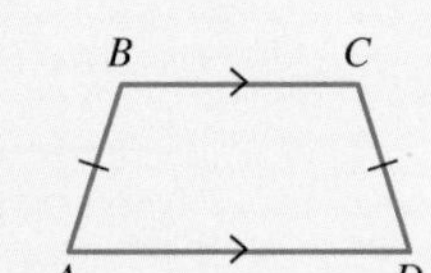

In $ABCD$, if $\overline{AD} \parallel \overline{BC}$ and $\angle A \cong \angle D$, then $\overline{AB} \cong \overline{CD}$.

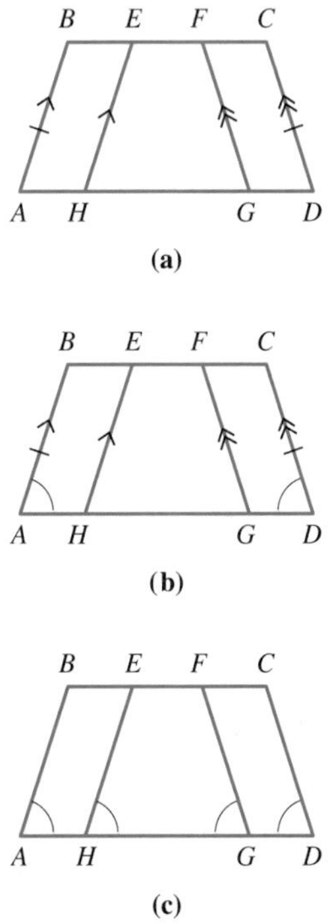

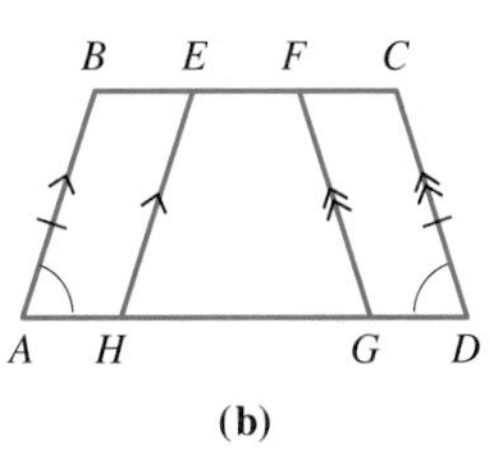

(b)

EXAMPLE 5.11 In isosceles trapezoid $ABCD$, $\overline{AB} \parallel \overline{EH}$ and $\overline{CD} \parallel \overline{FG}$ [Figure 5.36(a)]. Prove that $EFGH$ is an isosceles trapezoid.

SOLUTION

$EFGH$ is a trapezoid because $\overline{EF} \parallel \overline{GH}$ and the other sides are not parallel. Because $ABCD$ is isosceles, $\angle A = \angle D$ [Figure 5.36(b)]. Because $\overline{AB} \parallel \overline{EH}$, we have $\angle A = \angle EHG$, and $\overline{CD} \parallel \overline{FG}$ implies that $\angle D = \angle FGH$ [Figure 5.36(c)]. Since $\angle A = \angle D$, we have $\angle EHG = \angle FGH$, so by Theorem 5.31, $EFGH$ is isosceles. ■

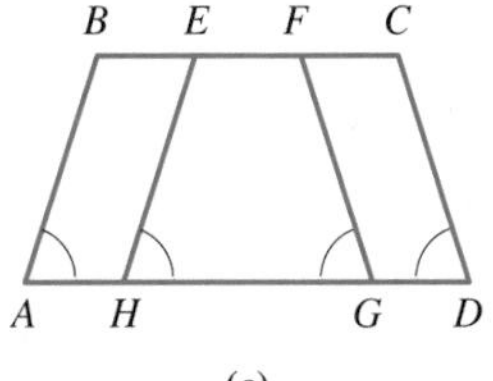

(c)

FIGURE 5.36

Attributes of Quadrilaterals

In Sections 5.3 and 5.4, we discussed many attributes of parallelograms, rhombuses, rectangles, squares, and trapezoids. Table 5.1 provides a summary of these attributes.

Figures \ Attributes	A diagonal forms congruent triangles	Both pairs of opposite sides are congruent	Both pairs of opposite angles are congruent	At least two opposite sides are congruent	Diagonals bisect each other	Diagonals are perpendicular	All sides are congruent	Diagonals bisect vertex angles	Diagonals are congruent	Has at least one right angle	Has at least two consecutive congruent angles	Has at least two consecutive supplementary angles	Consecutive angles are supplementary	Has at least two congruent sides
Parallelogram	X	X	X	X	X							X	X	X
Rhombus	X	X	X	X	X	X	X	X				X	X	X
Rectangle	X	X	X	X	X				X	X	X	X	X	X
Square	X	X	X	X	X	X	X	X	X	X	X	X	X	X
Trapezoid												X		
Isosceles Trapezoid				X					X		X	X		X

TABLE 5.1

Solution of Applied Problem

If the frame is to have four 90° angles, then it must be a rectangle. Because opposite sides are congruent, the frame is a parallelogram. To verify that $ABCD$ is a rectangle, Harry can find AC and BD. If the lengths of these diagonals are the same, then the frame is a rectangle and thus has four "square" corners. He could also verify that each angle is 90° by using the Pythagorean theorem to calculate what the length of a diagonal should be.

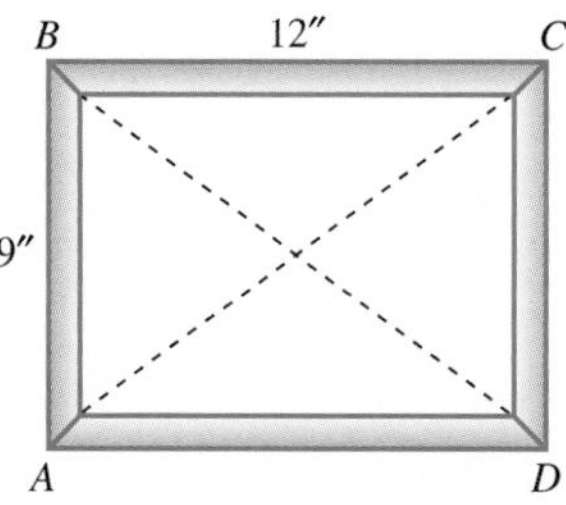

GEOMETRY AROUND US The structures of most radio towers include isosceles trapezoids. To increase the strength and stability of the towers, the (congruent) diagonals of the trapezoids are included.

PROBLEM SET 5.4

EXERCISES/PROBLEMS

Use the following figure for Exercises 1–10.

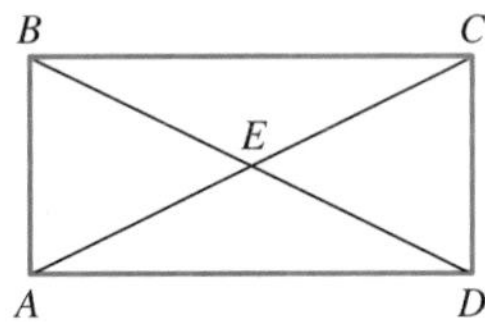

Determine whether each statement is true or false for rectangle ABCD. If true, explain why.

1. $\overline{AC} \perp \overline{BD}$
2. $\overline{AD} \parallel \overline{BC}$
3. $AE = CE$
4. $AE = BE$
5. $AC = BD$
6. $\angle BEC = 90°$
7. $\angle BAE = \angle DAE$
8. $\triangle BEC \cong \triangle AED$
9. $\triangle ABD \cong \triangle BAC$
10. $\triangle ABC \cong \triangle CDA$
11. A rectangle has a diagonal of 15 cm. The width of the rectangle is 3 cm less than its length. Find the area of the rectangle.
12. The length of a rectangle is one inch less than twice its width. The diagonal of the rectangle is two inches more than its length. Find the area of the rectangle.

Use the following figure for Exercises 13–22.

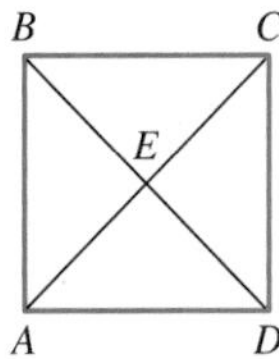

Determine whether each statement is true or false for square ABCD. If true, explain why.

13. $\overline{AC} \perp \overline{BD}$
14. $\overline{AB} \perp \overline{BC}$
15. $BE = ED$
16. $BE = AE$
17. $\angle ABD = \angle CDB$
18. $\angle BCA = \angle DCA$
19. $\triangle ABE \cong \triangle BCE$
20. $\triangle ABD \cong \triangle CDA$
21. $\angle BDA = 45°$
22. $\triangle AED$ is isosceles.
23. The length of a diagonal of a square is 16 mm. Find the length of a side of the square. Round to the nearest hundredth.
24. The side length of a square is 23 cm. Find the length of a diagonal of the square. Round to the nearest hundredth.

Use the following figure for Exercises 25–32.

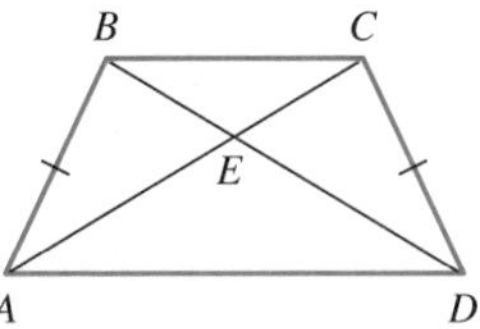

Determine whether the statement is true or false for trapezoid ABCD. If true, explain why.

25. $\angle BAD = \angle CDA$
26. $\angle BAC = \angle DAC$
27. $\triangle BAD \cong \triangle CDA$
28. $\triangle BAC \cong \triangle DCA$
29. $BD = CA$
30. $BE = DE$
31. $BE = CE$
32. $\angle ABD = \angle DCA$

33. Find the length of the legs in the following isosceles trapezoid. Round to the nearest hundredth.

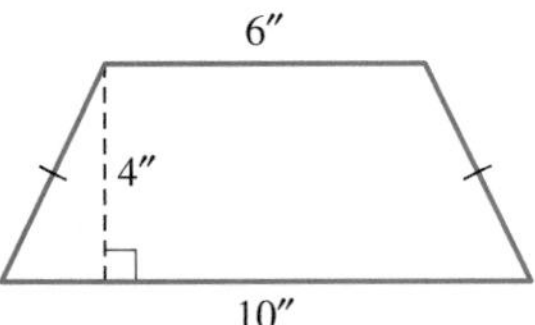

34. Find the area of the following isosceles trapezoid. Round to the nearest tenth.

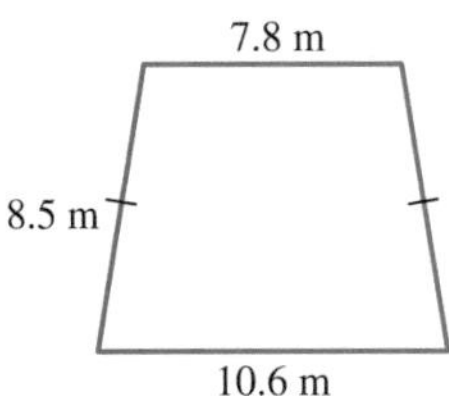

35. In isosceles trapezoid $EFGH$, $EF = 20''$, $FG = 25''$, and $\angle H = 60°$. Find EH.

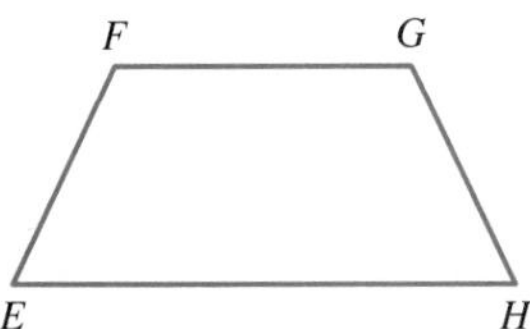

36. $PQRS$ is an isosceles trapezoid. If $\angle TRS = 100°$, $\angle T = 35°$, and $PS = TS$, find the measures of the numbered angles.

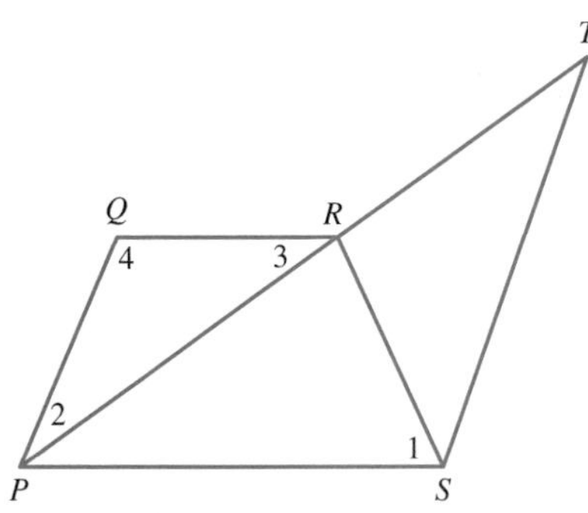

37. In isosceles trapezoid $WXYZ$, $XP = 5$ in., $YZ = 12$ in., $\overline{XZ} \perp \overline{WX}$, and $\overline{WY} \perp \overline{YZ}$ Find WZ.

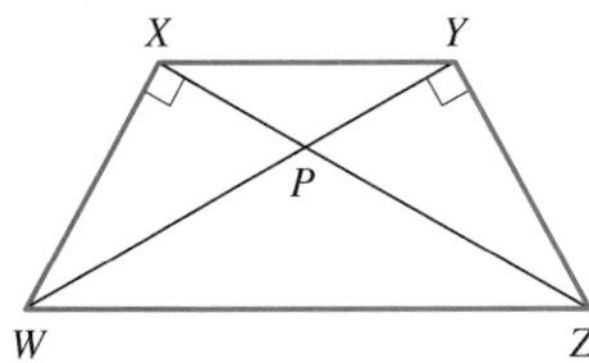

38. In the following figure, $BCDE$ is a rectangle and $ABDE$ is a parallelogram. Find the measures of $\angle 1$ and $\angle 2$.

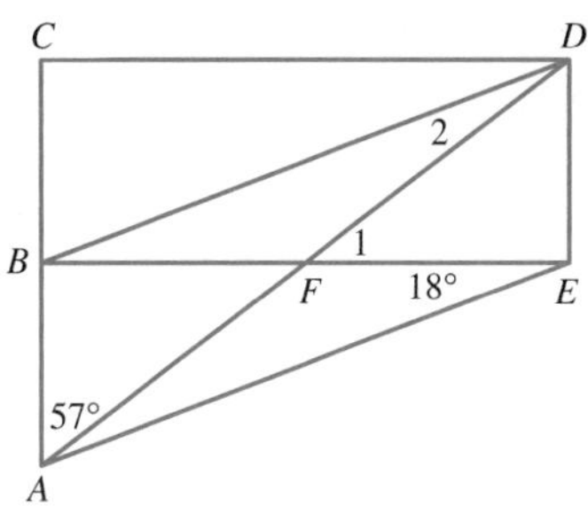

PROOFS

39. Prove: A square is a rhombus with one right angle.

40. Prove: A square is a rectangle with two adjacent sides congruent.

41. Prove: A square is a rhombus with congruent diagonals.

42. Prove: A square is a rectangle with perpendicular diagonals.

43. Prove: A parallelogram with one right angle is a rectangle (Theorem 5.27).

44. Prove: The diagonals of an isosceles trapezoid are congruent.

45. Prove: If the diagonals of a trapezoid are congruent, then it is an isosceles trapezoid.

46. In parallelogram $ABCD$, $\overline{BP}$ and $\overline{CQ}$ are altitudes. Prove that $PBCQ$ is a rectangle.

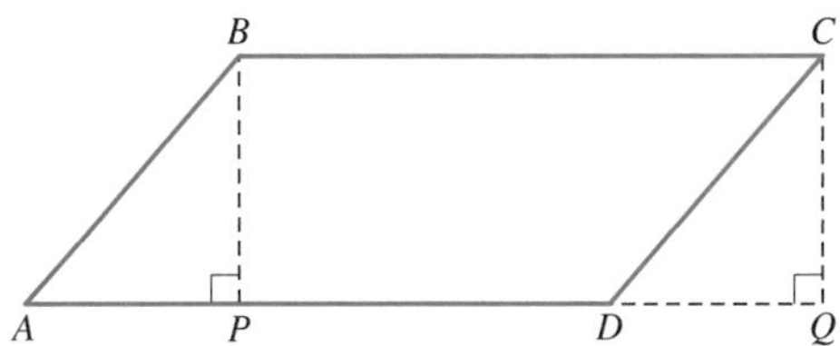

APPLICATIONS

47. While building the frame for a new door, you measure to determine if the frame is a rectangle. For each of the following situations, explain whether you can conclude the frame is a rectangle.

a. You measure and find opposite sides are the same length.

b. You measure and determine both diagonals are the same length.

c. You measure one angle and determine it is a right angle.

48. An antenna is erected at the center of a square, flat roof. Guy wires are attached at each corner of the roof as shown. Why are the lengths of the four guy wires the same?

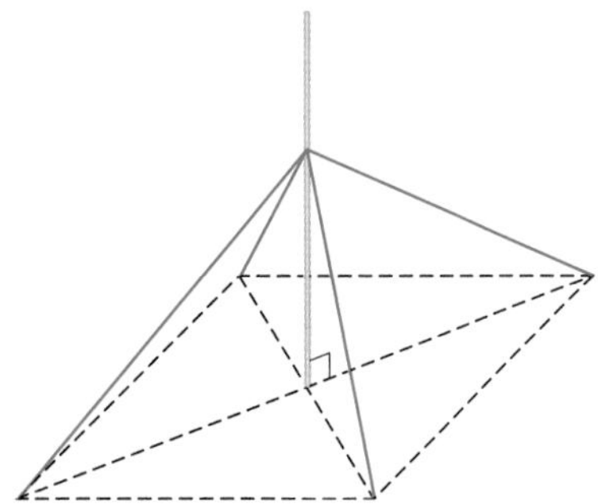

EXTENDED PROBLEMS

49. a. Quadrilateral $ABCD$ is a kite. Draw several other examples of kites. Then refer to the attributes of quadrilaterals given in Table 5.1. Which of these attributes seem to hold true for kites? Are there additional attributes that hold for kites?

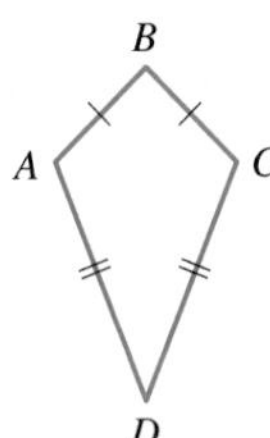

b. For each attribute you said held true for kites in part (a), prove your assertion.

50. Optical illusions capture our interest and play tricks on our minds. Research four famous optical illusions such as Schroder's staircase, the tribar, Zollner's illusion, Poggendorff's illusion, the parallelogram illusion, Orbison's illusion, and the irradiation illusion. Sketch a picture of each and describe how parallel lines and properties of quadrilaterals are used in creating the illusions.

51. We have established that if two triangles are congruent, then all six of their corresponding parts are congruent. In order to determine whether two triangles are congruent, however, we need not compare all six corresponding parts. In a similar way, two quadrilaterals are congruent if all eight of their corresponding parts are congruent. That is, quadrilateral $ABCD$ is congruent to quadrilateral $WXYZ$ if and only if all of the following statements are true:

$$\angle A \cong \angle W, \angle B \cong \angle X,$$
$$\angle C \cong \angle Y, \angle D \cong \angle Z,$$
$$\overline{AB} \cong \overline{WX}, \overline{BC} \cong \overline{XY},$$
$$\overline{CD} \cong \overline{YZ}, \text{and } \overline{DA} \cong \overline{ZW}.$$

Explore what minimal information we must have about two quadrilaterals in order to conclude that they are congruent. Does SSS or SAS congruence hold for quadrilaterals? Are there congruence theorems specific to quadrilaterals? For example, is there such a thing as SSSS congruence for quadrilaterals? Describe at least three different ways to determine whether quadrilaterals are congruent without comparing all eight corresponding parts.

5.5 GEOMETRIC CONSTRUCTIONS INVOLVING PARALLEL LINES

Applied Problem

In drafting, we can divide a segment $\overline{AB}$ into six congruent segments using a ruler as follows.

We construct a segment perpendicular to $\overline{AB}$ at B. Then we align a ruler as shown to the right. When we draw perpendicular segments to $\overline{AB}$ from marks at 1″, 2″, 3″, 4″, and 5″ on the ruler, $\overline{AB}$ is divided into six congruent segments. Why does this method work?

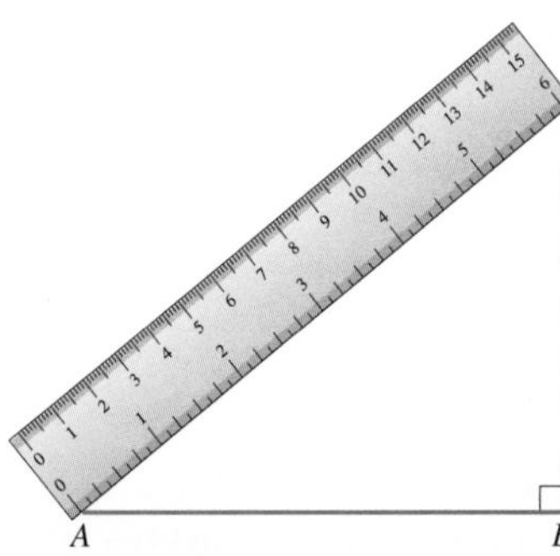

In Section 4.4, we gave several basic geometric constructions together with their justifications. In this section, we will use those basic constructions together with the ones that follow to construct parallel lines and quadrilaterals satisfying various conditions. These constructions will rely heavily on theorems from the first four sections of this chapter.

Constructing Parallel Lines

Construction 7. *To construct a line, through a point, parallel to a given line* [Figure 5.37(a)].

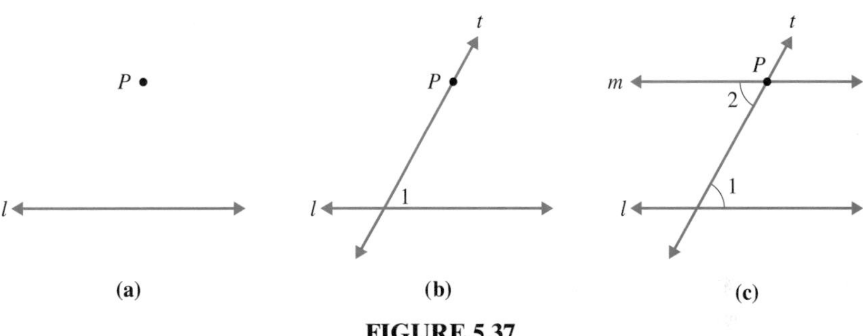

FIGURE 5.37

Construction Draw any line t through P that intersects l [Figure 5.37(b)]. Then copy $\angle 1$ to form $\angle 2$ with line m and line t as shown in Figure 5.37(c). Line m is parallel to line l.

Justification $\angle 1 \cong \angle 2$ by construction. Because they form a pair of alternate interior angles with lines l and m and transversal t, lines l and m are parallel. (NOTE: We could have also constructed congruent corresponding angles.) ■

Notice that rather than draw any line t through P, we could have constructed line t to be perpendicular to l. Then we could construct m through P perpendicular to t (Figure 5.38).

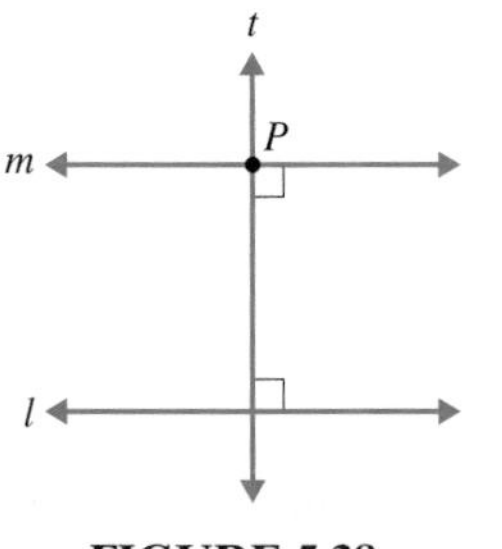

FIGURE 5.38

Constructing Quadrilaterals

Now, we can apply our seven basic constructions to construct a variety of quadrilaterals.

EXAMPLE 5.12 Construct a rectangle given the lengths a and b of its sides [Figure 5.39(a)].

a

b

FIGURE 5.39(a)

SOLUTION

STEP 1 Construct two lines, m and n, perpendicular at A [Figure 5.39(b)].

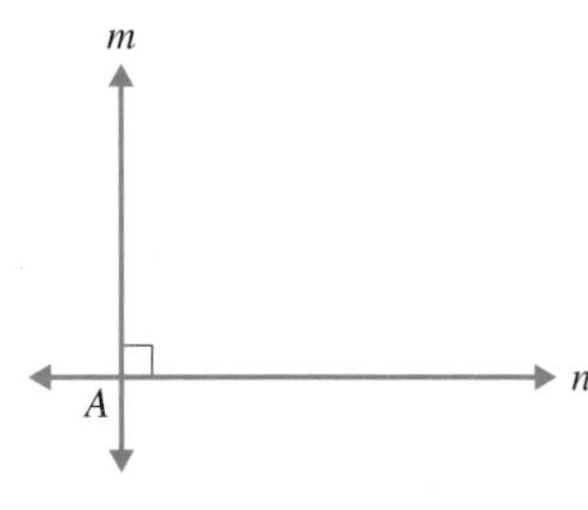

FIGURE 5.39(b)

STEP 2 Copy the lengths a and b, as illustrated in Figure 5.39(c).

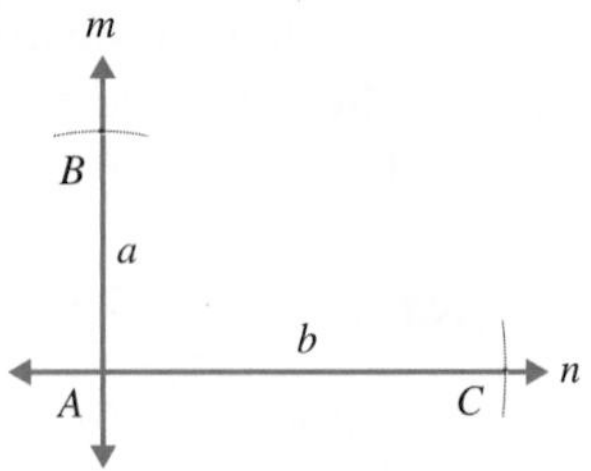

FIGURE 5.39(c)

STEP 3 From point B, swing an arc of length b to the right of m, and from C, swing an arc of length a above n to form point D [Figure 5.39(d)].

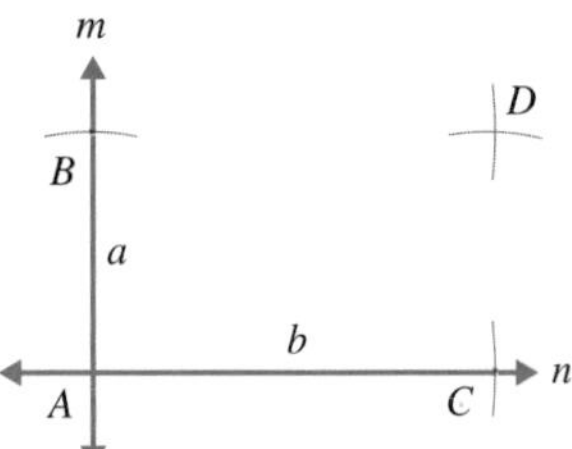

FIGURE 5.39(d)

Draw $\overline{BD}$ and $\overline{CD}$. Then $ABCD$ is a rectangle [Figure 5.39(e)].

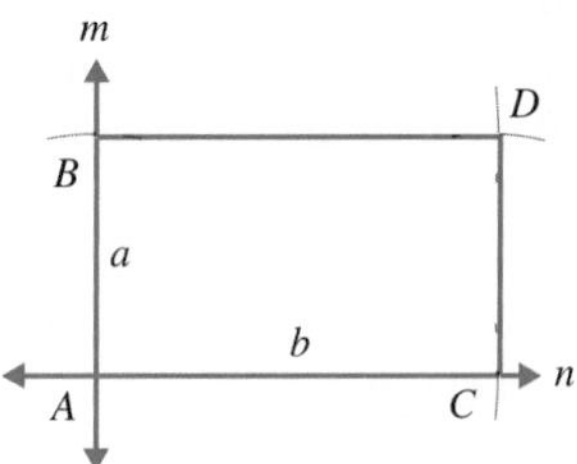

FIGURE 5.39(e)

Justification $ABDC$ is a quadrilateral with opposite sides congruent and at least one right angle (at A). By Theorems 5.17 and 5.27, $ABDC$ is a rectangle. ■

EXAMPLE 5.13 Construct a rhombus given the lengths a and b of its two diagonals [Figure 5.40(a)].

FIGURE 5.40(a)

SOLUTION

STEP 1 Construct a line segment $\overline{AB}$ of length b [Figure 5.40(b)].

FIGURE 5.40(b)

STEP 2 Construct the perpendicular bisector l of $\overline{AB}$. Let M represent the midpoint of $\overline{AB}$ [Figure 5.40(c)].

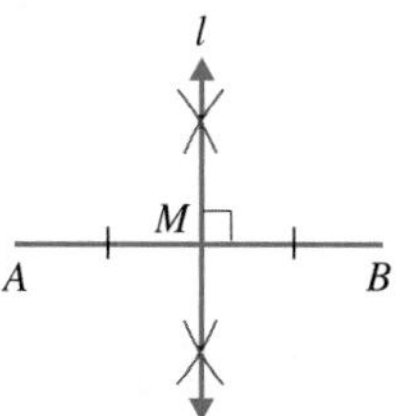

FIGURE 5.40(c)

STEP 3 Construct the perpendicular bisector of the original segment of length a [Figure 5.40(d)].

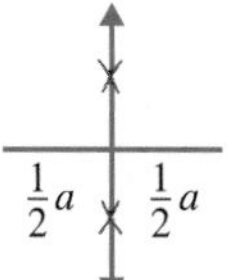

FIGURE 5.40(d)

Then mark off the points C and D on line l so that $CM = DM = \frac{1}{2}a$, as shown in Figure 5.40(e).

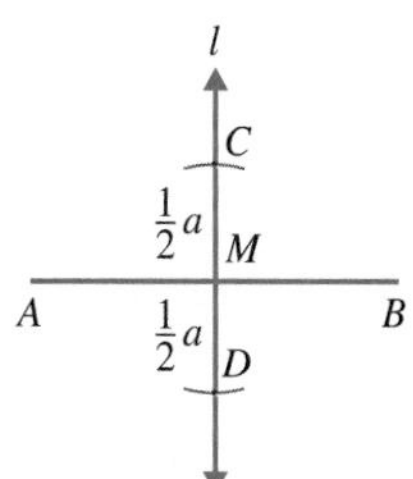

FIGURE 5.40(e)

STEP 4 Draw the segments $\overline{AC}$, $\overline{CB}$, $\overline{BD}$, and $\overline{AD}$. Then $ACBD$ is a rhombus [Figure 5.40(f)].

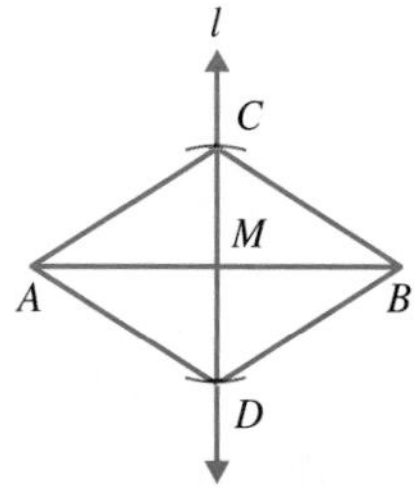

FIGURE 5.40(f)

Justification We constructed $\overline{AB}$ and $\overline{CD}$ to be perpendicular to each other where M is the midpoint of each segment. By Theorems 5.23 and 5.24, a quadrilateral is a rhombus if and only if its diagonals are perpendicular bisectors of each other. ■

Subdividing a Line Segment

In Chapter 4, we studied a construction to bisect a segment. By repeating that process, we could divide a line segment into four congruent segments, eight congruent segments, and so on. The following theorem shows how to divide a segment into any number of congruent subsegments.

THEOREM 5.32

If three parallel lines form congruent segments on one transversal, then they form congruent segments on any transversal.

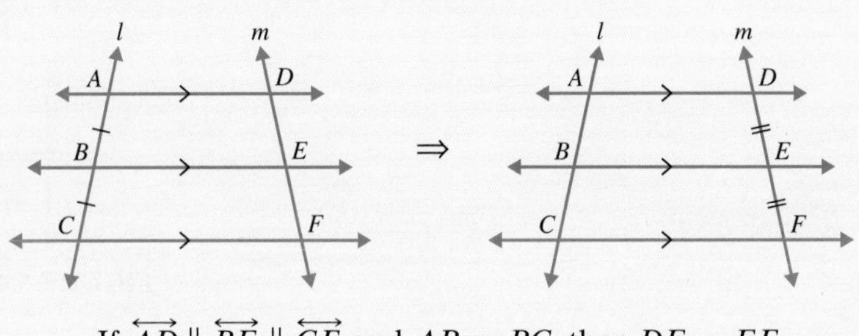

If $\overleftrightarrow{AD} \parallel \overleftrightarrow{BE} \parallel \overleftrightarrow{CF}$ and $AB = BC$, then $DE = EF$.

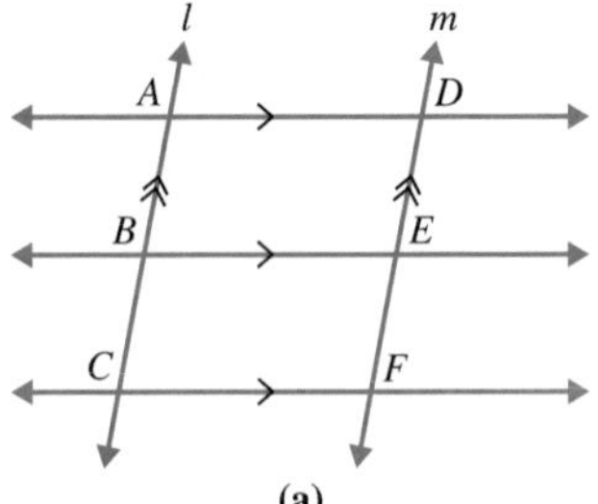

(a)

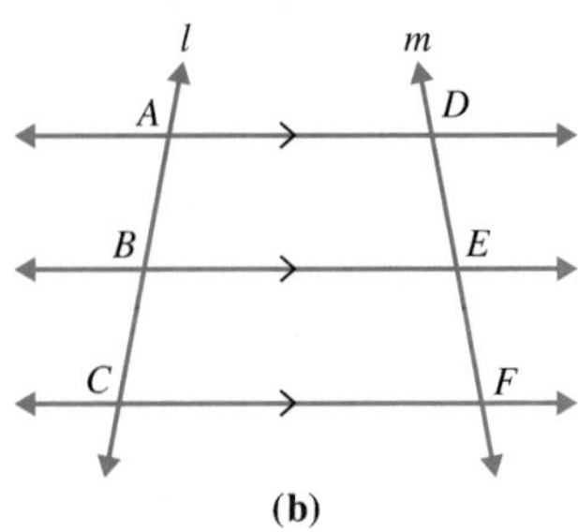

(b)

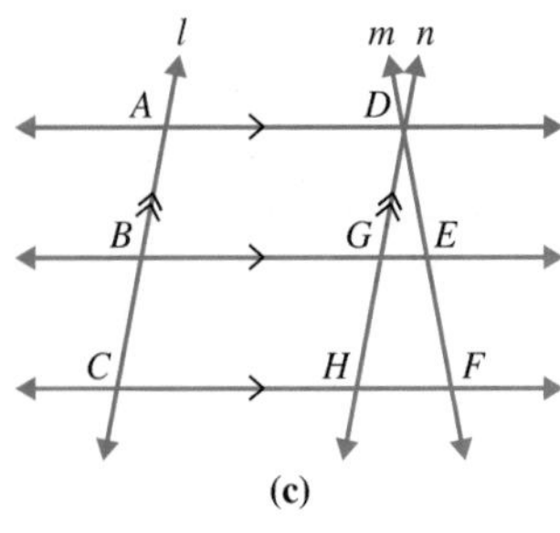

(c)

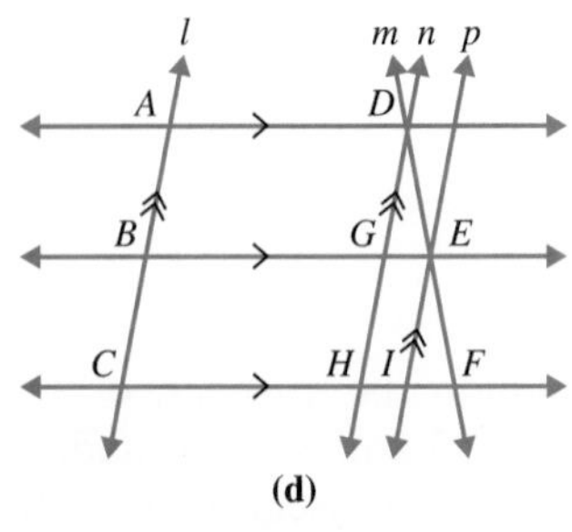

(d)

FIGURE 5.41

Given $\overline{AD} \parallel \overline{BE} \parallel \overline{CF}$ and $AB = BC$

Prove $DE = EF$

Proof There are two cases to consider. Case 1: l is parallel to m. Case 2: l is not parallel to m.

Case 1 l is parallel to m [Figure 5.41(a)].

Because $l \parallel m$ and $\overline{AD} \parallel \overline{BE}$, we can conclude that $ABED$ is a parallelogram. As a consequence, $AB = DE$ because opposite sides of a parallelogram are congruent. In a similar manner, $BC = EF$. Because it is given that $AB = BC$, we can conclude from our three equations that $DE = EF$.

Case 2 l is *not* parallel to m [Figure 5.41(b)].

An outline for the proof follows.

Subgoal 1 Let n be the line through D that is parallel to l, intersecting $\overleftrightarrow{BE}$ at G and intersecting $\overleftrightarrow{CF}$ at H [Figure 5.41(c)]. Use the result from Case 1 to show that $DG = GH$.

Subgoal 2 Let p be the line through E that is parallel to l and n, intersecting $\overleftrightarrow{CF}$ at I [Figure 5.41(d)]. Show that $BEIC$ is a parallelogram and hence, $BC = EI$.

Subgoal 3 Prove that $\triangle DGE \cong \triangle EIF$ using ASA. By corresponding parts, we can conclude that $DE = EF$. ■

The next construction uses this theorem to illustrate a method for dividing a segment into any number of congruent segments. It completes our collection of basic constructions.

Construction 8. *To divide a given segment into a specified number of congruent segments.*

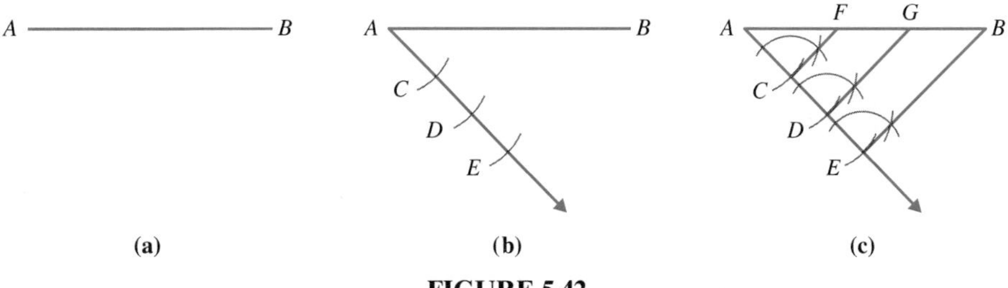

FIGURE 5.42

Construction We will construct three congruent segments on $\overline{AB}$ [Figure 5.42(a)].

STEP 1 Draw any ray with endpoint at A, but not containing $\overline{AB}$. Then mark off three congruent segments of any length on the ray as illustrated in Figure 5.42(b).

STEP 2 Draw $\overline{BE}$ and construct angles at C and D congruent to $\angle AEB$. Using these angles, draw lines through C and D that are parallel to $\overline{BE}$ [Figure 5.42(c)]. The points where the lines intersect $\overline{AB}$ will divide $\overline{AB}$ into the three congruent segments $\overline{AF}$, $\overline{FG}$, and $\overline{GB}$.

Justification Because we constructed $\angle ACF \cong \angle ADG \cong \angle AEB$, we have $\overline{CF} \parallel \overline{DG} \parallel \overline{EB}$ by congruent corresponding angles. We also know that $AC = CD = DE$, by construction. By Theorem 5.32, if parallel lines form congruent segments on one transversal, then they form congruent segments on any transversal. This means it must also be true that $AF = FG = GB$. ■

Solution of Applied Problem

When we draw the segments perpendicular to $\overline{AB}$, we obtain the following figure. Because all of these segments are perpendicular to $\overline{AB}$, they are parallel. The ruler acts as a transversal. Because the segments on the ruler are congruent, the segments on another transversal, like $\overline{AB}$, will also be congruent.

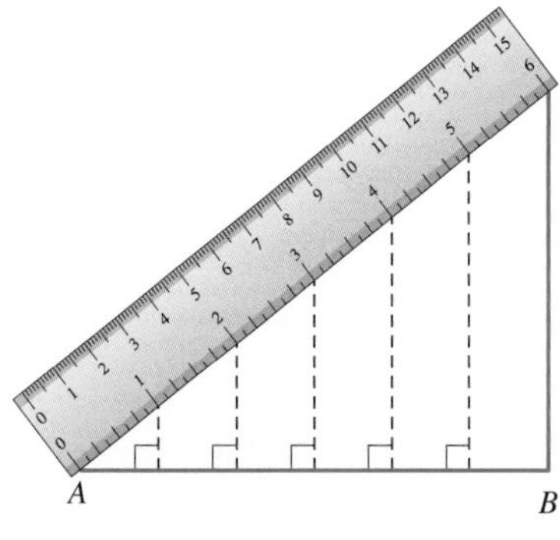

GEOMETRY AROUND US Parallel lines in the real world appear to get closer together as they get farther away. For example, railroad tracks appear to meet at a point far off in the distance, called the **vanishing point**. This phenomenon is the basis of a technique called **one-point perspective drawing**. To sketch a right square prism using one-point perspective drawing, draw a square and select a vanishing point. From each vertex of the square, draw a segment to the vanishing point. Add a horizontal and vertical segment for depth and then erase nonessential segments.

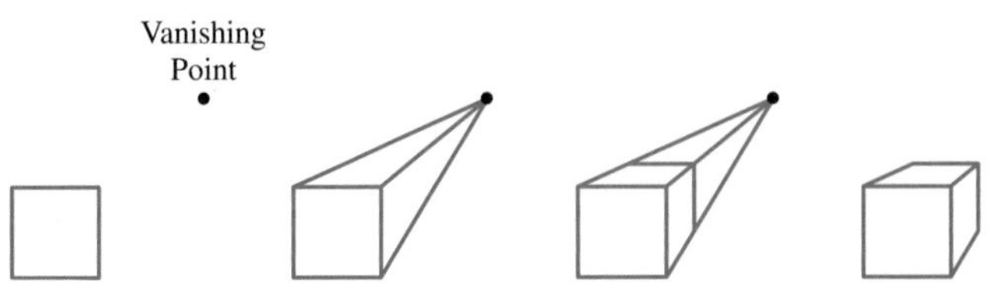

PROBLEM SET 5.5

EXERCISES/PROBLEMS

1. Construct a line through P that is parallel to line l.

2. Trace $\triangle ABC$. Then construct a line through B parallel to AC, a line through A parallel to $\overline{BC}$, and a line through C parallel to $\overline{AB}$. Extend each pair of lines so they meet at points D, E, and F. How does the area of $\triangle DEF$ compare to the area of $\triangle ABC$? Justify your answer.

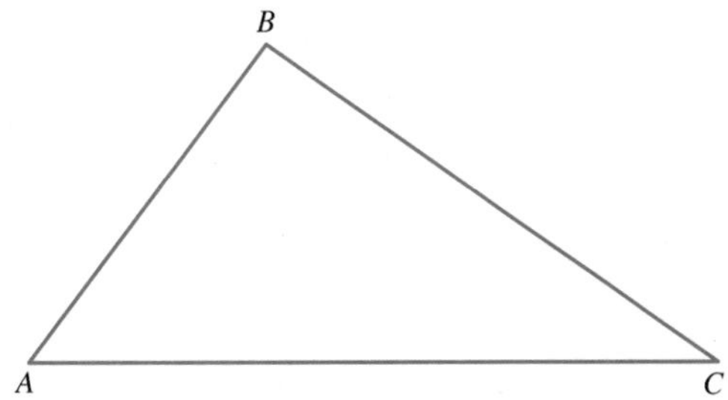

3. Construct a square given one of its sides a.

$\underline{\hspace{4cm}a\hspace{4cm}}$

4. Construct a rectangle given two adjacent sides a and b.

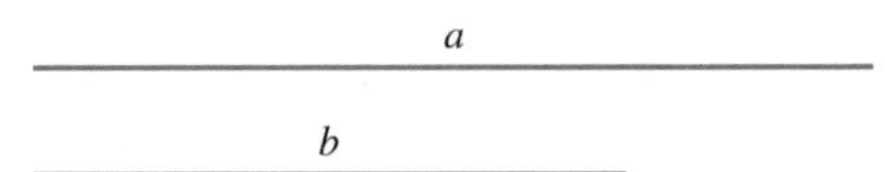

5. Construct a rhombus given one of its sides a and one of its angles A.

$\underline{\hspace{4cm}a\hspace{4cm}}$

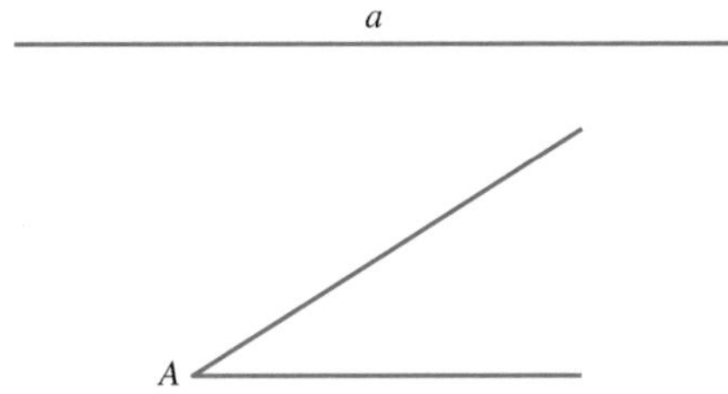

6. Construct a parallelogram given two adjacent sides, a and b, and the included angle A.

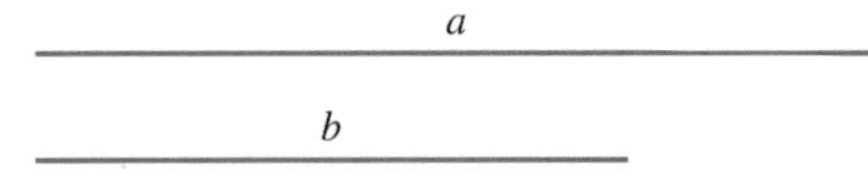

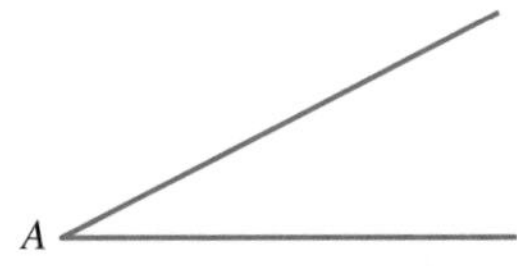

7. Construct a square given one of its diagonals a.

a

8. Construct a rectangle given one of its sides a and one of its diagonals b.

a

b

9. Construct an isosceles trapezoid given one base a and an angle A at one end of the base. (There are many possibilities.)

a

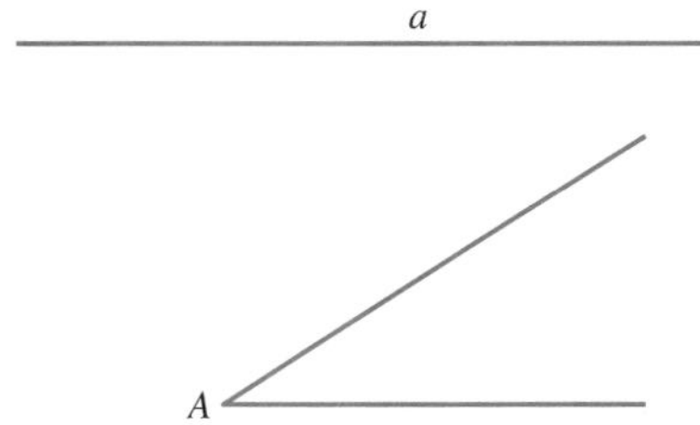

10. Construct an isosceles trapezoid with a base angle of 30°, one base of length b, and legs of length a.

b

a

11. Construct a parallelogram with adjacent sides having the lengths as given and with a vertex angle of 60°.

a

b

12. Construct a rhombus with a vertex angle of 60° and a diagonal of length d.

d

13. Construct a rhombus given the length of a diagonal and the angle between that diagonal and a side of the rhombus.

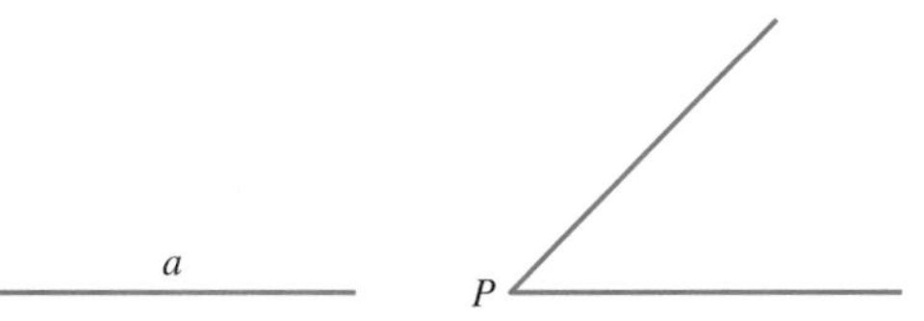

14. Construct a rhombus whose diagonals have the following lengths.

a

b

15. A parallelogram has one side of length a. Its diagonals have lengths b and c. Construct the parallelogram.

a

b

c

16. A parallelogram has a vertex angle of 45° and side lengths a and b. Construct the parallelogram.

a

b

17. Divide segment $\overline{AB}$ into five congruent segments using a compass and straightedge.

A ——— B

18. Divide segment $\overline{PQ}$ into six congruent segments using a compass and straightedge.

P ——— Q

19. Using only a compass and straightedge, divide segment $\overline{MN}$ into two segments such that one segment is twice as long as the other.

M ——— N

20. Using only a compass and straightedge, divide segment $\overline{ST}$ into three segments whose lengths are in the ratio 1:2:3.

S ——— T

21. In the figure, T is the midpoint of $\overline{PS}$, $RU = 5.8$ cm, and $SQ = 7.4$ cm. Find the area of $\triangle PRS$.

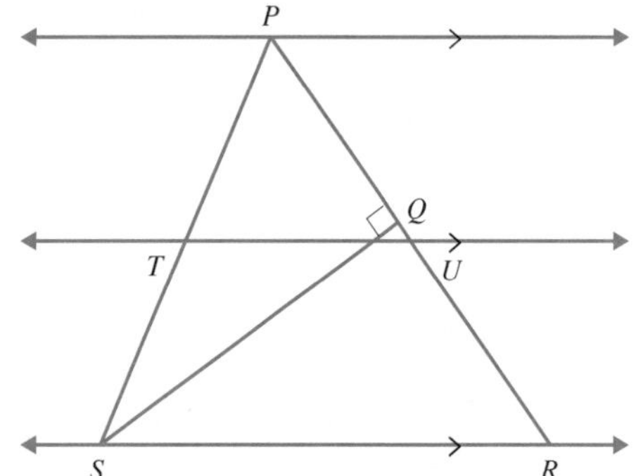

22. Find the value of x in the following figure.

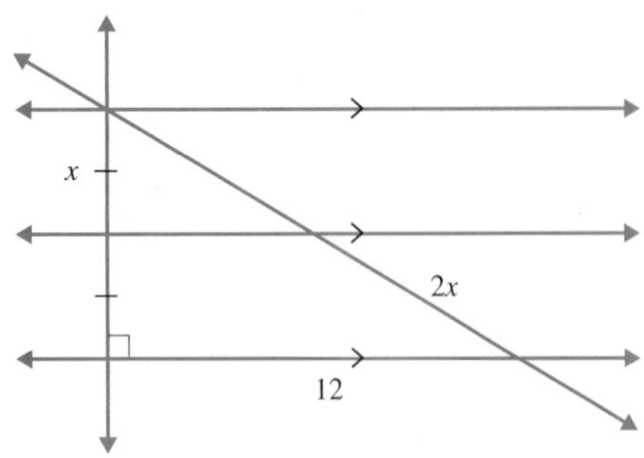

23. Trace $\triangle ABC$. Using a compass and straightedge, divide $\triangle ABC$ into two triangles with the same area.

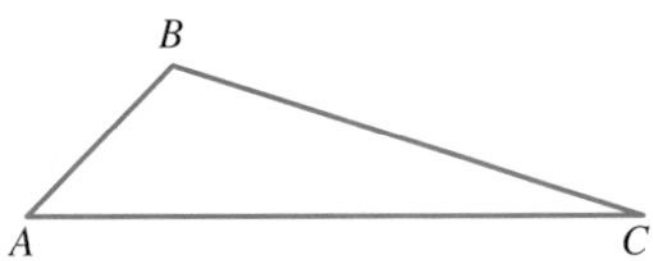

24. Trace $\triangle PQR$. Using a compass and straightedge, divide $\triangle PQR$ into five triangles, all with the same area.

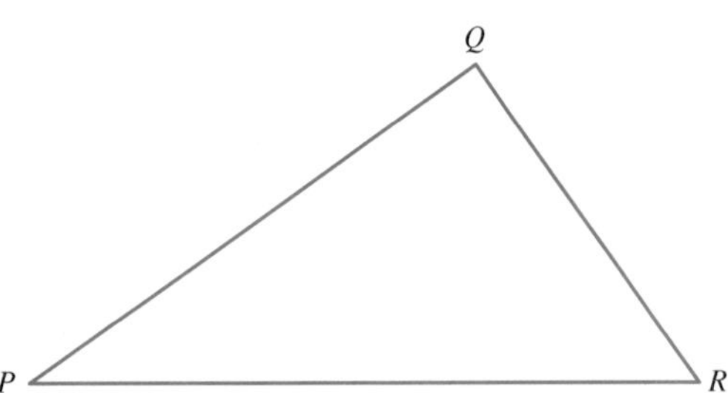

PROOFS

25. Justify your construction in Exercise 6.
26. Justify your construction in Exercise 7.
27. Justify your construction in Exercise 8.
28. Justify your construction in Exercise 19.
29. Justify your construction in Exercise 12.
30. Justify your construction in Exercise 13.
31. Justify your construction in Exercise 15.
32. Justify your construction in Exercise 24.

APPLICATIONS

33. The following figure shows a type of bridge truss called a **Baltimore truss**. If members $\overline{SC}$, $\overline{RU}$, and $\overline{QD}$ must be parallel and if $SR = RQ$, what other members must have the same length by Theorem 5.32?

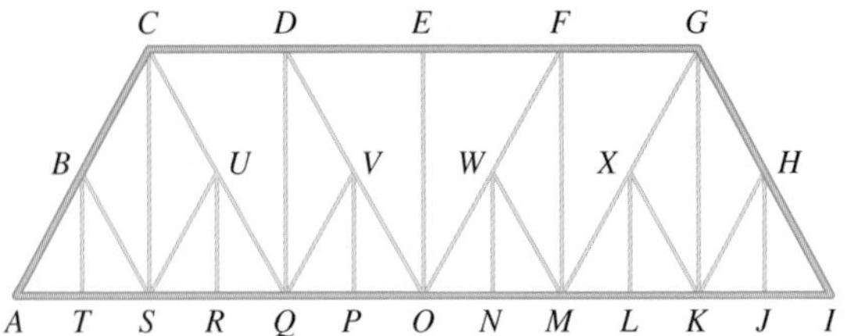

34. While constructing a deck, you place seven parallel beams 18 in. apart. To add support, you want to position another angled beam so that it attaches to all seven parallel beams. You measure and find $AB = 20.9$ in. How long should the angled beam be?

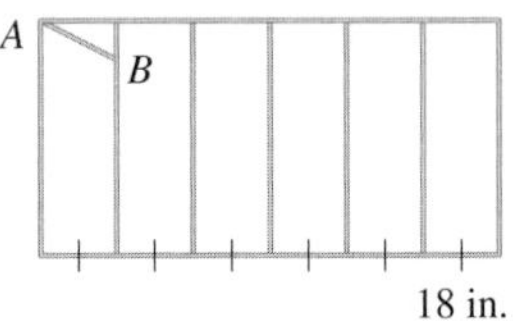

35. Draftpersons sometimes use the following method to construct parallel lines. A line is to be constructed through F parallel to $\overline{AB}$.

From a point C on $\overline{AB}$, swing arc FD. Using the same compass setting, swing an arc from point D through C. Set the compass to length DF and swing an arc from C to locate point E as shown in the following figure.

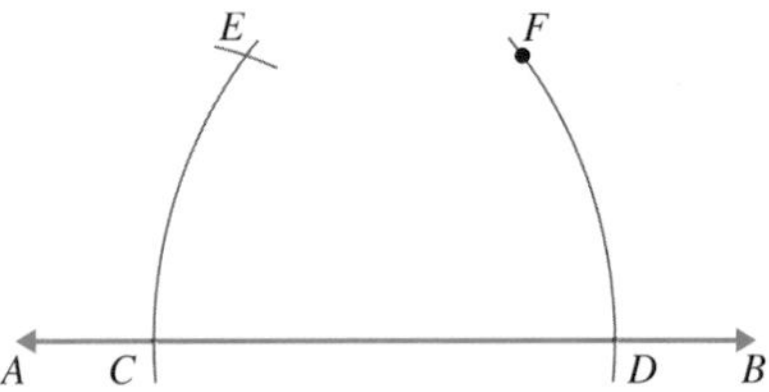

A line drawn through points E and F is parallel to $\overline{AB}$.

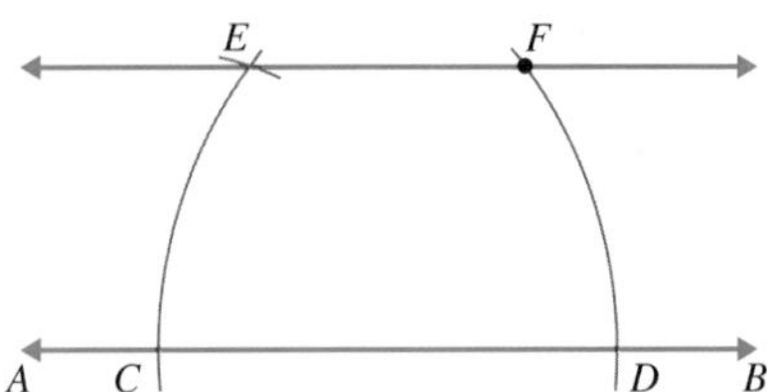

Explain why this construction produces parallel lines.

EXTENDED PROBLEMS

36. We can observe an interesting result called **Aubel's theorem** in the following construction.

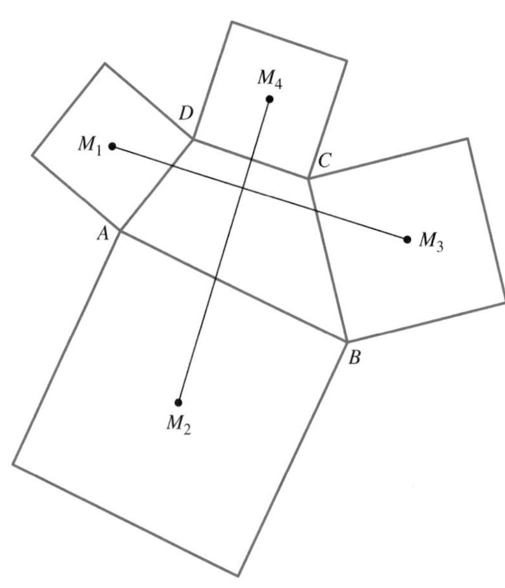

1. Draw any quadrilateral *ABCD*.
2. Using a compass and straightedge, construct a square on each side of quadrilateral *ABCD*.
3. Locate the midpoint of each square: M_1, M_2, M_3, M_4. Connect the midpoints of opposite squares with line segments.
   - **a.** Draw a quadrilateral like the one shown and perform the steps in (1), (2), and (3). Use your ruler to measure the lengths of segments $\overline{M_1M_3}$ and $\overline{M_2M_4}$. Use your protractor to measure the angles formed where $\overline{M_1M_3}$ and $\overline{M_2M_4}$ intersect.
   - **b.** Repeat the construction for two more quadrilaterals. Include one quadrilateral for which the squares overlap. Measure the segments and angles described in part (a).
   - **c.** What conclusion can you draw?

37. A special case of Aubel's theorem (see problem 36) is **Thébault's theorem**. Explore this theorem by performing the following constructions.

1. Draw a parallelogram *ABCD*.
2. Using a compass and straightedge, construct an external square on each side of parallelogram *ABCD*.
3. Locate the midpoint of each square: M_1, M_2, M_3, M_4. Connect consecutive midpoints with line segments.
   - **a.** Perform the construction on three different parallelograms including a square, a rectangle that is not a square, and a parallelogram that is not a rectangle.
   - **b.** For each construction, use a ruler to measure $\overline{M_1M_2}$, $\overline{M_2M_3}$, $\overline{M_3M_4}$, and $\overline{M_4M_1}$. Use your protractor to measure the vertex angles of the quadrilateral $M_1M_2M_3M_4$.
   - **c.** What conclusions can you draw?

38. We can represent numbers using line segments. For example, we can model both the number 3 and the number 2 geometrically by constructing line segments that are 3 units long and 2 units long, respectively. We can represent arithmetic operations such as $3 + 2$, $3 - 2$, 3×2, and $3 \div 2$ using line segments and compass and straightedge constructions. Research how to represent arithmetic operations using geometric constructions. Construct two line segments of different lengths. Show how to construct a segment that corresponds to the sum, difference, product, and quotient of the two segments. Include a description of each construction.

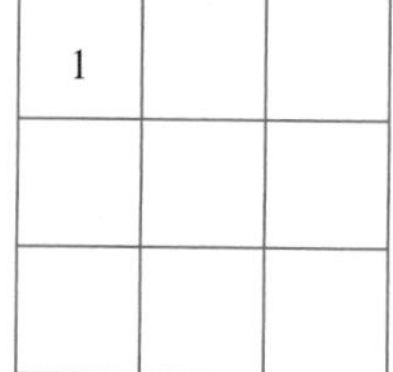

Solution of Initial Problem

We can arrange the numbers 1 through 9 in a 3 by 3 square array so that the sum of the numbers in each row, column, and diagonal is 15. Show that 1 cannot be in one of the corners.

Strategy: Use Indirect Reasoning

Suppose that 1 could be in a corner as shown. Then the row, column, and diagonal containing 1 must each have a sum of 15. Therefore, there must be *three* pairs of numbers from 2 to 9 whose sum is 14. However, there are only *two* such pairs, $5 + 9 = 14$ and $6 + 8 = 14$. Therefore, by indirect reasoning, it is impossible to have 1 in a corner.

Additional Problems Where the Strategy "Use Indirect Reasoning" Is Useful

1. For right triangle $\triangle ABC$ with sides a, b, c as shown, if c^2 is a perfect square, then the lengths a and b cannot both be odd.

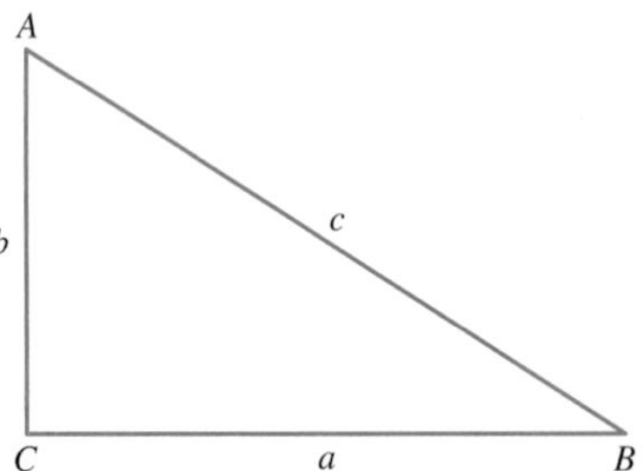

2. Given $\triangle ABC$ as shown, prove that there is only one altitude $\overline{BD}$ to $\overline{AC}$.

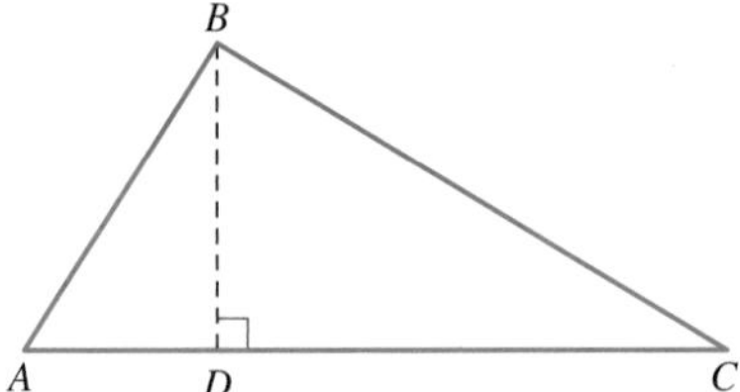

Writing for Understanding

1. Describe at least five examples in which properties of parallel lines are used in the real world. Explain how the parallel lines serve a useful function.
2. If asked to show that two lines are parallel, how might you do it? Describe as many ways as you can. Does one method have an advantage over other methods?
3. The "Geometry Around Us" at the end of Section 5.2 discusses reflectors. Do additional research on reflectors, such as the paraboloid. Summarize your findings, pointing out at least three important applications of the use of reflectors in daily life.
4. The "Geometry Around Us" at the end of Section 5.4 contains a discussion about towers. Do some research on the construction of at least three types of towers, focusing on the role geometry plays. Summarize your findings.

PEOPLE IN GEOMETRY

George Riemann (1826–1866) developed a non-Euclidean geometry by modifying the Parallel Postulate. He supposed that through any given point on a plane, there are no lines parallel to a given line. The resulting system could be modeled as the geometry on the surface of a sphere, where every line is a great circle and line segments are portions of great circles. In this realm, it can be shown that in every triangle, the sum of the angle measures greater than 180 degrees.

Riemann came from a poor family, and this poverty contributed to his early death at the age of 40. As a lecturer at the University of Göttingen, he received no salary and depended on contributions from his students. Yet it was during this time that he developed stunning new methods that cast all of geometry in a new light. Albert Einstein later used Riemann's mathematical descriptions of curved space in the theory of general relativity.

CHAPTER REVIEW

For each section of this chapter, you will find a list of vocabulary and notation, questions to assess your understanding of key concepts, and review problems similar to the problems you worked in your homework. Review each item in the *Vocabulary/Notation* list, mentally and, if necessary, refer back to the indicated page and write a definition. Then answer the *Concept Check Questions*, looking back at the section if you need help. Work the *Review Problems* as practice before you move on to the *Chapter Test*. Answers to the *Review Problems* and *Chapter Test* can be found at the back of the book.

SECTION 5.1 Indirect Reasoning and the Parallel Postulate

Vocabulary/Notation

Indirect reasoning (proof) 241
Skew lines 241
Transversal 242
Alternate interior angles 242
Alternate exterior angles 242
Interior angles on the same side of the transversal 242
Exterior angles on the same side of the transversal 242
Corresponding angles 242
Elliptic geometry 244
Hyperbolic geometry 244

Concept Check Questions

1. What must we assume to apply the technique of indirect reasoning to prove a theorem? 241
2. How are skew lines and parallel lines different? 241
3. When two lines are cut by a transversal, how many pairs of alternate interior angles are formed? 242
4. When two lines are cut by a transversal, how are corresponding angles positioned relative to the lines and the transversal? 242
5. When two lines are cut by a transversal, which pairs of angles, if congruent, guarantee that the lines are parallel? 242–243
6. If two parallel lines and a transversal form congruent interior angles on the same side of the transversal, how are the parallel lines and the transversal related? 243
7. What does the parallel postulate guarantee? 244

Review Problems

Use the following figure for review problems 1 through 3. Assume $m \parallel n$ and that $\angle 1 = 90°$, $\angle 13 = 130°$, and $\angle 11 = 75°$.

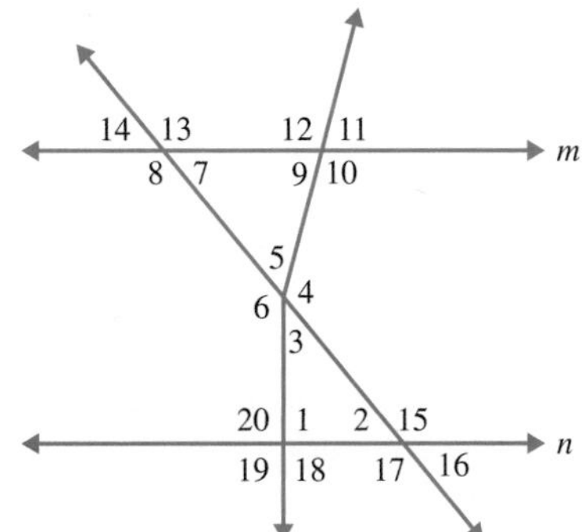

1. Determine whether each statement is true or false. If true, explain why.
   a. $\angle 1$ and $\angle 10$ are supplementary
   b. $\angle 13$ and $\angle 17$ are congruent.
   c. $\angle 8$ and $\angle 16$ are congruent.
   d. $\angle 2$ and $\angle 13$ are supplementary.
2. a. Name two pairs of alternate interior angles.
   b. Name two pairs of alternate exterior angles.
   c. Name four pairs of corresponding angles.
3. Find the measure of each numbered angle.
4. Prove: If two parallel lines are cut by a transversal, then they form a pair of supplementary exterior angles on the same side of the transversal.

SECTION 5.2 Important Theorems Based on the Parallel Postulate

Vocabulary/Notation

AAS Congruence 252
HA Congruence 253

Concept Check Questions

1. How can we use parallel line concepts to prove that the sum of the angle measures in a triangle is 180°? 250–251
2. How is the measure of an exterior angle of a triangle related to the sum of the measures of the two nonadjacent interior angles? 251
3. If a point is on the bisector of an angle, what does the Angle Bisector Theorem guarantee? 253
4. If a point is equidistant from the sides of an angle, what does the Angle Bisector Theorem guarantee? 253

Review Problems

1. Name three exterior angles of $\triangle ADB$ in the following figure.

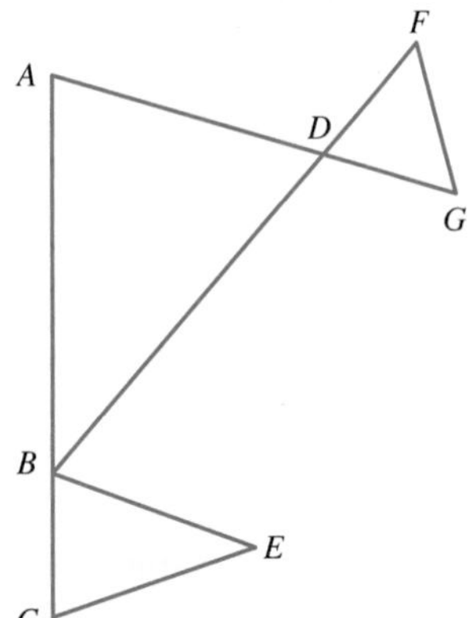

2. Find the value of x.

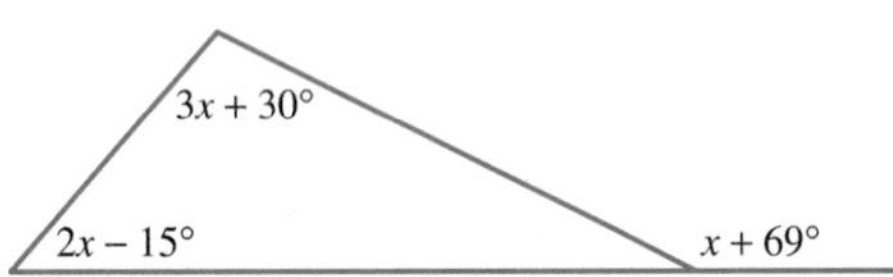

3. In the following figure with right angles as marked, $\overline{BE}$ bisects $\angle AED$. Determine whether each statement is always true or may be false. If true, explain why.

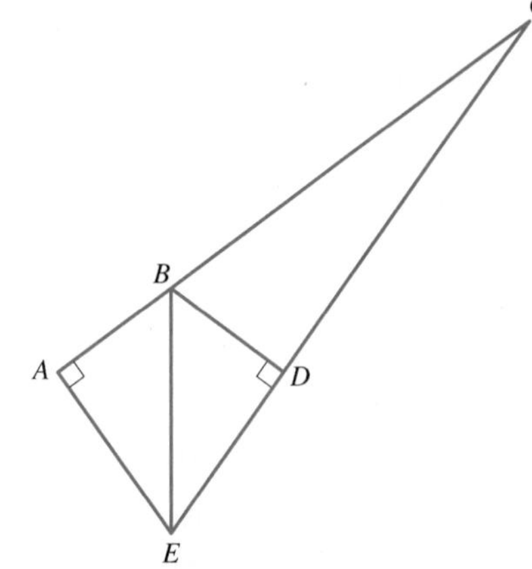

   a. $\triangle BAE \cong \triangle BDE$
   b. $\angle DBC = \angle DBE + \angle BED$
   c. $\angle ABE + \angle EBD = \angle BAE + \angle BCD$

4. Given that points A, B, D, and E in the following figure are collinear and $\overline{BC} \cong \overline{DC}$, prove $\angle CBA \cong \angle CDE$.

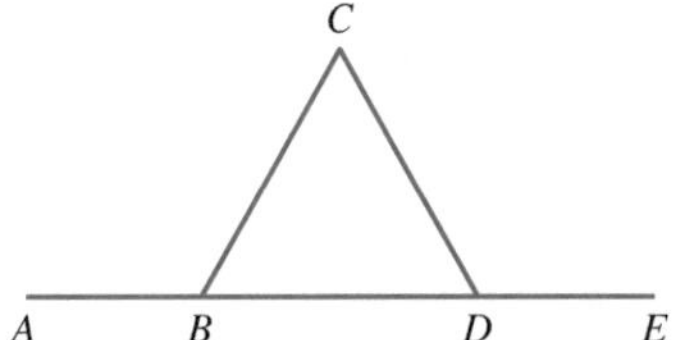

SECTION 5.3 Parallelograms and Rhombuses

Concept Check Questions

1. How can we use parallel line properties to prove that the diagonal of a parallelogram forms two congruent triangles? 257
2. What properties of parallelograms immediately follow from the fact that the diagonal of a parallelogram forms two congruent triangles? 258
3. What must be true about the sides of a quadrilateral in order to conclude that the quadrilateral is a parallelogram? 259
4. What must be true about the angles of a quadrilateral in order to conclude that the quadrilateral is a parallelogram? 259
5. Why can we conclude that a quadrilateral is a parallelogram if its diagonals bisect each other? 262
6. Why is every rhombus also a parallelogram? 262–263
7. What properties does a rhombus have that other types of parallelograms do not have? 264–265

Review Problems

1. $ABCD$ is a parallelogram. List four pairs of congruent segments in the figure.

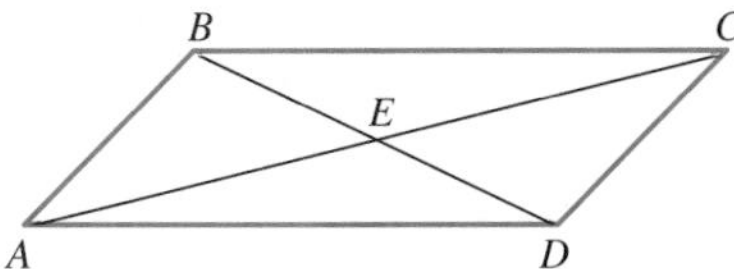

2. If two consecutive angles of a parallelogram measure $4n - 2°$ and $n - 3°$, find the measure of each angle in the parallelogram.
3. If the diagonals of a rhombus measure 18 in. and 25 in., find the area of the rhombus.
4. Find the measures of the angles $\angle WXY$ and $\angle XYZ$ in rhombus $WXYZ$.

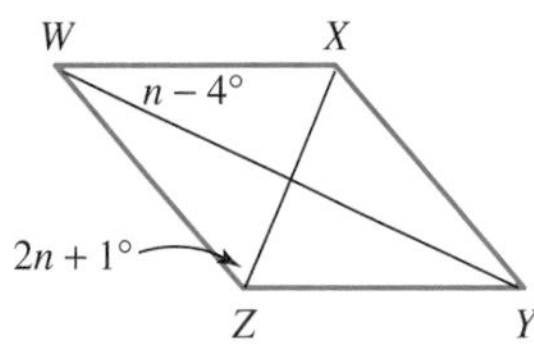

5. Given rhombus $PQRS$ with midpoints $A, B, C,$ and D of sides $\overline{PQ}, \overline{QR}, \overline{RS},$ and $\overline{SP}$, respectively, prove $ABCD$ is a parallelogram.

SECTION 5.4 Rectangles, Squares, and Trapezoids

Concept Check Questions

1. How can we use the definition of a rectangle to show that a rectangle is also a parallelogram? 270
2. If a parallelogram has one right angle, what must be true about the other three angles? 271
3. If a quadrilateral has congruent diagonals, is the quadrilateral a rectangle? 271–272
4. A square can be classified as what other three types of polygons? 272
5. If the base angles of a trapezoid are congruent, then what is true about the nonparallel sides? 274

Review Problems

1. Suppose $ABCD$ has congruent diagonals. What type of quadrilateral might it be?
2. What types of quadrilaterals with at least one right angle have diagonals that bisect each other?
3. In the following figure, $BCDE$ is a rectangle and $ABDE$ is a parallelogram. If $AE = 12$ cm and $\angle CEB = 30°$, find the exact lengths CF, CB, and CD.

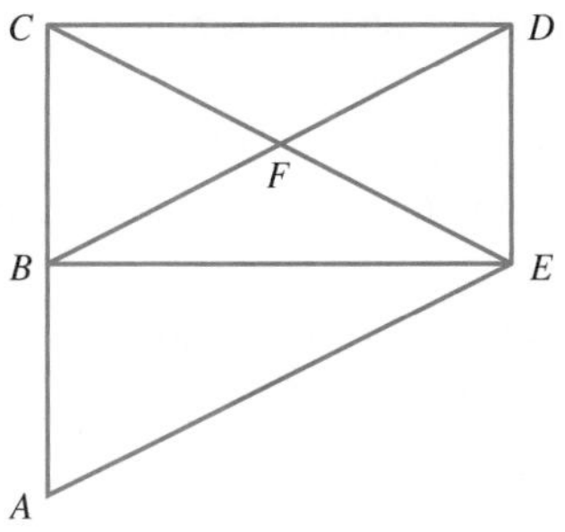

4. In the following figure, $ADFE$ is a square with side length 9 cm and $EABC$ is an isosceles trapezoid with base $AB = 9$ cm. Find the area of trapezoid $EABC$. Round to the nearest hundredth.

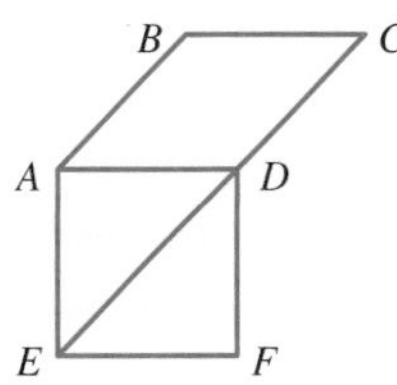

5. An isosceles trapezoid has legs of length 7 cm and base angles that measure 60°. If the shorter base of the trapezoid measures 15 cm, find the length of the longer base.

SECTION 5.5 Geometric Constructions Involving Parallel Lines

Concept Check Questions

1. What property of parallel lines was used to construct a line that is parallel to a given line and that passes through a given point? 279
2. What rectangle attributes can we use to construct a rectangle using a compass and a straightedge? 279–280
3. How can we use a compass and straightedge to divide a line segment into 2, 4, 8, 16, or any number of congruent segments that is a power of 2? 282
4. What property of parallel lines led to a construction that divides a segment into any number of congruent segments? 283

Review Problems

1. Construct a line through Q parallel to line m.

2. Construct a square given one of its diagonals d.

d

3 Construct a rectangle given one of its sides a and one of its diagonals d.

a

d

4. Construct a parallelogram with a 30° vertex angle and one side twice as long as the other.
5. Construct a rhombus given one of its sides a and one of its angles A. Justify the construction.

a A

6. Draw a line segment and divide it into two segments in which one segment is four times as long as the other.

CHAPTER 5 TEST

TRUE-FALSE

Mark as true any statement that is always true. Mark as false any statement that is never true or that is not necessarily true. Be able to justify your answers.

1. If two parallel lines are cut by a transversal, the alternate exterior angles are congruent.
2. If two lines are cut by a transversal forming a pair of supplementary corresponding angles, then the lines are parallel.
3. The angle sum in a quadrilateral is 180°.
4. A diagonal of an isosceles trapezoid forms two congruent triangles.
5. A quadrilateral is a rhombus if and only if its opposite angles are congruent.
6. A quadrilateral is a parallelogram if and only if two pairs of sides are parallel and congruent.
7. If the diagonals of a quadrilateral bisect each other, then it is a parallelogram.
8. If the diagonals of a quadrilateral are congruent, then it is a rectangle.
9. If a rhombus has a side of length 10 cm and a diagonal of length 16 cm, then the other diagonal measures 12 cm.
10. In any $\triangle ABC$, if P is a point on the bisector of $\angle ABC$, then $PA = PC$.

EXERCISES/PROBLEMS

11. In a parallelogram, one angle has measure 20°. Find the measures of the other three angles.

12. In rectangle $ABCD$, $AB = 1$ and $\angle BDA = 30°$. Find the measures of $\angle CBD$, BD, AC, and AD.

13. What type of quadrilateral is $ABCD$? Be as specific as you can. Use only the angle measures shown. That is, do not make your decision based on what the shape "appears" to be.

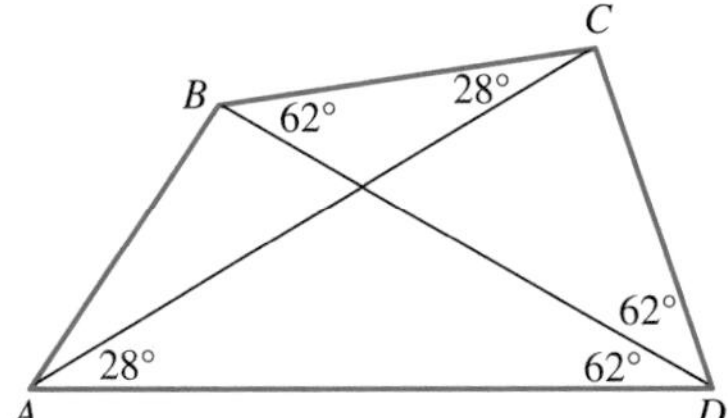

14 A rectangle has a length of 45 yards. The diagonal of the rectangle measures 3 yards more than twice the width. Find the width of the rectangle.

Refer to the following figure for questions 15–17.

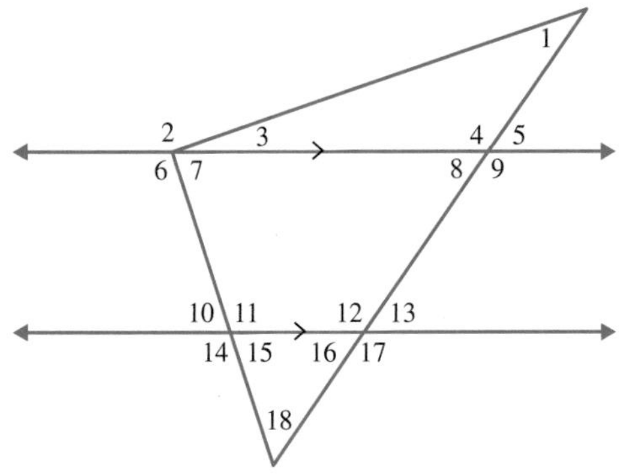

15. Identify four pairs of alternate interior angles in the figure.

16. Identify six pairs of corresponding angles in the figure.

17. Find the measures of $\angle 1$, $\angle 5$, and $\angle 10$, given that $\angle 12 = 140°$, $\angle 18 = 65°$, and $\angle 2 = 170°$.

18. Using only your compass and straightedge, divide $\overline{AB}$ into two parts such that the length of one segment is three times the length of the other segment.

19. Using only your compass and straightedge, construct a parallelogram with a diagonal of length d, one side of length a, and an angle between the side of length a and diagonal of length d that measures exactly 30°.

a ____________ d ____________

20. Using only your compass and straightedge, construct a rhombus whose diagonals have the following lengths.

a ____________ b ____________

PROOFS

21. In square $ABCD$, prove $\triangle ADE \cong \triangle ABE$.

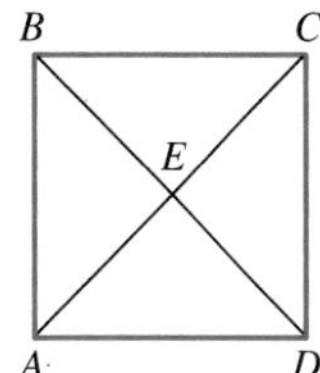

22. In parallelogram $ABCD$, E and F are the midpoints of $\overline{AB}$ and $\overline{CD}$, respectively. Prove that $AECF$ is a parallelogram.

APPLICATIONS

23. Plans for a patio in the shape of an isosceles trapezoid have dimensions as shown.

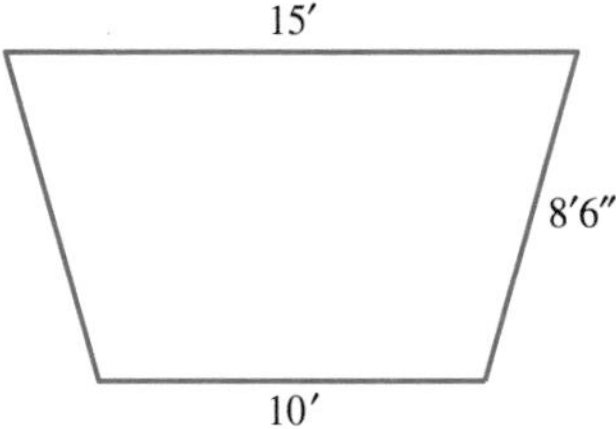

Workers will pour a layer of concrete 4 inches thick. Calculate the number of cubic feet of concrete that they will need for the project. Round your answer up to the nearest cubic foot.

24. A balance scale such as the one shown uses a parallel linkage. Explain why the pans of the scale remain horizontal, even when unequal weights are placed on the pans.

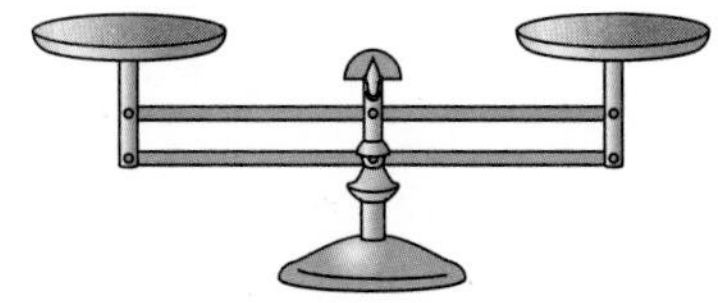

CHAPTER 6

SIMILARITY

Euclid's *The Elements* contained the following construction: "to cut a line segment in extreme and mean ratio." In modern terms, this means that a point B divides $\overline{AC}$ so that $\frac{AB}{BC} = \frac{BC}{AC}$.

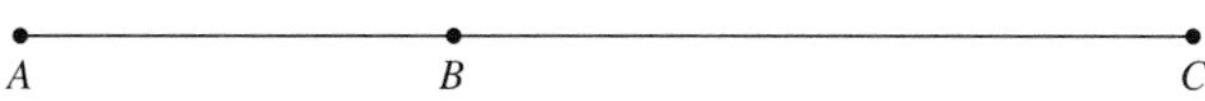

In this construction, the ratio $\frac{BC}{AB} = \frac{(1 + \sqrt{5})}{2}$, the golden ratio. The Greek historian Herodotus related that the Egyptians knew about this famous ratio before the Greeks. The designers of the famous Pyramid of Giza chose the dimensions of the square pyramid so that the square whose side is the height of the pyramid has the same area as any one of its triangular faces. A computation shows that the ratio of the altitude of a face to half the length of the base of a triangular face is the golden ratio.

Problem-Solving Strategies

1. Draw a Picture
2. Guess and Test
3. Use a Variable
4. Look for a Pattern
5. Make a Table
6. Solve a Simpler Problem
7. Look for a Formula
8. Use a Model
9. Identify Subgoals
10. Use Indirect Reasoning
11. *Solve an Equation*

Strategy 11: *Solve An Equation*

Often, when you apply the Use a Variable or Look for a Formula strategies to solve a problem, the representation of the problem will result in an equation. After you have solved the equation using techniques from algebra, you should substitute the answer in the *original problem* as a check.

Initial Problem

A photographer takes aerial photographs from a plane. A dam that is known to be 100 m long is 3 cm long in a photograph. A second picture taken from the same height shows another dam to be 5.7 cm long. How long is the second dam?

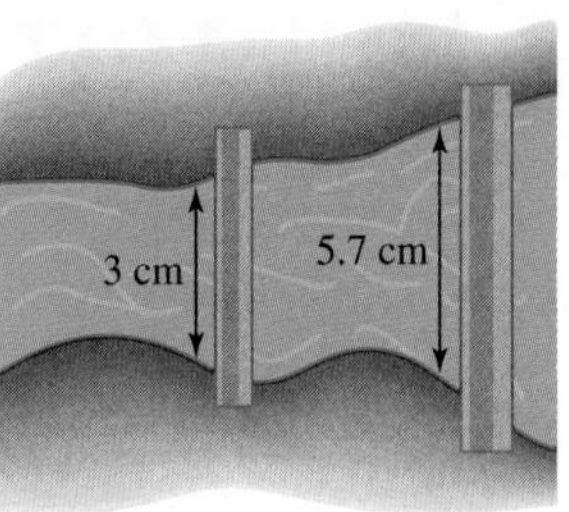

Clues

The Solve an Equation strategy may be appropriate when

- A variable has been introduced.
- The phrase "is," "is equal to," or "is proportional to" appears in a problem.
- The stated conditions can easily be represented in an equation.

A solution of this Initial Problem is on page 350.

Copy the following quadrilateral. Use paper folding or a compass and straightedge to find the midpoint of each side. Label consecutive midpoints as $A, B, C,$ and D respectively. Draw $ABCD$. $ABCD$ is called the midquad of the original quadrilateral. Use a ruler to measure each side of $ABCD$ and a protractor to measure each vertex angle in $ABCD$. What do you observe? What type of quadrilateral does $ABCD$ appear to be?

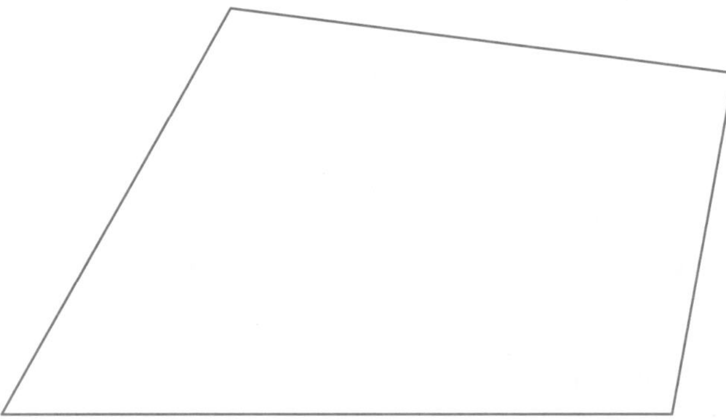

Draw the midquad for each of the following quadrilaterals. Describe each midquad as completely as possible.

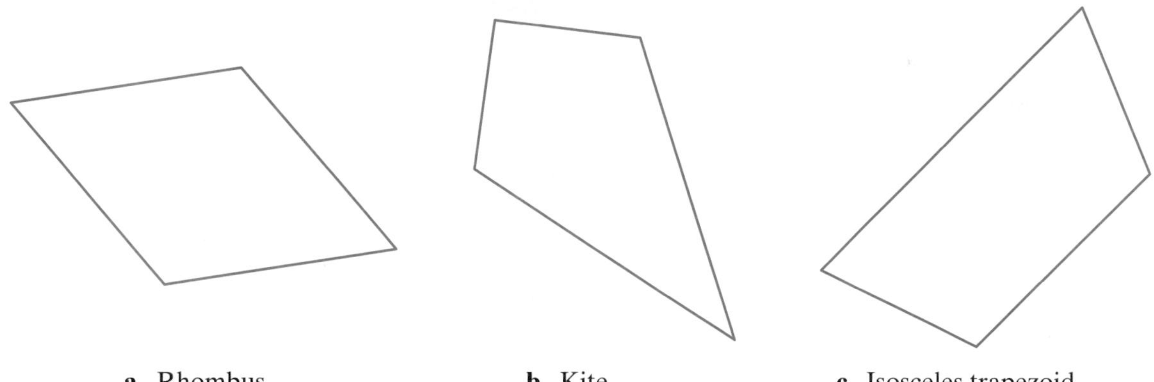

a. Rhombus **b.** Kite **c.** Isosceles trapezoid

1. Predict what the midquad of a rectangle will be. Draw a rectangle and its midquad to show that your prediction is reasonable.
2. Predict what the midquad of a square will be. Draw a square and its midquad to show that your prediction is reasonable.
3. If the midquad of a quadrilateral is a square, predict what specific shape the quadrilateral may have. Draw a figure (or figures) to show that your prediction is reasonable.
4. If the midquad of a quadrilateral is a rhombus, predict what specific shape the quadrilateral may have. Draw a figure (or figures) to show that your prediction is reasonable.
5. If the midquad of a quadrilateral is a rectangle, predict what specific shape the quadrilateral may have. Draw a figure (or figures) to show that your prediction is reasonable.

Introduction

Geometric figures that have the same shape, but not necessarily the same size, are said to be **similar**. Similarity is used in making maps, scale drawings, enlargements of photos, and indirect measurements of distance. For example, to determine the height of a building, we could compare the shadow of the building with the shadow of a meter stick and then use the fact that similar triangles are formed to calculate the height of the building.

We present the concepts of ratio and proportion in Section 6.1 and we use them in Sections 6.2 and 6.3, which contain the ideas central to the study of similar triangles. Section 6.4 introduces trigonometric relationships within right triangles. Section 6.5 introduces laws of trigonometry and applies them to solve geometry problems.

6.1 RATIO AND PROPORTION

Applied Problem

A computer-generated model of a car is to be sculpted into clay. The dimensions of the car are 400 cm long by 160 cm wide by 120 cm high. If the ratio of the dimensions of the model to the dimensions of the car is to be 1 to 4, what should be the dimensions of the model?

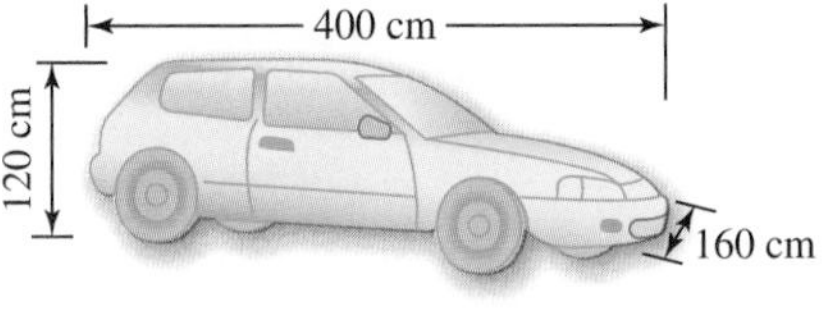

Ratios

The solution to the applied problem requires an understanding of the concepts of ratio and proportion. A **ratio** is an ordered pair of numbers. The ratio of the number a to the number b is represented in four common ways:

$$a \text{ to } b \quad a:b \quad \frac{a}{b} \quad a/b$$

(NOTE: The b in the ratio a to b can be zero. For example, the ratio of the number of stars to the number of circles on an American flag is 50 to 0. But, in this book, we will assume that $b \neq 0$; that is, denominators are nonzero.)

We could also write the ratio 1 to 4 in the preceding applied problem as $1:4$, $\frac{1}{4}$, or 1/4. When doing calculations with ratios, we express a ratio as a quotient of two numbers. For example, the ratio $\pi : 4.2$ would be written $\frac{\pi}{4.2}$.

EXAMPLE 6.1 For the rectangle in Figure 6.1, calculate the following ratios.

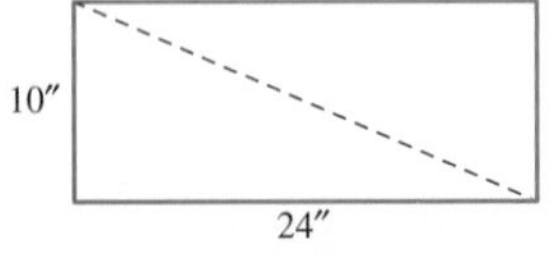

FIGURE 6.1

a. length to width

b. width to length

c. length of a diagonal to the perimeter

d. perimeter to area

SOLUTION

a. length : width $= 24 \text{ in.} : 10 \text{ in.} = \dfrac{24 \text{ in.}}{10 \text{ in.}} = \dfrac{12}{5}$

b. width : length $= 10 \text{ in.} : 24 \text{ in.} = \dfrac{10 \text{ in.}}{24 \text{ in.}} = \dfrac{5}{12}$

c. diagonal : perimeter
$$= \sqrt{10^2 + 24^2} \text{ in.} : (20 + 48) \text{ in.}$$
$$= 26 \text{ in.} : 68 \text{ in.}$$
$$= \frac{26 \text{ in.}}{68 \text{ in.}}$$
$$= \frac{13}{34}$$

d. perimeter : area
$$= 68 \text{ in.} : 240 \text{ in}^2$$
$$= \frac{68 \text{ in.}}{240 \text{ in}^2}$$
$$= \frac{17 \text{ in.}}{60 \text{ in}^2}$$ ■

Notice that we expressed each ratio in Example 6.1 in simplest form. Also notice that in parts (a)–(c), the units in the numerator and denominator were the same. In such cases, the units "cancel" and we express the ratio without units.

Ratios that include two different units of measures are called **rates**. Example 6.1 (d) showed an example of a rate. Other examples of rates are 12 miles per hour (12 miles: 1 hour), 5 percent per year (5 percent : 1 year), and \$3.69 per pound (\$3.69 : 1 pound). Applications often involve rates that must be converted into equivalent rates. As we showed in Section 2.5, we can use dimensional analysis to convert from one rate to another.

Proportions

When two ratios are involved, we often have a proportion. A **proportion** is an equation stating that two ratios are equal.

EXAMPLE 6.2 Show that the following are proportions.

a. $\dfrac{1}{2} = \dfrac{2}{4}$

b. $\dfrac{3.2}{4.8} = \dfrac{2}{3}$

c. $\pi : 3\pi = 1 : 3$

d. $\dfrac{7x}{5x} = \dfrac{7}{5}$

SOLUTION

a. $\dfrac{1}{2} = \dfrac{2}{4}$ because $\dfrac{1}{2} = \dfrac{(2)(1)}{(2)(2)} = \dfrac{2}{4}$

b. $\dfrac{3.2}{4.8} = \dfrac{2}{3}$ because $\dfrac{3.2}{4.8} = \dfrac{(2)(1.6)}{(3)(1.6)} = \dfrac{2}{3}$

c. $\pi : 3\pi = 1 : 3$ because $\dfrac{\pi}{3\pi} = \dfrac{(1)(\pi)}{(3)(\pi)} = \dfrac{1}{3}$

d. $\dfrac{7x}{5x} = \dfrac{7}{5}$ because $\dfrac{7x}{5x} = \dfrac{(7)(x)}{(5)(x)} = \dfrac{7}{5}$ ■

When a proportion is written as $\dfrac{a}{b} = \dfrac{c}{d}$, the numbers a, b, c, and d are called **terms**. The two numbers a and d are called the **extremes**, and b and c are called the **means**.

Next we show how the means and extremes of a proportion can be used in a convenient way to determine if two ratios form a proportion.

EXAMPLE 6.3 Determine if $\frac{10}{15} = \frac{8}{12}$ is a proportion.

SOLUTION

Rewrite $\frac{10}{15}$ and $\frac{8}{12}$ using the common denominator (15)(12). Thus, $\frac{10}{15} = \frac{(10)(12)}{(15)(12)}$ and $\frac{8}{12} = \frac{(15)(8)}{(15)(12)}$. Because the two fractions $\frac{(10)(12)}{(15)(12)}$ and $\frac{(15)(8)}{(15)(12)}$ have a common denominator, we need only determine if the numerators are equal. Because $(10)(12) = 120$ and $(15)(8) = 120$, we have $\frac{10}{15} = \frac{8}{12}$. ■

Notice that the numbers we compared were (10)(12), the product of the extremes, and (15)(8), the product of the means. We summarize this technique next.

THEOREM 6.1 Cross-Multiplication Theorem

$\frac{a}{b} = \frac{c}{d}$ if and only if the product of the means equals the product of the extremes.

That is,

$$\frac{a}{b} = \frac{c}{d} \quad \text{if and only if} \quad ad = bc$$

The proof of this theorem consists of two parts:

Subgoal 1 Prove that if $ad = bc$, then $\frac{a}{b} = \frac{c}{d}$.

Subgoal 2 Prove that if $\frac{a}{b} = \frac{c}{d}$, then $ad = bc$.

We will prove subgoal 1 here and leave subgoal 2 for the problem set.

Prove If $ad = bc$, then $\frac{a}{b} = \frac{c}{d}$.

Proof If $ad = bc$, then we can divide both sides of the equation by the same nonzero number, bd, to obtain $\frac{ad}{bd} = \frac{bc}{bd}$. Simplifying, we have $\frac{a}{b} = \frac{c}{d}$. ■

Using Theorem 6.1 this we can see that $\frac{3}{5} = \frac{6}{10}$ because $(3)(10) = (5)(6)$. Notice that we can interchange various terms of this proportion to form new proportions. So we can write

$\frac{10}{5} = \frac{6}{3}$ if the extremes of $\frac{3}{5} = \frac{6}{10}$ are interchanged

$\frac{3}{6} = \frac{5}{10}$ if the means of $\frac{3}{5} = \frac{6}{10}$ are interchanged

$\frac{5}{3} = \frac{10}{6}$ if both the ratios in $\frac{3}{5} = \frac{6}{10}$ are inverted

This discussion motivates the following theorem, which we can prove using Theorem 6.1.

> **THEOREM 6.2**
>
> $\frac{a}{b} = \frac{c}{d}$ if and only if $\frac{d}{b} = \frac{c}{a}$ (Exchange the extremes.)
>
> $\frac{a}{b} = \frac{c}{d}$ if and only if $\frac{a}{c} = \frac{b}{d}$ (Exchange the means.)
>
> $\frac{a}{b} = \frac{c}{d}$ if and only if $\frac{b}{a} = \frac{d}{c}$ (Invert each ratio.)

EXAMPLE 6.4 Determine if the following are proportions.

a. $\frac{21}{35} = \frac{27}{45}$ **b.** $\frac{21}{27} = \frac{35}{45}$ **c.** $\frac{45}{35} = \frac{27}{21}$

SOLUTION

a. $(21)(45) = 945$ and $(35)(27) = 945$. Thus, $\frac{21}{35} = \frac{27}{45}$ is a proportion by the Cross-Multiplication Theorem. To work this problem in another way, we could simplify both ratios.

$$\frac{21}{35} = \frac{(3)(7)}{(5)(7)} = \frac{3}{5} \quad \text{and} \quad \frac{27}{45} = \frac{(3)(9)}{(5)(9)} = \frac{3}{5}$$

Because both ratios equal $\frac{3}{5}$, they form a proportion.

b. From part (a), we know that $\frac{21}{35} = \frac{27}{45}$ is a proportion. By interchanging the means, we obtain $\frac{21}{27} = \frac{35}{45}$, which is a proportion by Theorem 6.2.

c. From part (a), we know that $\frac{21}{35} = \frac{27}{45}$. By exchanging the extremes we have $\frac{45}{35} = \frac{27}{21}$, which is a proportion by Theorem 6.2. ■

The proportion in (c) also could have been obtained from the one in (b) by inverting each ratio. We also could have used cross-multiplication in both parts (b) and (c).

Solving For a Missing Term in a Proportion

We can use the Cross-Multiplication Theorem to solve for a missing term in a proportion. The following example shows how this can be done.

EXAMPLE 6.5 Solve for the variable in each proportion.

a. $\frac{14}{17} = \frac{42}{y}$ **b.** $\frac{x}{7} = \frac{31}{42}$

SOLUTION

a. $\frac{14}{17} = \frac{42}{y}$ if and only if $14y = 17 \cdot 42$, or $y = \frac{17 \cdot 42}{14} = 51$.

b. $\frac{x}{7} = \frac{31}{42}$ if and only if $42x = 7 \cdot 31$, or $x = \frac{7 \cdot 31}{42} = \frac{31}{6}$. ■

We can solve many problems in mathematics and in geometry, in particular, using proportions.

EXAMPLE 6.6 A rectangular snapshot measures 3.5 inches tall by 5 inches wide. An enlargement of the photo will be made, where the larger photo will have a width of 18 inches. If none of the original picture will be cropped (left out), the dimensions of the enlargement are proportional to the dimensions of the original photo. What will be the height of the larger photograph (Figure 6.2)?

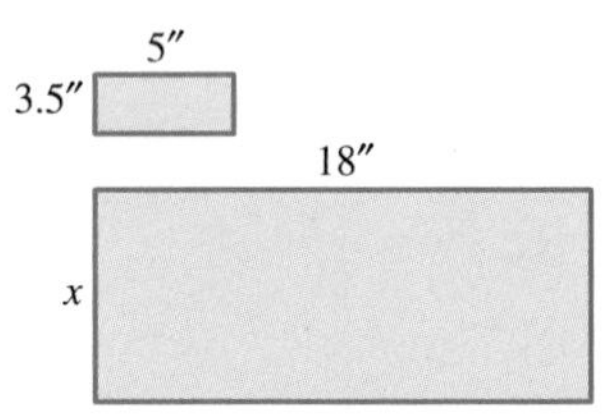

FIGURE 6.2

SOLUTION

We can use the proportion

$$\frac{\text{height of small photo}}{\text{width of small photo}} = \frac{\text{height of large photo}}{\text{width of large photo}}$$

So we have $\frac{3.5 \text{ inches}}{5 \text{ inches}} = \frac{x \text{ inches}}{18 \text{ inches}}$, or $\frac{3.5}{5} = \frac{x}{18}$.

Thus, $5x = (3.5)(18) = 63$ or $x = \frac{63}{5} = 12.6$ inches. Therefore, the height of the new photograph will be 12.6 inches. ■

The Geometric Mean

The **arithmetic average** or **arithmetic mean** of two numbers a and c is $\frac{a + c}{2}$. The **geometric mean** or **mean proportional** of a and c is, analogously, $\sqrt{ac}$ for positive a, c. Consider the proportion $\frac{a}{b} = \frac{b}{c}$. Here $ac = b^2$, or $b = \sqrt{ac}$. We state this result as follows.

> **THEOREM 6.3**
>
> If $\frac{a}{b} = \frac{b}{c}$, then b is the geometric mean (mean proportional) of a and c.

The geometric mean appears often in geometry. In particular, there is an application of it in the next section.

EXAMPLE 6.7 Find the geometric mean of 3 and 27.

SOLUTION

Using Theorem 6.3, we can find the geometric mean of 3 and 27 by solving the equation $\frac{3}{x} = \frac{x}{27}$. Here, $x^2 = 81$. Thus, $x = 9$ is the geometric mean. ■

Solution of Applied Problem

The dimensions of the car were 400 cm by 160 cm by 120 cm.

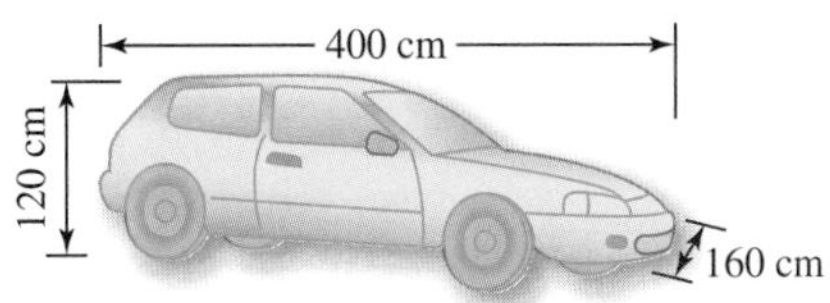

If the ratio of the dimensions of the clay model to the dimensionals of the actual car was to be 1:4 and we let l be the length of the model, w its width, and h its height, then these are the appropriate ratios:

$$\frac{1}{4} = \frac{l}{400} = \frac{w}{160} = \frac{h}{120}$$

Solving $\frac{1}{4} = \frac{l}{400}$, we get $4l = 400$, so $l = 100$ cm. In a similar manner, $w = 40$ cm and $h = 30$ cm.

GEOMETRY AROUND US The **golden ratio**, also called the **divine proportion**, was known to the Pythagoreans in 500 B.C.E. This ratio, the number $\frac{1 + \sqrt{5}}{2}$, occurs frequently in nature, art, and architecture. The Parthenon in Athens can be surrounded by a rectangle whose sides have length and width in the golden ratio. The Greeks believed that a rectangle with sides in this ratio had the most aesthetically pleasing shape.

PROBLEM SET 6.1

EXERCISES/PROBLEMS

Represent each ratio in Exercises 1–4 as a fraction in simplest form.

1. 16 to 64

2. 30 to 75

3. 82.5 to 16.5

4. $4\frac{1}{2}$ to $2\frac{1}{3}$

5. Show that each of the following is a proportion.

a. $\frac{11}{24} = \frac{66}{144}$

b. $14x : 91x = 4 : 26$

c. $\frac{3.8}{9.5} = \frac{26.6}{66.5}$

6. Show that each of the following is a proportion.

a. $\frac{6}{7} = \frac{35.4}{41.3}$

b. $15 : 60 = 18 : 72$

c. $\frac{4}{84} = \frac{15y}{315y}$

7. Determine if $18 : 51 = 3 : 8.5$ is a proportion. Justify your answer.

8. Determine if $\frac{0.9}{4.2} = \frac{96}{289}$ is a proportion.

Solve for the unknown in Exercises 9–20.

9. $\frac{57}{95} = \frac{18}{n}$

10. $2:130 = x:5$

11. $n/70 = 6/21$

12. $\frac{r}{84} = \frac{3}{14}$

13. $\frac{7}{5} = \frac{42}{s}$

14. $\frac{12}{t} = \frac{18}{45}$

15. $\frac{3}{5}:6 = d:25$

16. $b/8 = 2.25/18$

17. $\frac{x}{100} = \frac{4.8}{1.5}$

18. $\frac{57.4}{39.6} = \frac{7}{4x}$ (Round to one decimal place.)

19. $\frac{5}{8} = \frac{x}{x+9}$

20. $\frac{x}{30-x} = \frac{2}{3}$

Find the geometric mean of each pair of numbers in Exercises 21–24.

21. 2, 18

22. 5, 7

23. π, 10

24. n, $4n$

PROOFS

25. If the radius of the base of a cylinder is doubled and the height of the cylinder is tripled, find the ratio of the volume of the original cylinder to the volume of the new cylinder. Prove your assertion.

26. If the length and width of a rectangle are both doubled, find the ratio of the area of the new rectangle to the area of the original rectangle. Prove your assertion.

Prove each of the statements in problems 27–37.

27. $\frac{a}{b} = \frac{c}{d}$ if and only if $\frac{d}{b} = \frac{c}{a}$

28. $\frac{a}{b} = \frac{c}{d}$ if and only if $\frac{a}{c} = \frac{b}{d}$

29. $\frac{a}{b} = \frac{c}{d}$ if and only if $\frac{b}{a} = \frac{d}{c}$

30. If $\frac{a+b}{b} = \frac{c+d}{d}$, then $\frac{a}{b} = \frac{c}{d}$.

31. If $\frac{a}{b} = \frac{c}{d}$, then $\frac{a+b}{b} = \frac{c+d}{d}$.

32. If $\frac{a}{b} = \frac{c}{d}$, then $\frac{a-b}{b} = \frac{c-d}{d}$.

33. If $\frac{a}{b} = \frac{c}{b}$, then $a = c$.

34. If $\frac{a}{b} = \frac{a+c}{b+d}$, then $\frac{a}{b} = \frac{c}{d}$.

35. If $\frac{a}{b} = \frac{c}{d}$, then $\frac{a}{b} = \frac{a+c}{b+d}$.

36. If $\frac{a}{b} = \frac{c}{d} = \frac{e}{f}$, then $\frac{a+c+e}{b+d+f} = \frac{a}{b}$.

37. State and prove a generalization of problem 36.

APPLICATIONS

38. A map is drawn to scale such that $\frac{1}{8}$ inch represents 65 feet. If the shortest route from your house to the grocery store measures $23\frac{7}{16}$ inches on the map, how many miles is it to the grocery store? Round to the nearest tenth.

39. **a.** If 1 inch on a map represents 35 miles, how many miles are represented by 3 inches? 10 inches? n inches?

b. Los Angeles is about 1000 miles from Portland, Oregon. About how many inches apart would Portland and Los Angeles be on this map? Round to the nearest tenth.

40. A man who weighs 175 lb on earth would weigh 28 lb on the moon. How much would his 30-lb dog weigh on the moon?

41. Suppose that you drive your car an average of 4460 miles every half-year. At the end of 2.75 years, how far will your car have gone?

42. When doctors test blood cholesterol levels, they sometimes calculate a cardiac risk ratio.

$$\text{Cardiac Risk Ratio} = \frac{\text{total cholesterol level}}{\text{high density lipoprotein level (HDL)}}$$

For women, a ratio between 3.0 and 4.5 is desirable. A woman's blood test yields an HDL cholesterol level of 60 mg/dL and a total cholesterol level of 225 mg/dL. What is her cardiac risk ratio, expressed as a one-place decimal? Is her ratio in the normal range?

43. For the tax year 2004, the Internal Revenue Service audited 65 of every 10,000 individual returns.

a. In a community in which 34,000 people filed 2004 returns, how many returns might be expected to be audited?

b. In 2003 only 57 returns per 10,000 were audited. How many more of the 34,000 returns would be expected to be audited for 2004 than for 2003?

44. Sheila is climbing a hill that has a 17° slope. For every 5 feet she gains in altitude, she travels about 16.37 horizontal feet. If at the end of her uphill climb she has traveled 1 mile horizontally, how much altitude has she gained? Round to the nearest tenth.

45. Underwater pressure increases proportionately with depth. In an aquarium that is 18 in. tall, the pressure on the bottom of the tank is 93.6 lb/ft^2. In a large viewing aquarium, a tank is 8 ft deep. Find the pressure on the bottom of this larger tank.

46. a. A baseball pitcher has pitched a total of 25 innings so far during the season and has allowed 18 runs. At this rate, how many runs, to the nearest hundredth, would he allow in 9 innings? This number is called the pitcher's earned run average (ERA).

b. Tim Hudson of the Oakland Athletics had an ERA of 3.53 in 2004, At that rate, how many runs would he be expected to allow in 100 innings pitched? Round your answer to the nearest whole number.

47. The "Spruce Goose," a wooden flying boat built for Howard Hughes, had the world's largest wing span (319 ft 11 in.), according to the Guinness Book of World Records. It flew only once, in 1947 for a distance of about 1000 yd, and is now housed in the Evergreen Aviation Museum in McMinnville, Oregon. Shelly wants to build a scale model of the Spruce Goose, which is 218 ft 8 in. long. If her model will be 20 in. long, what will its wing span be? Round to the nearest whole number.

48. The aspect ratio of a TV screen is the ratio of its width to its height. The usual aspect ratio is 4 : 3. The size of a TV screen is specified by the length of its diagonal. Find the width and height of a 27-inch TV screen with an aspect ratio of 4 : 3.

49. A fashion doll popular with young girls for the last few decades is $11\frac{1}{2}$ inches tall and has a waist measurement of about 3 inches. Some concern has been expressed that this doll does not reflect realistic body proportions. Can you explain why? Suppose a "life-size" model of the same proportions with height 5 feet 6 inches were constructed. What would its waist measurement be? Round to the nearest whole number.

50. Many tires come with 13/32 inch of tread on them. The first 2/32 inch wears off quickly (say during the first 1000 miles). From then on the tire wears uniformly and more slowly. A tire is considered "worn out" when only 2/32 inch of tread is left.

a. How many 32nds of an inch of useable tread does a tire have after 1000 miles?

b. A tire has traveled 20,000 miles and has 5/32 inch of tread remaining. At this rate, how many total miles should the tire last before it is considered worn out?

51. Two professional drag racers are speeding down a quarter-mile track. The lead driver is traveling 1.738 feet for every 1.67 feet that the trailing car travels, and the trailing car is going 198 mph. How fast in mph is the lead car traveling? Round to the nearest whole number.

52. According to the "big-bang" hypothesis, the universe was formed approximately 10^{10} years ago. The analogy of a 24-hour day is often used to put the passage of this amount of time into perspective. Imagine that the universe was formed at midnight 24 hours ago. If the earth was formed approximately 10^5 years ago, to what time in the 24-hour day does this correspond?

EXTENDED PROBLEMS

53. We can relate the golden ratio to the lengths of line segments as follows. If we start with a line segment $\overline{AB}$ and locate point C on the segment such that $\frac{AB}{AC} = \frac{AC}{CB}$, then the value of each of these ratios is the golden ratio. Notice that AC is the geometric mean of AB and CB.

a. Another way to state the golden ratio as it relates to the lengths of line segments is $\frac{CB}{AC - CB} = \frac{AC}{CB}$. Show that this is equivalent to $\frac{AB}{AC} = \frac{AC}{CB}$. Then show that $\frac{AB}{AC} = \frac{AC}{CB}$ can be simplified to $\left(\frac{AC}{CB}\right)^2 = \frac{AC}{CB} + 1$. Let $x = \frac{AC}{CB}$ and use the quadratic formula to solve for x. What do you observe?

b. A **golden rectangle** is a rectangle such that the ratio of the long side to the short side is the golden ratio, $\frac{1 + \sqrt{5}}{2}$.

Designers and manufacturers sometimes utilize the aesthetic appeal of the golden rectangle in packaging and construction. Gather five rectangular items from your home and measure the dimensions. In each case, calculate the ratio of the long side to the short side and compare it to the golden ratio.

c. Research the golden ratio and describe how it is used in art and architecture. What famous structures were designed with the golden ratio in mind? When was the golden ratio first studied and by whom?

54. We can construct a golden rectangle with compass and straightedge as follows. Use the procedure given to construct a golden rectangle. Then measure the length and width of the rectangle you constructed and compare their ratio to the golden ratio.

a. Construct a square $ABCD$ of any size.

b. Construct the perpendicular bisector of $\overline{AD}$ and label the midpoint M.

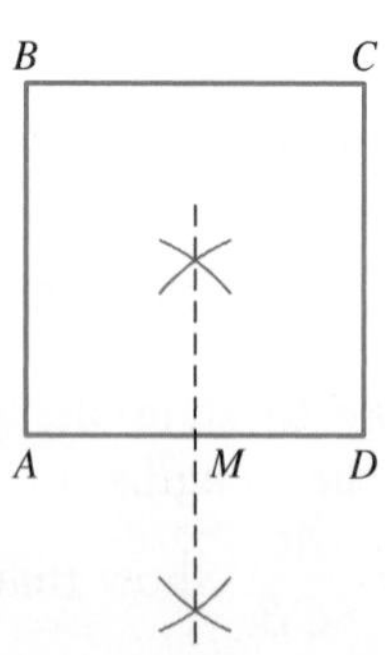

c. Use your compass to draw an arc with radius $\overline{MC}$ and with center M. Extend $\overline{AD}$ so that it intersects the arc at P.

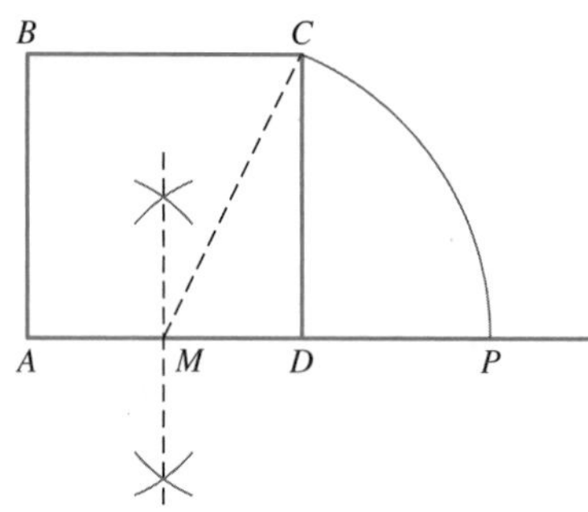

d. Construct a segment perpendicular to $\overline{AP}$ at P. Extend $\overline{BC}$ to meet this segment at Q.

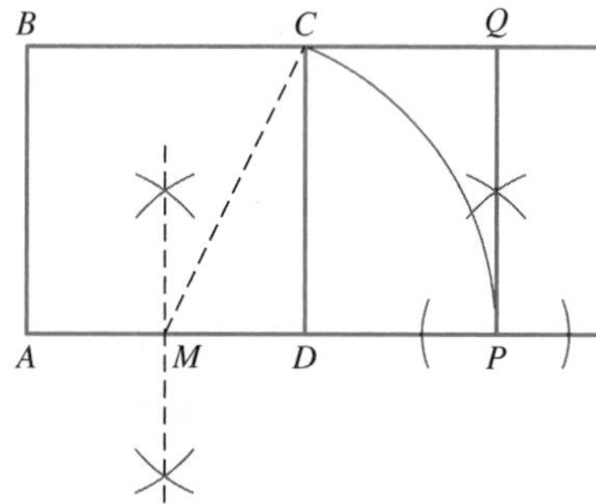

Now rectangle $ABQP$ is a golden rectangle. We could also say that D divides $\overline{AP}$ so that AD is the mean proportional of AP and DP.

55. The harmonic mean of two numbers a and b is defined as $\frac{2ab}{a + b}$.

a. Select five pairs of numbers and calculate the harmonic mean, arithmetic mean, and geometric mean for each pair.

b. Research the harmonic mean, arithmetic mean, and geometric mean, and investigate the following questions. How are the harmonic mean, arithmetic mean and geometric mean of two numbers related and under what circumstances would it be appropriate to use each of them? How is the harmonic mean related to the volume and surface area of cylinders? How can we construct the three different means using a compass and straightedge given two segments of length a and b?

6.2 SIMILAR TRIANGLES

Applied Problem

Campers walking along the south side of a river want to fell a tree tall enough so that they can walk on the tree to get across the river. How can they find the width of the river at its narrowest point without swimming across the river?

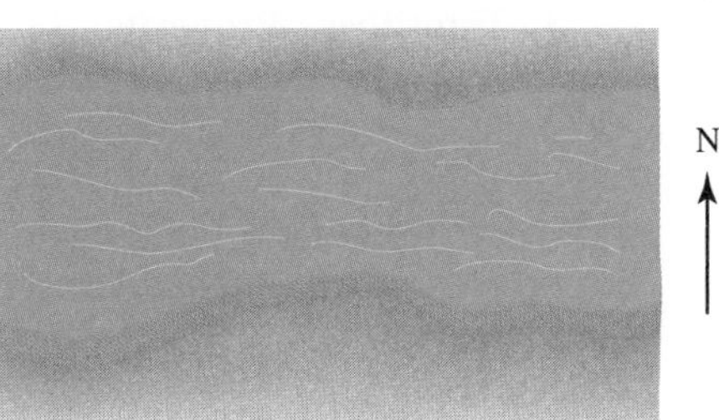

Similar Triangles

Recall that geometric shapes are similar if they have the same shape, though not necessarily the same size. For example, all squares are similar, as are all circles. However, a right triangle and an equilateral triangle do not have the same shape and so are not similar. In this section, we focus our attention on similar triangles because we often use them for finding measurements indirectly, that is, without actually measuring the distance. We will use similar triangles to solve the applied problem just stated.

DEFINITION Similar Triangles

$\triangle ABC$ is **similar** to $\triangle DEF$ (written $\triangle ABC \sim \triangle DEF$) under the correspondence $A \leftrightarrow D, B \leftrightarrow E, C \leftrightarrow F$ if and only if

1. All three pairs of corresponding angles are congruent.
2. All pairs of corresponding sides are proportional.

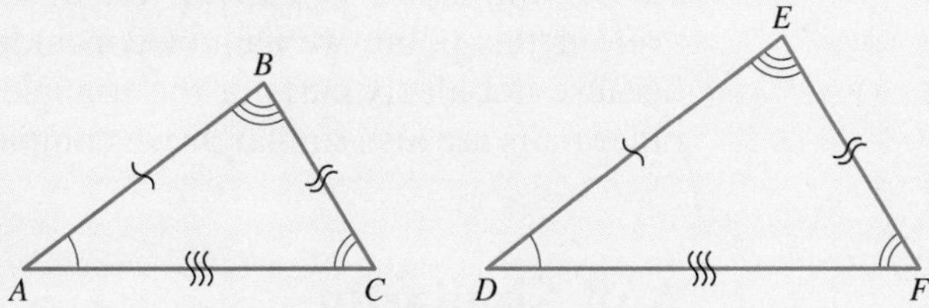

$\triangle ABC \sim \triangle DEF$ if and only if $\angle A \cong \angle D$, $\angle B \cong \angle E$, $\angle C \cong \angle F$, and

$$\frac{AB}{DE} = \frac{BC}{EF} = \frac{AC}{DF}.$$

Notice that we use curved marks to identify the pairs of corresponding sides that are proportional.

Congruent triangles are similar because their corresponding angles are congruent and the corresponding sides are congruent (hence, they are proportional in the ratio of $1:1$). Also, as with congruence, the following relationships hold for similarity:

1. **Reflexive Property**: $\triangle ABC \sim \triangle ABC$ for every $\triangle ABC$.
2. **Symmetric Property**: If $\triangle ABC \sim \triangle DEF$, then $\triangle DEF \sim \triangle ABC$.
3. **Transitive Property**: If $\triangle ABC \sim \triangle DEF$ and $\triangle DEF \sim \triangle GHI$, then $\triangle ABC \sim \triangle GHI$.

EXAMPLE 6.8 Find all pairs of triangles in Figure 6.3 that satisfy the definition of similar triangles. Also, identify other pairs that *appear* to be similar.

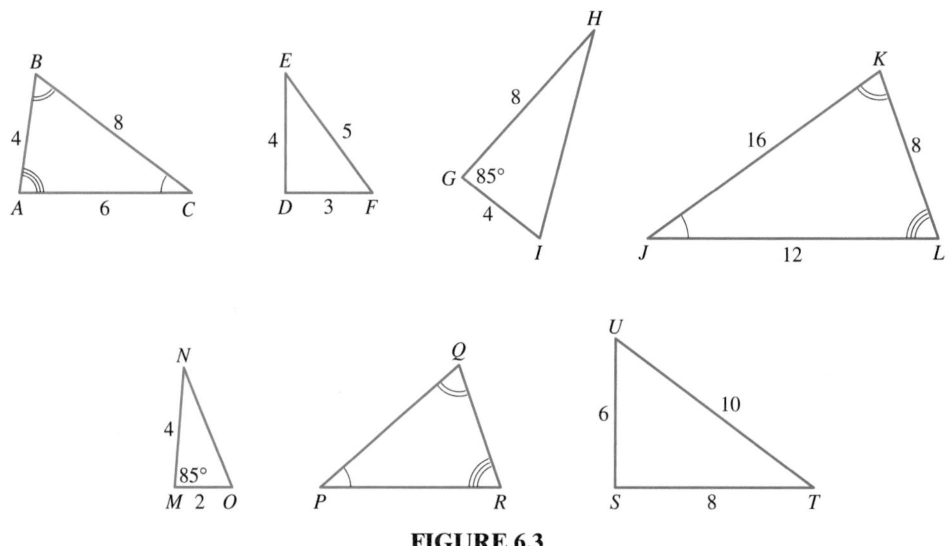

FIGURE 6.3

SOLUTION

$\triangle ABC$ and $\triangle LKJ$ are similar because (a) corresponding angles are congruent ($\angle A \cong \angle L$, $\angle B \cong \angle K$, $\angle C \cong \angle J$) and (b) corresponding sides are proportional in the ratio of 1 to 2 ($AC : LJ = BC : KJ = AB : LK = 1:2$). Other pairs (such as $\triangle DEF$ and $\triangle STU$ or $\triangle GHI$ and $\triangle MNO$) appear to have the same shape. However, at this point, we cannot conclude that they are similar from the definition because not all six parts of the triangles are known. We will be able to verify that other pairs are also similar as we complete this section. ■

AAA Similarity

In Figure 6.3, all three of the corresponding angles of $\triangle LKJ$ and $\triangle RQP$ are congruent and the two triangles appear to have the same shape, or to be similar. The fact that two triangles having congruent corresponding angles have the same shape is stated next in the Angle-Angle-Angle Similarity Postulate, abbreviated as AAA Similarity.

POSTULATE 6.1 AAA Similarity Postulate

Two triangles are similar if and only if three angles of one triangle are congruent, respectively, to three angles of the other triangle.

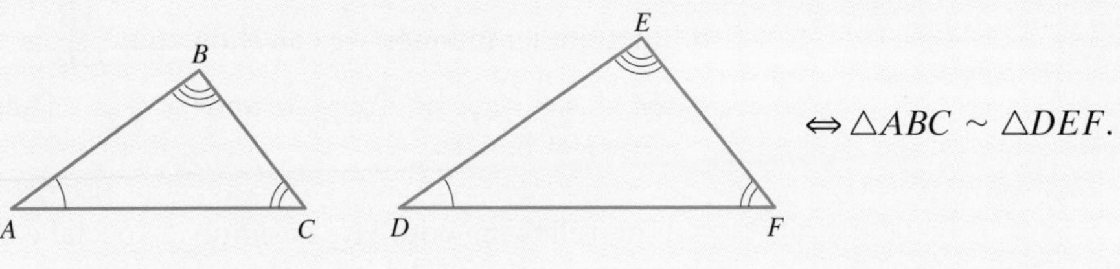

$\triangle ABC \sim \triangle DEF$ if and only if $\angle A \cong \angle D$, $\angle B \cong \angle E$, and $\angle C \cong \angle F$.

Now suppose that in $\triangle ABC$ and $\triangle DEF$ in Figure 6.4 we know $\angle A = \angle D = 30°$ and $\angle B = \angle E = 80°$. Then $\angle A + \angle B = \angle D + \angle E = 110°$.

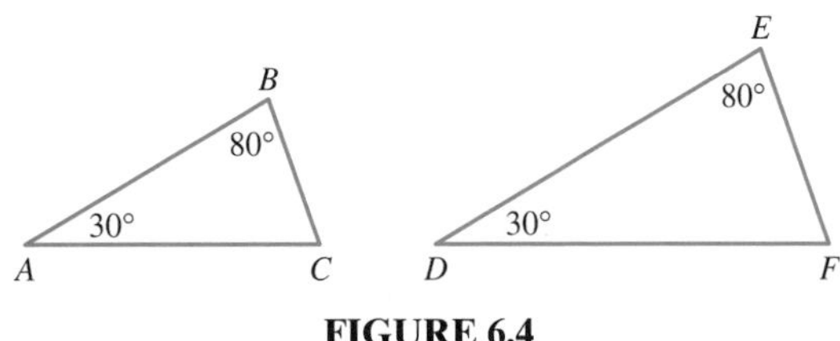

FIGURE 6.4

Moreover, because the sum of the measures of the angles in a triangle is 180°, the third angles must each equal $180° - 110° = 70°$. As a consequence, all three pairs of corresponding angles are congruent. Therefore, $\triangle ABC \sim \triangle DEF$ by the AAA Similarity Postulate. The following theorem, the Angle-Angle Similarity Theorem (AA Similarity), summarizes this result.

THEOREM 6.4 AA Similarity Theorem

Two triangles are similar if two angles of one triangle are congruent, respectively, to two angles of the other triangle.

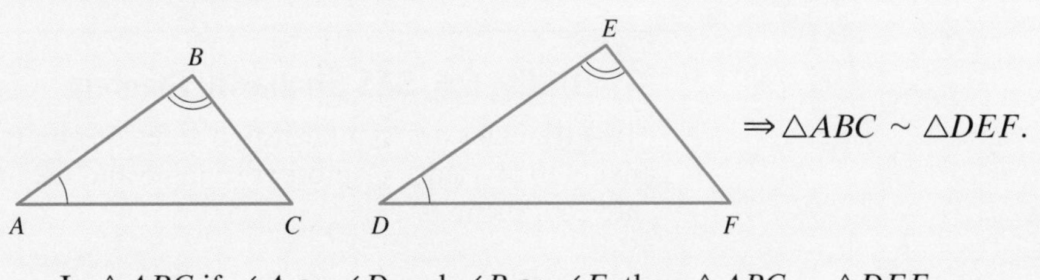

In $\triangle ABC$ if $\angle A \cong \angle D$ and $\angle B \cong \angle E$, then $\triangle ABC \sim \triangle DEF$.

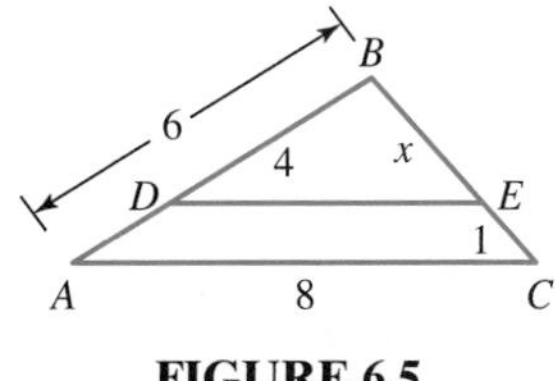

FIGURE 6.5

EXAMPLE 6.9 In $\triangle ABC$, $\overline{DE}$ is parallel to $\overline{AC}$ (Figure 6.5). If $AB = 6$, $CE = 1$, $AC = 8$, $BE = x$, and $BD = 4$, find **a.** DE and **b.** BC.

SOLUTION

First, $\overline{DE} \parallel \overline{AC}$ and $\angle C$ and $\angle BED$ are corresponding angles; hence, they are congruent. Also, because $\angle B$ is common to $\triangle ABC$ and $\triangle DBE$, we can conclude that $\triangle ABC \sim \triangle DBE$ by AA Similarity.

a. Because $\triangle ABC \sim \triangle DBE$, we have $\frac{AB}{DB} = \frac{AC}{DE}$, or $\frac{6}{4} = \frac{8}{DE}$. Therefore, $DE = \frac{32}{6} = 5\frac{1}{3}$.

b. In a similar manner, we can show that $\frac{AB}{DB} = \frac{BC}{BE}$, or $\frac{6}{4} = \frac{1 + x}{x}$. It follows that $6x = 4 + 4x$, or $2x = 4$ and $x = 2$. Thus, $BE = 2$ and $BC = x + 1 = 2 + 1 = 3$. ■

The following corollary is simply a special case of the AA Similarity Theorem, where one of the angles is a right angle.

COROLLARY 6.5

Two right triangles are similar if an acute angle of one triangle is congruent to an acute angle of the other triangle.

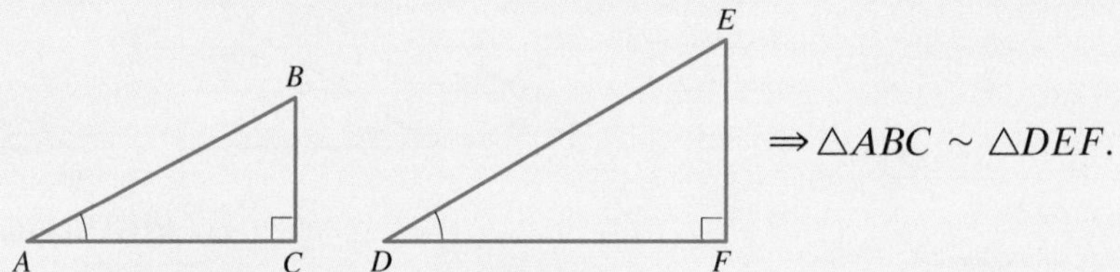

In $\triangle ABC$ and $\triangle DEF$, if $\angle C = \angle F = 90°$ and $\angle A \cong \angle D$, then $\triangle ABC \sim \triangle DEF$.

SAS Similarity

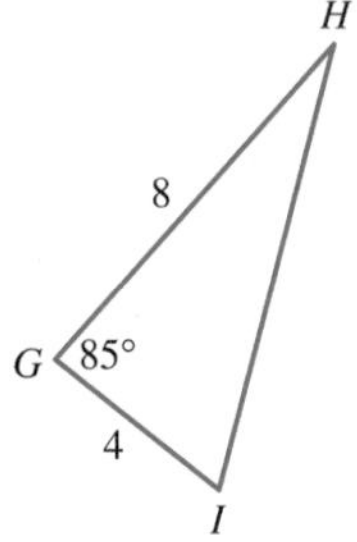

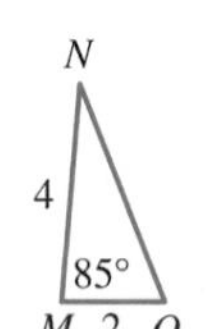

FIGURE 6.6

$\triangle GHI$ and $\triangle MNO$ from Figure 6.3, reproduced here in Figure 6.6, appear to be similar. Note that we know only that two corresponding sides are proportional and the included angles are congruent respectively. This suggests the following result, whose proof will be omitted. This suggests the following result, the Side-Angle-Side Similarity Theorem (SAS Similarity), whose proof will be omitted.

THEOREM 6.6 SAS Similarity Theorem

Two triangles are similar if two sides of one triangle are proportional, respectively, to two sides of another triangle and the angles included between the sides are congruent.

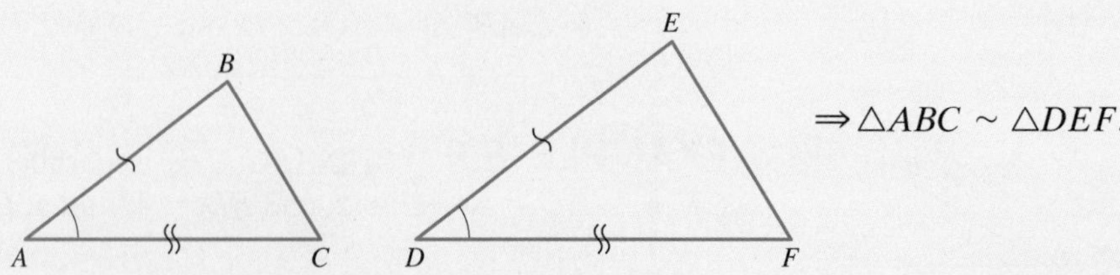

In $\triangle ABC$ and $\triangle DEF$, if $\frac{AB}{DE} = \frac{AC}{DF}$ and $\angle A \cong \angle D$, then $\triangle ABC \sim \triangle DEF$.

EXAMPLE 6.10 In Figure 6.7(a), $AB = 7$ cm, $BC = 10$ cm, and $AC = 14$ cm. Suppose $DB = 5$ cm and $BE = 3.5$ cm,

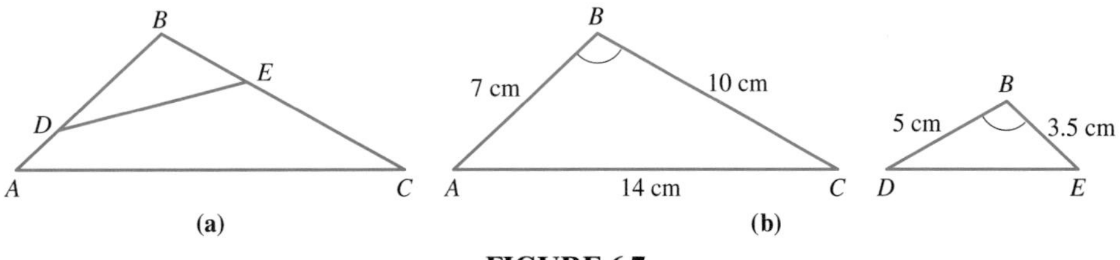

FIGURE 6.7

a. State which two triangles are similar.

b. Find ED.

SOLUTION

Note that $\triangle ABC$ and $\triangle DBE$ share $\angle B$, so they have one angle congruent. When one triangle is drawn inside another as in this case, it is sometimes helpful to redraw the figure, separating the two triangles [Figure 6.7(b)].

a. Because $BC = 2(BD)$ and $AB = 2(EB)$, we have $\frac{BC}{BD} = \frac{AB}{EB} = \frac{2}{1}$. Also, $\angle B \cong \angle B$. So $\triangle ABC \sim \triangle EBD$ by SAS Similarity.

b. By corresponding parts of similar triangles, we have $\frac{AC}{ED} = \frac{AB}{EB}$, so $\frac{14}{ED} = \frac{7}{3.5} = \frac{2}{1}$. Therefore, $2(ED) = 14$ or $ED = 7$ cm. ■

Notice that in Theorem 6.6, the proportion $\frac{AB}{DE} = \frac{AC}{DF}$ compared lengths between the two triangles. However, since $\frac{AB}{AC} = \frac{DE}{DF}$, Theorem 6.6 also holds when the ratios in the proportion compare lengths within each triangle.

The next corollary, the Leg-Leg Similarity Corollary, abbreviated as LL Similarity, is a special case of the SAS Similarity Theorem.

COROLLARY 6.7 LL Similarity

Two right triangles are similar if the legs of one triangle are proportional respectively to the legs of the other triangle.

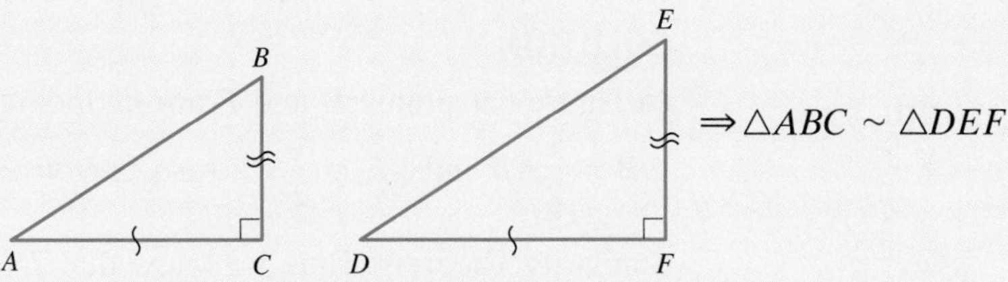

In $\triangle ABC$ and $\triangle DEF$, if $\angle C = \angle F = 90°$ and $\frac{AC}{DF} = \frac{BC}{EF}$, then $\triangle ABC \sim \triangle DEF$.

SSS Similarity

$\triangle DEF$ and $\triangle STU$ from Figure 6.3 are reproduced in Figure 6.8, although arranged differently.

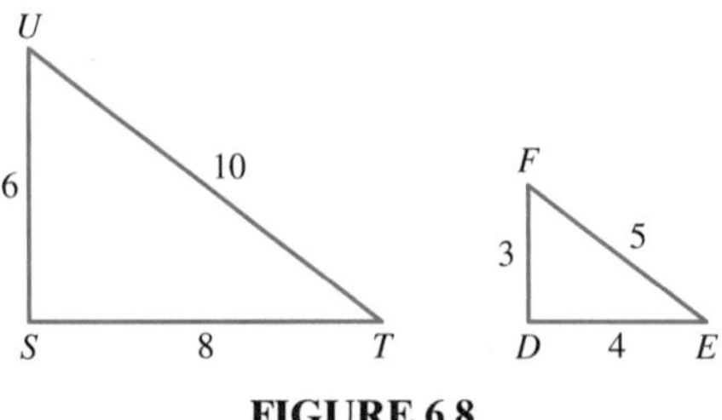

FIGURE 6.8

Although only the fact that their corresponding sides are proportional can be inferred from the figure, the two triangles appear to be similar. This suggests the following result. We will omit the proof of this theorem, the Side-Side-Side Similarity Theorem, which is abbreviated as SSS Similarity.

THEOREM 6.8 SSS Similarity Theorem

Two triangles are similar if three sides of one triangle are proportional to three sides of the other triangle.

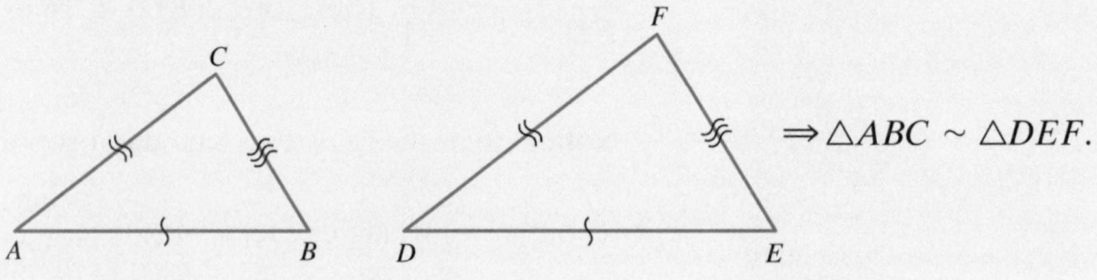

In $\triangle ABC$ and $\triangle DEF$, if $\frac{AB}{DE} = \frac{AC}{DF} = \frac{BC}{EF}$, then $\triangle ABC \sim \triangle DEF$.

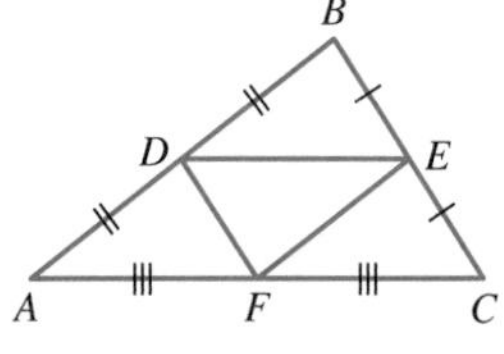

FIGURE 6.9

EXAMPLE 6.11 In Figure 6.9, D, E, and F are midpoints of the sides of $\triangle ABC$. Prove that $\triangle ABC \sim \triangle EFD$ (Notice the correspondence of the vertices in this similarity statement).

SOLUTION

In Figure 6.9, points D and E are midpoints of sides $\overline{AB}$ and $\overline{BC}$, respectively. So $DB = \frac{1}{2}AB$ and $BE = \frac{1}{2}BC$. Also, $\angle B \cong \angle B$. Thus, $\triangle BDE \sim \triangle BAC$ by the SAS Similarity Theorem. Thus, we know that $DE = \frac{1}{2}AC$. In a similar manner, we can show that, $DF = \frac{1}{2}CB$, and $EF = \frac{1}{2}AB$. Therefore, $\triangle ABC \sim \triangle EFD$ by the SSS Similarity Theorem because the corresponding sides have the same ratio. ■

Solution of Applied Problem

Sight point A opposite C. Then measure from C to point B, where $\overline{BC}$ is perpendicular to $\overline{AC}$. Find point E collinear with A and C and a convenient distance (say 40 ft) from C. Then find D, where $\overline{ED}$ is perpendicular to $\overline{AE}$ and A, B, and D are collinear. Measure ED. Then $\triangle ABC \sim \triangle ADE$ by AA Similarity. Thus, if the width of the river is x, then $\frac{11}{33} = \frac{x}{40 + x}$, or $440 + 11x = 33x$. Therefore, $440 = 22x$ and $x = 20$, so the river is 20 feet wide.

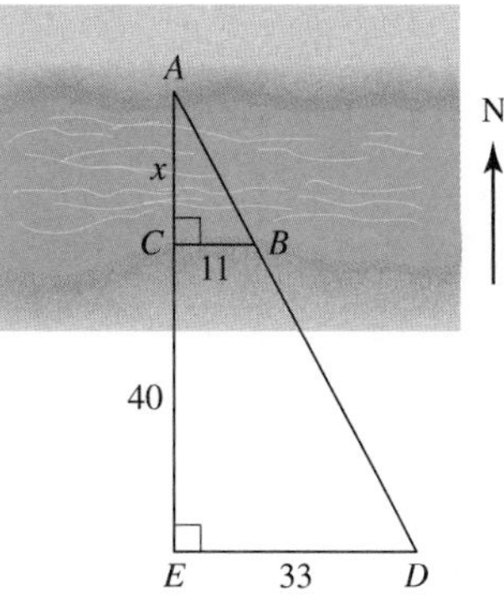

GEOMETRY AROUND US

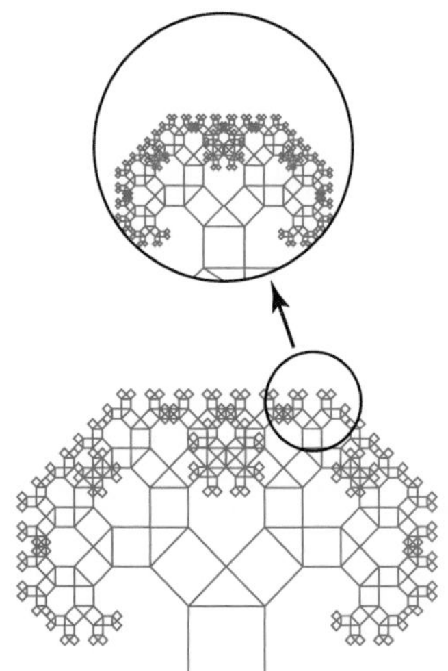

Many synthetic structures have common geometric shapes such as circles, triangles, and rectangles, but many naturally occurring forms, such as clouds, mountains, lightning, trees, and coastlines, cannot be adequately represented by these shapes. A new branch of geometry uses the idea of fractional dimension, or fractal, to more accurately describe complex forms.

A fractal has an interesting property called self-similarity. That is, the shape of a portion of the object, if magnified, looks like the original object.

PROBLEM SET 6.2

EXERCISES/PROBLEMS

In Exercises 1–4, determine which pairs of triangles are similar, note the correspondence of the vertices, and write an appropriate similarity statement. Also, explain why the triangles are similar.

1.

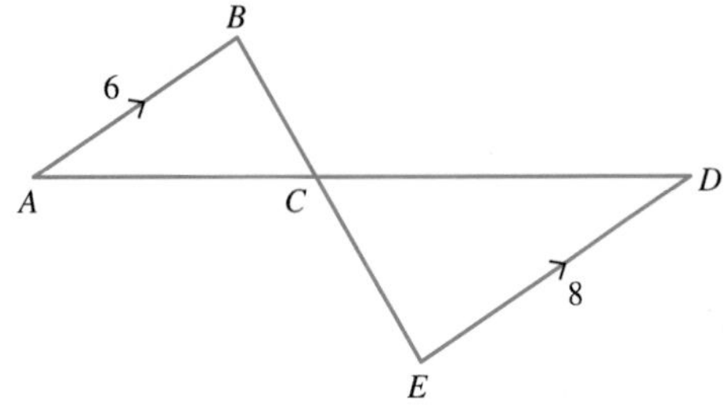

2.

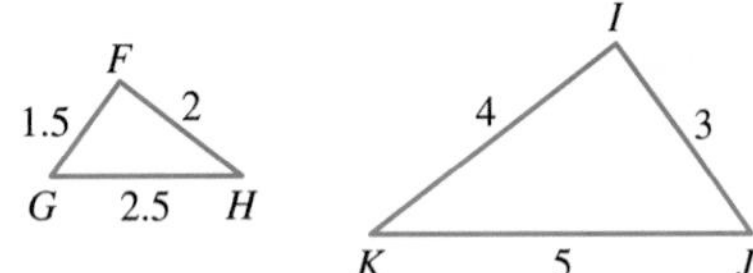

3.

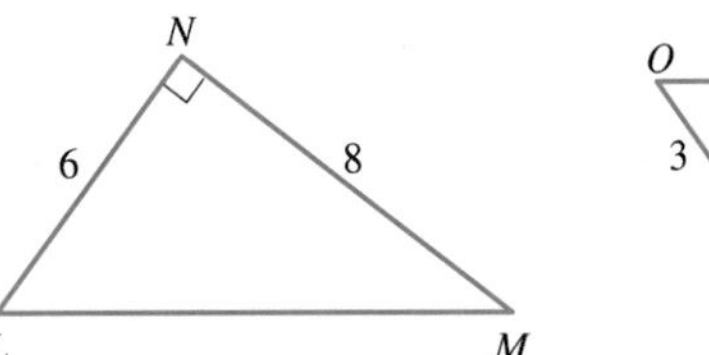

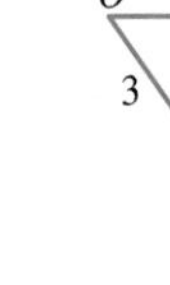

4.

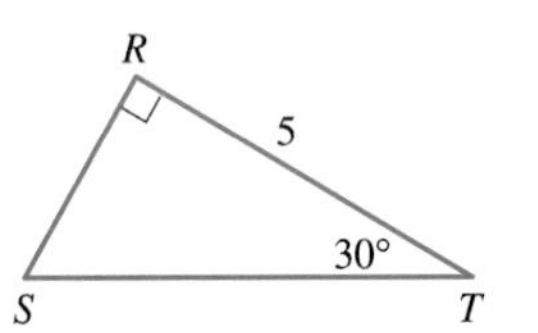

Pairs of similar triangles are shown in Exercises 5–12. Find the missing measures. Give exact answers rather than decimal approximations.

5.

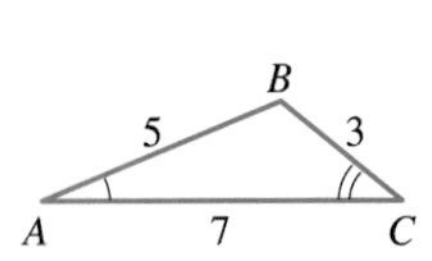

6.

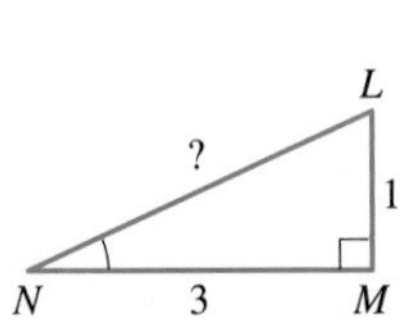

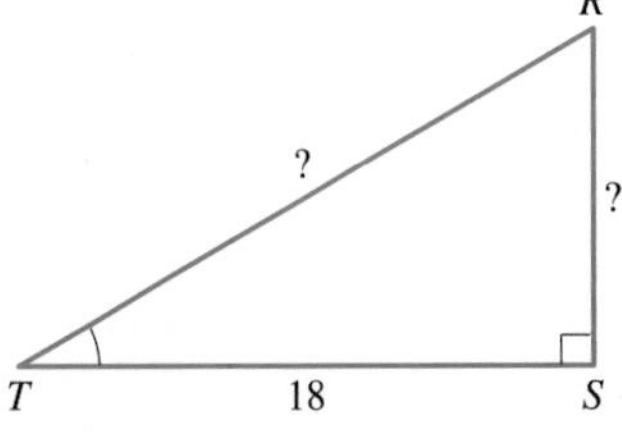

7.

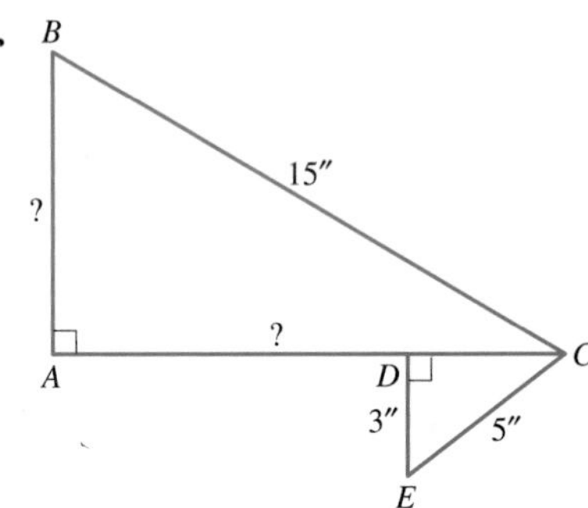

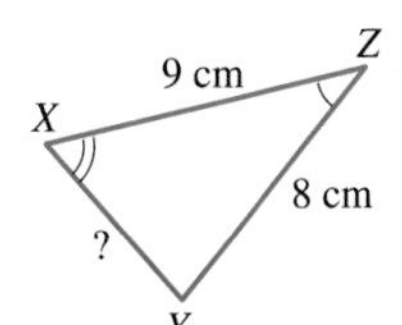

8.

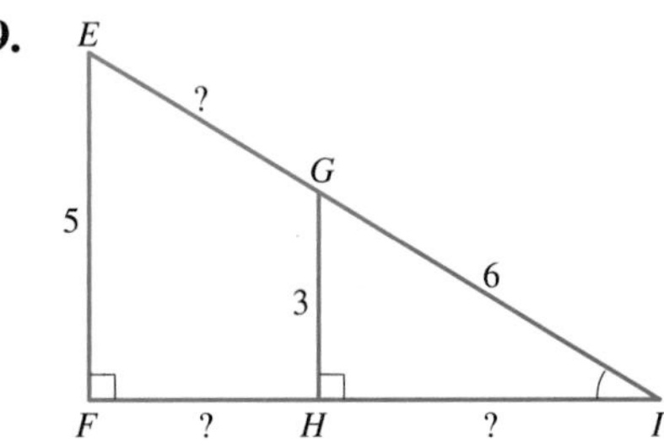

9.

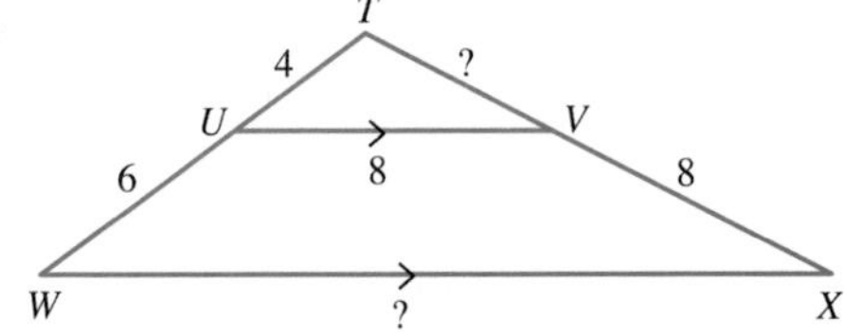

10.

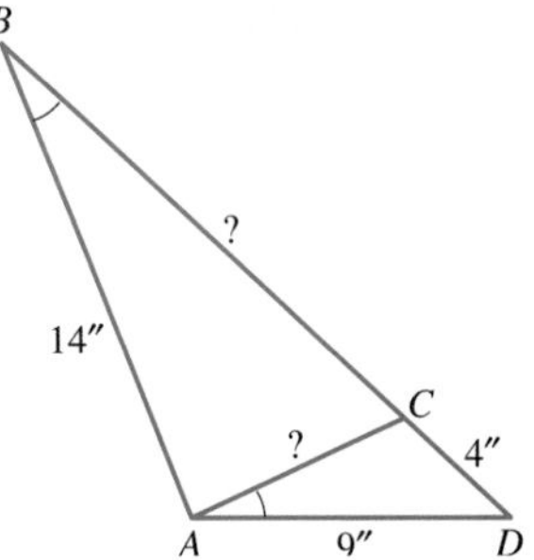

11.

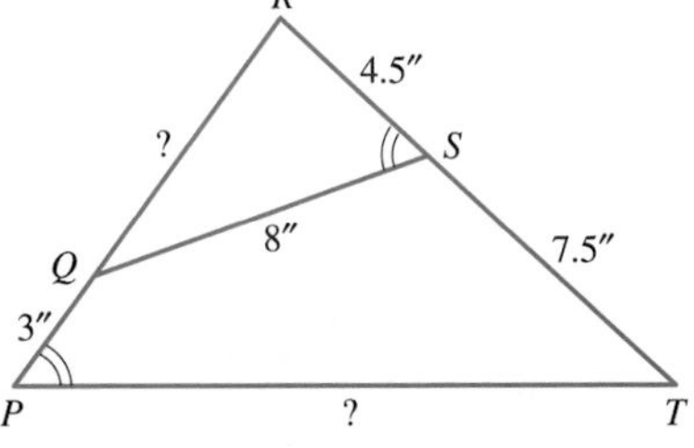

12.

13. Compute the ratios requested in the following pairs of similar figures.

a. Find the ratios of base : base, height : height, and area : area for the following pair of similar triangles.

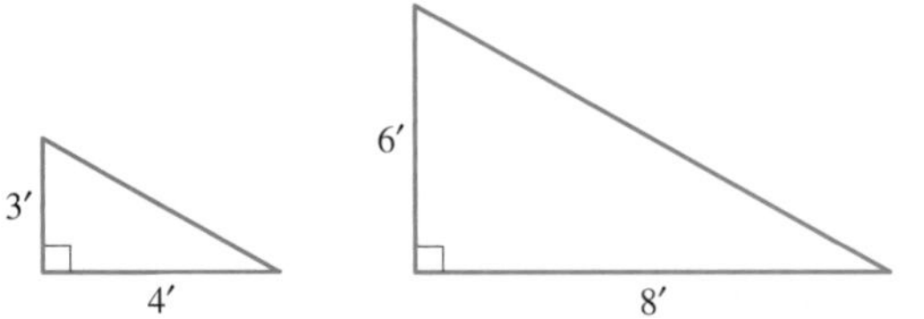

b. Find the ratios of base : base, height : height, and area : area for the following pair of similar triangles.

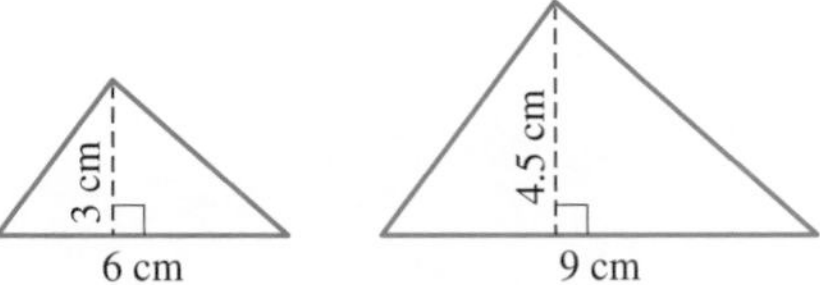

c. Do you see a pattern in the relationship between the ratios of the dimensions and the ratios of the areas in each pair of similar triangles? Test your conjecture by drawing several other pairs of similar figures and calculating these same ratios.

14. Compute the ratios requested in the following pairs of similar figures.

a. Find the ratios of base : base, height : height, and area : area for the following pair of similar rectangles.

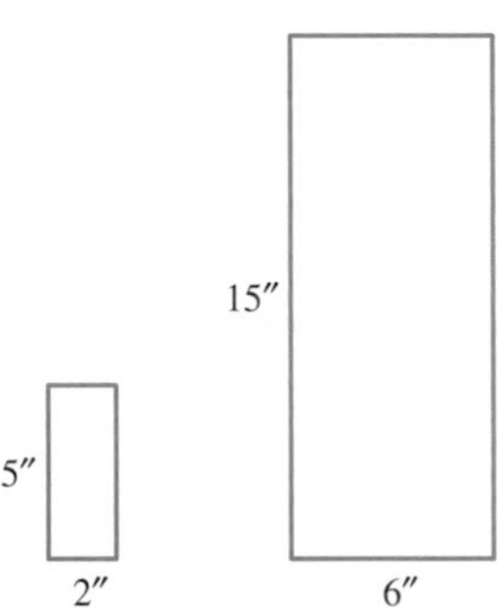

b. Find the ratios of base : base, height : height, and area : area for the following pair of similar rectangles.

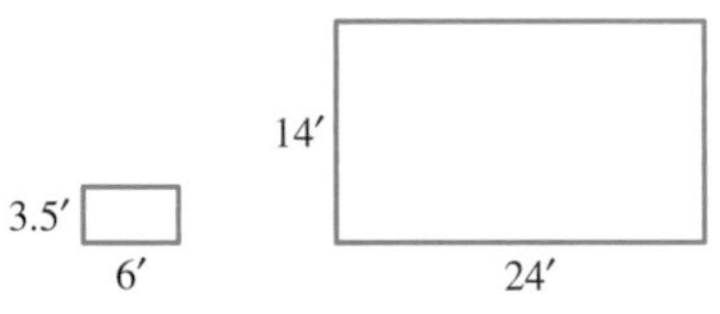

c. Do you see a pattern in the relationship between the ratios of the dimensions and the ratios of the areas in each pair of similar rectangles? Test your conjecture by drawing several other pairs of similar rectangles and calculating these same ratios.

PROOFS

15. Prove that all equilateral triangles are similar to each other.

16. Prove that all isosceles right triangles are similar to each other.

17. Prove that two isosceles triangles are similar if a base angle of one is congruent to a base angle of the other.

18. Prove that two isosceles triangles are similar if their vertex angles are congruent.

19. Let $ABCD$ be an isosceles trapezoid with diagonals intersecting in point E. Prove that $\triangle AED \sim \triangle CEB$.

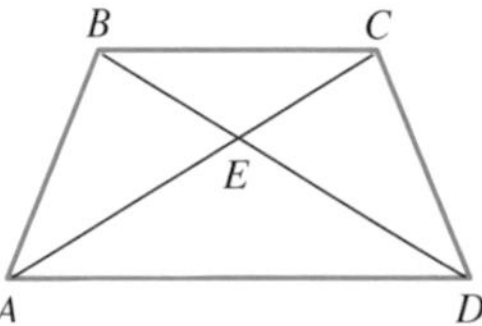

20. Prove that the diagonals of a trapezoid divide each other into proportional segments.

21. Prove that in similar triangles the ratio of the lengths of two corresponding medians is equal to the ratio of the lengths of any pair of corresponding sides.

22. Prove that in similar triangles, the ratio of the lengths of two corresponding altitudes is equal to the ratio of the lengths of their corresponding sides.

23. Prove that in similar triangles, the ratio of their areas is the square of the ratio of the lengths of their corresponding sides.

24. Prove that the perimeters of two similar triangles have the same ratio as the ratio of the lengths of any pair of corresponding sides.

25. The two right triangles shown are similar. Suppose that each triangle is rotated about a vertical axis, the dotted line in each case. Two right circular cones would be formed. Prove that the ratio of the volumes of these two cones is the cube of the ratio of the lengths of any two corresponding sides.

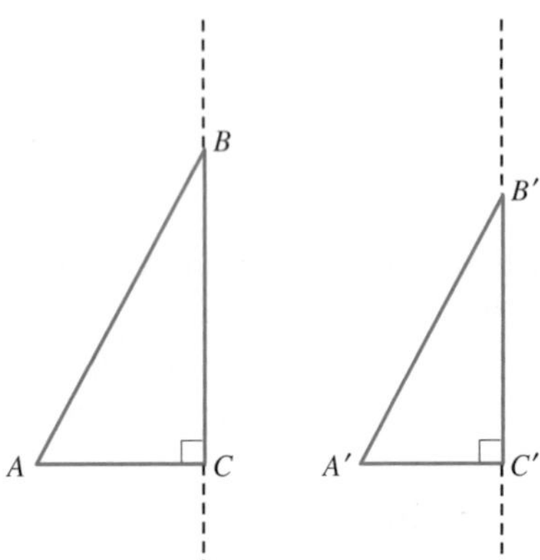

26. Given: $\angle A$ is supplementary to $\angle DEC$.

Prove: $\dfrac{BE}{AB} = \dfrac{BD}{BC}$.

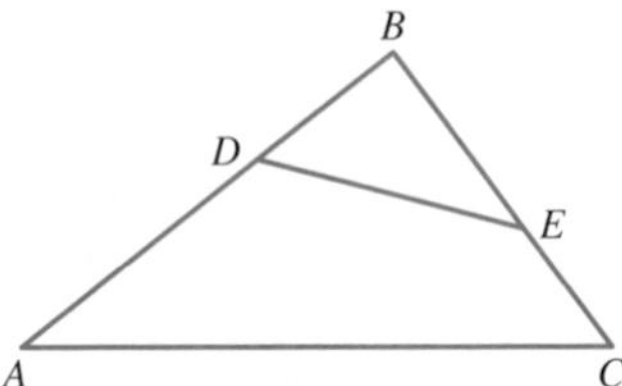

27. Prove that two triangles are similar if the lines containing their corresponding sides are perpendicular. That is, for the figure shown, prove that if $\overline{AB} \perp \overline{A'B'}$, $\overline{BC} \perp \overline{B'C'}$, and $\overline{AC} \perp \overline{A'C'}$ then $\triangle ABC \sim \triangle A'B'C'$.

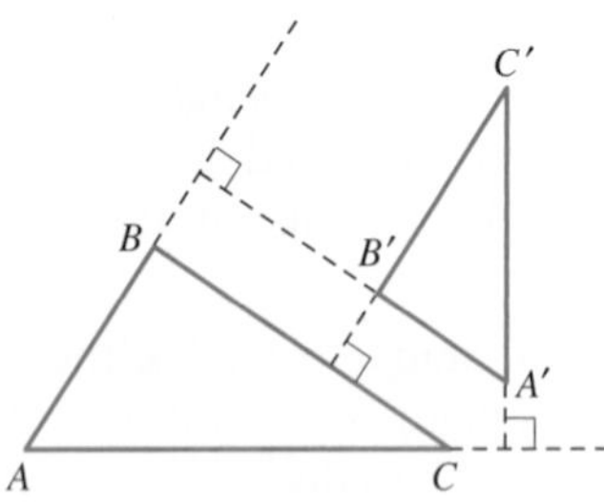

28. Show that $\triangle ABC$ is similar to $\triangle A'B'C'$ where x represents the measure of the angles.

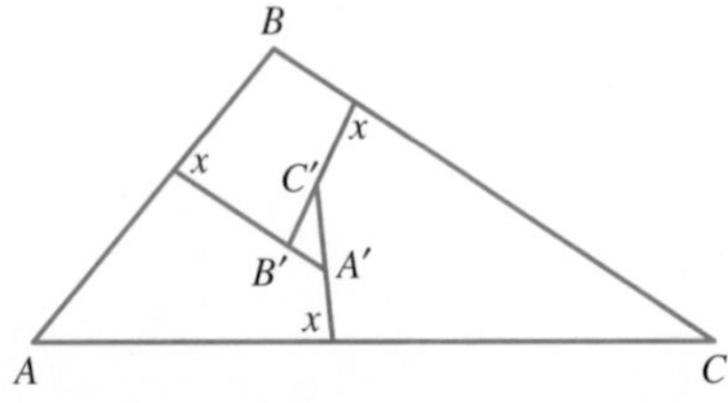

29. Prove that $\triangle ABC \sim \triangle A'B'C'$ if their corresponding sides are parallel to each other.

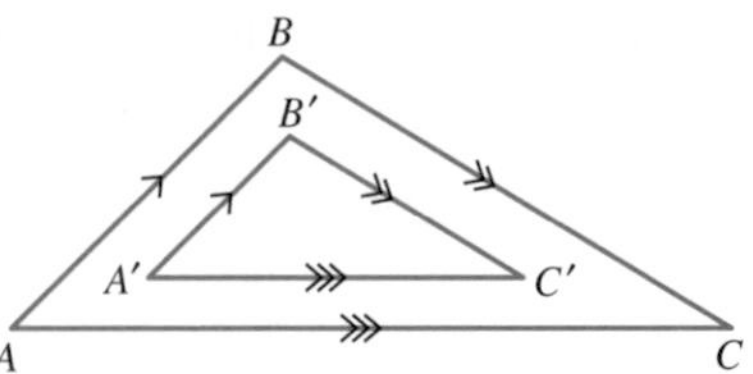

APPLICATIONS

30. Using the measurements given, find the distance, PQ, across the river.

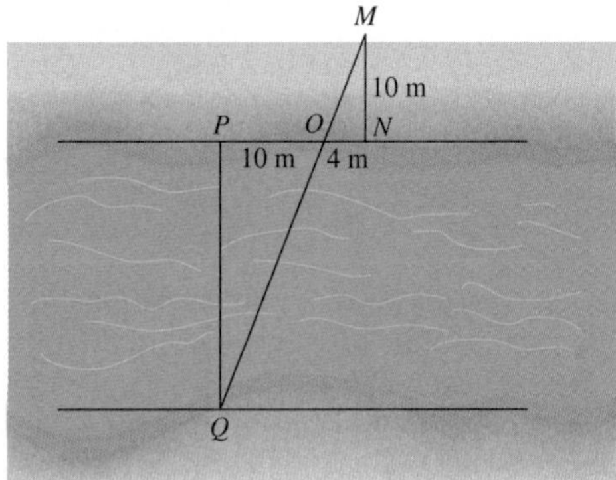

31. At a particular time, a tree casts a shadow 29 m long on horizontal ground. At the same time, a vertical pole 3 m high casts a shadow 4 m long. Calculate the height of the tree to the nearest meter.

32. A farmer wishes to measure the distance, EC, across a lake. He stands on dry ground at A and sights point B, which is in line with C on the other side of the lake. He then makes $\overline{DB}$ parallel to $\overline{EC}$. He finds that $EA = 105$ m, $AD = 45$ m, and $DB = 30$ m. How wide is the lake to the nearest meter?

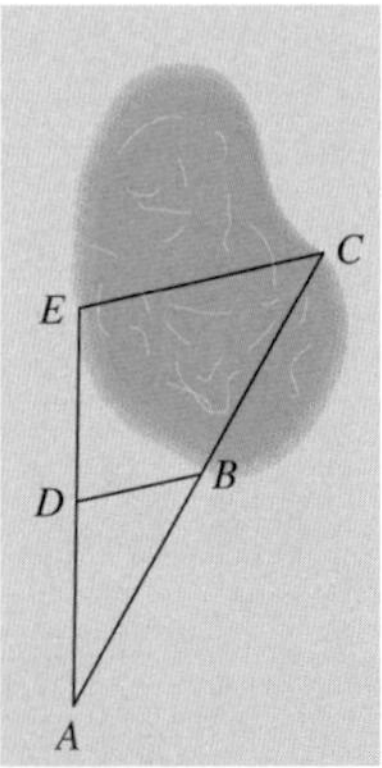

33. A folding workbench is constructed as shown and can be adjusted to various heights. If $AE = BE$ and $DE = CE$, explain why the workbench surface will always be parallel to the ground.

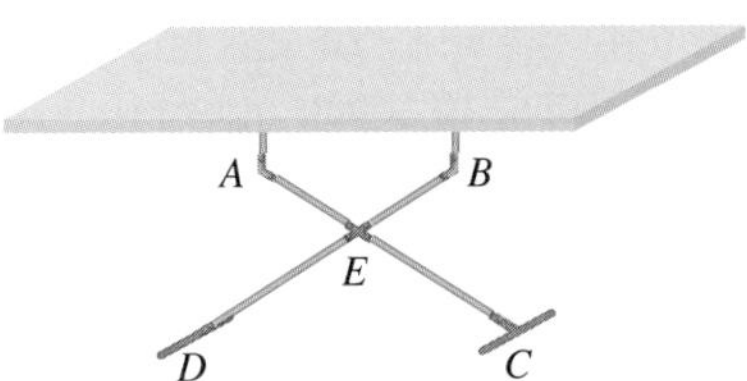

34. A boy and his friend wish to calculate the height of a flagpole. One boy holds a yardstick vertically at a point 40 feet from the base of the flagpole, as shown. The other boy backs away from the pole to a point where he sights the top of the pole over the top of the yardstick. If his position is 1 foot 9 inches from the yardstick and his eye level is 2 feet above the ground, find the height of the flagpole. Round to the nearest hundreth.

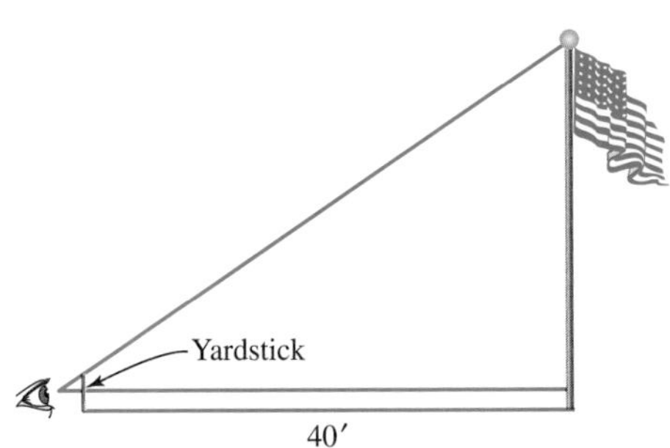

35. A new department store opens on a hot summer day. Kathy's job is to stand outside the main entrance and hand out a special supplement to incoming customers. The building is 30 feet tall and casts a shadow 8 feet wide in front of the store. If Kathy is 5 feet 6 inches tall, how far away from the building can she stand and still be in the shade? Round to the nearest hundredth.

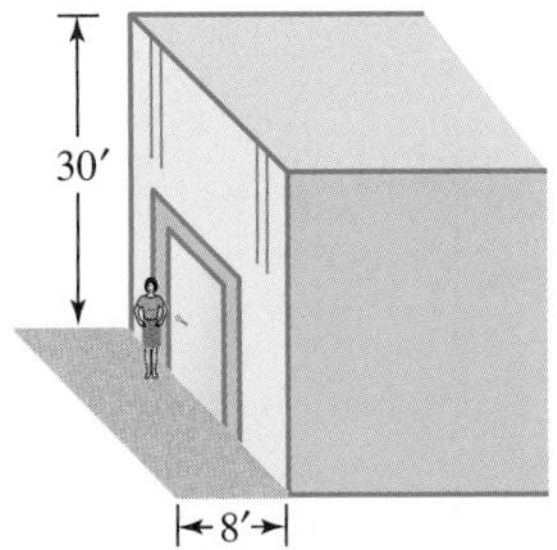

36. A water tank in the shape of an inverted cone has a radius of 5 m and a height of 15 m. What is the volume of water in the tank when the water is 6 m deep? Round to the nearest whole number.

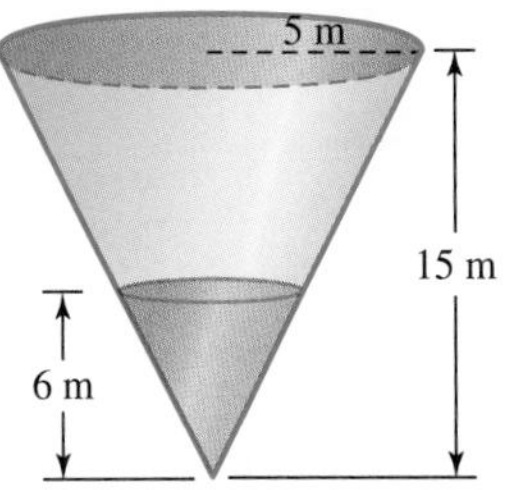

37. A paper cup is in the shape of a right circular cone with a diameter of 8 cm and a height of 12 cm.

a. Suppose the cup is filled with water to a depth of 8 cm. Calculate the volume of water in the cup. Round to the nearest whole number. (HINT: Use similar triangles.)

b. To what depth must the cup be filled in order to be half full of water? Round to the nearest tenth.

38. Sawdust is dropping from a conveyor belt onto a pile in the approximate shape of a right circular cone. At one time the height of the pile is 5 feet and the circumference of the base is 25 feet. Later the circumference of the base of the pile is measured as 62 feet. What is the height of the sawdust pile at the latter time and how many cubic feet of sawdust are in the pile? (HINT: Use similar triangles.) Round to the nearest tenth.

39. Tom and Carol are playing a shadow game. Tom is 6 feet tall and Carol is 5 feet tall. If Carol stands at the "shadow top" of Tom's head their combined shadows total 15 feet. How long is each shadow?

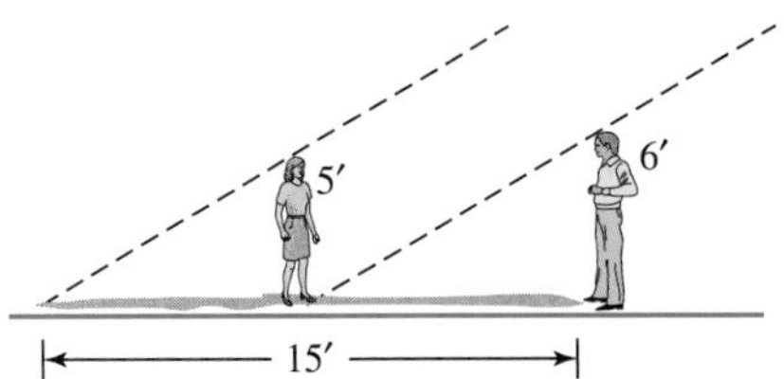

40. Another method of determining the height of an object is illustrated next. A pan of water is located at point B. Name the two similar triangles in the diagram and explain why they are similar. Find the height of the tree if a person 1.5 m tall sees the top of the tree when the pan of water is 18 m from the base of the tree and the person is 1 m from the pan of water.

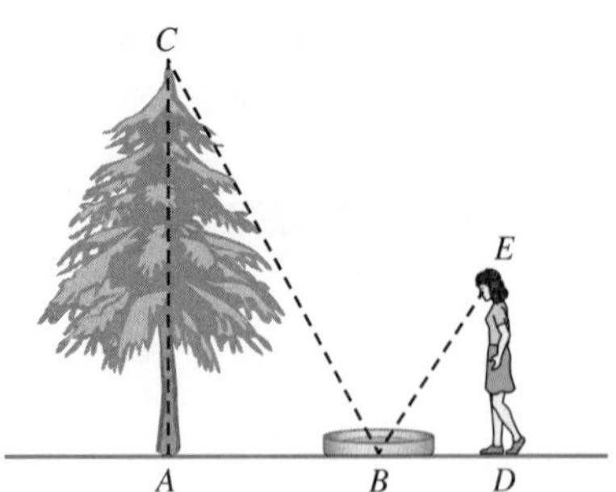

41. The distance from the lens to the mm in a 35 mm camera is 35 mm. The height of the film negative is 24 mm.

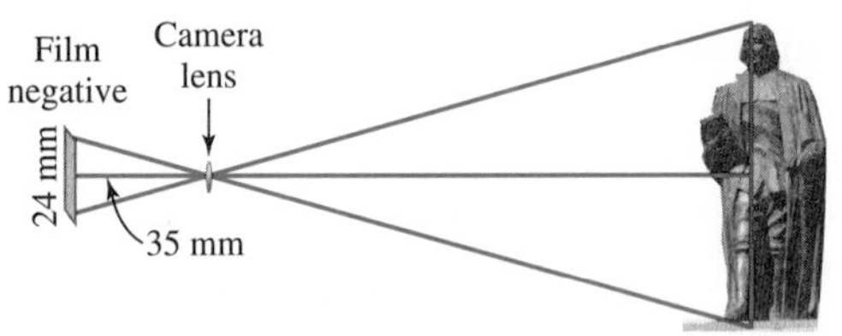

a. If you would like to photograph a 6-foot-tall statue so the height fills the negative, how far away from the statue must the camera be?

b. When the photo is printed, the top and bottom 0.5 mm of the image from the negative is not printed. How much of the head will be cropped from the print?

EXTENDED PROBLEMS

42. The Pythagorean theorem is often interpreted in terms of areas of squares as shown in the following figure.

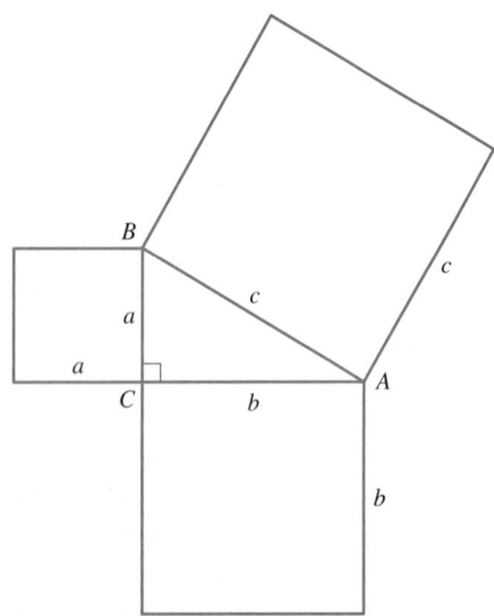

The equation $a^2 + b^2 = c^2$ means the sum of the areas of the squares drawn on the legs of the right triangle equals the area of the square drawn on the hypotenuse.

In fact, we can generalize the theorem to mean that if similar figures are drawn on each side of the right triangle, the sum of the areas of the two smaller figures equals the area of the larger figure. Verify that this is the case for each figure that follows.

a.

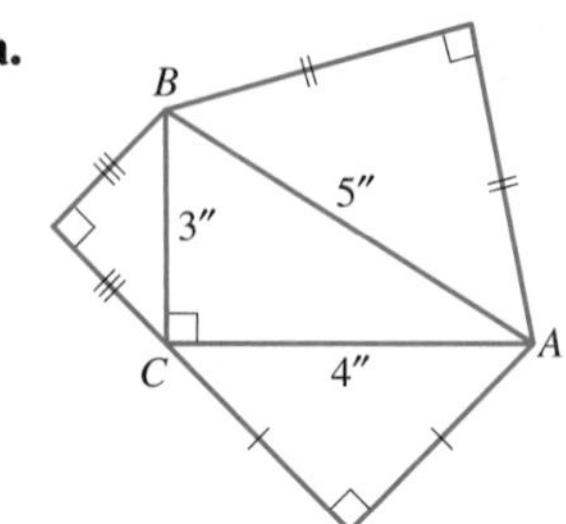

b.

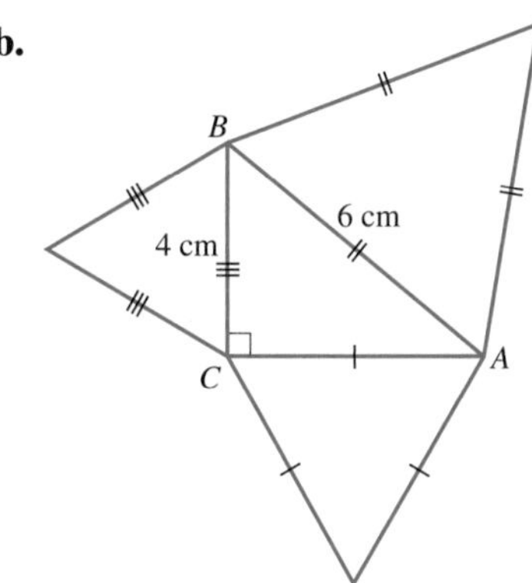

c.

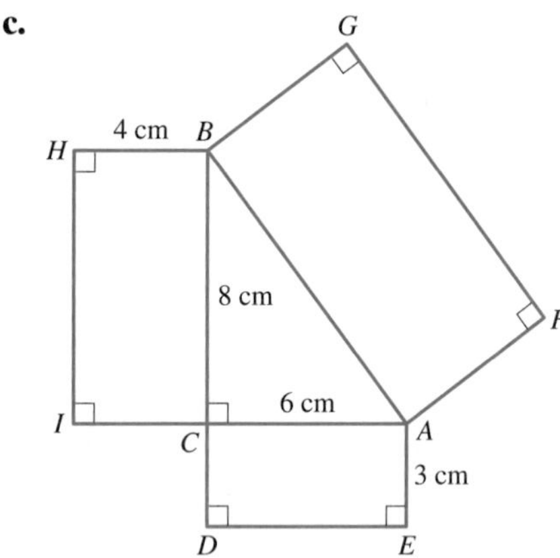

d.

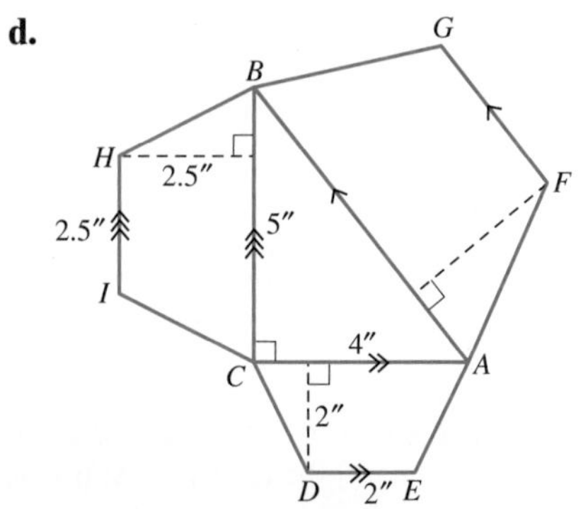

e. Create an example of this generalization by drawing similar polygons on each side of the right triangle. Choose a polygon with more than four sides and show the sum of the areas of the two smaller polygons is equal to the area of the larger polygon.

43. The ancient Greeks used shadows and similar triangles to estimate heights. The Greek mathematician Thales used this technique to estimate the height of the Great Pyramid. Research how Thales used similar triangles and shadows to obtain his estimate. Draw and label a sketch that shows the similar triangles Thales used. How close was Thales' estimate of the height of the Great Pyramid?

44. a. The following sequence of shapes composed of equilateral triangles produces a shape called the **Sierpinski triangle** if continued indefinitely.

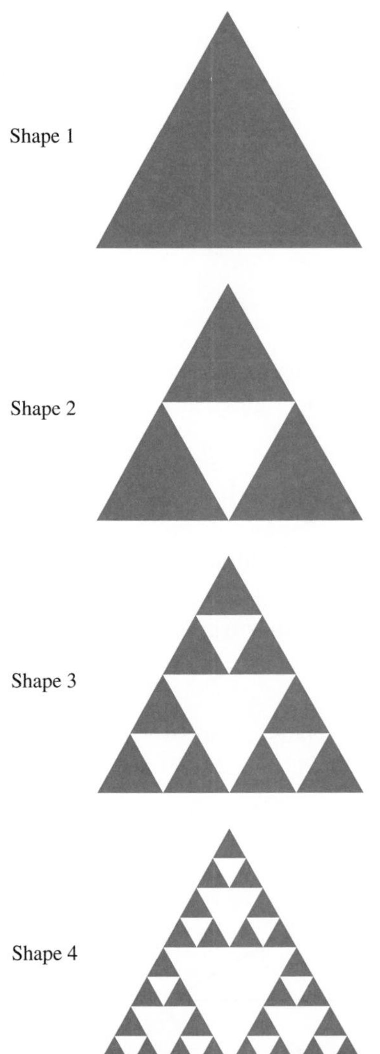

Each white triangle has as its vertices the midpoints of the sides of a larger blue triangle. Notice how a portion of the curve at one stage is similar to a portion of the curve at another stage. Suppose the first equilateral triangle has sides of length 1. Find the total perimeter and area of the blue triangles in each of the first four Sierpinski triangles. What are the total perimeter and total area of the blue triangles in the nth shape?

b. The following sequence of shapes composed of equilateral triangles produces a shape called the Koch snowflake if continued indefinitely.

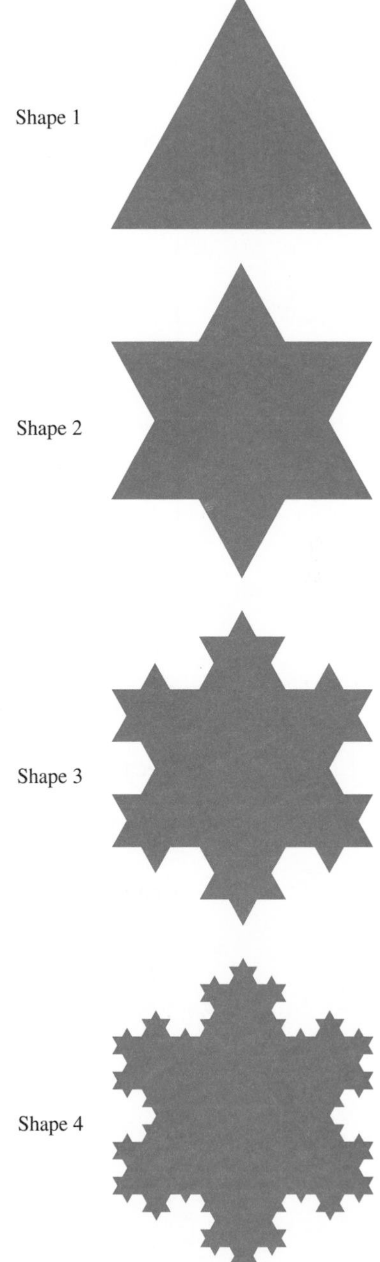

The sequence begins with an equilateral triangle. Next, equilateral triangles are constructed on the middle third of each side of the original triangle. Notice how a portion of the curve at one stage is similar to a portion of the curve at another stage. Suppose the first equilateral triangle has sides of length 1. Find the total perimeter and area of each of the first four shapes. Research the Koch snowflake. What happens to the perimeter and the area if the process is continued indefinitely? What are the perimeter and area of the nth shape?

c. Nature is full of shapes that resemble themselves at different scales. This quality is called **self-similarity**. Although natural snowflakes are incredibly complex, we can use fractal snowflakes to model natural snowflakes due to their self-similarity. Research fractals and self-similarity. Summarize your findings and list five other natural objects that we can model with fractals.

6.3 APPLICATIONS OF SIMILARITY

Applied Problem

A plastic kite without its cross-pieces may be ordered from a mail-order catalog and must be packed and mailed in a rectangular envelope. How can the kite shown be easily folded to fit inside a standard 10″ by 13″ mailer?

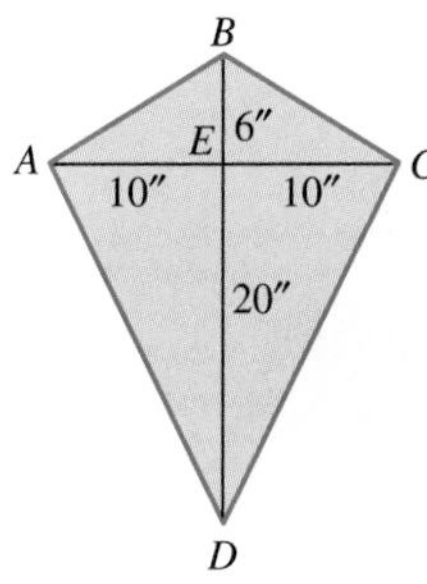

In this section, we will use similar triangles to derive some properties of geometric figures. We will be applying all of the postulates and theorems about similarity that we developed in the last section.

Mean Proportional in a Right Triangle

Recall that b is the mean proportional between a and c if $\frac{a}{b} = \frac{b}{c}$.

EXAMPLE 6.12 Name three pairs of similar triangles in Figure 6.10.

SOLUTION

$\overline{QS}$ divides $\triangle PQR$ into two $30°-60°$ right triangles. So by AA Similarity, we have $\triangle PQR \sim \triangle PSQ$, $\triangle PQR \sim \triangle QSR$, and $\triangle PSQ \sim \triangle QSR$.

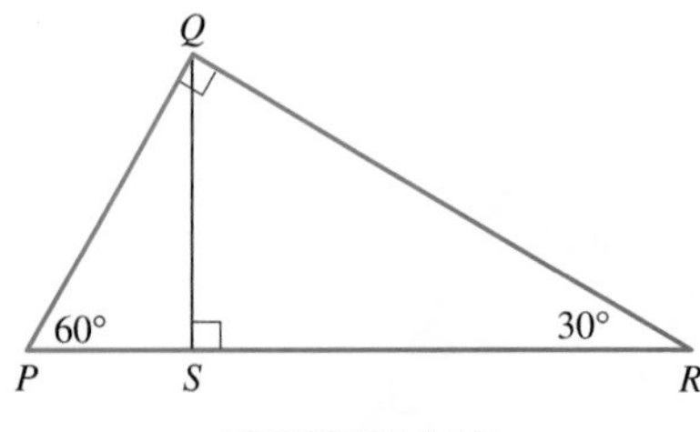

FIGURE 6.10

■

The fact that the two smaller triangles are similar to each other leads to the following result about the altitude to the hypotenuse in a right triangle.

THEOREM 6.9

In a right triangle, the altitude to the hypotenuse is the mean proportional between the two segments formed by the altitude on the hypotenuse.

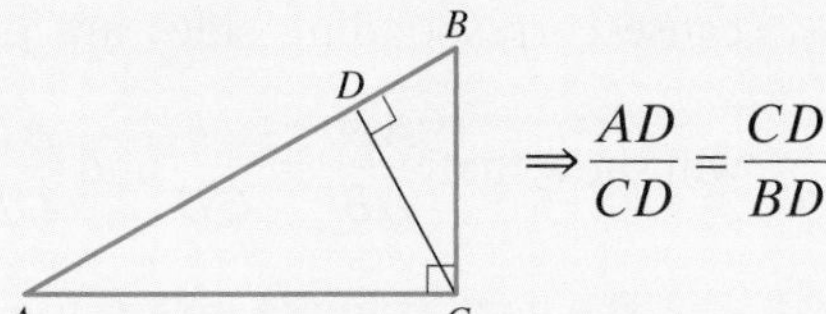

$$\Rightarrow \frac{AD}{CD} = \frac{CD}{BD}$$

In right triangle $\triangle ABC$, if $\overline{CD}$ is the altitude to $\overline{AB}$, then $\dfrac{AD}{CD} = \dfrac{CD}{BD}$.

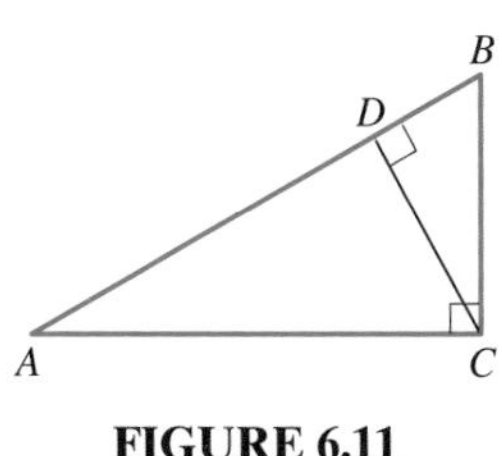

FIGURE 6.11

Given Right $\triangle ABC$ with $\overline{CD}$ perpendicular to $\overline{AB}$ (Figure 6.11)

Prove $\dfrac{AD}{CD} = \dfrac{CD}{BD}$

Proof Because $\angle A \cong \angle A$ and $\angle ADC \cong \angle ACB = 90°$, we can conclude that $\triangle ADC \sim \triangle ACB$ by AA Similarity. Also, $\triangle ACB \sim \triangle CDB$ by AA Similarity because $\angle ACB \cong \angle CDB$ and $\angle B \cong \angle B$. Thus, we have $\triangle ADC \sim \triangle ACB$ and $\triangle ACB \sim \triangle CDB$, so, by the transitive property, we have that $\triangle ADC \sim \triangle CDB$. Because corresponding sides of $\triangle ADC$ and $\triangle CDB$ are proportional, it follows that $\dfrac{AD}{CD} = \dfrac{CD}{BD}$. ■

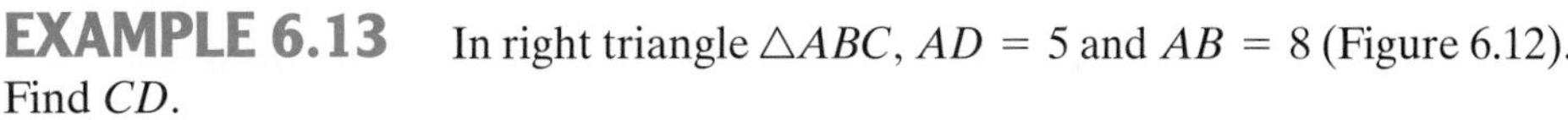

EXAMPLE 6.13 In right triangle $\triangle ABC$, $AD = 5$ and $AB = 8$ (Figure 6.12). Find CD.

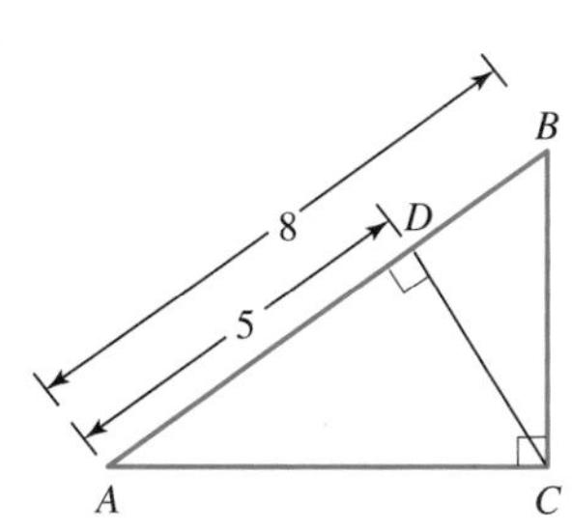

FIGURE 6.12

SOLUTION

First, $BD = AB - AD = 8 - 5 = 3$. Next, by Theorem 6.9, $\dfrac{AD}{CD} = \dfrac{CD}{BD}$, so $\dfrac{5}{CD} = \dfrac{CD}{3}$. Thus, $(CD)^2 = 15$, or $CD = \sqrt{15}$. ■

Side Splitting Theorem

We can prove the next theorem, which is a generalization of Example 6.9 in Section 6.2, using AA Similarity.

THEOREM 6.10 Side Splitting Theorem

A line parallel to one side of a triangle forms a triangle similar to the original triangle and divides the other two sides of the triangle into proportional corresponding segments.

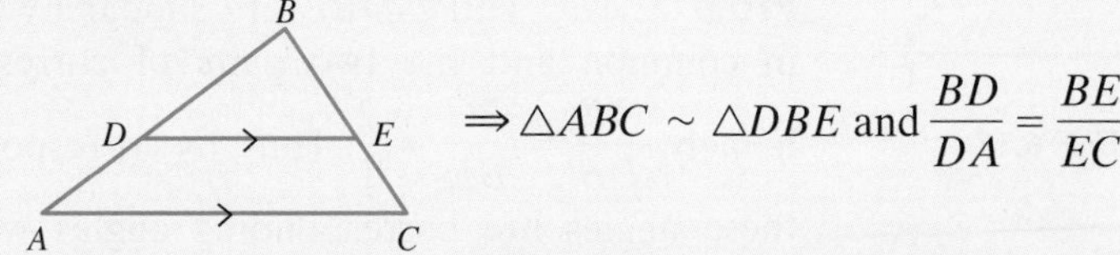

$$\Rightarrow \triangle ABC \sim \triangle DBE \text{ and } \frac{BD}{DA} = \frac{BE}{EC}$$

In $\triangle ABC$, if $\overline{DE} \parallel \overline{AC}$, then $\triangle ABC \sim \triangle DBE$ and $\dfrac{BD}{DA} = \dfrac{BE}{EC}$.

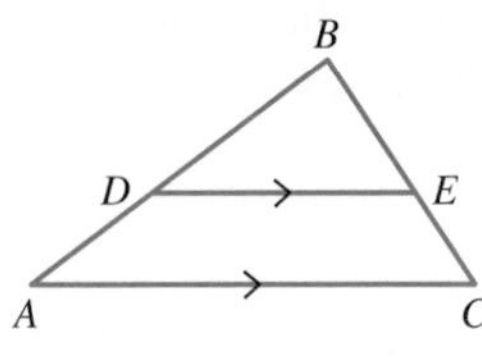

FIGURE 6.13

Given $\triangle ABC$ with $\overline{DE}$ parallel to $\overline{AC}$ (Figure 6.13)

Prove $\triangle ABC \sim \triangle DBE$ and $\dfrac{BD}{BA} = \dfrac{BE}{BC}$

Proof $\angle BDE \cong \angle BAC$ and $\angle BED \cong \angle BCA$ because parallel lines form congruent corresponding angles. Therefore, $\triangle ABC \sim \triangle DBE$ by AA Similarity. Because corresponding sides are proportional, $\dfrac{BD}{BA} = \dfrac{BE}{BC}$. In the problem set, you will show that $\dfrac{AD}{AB} = \dfrac{CE}{CB}$ and $\dfrac{AD}{DB} = \dfrac{CE}{EB}$. ■

The next example makes use of the Side Splitting Theorem.

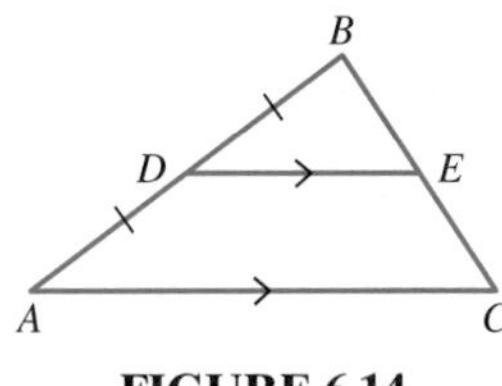

FIGURE 6.14

EXAMPLE 6.14 In $\triangle ABC$, $\overline{DE}$ is parallel to $\overline{AC}$ and D is the midpoint of $\overline{AB}$ (Figure 6.14). Show that E is the midpoint of $\overline{CB}$.

SOLUTION

By the Side Splitting Theorem, $\dfrac{BD}{AD} = \dfrac{BE}{CE}$. Because $\dfrac{BD}{AD} = 1$, it follows that $\dfrac{EB}{CE} = 1$, so $EB = CE$ and E is the midpoint of $\overline{CB}$.

The Midsegment Theorem

In a triangle, a line segment that joins the midpoints of two sides of the triangle is called a **midsegment** of the triangle. The next theorem shows two properties of midsegments.

> **THEOREM 6.11 Midsegment Theorem**
>
> A midsegment of a triangle is parallel to the third side and is half its length.
>
> $\Rightarrow \overline{DE} \parallel \overline{AC}$ and $DE = \dfrac{1}{2}AC$
>
> In $\triangle ABC$, if $AD = DB$ and $BE = EC$, then $\overline{DE} \parallel \overline{AC}$ and $DE = \dfrac{1}{2}AC$.

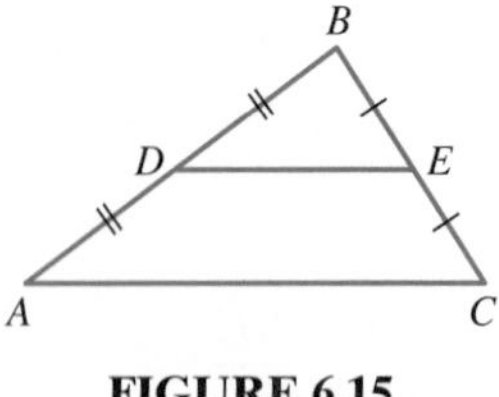

FIGURE 6.15

Given $\triangle ABC$ with midsegment $\overline{DE}$ (Figure 6.15)

Prove $\overline{DE} \parallel \overline{AC}$ and $DE = \dfrac{1}{2}AC$

Proof $\triangle ABC$ and $\triangle DBE$ are similar by SAS Similarity because they have $\angle B$ in common and the two pairs of corresponding sides have a common ratio, namely, $\dfrac{AB}{DB} = \dfrac{BC}{BE} = \dfrac{2}{1}$. Thus, by corresponding parts, $\angle BAC \cong \angle BDE$. Because these angles are corresponding angles with respect to $\overline{DE}$ and $\overline{AC}$, we have $\overline{DE} \parallel \overline{AC}$. Also, the ratio of DE to AC is 1:2, so $DE = \dfrac{1}{2}AC$. ■

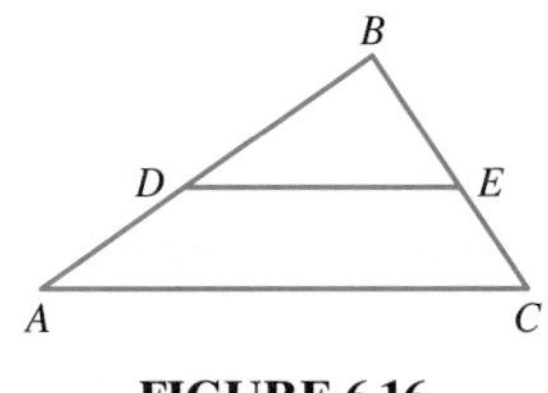

FIGURE 6.16

Although we proved the Midsegment Theorem using similarity, we could also prove it using triangle congruence results. However, the congruence proof is more complex. Not only is the similarity proof simpler, but also, it can be generalized to the situation where D and E are not necessarily midpoints of $\overline{AB}$ and $\overline{BC}$, respectively. In Figure 6.16, if $\frac{BD}{BA} = \frac{BE}{BC}$, then $DE \parallel AC$ and $\frac{DE}{AC} = \frac{BD}{BA}$. For example, if $\frac{BD}{BA} = \frac{BE}{BC} = \frac{1}{3}$, then $\overline{DE} \parallel \overline{AC}$ and $DE = \frac{1}{3}AC$. The proof of this result is analogous to the proof of the Midsegment Theorem.

Now suppose that $ABCD$ is *any* quadrilateral and E, F, G, H are midpoints of its sides. [Figure 6.17(a)]. When we draw $\overline{BD}, \overline{EH}$, and $\overline{FG}$, $\triangle ABD$ and $\triangle CBD$ have midsegments $\overline{EH}$ and $\overline{FG}$, respectively [Figure 6.17(b)].

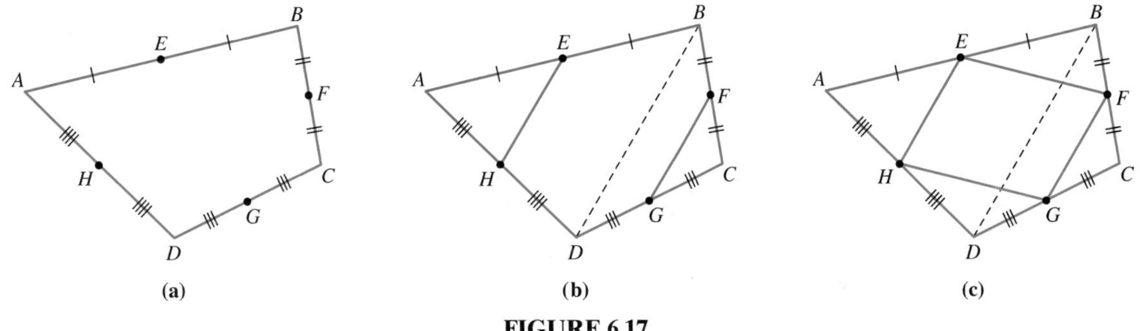

FIGURE 6.17

Using the Midsegment Theorem, we find that $\overline{EH} \parallel \overline{BD}$ and $\overline{FG} \parallel \overline{BD}$, so $\overline{EH} \parallel \overline{FG}$. Also, $EH = FG$ because they are each equal to $\frac{1}{2}BD$ in Figure 6.17(c). Therefore, by the Midsegment Theorem, $EFGH$ is a parallelogram, because it is a quadrilateral with one pair of opposite sides parallel and congruent. Because $EFGH$ is a quadrilateral joining the midpoints of the sides of $ABCD$, we call it the **midquad** of $ABCD$. Thus, we have the following surprising result.

COROLLARY 6.12 Midquad Theorem

If $ABCD$ is any quadrilateral and E, F, G, H are midpoints as shown, then $EFGH$ is a parallelogram.

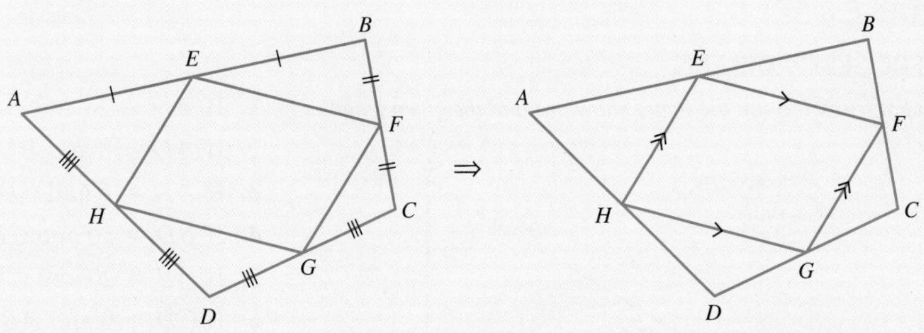

In $ABCD$, if $AE = BE$, $BF = CF$, $CG = DG$, and $AH = DH$, then $EFGH$ is a parallelogram.

Solution of Applied Problem

If the midpoints of the sides are joined as shown, a parallelogram is formed, by the Midquad Theorem. Since the diagonals of a kite are perpendicular, the sides of the parallelogram are perpendicular. Thus, a rectangle of 10″ by 13″ is formed. Furthermore, if the kite is folded on the dashed lines, each triangle outside the rectangle folds exactly onto the rectangle where points A, B, C, and D meet at point E.

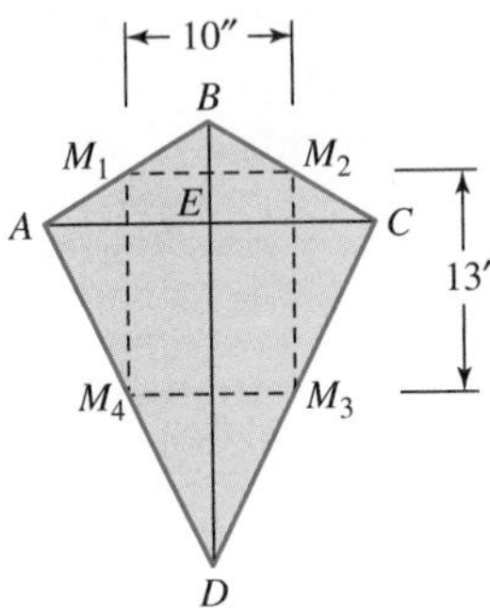

GEOMETRY AROUND US

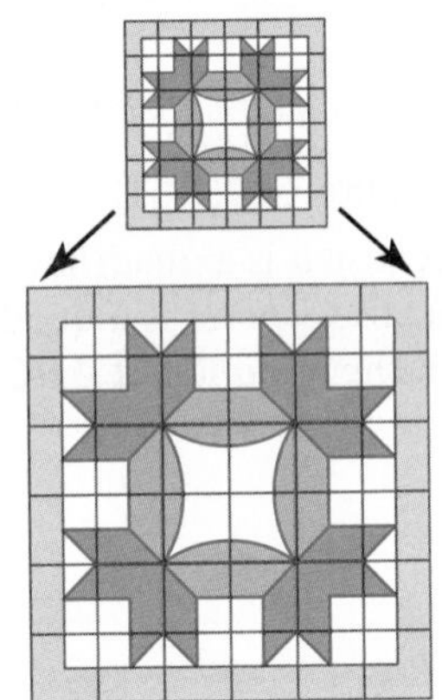

Patterns for quilting, crafts, woodworking, and so on must frequently be enlarged from the size printed in a book or instruction sheet. The larger figure must be similar to the original figure. Such an enlargement is often made using square grids. The pattern is printed on a grid of small squares, and the larger figure is made by duplicating the figure, square by square, on a grid of larger squares.

PROBLEM SET 6.3

EXERCISES/PROBLEMS

For Exercises 1–6 use the figure shown to find the given length.

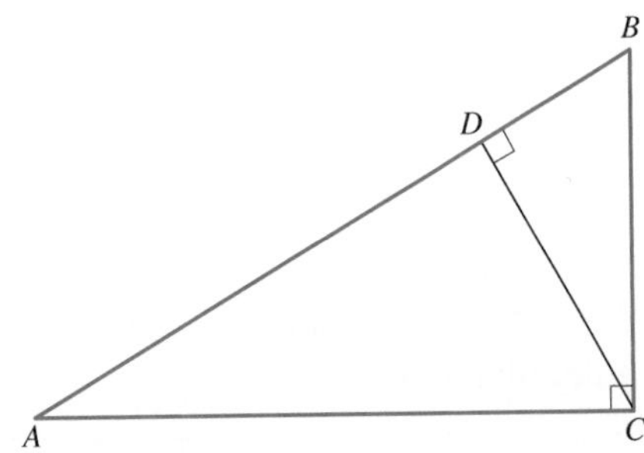

1. If $AD = 24$ and $BD = 9$, find CD.
2. If $AD = 8$ and $BD = 4$, find CD.
3. If $AD = 6$ and $CD = 4$, find BD.
4. If $BD = 10$ and $CD = 15$, find AD.
5. If $AB = 20$ and $CD = 6$, find BD.
6. If $CD = 9$ and $AB = 30$, find AD.

For Exercises 7–12, find the given length if $\overline{BC} \parallel \overline{DE}$.

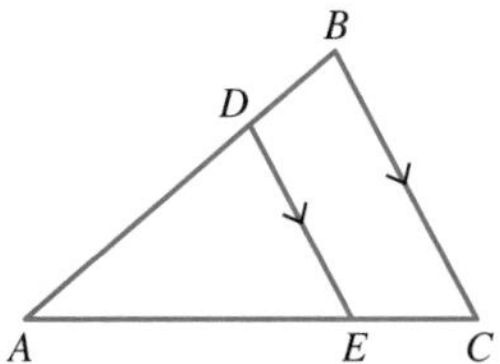

7. If $AE = 3$, $AB = 5$, and $AD = 4$, find AC.

8. If $AD = 5$, $AB = 8$, and $AC = 7$, find AE.

9. If $BD = 3$, $EC = 4$, and $AC = 9$, find AD.

10. If $AD = EC$, $DB = 4$, and $AE = 9$, find AD.

11. If $DE = 5$, $BC = 7$, and $AD = 4$, find AB.

12. If $ED = 7$, $BC = 10$, and $AE = 2 + EC$, find AE.

For Exercises 13–16, use the figure shown, where parallel lines are marked.

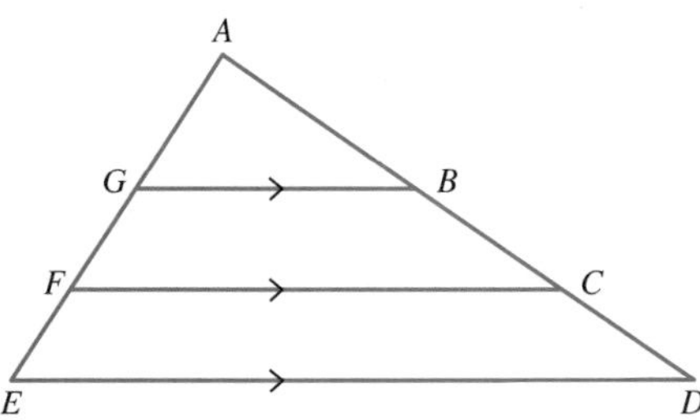

13. If $AB = 4$, $CD = 3$, $AG = 3$, and $GF = 2$, find EF and BC.

14. If $BG = 5$, $DE = 15$, and $AD = 12$, find BD.

15. If $AB = x$, $AG = x - 2$, $EG = 7$, and $BD = 10.5$, find AB and AG.

16. If $BG = x$, $AB = x + 2$, $BC = x + 3$, $DE = 18$, and $CD = x - 1$, find BG and AD.

17. In $\triangle PQR$, M and N are midpoints of $\overline{PQ}$ and $\overline{QR}$, respectively. In addition, $PQ = 20$ cm and $PR = 12$ cm. Find MN in two different ways.

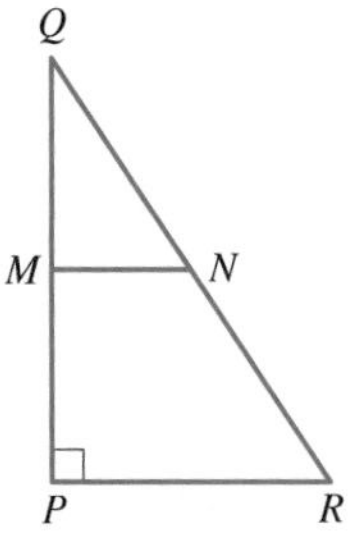

18. In rectangle $ABCD$, F and I are the midpoints of $\overline{AB}$ and $\overline{CD}$, respectively. Segments $\overline{BE}$, $\overline{FI}$, and $\overline{CE}$ are drawn. Congruent segments are indicated. If $BG = 10$, $CH = 15$, and $ID = 8$, find GH.

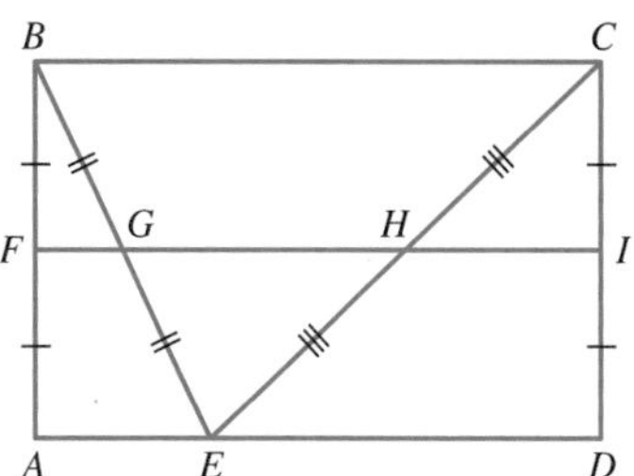

19. In trapezoid $ABCD$, M and N are midpoints of $\overline{AB}$ and $\overline{CD}$, respectively. If $MN = 15$ cm and $BE = 7$ cm, find the area of $ABCD$.

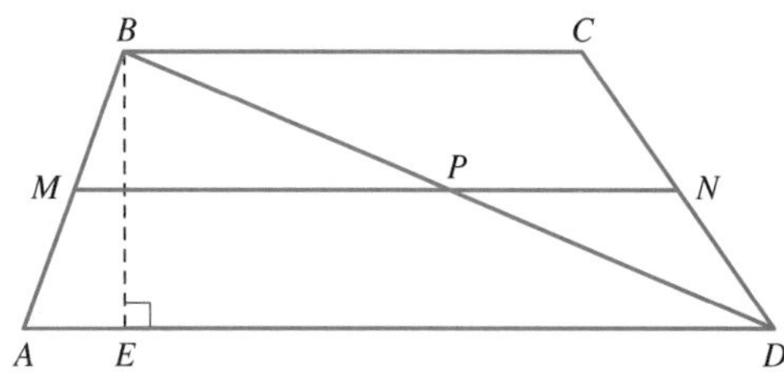

20. In parallelogram $PQRS$, points A, B, C, and D are midpoints of the sides. If the area of $\triangle AQB$ is 5 in^2, find the area of hexagon $PABRCD$.

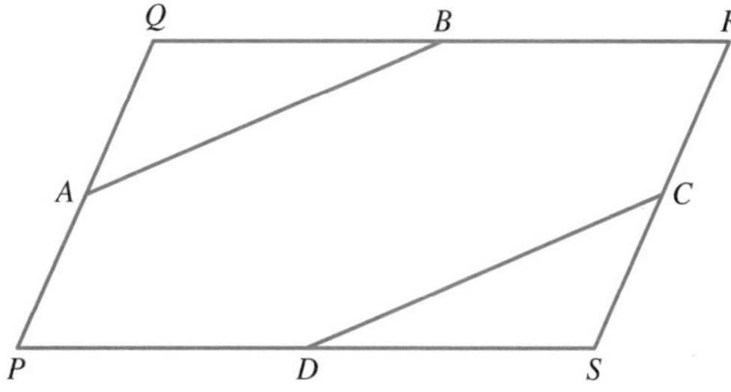

21. In the figure shown, $\angle A \cong \angle C$, M and N are midpoints of sides $\overline{AB}$ and $\overline{BC}$, and $MNPQ$ is a rectangle. Show that $\triangle MQA \cong \triangle NPC$.

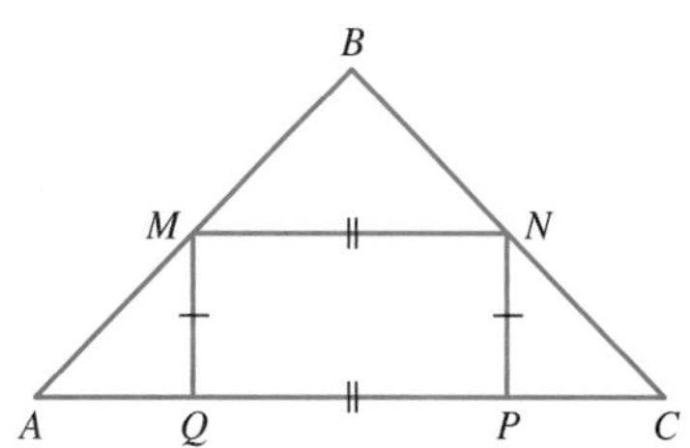

Find the area of $MNPQ$, where $MN = 14$ m and $BC = 22$ m. Round to the nearest hundredth.

22. In equilateral $\triangle STU$, $ST = 10$ and V and W are midpoints of $\overline{ST}$ and $\overline{TU}$, respectively. What is the area of $\triangle TVW$ if $\overline{TX} \perp \overline{SU}$?

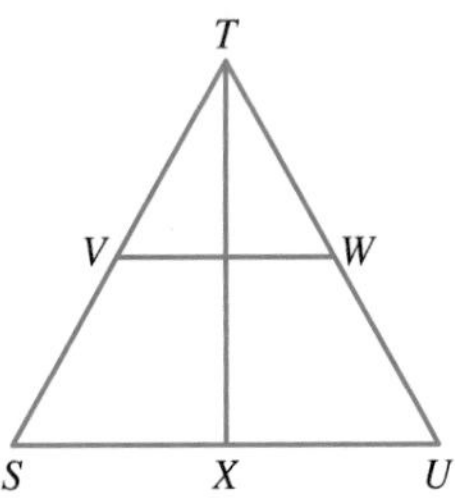

PROOFS

23. Explain how the figure shown can be used to find $\sqrt{a}$ for a given length a. (HINT: Find x.)

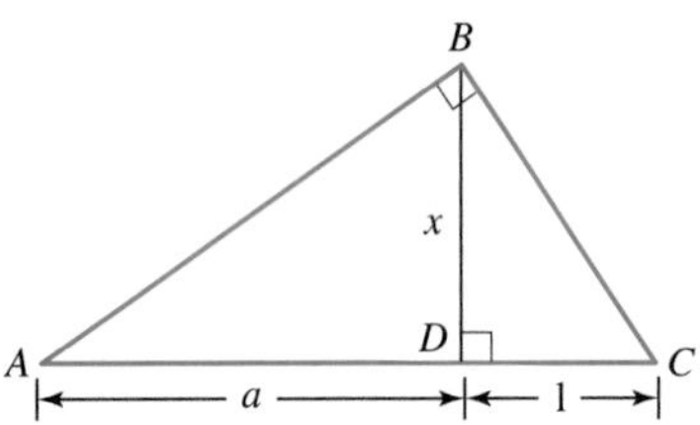

24. In $\triangle ABC$, prove that $\dfrac{AD}{AC} = \dfrac{AC}{AB}$.

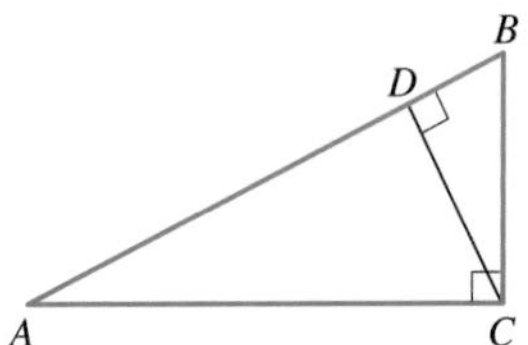

25. In $\triangle ABC$, $\overline{DE} \parallel \overline{AC}$. Prove that $\dfrac{AD}{DB} = \dfrac{CE}{EB}$.

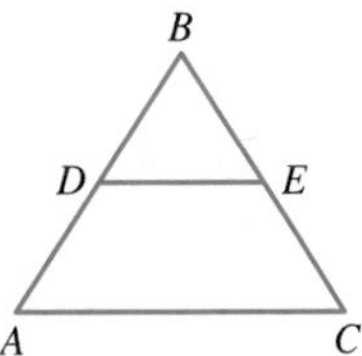

26. In $\triangle ABC$, $\overline{DE} \parallel \overline{AC}$. Prove that $\dfrac{AD}{AB} = \dfrac{CE}{CB}$.

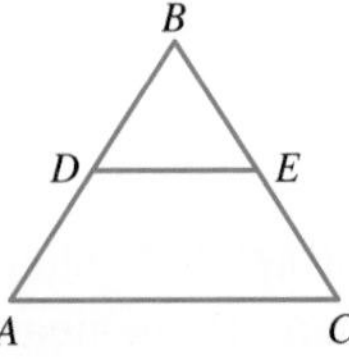

27. Prove that the midpoint of the hypotenuse of a right triangle is equidistant from its vertices.

28. Given: $(AC)^2 = AB \cdot AD$

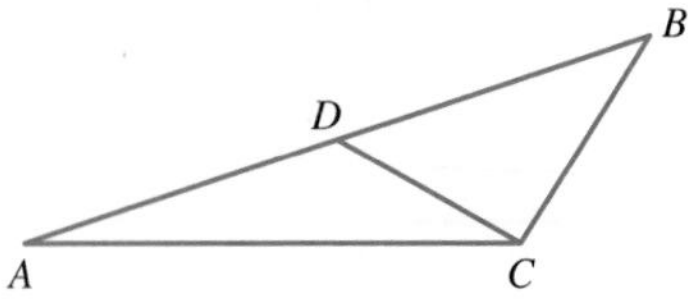

Prove: $\triangle ABC$ is similar to $\triangle ACD$.

29. $\overline{BE}$ bisects $\angle ABC$ and $\dfrac{AB}{BE} = \dfrac{BD}{BC}$. Prove that $\angle A = \angle D$.

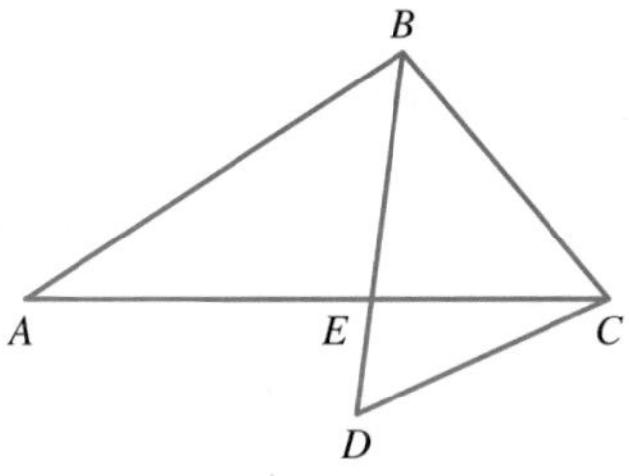

30. Prove that if three parallel lines are cut by intersecting transversals, then the corresponding segments of the transversals are proportional. That is, show that $AD : DE : EG = BC : CE : EF$.

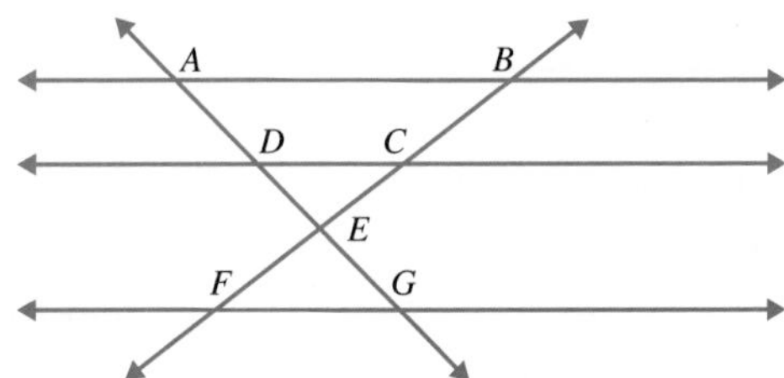

31. Using Theorem 6.9 and the following figure, prove the Pythagorean theorem.

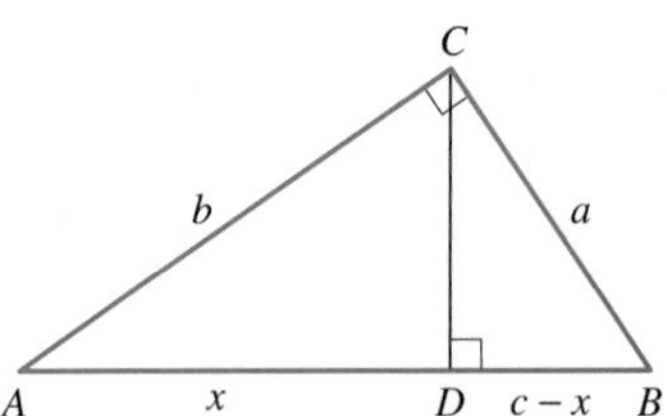

32. In $\triangle ABC$, $\dfrac{DB}{AB} = \dfrac{EB}{CB}$. Prove that $\overline{DE}$ is parallel to $\overline{AC}$.

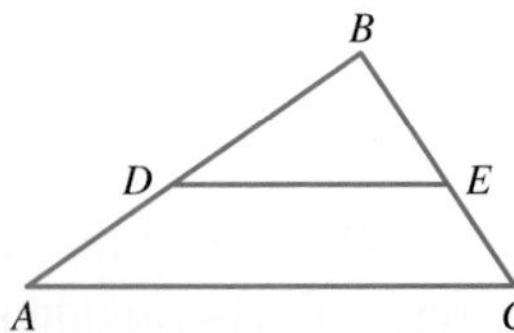

33. Prove that the midsegment of an equilateral triangle forms another equilateral triangle.

34. Prove that the quadrilateral formed by joining the midpoints of adjacent sides of a rectangle is a rhombus.

35. Prove: If the midpoints of the sides of a rhombus are connected consecutively, the figure formed is a rectangle.

36. Prove: If the midpoints of the sides of a kite are connected consecutively, the figure formed is a rectangle.

37. Prove: If the midpoints of the sides of an isosceles trapezoid are connected consecutively, the figure formed is a rhombus.

38. In the next figure, M is the midpoint of $\overline{AB}$ and N is the midpoint of $\overline{BC}$. If $MN = NP$, prove that $AMPC$ is a parallelogram.

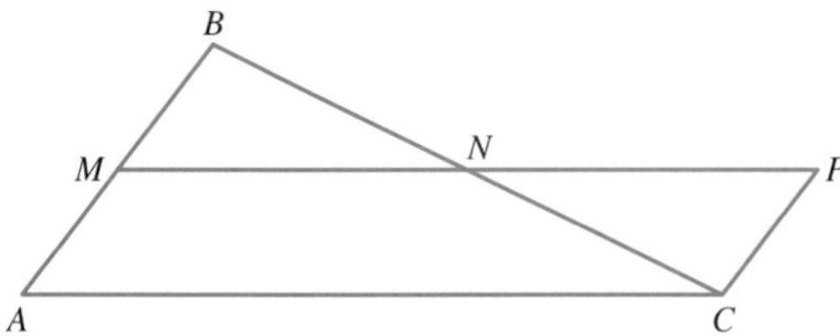

39. In $\triangle ABC$, points D, E, and F are midpoints. Prove that the four smaller triangles are congruent.

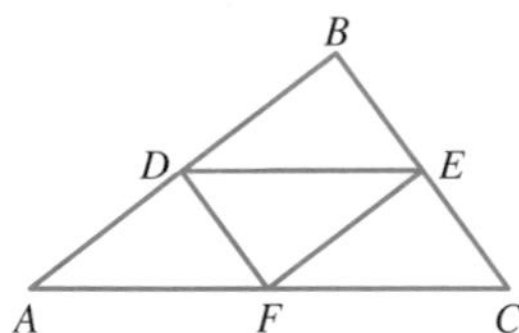

APPLICATIONS

40. A quilt piece looks like the figure shown, where $\triangle ABC$ is an equilateral triangle and points M, N, O, P, Q, and R are midpoints.

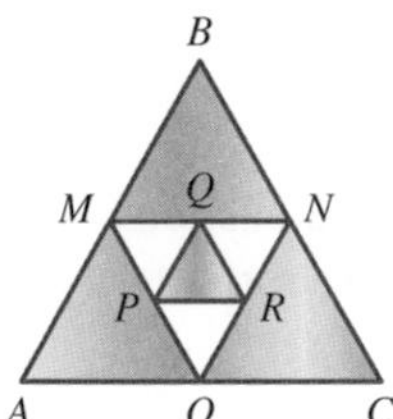

a. Which triangles are congruent and why?

b. How does the area of the shaded triangles compare to the area of the unshaded triangles?

41. A single piece of rectangular cardboard can be folded around a 5″ by 7″ photo and taped along the seams to create a protective cover suitable for mailing. If the cardboard is folded as shown with no gap or overlap, find the dimensions of the original cardboard. Round to the nearest hundredth. (HINT: Use Theorem 6.9.)

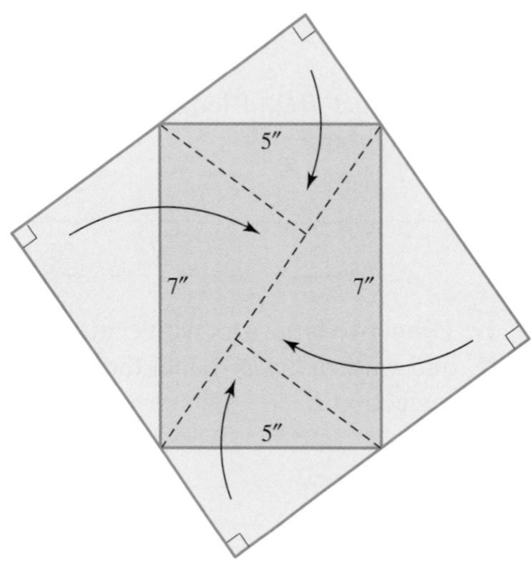

EXTENDED PROBLEMS

42. The midquad of any quadrilateral is a parallelogram. Explore other relationships between a quadrilateral and its midquad as follows.

a. Construct a square and its midquad. Find and compare the perimeter of the midquad and the lengths of the diagonals of the square.

b. Construct a rectangle and its midquad. Compare the perimeter of the midquad and the lengths of the diagonals of the rectangle.

c. Make a conjecture about the relationship between the diagonals of a quadrilateral and the perimeter of the midquad of the quadrilateral. Prove your conjecture.

d. Construct a square and its midquad. Find and compare the areas. Repeat with a rectangle and its midquad, a rhombus and its midquad, and an isosceles trapezoid and its midquad. How does the area of the midquad compare to the original quadrilateral?

43. One method of constructing the geometric mean of a and b, for $a \geq b$, is described next.

a $\qquad$ b

i. Construct $\overline{AC}$ of length a.

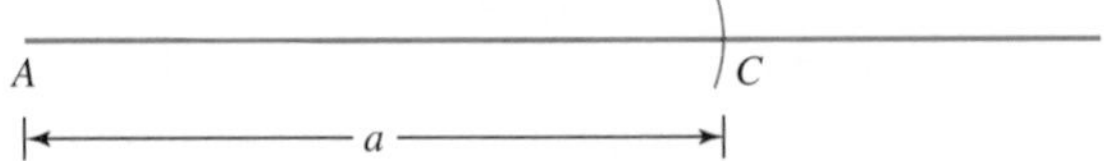

ii. Locate B such that $\overline{BC}$ has length b and B is between A and C.

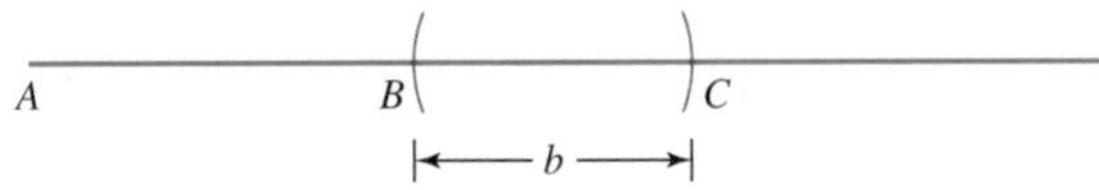

iii. Construct $\overline{BD}$ of length a, where D is to the right of C.

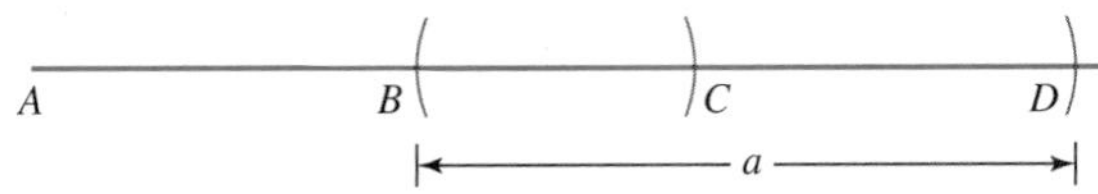

iv. Draw two large arcs with centers A and D, and with radius a. Label their point of intersection E.

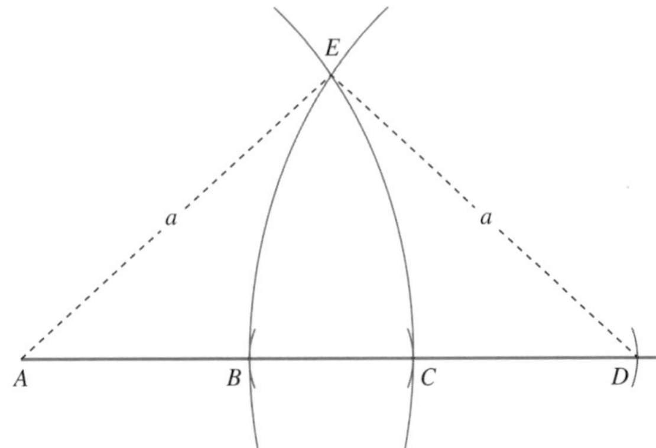

v. The geometric mean of a and b is x, where $x = EB$.

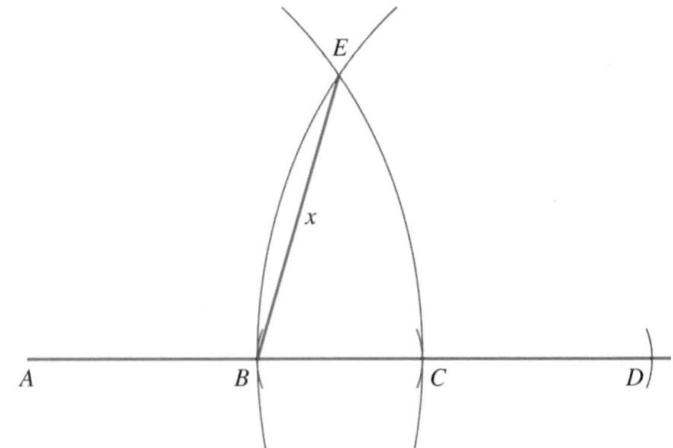

a. Construct the geometric mean of a and b, where the following segments have lengths a and b.

a b

b. If the segment shown, next has length 1. construct the geometric mean between 1 and 2. What is the length you have constructed?

1

c. Prove that the length x in the construction described is in fact the geometric mean between a and b. (HINT: Consider $\triangle AEC$ and $\triangle ECB$.)

44. The midsegment of a trapezoid is the segment that connects the midpoints of its nonparallel legs.

a. Construct three different trapezoids including one isosceles trapezoid and one trapezoid with a right angle. For each trapezoid, construct the midsegment. Measure the lengths of the bases and the midsegment. Then find the arithmetic mean of the lengths of the bases. What do you notice?

b. Prove that for any trapezoid, the midsegment is parallel to the bases of the trapezoid and its length is the arithmetic mean of the lengths of the bases.

6.4 USING RIGHT TRIANGLE TRIGONOMETRY TO SOLVE GEOMETRY PROBLEMS

Applied Problem

An approach ramp is being planned to lead to a suspension bridge. Given the dimensions in the figure, at what angle from the horizontal should the ramp be sloped?

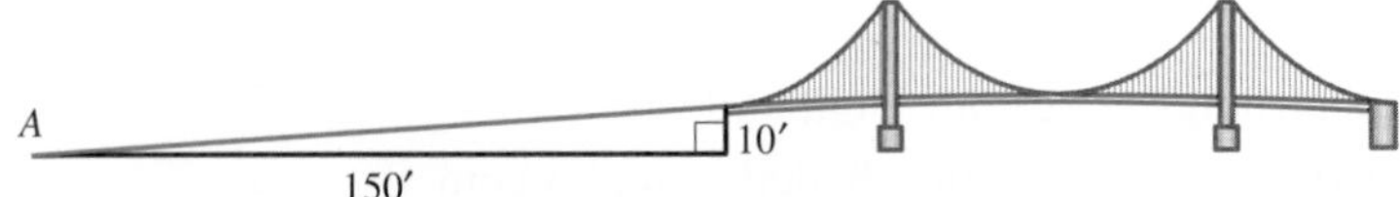

The Tangent Ratio

We can solve this section's applied problem using methods from right triangle trigonometry. **Trigonometry**, as its name suggests, is the study of the measures (*metry*) of triangles (*triang*). Consider the right triangles $\triangle ABC$ and $\triangle DEF$ in Figure 6.18, where $\angle A \cong \angle D$.

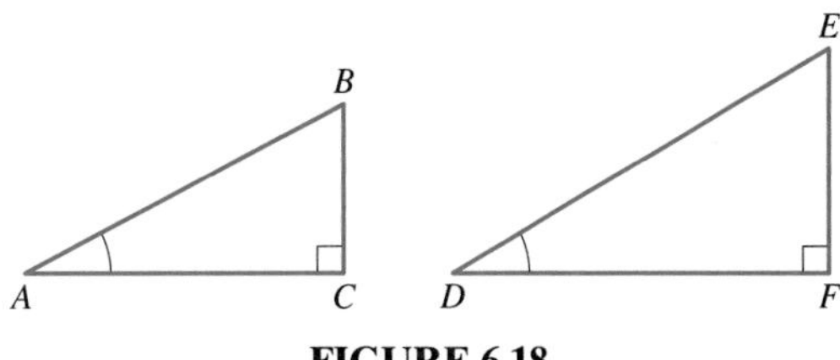

FIGURE 6.18

By AA Similarity, $\triangle ABC \sim \triangle DEF$. Hence, the following ratios are equal:

$$\frac{BC}{AC} = \frac{EF}{DF} \qquad \frac{BC}{AB} = \frac{EF}{DE} \qquad \frac{AC}{AB} = \frac{DF}{DE}$$

This suggests that associated with any acute angle in a right triangle (such as $\angle A$ in Figure 6.18), there are three unique ratios that depend only on the measure of the angle, *not* on the size of the triangle. As shown in Figure 6.19, these ratios are $\frac{a}{b}$, $\frac{a}{c}$, and $\frac{b}{c}$. Notice how lowercase letters are opposite their uppercase counterparts in the triangle.

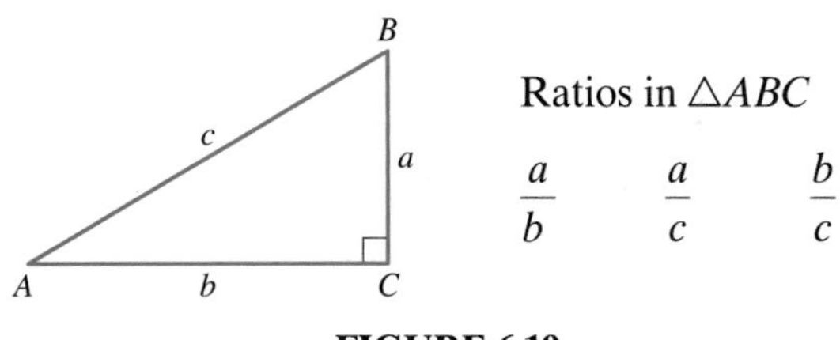

FIGURE 6.19

These ratios form the basis of right triangle trigonometry and are given special names. We study the first of these ratios, the tangent ratio, next. In right triangle $\triangle ABC$, as shown in Figure 6.19, the **tangent** of acute $\angle A$, written **tan *A***, is the ratio $\frac{a}{b}$. That is, the tangent of $\angle A$, is the ratio of the length of the leg *opposite* $\angle A$ (called opposite side) to the length of the leg *adjacent* to $\angle A$ (called the adjacent side).

DEFINITION Tangent of an Acute Angle

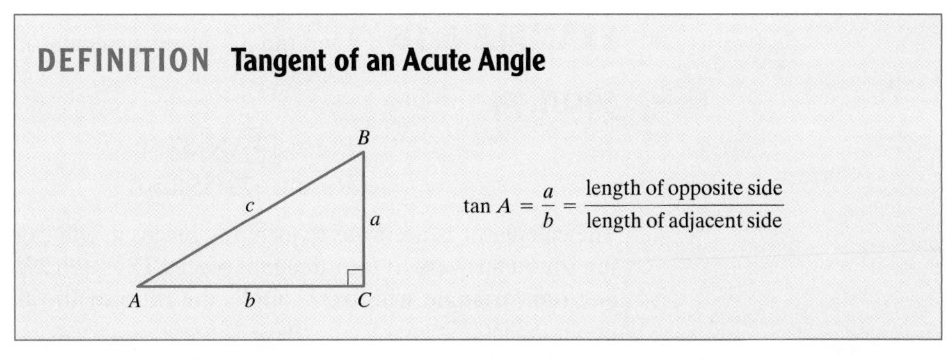

EXAMPLE 6.15 Calculate the following.

a. tan 60° **b.** tan 30° **c.** tan 45°

SOLUTION

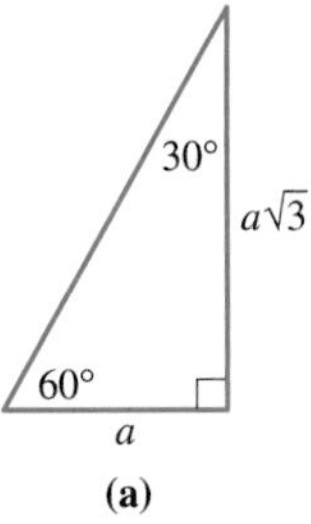

(a)

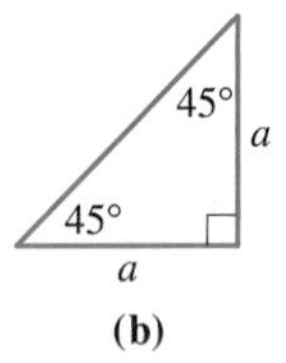

(b)

FIGURE 6.20

a. Recall that in a 30°–60° right triangle, if the length of the shorter leg is a, then the longer leg has length $a\sqrt{3}$ [Figure 6.20(a)]. We can find tan 60° as follows.

$$\tan 60^\circ = \frac{\text{opp}}{\text{adj}} = \frac{a\sqrt{3}}{a} = \sqrt{3} \approx 1.7321$$

The exact value of tan 60° is $\sqrt{3}$ and an approximate value is 1.7321.

b. We also can find tan 30° using the triangle in Figure 6.20(a), as follows:

$$\tan 30^\circ = \frac{\text{opp}}{\text{adj}} = \frac{a}{a\sqrt{3}} = \frac{1}{\sqrt{3}} \approx 0.5774$$

c. In a 45°–45° right triangle, both legs have the same length a [Figure 6.20(b)]. Thus, we have

$$\tan 45^\circ = \frac{\text{opp}}{\text{adj}} = \frac{a}{a} = 1$$

■

Figure 6.21 shows a right triangle with one leg of length 1. The other leg has length a, where $a > 0$ and a can be arbitrarily large. Therefore, $\tan A = \frac{a}{1} = a$. This shows that the tangent of an acute angle must be greater than zero and may be arbitrarily large.

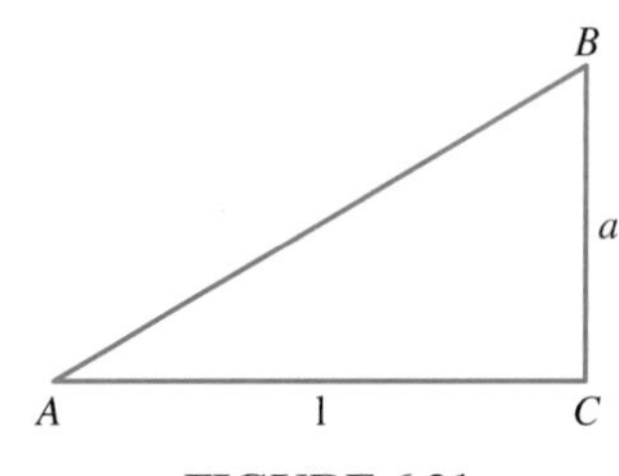

FIGURE 6.21

You can use a graphing or scientific calculator to find the tangent of any angle. Because there are various ways to measure angles other than using degrees, be sure that your calculator is set for degree measure.

EXAMPLE 6.16 Find tan 37° using a calculator.

SOLUTION

Press: [TAN] 37

Result: .7535540501

The calculator is accurate to as many places as the calculator can display, but we often round answers to four decimal places. Thus, tan 37° ≈ 0.7536. This means that in any right triangle with a 37° angle, the ratio of the shorter leg to the longer leg is about 0.7536.

■

(NOTE: The sequence of keystrokes on your calculator and number of decimal places shown may differ from the ones given in this book. Consult the manual for your calculator to be sure of the correct sequence.)

We can use the tangent ratio to solve problems involving right triangles.

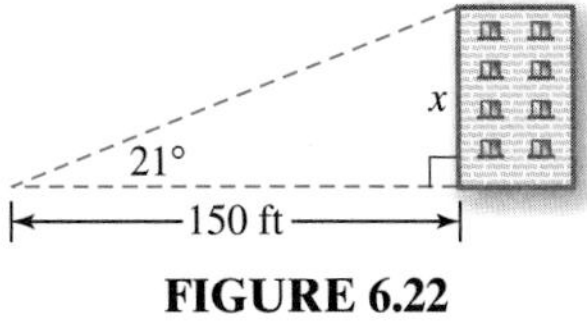

FIGURE 6.22

EXAMPLE 6.17 At a horizontal distance of 150 feet from the base of a building, the line of sight to the top of the building makes an angle of 21° with level ground (Figure 6.22). That is, the angle of elevation to the top of the building is 21°. About how tall is the building? Assume that the building is perpendicular to the ground.

SOLUTION

Let x represent the height of the building. Then $\tan 21° = \frac{\text{opp}}{\text{adj}} = \frac{x}{150}$, or $x = 150 \tan 21°$. On a calculator, we can find 150 tan 21° as follows:

$$x = 150 \boxed{\text{TAN}} 21 = 57.57960526$$

So the building is about 58 feet tall. ■

The Sine and Cosine Ratios

The sine and cosine ratios compare the lengths of the legs of a right triangle to the length of its hypotenuse. In the right triangle $\triangle ABC$, as shown in Figure 6.19, the **sine** of $\angle A$, written **sin *A***, is the ratio $\frac{a}{c}$. That is, the sine of $\angle A$ is the ratio of the length of the leg *opposite* $\angle A$ to the length of the *hypotenuse.*

DEFINITION Sine of an Acute Angle

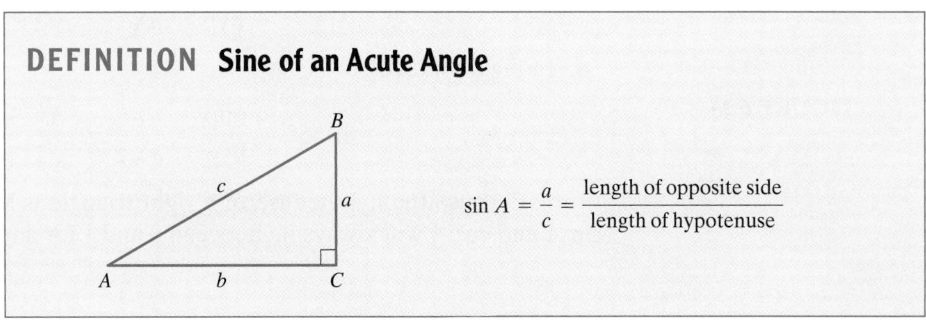

In the right triangle $\triangle ABC$, as shown in Figure 6.19, the **cosine** of $\angle A$, written **cos *A***, is the ratio $\frac{b}{c}$. That is, the cosine of $\angle A$ is the ratio of the length of the leg *adjacent* to $\angle A$ to the length of the *hypotenuse.*

DEFINITION Cosine of an Acute Angle

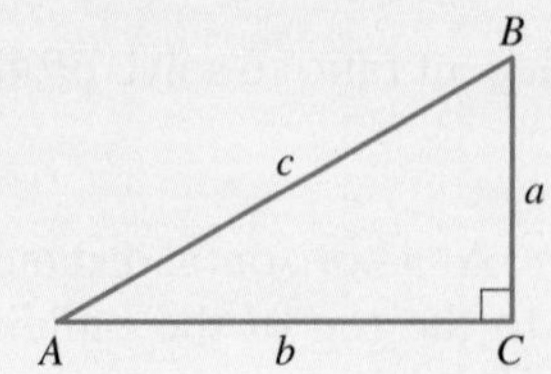

$$\cos A = \frac{b}{c} = \frac{\text{length of adjacent side}}{\text{length of hypotenuse}}$$

(NOTE: Although it is common to speak of the tangent, sine, or cosine of an *angle*, it is also customary to find the tangent, sine, or cosine of the *number* of degrees, in this book that is the measure of an angle. Ratios such as the tangent, sine, and cosine are referred to as **trig ratios**.)

EXAMPLE 6.18 Calculate the following, giving both an exact and an approximate answer.

a. sin 60° **b.** sin 30° **c.** cos 45° **d.** sin 45°

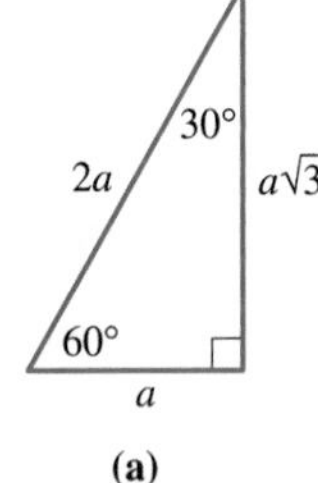

(a)

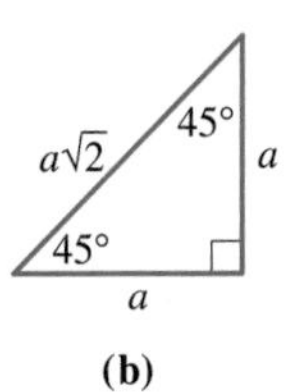

(b)

FIGURE 6.23

SOLUTION

a. In Figure 6.23(a),

$$\sin 60° = \frac{\text{opp}}{\text{hyp}} = \frac{a\sqrt{3}}{2a} = \frac{\sqrt{3}}{2} \approx 0.8660$$

b. In Figure 6.23(a),

$$\sin 30° = \frac{\text{opp}}{\text{hyp}} = \frac{a}{2a} = 0.5$$

c. In Figure 6.23(b),

$$\cos 45° = \frac{\text{adj}}{\text{hyp}} = \frac{a}{a\sqrt{2}} = \frac{1}{\sqrt{2}} \approx 0.7071$$

d. In Figure 6.23(b),

$$\sin 45° = \frac{\text{opp}}{\text{hyp}} = \frac{a}{a\sqrt{2}} = \frac{1}{\sqrt{2}} \approx 0.7071$$ ■

(NOTE: Because the hypotenuse of a right triangle is longer than either of the legs, sin A and cos A will always be between 0 and 1 for any acute angle.)

EXAMPLE 6.19 Find sin 37° and cos 37° using a calculator.

SOLUTION

Press: SIN 37
Result: 0.601815023
Press: COS 37
Result: 0.79863551

So, to four decimal places, sin 37° ≈ 0.6018 and cos 37° ≈ 0.7986. ■

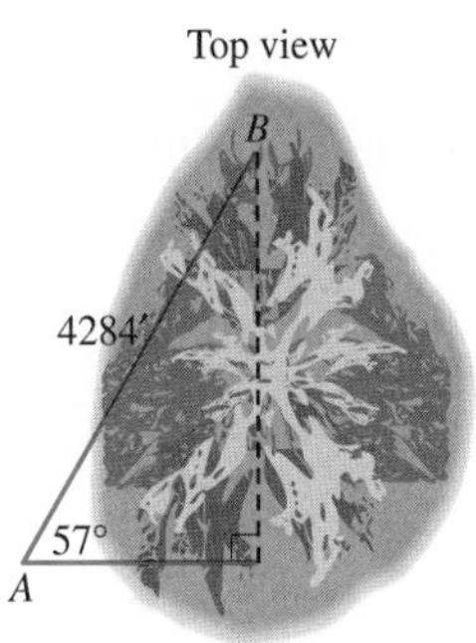

FIGURE 6.24

EXAMPLE 6.20 To determine the feasibility of constructing a tunnel through a mountain, surveyors want to measure the north-south distance BC through the mountain. They found that $AB = 4284$ ft, and $\angle A = 57°$, and $\angle C$ was a right angle based on compass measurement (Figure 6.24). How long will the tunnel be?

SOLUTION

$\sin A = \dfrac{\text{opp}}{\text{hyp}} = \dfrac{BC}{AB}$. Thus, $\sin 57° = \dfrac{BC}{4284}$.

So $BC = 4284 \sin 57°$ or about 3593 ft. ■

Examples 6.16 and 6.19 show how to find the tangent, sine, and cosine of given angles (numbers). On the other hand, we can also use a calculator to find the approximate measure of an angle given one of its trig ratios, as shown next.

Inverse Trigonometric Functions

Now we investigate inverse trigonometric functions that are useful for solving applications involving angles of triangles.

EXAMPLE 6.21 Find an approximate value for $\angle A$ in each of the following, using inverse trig function calculator keys.

a. $\tan A = 0.8391$ **b.** $\sin A = 0.3581$

SOLUTION

We must find the angle that has a tangent of 0.8391 or a sine of 0.3581.

a. Press: [TAN^{-1}] .8391

Result: 40.0000124

Therefore, $A \approx 40°$.

This means that an angle which has a tangent of 0.8391 is about 40°.

b. Press: [SIN^{-1}] .3581

Result: 20.98355628

Therefore, $A \approx 21°$. ■

(NOTE: On most calculators, [TAN^{-1}] is the second function on the [TAN] key. Thus, you must press the second function key *before* pressing the [TAN] key. Again, the sequence of keystrokes for your calculator may differ. Consult your manual to be sure).

Example 6.21 used the [TAN^{-1}] and [SIN^{-1}] keys to find an angle given its tangent and sine, respectively. These keys, together with the [COS^{-1}] key, are called **inverse trig function keys** because we can use them to find the measure of an angle given its corresponding trig ratio. Some calculators have an [ARCTAN] or use the two keys [INV] [TAN] in place of [TAN^{-1}].

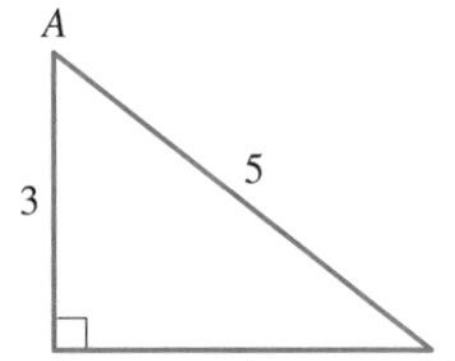

FIGURE 6.25

EXAMPLE 6.22 A familiar right triangle is the 3-4-5 triangle as shown in Figure 6.25. Find the measures of $\angle A$ and $\angle B$.

SOLUTION

Using the tangent ratio and the given side lengths, we have $\tan B = \dfrac{3}{4} = 0.75$. To find $\angle B$ using the calculator, we find [TAN^{-1}] .75, which gives a result of about 36.8699. Therefore, $\angle B \approx 36.9°$ and $\angle A \approx 90° - 36.9° = 53.1°$. ■

Relating the Sine, Cosine, and Tangent Ratios

The trig ratios are interrelated in two interesting ways. First, consider all three ratios $\frac{a}{b}, \frac{a}{c}$, and $\frac{b}{c}$ for the triangle in Figure 6.26.

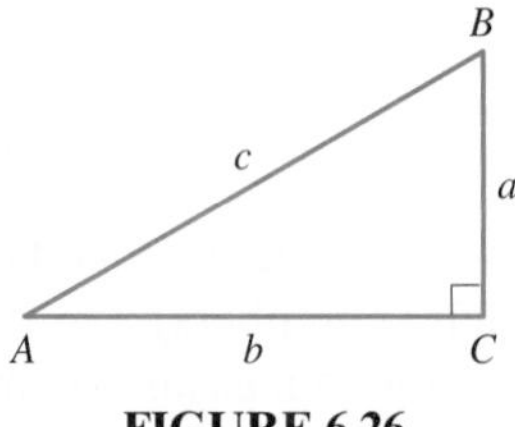

FIGURE 6.26

Notice that $\frac{a}{c} \div \frac{b}{c} = \left(\frac{a}{c}\right)\left(\frac{c}{b}\right) = \frac{a}{b}$.

Because $\frac{a}{c} = \sin A$, $\frac{b}{c} = \cos A$, and $\frac{a}{b} = \tan A$, we can conclude that $\sin A \div \cos A = \tan A$.

THEOREM 6.13

In a right triangle with acute angle A, $\sin A$ divided by $\cos A$ is $\tan A$.

$$\frac{\sin A}{\cos A} = \tan A$$

Because the trig ratios pertain to right triangles, we can apply the Pythagorean theorem to derive another relationship among these ratios. For $\triangle ABC$ in Figure 6.26, we have $a^2 + b^2 = c^2$. Dividing both sides of this equation by c^2 yields

$$\frac{a^2}{c^2} + \frac{b^2}{c^2} = \frac{c^2}{c^2} \quad \text{or} \quad \left(\frac{a}{c}\right)^2 + \left(\frac{b}{c}\right)^2 = 1$$

Because $\frac{a}{c} = \sin A$ and $\frac{b}{c} = \cos A$, we have $(\sin A)^2 + (\cos A)^2 = 1$. It is customary to write $(\sin A)^2$, $(\cos A)^2$, and $(\tan A)^2$ as $\sin^2 A$, $\cos^2 A$, and $\tan^2 A$, respectively. Thus, we have the following result.

THEOREM 6.14

In a right triangle with acute angle A, the sum of $\sin^2 A$ and $\cos^2 A$ is 1.

$$\sin^2 A + \cos^2 A = 1$$

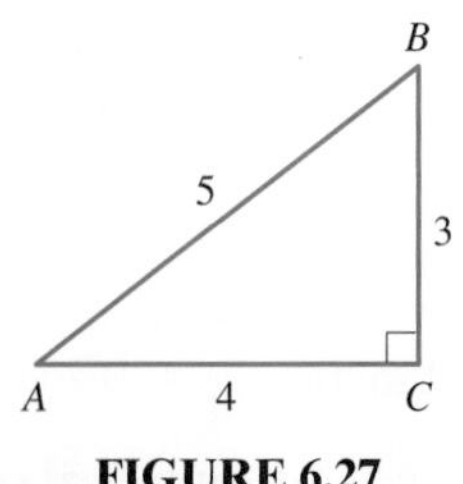

FIGURE 6.27

EXAMPLE 6.23 Verify Theorems 6.13 and 6.14 in the case of a 3-4-5 right triangle (Figure 6.27).

SOLUTION

First, $\sin A = \frac{3}{5}$, $\cos A = \frac{4}{5}$, and $\tan A = \frac{3}{4}$. So

$$\frac{\sin A}{\cos A} = \frac{\frac{3}{5}}{\frac{4}{5}} = \left(\frac{3}{5}\right)\left(\frac{5}{4}\right) = \frac{3}{4} = \tan A$$

This verifies Theorem 6.13. We also have

$$\sin^2 A + \cos^2 A = \left(\frac{3}{5}\right)^2 + \left(\frac{4}{5}\right)^2 = \frac{9}{25} + \frac{16}{25} = \frac{25}{25} = 1$$

This verifies Theorem 6.14. ■

EXAMPLE 6.24 Given that $\sin A = 0.5736$, find $\cos A$.

SOLUTION

Because $\sin^2 A + \cos^2 A = 1$, we have $\cos^2 A = 1 - \sin^2 A$, or $\cos A = \sqrt{1 - \sin^2 A}$. Thus, $\cos A = \sqrt{1 - (0.5736)^2}$, or $\cos A \approx 0.8191$. ■

Solving Right Triangles

If we know the lengths of any two of the sides of a right triangle, we can find the length of the third side using the Pythagorean theorem. Moreover, we can also find the measures of the two acute angles using the sine or cosine ratios. In a similar manner given the measure of one angle and the length of one side of a right triangle, we can find the lengths of the other sides using trigonometry. The next two examples illustrate the technique known as **solving right triangles**, which is used to find the measures of all of the angles and sides of a right triangle.

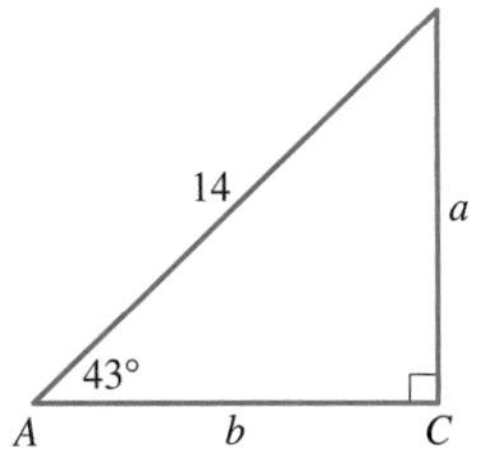

EXAMPLE 6.25 Solve the right triangle $\triangle ABC$ in Figure 6.28.

FIGURE 6.28

SOLUTION

First, $\angle B = 90° - 43° = 47°$. Then, since $\sin 43° = \frac{a}{14}$, we have $a = 14 \sin 43° \approx 9.547977041$. Then, since $\sin 47° = \frac{b}{14}$, we have $b = 14 \sin 47° \approx 10.23895182$. In summary,

$$a \approx 9.55 \qquad b \approx 10.24 \qquad c = 14$$

$$\angle A = 43° \qquad \angle B = 47° \qquad \angle C = 90°$$ ■

Notice that we also could have found both a and b using the cosine as follows:

$$\cos 43° = \frac{b}{14}, \text{ so } b = 14 \cos 43° \approx 10.23895182$$

$$\cos 47° = \frac{a}{14}, \text{ so } a = 14 \cos 47° \approx 9.547977041$$

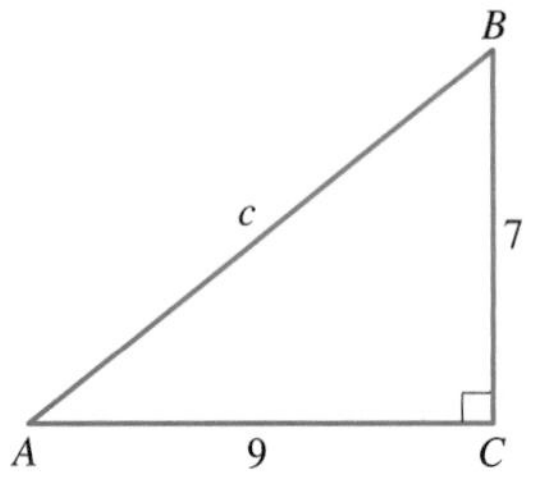

FIGURE 6.29

EXAMPLE 6.26 Solve the right triangle $\triangle ABC$ in Figure 6.29.

SOLUTION

By the Pythagorean theorem, $c^2 = 7^2 + 9^2$. Therefore, $c = \sqrt{49 + 81} \approx 11.4$. Also, $\tan A = \frac{7}{9}$, so $\angle A = \tan^{-1}\left(\frac{7}{9}\right) \approx 37.87498365 \approx 38°$. Hence, $\angle B \approx 90° - 38° = 52°$. In summary, we have the following:

$$\begin{array}{ccc} a = 7 & b = 9 & c \approx 11.4 \\ \angle A \approx 38° & \angle B \approx 52° & \angle C = 90° \end{array}$$

Using Trig Ratios to Calculate Area

We may use trigonometry to determine the area of a triangle or another geometric figure. The next example illustrates how we can calculate the height and area of a triangle by choosing an appropriate trig ratio.

EXAMPLE 6.27 Determine the area of $\triangle ABC$ in Figure 6.30, given that $\angle A = 50°$, $AB = 35$ mm, and $AC = 75$ mm.

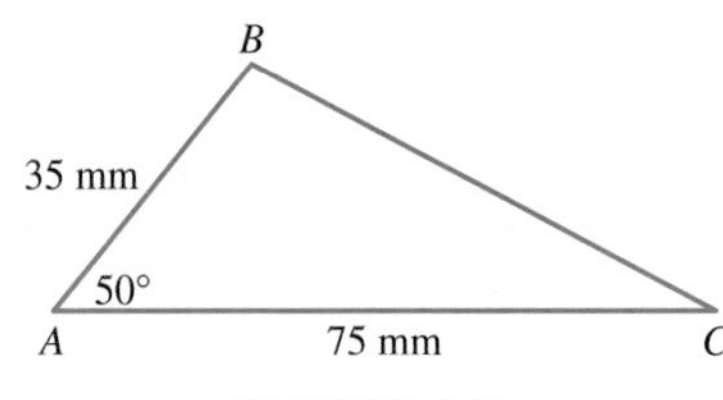

FIGURE 6.30

SOLUTION

We will use the fact that the area of a triangle is given by $A = \frac{1}{2}bh$, where $b = 75$ mm in $\triangle ABC$. We must determine the height h of the triangle (Figure 6.31) in order to calculate the area of $\triangle ABC$.

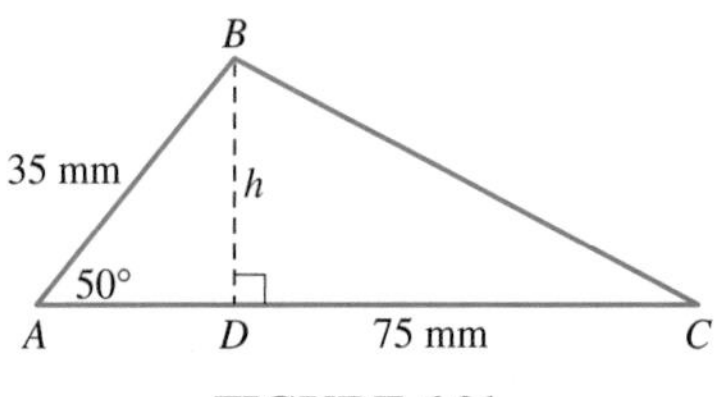

FIGURE 6.31

$\triangle ABD$ is a right triangle in which the hypotenuse is 35 mm long and the measure of the side opposite the 50° angle is h. We use the sine ratio to determine h.

$$\sin 50° = \frac{h}{35}, \text{ so } h = 35 \sin 50° \approx 26.81 \text{ mm}$$

Now we can calculate the area of $\triangle ABC$ as $A = \frac{1}{2}bh = \frac{1}{2}(75)(26.81) \approx 1005 \text{ mm}^2$.

Solution of Applied Problem

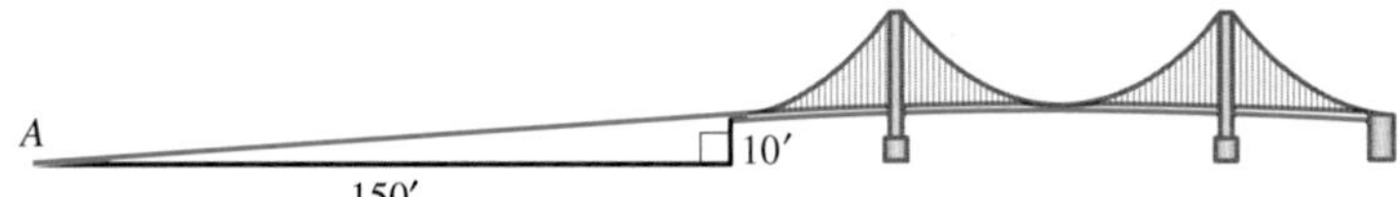

In the figure, $\tan A = \dfrac{\text{opp}}{\text{adj}} = \dfrac{10}{150} = \dfrac{1}{15}$, so $A = \tan^{-1}\left(\dfrac{1}{15}\right) \approx 3.814074834 \approx 3.8°$.
So the ramp has an angle of inclination of approximately 3.8°.

GEOMETRY AROUND US Modern house construction involves the use of a framework called a roof truss system. Trusses are constructed at a factory and then delivered to the building site and lifted onto the tops of the walls. As illustrated in the figure, right triangles play a critical role in the construction of trusses.

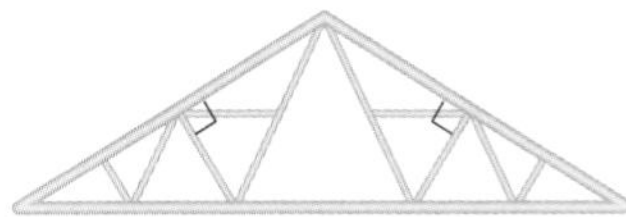

PROBLEM SET 6.4

EXERCISES/PROBLEMS

Use the right triangle shown for Exercises 1–6.

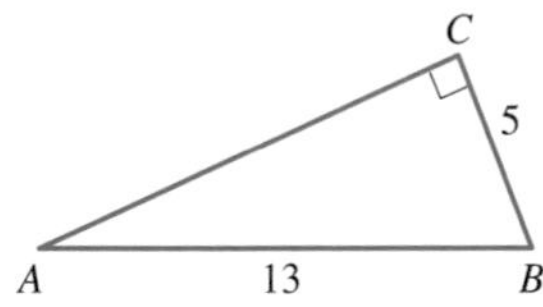

Find the following. Leave answers in fraction form.

1. $\sin A$
2. $\cos A$
3. $\tan A$
4. $\sin B$
5. $\cos B$
6. $\tan B$

Use this figure for Exercises 7–24.

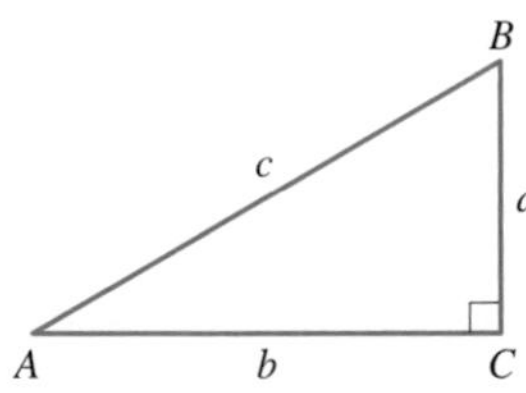

Find the following. Round lengths to the nearest hundredth and angle measures to the nearest tenth.

7. $\tan A$ if $a = 7$ and $b = 5$
8. $\sin B$ if $a = 5$ and $c = 12$
9. $\cos A$ if $b = 7$ and $c = 11$
10. $\tan B$ if $a = 5$ and $c = 13$
11. $\cos B$ if $a = b$
12. $\sin A$ if $a = 5$ and $b = 12$
13. a if $\sin A = 0.503$ and $c = 10$
14. b if $\cos A = 0.7$ and $c = 12$
15. c if $\tan A = 0.6$ and $a = 5$
16. b if $\sin B = 0.5$ and $a = 12$
17. $\angle A$ if $a = 5$ and $b = 7$
18. $\angle B$ if $a = 7$ and $c = 9$
19. $\angle A$ if $b = \pi$ and $c = \sqrt{21}$
20. $\angle B$ if $a = \sqrt{2}$ and $b = \sqrt{3}$
21. $\cos B$ if $b = a\sqrt{3}$
22. $\tan A$ if $c = 3a$
23. c if $\angle B = 70°$ and $b = 0.8$
24. b if $\angle A = 27.3°$ and $c = 5$

Solve each of the right triangles in Exercises 25–28. Round to the nearest tenth.

25.

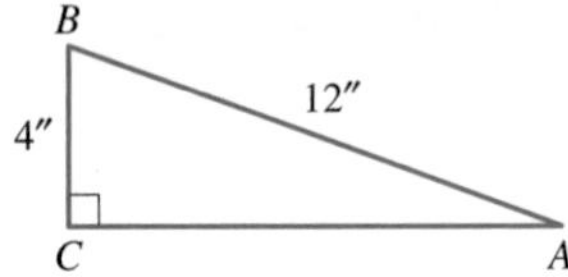

26.

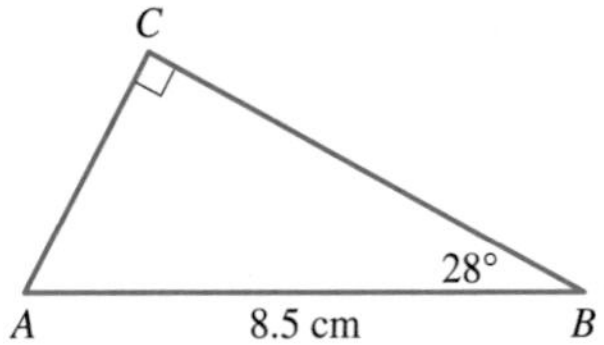

27.

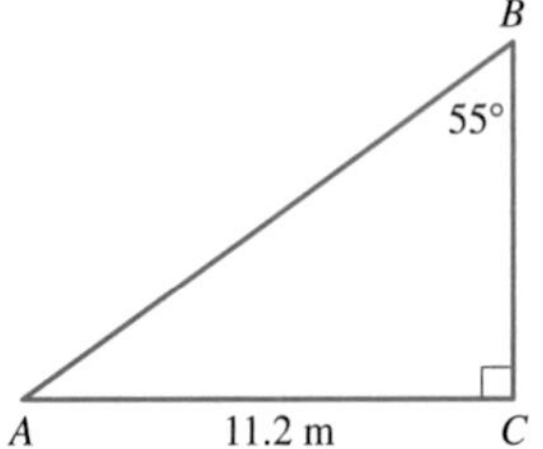

28.

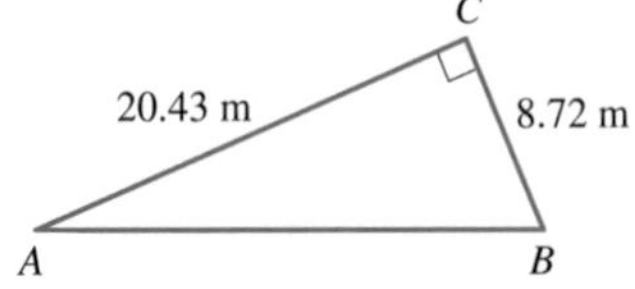

Verify Theorems 6.13 and 6.14 for Exercises 29–32.

29. $\angle A = 45°$

30. $\angle A = 13°$

31. $\angle A = 30°$

32. $\angle A = 49°$

In Exercises 33–40, solve for the missing parts of the following right triangle using the information given. Round to the nearest tenth.

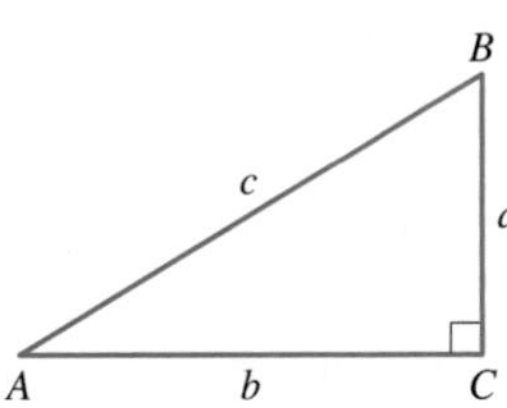

33. $c = 10$ and $\angle A = 30°$

34. $b = 8$ and $\angle B = 45°$

35. $a = 24$ and $b = 10$

36. $c = 7$ and $b = 5$

37. $\angle A = 23°$ and $b = 4$

38. $a = 9$ and $\angle B = 41°$

39. $\angle B = 12°$ and $c = 10.3$

40. $c = 21.7$ and $\angle A = 49°$

Find the area of each triangle in Exercises 41–44. Round to the nearest tenth.

41.

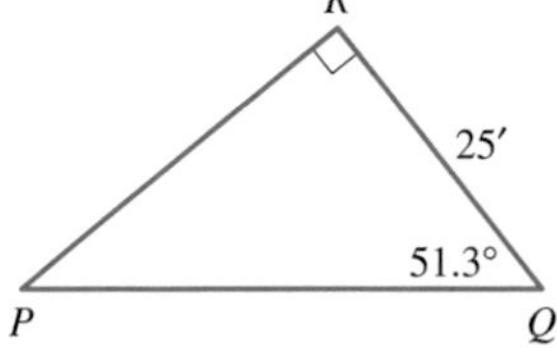

42.

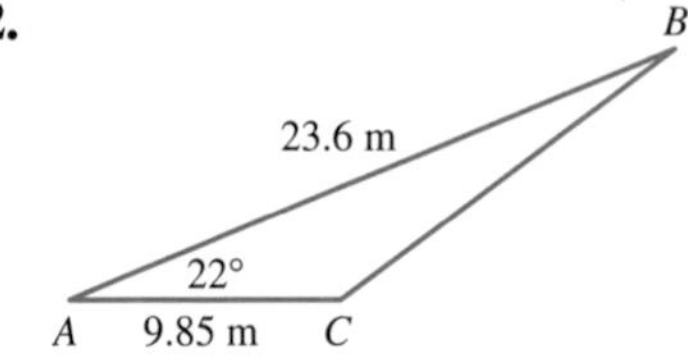

43.

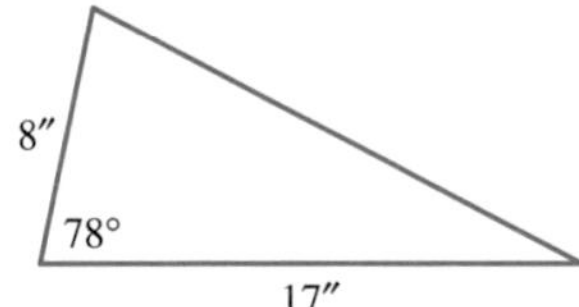

44.

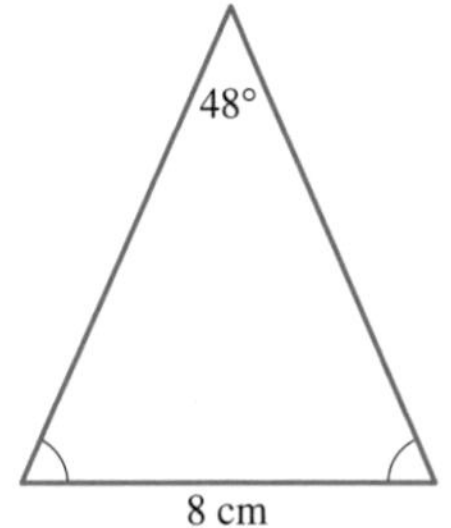

Find the area of each quadrilateral in Exercises 45–46.

45.

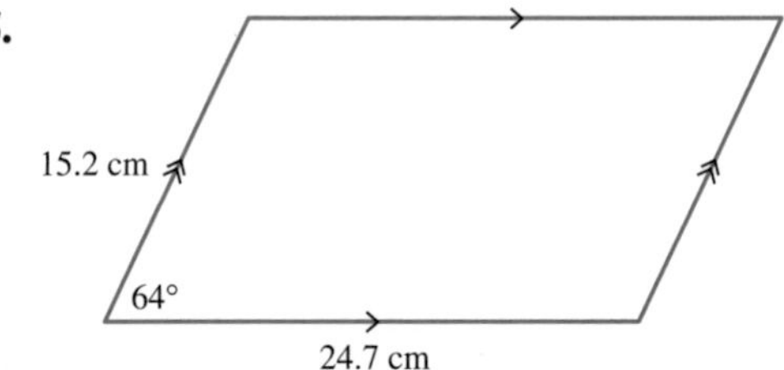

46.

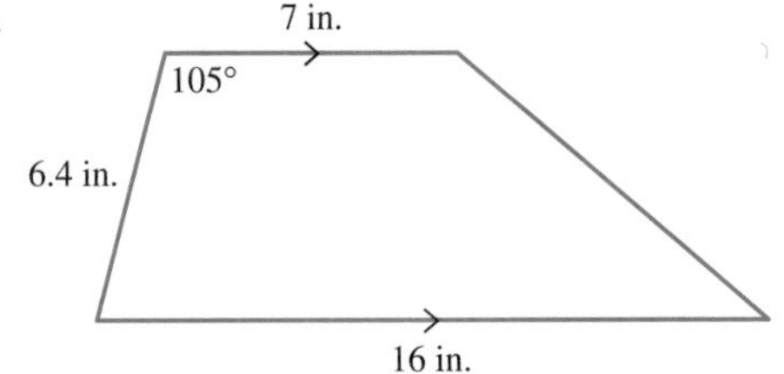

PROOFS

47. Prove: $\sin A = (\cos A)(\tan A)$

48. Prove: If A and B are the acute angles in a right triangle, then $\sin A = \cos B$.

49. Prove: If A is an acute angle in an isosceles right triangle, then $\sin A = \cos A$.

50. Show that $\sin A = \cos(90° - A)$.

51. Show that $\cos A = \sin(90° - A)$.

Three other trig ratios are defined as follows:

$$\cot A = \frac{\text{adjacent}}{\text{opposite}} \quad \text{(the \textbf{cotangent} ratio)}$$

$$\sec A = \frac{\text{hypotenuse}}{\text{adjacent}} \quad \text{(the \textbf{secant} ratio)}$$

$$\csc A = \frac{\text{hypotenuse}}{\text{opposite}} \quad \text{(the \textbf{cosecant} ratio)}$$

Using these definitions, prove the following:

52. $\cot A = \dfrac{1}{\tan A}$

53. $\sec A = \dfrac{1}{\cos A}$

54. $\csc A = \dfrac{1}{\sin A}$

55. $\cot A = \dfrac{\cos A}{\sin A}$

56. $\cot A = \dfrac{\csc A}{\sec A}$

57. $\csc^2 A - \cot^2 A = 1$

58. $\sec^2 A - \tan^2 A = 1$

APPLICATIONS

59. A hill makes an angle of 38° with the horizontal. A 150-foot-long fence is built from the top of the hill straight down to the base of the hill. How high is the hill? Round to the nearest whole number.

60. A ramp leading to a freeway overpass is 300 feet long and rises 20 feet. What is the angle of inclination of the ramp (the acute angle between the ramp and the horizontal)? Round to the nearest tenth.

61. A guy wire is attached to the top of a radio antenna and anchored to the ground at a point 20 feet from the base of the antenna. If the wire makes an angle of 65° with the level ground, how high is the antenna? Round to the nearest hundredth.

62. A surveyor at point P wishes to measure the distance PQ in the drawing that follows. The surveyor sights point Q, makes a right angle at P, steps off 100 ft to R, and again sights point Q. $\angle PRQ$ is determined to be 64°. Find the distance PQ. Round to the nearest whole number.

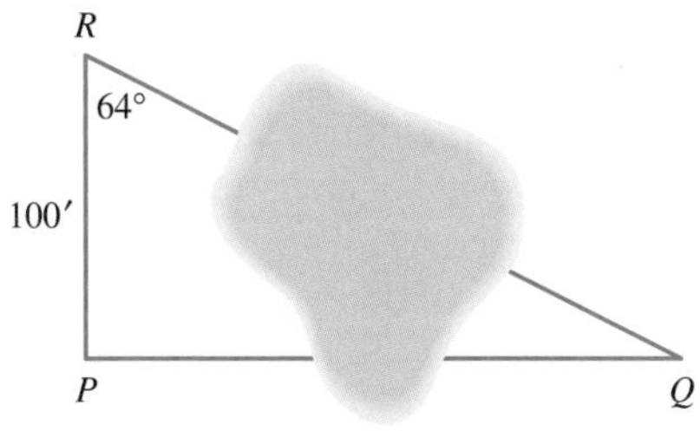

63. A girl flying a kite lets out all 120 feet of string and ties the string to an object on the ground. If the string forms an angle of 42° with level ground, how high is the kite at this moment? Round to the nearest hundredth.

64. A forester standing 50 feet from the base of a Douglas fir measures the angle of elevation to the top of the tree as 71.5°. Find the height of the tree. Round to the nearest whole number.

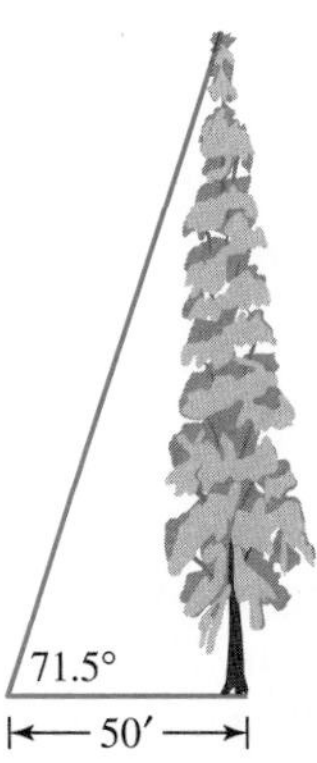

65. A conveyor that is used to put bales of hay into storage is 100 feet long. The recommended angle of elevation for the conveyor is 30°.

a. To what height can the hay be moved?

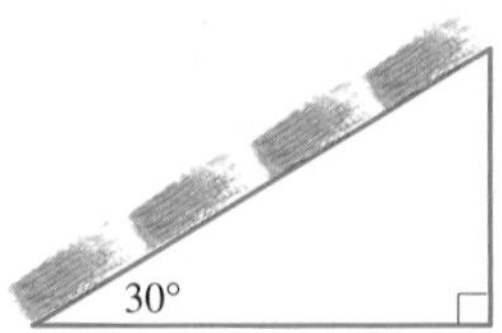

b. If the conveyor is positioned at an angle of 28°, to what height can the hay be moved? Round to the nearest tenth.

66. A truck passes a sign that reads 8 PERCENT GRADE NEXT 10 MILES. What is the angle of elevation of the road? Round to the nearest tenth. (NOTE: Percent grade is the slope of the road expressed as a percentage.)

67. A surveyor with a transit stands 100 ft away from a flagpole. The angle of elevation to the top of the flagpole is measured as 21°45′, and the angle of depression to the base of the flagpole is measured as 1°22′. Find the height of the flagpole to the nearest tenth.

68. A detective discovered that a bullet passed through a window at a point 4′2″ from the floor. It lodged in a wall 14′8″ from the hole in the window and 7′ up on the wall. At what angle from the horizontal was the bullet fired? Round to the nearest tenth.

69. To determine the height of a building, a surveyor stands at a point P and measures the angle of elevation to the top of the building as 19.42°. From a point Q, 60 m further away from the building, the angle of elevation to the top of the building is 16.8°. Find the height of the building to the nearest meter. Round to the nearest whole number.

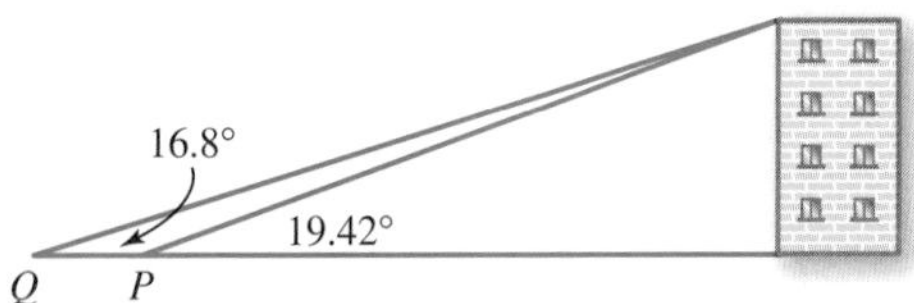

EXTENDED PROBLEMS

70. Due to the movement of the earth around the sun, our perception of the position of a nearby star relative to other distant stars changes. This apparent motion is called **stellar parallax**. By using stellar parallax and trigonometry, we can determine the distance to a star. Research stellar parallax and write a description of how astronomers use stellar parallax to measure the distance to stars. When was stellar parallax first used to estimate the distance of a star? Explain and draw a diagram to show how to use trigonometry and stellar parallax to find the distance to the star Alpha Centauri.

71. **Spherical geometry** is the study of figures on the surface of a sphere. In spherical geometry, triangles can have more than one right angle. A **spherical triangle** is made up of sections of three circles, whose centers are the center of the sphere. A version of the Pythagorean theorem for spherical geometry involves trigonometry. Research the spherical Pythagorean theorem. Using an inflated balloon and a marker, draw a spherical triangle, labeling each side and each angle. Verify the spherical Pythagorean theorem for a spherical triangle made up of the equator and two lines of longitude, which are all circles that have as their centers the center of the earth. A line of longitude also passes through the north and south poles.

72. The golden ratio was introduced in the Geometry Around Us in Section 6.1 and is denoted ϕ, where $\phi = \dfrac{1 + \sqrt{5}}{2}$. A **golden triangle** is an acute isosceles triangle like the one shown, where $\dfrac{c}{a} = \phi$.

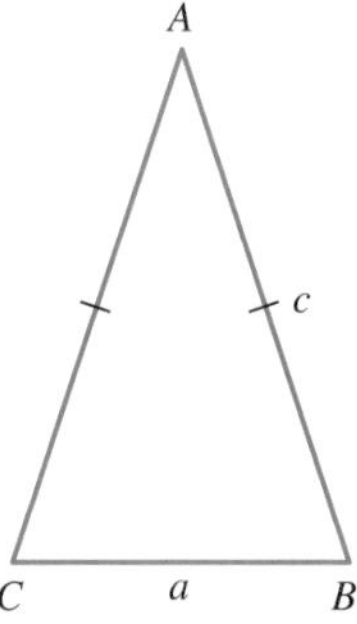

A **golden gnomon** is an obtuse isosceles triangle like the one shown, where $\dfrac{c}{a} = \dfrac{1}{\phi}$.

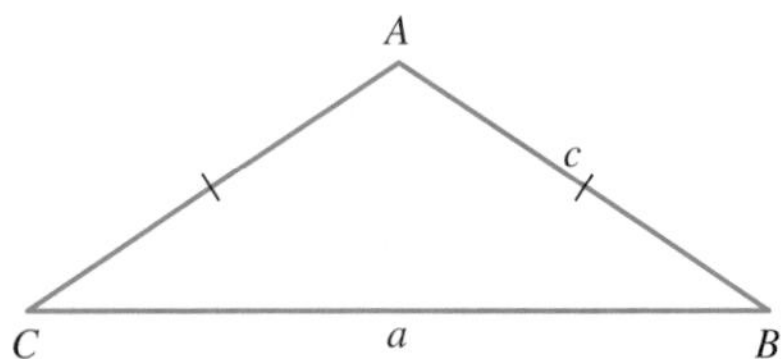

a. Use trigonometry to determine the vertex angle measures of a golden triangle. Use a protractor to draw and label an example of a golden triangle.

b. Use trigonometry to determine the vertex angle measures of a golden gnomon. Use a protractor to draw and label an example of a golden gnomon.

c. For the golden triangle drawn in part (a), bisect one of the base angles and verify that the small triangle formed by the angle bisector is a golden triangle and is similar to the original triangle. If you continue to create smaller similar triangles by bisecting one base angle of the previously created smaller triangle and then connect vertices in a certain way, you will form a **golden spiral**. Research the golden spiral, describe how it is created, and give examples of where it is found in nature. Draw and label a large golden spiral on a piece of paper and list the similar triangles.

6.5 USING LAWS OF TRIGONOMETRY TO SOLVE GEOMETRY PROBLEMS

Applied Problem

A surveyor estimates the distance across a lake by walking 75 meters from one end of the lake to a marker, turning 100°, and walking 125 meters to the other end of the lake. To the nearest meter, approximately how wide is the lake?

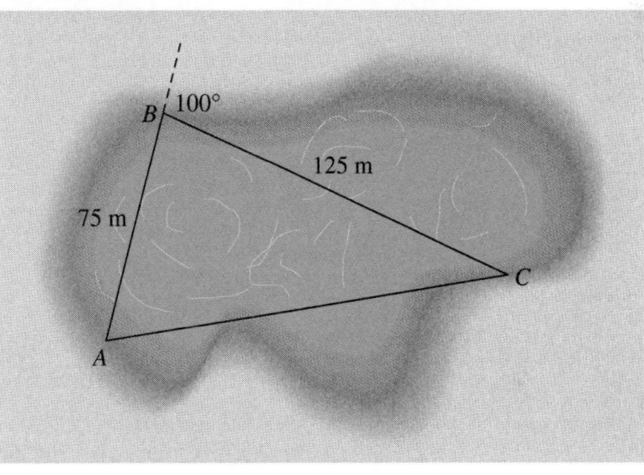

In Section 6.4, we used the trig functions sine, cosine, and tangent to solve geometry problems involving right triangles. Next we will derive two famous laws of trigonometry that we can use to solve geometry problems that are not limited to right triangles. Triangles that have no right angles are called **oblique triangles**. Oblique triangles can be acute or obtuse. In this section, we will limit our discussion to acute triangles.

The Law of Sines

In Section 6.4, we saw that we could use trigonometry to calculate the height of a triangle. We next use the height of a triangle to develop a relationship between the sides and angles of a triangle called the Law of Sines. Consider $\triangle ABC$ in Figure 6.32 where $\overline{BD}$ is the perpendicular line segment from B to side $\overline{AC}$ and where side lengths are as shown.

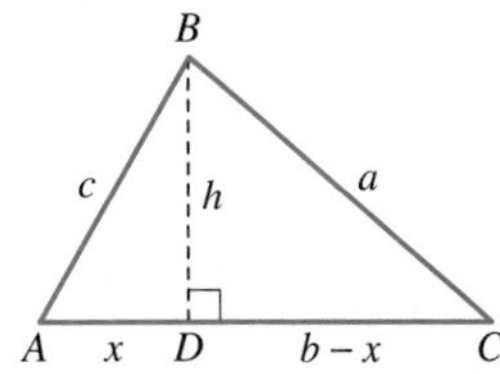

FIGURE 6.32

We can calculate the height h of $\triangle ABC$ using $\triangle ABD$ and $\angle A$, as follows.

$$\sin A = \frac{\text{opp}}{\text{hyp}} = \frac{h}{c}, \text{ so } h = c \sin A$$

In a similar manner, using $\triangle CBD$, we can express the height of $\triangle ABC$ as $h = a \sin C$. Thus, we have expressed h in two ways, so

$$c \sin A = a \sin C, \text{ or}$$

$$\frac{\sin A}{a} = \frac{\sin C}{c}$$

In a similar manner, we can show that $\frac{\sin A}{a} = \frac{\sin B}{b}$. Combining these last two equations produces what is known as the **Law of Sines**.

THEOREM 6.15 The Law of Sines

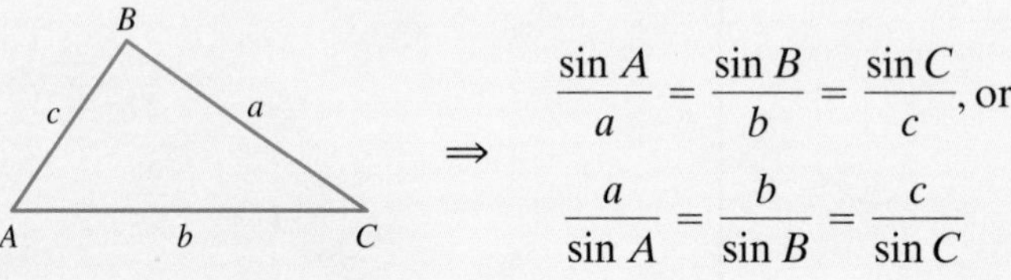

$$\Rightarrow \quad \frac{\sin A}{a} = \frac{\sin B}{b} = \frac{\sin C}{c}, \text{ or} \quad \frac{a}{\sin A} = \frac{b}{\sin B} = \frac{c}{\sin C}$$

We have seen that we may use the trig ratios to find the length of a side or the measure of an angle in a right triangle. We may use the Law of Sines to find the length of a side or the measure of an angle in an acute triangle. To solve a triangle using the Law of Sines, we must know the measure of at least one side and the measures of any two angles, or we must know the measures of any two sides and the measure of an angle opposite one of them.

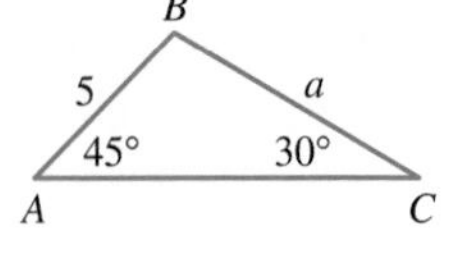

FIGURE 6.33

EXAMPLE 6.28 Find side length a in the triangle in Figure 6.33.

SOLUTION

According to the Law of Sines, we have

$$\frac{a}{\sin 45°} = \frac{5}{\sin 30°}, \text{ or } a = \frac{5 \sin 45°}{\sin 30°}$$

Using our results in Example 6.18 (b) and (d), we have

$$a = \frac{5 \sin 45°}{\sin 30°} = \frac{5\left(\frac{1}{\sqrt{2}}\right)}{\frac{1}{2}} = 5\left(\frac{2}{\sqrt{2}}\right) = 5\sqrt{2}$$

■

EXAMPLE 6.29 To estimate the height of a bridge over a stream, a hiker stands at point A on the bridge and measures the angle of depression to a rock in the stream as 65° (Figure 6.34). Then she walks 50 feet across the bridge to point B and measures the angle of depression to the same rock to be 42°. How high is the bridge above the water?

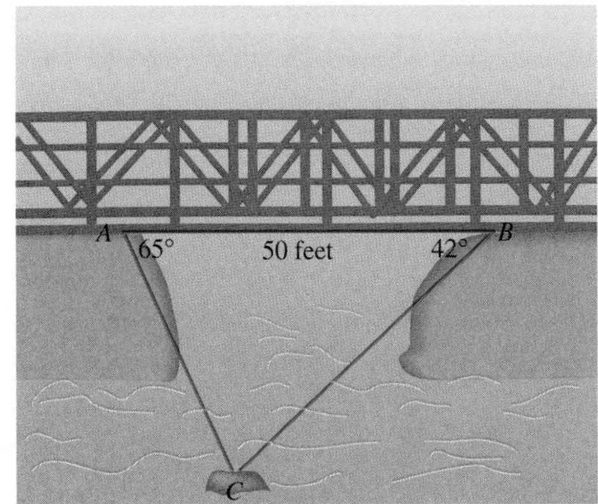

FIGURE 6.34

SOLUTION

We would like to find the altitude of the triangle from vertex C as shown in Figure 6.35, but we do not have enough information yet.

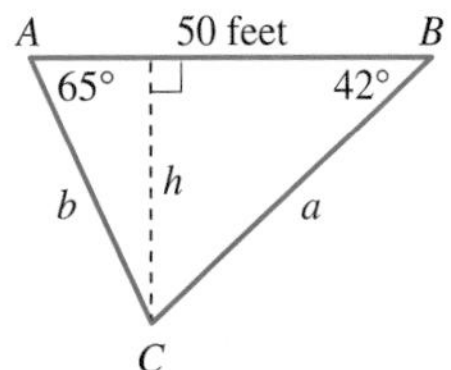

FIGURE 6.35

$\triangle ABC$ is an acute triangle, so we can use the Law of Sines to find the length of one of the other sides. The third angle in the triangle is $180° - 65° - 42° = 73°$ (Figure 6.36).

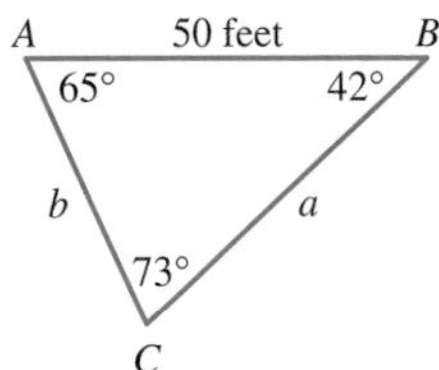

FIGURE 6.36

Next we use the Law of Sines to find side length a.

$$\frac{a}{\sin 65°} = \frac{50}{\sin 73°}, \text{ so } a = \frac{50 \sin 65°}{\sin 73°} \approx 47.39 \text{ feet}$$

Now we can find the altitude of the triangle from vertex C, which is the height of the bridge above the water (Figure 6.37).

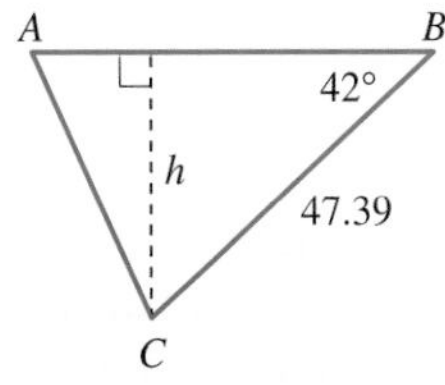

FIGURE 6.37

Using the sine ratio, we have $\sin 42° = \frac{\text{opp}}{\text{hyp}} = \frac{h}{47.39}$, so $h = 47.39 \sin 42° \approx$ 31.71 feet.

Therefore, the bridge is approximately 32 feet above the water. ■

Law of Cosines

We have seen that the Law of Sines provides a way to solve for angles or sides in an acute triangle. Another trigonometric law that we can use to solve acute triangles is called the Law of Cosines. We next explore that relationship between the sides and angles of a triangle.

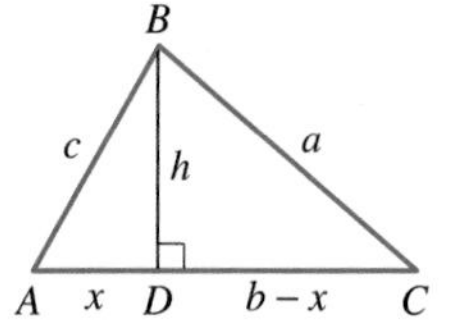

FIGURE 6.38

Consider $\triangle ABC$ in Figure 6.38, where $\overline{BD} \perp \overline{AC}$. The altitude $\overline{BD}$ divides $\triangle ABC$ into two right triangles. By the Pythagorean theorem, we have

$c^2 = x^2 + h^2$, which is equivalent to $h^2 = c^2 - x^2$, and

$a^2 = (b - x)^2 + h^2$, which is equivalent to $h^2 = a^2 - (b - x)^2$.

The two preceding equations on the right lead to

$c^2 - x^2 = a^2 - (b - x)^2$, which is equivalent to $c^2 - x^2 = a^2 - b^2 + 2bx - x^2$.

Simplifying the preceding equation on the right yields

$$a^2 = b^2 + c^2 - 2bx$$

Notice that in Figure 6.38, $\cos A = \frac{x}{c}$, so $c \cos A = x$. Substituting $c \cos A$ for x in the preceding equation, we have

$$a^2 = b^2 + c^2 - 2bc \cos A.$$

We can derive the following two equations in a similar fashion.

$$b^2 = a^2 + c^2 - 2ac \cos B$$

$$c^2 = a^2 + b^2 - 2ab \cos C.$$

These last three equations make up what is called the **Law of Cosines**, which we summarize next.

THEOREM 6.16 The Law of Cosines

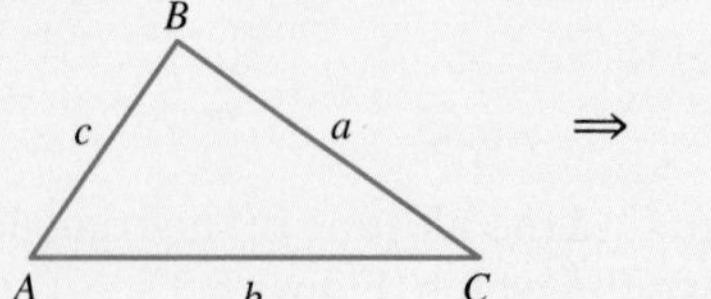

$\Rightarrow$

$$a^2 = b^2 + c^2 - 2bc \cos A$$

$$b^2 = a^2 + c^2 - 2ac \cos B$$

$$c^2 = a^2 + b^2 - 2ab \cos C$$

Like the Law of Sines, the Law of Cosines may be used to solve acute triangles. To use the Law of Cosines to solve a triangle, we must know the measures of two sides and the included angle of the triangle, or we must know the measures of the three sides of the triangle.

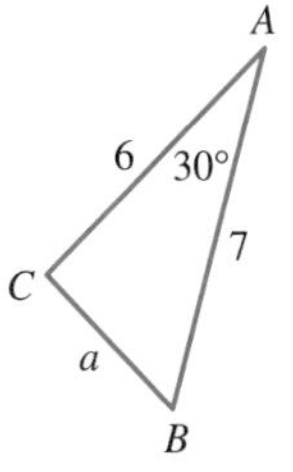

FIGURE 6.39

EXAMPLE 6.30 Find side length a in the triangle in Figure 6.39.

SOLUTION

In $\triangle ABC$ we are given $b = 6$, $c = 7$, and $\angle A = 30°$. According to the Law of Cosines,

$$a^2 = b^2 + c^2 - 2bc\cos A$$
$$a^2 = 6^2 + 7^2 - 2(6)(7)\cos 30°$$
$$a^2 = 12.253866082$$

Therefore, $$a \approx \sqrt{12.253866082} \approx 3.5$$ ■

Thus far, we have used the Law of Sines and the Law of Cosines to find the length of a side or a height in a triangle. The next example shows how we may use the Law of Cosines to determine the measure of an angle in a triangle.

EXAMPLE 6.31 An awning kit comes with identical sets of three metal rods to be assembled at points A, B, and C as shown in Figure 6.40. If the wall-mounted rod is 3 feet long, and a support rod 2.08 feet long is attached 2.43 feet from the end of the longest rod, what angle will the awning make with the wall?

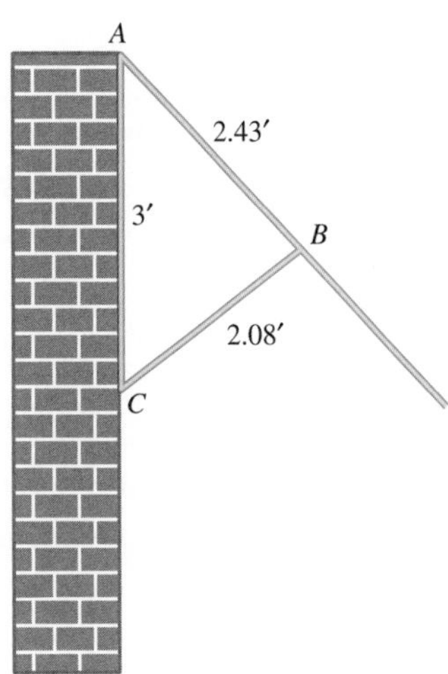

FIGURE 6.40

SOLUTION

The angle that the awning makes with the wall is $\angle CAB$ or $\angle A$ in $\triangle ABC$. Because we know the lengths of all three sides of the triangle, we use the Law of Cosines. According to the Law of Cosines,

$$a^2 = b^2 + c^2 - 2bc\cos A$$
$$2.08^2 = 3^2 + 2.43^2 - 2(3)(2.43)\cos A$$
$$\frac{2.08^2 - 3^2 - 2.43^2}{-2(3)(2.43)} = \cos A$$
$$0.725548697 \approx \cos A$$

To find the measure of $\angle A$, we now use the inverse cosine function.

$$A = \cos^{-1}(0.725548697) \approx 43.5°$$

Therefore, the awning makes a 43.5° angle with the wall. ■

Solution of Applied Problem

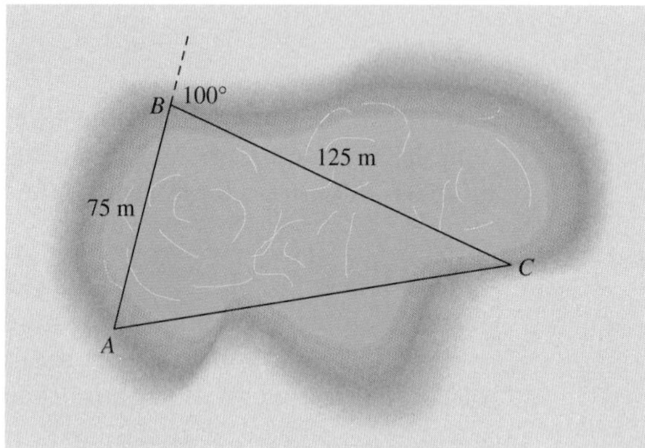

The distance across the lake is side length b in the figure. We must first find the measure of an angle in the triangle. We know that $\angle ABC = 180° - 100° = 80°$. Because we know the measures of two sides and their included angle, we will use the Law of Cosines to find b.

$$b^2 = a^2 + c^2 - 2ac \cos B$$
$$b^2 = 125^2 + 75^2 - 2(125)(75) \cos 80°$$
$$b^2 \approx 17{,}994.0967$$
$$b \approx \sqrt{17{,}994.0967} \approx 134.14 \text{ m}$$

So the lake is approximately 134 m wide.

GEOMETRY AROUND US **Triangulation** is a process by which distances are calculated using measured angles and a measured length in a triangle. First, the distance between two reference points, called the base line, is measured. Then, using a third reference point to complete a triangle, the angles between the base line and the sides of the triangle are measured. By using the Law of Sines or the Law of Cosines, we can calculate the lengths of the other two sides of the triangle. Triangulation is used in astronomy, navigation, and surveying. One famous use of this method is the Retriangulation of Great Britain, carried out from 1935 to 1962. More than 6000 pillars were placed across Great Britain as vertices of triangles. Very accurate maps of Great Britain resulted from the retriangulation.

PROBLEM SET 6.5

EXERCISES/PROBLEMS

In problems 1–6, use the Law of Sines to find the indicated side length in $\triangle ABC$. Round to the nearest tenth.

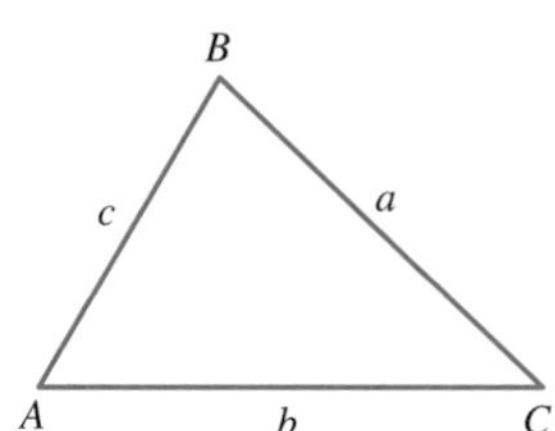

1. If $\angle A = 27°$, $\angle C = 70°$, and $a = 10$, find c.
2. If $\angle A = 33°$, $\angle C = 42°$, and $a = 23$, find c.
3. If $\angle B = 19°$, $\angle C = 73°$, and $c = 11$, find b.
4. If $\angle B = 52°$, $\angle C = 59°$, and $c = 25$, find b.
5. If $\angle A = 85°$, $\angle B = 12°$, and $b = 9.5$, find a.
6. If $\angle A = 24°$, $\angle B = 83°$, and $b = 19.7$, find a.

In problems 7–10, use the Law of Sines to find the indicated angle measure in $\triangle ABC$. Round to the nearest tenth.

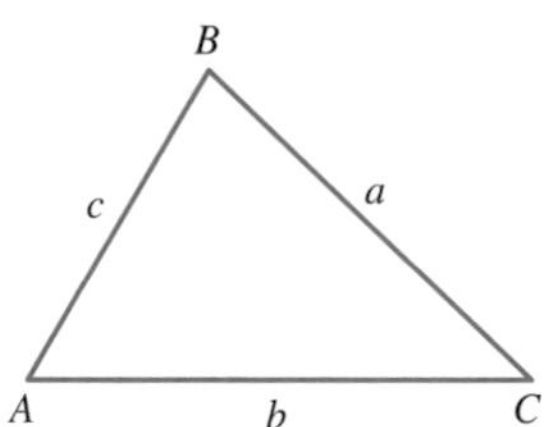

7. If $\angle A = 33°$, $a = 14$, and $b = 8.5$, find $\angle B$.
8. If $\angle B = 76°$, $b = 19.7$, and $c = 4.6$, find $\angle C$.
9. If $\angle C = 61.3°$, $a = 43$, and $c = 44.2$, find $\angle A$.
10. If $\angle A = 52.3°$, $a = 5.44$, and $b = 4.57$, find $\angle B$.

In problems 11–16, use the Law of Sines and the given information to solve $\triangle ABC$. Round approximate answers to the nearest tenth.

11. $\angle A = 89.5°$, $\angle C = 37°$, and $a = 7\sqrt{3}$
12. $\angle B = 47.6°$, $\angle C = 61°$, and $c = 10\sqrt{2}$
13. $\angle B = 68°$, $b = 91.1$, and $c = 67$
14. $\angle A = 83.2°$, $a = 11$, and $b = 9.6$
15. $\angle C = 73°$, $b = \sqrt{29}$, and $c = 7.2$
16. $\angle B = 87°$, $b = 18.21$, and $c = 17.45$

In problems 17–22, use the Law of Cosines to find the indicated side length in $\triangle ABC$. Round to the nearest tenth.

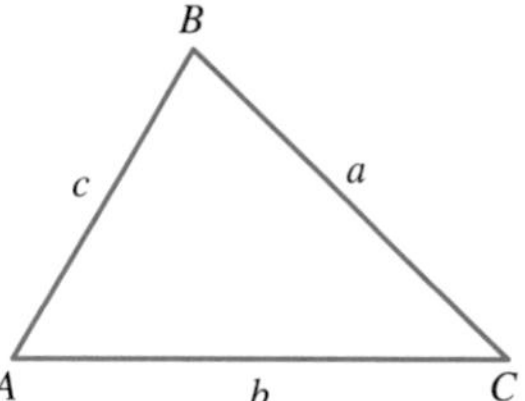

17. If $\angle A = 85°$, $b = 36.2$, and $c = 54.6$, find a.
18. If $\angle A = 79°$, $b = 8.55$, and $c = 10.47$, find a.
19. If $\angle B = 51°$, $a = 47$, and $c = 47$, find b.
20. If $\angle B = 75°$, $a = 16$, and $c = 32$, find b.
21. If $\angle C = 60.26°$, $a = 18.75$, and $b = 21.30$, find c.
22. If $\angle C = 68.35°$, $a = 19.375$, and $b = 15.4$, find c.

In problems 23–26, use the Law of Cosines to find the indicated angle measure in $\triangle ABC$. Round to the nearest tenth.

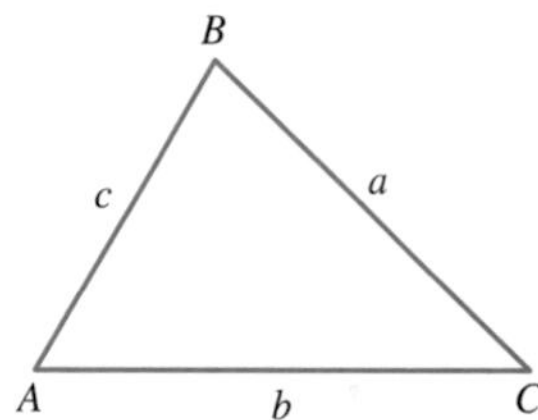

23. If $a = 14$, $b = 8$, and $c = 11$, find $\angle B$.
24. If $a = 22$, $b = 21$, and $c = 24$, find $\angle A$.
25. If $a = 2.3$, $b = 1.9$, and $c = 2.4$, find $\angle C$.
26. If $a = 15.67$, $b = 12.33$, and $c = 14.92$, find $\angle B$.

In problems 27–32, use the Law of Cosines and the given information to solve $\triangle ABC$. Round approximate answers to the nearest tenth.

27. $\angle A = 85°$, $b = 36.2$, and $c = 54.6$
28. $\angle B = 79°$, $a = 8.55$, and $c = 10.47$
29. $a = 55$, $b = 67$, and $c = 43$
30. $a = 19$, $b = 11$, and $c = 16$
31. $a = 75.6$, $b = 92.3$, and $c = 69.9$
32. $a = 100.34$, $b = 100.55$, and $c = 26.4$

In problems 33–36, do the following:

- **a.** State whether the Law of Sines or the Law of Cosines must be used to find the indicated measurement, and
- **b.** Find the indicated measurement. Round to the nearest tenth.

33. Find b.

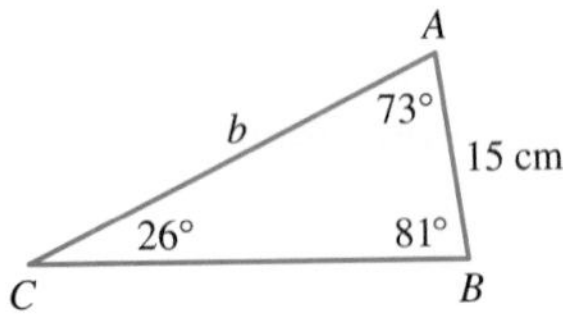

34. Find a.

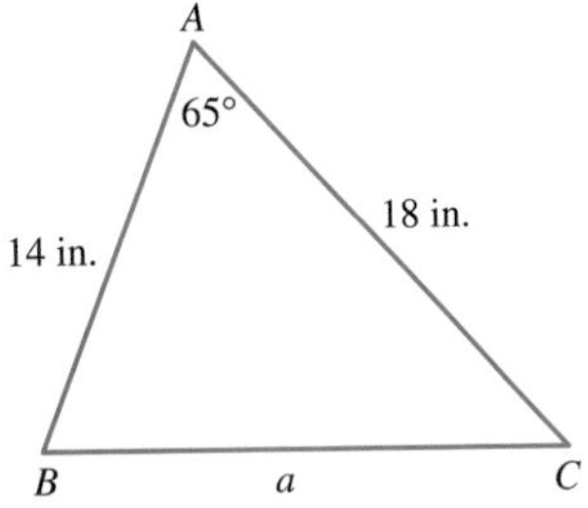

35. Find $\angle A$.

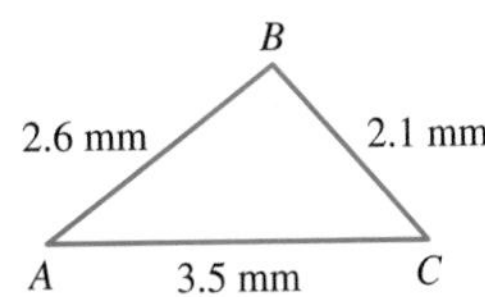

36. Find $\angle B$.

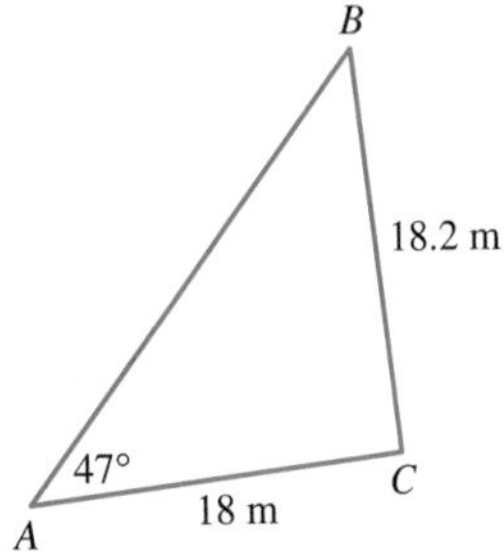

37. Find the area of a regular heptagon with side length 10 inches. Round to the nearest hundredth.

38. Find the perimeter of the following regular pentagon. Round to the nearest tenth.

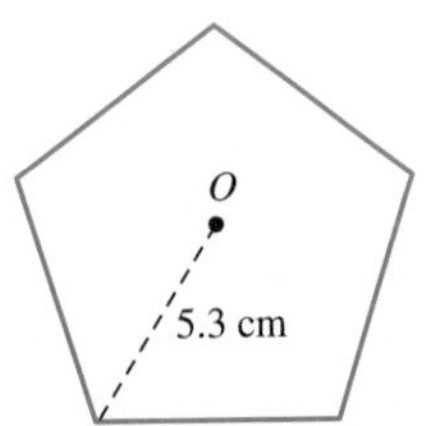

PROOFS

39. Consider isosceles $\triangle ABC$.

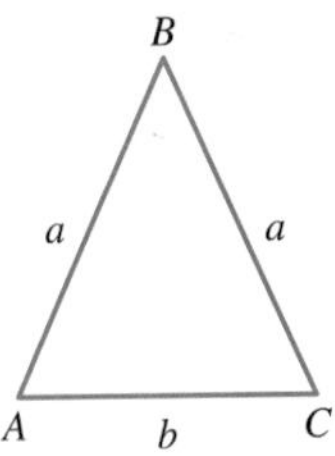

Use the Law of Cosines to prove

$$b^2 = 2a^2(1 - \cos B)$$

40. Consider isosceles $\triangle ABC$.

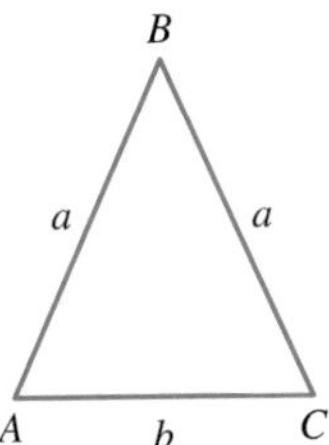

Use the Law of Cosines to prove

$$\cos A = \frac{b}{2a}$$

41. Consider $\triangle ABC$ as shown next.

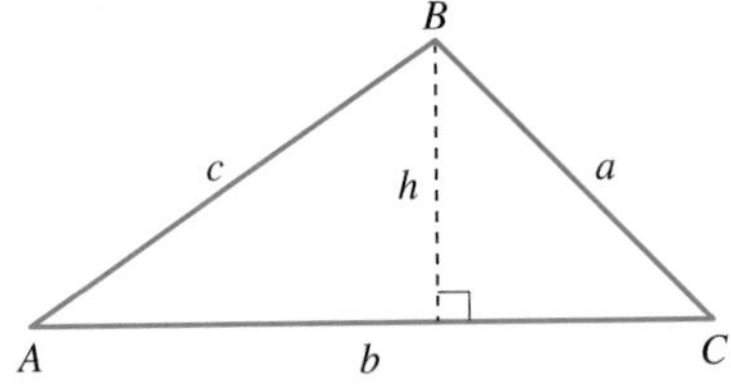

Use right triangle trigonometry, the result that $\sin^2 x + \cos^2 x = 1$, and the Law of Cosines to prove

$$\text{Area of } \triangle ABC = \frac{1}{2}bc\sqrt{1 - \left(\frac{b^2 + c^2 - a^2}{2bc}\right)^2}$$

42. In this text, we have discussed several ways to determine the area of a triangle, including $A = \frac{1}{2}bh$ and Heron's formula. We can also express the area of a triangle in terms of one side length and the three angle measures. Consider $\triangle ABC$ as shown next.

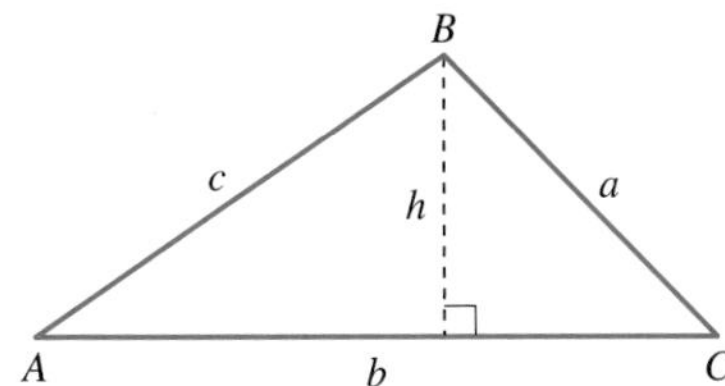

Use the Law of Sines to prove

$$\text{Area of } \triangle ABC = \frac{b^2 \sin A \sin C}{2 \sin B}$$

APPLICATIONS

43. Construction began on the Leaning Tower of Piza in 1173. Before completion, the tower began to tilt as a result of the unstable soil on which it was being erected. The original height of the tower was 58.36 meters. If an observer 33 meters from the center of the tower finds the angle of elevation to the top of the tower to be 65°, what is the acute angle the tower makes with the ground? Round to the nearest whole number.

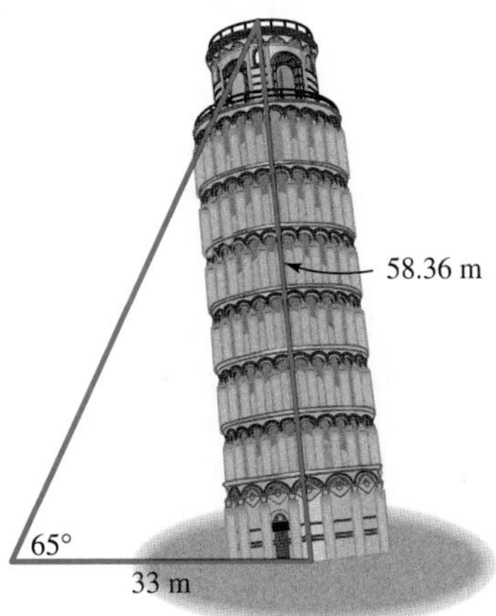

44. While whale watching off the Oregon coast, two observers standing 65 feet apart on the deck of a tour boat estimate the angle from the boat to the whale to be 35° and 85° respectively. How far is the whale from each observer? Round to the nearest whole number.

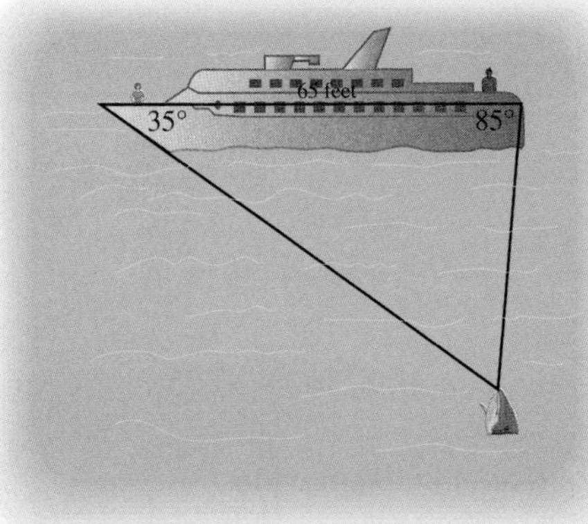

45. Two airplanes take off from the same airport. At a certain time, one plane is 250 miles from the airport and the other is 300 miles from the airport. If the angle between them is 75°, how far apart are the planes? Round to the nearest whole number.

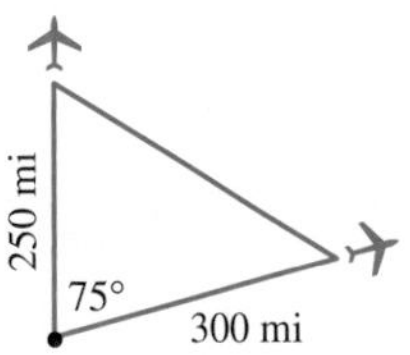

46. Two fishing boats head out to sea from a dock at the same time. Boat A travels 30 mph while Boat B travels 43 mph. After 1 hour and 15 minutes, both boats stop and drop anchor when positioned as shown in the following figure. How far apart are the boats? Round to the nearest whole number.

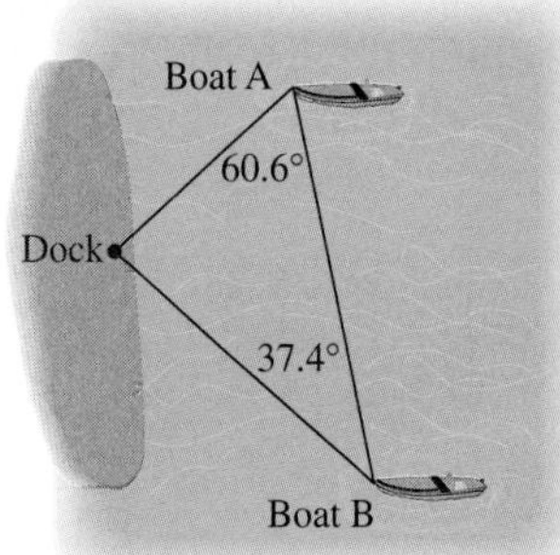

47. One observer estimates the angle of elevation to the basket of a hot air balloon to be 55°, while another observer 100 yards away estimates the angle of elevation to be 36°. How high off the ground is the basket of the hot air balloon? Round to the nearest whole number.

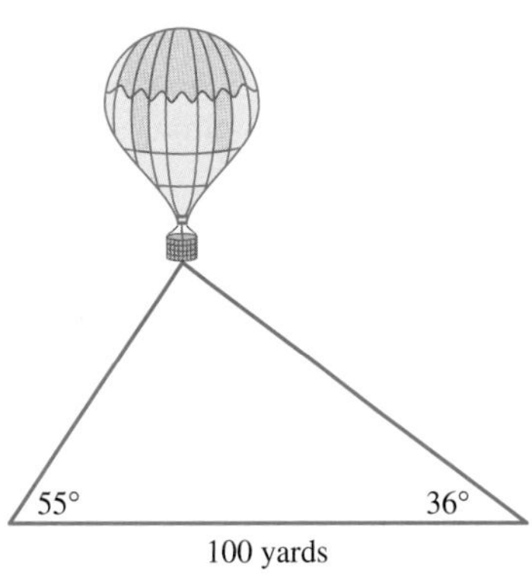

48. A log bridge over a canyon is 15 feet long. The angles of depression from each side of the log to a weed on the canyon floor are 84° and 86°, respectively. How high off the ground is the bridge? Round to the nearest whole number.

49. **Velocity** is the speed of an object in a particular direction. An airplane flies at a velocity of 450 mph due north. The wind blows at a velocity of 80 mph in the direction S 85° E. Find the resultant velocity (speed and direction) of the airplane. Round to the nearest hundredth. (HINT: See Problem 29 in Section 2.1 for a discussion of bearings.)

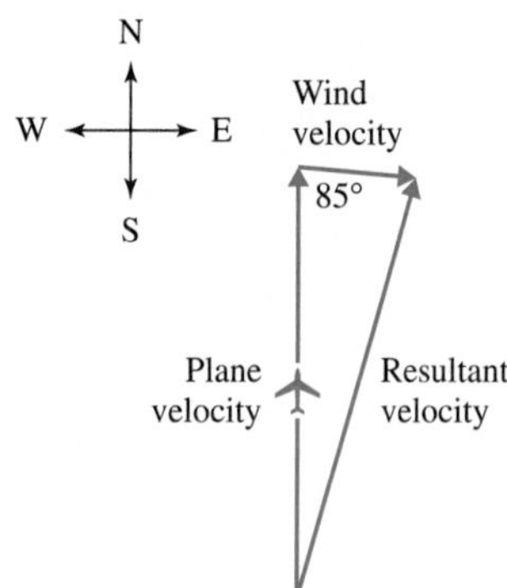

EXTENDED PROBLEMS

50. Problem 41 used the Law of Cosines to derive a formula for the area of a triangle that requires knowing only the lengths of its three sides. Heron's formula, which was stated in problem 37 in Section 3.1 as Area $= \sqrt{s(s-a)(s-b)(s-c)}$ where $s = \dfrac{a+b+c}{2}$, also can be used to find area of a triangle when only the three side lengths are known. Research Heron of Alexandria and summarize his contributions to mathematics. What inventions are attributed to him? What other formula is a special case of Heron's formula?

51. In the Olympics and other athletic events, the **Leica surveying system** is now used to measure distances using lasers. It has become increasingly popular due to its speed and accuracy. Research this surveying system and describe how measurements are taken and how trigonometry is used to find distances. In which Olympic games is the Leica system currently used? Also investigate what measurement technique was used previously and how the accuracy of the earlier system and the Leica system compare.

52. Trigonometry is one of the earliest recorded forms of mathematics. Research the history of trigonometry. Where and when did it originate? Create a time line of important trigonometric developments and identify who is given credit for them. In particular, be sure to summarize the contributions of Hipparchus and Ptolemy and list three of the earliest uses for trigonometry. Until calculators became widely available, trigonometric tables were widely used. Print out a trigonometric table and explain how to use it.

Solution of Initial Problem

A photographer takes aerial photographs from a plane. A dam that is known to be 100 m wide appears to be 3 cm long in a photograph. A second picture taken from the same height shows another dam to be 5.7 cm long. How wide is the second dam?

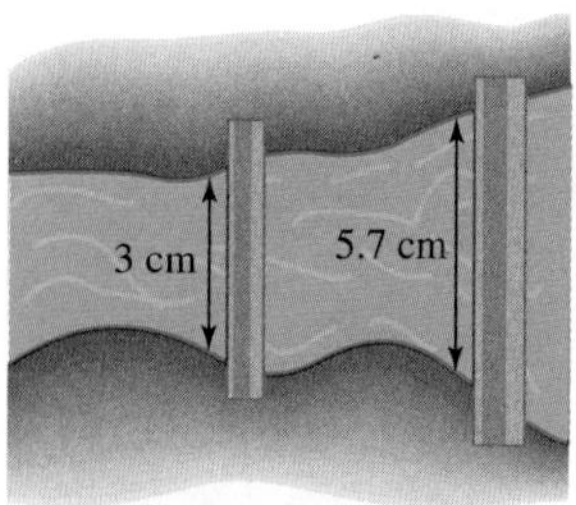

The ratio $\dfrac{3\text{ cm}}{100\text{ m}} = \dfrac{5.7\text{ cm}}{x\text{ m}}$ leads to the equation $3x = 100(5.7)$. Solving for x, we have $x = \dfrac{570}{3} = 190$. Thus, the second dam is 190 m wide.

Additional Problems Where the Strategy "Solve an Equation" Is Useful

1. One leg of a right triangle is 2 cm less than twice as long as the other leg. The hypotenuse of the triangle is 2 cm more than twice as long as the shorter leg. Find the lengths of the sides of the triangle.
2. Find the length of diagonal AC in the rhombus $ABCD$ given that $AD = 5$ in. and $\angle B = 100°$. Round your answer to two decimal places.

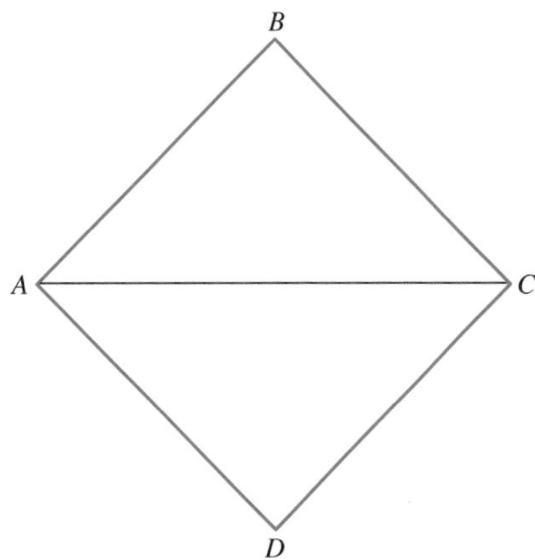

Writing for Understanding

1. Describe at least five ways in which ratios and proportions are used in everyday life. Write one problem illustrating each application that you mentioned.
2. Find at least five examples of the use of similar figures in the real world. Explain how you know that the figures are similar and how the concept of similarity is useful in each situation.
3. Research and write a brief article on the golden ratio (also called the divine proportion) as discussed in the "Geometry Around Us" at the end of Section 6.1.
4. Research and write a brief article on the geometry of fractals as discussed in the "Geometry Around Us" at the end of Section 6.2.

PEOPLE IN GEOMETRY

Benoit Mandelbrot (b. 1924) once said, "Clouds are not spheres, mountains are not cones, coastlines are not circles, and bark is not smooth, nor does lightning travel in a straightline." This is how he described the inspiration for fractal geometry, a new field of mathematics that finds order in chaotic, irregular shapes and processes. Mandelbrot was largely responsible for developing the mathematics of fractal geometry while at IBM's Watson Research Center. Before Mandelbrot, other mathematicians had considered the requisite mathematics, but they described it as "pathological," giving rise to "mathematical monstrosities." Mandelbrot once said, "The question I raised in 1967 is, 'How long is the coast of Britain,' and the correct answer is 'it all depends.' It depends on the size of the instrument used to measure length. As the measurement becomes increasingly refined, the measured length will increase. Thus, the coastline is of infinite length in some sense."

CHAPTER REVIEW

For each section of this chapter, you will find a list of vocabulary and notation, questions to assess your understanding of key concepts, and review problems similar to the problems you worked in your homework. Review each item in the *Vocabulary/Notation* list mentally, and, if necessary, refer back to the indicated page and write a definition. Then answer the *Concept Check Questions*, looking back at the section if you need help. Finally, work the *Review Problems* as practice before you move on to the *Chapter Test*. Answers to the *Review Problems* and *Chapter Test* can be found at the back of the book.

SECTION 6.1 Ratio and Proportion

Vocabulary/Notation

Similar 298
Ratio 298
Rate 299
Proportion 299
Terms 299
Extremes 299
Means 299
Arithmetic average (arithmetic mean) 302
Geometric mean (mean proportional) 302

Concept Check Questions

1. What are four different ways to represent the ratio of the number a to the number b? 298
2. Why are ratios expressed without units? 299
3. What is the difference between a ratio and a rate? 299
4. How are the means and extremes of a proportion related? 299
5. How are the means and extremes used to determine whether two ratios form a proportion? 300
6. In what ways can the terms of a proportion be interchanged to form a new proportion? 301
7. How can a missing term in a proportion be determined? 301–302
8. How is the arithmetic mean similar to the geometric mean? 302

Review Problems

1. Represent the ratio 4 : 5 in three other ways.
2. Represent the ratio 10 : 75 as a fraction in simplest form.
3. Determine whether $\frac{18}{32} = \frac{27}{48}$ is a proportion.
4. Scale drawings will be created for the Eiffel Tower which is 986 feet tall and the Empire State Building which is 1250 feet tall. If the Empire State Building is drawn 5.6 cm long, how long should the Eiffel Tower be drawn? Round to the nearest tenth.
5. Find the geometric mean of the numbers 6.2 and 24.8.
6. Prove Subgoal 2 of Theorem 6.1: If $\frac{a}{b} = \frac{c}{d}$, then $ad = bc$.

SECTION 6.2 Similar Triangles

Vocabulary/Notation

Similar triangles 307
Reflexive Property 308
Symmetric Property 308
Transitive Property 308
AAA Similarity 309
AA Similarity 309
SAS Similarity 310
LL Similarity 311
SSS Similarity 312

Concept Check Questions

1. If two triangles are similar, what is true about pairs of corresponding angles? 307
2. If two triangles are similar, what is true about pairs of corresponding sides? 307

3. What property guarantees every triangle is similar to itself? 308
4. If two triangles are each similar to a third triangle, what property guarantees the two triangles are similar to each other? 308
5. If two angles of one triangle are congruent to two angles of another triangle, why can we conclude the two triangles are similar? 309
6. If one angle of one triangle is congruent to one angle in another triangle, what else must be true in order to use the SAS Similarity Theorem to conclude the triangles are similar? 310
7. Why is LL Similarity a special case of SAS Similarity? 311
8. If no angle information can be determined about two triangles, what must be true about the sides in order to conclude the triangles are similar? 312

Review Problems

1. If $\triangle ABC \sim \triangle PQR$, list three pairs of congruent angles and two pairs of corresponding sides that are proportional to $\frac{AB}{PQ}$.
2. If $\triangle ABC \sim \triangle PQR$ and $\triangle PQR \sim \triangle MNO$, fill in the blank using the given property.
   a. $\triangle MNO \sim \triangle$_______by the reflexive property.
   b. $\triangle ABC \sim \triangle$_______by the transitive property.
3. Determine whether the following triangles are similar. If similar, give an appropriate similarity statement and explain why the triangles are similar.

a.

D F E: 5″, 3.1″, 6.8″

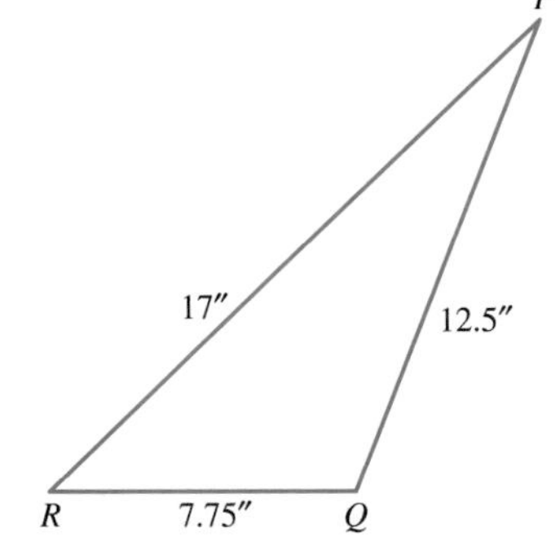

b.

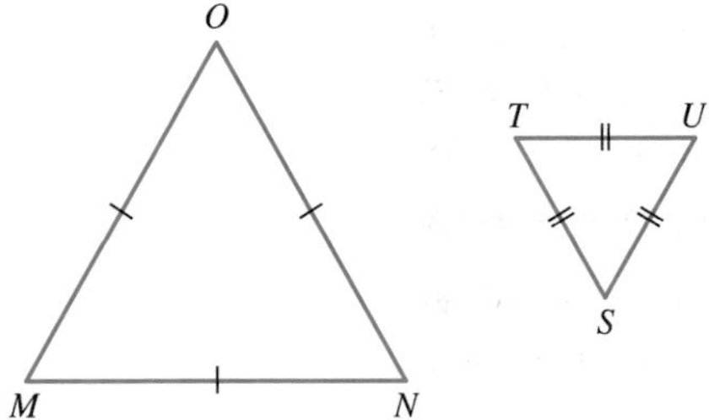

c.

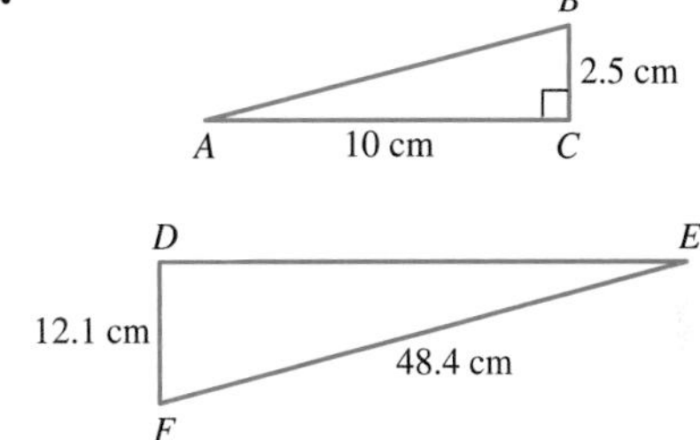

d.

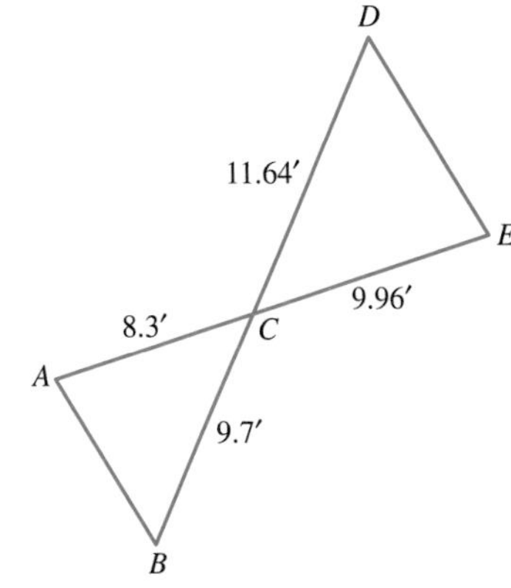

4. In the following figure, $AE \parallel BD$. Find ED.

A, B, C, D, E; 14.25 cm, 9.5 cm, 15 cm

5. To measure the distance across a lake, you mark points A and B on opposite ends of the lake. From point B, you walk 65 yards and place a stake at point C then continue 20 yards to point D. From point D, you walk parallel to $\overline{AB}$ and locate point E such that E, C, and A are collinear. If $DE = 42.5$ yards, what is the distance across the lake?

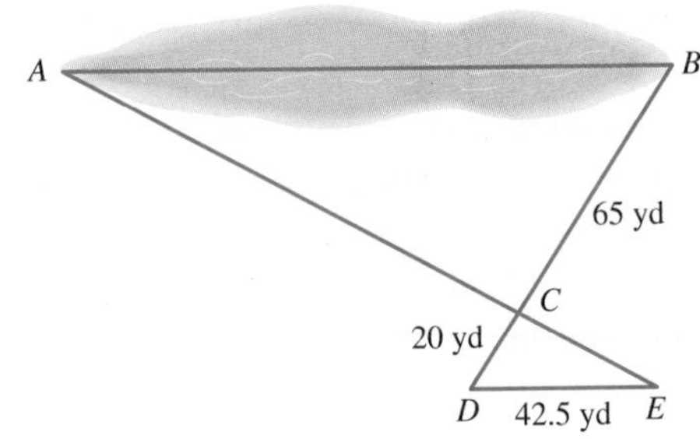

SECTION 6.3 Applications of Similarity

Vocabulary/Notation

Midsegment 322

Midquad 323

Concept Check Questions

1. How is the altitude to the hypotenuse related to the two segments formed by the altitude on the hypotenuse? 321
2. In what situation can the Side Splitting Theorem be used? 321
3. What are two properties of the midsegment of a triangle? 322
4. How is the midquad of a quadrilateral formed? 323
5. The midquad of any quadrilateral is what kind of polygon? 323

Review Problems

1. Use the following figure to find the indicated lengths. Round to the nearest hundredth.

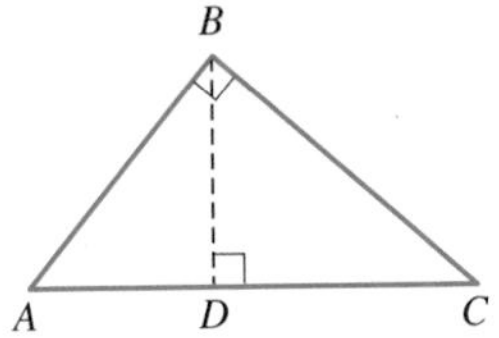

   a. If $AD = 9$ and $DC = 11$, find BD and BC.
   b. If $BD = 4$ and $AD = 2.5$, find DC and AB.

2. Use the following figure to find the indicated lengths.

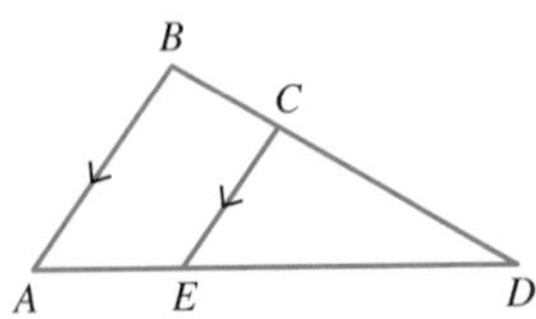

   a. If $AE = 4$, $CD = 7$, and $BC = 2.5$, find DE.
   b. If $BD = 16$, $ED = 18$, and $AE = 14$, find BC.

3. If $\overline{CE} \parallel \overline{BF} \parallel \overline{AG}$, find x and AG.

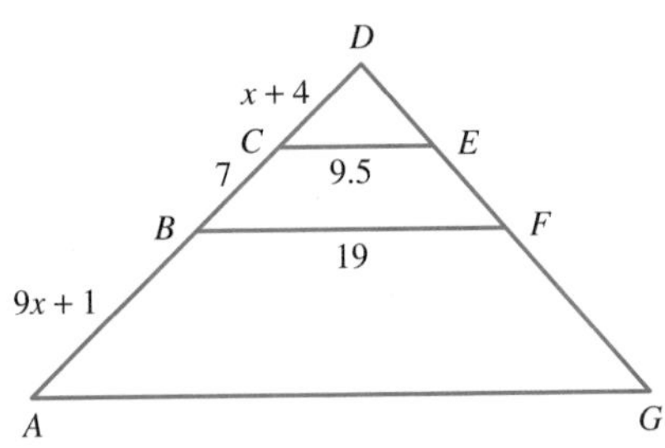

4. Find the perimeter and area of the midquad of a 10 cm by 5 cm rectangle.
5. In $\triangle ABC$, $\overline{DE}$ is the midsegment and $\triangle BDE \cong \triangle BED$. Prove $ADEC$ is an isosceles trapezoid.

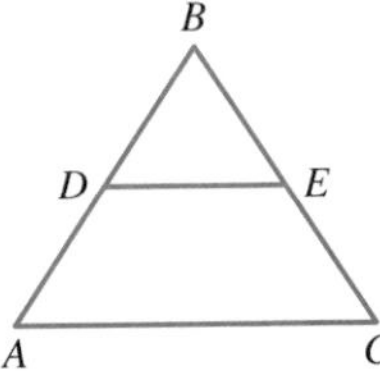

6. Prove: If the midpoints of the sides of a square are connected consecutively, the figure formed is a square.

SECTION 6.4 Using Right Triangle Trigonometry to Solve Geometry Problems

Vocabulary/Notation

Trigonometry 329
Tangent 329
tan A 329
Sine 331
sin A 331
Cosine 331
cos A 331
Trig ratios 332
Inverse trig function keys 333
Solving right triangles 335

Concept Check Questions

1. What is true about the ratio of a pair of corresponding sides in any pair of similar right triangles? 329
2. How is the tangent of an acute angle in a right triangle defined? 329
3. If we use a calculator to find the tangent of an acute angle, how do we interpret the results? 330
4. Why are the sine of an acute angle and the cosine of an acute angle always between 0 and 1? 332
5. What is the result of using an inverse trigonometric function? 333
6. How are the sine, cosine, and tangent of an acute angle related? 334
7. How do we use the Pythagorean theorem to derive a relationship between the sine and cosine of an acute angle? 334

Review Problems

1. Solve the right triangle. Round to the nearest whole number.

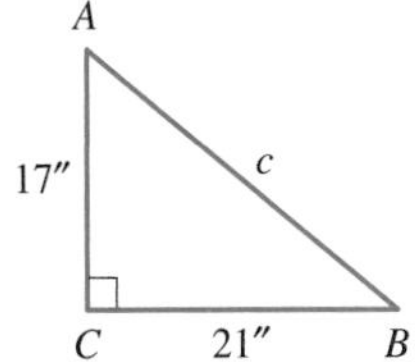

2. Use the following figure.

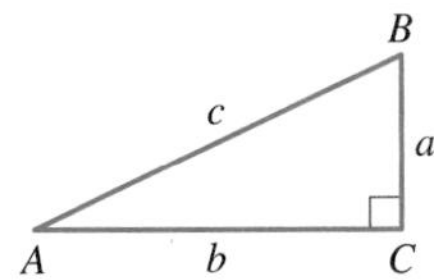

   a. Find the exact value of tan A if $c = 5$ and $b = 2$.
   b. Find c if $\angle B = 22°$ and $b = 16.3$. Round to the nearest hundredth.

3. Find the exact value of each of the following.
   a. sin 45°
   b. cos 30°
   c. tan 60°
4. Find the area of the triangle. Round to the nearest hundredth.

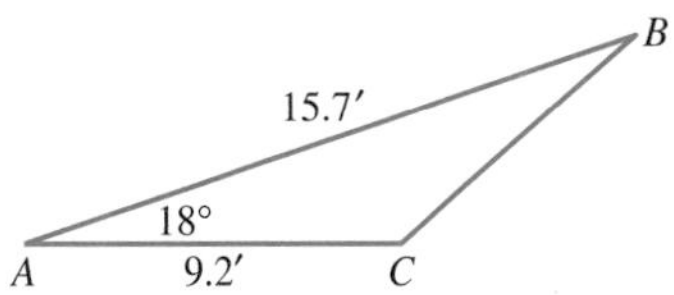

5. An escalator is 508 feet long and the angle it forms with the horizontal is 32°. What is the vertical distance traveled if a passenger rides from the bottom of the escalator to the top? Round to the nearest tenth.

SECTION 6.5 Using the Laws of Trigonometry to Solve Geometry Problems

Vocabulary/Notation

Oblique triangle 341
Law of Sines 342
Law of Cosines 344

Concept Check Questions

1. Under what conditions can we use the Law of Sines to solve a triangle? 342
2. Under what conditions can we use the Law of Cosines to solve a triangle? 344

Review Problems

1. Use the Law of Sines to solve for the missing measurements in $\triangle ABC$ if $\angle A = 38°$, $\angle C = 54°$, and $a = 18$ cm. Round approximate answers to the nearest hundredth.
2. Use the Law of Cosines to find the missing measurements in $\triangle ABC$ if $\angle A = 73.2°$, $b = 13.54$ mm, and $c = 15.32$ mm. Round to the nearest hundredth.
3. The plans for a Fink roof truss as shown next specify that $GC = 7'$, $DG = 3.5'$, and $\angle C = 29°$. Find $\angle GDC$ and DC. Round to the nearest tenth.

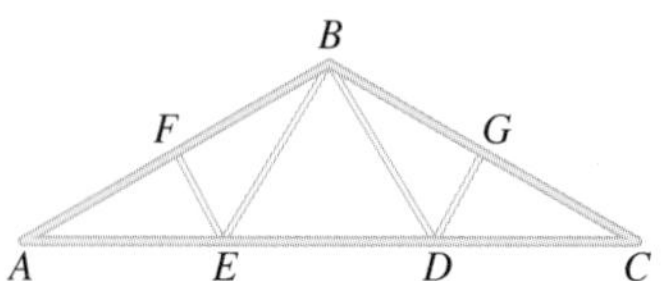

4. Find the angle measures of a triangle with side lengths 25 feet, 19 feet, and 22 feet. Round answers to the nearest tenth.
5. An airplane flies due south at a speed of 300 mph. The wind blows N 85° W at 65 mph. Find the resultant speed and direction of the plane. Round to the nearest tenth.

CHAPTER 6 TEST

TRUE-FALSE

Mark as true any statement that is always true. Mark as false any statement that is never true or that is not necessarily true. Be able to justify your answers.

1. If $\frac{a}{b} = \frac{c}{d}$, then $\frac{a}{c} = \frac{d}{b}$.
2. In a proportion, the product of the means equals the product of the extremes.
3. Two triangles are similar if two sides of one triangle are proportional to two sides of the other triangle.
4. A midsegment of a triangle is perpendicular to one of the sides of the triangle.
5. If $\angle A$ is an acute angle that measures less than 45°, then $\sin A < \cos A$.
6. Any two right isosceles triangles are similar.
7. If four angles of one quadrilateral are congruent, respectively, to four angles of another quadrilateral, then the quadrilaterals are similar.
8. In a right triangle, the altitude to the hypotenuse creates two right triangles, each of which is similar to the original triangle.
9. If the ratio of the lengths of the sides of two equilateral triangles is 3:5, then the ratio of their areas is 9:25.
10. The cosine ratio is the reciprocal of the sine ratio.
11. We can use the Law of Sines to solve a triangle when we know the lengths of two sides and the measure of the included angle.
12. It is possible to find the measures of the vertex angles in a triangle if we know the measures of the three sides.

EXERCISES/PROBLEMS

13. Find x in the proportion $\frac{x}{24} = \frac{5}{16}$.
14. Find the geometric mean of 7 and 13.
15. Suppose $\triangle ABC \sim \triangle DEF$, $AB = 5$, $BC = 9$, and $DE = 35$. Find EF.
16. a. Draw a right triangle with sides of lengths 5 units, 12 units, and 13 units. Verify that this triangle is a right triangle.
    b. Determine the sine and cosine of each acute angle in the triangle.
    c. Verify that the relationship $\sin^2 A + \cos^2 A = 1$ holds for each acute angle in the triangle.
17. Square $ABCD$ has $AC = 20$ cm. Find the perimeter of the quadrilateral $EFGH$ formed by joining the midpoints of $ABCD$.

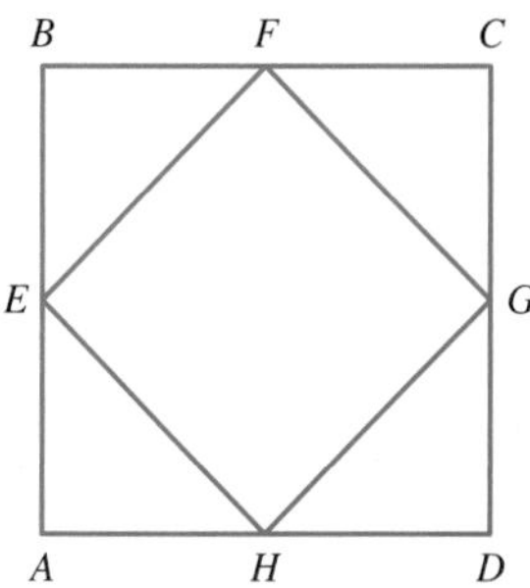

18. In a right triangle, one leg has length 10 and its opposite angle is 48°. Find the length of the hypotenuse to the nearest tenth.

19. Find x in the figure where $\overline{BC} \parallel \overline{DE}$.

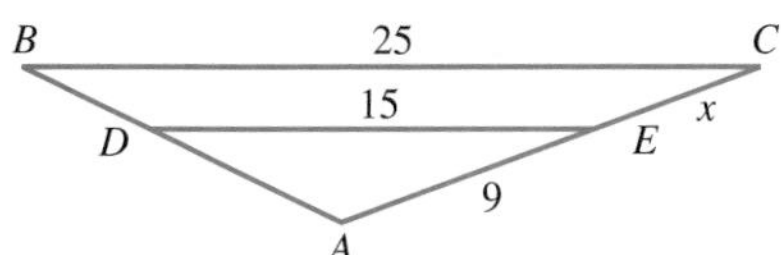

20. Find two different pairs of similar triangles in the figure where $DE \parallel AC$. Justify your answers.

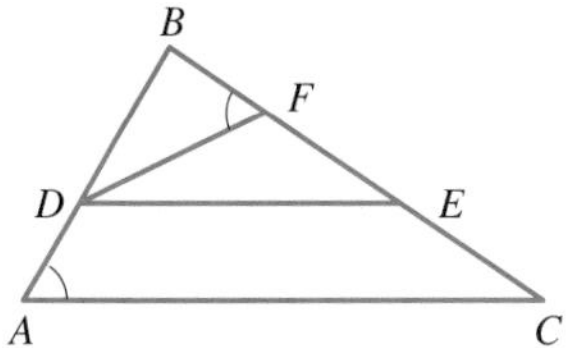

21. Find the area of the parallelogram shown. Round your answer to the nearest whole number.

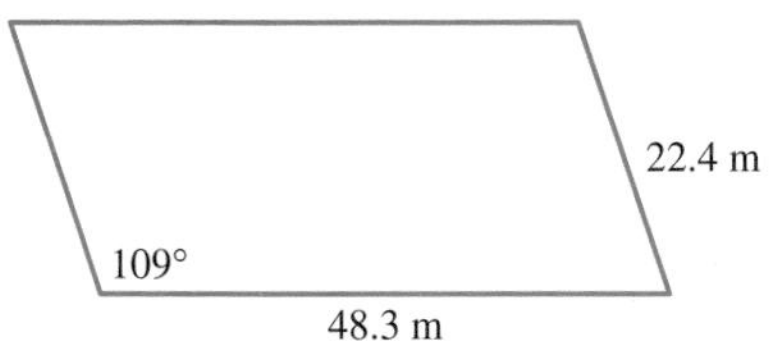

22. One group touring Washington D.C. leaves Washington Circle and walks 7 blocks along New Hampshire Avenue. Another group leaves Washington Circle and walks 5.5 blocks along Pennsylvania Avenue. How far apart are the groups? Round to the nearest tenth.

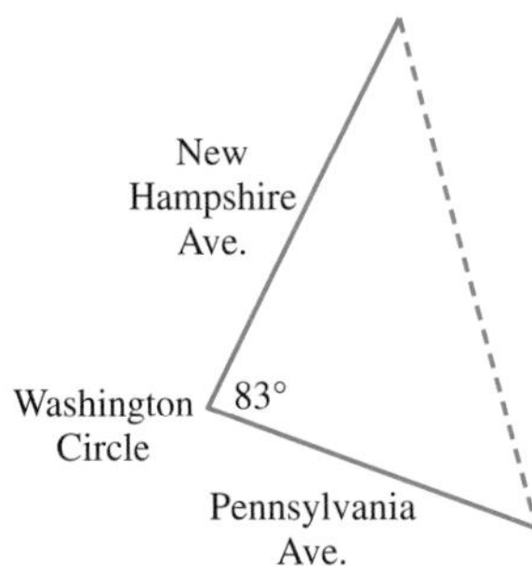

23. Solve for the missing parts of the triangle shown. Round to the nearest tenth.

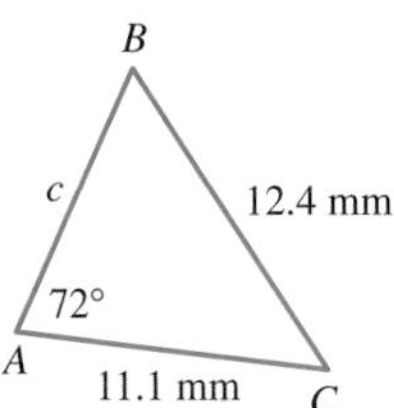

24. Solve for the missing parts of the triangle shown, given that $b = 55$ and $c = 73$. Round angle measures to the nearest tenth.

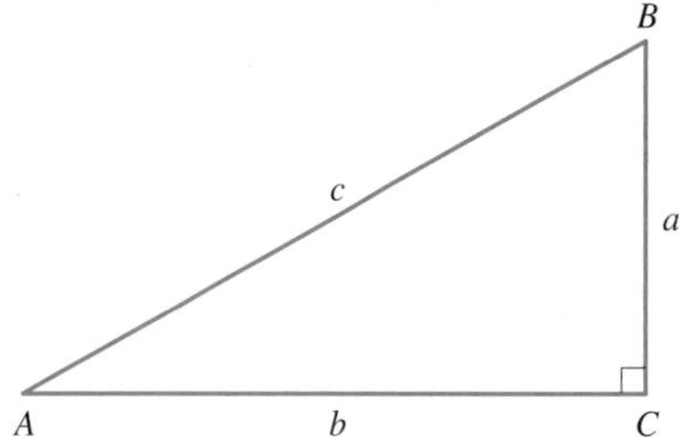

PROOFS

25. **a.** If the height of a right circular cone is doubled and the radius of the cone is cut in half, what is the ratio of the volume of the new cone to the volume of the original cone?

b. Prove your assertion in part (a)

26. Two parallel lines are crossed by two different transversals that are not parallel. The two transversals intersect in a point that is between the two parallel lines. Prove that the two triangles formed are similar.

APPLICATIONS

27. An observer in a lighthouse is 70 feet above sea level. She sees a whale spout and measures the angle of depression to the whale as 16.2°. What was the horizontal distance from the whale to the lighthouse when the woman saw it? Round to the nearest whole number.

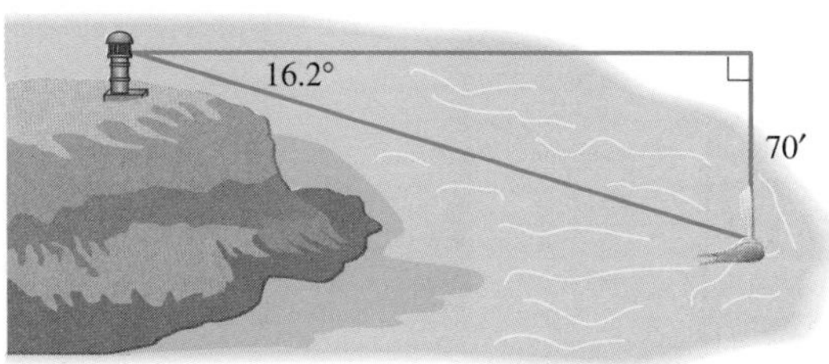

28. A wire is strung between two posts as shown. Find the length of the wire from A to B to E to F. Round to the nearest tenth.

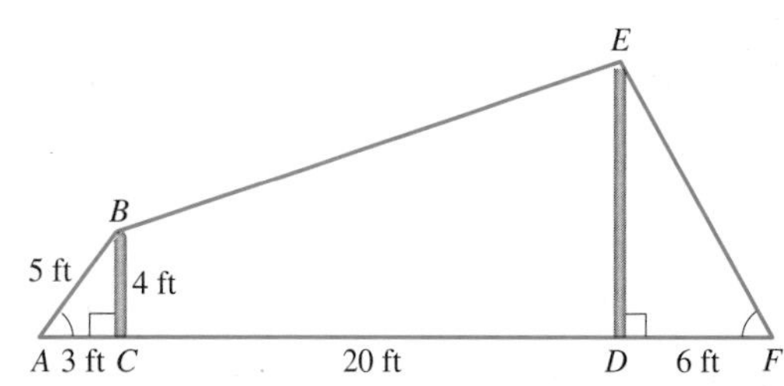

29. The crankshaft is part of a car's engine that translates back-and-forth motion from a piston into rotational motion that turns the wheels. The crankshaft is connected to a piston by a connecting rod as shown in the following figure. If the connecting rod measures 9 cm, find RP when $\angle SRP = 80°$. Round to the nearest hundredth.

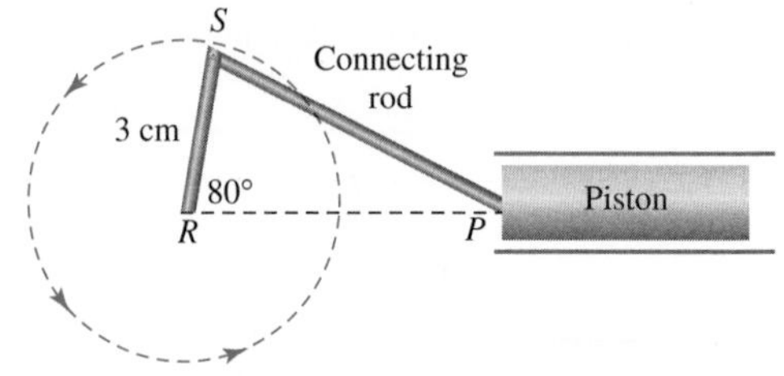

CHAPTER 7

CIRCLES

A **cycloid** is the path traced out by a point P on the circumference of a circle as the circle rolls (without slipping) along a horizontal line l. A portion of a cycloid is shown next.

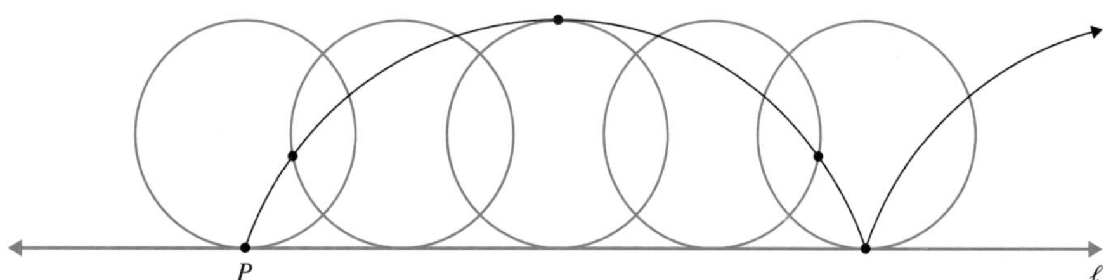

During the 17th century, a number of arguments among prominent mathematicians of the time centered on this curve. Galileo, Pascal, Descartes, Johann and Jakob Bernoulli, Leibniz, Newton, and others studied the mechanical properties of the cycloid. Among their discoveries were the following:

1. An inverted cycloid is the curve of equal descent. That is, if two marbles at different points on an inverted cycloid are released at the same time, they will reach the bottom at exactly the same time.

2. An inverted cycloid is also the curve of *fastest* descent. That is, if a marble at one point is to roll down to another point in the shortest amount of time, the path it should follow is, surprisingly, not a straight line, but an inverted cycloid.

Problem-Solving Strategies

1. Draw a Picture
2. Guess and Test
3. Use a Variable
4. Look for a Pattern
5. Make a Table
6. Solve a Simpler Problem
7. Look for a Formula
8. Use a Model
9. Identify Subgoals
10. Use Indirect Reasoning
11. Solve an Equation
12. *Use Cases*

Strategy 12: *Use Cases*

We can solve many problems more easily by separating the problem into various cases. For example, to determine if a certain property is true for all triangles, it may be easier to first examine all right triangles for this property, then all acute triangles, and, finally, all obtuse triangles.

Initial Problem

The circumscribed circle of a triangle is the unique circle that contains all three vertices of the triangle. For a given triangle, develop a simple test to determine if the center of the circumscribed circle is inside, on, or outside the triangle.

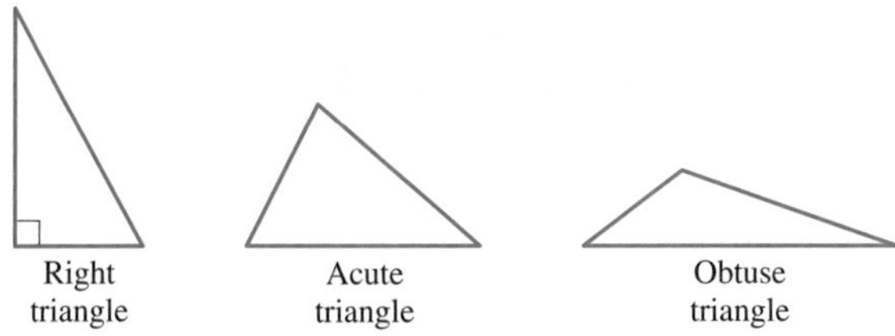

Clues

The Use Cases strategy may be appropriate when

- You can separate a problem into several distinct categories.
- You can generalize investigations in specific cases.

A solution of this Initial Problem is on page 410.

Nap△le△n's Triangle

The following construction has been attributed to Napoleon Bonaparte (1769–1821), who ruled as the Emperor of France and is said to have excelled at mathematics.

1. Draw any triangle. Use a compass and straightedge to construct an equilateral triangle on each side of the triangle.

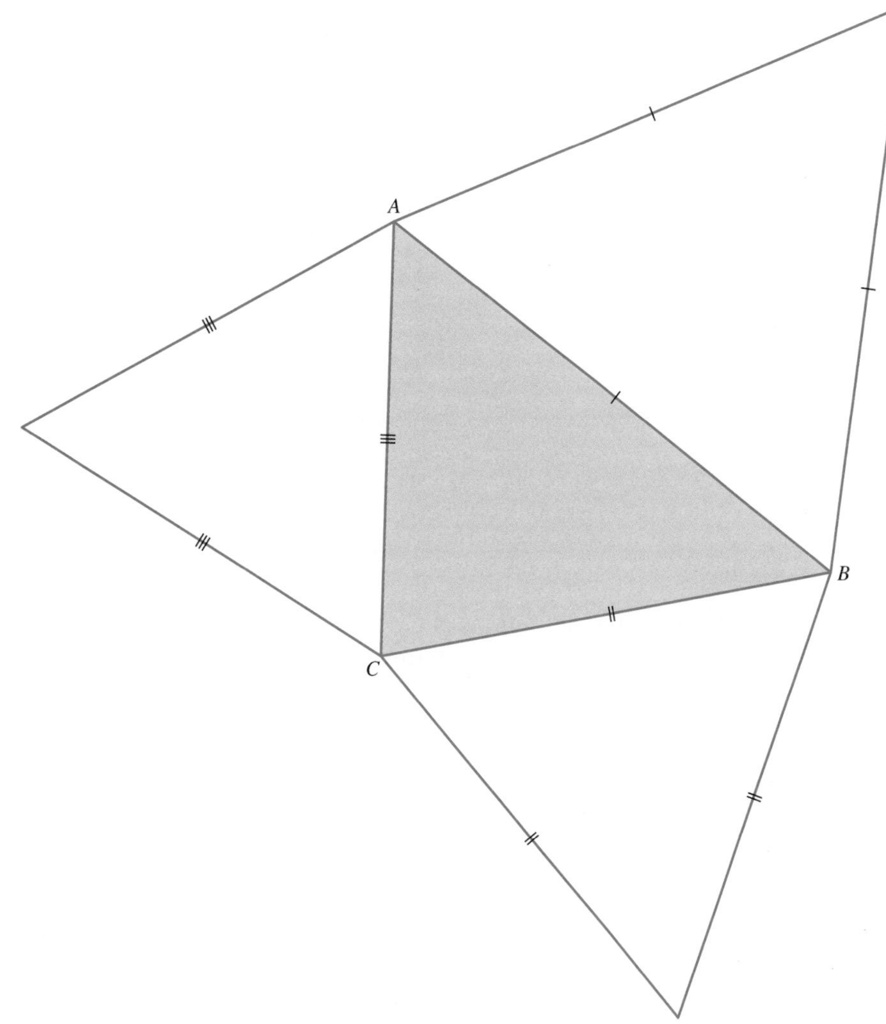

2. For each of the equilateral triangles, construct the median to each side. What do you notice about the three medians in each triangle?
3. Label the points of intersection of the medians in the three triangles D, E, and F. Draw $\triangle DEF$. Measure the side lengths and the vertex angles of $\triangle DEF$. Describe $\triangle DEF$ as completely as possible.
4. Draw a different $\triangle ABC$ and repeat the construction described in (1)–(3) to form a new $\triangle DEF$. Describe the $\triangle DEF$ as completely as possible. Is your description of $\triangle DEF$ the same as in (3)?
5. Repeat the construction on a third triangle at least as large as the original $\triangle ABC$. But this time, rather than constructing equilateral triangles on the outside of $\triangle ABC$, construct equilateral triangles oriented toward the inside of $\triangle ABC$. What type of triangle is $\triangle DEF$? Is it the same as $\triangle DEF$ in (3) and (4)?
6. Write a statement of what you think Napoleon's theorem says about any triangle.

Introduction

This chapter contains a study of circles and portions of circles as well as special lines and angles associated with circles. This material is particularly rich in that its organization ties together many ideas and consequently helps one see many interesting mathematical connections.

7.1 CENTRAL ANGLES AND INSCRIBED ANGLES

Applied Problem

A lawn sprinkler sprays water over a radius of 30 feet. If the sprinkler is set to turn through an angle of 110°, calculate the area that will be watered by the sprinkler.

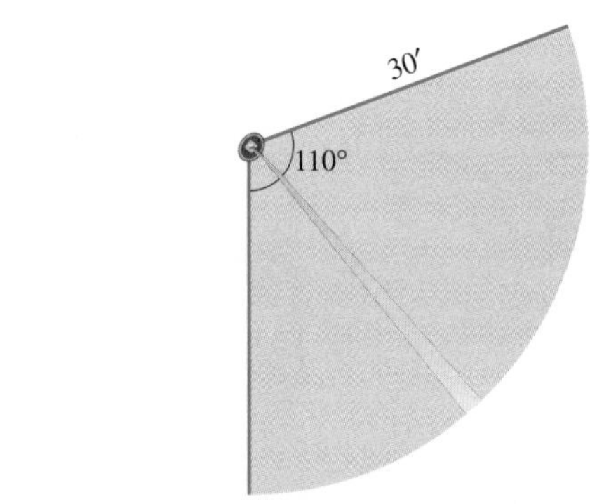

We gave the definitions of circle, radius, and diameter in Chapter 2. In this section, we introduce some new terms associated with circles.

Arcs of a Circle

An **arc** $\widehat{AB}$ of a circle consists of the points A and B of a circle together with the portion of the circle contained between the two points as shown in Figure 7.1. Points A and B are called the **endpoints** of the arc [Figure 7.1(a)]. A **semicircle** is an arc whose endpoints are the endpoints of a diameter [$\widehat{AB}$ in Figure 7.1(b)]. An arc of a circle that is shorter than a semicircle is called a **minor arc** of the circle [$\widehat{AB}$ in Figure 7.1(c)].

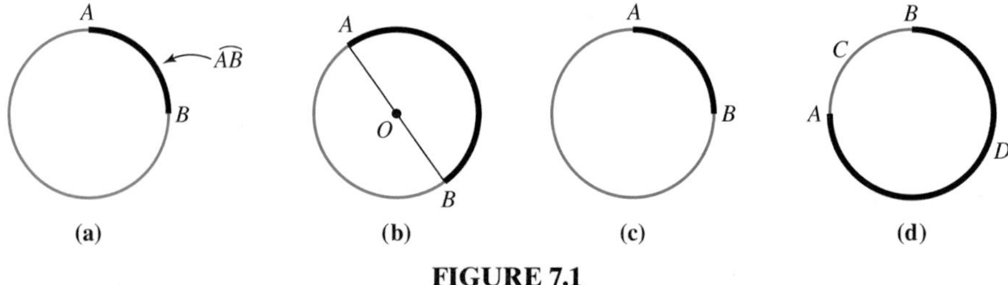

FIGURE 7.1

An arc that is longer than a semicircle is called a **major arc** [$\widehat{ADB}$ in Figure 7.1(d)].

Notice that due to possible ambiguity, we may also name arcs using a third letter on the arc. In Figure 7.1(d), the major arc is named $\overset{\frown}{ADB}$ or $\overset{\frown}{BDA}$ and the minor arc is named $\overset{\frown}{ACB}$. If there are only two letters, we agree that a minor arc is being named, so $\overset{\frown}{AB}$ describes the same arc as $\overset{\frown}{ACB}$. Arcs can be added or subtracted much like line segments. For example, in Figure 7.1(d), $\overset{\frown}{AB} = \overset{\frown}{AC} + \overset{\frown}{CB}$ and $\overset{\frown}{ADB} - \overset{\frown}{DB} = \overset{\frown}{AD}$.

Central Angles in a Circle

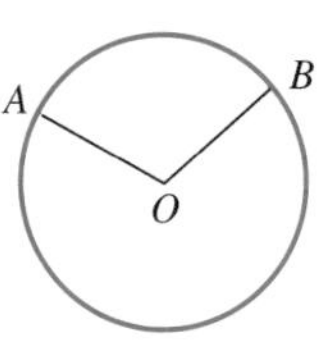

FIGURE 7.2

An angle whose vertex is the center of a circle and whose sides are radii is called a **central angle**. If O is the center of a circle containing points A and B, then $\angle AOB$ is a central angle (Figure 7.2). When the sides of an angle such as $\angle AOB$ intersect a circle at the two points A and B as in Figure 7.2, we say that $\angle AOB$ **intercepts** the arc $\overset{\frown}{AB}$. The measure of an arc of a circle is related to the central angle that intercepts it.

DEFINITION Measure of an Arc

The measure of an arc is the measure of the central angle that intercepts the arc.

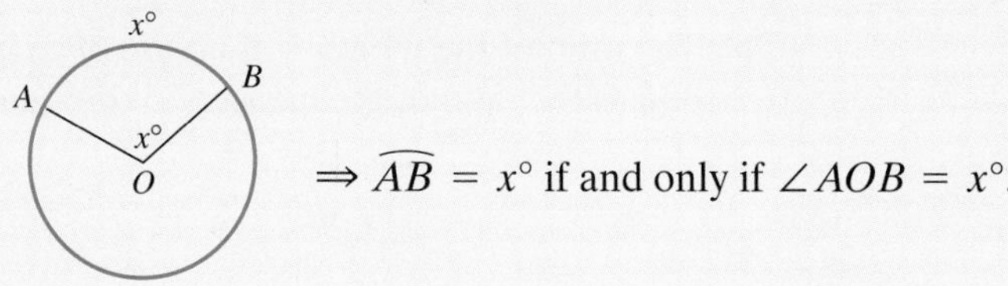

$\Rightarrow \overset{\frown}{AB} = x°$ if and only if $\angle AOB = x°$.

The measure of a semicircle is 180°. This means the measure of a minor arc will be less than 180° and the measure of a major arc is greater than 180° but less than 360°. Just as in the case of angle measure, the measure of $\overset{\frown}{AB}$ is written as $m(\overset{\frown}{AB})$ or simply $\overset{\frown}{AB}$ when the context is clear.

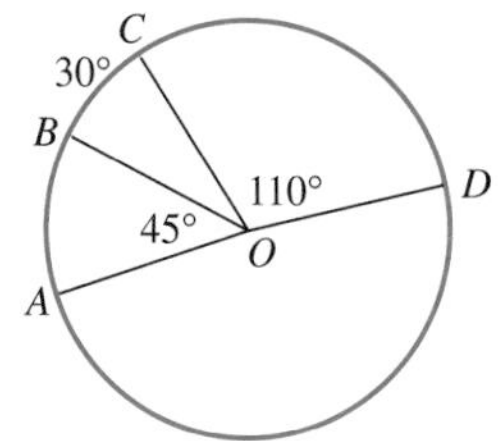

FIGURE 7.3

EXAMPLE 7.1 Use circle O in Figure 7.3 to find the measures of the following.

a. $\overset{\frown}{AB}$ **b.** $\angle BOC$ **c.** $\overset{\frown}{ACD}$ **d.** $\overset{\frown}{AD}$

SOLUTION

a. $\overset{\frown}{AB} = 45°$ because $\angle AOB = 45°$

b. $\angle BOC = 30°$ because $\overset{\frown}{BC} = 30°$

c. $\overset{\frown}{ACD} = \overset{\frown}{AB} + \overset{\frown}{BC} + \overset{\frown}{CD} = 45° + 30° + 110° = 185°$

d. $\overset{\frown}{ACD} + \overset{\frown}{AD} = 360°$. Therefore, $\overset{\frown}{AD} = 360° - \overset{\frown}{ACD} = 360° - 185° = 175°$. ■

Length of an Arc of a Circle

We have seen that the degree measure of an arc of a circle is defined to be the same as the degree measure of the central angle that intercepts it. We can use the degree measure of an arc to describe the fraction of a circle an arc comprises. For example, an arc of 90° would be equivalent to one quarter of a circle. But two arcs of 90°, not in the same circle, will not necessarily be the same *length*. The length of an arc of a circle depends on both the central angle *and* the radius of the circle.

EXAMPLE 7.2 A clock has a pendulum 32 in. long. If the pendulum swings through an arc of 12°, how far does the pendulum travel?

SOLUTION

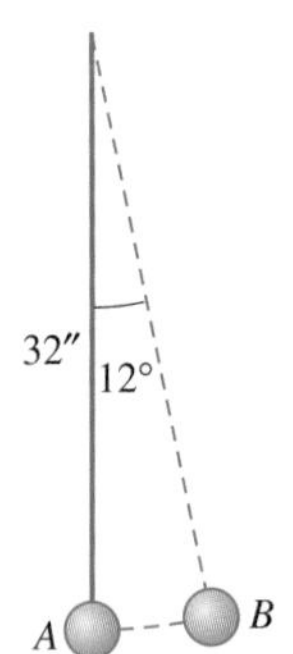

FIGURE 7.4

The pendulum swings through an arc of a circle as shown in Figure 7.4. The central angle of 12° means that $\widehat{AB}$ is $\frac{12°}{360°} = \frac{1}{30}$ of a circle with radius 32 in. So the length of arc $\widehat{AB}$ is $\frac{1}{30}$ of the circumference of a circle with radius 32 in. Therefore,

$$\widehat{AB} = \frac{1}{30}(2\pi r) = \frac{1}{30}(2\pi)(32 \text{ in.}) \approx 6.7 \text{ in.}$$

■

We can generalize this example to yield the next postulate.

POSTULATE 7.1 Length of an Arc

The ratio of the length l, of an arc of a circle to the circumference, C, of the circle equals the ratio of the measure of the central angle of the arc, $x°$, to 360°.

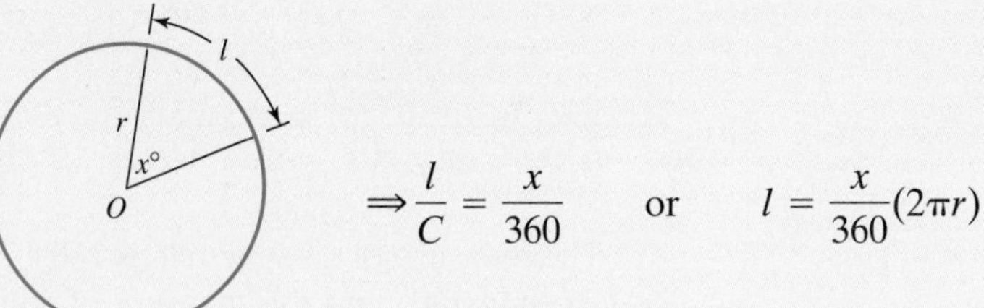

$$\Rightarrow \frac{l}{C} = \frac{x}{360} \quad \text{or} \quad l = \frac{x}{360}(2\pi r)$$

Area of a Sector of a Circle

We can also use the measure of a central angle to calculate the area of a portion of a circular region called a sector. A **sector** of a circle is a region bounded by two radii of the circle and their intercepted arc.

EXAMPLE 7.3 Find the area of the sector shaded in Figure 7.5.

SOLUTION

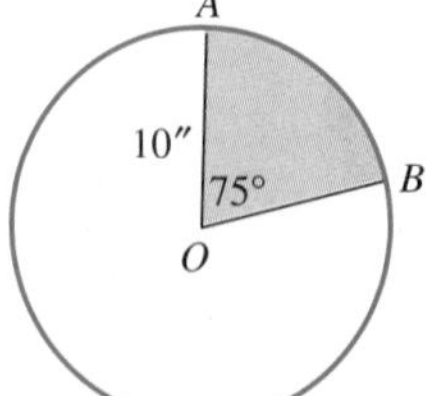

FIGURE 7.5

The region bounded by $\overline{OA}$, $\overline{OB}$, and $\widehat{AB}$ is a sector of circle O. Its area is some fraction of the area of circle O. Because $\angle AOB = 75°$, the area of the sector should be $\frac{75°}{360°} = \frac{5}{24}$ of the area of circle O. Therefore, the area of the shaded region will be $\left(\frac{5}{24}\right)\pi\,(10 \text{ in.})^2 \approx 65.4 \text{ in}^2$.

■

We can generalize this example to yield the following postulate.

POSTULATE 7.2 Area of a Sector

The ratio of the area of a sector of a circle to the area of the circle is equal to the ratio of the measure of the central angle of the sector, $x°$, to 360°.

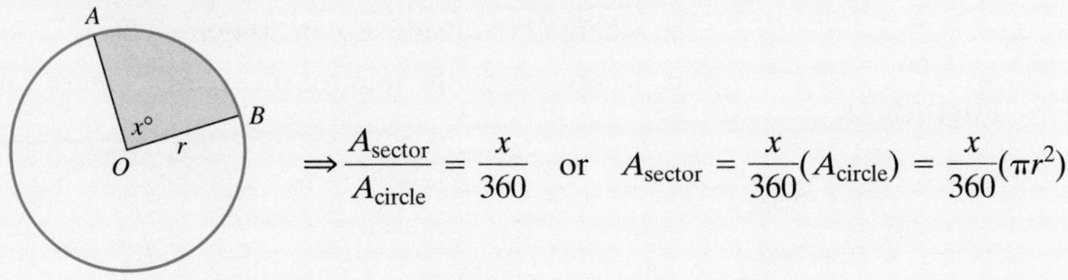

$$\Rightarrow \frac{A_{\text{sector}}}{A_{\text{circle}}} = \frac{x}{360} \quad \text{or} \quad A_{\text{sector}} = \frac{x}{360}(A_{\text{circle}}) = \frac{x}{360}(\pi r^2)$$

Inscribed Angles in a Circle

A central angle has the center of the circle as its vertex. Next we will study an angle whose vertex is a point of the circle. An angle is an **inscribed angle** in a circle if its vertex is on the circle and its sides each intersect the circle in another point. In Figure 7.6, $\angle ABC$ is an inscribed angle in the circle shown and $\widehat{AC}$ is the arc that $\angle ABC$ intercepts. As was the case with a central angle, the measure of an inscribed angle is related to the measure of the arc it intercepts. The next example shows how.

FIGURE 7.6

EXAMPLE 7.4 Find the measure of $\angle BAC$ in circle O (Figure 7.7).

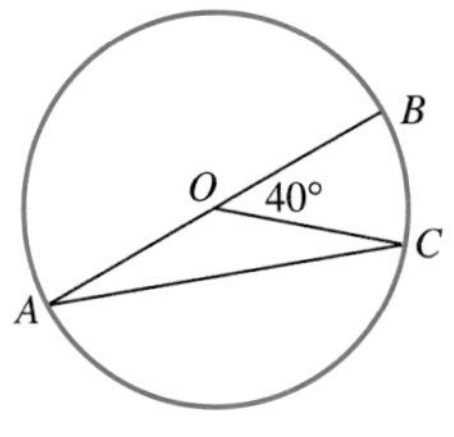

FIGURE 7.7

SOLUTION

We know that $\widehat{BC} = 40°$ because $\angle BOC = 40°$. $\triangle AOC$ is isosceles because $OA = OC$, so $\angle OAC = \angle OCA$. By the Exterior Angle Theorem, $\angle OAC + \angle OCA = 40°$. Therefore, $\angle OAC = 20°$. Notice that this means that $\angle BAC = 20° = \frac{1}{2}\widehat{BC}$.

Example 7.4 suggests the following theorem.

THEOREM 7.1 Inscribed Angle Theorem

The measure of an inscribed angle in a circle is equal to half the measure of its intercepted arc.

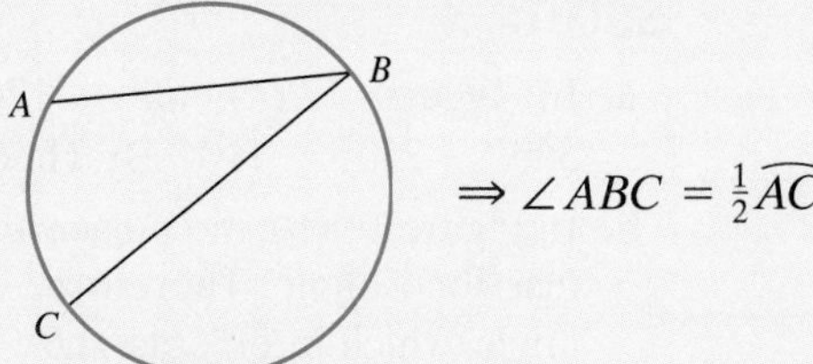

$$\Rightarrow \angle ABC = \tfrac{1}{2}\widehat{AC}$$

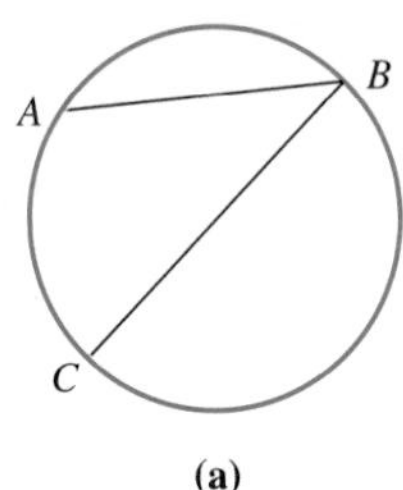

(a)

FIGURE 7.8

Given $\angle ABC$ inscribed in a circle [Figure 7.8(a)]

Prove $\angle ABC = \frac{1}{2}\widehat{AC}$

Proof There are three cases to consider:

a. where the center O is on $\overline{CB}$ [Figure 7.8(b)]

b. where O is outside $\angle ABC$ [Figure 7.8(c)]

c. where O is in the interior of $\angle ABC$ [Figure 7.8(d)]

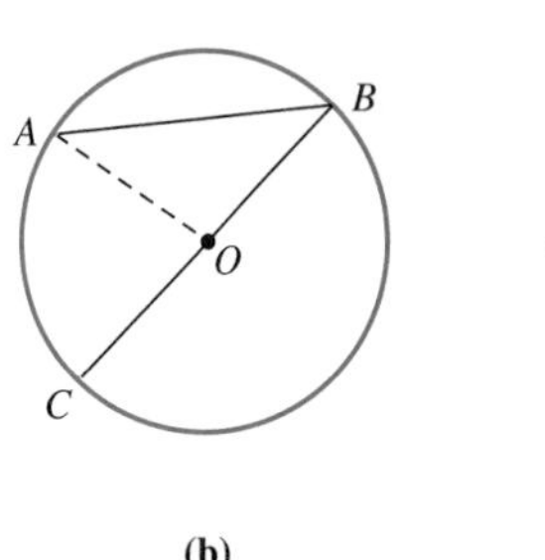

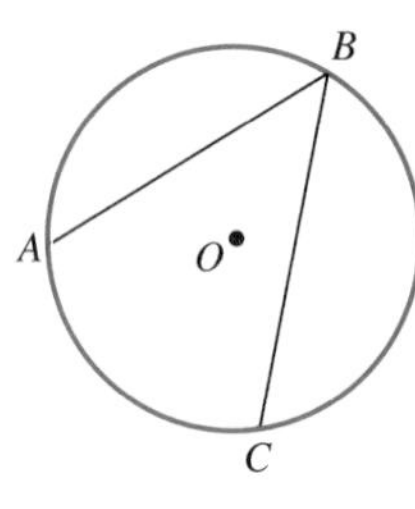

(b) (c) (d)

FIGURE 7.8

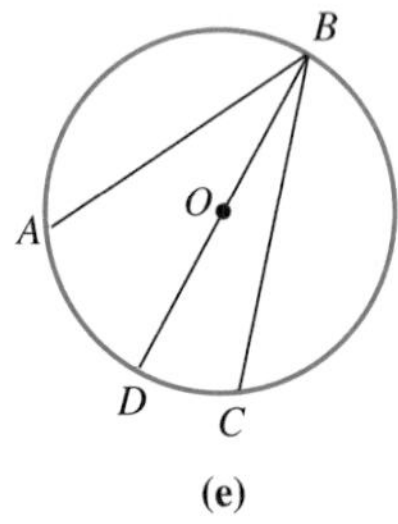

(e)

FIGURE 7.8

The proofs of part **a.**, which is analogous to the solution of Example 7.4 if radius $\overline{OA}$ is drawn, and part **b.** are requested in the problem set. We will prove the case where O is in the interior of $\angle ABC$ [Figure 7.8 (d)]. Draw the diameter $\overline{BD}$ [Figure 7.8(e)]. Then $\angle ABD = \frac{1}{2}\widehat{AD}$ and $\angle CBD = \frac{1}{2}\widehat{DC}$. Therefore,

$$\angle ABC = \angle ABD + \angle CBD = \tfrac{1}{2}\widehat{AD} + \tfrac{1}{2}\widehat{DC} = \tfrac{1}{2}(\widehat{AD} + \widehat{DC}) = \tfrac{1}{2}\widehat{AC}$$ ■

The next example shows how to use the Inscribed Angle Theorem to find missing angle measures.

EXAMPLE 7.5

a. Find x in Figure 7.9(a).

b. Find y in Figure 7.9(b) where O is the center of the circle.

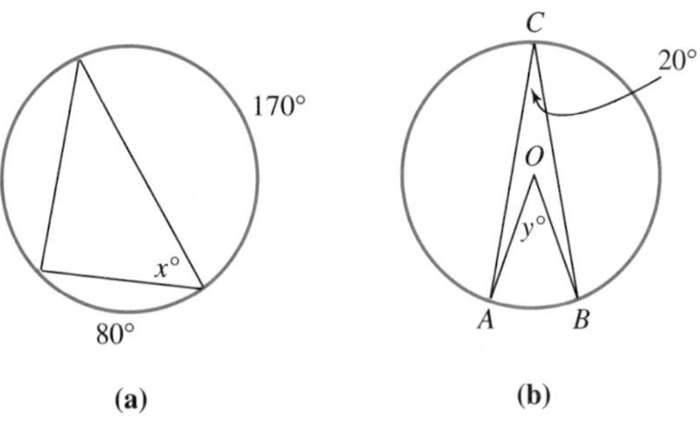

(a) (b)

FIGURE 7.9

SOLUTION

a. In Figure 7.9(a), $80° + 170° = 250°$, so the unlabeled arc has measure $360° - 250° = 110°$. By Theorem 7.1, $x = \frac{1}{2}(110°) = 55°$.

b. In Figure 7.9(b), the measure of $\widehat{AB}$ is 40° by Theorem 7.1 because $\angle ACB$ is an inscribed angle. Therefore, $y = 40°$ because it is the measure of the central angle, which intercepts $\widehat{AB}$. ■

The next two corollaries about inscribed angles follow from Theorem 7.1.

COROLLARY 7.2

Inscribed angles that intercept the same arc (or congruent arcs) are congruent.

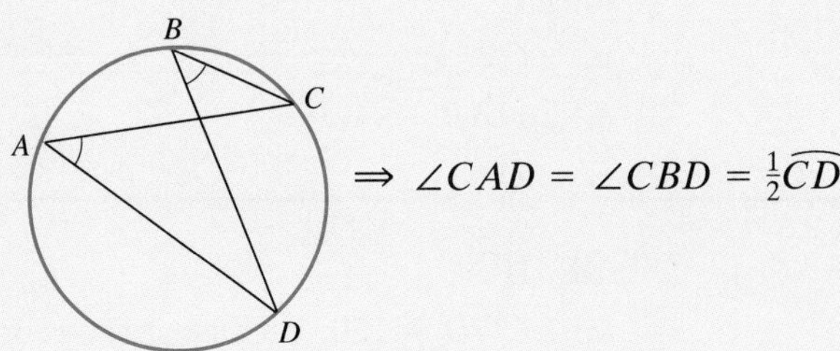

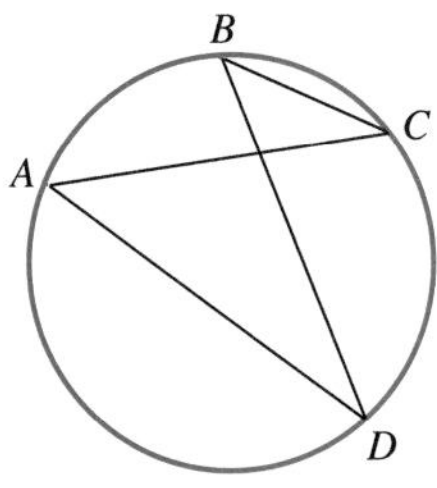

FIGURE 7.10

Given Inscribed angles $\angle CAD$ and $\angle CBD$ (Figure 7.10)

Prove $\angle CAD = \angle CBD$

Proof By the Inscribed Angle Theorem, we have $\angle CAD = \frac{1}{2}\widehat{CD}$ and $\angle CBD = \frac{1}{2}\widehat{CD}$. Therefore, $\angle CAD = \angle CBD$. ■

COROLLARY 7.3

An angle is inscribed in a semicircle if and only if it is a right angle.

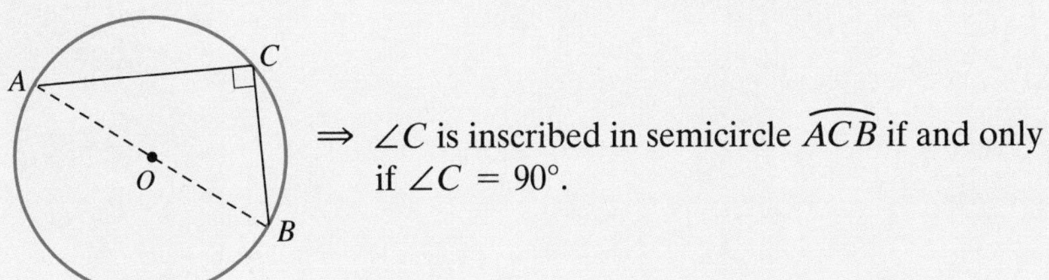

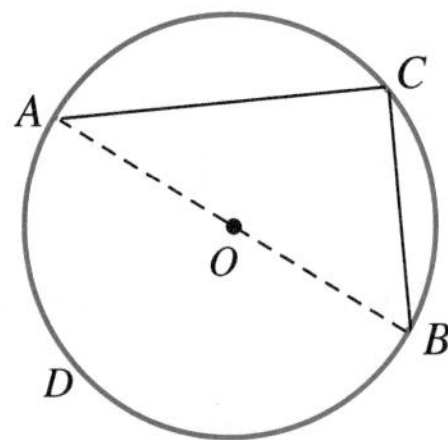

FIGURE 7.11

Given $\angle ACB$ inscribed in a circle (Figure 7.11)

Prove

a. If $\widehat{ACB}$ is a semicircle, then $\angle C$ is a right angle.

b. If $\angle C$ is a right angle, then $\widehat{ACB}$ is a semicircle.

Proof

a. If $\widehat{ACB}$ is a semicircle, then $\widehat{ADB}$ is a semicircle. Because an arc that is a semicircle has measure 180°, $\angle C = \frac{1}{2}(180°) = 90°$.

b. If $\angle C$ is a right angle, then it intercepts an arc whose measure is $2(90°) = 180°$. Thus, $\widehat{ADB}$ is a semicircle and $\widehat{ACB}$ must also be a semicircle. ■

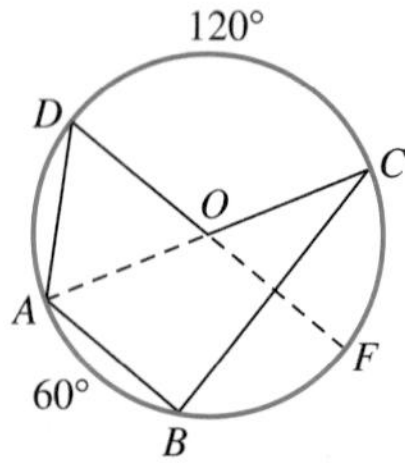

FIGURE 7.12

EXAMPLE 7.6 Find the measures of the following angles in Figure 7.12, where O is the center of the circle and $\overline{AC}$ and $\overline{DF}$ are diameters.

a. $\angle COD$

b. $\angle B$

c. $\angle C$

d. $\angle DAB$

e. $\angle D$

SOLUTION

a. $\angle COD = 120°$ because it is a central angle and $\widehat{DC} = 120°$.

b. Because AC is a diameter, $\angle B = 90°$ by Corollary 7.3.

c. $\angle C = \frac{1}{2}\widehat{AB} = \frac{1}{2}(60°) = 30°$ by the Inscribed Angle Theorem.

d. Because $\widehat{CD} = 120°$ and $\widehat{BFC} = 180° - 60° = 120°$, we have $\widehat{DCB} = 120° + 120° = 240°$. Thus, $\angle DAB = \frac{1}{2}(240°) = 120°$ by the Inscribed Angle Theorem.

e. Because $\angle COD = 120°$ and $\angle DOA = 180° - 120° = 60°$, we have $\widehat{AD} = 60°$. It follows that $\widehat{BF} = 180° - 60° - 60° = 60°$. Because $\widehat{ABF} = 60° + 60° = 120°$, we have $\angle D = \frac{1}{2}(120°) = 60°$ by the Inscribed Angle Theorem. ■

Solution of Applied Problem

The region watered by the sprinkler is a sector of a circle. Because the central angle of the sector is 110° and the radius of the circle is 30 ft, the irrigated area is $A = \frac{110}{360}\pi(30\text{ ft})^2 \approx 864\text{ ft}^2$.

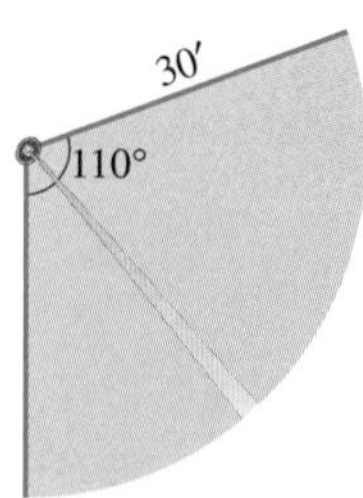

GEOMETRY AROUND US Circle graphs such as the one shown appear frequently in newspapers to depict relative amounts of a whole. In a circle graph, the area of each sector is proportional to the fraction or percentage that it represents. The graph shown gives the percentage of the population of the United States living in each region as of July 1, 2005. The population living in the northeast region is approximately half the population living in the south region because the area of the northeast sector is about half the area of the south sector. The percentage of the population living in the midwest is about the same as the population living in the west because the area of each of those sectors is about the same. To construct a circle graph, the percentages are calculated first and are used to determine the central angle for each sector.

United States Population by Region, 2005

West 23.0%
Northeast 18.4%
Midwest 22.3%
South 36.3%

PROBLEM SET 7.1

Refer to the following figure for Exercises 1–4. Segment $\overline{AC}$ is a diameter.

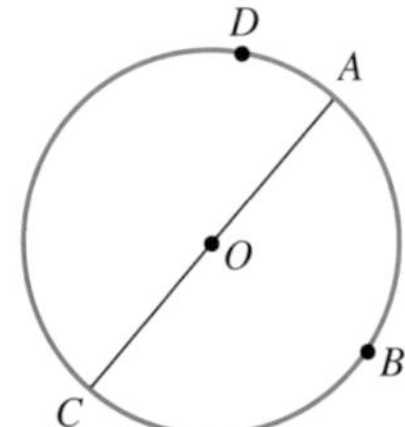

1. Name five minor arcs.
2. Name six major arcs.
3. Name two arcs that can be added to create semicircle $\overset{\frown}{ABC}$.
4. Name an arc that can be subtracted from major arc $\overset{\frown}{DBC}$ to create a semicircle.

Refer to the following figure for Exercises 5–14. Segment $\overline{AC}$ is a diameter.

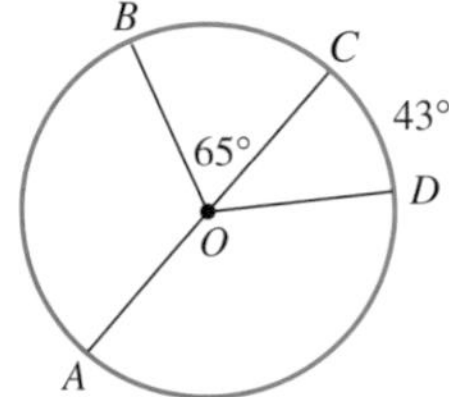

5. What kind of angle is $\angle BOC$ with respect to circle O?
6. What arc is intercepted by $\angle BOA$?
7. What angle intercepts $\overset{\frown}{CD}$?
8. What angle intercepts $\overset{\frown}{AD}$?
9. What is the measure of $\angle AOD$?
10. What is the measure of $\angle COD$?
11. What is the measure of $\overset{\frown}{BC}$?
12. What is the measure of $\overset{\frown}{AD}$?
13. What is the measure of $\overset{\frown}{ADB}$?
14. What is the measure of $\overset{\frown}{CDB}$?

Refer to the following figure for Exercises 15–18. Segment $\overline{AC}$ is a diameter, $\overset{\frown}{BC} = 115°$ $\angle DOC = 25°$, and $\overline{OA} = 8$ cm.

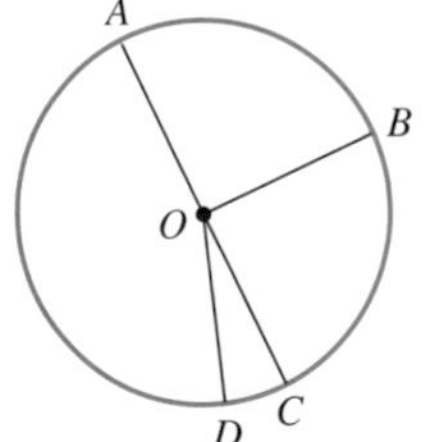

15. Find the lengths (in cm) of $\overset{\frown}{AC}$ and $\overset{\frown}{DC}$. Round to the nearest hundredth.
16. Find the lengths (in cm) of $\overset{\frown}{BC}$ and $\overset{\frown}{AD}$. Round to the nearest hundredth.
17. Find the area of sector AOB. Round to the nearest hundredth.
18. Find the area of sector DOC. Round to the nearest hundredth.

Refer to the following figure for Exercises 19–30.

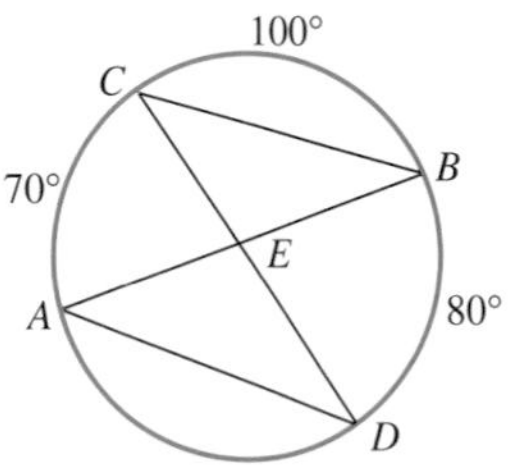

19. Name two inscribed angles that intercept $\overset{\frown}{BD}$.
20. Name two inscribed angles that intercept $\overset{\frown}{CA}$.
21. Name an angle that intercepts $\overset{\frown}{CB}$ and indicate whether the angle is inscribed, central, or neither.
22. Name an angle that intercepts $\overset{\frown}{DA}$ and indicate whether the angle is inscribed, central, or neither.
23. What is the measure of $\angle C$?
24. What is the measure of $\angle A$?
25. What is the measure of $\angle B$?
26. What is the measure of $\angle D$?
27. What is the measure of $\angle CEB$?
28. What is the measure of $\angle CEA$?
29. What is the measure of $\overset{\frown}{AD}$?
30. What is the measure of $\overset{\frown}{BDA}$?

Refer to the following figure for Exercises 31–36, where $\angle D = 30°$ and $\overline{AB} \parallel \overline{CD}$.

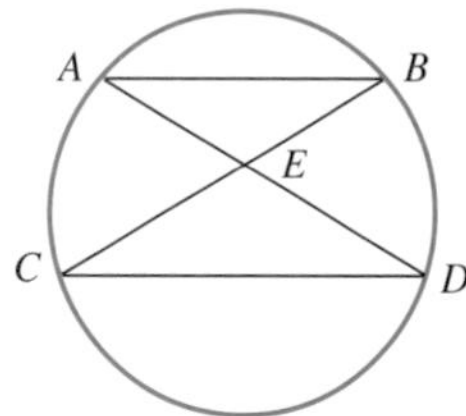

31. What are the measures of $\angle A$, $\angle B$, and $\angle C$?

32. Why is $\triangle ABE$ isosceles?

33. What is the measure of $\angle AEB$?

34. What is the measure of $\overparen{AC}$?

35. Why can we not necessarily conclude that $\overparen{AB} = \angle AEB$?

36. If $\overparen{CD} = 2\,\overparen{AB}$, what is the measure of $\overparen{CD}$?

37. In circle O, if $\angle BAC = 40°$, what is the measure of $\angle BOC$?

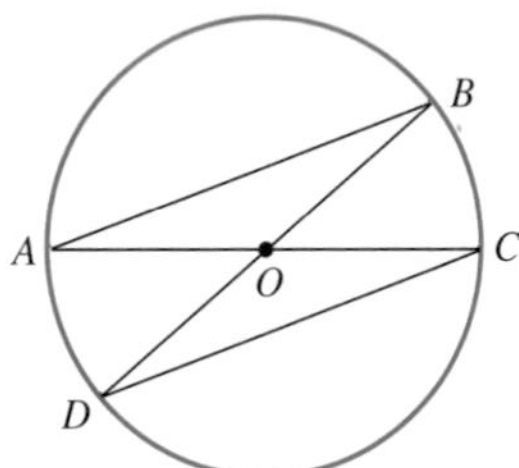

38. In circle O, if $\overparen{BC} = 38°$, what is the measure of $\overparen{AB}$?

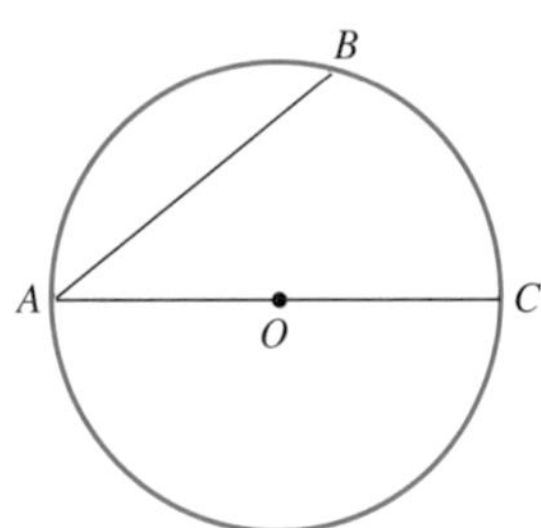

39. If $AB = 36''$ and $BC = 15''$, what is the radius of circle O?

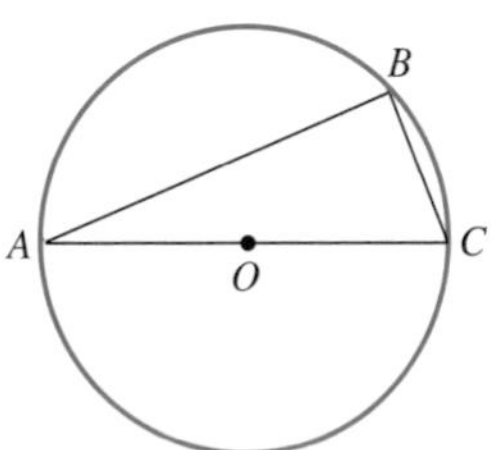

40. In the figure shown, the measure of $\overparen{AB}$ is twice the measure of $\overparen{BC}$ and the measure of $\overparen{BC}$ is 40° less than the measure of $\overparen{AC}$. Find $\angle A$.

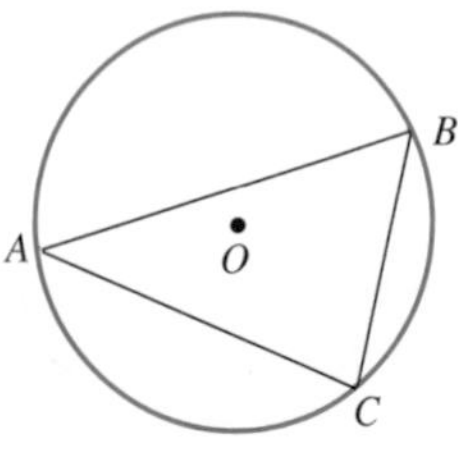

41. If $\overparen{AB} = 110°$, $\overparen{BC} = 80°$, and $\overparen{CD} = 50°$, what is the measure of $\angle B$?

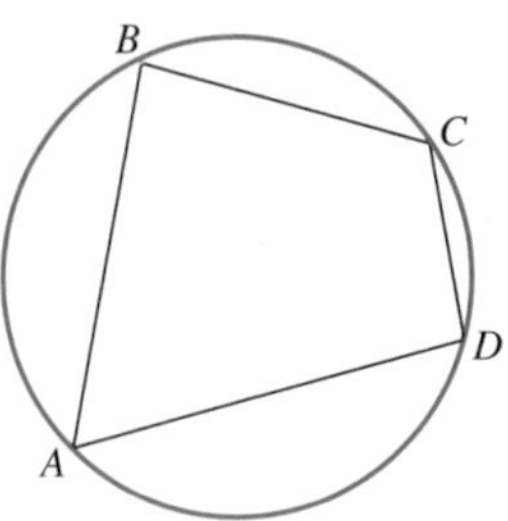

42. In circle O, $\angle PRS = 50°$ and $\overparen{SR} = \frac{2}{3}\overparen{PQ}$. Find the measure of $\angle QPR$.

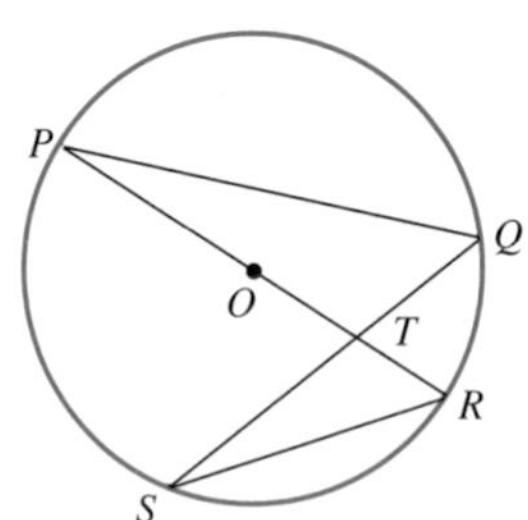

43. For quadrilateral $PQRS$, find PS. Round to the nearest hundredth.

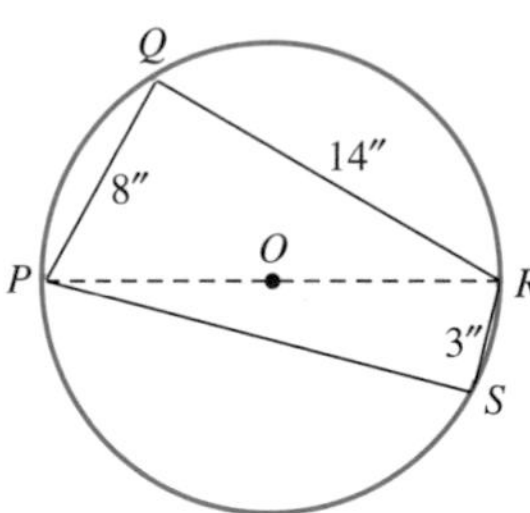

44. In circle O, $\overparen{AB} = 130°$. Find the measure of $\angle OBC$.

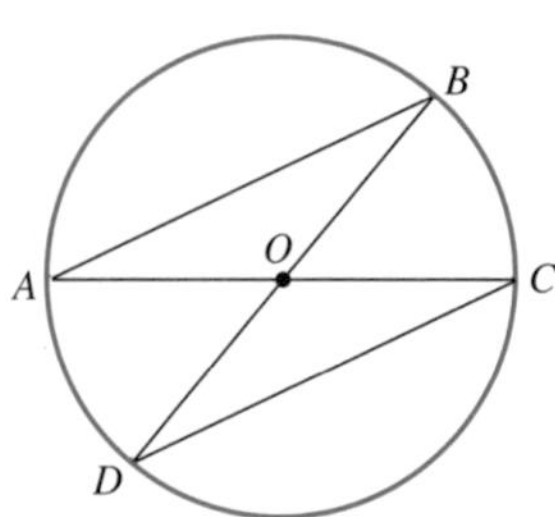

45. In circle O, $\overparen{AE} = 120°$ and $\overparen{CD} = 42°$. Find the measures of each of the five angles in pentagon $ABCDE$.

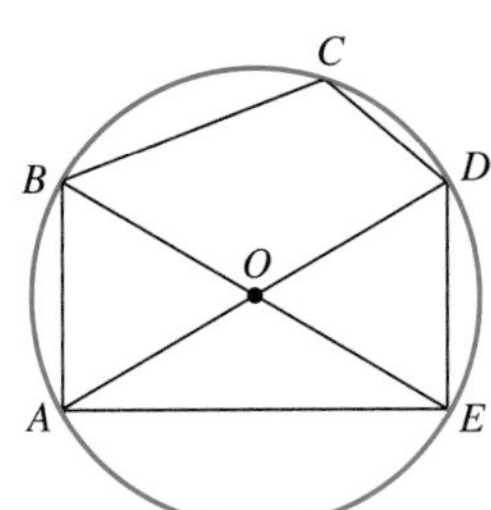

46. Find the measure of each angle of the star shown in the figure.

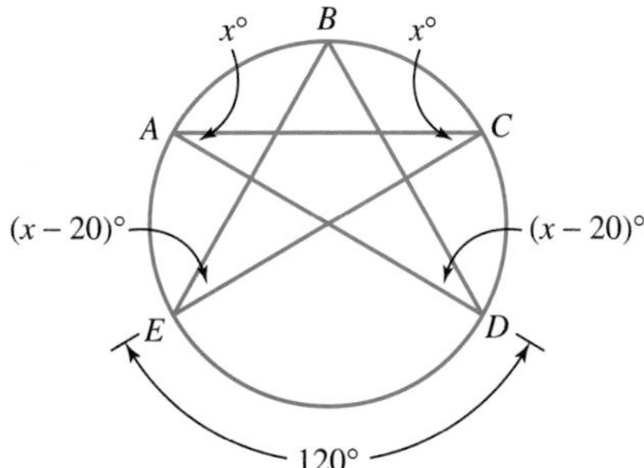

PROOFS

47. In the figure given, $ABCD$ is a rectangle. Prove that $\overline{AC}$ and $\overline{BD}$ are diameters of the circle.

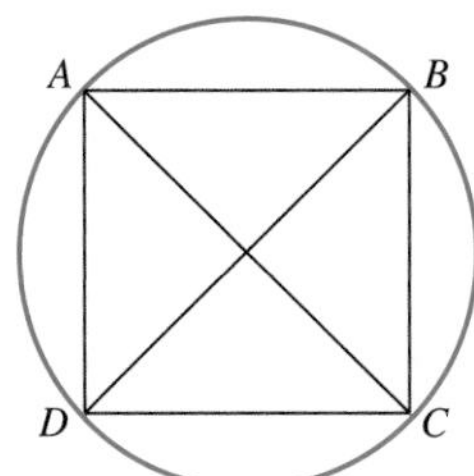

48. In the figure shown, $\overline{AB} \parallel \overline{CD}$. Prove that $\triangle CED$ and $\triangle AEB$ are isosceles triangles.

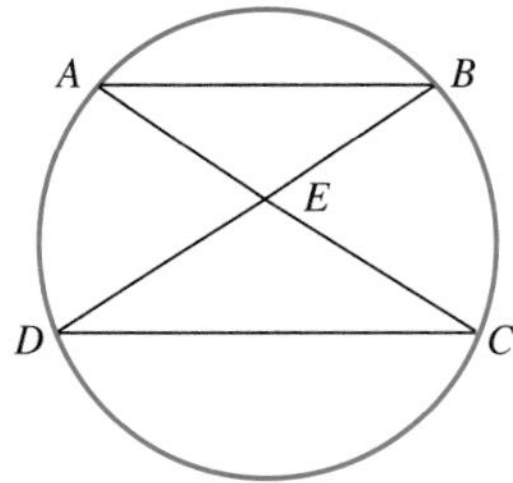

49. In the figure shown, $\overline{BD}$ is a diameter. Prove that $\angle ADC$ and $\angle ABC$ are supplementary angles.

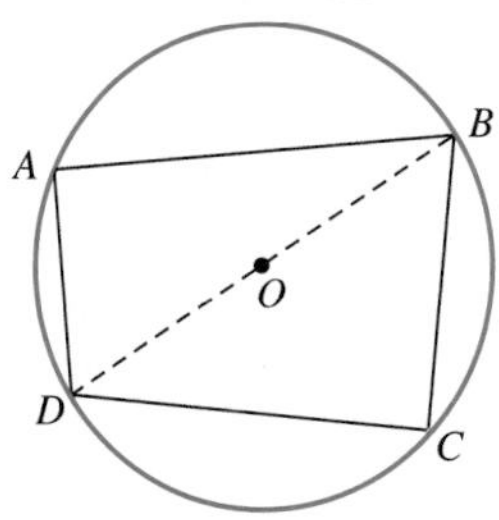

50. In the figure shown, $\overline{AB} \parallel \overline{CD}$. Prove that $ABCD$ is an isosceles trapezoid.

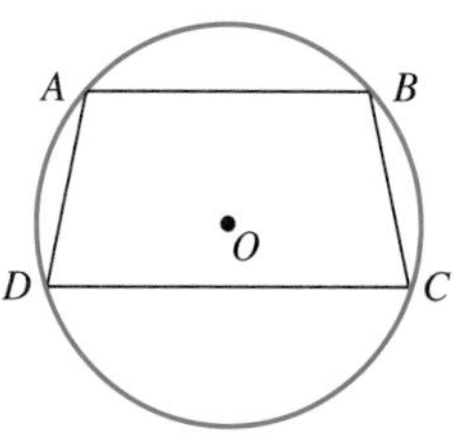

51. Prove Theorem 7.1, case (a): Show $\angle ABC = \frac{1}{2}\widehat{AC}$.

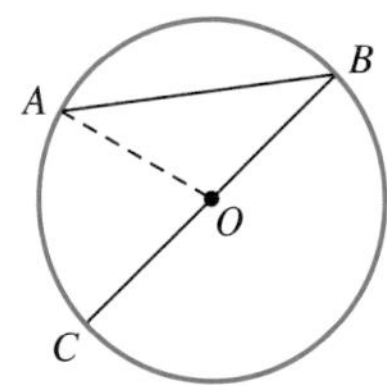

52. Prove Theorem 7.1, case (b): Show $\angle ABC = \frac{1}{2}\widehat{AC}$.

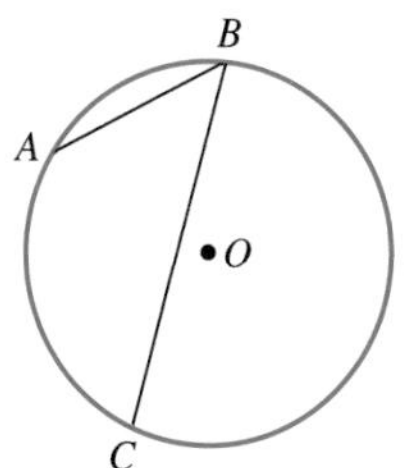

53. In the following figure, $\overline{BD}$ is a diameter of the circle. Let $BD = d$.

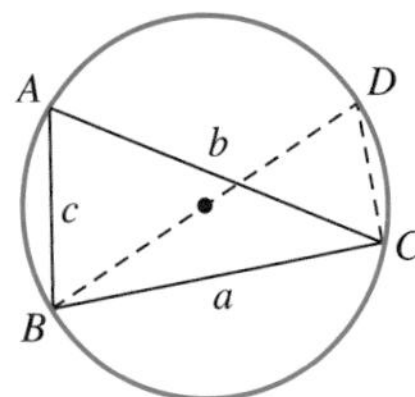

Use the Law of Sines and the Inscribed Angle Theorem to prove

$$\frac{a}{\sin A} = \frac{b}{\sin B} = \frac{c}{\sin C} = d$$

APPLICATIONS

54. The minute hand of a clock has a length of 5 inches. Find the length (in inches) of the arc the minute hand sweeps out in 22 minutes. Round to the nearest hundredth.

55. The pendulum of a clock is 40 inches long and swings through an arc of 8°. Find the length (in inches) of arc that the pendulum traces out. Round to the nearest hundredth.

56. A windshield wiper is 18 inches long and has a blade 12 inches long. If the wiper sweeps through an angle of 130°, how large an area does the wiper blade clean? Round your answer to the nearest whole number.

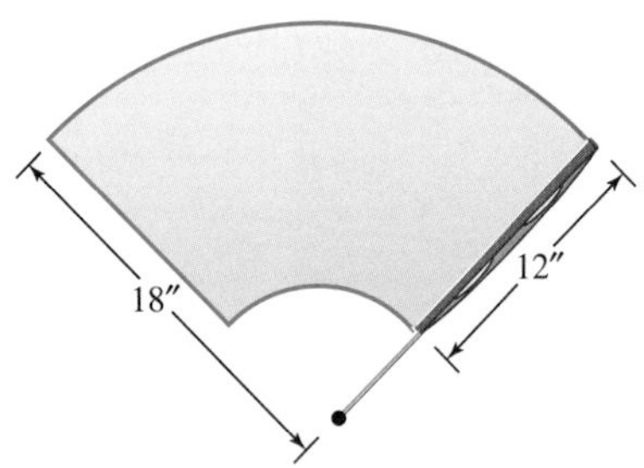

57. A wheel of cheese is 22 inches in diameter and 4 inches thick. If a wedge of cheese with a central angle of 14° is cut from the wheel, find the volume of the cheese wedge. Round to the nearest hundredth.

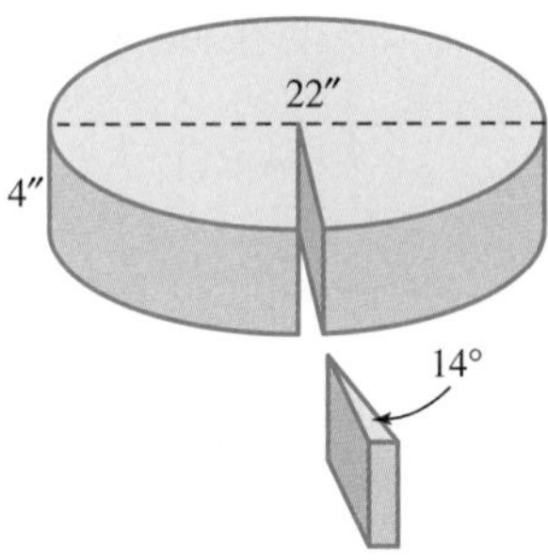

58. A paper cup has the shape shown (called a frustum of a right circular cone).

If the cup is sliced open and flattened, the side of the cup has the following shaded shape.

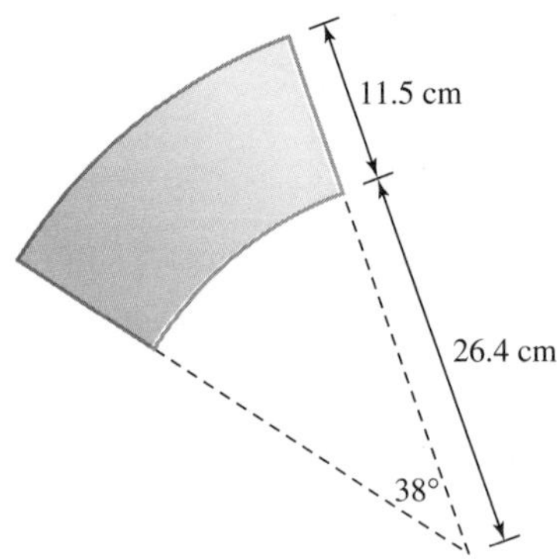

Use the dimensions given to calculate the number of square meters of paper used in the construction of the sides of 10,000 of these cups. Round your answer to the nearest square meter.

59. A draftsperson might use a right triangle like the one shown to find the center of a given circle. How can that be done? Explain why the technique works.

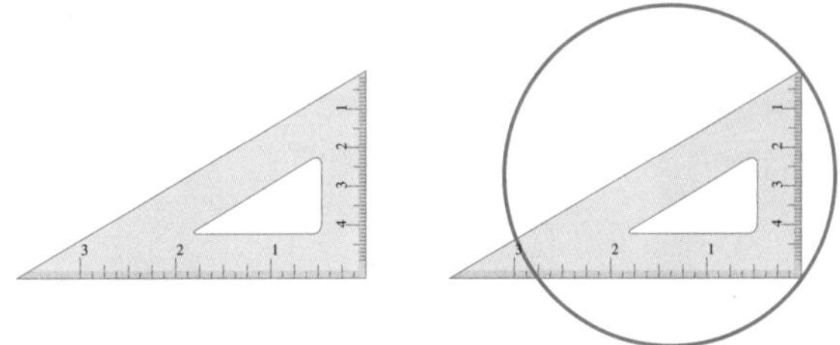

EXTENDED PROBLEMS

60. **Parallels** are circles on the Earth in an east-west direction, parallel to the equator. We use parallels to measure how far north or south of the equator a particular place is. That distance is specified in degrees of **latitude** N or S. For example, Anchorage, Alaska, has a latitude of approximately 61°N as shown in the following figure.

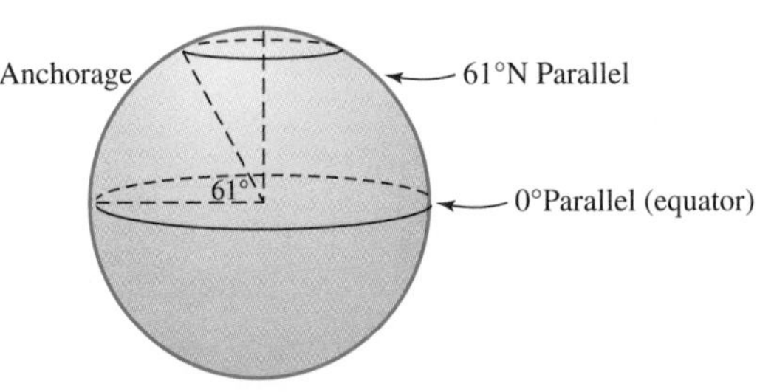

Meridians are north-south circles through the poles that measure distance east or west of the Prime Meridian. By using the latitudes of two points that have the same **longitude**, or that lie on the same meridian, we can calculate the distance between them.

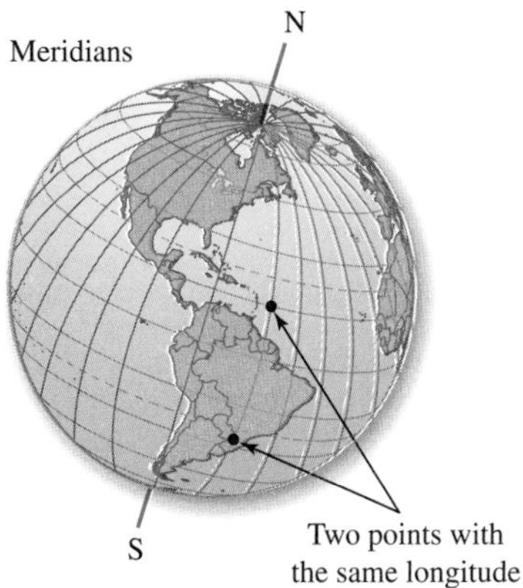

Research the latitude and longitude of various cities around the world. Locate three pairs of cities that have the same longitude and use the fact that the radius of the earth is approximately 3960 miles to calculate the distance between each pair of cities.

61. One **radian** is the measure of the central angle that intercepts an arc with length equal to the radius of the circle. Angles are traditionally labeled using the Greek letter θ. Thus, in the following figure, $\theta = 1$ radian.

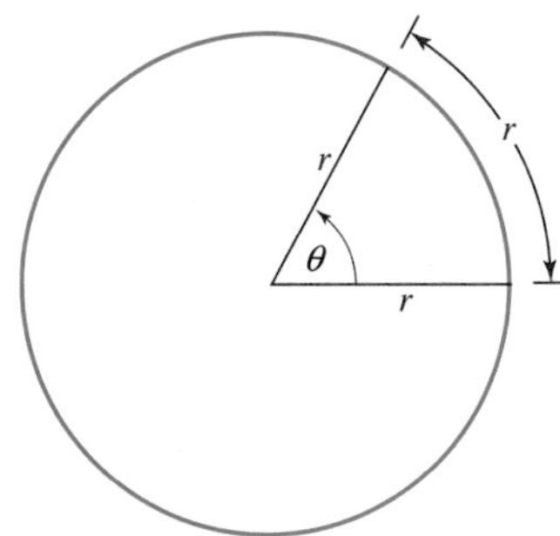

It follows that the measure (in radians) of any central angle θ is the ratio of the length of the arc intercepted by the angle and the radius of the circle. Let s represent the arc length and r represent the radius of the circle. Then the measure (in radians) of the angle is $\theta = \frac{s}{r}$.

a. Use this definition of the radian measure of an angle to find the radian measure of one full revolution. Rewrite the formula for the length of an arc and the area of a sector of a circle in terms of radian measure.

b. Research the history of radian measure. Who is credited with its development? In what situations are radians preferred to degrees and why?

62. In 1889, Joseph Louis Bertrand studied a problem (which came to be called Bertrand's paradox) involving segments, called chords, whose endpoints are on a circle. This problem is stated as follows: What is the probability that a random chord in a circle will have a length longer than the side of an equilateral triangle inscribed in the circle? (NOTE: Chords are studied in Section 7.2.)

a. Construct a circle and an inscribed equilateral triangle, that is, a triangle in which all three vertices lie on the circle. Measure and record the side length of the equilateral triangle. Draw 25 chords randomly in the circle, and measure and compare their lengths to the side length of the equilateral triangle. Determine the experimental probability that the length of a chord is greater than the length of the side of the triangle by dividing the number of times the length of the chord was longer than the side of the triangle by 25. Repeat the process with another circle. What do you conclude?

b. Research Bertrand's paradox. Explain why the problem is called a paradox and describe each of the three interpretations of the solution.

7.2 CHORDS OF A CIRCLE

Applied Problem

Some large circles turned up mysteriously in farmers' fields in Iowa and England. Conjectures concerning their origin include UFOs, freak wind storms, and magnetic fields. If the researchers who examined the fields needed to find the centers of the circles, how could they do it?

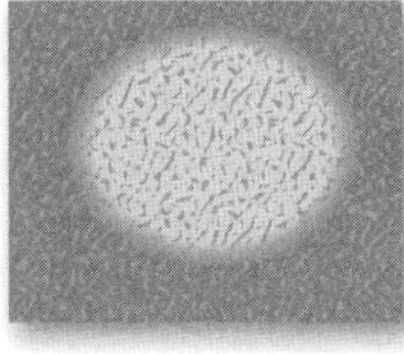

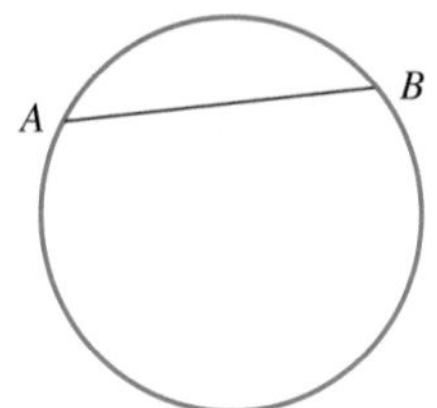

FIGURE 7.13

Radii and diameters are line segments that contain the centers of their circles. A **chord** is any line segment whose endpoints are on a circle. In Figure 7.13, $\overline{AB}$ is a chord of the circle. Notice that, according to the definition, diameters are chords that contain the center of the circle.

Perpendicular Bisector of a Chord

In Chapter 4, we showed that if a point is equidistant from the endpoints of a line segment, then it is on the perpendicular bisector of the segment. The next theorem and its corollary show how we can use this result to locate the center of a circle.

THEOREM 7.4

The perpendicular bisector of a chord contains the center of the circle.

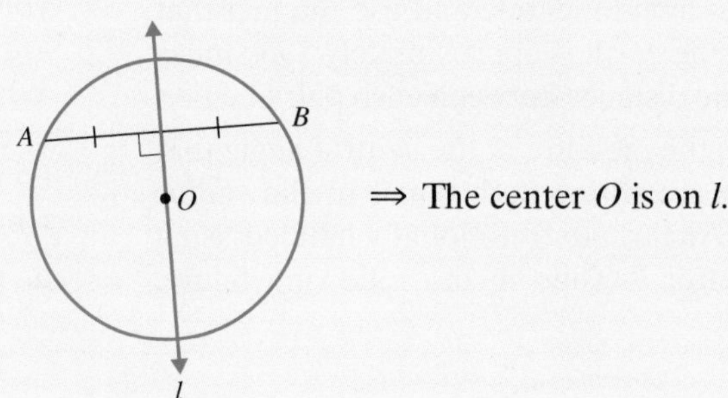

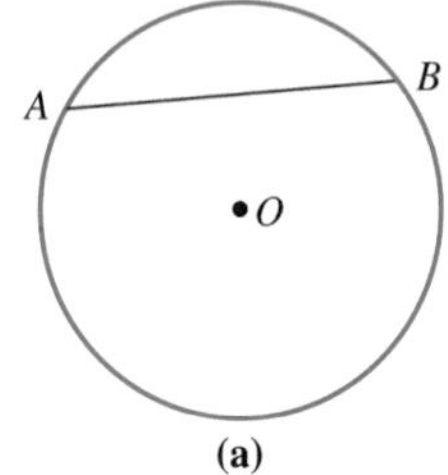

(a)

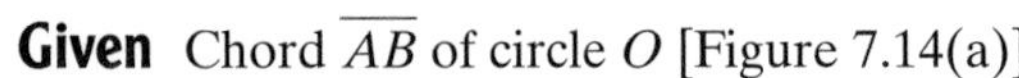

Given Chord $\overline{AB}$ of circle O [Figure 7.14(a)]

Prove O lies on the perpendicular bisector of $\overline{AB}$.

Proof Consider $\overline{AO}$ and $\overline{BO}$. Because $\overline{AO}$ and $\overline{BO}$ are radii, we have $AO = BO$ [Figure 7.14(b)]. This means that O is equidistant from A and B. Therefore, O is on l, the perpendicular bisector of $\overline{AB}$ [Figure 7.14(c)]. ■

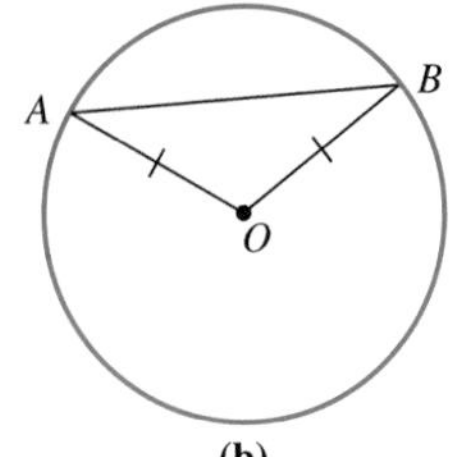

(b)

The next corollary extends Theorem 7.4 and states a relationship between the perpendicular bisectors of two different chords.

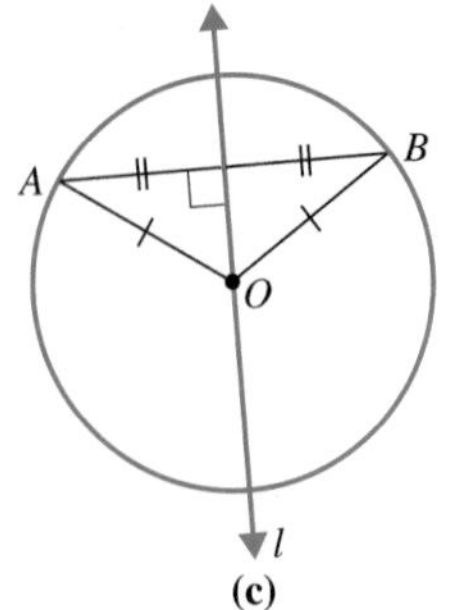

(c)

FIGURE 7.14

COROLLARY 7.5

The intersection of the perpendicular bisectors of any two nonparallel chords of a circle is the center of the circle.

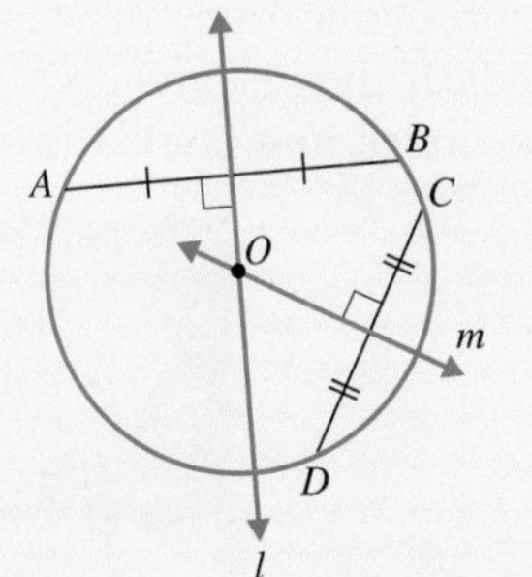

⇒ l and m intersect at the center O.

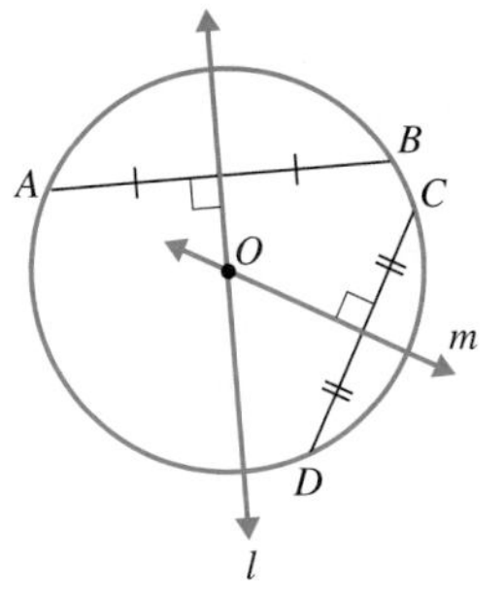

FIGURE 7.15

Given Two nonparallel chords $\overline{AB}$ and $\overline{CD}$ and their perpendicular bisectors l and m (Figure 7.15).

Prove The lines l and m intersect in the center.

Proof Because the chords are not parallel, neither are their perpendicular bisectors. Thus, the perpendicular bisectors intersect in one point. By Theorem 7.4, this point must be the center because both lines contain the center. ■

Corollary 7.5 provides a practical method for locating the center of a given circle. If we draw any two nonparallel chords and construct their perpendicular bisectors, the center of the circle will be the point of intersection of the perpendicular bisectors. We will use this method in Section 7.4. The following corollary also can be proved using Theorem 7.4.

COROLLARY 7.6

If two circles, O and O', intersect in two points A and B, then the line containing O and O' is the perpendicular bisector of $\overline{AB}$.

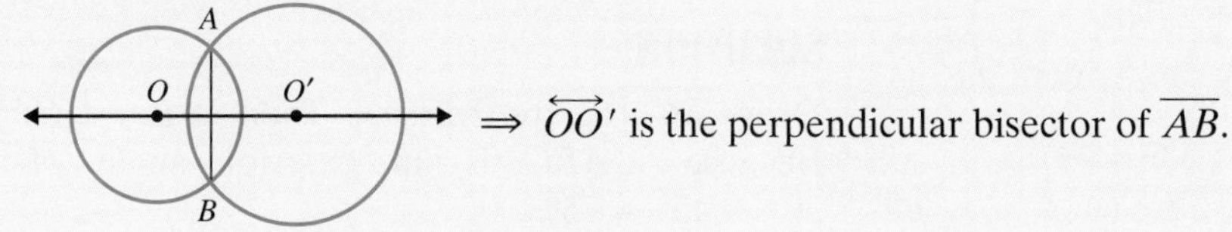

(NOTE: The line, such as $\overleftrightarrow{OO'}$, connecting the centers of two circles is called the **line of centers**.)

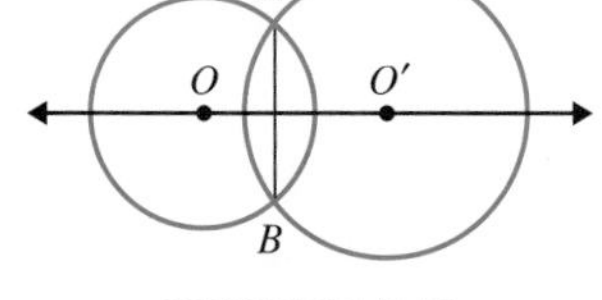

FIGURE 7.16

Given Circles O and O' intersecting in points A and B (Figure 7.16)

Prove $\overleftrightarrow{OO'}$ is the perpendicular bisector of $\overline{AB}$.

Proof By Theorem 7.4, the perpendicular bisector of $\overline{AB}$ contains the centers of both circle O and O'. Thus, since $\overline{AB}$ is a chord of each circle, the line of centers $\overleftrightarrow{OO'}$ is the perpendicular bisector of $\overline{AB}$. ■

EXAMPLE 7.7 Show three different ways to construct the perpendicular bisector of the chord $\overline{AB}$ in the circle with center O [Figure 7.17(a)].

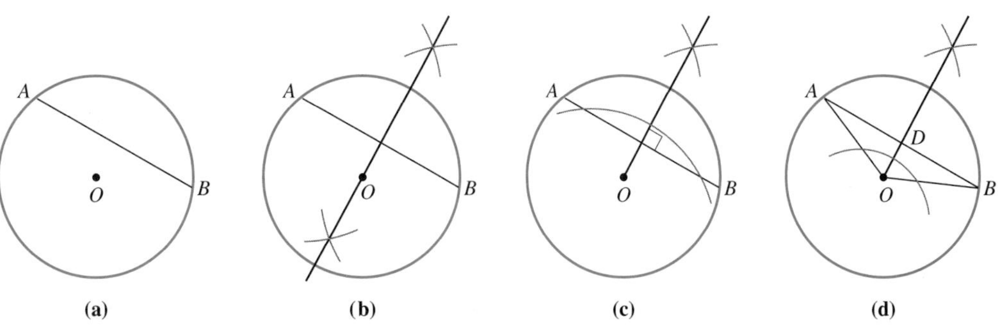

FIGURE 7.17

SOLUTION

1. Construct the perpendicular bisector of $\overline{AB}$ using Construction 6 of Section 4.4 [Figure 7.17(b)].
2. Construct a perpendicular from O to $\overline{AB}$ using Construction 5 of Section 4.4 [Figure 7.17(c)]. There is a unique perpendicular from a point to a line, and by Theorem 7.4, O lies on the perpendicular bisector of $\overline{AB}$. So this segment is the perpendicular bisector of $\overline{AB}$.
3. Bisect $\angle AOB$ using Construction 3 of Section 4.4 [Figure 7.17(d)]. Because $\triangle AOB$ is isosceles, $\overline{OD}$ is the perpendicular bisector of $\overline{AB}$. ■

Measures of Angles Formed by Chords

Next consider the following example that examines the measures of angles formed by chords.

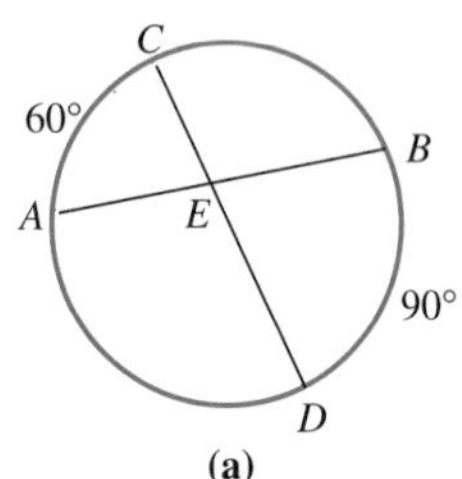

(a)

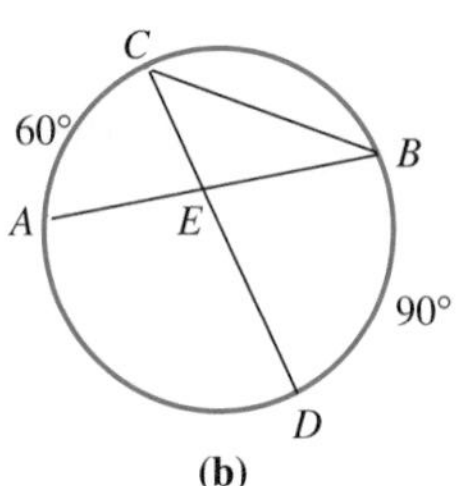

(b)

FIGURE 7.18

EXAMPLE 7.8 In Figure 7.18(a), $\widehat{AC} = 60°$ and $\widehat{BD} = 90°$. Find the measure of $\angle AEC$.

SOLUTION

Draw $\overline{BC}$ [Figure 7.18(b)]. Then $\angle B = \frac{1}{2}(60°) = 30°$, and $\angle C = \frac{1}{2}(90°) = 45°$. Because $\angle AEC$ is an exterior angle of $\triangle BCE$, $\angle AEC = \angle B + \angle C = 30° + 45° = 75°$. ■

The technique used to find $\angle AEC$ in Example 7.8 is generalized to any pair of intersecting chords in the next theorem.

THEOREM 7.7

If two chords intersect, then the measure of any one of the vertical angles formed is equal to half the sum of the measures of the two arcs intercepted by the two vertical angles.

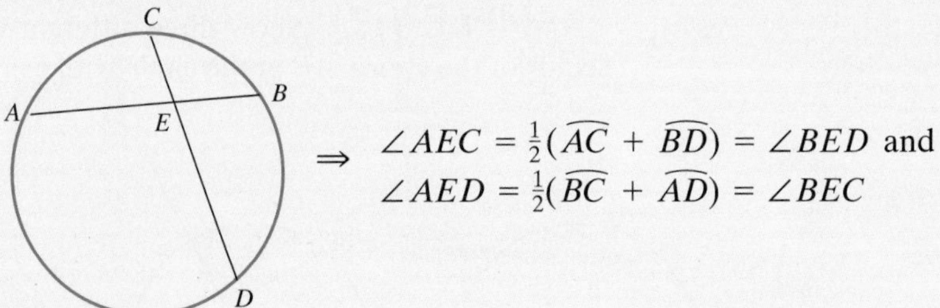

$\angle AEC = \frac{1}{2}(\widehat{AC} + \widehat{BD}) = \angle BED$ and
$\angle AED = \frac{1}{2}(\widehat{BC} + \widehat{AD}) = \angle BEC$

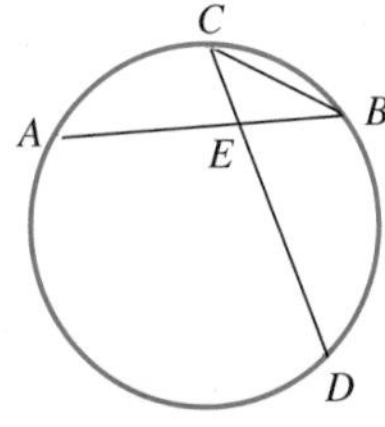

FIGURE 7.19

Given Chords $\overline{AB}$ and $\overline{CD}$ intersecting at E (Figure 7.19)

Prove $\angle AEC = \frac{1}{2}(\widehat{AC} + \widehat{BD})$

Proof In Figure 7.19, $\angle C = \frac{1}{2}\widehat{BD}$ and $\angle B = \frac{1}{2}\widehat{AC}$ by the Inscribed Angle Theorem. Because $\angle AEC$ is an exterior angle to $\triangle BCE$, $\angle AEC = \angle C + \angle B$ by the Exterior Angle Theorem. Therefore,

$$\angle AEC = \tfrac{1}{2}\widehat{AC} + \tfrac{1}{2}\widehat{BD} = \tfrac{1}{2}(\widehat{AC} + \widehat{BD})$$ ■

Notice that $\angle AEC = \angle BED$ because they are vertical angles. But this does *not* mean that $\widehat{AC}$ must be congruent to $\widehat{BD}$.

Measures of Segments of Chords

The next theorem develops a relationship between the four segments formed by intersecting chords.

THEOREM 7.8

If two chords of a circle intersect, the product of the lengths of the two segments formed on one chord is equal to the product of the lengths of the two segments formed on the other chord.

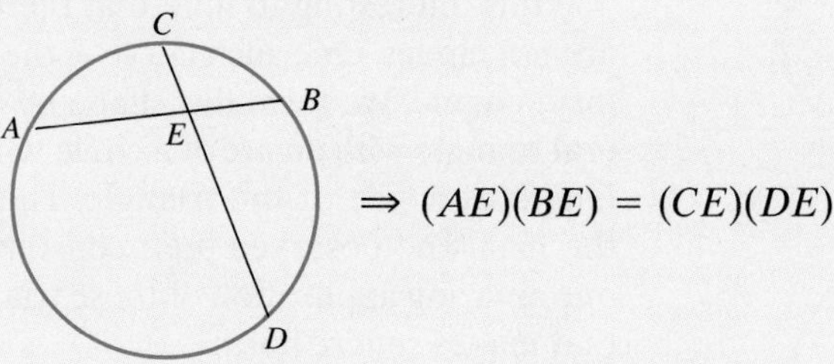

$\Rightarrow (AE)(BE) = (CE)(DE)$

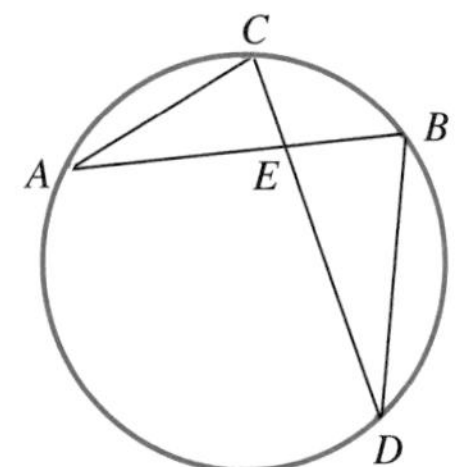

FIGURE 7.20

Given Chords $\overline{AB}$ and $\overline{CD}$ intersecting at E (Figure 7.20)

Prove $(AE)(BE) = (CE)(DE)$

Plan Show $\triangle AEC \sim \triangle DEB$. Then use C.P.

PROOF

Statements	Reasons
1. $\angle AEC \cong \angle DEB$	**1.** Vertical angles
2. $\angle C = \angle B = \frac{1}{2}\widehat{AD}$	**2.** Inscribed Angle Theorem
3. $\triangle AEC \sim \triangle DEB$	**3.** AA Similarity
4. $\dfrac{AE}{CE} = \dfrac{DE}{BE}$	**4.** Corresponding parts
5. $(AE)(BE) = (CE)(DE)$	**5.** Cross multiplication

■

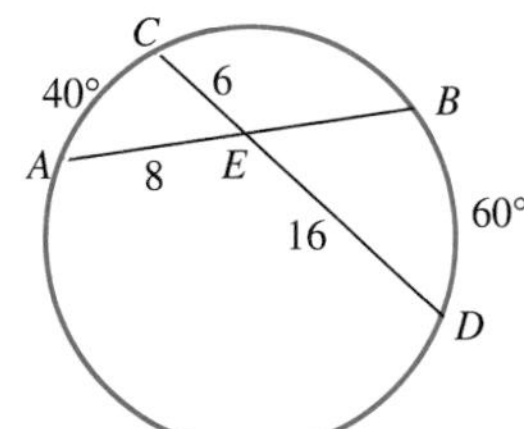

FIGURE 7.21

EXAMPLE 7.9 In Figure 7.21, find **a.** $\angle AEC$ and **b.** EB.

SOLUTION

a. $\angle AEC = \frac{1}{2}(40° + 60°) = 50°$ by Theorem 7.7.

b. By Theorem 7.8, $8(EB) = 6(16)$, so $EB = 12$. ■

Solution of Applied Problem

By Corollary 7.5, if we draw any two nonparallel chords, their perpendicular bisectors will intersect in the center of the circle. Thus, we would draw two nonparallel chords of the circle on a photo or scale drawing for which we wish to find the center and construct their perpendicular bisectors. The point of intersection of the perpendicular bisectors is the center of the circle.

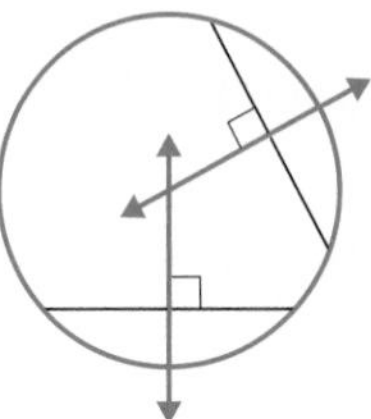

GEOMETRY AROUND US

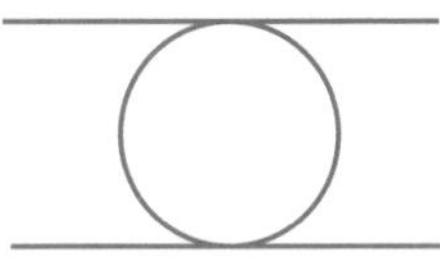

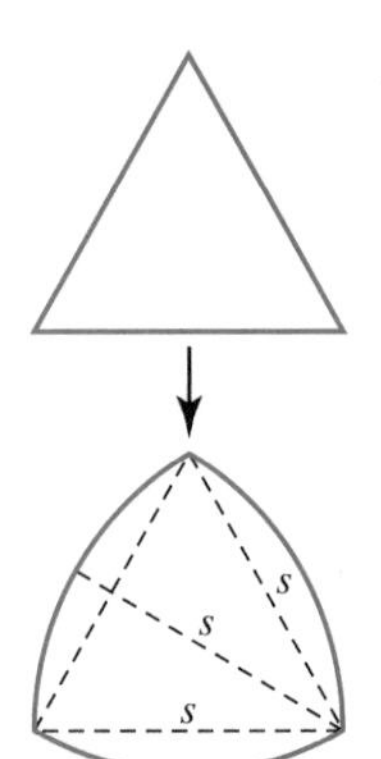

A circle is a **curve of constant width**. That is, if a circle rolls in a channel of parallel lines, the circle will always touch the top and bottom of the channel as shown.

It is interesting to note that there are many other curves of constant width that are not circles. One such curve is the shape of the rotor found in the Wankel (or rotary) engine. We form this shape by connecting every pair of vertices of an equilateral triangle with an arc of a circle whose center is at a vertex and whose radius is the length of a side of the triangle. These shapes are called **Reuleaux triangles** after the man who observed their constant width property. An English engineer devised one amazing application of these triangles, using a Reuleaux triangle to design a drill that makes *square* holes.

PROBLEM SET 7.2

EXERCISES/PROBLEMS

1. In the following circle, $\overline{EF}$ is the perpendicular bisector of $\overline{CD}$, and $\overline{GH}$ is the perpendicular bisector of $\overline{AB}$. How do GH and EF compare? What is the measure of $\widehat{FGE}$?

2. In circle O, $\overline{AC}$ and $\overline{BD}$ are diameters and $\overline{AC} \perp \overline{BD}$. Find AB, $\angle OBA$, and OF. For approximate answers, round to the nearest hundredth.

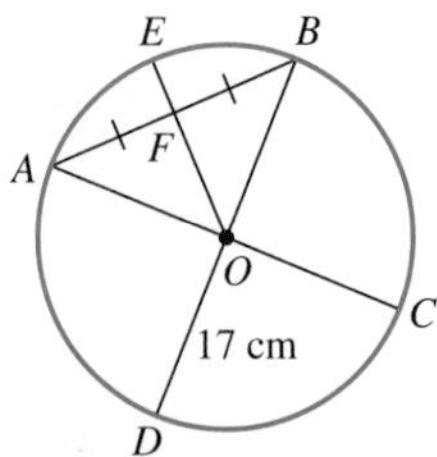

For problems 3–6, refer to the following figure.

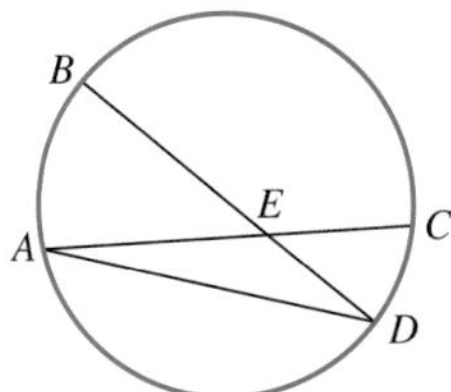

3. If $\overset{\frown}{AD} = 125°$ and $\overset{\frown}{BC} = 173°$, find $\angle BEC$.
4. If $\angle AEB = 35°$ and $\overset{\frown}{AB} = 24°$, find $\overset{\frown}{CD}$.
5. If $\angle CAD = 30°$ and $\angle BDA = 40°$, find $\angle CED$.
6. If $\angle CED = 41°$ and $\angle DAC = 25°$, find $\overset{\frown}{AB}$.

For problems 7–10, refer to the following figure.

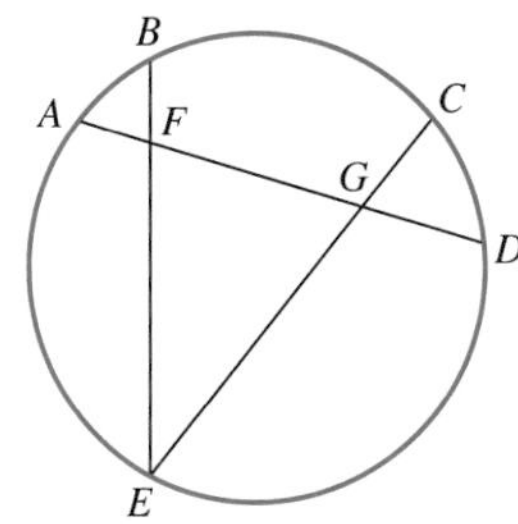

7. If $BF = 7''$, $EF = 18''$, and $FD = 15''$, find AF.
8. If $AF = 2$ cm, $GD = 3$ cm, $GC = 4$ cm, and $EG = 9$ cm, find FG.
9. If $EF = 3.9FB$, $AF = 5.2$ cm, and $FD = 12$ cm, find BF.
10. If $AF = 3$ mm, $GD = 4$ mm, $EF = 3BF$, and $FG = 2BF$, find EF. Round to the nearest tenth.

For problems 11–14, refer to the following figure.

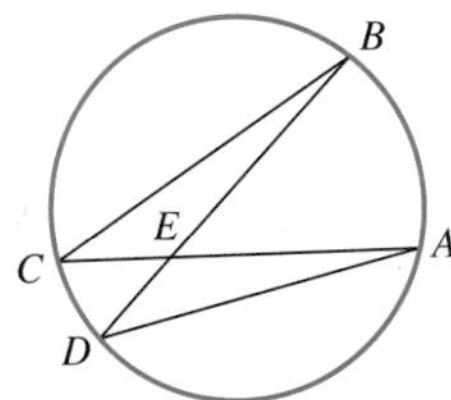

11. If $BE = 5$ cm, $AE = 4$ cm, and $CE = 2$ cm, find DE.
12. If $AE = 9.8$ m, $CE = 5.7$ m, and $DE = 3.8$ m, find BE.
13. If $BE = 2(EC)$, $AC = 12''$, and $DE = 3.5''$, find CE.
14. If $AC = 10$ cm, $BE = 8$ cm, and $DE = 3$ cm, find AE.

For problems 15–18 refer to the following figure.

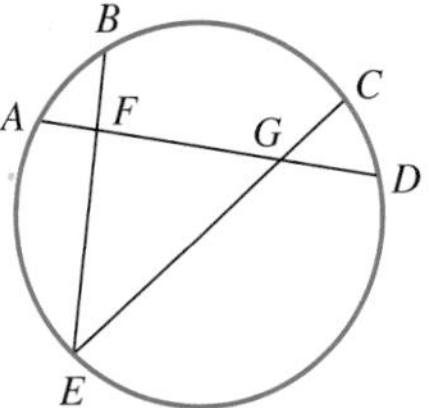

15. If $\angle BEC = 42°$, $\overset{\frown}{CD} = 30°$, and $\overset{\frown}{AE} = 95°$, find $\angle AFE$.
16. If $\angle AFB = 70°$, $\overset{\frown}{AE} = 80°$, $\overset{\frown}{ED} = 120°$, and $\overset{\frown}{CD} = 15°$, find $\overset{\frown}{BC}$.
17. If $\overset{\frown}{AE} = 83°$, $\angle CGD = 62°$, $\angle BFG = 123°$, and $\overset{\frown}{ED} = 5\overset{\frown}{AB}$, find $\overset{\frown}{ED}$, $\overset{\frown}{BC}$, and $\overset{\frown}{CD}$.
18. If $\angle BEC = 71°$, $\overline{EF} \cong \overline{EG}$, $\overset{\frown}{AB} = 18°$, and $\overset{\frown}{AE} = 3\overset{\frown}{CD}$, find $\overset{\frown}{AE}$, $\overset{\frown}{CD}$, and $\overset{\frown}{ED}$.

For problems 19–22, refer to the following figure.

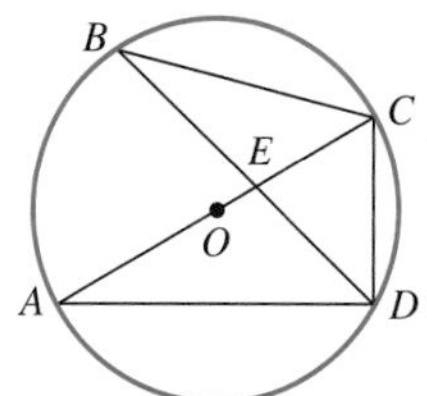

19. If $AD = 15$ cm, $CD = 8$ cm, $BE = 10$ cm, and $DE = 6$ cm, find AE.
20. If $\angle ADB = 65°$ and $\angle BEC = 80°$, find $\overset{\frown}{AD}$.
21. If $AE = 14'$, $BE = 12'$, $DE = 7'$, and $CD = 12'$, find AD.
22. If $\angle CBD = 30°$, $CD = 6''$, $BE = 5''$, and $ED = 3''$, find CE.
23. Use a compass and straightedge to find the center of the following circle.

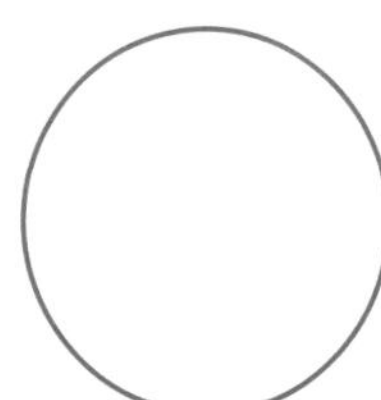

24. Use a compass and straightedge to find the center of the circle to which the following arc belongs. Check your answer by drawing the circle with your compass.

25. A **cyclic quadrilateral** is a quadrilateral such that each of the four vertices is contained in a circle. Use a compass and straightedge to construct the circle that contains the vertices of cyclic quadrilateral *ABCD*.

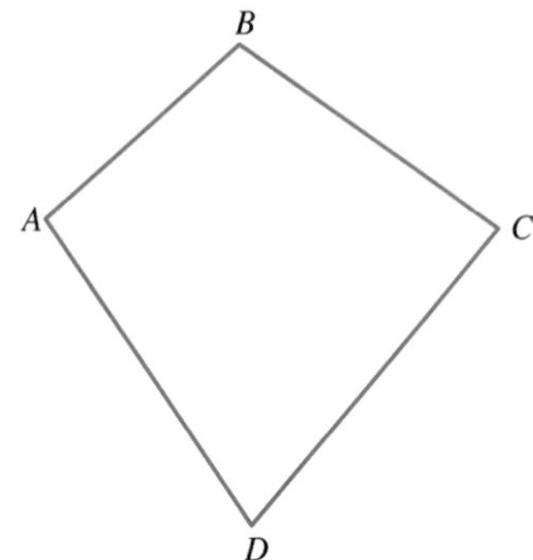

26. For quadrilateral *ABCD*, use a compass and straightedge to construct the angle bisector of each vertex angle. Construct the circle that contains the intersection of the angle bisectors for angles *A* and *B*, angles *B* and *C*, angles *C* and *D*, and angles *D* and *A*.

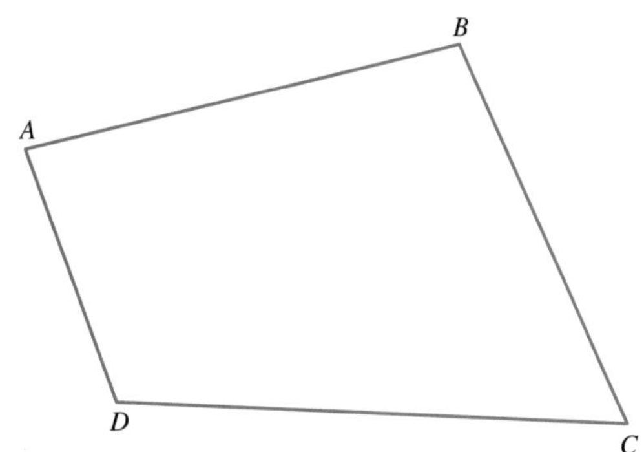

PROOFS

27. Prove that in the same or congruent circles, congruent chords intercept congruent arcs.
28. Prove that in the same or congruent circles, congruent arcs have congruent chords.
29. Prove that in the same or congruent circles, congruent central angles have congruent chords.
30. Prove that in the same or congruent circles, congruent chords determine congruent central angles.
31. Prove that two arcs are congruent if they are contained in congruent circles and have the same central angle.
32. Prove that in a circle, parallel lines intercept congruent arcs.
33. Prove that in the same or congruent circles, congruent chords are equidistant from the center.
34. Given circle *O* with diameter $\overline{AB}$ and a point *C* in the exterior of the circle, determine if $\angle ACB$ is acute, right, or obtuse, and prove your assertion.
35. Prove that if a kite is inscribed in a circle, then its two congruent angles must be right angles.
36. Prove that we can inscribe a quadrilateral in a circle if and only if the quadrilateral's opposite angles are supplementary.
37. Prove that a line drawn from the center of a circle through the midpoint of a chord is the perpendicular bisector of the chord.
38. Prove that a line containing the center of a circle and perpendicular to a chord contains the midpoint of the chord.
39. Prove that the perpendicular to a chord bisects the arcs determined by the chord.
40. A proof of the Law of Cosines based on the following figure is attributed to Roger Nelson, professor of mathematics at Lewis and Clark College in Portland, Oregon. In the figure, *A* is the center of the circle, and $\angle C$ is acute. Use the following figure, in which *a* represents *BC*, *b* represents *AC*, and *c* represents *AB*, to prove $c^2 = a^2 + b^2 - 2ab\cos C$. (HINT: Use Theorem 7.8.)

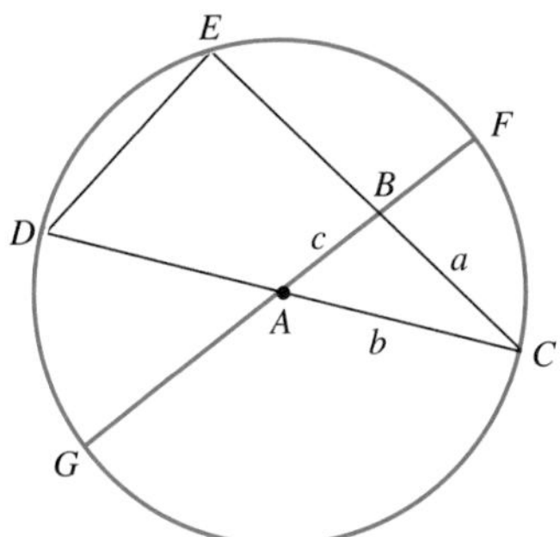

APPLICATIONS

41. Wally, Jane, and Steve decide to get up a game of wastepaper basketball. They are seated at their desks in positions *W*, *J*, and *S*. Where should they place the wastepaper basket so that each person has to throw the same distance to the basket?

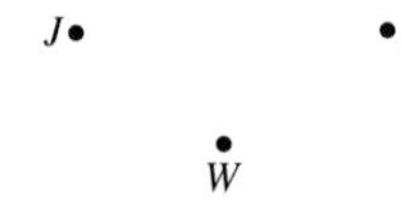

42. The bridge shown has dimensions as given where the arch measures 18′ at its highest point. The arch of the bridge is an arc of a circle, though *not* a semicircle. Find the radius of the circle that contains the arc.

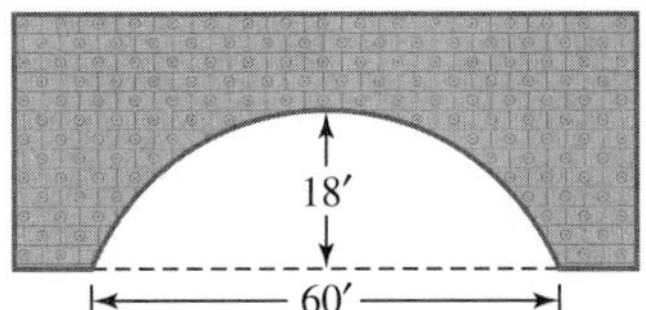

43. A piece of a broken wheel is shown. It was taken to a machine shop to be replaced with a new "whole" wheel. Find the radius of the wheel if $AC = 10$ cm, $BD = 3$ cm, and D is the midpoint of $\overline{AC}$. Round to the nearest hundredth.

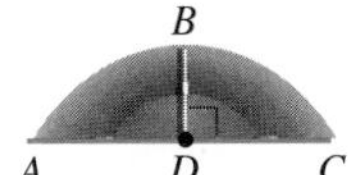

EXTENDED PROBLEMS

44. A conic section is a curve formed from the intersection of a plane and a right circular cone. An ellipse is formed when a plane intersects a right circular cone and the plane does not intersect the cone's base. A hyperbola is formed when a plane intersects two right circular cones as shown in the following figure.

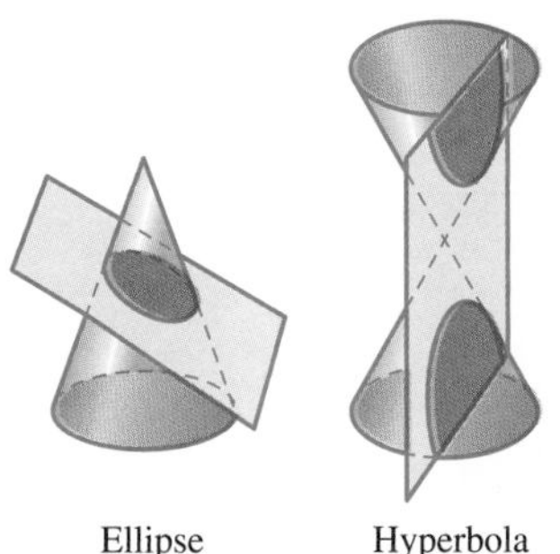

We can create two conic sections, a hyperbola and an ellipse, using a circle and paper-folding techniques. Research paper-folding methods that use a circle to create a hyperbola and an ellipse. Describe the set of points that make up each of the conic sections and list the steps used in the paper-folding techniques. Perform the folds to create a hyperbola and an ellipse. Then create another hyperbola and ellipse by selecting different initial points and different-sized circles. What causes a hyperbola to be wide or narrow? What causes an ellipse to be long and skinny or very round? Refer back to Problem 44 in Section 2.4 and find other applications of ellipses and hyperbolas.

45. In the "Geometry Around Us" at the end of Section 7.2, we discussed two curves of constant width, the circle and the Reuleaux triangle. Research other curves of constant width. Investigate how to create Reuleaux curves using a regular pentagon and a regular heptagon (seven-sided polygon) and make a few examples. How is the perimeter of a Reuleaux triangle related to the circumference of the circle with radius r used to form the triangle? How is the perimeter of a Reuleaux pentagon related to the circumference of the circle used to create it? How is the perimeter of a Reuleaux heptagon related to the circumference of the circle used to create it? Why are wheels generally not made of Reuleaux polygons?

46. An **arbelos** is the shaded region made up of semicircles in the following figure. The sum of the diameters of the two smaller semicircles is equal to the diameter of the large semicircle.

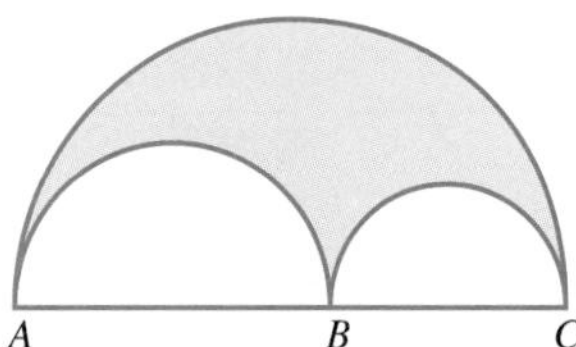

a. If the diameter of the large semicircle is 1 unit, find the arc length $\widehat{AB} + \widehat{BC}$.

b. If the radius of the large semicircle is r and the radii of the smaller semicircles are r_1 and r_2, find an expression for the area of the arbelos in terms of r_1 and r_2.

c. In the following figure, $\overline{BD} \perp \overline{AC}$, the radius of the large semicircle is r, and the radii of the smaller semicircles are r_1 and r_2. Find an expression for BD in terms of r_1 and r_2. Then express the area of the arbelos in terms of BD.

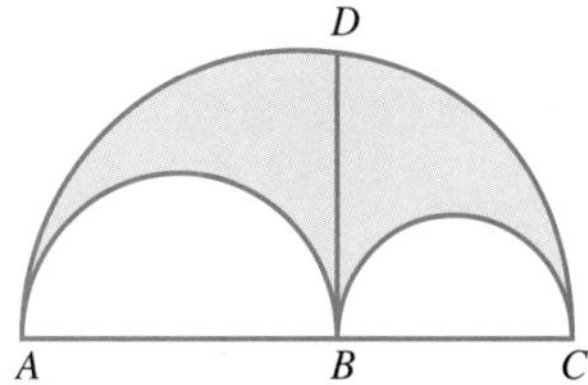

d. Research the arbelos and discuss the significance of the name. Who first studied the arbelos? Describe two special properties of the arbelos.

7.3 SECANTS AND TANGENTS

Applied Problem

Three planets are aligned as shown. The diameter of the smallest planet is 3000 miles and the diameter of the planet in the middle is 8000 miles. Given the other dimensions in the figure, what is the diameter of the largest planet?

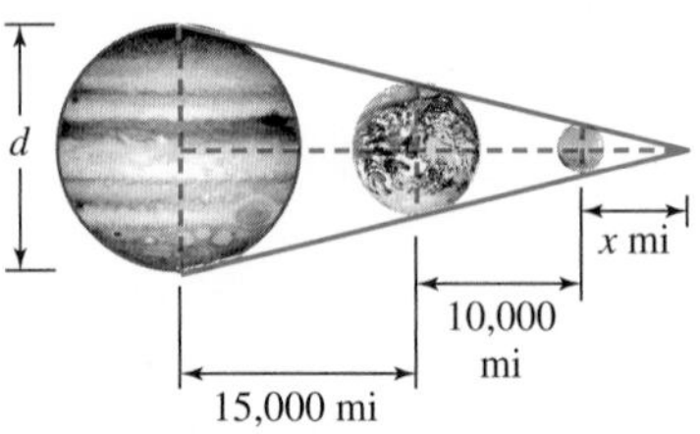

In the previous section, we discussed chords of circles and associated measures. This section presents similar results about secants and tangents.

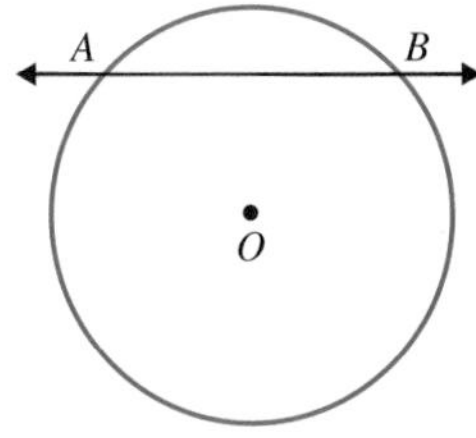

FIGURE 7.22

Secant Lines

A **secant line** is a line that intersects a circle in two points. In Figure 7.22, $\overleftrightarrow{AB}$ is a secant line for circle O. The next example shows how we can find the measure of the angle formed by two secants intersecting outside of a circle.

EXAMPLE 7.10 Find the measure of $\angle AEC$ in Figure 7.23 where $\widehat{AC} = 80°$ and $\widehat{BD} = 20°$.

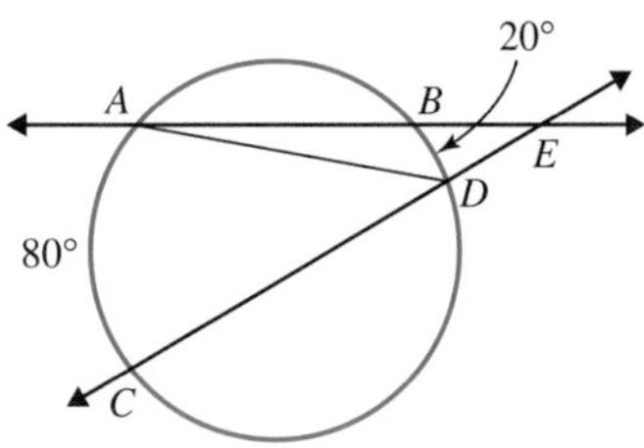

FIGURE 7.23

SOLUTION

Because $\widehat{AC} = 80°$, $\angle ADC = \frac{1}{2}(80)° = 40°$. Also, $\widehat{BD} = 20°$ implies that $\angle DAB = \frac{1}{2}(20)° = 10°$. So by the Exterior Angle Theorem, $\angle ADC = \angle DAB + \angle AEC$ and $\angle AEC = \angle ADC - \angle DAB = 40° - 10° = 30°$. ■

We generalize this example in the next theorem.

THEOREM 7.9

If two secants intersect outside a circle, the measure of the acute angle formed is half the difference of the measures of the intercepted arcs.

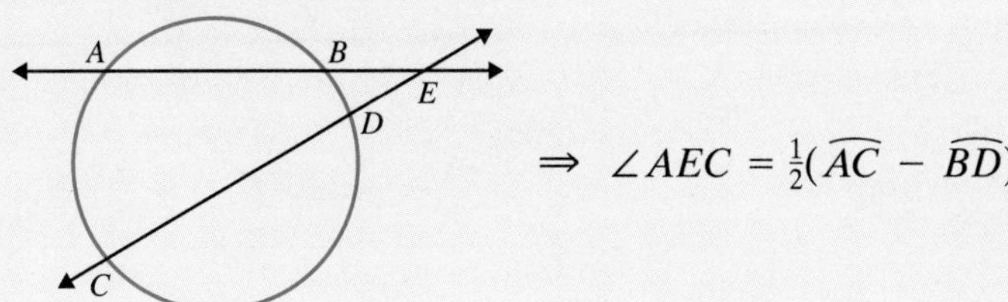

Given Secants $\overleftrightarrow{AB}$ and $\overleftrightarrow{CD}$ intersecting outside a circle at point E [Figure 7.24(a)].

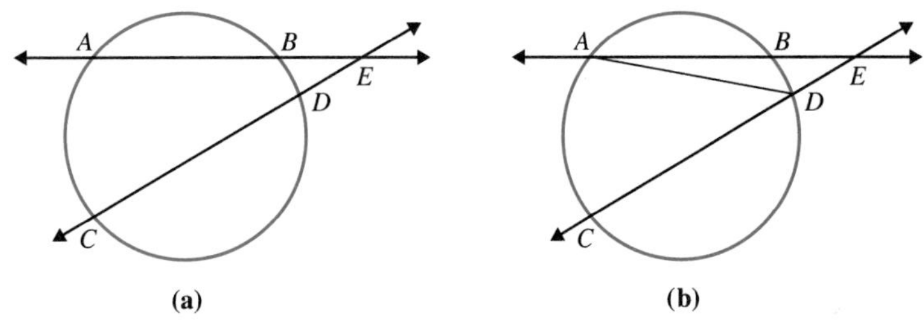

FIGURE 7.24

Prove $\angle AEC = \frac{1}{2}(\overset{\frown}{AC} - \overset{\frown}{BD})$

Proof Draw $\overline{AD}$ [Figure 7.24(b)].

Statements	Reasons
1. $\angle ADC = \frac{1}{2}\overset{\frown}{AC}$	**1.** Inscribed Angle Theorem
2. $\angle DAB = \frac{1}{2}\overset{\frown}{BD}$	**2.** Inscribed Angle Theorem
3. $\angle ADC = \angle DAB + \angle AEC$	**3.** Exterior Angle Theorem
4. $\angle AEC = \angle ADC - \angle DAB$	**4.** Subtraction
5. $\angle AEC = \frac{1}{2}\overset{\frown}{AC} - \frac{1}{2}\overset{\frown}{BD} = \frac{1}{2}(\overset{\frown}{AC} - \overset{\frown}{BD})$	**5.** Substitution/Simplification from Steps 1, 2, and 4 ■

In the last section we found a relationship between the lengths of the segments formed by intersecting chords (Theorem 7.8). A similar relationship exists between the lengths of segments formed by intersecting secants.

THEOREM 7.10

If two secants intersect outside a circle, then the product of the lengths of two segments formed on one secant is equal to the product of the lengths of the corresponding segments on the other secant.

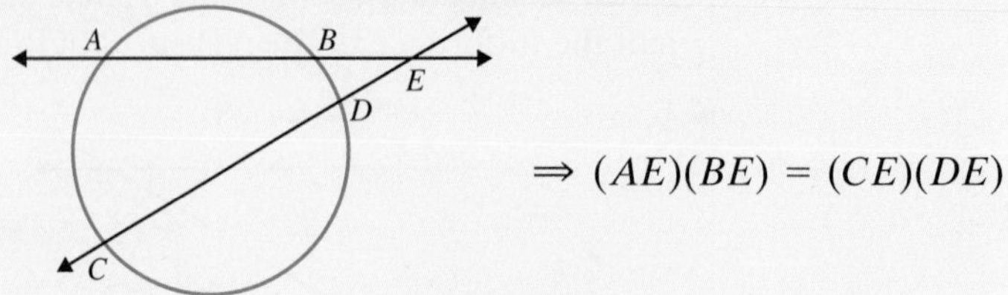

$\Rightarrow (AE)(BE) = (CE)(DE)$

Given Secants $\overleftrightarrow{AB}$ and $\overleftrightarrow{CD}$ intersecting outside a circle at point E

Prove $(AE)(BE) = (CE)(DE)$

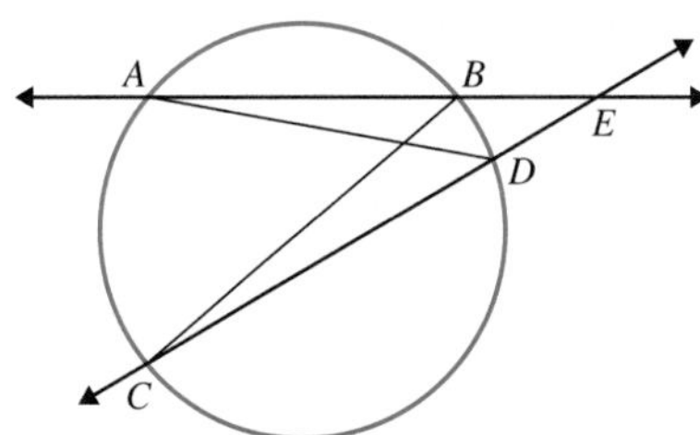

FIGURE 7.25

Proof Draw $\overline{AD}$ and $\overline{BC}$. Consider $\triangle AED$ and $\triangle CEB$ in Figure 7.25. Because E is common to both triangles and $\angle ABC$ and $\angle CDA$ intercept the same arc, we have $\angle ABC = \angle CDA$. Now, $\angle ADE = \angle CBE$ because they are supplements of the congruent angles $\angle CDA$ and $\angle ABC$, respectively. Thus, we can conclude that $\triangle AED \sim \triangle CEB$ by AA Similarity. Therefore, $\frac{AE}{DE} = \frac{CE}{BE}$, or $(AE)(BE) = (CE)(DE)$. ■

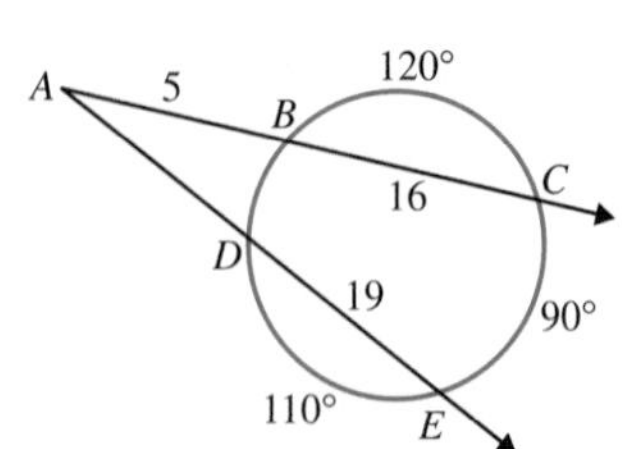

FIGURE 7.26

EXAMPLE 7.11 In Figure 7.26, find the measures of the following.

a. $\angle CAE$ **b.** $\overline{AD}$

SOLUTION

a. $\widehat{BD} = 360° - (120° + 110° + 90°) = 40°$. Thus, by Theorem 7.9,

$$\angle CAE = \frac{1}{2}(90° - 40°) = 25°$$

b. By Theorem 7.10, $(AE)(AD) = (AC)(AB)$. Let x represent AD.

$$(19 + x)(x) = (5 + 16)(5)$$
$$19x + x^2 = 105$$
$$x^2 + 19x - 105 = 0$$

Recall that the solutions of a quadratic equation of the form $ax^2 + bx + c = 0$ are given by the quadratic formula:

$$x = \frac{-b \pm \sqrt{b^2 - 4ac}}{2a}$$

Using the quadratic formula, we have

$$x = \frac{-19 \pm \sqrt{19^2 - 4(-105)}}{2}$$

Because x must be positive, we have

$$x = \frac{-19 + \sqrt{781}}{2} \approx 4.47 \text{ in.}$$

■

Tangent Lines

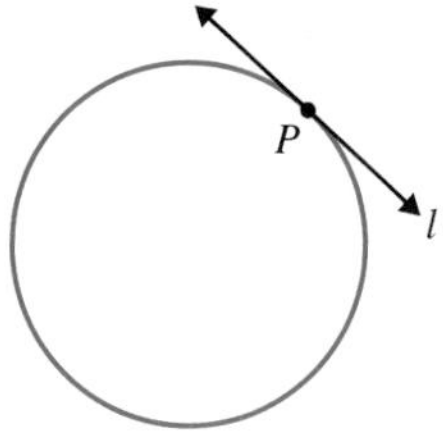

FIGURE 7.27

Chords and secants each intersect a circle in exactly two points. A **tangent line** is a line that intersects a circle in exactly one point. This point is called the **point of tangency**. In Figure 7.27, l is a tangent line and P is the point of tangency. Notice that all points on a tangent line except the point of tangency lie outside the circle. Rays and line segments are said to be tangent to a circle if they lie on a line tangent to the circle and if they intersect the circle in one point.

Suppose a radius of a circle is drawn to the point of tangency (Figure 7.28). Notice that $\overline{OA}$ appears to be perpendicular to the tangent line $\overleftrightarrow{AB}$. The next theorem verifies this fact. This theorem will be useful when we want to construct a tangent line to a point on a circle.

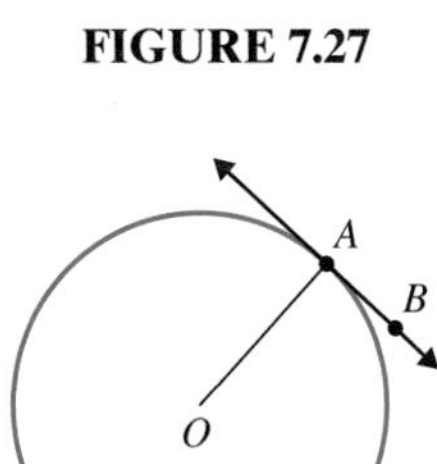

FIGURE 7.28

THEOREM 7.11

A radius or diameter of a circle is perpendicular to a tangent line at its point of tangency.

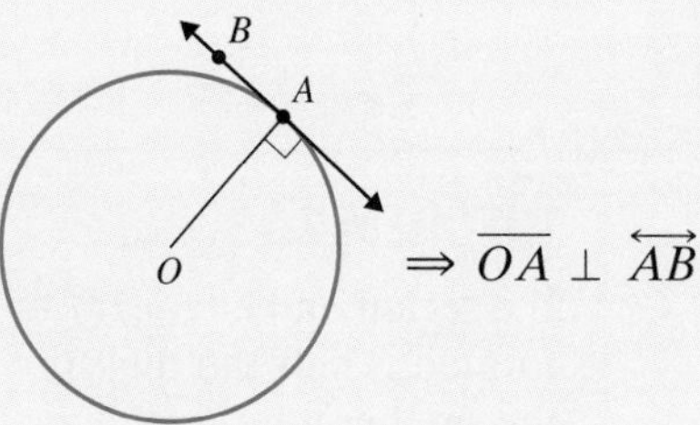

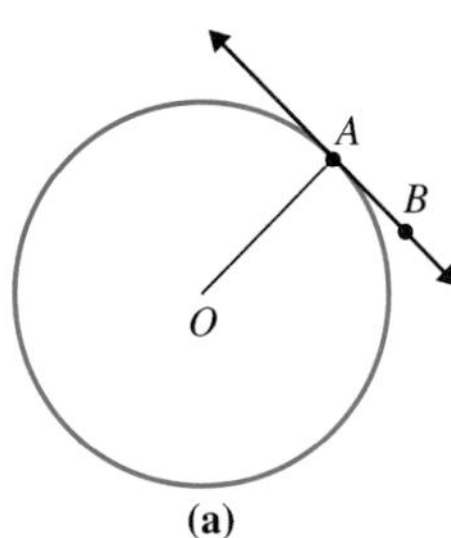

(a)

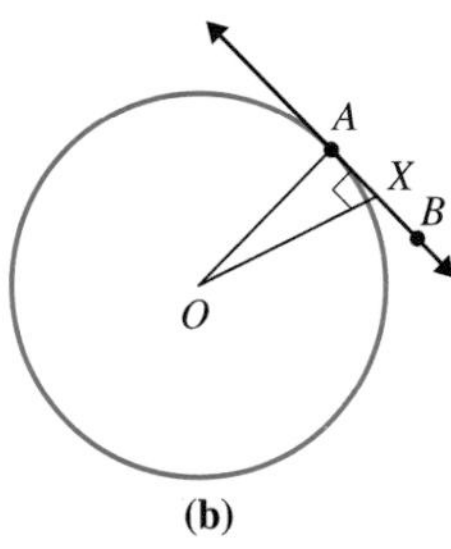

(b)

FIGURE 7.29

Given Circle O with radius $\overline{OA}$ and tangent line $\overleftrightarrow{AB}$ [Figure 7.29(a)]

Prove $\overline{OA} \perp \overleftrightarrow{AB}$

Plan We will prove this result indirectly.

Proof Suppose $\angle OAB$ is *not* a right angle. Then construct the perpendicular $\overline{OX}$ to the tangent line [Figure 7.29(b)], where X is on the tangent line $\overleftrightarrow{AB}$. Then, because $\triangle OAX$ is a right triangle with hypotenuse $\overline{OA}$, we have $OA > OX$. However, because X is outside the circle, it must be true that $OX > OA$. Thus, we have a contradiction, so $\angle OAB$ must be a right angle. ■

Angles Formed By Tangent Lines

The next three theorems describe results involving tangents, secants, and angle measures based on the measure of intercepted arcs. As you proceed, observe how each succeeding theorem uses the previous result.

THEOREM 7.12

The measure of an angle formed when a chord intersects a tangent line at the point of tangency is half the measure of the arc intercepted by the chord and the tangent line.

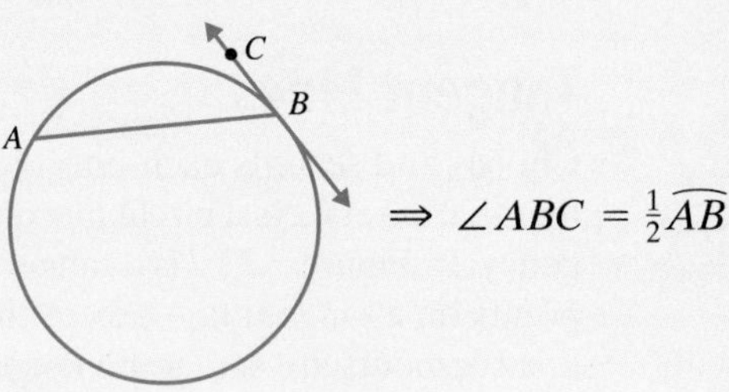

$\Rightarrow \angle ABC = \frac{1}{2}\overparen{AB}$

Given Chord $\overline{AB}$ and tangent line $\overleftrightarrow{CB}$ [Figure 7.30(a)]

Prove $\angle ABC = \frac{1}{2}\overparen{AB}$

Proof Consider $\overparen{AB}$, tangent line $\overleftrightarrow{CB}$, and diameter $\overline{BD}$ [Figure 7.30(b)]. Because, by Theorem 7.11, a diameter is perpendicular to a tangent line at its point of tangency, $\angle CBD = 90° = \frac{1}{2}\overparen{DAB}$. Also, $\angle ABD = \frac{1}{2}\overparen{AD}$ because it is an inscribed angle. Therefore, we have

$$\angle ABC = \angle CBD - \angle ABD = \tfrac{1}{2}\overparen{DAB} - \tfrac{1}{2}\overparen{AD} = \tfrac{1}{2}(\overparen{DAB} - \overparen{AD}) = \tfrac{1}{2}\overparen{AB} \quad \blacksquare$$

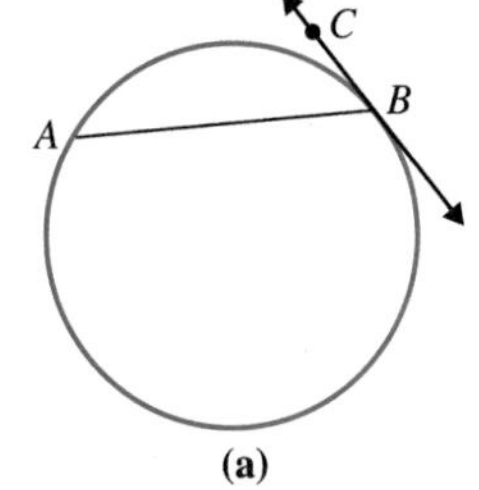

(a)

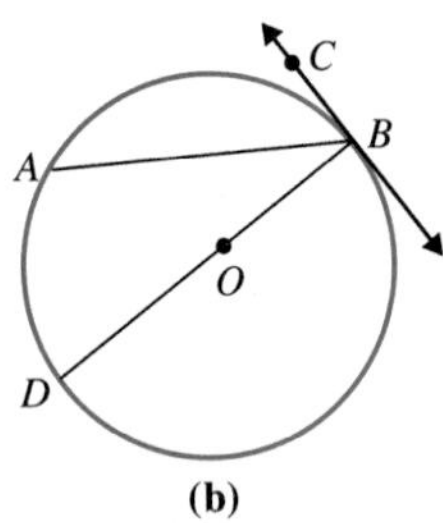

(b)

FIGURE 7.30

THEOREM 7.13

If a secant and a tangent line intersect outside a circle, the measure of the angle formed is half the measure of the larger intercepted arc minus the measure of the smaller intercepted arc.

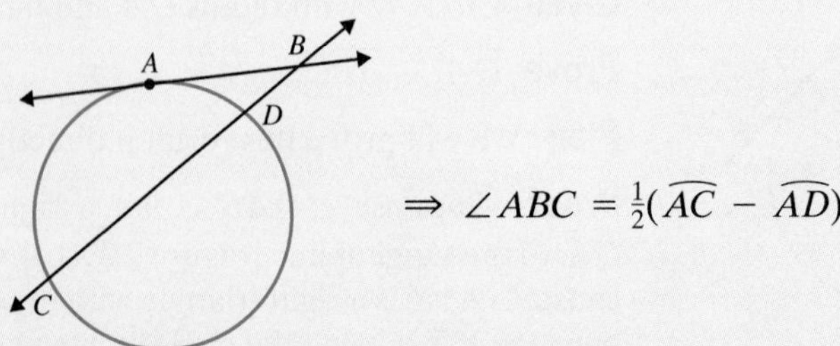

$\Rightarrow \angle ABC = \frac{1}{2}(\overparen{AC} - \overparen{AD})$

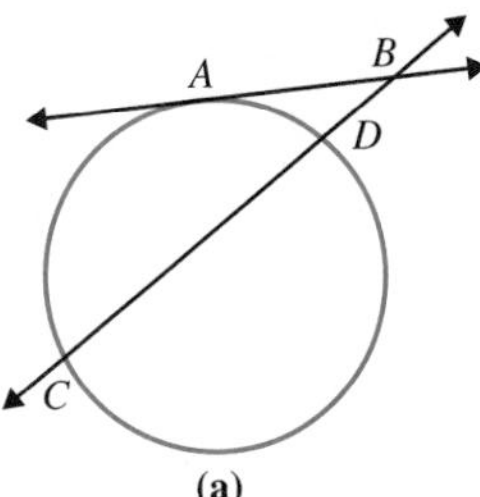

(a)

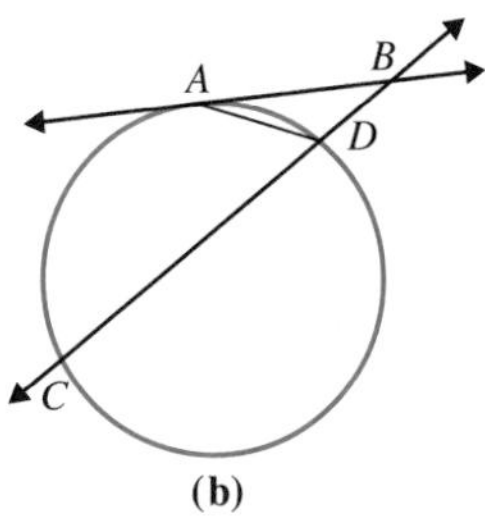

(b)

FIGURE 7.31

Given Secant line $\overleftrightarrow{CD}$ and tangent line $\overleftrightarrow{AB}$, where A is the point of tangency [Figure 7.31(a)]

Prove $\angle ABC = \frac{1}{2}(\widehat{AC} - \widehat{AD})$

Proof Draw $\overline{AD}$ [Figure 7.31(b)].

Statements	Reasons
1. $\angle ADC = \frac{1}{2}\widehat{AC}$	**1.** Inscribed Angle Theorem
2. $\angle BAD = \frac{1}{2}\widehat{AD}$	**2.** Theorem 7.12
3. $\angle ADC = \angle BAD + \angle ABC$	**3.** Exterior Angle Theorem
4. $\angle ABC = \angle ADC - \angle BAD$	**4.** Subtraction
5. $\angle ABC = \frac{1}{2}\widehat{AC} - \frac{1}{2}\widehat{AD} = \frac{1}{2}(\widehat{AC} - \widehat{AD})$	**5.** Substitution/Simplification ■

We can use Theorem 7.13 to prove a similar result about two tangent lines.

THEOREM 7.14

If two tangent lines intersect outside a circle, the measure of the angle formed is half the measure of the larger intercepted arc minus the measure of the smaller intercepted arc.

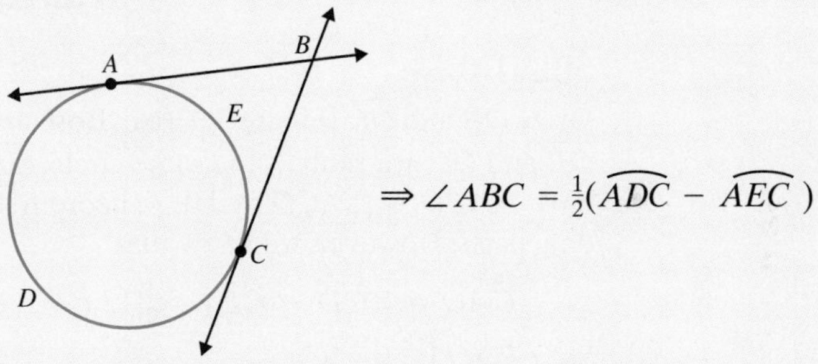

Given Tangent lines $\overleftrightarrow{AB}$ and $\overleftrightarrow{BC}$ intersecting outside a circle where A and C are the points of tangency [Figure 7.32(a)]

Prove $\angle ABC = \frac{1}{2}(\widehat{ADC} - \widehat{AC})$

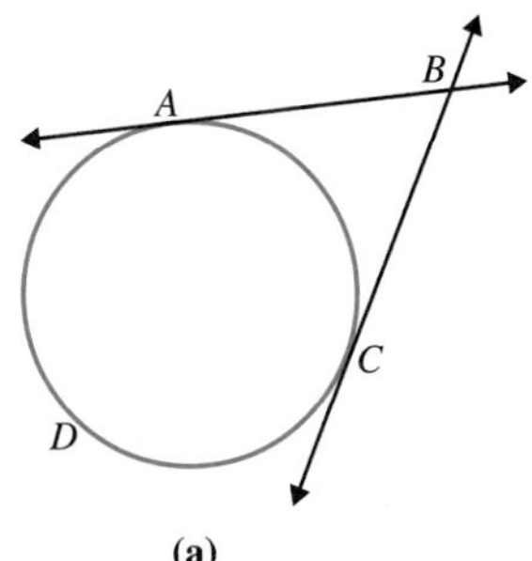

(a)

FIGURE 7.32

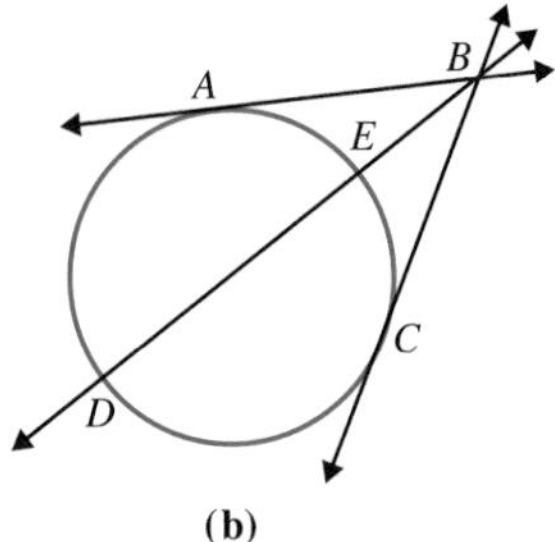

(b)

FIGURE 7.32

Proof Draw $\overleftrightarrow{BD}$ and label E as the point of intersection of $\overleftrightarrow{BD}$ and the circle [Figure 7.32(b)]. Using Theorem 7.13, we have $\angle ABD = \frac{1}{2}\widehat{AD} - \frac{1}{2}\widehat{AE}$ and $\angle DBC = \frac{1}{2}\widehat{DC} - \frac{1}{2}\widehat{EC}$. Because $\angle ABC = \angle ABD + \angle DBC$, we have the following:

$$\begin{aligned}\angle ABC &= \tfrac{1}{2}\widehat{AD} - \tfrac{1}{2}\widehat{AE} + \tfrac{1}{2}\widehat{DC} - \tfrac{1}{2}\widehat{EC} \\ &= \tfrac{1}{2}\widehat{AD} + \tfrac{1}{2}\widehat{DC} - \tfrac{1}{2}\widehat{AE} - \tfrac{1}{2}\widehat{EC} \\ &= \tfrac{1}{2}(\widehat{AD} + \widehat{DC}) - \tfrac{1}{2}(\widehat{AE} + \widehat{EC}) \\ &= \tfrac{1}{2}\widehat{ADC} - \tfrac{1}{2}\widehat{AEC}\end{aligned}$$

■

EXAMPLE 7.12 In Figure 7.33(a), if $\widehat{CD} = 120°$, find $\angle BAO$.

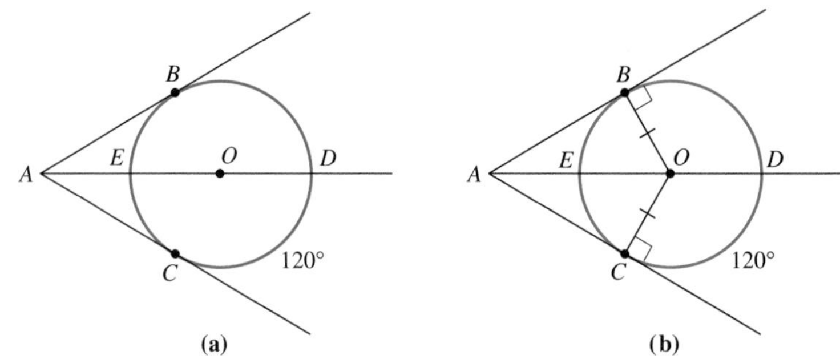

FIGURE 7.33

SOLUTION

Draw $\overline{OB}$ and $\overline{OC}$ [Figure 7.33(b)]. Both are perpendicular to tangent lines. Because $\overline{OB}$ and $\overline{OC}$ are radii of the same circle, $\overline{OB} \cong \overline{OC}$. Also, $\overline{AO} \cong \overline{AO}$. Therefore, we have $\triangle ABO \cong \triangle ACO$ by HL (Theorem 4.3). So $\angle BAO \cong \angle CAO$ by corresponding parts. However, we know that

$$\angle CAO = \tfrac{1}{2}(\widehat{CD} - \widehat{CE}) = \tfrac{1}{2}[120° - (180° - 120°)] = \tfrac{1}{2}(120° - 60°) = 30°$$

So $\angle BAO$ is also 30°. ■

Segments on Tangent Lines

Notice that because $\triangle ABO \cong \triangle ACO$ in Example 7.12, $\overline{AB} \cong \overline{AC}$. We generalize this result next. We leave the proof of the corollary for the problem set.

COROLLARY 7.15

If two tangent lines are drawn to a circle from the same point in the exterior of the circle, the distances from the common point to the points of tangency are equal.

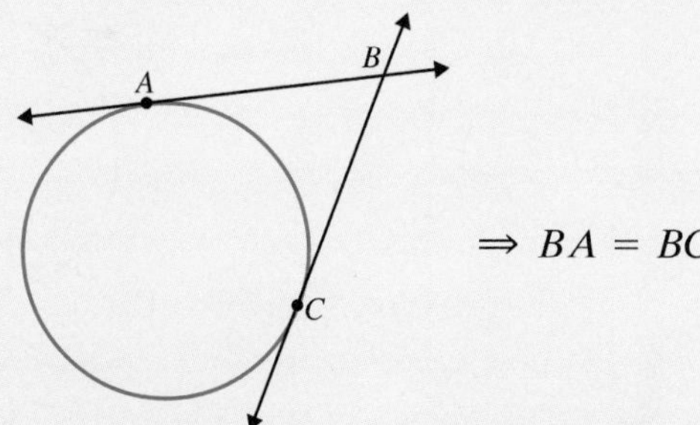

$\Rightarrow BA = BC$

The next example shows an application of Corollary 7.15 to finding lengths of segments in a triangle.

EXAMPLE 7.13 In Figure 7.34(a), the sides of $\triangle ABC$ are each tangent to circle O. We say that circle O is **inscribed** in $\triangle ABC$. If P, Q, and R are points of tangency, find AR and BP.

SOLUTION

Let $AR = x$ and $BP = y$. Because $\overline{AP}$ and $\overline{AR}$ are tangents to circle O from a common point, $AR = AP = x$, by Corollary 7.15. Likewise, $BQ = y$ [Figure 7.34(b)]. Because $AC = 4$ in., we have $RC = 4 - x$ in. Because $BC = 3$ in., we have $QC = 3 - y$ in. [Figure 7.34(c)]. Because $\overline{QC}$ and $\overline{RC}$ are tangents to circle O from point C, by Corollary 7.15 we have $QC = RC$. So $4 - x = 3 - y$, or

$$x - y = 1 \tag{1}$$

Because $AB = 5$ in., we have

$$x + y = 5 \tag{2}$$

Solving the system of equations (1) and (2), we get

$$\begin{array}{r} x - y = 1 \\ \underline{x + y = 5} \\ 2x \quad\; = 6 \end{array}$$

or $x = 3''$ and $y = 2''$. So $AR = 3''$ and $BP = 2''$. ■

(a)

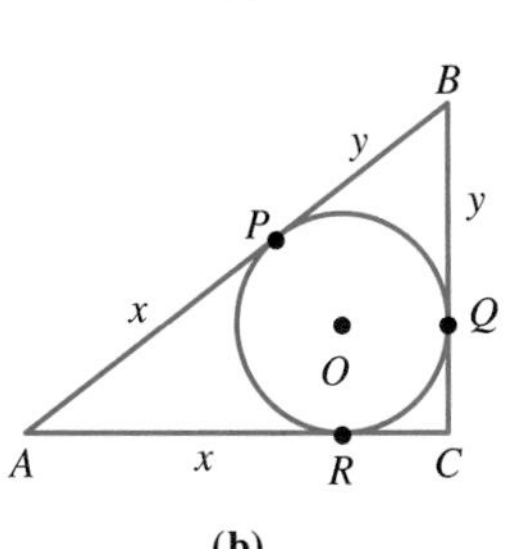

(b)

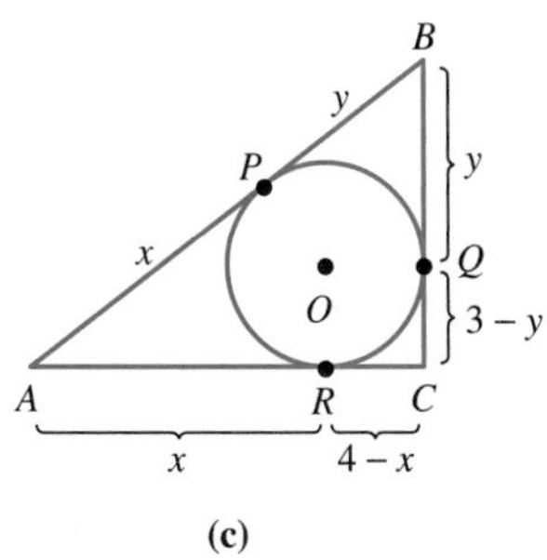

(c)

FIGURE 7.34

THEOREM 7.16

If we draw tangent and secant lines to a circle from the same point in the exterior of the circle, the length of the tangent segment is the mean proportional between the length of the external secant segment and the length of the secant.

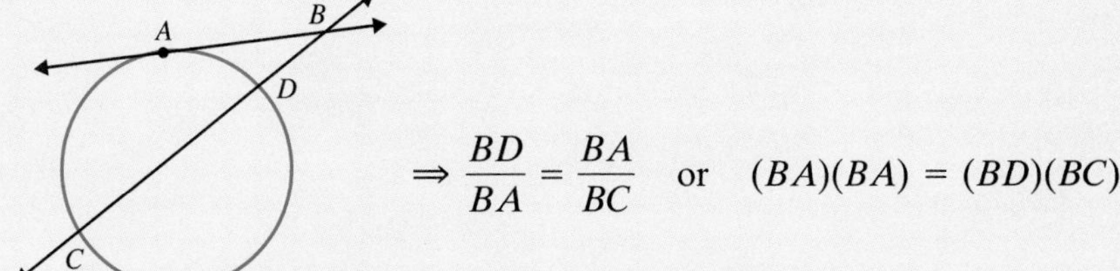

Given Tangent $\overline{AB}$ and secant $\overline{BC}$ [Figure 7.35(a)]

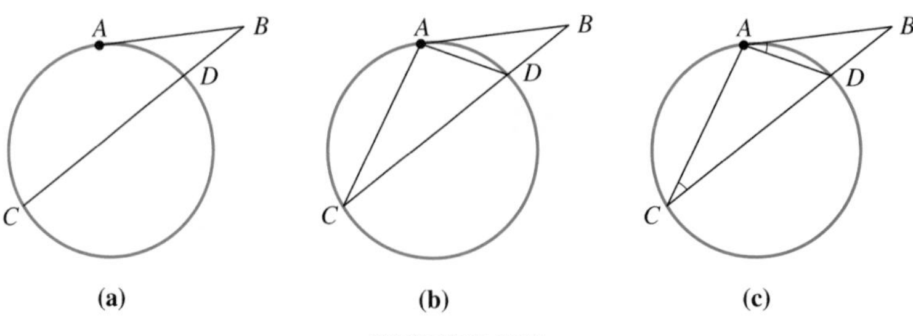

FIGURE 7.35

Prove $\frac{BD}{BA} = \frac{BA}{BC}$

Subgoal Show that $\triangle BDA \sim \triangle BAC$ [Figure 7.35(b)]

PROOF

Statements	Reasons
1. $\angle B \cong \angle B$	1. Reflexive Property
2. $\angle C = \frac{1}{2}\widehat{AD}$	2. Inscribed Angle Theorem
3. $\angle BAD = \frac{1}{2}\widehat{AD}$	3. Theorem 7.12
4. $\angle BAD = \angle C$	4. Substitution [Figure 7.35(c)]
5. $\triangle BDA \sim \triangle BAC$	5. AA Similarity
6. $\frac{BD}{BA} = \frac{BA}{BC}$	6. Corresponding sides in similar triangles are proportional.

We have discussed many results about angles and segments in this chapter. Table 7.1 summarizes many of the major definitions, theorems, and postulates that relate to angle measures and relationships between the lengths of segments in circles.

Figure	Angle Measure	Segment Relationships
Central angle 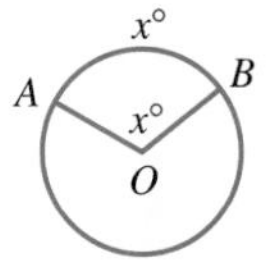	$\angle AOB = \widehat{AB}$	None
Inscribed angle 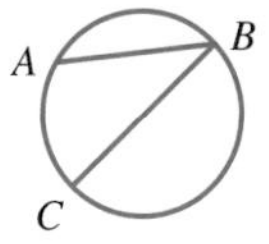	$\angle ABC = \frac{1}{2}\widehat{AC}$	None
Intersecting chords 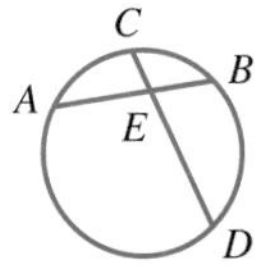	$\angle AEC = \frac{1}{2}(\widehat{AC} + \widehat{BD})$	$(AE)(BE) = (CE)(DE)$
Intersecting secants 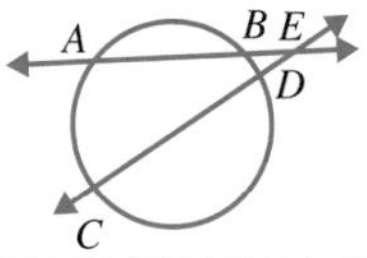	$\angle AEC = \frac{1}{2}(\widehat{AC} - \widehat{BD})$	$(AE)(BE) = (CE)(DE)$
Intersecting tangent and radius 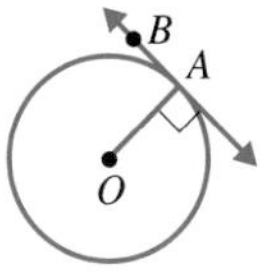	$\angle OAB = 90°$	None
Intersecting tangent and chord 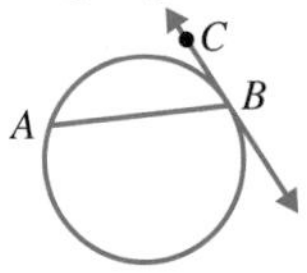	$\angle ABC = \frac{1}{2}\widehat{AB}$	None
Intersecting tangent and secant 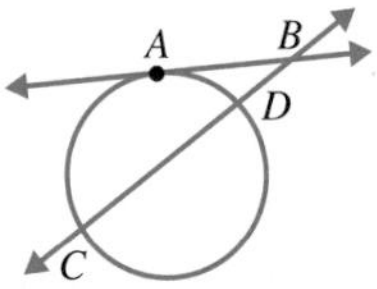	$\angle ABC = \frac{1}{2}(\widehat{AC} - \widehat{AD})$	$\frac{BD}{BA} = \frac{BA}{BC} \Rightarrow (BA)(BA) = (BD)(BC)$
Two intersecting tangents 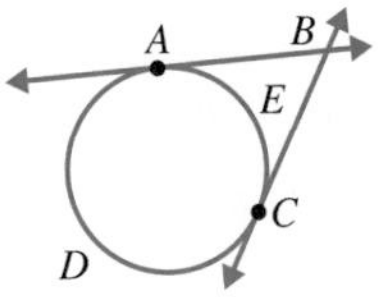	$\angle ABC = \frac{1}{2}(\widehat{ADC} - \widehat{AEC})$	$AB = CB$

TABLE 7.1

Solution of Applied Problem

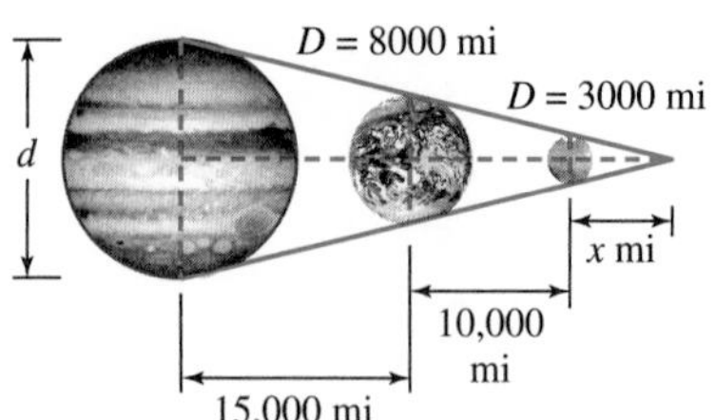

All three triangles formed are similar. Therefore, we have $\frac{x}{3000} = \left(\frac{x + 10{,}000}{8000}\right)$, which yields $8000x = 3000(x + 10{,}000)$, or $x = 6000$ mi. Comparing the smallest triangle with the largest one, we have $\frac{x}{3000} = \frac{25{,}000 + x}{d}$, or $\frac{6000}{3000} = \frac{31{,}000}{d}$. So $d = 15{,}500$ mi.

GEOMETRY AROUND US A rainbow occurs when drops of water in rain or mist reflect and refract (bend) rays of sunlight. As a ray enters a raindrop, it is refracted and separated into colors. Then it is reflected and bent again as it leaves the raindrop. Each color of the spectrum is bent through a different angle, for example, 42° for red and 40° for violet.

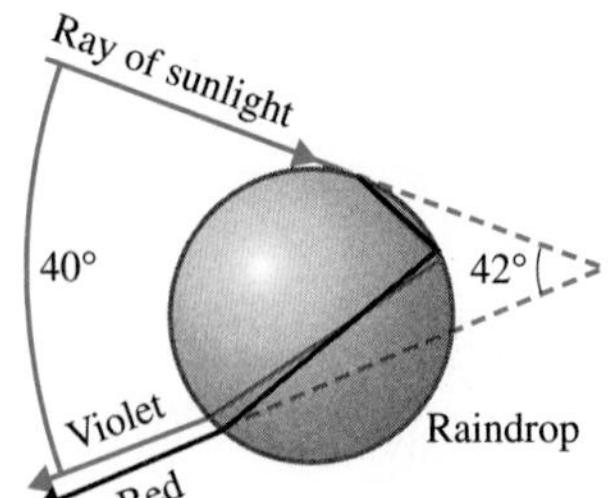

PROBLEM SET 7.3

EXERCISES/PROBLEMS

Refer to the following figure for Exercises 1–6.

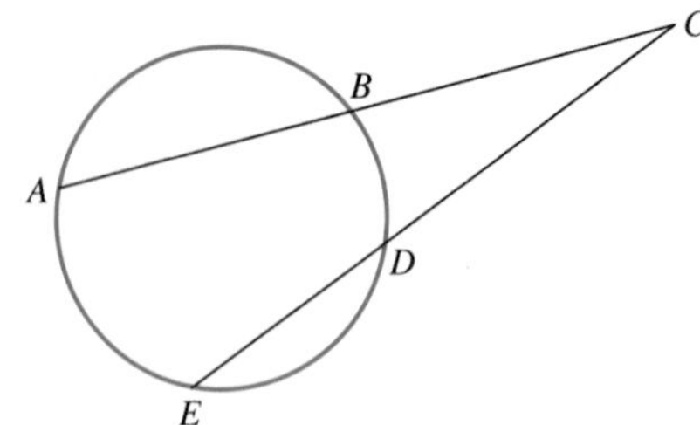

1. If $\widehat{AE} = 90°$ and $\widehat{BD} = 38°$, find the measure of $\angle C$.
2. If $\angle C = 25°$ and $\widehat{BD} = 34°$, find the measure of $\widehat{AE}$.
3. If $AC = 10$ cm, $AB = 4$ cm, and $DE = 3$ cm, find EC. Round to the nearest hundredth.
4. If $AC = 10$ cm, $AB = 4$ cm, and $DC = 3$ cm, find EC.
5. If $DE = 6$ in., $DC = 9$ in., and $\overline{AB}$ is 8 in. shorter than $\overline{BC}$, find AB. Round to the nearest hundredth.
6. If the length of $\overline{DC}$ is 2 cm more than the length of $\overline{DE}$, $AB = 3$ cm, and $BC = 7$ cm, find DC. Round to the nearest hundredth.

Refer to the following figure for Exercises 7–10. Point O is the center of the circle.

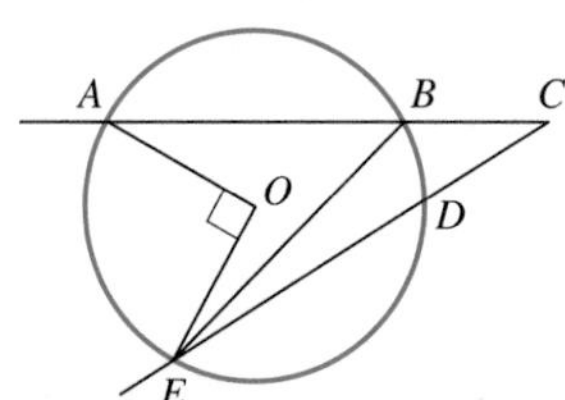

7. If $\angle OED = \angle ABE$ and BE bisects $\angle OED$, what is the measure of $\overset{\frown}{BD}$ and $\angle C$?

8. If $\angle C = 32°$, what is the measure of $\angle BEC$?

9. If $AB = 10$, $BC = 6$, and $CD = 7$, what is DE? Round to the nearest hundredth.

10. If $AB = 12$, $AC = 20$, and $CE = 18$, what are DE and CD? Round to the nearest hundredth.

Refer to the following figure for Exercises 11–14, where $\overleftrightarrow{ST}$ is tangent to the circle at P.

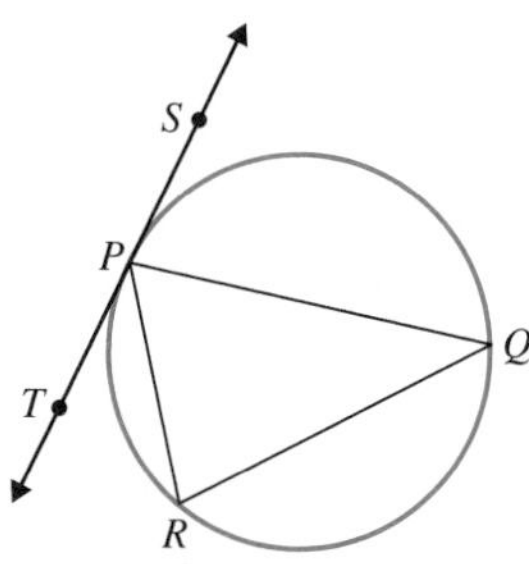

11. If $\overset{\frown}{PRQ} = 230°$, find the measure of $\angle SPQ$.

12. If $\angle PQR = 24°$, find the measure of $\angle TPR$.

13. If $\angle TPR = 28°$ and $\angle RPQ = 44°$, find $\overset{\frown}{PQ}$.

14. If $\angle PRQ = 70°$ and $\overset{\frown}{RQ} = 60°$, find the measure of $\angle TPR$.

Refer to the following figure for Exercises 15 and 16 where $\overline{UP}$ is tangent to the circle at P.

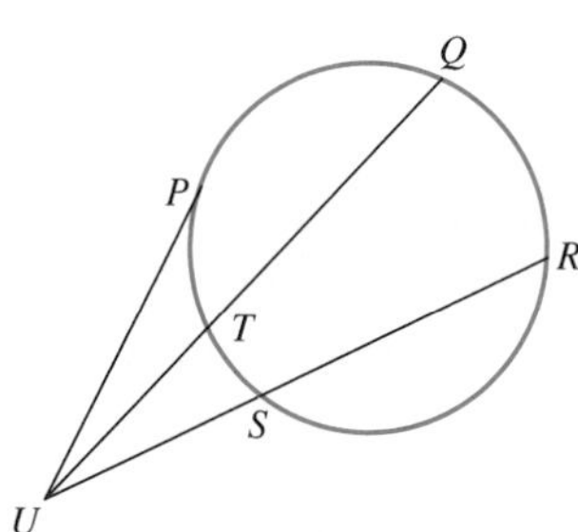

15. If $\angle SUP = 24°$ and $\overset{\frown}{PTS} = 40°$, find the measure of $\overset{\frown}{PQR}$.

16. If $\overset{\frown}{PQ} = 55°$ and $\overset{\frown}{PT} = 31°$, find the measure of $\angle TUP$.

Refer to the following figure for Exercises 17–22, where l and m are tangent to circle O and A and B are points of tangency.

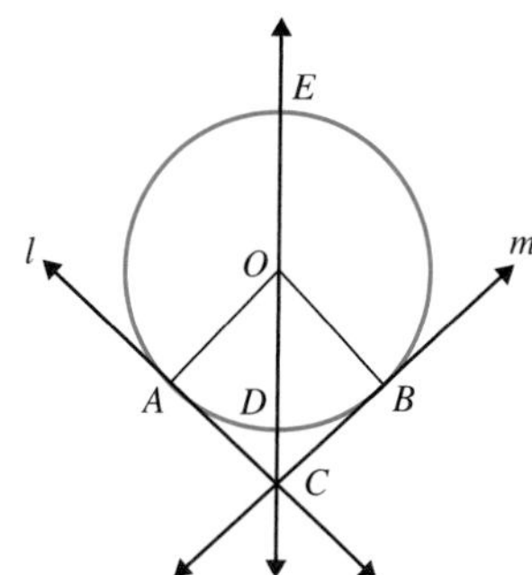

17. What is the measure of $\angle CBO$?

18. What is the measure of $\angle CAO$?

19. Why is $\triangle ACO \cong \triangle BCO$?

20. If $CD = 2$ and $BC = 6$, what is DE?

21. If $\angle AOD = 30°$, what is the measure of $\overset{\frown}{AE}$?

22. If $\overline{OD}$ bisects $\angle AOB$ and $\angle EOB = 150°$, find $\overset{\frown}{AED}$ and $\angle ACB$ in two different ways.

23. In the following figure, $\overline{AB}$ is tangent to the circle with center O. If $OA = 4$ cm and $OB = 12$ cm, find AB. Round to the nearest hundredth.

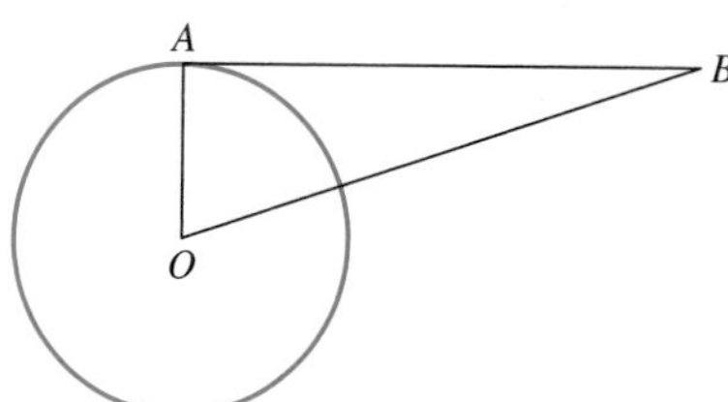

24. $\overline{XY}$ is tangent to the circle O. If $OZ = ZX$ and $XY = 6$ in., find the radius of the circle. Round to the nearest hundredth.

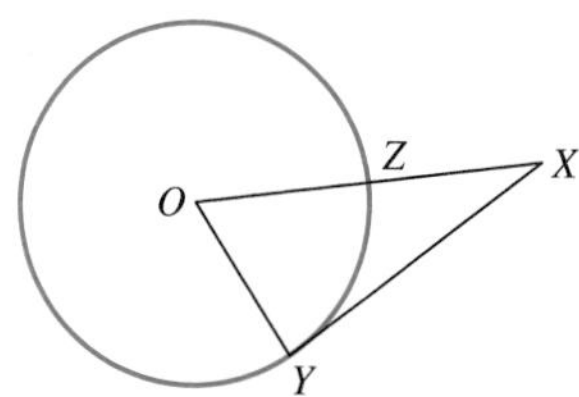

25. $\triangle ABC$ is isosceles with the inscribed circle as shown. $AB = BC = 12$ ft and $AC = 7$ ft. Find AP and BQ.

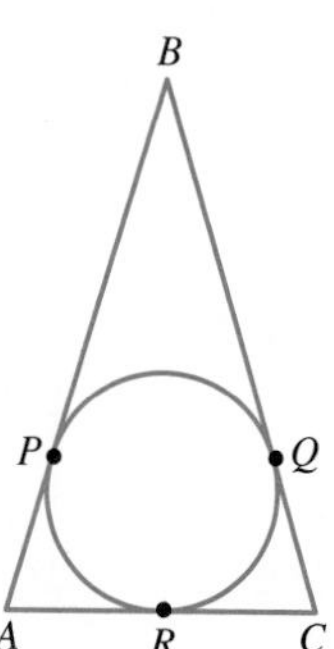

26. Given the following $\triangle ABC$, with $AB = 9$ in., $BC = 7$ in., and $AC = 10$ in., and its inscribed circle, find BX and AZ.

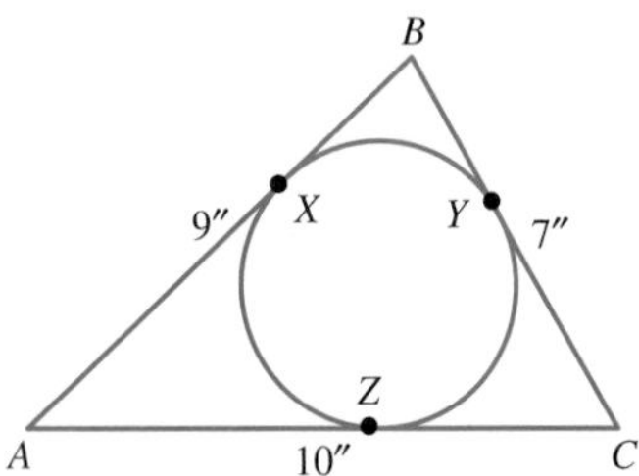

27. If $\angle A = 48°$, $\angle B = 72°$, $\angle C = 60°$, and the measures of the arcs are as indicated in the following figure, find the measures of $\widehat{EF}$, $\widehat{GH}$, and $\widehat{DI}$.

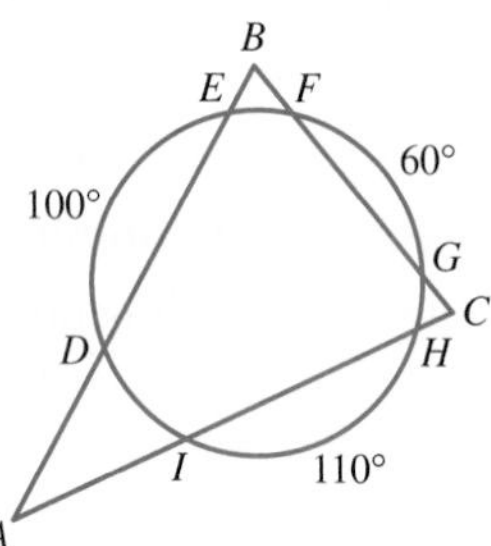

28. If $\angle A = 45°$, $\angle B = 39°$, and the measures of the arcs are as indicated in the figure, find the measures of $\widehat{DE}$, $\widehat{EF}$, and $\widehat{GC}$.

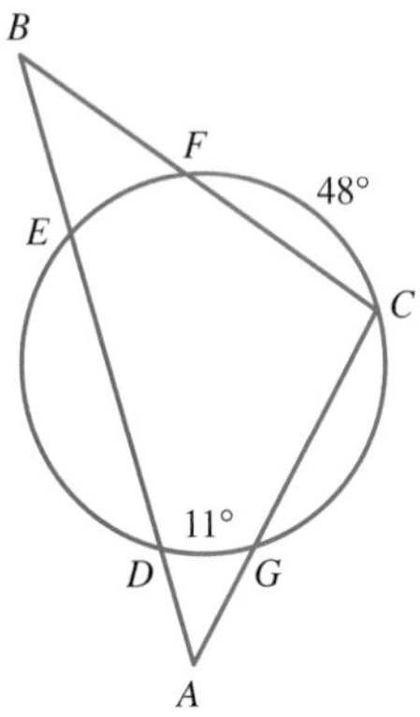

29. In the following figure, the two circles have the same center and the sides of the square are tangent to the smaller circle. If the radius of the larger circle is 1 unit, find the exact area of the shaded region.

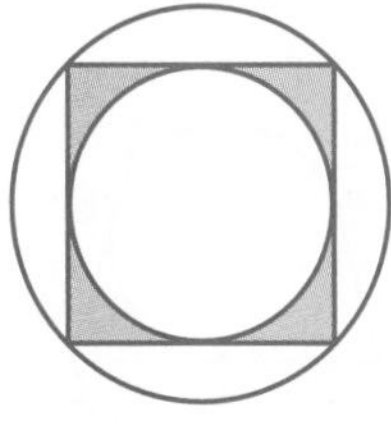

30. In the following figure, the two circles have the same center and the sides of the square are tangent to the smaller circle. If the radius of the smaller circle is 1 unit, find the exact area of the shaded region.

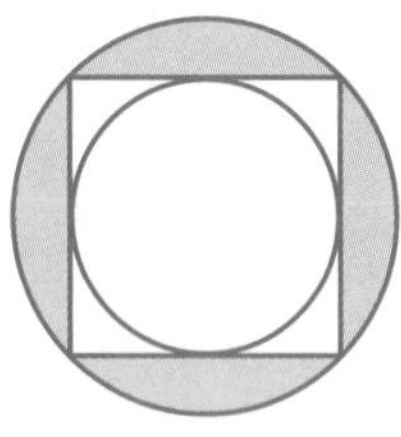

PROOFS

31. Prove Corollary 7.15: If two tangent lines are drawn to a circle from the same point in the exterior of the circle, the distances from the common point to the points of tangency are equal.

32. Prove that if a tangent line and a secant are parallel, then the arcs they intercept are congruent.

33. Prove that a line perpendicular to a radius at the point where the radius intersects the circle is a tangent line.

34. Given are two circles with centers O and O' and two common tangents from P.

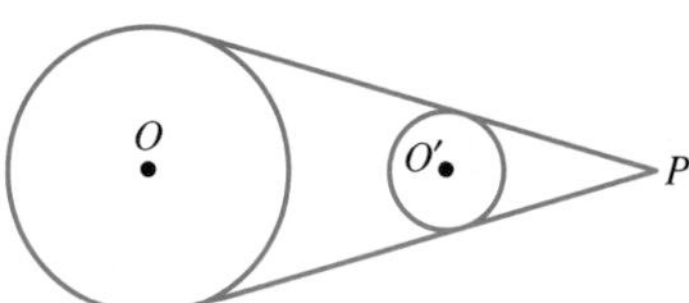

Prove that O, O', and P are collinear.

35. Prove that if two secants of a circle are parallel, then the arcs between them are congruent.

36. Two circles are **externally tangent** if they intersect in one point, have a common tangent line at the point of intersection and each circle is in the exterior of the other, except for the point in common, as shown next.

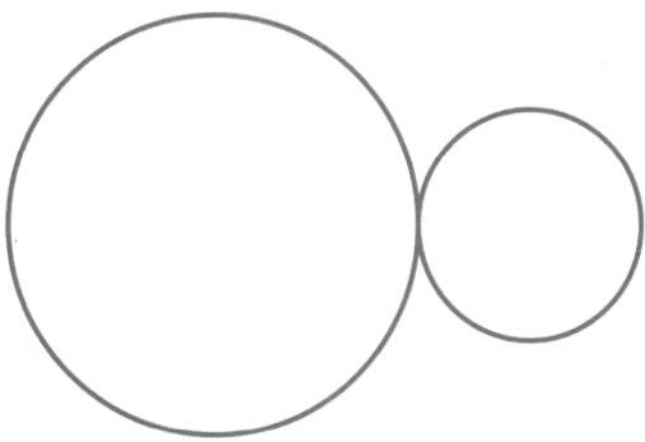

Prove that if two circles are externally tangent, then their centers and their point of intersection are collinear.

37. Two circles are **internally tangent** if they intersect in one point, have a common tangent line at the point of intersection, and one circle is in the interior of the other, except for the point in common, as shown next.

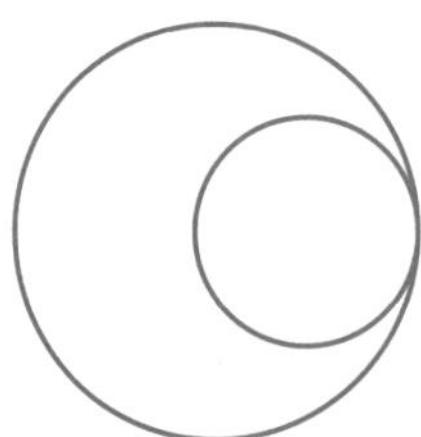

Prove that if two circles are internally tangent, then their centers and their point of intersection are collinear.

38. In the following figure, the small circle and the large circle have the same center O. Chord $\overline{AB}$ is tangent to the smaller circle at C. Prove that the area between the large and small circles is the same as the area of the circle that has $\overline{AB}$ as a diameter.

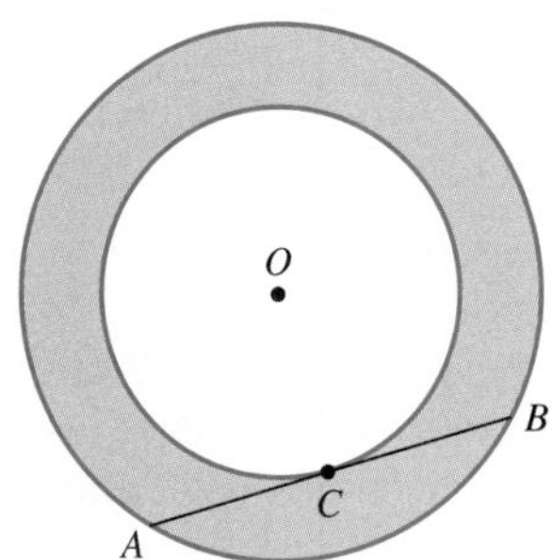

APPLICATIONS

39. In the corner of a fenced yard is a small triangular storage area closed off from the rest of the yard by a gate as shown. If the dimensions of the storage area are as shown, what is the diameter of the largest circular trash can that could be stored inside?

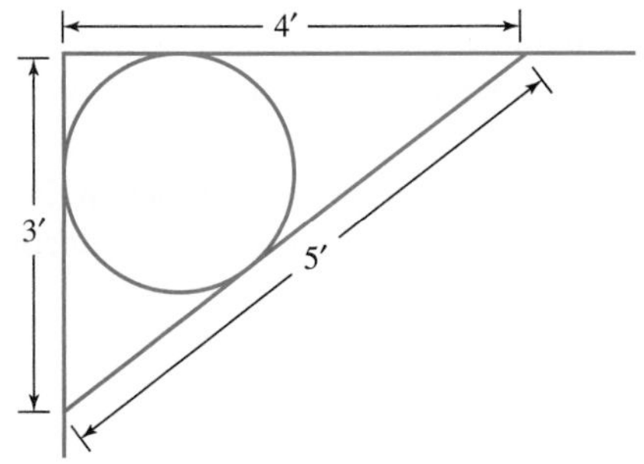

40. Suppose a person is standing at an altitude h above sea level, as shown in the figure. The distance d the person can see is the distance to the horizon. We can calculate it using the radius r of the earth and a property of circles.

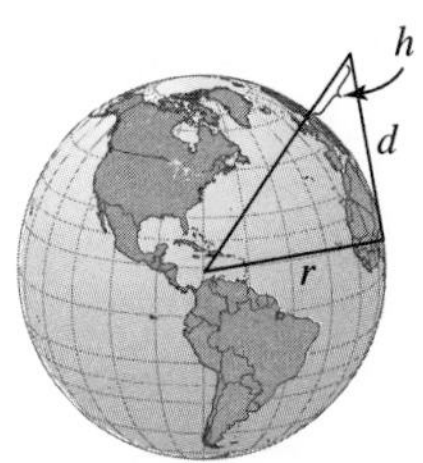

a. Express the distance, d, in terms of h and r.

b. Use 3960 miles as the radius of the earth, and determine the distance to the horizon for an observer at an altitude of 590 feet. Round your answer to the nearest whole number.

c. Use 3960 miles as the radius of the earth, and determine the distance to the horizon for a pilot flying at 12,000 feet. Round your answer to the nearest mile.

41. A **total solar eclipse** occurs when the moon passes between the earth and the sun, and the darkest shadow cast by the moon, called the **umbra**, hits the surface of the earth. If the umbra does not hit the surface, as shown in the following figure, then a total solar eclipse is not possible. In other words, for a total solar eclipse to occur, point P must lie inside the circle for the earth.

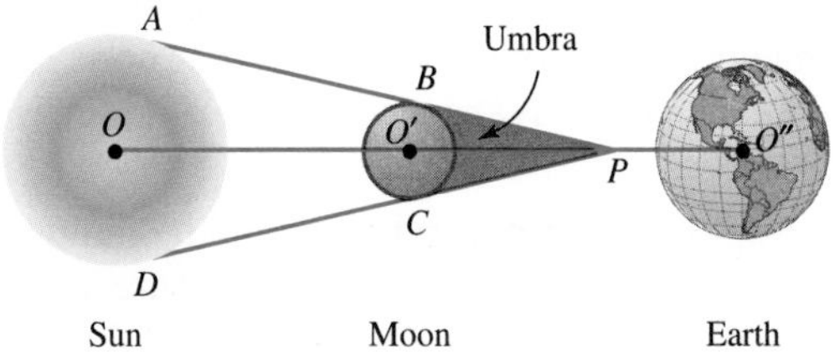

Assume the diameter of the sun is 870,000 miles, the diameter of the moon is 2160 miles, the diameter of the earth is 7920 miles, and the distance from the center of the sun to the center of the earth is approximately 93,000,000 miles. The distance from the moon to the earth varies, but the maximum distance from the center of the moon to the center of the earth is 252,700 miles, and is called the **lunar apogee**. How far is P from the center of the earth during lunar apogee? Round to the nearest thousand. Can there be a total solar eclipse during lunar apogee? Explain.

EXTENDED PROBLEMS

42. There are several variations of the classic **circle-packing problem**, one of which involves packing identical circles into a larger circle so that none of the circles overlap and the radius of the smaller circles is as large as possible. Let N represent the number of small circles to be packed into a larger circle. The following figure shows a solution to the circle-packing problem for $N = 2$.

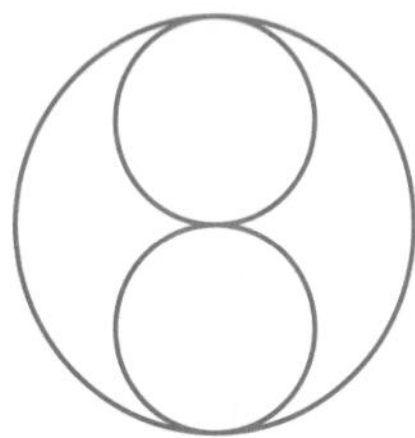

a. For a circle with radius 1 inch, determine the largest radius of two identical circles that can be packed into the circle.

b. For a circle with radius 1 inch, determine the largest radius of three identical circles that can be packed into the circle.

c. Research the circle-packing problem. How is an optimal packing defined? For what values of N has an optimal packing been found? Summarize your findings and include figures that show an optimal circle packing for the cases where $N = 3$ through 9.

43. René Descartes developed a formula related to four circles of various sizes that are mutually tangent, that is, four circles such that none overlap and each circle is tangent to three other circles internally or externally.

a. Explore this problem by finding the radius of the smallest circle r in the following figure, where the radii of the other three circles are as shown.

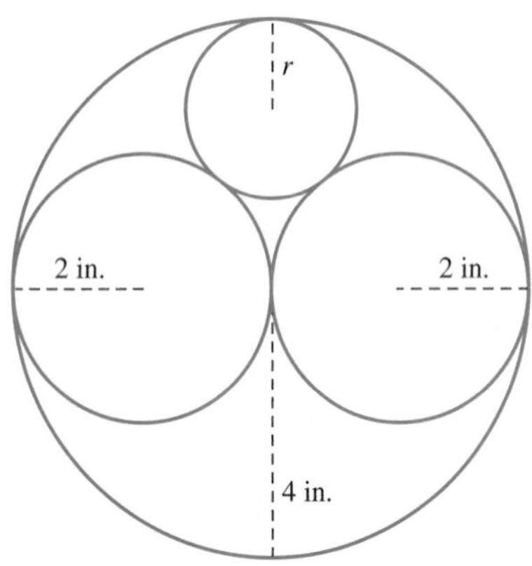

b. The **curvature** of a circle with radius r is defined as $\frac{1}{r}$. Descartes's formula gives a relationship between the curvatures of the four circles. Research Descartes's formula and the curvature of a circle. What does the sign of the curvature of a circle mean? State Descartes's formula and use it to verify the radius you found in part (a).

44. A **hypocycloid** is a curve traced out by a fixed point on a circle as it rolls around the inside of and remains tangent to a larger circle. The following figure shows one example of a hypocycloid. Because three curves are formed, this hypocycloid is called a **three-cusped hypocycloid**.

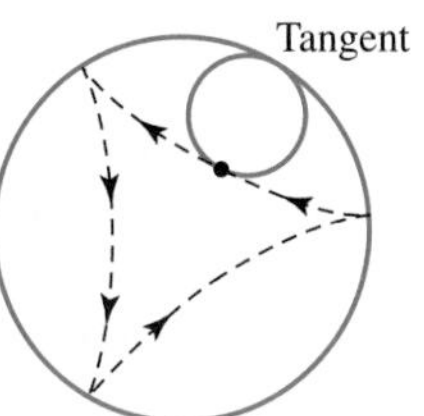

The radii of the two circles determine the shape of the hypocycloid.

a. Construct a circle of radius 9 cm on heavy card-stock paper and carefully remove the circle leaving the paper with a 9 cm radius circular hole. This paper will serve as a template in which smaller circles will rotate. From another piece of card-stock paper, cut out circles of radii 4.5 cm, 3 cm, 2.25 cm, and 1.8 cm. Mark one point on the circumference of each circle. Lay the circle template on a clean sheet of paper and carefully rotate one of the smaller circles around the inside of the 9 cm radius circle. Trace the curve mapped out by the point you marked. Repeat the process using a clean sheet of paper for each of the smaller circles. How is the curve related to the radii of the two circles? In particular, what happens if the ratio of the large circle's radius to the small circle's radius is 2, 3, 4, or 5?

b. Research the hypocycloid, when it was first studied, and by whom. For what purpose was it investigated?

7.4 CONSTRUCTIONS INVOLVING CIRCLES

Applied Problem

A woodworker needs to make a circular cut at a corner as shown. The arc should be tangent to each edge of the board. How can she draw the arc?

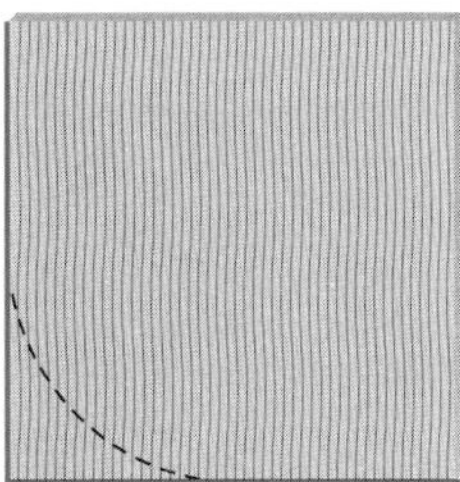

We can also use the basic constructions presented in Sections 4.4 and 5.5 to perform constructions involving circles. In this section, we construct particular kinds of circles and tangents.

Constructing the Circumscribed Circle of a Triangle

The **circumscribed circle** of a triangle is the unique circle that contains the vertices of the triangle. Next we will show how we can construct the circumscribed circle of a triangle with a compass and straightedge.

Construction 9. *To construct the circumscribed circle of a triangle.*

Procedure Consider $\triangle ABC$ [Figure 7.36(a)]. Construct the perpendicular bisectors of any two sides, say $\overline{AC}$ and $\overline{AB}$, and label their points of intersection P [Figure 7.36(b)]. Using P as center and AP as radius, construct the circumscribed circle. The point P, the point of intersection of the perpendicular bisectors of the sides of the triangle, is called the **circumcenter** of the triangle.

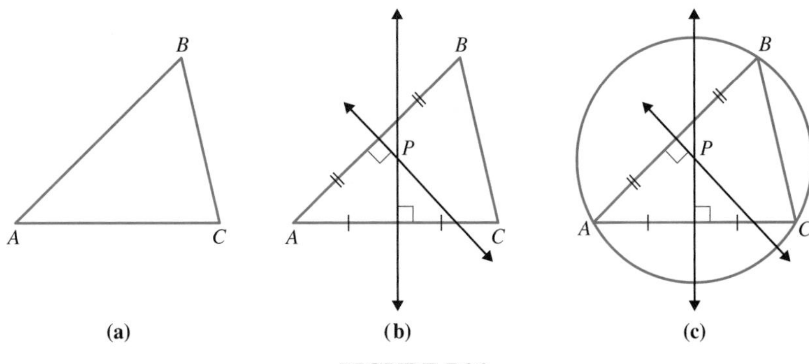

FIGURE 7.36

Justification Because P is on the perpendicular bisector of $\overline{AC}$, by Theorem 4.10, it is equidistant from A and C, so $AP = CP$. In a similar manner, P is equidistant from A and B, hence, $AP = BP$. Because $AP = CP$ and $AP = BP$, we can conclude that $CP = BP$. This says that P is equidistant from C and B. Thus, we can conclude that P is on the perpendicular bisector of $\overline{BC}$. Because $AP = BP = CP$, we can construct the circumscribed circle with center P and radius AP around $\triangle ABC$ [Figure 7.36(c)]. ■

This justification leads to the following theorem.

THEOREM 7.17 Circumcenter of a Triangle

The perpendicular bisectors of the sides of a triangle intersect in a single point, the circumcenter.

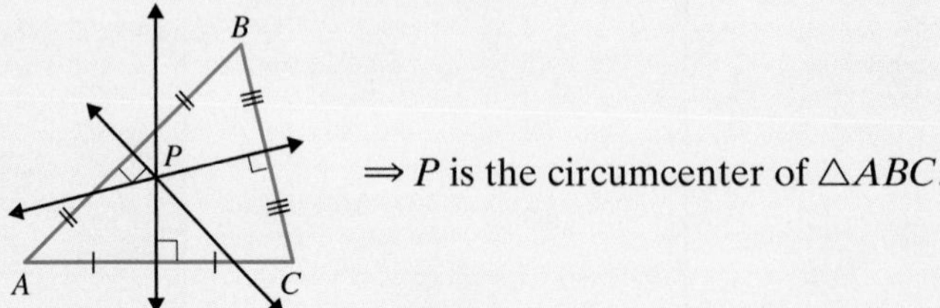

Constructing the Orthocenter of a Triangle

The altitudes of a triangle also intersect in a common point called the **orthocenter** of the triangle. Although this result may not seem surprising, the proof of this fact follows from the previous discussion!

Construction 10. *To construct the orthocenter of a triangle.*

Procedure Consider $\triangle ABC$ [Figure 7.37(a)]. Construct any two altitudes in $\triangle ABC$. They intersect in P, the orthocenter of the triangle [Figure 7.37(b)].

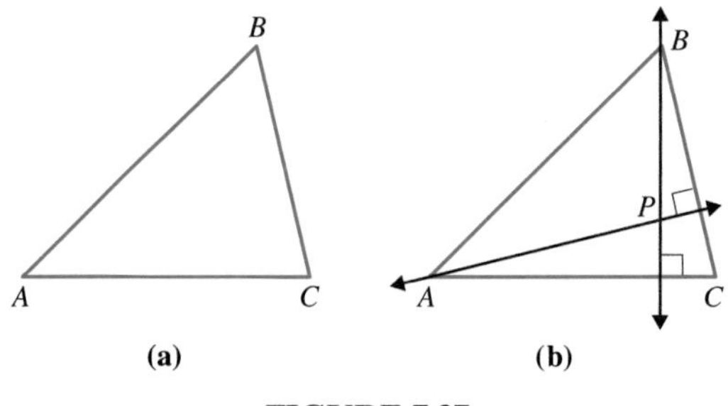

FIGURE 7.37

Justification Using the triangle in Figure 7.37(b), draw lines through the vertices, making each one parallel to the opposite sides of $\triangle ABC$ [Figure 7.37(c)].

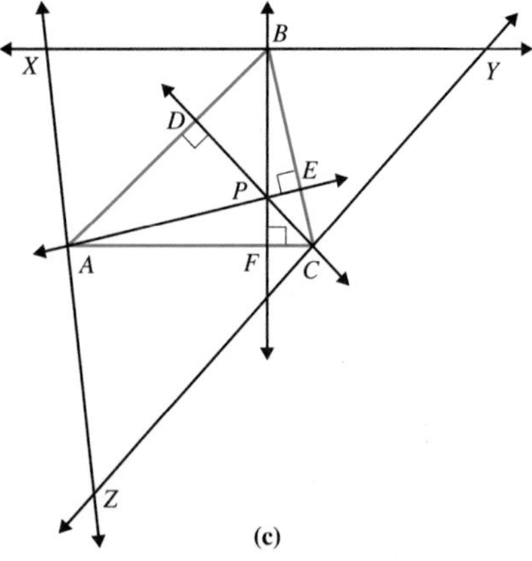

FIGURE 7.37

Now we will prove that $\overline{BF}$, $\overline{AE}$, and $\overline{CD}$ are perpendicular bisectors of $\overline{XY}$, $\overline{XZ}$, and $\overline{YZ}$, respectively. Because $\overline{AC} \parallel \overline{XY}$ and $\overline{BF} \perp \overline{AC}$, we can conclude that (1) $\overline{BF} \perp \overline{XY}$. Also, because the opposite sides are parallel, $XBCA$ and $BYCA$ are both parallelograms. Because opposite sides of parallelograms are congruent, we have $XB = CA$ and $BY = CA$, or that (2) $XB = BY$. From (1) and (2), it follows that $\overline{BF}$ is the perpendicular bisector of $\overline{XY}$ (keep in mind that $\overline{BF}$ is also the altitude from vertex B). In a similar manner we can show that $\overline{AE}$ is the perpendicular bisector of $\overline{XZ}$, and $\overline{CD}$ is the perpendicular bisector of $\overline{YZ}$. Thus, by Construction 9, $\overline{BF}$, $\overline{AE}$, and $\overline{CD}$ intersect in the point P. Because these segments are also the altitudes of $\triangle ABC$, the altitudes intersect in a point. ■

This justification provides a proof of the next theorem.

THEOREM 7.18 Orthocenter of a Triangle

The altitudes of a triangle intersect in a single point, the orthocenter.

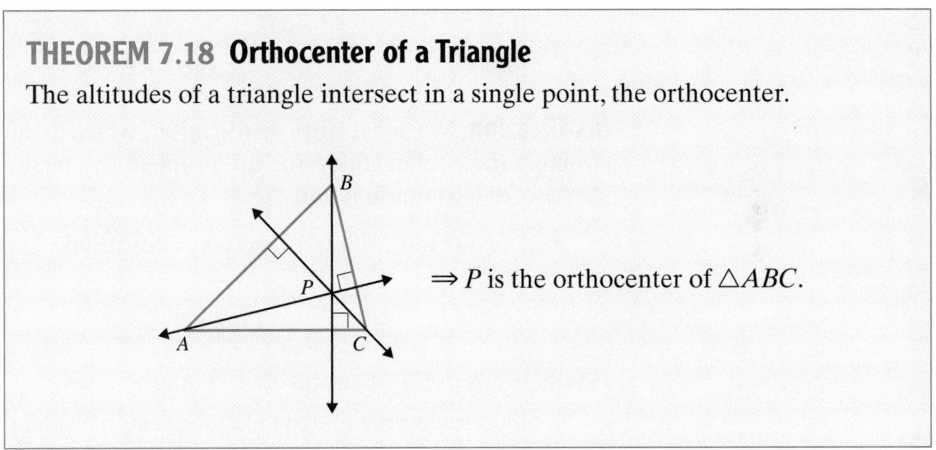

The orthocenter does not serve as a center of a circle related to the original triangle. However, if P is the orthocenter of $\triangle ABC$, the points P, A, B, and C form an **orthocentric set** which has the property that the triangle formed by any three of the four points in the set has the fourth point as its orthocenter.

Constructing the Inscribed Circle of a Triangle

In every triangle, we may construct a unique circle, called the **inscribed circle**, that is tangent to all three sides of the triangle. The center of the inscribed circle, the **incenter**, is the intersection of the angle bisectors of the triangle.

Construction 11. *To construct the inscribed circle of a triangle.*

Procedure Consider $\triangle ABC$ [Figure 7.38(a)]. Construct any two angle bisectors of $\triangle ABC$. They meet in the incenter, that is, the point that is equidistant from all three sides of the triangle [Figure 7.38(b)].

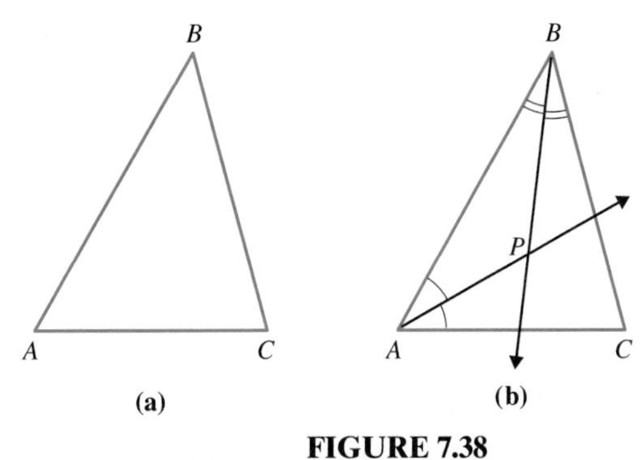

FIGURE 7.38

Next, using Construction 3 in Section 4.4, construct a perpendicular from P to $\overline{AC}$, and call the point of intersection D [Figure 7.38(c)]. Using PD as a radius, construct the inscribed circle [Figure 7.38(d)].

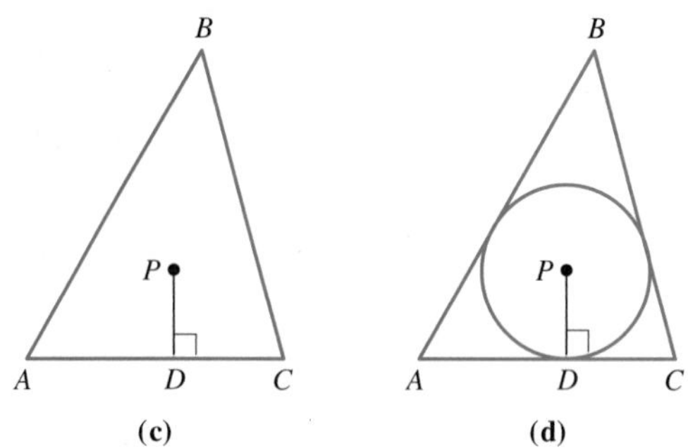

FIGURE 7.38

Justification We leave this justification, which is analogous to the one for constructing the circumcenter, for the problem set. First, show that the angle bisectors intersect in a common point. ■

The next result follows from this justification.

THEOREM 7.19 Incenter of a Triangle

The angle bisectors of a triangle intersect in a single point, the incenter.

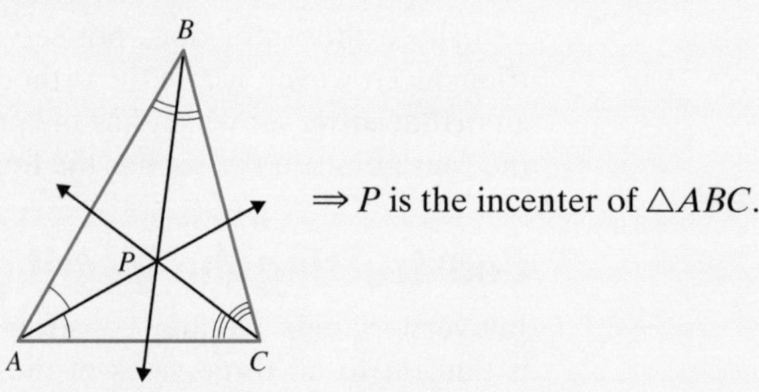

Constructing the Centroid of a Triangle

It is interesting to note that the medians of any triangle are also concurrent; their point of intersection is called the **centroid** of the triangle. The centroid is the **center of gravity**, or balance point, of a triangle.

Construction 12. *To construct the centroid of a triangle.*

Procedure Consider $\triangle ABC$ [Figure 7.39(a)]. Construct the midpoints of any two sides of $\triangle ABC$ and draw the medians [Figure 7.39(b)]. Their intersection, point P, is the centroid of $\triangle ABC$.

Justification We will show that the medians are concurrent. Consider $\triangle ABC$ with medians from A and C [Figure 7.39(b)]. Draw $\overrightarrow{BP}$, where O is the point where $\overrightarrow{BP}$ crosses $\overline{AC}$ [Figure 7.39(c)]. Let D be the point on $\overrightarrow{BP}$, where P is the midpoint

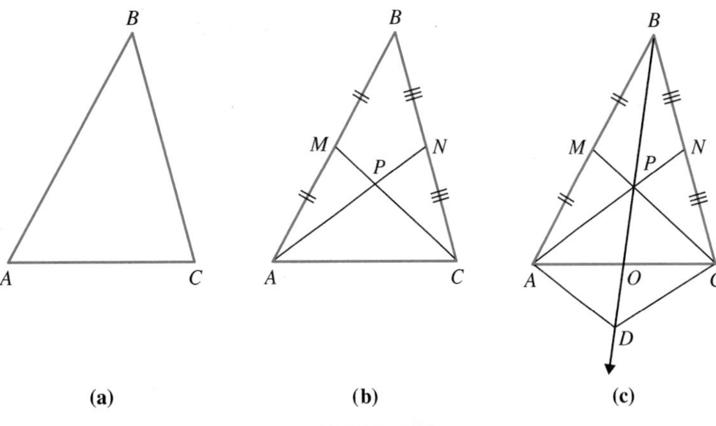

FIGURE 7.39

of $\overline{BD}$. Next we will show that $APCD$ is a parallelogram. Because the diagonals of a parallelogram bisect each other, this will show that O is the midpoint of $\overline{AC}$; thus, $\overline{BO}$ is the third median through P. Since $\overline{MP}$ joins the midpoints of two sides of $\triangle ABD$, $\overline{MP}$ is parallel to $\overline{AD}$ by the Midsegment Theorem. Also, $\overline{PC}$ is parallel to $\overline{AD}$ since $\overline{PC}$ is an extension of $\overline{MP}$. In a similar manner, $\overline{NP}$ is parallel to $\overline{DC}$ and, hence, $\overline{AP}$ is parallel to $\overline{DC}$. Thus, $APCD$ is a parallelogram since opposite sides are parallel. ■

We can also use the preceding justification to show that the point P divides each median into segments whose lengths are in a ratio of 2:1. We leave the proof of that fact for the problem set. We summarize the preceding discussion in the next theorem.

THEOREM 7.20 Centroid of a Triangle

The medians of a triangle intersect in a single point, the centroid, which is two-thirds of the way from any vertex to the other endpoint of the median from that vertex.

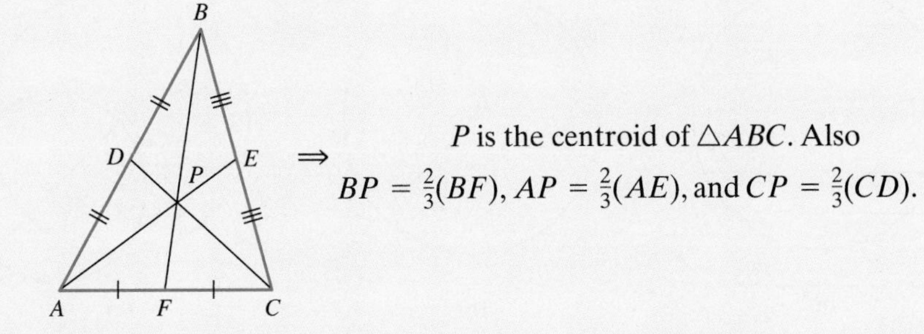

Table 7.2 summarizes the four centers we have constructed for a triangle.

Name	Point of Intersection	Significance
Circumcenter	Perpendicular bisectors	Center of circumscribed circle
Orthocenter	Altitudes	Forms orthocentric set together with vertices
Incenter	Angle bisectors	Center of inscribed circle
Centroid	Medians	Center of gravity

TABLE 7.2

Constructing Tangents to Circles

Next we will look at constructions involving tangents to a circle. Theorem 7.11 provides a method for constructing tangents to a given circle.

Construction 13. *To construct a tangent to a circle at a point on the circle.*

Procedure To construct a tangent line at point P of circle O [Figure 7.40(a)], first draw $\overrightarrow{OP}$ [Figure 7.40(b)]. Then, at P, construct a line, l, perpendicular to $\overrightarrow{OP}$ using Construction 4 in Section 4.4 [Figure 7.40(c)].

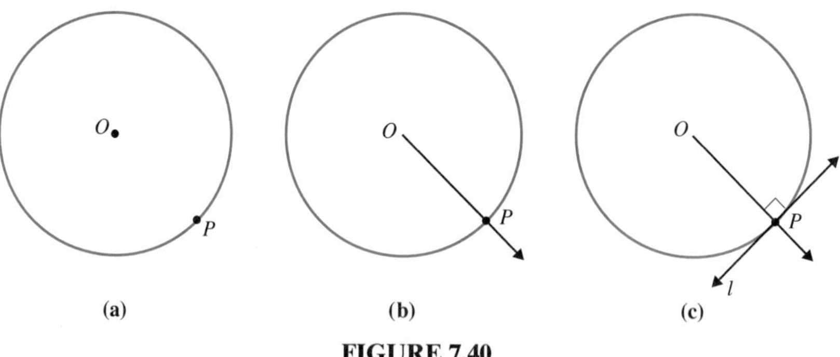

FIGURE 7.40

Justification By Theorem 7.11, l is tangent to the circle at P. ■

Next, we demonstrate the construction of a tangent line to a circle from a point outside the circle.

Construction 14. *To construct a tangent to a circle from a point outside the circle.*

Procedure Consider a circle with center O and a point P outside the circle [Figure 7.41(a)]. Draw $\overline{OP}$ [Figure 7.41(b)].

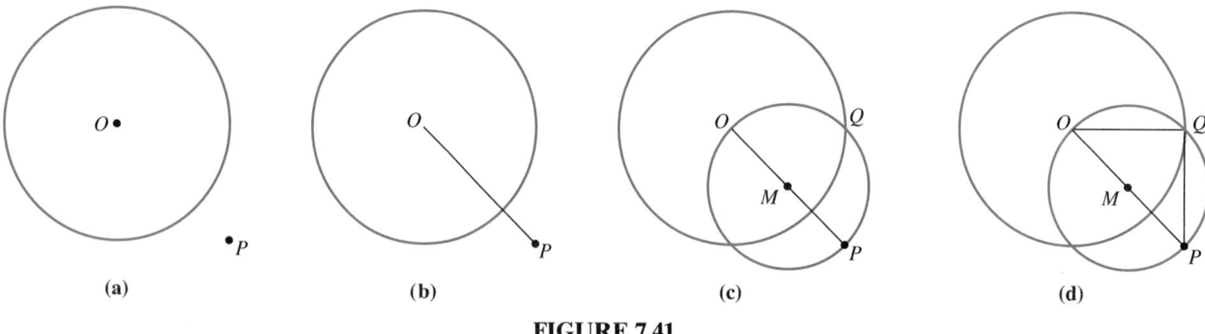

FIGURE 7.41

Construct the midpoint M of $\overline{OP}$, and construct the circle with center M and radius $MP = MO$ [Figure 7.41(c)]. Let Q be one of the points of intersection of the two circles [Figure 7.41(c)]. Draw $\overline{PQ}$ [Figure 7.41(d)]. $\overline{PQ}$ will be tangent to circle O.

Justification Because $\angle OQP$ is inscribed in a semicircle, it is a right angle, so $\overleftrightarrow{QP}$ is tangent to circle O at Q. ■

If two circles do not intersect and have no parts of their interiors in common, there are two types of tangents that we can construct. When the circles are on opposite sides of a common tangent line, the line is called a **common internal tangent**

[Figure 7.42(a)]. When the circles are on the same side of the tangent line, the line is called a **common external tangent** [Figure 7.42(b)].

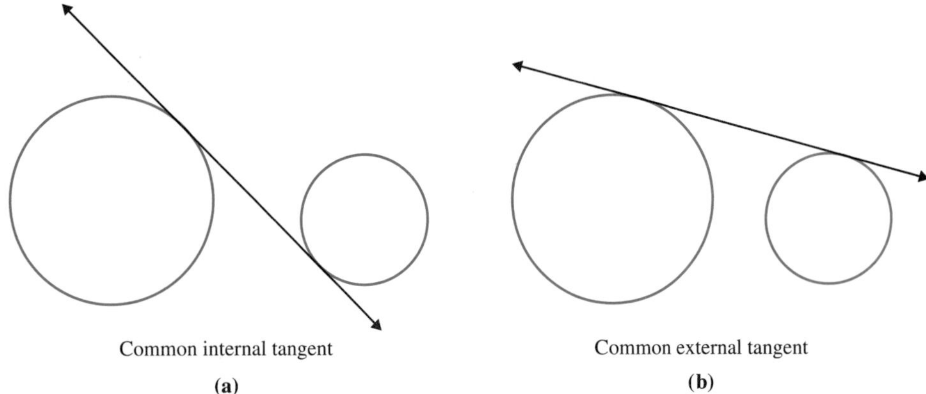

FIGURE 7.42

The following discussion shows how to construct these two types of tangent lines.

Construction 15. *To construct a common internal tangent to two circles.*

Procedure Consider circles O and P [Figure 7.43(a)].

FIGURE 7.43

STEP 1 Draw $\overline{OP}$, giving points of intersection Q and R [Figure 7.43(b)].

STEP 2 Using O as center, construct a circle whose radius is $OQ + PR$ [Figure 7.43(c)].

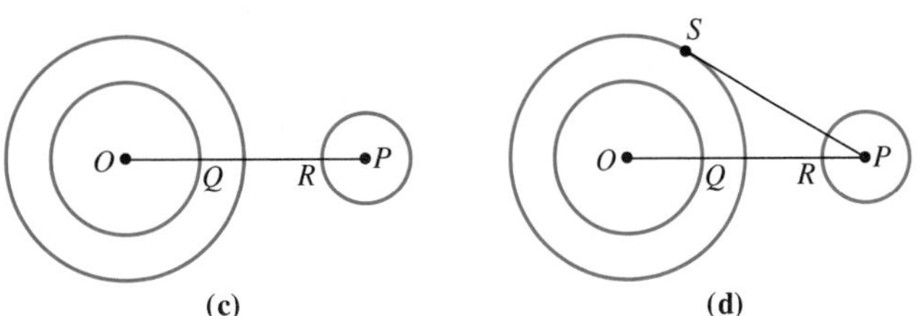

FIGURE 7.43

STEP 3 Construct a tangent line from P to the new circle at S using Construction 14 [Figure 7.43(d)].

STEP 4 Draw $\overline{OS}$, calling T the point where $\overline{OS}$ intersects the original circle with center O [Figure 7.43(e)].

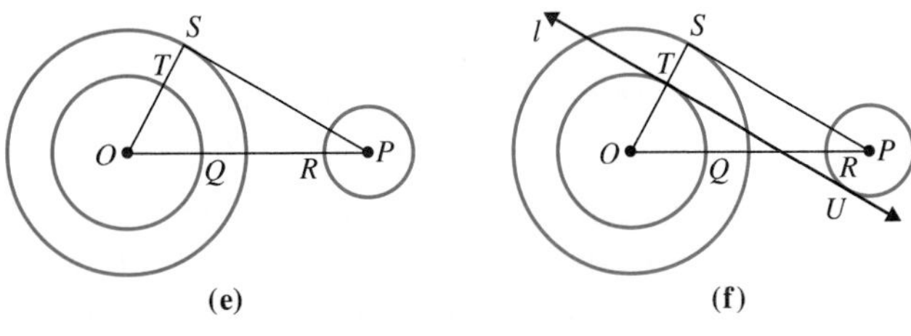

FIGURE 7.43

STEP 5 Construct a line, l, through T, parallel to $\overline{SP}$ [Figure 7.43(f)]. Line l is the common internal tangent to the original two circles.

Justification Because $l \parallel \overline{SP}$ and $\overline{SP} \perp \overline{OS}$, we have $l \perp \overline{OT}$ at T. Hence, l is tangent to the original circle at T. Let U be the point on l such that $\overline{PU} \perp l$. Then $PSTU$ is a rectangle because it has four right angles. Therefore, $PU = ST$, because they are opposite sides of a rectangle. Also $ST = PR$ by construction. Thus, PU is a radius perpendicular to l at U. Therefore, l is tangent to circle P at U. ■

Construction 16. *To construct a common external tangent to two circles.*

Procedure Consider circles O and P [Figure 7.44(a)].

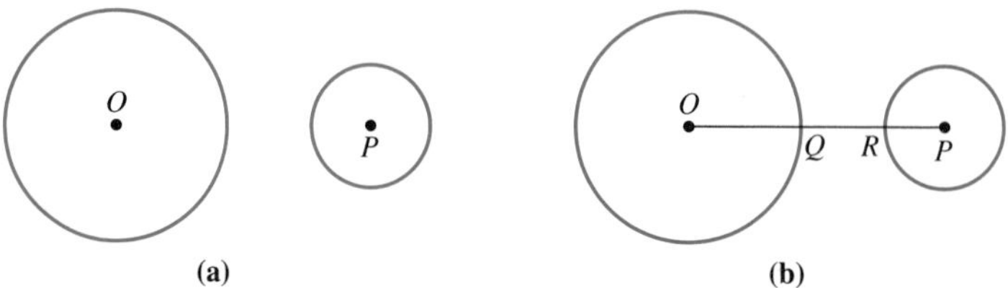

FIGURE 7.44

STEP 1 Draw $\overline{OP}$, yielding points Q and R on circles O and P, respectively [Figure 7.44(b)].

STEP 2 Using O as center, construct a circle having radius $OQ - PR$ [Figure 7.44(c)].

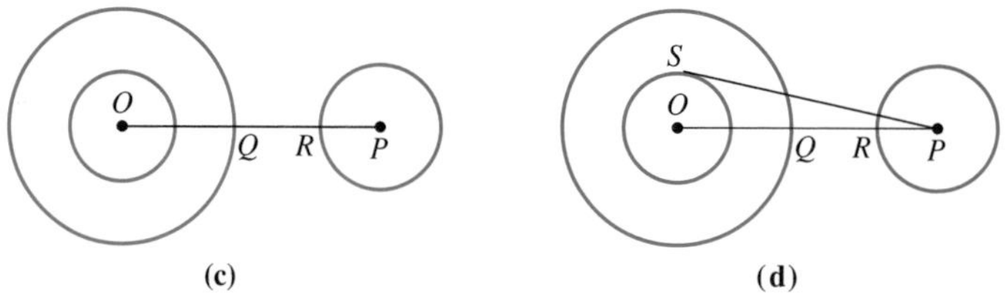

FIGURE 7.44

STEP 3 Construct a tangent line from P to the new circle at S, using Construction 14 [Figure 7.44(d)].

STEP 4 Construct perpendiculars to $\overline{SP}$ at S and P, forming points T and U [Figure 7.44(e)].

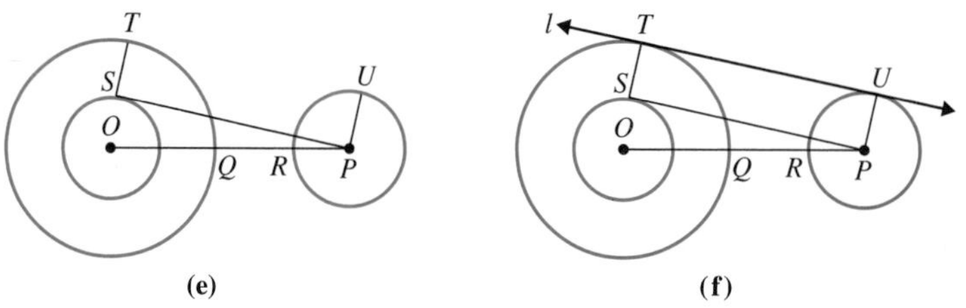

FIGURE 7.44

STEP 5 Draw line l through T and U [Figure 7.44(f)]. Then line l is the common external tangent to the original two circles.

Justification By construction, $TS = UP$ and $\overline{TS} \parallel \overline{UP}$, because they are both perpendicular to $\overline{SP}$. Therefore, $PSTU$ is a parallelogram. In addition, $PSTU$ contains a right angle; hence, it is a rectangle. Thus, l is a common tangent since it is perpendicular to radii $\overline{OT}$ and $\overline{PU}$. ■

Solution of Applied Problem

First, we must specify a desired radius for the circular arc. Using that radius, we can make an arc with the center at the corner C, giving points of intersection A and B. We can construct square with A, B, and C as vertices. The fourth vertex, D, will be the center of the circle containing the desired arc. We can now draw the arc using the same radius.

Because $ADBC$ is a square, $AD \perp AC$ and $BD \perp CB$. Thus, the arc is tangent to the edges of the board, as required.

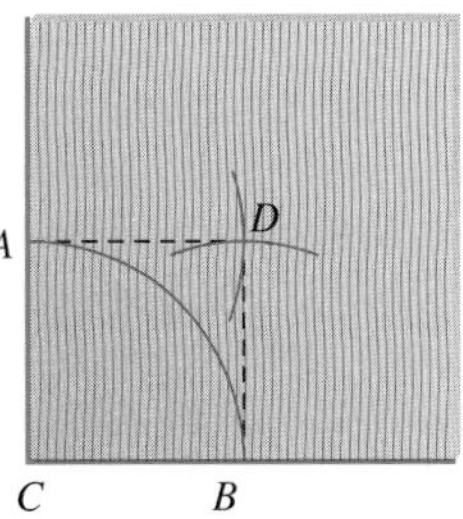

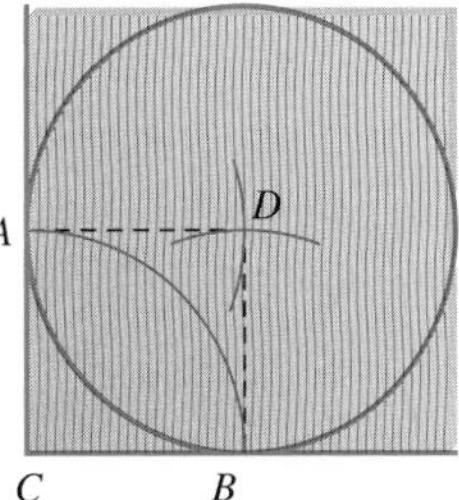

GEOMETRY AROUND US

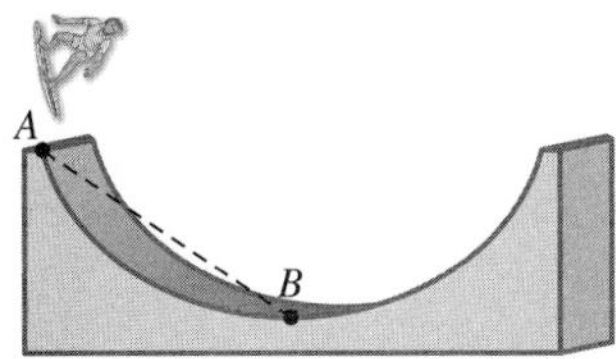

The cross section of an optimal skateboard ramp is in the shape of an inverted cycloid, called a **brachistochrone**. The "halfpipe" used in Olympic snowboarding events and some BMX competitions is also based on this curve. The brachistochrone is the curve of fastest descent. Although it might seem that a straight line ramp would be the fastest path from A to B, the inverted cycloid allows a skater or snowboarder to get from A to B in the shortest time of any curve.

PROBLEM SET 7.4

EXERCISES/PROBLEMS

For Exercises 1–3, trace each triangle and construct its circumscribed circle.

1.

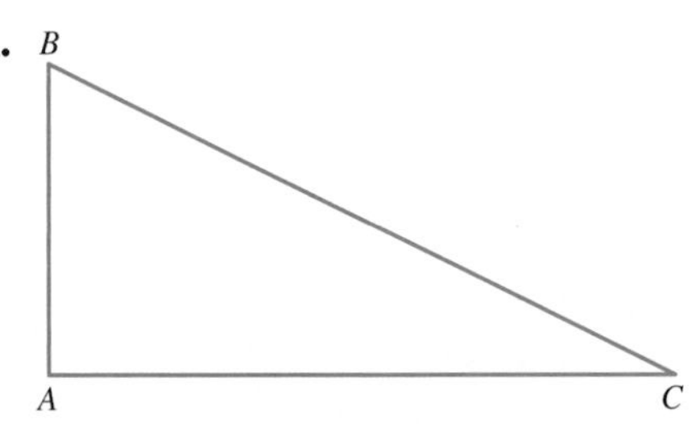

2.

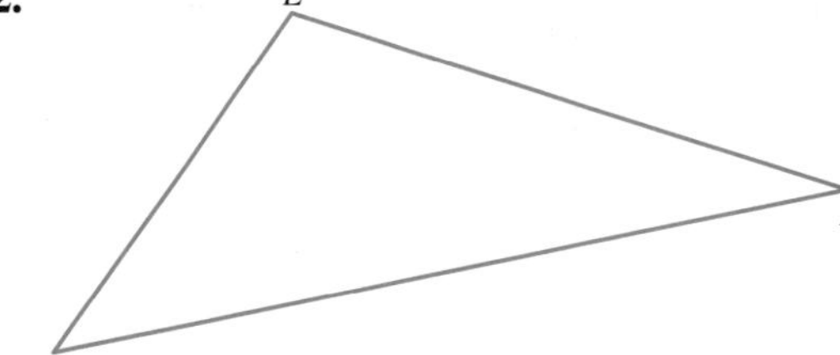

3.

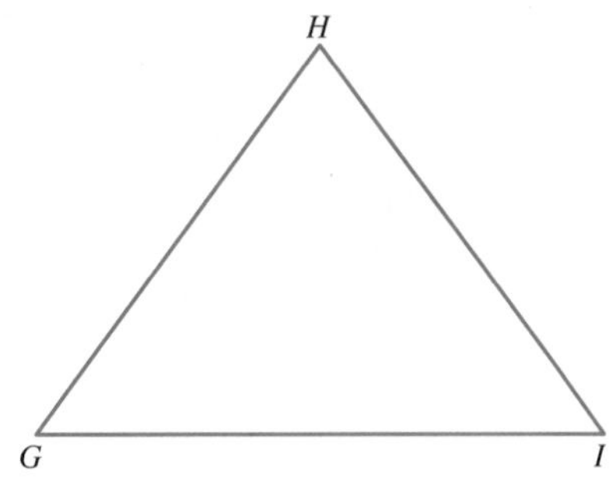

4. What did you notice about the circumcenter in Exercises 1–3? Was it always inside the triangle? Try a few more examples to test your conjecture.

5. Trace $\triangle PQR$ and locate its orthocenter using your compass and straightedge.

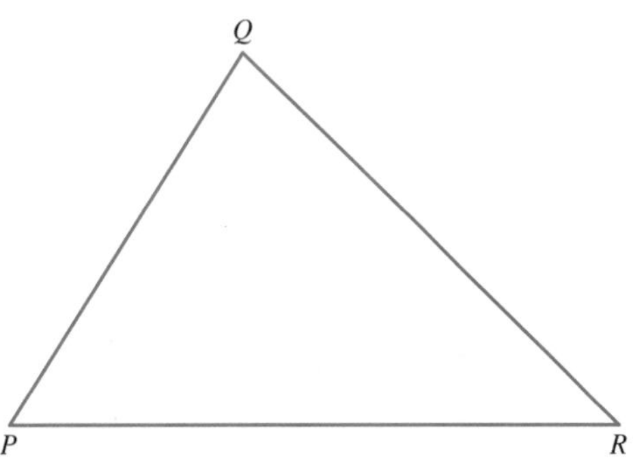

6. Trace $\triangle PQR$ and locate its orthocenter using your compass and straightedge.

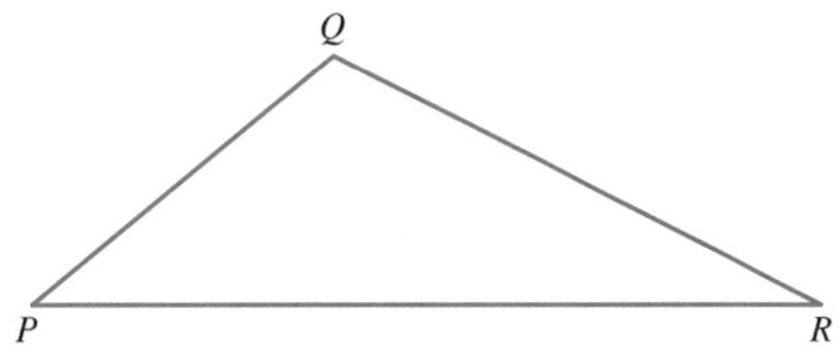

7. Use a compass and straightedge to construct point D to complete an orthocentric set.

B•

A•

Then copy the four points. Connect points D, B, and C to form a triangle and construct the orthocenter using a compass and straightedge. What do you notice?

8. Use a compass and straightedge to construct point C so that point D is the orthocenter of $\triangle ABC$. (HINT: Think about the points that form an orthocentric set.)

A•

•D

B•

For Exercises 9–11, trace each triangle and construct its inscribed circle.

9.

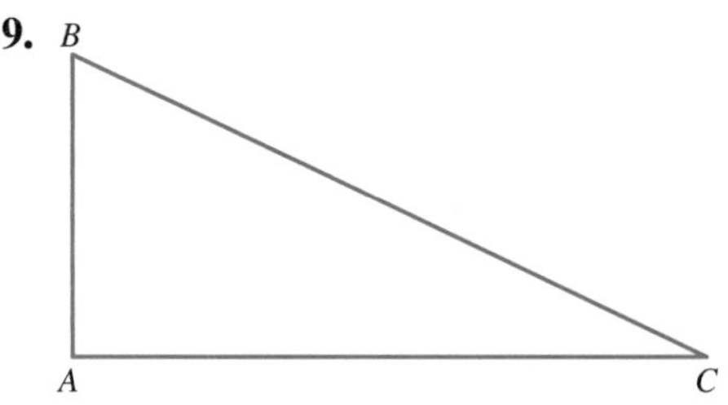

10.

E

F

D

11.

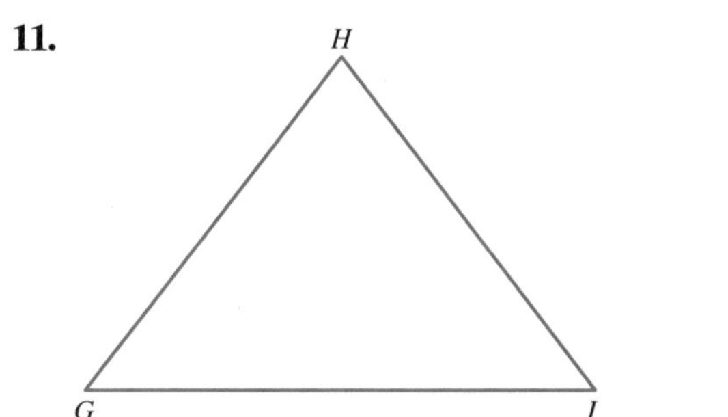

12. Why will the incenter of a triangle always be inside the triangle? Explain.

13. Trace $\triangle ABC$ and find its centroid using your compass and straightedge. By measuring, verify that the centroid is two-thirds of the way from each vertex to the midpoint of the opposite side.

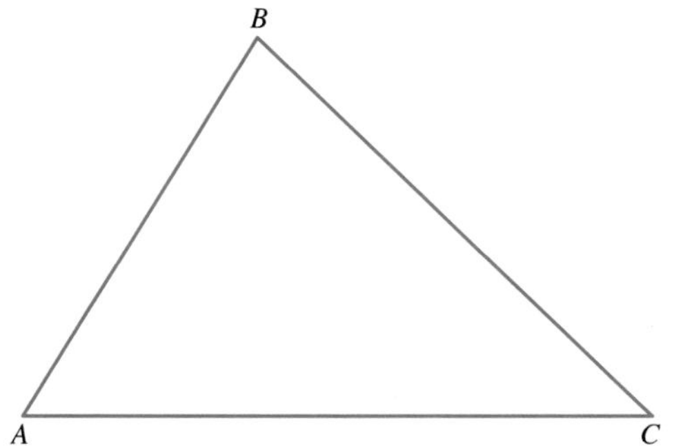

14. Trace $\triangle ABC$ and find its centroid using your compass and straightedge. By measuring, verify that the centroid is two-thirds of the way from each vertex to the midpoint of the opposite side.

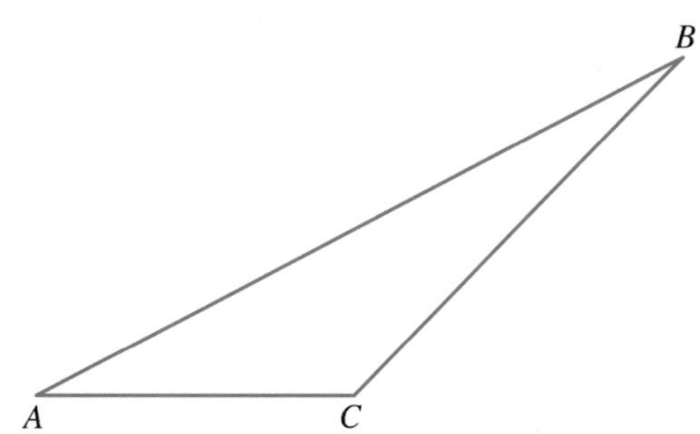

15. Use a compass and straightedge to construct point C so that point P is the centroid of $\triangle ABC$.

16. Use a compass and straightedge to construct point C so that point P is the centroid of $\triangle ABC$.

•*P*

17. Construct a tangent line to circle O at P.

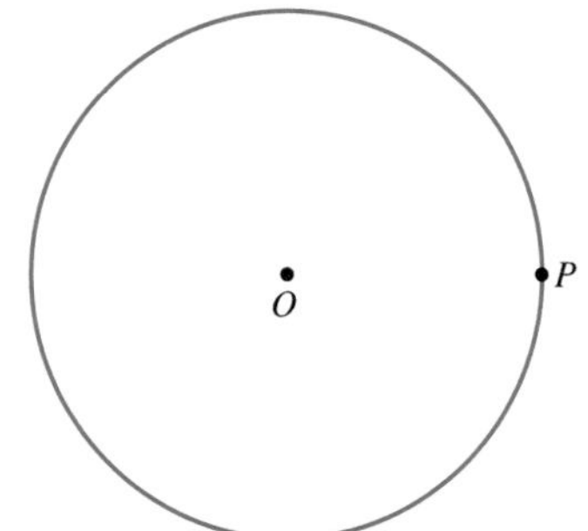

18. Construct a tangent line from P to the circle O.

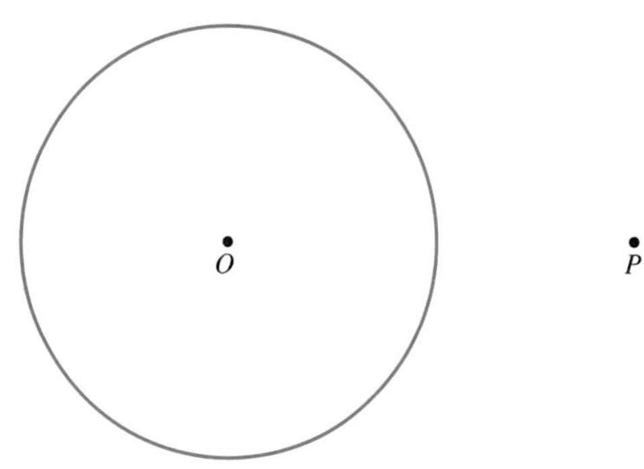

19. Construct a common internal tangent to circles O and P.

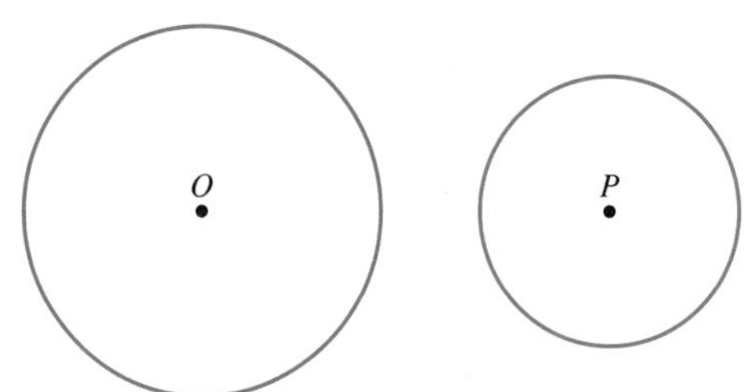

20. Construct a common external tangent to circles O and P.

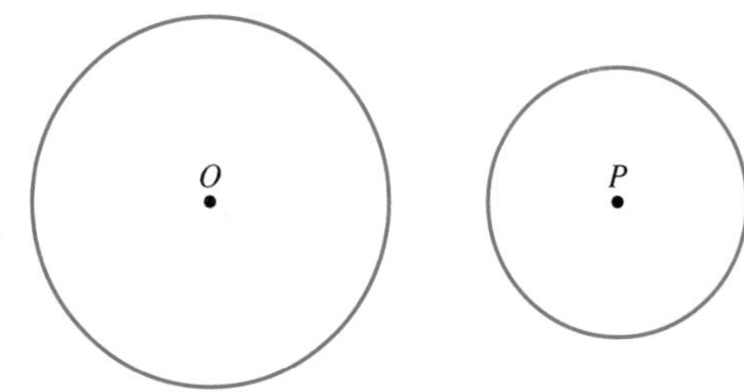

21. Draw a circle and construct a square inscribed in your circle. (HINT: Recall that an angle inscribed in a semicircle is a right angle.)

22. Draw a circle and construct a square circumscribed around your circle. (HINT: Recall that a radius is perpendicular to the tangent line at the point of tangency.)

23. Draw a circle and construct a regular octagon circumscribed about your circle.

24. Draw a circle and construct a regular hexagon circumscribed about your circle.

25. Construct a circle having a central angle of 60°. Construct the tangents to the circle where the sides of the angle intersect the circle. Draw the quadrilateral formed by the two tangents and the radii.

Find the measures of the other three vertex angles of the quadrilateral. Name the quadrilateral.

26. Construct a central angle of 90° whose legs are two radii of a circle. Construct a tangent line at each of the endpoints of the radii where they intersect the circle. Give the most complete description possible of the quadrilateral formed by the two radii and the portions of the tangent lines contained between the circle and their point of intersection.

27. The **Euler line** of a triangle is the line that contains the circumcenter, orthocenter, and centroid of the triangle. Draw a triangle. Show by construction that these three points are collinear and draw the triangle's Euler line.

28. The "Geometry Investigation" in Chapter 4 involved the construction of a regular hexagon inscribed in a circle. Use the following method to construct a regular pentagon inscribed in a circle.

    a. Draw circle O with diameter $\overline{AB}$.

    b. Construct the perpendicular bisector of $\overline{AB}$ and label as C one of its points of intersection with the circle.

    c. Construct the midpoint of $\overline{BO}$ and label it as point D.

    d. Swing an arc of length DC from D to intersect $\overline{AB}$. Label this point E.

    e. Swing five arcs of length CE along circle O to form the vertices of the regular pentagon.

PROOFS

29. Prove Theorem 7.19: The angle bisectors of a triangle intersect in a single point, the incenter.

30. Prove Theorem 7.17: The perpendicular bisectors of the sides of a triangle intersect in a single point, the circumcenter.

31. Justify Construction 11: To construct the inscribed circle of a triangle.

32. Prove Theorem 7.20: The medians of a triangle intersect in a single point, the centroid, which is two-thirds of the way from any vertex to the other endpoint of the median from that vertex.

33. Construct the tangent lines at the endpoints of a diameter of a circle. Prove that the two tangent lines are parallel.

34. Prove that the orthocenter of a triangle forms an orthocentric set with the vertices of the triangle.

35. Napoleon's theorem was the focus of the "Geometry Investigation" in this chapter. Consider equilateral triangles constructed on each side of $\triangle ABC$.

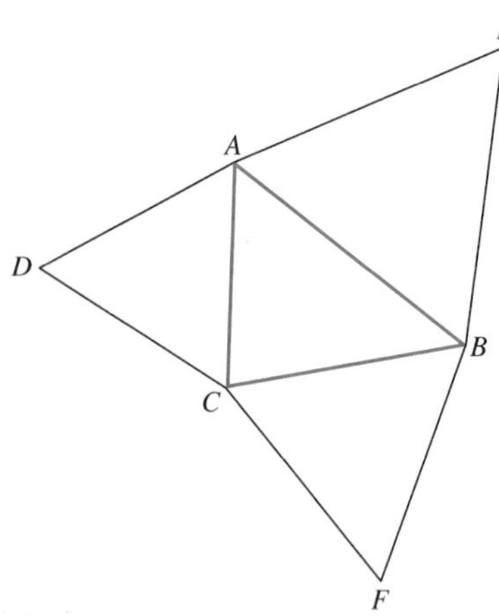

Napoleon's theorem states that the triangle formed by the centroids of the equilateral triangles is itself an equilateral triangle. Let G, H, and I be the centroids of $\triangle DAC$, $\triangle EBA$, and $\triangle FCB$, respectively. We can use the Law of Cosines to show that $\triangle GHI$ is equilateral. We will proceed as follows to prove the theorem.

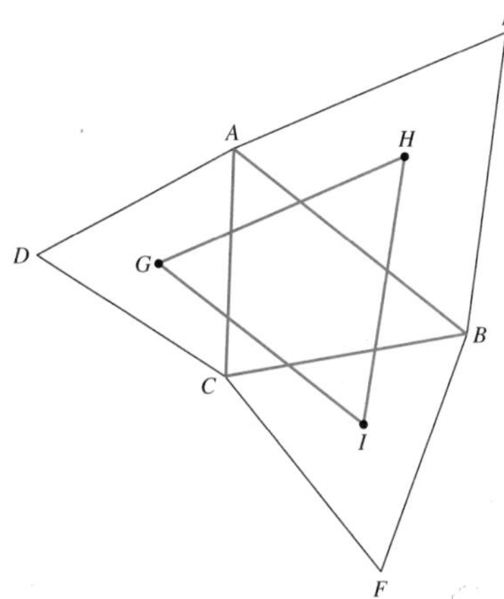

a. Draw segment $\overline{CE}$ and consider $\triangle CBE$ and $\triangle CAE$.

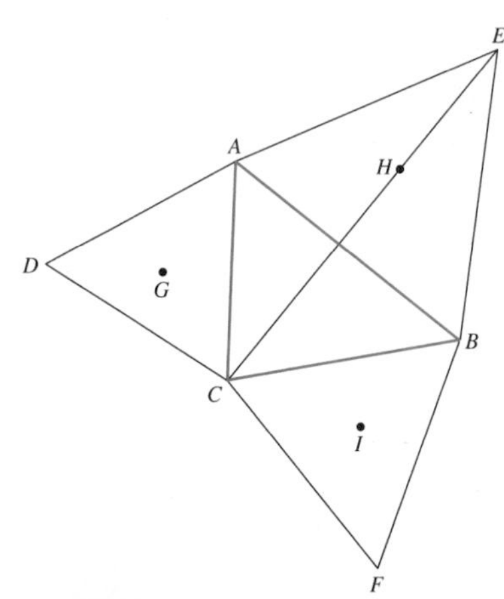

Use the Law of Cosines to express CE in two different ways.

b. Use the fact that $\triangle EBA$ is equilateral to rewrite the expressions from part (a) in terms of the lengths of sides of $\triangle ABC$.

c. Draw in $\overline{BI}$ and $\overline{BH}$. Apply the Law of Cosines to $\triangle HBI$ to write an expression for HI.

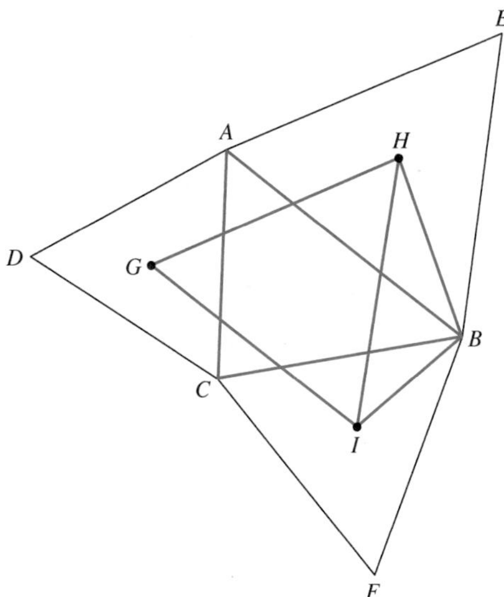

d. Recall that H and I are centroids of the equilateral triangles $\triangle AEB$ and $\triangle BFC$, respectively. Use Theorem 7.20 to express BH and BI in terms of AB and CB, respectively, and rewrite the expression for HI in part (c) in terms of the sides of $\triangle ABC$.

e. Draw in $\overline{AG}$ and $\overline{AH}$. Apply the Law of Cosines to $\triangle GAH$ and write an expression for GH.

f. Using Theorem 7.20 and the fact that G and H are centroids, rewrite the expression for GH in part (e) in terms of the sides of $\triangle ABC$.

g. Substitute results from parts (d) and (f) into part (b) to show that $HI = GH$.

We can use similar argument to show $GH = GI$ or $HI = GI$. Thus, $\triangle GHI$ is equilateral.

APPLICATIONS

36. A pulley-belt system is often used to transfer motion to another location in an engine. In the following pulley-belt configuration, the radius of the large pulley is $R = 8$ inches, the radius of the small pulley is $r = 5$ inches, and the distance between the centers of the pulleys is $D = 16$ inches. Find the length of the belt surrounding the pulley system. Assume that the radii are drawn to a point of tangency on the belt. Round to the nearest whole number.

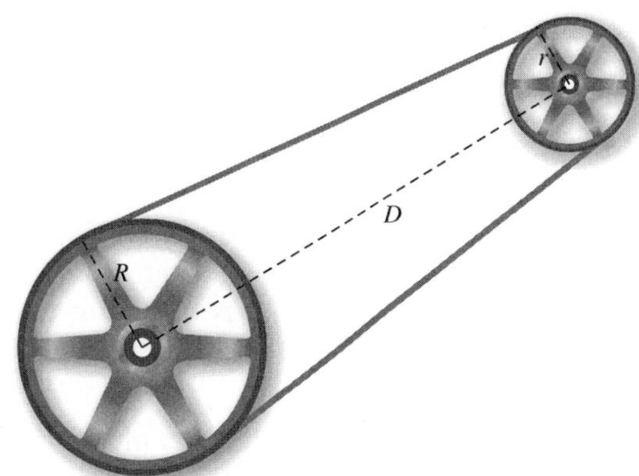

37. To reverse the direction of a pulley in an engine, the following pulley-belt system is used. The radius of the large pulley is $R = 7$ cm, the radius of the small pulley is $r = 2$ cm, and the distance between the centers of the pulleys is $D = 15$ cm. Find the length of the belt surrounding the pulley system. Assume that the radii are drawn to a point of tangency on the belt. Round to the nearest whole number.

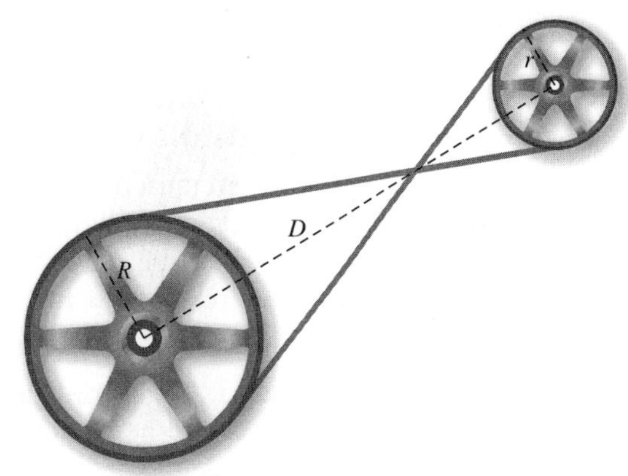

EXTENDED PROBLEMS

38. Explore three-pulley systems and print a photo or diagram of a pulley system in an engine.

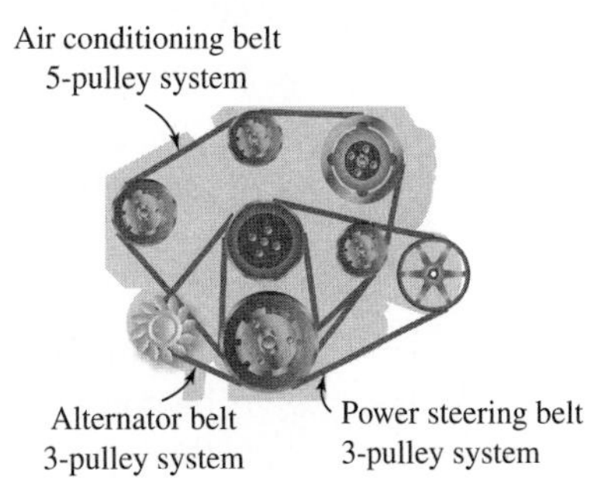

Measure the pulley diameters, the distance between the centers of the pulleys, and the angles formed by the lines between the centers. Use properties of right triangles and tangent lines to derive a formula for the length of the belt.

39. An **excircle** is a circle outside a triangle that is tangent to one side of the triangle and also tangent to the extensions of the other two sides of the triangle. Explore properties of the excircles of a triangle by doing the following.

a. Draw $\triangle ABC$ and extend the sides of the triangle to form the exterior angles of the triangle.

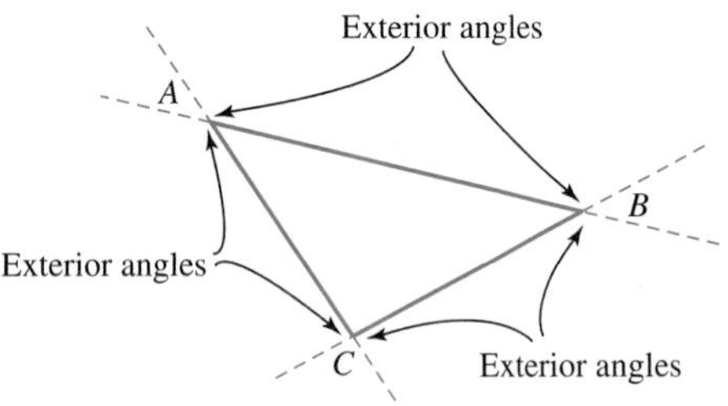

Construct the bisector of each exterior angle. What must be true about the points of intersection of pairs of these angle bisectors according to the Angle Bisector Theorem? How is an excircle of a triangle related to the exterior angle bisectors?

b. Research the excircle. How are the three centers of the excircles and the incenter of the triangle related?

40. Karl Feuerbach proved a theorem that states that for any triangle, one circle contains the following nine points: the three midpoints of the sides, the three feet of the altitudes, and the midpoints of the three segments joining the vertices of the triangle to its orthocenter.

a. Draw a triangle, construct the nine points described here, and find a circle that fits the nine points.

b. Research the history of the nine-point circle. Is a single person credited with its discovery? How is the nine-point circle related to the incenter of the triangle? How is the nine-point circle related to the three excircles of a triangle? Describe two other interesting facts about the nine-point circle.

Solution of Initial Problem

The circumscribed circle of a triangle is the unique circle that contains all three of the vertices of the triangle. For a given triangle, develop a simple test to determine if the center of a circumscribed circle is inside, on, or outside the triangle. The center of the circle is the intersection of the perpendicular bisectors of the sides of the triangle.

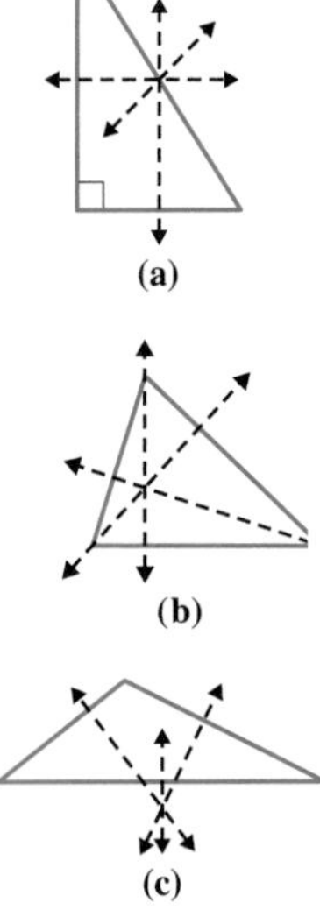

FIGURE 7.45

Case 1 Right triangle [Figure 7.45(a)]. In this case, the midpoint of the hypotenuse is the circumcenter hence, it is on the triangle.

Case 2 Acute triangle [Figure 7.45(b)]. In this case, the circumcenter is inside the triangle.

Case 3 Obtuse triangle [Figure 7.45(c)]. Here the circumcenter is outside the triangle, opposite the obtuse angle.

Hence, the center is on, inside, or outside a triangle, depending on whether the triangle is right, acute, or obtuse, respectively.

Additional Problems Where the Strategy "Use Cases" Is Useful

1. Determine whether the centroid of a triangle will always lie inside the triangle.
2. The measures of the angles of a triangle are consecutive integers. Show that the measures of two of the angles must be odd.

Writing for Understanding

1. The many angles involving circles and lines that we discussed in this chapter can be confusing to a beginning geometry student. Describe how you might classify the types of angles and determine their measures.

2. Choose one construction involving a circle from this chapter. Describe carefully the steps involved in that construction and discuss where such a construction might be useful.
3. The Wankel rotary engine, which has been used in the Mazda RX-8, was mentioned in the "Geometry Around Us" at the end of Section 7.2. Do some research on how circles and other curves of constant width, such as Reuleaux triangles, are important in the design of car engines. Summarize your findings.
4. One use of a brachistochrone is described in the "Geometry Around Us" at the end of Section 7.4. Do some research on this curve, focusing on its applications and the history of the "brachistochrone problem." Summarize your findings.

PEOPLE IN GEOMETRY

Carl Friedrich Gauss (1777–1855), according to the historian E. T. Bell, "lives everywhere in mathematics." His contributions to geometry, number theory, and analysis were deep and wide ranging. Yet he also made crucial contributions in applied mathematics. When the tiny planet Ceres was discovered in 1800, Gauss developed a technique for calculating its orbit, based on meager observations of its direction from earth at several known times. Gauss contributed to the modern theory of electricity and magnetism, and, with the physicist W. E. Weber, constructed one of the first practical electric telegraphs. In 1807, he became director of the astronomical observatory at Göttingen, where he served until his death. At age 18, Gauss devised a method for constructing a 17-sided regular polygon, using only a compass and straightedge. Then, remarkably, he derived a general rule that predicted which regular polygons are likewise constructible.

CHAPTER REVIEW

For each section of this chapter, you will find a list of vocabulary and notation, questions to assess your understanding of key concepts, and review problems similar to the problems you worked on your homework. Review each item in the *Vocabulary/Notation* list, mentally and, if necessary, refer back to the indicated page and write a definition. Then answer the *Concept Check Questions*, looking back at the section if you need help. Work the *Review Problems* as practice before you move on to the *Chapter Test*. Answers to the *Review Problems* can be found at the back of the book.

SECTION 7.1 Central Angles and Inscribed Angles

Vocabulary/Notation

Arc 362
Endpoints of an arc 362
Semicircle 362
Minor arc 362
Major arc 362
Central angle 363
Intercepts 363
Measure of an arc 363
Length of an arc 364
Sector 364
Area of a sector 365
Inscribed angle 365

Concept Check Questions

1. How is the measure of an arc related to the central angle that intercepts the arc? 363
2. On what two measurements does the length of an arc depend? 364
3. What is the difference between a central angle and an inscribed angle? 365
4. How is the measure of an arc related to an inscribed angle that intercepts the arc? 365

Review Problems

1. In circle O, $\overline{AC}$ is a diameter.

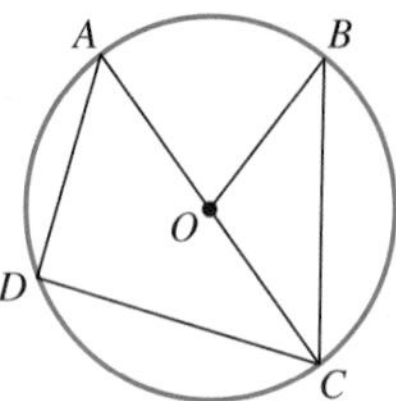

   a. What kind of angle is $\angle ACD$ with respect to circle O?
   b. What kind of angle is $\angle BOC$ with respect to circle O?
   c. How do the measures of $\angle AOB$ and $\angle ACB$ compare?
   d. What is the measure of $\widehat{ABC}$?
   e. What is the measure of $\angle D$?

2. Use circle O with diameter $\overline{BD}$ to find the measures of $\angle DOC$ and $\widehat{AB}$.

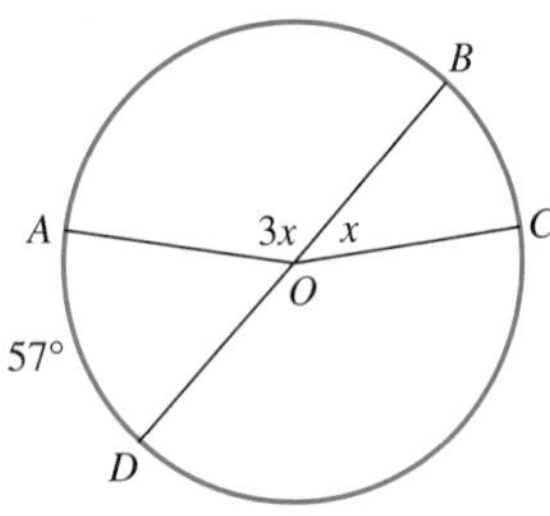

3. Find the measures of $\angle A$, $\angle C$, $\angle D$, $\widehat{AB}$, and $\widehat{AD}$ in the circle shown.

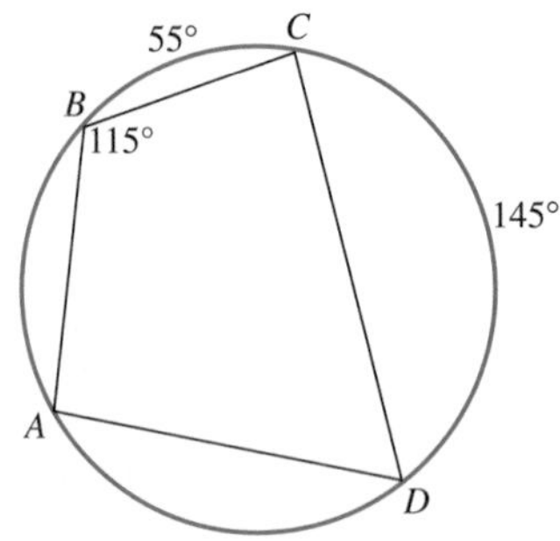

4. To lay out the pattern for a curved patio, you tie a 30-foot-long string to a stake in the ground at point C and walk from point A to point B along the curved path shown, holding the string taut.

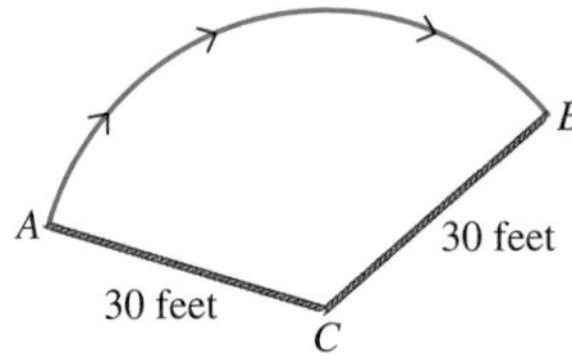

   a. If you walk a total of 54 feet, find the area of the patio. Round to the nearest whole number.
   b. If you want the area of the patio to be 550 square feet, how far should you walk along the curved path? Round to the nearest tenth.

SECTION 7.2 Chords of a Circle

Vocabulary/Notation

Chord 374

Line of centers 375

Concept Check Questions

1. How is the perpendicular bisector of any chord in a circle related to the center of the circle? 374
2. How can you locate the center of a circle 374
3. When two chords in a circle intersect, how are the vertical angles formed related to the intercepted arcs? 376
4. What is the relationship between the four segments formed by two intersecting chords in a circle? 377

Review Problems

1. Circle O has a radius of 7 cm. Circle O' has a radius of 9 cm. If the circles intersect in points A and B and $AB = 11$ cm, find OO'. Round to the nearest hundredth.

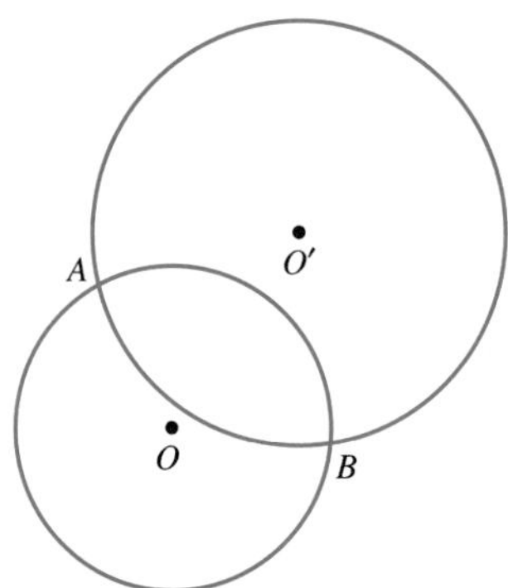

2. If $\widehat{AD} = 95°$ and $\angle BDC = 23°$, find $\angle AEB$.

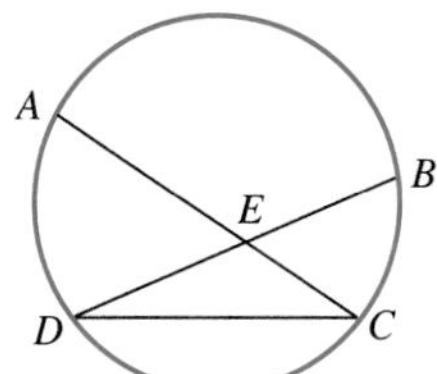

3. If $AF = CG$, use the given measurements to find the exact value of BF.

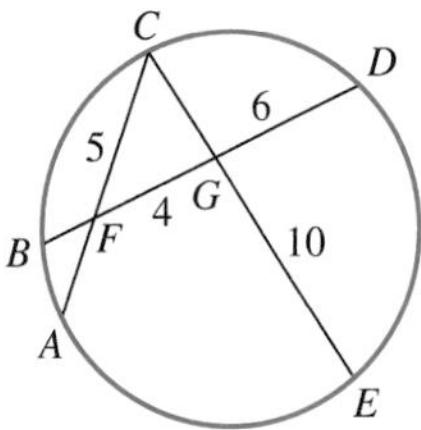

4. In circle O, $\widehat{AB} = \widehat{BC}$. Prove $\triangle ABC$ is isosceles.

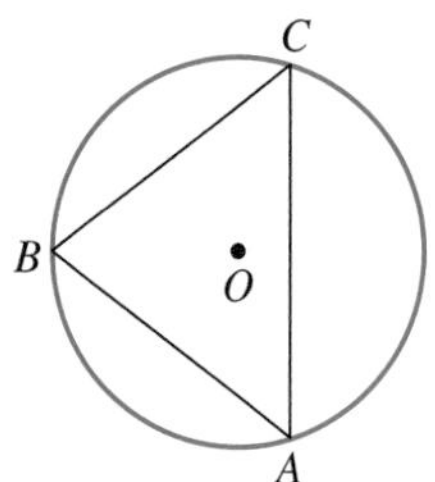

5. Prove that there is only one circle determined by three noncollinear points.

SECTION 7.3 Secants and Tangents

Vocabulary/Notation

Secant line 382 Tangent line 385 Point of tangency 385

Concept Check Questions

1. When two secant lines intersect outside a circle, how is the measure of the angle of intersection related to the arcs intercepted? 383
2. What is the relationship between the segments formed when two secant lines intersect outside of a circle? 384
3. How is a tangent line different from a secant line? 385
4. Which points on the tangent line to a circle lie outside the circle? 385
5. Under what conditions is a ray or a line segment said to be tangent to a circle? 385
6. If the diameter or the radius of a circle intersects a tangent line at the point of tangency, what angle is formed? 385
7. When a chord and a tangent line intersect at the point of tangency, the measure of which angle is equal to half the measure of the intercepted arc? 386
8. How is the angle formed by the intersection of a tangent line and a secant line related to the two arcs intercepted? 386
9. How is the angle formed by the intersection of two tangent lines related to the two arcs they intercept? 387
10. If two tangent lines intersect at a point, what must be true about the distance from the intersection point to each point of tangency? 388

Review Problems

1. In circle O, if $\overset{\frown}{GF} = 2\overset{\frown}{FE}$, $\overset{\frown}{GB} = 105°$, $\overset{\frown}{CD} = 25°$, and $\angle BAD = 35°$, find $\overset{\frown}{ED}$, $\overset{\frown}{FE}$, and $\overset{\frown}{BC}$.

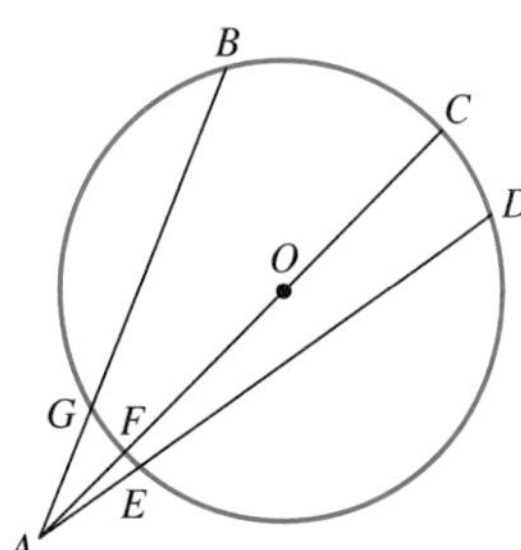

2. The radius of circle O is $\sqrt{73}$ cm and $BE = 8\sqrt{3}$ cm. Find DE.

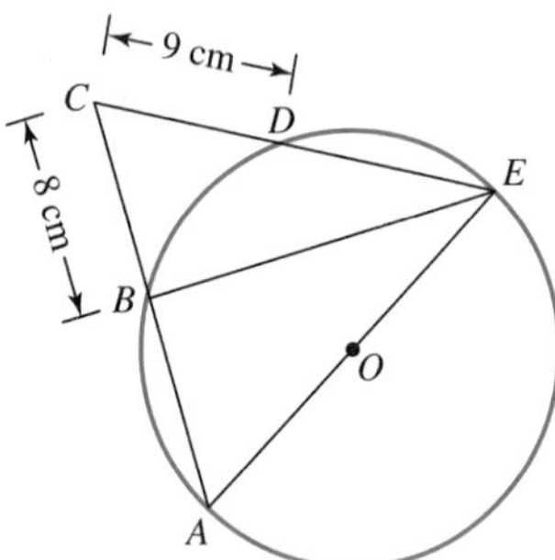

3. $\overline{AB}$ is tangent to circle O at C and $\angle ECD = 42.5°$. Find the measures of $\angle ACE$, $\angle BCD$, and $\overset{\frown}{CD}$.

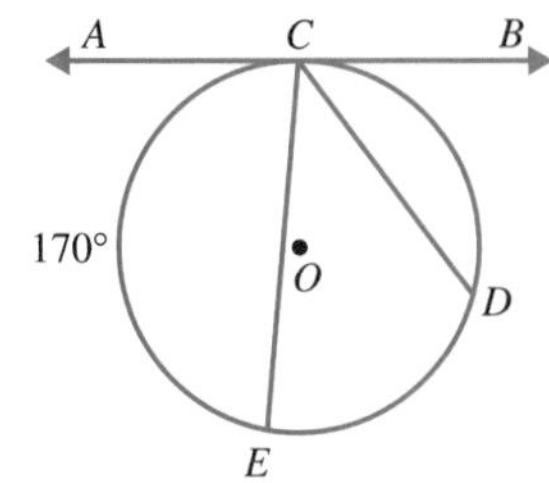

4. In circle O, $\angle ABC = 67.38°$ $BC = 7.5$ cm, and $OE = 5$ cm.

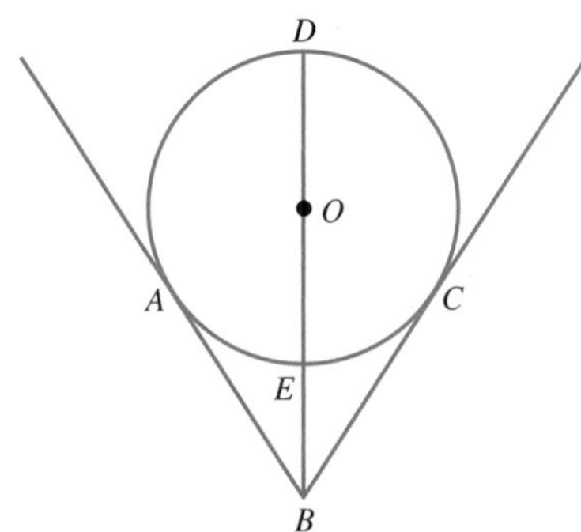

   a. Find $\overset{\frown}{DC}$ and $\overset{\frown}{AE}$.

   b. Find BE. Round to the nearest hundredth.

5. If, in a certain circle, the midpoints of all chords with a given length were connected, what figure would be formed? Justify your answer.

SECTION 7.4 Constructions Involving Circles

Vocabulary/Notation

Circumscribed circle 397
Circumcenter 397
Orthocenter 398
Orthocentric set 399
Inscribed circle 399
Incenter 399
Centroid 400
Center of gravity 400
Common internal tangent 402
Common external tangent 403

Concept Check Questions

1. How is the circumscribed circle for a triangle constructed? 397
2. What is the relationship between the three vertices of a triangle and the orthocenter of the triangle? 399
3. How are the sides of a triangle related to its inscribed circle? 399
4. How is the incenter of a triangle constructed? 399–400
5. The intersection of the medians of a triangle locates what special point? 400
6. How can you construct the tangent to a circle? 402
7. What types of tangents can be constructed to two circles that do not have any points in common? 402–403

Review Problems

1. Use a compass and straightedge to construct the circumscribed circle of the following triangle.

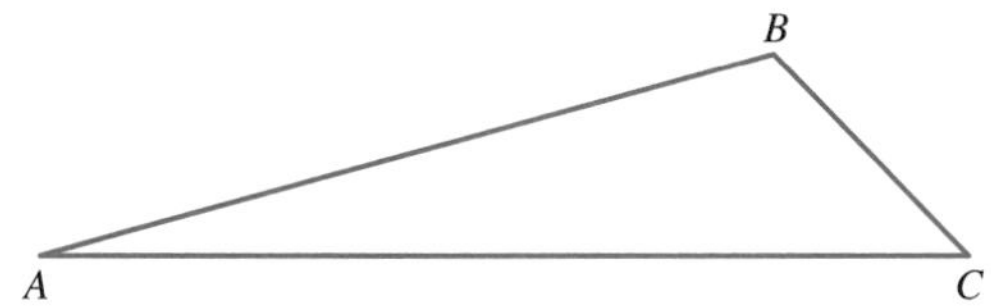

2. **a.** Construct the orthocenter of $\triangle ABC$ using a compass and straightedge. Label the orthocenter D.

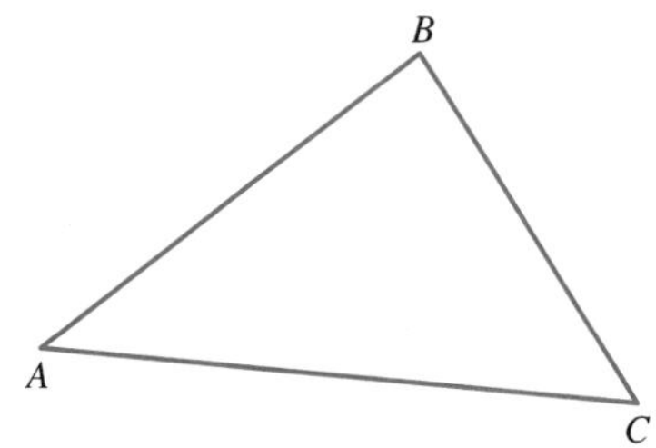

   b. Using points A, B, C, and D from part (a), can you determine the orthocenters of $\triangle ABD$, $\triangle ACD$, and $\triangle BCD$?

3. Use a compass and straightedge to locate point C so that point P is the incenter of $\triangle ABC$.

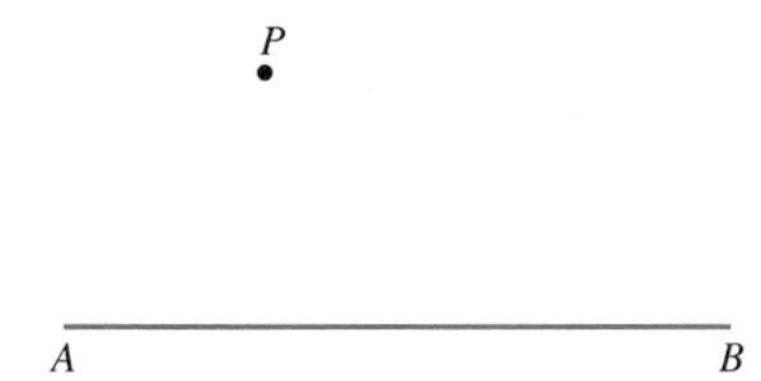

4. Use a compass and straightedge to construct the centroid of $\triangle ABC$.

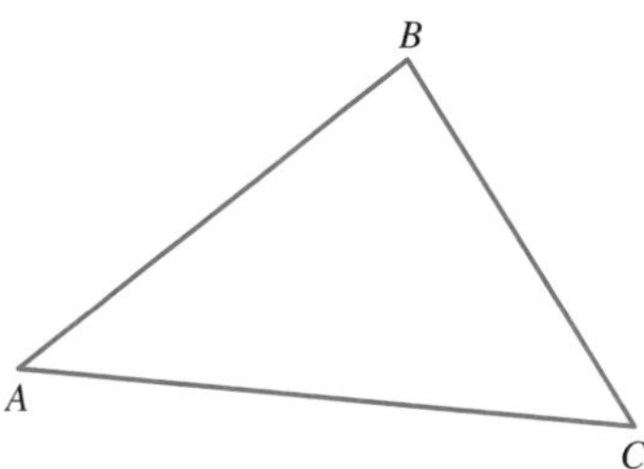

5. Construct a square and then construct a circle O inscribed in the square.
6. Use a compass and straightedge to construct
   a. the common external tangent to the following circles, and
   b. the common internal tangent to the circles.

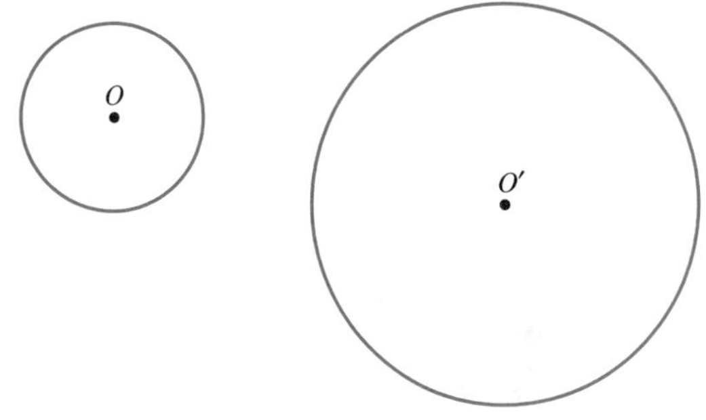

CHAPTER 7 TEST

TRUE-FALSE

Mark as true any statement that is always true. Mark as false any statement that is never true or that is not necessarily true. Be able to justify your answers.

1. If a right triangle is inscribed in a circle, then the hypotenuse of the triangle must be a diameter of the circle.
2. The incenter of an obtuse triangle lies outside the triangle.
3. The circumcenter of an obtuse triangle lies outside the triangle.
4. If the radius of a circle is 2 in., then the longest possible chord has length 6 in.
5. The measure of an inscribed angle in a circle equals the measure of its intercepted arc.
6. If two central angles of a circle are congruent, then the arcs they intercept are congruent.
7. If the area of a sector of a circle is 2π square inches and the radius of the circle is 4 inches, then the central angle of the sector is 45°.
8. If $\overleftrightarrow{AB}$ is tangent to circle O at point P, then $\angle APO = 90°$.
9. If circle O_1 and circle O_2 intersect in points A and B, then $\overline{O_1O_2}$ is the perpendicular bisector of $\overline{AB}$.
10. If chord $\overline{AB}$ of circle O measures 8 cm and chord $\overline{PQ}$ of circle O measures 5 cm, then the measure of $\angle AOB$ is less than the measure of $\angle POQ$.

EXERCISES/PROBLEMS

11. Using the figure shown, give one example of each of the following. (NOTE: O is the center of the circle.)

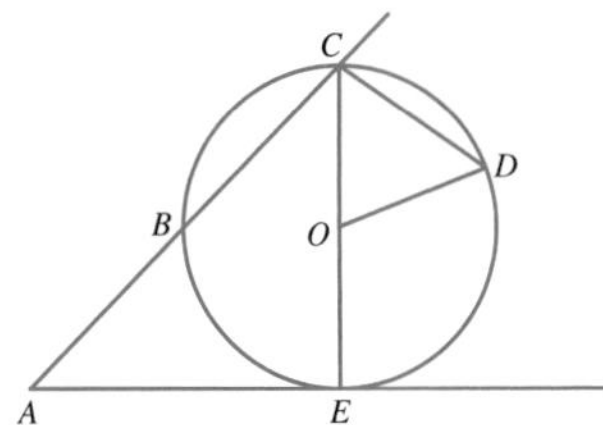

a. A secant line

b. An inscribed angle

c. A major arc

d. A right angle

e. A chord

For problems 12–14, refer to the following figure, where $\widehat{AC} = 126°$, $\angle BAC = 35°$, *and* $\widehat{AD} = 140°$.

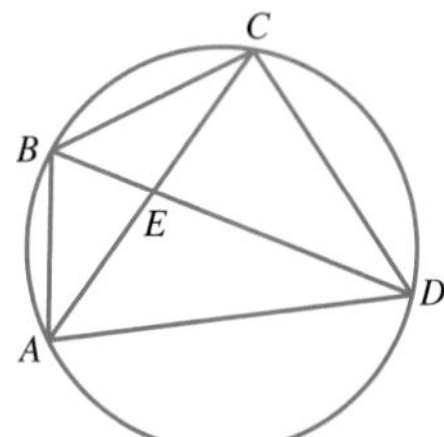

12. Find the measure of $\widehat{AB}$.

13. Find the measure of $\angle CAD$.

14. Find the measure of $\angle BEC$.

15. a. Imagine the largest square plug that fits into a circular hole. How well does the plug fit? That is, what percentage of the circular hole does the square plug occupy?

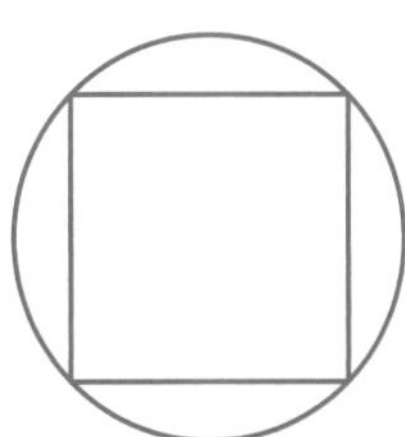

b. Now imagine the largest circular plug that fits into a square hole. How well does this plug fit? That is, what percentage of the square hole does the circular plug occupy?

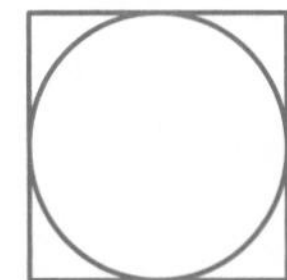

c. Which of the two plugs described in (a) and (b) is a better fit?

16. For the figure shown, determine the measures of each of the following, where $\widehat{CE} = 14°$ and $\angle ADB = 8°$

a. $\widehat{AB}$ **b.** $\angle EDF$

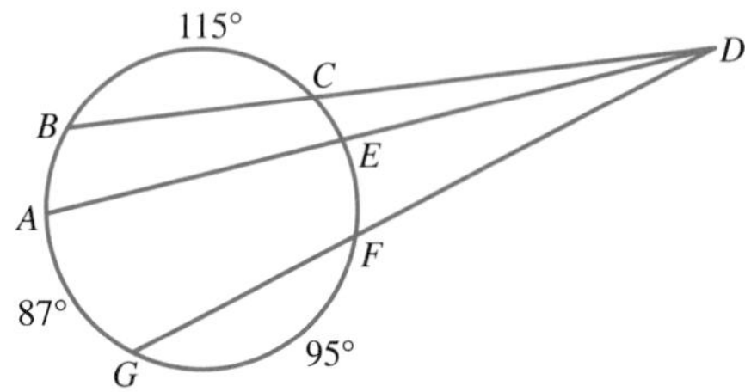

17. In the following figure, $PT = 12$ ft and $TR = 5$ ft. If $QS = 19$ ft, find TS and QT.

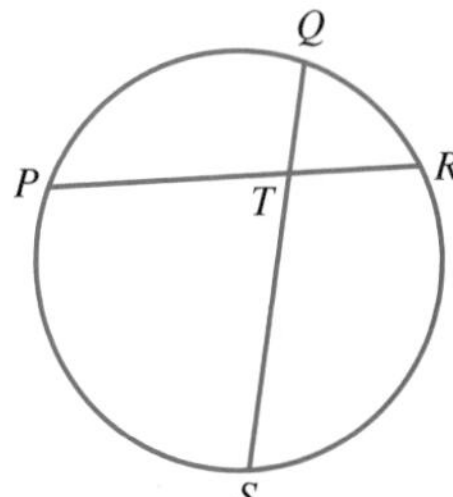

18. In the figure shown, $\overleftrightarrow{AB}$ and $\overleftrightarrow{BC}$ are tangent to circle O, and A and C are the points of tangency. For $AB = 7$ and $EB = 3$, find the exact value of DE.

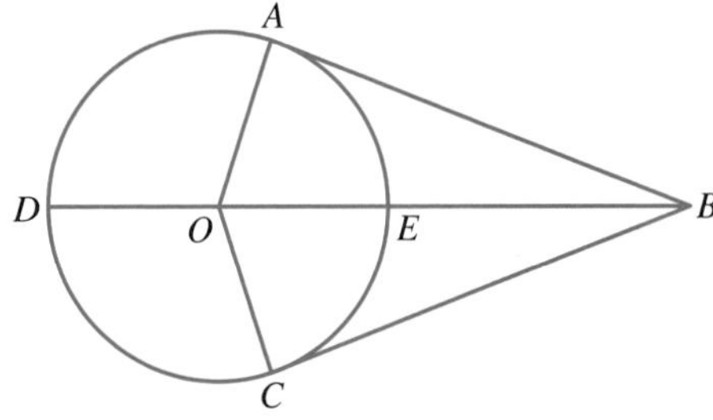

19. Trace $\triangle ABC$ and use your compass and straightedge to construct its inscribed circle.

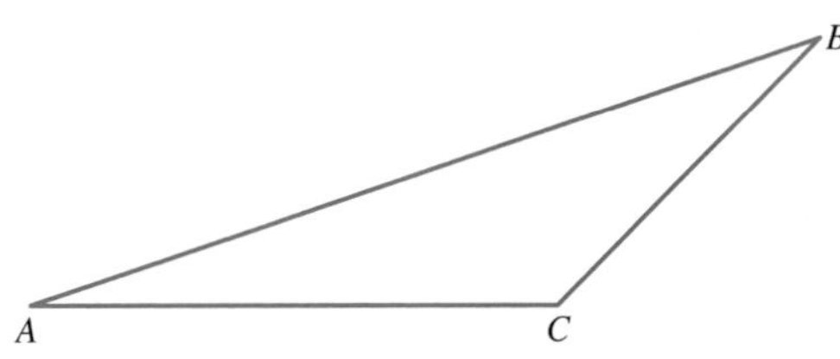

20. Use your compass and straightedge to construct a tangent line to circle O from point P.

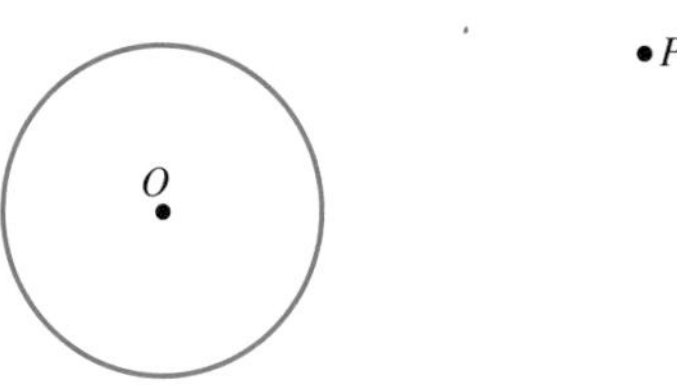

PROOFS

21. Prove that a line drawn through the midpoint of an arc and the center of its circle is perpendicular to the chord determined by the arc.

22. In the following figure, O is the center of the circle. The measure of $\angle OBC$ is twice the measure of $\angle ODC$.

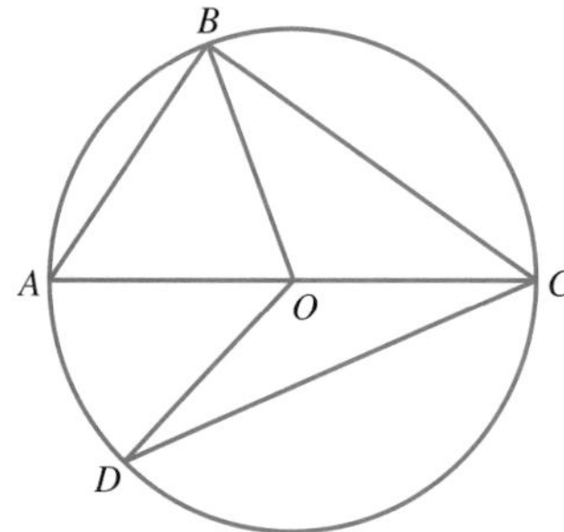

   a. Prove that the measure of $\angle AOB$ is four times the measure of $\angle OCD$.

   b. Prove that $\angle AOD$ and $\angle ABO$ are complementary.

APPLICATIONS

23. The maximum safe speed on a curve is related to the curvature of the road. Suppose a road has a curve as shown, where the curve is an arc of a circle. The **radius of curvature** is defined as the radius of a circle that would contain the arc. Use the dimensions shown to find the radius of curvature r. Round to the nearest whole number.

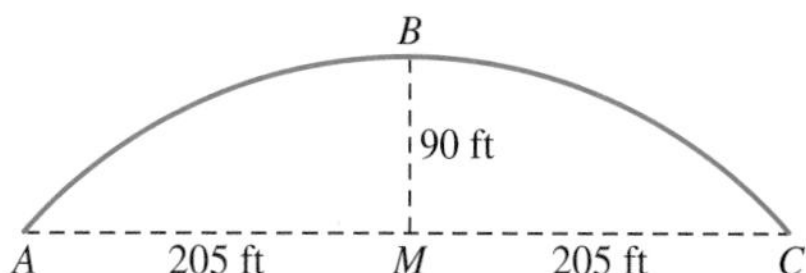

24. A baseball diamond is located on a field that is a sector of a circle as shown. To calculate the amount of fertilizer to be used on the grass, we must determine the area. Use the dimensions shown to find the area, to the nearest hundred square feet.

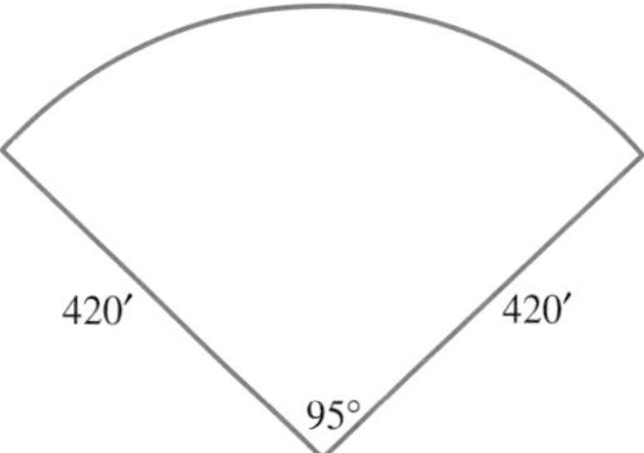

CHAPTER 8

COORDINATE GEOMETRY

The subjects of algebra and geometry evolved on parallel tracks until the French philosopher and mathematician René Descartes (1596–1650) developed a method of linking them. This integration of algebra and geometry made possible the development of calculus. Because of his important contribution to the evolution of mathematics, Descartes has been called the father of modern mathematics. The coordinate system used in analytic geometry is called the Cartesian coordinate system in his honor. The following timeline identifies some of the significant geometric contributions up to the time of Descartes.

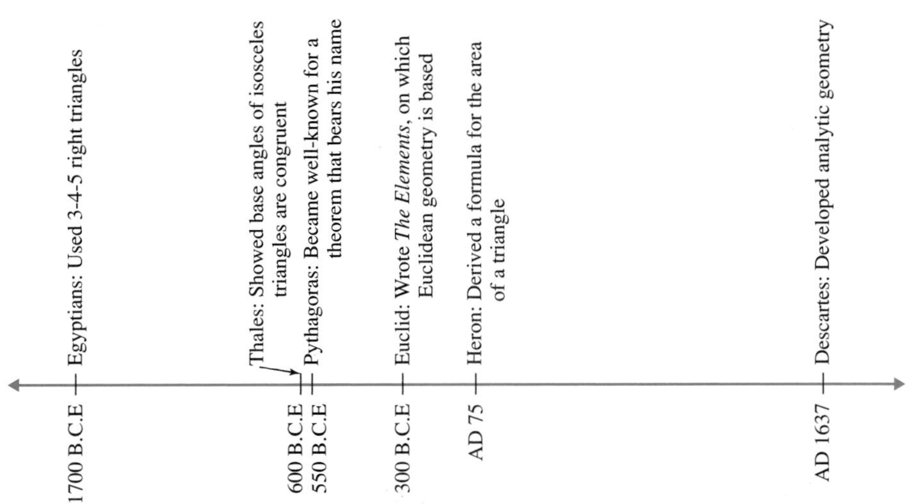

Problem-Solving Strategies

1. Draw a Picture
2. Guess and Test
3. Use a Variable
4. Look for a Pattern
5. Make a Table
6. Solve a Simpler Problem
7. Look for a Formula
8. Use a Model
9. Identify Subgoals
10. Use Indirect Reasoning
11. Solve an Equation
12. Use Cases
13. *Use Coordinates*

Strategy 13: *Use Coordinates*

We may simplify many two-dimensional geometry problems by assigning coordinates to points on a figure. To accomplish this, we introduce a coordinate system of some type. Coordinate systems are also useful in three dimensions and on curved surfaces, such as a sphere.

Initial Problem

A treasure has been hidden in a desert location. A map shows $\triangle ABC$, where B is 50 miles northeast of A and C is 80 miles east of A. You are told that the treasure is at the midpoint of the median from A (Figure 8.1). Locate the treasure.

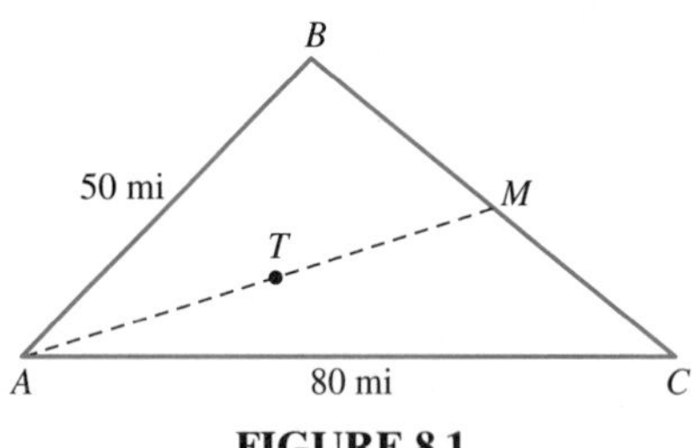

FIGURE 8.1

Clues

The Use Coordinates strategy may be appropriate when

- You can represent a problem using two variables.
- You cannot easily solve a geometry problem by traditional Euclidean methods.
- You must represent lines, circles, or other curves algebraically.
- A problem involves slope, parallel lines, perpendicular lines, and so on.
- The location of a geometric shape with respect to other shapes is important.
- A problem involves maps.

A solution of this Initial Problem is on page 469.

Triangle AREA Using Coordinates

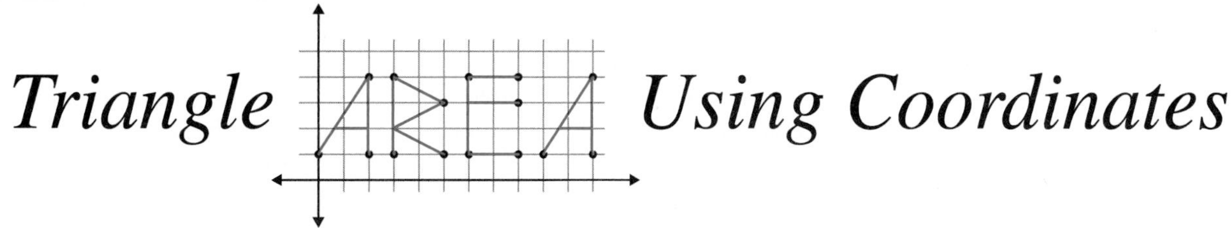

We have seen how to find the area of a triangle using the length of one side and the length of the altitude to that side. If we know the lengths of the three sides of a triangle, we can find its area using Heron's formula. (See Section 3.1.) A third method requires only that we know the coordinates of the vertices of the triangle.

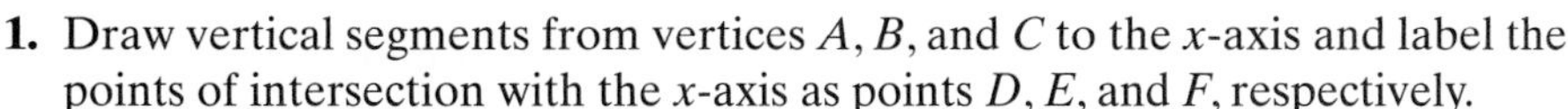

Consider $\triangle ABC$ with vertices $A(2,3)$, $B(4,6)$, and $C(9,8)$, as shown in the figure to the right.

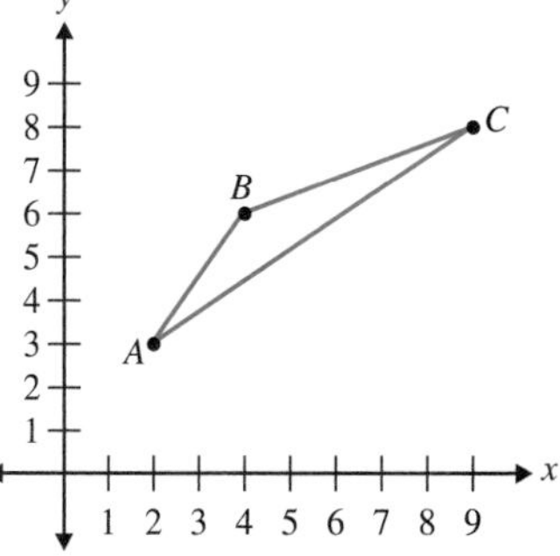

1. Draw vertical segments from vertices A, B, and C to the x-axis and label the points of intersection with the x-axis as points D, E, and F, respectively.
2. Find the lengths DA, EB, and FC.
3. Find the lengths DE, EF, and DF.
4. Use these lengths to find the areas of trapezoids $DABE$, $EBCF$, and $DACF$.
5. Using a combination of these three areas, find the area of $\triangle ABC$.

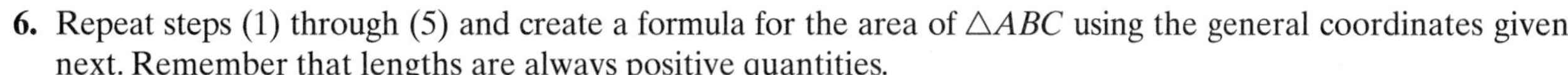

6. Repeat steps (1) through (5) and create a formula for the area of $\triangle ABC$ using the general coordinates given next. Remember that lengths are always positive quantities.

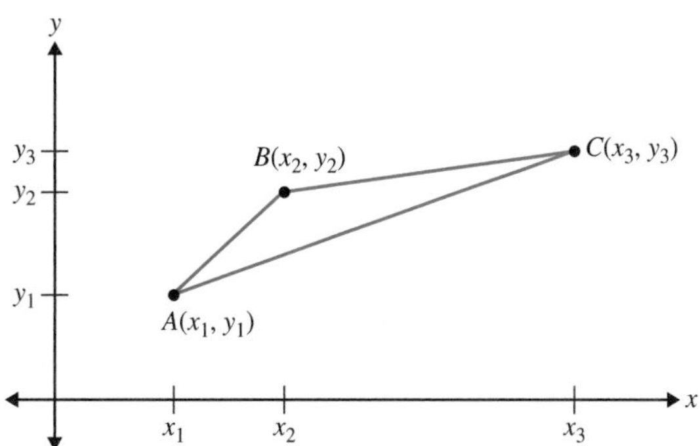

Area of $\triangle ABC$ = ____

7. Substitute the coordinates $A(2,3)$, $B(4,6)$, and $C(9,8)$ into your area formula from (6) and verify that this formula gives the area found in (5).
8. Plot the following points to form two triangles and use the formula from (6) to find the area of each triangle. Use the formula Area $= \frac{1}{2}bh$ to check your results.

   a. $A(1,2)$, $B(3,6)$, and $C(8,2)$ **b.** $A(2,3)$, $B(2,7)$, and $C(9,8)$

9. Create your own triangle by selecting vertices with coordinates that satisfy $x_3 \geq x_2 \geq x_1 \geq 0$ and $y_3 \geq y_2 \geq y_1 \geq 0$. Find its area using the formula you created and verify the area another way. We develop a more general area formula using coordinates in the Section 8.1 problem set.

Introduction

In this chapter, we study geometry using the coordinate plane. By imposing a coordinate system on the plane, we are able to prove many useful results about lines, polygons, circles, and so on. In Section 8.1, we initiate the study of coordinates and the concept of distance. We then use slope in Section 8.2 to derive results about parallel and perpendicular lines. In Section 8.3, we apply the study of coordinates to lines and circles. Last, in Section 8.4, we prove many interesting geometric results from a coordinate point of view.

8.1 COORDINATES AND DISTANCE IN THE PLANE

Applied Problem

A lot has corners with coordinates as shown, where measurements are given in feet. Find the perimeter of the lot.

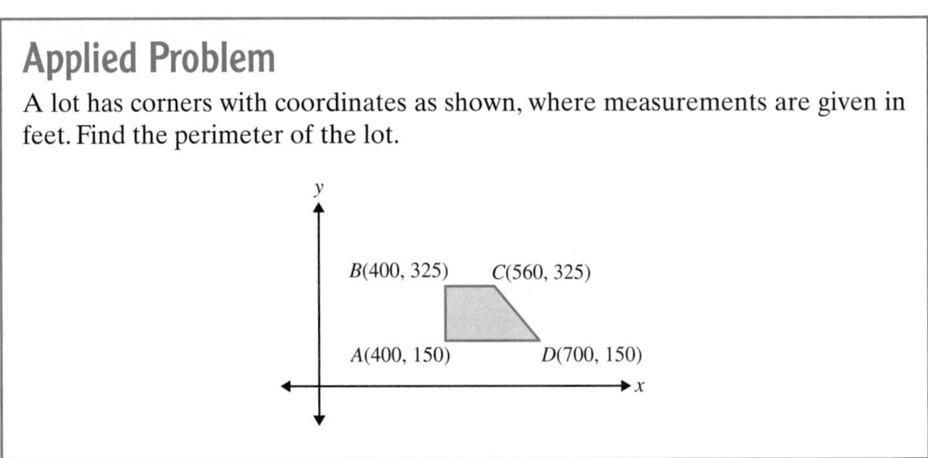

Coordinates

Suppose we choose two perpendicular real number lines in the plane and use their intersection point as a reference point, O, called the **origin** [Figure 8.2(a)]. The horizontal line is called the ***x*-axis** and the vertical line is called the ***y*-axis** (plural of axis is **axes**). The plane with an (x, y)-coordinate system on it is called a **coordinate plane**. Each point P on the coordinate plane is identified by an **ordered pair** (x, y). The ordered pair (x, y) is also denoted $P(x, y)$ [Figure 8.2(b)].

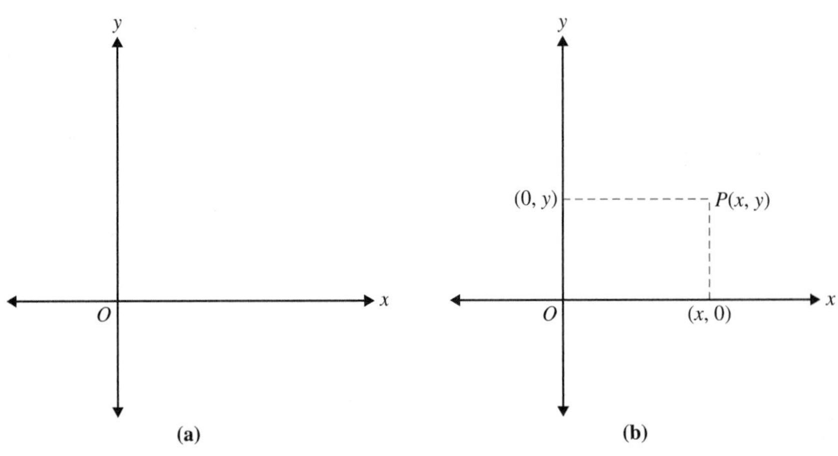

FIGURE 8.2

In the ordered pair (x, y) pictured in Figure 8.2(b), the first number, x, is called the ***x*-coordinate** and the second number, y, is called the ***y*-coordinate**. The x-coordinate locates the position of a point in relation to the y-axis. That is, it specifies the directed distance of the point from the y-axis. In a similar manner, the y-coordinate specifies the directed distance of the point from the x-axis.

In Figure 8.3(a), several points are graphed and their coordinates are shown.

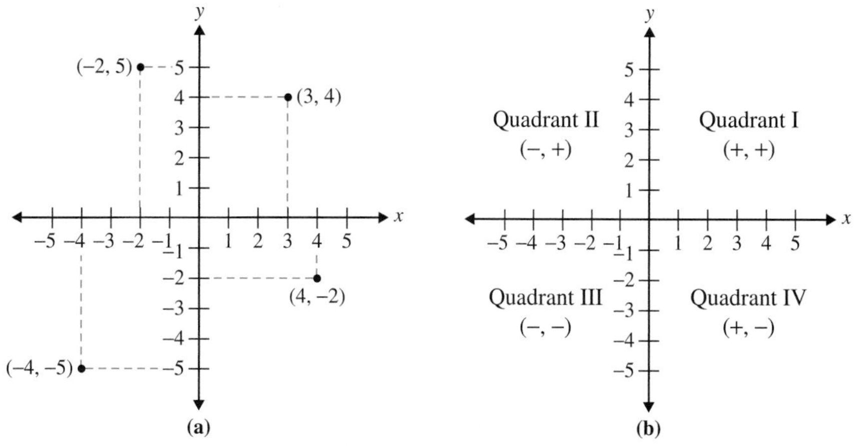

FIGURE 8.3

Notice how signs on the coordinates in the ordered pairs in Figure 8.3(a) are determined by the location of the points relative to the x- and y-axes. The two axes divide the plane into four **quadrants**. These quadrants are named in Figure 8.3(b) and the signs of the coordinates in each quadrant are indicated.

EXAMPLE 8.1 Plot the points with the following coordinates: $A(-2, 4)$, $B(4, 3)$, $C(5, -2)$, and $D(-4, -3)$. Then draw the following segments: $\overline{AB}$, $\overline{BC}$, $\overline{CD}$, and $\overline{AD}$.

SOLUTION

See Figure 8.4.

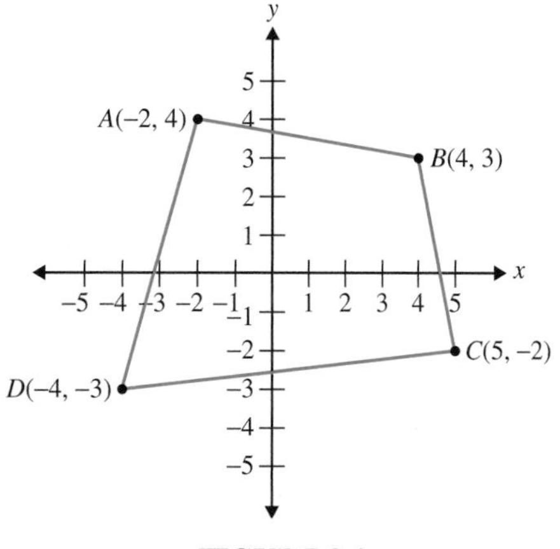

FIGURE 8.4

■

Distance

We can determine distances between points in the coordinate plane using the coordinates of the points and the Pythagorean theorem.

EXAMPLE 8.2 Find the distance between $P(-2, 1)$ and $Q(4, 3)$.

SOLUTION

Consider the points as graphed in Figure 8.5(a). Draw in the line segments as shown in Figure 8.5(b) to form a right triangle.

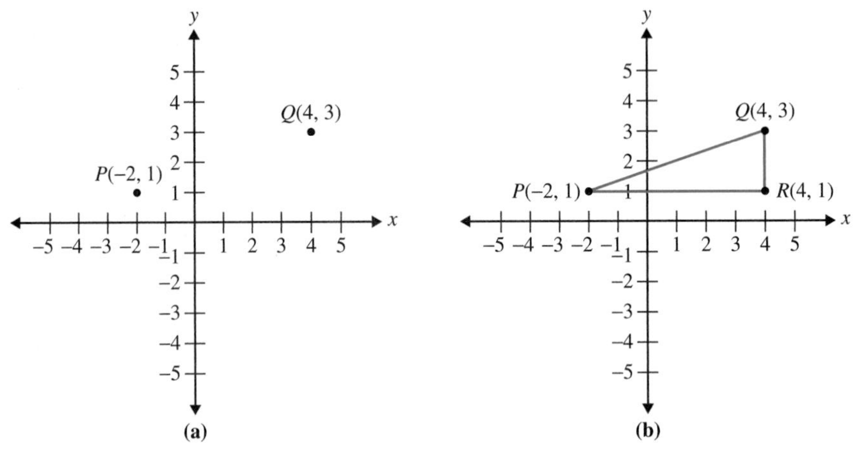

FIGURE 8.5

By the Pythagorean theorem, we have $(PR)^2 + (QR)^2 = (PQ)^2$. Since $PR = 4 - (-2) = 6$ and $QR = 3 - 1 = 2$, we have $6^2 + 2^2 = 40 = (PQ)^2$. Therefore, $PQ = \sqrt{40} = 2\sqrt{10}$. ■

We generalize the method in Example 8.2 in the following theorem, whose proof we leave for the problem set.

THEOREM 8.1 Distance Formula

If $P(x_1, y_1)$ and $Q(x_2, y_2)$ are two points in the coordinate plane, then

$$PQ = \sqrt{(x_2 - x_1)^2 + (y_2 - y_1)^2}$$

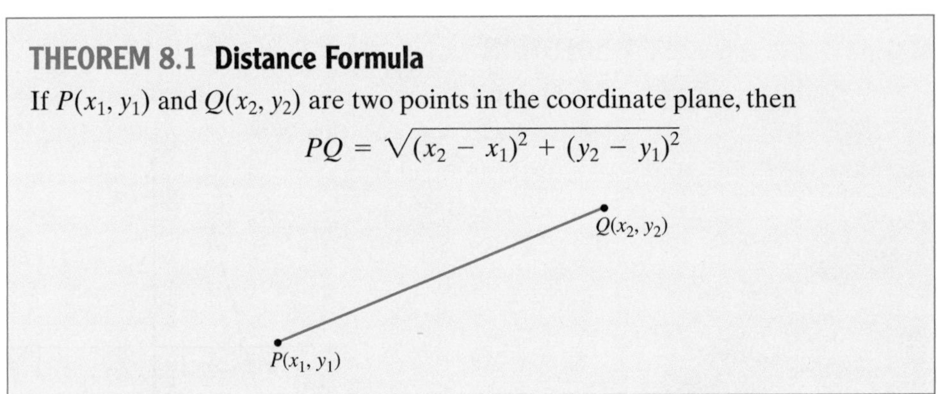

Since $(x_2 - x_1)^2 = (x_1 - x_2)^2$ and $(y_2 - y_1)^2 = (y_1 - y_2)^2$, we can also find PQ using $(x_1 - x_2)^2$ and $(y_1 - y_2)^2$.

We can use the coordinates of the vertices of a geometric figure together with the distance formula to classify a polygon by type. The following example illustrates this process.

EXAMPLE 8.3 The vertices of a quadrilateral $ABCD$ are $A(-7, -3)$, $B(0, 3)$, $C(5, 0)$, and $D(2, -5)$. Describe this quadrilateral as specifically as possible.

SOLUTION

$ABCD$ is graphed in Figure 8.6.

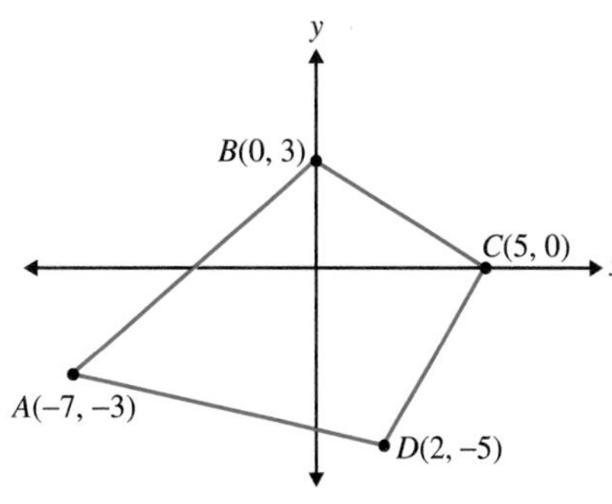

FIGURE 8.6

Using the distance formula, we can calculate the lengths of the sides of the quadrilateral.

$$AB = \sqrt{(-7 - 0)^2 + (-3 - 3)^2} = \sqrt{49 + 36} = \sqrt{85}$$
$$BC = \sqrt{(0 - 5)^2 + (3 - 0)^2} = \sqrt{25 + 9} = \sqrt{34}$$
$$CD = \sqrt{(5 - 2)^2 + [0 - (-5)]^2} = \sqrt{9 + 25} = \sqrt{34}$$
$$AD = \sqrt{(-7 - 2)^2 + [-3 - (-5)]^2} = \sqrt{81 + 4} = \sqrt{85}$$

Because $AB = AD$ and $BC = CD$, $ABCD$ is a kite. ■

We can also use the distance formula to locate points on a geometric figure placed on a coordinate system.

EXAMPLE 8.4 A circle has a radius of 5 and has its center at the point $O(0, 0)$ (Figure 8.7).

a. Show that the point $P(3, 4)$ lies on this circle.

b. Find coordinates of six other points on the circle.

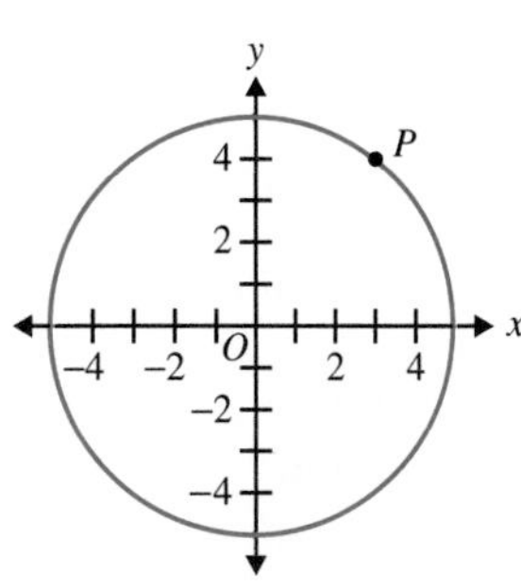

FIGURE 8.7

SOLUTION

a. The circle consists of every point whose distance from the center O is 5. Using the distance formula, we have

$$OP = \sqrt{(3-0)^2 + (4-0)^2} = \sqrt{9+16} = \sqrt{25} = 5$$

Thus, point P lies on the circle.

b. The symmetry of the graph allows us to use the given point $P(3, 4)$ to find other points on the circle that lie in Quadrants II, III, and IV: $(-3, 4)$, $(-3, -4)$, and $(3, -4)$. To locate other points on the circle, we must find other points at a distance of 5 from the center O. For example, if we choose $x = 4$, we have

$$\begin{aligned} \sqrt{(4-0)^2 + (y-0)^2} &= 5 \\ \sqrt{16 + y^2} &= 5 \\ 16 + y^2 &= 5^2 \\ 16 + y^2 &= 25 \\ y^2 &= 9 \\ y &= \pm 3 \end{aligned}$$

This result gives two more points on the circle: $(4, 3)$ and $(4, -3)$. There are also points on the coordinate axes that lie on the circle, such as $(0, 5)$. Figure 8.8 shows these six points. There are infinitely many other points on the circle that we could have chosen as examples. For example, $(2, \sqrt{21})$ is one other point on the circle because $\sqrt{2^2 + (\sqrt{21})^2} = \sqrt{4+21} = \sqrt{25} = 5$.

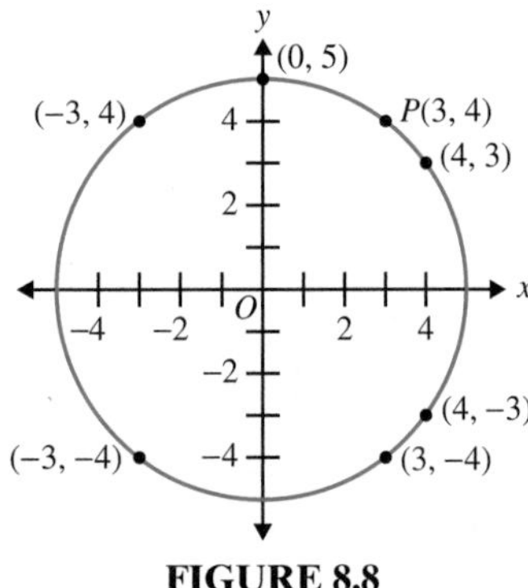

FIGURE 8.8

Test For Collinearity

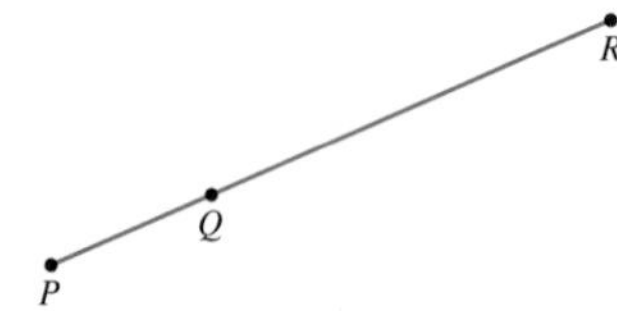

FIGURE 8.9

Recall that three points are collinear if there is a line that contains all three points. Notice that points P, Q, and R in Figure 8.9, shown left, are collinear and $PQ + QR = PR$. This suggests the following result, whose proof we leave for the problem set.

THEOREM 8.2 Test for Collinearity

The points P, Q, and R are collinear, with Q between P and R, if and only if $PQ + QR = PR$.

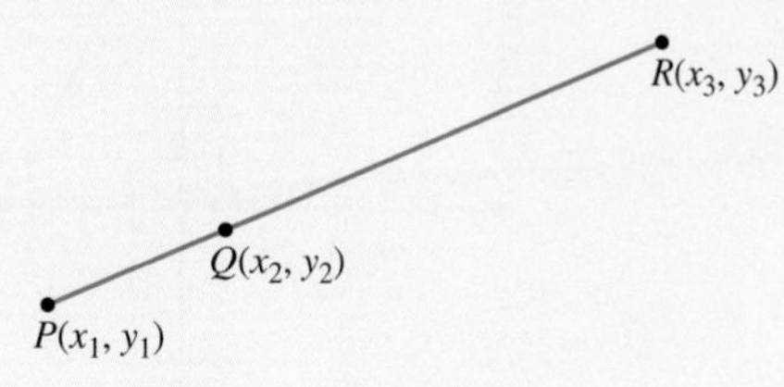

EXAMPLE 8.5 Determine whether the points $A(-2, 6)$, $B(1, 2)$, and $C(7, -6)$ are collinear (Figure 8.10).

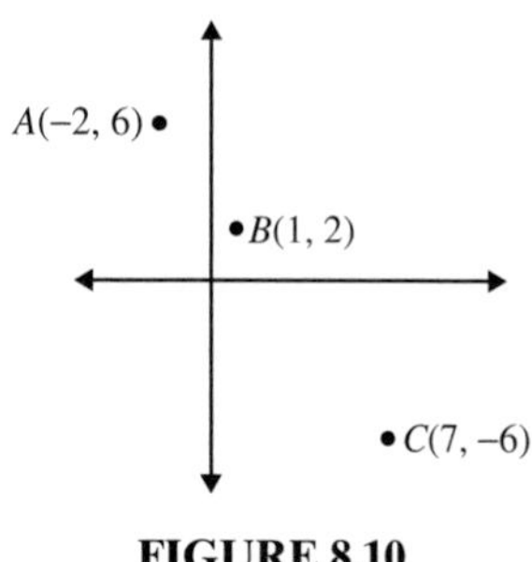

FIGURE 8.10

SOLUTION

Although the points in the figure appear to be collinear, we cannot say for certain without further investigation.

$$AB = \sqrt{[1-(-2)]^2 + (2-6)^2} = \sqrt{9+16} = \sqrt{25} = 5$$
$$AC = \sqrt{[7-(-2)]^2 + (-6-6)^2} = \sqrt{81+144} = \sqrt{225} = 15$$
$$BC = \sqrt{(7-1)^2 + (-6-2)^2} = \sqrt{36+64} = \sqrt{100} = 10$$

Because $5 + 10 = 15$, we have $AB + BC = AC$. This shows that A, B, and C are collinear and that B is between A and C. ■

Midpoint of a Segment

Using the distance formula, we can also find the midpoint of any line segment, as the next example shows.

EXAMPLE 8.6 Find the midpoint of the segment whose endpoints are $A(3, 1)$ and $B(1, 5)$ [Figure 8.11(a)].

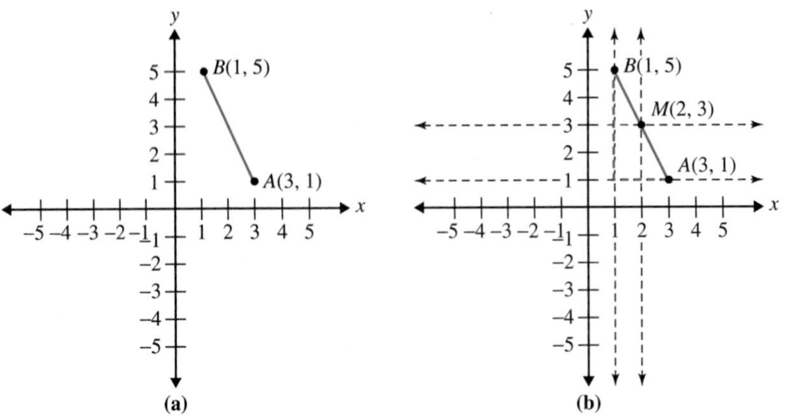

FIGURE 8.11

SOLUTION

Notice that in Figure 8.11(b), the vertical lines through $(1, 0)$ and $(2, 0)$ are parallel. By the Side Splitting Theorem, the vertical line through $(2, 1)$ bisects $\overline{AB}$. In a similar fashion, the horizontal line through $(1, 3)$ bisects $\overline{AB}$. Thus, $M(2, 3)$ is the midpoint of $\overline{AB}$. ■

We generalize this technique of finding midpoints as follows. We leave its proof for the problem set.

THEOREM 8.3 Midpoint Formula

If $P(x_1, y_1)$ and $Q(x_2, y_2)$ are endpoints of a line segment, then the midpoint of $\overline{PQ}$ is $M\left(\frac{x_1 + x_2}{2}, \frac{y_1 + y_2}{2}\right)$.

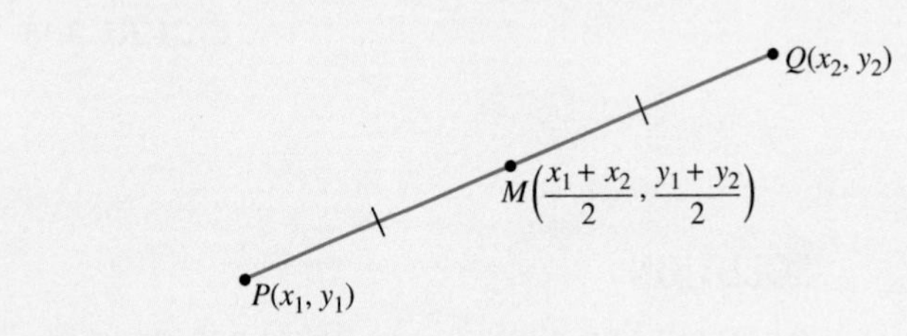

EXAMPLE 8.7 Find the coordinates of center O for the circle shown in Figure 8.12.

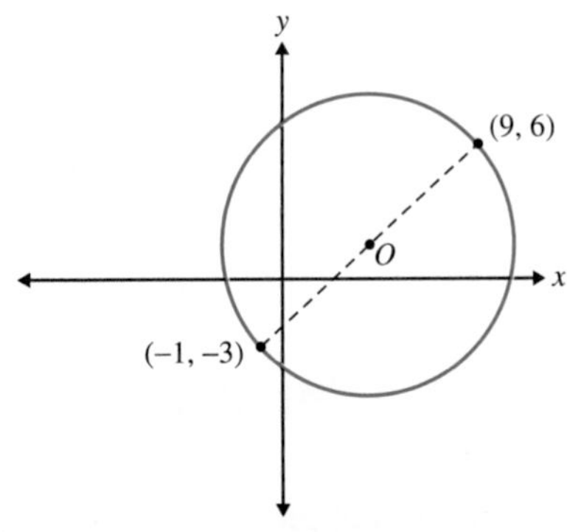

FIGURE 8.12

SOLUTION

Since the segment joining (9, 6) and (−1, −3) is a diameter of the circle, O is the midpoint of the segment. The coordinates of O are $\left(\frac{-1 + 9}{2}, \frac{-3 + 6}{2}\right) = \left(4, \frac{3}{2}\right)$. ■

Solution of Applied Problem

$AB = 325\text{ ft} - 150\text{ ft} = 175\text{ ft}$, $BC = 160\text{ ft}$, and $AD = 300$ ft. Using the distance formula, we find

$$CD = \sqrt{(700 - 560)^2 + (150 - 325)^2} = \sqrt{140^2 + (-175)^2}$$

This is approximately 224 ft. So the perimeter of the lot is approximately $175\text{ ft} + 160\text{ ft} + 300\text{ ft} + 224\text{ ft} = 859\text{ ft}$.

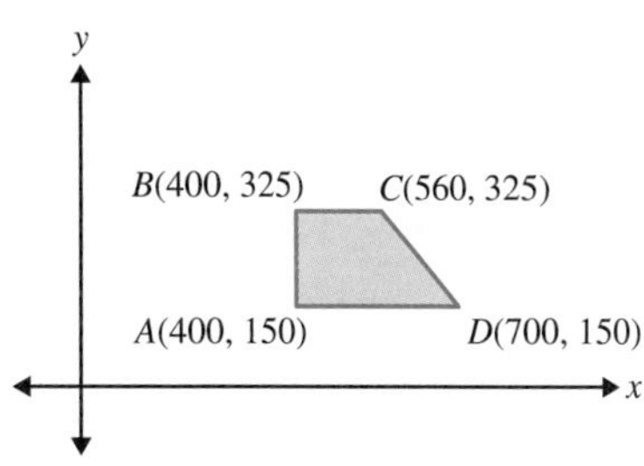

GEOMETRY AROUND US Coordinates are essential to graphics programming on any computer. Positions of pixels on the screen are identified by horizontal and vertical components, or x- and y-coordinates. One possible graphics screen is shown. A popular screen measures 1024 pixels horizontally by 768 pixels vertically. Notice that the screen is oriented so that the origin is at the upper left-hand corner and all distances from that point are taken to be positive. The highlighted pixel is at the position described by (825, 450).

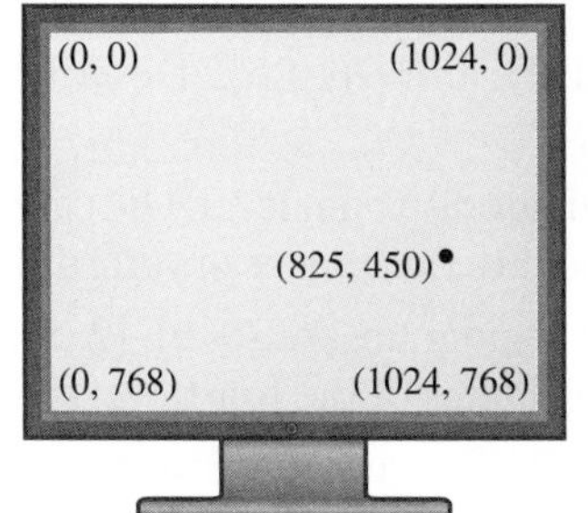

PROBLEM SET 8.1

EXERCISES/PROBLEMS

1. Plot the following sets of points on a coordinate system.
   a. (3, 2), (−3, 2), (− 3, −2), (3, −2)
   b. (0, 4), (4, 0), (3, −5), (−2, 4)
2. Plot the following sets of points on a coordinate system.
   a. (−3, −4), (−1, 5), (2, −4), (3, 6)
   b. (−1, −1), (−4, −4), (3, 3), (5, 5)
3. Find the distance between the given pairs of points.
   a. (0, 0), (3, 4)
   b. (−1, 2), (3, 5)
   c. (−1, −3), (−5, −3)
   d. (3, −4), (3, 5)
4. Find the distance between the given pairs of points.
   a. (0, 0), (5, 12)
   b. (3, −2), (−3, 2)
   c. (−1, −3), (−1, 3)
   d. (−4, −4), (3, 3)
5. A quadrilateral $ABCD$ has vertices $A(-1, 1)$, $B(2, 4)$, $C(6, 1)$, and $D(3, -2)$. Use the distance formula to verify that $ABCD$ is a parallelogram.

6. A quadrilateral $PQRS$ has vertices $P(0, -3)$, $Q(-1, 2)$, $R(4, 1)$, and $S(5, -4)$. Use the distance formula to verify that $PQRS$ is a rhombus.

7. Use the distance formula to determine if points P, Q, and R are collinear.

a. $P(-1, 4)$, $Q(-2, 3)$, and $R(-4, 1)$
b. $P(-2, 1)$, $Q(3, 4)$, and R $(12, 10)$
c. $P(-2, -3)$, $Q(2, -1)$, and R $(10, 3)$

8. Use the distance formula to determine if points P, Q, and R are collinear.

a. $P(0, 4)$, $Q(-6, 0)$, $R(6, 9)$
b. $P(-4, -19)$, $Q(2, -1)$, $R(5, 8)$
c. $P(-14, -10)$, $Q(-8, -7)$, $R(6, 0)$

9. The distance from $S(5, 2)$ to $T(2n, n)$ is $\sqrt{2}$ units. Find the value of n.

10. The length of $\overline{AB}$ is $2\sqrt{10}$, where A has coordinates $(-3, 4)$ and B has coordinates $(-1, k)$. Find the value of k.

11. Which of the given points are collinear with $A(-3, 0)$ and $B(1, 3)$?

a. $(6, 6)$ **b.** $(0, 2)$ **c.** $(-7, -3)$ **d.** $(9, 9)$

12. Which of the given points are collinear with $A(-2, -3)$ and $B(4, -1)$?

a. $(1, -2)$ **b.** $(7, 1)$ **c.** $(12, 2)$ **d.** $(-8, -5)$

13. The endpoints of a segment are given. Find the coordinates of the midpoint of the segment.

a. $(0, 2)$ and $(-3, 2)$ **b.** $(-5, -1)$ and $(3, 5)$
c. $(-2, 3)$ and $(-3, 6)$ **d.** $(3, -5)$ and $(3, 7)$
e. $(1, 5)$ and $(3, 9)$ **f.** $(6, -2)$ and $(-3, 5)$

14. The coordinates of two points are given. If M is the midpoint of $\overline{AB}$, find the coordinates of the third point.

a. $A(-2, 4)$, $B(-1, 10)$
b. $A(-1, -3)$, $B(5, 12)$
c. $A(3, -5)$, $B(3, 7)$
d. $A(2, 6)$, $M(4, -3)$
e. $A(1, -3)$, $M(5, 2)$
f. $M(-2, -5)$, $B(3, -4)$

15. Draw triangles that have vertices with the given coordinates. Describe each triangle as scalene, isosceles, equilateral, acute, right, and/or obtuse. Be as complete as possible in your descriptions and justify your answer.

a. $(0, 0)$, $(0, -5)$, $(5, -5)$
b. $(-3, 1)$, $(1, 3)$, $(5, -5)$
c. $(-2, -1)$, $(2, 2)$, $(6, -1)$
d. $(-4, -2)$, $(-1, 3)$, $(4, -2)$

16. Draw triangles that have vertices with the given coordinates. Describe each triangle as scalene, isosceles, equilateral, acute, right, and/or obtuse. Be as complete as possible in your descriptions and justify your answer.

a. $(2, -3)$, $(5, -1)$, $(8, 4)$
b. $(1, -6)$, $(5, 1)$, $(9, -6)$
c. $(-4, 0)$, $(-3, 4)$, $(1, 3)$
d. $(-7, 3)$, $(-4, -7)$, $(-1, 4)$

17. Given are the vertices of triangles. Use the distance formula to determine if each triangle is a right triangle.

a. $(-2, 3)$, $(5, 5)$, $(7, -2)$
b. $(0, -6)$, $(1, 0)$, $(6, -7)$
c. $(-7, -5)$, $(-4, 8)$, $(3, 5)$

18. Given are the vertices of triangles. Use the distance formula to determine if each triangle is a right triangle.

a. $(-2, 5)$, $(0, -1)$, $(12, 3)$
b. $(2, 3)$, $(-2, -3)$, $(-6, 1)$
c. $(-3, -2)$, $(5, -2)$, $(1, 2)$

19. Use the distance formula to determine whether $PQRS$ is a rectangle for points $P(1, 1)$, $Q(4, 2)$, $R(2, 5)$, and $S(-1, 4)$. (HINT: Use the diagonals of $PQRS$.)

20. What general type of quadrilateral is $ABCD$, for $A(0, 0)$, $B(-4, 3)$, $C(-1, 7)$, and $D(10, 5)$? Describe it as completely as possible.

21. Two vertices of an equilateral triangle $\triangle ABC$ are $A(0, 0)$ and $B(0, 10)$. Find the coordinates of vertex C.

22. Three vertices of a parallelogram are $(0, 0)$, $(2, 5)$, and $(8, 0)$. Find the coordinates of the fourth vertex.

23. Draw quadrilateral $ABCD$ whose vertices are $A(3, 0)$, $B(6, 6)$, $C(6, 9)$, and $D(0, 6)$. Divide each of the coordinates by 3 and graph the new quadrilateral $A'B'C'D'$. For example, A' has coordinates $(1, 0)$. How do the lengths of corresponding sides compare?

24. Draw quadrilateral $ABCD$ whose vertices are $A(-8, 0)$, $B(0, -16)$, $C(8, 4)$, and $D(4, 12)$. Divide each of the coordinates by 4 to get the coordinates of a new quadrilateral $A'B'C'D'$. For example, A' has coordinates $(-2, 0)$. Graph quadrilateral $A'B'C'D'$. How do the lengths of corresponding sides of the two quadrilaterals compare?

25. Draw $\triangle ABC$ for points $A(2, 0)$, $B(-1, 2)$, and $C(0, 0)$. Multiply each of the coordinates by 2 and graph $\triangle A'B'C'$. For example, A' has coordinates $(4, 0)$.

a. How do the lengths of corresponding sides compare?

b. How do the areas of the two triangles compare?

26. Draw quadrilateral $ABCD$ for $A(4, -2)$, $B(4, 2)$, $C(-2, 2)$, and $D(-2, -2)$. Multiply each coordinate by 3 and graph the resulting quadrilateral $A'B'C'D'$. For example, A' has coordinates $(12, -6)$.

a. How do the perimeters of $ABCD$ and $A'B'C'D'$ compare?

b. How do the areas of $ABCD$ and $A'B'C'D'$ compare?

27. Draw quadrilateral $ABCD$ for $A(-9, 3)$, $B(-6, -6)$, $C(3, -3)$, and $D(0, 6)$.

a. Describe $ABCD$ as completely as possible.

b. Calculate the area of $ABCD$.

c. Multiply each coordinate of $ABCD$ by $\frac{1}{3}$ to get the coordinates of a new quadrilateral $A'B'C'D'$. Graph quadrilateral $A'B'C'D'$. How does the area of $A'B'C'D'$ compare with the area of $ABCD$?

28. Draw quadrilateral $ABCD$ for $A(-8, 0)$, $B(-6, 4)$, $C(6, 4)$, and $D(8, 0)$.

a. Describe $ABCD$ as completely as possible.

b. Calculate the area of $ABCD$.

c. Multiply each coordinate of $ABCD$ by $\frac{1}{2}$ and graph the resulting quadrilateral $A'B'C'D'$. How does the area of $A'B'C'D'$ compare with the area of $ABCD$?

29. Find the coordinates of six points on a circle centered at the origin with radius 13.

30. Find the coordinates of six points on a circle centered at the origin with radius 10.

31. We have seen that the location of a point in the coordinate plane can be described by an ordered pair (x, y). We may generalize Cartesian coordinates to three-dimensional space so that the location of a point in space is described by an ordered triple (x, y, z). The three coordinates describe the location of the point relative to three axes as shown. Point P with coordinates (x, y, z) is plotted by going x units along the x-axis, y units along the y-axis, and z units along the z-axis as shown. Plot the following points in three-dimensional space.

a. $(2, 1, 3)$ **b.** $(-2, 1, 0)$ **c.** $(3, -1, -2)$

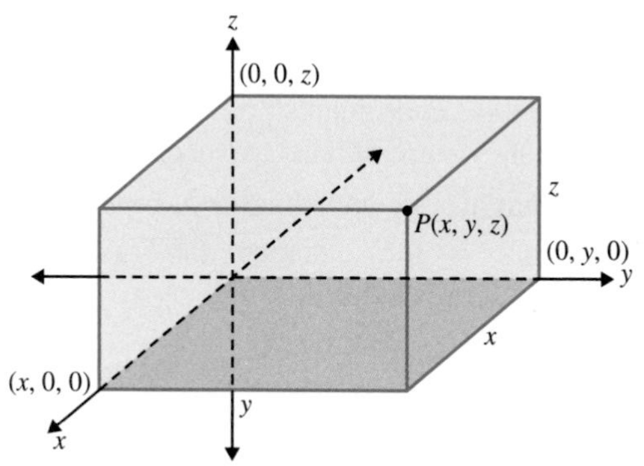

32. The three coordinate axes taken in pairs determine three coordinate planes:

The (horizontal) xy-plane, where $z = 0$
The (vertical) yz-plane, where $x = 0$
The (vertical) xz-plane, where $y = 0$

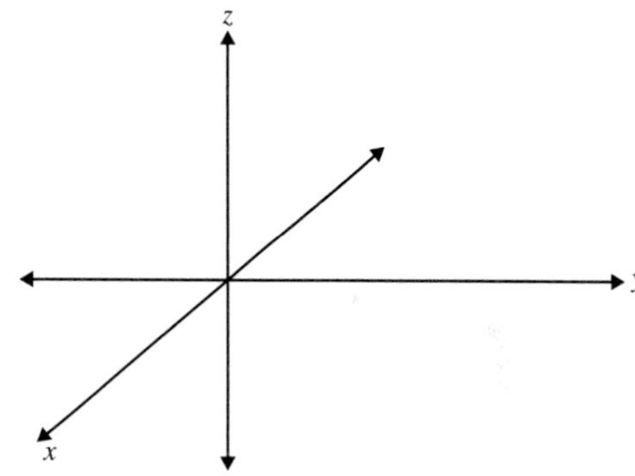

These three planes divide space into eight regions, called **octants**. Looking above the xy-plane, the front right octant is Octant 1. Octants 2, 3, and 4 are found by going counterclockwise through the other upper-level octants. Octant 5 is directly below Octant 1, 6 is below 2, 7 is below 3, and 8 is below 4. Given the following points, indicate in which octant each is found.

a. $(3, 2, 1)$

b. $(-3, -3, 1)$

c. $(-1, 2, -3)$

d. $(-5, -3, -2)$

e. $(6, -3, 5)$

f. $(8, 4, -2)$

33. In Octant 1, the x-, y-, and z-coordinates are all positive. Characterize the coordinates in the remaining seven octants.

34. We can generalize the distance formula to three-dimensional space. Let P have coordinates (x_1, y_1, z_1) and Q have coordinates (x_2, y_2, z_2). The faces of the prism shown are parallel to the coordinate planes.

a. What are the coordinates of point R?

b. What is the distance PR? (HINT: Use the Pythagorean theorem.)

c. What is the distance QR?

d. What is the distance PQ?

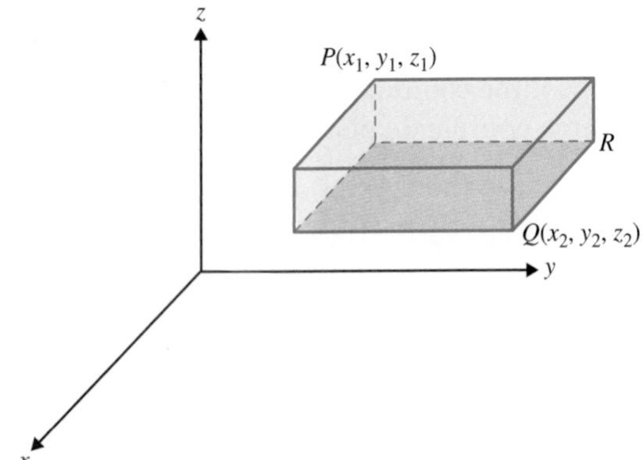

35. Use the distance formula in three dimensions to find the distance between the following pairs of points.

a. $(1, 2, 3)$ and $(2, 3, 1)$

b. $(-1, 0, 5)$ and $(6, 2, -1)$

36. We can generalize the midpoint formula to points in three-dimensional space. The midpoint of $\overline{PQ}$ for $P(x_1, y_1, z_1)$ and $Q(x_2, y_2, z_2)$ is

$$M\left(\frac{x_1 + x_2}{2}, \frac{y_1 + y_2}{2}, \frac{z_1 + z_2}{2}\right).$$

a. Find the midpoint of the segment having endpoints $(2, 0, -3)$ and $(6, 4, -7)$.

b. If the midpoint of $\overline{PQ}$ is $M(1, -1, 3)$ and Q has coordinates $(4, -3, 0)$, find the coordinates of P.

c. The endpoints of a diameter of a sphere are $(1, 1, 4)$ and $(2, -1, 5)$. Find the coordinates of the center of the sphere.

The following discussion pertains to upcoming problems.

In Chapter 6, we defined the sine and cosine of an acute angle as the ratio of two sides of a right triangle. If we use coordinates to define the sine and cosine of an angle in another way, we can calculate sines and cosines of right angles and obtuse angles. Consider a circle with radius 1 that is centered at the origin. This circle is called a **unit circle**. Consider point P on the circle and in Quadrant I as shown next.

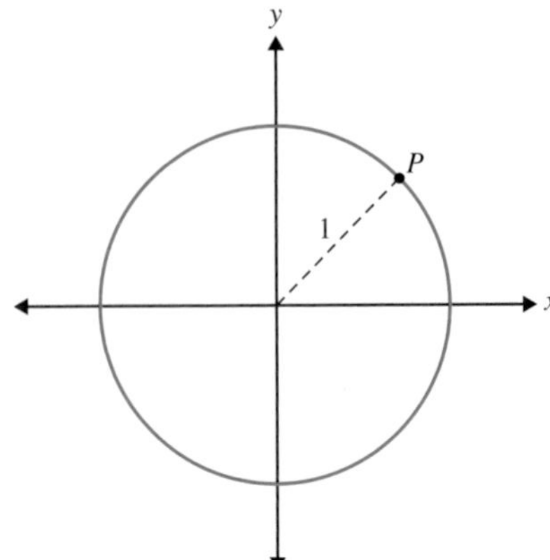

We can find the coordinates of point P by considering the right triangle formed by the x-axis, a vertical segment through P, and the radius of the unit circle. The radius that connects point P to the origin forms an angle with the positive x-axis. This angle will be denoted as $\angle A$. It turns out that the sine and cosine of $\angle A$ are related to the coordinates of point P.

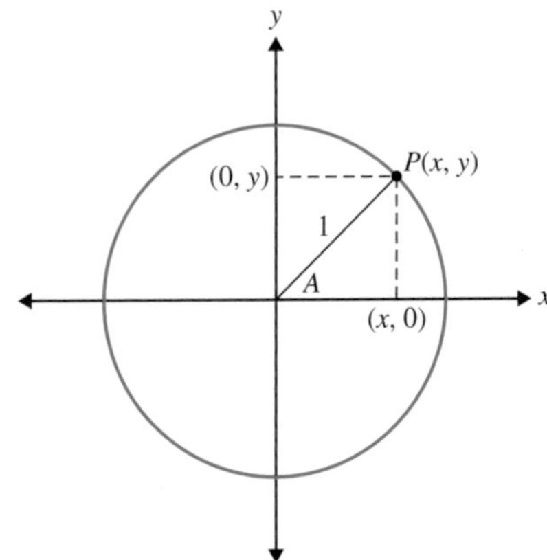

Using right triangle trigonometry, we see

$$\sin A = \frac{\text{opposite}}{\text{hypotenuse}} = \frac{y}{1}$$

and

$$\cos A = \frac{\text{adjacent}}{\text{hypotenuse}} = \frac{x}{1}$$

so $y = \sin A$ and $x = \cos A$. The coordinates of point P are $(x, y) = (\cos A, \sin A)$. Thus, we can define the cosine of an angle as the x-coordinate of a point on the unit circle and the sine of an angle as the y-coordinate of a point on the unit circle. This definition holds when point P is in

Quadrant I and $\angle A$ is acute, as is the case here. It also holds when P lies on one of the coordinate axes or in one of the other three quadrants. Use this unit circle definition of the sine and cosine of an angle to complete problems 37–44, proofs 46, 50, 53 and 54.

37. Draw a unit circle centered at the origin and draw $\angle A$ with the given measure. Remember that $\angle A$ is measured from the positive x-axis so that one side of $\angle A$ lies on the positive x-axis. Label as P the point where the other side of $\angle A$ intersects the unit circle. Find the coordinates of point P on the unit circle given each of the following measures for $\angle A$.

a. $\angle A = 30°$ **b.** $\angle A = 90°$

c. $\angle A = 135°$ **d.** $\angle A = 0°$

38. Draw a unit circle centered at the origin and draw $\angle A$ with the given measure. Remember that $\angle A$ is measured from the positive x-axis. Label the point where one side of $\angle A$ intersects the unit circle as P. Find the coordinates of point P on the unit circle given each of the following measures for $\angle A$.

a. $\angle A = 60°$ **b.** $\angle A = 180°$

c. $\angle A = 210°$ **d.** $\angle A = 270°$

39. Use the unit circle definition of the sine and cosine of an angle to evaluate sin A and cos A for each of the following. Use your calculator to verify your answers.

a. $\angle A = 30°$ **b.** $\angle A = 90°$

c. $\angle A = 135°$ **d.** $\angle A = 0°$

40. Use the unit circle definition of the sine and cosine of an angle to evaluate sin A and cos A for each of the following. Use your calculator to verify your answers.

a. $\angle A = 60°$ **b.** $\angle A = 180°$

c. $\angle A = 210°$ **d.** $\angle A = 270°$

41. a. Use the unit circle definition of the sine and cosine of an angle to find the exact values for sin 60° and sin 120°. Use your calculator to verify your answers.

b. Use the unit circle definition of the sine and cosine of an angle to find the exact values for sin 45° and sin 135°. Use your calculator to verify your answers.

c. Based on the results from parts (a) and (b), how is sin A related to sin $(180° - A)$?

42. a. Use the unit circle definition of the sine and cosine of an angle to find the exact values for cos 60° and cos 120°. Use your calculator to verify your answers.

b. Use the unit circle definition of the sine and cosine of an angle to find the exact value for cos 45° and cos 135°. Use your calculator to verify your answers.

c. Based on the results from parts (a) and (b), how is cos A related to cos $(180° - A)$?

43. We introduced the Law of Cosines in Section 6.5. We can now apply the Law of Cosines to obtuse angles. Use the Law of Cosines and a calculator to find the indicated measurement in $\triangle ABC$. Round to the nearest tenth.

a. If $\angle C = 115°$, $a = 10.1$, and $b = 12$, find c.

b. If $\angle A = 170°$, $b = 18.2$, and $c = 25.3$, find a.

44. Use the Law of Cosines and a calculator to find the indicated measurement in $\triangle ABC$. Round to the nearest tenth.

a. If $\angle B = 126°$, $a = 5.47$, and $c = 8.56$, find b.

b. If $\angle A = 97°$, $b = 10.91$, and $c = 35.21$, find a.

45. In the "Geometry Investigation" at the beginning of this chapter, you found a coordinate formula for the area of a triangle in one special case. We will explore that formula in more detail next. Points $A(x_1, y_1)$, $B(x_2, y_2)$, and $C(x_3, y_3)$ form the vertices of a triangle, where $x_3 > x_2 > x_1 > 0$ and $y_3 > y_2 > y_1 > 0$. Vertical segments have been drawn from each vertex to the x-axis. Points where the segments from A, B, and C intersect the x-axis are labeled D, E, and F, respectively, as shown in the following figure.

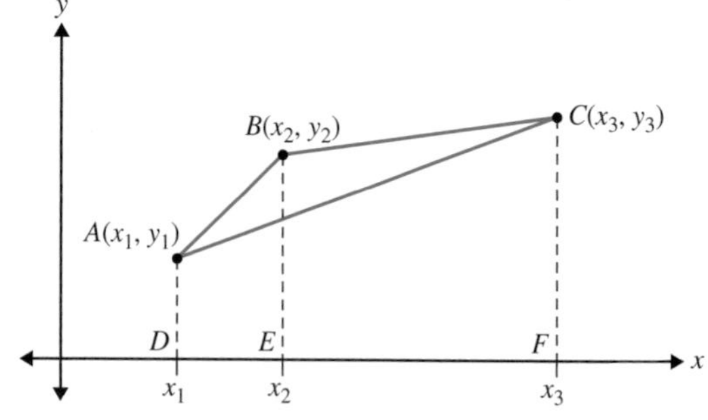

a. The area of $\triangle ABC$ = area of $DABE$ + area of $EBCF$ − area of $DACF$. Show that this relationship leads to the following coordinate area formula.

$$A = \frac{1}{2}[x_1(y_3 - y_2) + x_2(y_1 - y_3) + x_3(y_2 - y_1)]$$

b. Use the formula from part (a) to find the area for the triangle with vertices $A(2, 3)$, $B(4, 6)$, and $C(9, 8)$. Compare your result to the area calculated using the area formula developed in the "Geometry Investigation." Are the results the same?

c. In the "Geometry Investigation," the coordinates for the vertices of $\triangle ABC$ were carefully selected. For coordinates $A(-5, -4)$, $B(4, -4)$, and $C(-5, 9)$ use the formula from part (a) to find the area of $\triangle ABC$. Find the area using the formula $\frac{1}{2}bh$. How are the two areas different? Why do the results differ? How can you adjust the formula from part (a) so that it gives the correct area of *any* triangle?

PROOFS

46. Prove the Law of Cosines is equivalent to the Pythagorean theorem if the angle measure is 90°. That is, show that

$$c^2 = a^2 + b^2 - 2ab\cos C$$

is the same as the Pythagorean theorem if $\angle C = 90°$.

47. In the development of the midpoint formula, P and Q have coordinates (x_1, y_1) and (x_2, y_2), respectively, and M is the point $\left(\frac{x_1 + x_2}{2}, \frac{y_1 + y_2}{2}\right)$. Use the distance formula to verify that P, M, and Q are collinear and that $PM = MQ$.

48. Verify the Test for Collinearity by proving parts (a) and (b).

a. If Q is on $\overline{PR}$, then $PQ + QR = PR$. (HINT: Draw a picture and add corresponding lengths.)

b. If $PQ + QR = PR$, then Q is on $\overline{PR}$. (HINT: Use indirect reasoning: Assume that Q is not on $\overline{PR}$. Draw a picture showing the various possibilities and compare lengths.)

49. Suppose that l is a line with an equation of the form $y = mx + b$. Show that if a point (x, y) in the plane satisfies the equation $y = mx + b$, then (x, y) is on line l. (HINT: Choose two points on l by giving x two values. Then show that the point $(x, mx + b)$ is collinear with the points you have chosen.)

50. A **cevian** is a segment from the vertex of a triangle to the opposite side. In $\triangle ABC$, $\overline{AD}$ is a cevian from vertex A to side $\overline{BC}$, and it divides $\overline{BC}$ into two segments with lengths m and n.

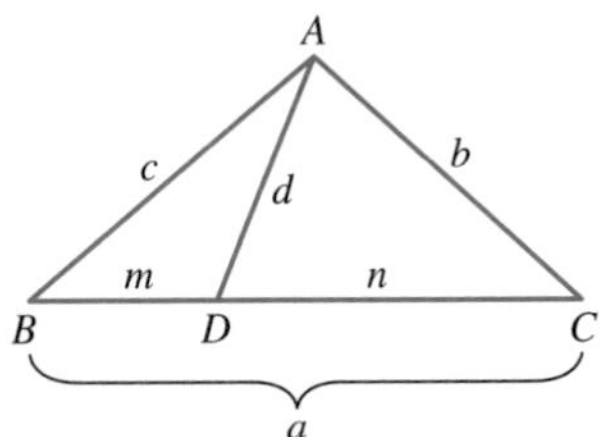

a. Use the fact that for any angle A, $\cos(180° - A) = -\cos A$ and apply the Law of Cosines to $\triangle BDA$ and $\triangle CDA$ to prove that in $\triangle ABC$ with cevian $\overline{AD}$,

$$mna + d^2a = c^2n + b^2m$$

b. Find AD in the following triangle in two ways. First use the relationship from part (a), substituting known values for m, n, a, b, and c. Then use the Law of Cosines to find the measure of $\angle B$ and AD. Are the results the same?

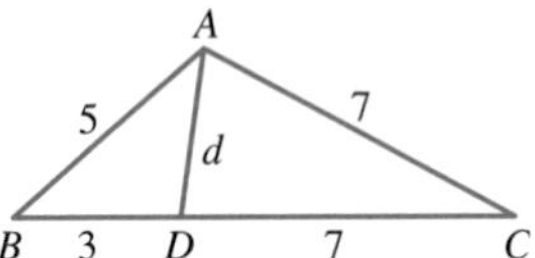

APPLICATIONS

51. A landscape designer plans to add five trees in a line along a retaining wall for added privacy. If she selects coordinates (0, 5) and (14, 0) for the outermost trees, and all the trees are equally spaced, find the coordinates of the positions of the remaining three trees.

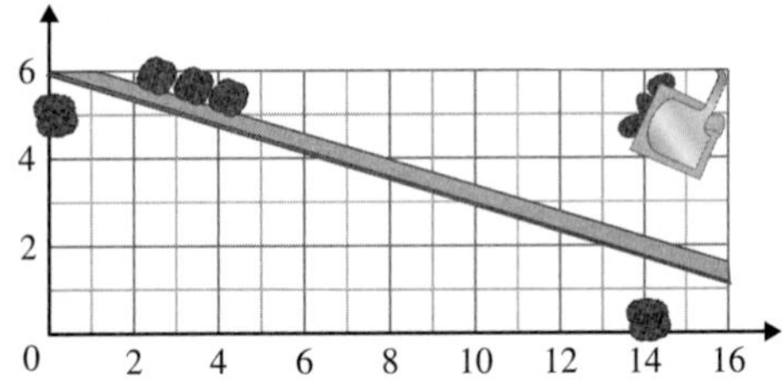

52. a. A coordinate system is set up so that (0, 0) is the center of the sun, (67,499,999, 62,512,000) the center of the earth, and (67,600,000, 62,728,916) the center of the moon. If all distances are in miles, find the distance from the center of the sun to the center of the earth and the distance from the center of the earth to the center of the moon. Round to the nearest whole number.

b. During a total solar eclipse, the centers of the sun, moon, and earth are collinear and the moon is located between the sun and the earth. Using the distances from part (a), give the coordinates of the centers of the moon and earth during a total solar eclipse if the sun is placed at the origin and the moon and earth are placed on the positive x-axis.

53. Recall from Section 6.5 that the velocity of an object is the speed of the object in a particular direction. A person in a rowboat would like to row from a dock to a ship anchored 5 km away. He rows at a velocity of 2.5 kph directly toward the ship. The velocity of the current is 3.6 kph directed as shown in the following figure. When the person has rowed for 30 minutes, how far from the ship will he be? Round to the nearest tenth.

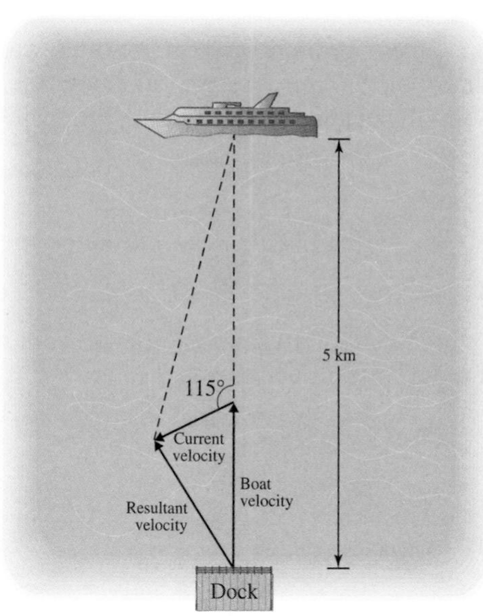

EXTENDED PROBLEMS

54. By defining the sine and cosine of an angle using the unit circle, we may apply the Law of Sines and the Law of Cosines to obtuse triangles. However, an interesting situation sometimes arises with the Law of Sines. Research the Law of Sines as it applies to obtuse triangles. Under what general conditions will the Law of Sines give two solutions? How can we use the Law of Sines to solve a triangle in the two-solution case? Include figures and an example that demonstrates conditions under which two solutions arise along with your explanation.

55. The **Universal Transverse Mercator** (UTM) grid is a coordinate system derived from a projection of the globe onto a two-dimensional surface. Research the UTM grid, print a Mercator projection map with the coordinate system labeled, and use coordinates to identify five locations on the map. Answer the following questions briefly. How is the Mercator projection created and what does it mean to say the Mercator projection is conformal? Into how many zones is the globe divided? How is each zone labeled? What do the coordinates of a location represent? How accurate is this type of map?

56. One way to describe the location of a point in the coordinate plane, also called the **rectangular coordinate system**, is to use an ordered pair that describes the distance of the point from the coordinate axes. We can also represent every point in the coordinate plane by specifying a distance from the origin, called the **pole**, and specifying an angle measured counterclockwise from the positive x-axis, called the **polar axis**. This (distance, angle) or (r, A) representation is called the **polar coordinates** of a point.

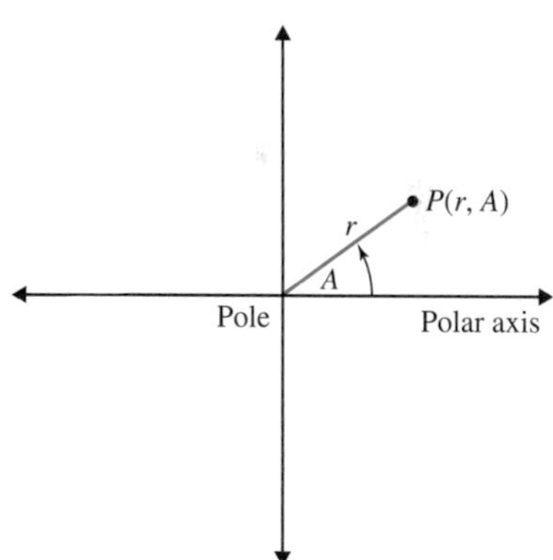

Research polar coordinates and explain why any one point has infinitely many polar coordinate representations. Give several examples to illustrate this concept. Describe how to convert from polar coordinates to rectangular coordinates. For example, if a point P has rectangular coordinates $(x, y) = (-2, 2)$, what polar coordinates (r, A) might be used to represent point P?

8.2 SLOPE

Applied Problem

A conveyor belt is used to move bundles of shingles to the roof of a house. If the conveyor belt is 60 feet long and makes an angle of 35° with the horizontal, what is the slope of the conveyor? That is, how many feet is a bundle of shingles elevated for every one foot it travels horizontally?

Slope

When graphing lines or line segments on a coordinate system, we often want to determine the steepness or slant of a line. For that purpose, we talk about the slope of a line. The slope of a line is a measure of the inclination of the line with respect to the horizontal.

DEFINITION Slope of a Line

Given the coordinates of any two points that determine a nonvertical line, the slope, m, of the line is the ratio of the change in y to the change in x.

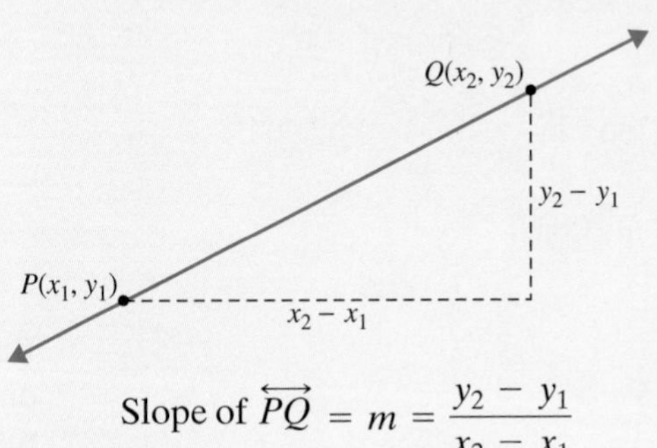

$$\text{Slope of } \overleftrightarrow{PQ} = m = \frac{y_2 - y_1}{x_2 - x_1}$$

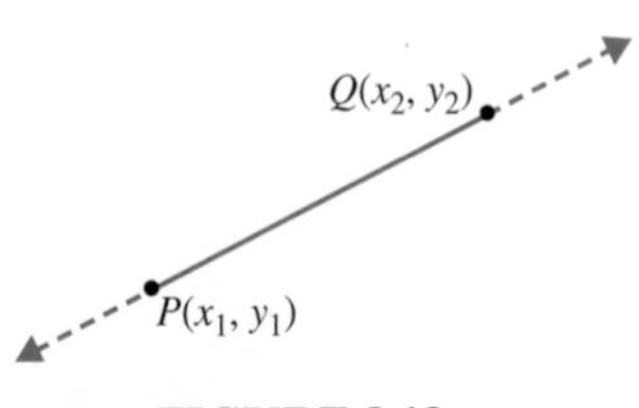

FIGURE 8.13

The difference $y_2 - y_1$ is called the **rise** and $x_2 - x_1$ is called the **run** of $\overleftrightarrow{PQ}$. Thus, we could also define the slope of a nonvertical line to be the ratio of its rise to its run. Notice that we can also write the slope of $\overleftrightarrow{PQ}$ as $\frac{y_1 - y_2}{x_1 - x_2}$ because $\frac{y_2 - y_1}{x_2 - x_1} = \frac{y_1 - y_2}{x_1 - x_2}$. The **slope of a line segment** $\overline{PQ}$ is the slope of $\overleftrightarrow{PQ}$ (Figure 8.13). If $\overleftrightarrow{PQ}$ is vertical, then $x_1 = x_2$ and the denominator in the slope definition, namely $x_2 - x_1$ is zero. Therefore, we say the slope of a vertical line is undefined.

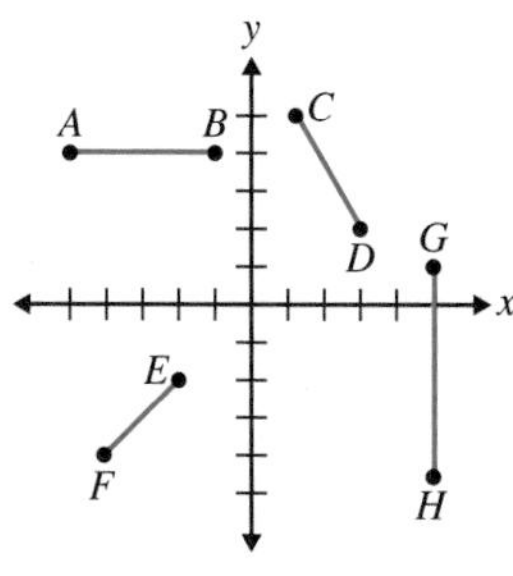

FIGURE 8.14

EXAMPLE 8.8 Find the slopes, if defined, of each of the line segments in Figure 8.14 with endpoints $A(-5,4)$, $B(-1,4)$, $C(1,5)$, $D(3,2)$, $E(-2,-2)$, $F(-4,-4)$, $G(5,1)$, and $H(5,-4)$.

SOLUTION

a. Slope of $\overline{AB} = \dfrac{4-4}{-5-(-1)} = \dfrac{0}{-4} = 0$

b. Slope of $\overline{CD} = \dfrac{5-2}{1-3} = \dfrac{3}{-2} = \dfrac{-3}{2}$

c. Slope of $\overline{EF} = \dfrac{-2-(-4)}{-2-(-4)} = \dfrac{2}{2} = 1$

d. Slope of $\overline{GH}$ is undefined because $\overline{GH}$ is vertical. ■

We can make several generalizations about slopes of lines by reviewing Example 8.8. First, notice that the slope of the horizontal line segment $\overline{AB}$ is zero. Because the y-coordinates of any two points on a horizontal line are equal, the slope of *any* horizontal line is zero. The slope of $\overline{CD}$ is negative. This is true of *any* line that falls from left to right. The slope of $\overline{EF}$ is positive which is true for *any* line that rises from left to right. Figure 8.15 gives a visual summary of slopes. We will discuss more generalizations regarding slopes in the problem set.

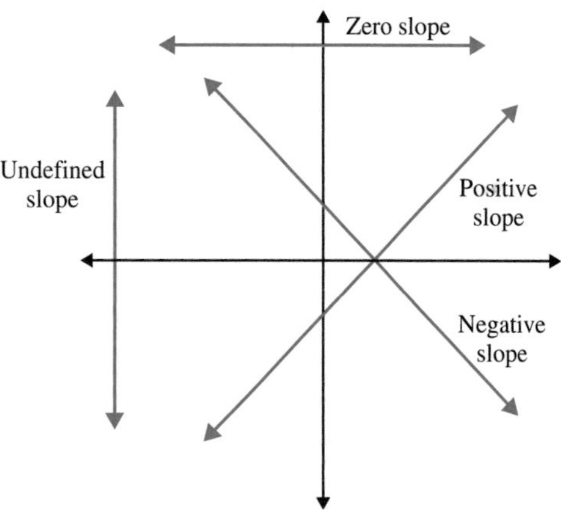

FIGURE 8.15

Collinearity and Slope

We have seen how to calculate the slope of a line or line segment given the coordinates of two points contained in it. If we know the coordinates of three different points on a line, we can determine the slope of that line by choosing any two of the three points and using the slope formula. In Figure 8.16, the three points P, Q, and R are collinear. The slopes of $\overline{PQ}$, $\overline{QR}$, and $\overline{PR}$ will be the same. That is, the choice of points does not affect the value of the slope. We will discuss this idea further in the problem set.

In Section 8.1, we developed a test for collinearity using the distance formula. An alternate form of the test, and one that is often easier to use, involves slope. We will show in the problem set that if the slope of $\overline{PQ}$ equals the slope of $\overline{QR}$, then P, Q, and R are collinear.

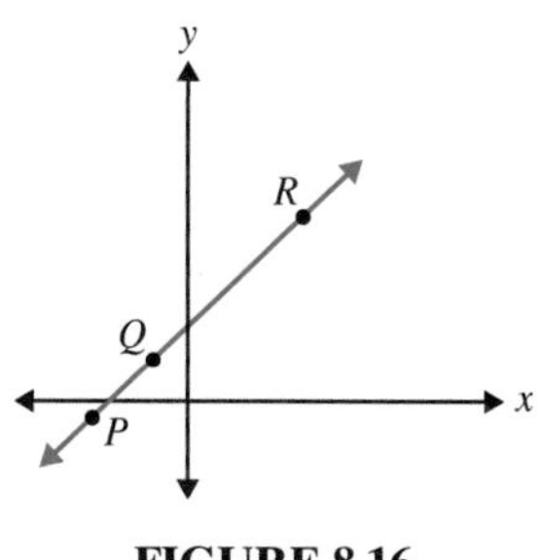

FIGURE 8.16

EXAMPLE 8.9 An experiment yielded the data points graphed in Figure 8.17. Determine whether the three points are collinear.

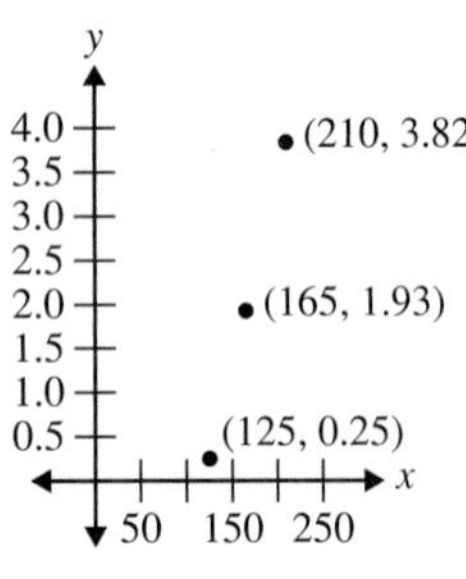

FIGURE 8.17

SOLUTION

If we label the three points $A(125, 0.25)$, $B(165, 1.93)$, and $C(210, 3.82)$, then we have

$$\text{Slope of } \overline{AB} = \frac{1.93 - 0.25}{165 - 125} = \frac{1.68}{40} = 0.042$$

$$\text{Slope of } \overline{BC} = \frac{3.82 - 1.93}{210 - 165} = \frac{1.89}{45} = 0.042$$

Because the slopes of $\overline{AB}$ and $\overline{BC}$ are the same, the three points are collinear. Notice that the two segments must share a common point, B in this case, for this test to work. ■

Slopes of Parallel and Perpendicular Lines

Because slope is a measure of the inclination of a line with respect to the horizontal, two (nonvertical) parallel lines must have the same slope. Also, vertical lines are parallel (Figure 8.18). The next theorem summarizes these relationships.

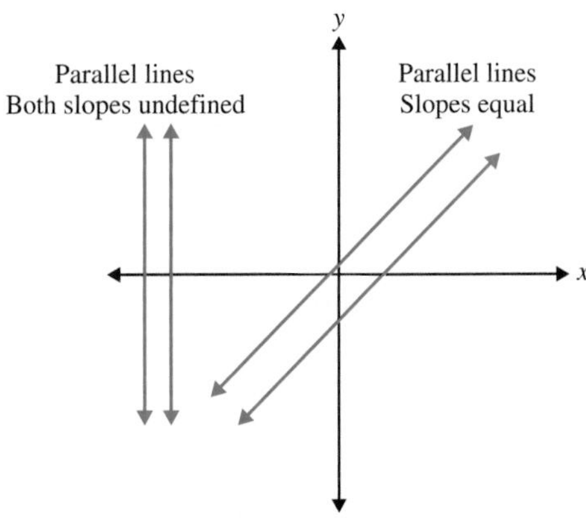

FIGURE 8.18

> **THEOREM 8.4 Slopes of Parallel Lines**
> Two lines in a coordinate plane are parallel if and only if
> **a.** their slopes are equal, or
> **b.** their slopes are undefined.

We can use slope to classify and to verify properties of particular geometric figures as the next example demonstrates.

EXAMPLE 8.10 For quadrilateral $ABCD$ with $A(-2, 5)$, $B(6, 3)$, $C(8, -2)$, and $D(0, 0)$,

a. Use slope to verify that $ABCD$ is a parallelogram.

b. Verify that the diagonals of $ABCD$ bisect each other.

SOLUTION

$ABCD$ is graphed in Figure 8.19.

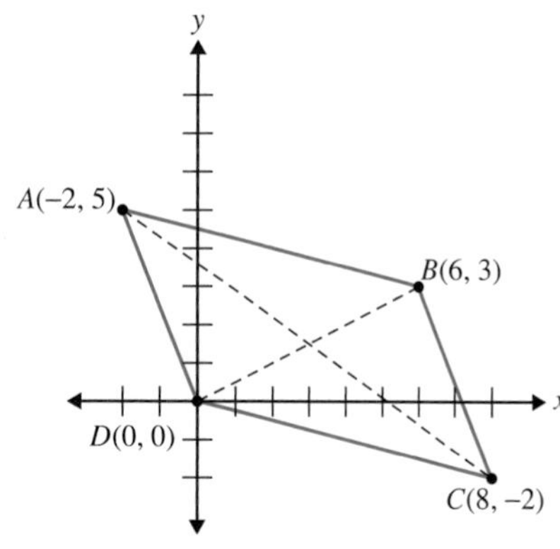

FIGURE 8.19

a. While the figure *appears* to be a parallelogram, we will verify that it is one by using slope.

$$\text{Slope of } \overline{AB} = \frac{3-5}{6+2} = \frac{-2}{8} = -\frac{-1}{4}$$

$$\text{Slope of } \overline{BC} = \frac{-2-3}{8-6} = \frac{-5}{2}$$

$$\text{Slope of } \overline{CD} = \frac{0+2}{0-8} = \frac{2}{-8} = \frac{-1}{4}$$

$$\text{Slope of } \overline{DA} = \frac{5-0}{-2-0} = \frac{5}{-2} = \frac{-5}{2}$$

Because both pairs of opposite sides are parallel, $ABCD$ is a parallelogram. (NOTE: We could also use the distance formula to verify that $ABCD$ is a parallelogram. However, the slope formula is often easier to use.)

b. To show that the diagonals bisect each other, we must show that the midpoint of diagonal $\overline{AC}$ is the same point as the midpoint of diagonal $\overline{BD}$.

$$\text{Midpoint of } \overline{AC} = \left(\frac{-2+8}{2}, \frac{5+(-2)}{2}\right) = \left(3, \frac{3}{2}\right)$$

$$\text{Midpoint of } \overline{BD} = \left(\frac{0+6}{2}, \frac{0+3}{2}\right) = \left(3, \frac{3}{2}\right)$$

Because their midpoints are the same, diagonals $\overline{AC}$ and $\overline{BD}$ must intersect at their midpoints. Therefore, the diagonals of $ABCD$ bisect each other. ■

Just as we are able to use slope to determine whether two lines are parallel, we can also use slope to determine whether two lines are perpendicular as the next example shows.

EXAMPLE 8.11 In Figure 8.20, prove that $\triangle ABC$ has a right angle at C. Then find the product of the slopes of $\overline{AC}$ and $\overline{BC}$.

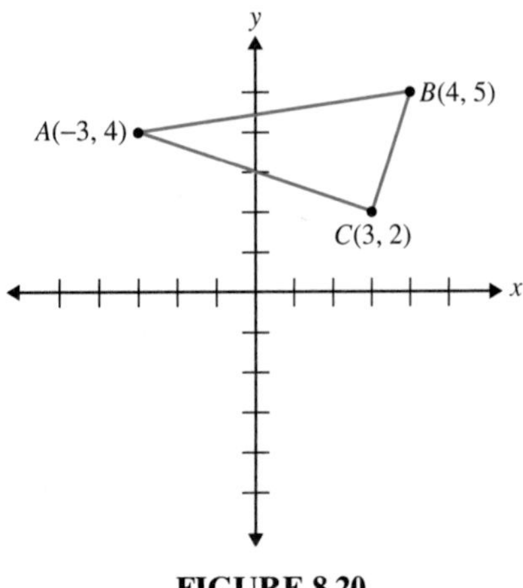

FIGURE 8.20

SOLUTION

To prove that $\triangle ABC$ is a right triangle, we will use the converse of the Pythagorean theorem (Test for Right Angles).

$AB = \sqrt{(4-(-3))^2 + (5-4)^2} = \sqrt{49+1} = \sqrt{50}$. Thus, $(AB)^2 = 50$.

$AC = \sqrt{[(-3)-3]^2 + (4-2)^2} = \sqrt{36+4} = \sqrt{40}$. Thus, $(AC)^2 = 40$.

$BC = \sqrt{(4-3)^2 + (5-2)^2} = \sqrt{1+9} = \sqrt{10}$. Thus, $(BC)^2 = 10$.

Because $(AC)^2 + (BC)^2 = 40 + 10 = 50 = (AB)^2$, we have that $\triangle ABC$ is a right triangle by the converse of the Pythagorean theorem. The slope of $\overline{AC}$ equals $\frac{4-2}{(-3)-3} = -\frac{1}{3}$ and the slope of $\overline{BC}$ is $\frac{5-2}{4-3} = 3$. Thus, the product of their slopes is -1. ■

In general, the product of the slopes of two perpendicular lines is -1, if neither line is vertical. We summarize this result in the next theorem.

> **THEOREM 8.5 Slopes of Perpendicular Lines**
> Two lines in a coordinate plane are perpendicular if and only if
> **a.** one line is horizontal and the other is vertical, or
> **b.** the product of their slopes equals -1.

Given Two perpendicular lines l_1, and l_2 through the origin, neither of which is vertical [Figure 8.21(a)].

Prove The product of their slopes is -1.

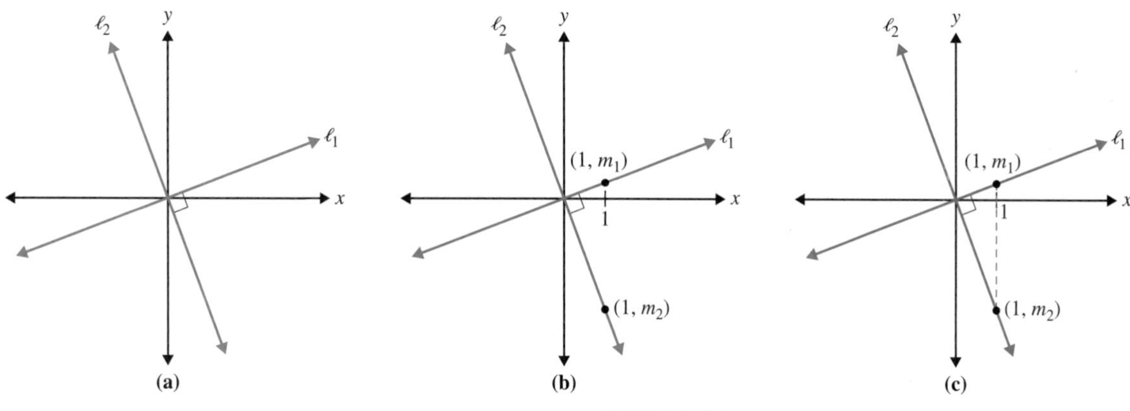

FIGURE 8.21

Proof Suppose line l_1 has slope m_1 and line l_2 has slope m_2. Then the equation of line l_1 is $y = m_1x$ and the equation of line l_2 is $y = m_2x$. For an x-coordinate of 1, line l_1 has a y-coordinate of $y = m_1(1) = m_1$ and line l_2 has a y-coordinate of $y = m_2(1) = m_2$, so the point $(1, m_1)$ is a point on l_1 and the point $(1, m_2)$ is a point on l_2 [Figure 8.21(b)]. These two points together with the origin form the vertices of a right triangle [Figure 8.21(c)]. Notice that the sides of this right triangle have lengths $\sqrt{1 + m_1^2}$ and $\sqrt{1 + m_2^2}$ and the hypotenuse has length $m_1 - m_2$. Applying the Pythagorean theorem to this triangle, we have

$$\left(\sqrt{1 + m_1^2}\right)^2 + \left(\sqrt{1 + m_2^2}\right)^2 = (m_1 - m_2)^2$$

$$1 + m_1^2 + 1 + m_2^2 = (m_1 - m_2)^2$$

$$2 + m_1^2 + m_2^2 = m_1^2 - 2m_1m_2 + m_2^2$$

$$2 = -2m_1m_2$$

$$-1 = m_1m_2$$

Therefore, the product of the slopes of l_1 and l_2 is -1, which is what we wanted to show. Although we proved this result only for the case when the two lines intersect at the origin, this result holds for any pair of perpendicular lines (neither of which is horizontal). ■

We can use slope to prove that segments are perpendicular or that geometric figures contain right angles, as the next example shows.

EXAMPLE 8.12 Prove that the diagonals of rhombus $PQRS$ in Figure 8.22 are perpendicular using a coordinate proof.

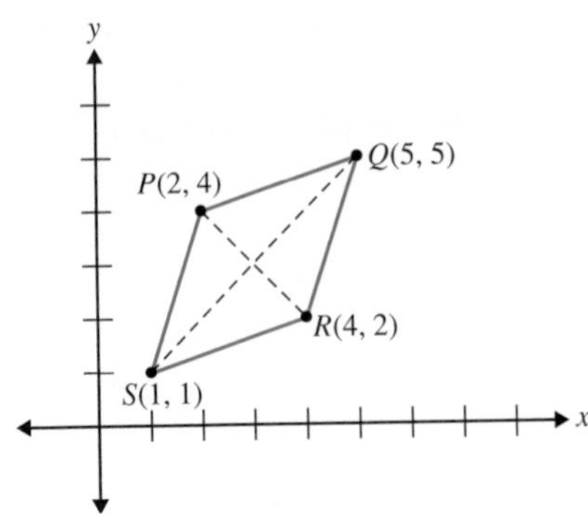

FIGURE 8.22

SOLUTION

The slope of diagonal $\overline{PR}$ is $\frac{4-2}{2-4} = -1$ and the slope of diagonal $\overline{QS}$ is $\frac{5-1}{5-1} = 1$. Since the product of these two slopes is -1, the diagonals are perpendicular. (NOTE: The result that the diagonals of *any* rhombus are perpendicular will be proved in Section 8.4) ■

Solution of Applied Problem

Using the given information, we can calculate the "rise" and the "run" in the right triangle shown.

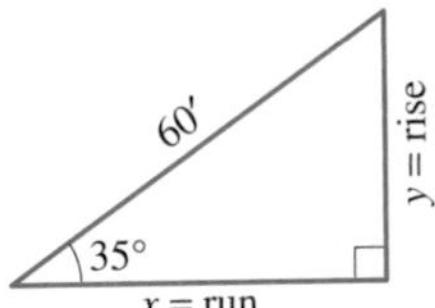

We know that $\sin 35^\circ = \frac{y}{60}$, so $y = 60 \sin 35^\circ \approx 34.4$ ft. In a similar manner, we know that $x = 60 \cos 35^\circ \approx 49.1$ ft. The slope of the conveyor is

$$m = \frac{\text{rise}}{\text{run}} = \frac{34.4 \text{ ft}}{49.1 \text{ ft}} \approx 0.70$$

So, for every one foot a bundle of shingles travels horizontally, it rises 0.7 ft. (NOTE: We can also find the slope m by using $\tan 35^\circ = 0.70$, because $\tan 35^\circ = \frac{y}{x}$.)

GEOMETRY AROUND US The steepness of a sloping roof is described by its pitch, which is similar to the slope of a line. A "1 in 2" pitch, for example, means that the roof rises one foot vertically for every two feet of horizontal distance that it runs. In other words, the roof has slope $\frac{1}{2}$. Other common roof pitches are 5 in 12 and 5 in 8.

PROBLEM SET 8.2

EXERCISES/PROBLEMS

1. Let $P(x_1, y_1) = (-4, 5)$ and $Q(x_2, y_2) = (6, -3)$. Use the definition of slope to find the slope of $\overline{PQ}$. Then let $P(x_1, y_1) = (6, -3)$ and $Q(x_2, y_2) = (-4, 5)$ and find the slope of $\overline{PQ}$. What do you notice?

2. Let $P(x_1, y_1) = (-2, -3)$ and $Q(x_2, y_2) = (3, 2)$. Use the definition of the slope to find the slope of $\overline{PQ}$. Then let $P(x_1, y_1) = (3, 2)$ and $Q(x_2, y_2) = (-2, -3)$ and find the slope of $\overline{PQ}$. What do you notice?

3. Find the slopes of the lines containing the following pairs of points.

a. $(3, 2)$ and $(5, 3)$

b. $(-2, 1)$ and $(-5, -3)$

c. $(3, -5)$ and $(-6, -5)$

d. $(4, -1)$ and $(4, 2)$

4. Find the slopes of the lines containing the following pairs of points.

a. $(9, 4)$ and $(9, 7)$

b. $(-6, -3)$ and $(-5, 8)$

c. $(3, -3)$ and $(-4, 10)$

d. $(1, -6)$ and $(7, -6)$

5. Classify the slope of each line shown as positive, negative, zero, or undefined.

a.

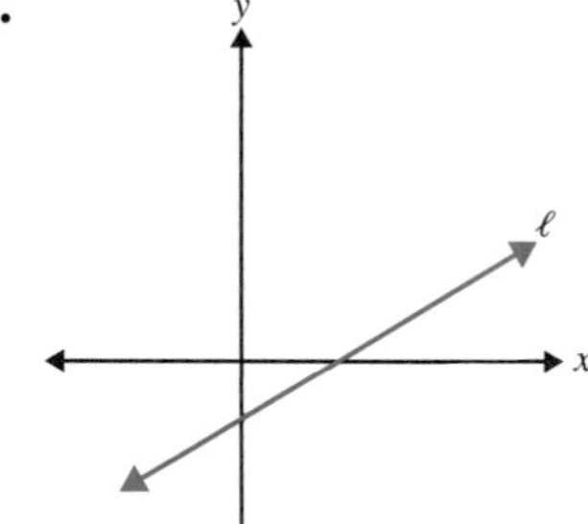

b.

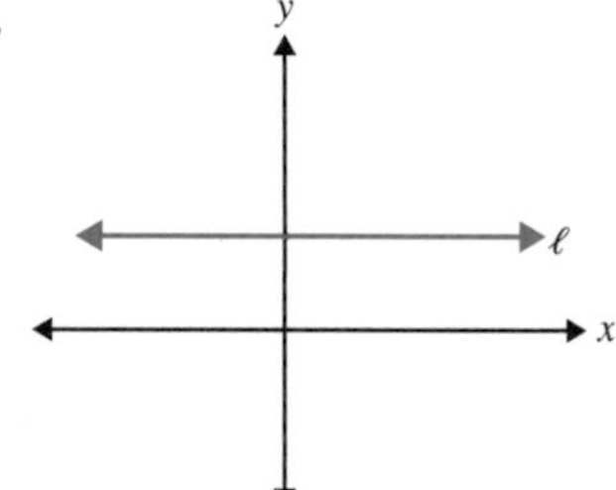

6. Classify the slope of each line shown as positive, negative, zero, or undefined.

a.

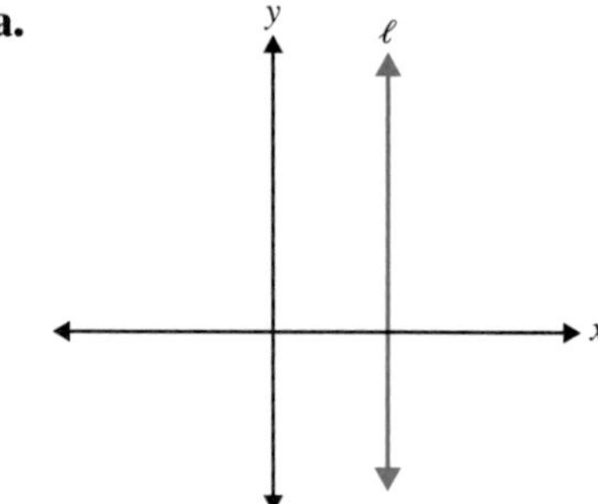

b.

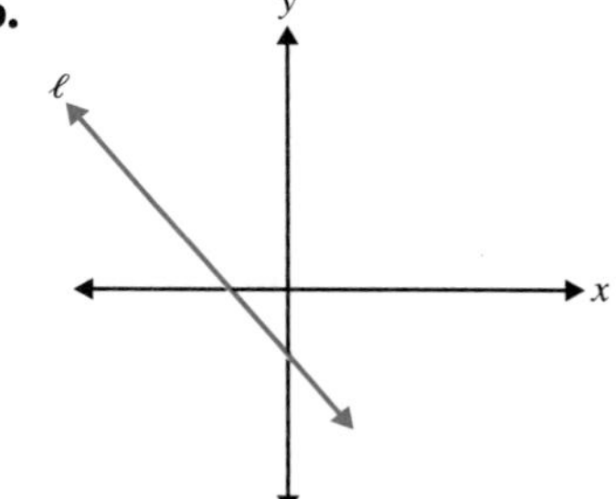

7. Given are the slopes of several lines. Indicate whether each line is horizontal or vertical, rises from left to right, or falls from left to right.

a. $-\frac{8}{9}$

b. 0

c. 13

d. undefined

8. Given are the slopes of several lines. Indicate whether each line is horizontal or vertical, rises from left to right, or falls from left to right.

a. $\frac{3}{4}$

b. undefined

c. 0

d. $-\frac{5}{6}$

9. Use slopes to determine if $A(3, -2)$, $B(1, 2)$, and $C(-3, 10)$ are collinear.

10. Use slopes to determine if $A(0, 7)$, $B(2, 11)$, and $C(-2, 1)$ are collinear.

11. Three of the four points given are collinear. Use slopes to determine which three points are collinear.

$A(-8, -11)$ $\quad B(4, -4)$ $\quad C(12, -1)$ $\quad D(-2, -8)$

12. Three of the four points given are collinear. Use slopes to determine which three points are collinear.

$A(2,3)$ $B(5,-2)$ $C(-2,-1)$ $D(6,7)$

13. Use slopes to determine if $\overline{AB} \parallel \overline{PQ}$.

a. $A(-1,0), B(4,5), P(3,9), Q(-2,4)$

b. $A(0,4), B(6,8), P(-4,-6), Q(2,-2)$

14. Determine if the following pairs of line segments are parallel.

a. The segment from $(0, 2)$ to $(1, 3)$ and the segment from $(2,0)$ to $(3,2)$

b. The segment from $(-5,3)$ to $(-1,2)$ and the segment from $(-3,1)$ to $(0,0)$

c. The segment from $(-1,1)$ to $(-1,4)$ and the segment from $(1,0)$ to $(1,4)$

d. The segment from $(-5,-3)$, to $(-3,-1)$ and the segment from $(-5,-4)$ to $(-2,-1)$

15. Give the slope of a line parallel to $\overline{PQ}$.

a. $P(3,2), Q(5,-7)$

b. $P(-3,1), Q(-1,9)$

16. Give the slope of a line parallel to $\overline{PQ}$.

a. $P(1,-5), Q(1,4)$

b. $P(-2,-3), Q(-5,-2)$

17. Use slopes to determine if the quadrilaterals with the given vertices are parallelograms.

a. $(1,4), (4,4), (5,1)$, and $(2,1)$

b. $(1,-1), (6,-1), (6,4)$, and $(1,-4)$

18. Use slopes to determine if the quadrilaterals with given vertices are parallelograms.

a. $(1,-2), (4,2), (6,2)$, and $(3,-2)$

b. $(-10,5), (-5,10), (10,-5)$, and $(5,-10)$.

19. Three vertices of a parallelogram are given. Find the coordinates of a fourth vertex.

a. $(-2,1), (-4,6), (8,1)$

b. $(-2,1), (0,3), (7,2)$

20. Three vertices of a rectangle are given. Find the coordinates of the fourth vertex.

a. $(-1,4), (3,5), (0,0)$

b. $(6,7), (9,-2), (-3,4)$

21. A quadrilateral $ABCD$ has vertices $A(-1,1), B(2,4), C(6, 1)$, and $D(3,-2)$. Use slopes to verify that $ABCD$ is a parallelogram.

22. A quadrilateral $ABCD$ has vertices $A(-3,2), B(3,4), C(6,0)$, and $D(-3,-3)$. Use slopes and the distance formula to verify that $ABCD$ is an isosceles trapezoid.

23. Give the slope of a line perpendicular to $\overline{AB}$.

a. $A(1,6), B(2,5)$

b. $A(4,2), B(-5,2)$

24. Give the slope of a line perpendicular to $\overline{AB}$.

a. $A(0,4), B(-6,-5)$

b. $A(-1,5), B(-1,3)$

25. In each part, determine which, if any, of $\overline{AB}$, $\overline{BC}$, and $\overline{CA}$ are perpendicular.

a. $A(3,3), B(3,9)$, and $C(12,9)$

b. $A(2,8), B(7,1)$, and $C(14,6)$

26. In each part, determine which, if any, of $\overline{AB}$, $\overline{BC}$, and $\overline{CA}$ are perpendicular.

a. $A(-8,-3), B(8,2)$, and $C(6,9)$

b. $A(-7,-2), B(-4,4)$, and $C(10,-3)$

27. Find a so that the slope of the line segment having endpoints $P(-11,4)$ and $Q(a,-3)$ is $-\frac{1}{3}$.

28. Find b so that the slope of the line segment having endpoints $P(5, 1)$ and $Q(-6, b)$ is perpendicular to the line segment having endpoints $(1,4)$ and $(-1,0)$.

29. One diagonal of a rhombus $ABCD$ contains the vertices $A(-4,-1)$ and $C(7,-2)$. Find the slope of the other diagonal.

30. One side of a rectangle $ABCD$ contains the vertices $A(9,-3)$ and $C(6,1)$. Find the slope of an adjacent side.

31. Use slopes to determine which, if any, of the triangles with the given vertices is a right triangle.

a. $(2,2), (8,6)$, and $(4,8)$

b. $(1,-1), (9,-4)$, and $(10,11)$

c. $(8,1), (6,3)$ and $(1,-6)$

32. Use slopes to determine which, if any, of the quadrilaterals with the given vertices is a rectangle.

a. $(0,0), (4,-2), (8,4)$, and $(6,6)$

b. $(-3,8), (0,12), (12,3)$, and $(9,-1)$

c. $(-10,-5), (-6,15), (14,11)$, and $(10,-9)$

33. Which of the following properties are true of the two kites having the given vertices?

i. $(0,0), (2,5), (0,8), (-2,5)$

ii. $(0,0), (2,3), (3,-2), (10,2)$

a. The diagonals are congruent.

b. The diagonals are perpendicular to each other.

c. The diagonals bisect each other.

d. The kite has two right angles.

34. Which of the following properties are true of the two trapezoids having the given vertices?

i. $(0,0), (16,0), (6,4), (10,4)$

ii. $(-4,0), (-4,5), (7,5), (9,0)$

a. The diagonals have the same length.

b. The diagonals are perpendicular to each other.

c. The diagonals bisect each other.

d. The trapezoid has two right angles.

PROOFS

35. Recall that a midsegment of a triangle is a segment joining the midpoints of two sides of the triangle. For $\triangle ABC$, with $A(-3,0)$, $B(0, 4)$, and $C(6, 2)$, verify that the midsegment connecting midpoints of $\overline{AB}$ and $\overline{BC}$ is parallel to and half the length of $\overline{AC}$.

36. Recall that we obtain the midquad of a quadrilateral by connecting consecutive midpoints of the sides of the quadrilateral. For quadrilateral $ABCD$, with $A(1, 1)$, $B(-2,2)$, $C(5, 5)$, and $D(7, 1)$, verify that the midquad is a parallelogram.

37. A quadrilateral $ABCD$ has vertices $A(-1,2)$, $B(3, 3)$, $C(7, 2)$, and $D(3, 1)$. Use slopes to verify that $ABCD$ is a rhombus.

38. Let P, Q, R, and S be any points on the line as shown. By drawing horizontal and vertical segments, draw right triangles $\triangle PQO$ and $\triangle RST$. Follow the given steps to prove that the slope of the line is independent of the pairs of points selected.

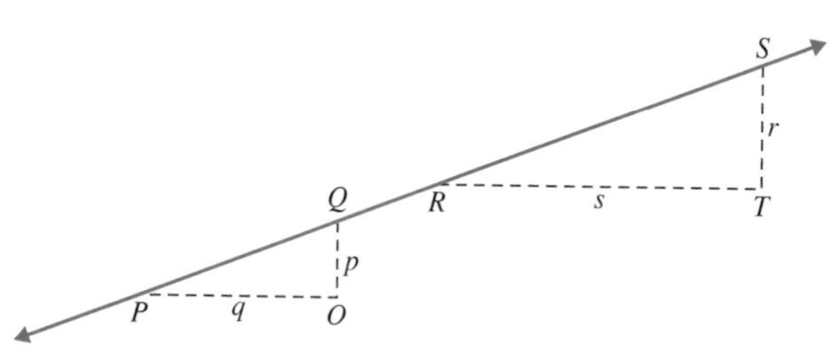

a. Show that $\triangle PQO \sim \triangle RST$.

b. Show that $\dfrac{p}{q} = \dfrac{r}{s}$.

c. Is the slope of $\overline{PQ}$ equal to the slope of $\overline{RS}$? Explain.

39. If the slope of $\overline{PQ}$ is equal to the slope of $\overline{PR}$, we can show that P, Q, and R are collinear. The following justification uses similar triangles.

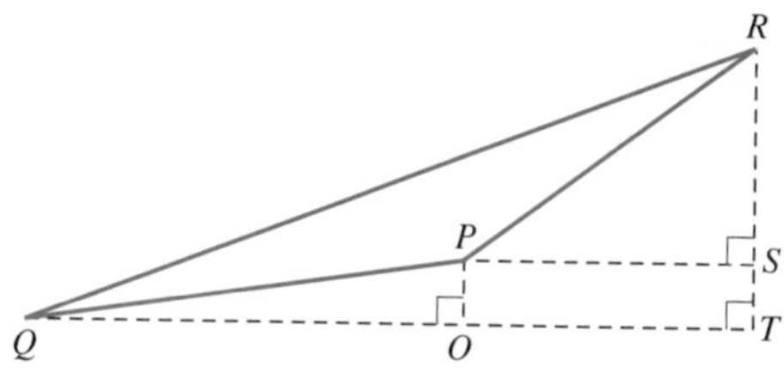

a. Horizontal lines are drawn through points P and Q and vertical lines through P and R, intersecting at points O, S, and T, as illustrated. If the slope of $\overline{PQ}$ and the slope of $\overline{PR}$ are equal, what two ratios of sides are equal?

b. Show that $\triangle QOP \sim \triangle PSR$.

c. Which sides are proportional?

d. Verify that $\dfrac{QO + PS}{PS} = \dfrac{PO + RS}{RS}$.

e. Show that $\triangle PSR \sim \triangle QTR$.

f. Which sides are proportional?

g. Verify that $QR = PR \cdot \dfrac{QT}{PS}$.

h. Verify that $QP + PR = PR \cdot \dfrac{QT}{PS}$.

i. Verify that points P, Q, and R are collinear.

40. Complete the proof of Theorem 8.5 by showing the following: If the product of the slopes of l and m is -1, then l and m are perpendicular. (HINT: Show that $\triangle OPQ$ is a right triangle.)

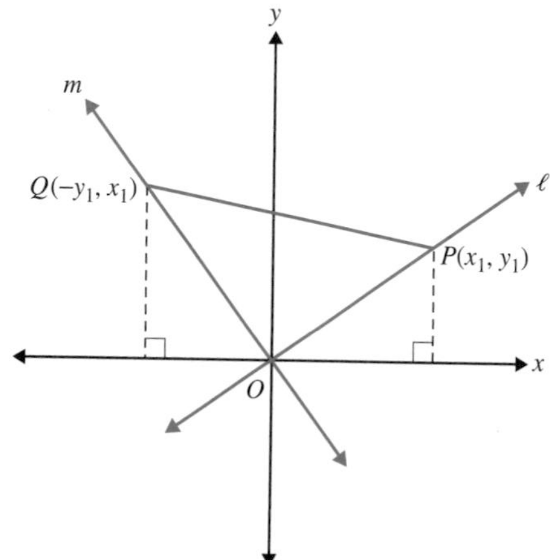

APPLICATIONS

41. Gutters on the eaves of a home transfer water runoff from the roof to the downspouts. Properly sloped gutters drop $\frac{1}{16}$ inch for every 1 foot in length.

a. Determine the vertical drop, in inches, for a gutter length of 25 feet. Round to the nearest hundredth.

b. If a certain length of gutter drops $\frac{5}{8}$ inch, find the length of the gutter in feet.

42. What is the slope of the roof shown? What is its pitch? (HINT: See the "Geometry Around Us" in this section.)

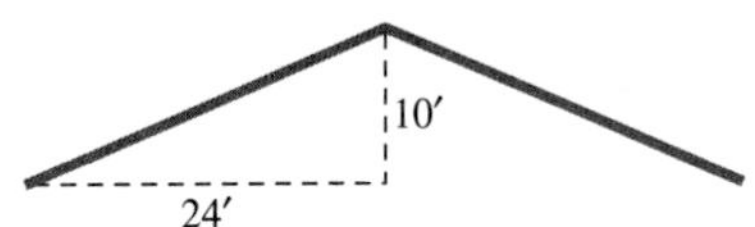

43. A road rises 11.2 feet in a horizontal distance of 140 feet. What is the slope of the road? What is its angle of inclination (the angle formed with the horizontal)?

44. The **percent grade** of a highway is the amount that the highway rises (or falls) in a given horizontal distance. For example, a highway with a four percent grade rises 0.04 mile for every 1 mile of horizontal distance.

a. How many feet does a highway with a six percent grade rise in 2.5 miles of horizontal distance?

b. How many feet would a highway with a six percent grade rise in 90 miles of horizontal distance if the grade remained constant?

c. How is percent grade related to slope?

45. A freeway ramp connects a highway to an overpass 10 meters above the ground. The ramp, which starts 150 meters from the overpass, is 150.33 meters long and 9 meters wide. What is the percent grade of the ramp to three decimal places?

46. A tennis court should be sloped side-to-side 1 inch in 10 feet if the court is made of asphalt or concrete. Gravity moves water from the court to a drainage system located along the side of the court. The drainage system should have a percent grade of one percent.

a. Give the percent grade for a tennis court made of concrete, rounded to the nearest hundredth, and determine the difference in height at the sides of a 36-foot wide court.

b. If the drainage system is located along a 120-foot fenceline, find the difference in the heights of the drainpipe at each end of the fenceline.

EXTENDED PROBLEMS

47. In Problem 44, we defined the percent grade of a highway. Research percent grade and how the Federal Highway Administration uses it. Is there a maximum percent grade for interstate highways? Where in the United States is the street, road, or highway with the largest percent grade, and what is the grade? List the locations and percent grades for the top five steepest streets. Create a scale drawing showing the slope of each street you listed.

48. In this section, we have determined the slope of a line using two points on the line. A primary focus of differential calculus is determining the slope of a curve at a point, or finding the derivative of the curve at a point. The derivative at a point is the slope of the line tangent to the curve at that point.

a. Research the derivative of a function. Explain how we can find the slope of a tangent line to a curve, when only one point is known, by using the slopes of secant lines. Demonstrate this method by constructing a circle on a coordinate system and a secant line through two points P and Q on the circle. Show how to use the slopes of several secant lines to estimate the slope of the tangent line at P.

b. Give examples of how the derivative is used in real life and create a timeline of the history and major advances in calculus.

49. It is dangerous and often impossible to measure the peak discharge of a river during a flood, so for many years the U.S. Geological Survey has estimated the peak discharge of a stream indirectly using the **slope-area method** developed by Robert Manning in the late 19th century. This method uses the slope of the river channel and the cross-sectional area at a point along the river. Research the slope-area method and define the terms reach length, fall, flow area, and average Manning. Explain how they are estimated. Describe the two different slopes used in this method. For what type of stream is this method suitable?

8.3 EQUATIONS OF LINES AND CIRCLES

Applied Problem

We can place a triangular metal plate of uniform thickness and density on a pair of coordinate axes so that it has vertices $A(0, 0)$, $B(2, 4)$, and $C(8, 0)$. Find the center of mass of the plate.

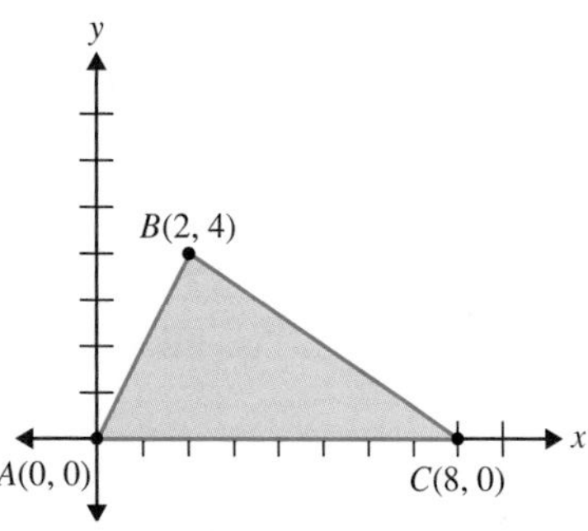

Equations of Lines

We can use the slope of a line to derive its equation as illustrated in the next example.

EXAMPLE 8.13 Given points $(-4, 2)$ and $(2, 5)$ on line l (Figure 8.23), find an equation that describes all of the points on l.

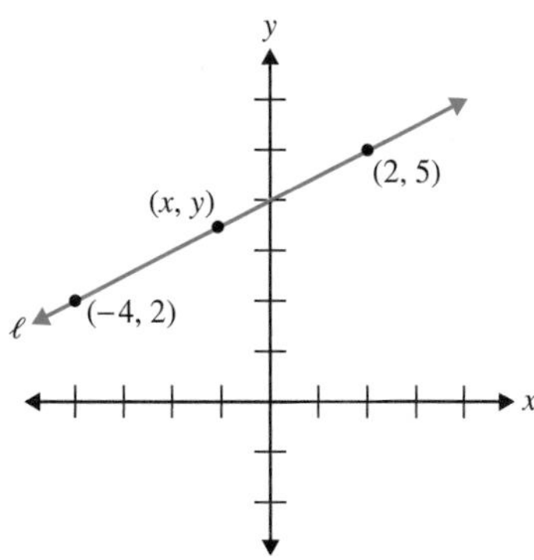

FIGURE 8.23

SOLUTION

Because $(-4, 2)$ and $(2, 5)$ are points on l, the slope of l is $\frac{5 - 2}{2 - (-4)} = \frac{3}{6} = \frac{1}{2}$. Let (x, y) represent any point on l (Figure 8.23). Because points $(-4, 2)$, $(2, 5)$, and (x, y) are collinear, the slope determined by *any* pair of them must be $\frac{1}{2}$. So for points $(-4, 2)$ and (x, y), we have the slope of l is $\frac{y - 2}{x - (-4)}$, which, in turn, must be equal to $\frac{1}{2}$. Thus, we have $\frac{y - 2}{x - (-4)} = \frac{1}{2}$, or

$$\textbf{(i)}\quad y - 2 = \frac{1}{2}[x - (-4)]$$

The previous equation is usually simplified and written in one of two other equivalent forms:

(ii) $y = \frac{1}{2}x + 4$
(iii) $x - 2y = -8$ ■

Equation (i), $y - 2 = \frac{1}{2}[x - (-4)]$, displays the slope, $\frac{1}{2}$ and the coordinates of one of the points, $(-4, 2)$, that we used to derive this equation. For this reason, equation (i) is called the **point-slope form** of line l. Equation (ii), $y = \frac{1}{2}x + 4$, displays the slope as the coefficient of x. If we substitute 0 for x in equation (ii), we obtain the equation $y = 4$. The point $(0, 4)$ is where line l intersects the y-axis. For this reason, $(0, 4)$, or simply 4, is called the y-intercept of this equation. In general, the **y-intercept** of a line is the point where the line intersects the y-axis. Because it shows the slope and the y-intercept of the line, equation (ii) is called the **slope-intercept form** of line l. Equation (iii) is called the **standard form** of line l.

THEOREM 8.6 Forms of Equations of Lines

Every nonvertical line in the plane can be expressed in either of the following two forms where m represents the slope of the line, b is the y-intercept of the line, and (x_1, y_1) is a point on the line.

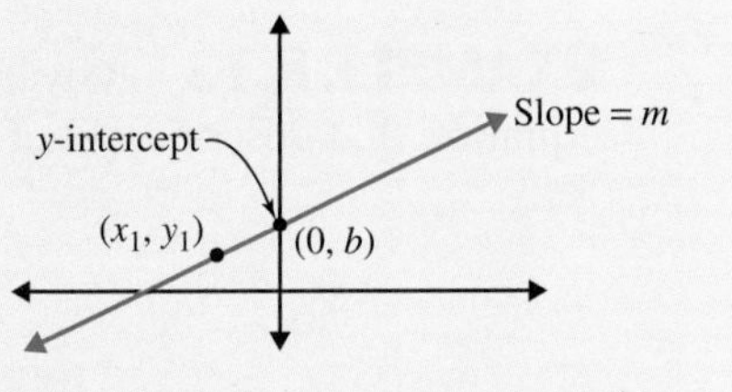

Point-slope form: $y - y_1 = m(x - x_1)$

Slope-intercept form: $y = mx + b$

EXAMPLE 8.14 Find the equation of the perpendicular bisector of $\overline{AB}$, for $A(-5, -2)$ and $B(1, 7)$ [Figure 8.24(a)].

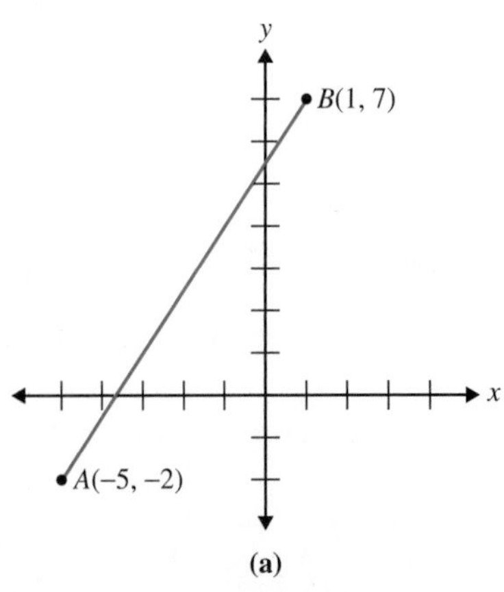

FIGURE 8.24(a)

SOLUTION

We will demonstrate two different approaches to solving this problem.

Method 1 First we find the midpoint of $\overline{AB}$ and the slope of a line perpendicular to $\overline{AB}$. The midpoint, M, of $\overline{AB}$ is $\left(\frac{-5+1}{2}, \frac{-2+7}{2}\right) = \left(-2, \frac{5}{2}\right)$ [Figure 8.24(b)]. The slope of $\overline{AB}$ is $\frac{7-(-2)}{1-(-5)} = \frac{9}{6} = \frac{3}{2}$, so the slope of any line perpendicular to $\overline{AB}$ is $-\frac{2}{3}$. We want the equation of the line through the point $\left(-2, \frac{5}{2}\right)$ having slope $-\frac{2}{3}$. Using the point-slope form, we obtain $y - \frac{5}{2} = -\frac{2}{3}(x+2)$, which simplifies to $y - \frac{5}{2} = -\frac{2}{3}x - \frac{4}{3}$, or $y = -\frac{2}{3}x + \frac{7}{6}$.

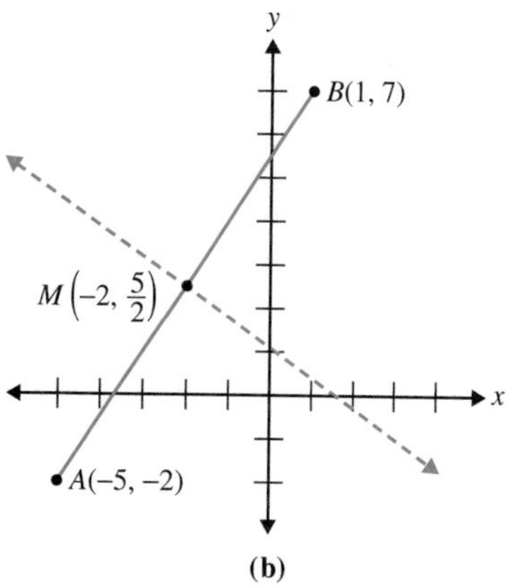

FIGURE 8.24(b)

Method 2 Next we use the distance formula and a property of points on the perpendicular bisector of a segment. We know that any point on the perpendicular bisector of a segment is equidistant from the endpoints of the segment. Let $P(x, y)$ be any point on the perpendicular bisector of $\overline{AB}$ [Figure 8.24(c)].

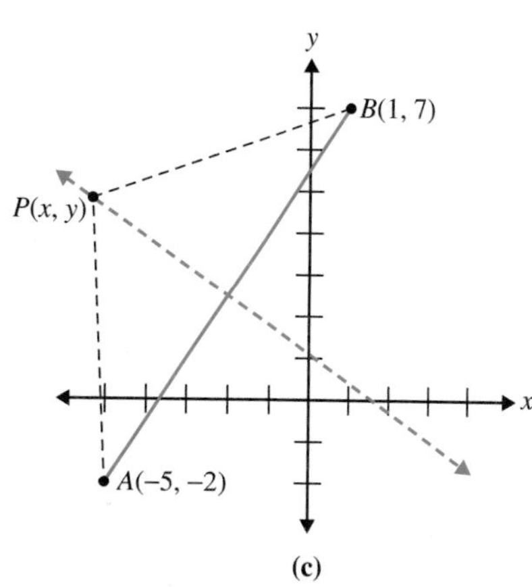

FIGURE 8.24(c)

Thus, $PA = PB$. Using the distance formula, we find that

$$\sqrt{(x+5)^2 + (y+2)^2} = \sqrt{(x-1)^2 + (y-7)^2}.$$

Squaring both sides of this equation and simplifying, we have

$$\begin{aligned} x^2 + 10x + 25 + y^2 + 4y + 4 &= x^2 - 2x + 1 + y^2 - 14y + 49 \\ 10x + 4y + 29 &= -2x - 14y + 50 \\ 12x + 18y &= 21 \\ 4x + 6y &= 7 \end{aligned}$$

This equation is equivalent to the one obtained using Method 1. Here the equation is in standard form whereas the one obtained by Method 1 is written in slope-intercept form. ■

Systems of Equations

Two or more equations, called a **system of equations**, are said to have a **simultaneous solution** if there is an ordered pair that is a solution of both the equations. For example, since the ordered pair (2, 3) is a solution of $3x + 4y = 18$ *and* $2x - y = 1$, it is a *simultaneous* solution of the equations. Using coordinate geometry, we can determine how many simultaneous solutions equations have, because these solutions will be points common to each of their graphs. The following example shows how to find the simultaneous solutions of systems of linear equations, that is, equations of lines.

EXAMPLE 8.15 Determine the number of simultaneous solutions of the following systems of linear equations.

a. $y = 3x - 2$ and $y = 8 - x$

b. $y = 2x + 7$ and $y = 2x - 3$

c. $y = x - 5$ and $2x = 2y + 10$

SOLUTION

a. The graph of $y = 3x - 2$ has slope 3 and the graph of $y = 8 - x$ has slope -1. Because their slopes are different, their graphs must intersect in a single point. Therefore, they have one simultaneous solution [Figure 8.25(a)].

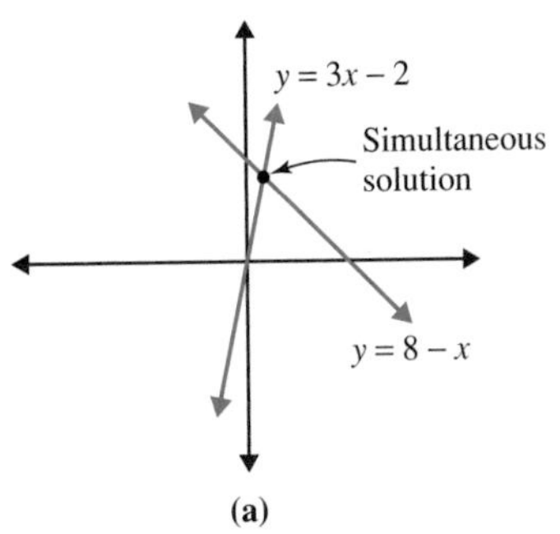

FIGURE 8.25(a)

b. The graphs of $y = 2x + 7$ and $y = 2x - 3$ have the same slope, namely 2, but *different* y-intercepts. Thus, they are parallel lines with no points in common, so the equations have no common solution [Figure 8.25(b)].

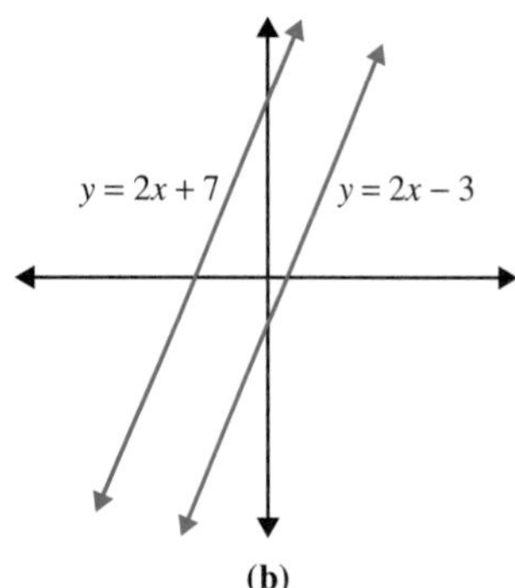

(b)

FIGURE 8.25(b)

c. The equations $y = x - 5$ and $2x = 2y + 10$ are two forms of the same equation. Therefore, their graphs are the same. Because every ordered pair that is a solution of one is also a solution of the other, every ordered pair on the graph is a simultaneous solution of both equations. So the equations have infinitely many solutions in common. ■

We summarize Example 8.15 in the following theorem.

THEOREM 8.7 Solutions of a System of Two Linear Equations

Two linear equations have 0, 1, or infinitely many simultaneous solutions if and only if the lines they represent are parallel, intersect in one point, or coincide, respectively.

EXAMPLE 8.16 Find the point M where the diagonals of $ABCD$ intersect (Figure 8.26).

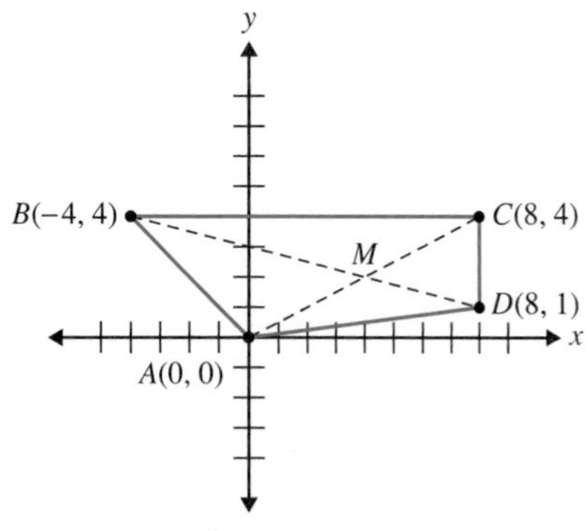

FIGURE 8.26

SOLUTION

We must find equations for the lines containing $\overline{AC}$ and $\overline{BD}$ and then solve a system of equations. The line containing $\overline{AC}$ has slope $m = \dfrac{4 - 0}{8 - 0} = \dfrac{1}{2}$. The y-intercept of this line is 0, so the equation of the line is $y = \dfrac{1}{2}x$.

The line containing $\overline{BD}$ has slope $m = \frac{1-4}{8-(-4)} = \frac{-3}{12} = -\frac{1}{4}$. Substituting this slope and the point $D(8, 1)$ into the point-slope form, we have

$$y - 1 = -\frac{1}{4}(x - 8)$$
$$y - 1 = -\frac{1}{4}x + 2$$
$$y = -\frac{1}{4}x + 3$$

The point of intersection satisfies both of these equations, so, it is the point of intersection of $y = \frac{1}{2}x$ and $y = -\frac{1}{4}x + 3$. Substituting $\frac{1}{2}x$ for y in the second equation, we obtain the following:

$$\frac{1}{2}x = -\frac{1}{4}x + 3$$
$$2x = -1x + 12 \qquad \text{Multiplying both sides by 4}$$
$$3x = 12$$
$$x = 4$$

Thus, one coordinate of the point of intersection is $x = 4$. Because $y = \frac{1}{2}x$ at the point of intersection, we know that $y = \frac{1}{2}(4) = 2$. Therefore, the point M at which the diagonals intersect is $(4, 2)$, which is consistent with the graph shown in Figure 8.26. ■

The method we used to solve the system of equations in Example 8.16 is called the **substitution method** because we solve for a variable in one equation and then substitute the value of that variable in the other equation. We will discuss two other methods for solving a system of equations in the problem set.

Equation of a Circle

We have seen in Example 8.14 how we can apply the distance formula to derive an equation of the perpendicular bisector of a line segment. We can also derive the equation of a circle using the distance formula.

EXAMPLE 8.17 Find the equation that describes all points of the plane at a distance of 5 from the point $(1, 2)$.

SOLUTION

The figure described is the circle with center $(1, 2)$ and radius 5 (Figure 8.27). If (x, y) is any point on the circle, then by the distance formula, we have $5 = \sqrt{(x-1)^2 + (y-2)^2}$. Squaring both sides of this equation, we obtain the equation $(x-1)^2 + (y-2)^2 = 25$, which is the equation of the circle. ■

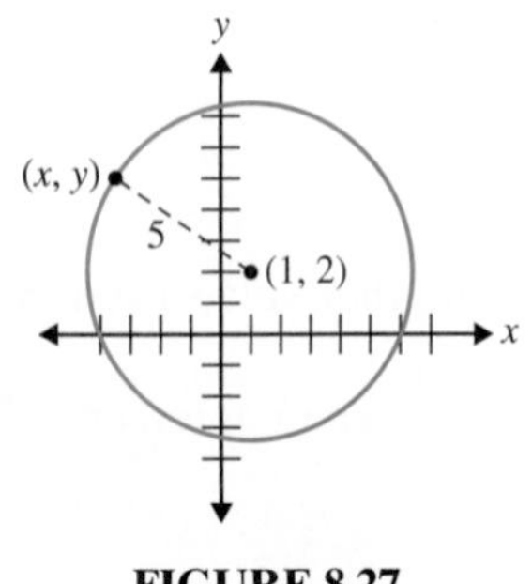

FIGURE 8.27

The form of the equation for a circle shown in Example 8.17 is called the **standard form** of the equation of a circle. We can modify this example to derive the equation of any arbitrary circle whose center is (h, k) and whose radius is r. We state this result next and leave its proof for the problem set.

THEOREM 8.8 Equation of a Circle
The circle with center (h, k) and radius r has the equation

$$(x - h)^2 + (y - k)^2 = r^2$$

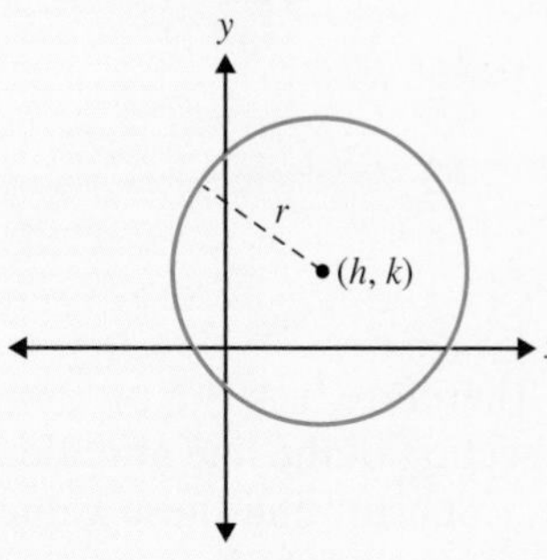

The next example shows how we can use the equation of a circle in conjunction with the equation of a line to find their points of intersection.

EXAMPLE 8.18 Determine where the circle with center $(-5, 3)$ and radius 6 intersects the line whose slope is 1 and whose y-intercept is 2.

SOLUTION

First we find the equations of the circle and the line. Their graphs are sketched in Figure 8.28. We see that there are two points of intersection.

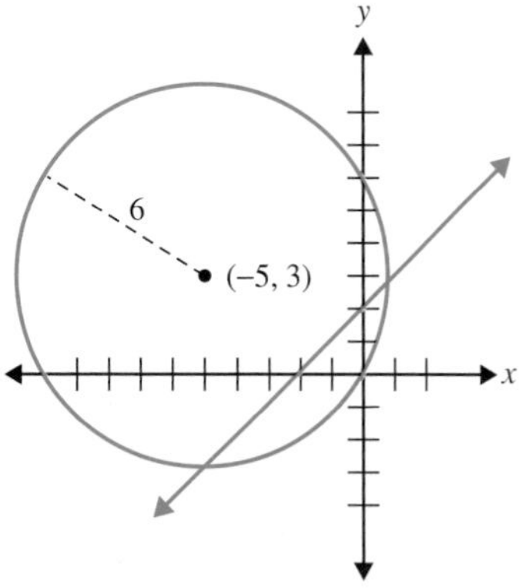

FIGURE 8.28

$$\text{Equation of the circle: } [x - (-5)]^2 + (y - 3)^2 = 6^2$$

$$\text{Equation of the line: } y = x + 2$$

To find the points of intersection, we must find the simultaneous solution of the system of equations $(x + 5)^2 + (y - 3)^2 = 36$ and $y = x + 2$.

We can substitute $x + 2$ for y in the equation of the circle and simplify the resulting equation.

$$\begin{aligned}
(x + 5)^2 + [(x + 2) - 3]^2 &= 36 \\
(x + 5)^2 + (x - 1)^2 &= 36 \\
x^2 + 10x + 25 + x^2 - 2x + 1 &= 36 \\
2x^2 + 8x + 26 &= 36 \\
2x^2 + 8x - 10 &= 0 \\
x^2 + 4x - 5 &= 0 \\
(x + 5)(x - 1) &= 0
\end{aligned}$$

Therefore, $x = -5$ or $x = 1$. These are the x-coordinates of the two points of intersection of the line and the circle.

Substituting these values for x in the equation of the line, we obtain the corresponding values of $y: y = (-5) + 2 = -3$ and $y = 1 + 2 = 3$. Thus, the points of intersection are $(-5, -3)$ and $(1, 3)$. ■

Solution of Applied Problem

The center of mass, or center of gravity, of the triangle is its "balance point." It is located at the point of intersection of the medians of the triangle, the centroid.

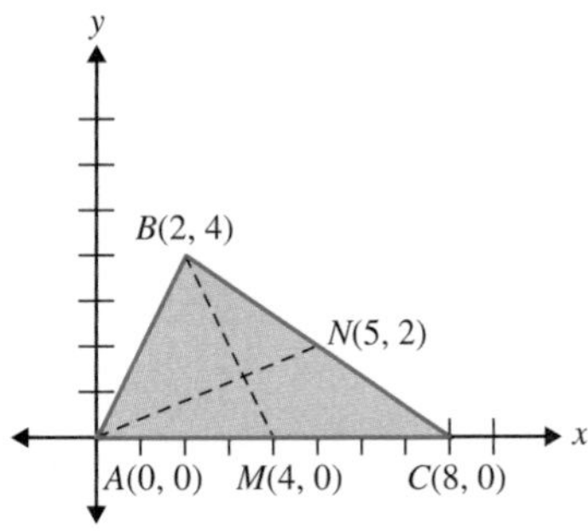

So we must find equations for two of the medians of $\triangle ABC$ and determine their simultaneous solution. The midpoint of $\overline{AC}$ is $M(4, 0)$ and the midpoint of $\overline{BC}$ is $N(5, 2)$. The slope of $\overline{MB}$ is $\frac{4 - 0}{2 - 4} = -2$. Thus, an equation of the line containing $\overline{MB}$ is $y - 0 = -2(x - 4)$, or $y = -2x + 8$. In a similar manner, we find that the equation of the line containing $\overline{AN}$ is $y = \frac{2}{5}x$. The center of mass is the simultaneous solution of $y = -2x + 8$ and $y = \frac{2}{5}x$. Solving this system of equations, we get $x = \frac{10}{3}$ and $y = \frac{4}{3}$.

GEOMETRY AROUND US In business, a cost function and a revenue function are often plotted on the same graph. The cost function gives the total cost to produce x items, and the revenue function gives the total revenue from the sale of x items. The point at which these two graphs intersect is significant—it is called the **break-even point**. When the graph of revenue lies above the graph of cost, the company makes a profit. If the revenue graph lies below the cost graph, the company loses money.

PROBLEM SET 8.3

EXERCISES/PROBLEMS

1. Show that point P lies on the line with the given equation.
   a. $y = 7x - 2$; $P(1, 5)$
   b. $-2x = 6y + 3$; $P(7.5, -3)$
2. Show that point P lies on the line with the given equation.
   a. $y = -2x + 5$; $P(-2, 9)$
   b. $3y = 4x + 2$; $P(4, 6)$
3. For each of the following equations, find three points whose coordinates satisfy the equation.
   a. $x - y = 4$ b. $2x - 3y = 6$
4. For each of the following equations, find three points whose coordinates satisfy the equation.
   a. $x = 3$ b. $x + 4y = 0$
5. Write each equation in the form $y = mx + b$. Identify the slope and y-intercept of the line.
   a. $2 + y = -3x$
   b. $-2x + y = -5$
   c. $2y = 5x + 6$
   d. $2x - 7y = 8$
6. Write each equation in the form $y = mx + b$. Identify the slope and y-intercept of the line.
   a. $2y = 6x + 12$
   b. $4y + 3x = 0$
   c. $8y - 2 = 0$
   d. $3x - 4y = 12$
7. Write the equation of the line, given its slope m and y-intercept b.
   a. $m = 3, b = 7$ b. $m = -1, b = -3$
8. Write an equation of the line, given its slope m and y-intercept b.
   a. $m = 2, b = 5$ b. $m = -3, b = -\frac{1}{4}$
9. Write the equation in slope-intercept form for the line containing these pairs of points.
   a. $(6, 3), (0, 2)$
   b. $(-4, 8), (3, -6)$
10. Write the equation in slope-intercept form for the line containing these pairs of points.
    a. $(-5, 2), (-3, 1)$
    b. $(4, 7), (10, 7)$
11. Tell whether the line described by each equation rises to the right, falls to the right, is vertical, or is horizontal.
    a. $y = 5x$
    b. $y = -\frac{1}{3}x - 1$
    c. $y = -10$
12. Tell whether the line described by each equation rises to the right, falls to the right, is vertical, or is horizontal.
    a. $x = -10$
    b. $y + x = 0$
    c. $2y - 3x = 0$
13. Write equations of the line that satisfies each of the following conditions.
    a. Vertical through $(0, 0)$
    b. Horizontal through $(2, 5)$
    c. Vertical through $(-9, 3)$
    d. Horizontal through $(-10, 7)$

14. Write an equation of the line that satisfies each of the following conditions.

a. Vertical through $(1, 3)$

b. Vertical through $(-5, -2)$

c. Horizontal through $(-3, 6)$

d. Horizontal through $(3, -3)$

15. Graph the line $3y = 4x + 3$, using the following steps.

a. Rewrite the equation in the $y = mx + b$ form.

b. What is its y-intercept? Plot the y-intercept.

c. What is its slope? Use the slope to find another point on the line and then graph the line.

16. Graph the line $-6 + 3y = 5x$, using the following steps.

a. Rewrite the equation in $y = mx + b$ form.

b. What is the y-intercept of the line? Plot the y-intercept.

c. What is the slope of the line? Use the slope to find another point on the line and then graph the line.

17. Graph the line described by each equation.

a. $y = 2x - 1$ **b.** $x = -2$ **c.** $2x = -6y$

18. Graph the line described by each equation.

a. $y = -3x + 2$ **b.** $y = 2x + 1$ **c.** $y = 3$

19. a. Graph and label the following lines on the same pair of coordinate axes.

i. $y = 2x$ **ii.** $y = 2x + 3$ **iii.** $y = 2x - 5$

b. What is the relationship among these lines?

c. Describe all lines of the form $y = cx + d$, where c is a fixed real number and d can be any real number.

20. a. Graph and label the following lines on the same pair of coordinate axes.

i. $y = 2x + 3$ **ii.** $y = -3x + 3$

iii. $y = \frac{1}{2}x + 3$

b. What is the relationship between these lines?

c. Describe all lines of the form $y = cx + d$, where d is a fixed real number and c can be any real number.

21. Write the equation of the line that passes through the given point and is parallel to the line whose equation is given.

a. $(5, -1)$; $y = 2x - 3$

b. $(1, -2)$; $y = -x - 2$

c. $(2, -5)$; $3x + 5y = 1$

d. $(-1, 0)$; $3x + 2y = 6$

22. Write the equation of the line that passes through the given point and is perpendicular to the line whose equation is given.

a. $(-2, 5)$; $y = -2x + 1$

b. $(6, 0)$; $y = 3x - 1$

c. $(1, -3)$; $2x + 4y = 6$

d. $(-5, -4)$; $3x - 2y = 8$

23. Find the equation of the perpendicular bisector of the segment whose endpoints are $(-3, -1)$ and $(6, 2)$.

24. Find equations of the perpendicular bisectors of the sides of the triangle whose vertices are $(0, 0)$, $(4, 0)$, and $(0, 3)$.

25. Find the equation of the line containing the median $\overline{AD}$ of $\triangle ABC$ with vertices $A(3, 7)$, $B(11, 2)$, and $C(1, 4)$.

26. Find the equation of the line containing altitude $\overline{PT}$ of $\triangle PRS$ with vertices $P(3, 5)$, $R(-1, 1)$, and $S(7, -3)$.

27. Use the substitution method to find the simultaneous solution of each system of linear equations, if a solution exists. If no solution exists, explain why not.

a. $y = 2x - 4$
$y = -5x + 17$

b. $4x + y = 8$
$5x + 3y = 3$

c. $x - 2y = 1$
$x = 2y + 3$

d. $x - y = 5$
$2x - 4y = 7$

28. Use the substitution method to find the simultaneous solution of each system of linear equations, if a solution exists. If no solution exists, explain why not.

a. $y = x - 1$
$y = -2x + 8$

b. $y = \frac{1}{2}x - 1$
$y = \frac{2}{3}x + \frac{5}{3}$

c. $2x - 4y = 7$
$5x + y = 1$

d. $y = \frac{2}{3}x + 2$
$3x - 9y = 14$

29. One method of solving a system of linear equations is called the **graphical method**. The lines are graphed and, if they intersect, the coordinates (or approximate coordinates) of the intersection point are determined. Graph the following pairs of linear equations to determine their simultaneous solutions, if any exist.

a. $y = 2x + 1$
$y = 2x + 4$

b. $x + 2y = 4$
$x + 2y = -2$

c. $-2x + 3y = 9$
$x + y = -2$

d. $-2x + 3y = 9$
$4x - 6y = -18$

30. Use the graphical method to find the simultaneous solution of each system of linear equations, if a solution exists.

a. $y = x + 1$
$y = 3x - 1$

b. $x - y = -8$
$2x + y = 2$

c. $y = -\frac{1}{2}x + 2$
$x + 2y = 7$

d. $4x - 6y = 19$
$-x + 3y = -7$

31. Another algebraic method of solving systems of equations is called the **elimination method** and involves eliminating one variable by adding or subtracting equivalent expressions. Consider the system

$$2x + y = 7$$
$$3x - y = 3$$

a. Add the left-hand sides of the equations and the right-hand sides. Because the original expressions were equal, the resulting sums are also equal. Notice that the variable y is eliminated.

b. Solve the equation you obtained in part (a).

c. Substitute this value of the variable into one of the original equations to find the value of the other variable.

32. Use the elimination method to solve the following systems of equations. (HINT: You may have to multiply one equation by some number to facilitate eliminating one of the variables.)

a. $5x + 3y = 17$
$2x - 3y = -10$

b. $4x - 4y = -3$
$7x + 2y = 6$

33. Rectangle $PQRS$ has vertices $P(-2, -1)$, $Q(-2, 4)$, $R(8, 4)$, and $S(8, -1)$.

a. Find equations of the lines containing the diagonals of the rectangle.

b. Use the substitution method to find the coordinates of the point of intersection of the diagonals.

34. Kite $ABCD$ has vertices $A(-5, -5)$, $B(-4, 3)$, $C(1, 4)$, and $D(2, -1)$.

a. Find equations of the lines containing the diagonals of the kite.

b. Use the substitution method to find the coordinates of the point of intersection of the diagonals.

35. Show that point P lies on the circle with the given equation.

a. $P(3, 4)$; $x^2 + y^2 = 25$

b. $P(-3, 5)$; $x^2 + y^2 = 34$

36. Show that point P lies on the circle with the given equation.

a. $P(-3, 7)$; $(x + 1)^2 + (y - 2)^2 = 29$

b. $P(\sqrt{5}, -3)$; $x^2 + (y + 5)^2 = 9$

37. Identify the center and radius of the circles whose equations are given.

a. $(x + 7)^2 + (y - 1)^2 = 121$

b. $(x - 5)^2 + (y - 8)^2 = 27$

38. Identify the center and radius of each circle whose equation is given.

a. $(x - 2)^2 + (y + 5)^2 = 64$

b. $(x + 3)^2 + (y - 4)^2 = 20$

39. Write the equation of the circle given by each of the following descriptions.

a. Center $(-1, -2)$ and radius $\sqrt{5}$

b. Center $(2, -4)$ and passing through $(-2, 1)$

c. Endpoints of a diameter at $(-1, 6)$ and $(3, -2)$

40. Write an equation of the circle satisfying each of the following conditions.

a. Center $(3, -4)$ and passing through $(2, -6)$

b. Center $(-2, 5)$ and radius 7

c. Endpoints of a diameter at $(-4, 5)$ and $(12, 7)$

41. Find the coordinates of the centroid of the triangle with vertices $A(-5, -1)$, $B(3, 3)$, and $C(5, -5)$.

42. Find the coordinates of the orthocenter of the triangle with vertices $A(-2, 0)$, $B(-1, 6)$, and $C(5, 0)$.

43. Find the coordinates of the circumcenter of the triangle with vertices $A(-5, -1)$, $B(3, 3)$, and $C(5, -5)$.

44. Find the equation of the circumscribed circle for the triangle whose vertices are $J(4, 5)$, $K(8, -3)$, and $L(-4, -3)$.

45. a. Use the graphical method to predict solutions to the system of equations given. How many solutions do you expect?

$$x^2 + y^2 = 1 \quad \text{and} \quad y = \frac{x}{2} + 1$$

b. Use the substitution method to solve the simultaneous equations.

46. Solve the following pairs of simultaneous equations.

a. $x^2 + y^2 = 4$
$(x - 2)^2 + (y - 2)^2 = 4$

b. $x^2 + y^2 = 4$
$y + x = 0$

47. Find the area of a triangle formed by the x-axis, y-axis, and the line $2x + 3y = 6$.

48. Find the area of the triangle formed by the x-axis, the line $y = 2x$, and the line $8x + y = 40$.

PROOFS

49. Prove Theorem 8.8: The circle with center (h, k) and radius r has the equation $(x - h)^2 + (y - k)^2 = r^2$.

50. We explored Napoleon's theorem in the "Geometry Investigation" for Chapter 7. That theorem involved constructing equilateral triangles on each side of a triangle, $\triangle ABC$. For the special case in which the vertices of $\triangle ABC$ are $A(0, 0)$, $B(0, 4)$, and $C(4\sqrt{3}, 0)$ and form a 30°-60°-90° triangle, prove Napoleon's theorem. That is, if equilateral triangles are constructed on each side of $\triangle ABC$, prove that the triangle formed by connecting the centroid of each equilateral triangle is an equilateral triangle.

51. Prove Napoleon's theorem for the special case in which the vertices of $\triangle ABC$ are $A(0, 0)$, $B(0, 6)$, and $C(6, 0)$ and form a 45°-45°-90° triangle. That is, if equilateral triangles are constructed on each side of $\triangle ABC$, prove that the triangle formed by connecting the centroid of each equilateral triangle is an equilateral triangle.

APPLICATIONS

52. A catering company will cater a reception for \$3.50 per person plus fixed costs of \$100.

a. Complete the following chart.

Number of People	1	30	50	75	100	n
Total Cost						

b. Write a linear equation representing the relationship between number of people, x, and total cost, y.

c. What is the slope of this line? What does it represent?

d. What is the y-intercept? What does it represent?

53. A cab company charges a fixed fee of \$0.60 plus \$0.50 per mile.

a. Find the cost of traveling 10 miles, and of traveling 25 miles.

b. Write an expression for the cost, y, of a trip of x miles.

54. A barrel in the form of a right circular cylinder having a diameter of three feet is to be constructed.

a. Write a linear equation that describes the volume of the barrel y as a function of its height x.

b. What is the slope of the line whose equation you wrote in (a)? What does it represent?

c. Write a linear equation that describes the surface area of the barrel y as a function of its height x.

d. What is the slope of the line whose equation you wrote in (c)? What does it represent?

55. Three sides of a lot must be fenced as shown. One end requires 60 feet of fence, and it must be fenced. The lengths of the other two congruent sides depend on the owner's wishes and finances. The type of fencing to be used costs \$7.25 per linear foot.

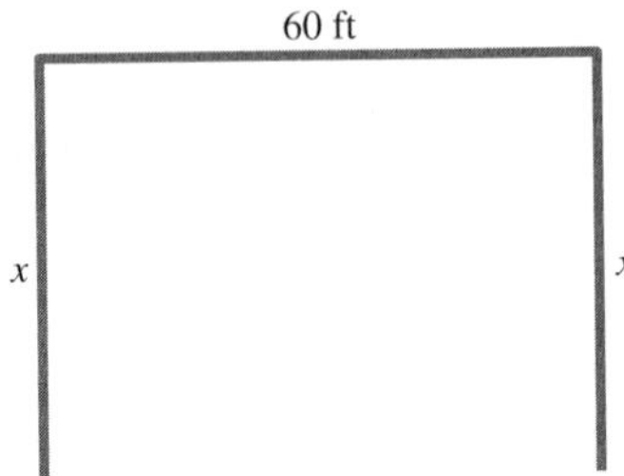

a. Write a linear equation that describes the total cost, y, in terms of the length of a side, x.

b. What is the slope of the line whose equation you wrote in (a)? What does it represent?

c. What is the y-intercept of the line whose equation you wrote in (a)? What does it represent?

d. If the owner has available \$1700 for this fencing project, what is the maximum length for the side of the fenced area?

56. a. A manufacturer can produce items at a cost of \$0.65 per item with an operational overhead of \$350. Let y represent the total production costs and x represent the number of items produced. Write an equation representing total production costs.

b. The manufacturer sells the items for \$1.15 per item. Let y represent the revenue and x represent the number of items sold. Write an equation representing total revenue.

c. How many items does the manufacturer need to produce and sell to break even?

57. An organization is planning a dance. The band will cost \$400 and advertising costs will be \$100. Food will be supplied at \$2 per person.

a. How many people need to attend the dance in order for the organization to break even if tickets are sold at \$7 each?

b. How many people need to attend the dance in order for the organization to break even if tickets are sold at \$6 each?

c. How many people need to attend the dance if the organization wants to earn \$400 profit and sells tickets for \$6 each?

EXTENDED PROBLEMS

58. A solution to a linear equation involving two variables is an ordered pair (x, y) that makes the linear equation true. A linear equation of the form $ax + by = c$, where $a, b,$ and c are constants, has infinitely many solutions. When plotted, those solutions form a line. **A linear inequality** is a statement of the form $ax + by \leq c, ax + by \geq c, ax + by < c,$ or $ax + by > c$. Like a linear equation, a linear inequality also has infinitely many solutions of the form (x, y). When plotted, those solutions fill a region in the coordinate plane. For example, consider the linear inequality $x + y \leq 6$. A solution to this inequality is any ordered pair (x, y) that makes the inequality true. Any point on the line $x + y = 6$ satisfies the inequality $x + y \leq 6$, and all the points on one side of the line also satisfy the inequality. To graph the solution set, first graph the line $x + y = 6$ and then select a convenient point on one side of the line and see if the point satisfies $x + y \leq 6$. If it does, shade that side of the line. If it does not, shade the other side of the line. A convenient point to choose is $(0, 0)$. When we substitute this point into the inequality $x + y \leq 6$, we get $0 + 0 \leq 6$, a true statement. The shaded half of the coordinate system that includes the point $(0, 0)$ is the solution set of the linear inequality as shown next.

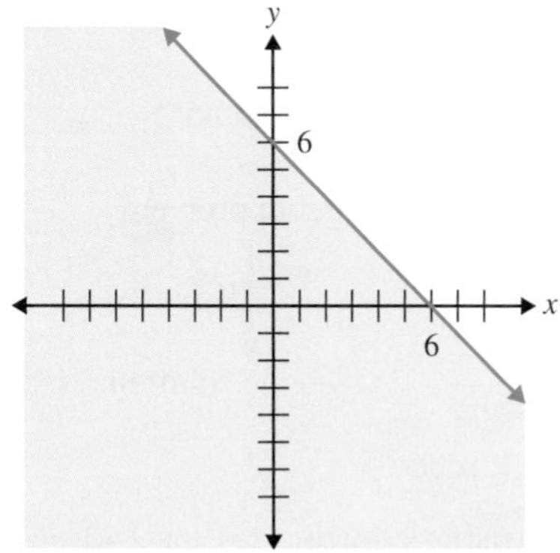

For a system of linear inequalities, graph each individual linear inequality and note where the shading overlaps. The solution set is the region that has been shaded by each linear inequality. For each of the following, shade the region that makes the system of linear inequalities true.

a. $x \geq 0, y \geq 0, x + y \leq 8$

b. $x \leq 6, y \geq -3, -x + 2y \leq 4$

59. In the Section 8.1 problem set, we introduced a coordinate system for three-dimensional space, as well as a formula for finding the distance between two points (x, y, z) in space.

a. A sphere is a set of points in space equidistant from a point called the center. Research the sphere and describe how many points uniquely determine a sphere.

b. What is the equation of a sphere centered at the origin? What is the equation of a sphere that is not centered at the origin?

c. How many ways can a line and sphere intersect? How can a plane and a sphere intersect? What is the name for the set of points that make up the intersection of a plane and a sphere when the plane passes through the center of the sphere?

60. At one time, it was thought that the earth revolved around the sun in a circular orbit. In 1609, Johannes Kepler published his discovery that the orbits of the planets are elliptical. While a circle is the set of points in a plane equidistant from a single point, an **ellipse** is a set of points in which the distance from two fixed points, called foci, to a point on the ellipse is constant. In the following figure, the two fixed points are labeled $(-c, 0)$ and $(c, 0)$.

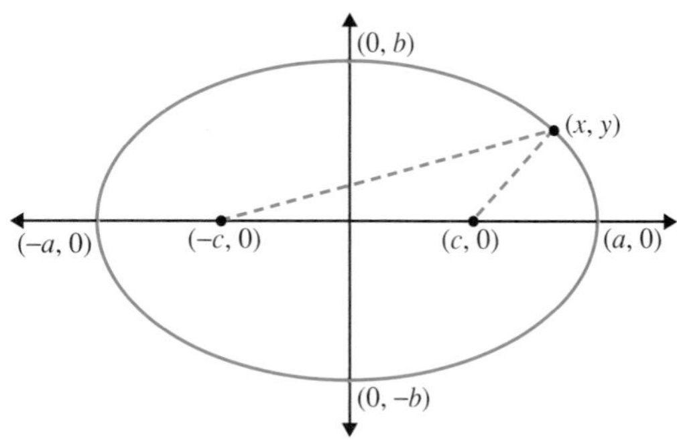

a. Research the equation of the ellipse and define the terms minor axis and major axis. How are the lengths a, b, and c, as shown in the figure, related? Use the distance formula to derive the equation of an ellipse centered at the origin.

b. Create the equations of three different ellipses, graph each, and label the intercepts, major axis, and minor axis of each.

c. Research distances related to the earth's orbit to determine where the sun is located within the elliptical orbit of the earth. Create a coordinate system showing the orbit of the earth around the sun. Center the orbit at the origin. Give the coordinates of the endpoints of the major axis, minor axis, and foci.

8.4 PROBLEM SOLVING USING COORDINATES

Applied Problem

Two lookouts posted in fire towers located 8 miles apart both spot the smoke from a fire. One observer determines the direction of the fire from her position as N 42° E, and the second sights the fire at N 30° W. Use the angles shown and the lines drawn to determine the location of the fire. (NOTE: This method of locating an object is called triangulation.)

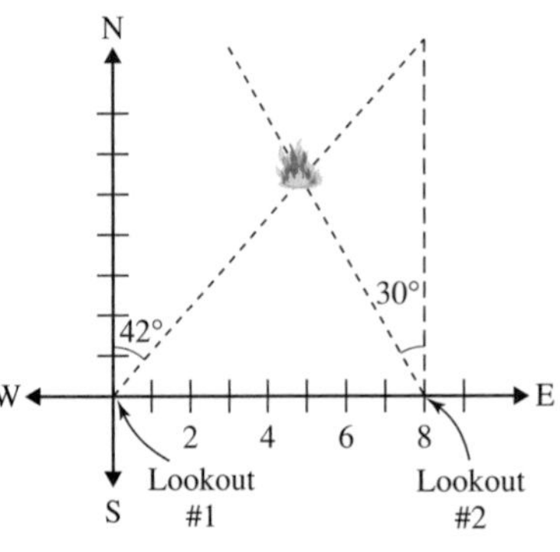

The most beautiful feature of coordinate geometry is that we can represent many common geometric shapes by equations. Thus, if we want to solve a problem in geometry, we may be able to represent the problem in algebraic terms, solve the problem using algebra, and then interpret the result geometrically (Figure 8.29).

Problem Solving Using Coordinate Geometry

Geometry problem —*Translate*→ Algebra problem

Algebra problem —*Solve*→ Algebraic solution

Algebraic solution —*Interpret*→ Geometric solution

Geometric solution —*Check*→ Geometry problem

FIGURE 8.29

We can solve some problems more easily using coordinates instead of the triangle congruence and similarity techniques used in Chapters 2–6. The following examples use familiar theorems to illustrate coordinate proofs. In doing such proofs, we should orient geometric figures as conveniently as possible in an effort to simplify computations.

Assigning Coordinates to Geometric Figures

When writing a proof using coordinate geometry, we locate the coordinate axes conveniently to simplify the computations. For example, we may choose to put one vertex of a rectangle or parallelogram at the origin and another vertex somewhere else on the x-axis because this orientation makes several of the coordinates of the vertices zero. We use variables for some of the coordinates of the vertices to write our proof for a general figure. The next example shows how we can assign coordinates to some common geometric figures.

EXAMPLE 8.19 Choose convenient coordinates for the vertices of each of the following.

a. rectangle **b.** a parallelogram

SOLUTION

a. Because the coordinate axes are perpendicular, we can place two sides of the rectangle on the x-axis and y-axis [Figure 8.30(a)]. Placing three vertices on the coordinate axes also makes four coordinates of the vertices zero, which can simplify computations when we determine slopes, distances, equations of sides, and so on. These calculations are usually required when we write a coordinate proof. Notice that vertices B and D were given coordinates $(0, b)$ and $(a, 0)$, respectively, to allow for the fact that sides $\overline{AB}$ and $\overline{AD}$ may not be congruent. Hence, $ABCD$ is a general rectangle. If we had used vertices $A(0, 0)$, $B(0, a)$, $C(a, a)$, and $D(a, 0)$, the rectangle would have been a square. Any result we subsequently proved using those coordinates would be true only for squares and might not be true for all rectangles in general.

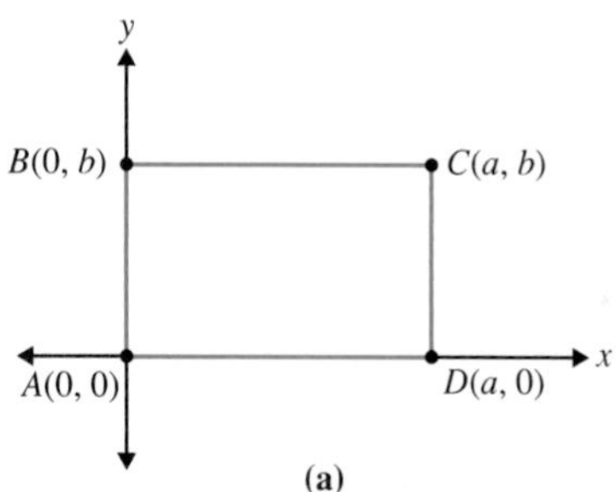

FIGURE 8.30

b. To represent a parallelogram, we can place one side along the x-axis with one vertex at the origin [Figure 8.30(b)]. Now we must determine coordinates for vertices B and C. We choose two new variable coordinates (b, c), for point B [Figure 8.30(c)]. The y-coordinate of point C must be c, the same as the y-coordinate of B, because $\overline{BC}$ is parallel to $\overline{AD}$. Moreover, because the slope of $\overline{AB}$ must equal the slope of $\overline{DC}$, the x-coordinate of C must be $a + b$ [Figure 8.30(c)].

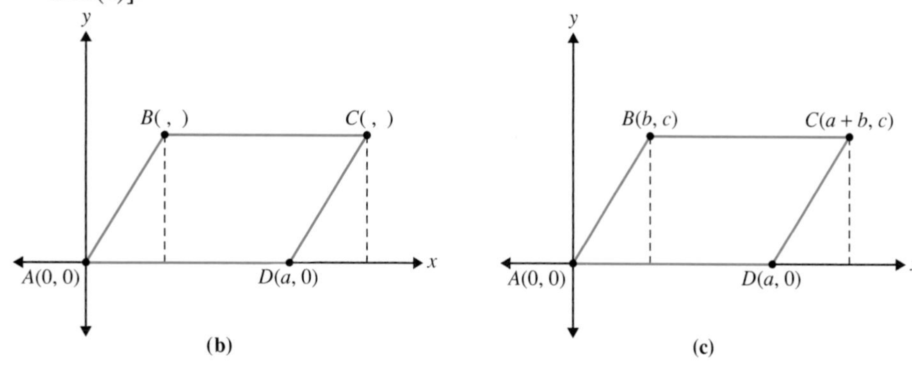

FIGURE 8.30

Keep in mind that the representations shown in Example 8.19 are not unique. Because the figures could be oriented differently on the coordinate axes, many other representations are possible. The ones shown, however, are convenient for many purposes.

Coordinate Proofs

When we use coordinates to prove a theorem about a geometric figure, there are two steps that we must complete.

1. *Assign appropriate coordinates to the figure.* As Example 8.19 illustrates, this means making certain that the figure is as general as necessary. For example, if we plan to prove a theorem about triangles in general, we should not draw a special case of a triangle such as a right triangle or an isosceles triangle. Placing sides on the coordinate axes and using zero for as many coordinates as possible will often simplify calculations.
2. *Use the assigned coordinates to write a proof.* Some of the tools we have at hand to write coordinate proofs are
   - **a.** Slope (useful for showing that segments are parallel or perpendicular)
   - **b.** The midpoint formula (useful for showing when segments are bisected, for example)
   - **c.** The distance formula (useful for showing that segments are congruent)
   - **d.** The equation of a line (useful for finding points of intersection)

The following examples show how we apply these two steps to construct a coordinate proof of a theorem. We proved each of the theorems earlier in the text using standard Euclidean techniques.

EXAMPLE 8.20 Use coordinates to prove that the diagonals of a rectangle are congruent.

SOLUTION

Using the steps mentioned in the preceding discussion, we must

1. assign coordinates to the vertices of a rectangle $ABCD$ (Figure 8.31), and
2. show that its diagonals are congruent, or that $\overline{AC} \cong \overline{BD}$ in Figure 8.31.

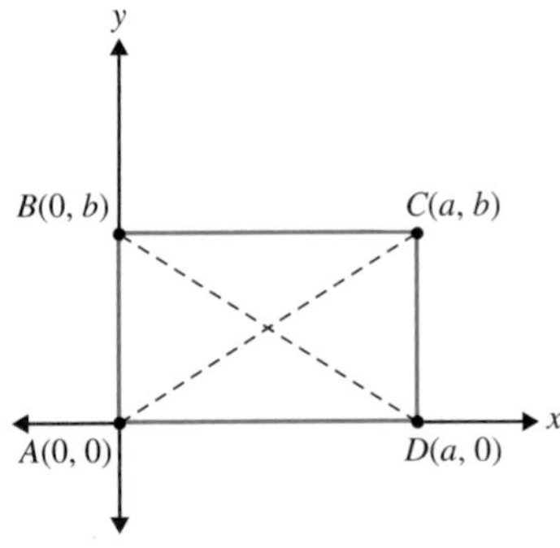

FIGURE 8.31

Using the distance formula, we have

$$AC = \sqrt{(a - 0)^2 + (b - 0)^2} = \sqrt{a^2 + b^2}$$

and

$$DB = \sqrt{(a - 0)^2 + (0 - b)^2} = \sqrt{a^2 + b^2}$$

Thus, $AC = BD$. Therefore, the diagonals are congruent. ■

EXAMPLE 8.21 Use coordinates to prove that the diagonals of a rhombus are perpendicular.

SOLUTION

1. First we must assign coordinates to a rhombus $ABCD$. Since a rhombus is a parallelogram, we can use the vertices for a parallelogram obtained in Example 8.19 and shown in Figure 8.32(a). The fact that this figure is a rhombus means that $AB = AD$, or $\sqrt{a^2 + b^2} = c$, or $a^2 + b^2 = c^2$. We will use this relationship in part (2).

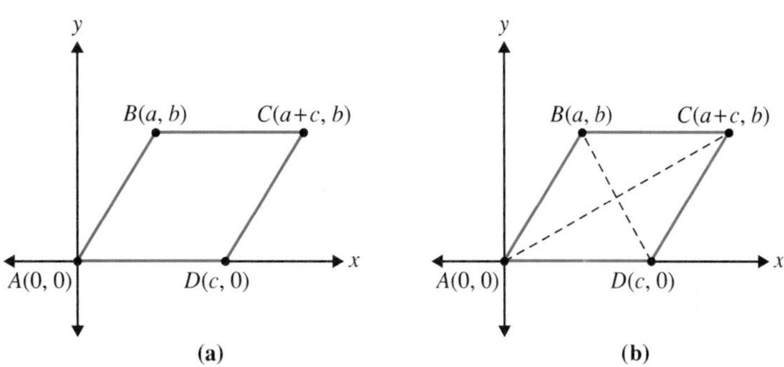

FIGURE 8.32

2. We must show that $\overline{AC} \perp \overline{BD}$ [Figure 8.32(b)]. We will show that the product of their slopes is -1.

$$\text{Slope of } \overline{AC} = \frac{b - 0}{(a + c) - 0} = \frac{b}{a + c}$$

$$\text{Slope of } \overline{BD} = \frac{b - 0}{a - c} = \frac{b}{a - c}$$

Thus, the product of their slopes is

$$\textbf{(i)} \quad \left(\frac{b}{a + c}\right)\left(\frac{b}{a - c}\right) = \frac{b^2}{a^2 - c^2}$$

Recall that in (1) we found that $a^2 + b^2 = c^2$, or $b^2 = c^2 - a^2$. Substituting this expression in place of b^2 in equation (i), we have that the product of the slopes is $\frac{b^2}{a^2 - c^2} = \frac{c^2 - a^2}{a^2 - c^2} = -1$. Therefore, $\overline{AC} \perp \overline{BD}$, that is, the diagonals of this general rhombus are perpendicular. ■

EXAMPLE 8.22 Use coordinates to prove that for any quadrilateral $ABCD$ whose sides have midpoints W, X, Y, and Z as shown in Figure 8.33(a), $WXYZ$ is a parallelogram.

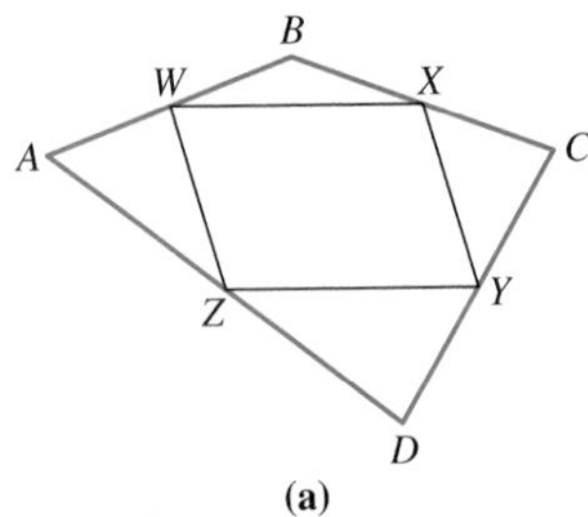

(a)

FIGURE 8.33

SOLUTION

1. First we place $ABCD$ on the coordinate plane. Point A is placed at the origin and $\overline{AD}$ is on the x-axis [Figure 8.33(b)].

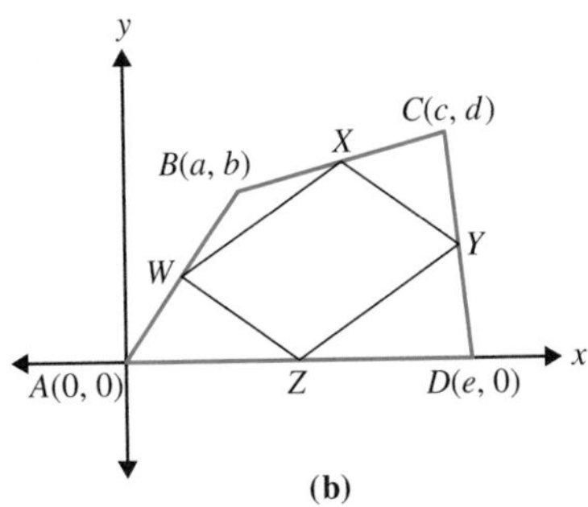

(b)

FIGURE 8.33

2. Now we must show that $WXYZ$ is a parallelogram. We will show $\overline{WX} \parallel \overline{ZY}$ and $\overline{WZ} \parallel \overline{XY}$. We find the coordinates of $WXYZ$ using the midpoint formula as follows:

$$W = \left(\frac{a+0}{2}, \frac{b+0}{2}\right) = \left(\frac{a}{2}, \frac{b}{2}\right) \qquad X = \left(\frac{a+c}{2}, \frac{b+d}{2}\right)$$

$$Y = \left(\frac{c+e}{2}, \frac{d+0}{2}\right) = \left(\frac{c+e}{2}, \frac{d}{2}\right) \quad Z = \left(\frac{e}{2}, 0\right)$$

$$\text{Slope of } \overline{WX} = \frac{\frac{b}{2} - \frac{b+d}{2}}{\frac{a}{2} - \frac{a+c}{2}} = \frac{\frac{b-b-d}{2}}{\frac{a-a-c}{2}} = \frac{\frac{-d}{2}}{\frac{-c}{2}} = \frac{d}{c}$$

$$\text{Slope of } \overline{ZY} = \frac{\frac{d}{2} - 0}{\frac{c+e}{2} - \frac{e}{2}} = \frac{\frac{d}{2}}{\frac{c+e-e}{2}} = \frac{\frac{d}{2}}{\frac{c}{2}} = \frac{d}{c}$$

Thus, $\overline{WX} \parallel \overline{ZY}$ because they have the same slope. We can show similarly that $\overline{WZ} \parallel \overline{XY}$. (we leave details for the problem set). Because the opposite sides of $WXYZ$ are parallel, it is a parallelogram. ■

Solution of Applied Problem

Find the equation of each line in slope-intercept form. The slope of Line 1 is $m = \tan 48° \approx 1.1106$. Line 1 has a y-intercept of 0. So Line 1 has equation $y = 1.1106x$. The slope of Line 2 is $m = -\tan 60° \approx -1.7321$. Using this slope and the point $(8, 0)$, we have

$$y - 0 = -1.7321\,(x - 8) \qquad \text{or} \qquad y = -1.7321x + 13.856$$

Solving these equations simultaneously, we find $x \approx 4.9$ and $y \approx 5.4$. Hence, the fire is at the point $(4.9, 5.4)$.

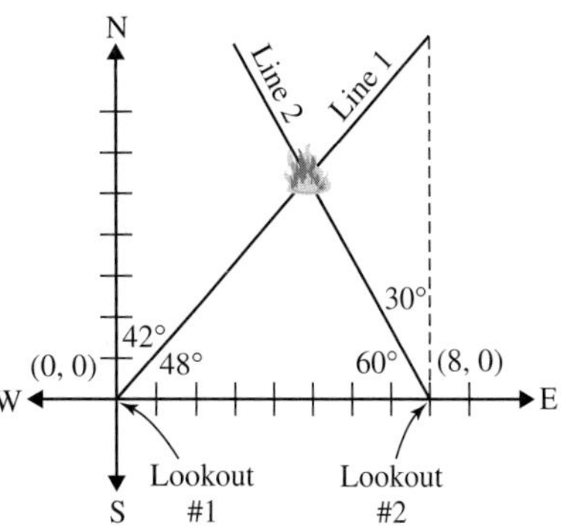

GEOMETRY AROUND US If you hold up one finger and look at it with your left eye closed, then open your left eye and close your right eye, the finger will seem to have moved. This apparent change in location is called **parallax**, and it can be used indirectly to calculate the distance to nearby stars. By observing the star from two opposite points in Earth's orbit (six months apart), astronomers can determine the angle between these two lines of sight. They can use this angle and the known distance between the two observation points to calculate the distance to the star. Astronomers have defined a parsec, a unit of distance, in terms of parallax. One **parsec** is defined to be the distance at which a star would have a parallax of one second of arc.

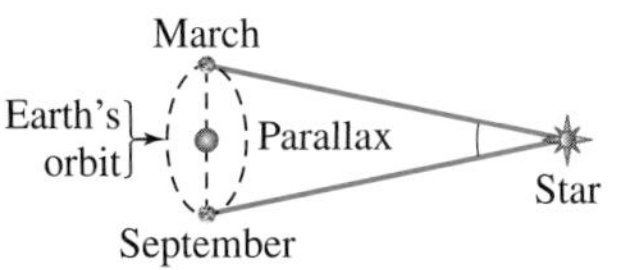

PROBLEM SET 8.4

EXERCISES/PROBLEMS

1. Two vertices of a trapezoid have been assigned coordinates in the following figure.

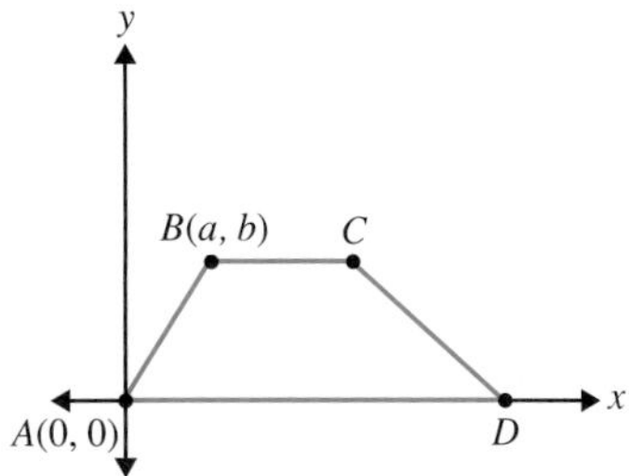

Determine appropriate coordinates for the remaining two vertices. Use the coordinates to verify that the resulting figure is a trapezoid and that it is not a special case.

2. In each of the following figures, two vertices of an isosceles trapezoid have been assigned coordinates.

a. **b.**

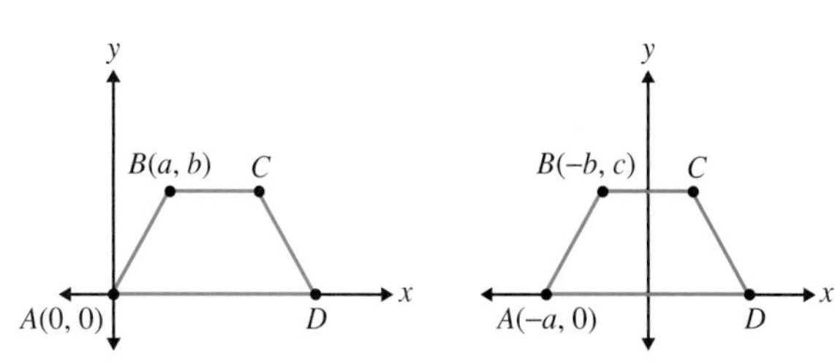

Determine appropriate coordinates for the remaining two vertices. Use the coordinates to verify that the resulting figure is an isosceles trapezoid in each case.

3. A theorem is to be proved about an isosceles triangle. Two vertices have been assigned coordinates as shown.

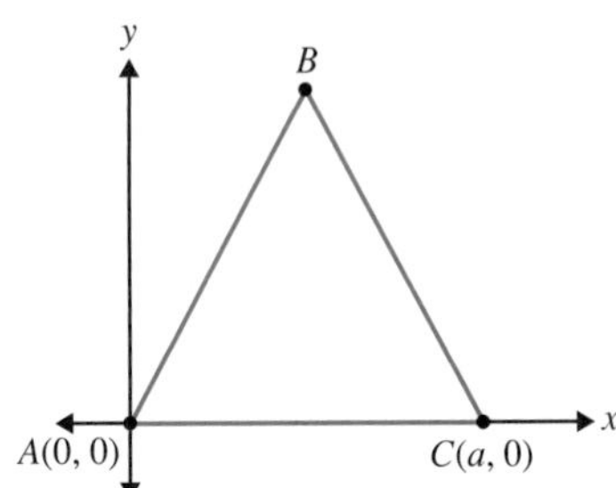

Determine appropriate coordinates for the third vertex. Use the coordinates to verify that the triangle is isosceles, but not necessarily equilateral.

4. A theorem involving right triangles is to be proved.

a. The sides have been positioned along the coordinate axes as shown.

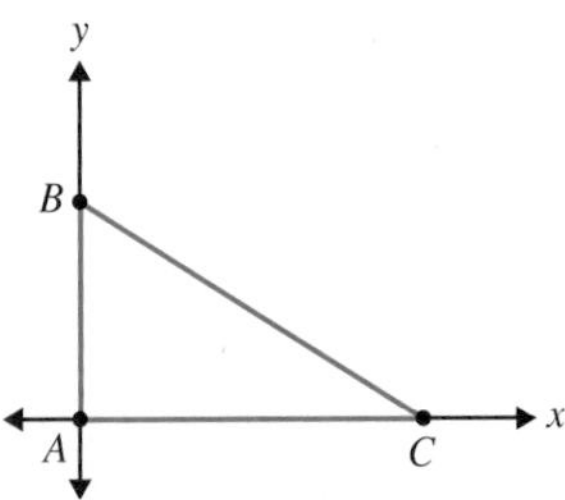

Label the coordinates of the vertices. Why is $\triangle ABC$ a right triangle?

b. Suppose that the vertices of the right triangle were assigned the coordinates $(0, 0)$, $(0, a)$, and $(a, 0)$. Would these coordinates be appropriate? Why or why not?

5. a. A kite has been placed on the coordinate system as shown next, and two vertices have been assigned coordinates.

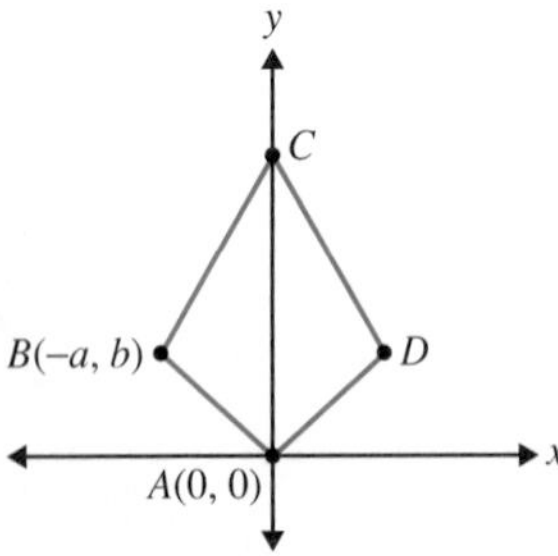

Determine appropriate coordinates for the remaining two vertices. Use the coordinates to verify that the resulting figure is a kite.

b. Suppose that vertices for a kite were given as $A(0, 0)$, $B(-a, a)$, $C(0, 2a)$, and $D(a, a)$. Would these coordinates be appropriate? Why or why not?

6. A student asked to draw a general quadrilateral and label the vertices drew the following figure.

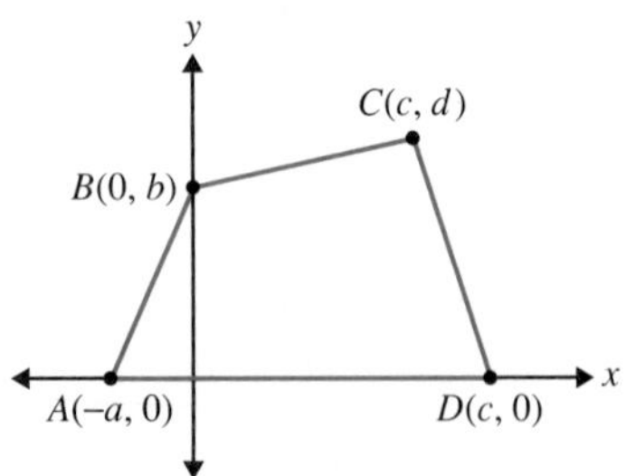

a. Why is this representation of a general quadrilateral not appropriate?

b. How would you change one vertex of the figure so that it is acceptable?

7. Complete the proof in Example 8.22 by showing that $\overline{WZ} \parallel \overline{XY}$.

8. For $\triangle ABC$ shown, write the equation of the median from vertex A.

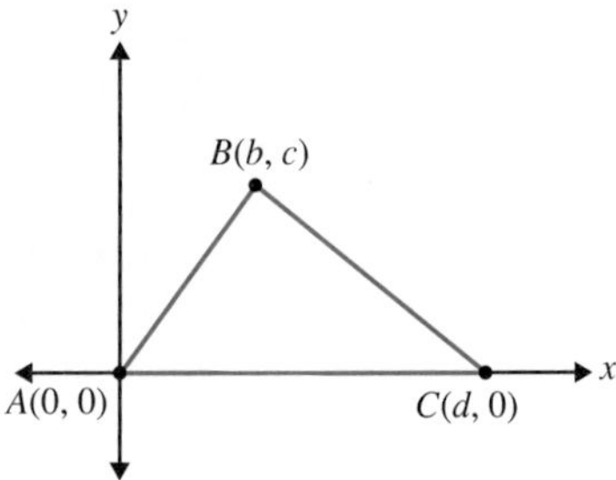

PROOFS

Use coordinates to write the following proofs.

9. Prove: If $ABCD$ is a square, then $\overline{AC} \perp \overline{BD}$.

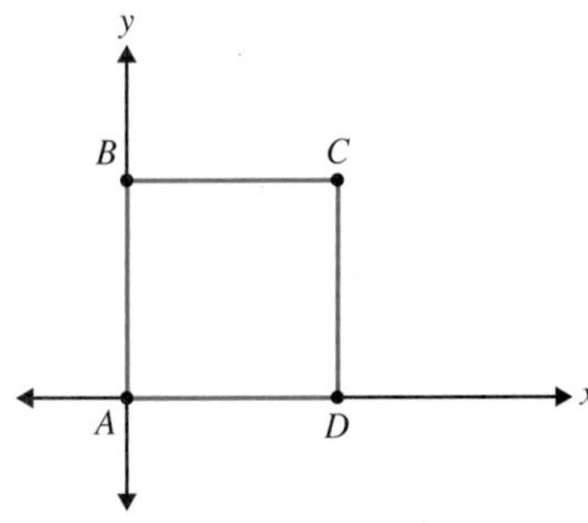

10. Prove: If $ABCD$ is a parallelogram, then $\overline{AC}$ and $\overline{BD}$ bisect each other.

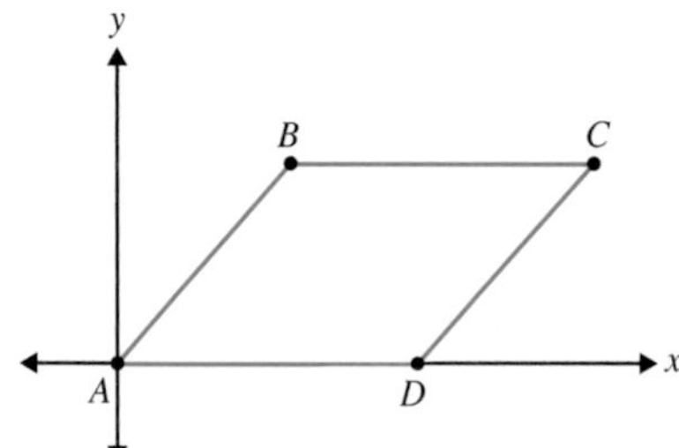

11. Prove: The opposite sides of a parallelogram are congruent.
12. Prove: The median from the vertex of an isosceles triangle is perpendicular to the base.
13. Prove: If M is the midpoint of $\overline{BC}$, where $\triangle ABC$ is a right triangle with $\angle A = 90°$, then $AM = BM = CM$.

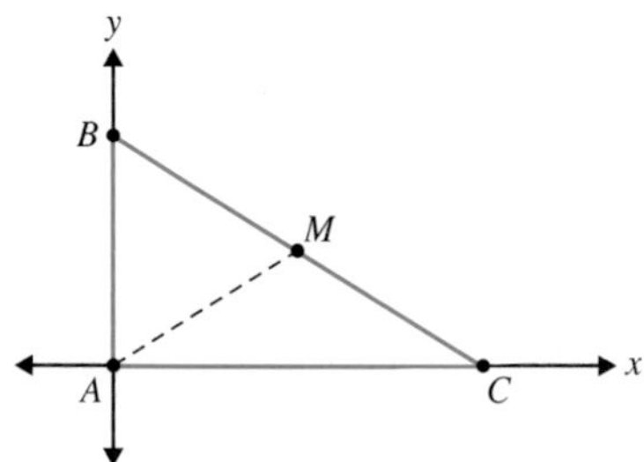

14. Prove: If M and N are midpoints of $\overline{AB}$ and $\overline{BC}$, respectively, then $\overline{MN} \parallel \overline{AC}$ and $MN = \frac{1}{2}AC$.

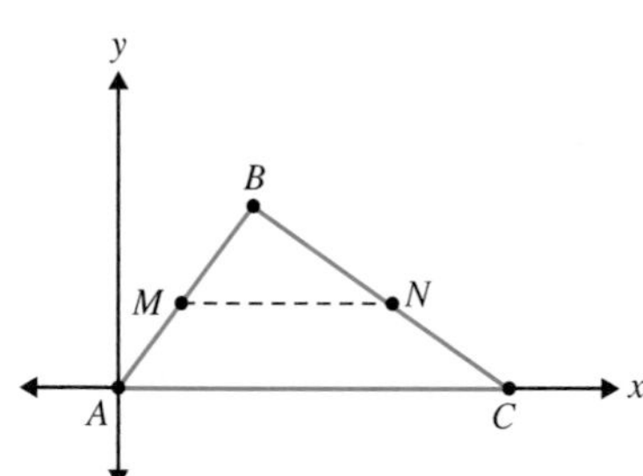

15. Prove: If $ABCD$ is an isosceles trapezoid, then $AC = BD$.

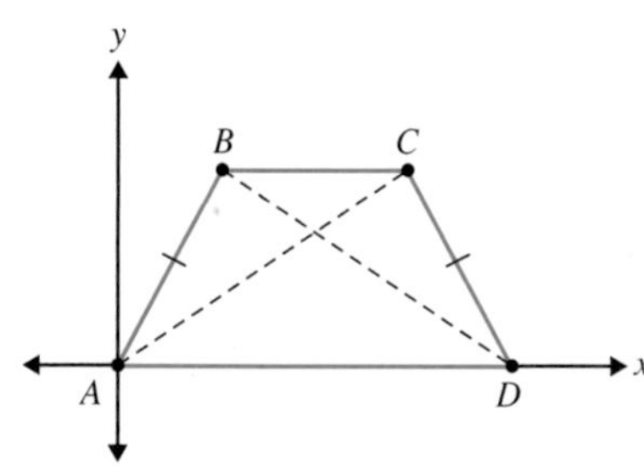

16. Prove: If M and N are midpoints of the legs of trapezoid $ABCD$, then $MN = \frac{1}{2}(AD + BC)$.

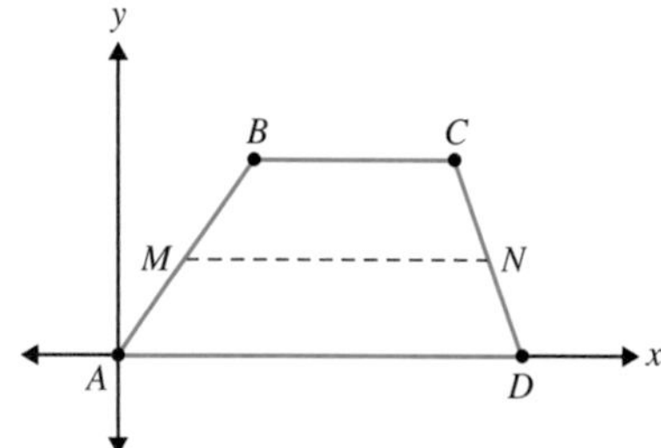

17. Prove: If $ABCD$ is a rectangle and M, N, O, and P are midpoints of the sides as indicated, then $MNOP$ is a rhombus.

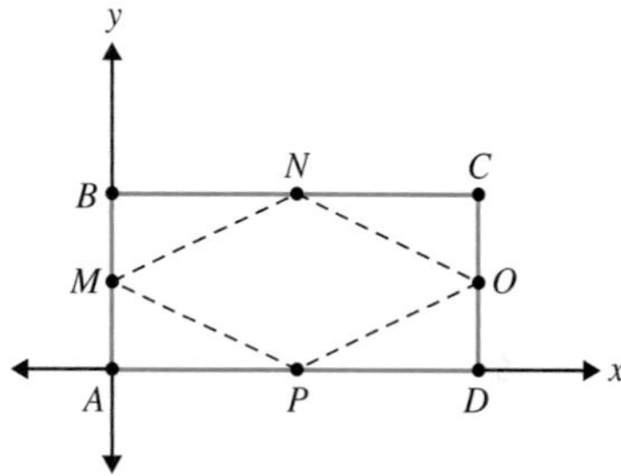

18. Prove: If $ABCD$ is a kite and M, N, O, and P are midpoints of the sides as indicated, then $MNOP$ is a rectangle.

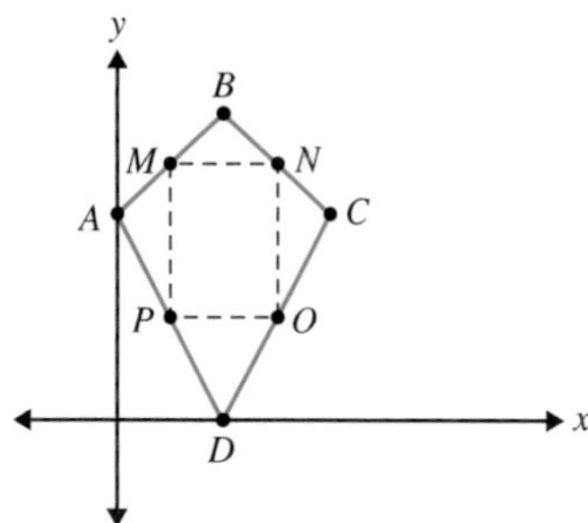

19. Prove: The segments connecting consecutive midpoints of the sides of a rhombus form a rectangle.
20. Prove: The segments connecting the midpoints of opposite sides of a quadrilateral bisect each other.
21. Prove: If M and N are midpoints of the congruent sides of an isosceles triangle as indicated, then $AN = CM$.

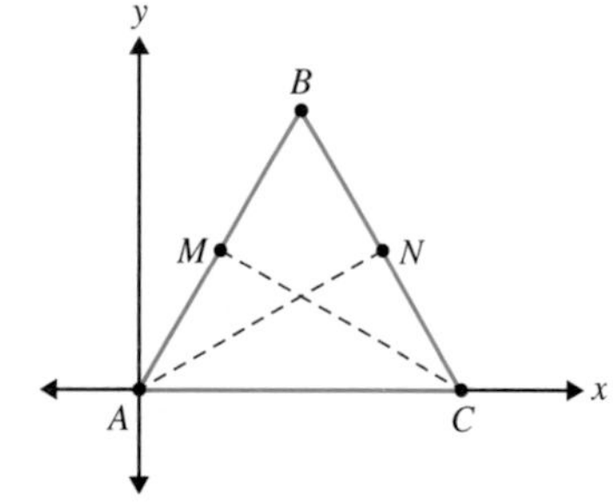

22. Prove: If AN and CM are altitudes to the congruent sides of an isosceles triangle, then $AN = CM$.

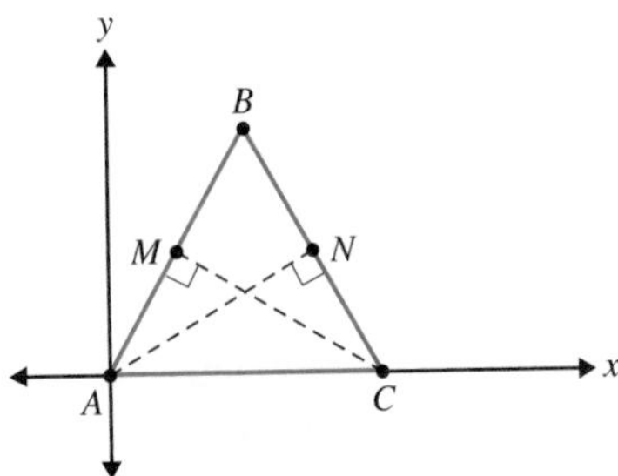

23. Prove: If the diagonals of a parallelogram are congruent, then it is a rectangle.
24. Prove: If the diagonals of a parallelogram are perpendicular, then it is a rhombus.
25. Prove: The perpendicular bisectors of the sides of a triangle are concurrent—this point is the circumcenter. (HINT: Write an equation for each of the three perpendicular bisectors. Then determine the point of intersection for each pair of lines.)
26. Prove: The altitudes of a triangle are concurrent—this point is the orthocenter. (HINT: Write an equation for each of the three altitudes. Then determine the point of intersection for each pair of lines.)
27. Prove: The medians of a triangle are concurrent—this point is the centroid. (HINT: Write an equation for each of the three medians. Then determine the point of intersection for each pair of lines.)
28. In the problem set of Section 3.1, an 8 by 8 square with an area of 64 square units was subdivided into two right triangles and two trapezoids as shown.

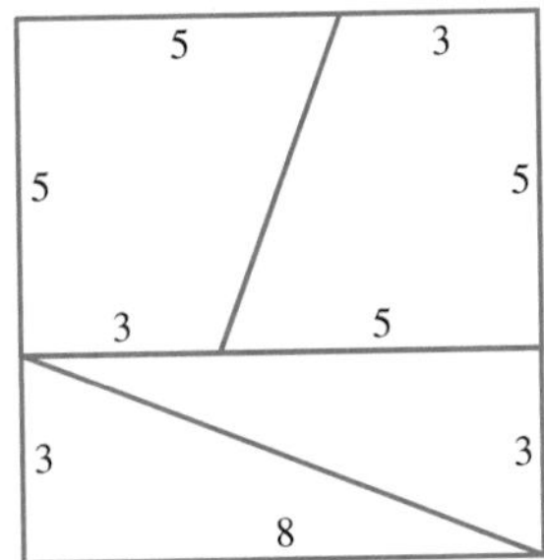

When the pieces were rearranged to form an 8 by 13 rectangle as shown, the area was calculated to be 65 square units.

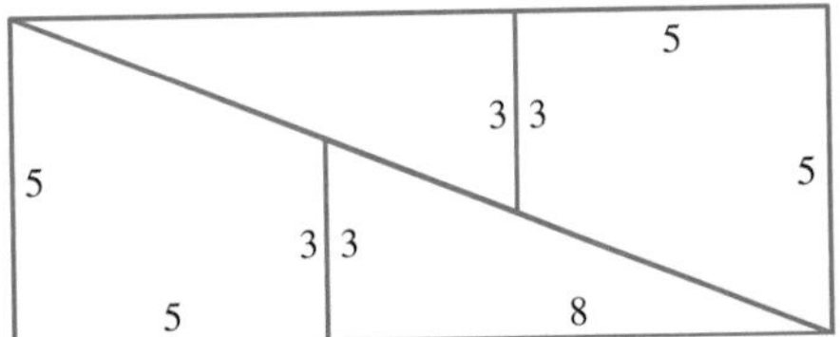

Explain this apparent contradiction by placing the rectangle in a coordinate system. Identify the coordinates of the vertices of each shape that makes up the rectangle. Calculate the slope of the diagonal of the rectangle and the slopes of the segments that make up the diagonal in the figure shown here. How would you explain the difference in the areas of the square and the rectangle?

EXTENDED PROBLEMS

29. Blaise Pascal (1623–1662), a French mathematician, physicist, and philosopher, proposed a theorem about a hexagon inscribed in a conic section. Pascal claimed that if the three pairs of opposite sides are extended until they intersect, the three points of intersection are collinear. A circle is a conic section. Research Pascal's theorem and then use a compass and a straightedge to draw a circle and construct any irregular hexagon inscribed in the circle. Use the straightedge to extend pairs of opposite sides until they intersect. Verify the three intersection points are collinear.
30. Pierre de Fermat considered extensions to the Pythagorean theorem. While there are infinitely many Pythagorean triples that satisfy the equation $x^2 + y^2 = z^2$, Fermat's famous Last Theorem states that $x^n + y^n = z^n$ has no nonzero integer solutions for x, y, and z when $n > 2$. Research Fermat's Last Theorem and summarize the progress made toward a proof since the discovery of the theorem. Has the theorem been proven? If so, when and by whom? What is the significance of Fermat's Last Theorem?
31. In Section 5.3, we defined a vector as a directed line segment and described it in terms of a direction angle and a magnitude. Vectors may also be represented using coordinates. A vector in the Cartesian coordinate system with its initial point placed at the origin, as shown next, is called a **position vector**. Position vectors are commonly specified in what is called component form or in unit vector form. Research both ways of specifying a vector.

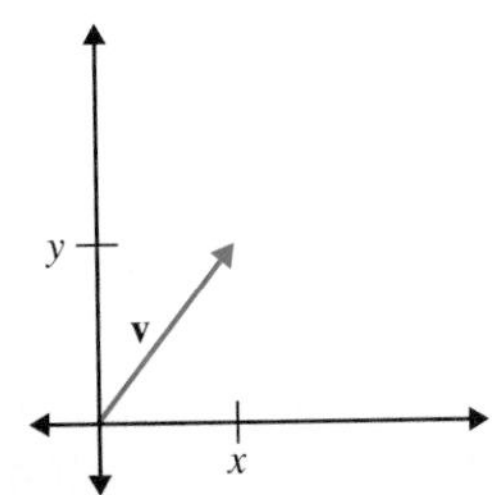

a. Answer each of the following questions about vectors and give an example to support your explanation. What are the components of a vector whose initial point is at the origin? What are the components of a vector whose initial point is not at the origin? How can we use the components of a vector to find the magnitude, or length, of the vector? Under what conditions are two vectors equivalent? How is the resultant vector found when two vectors in component form are added?

b. What is a unit vector? What are the components of the unit vector whose direction is along the x-axis? What are the components of the unit vector whose direction is along the y-axis? Show how a vector can be written using unit vectors. How can the resultant vector be found when two vectors in unit vector form are added?

Solution of Initial Problem

A treasure has been hidden in a desert location. A map shows $\triangle ABC$, where B is 50 miles NE of A and C is 80 miles east of A [Figure 8.34(a)]. You are told that the treasure is at the midpoint of the median from A. Locate the treasure.

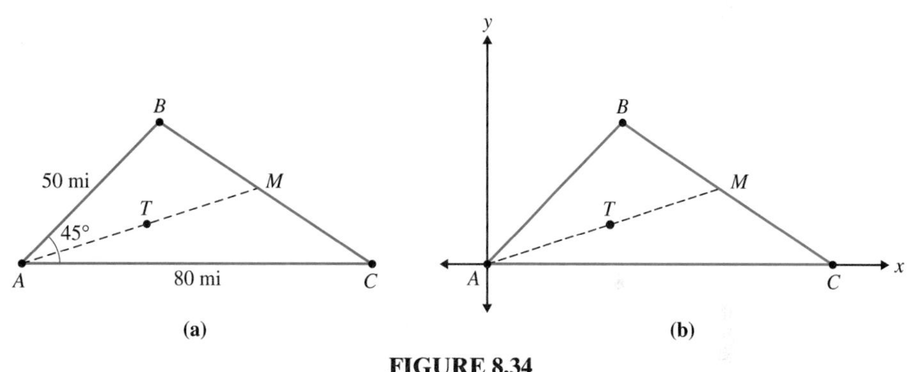

FIGURE 8.34

Place A at $(0, 0)$ and $\overline{AC}$ on the x-axis [Figure 8.34(b)]. The coordinates of C are $(80, 0)$. Because B is 50 miles NE of A and $\angle A = 45°$, B's coordinates are given by $(25\sqrt{2}, 25\sqrt{2})$. The midpoint M of $\overline{BC}$ is $\left(\frac{80 + 25\sqrt{2}}{2}, \frac{25\sqrt{2}}{2}\right)$. Point T, the midpoint of $\overline{AM}$, has coordinates $\left(\frac{80 + 25\sqrt{2}}{4}, \frac{25\sqrt{2}}{4}\right)$ [Figure 8.34(c)].

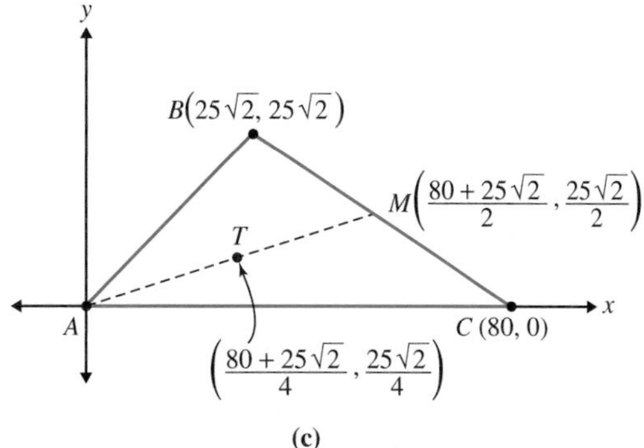

FIGURE 8.34

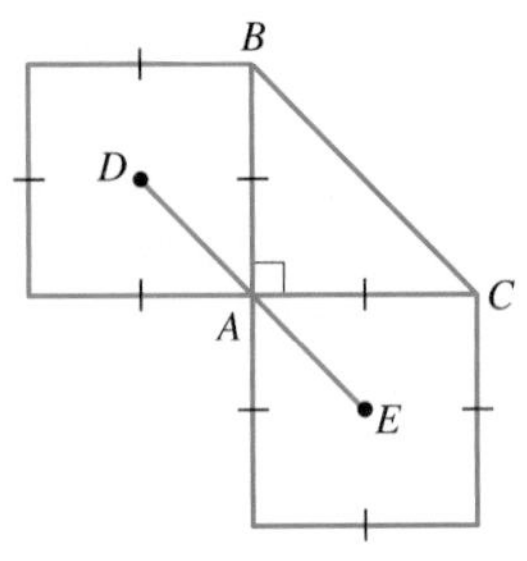

FIGURE 8.35

Additional Problems Where the Strategy "Use Coordinates" Is Useful

1. An explorer leaves base camp and travels 4 km north, 3 km west, and 2 km south. How far is he from the base camp?
2. Prove that $\overline{BC}$ is parallel to $\overline{DE}$ where D and E are the centers of the squares (Figure 8.35). (HINT: Use slopes.)

Writing for Understanding

1. A student claims that it does not make sense to say that a vertical line has "no slope." He says it seems more reasonable to say that a horizontal line has no slope, because it is flat. How would you explain the distinction, in terms of slope, between horizontal and vertical lines so that it makes sense to this student?
2. Give a geometric interpretation of what happens when you solve a system of two linear equations simultaneously. Be sure to include all possible outcomes. How do the possibilities change when there are more than two equations? When at least one of the equations is not linear?
3. Select two problems in Section 8.4. Choose one whose proof is easier to do using coordinates than using triangle congruence principles as in Chapter 5, and choose a second that is easier to do using triangle congruence than coordinates. Based on these two examples, state which method you prefer and why.
4. The "Geometry Around Us" at the end of Section 8.2 relates slope to the pitch of a roof. Do some research on the pitch of a roof, focusing on the advantages of one pitch over another, taking into account such factors as space, construction materials, the climate, and so on. Summarize your findings.

PEOPLE IN GEOMETRY

Maria Agnesi's (1718–1799) mathematical fame rests on the highly regarded *Instituzioni Analitiche*, a 1020-page, two-volume presentation of algebra, analytic geometry, and calculus. Published in 1748, it brought order and clarity to the mathematics invented by Descartes, Newton, Leibniz, and others in the 17th century. Maria was the eldest of 21 children in a wealthy Italian family. She was a gifted child, with an extraordinary talent for languages. Her parents encouraged her to excel, and she received the best schooling available. Maria began work on *Instituzioni Analitiche* at age 20 and finished it 10 years later, supervising its printing on presses installed in her home. After its publication, she was appointed honorary professor at the University of Bologna. Instead, Maria decided to dedicate her life to charity and religious devotion, and she spent the last 45 years of her life caring for the sick, aged, and indigent.

CHAPTER REVIEW

For each section of this chapter, you will find a list of vocabulary and notation, questions to assess your understanding of key concepts, and review problems similar to the problems you worked in your homework. Review each item in the *Vocabulary/Notation* list mentally, and, if necessary, refer back to the indicated page and write a definition. Then answer the *Concept Check Questions*, looking back at the section if you need help. Work the *Review Problems* as practice before you move on to the *Chapter Test*. Answers to the *Review Problems* and the *Chapter Test* can be found at the back of the book.

SECTION 8.1 Coordinates and Distance in the Plane

Vocabulary/Notation

Origin 422
x-axis 422
y-axis 422
Axes 422
Coordinate plane 422
Ordered pair 422
x-coordinate 423
y-coordinate 423
Quadrants 423

Concept Check Questions

1. How is each point in the coordinate plane represented? 422
2. How are the signs of the coordinates of a point determined by the quadrant in which the point lies? 423
3. How can we find the distance between two points? 424
4. What must be true about the distances between three collinear points? 427
5. How do we use the Side Splitting Theorem to derive the midpoint formula? 428

Review Problems

1. For each of the following points, indicate in which quadrant or on which axis each point is located.
   a. $(-5, -8)$
   b. $(3, -6)$
   c. $(0, 9)$
   d. $(-12, 17)$
   e. $(-7, 0)$
   f. $(1, 4)$
2. Consider points $A(-5, 6)$ and $B(7, 1)$. Determine the coordinates of point C so that A, B, and C are collinear, point B is between A and C, and $AB = 2BC$.
3. Quadrilateral $ABCD$ has vertices $A(-11, -7)$, $B(-2, 0)$, $C(9, -3)$, and $D(0, -10)$.
   a. Describe $ABCD$ as completely as possible.
   b. Find the coordinates of the midpoint of the diagonals of $ABCD$.
4. Determine whether points $A(-4, 2)$, $B(-2, 6)$, and $C(1, 12)$ are collinear. Justify your response.
5. Suppose that $\triangle ABC$ has vertices $A(a, b)$, $B(c, d)$, and $C(e, f)$. If D is the midpoint of $\overline{AB}$ and E is the midpoint of $\overline{BC}$, prove that $DE = \frac{1}{2}AC$.

SECTION 8.2 Slope

Vocabulary/Notation

Slope of a line 436
Rise 436
Run 436
Slope of a line segment 436

Concept Check Questions

1. How do we measure the steepness of a line? 436
2. Why do we say the slope of a vertical line is undefined? 436
3. What does the graph of a line with positive slope look like? 437
4. How can we use slope to show three points are collinear? 437
5. How are the slopes of parallel lines and perpendicular lines related? 438–439
6. How are the slopes of perpendicular lines related? 441

Review Problems

1. The slope of the line through $(-3, 12)$ and $(x, 7)$ is $\frac{4}{5}$. Find the value of x.
2. The slopes of several lines are given. For each, give the slope of another line that is parallel, and the slope of a line that is perpendicular, to the given line.
   a. 9
   b. 0
   c. $\frac{2}{7}$
   d. $-\frac{5}{3}$
   e. undefined
3. A quadrilateral has coordinates $A(-6, 2)$, $B(-4, -6)$, $C(12, -2)$, and $D(10, 6)$. Describe quadrilateral $ABCD$ as completely as possible.
4. Three of the four coordinates of a quadrilateral are $A(0, 0)$, $C(7, 4)$, and $D(11, 0)$.
   a. Find the coordinates of point B so that B is in Quadrant I and $ABCD$ is an isosceles trapezoid.
   b. Find the coordinates of point B so that B is in Quadrant IV and $ABDC$ is a parallelogram.
5. Let point A have coordinates (a, b) and point B have coordinates (c, d). If M is the midpoint of $\overline{AB}$, prove A, M, and B are collinear.

SECTION 8.3 Equations of Lines and Circles

Vocabulary/Notation

Concept Check Questions

1. How can we use the slope of a line to derive its equation? 447–448
2. In what three equivalent ways can we express the equation of a line? 447–448
3. What must be true about a simultaneous solution for a system of equations? 450
4. How can we use the slopes and y-intercepts of two lines to determine whether a system of linear equations has zero, one, or infinitely many simultaneous solutions? 450–451
5. What steps must we take to solve a system of equations using the substitution method? 452
6. How can we determine the center and radius of a circle from the standard form of the equation of the circle? 453

Review Problems

1. For each of the following lines, identify the slope and the y-intercept. Graph the line.
   a. $y = -\frac{3}{4}x + 6$
   b. $x = 4y - 12$
   c. $\frac{y}{5} = 2 - x$
   d. $-3x + 3y = 21$
2. a. Write an equation of the horizontal line through the point $(-12, 5)$.
   b. Write an equation of the vertical line through the point $(4, -7)$.
   c. Write an equation of the line in slope-intercept form that passes through the point $(2, -4)$ and that is perpendicular to the line $2x - 3y = 12$.
3. Use the substitution method to find the simultaneous solution(s), if any, to the system of equations. If no solution exists, explain why not.
   a. $-5x + 4y = 24$ and $x + y = -3$
   b. $\frac{-3}{2}x + y = 3$ and $2y = 3x - 4$
4. Find an equation of the perpendicular bisector of $\overline{AB}$ if the coordinates of the endpoints of the segment are $A(0, 0)$ and $B(2, 5)$.
5. Write an equation of the circle having center $(2, 4)$ and passing through the point $(-5, 3)$.

SECTION 8.4 Problem Solving Using Coordinates

Concept Check Questions

1. Which locations in a coordinate system are convenient choices when placing vertices of geometric figures? 460–461
2. Are coordinate representations of geometric figures unique? 462
3. What are the two steps that we must complete when using coordinates to prove a theorem? 462
4. What are some of the algebraic tools available when writing coordinate proofs? 462

Review Problems

1. Suppose that rhombus $ABCD$ is centered at the origin and the coordinates of A are $(-a, 0)$ as shown. Determine appropriate coordinates of the other vertices of $ABCD$.

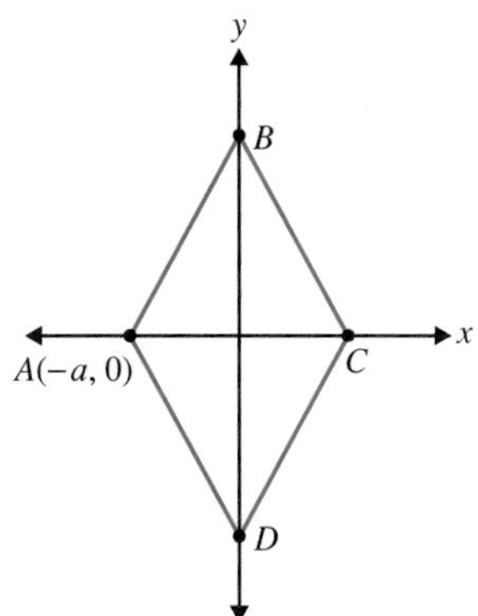

2. Use the coordinates from Problem 1 to verify that quadrilateral $ABCD$ is a rhombus. That is, verify that your choices of coordinates make $ABCD$ a rhombus.
3. Prove: If $\triangle ABC$ is isosceles with $\overline{AB} = \overline{BC}$ and if $\overline{BM}$ is the median to side $\overline{AC}$, then $\overline{BM} \perp \overline{AC}$.

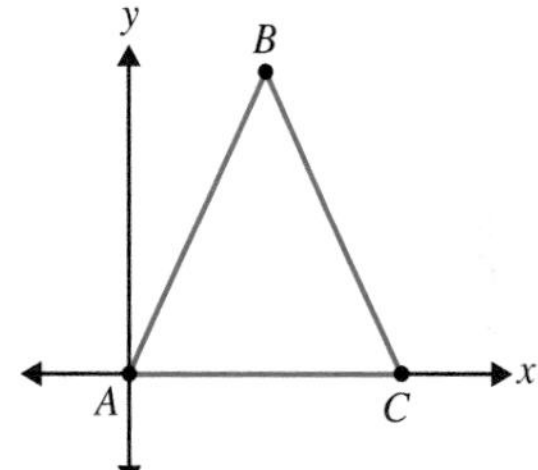

4. Prove: If $ABCD$ is an isosceles trapezoid and M, N, O, and P are midpoints of the sides $\overline{AB}$, $\overline{BC}$, $\overline{CD}$, and $\overline{AD}$ respectively, then $MNOP$ is a rhombus.

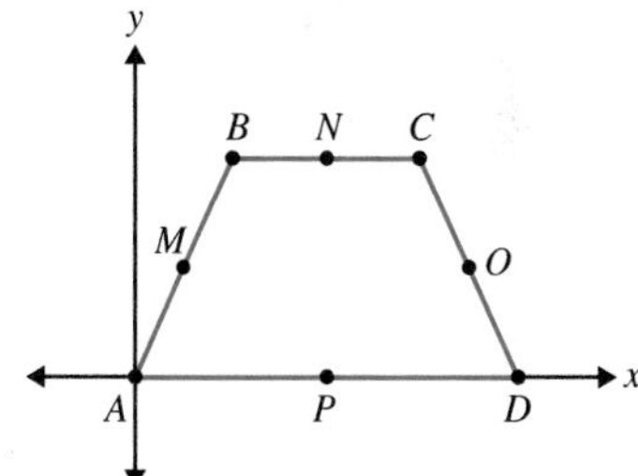

CHAPTER 8 TEST

TRUE-FALSE

Mark as true any statement that is always true. Mark as false any statement that is never true or that is not necessarily true. Be able to justify your answers.

1. If the slope of a line is 2, then the slope of any line perpendicular to it is −2.
2. If points A, B, and C are collinear, then $AB + BC = AC$.
3. If points A, B, and C are collinear and lie on a line that is not vertical, then the slope of $\overline{AB}$ equals the slope of $\overline{AC}$.
4. If two line segments bisect each other, then they intersect at their midpoints.
5. Given the coordinates of the vertices of $ABCD$, you can use the slope formula to determine whether it is a rhombus.
6. A system of two linear equations has exactly one simultaneous solution.
7. A horizontal line has a slope of 0.
8. The line $ax + by = c$ (where $b \neq 0$) has a slope of $\frac{a}{b}$.

9. If each coordinate of each vertex of $\triangle ABC$ is doubled, the area of the resulting triangle will be twice the area of $\triangle ABC$.

10. The graph of the line $y = 2 - 5x$ has y-intercept 2.

EXERCISES/PROBLEMS

11. For the points $A(3, -2)$ and $B(9, 4)$, find the following.

a. The midpoint of $\overline{AB}$

b. The exact distance AB

c. The slope of $\overline{AB}$

d. The equation of the line containing $\overline{AB}$

12. Determine whether the three points $A(-30, 400)$, $B(-16, 232)$, and $C(20, -200)$ are collinear. Explain.

13. Sketch the graphs of the following equations.

a. $4x - 3y = 6$

b. $y = -4x$

c. $x = 2$

d. $x^2 + y^2 = 9$

e. $(x + 5)^2 + (y - 2)^2 = 36$

14. Solve each of the following systems of equations. If there is no solution, explain why not.

a. $3x - y = 11$ and $5x + 4y = 24$

b. $y = -x + 5$ and $3x + 3y = 13$

c. $5x + 3y = 15$ and $(x - 4)^2 + (y - 3)^2 = 4$

15. Write an equation of the perpendicular bisector of $\overline{AB}$ for $A(1, -6)$ and $B(5, 2)$.

16. The coordinates of three vertices of parallelogram $ABCD$ are shown. Determine the coordinates of B.

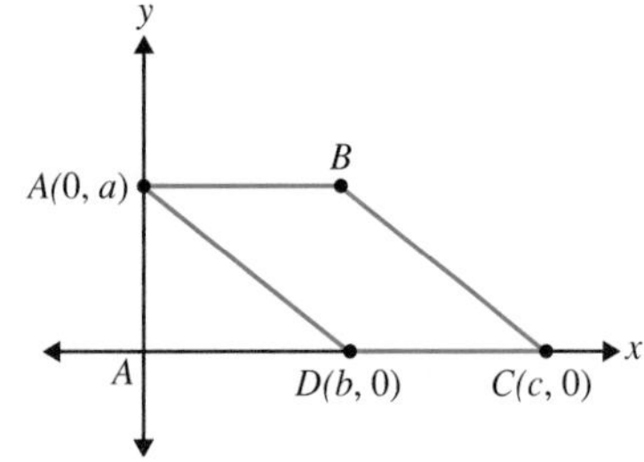

17. Quadrilateral $ABCD$ has vertices $A(-1, 1)$ $B(-2, 6)$, $C(3, 7)$, and $D(4, 2)$. Use these coordinates to verify that the diagonals of $ABCD$ are congruent.

18. A circle has center $O(-1, 4)$ and passes through the point $(5, 0)$. Write an equation of the circle.

19. Quadrilateral $ABCD$ has vertices $A(-5, 4)$, $B(8, 6)$, $C(12, 0)$, and $D(-1, -2)$. What type of quadrilateral is $ABCD$? Justify your answer.

20. The coordinates of the midpoints of the sides of a triangle are $M_1(6, 5)$, $M_2(5, -1)$ and $M_3(-1, 5)$. Find the coordinates of the vertices of the triangle.

PROOFS

21. Use coordinates to prove that if M_1, M_2 and M_3 are midpoints of the sides of equilateral $\triangle ABC$, then $\triangle M_1M_2M_3$ is an equilateral triangle.

22. Use coordinates to prove that if the diagonals of a parallelogram are perpendicular, then the parallelogram is a rhombus.

APPLICATIONS

23. A highway rises 220 feet in 0.8 mile. What is the slope of the road? What is the percent grade?

24. A copper wire has a diameter of 0.2 cm.

a. If x represents the length of the wire in cm and y represents the total volume of the wire, write a linear equation expressing y in terms of x.

b. What is the slope of the line whose equation you wrote in (a)? What does it mean?

c. The density of copper is 8.94 g/cm^3. Again, let x represent the length of the wire and now let y represent the total mass of the wire. Write a linear equation expressing y in terms of x.

d. What is the slope of the line whose equation you wrote in (c)? What does it mean?

CHAPTER 9

TRANSFORMATION GEOMETRY

Coordinate geometry provides one alternative way to verify relationships in geometry. Another, more modern method, transformation geometry, involves a more dynamic approach. Here, we study properties of geometric figures with respect to translations, rotations, reflections, and size transformations. The art of M. C. Escher uses the concepts associated with transformation geometry. In his *Circle Limit 1* shown next, Escher pictures flying figures that can be rotated to coincide with each other, figures that can be reflected and reduced in size to another, and so on. We can also describe many objects in nature using the language of transformation geometry. For example, the silhouette of a fir tree has reflection symmetry with respect to a vertical line through the center of its trunk, a starfish has 72° rotational symmetry because it rotates to itself after a turn equal to one-fifth of 360°, and so on.

Problem-Solving Strategies

1. Draw a Picture
2. Guess and Test
3. Use a Variable
4. Look for a Pattern
5. Make a Table
6. Solve a Simpler Problem
7. Look for a Formula
8. Use a Model
9. Identify Subgoals
10. Use Indirect Reasoning
11. Solve an Equation
12. Use Cases
13. Use Coordinates
14. *Use Symmetry*

Strategy 14: *Use Symmetry*

In Chapter 2, we observed symmetry in many geometric shapes. If a figure has reflection symmetry, we know that angles and sides that reflect onto corresponding angles and sides are congruent. We can deduce similar information from figures with rotation symmetry. By using a reflection or rotation, we can transform the following problem into a simpler, equivalent problem using symmetry.

Initial Problem

Houses A and B are to be connected to a television cable line, l, at a transformer at some point P. Where should P be located so that the sum of the distances, $AP + PB$, is as small as possible?

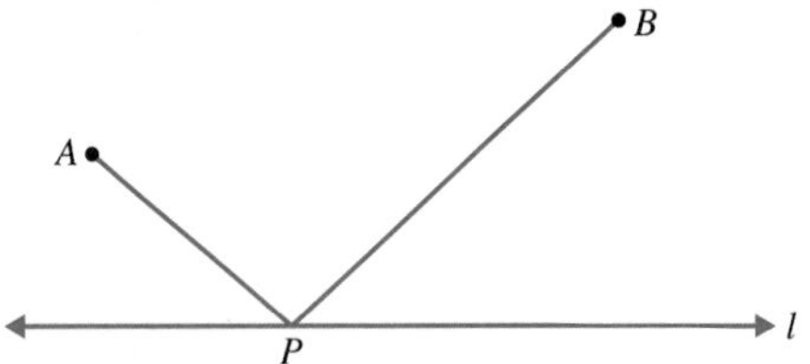

Clues

The Use Symmetry strategy may be appropriate when

- Geometry problems involve transformations.
- Interchanging values does not change the representation of the problem.
- Symmetry limits the number of cases that we need to consider.
- Pictures or algebraic procedures appear to be symmetric.

A solution of the Initial Problem is on page 519.

Tessellation Art

A tessellation or tiling is an arrangement of congruent figures in which the figures completely cover the plane with no gaps or overlaps. We can form tessellations using polygons as tiles or tiles derived from polygon regions. Follow the steps below to create a tessellation starting with square $ABCD$.

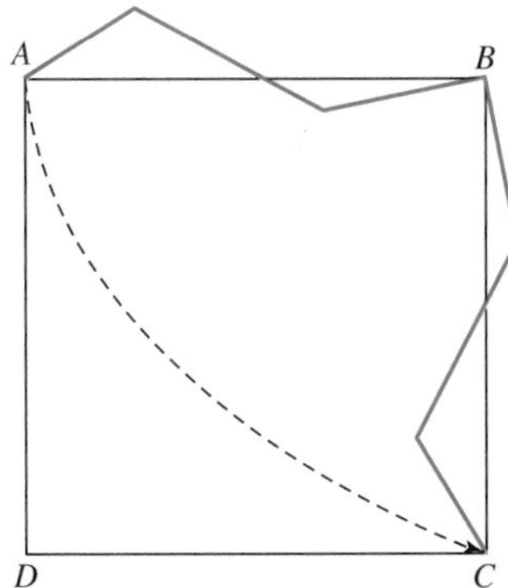

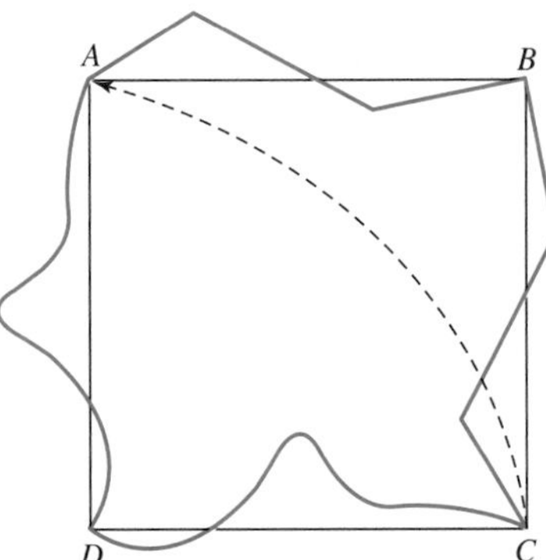

I. Draw a curve from vertex A to vertex B. Trace the curve and rotate it so that A rotates onto C and B stays fixed.

II. Draw a curve from vertex C to vertex D. Trace the curve and rotate it so that C rotates onto A and D stays fixed.

III. Erase the original square and add decorations to the shape. Make at least eight copies of this tile and arrange them to tessellate the plane.

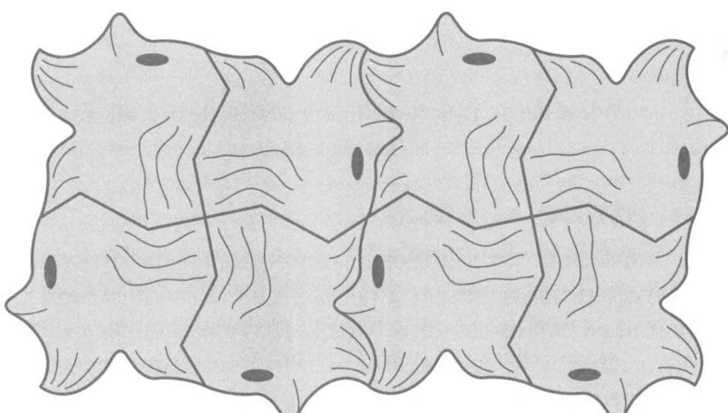

1. Draw a 2-inch by 2-inch square and follow the preceding instructions to create your own tessellation. Slide each piece into place and describe any other ways in which you must move the pieces to fit together. Must you flip over any pieces? Must you rotate any pieces?
2. For each of the following parts, make at least eight copies of the shape and fit them together to tessellate the plane. Describe how you must slide, rotate, or flip the pieces to make them fit together.

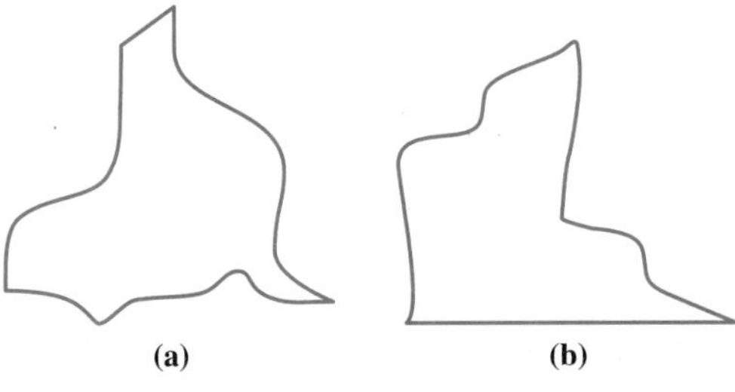

Introduction

We studied reflection and rotation symmetries in Chapter 2. In this chapter, we extend the notion of symmetry to a study of geometric figures using transformations. In Section 9.1, we study translations (slides), rotations (turns), reflections (flips), and combinations of them. In Section 9.2, we study similitudes, transformations that preserve shapes of objects. Section 9.3 is devoted to solving problems using translations, rotations, reflections, and glide reflections.

9.1 ISOMETRIES AND CONGRUENCE

Applied Problem

An astronomer is constructing a star map to show the position of the constellation Cassiopeia at different times during one evening. The following map shows its position with respect to the North Pole at 9:00 P.M.

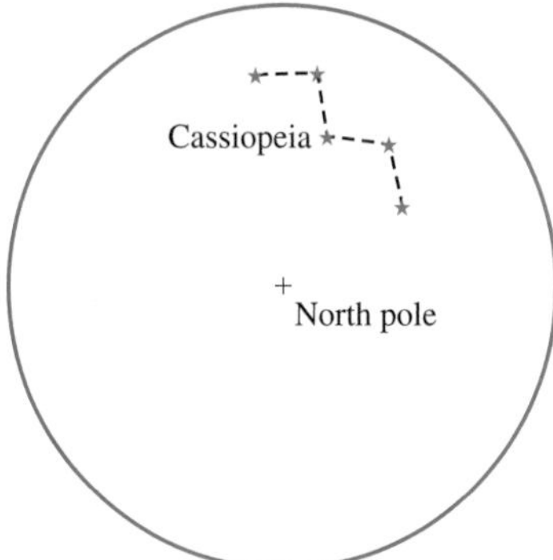

What is the position of the constellation at 3:00 A.M. the next morning?

Transformations

A translation, or slide, is one example of a transformation. To think of a slide, imagine a **directed line segment**, that is, a line segment where one end of the segment has been designated by an arrowhead. We call such a directed line segment a **vector**. Associated with each vector is a distance (the length of the vector) and a direction (the measure of the angle the vector makes with the positive x-axis) [Figure 9.1(a)]. Now imagine moving every point in a plane the same distance and in the same direction as indicated by the vector v in Figure 9.1(a) [Figure 9.1(b)].

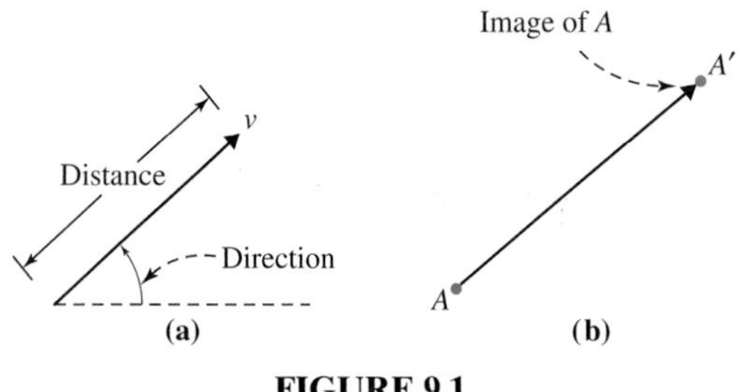

FIGURE 9.1

A one-to-one correspondence that assigns every point A in a plane to another point A' in the plane is called a **transformation**. Point A' is called the **image** of A and

A is called **preimage** of A'. In the case of a transformation described by a vector, the image of A is at the end of the arrow's tip on point A' [Figure 9.1(b)]. We can denote such a vector as $\overrightarrow{AA'}$. (NOTE: Although the arrow written above two letters has represented a ray thus far, in this chapter this notation will represent a vector of finite length unless stated otherwise. In this context, an arrow such as the one in Figure 9.1(b) also has the *finite* length AA', whereas a ray has infinite length.) We may assign one or many of the points to themselves under a transformation. For example, in a rotation, we will show that the center corresponds to itself.

We will study several types of transformations in this chapter, including translations, rotations, and reflections, as well as combinations of these transformations. We will emphasize their properties and how we can use them to solve problems.

Translations

Every vector v defines a transformation in that it assigns to every point A in a plane, a point A' with a distance and direction determined by v [Figure 9.1(b)]. This type of transformation is called a **translation**. We can informally think of a translation of the plane as a slide, sliding all the points of the plane the same distance and same direction.

EXAMPLE 9.1 Describe the image of $\triangle ABC$ under the translation determined by the vector v in Figure 9.2(a).

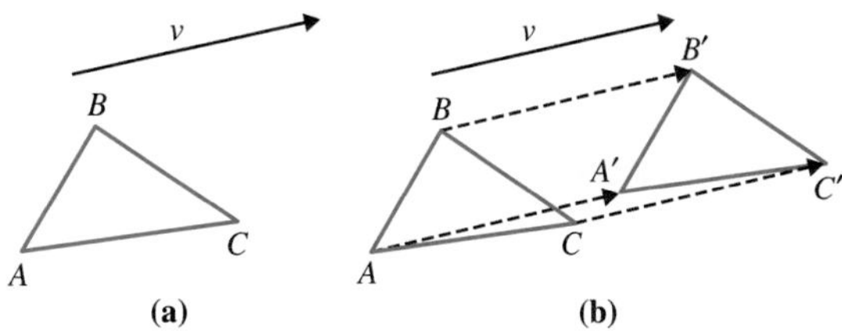

FIGURE 9.2

SOLUTION

Because the vector v takes all points the same distance and direction, $\triangle ABC$ is assigned to $\triangle A'B'C'$ as shown in Figure 9.2(b). Notice that vectors $\overrightarrow{AA'}$, $\overrightarrow{BB'}$, and $\overrightarrow{CC'}$ all have the same length and direction as the vector v. ■

In Figure 9.2(b), it appears that $\triangle ABC \cong \triangle A'B'C'$. We can prove this is true by studying Figure 9.3. The translation that is determined by vector v forms the parallelogram $ABB'A'$ because $\overline{AA'}$ and $\overline{BB'}$ are congruent and parallel. Therefore, $\overline{AB} \cong \overline{A'B'}$ because they are opposite sides of a parallelogram. In a similar manner $\overline{BC} \cong \overline{B'C'}$ and $\overline{AC} \cong \overline{A'C'}$. Thus, $\triangle ABC \cong \triangle A'B'C'$ by SSS Congruence.

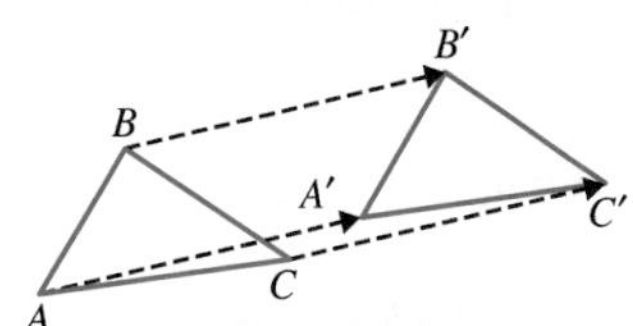

FIGURE 9.3

By corresponding parts, $\angle ABC \cong \angle A'B'C'$, $\angle BAC \cong \angle B'A'C'$, and $\angle ACB \cong \angle A'C'B'$.

THEOREM 9.1 Properties of Translations

1. Translations take lines to lines, rays to rays, and line segments to line segments.
2. Translations preserve distance.

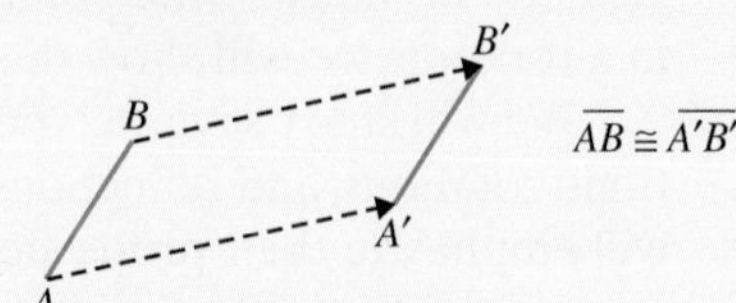

3. Translations preserve angle measure.

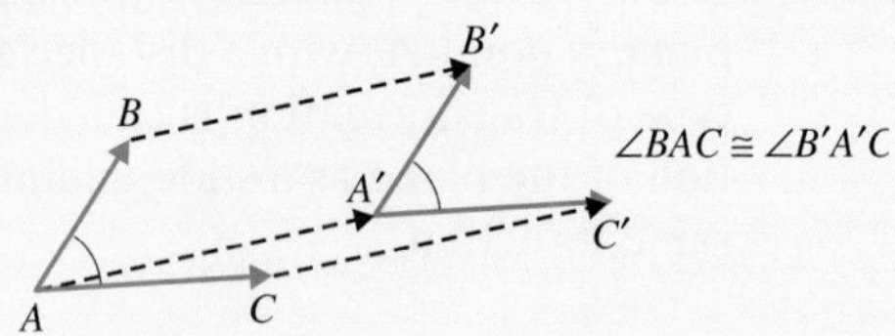

4. Translations preserve perpendicularity.
5. Translations preserve parallelism.

Proof

Part 1 Because the proof of this result requires sophisticated ideas that are beyond the scope of this book, we will accept this result as we do postulates.

Part 2 We proved this result in the paragraph preceding Theorem 9.1.

Part 3 This proof also was given in the paragraph preceding Theorem 9.1.

Part 4 Because perpendicular lines form right angles and angle measure is preserved by part 3, it follows that perpendicularity is preserved.

Part 5 If two parallel lines are cut by a transversal they form congruent alternate interior angles. By part 1, the lines go to lines under translations. By part 3, angle measure is preserved. Therefore, the alternate interior angles of the new set of lines formed by the transversal are congruent. So the lines are parallel by Theorem 5.1. ■

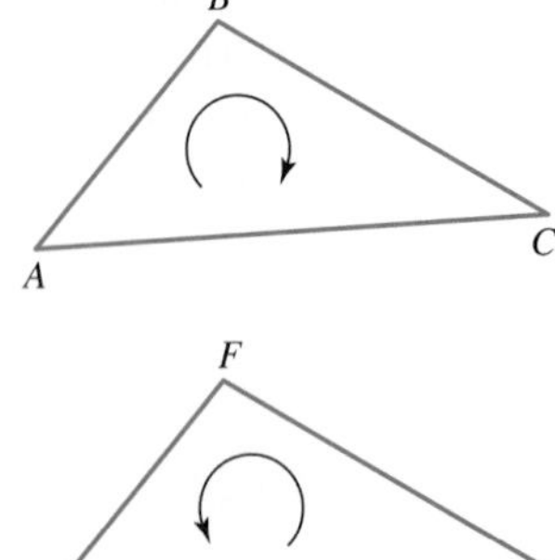

FIGURE 9.4

Part 5 of Theorem 9.1 states that translations preserve parallelism. That is, if two lines are parallel, their images are also parallel. However, as suggested in Figure 9.3, under a translation, the image of any line is parallel to the original line. For example, $\overline{AB}$ and $\overline{A'B'}$ in Figure 9.3 are parallel.

Another important property of transformations is orientation. $\triangle ABC$ is said to have **clockwise orientation** if its vertices when read A, B, C are pictured in a clockwise orientation. Otherwise, the triangle is said to have a **counterclockwise orientation**. In Figure 9.4, $\triangle ABC$ has clockwise orientation and $\triangle DEF$ has counterclockwise orientation. Notice that $\triangle ACB$ has a counterclockwise orientation.

Check the orientation of a triangle and its image under several translations and you will find that translations preserve orientation of triangles (and all other shapes). Later in this section, we will study transformations that do not preserve orientation.

We can use coordinates to illustrate the effects of a translation as shown next.

EXAMPLE 9.2 A vector v, which represents a translation, is graphed on the coordinate system in Figure 9.5(a).

a. A triangle has vertices $A(-1, 1)$, $B(4, -2)$, and $C(3, 6)$. Find the image of $\triangle ABC$ under the translation described by v. That is, find $\triangle A'B'C'$.

b. Verify that $\triangle ABC \cong \triangle A'B'C'$ and that the translation preserves orientation.

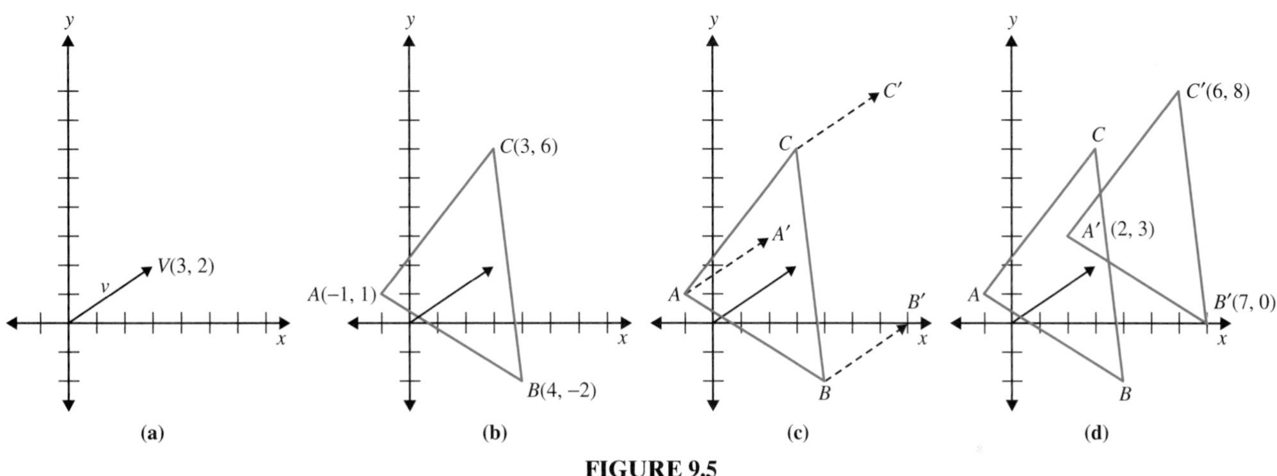

FIGURE 9.5

SOLUTION

a. $\triangle ABC$ is graphed in Figure 9.5(b). Note that $\triangle ABC$ has a counterclockwise orientation. The effect of vector v will be to translate each point of $\triangle ABC$ three units to the right and two units up. If we apply the translation to each vertex of $\triangle ABC$, we get the three points A', B', and C' shown in Figure 9.5(c). Connecting the vertices, we finally obtain $\triangle A'B'C'$, the image of $\triangle ABC$ in Figure 9.5(d).

b. We can use the distance formula and SSS Congruence to verify that $\triangle ABC \cong \triangle A'B'C'$.

In $\triangle ABC$,

$$\begin{aligned} AB &= \sqrt{(4+1)^2 + (-2-1)^2} = \sqrt{34} \\ BC &= \sqrt{(3-4)^2 + (6+2)^2} = \sqrt{65} \\ AC &= \sqrt{(3+1)^2 + (6-1)^2} = \sqrt{41} \end{aligned}$$

In $\triangle A'B'C'$,

$$\begin{aligned} A'B' &= \sqrt{(7-2)^2 + (0-3)^2} = \sqrt{34} \\ B'C' &= \sqrt{(6-7)^2 + (8-0)^2} = \sqrt{65} \\ A'C' &= \sqrt{(6-2)^2 + (8-3)^2} = \sqrt{41} \end{aligned}$$

Because $AB = A'B'$, $BC = B'C'$, and $AC = A'C'$, we have $\triangle ABC \cong \triangle A'B'C'$ by SSS Congruence. We have shown that the translation determined by the vector v takes $\triangle ABC$ to congruent triangle $\triangle A'B'C'$. Also, because both triangles have counterclockwise orientation, the translation preserves orientation. ■

Rotations

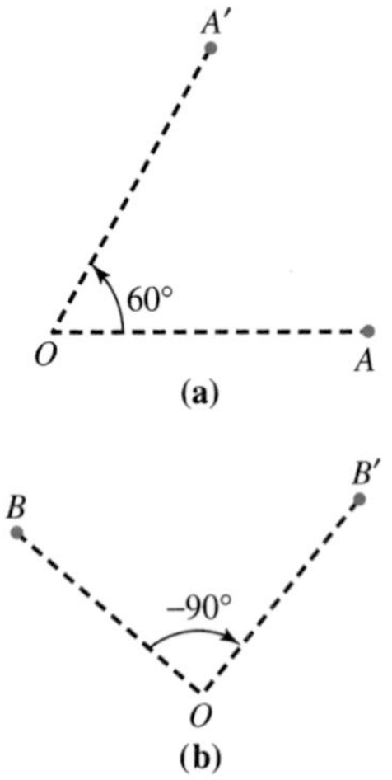

FIGURE 9.6

Another type of transformation is a rotation, or a turn. A rotation is determined by a point, O, and a directed angle. A **directed angle** is an angle where one side is identified as the initial side and a second side is the terminal side. An angle can be directed either clockwise or counterclockwise. In Figure 9.6(a), point A is rotated counterclockwise 60° around O to A'. Here $\overline{OA}$ is the initial side of $\angle AOA'$ and $\overline{OA'}$ is its terminals side. Angles that are directed counterclockwise are assigned a positive measure. Thus, we say that the measure of directed angle $\angle AOA'$ is 60°. In Figure 9.6(b), point B is rotated clockwise 90° around O to B', and, so directed angle $\angle BOB'$ has measure $-90°$ to indicate its clockwise rotation. The point O is called the **center of rotation** in each case.

Notice that under these rotations, $OA = OA'$ and $OB = OB'$. Also, the orientation of a directed angle is indicated by the way the angle is *named*. For example, the directed angle $\angle AOA'$ in Figure 9.6(a) is directed counterclockwise, and, hence, is positive, whereas the directed angle $\angle A'OA$ has a clockwise direction, hence, its measure is $-60°$.

Based on the directed angles in Figure 9.6, we can state the following formal definition of a **rotation**. The image of a point X under the rotation determined by the directed angle $\angle AOB$ in Figure 9.7(a) is the point X' where

1. $OX = OX'$
2. $\angle XOX' = \angle AOB$ as *directed* angles [Figure 9.7(b)]

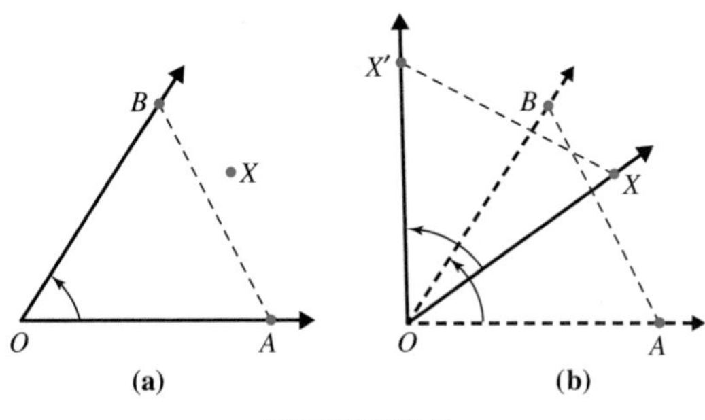

FIGURE 9.7

Once a rotation is defined by its center and its directed angle (whose vertex is the center of rotation), the image of every point in the plane is determined. That is, we can find the image A' of any point A in the plane.

The center, O, of a rotation always corresponds to itself. Thus, we call it a fixed point. In general, a point A is called a **fixed point** under a transformation if A and its image A' are the same point. The transformation that leaves all points fixed is called the **identity transformation**. The translation determined by the zero vector is the identity transformation. All other translations have no fixed points. The rotation whose directed angle is 360° is also the identity transformation since each point corresponds to itself.

Next we consider the effect a rotation has on a triangle.

EXAMPLE 9.3 Describe the image of $\triangle ABC$ under the rotation with center O and directed angle $\angle XOX'$, as shown in Figure 9.8(a).

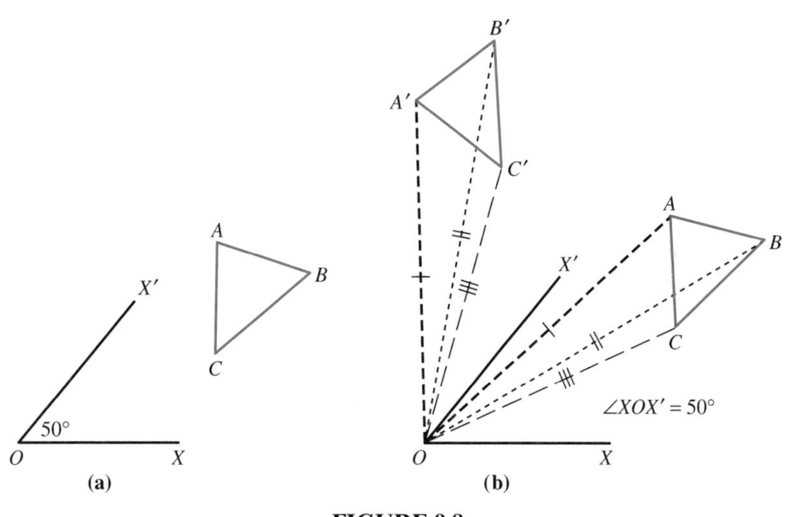

FIGURE 9.8

SOLUTION

The respective images A', B', and C' are shown in Figure 9.8(b). Notice that $OA = OA'$, $OB = OB'$, and $OC = OC'$. Also, $\angle AOA' = \angle BOB' = \angle COC' = 50°$. ■

It appears that $\triangle A'B'C' \cong \triangle ABC$ but it has been rotated 50° counterclockwise around O. Due to the nature of the rotation in Figure 9.8(b), we can observe the following:

1. $OA = OA'$
2. $OB = OB'$
3. $\angle AOA' = \angle BOB' = 50°$

If we subtract $\angle AOB'$ from $\angle AOA'$ and $\angle BOB'$, we have

4. $\angle A'OB' = \angle AOB$

Thus, by SAS Congruence, we can conclude that $\triangle AOB \cong \triangle A'OB'$. Therefore, $AB = A'B'$. In a similar manner, $AC = A'C'$ and $BC = B'C'$. Using SSS Congruence, we have that $\triangle A'B'C' \cong \triangle ABC$. This shows, by example, that a rotation takes a triangle to a congruent triangle. Using methods analogous to those we used for translations, we can prove the following theorem.

THEOREM 9.2 Properties of Rotations

1. Rotations take lines to lines, rays to rays, and line segments to line segments.
2. Rotations preserve distance.

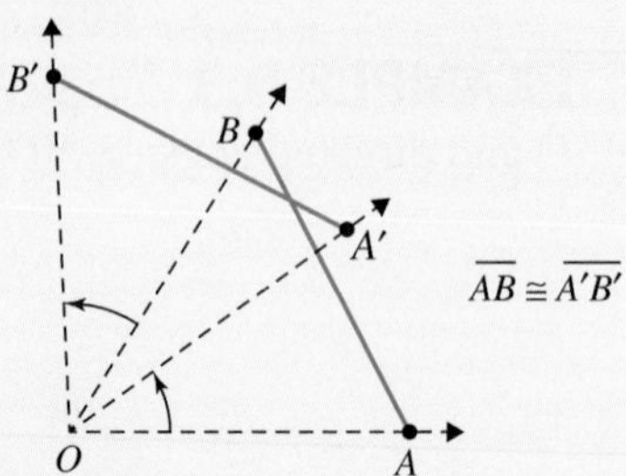

3. Rotations preserve angle measure.

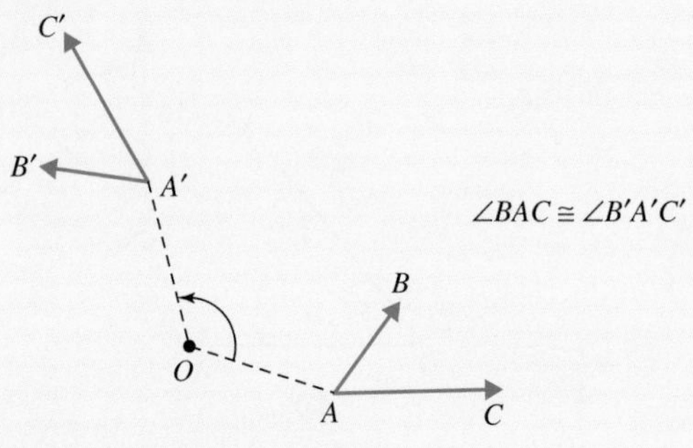

4. Rotations preserve perpendicularity.
5. Rotations preserve parallelism.

We leave the verification of properties 2–5 for the problem set. Several examples should make it clear that rotations preserve orientation just as translations do.

Although Theorems 9.1 and 9.2 show that translations and rotations have many similar properties, they differ in some ways. Except for the zero translation, translations have no fixed points. However, rotations always have at least one fixed point, namely the center. Also, the image of a line under a translation is parallel to the original line. However, this is not generally true in a rotation. For example, in Figure 9.8(b), the image of $\overline{AB}$, namely $\overline{A'B'}$, is not parallel to $\overline{AB}$.

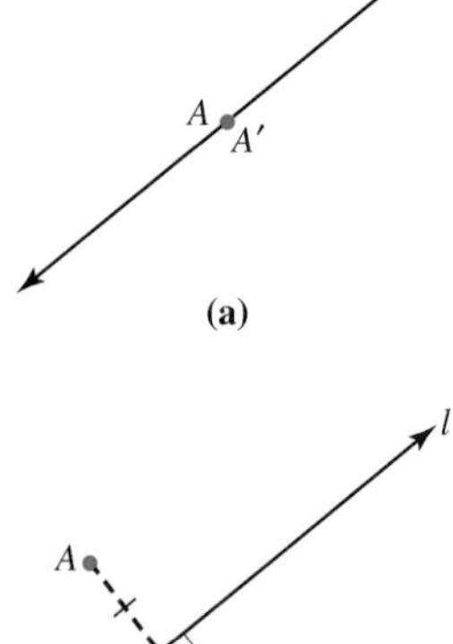

FIGURE 9.9

Reflections

In Section 2.2, we observed that some geometric figures were symmetric with respect to a line. This idea is associated with a type of transformation called a reflection. A **reflection with respect to line *l*** is defined by describing the location of the image of each point of the plane as follows:

1. If A is a point on l, then $A = A'$ (that is, a point on the line of reflection is its own image) [Figure 9.9(a)].
2. If A is not on l, then l is the perpendicular bisector of $\overline{AA'}$ [Figure 9.9(b)].

Next we consider the effect of a reflection on a triangle.

EXAMPLE 9.4 Describe the image of $\triangle ABC$ under the reflection with respect to line l as shown in Figure 9.10(a).

SOLUTION

Line l will be the perpendicular bisector of $\overline{AA'}$. So we construct a line through A that is perpendicular to l at point P, then mark off $\overline{A'P} \cong \overline{AP}$ [Figure 9.10(b)]. Then we find points B' and C' in the same way. The respective images A', B', and C' of A, B, and C are shown in Figure 9.10(c). Here, again, it appears that $\triangle A'B'C' \cong \triangle ABC$. However, notice that while the orientation of $\triangle ABC$ is clockwise, the orientation of $\triangle A'B'C'$ is counterclockwise.

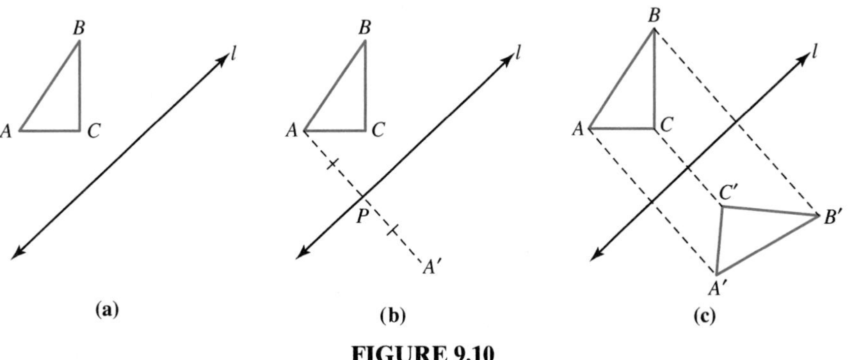

FIGURE 9.10

It can be proved that $\triangle A'B'C' \cong \triangle ABC$. This result is a consequence of the next theorem whose proof we leave for the problem set.

THEOREM 9.3 Properties of Reflections

1. Reflections take lines to lines, rays to rays, and line segments to line segments.
2. Reflections preserve distance.

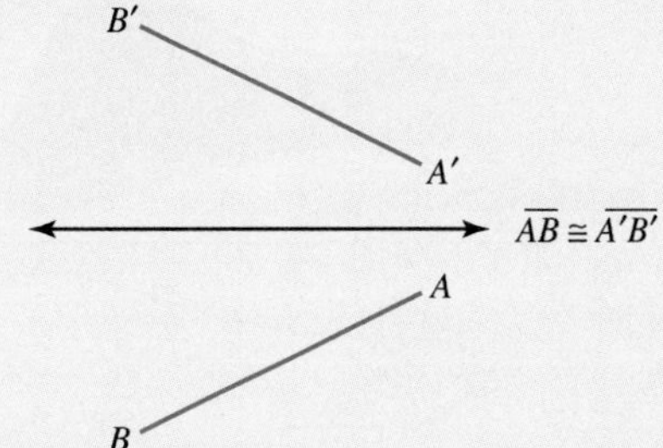

3. Reflections preserve angle measure.

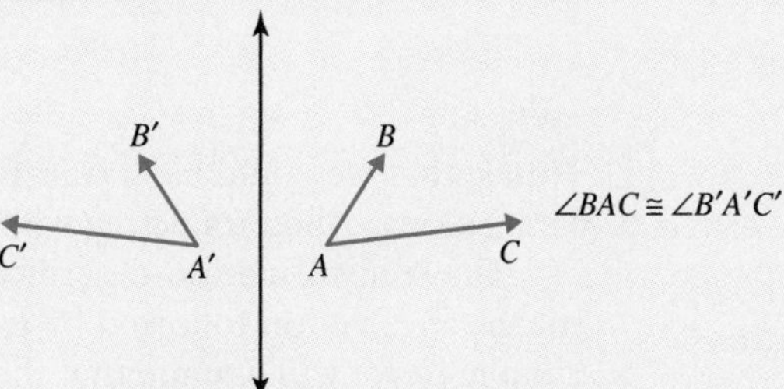

4. Reflections preserve perpendicularity.
5. Reflections preserve parallelism.

Notice that $\overline{AB}$ is *not* parallel to its image $\overline{A'B'}$ in Figure 9.10(c). Therefore, reflections are similar to rotations in that the image of a line is not necessarily parallel to the original line. In Figure 9.10(c), we observed that $\triangle ABC$ has clockwise orientation and $\triangle A'B'C'$ has counterclockwise orientation. Thus, reflections differ from both translations and rotations since orientation is not preserved.

Glide Reflections

To motivate the final type of transformation in this section, consider the footprints in the sand pictured in Figure 9.11.

FIGURE 9.11

It is impossible to translate or rotate the left foot to the right foot since the feet have different orientations. Also, we cannot find any reflection line to reflect one foot onto the other. However, as Figure 9.12 shows, we can obtain one foot from the other by performing a reflection followed by a translation.

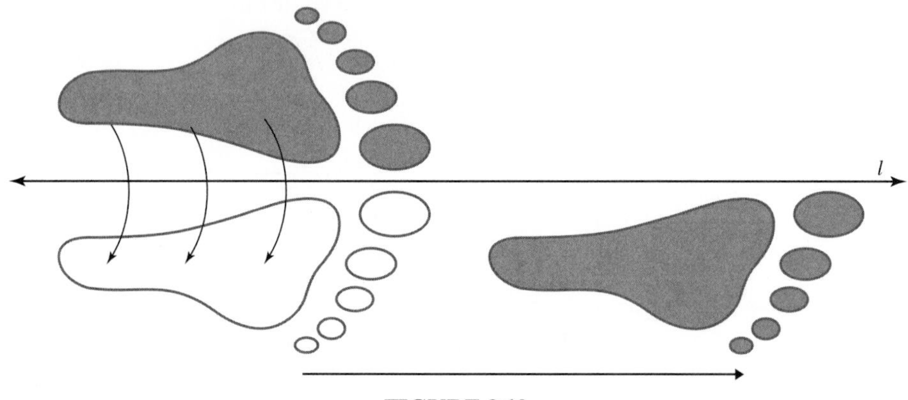

FIGURE 9.12

Notice that we could have transformed the left foot to the right foot by first translating along l, then reflecting with respect to line l.

The transformation described in Figure 9.12 is called a **glide reflection** which means a reflection followed by a translation that is not the identity translation. For simplicity, we will assume that the translation is in a direction parallel to the line of reflection. However, we can show that this assumption is not necessary. Because a glide reflection is composed of a reflection and a translation, it has all the properties common to both of these, as stated in the next theorem.

> **THEOREM 9.4 Properties of Glide Reflections**
> 1. Glide reflections take lines to lines, rays to rays, and line segments to line segments.
> 2. Glide reflections preserve distance.
> 3. Glide reflections preserve angle measure.
> 4. Glide reflections preserve perpendicularity.
> 5. Glide reflections preserve parallelism.

Notice that a glide reflection reverses the orientation of geometric figures because the reflection reverses orientation, but the translation that we used in combination with the reflection does not. The results listed in Theorem 9.4 follow immediately due to the fact that glide reflections are combinations of a translation and a reflection, each of which satisfies 1 to 5 in the theorem.

We can use coordinates to describe the effect of a glide reflection, as shown in the next example.

EXAMPLE 9.5 $\triangle ABC$ with vertices $A(1,0)$, $B(2,3)$, and $C(5,1)$ is shown in Figure 9.13(a).

a. Find the image of $\triangle ABC$ under the glide reflection defined by the translation from $(0,0)$ to $(0,-2)$, followed by a reflection with respect to the y-axis.

b. If a point has coordinates (s,t), what are the coordinates of its image under this glide reflection?

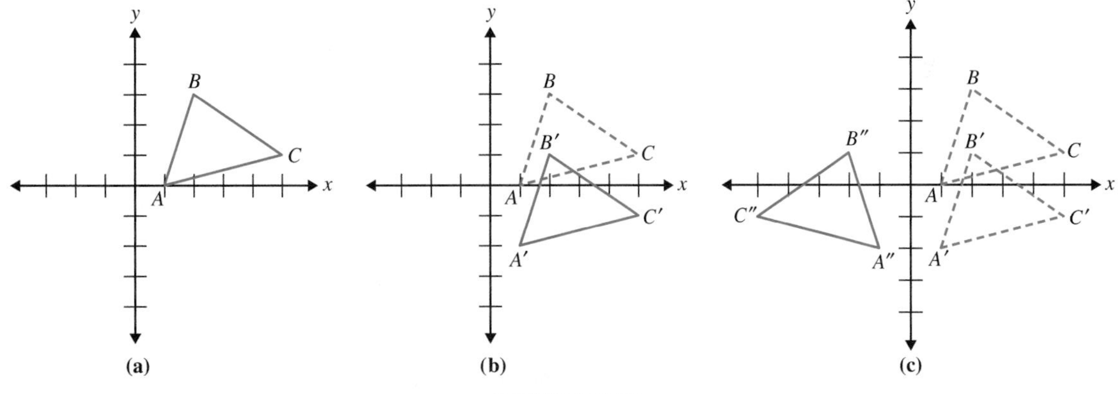

FIGURE 9.13

SOLUTION

a. The translation from $(0,0)$ to $(0,-2)$ shifts each point of the plane down two units. The image $\triangle A'B'C'$ of $\triangle ABC$ under this translation is shown in Figure 9.13(b). A reflection with respect to the y-axis flips $\triangle A'B'C'$ across the y-axis to yield $\triangle A''B''C''$ [Figure 9.13(c)]. Thus, $\triangle A''B''C''$ is the image of $\triangle ABC$ under this glide reflection.

b. Consider the images of points A, B, and C under this glide reflection. The points and their images are listed in Table 9.1.

Point	Image
$A(1, 0)$	$A''(-1, -2)$
$B(2, 3)$	$B''(-2, 1)$
$C(5, 1)$	$C''(-5, -1)$
(s, t)	$(?, ?)$

TABLE 9.1

In each case, the y-coordinate of the original point is decreased by 2 because of the downward translation. The x-coordinate of each image point is the opposite of the x-coordinate of its preimage because of the reflection with respect to the y-axis. Therefore, the image of the point (s, t) is $(-s, t - 2)$. ■

Isometries

All of the transformations we have discussed share the property that they preserve distance. A transformation that preserves distance is called an **isometry**—iso means same and metry means measure. That is, for any pair of points A and B and their images A' and B', we have $AB = A'B'$. Translations, rotations, reflections, and glide reflections are isometries because they all preserve distance. Although we will not prove it in this book, it can be proven that these are the *only* isometries. That is, if there is a transformation that preserves distance, then it must be one of the four types that we have studied.

Because isometries preserve distance, they also preserve angle measure, perpendicularity, and parallelism, and they take lines to lines, rays to rays, and so on. In fact, we could have started this chapter by proving these properties about isometries. Then, when we showed that translations, rotations, and reflections preserved distance, they would automatically satisfy all the other properties of isometries.

In Chapter 4, we studied the concept of congruence of triangles. Because isometries preserve distance, the image of one triangle must be congruent to the original triangle by the SSS Congruence Postulate. This idea motivates the next theorem.

THEOREM 9.5 Congruent Triangles and Isometries

Two triangles are congruent if and only if there is an isometry that takes one triangle to the other.

In general, two polygons are congruent if there is a one-to-one correspondence between the vertices of one polygon and the vertices of the other such that the corresponding angles and the corresponding sides are congruent. Because we can subdivide polygons into triangles by connecting vertices with diagonals, we can extend congruence to polygons by using triangle congruence. We summarize this result next.

THEOREM 9.6 Congruent Polygons and Isometries

Two polygons are congruent if and only if there is an isometry that takes one polygon to the other.

The value in studying isometries is that we can use isometries to decide if *any* shapes, including those that are not polygons, are congruent. In general, we say that two shapes are **congruent** if and only if there is an isometry that takes one to the other. For example, the shaded footprints in Figure 9.11 are congruent because we can use a glide reflection to take one footprint to the other.

Solution of Applied Problem

In six hours, the constellation will make one quarter turn about the North Pole. So the astronomer can draw the constellation's new position by finding the image of Cassiopeia after a 90° rotation. The new position is shown next.

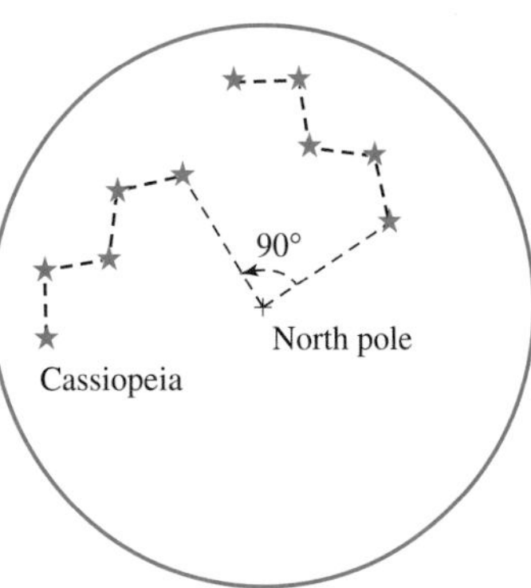

GEOMETRY AROUND US We can apply the four isometries described in this section again and again to create a repeated pattern that is pleasing to the eye. For example, patterns on wallpaper, pottery, and fabric often exhibit translations, reflections, rotations, or glide reflections. The pot and strip of wallpaper shown use a translation and glide reflection in their patterns.

PROBLEM SET 9.1

EXERCISES/PROBLEMS

1. Draw $\overrightarrow{PQ}$ and $\overline{AB}$ as illustrated next.

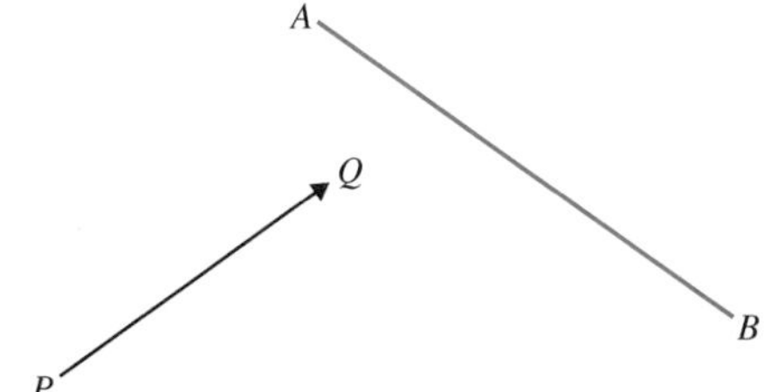

With a sheet of tracing paper on top, trace $\overline{AB}$ and point P. Slide the tracing paper (without turning) so that the traced point P moves to point Q. Make impressions of points A and B by pushing your pencil tip down. Label these impressions A' and B'. Draw $\overline{A'B'}$, the translation image of $\overline{AB}$. Does $\overline{A'B'}$ appear to be parallel to $\overline{AB}$? Should it be? Explain.

2. Using tracing paper, find the image, $\triangle A'B'C'$, of $\triangle ABC$ under the translation $\overrightarrow{PQ}$.

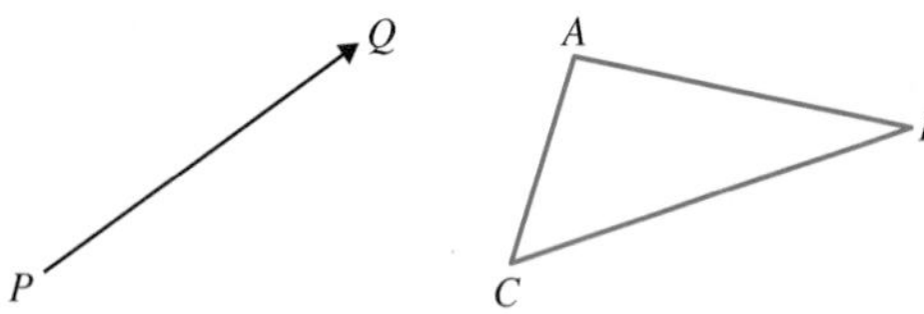

Does $\triangle A'B'C'$ appear to be congruent to $\triangle ABC$? Should it be? Explain.

3. a. On graph paper, draw three vectors that describe the translation that moves points 3 units down and 4 units to the right.

b. Draw a vector that describes the translation that maps $\triangle RST$ to $\triangle R'S'T'$.

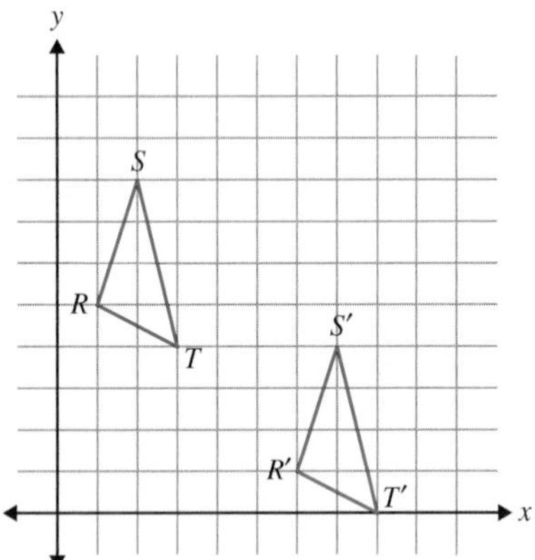

4. a. On graph paper, draw three vectors that describe the translation that moves points 5 units up and 3 units to the left.

b. Draw a vector that describes the translation that maps $\triangle RST$ to $\triangle R'S'T'$.

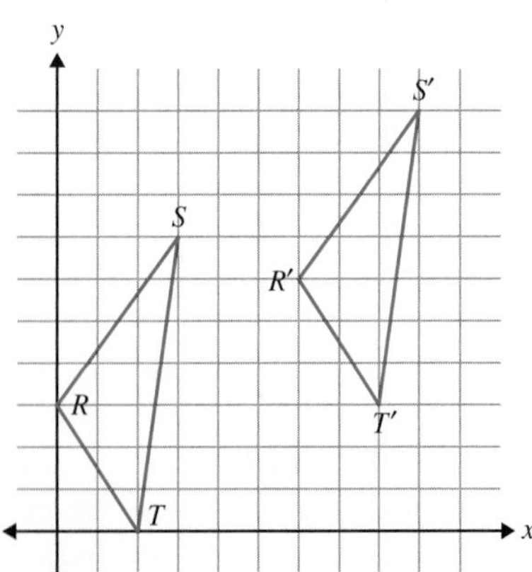

5. Find the coordinates of A' and B' that are the images of A and B under the translations shown in parts (a), (b), (c), and (d).

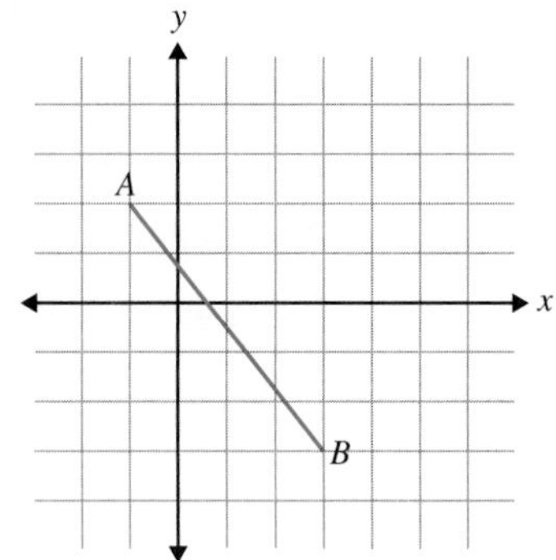

a.

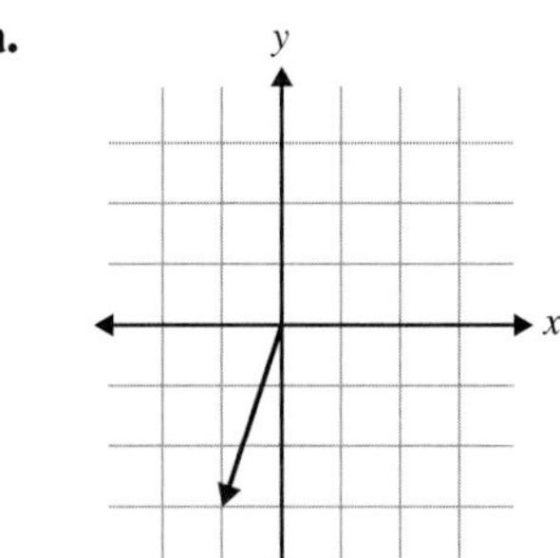

b.

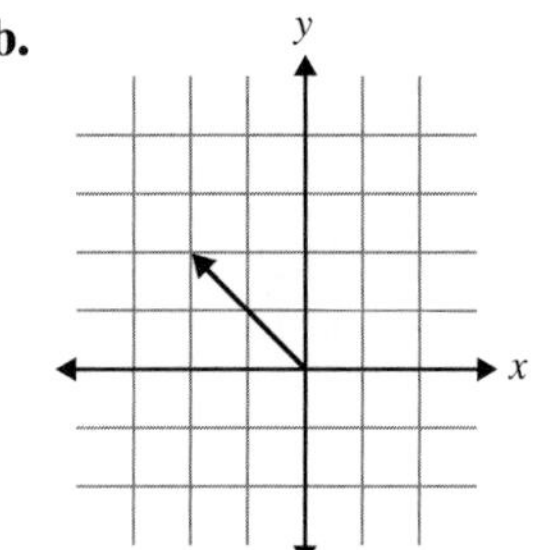

c.

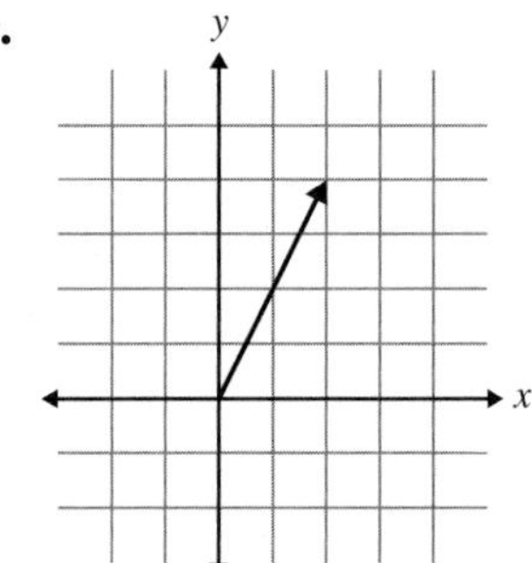

d.

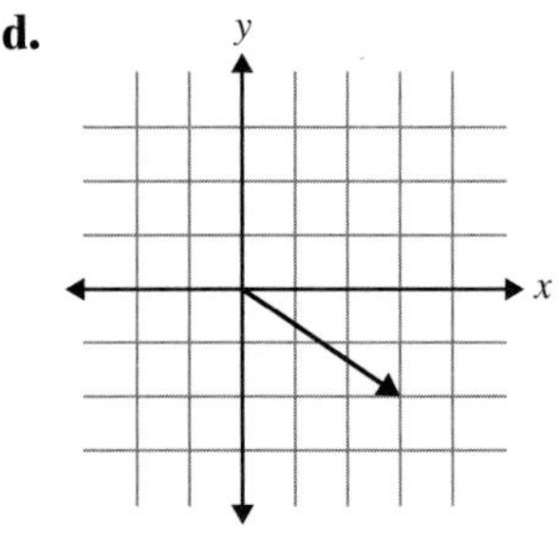

6. Find the coordinates of A' and B' that are the images of A and B under the translations listed in parts (a), (b), (c), and (d).

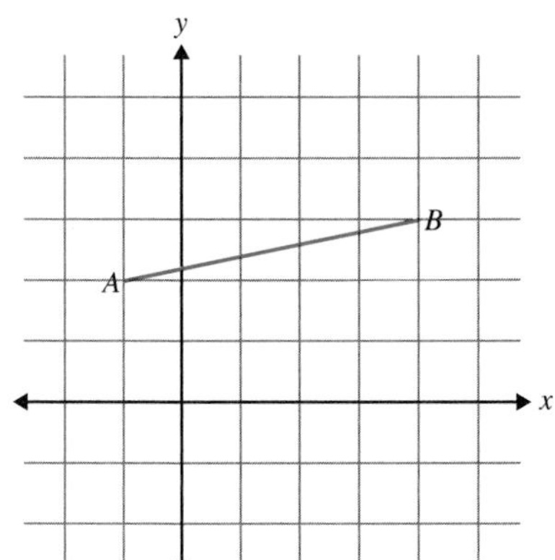

a.

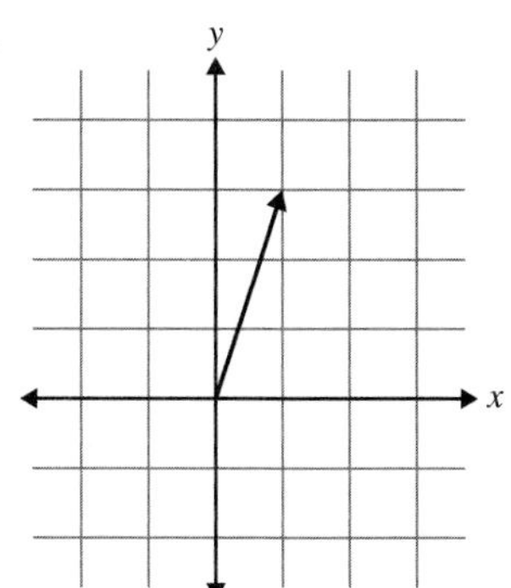

b.

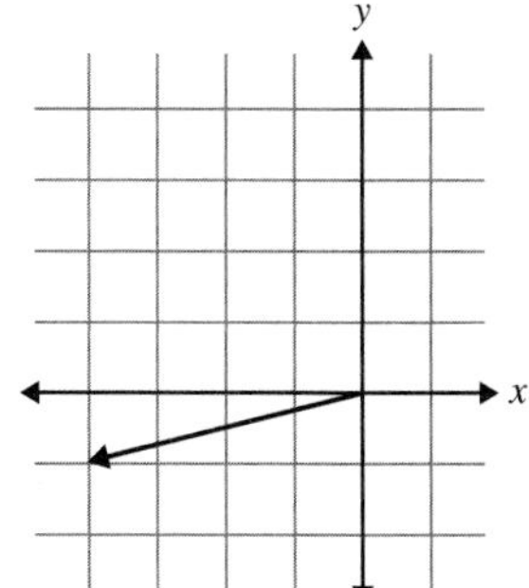

c.

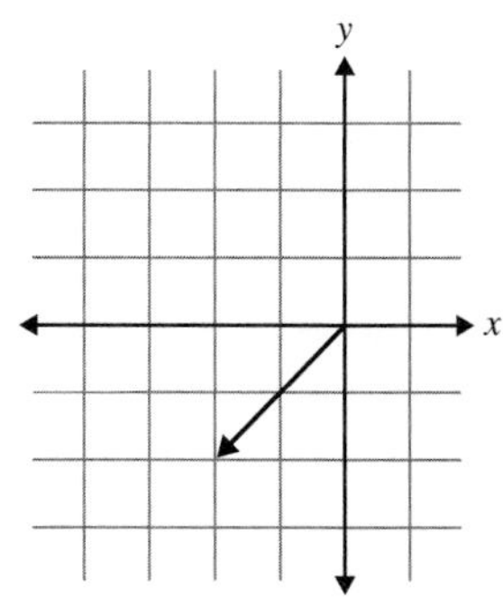

d.

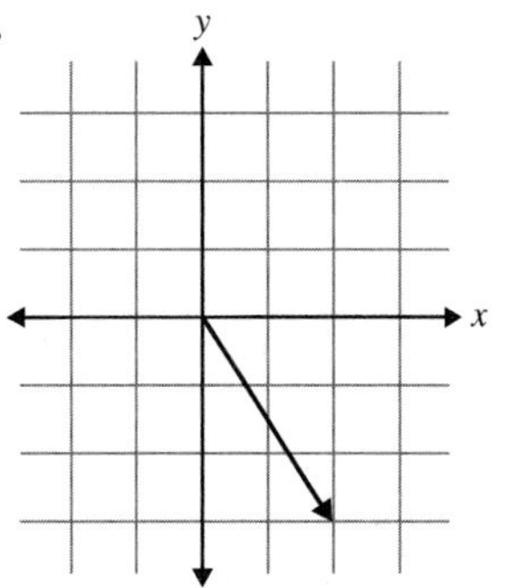

7. Using tracing paper, find the image $\triangle A'B'C'$ of $\triangle ABC$ under the translation $\overrightarrow{PQ}$.

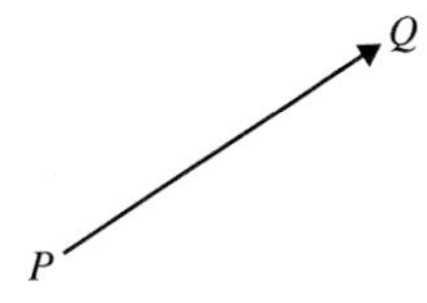

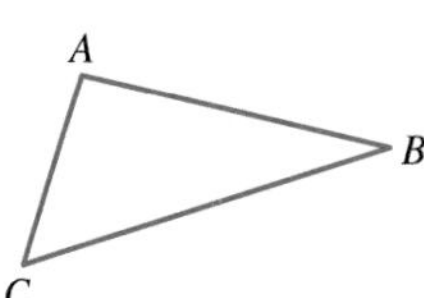

Then find the image $\triangle A''B''C''$ of $\triangle A'B'C'$ under the translation $\overrightarrow{RS}$. Does $\triangle A''B''C''$ appear to be congruent to $\triangle ABC$? Should it be? Explain.

8. Using tracing paper, find the image $\triangle A'B'C'$ of $\triangle ABC$ under the translation $\overrightarrow{PQ}$.

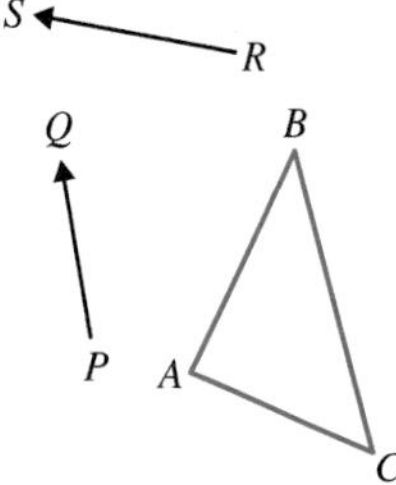

Then find the image $\triangle A''B''C''$ of $\triangle A'B'C'$ under the translation $\overrightarrow{RS}$. Does $\triangle A''B''C''$ appear to be congruent to $\triangle ABC$? Should it be? Explain.

9. Use a compass and straightedge to construct the image of $\overline{AB}$ under the translation $\overrightarrow{PQ}$ (HINT: Use the fact that $AA'B'B$ must be a parallelogram.)

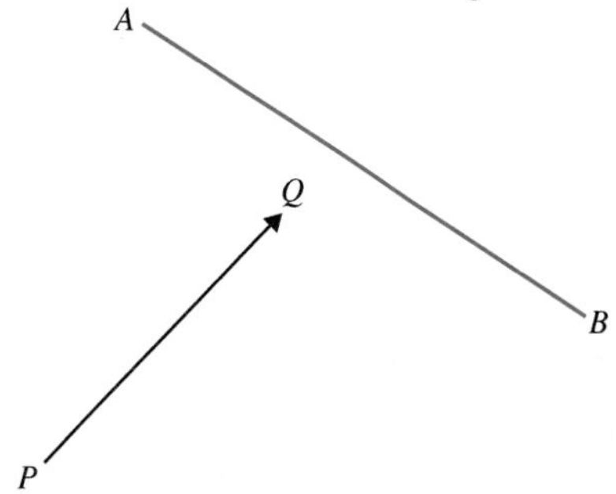

10. Use a compass and straightedge to construct the image of $\triangle ABC$ under the translation $\overrightarrow{PQ}$.

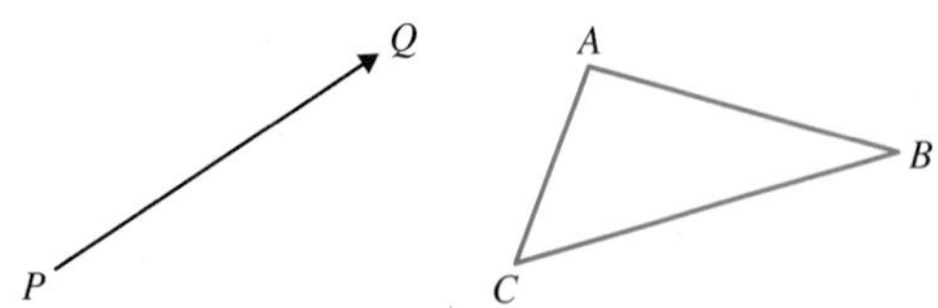

11. The coordinates of $\triangle ABC$ are $A(-1,1)$, $B(3,3)$, and $C(4,-2)$. Find the coordinates of points A', B', and C' under the translations that take

a. $P(0,0)$ to $Q(1,3)$

b. $P(0,-1)$ to $Q(4,-2)$

c. $P(-1,-2)$ to $Q(-2,2)$

12. The coordinates of $\triangle ABC$ are $A(-4,-2)$, $B(0,1)$, and $C(3,-1)$. Find the coordinates of points A', B', and C' under the translations that take

a. $P(0,0)$ to $Q(2,5)$

b. $P(0,3)$ to $Q(2,-2)$

c. $P(-3,-1)$ to $Q(0,4)$

13. Find the 90° counterclockwise rotation of point P around point O.

a. P O

b. P O

c. P O

14. Find the rotation image of $\overline{AB}$ around point O for each of the following directed angles.

a. −90° A O B

b. 90° A B O

c. 180° A O B

15. Find the measure of each of the following directed angles $\angle ABC$.

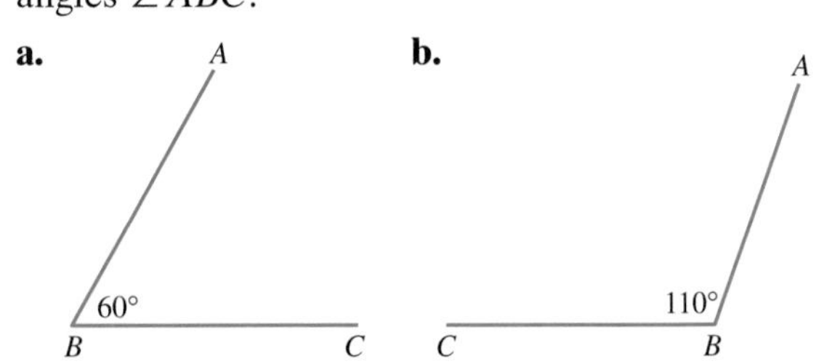

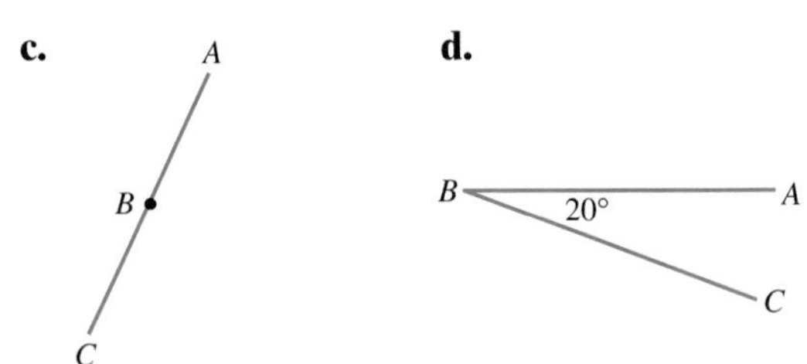

16. Using a protractor, draw and label $\angle ABC$ to represent the following directed angles.

a. 40° **b.** −60° **c.** 110° **d.** −150°

17. A protractor and tracing paper may be used to find rotation images. For example, find the image of point A under a rotation of −50° about center O by following these steps

A O

i. Draw a ray from O through A.

ii. With ray $\overrightarrow{OA}$ as the initial side, use your protractor to draw a directed angle $\angle AOB$ of −50°.

iii. Place tracing paper on top and trace A.

iv. Keep O fixed and turn the tracing paper until A is on ray $\overrightarrow{OB}$. Make an imprint with your pencil for A'.

18. Using a protractor and tracing paper, find the rotation image of A about O for the following directed angles.

a. $75°$ **b.** $-90°$ **c.** $-130°$ **d.** $180°$

A

O

19. Using a protractor, find the image, $\triangle A'B'C'$, of $\triangle ABC$ under the rotation about point O of $90°$. Are the sides of $\triangle ABC$ parallel to the corresponding sides of $\triangle A'B'C'$? Should they be? Explain.

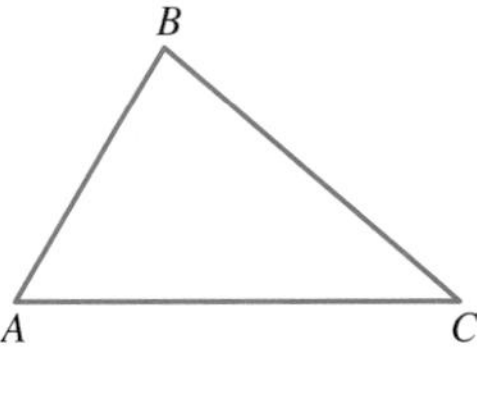

O

20. Using a protractor, find the image, $\triangle A'B'C'$, of $\triangle ABC$ under the rotation around O of $70°$. Then find the rotation image, $\triangle A''B''C''$, of $\triangle A'B'C'$, under the rotation of $110°$ around O. Is there a rotation around O that takes $\triangle ABC$ to $\triangle A''B''C''$? If so, describe it. Are the sides of $\triangle ABC$ parallel to the corresponding sides of $\triangle A''B''C''$? Should they be? Explain.

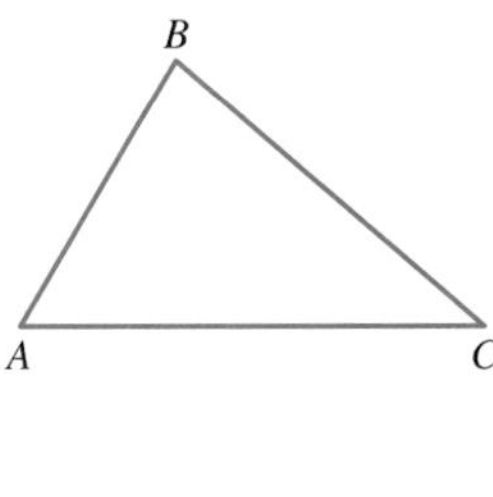

21. Using a compass, straightedge, and/or protractor, find the $-60°$ rotation image of $\triangle ABC$ around O.

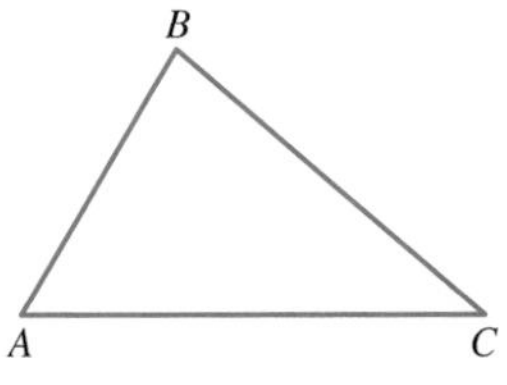

O

22. Using a compass, straightedge, and/or protractor, find the image of $\triangle ABC$ after it is rotated $40°$ around O.

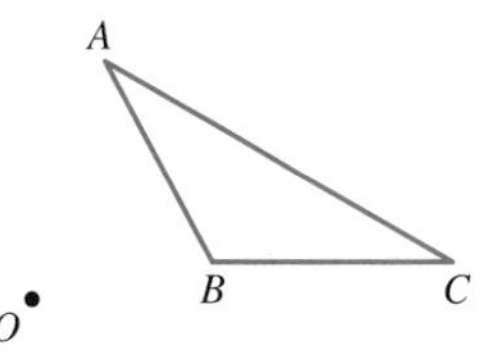

23. Using a compass, straightedge, and/or protractor, find the image, $\triangle A'B'C'$, of $\triangle ABC$ under the rotation of $-70°$ about point O. Then, using tracing paper, find the translation image $\triangle A''B''C''$ of $\triangle A'B'C'$ under the translation from P to Q. (NOTE: Use the location of P and Q as given, not as they would appear after a rotation.)

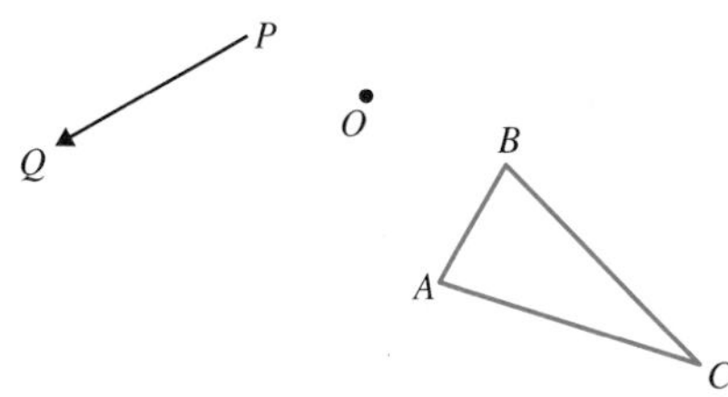

24. Using a compass and straightedge, find the image, $\triangle A'B'C'$, of $\triangle ABC$ under the rotation of $90°$ about point O. Then, using tracing paper, find the translation image $\triangle A''B''C''$ of $\triangle A'B'C'$ under the translation from P to Q. (NOTE: Use the location of P and Q as given, not as they would be after the rotation.)

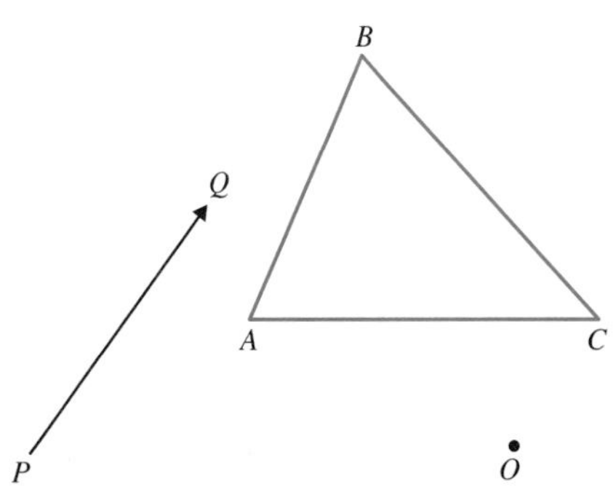

25. Give the coordinates of the images of the following points under the $-90°$ rotation around the origin.

a. $(1, 5)$ **b.** $(-1, 3)$ **c.** $(-2, 4)$
d. $(-3, -1)$ **e.** $(5, -2)$ **f.** (x, y)

26. Give the coordinates of the images of the following points under the rotation of $180°$ around the origin.

a. $(3, -1)$ **b.** $(-6, -3)$ **c.** $(-4, 2)$ **d.** (x, y)

27. Find the reflection of point A with respect to each of the given lines.

a. 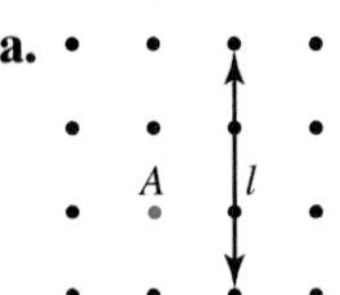

b. 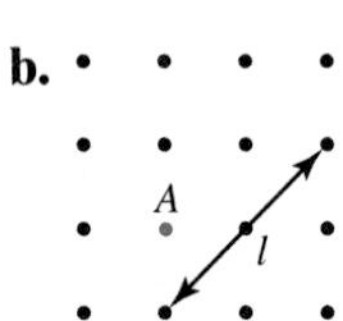

c. 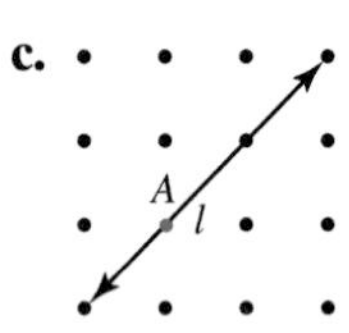

28. Find the image $\overline{A'B'}$ of the segment $\overline{AB}$ under a reflection with respect to each of the given lines l.

a. 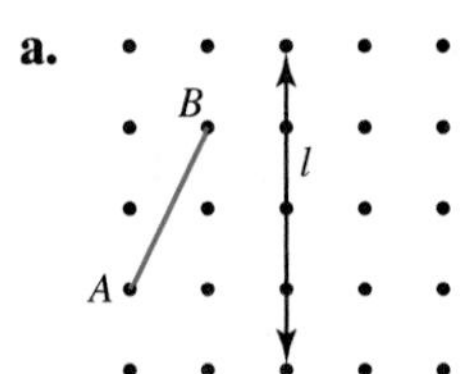

b. 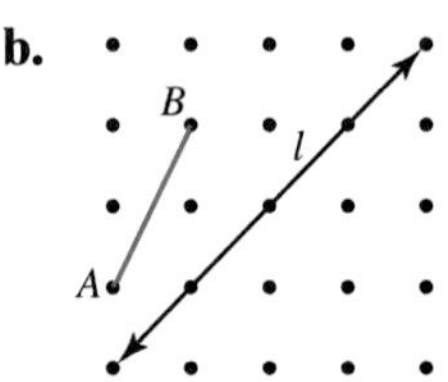

c. 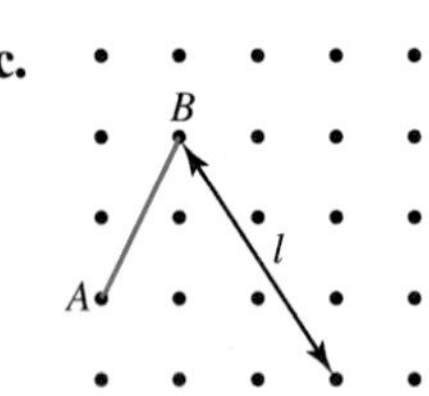

29. The reflection image of P can be found using tracing paper.

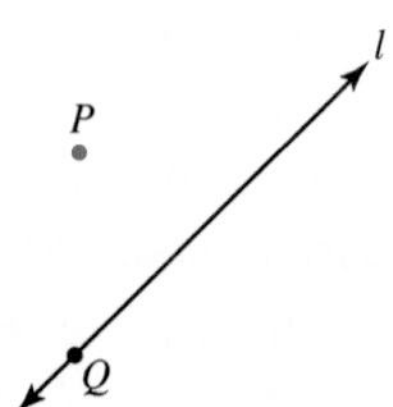

i. Choose a point Q on line l.

ii. Trace line l and points P and Q on your tracing paper.

iii. Flip your tracing paper over, matching line l and point Q.

iv. Make an impression for P'.

a. What kind of triangle is $\triangle PP'Q$? Explain.

b. If M is the point where $\overline{PP'}$ intersects l, what kind of triangle is $\triangle PMQ$? Explain.

30. Find the reflection image of Q and P with respect to line l. Describe the figure $PP'Q'Q$ as completely as possible. Explain.

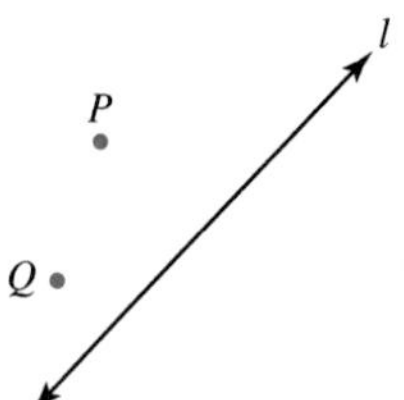

31. Using a compass and straightedge, find the reflection image of $\triangle ABC$ with respect to line l. Describe $ABCB'$ as completely as possible. Explain.

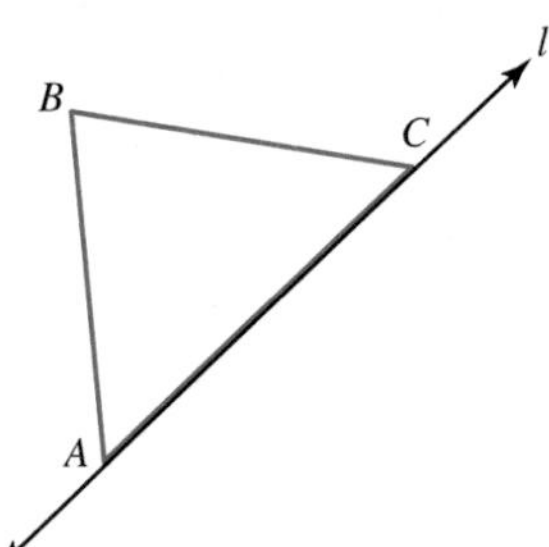

32. Using a compass and straightedge, find the reflection image of $\triangle ABC$ with respect to line l. Does the image appear to be congruent to $\triangle ABC$? Does it have the same orientation?

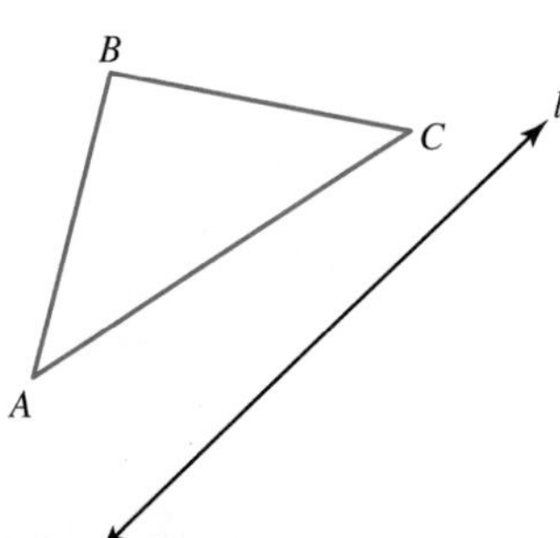

33. The reflection image of A with respect to line l can be constructed with a compass and straightedge. Use the fact that line l is the perpendicular bisector of $\overline{AA'}$. Recall that if point P is the intersection of $\overline{AA'}$ and line l, then $\overline{AA'} \perp l$ and $AP = PA'$. Using a compass and straightedge, find A'.

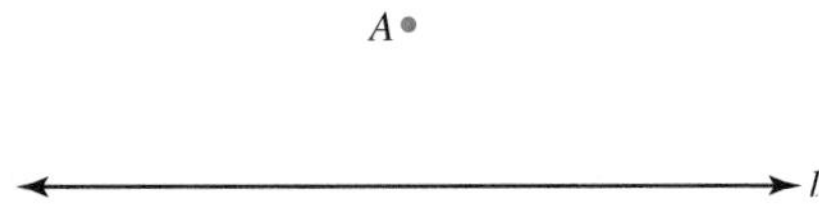

34. Using a compass and straightedge, find the reflection image of $\overline{AB}$ with respect to line l.

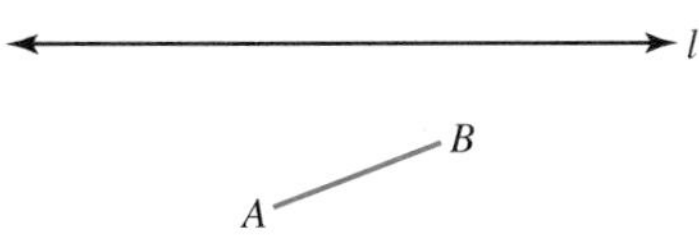

35. **a.** Graph $\triangle ABC$ with $A(2, 1)$, $B(3, -5)$, and $C(6, 3)$, and its image with respect to the reflection across the y-axis.
b. What are the coordinates of the points A, B, and C under the reflection with respect to the y-axis?
c. If point P has coordinates (a, b), what are the coordinates of its image under the reflection with respect to the y-axis?

36. **a.** Graph $\triangle ABC$, with $A(3, 1)$, $B(4, 3)$, and $C(5, -2)$, and its image under the reflection with respect to the line $y = x$.
b. What are the coordinates of the images of points A, B, and C under the reflection with respect to the line $y = x$.
c. If a point P has coordinates (a, b), what are the coordinates of its image under the reflection with respect to the line $y = x$.

37. Using a compass and straightedge, find the image of $\overline{AB}$ under the glide reflection composed of a reflection with respect to line l followed by a translation from X to Y.

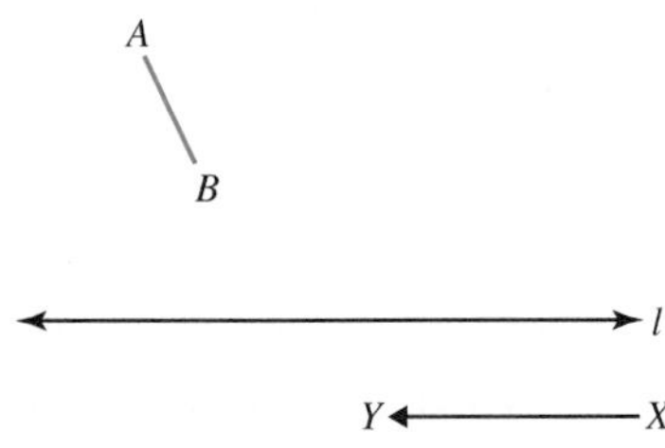

38. Using a compass and straightedge, find the image of $\overline{AB}$ under the glide reflection composed of the translation from X to Y followed by a reflection with respect to the line l.

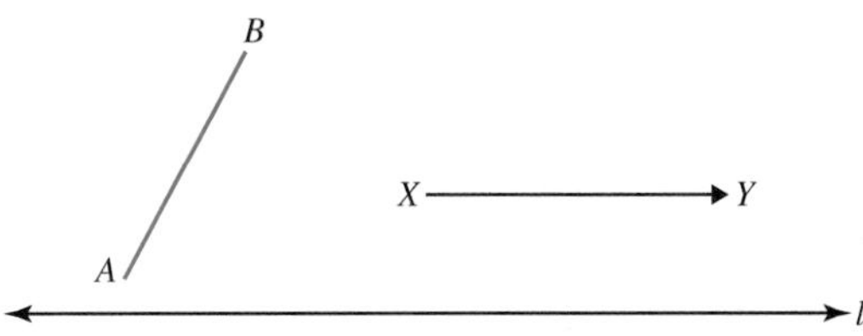

39. **a.** Graph the triangle with vertices $A(1, 2)$, $B(3, 5)$, and $C(6, 1)$. Graph the image of $\triangle ABC$ under the glide reflection described by the translation from $P(0, 0)$ to $Q(-3, -5)$ followed by the reflection with respect to the y-axis.
b. What are the coordinates of the images of points A, B, and C under this glide reflection?
c. If a point has coordinates (a, b), what are the coordinates of its image under this glide reflection?

40. **a.** Graph $\triangle ABC$ with $A(-2, 1)$, $B(0, 3)$, and $C(3, -2)$, and its image under the glide reflection described by the translation from $P(1, 1)$ to $Q(1, 6)$ followed by the reflection with respect to the y-axis.
b. What are the coordinates of the images of points A, B, and C under this glide reflection?
c. If a point has coordinates (a, b), what are the coordinates of its image under this glide reflection?

41. In each part, identify the type of transformation that maps rectangle $ABCD$ onto rectangle $A'B'C'D'$.

a.

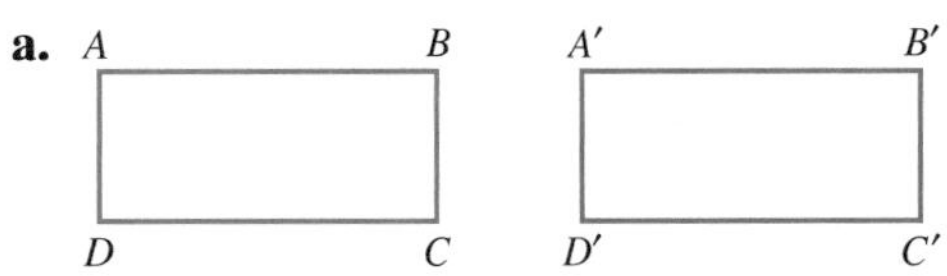

b.

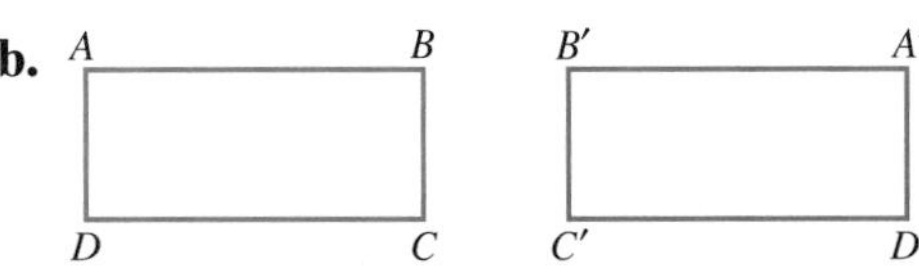

42. In each part, identify the type of transformation that maps rectangle $ABCD$ onto rectangle $A'B'C'D'$.

a.

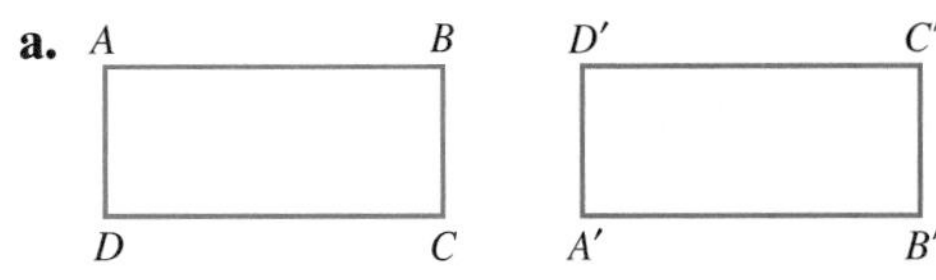

b.

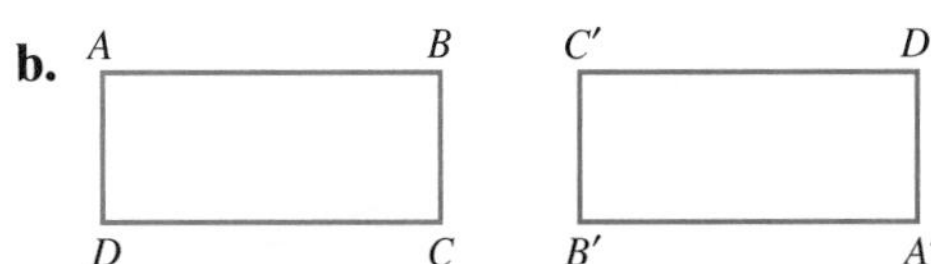

43. Trace the following pair of figures. Use a compass and straightedge to draw the reflection line that takes $\triangle ABC$ to $\triangle A'B'C'$.

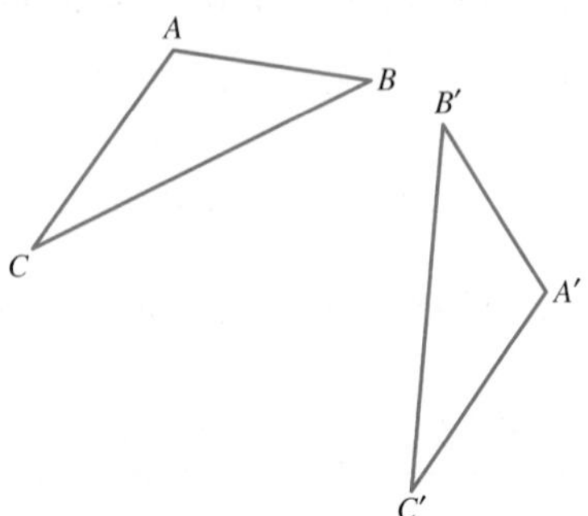

44. Trace the following pair of figures. Use a compass and straightedge to draw the reflection line that takes $\triangle ABC$ to $\triangle A'B'C'$.

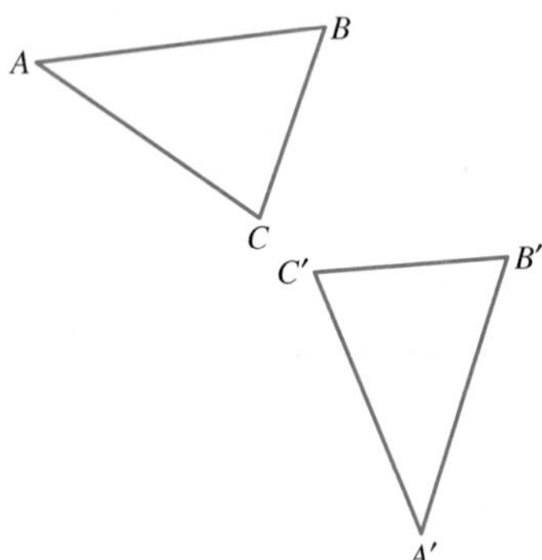

45. Determine the type of isometry that maps the shape on the left onto the shape on the right.

a.

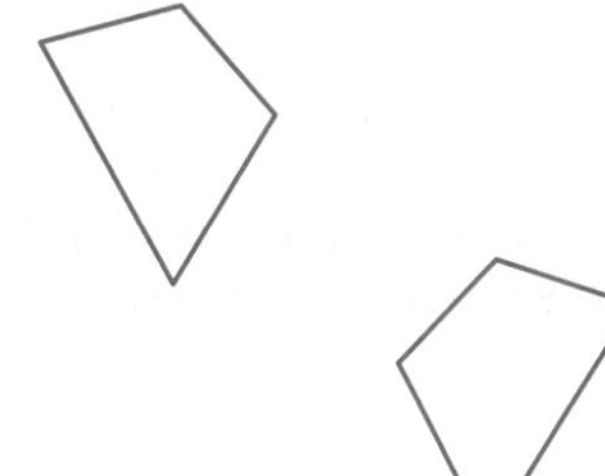

b.

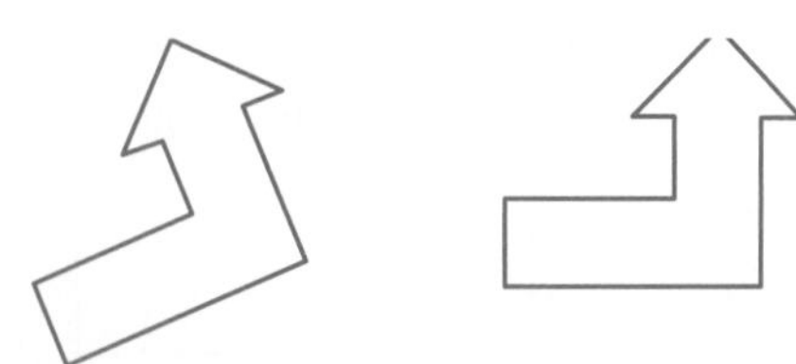

c.

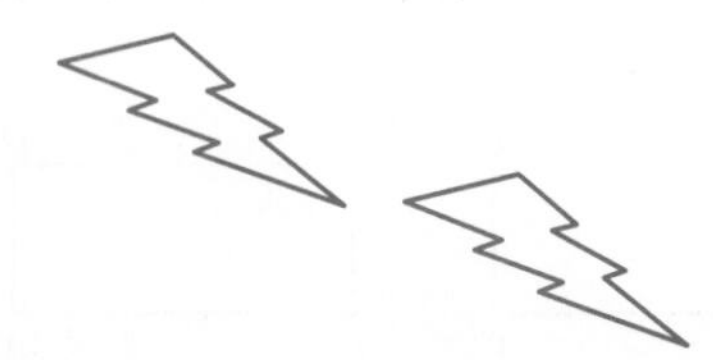

46. Determine the type of isometry that maps the shape on the left onto the shape on the right.

a.

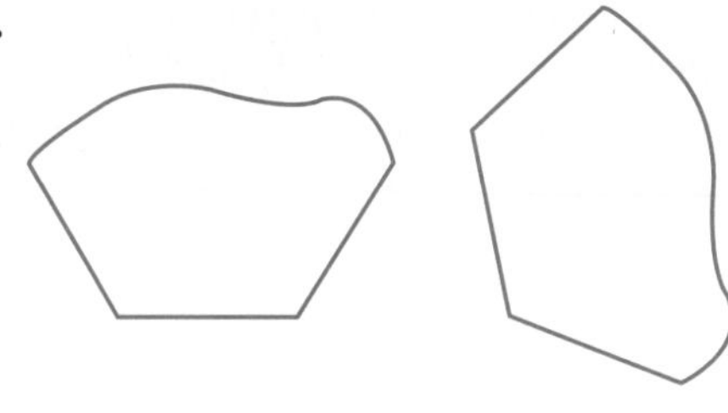

b.

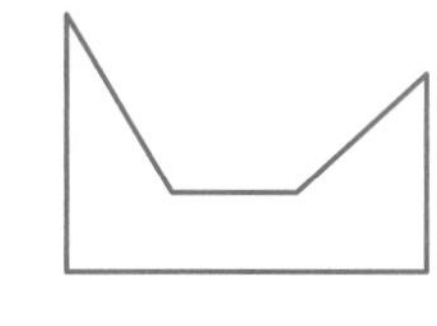

c.

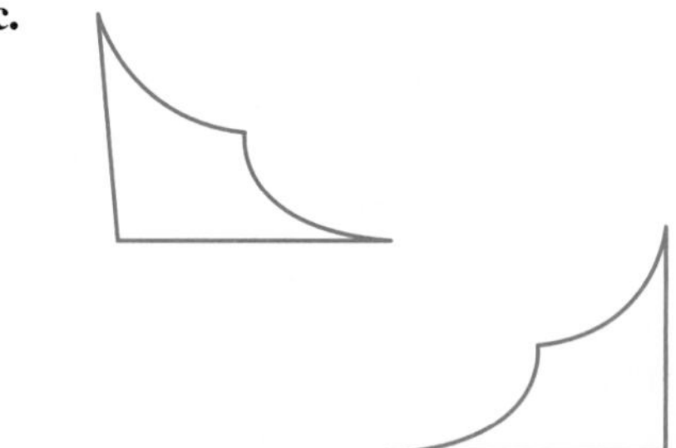

47. Several reflections are described next. In each part, draw figures and their images to find the reflection line. The notation $(x, y) \to (x, 2 - y)$ means that the image of (x, y) is $(x, 2 - y)$.

a. $(x, y) \to (x, 2 - y)$.

b. $(x, y) \to (x, 8 - y)$

c. $(x, y) \to (x, -4 - y)$

48. We could express the results of the rotation of 90° around the origin applied to (4, 2) in the following way: (4, 2) rotates 90° around the origin to (−2, 4). Complete the following statements.

a. (x, y) rotates 180° around the origin to (?, ?).

b. (x, y) rotates 270° around the origin to (?, ?).

c. (x, y) rotates 360° around the origin to (?, ?).

49. In the "Geometry Investigation" for Chapter 9, we created tessellations using rotations. We can also create tessellations using translations.

a. Given square $ABCD$, a curve has been drawn from A to B and from B to C.

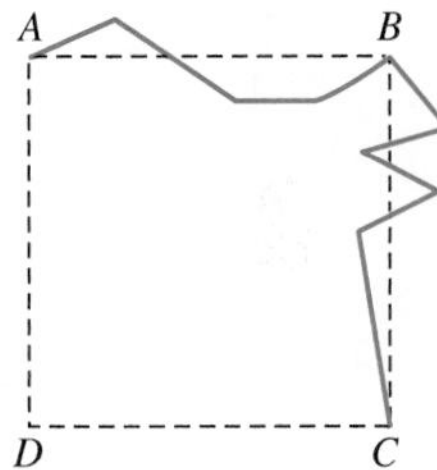

To create a tessellation based on translations, draw the image of the curve from A to B under the translation $\overrightarrow{AD}$. Then draw the image of the curve from B to C under the translation $\overrightarrow{BA}$. Erase the original square and make eight copies of the tile. List all isometries used to position the tiles to form a tessellation.

b. Make your own tessellation based on translations using a square.

50. a. Given regular hexagon $ABCDEF$, a curve has been drawn from A to B, from B to C, and from C to D.

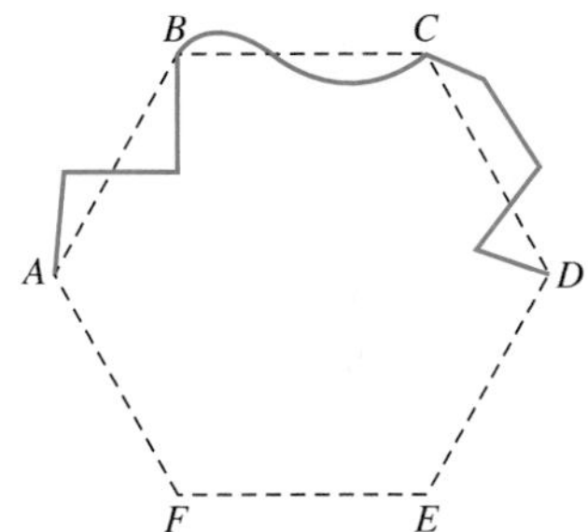

To create a tessellation based on translations, draw the image of the curve from A to B under the translation $\overrightarrow{AE}$. Then draw the image of the curve from B to C under the translation $\overrightarrow{BF}$. Last, draw the image of the curve from C to D under the translation $\overrightarrow{CA}$. Erase the original hexagon and make eight copies of the tile. List all isometries used to position the tiles to form a tessellation.

b. Make your own tessellation based on translations using a regular hexagon.

PROOFS

51. Prove property 2 of Theorem 9.2: Rotations preserve distance.

52. Prove property 3 of Theorem 9.2: Rotations preserve angle measure.

53. Prove property 4 of Theorem 9.2: Rotations preserve perpendicularity.

54. Prove property 5 of Theorem 9.2: Rotations preserve parallelism.

55. Prove property 2 of Theorem 9.3: Reflections preserve distance.

56. Prove property 3 of Theorem 9.3: Reflections preserve angle measure.

57. Prove property 4 of Theorem 9.3: Reflections preserve perpendicularity.

58. Prove property 5 of Theorem 9.3: Reflections preserve parallelism.

59. Prove property 2 of Theorem 9.4: Glide reflections preserve distance.

60. Prove property 3 of Theorem 9.4: Glide reflections preserve angle measure.

APPLICATIONS

61. Each wallpaper border shown illustrates at least one, and possibly all, of the isometries in this section. For each border, list all possible isometries that can take the design onto itself.

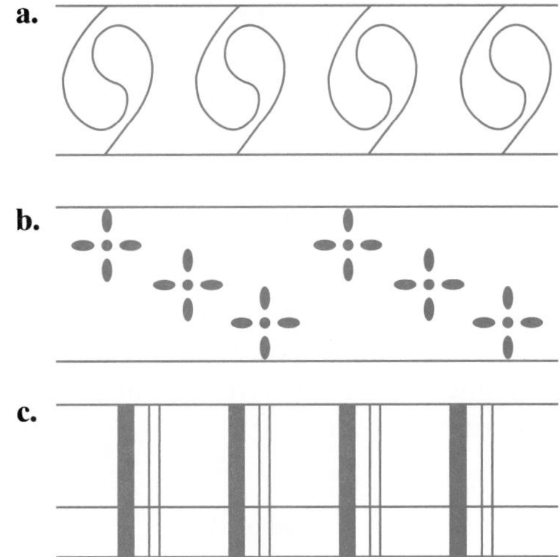

62. Each wallpaper border shown illustrates at least one, and possibly all, of the isometries in this section. For each border, list all possible isometries that can take the design onto itself.

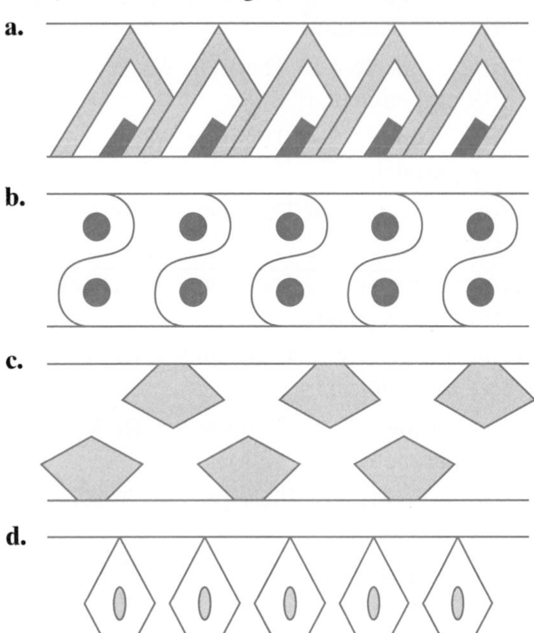

63. An unfinished quilt block is shown next. Complete the quilt block by drawing the images of the shaded figures after a rotation of 90° about point O, after a rotation of 180° about point O, and after a rotation of 270° about point O.

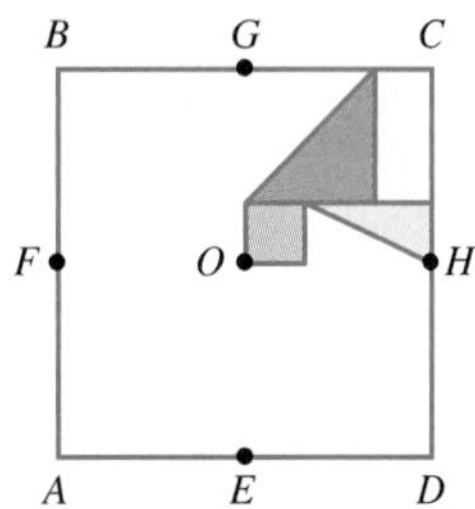

64. An unfinished quilt block is shown next. Complete the quilt block by drawing the image of the shaded triangle after performing the translation $\overrightarrow{CG}$. Then draw the images of the two triangles after a reflection with respect to $\overline{FH}$.

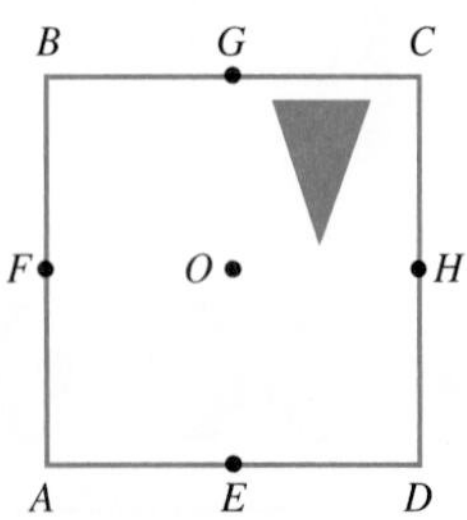

65. In the game of chess, the knight's move is two squares away in any row or column and then one square at a right angle to the line of the first move. Given the knight represented by the dot in the position shown, find all possible moves for that piece and describe those moves in terms of transformations.

66. **Tetris** is a game involving tetrominos, which are shapes made up of four connected squares. Players use translations and rotations to place playing pieces on the playing space without gaps or overlaps. The next figure shows an example of three tetromino pieces placed on the game board.

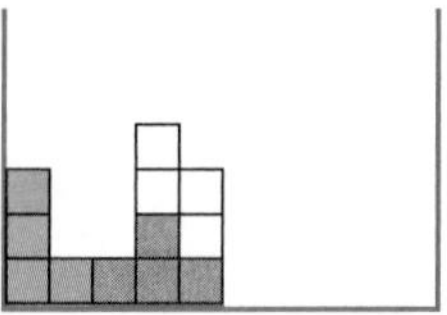

For each of the following tetromino pieces, give the positive angle measure of the rotation required to rotate the piece into position as shown in the preceding figure. The center of rotation is indicated on each piece.

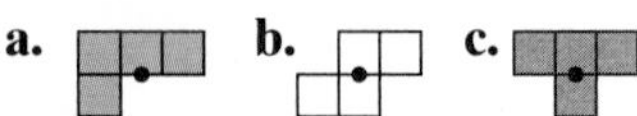

67. **Tic-Tac-Toe** is a game played on a 3 by 3 grid in which players take turns placing an X or an O in the grid. To win, a player must place three of their symbols along a line horizontally, vertically, or diagonally. One possible win by player O is shown.

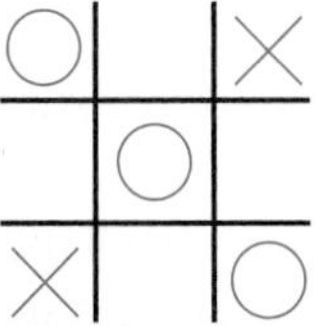

a. A player makes a move by placing a symbol in the grid. One move in Tic-Tac-Toe is equivalent to another if there is an isometry that can take the first grid pattern to the other. Consider an empty grid and the first move of the game by player O. There are nine positions in which the player can place an O. Draw all possible moves and indicate which moves are equivalent. How many unique (nonequivalent) moves are there for player O?

b. Once player O makes a move, how many unique moves are possible for player X?

EXTENDED PROBLEMS

68. A **Frieze pattern** is a pattern that continues in one direction. An example of a Frieze pattern is a wallpaper border. (See problems 61 and 62.) We create Frieze patterns by starting with a basic design and using a translation to create a repeated pattern. We may also use other isometries to create a Frieze pattern. We can classify all Frieze patterns as one of seven types. There is a classification system used in **crystallography**, the study of crystals, that we can use to classify Frieze patterns. Research the four-symbol crystallographic classification system of Frieze patterns and explain why each classification begins with the letter p. Explain what the letter or number in each position means. Draw an example of each of the seven types of patterns and give the classification of each.

69. M. C. Escher (1898–1972) was a Dutch artist whose work frequently incorporated transformation geometry. For one example of Escher's artwork, see *Circle Limit I* at the start of this chapter. Research the art of M. C. Escher and find examples of Escher works that display each of the following isometries: translation, reflection, rotation, and glide reflection.

70. Word processors on every computer have some sort of shape or object creator. Once a shape has been created, the user can edit the shape by translating it, flipping it, or rotating it. In other words, the program uses isometries to move the object to the desired location and give it the desired orientation. Explore the object creation feature of a word processor or drawing program on your computer and use it to create a rectangle that is not a square.

a. Use the object manipulation feature of the software to translate the rectangle on the page and verify Theorem 9.1.

b. Rotate the rectangle and determine the location of the center of rotation.

c. Flip the rectangle horizontally. Where is the line over which the rectangle has been reflected? Repeat for a vertical flip.

9.2 SIMILITUDES AND SIMILARITY

Applied Problem

The scale drawing shown represents the basic floor plan in the design of a home.

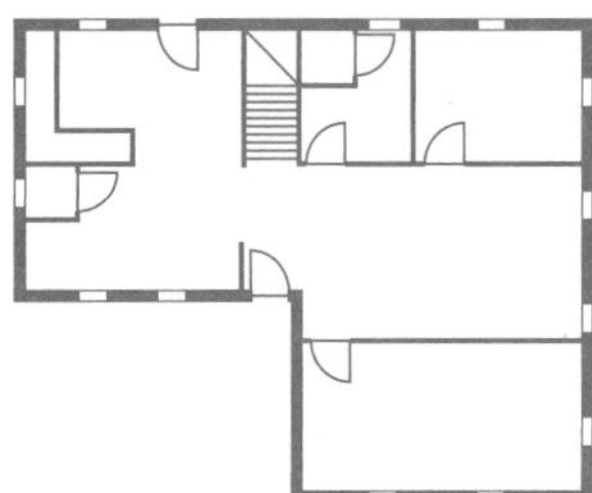

A draftsperson needs to create an enlargement of this floor plan that has four times the area of the plan shown but retains the same shape. How can this be done?

Just as we generalize congruence using the concept of isometry, we can generalized similarity of triangles to any geometric figures.

O

$k = 1.5$

(a)

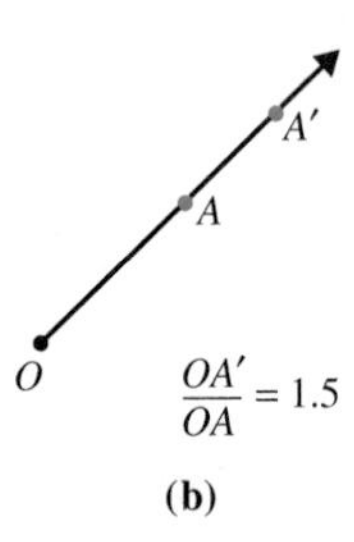

(b)

FIGURE 9.14

Size Transformations

Congruence is related to the sizes and shapes of geometric figures, but similarity is related to only the shapes of figures. We begin our study of similarity using transformations by enlarging and reducing the sizes of figures proportionally.

A **size transformation** of the plane is defined as follows. Let O be any point (called the **center of the size transformation**) and k be a positive real number (called the **scale factor**). The image of any point A under the size transformation with center O and scale factor k is the point A' where

1. A' is on ray $\overrightarrow{OA}$.
2. $\dfrac{OA'}{OA} = k$ [Figure 9.14(b)]

If $k = 1.5$ and O and A are given as in Figure 9.14(a), then A' is found as shown in Figure 9.14(b).

EXAMPLE 9.6 In Figure 9.15(a), find the image of point A for the following size transformations.

a. The size transformation with center O and scale factor 3
b. The size transformation with center O and $k = 0.5$
c. The size transformation with center O and scale factor 1

SOLUTION

a. To find the image of A, we extend $\overline{OA}$ through A and mark off point A' where $OA' = 3(OA)$ [Figure 9.15(b)].
b. In Figure 9.15(c), the image is A' because $OA' = 0.5(OA)$ and A' is on ray $\overrightarrow{OA}$.
c. In Figure 9.15(d), the image is $A' = A$ because $OA = OA'$. This size transformation is called the **identity size transformation**.

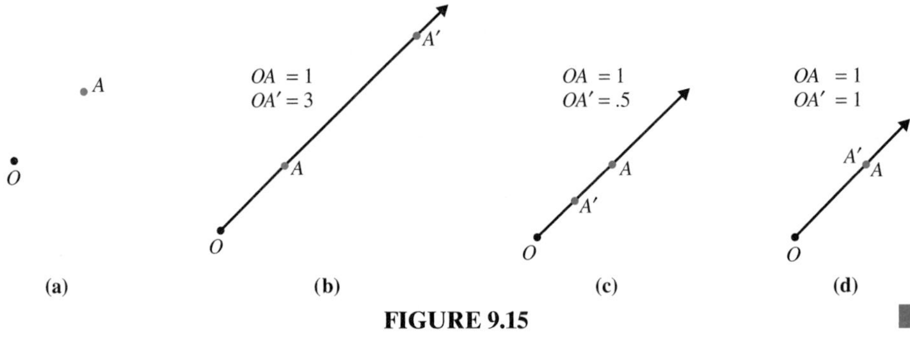

FIGURE 9.15

Size transformations have several properties similar to those of isometries. The next example illustrates some of them.

EXAMPLE 9.7 In Figure 9.16(a), find the image of $\triangle ABC$ for the size transformation with center O and scale factor 2. Is $\triangle ABC \sim \triangle A'B'C'$?

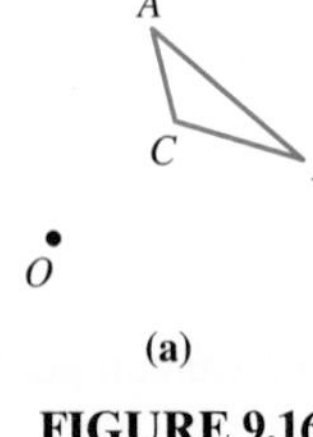

(a)

FIGURE 9.16

SOLUTION

The image is $\triangle A'B'C'$ [Figure 9.16(b)]. By the definition of a size transformation, we know that $\dfrac{OA'}{OA} = \dfrac{OB'}{OB} = 2$. Also $\angle AOB \cong \angle A'OB'$ because A and A' are

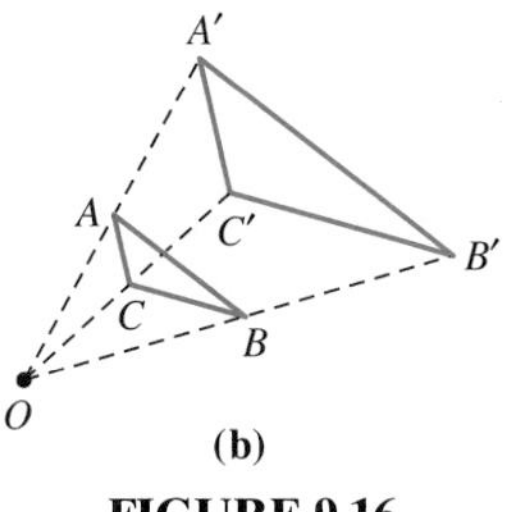

(b)

FIGURE 9.16

collinear, as are B and B'. So, by SAS Similarity, $\triangle OAB \sim \triangle OA'B'$. Thus $\dfrac{A'B'}{AB} = \dfrac{OA'}{OA} = 2$. By the same reasoning, we can show that $\triangle OAC \sim \triangle OA'C'$ and $\triangle OCB \sim \triangle OC'B'$. Therefore, we have $\dfrac{A'C'}{AC} = 2$ and $\dfrac{B'C'}{BC} = 2$, respectively. As a consequence, $\triangle ABC \sim \triangle A'B'C'$ by SSS Similarity. It follows that $\angle A \cong \angle A'$, $\angle B \cong \angle B'$, and $\angle C \cong \angle C'$. Moreover, it can be shown that the respective sides of the two triangles are parallel. ■

This example suggests some of the results stated in the following theorem. We leave the proofs of properties 3–6 for the problem set.

THEOREM 9.7 Properties of Size Transformations

1. Size transformations take lines to lines, rays to rays, and line segments to line segments.
2. Size transformations preserve orientation.
3. Size transformations preserve ratios of distances.
4. Size transformations preserve angle measure.
5. Size transformations preserve perpendicularity.
6. Size transformations take lines to parallel lines.

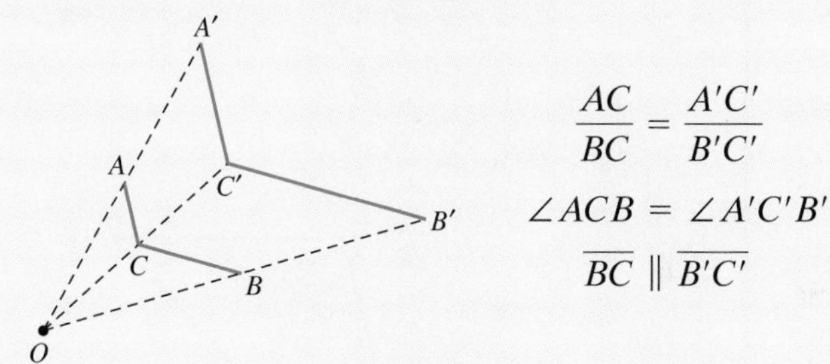

$$\frac{AC}{BC} = \frac{A'C'}{B'C'}$$

$$\angle ACB = \angle A'C'B'$$

$$\overline{BC} \parallel \overline{B'C'}$$

EXAMPLE 9.8 Is there a size transformation that takes $\triangle RST$ in Figure 9.17(a) to $\triangle R'S'T'$? If so, find its center and scale factor. If not, explain why not.

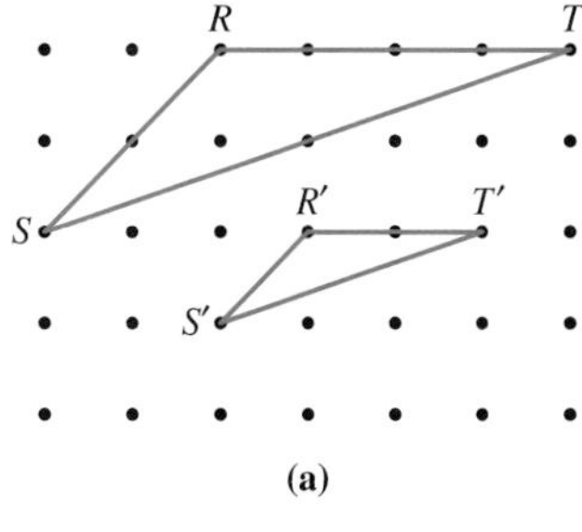

(a)

FIGURE 9.17

SOLUTION

If such a size transformation exists, then $\triangle RST \sim \triangle R'S'T'$. By comparing lengths of the sides, we have

$$\frac{R'T'}{RT} = \frac{S'R'}{SR} = \frac{S'T'}{ST} = \frac{1}{2}$$

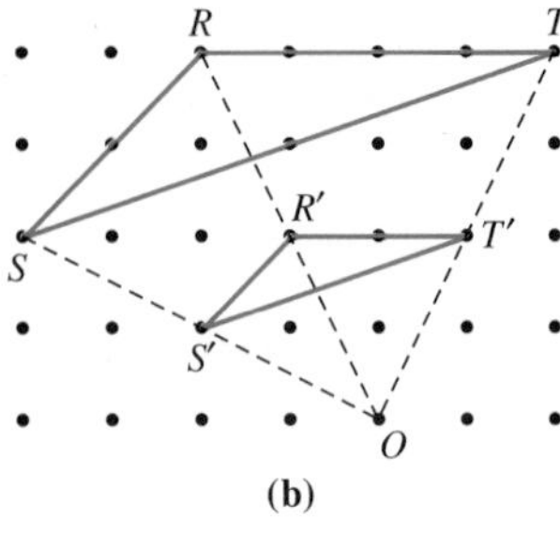

(b)

FIGURE 9.17

So $\triangle RST \sim \triangle R'S'T'$ by SSS Similarity. Therefore, we *may* be able to find a size transformation that takes $\triangle RST$ to $\triangle R'S'T'$. Since the corresponding sides are parallel, such a size transformation should exist. The center of the size transformation is collinear with S and S', with R and R', and with T and T'. The lines determined by $\overline{SS'}$, $\overline{RR'}$, and $\overline{TT'}$ intersect at the center O [Figure 9.17(b)]. Thus, the size transformation with center O and scale factor $\frac{1}{2}$ takes $\triangle RST$ to $\triangle R'S'T'$. ■

Similitudes

In Section 9.1, we found that we can define congruence in terms of isometries. As suggested by Examples 9.7 and 9.8, a size transformation takes a figure to a similar figure. However, consider $\triangle ABC$ and $\triangle DEF$ [Figure 9.18(a)].

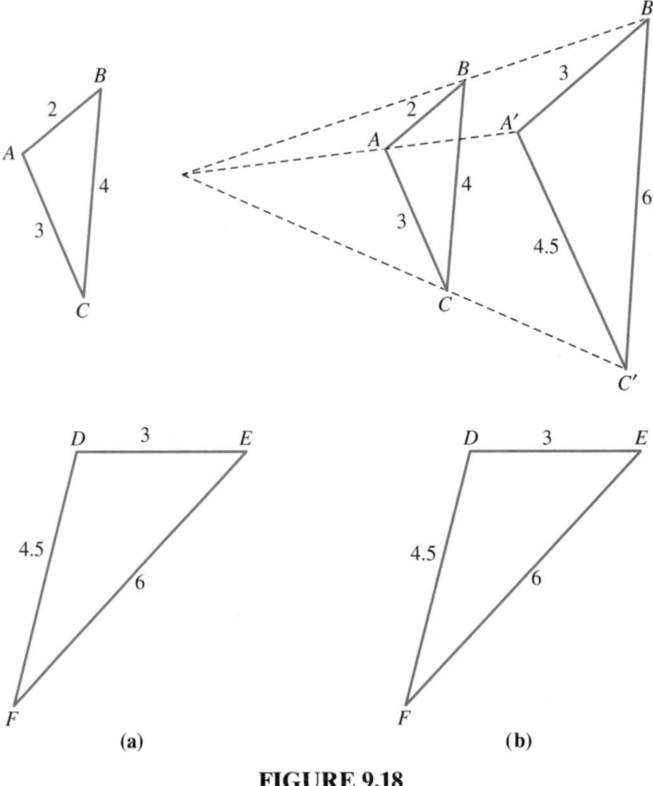

FIGURE 9.18

Although the two triangles are similar by SSS Similarity, no size transformation will take $\triangle ABC$ to $\triangle DEF$ because their corresponding sides are not parallel as required by part 6 of Theorem 9.7. However, a size transformation with scale factor 1.5 will take $\triangle ABC$ to $\triangle A'B'C'$ where $\triangle A'B'C' \cong \triangle DEF$ [Figure 9.18(b)]. Then we can find an isometry (here a rotation) to take $\triangle A'B'C'$ to $\triangle DEF$ since they are congruent. A **similitude** is defined to be a combination of a size transformation and an isometry. This discussion motivates the following result.

> **THEOREM 9.8 Similar Triangles and Similitudes**
>
> $\triangle ABC \sim \triangle A'B'C'$ if and only if there is a similitude that takes $\triangle ABC$ to $\triangle A'B'C'$.

EXAMPLE 9.9 Suppose $\triangle ABC \sim \triangle DBE$ [Figure 9.19(a)]. Find the similitude that takes $\triangle ABC$ to $\triangle DBE$. (NOTE: We can express any similitude as the composition of a size transformation followed by an isometry in infinitely many ways depending on what point is chosen as the center of the size transformation. Thus, although there is only one similitude, there are many different ways to describe it.)

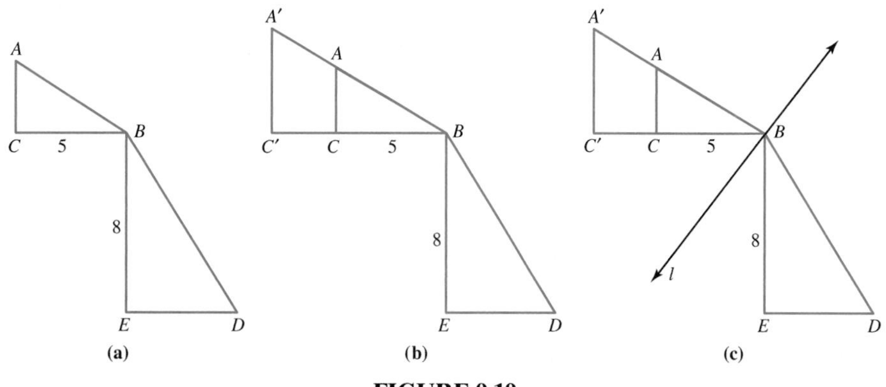

FIGURE 9.19

SOLUTION

Because the ratio $\frac{BE}{BC}$ is $\frac{8}{5}$, the scale factor of our size transformation must be $\frac{8}{5}$. For convenience, we choose the center of the size transformation to be B (however, we could have used any other point in the plane). Then the image of $\triangle ABC$ under the size transformation with center B and scale factor $\frac{8}{5}$ is $\triangle A'BC'$ [Figure 9.19(b)]. Because $\triangle A'BC'$ and $\triangle DBE$ have opposite orientations, the isometry that takes $\triangle A'BC'$ to $\triangle DBE$ must be a reflection or a glide reflection. Notice that the reflection with respect to the line l that bisects $\angle C'BE$ takes $\triangle A'BC'$ to $\triangle DBE$ [Figure 9.19(c)]. Thus, the size transformation with center B and with scale factor $\frac{8}{5}$ followed by the reflection with respect to line l is a similitude that takes $\triangle ABC$ to $\triangle DBE$. ■

Two polygons are defined to be **similar polygons** if and only if there is a correspondence between the polygons such that the corresponding angles are congruent and the corresponding sides are proportional.

EXAMPLE 9.10 Prove that any two squares are similar.

SOLUTION

Let $ABCD$ and $EFGH$ be two squares with sides of length s and t [Figure 9.20(a)].

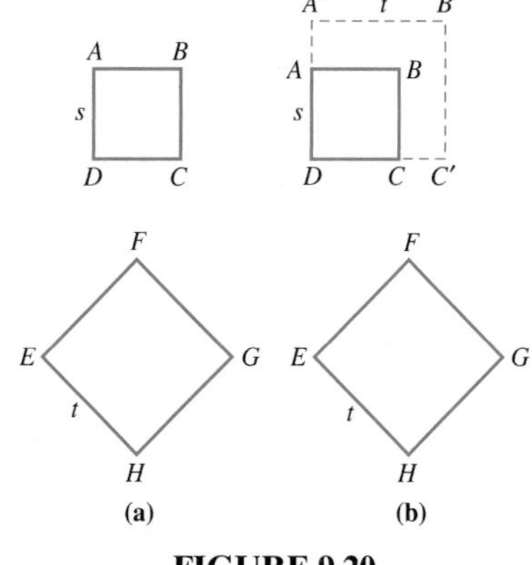

FIGURE 9.20

Any size transformation with scale factor $\frac{t}{s}$ takes $ABCD$ to a square $A'B'C'D'$ where $A'B'C'D' \cong EFGH$. One such size transformation with center D is shown in Figure 9.20(b). Because the squares $A'B'C'D'$ and $EFGH$ are congruent, there must be an isometry that takes $A'B'C'D'$ to $EFGH$. Thus, the combination of the size transformation and the isometry produces a similitude that takes $ABCD$ to $EFGH$. Since similitudes preserve angle measure and proportionality of side lengths, the squares are similar. ■

This example suggests that all equilateral triangles are similar also. The following general result about polygons holds.

THEOREM 9.9 Similar Polygons and Similitudes

Two polygons are similar if and only if there is a similitude that takes one to the other.

Just as with isometries, the value in studying similitudes is that we can use similitudes to decide if *any* shapes, including those that are not polygons, are similar. In general, we say that two shapes are **similar shapes** if and only if there is a similitude that takes one to the other.

Solution of Applied Problem

If the ratio of the sides of two similar figures is a:b, then the ratio of their areas is $a^2 : b^2$. Thus, if the dimensions of the sides of the floor plan are doubled, that is, $b = 2a$, the area of the new floor plan would be four times the area of the original plan since $b^2 = (2a)^2 = 4a^2$.

GEOMETRY AROUND US

Computer graphics programs allow the user to manipulate a geometric figure drawn on the screen using transformations such as those described in this chapter. One graphics package has options such as Flip Horizontal, Flip Vertical, and Rotate. Users can also perform a size transformation on a figure using a Scaling command and can perform a translation by moving a figure using the mouse. An example of the use of the Flip Vertical Option is shown here.

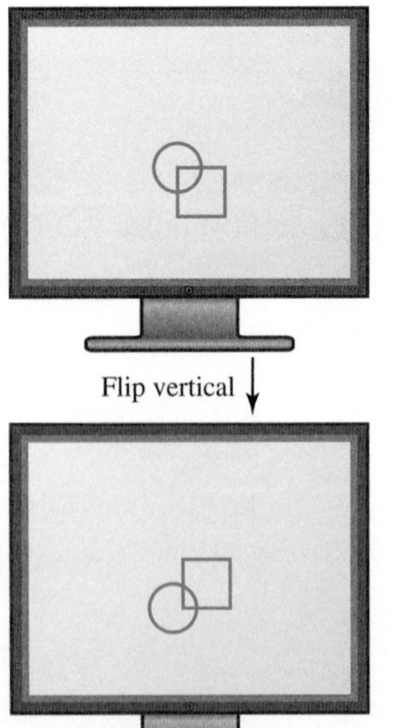

PROBLEM SET 9.2

EXERCISES/PROBLEMS

1. Trace O and A. Using a ruler, find A' for a size transformation with center O having the following scale factors.

   a. 4 **b.** 3.75 **c.** $2\frac{5}{8}$ **d.** $\frac{2}{3}$

2. Trace O and A. Using a compass and straightedge, find A' for a size transformation with center O having the following scale factors.

   a. 3.5 **b.** $\frac{3}{4}$ **c.** 1 **d.** 2.25

3. On the square lattice portions shown, find the image of point P under the size transformation with center O and scale factor 3.

   a.

   b.

   c.

4. On the square lattice portions shown, find the image of $\overline{AB}$ under the size transformation with center O and scale factor 2.

   a.

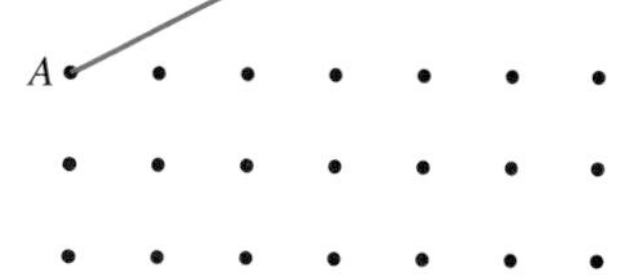

   b.

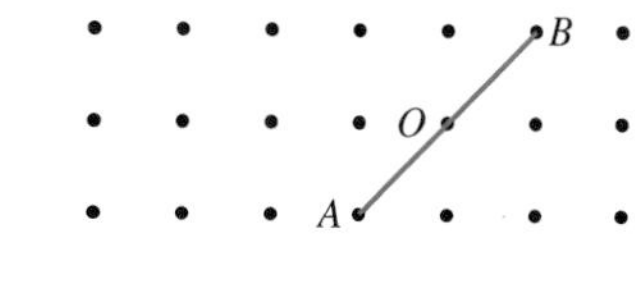

   c.

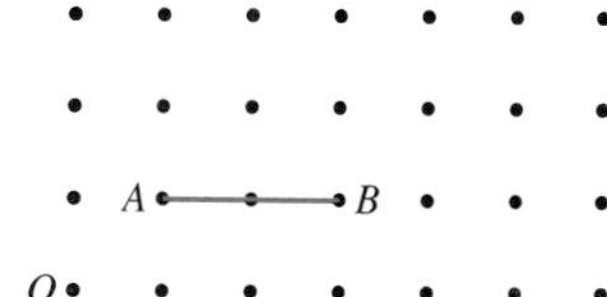

5. On the square lattice portions shown, find the image of the circle under a size transformation with center O and scale factor as given.

   a. $k = 2$

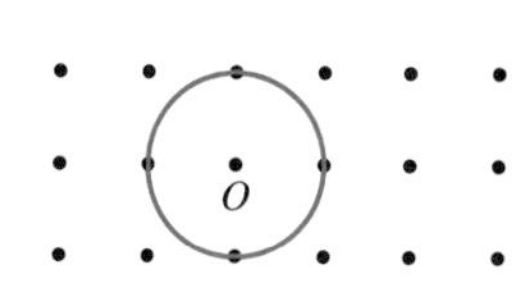

   b. $k = \frac{1}{2}$

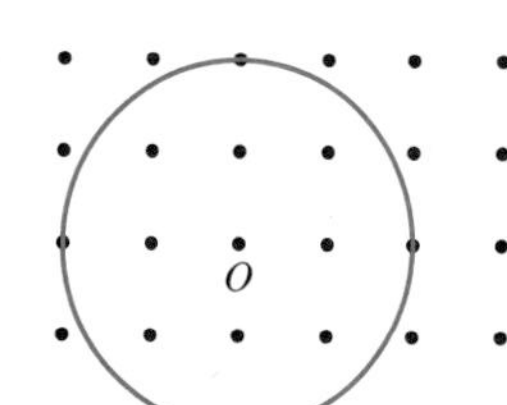

6. On the square lattice portions shown, find the image of parallelogram $ABCD$ under a size transformation with center O and scale factor as given.

a. $k = 3$

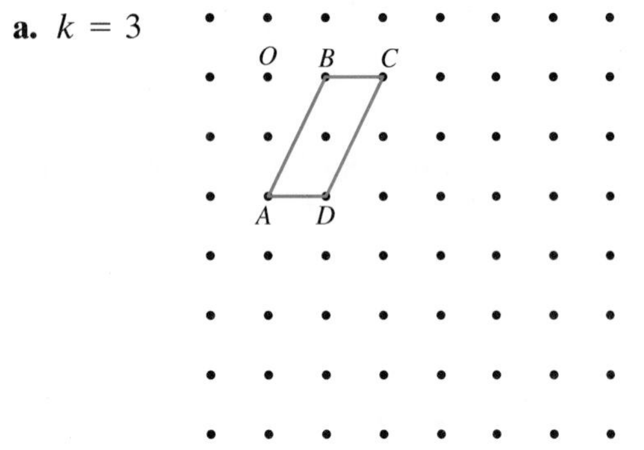

b. $k = 0.5$

7. a. Find the image of $\angle ABC$ under the size transformation with center O and scale factor 2.

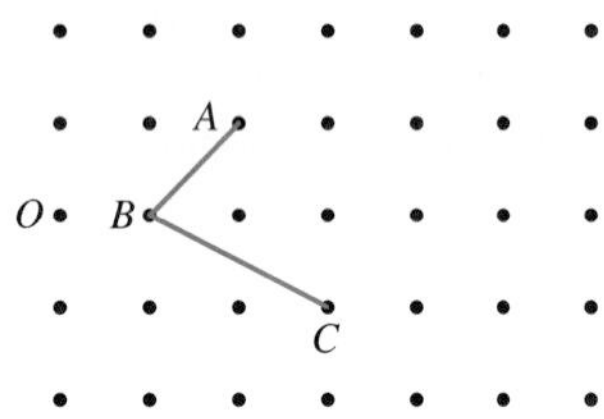

b. How do the measures of $\angle ABC$ and $\angle A'B'C'$ compare?

8. a. Find the image of $\triangle ABC$ under the size transformation with center O and scale factor 3.

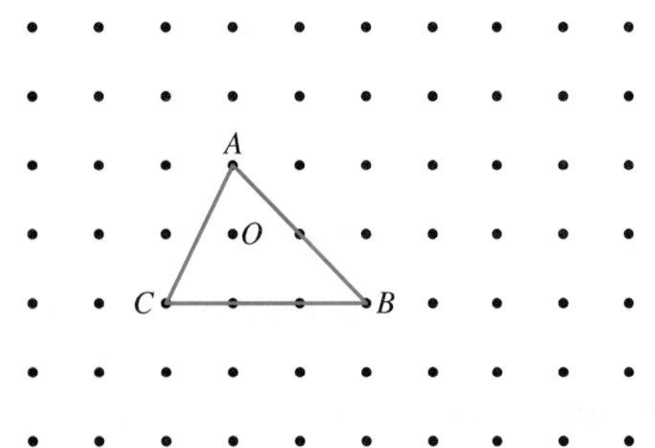

b. Find the lengths of $\overline{AB}$ and $\overline{A'B'}$. What is the ratio $\frac{A'B'}{AB}$?

c. Find the lengths of $\overline{BC}$ and $\overline{B'C'}$. What is the ratio $\frac{B'C'}{BC}$?

d. Find the lengths of $\overline{AC}$ and $\overline{A'C'}$. What is the ratio $\frac{A'C'}{AC}$?

e. In addition to the ratio of their lengths, what other relationship exists between a line segment and its image under the size transformation?

9. Given $\triangle ABC$ with vertices $A(0,0)$, $B(4,0)$, and $C(4,3)$, find the following:

a. The image of $\triangle ABC$ after the size transformation with center A and with scale factor 2

b. The ratio of the area of $\triangle A'B'C'$ to the area of $\triangle ABC$

10. Given $\triangle ABC$ with vertices $A(0, 0)$, $B(4, 0)$, and $C(4, 3)$, find the following:

a. The image of $\triangle ABC$ after the size transformation with center C and scale factor $\frac{1}{2}$

b. The ratio of the perimeter of $\triangle A'B'C'$ to the perimeter of $\triangle ABC$

11. Trace the following circle. Then draw the image of the circle under a size transformation with center O having scale factor 3. Compare the circumference and area of the circle with those of its size transformation image.

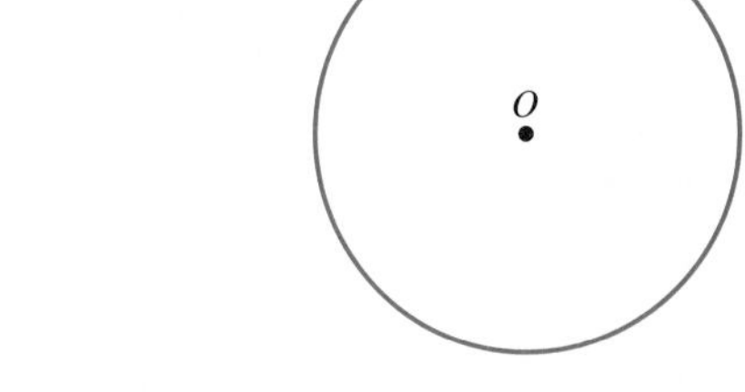

12. Trace the following circle. Then draw the image of the circle under a size transformation with center A having scale factor 0.5. Compare the circumference and area of the circle with those of its size transformation image.

13. **a.** Find the center of the size transformation that maps P to P' and Q to Q'.

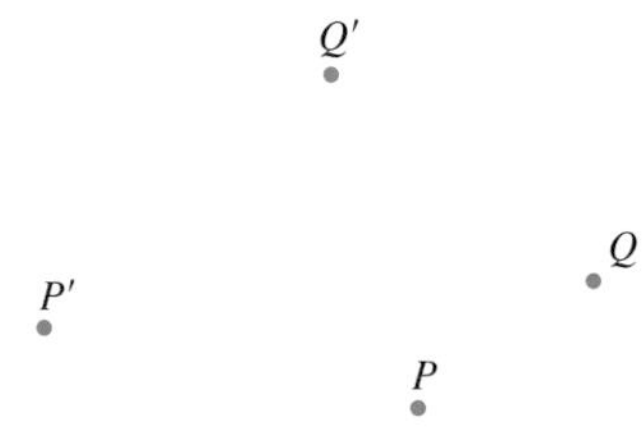

b. Give an estimate for the scale factor of the size transformation.

14. **a.** Find the center of the size transformation that maps P to P' and Q to Q'.

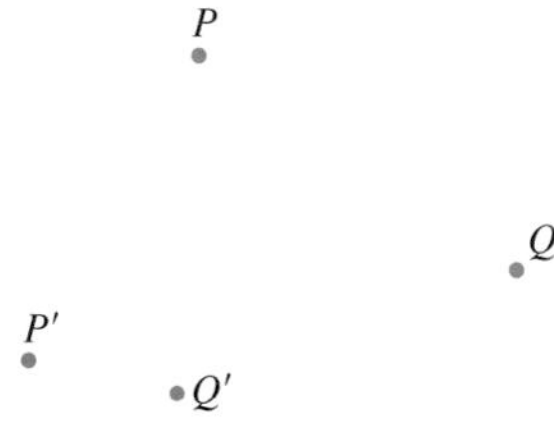

b. Give an estimate for the scale factor of the size transformation.

15. Determine if $\triangle A'B'C'$ is a size transformation image of $\triangle ABC$ for $A'(0,2)$, $B'(4,4)$, $C'(8,0)$, $A(0,1)$, $B(2,2)$, and $C(4,0)$. If yes, find the center and scale factor. If no, explain why not.

16. Determine if $\triangle A'B'C'$ is a size transformation image of $\triangle ABC$ for $A'(5,3)$, $B'(1,5)$, $C'(9,5)$, $A(5,0)$, $B(3,1)$, and $C(7,2)$. If yes, find the center and scale factor. If no, explain why not.

17. A quadrilateral $ABCD$ has vertices $A(0,0)$, $B(1,3)$, $C(4,2)$, and $D(5,0)$.

a. Sketch the image of $ABCD$ under the size transformation with center A and scale factor 2 followed by a reflection with respect to the y-axis.

b. What are the coordinates of A'', B'', C'', and D''?

c. If a point has coordinates (a, b), what are the coordinates of its image under this similitude?

18. A quadrilateral $ABCD$ has vertices $A(0,0)$, $B(1,3)$, $C(4,2)$, and $D(5,0)$.

a. Sketch the image of $ABCD$ under the size transformation with center A and scale factor 3 followed by a translation of 2 units to the right and 1 unit up.

b. What are the coordinates of A'', B'', C'', and D''?

c. If a point has coordinates (a, b), what are the coordinates of its image under this similitude?

19. A quadrilateral $ABCD$ has vertices $A(0,0)$, $B(1,3)$, $C(4,2)$, and $D(5,0)$.

a. Sketch the image of $ABCD$ under the rotation of 180° around A followed by the size transformation with center A and scale factor $\frac{1}{2}$.

b. What are the coordinates of A'', B'', C'', and D''?

c. If a point has coordinates (a, b), what are the coordinates of its image under this similitude?

20. A quadrilateral $ABCD$ has vertices $A(0,0)$, $B(1,3)$, $C(4,2)$, and $D(5,0)$.

a. Sketch the image of $ABCD$ under a rotation of $-90°$ around point A followed by the size transformation with center A and scale factor 2.

b. What are the coordinates of A'', B'', C'', and D''?

c. If a point has coordinates (a, b), what are the coordinates of its image under this similitude?

21. Find the similitude that takes $\triangle ABC$ to $\triangle A''B''C''$ for $A(0,0)$, $B(0,2)$, $C(2,0)$, $A''(-3,0)$, $B''(-3,9)$, and $C''(6,0)$. (HINT: First find a size transformation that takes $\triangle ABC$ to some $\triangle A'B'C'$ that is congruent to $\triangle A''B''C''$.)

22. Find the similitude that takes $\triangle ABC$ to $\triangle A''B''C''$ for $A(1,0)$, $B(1,3)$, $C(2,0)$, $A''(-3,0)$, $B''(-3,9)$, and $C''(-6,0)$.

23. Find the similitude that takes $\triangle ABC$ to $\triangle A''B''C''$. Describe the similitude as completely as possible.

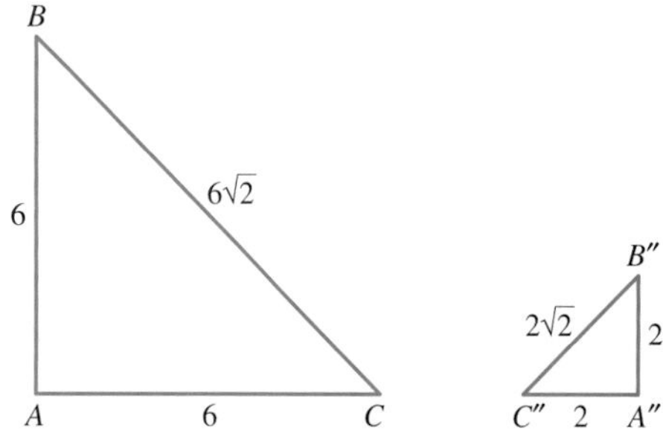

24. Find the similitude that takes $\triangle ABC$ to $\triangle A''B''C''$. Describe the similitude as completely as possible.

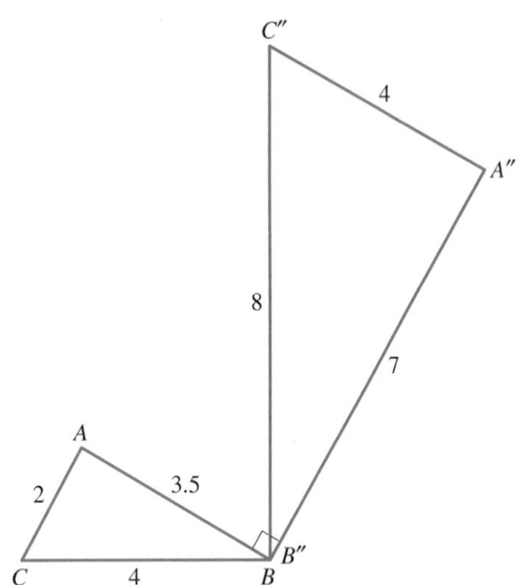

PROOFS

25. Describe in words how you would prove that equilateral $\triangle ABC$ is similar to equilateral $\triangle DEF$.

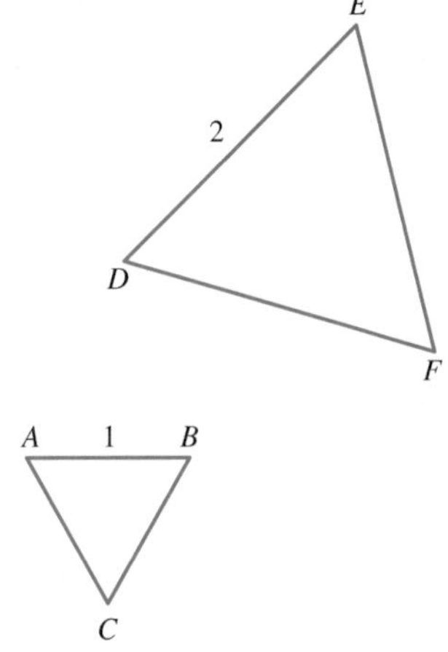

26. Prove that any circle is similar to any other circle.

27. Prove part 3 of Theorem 9.7: Size transformations preserve ratios of distances.

28. Prove part 4 of Theorem 9.7: Size transformations preserve angle measure.

29. Prove part 5 of Theorem 9.7: Size transformations preserve perpendicularity.

30. Prove part 6 of Theorem 9.7: Size transformations take lines to parallel lines.

APPLICATIONS

31. Oil from a drum leaks into water surrounding the drum, creating a circular oil slick 100 square feet in area. Find the scale factor of the size transformation of the oil slick when it is 450 square feet in area. Round to the nearest hundredth. Where is the center of the size transformation?

32. A light source in a movie projector illuminates a piece of film and projects a 19-foot-tall image of a rectangular building onto a theater screen. If the image on the film is 14.2 mm wide and 23 mm tall, what is the scale factor and how wide is the building on the screen? Round to the nearest whole number.

33. A **pantograph**, shown next, is a drawing tool made up of hinged rods. A pantograph is used to copy, enlarge, or reduce the size of a figure.

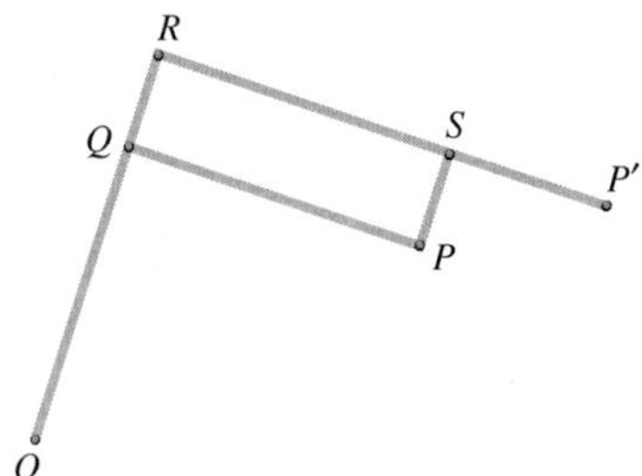

The pantograph is constructed so that quadrilateral $QRSP$ is a parallelogram. Point O is a fixed point, the center of the size transformation. Point P is used to trace a figure while a pen placed at P' draws an enlarged figure. If $OQ = 18$ inches and $QR = 6$ inches, find the scale factor of the size transformation that takes the original figure to the enlarged figure drawn by the pantograph.

34. The following quilt was created using similitudes. Several points have been labeled.

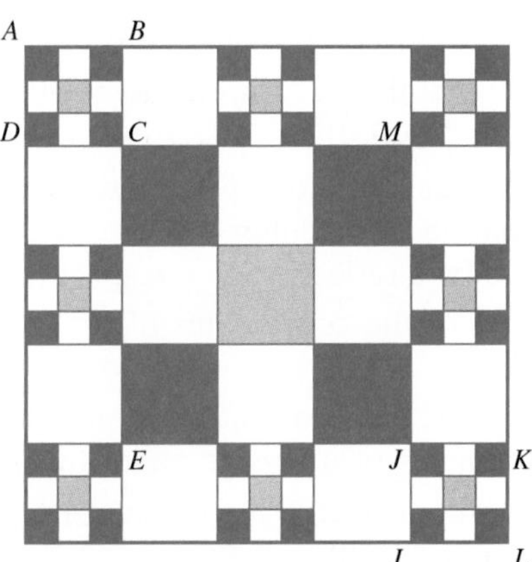

a. Describe as completely as possible a similitude that takes block $ABCD$ to $CMJE$.

b. Describe a similitude that takes block $ABCD$ to $LKJI$.

35. An unfinished quilt is shown next. Follow the directions to transform quilt block $ABNL$ and color the image.

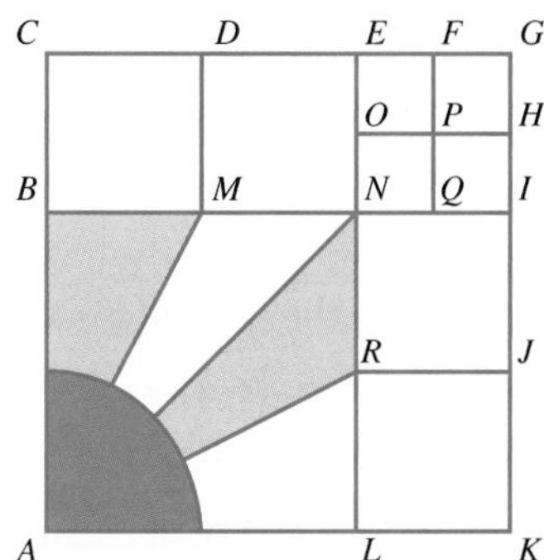

a. Perform a size transformation on $ABNL$ with scale factor $\frac{1}{2}$ and center A followed by translation $\overrightarrow{AR}$. Color the quilt block.

b. Perform a size transformation on $ABNL$ with scale factor $\frac{1}{4}$ and center N, a 90° rotation with center of rotation N followed by translation $\overrightarrow{NO}$. Color the quilt block.

EXTENDED PROBLEMS

36. Architects cannot create full-size designs of buildings. To be practical, they must draw designs to scale. Research scale drawings and determine if there is a typical scale factor used for residential drawings. Create a scale drawing of your home. Select an appropriate scale factor. Your drawing should include all walls, doorways, and windows.

37. **Projective geometry** is a branch of geometry developed in the early 19th century by Gérard Desargues (1591–1661). It is the study of geometric figures that remain unchanged under a **projection**, which is a transformation of a figure from one plane to another. Research projective geometry and describe its basic elements. List the axioms of the system. Are lengths, areas, and angles unchanged under projection? What properties or figures are unaffected under a projection? Do parallel lines exist in projective geometry? Explain what duality means.

38. In a word processor or a computer drawing program, a shape or object can be scaled larger or smaller. Create a shape in a drawing program on a computer. Copy the shape and place the copy on top of the original. Then use the scaling feature of the program to double the size of the shape. Use the original shape and the scaled shape to determine the center of the size transformation. Repeat the process, but this time use a scaled-down version of the original shape. Is the center of the size transformation the same?

9.3 PROBLEM SOLVING USING TRANSFORMATIONS

Applied Problem

Towns A and B are on opposite sides of a river. The towns are to be connected with a bridge, $\overline{CD}$, perpendicular to the river, so that the distance $AC + CD + DB$ is as small as possible. Where should the bridge be located?

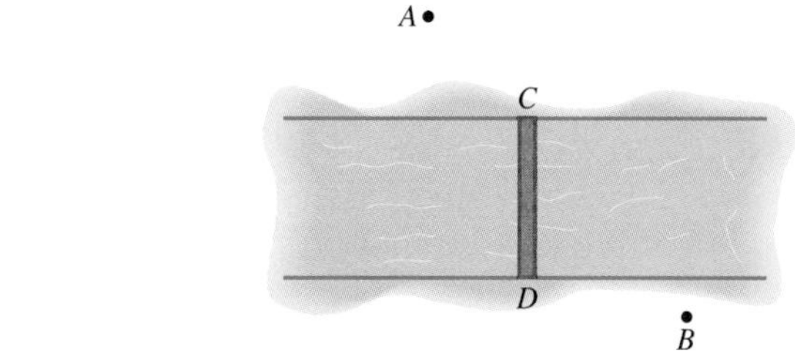

We can apply our knowledge of transformations to solve problems. The following examples are of two types. First, we examine proofs of some familiar results using transformations. We present these to reinforce the idea that it is acceptable to prove a result using any of several techniques. Second, we look at examples of problems that may be difficult to prove using techniques other than those of transformation geometry.

Transformation Proofs

In Chapter 5, we used congruent triangles to verify properties of quadrilaterals. The following examples show how we can verify properties using transformations.

EXAMPLE 9.11 Use isometries to show that the base angles of an isosceles triangle are congruent.

SOLUTION

Let $\triangle ABC$ be isosceles with $\overline{AB} \cong \overline{AC}$ [Figure 9.21(a)]. Reflect $\triangle ABC$ with respect to $\overleftrightarrow{AC}$ forming $\triangle AB'C$ [Figure 9.21(b)]. As a result, we know that $\angle ABC \cong \angle AB'C$ and $\angle BAC \cong \angle B'AC$, since reflections preserve angle measure. Next, consider the rotation with center A and directed angle $\angle B'AC$. Since $\angle B'AC \cong \angle BAC$, and since $\overline{AB} \cong \overline{AC} \cong \overline{AB'}$, we know that this rotation maps $\triangle AB'C$ to $\triangle ACB$. Hence, $\angle AB'C \cong \angle ACB$. Combining this with our preceding observation that $\angle AB'C \cong \angle ABC$, we have $\angle ABC \cong \angle ACB$.

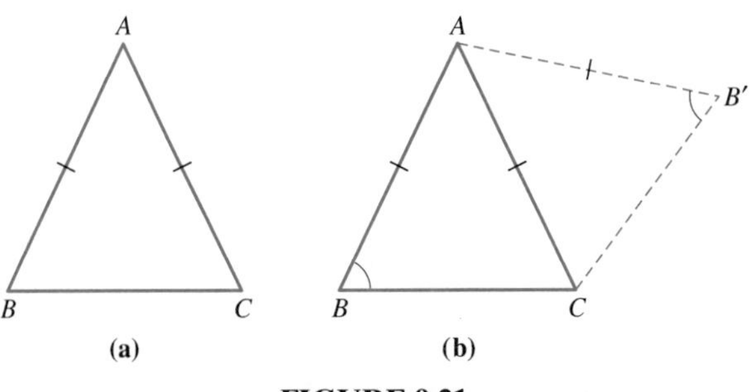

FIGURE 9.21

A particular type of rotation, called a half-turn, is especially useful in verifying properties of polygons. A **half-turn** is a rotation through 180° with any point, O, as the center. Figure 9.22, shows a half-turn image of $\triangle ABC$, with point O as the center

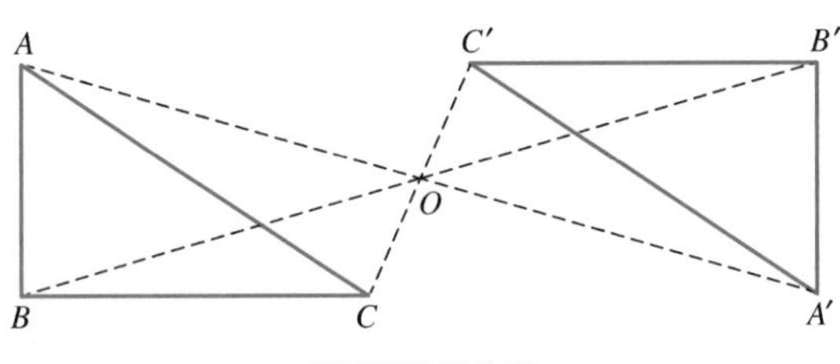

FIGURE 9.22

EXAMPLE 9.12 Using transformations, show that if the diagonals of a quadrilateral $ABCD$ bisect each other, then it is a parallelogram.

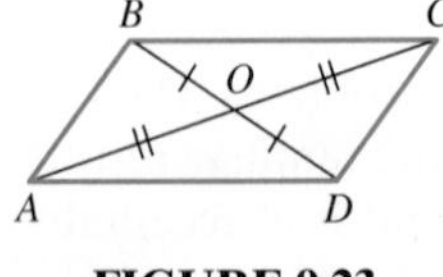

FIGURE 9.23

SOLUTION

Let $ABCD$ have diagonals $\overline{AC}$ and $\overline{BD}$ that bisect each other at O (Figure 9.23). Consider the half-turn with center O. Because the diagonals bisect each other, the half-turn takes A to C and B to D, so it must take $\overline{AB}$ to $\overline{CD}$. Therefore, $\overline{AB} \cong \overline{CD}$ because a rotation preserves length. In a similar manner, $\overline{AD} \cong \overline{CB}$. Thus, $ABCD$ is a parallelogram because it has two pairs of opposite sides that are congruent. ■

Theorem 8.5 states that two lines are perpendicular if and only if the product of their slopes is -1. The next example illustrates how using a rotation can lead to a simpler proof.

EXAMPLE 9.13

Given Two perpendicular lines l and m, neither of which is vertical [Figure 9.24(a)]

Prove The product of the slopes of l and m is -1

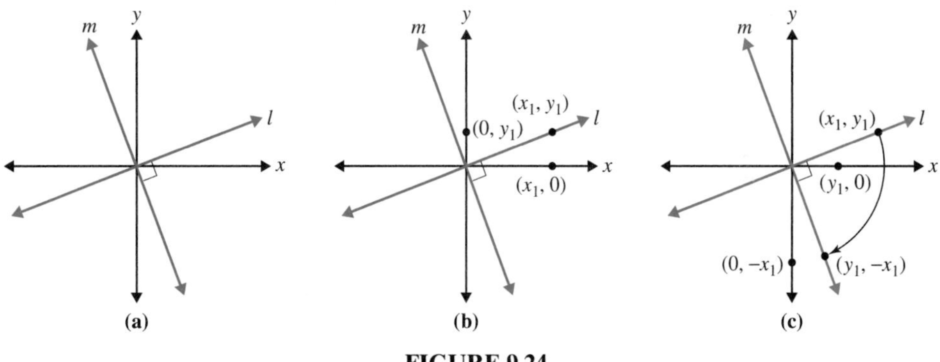

FIGURE 9.24

SOLUTION

Proof Let $(x_1, 0), (0, y_1), (x_1, y_1)$ be three points as shown in Figure 9.24(b), where (x_1, y_1) is a point on l. Imagine that line l is rotated 90° counterclockwise to line m [Figure 9.24(c)]. The point (x_1, y_1) would correspond to the point $(y_1, -x_1)$ on m. Using points $(0, 0)$ and (x_1, y_1), we find that the slope of l is $\frac{y_1}{x_1}$. Using points $(0, 0)$ and $(y_1, -x_1)$, we find that the slope of m is $\frac{-x_1}{y_1}$. Therefore, the product of the slopes of l and m is $\frac{y_1}{x_1} \cdot \frac{-x_1}{y} = -1$. ■

The next example gives a transformation proof of a result proved in Chapter 6 using similar triangles.

EXAMPLE 9.14

In $\triangle ABC$ in Figure 9.25, $\overline{DE}$ divides $\overline{AB}$ and $\overline{CB}$ proportionally as indicated. Prove that $\overline{DE} \parallel \overline{AC}$.

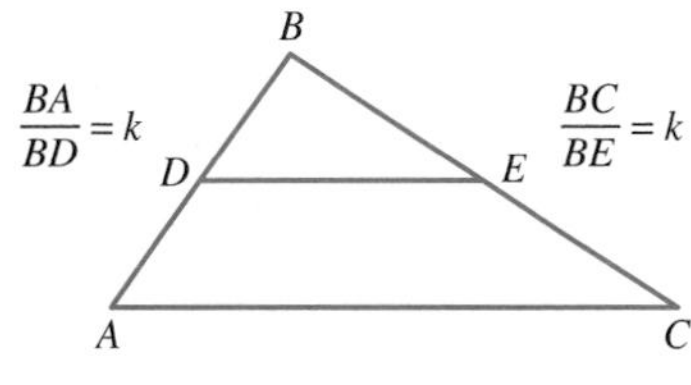

FIGURE 9.25

SOLUTION

Consider the size transformation with center B and scale factor k. Because $\frac{BA}{BD} = k$, we know that the size transformation takes D to A. In a similar fashion, the size transformation takes E to C. Therefore, the size transformation takes $\overline{DE}$ to $\overline{AC}$. Because size transformations take line segments to parallel line segments, we have $\overline{DE} \parallel \overline{AC}$. ■

Applications of Transformations

Transformations can also be useful in solving certain applied problems. Consider the following problem related to shots on a pool table.

EXAMPLE 9.15 In the game of pool shown in Figure 9.26(a), ball A is blocking a direct shot from the cue ball C to the object ball B. Assuming there is no "spin" on the cue ball, where on the upper rail should a player aim ball C to hit ball B?

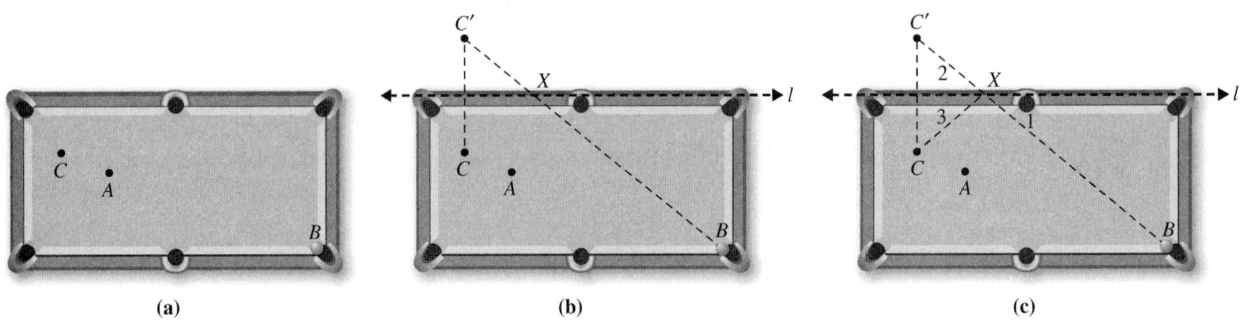

FIGURE 9.26

SOLUTION

Consider line l defined by the edge of the upper rail [Figure 9.26(b)]. Reflect C across l to C' and draw $\overline{C'B}$. The point X where $\overline{C'B}$ intersects l is the point where ball C should be hit [Figure 9.26(b)]. We will use angle measures to prove that X is the correct point [Figure 9.26(c)]. First, $\angle 1$ is the correct angle to hit B from X. Next, $\angle 1 = \angle 2$ because they are vertical angles. Finally, $\angle 2 = \angle 3$ because reflections preserve angle measure and $\angle 3$ was reflected to $\angle 2$. Thus, when ball C hits l at X forming $\angle 3$, it will rebound forming $\angle 1$ and hit ball B. ■

Other examples of pool shot paths caroming or banking off several cushions appear in the problem set.

Solution of Applied Problem

No matter where the bridge is located, CD, the width of the river, will be a constant in the sum $AC + CD + DB$. Hence, we wish to minimize the sum $AC + DB$. Consider a translation in a direction from B toward the river and perpendicular to it for a distance equal to the width of the river, d. Let B' be the image of B under this translation. (See figure.) Then $\overline{AB'}$ is the shortest path from A to B'. Let C be the intersection of $\overline{AB'}$ and m, a line that represents one side of the river.

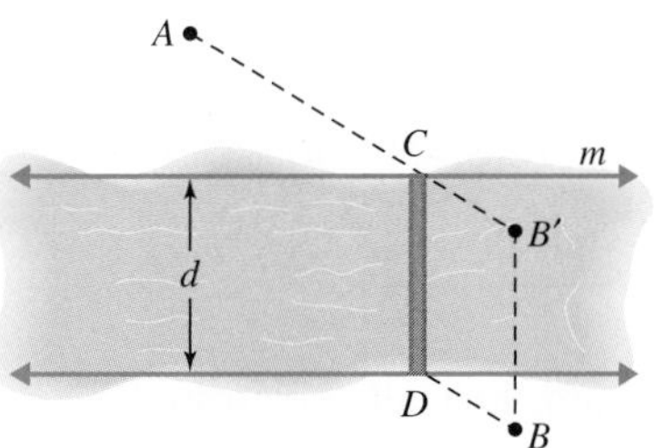

Let D be opposite C on the other side of the river, where $\overline{CD} \perp m$. That is, $\overline{CD} \parallel \overline{B'B}$ and $CD = B'B$. Hence, $BB'CD$ is a parallelogram because the opposite sides $\overline{CD}$ and $\overline{B'B}$ are congruent and parallel. Thus, the sum $AC + CD + DB$ is as small as possible, since $\overline{AB'}$ was the shortest path.

GEOMETRY AROUND US The German mathematician A. F. Moebius created a surface with unique properties. To make a Moebius strip, start with a rectangular strip $ABCD$. Twist the strip once and tape the two ends together to form a "twisted loop." Then, for a surprising result, draw a continuous line down the middle of "one side" of your strip.

Moebius strips are among the shapes studied in a branch of mathematics called topology. **Topology** focuses on those features of geometric shapes that are not changed when the shape is stretched or deformed. One practical application of the Moebius strip is the manufacture of conveyor belts, fan belts, and so on, which wear more evenly.

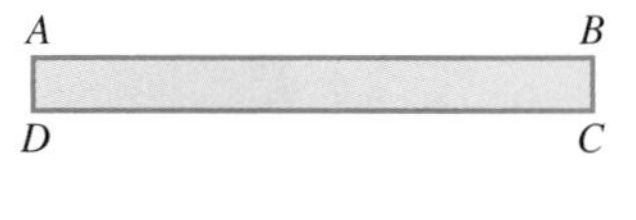

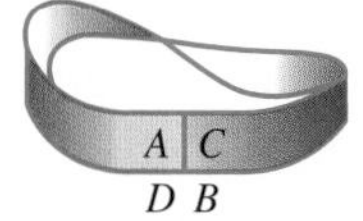

PROBLEM SET 9.3

EXERCISES/PROBLEMS

1. $ABCD$ is a square and points E, F, G, and H are midpoints of its sides. List four reflections that map the square onto itself.

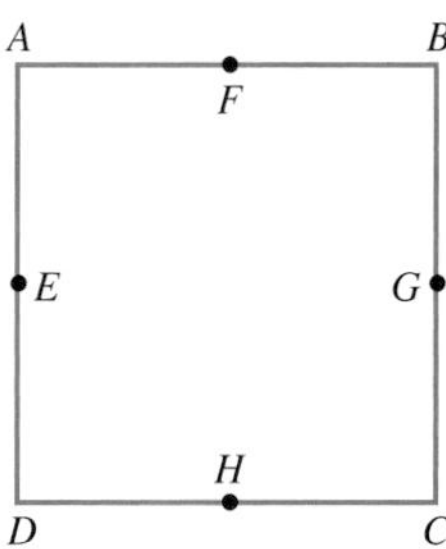

2. *ABCDEF* is a regular hexagon with center *O*. List three reflections and three rotations that map the hexagon onto itself.

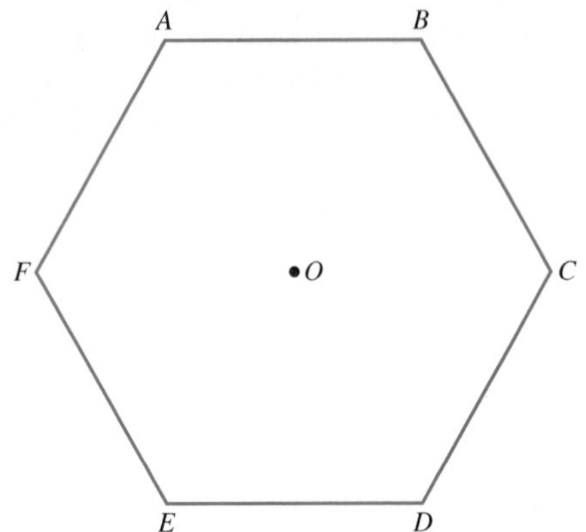

3. *ABCD* is a parallelogram. List all the isometries that map *ABCD* to itself.

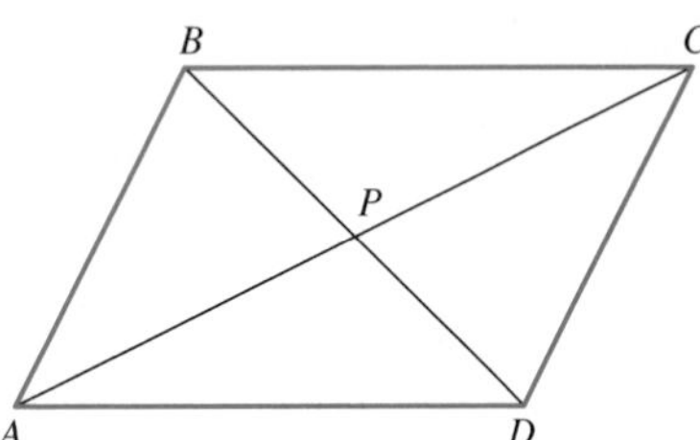

4. *ABCD* is a kite. List all the isometries that map *ABCD* onto itself.

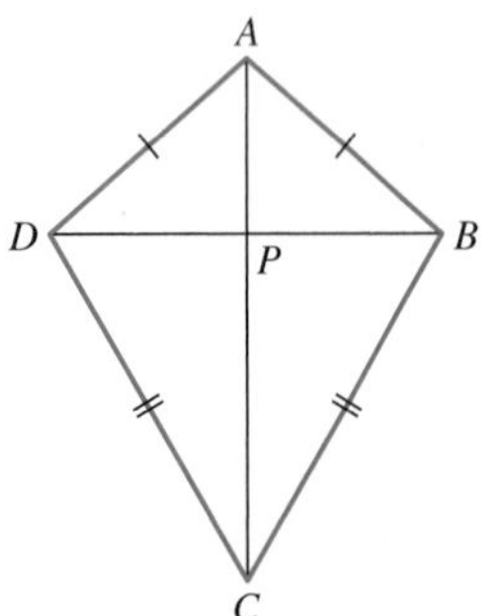

5. *ABCD* is a rhombus.

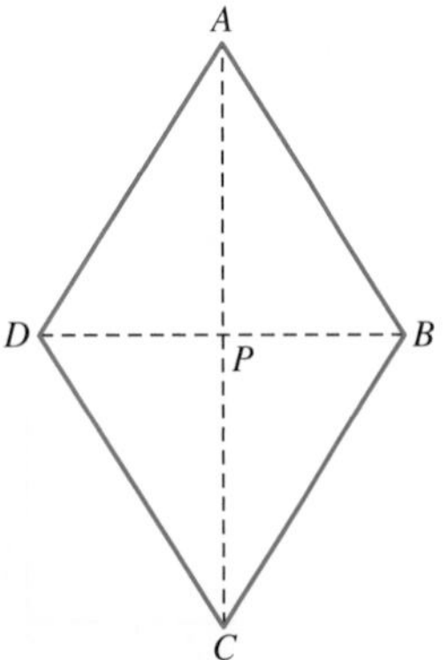

a. List all the isometries that map *ABCD* onto itself.

b. How is your solution related to problems 3 and 4?

6. *ABCD* is an isosceles trapezoid. Are there two isometries that map *ABCD* onto itself? Explain.

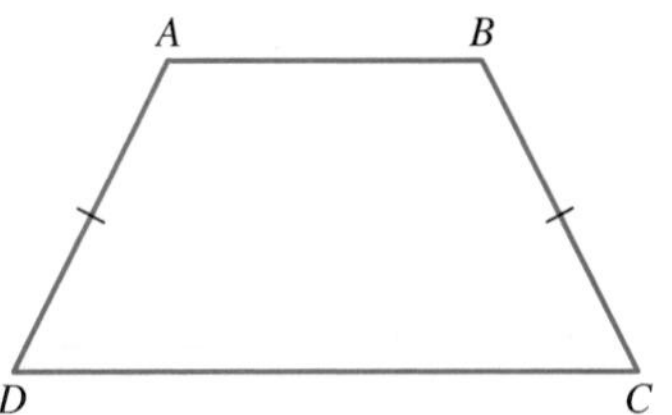

7. Suppose lines *l* and *m* as shown are parallel. Reflect $\triangle ABC$ over *l* and then reflect its image $\triangle A'B'C'$ over *m* to $\triangle A''B''C''$.

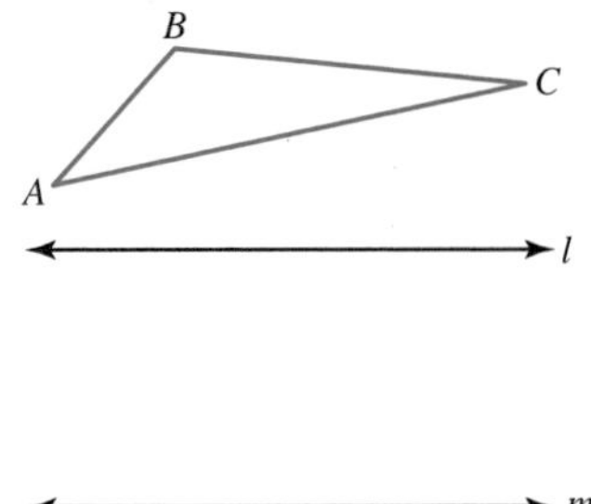

a. Is $\triangle ABC \cong \triangle A''B''C''$? Explain.

b. Do $\triangle ABC$ and $\triangle A''B''C''$ have the same orientation?

c. Will any points of the plane be fixed points after the two reflections have been performed?

d. Is the result of two reflections an isometry? Explain.

e. We stated in Section 9.1 that an isometry must be either a translation, a rotation, a reflection, or a glide reflection. What type of isometry will be the result of two reflections across parallel lines? Explain.

8. Suppose lines *l* and *m* intersect at *O*. Reflect $\triangle ABC$ over *l* and then reflect its image $\triangle A'B'C'$ over *m* to $\triangle A''B''C''$.

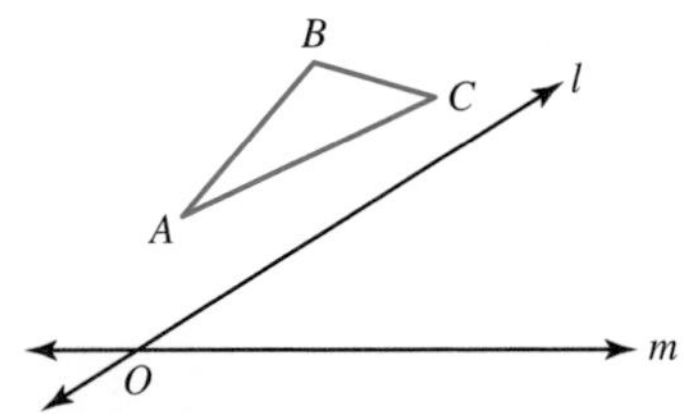

a. Is $\triangle ABC \cong \triangle A''B''C''$? Explain.

b. Do $\triangle ABC$ and $\triangle A''B''C''$ have the same orientation?

c. Will any points of the plane be fixed points after the two reflections have been performed? Explain.

d. Is the result of two reflections an isometry? Explain.

e. We stated in Section 9.1 that an isometry must be either a translation, a rotation, a reflection, or a glide reflection. What type of isometry will be the result of two reflections over intersecting lines? Explain.

PROOFS

9. $ABCD$ is a square. Point P is the intersection of the diagonals. Determine which of the transformations in (a)–(d) can be used to map the square onto itself.

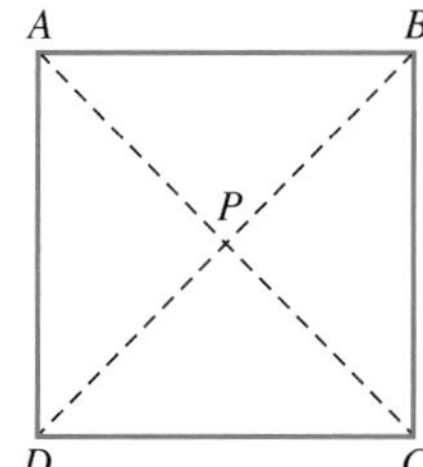

a. The rotation around P of 90°

b. The reflection with respect to $\overline{AC}$

c. The reflection with respect to $\overline{BD}$

d. The rotation around P of 270°

e. Which of the transformations in (a)–(d) can be used to prove that the diagonals of a square are congruent? Explain.

10. $ABCD$ is a kite and point E is the intersection of its diagonals. Which of the following transformations in (a)–(d) can be used to prove that $\triangle ABC \cong \triangle ADC$ using the transformation definition of congruence? Explain.

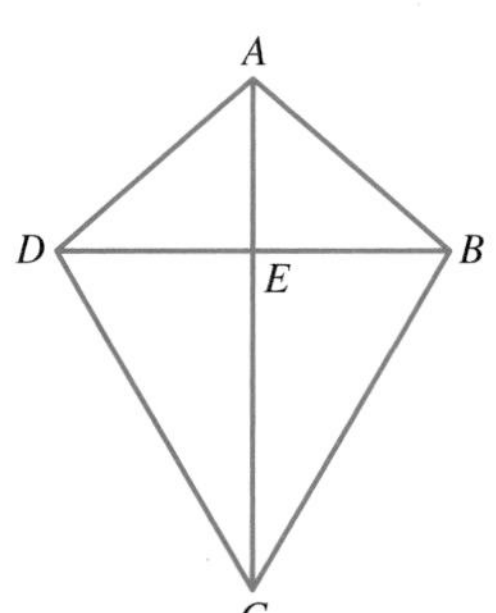

a. The rotation around E of 180°

b. The translation from B to E

c. The rotation around E of 90°

d. The reflection with respect to $\overleftrightarrow{AC}$

11. $ABCDE$ is a regular pentagon with center O. Points F, G, H, I, and J are the midpoints of the sides. List all the reflections and all the rotations that map the pentagon onto itself. Then show how to prove that all the diagonals of a regular pentagon are congruent.

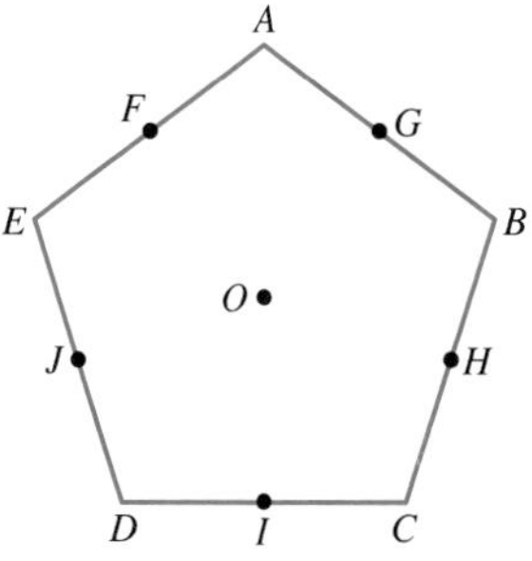

12. In $ABCD$, $AP = CP$ and $\overline{AC} \perp \overline{BD}$.

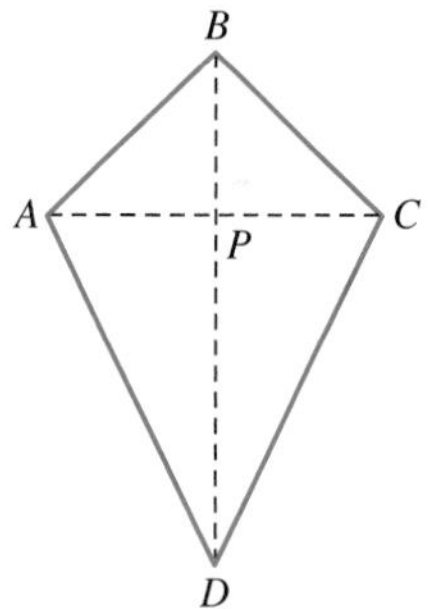

Use transformations to prove that

a. $AB = BC$

b. $AD = CD$

c. $ABCD$ is a kite.

13. In $ABCD$, $AP = CP$, $BP = DP$, and $\overline{AC} \perp \overline{BD}$. Using transformations, prove that $ABCD$ is a rhombus.

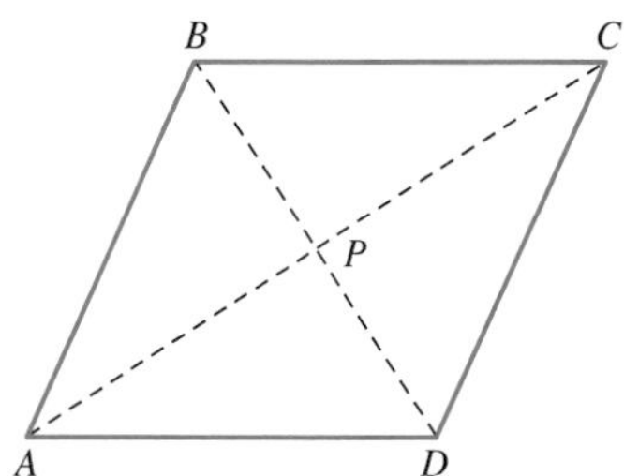

14. In $ABCD$, $AP = CP = BP = DP$ and $\overline{AC} \perp \overline{BD}$. Using transformations, prove that $ABCD$ is a square.

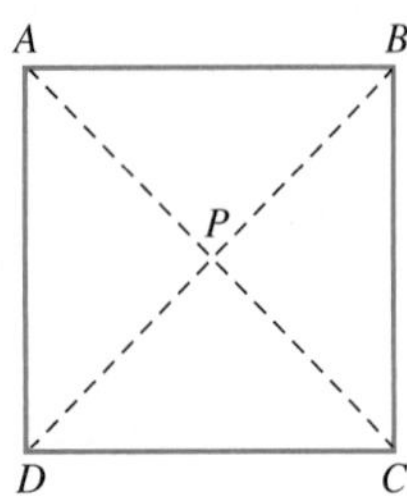

15. $ABCD$ is a parallelogram. Show that diagonal $\overline{AC}$ creates two congruent triangles using the transformation definition of congruence. (HINT: Use a rotation of 180° around the midpoint of $\overline{AC}$ and the fact that 180° rotations take a line to a parallel line.)

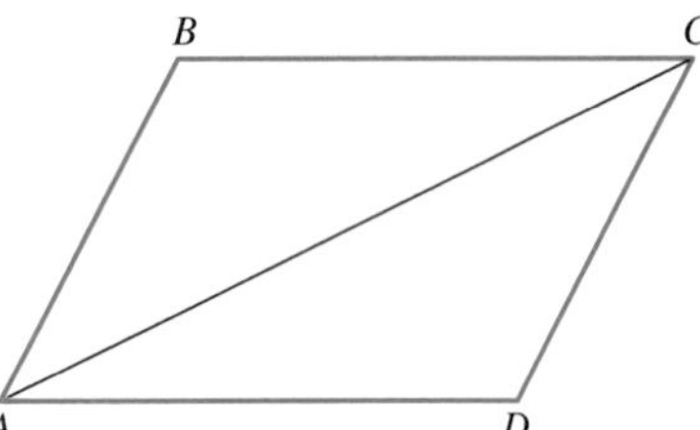

16. Use the results of problem 15 to show that the following statements are true.

a. Opposite angles of a parallelogram are congruent.

b. Opposite sides of a parallelogram are congruent.

17. Suppose B is equidistant from A and C as shown. Using a transformation proof, show that B is on the perpendicular bisector of $\overline{AC}$. (HINT: Let P be a point on $\overline{AC}$ so that $\overline{BP}$ is the angle bisector of $\angle ABC$. Then use a reflection with respect to the line $\overleftrightarrow{BP}$)

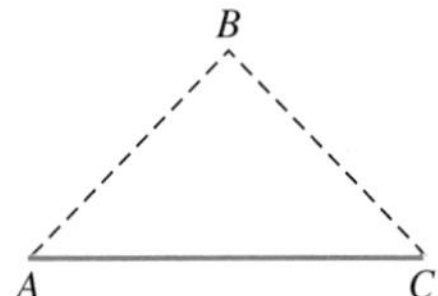

18. Let $ABCD$ be a kite with $AB = BC$ and $AD = DC$. Explain how problem 17 shows that the following statements are true.

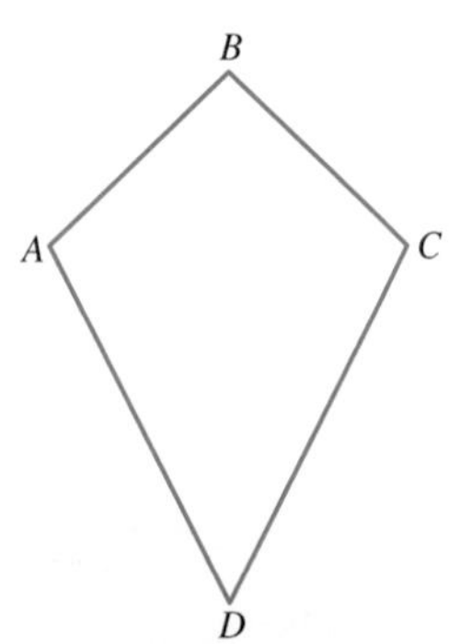

a. The diagonals of a kite are perpendicular.

b. A kite has reflection symmetry.

19. Suppose B is on the perpendicular bisector l of $\overline{AC}$. Use the reflection across l to show that $AB = CB$ (i.e., that B is equidistant from A and C).

20. Suppose that P is on the angle bisector l of $\angle ABC$ where $\overline{AP} \perp \overline{AB}$ and $\overline{CP} \perp \overline{CB}$. Use the reflection across l to show that $AP = CP$ (i.e., that P is equidistant from A and C).

21. Suppose $ABCD$ is a parallelogram. Using transformations, show that the diagonals $\overline{AC}$ and $\overline{BD}$ bisect each other. (HINT: Let P be the midpoint of $\overline{AC}$ and show that under a rotation of 180° about P, the image of B is D.)

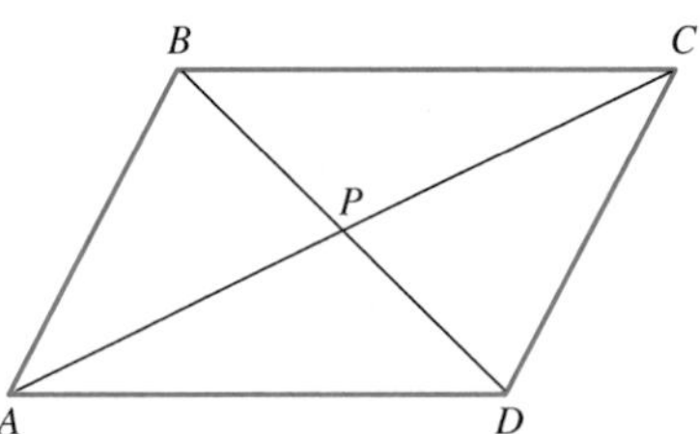

22. Suppose $ABCD$ is a parallelogram and P is the intersection of the diagonals. How does problem 21 show that a parallelogram has rotation symmetry?

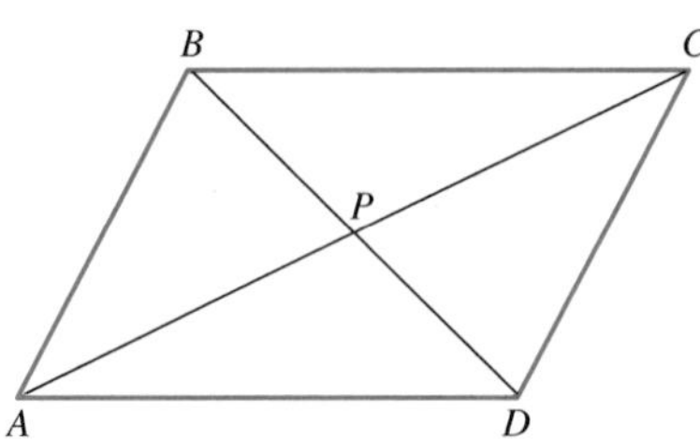

23. Points P, Q, and R are the midpoints of the sides of $\triangle ABC$.

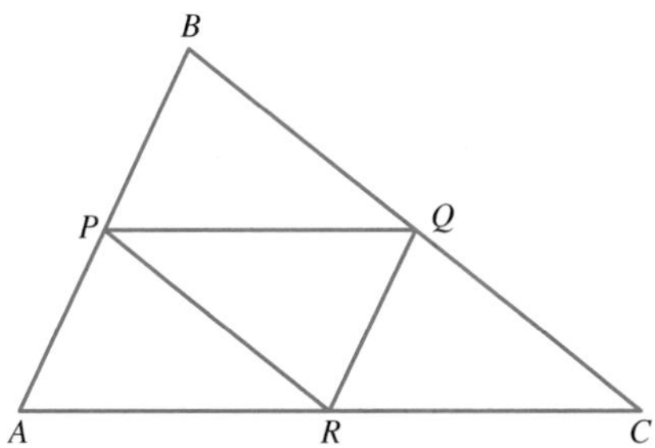

a. Show that a size transformation with center A and scale factor 2 takes $\triangle APR$ to $\triangle ABC$.

b. What is the image of $\triangle PBQ$ under the size transformation with center B and scale factor 2?

c. Find a size transformation that takes $\triangle QCR$ to $\triangle BCA$.

24. Show how parts (a), (b), and (c) in problem 23 can be used to prove that $PBQR$ is a parallelogram.

25. The following sequence of diagrams suggests a transformational proof of the Pythagorean theorem. Explain why this works.

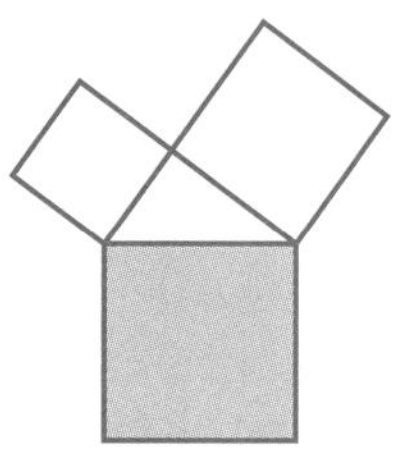

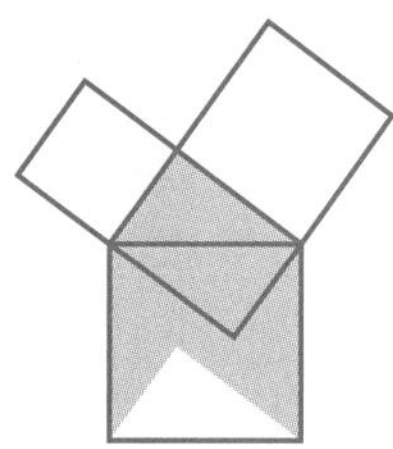

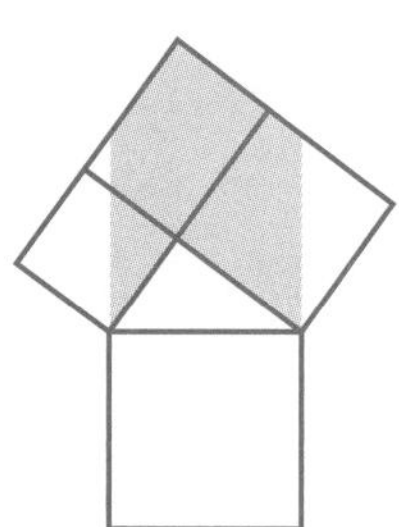

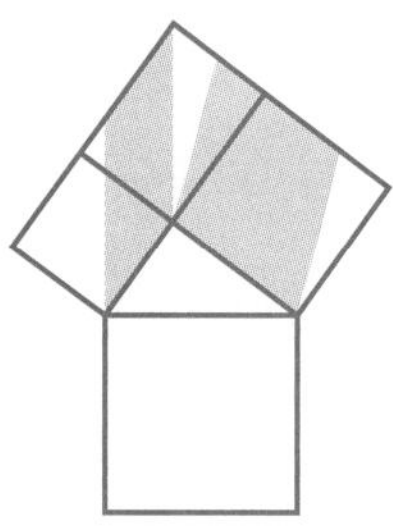

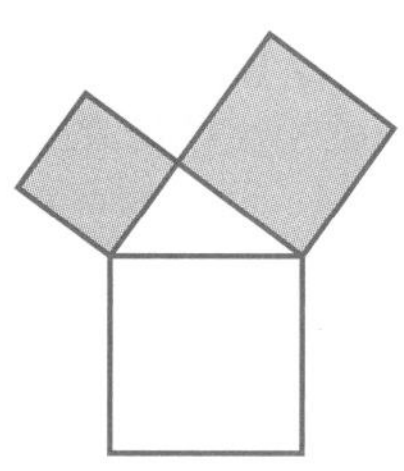

APPLICATIONS

26. On the billiards table, ball A is to carom off two of the rails (sides) and strike ball B. Draw a path for a successful shot.

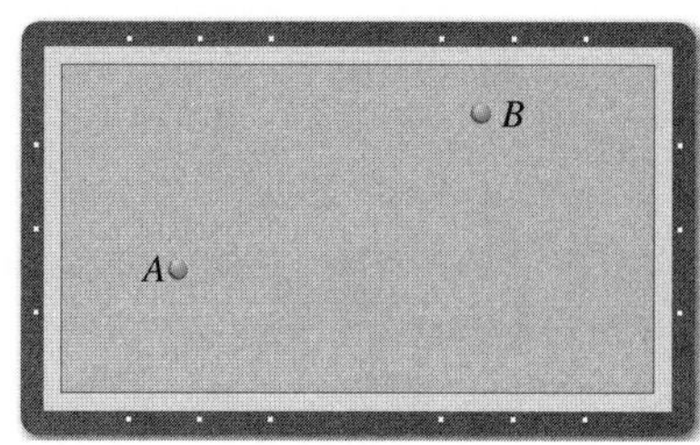

27. On the billiards table, ball A is to carom off three rails, then strike ball B. Draw a path for a successful shot.

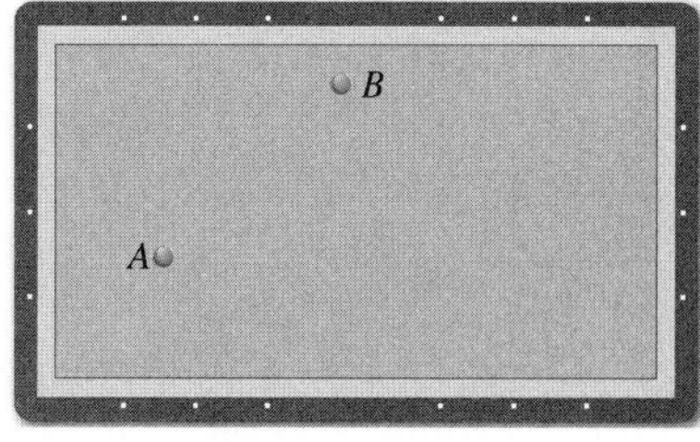

EXTENDED PROBLEMS

28. Every real number is part of a larger system of numbers called the **complex numbers**. All complex numbers are of the form $a + bi$, where a and b are real numbers. We can use a coordinate system, called the **complex plane**, to plot complex numbers. The x-axis is the real axis while the y-axis is the imaginary axis. Associated with each complex number is a distance r from the origin r and an angle A measured from the positive real axis. For example, the complex number $5 + 6i$ appears as a point in the complex plane as shown.

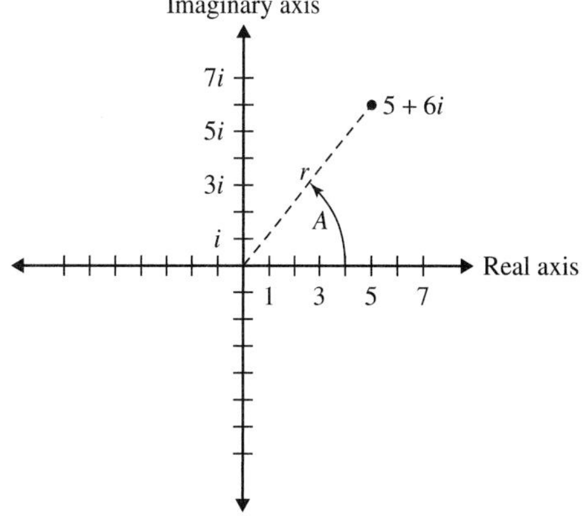

For the complex number $a + bi$, we can find the distance r using the Pythagorean theorem. For the number $5 + 6i$, we have

$$r = \sqrt{a^2 + b^2}$$
$$= \sqrt{5^2 + 6^2}$$
$$= \sqrt{61}$$

We can calculate the measure of angle A by noting that for the complex number $a + bi$,

$$\tan A = \frac{b}{a}$$

Thus, for the complex number $5 + 6i$, we have $\tan A = \frac{6}{5}$, so $A = \tan^{-1}\left(\frac{6}{5}\right) \approx 50.2°$.

Verify this calculation by measuring the angle in the figure with a protractor. If we calculate powers of a complex number and plot the resulting complex numbers, an interesting pattern involving symmetry and transformation geometry emerges.

a. Plot the complex number $1 + i$ in the complex plane. Find r and $\angle A$. Find $(1 + i)^2$, plot the resulting complex number in the complex plane, and find $\angle A$ for $(1 + i)^2$. Repeat this process for the first six powers of the complex number $1 + i$. Describe the symmetry you see in the graph in terms of size transformations and rotations.

b. We discussed polar coordinates in Section 8.1. Research complex numbers and explain how a complex number of the form $a + bi$ can be written in polar form. What is DeMoivre's theorem and how does it verify the symmetry you discovered in part (a)?

29. Geometric optics includes the study of objects reflected in mirrors. For the case of a flat mirror, an object placed a distance d in front of the mirror will have an image that appears to be the same distance away from the mirror and on the opposite side of the mirror. The image is also oriented the same direction as the original object and is the same size as the original object, as shown in the following figure.

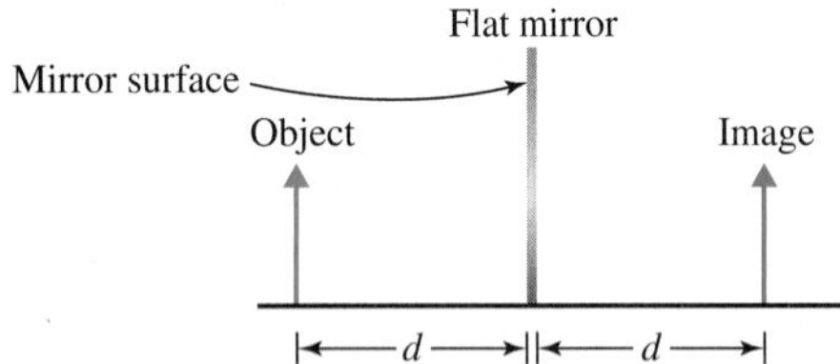

Image location, orientation, and size can change if the mirror is not flat. Consider the case of a convex mirror in which C is the center of a sphere from which the mirror is cut (C is called the **center of curvature**) and F is the focal point. Both C and F are located along a principal axis as shown in the following figure. The **focal point** is the point at which rays of light traveling parallel to the principal axis will meet after reflection.

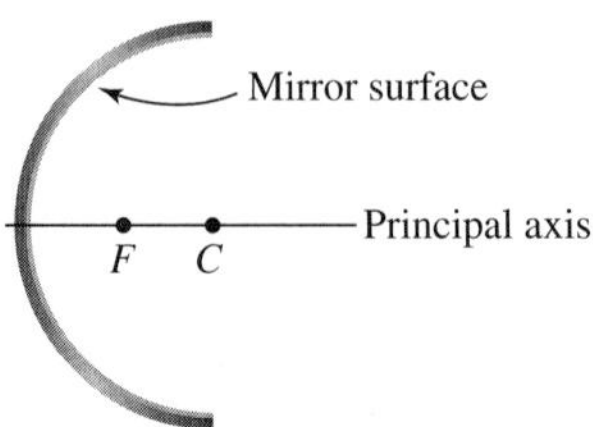

Research reflective properties of a convex mirror and determine what happens to the size, orientation, and location of an image if the object is placed as follows.

a. The object is between the mirror and the focal point.

b. The object is at the focal point.

c. The object is between the focal point and the center of curvature.

d. The object is at the center of curvature.

e. The object is beyond the center of curvature.

30. We discussed the Moebius strip in the "Geometry Around Us" in Section 9.3.

a. Create a Moebius strip. Draw a line down the middle of one side of the strip. What happens? Predict what will happen if you cut the Moebius strip along the line you drew. Cut the strip along that line and describe what happens.

b. Create another Moebius strip. Draw a line one-third of the way from one edge of the Moebius strip and cut along the line. Describe what happens.

c. Construct a strip with two half twists. Is it a Moebius strip? How can you tell?

d. Research the Moebius strip and its properties. What are some other interesting results that occur when a Moebius strip or a combination of strips is cut? What is one practical application of the Moebius strip? How did M. C. Escher incorporate the Moebius strip into his artwork?

Solution of Initial Problem

Houses A and B are to be connected to a television cable line, l, at a transformer at some point P. Where should P be located so that the sum of the distances, $AP + PB$, is as small as possible?

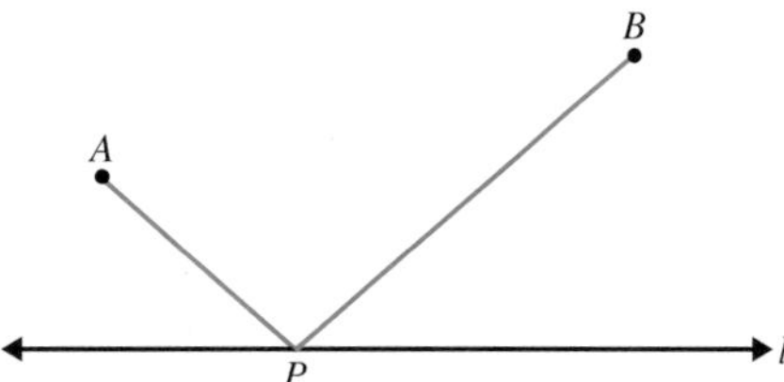

Reflect point B across line l to point B'. Then draw $\overline{AB'}$ and $\overline{BB'}$, forming points of intersection P and Q, respectively.

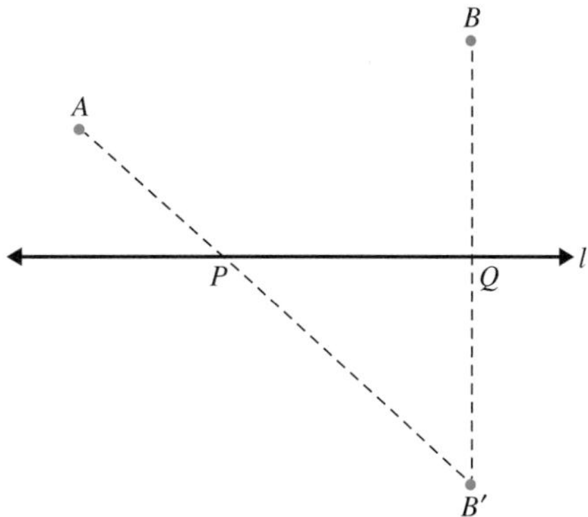

Next draw the perpendicular $\overline{AR}$ and the segment $\overline{PB}$.

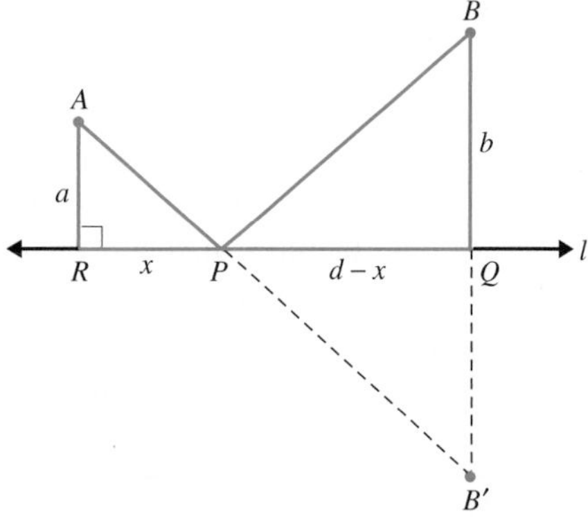

Because B' is the reflection image of B, we have $\angle BPQ \cong \angle B'PQ$. Also, $\angle APR \sim \angle B'PQ$ since they are vertical angles. Therefore, $\angle APR \cong \angle BPQ$. Thus, $\triangle APR \sim \triangle BPQ$ by AA Similarity. Let $RP = x$ and $RQ = d$. Thus, $PQ = d - x$. Because corresponding parts of similar triangles are proportional, we have

$$\frac{x}{d - x} = \frac{a}{b}$$

so that

$$bx = a(d - x), \text{ or}$$

$$x = \frac{ad}{a + b}$$

where $d = RQ$. So the transformer P should be located at a distance $\dfrac{ad}{a + b}$ from R along $\overline{RQ}$.

Additional Problems Where the Strategy "Use Symmetry" Is Useful

1. A 4-by-4 quilt of squares shown has reflection symmetry.

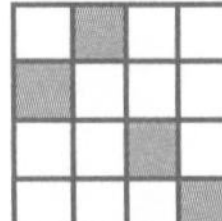

Explain how to find all such quilts having reflection symmetry that are made up of 12 light squares and 4 dark squares.

2. Two vertices of an isosceles trapezoid are given. What are two possible sets of coordinates for the other vertices R and S?

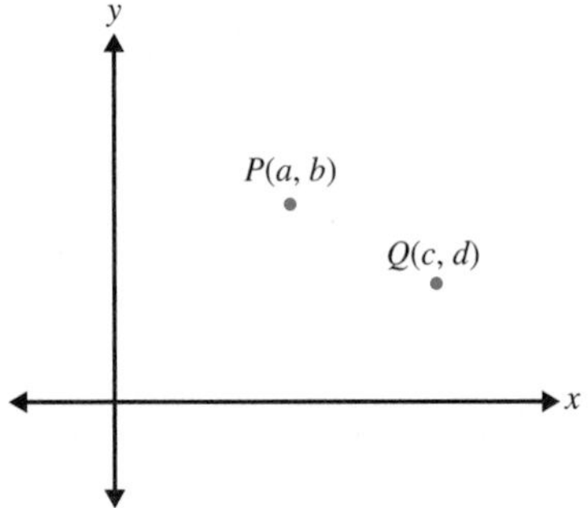

Writing for Understanding

1. Discuss all possible outcomes when two isometries are performed in sequence. For example, what is the result when a reflection is followed by a reflection?
2. The relationship of the size of a photograph of an object to the size of the object is an example of a size transformation. Describe several other applications of size transformations in everyday life.
3. Designs based on symmetries and patterns based on isometries and similitudes are found in many cultures as mentioned in the "Geometry Around Us" in Section 9.1. Find pictures of such designs in pottery, weavings, etc., from a culture other than your own. Describe how transformations are used in the artwork in the pictures.
4. Do research on tessellations created by M. C. Escher. Select three of your favorite ones and explain how transformations play a role in the tessellation.

PEOPLE IN GEOMETRY

H. S. M. Coxeter (1907–2003) is known for his research and expositions of geometry. He was the author of 11 books, including *Introduction to Geometry, Projective Geometry, Non-Euclidean Geometry, The Fifty-Nine Icosahedra*, and *Mathematical Recreations and Essays.* He contended that Russia, Germany, and Austria are the countries that do the best job of teaching geometry in the schools, because they regard it as a subject worth studying: "In English-speaking countries, there was a long tradition of dull teaching of geometry. People thought that the only thing to do in geometry was to build a system of axioms and see how you would go from there. So children got bogged down in this formal stuff and didn't get a lively feel for the subject. That did a lot of harm."

CHAPTER REVIEW

For each section of this chapter, you will find a list of vocabulary and notation, questions to assess your understanding of key concepts, and review problems similar to the problems you worked in your homework. Review each item in the *Vocabulary/Notation* list mentally, and, if necessary, refer back to the indicated page and write a definition. Then answer the *Concept Check Questions*, looking back at the section if you need help. Work the *Review Problems* as practice before you move on to the *Chapter Test*. Answers to the *Review Problems* and *Chapter Test* can be found at the back of the book.

SECTION 9.1 Isometries and Congruence

Vocabulary/Notation

Directed line segment 478
Vector 478
Transformation 478
Image 478
Preimage 479
Translation (slide) 479
Clockwise orientation 480
Counterclockwise orientation 480
Directed angle 482
Center of rotation 482
Rotation (turn) 482
Fixed point 482
Identity transformation 482
Reflection (flip) with respect to a line 484
Glide reflection 486
Isometry 488
Congruent shapes 488

Concept Check Questions

1. What properties are associated with a vector? 478
2. How can we use a vector to translate a point to its image? 478–479
3. What are the five properties of translations? 480
4. How can the orientation of a figure be described? 480
5. In a rotation, which point is always a fixed point? 482
6. What is the relationship between a point and its image after a reflection with respect to a line? 484
7. Which two transformations make up a glide reflection? 486
8. Which transformations are isometries? 488
9. How can congruence be defined in terms of isometries? 488–489

Review Problems

1. **a.** Find the coordinates of the images of the vertices of quadrilateral $ABCD$ for $A(-2,0)$, $B(0,2)$, $C(3,0)$, and $D(-1,-2)$ under a translation from $P(0,0)$ to $Q(-1,-2)$. Graph quadrilateral $ABCD$ and its image.
   b. Given a point (x,y), what are the coordinates of its image under this translation?
2. Give the coordinates of the images of the following points under the rotation of 90° around the origin.
   a. $(2,3)$ **b.** $(-1,3)$ **c.** $(-1,4)$
   d. $(-4,-2)$ **e.** $(2,-4)$ **f.** (x,y)
3. Consider quadrilateral $ABCD$ with $A(-2,0)$, $B(0,2)$, $C(3,0)$, and $D(-1,-2)$.
   a. Find the coordinates of the image of quadrilateral $ABCD$ under a rotation of 180° around the origin.
   b. Find the coordinates of the image of quadrilateral $ABCD$ after a reflection over the x-axis.
4. Use a compass and straightedge to find the reflection image of $\triangle ABC$ with respect of line l.

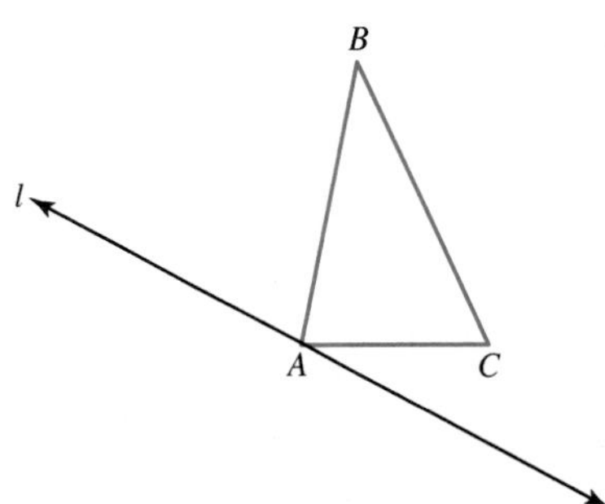

5. Find the image of each of the following points under the glide reflection described by the translation from $P(2,3)$ to $Q(-3,5)$ followed by the reflection with respect to the y-axis.
   a. $(6,10)$ **b.** $(-5,1)$
   c. $(-4,-8)$ **d.** (x,y)

SECTION 9.2 Similitudes and Similarity

Vocabulary/Notation

Size transformation 500
Center of a size transformation 500
Scale factor 500
Identity size transformation 500
Similitude 502
Similar polygons 503
Similar shapes 504

Concept Check Questions

1. What information is required in order to carry out a size transformation? 500
2. How can the image of a point under a size transformation be found? 500
3. For what size transformation will the original point and its image be the same? 500
4. What properties are preserved under a size transformation? 501
5. How can we determine the center of a size transformation? 501–502
6. How can we define similarity in terms of a similitude? 502
7. How can we determine the scale factor of a size transformation? 501

Review Problems

1. Find the image of $\triangle ABC$ for the following size transformations.

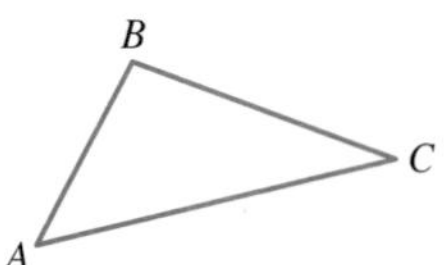

   a. Scale factor 2, center A
   b. Scale factor $\frac{1}{3}$, center C

2. a. Find the center and scale factor of the size transformation that takes $ABCD$ to $A'B'C'D'$.

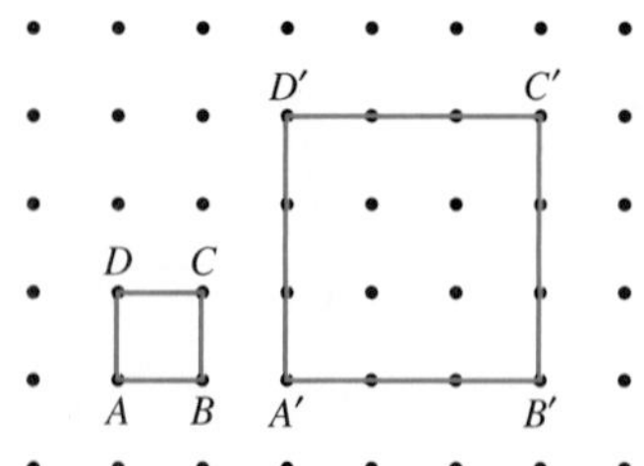

   b. Find each of the following ratios: $\frac{A'B'}{AB}$, $\frac{B'C'}{BC}$, $\frac{C'D'}{CD}$, and $\frac{D'A'}{DA}$.
   c. Find the ratio $\frac{\text{Area of } A'B'C'D'}{\text{Area of } ABCD}$.

3. Given $\triangle ABC$ with $A(0, 0)$, $B(4, 1)$, and $C(2, 5)$, find the coordinates of $\triangle A'B'C'$ after a size transformation with center $O(-3, 2)$ and scale factor 2. Then find the coordinates of $\triangle A''B''C''$ after a reflection with respect to the y-axis. Is $\triangle ABC$ similar to $\triangle A''B''C''$? Explain.

4. $ABCD$ is a square with side length 4 inches. Find a similitude that takes $ABCD$ to $A''B''C''D''$. Describe the similitude as completely as possible.

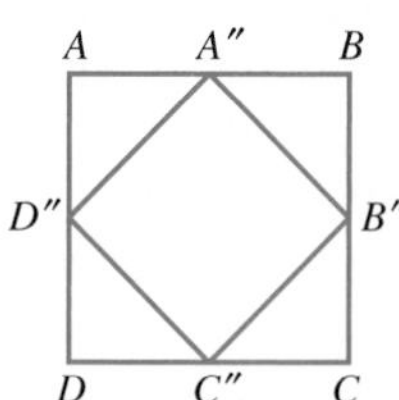

5. Describe how to prove that any regular hexagon $ABCDEF$ is similar to any other regular hexagon $GHIJKL$.

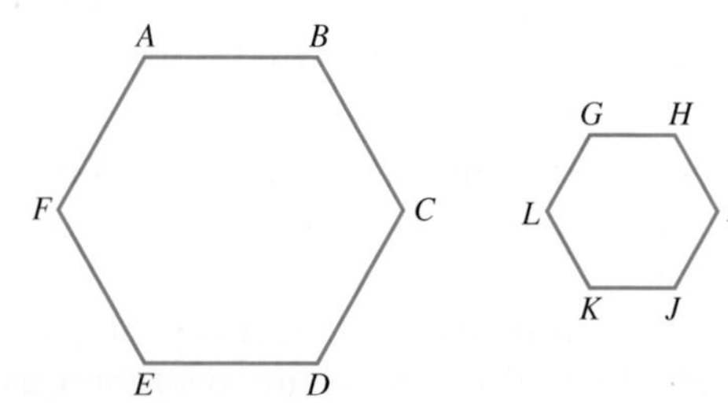

SECTION 9.3 Problem Solving Using Transformations

Vocabulary/Notation

Half-turn 510

Concept Check Questions

1. What types of transformations might be useful when creating a proof that requires congruent figures? 510–511

2. What types of transformations might be useful when creating a proof that requires similar figures? 511–512

Review Problems

1. $ABCD$ is a rectangle and points E, F, G, and H are midpoints of its sides. Point O is the center of $ABCD$. List all the isometries that map $ABCD$ to itself.

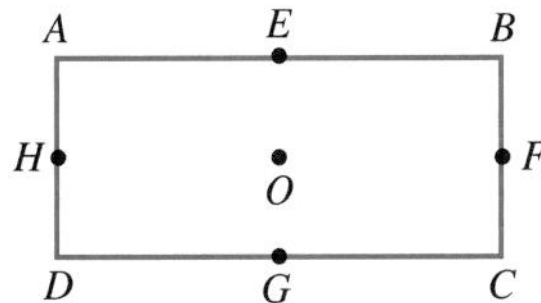

2. In isosceles trapezoid $ABCD$, $\overline{AD} \parallel \overline{BC}$ and $AB = CD$. Use a transformation to prove $\angle A \cong \angle D$.

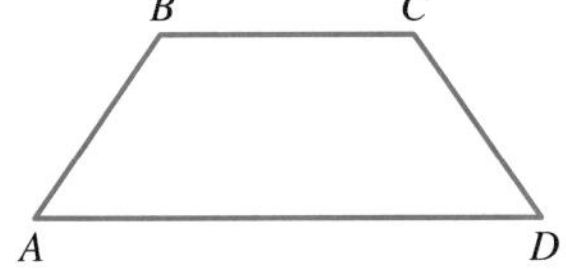

3. On a miniature golf course, a ball at point A is to be putted into a hole at point B. Is it possible to hit the ball into the hole in one shot, allowing for caroms off sides? If so, draw the path of the ball. If not, explain why it is impossible.

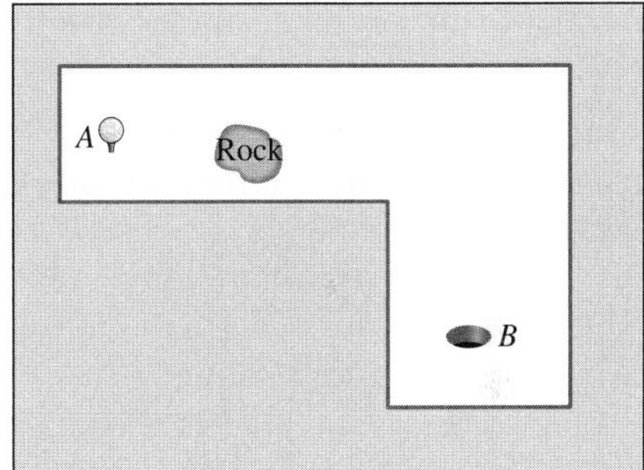

CHAPTER 9 TEST

TRUE-FALSE

Mark as true any statement that is always true. Mark as false any statement that is never true or that is not necessarily true. Be able to justify your answers.

1. Every rotation has at least one fixed point.
2. Every reflection has at least one fixed point.
3. A size transformation takes a line segment to a parallel line segment.
4. A glide reflection takes a line segment to a parallel line segment.
5. If the perimeter of $\triangle ABC$ is 20 cm and $\triangle A'B'C'$ is the image of $\triangle ABC$ under any isometry, then the perimeter of $\triangle A'B'C'$ is 20 cm.
6. A reflection preserves the orientation of a triangle.
7. The transformation that takes each point (a, b) to the point $(a - 3, b)$ is a translation.
8. The transformation that takes each point (a, b) to the point $(a, -b)$ is a reflection.
9. If two triangles are congruent and have the same orientation, then there is a translation that takes one triangle to the other.
10. If two polygons are similar and have the same orientation, then there is a size transformation that takes one polygon to the other.

EXERCISES/PROBLEMS

11. a. Find coordinates of the vertices of $\triangle A'B'C'$, the image of $\triangle ABC$ under the translation $\overrightarrow{PQ}$.

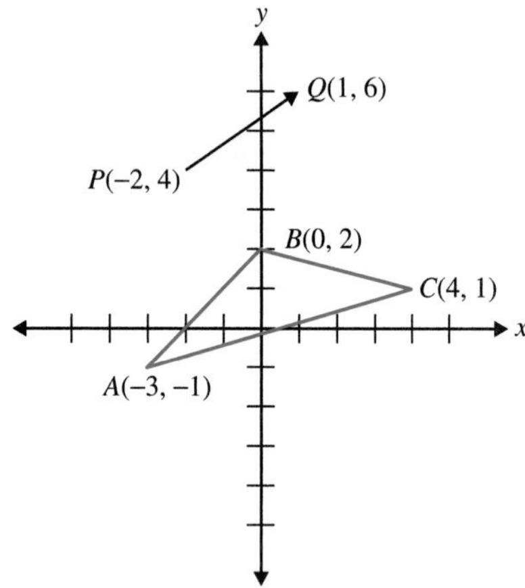

b. If a point has coordinates (a, b), what are the coordinates of its image under the translation $\overrightarrow{PQ}$?

12. Use your protractor and a ruler to draw the image of the figure shown under a rotation of $-40°$ about center O.

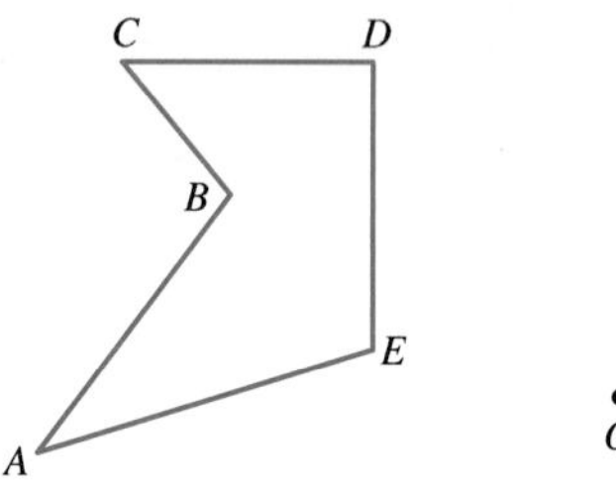

13. Using a compass and straightedge, find the reflection image of $\triangle ABC$ with respect to line l.

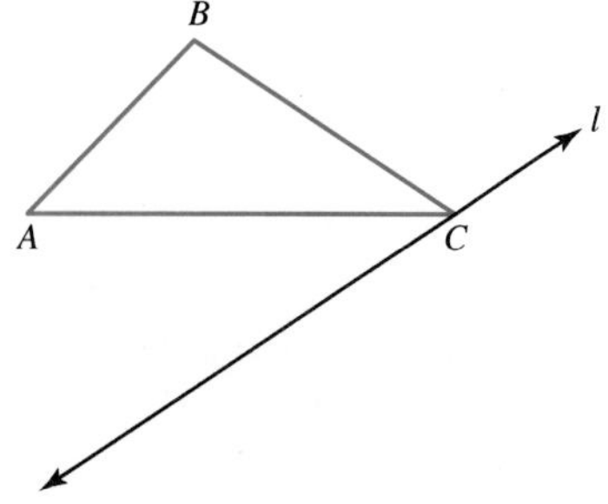

14. Using a protractor and ruler, find the image of $ABCD$ under the glide reflection that consists of the translation from A to B followed by the reflection with respect to line l.

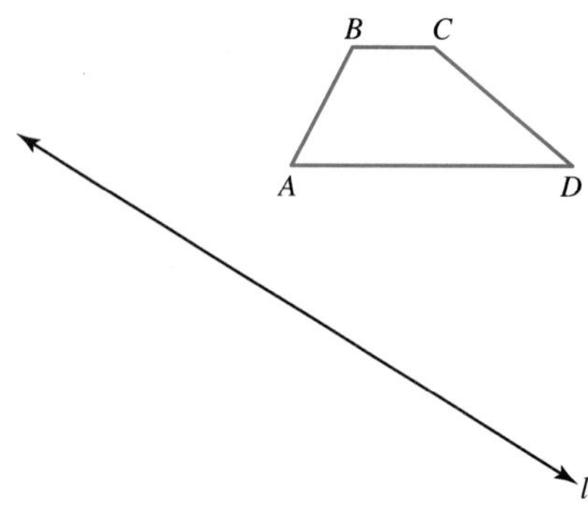

15. On the square lattice shown, draw the image of $\overline{AB}$ under a rotation of 270° about point O.

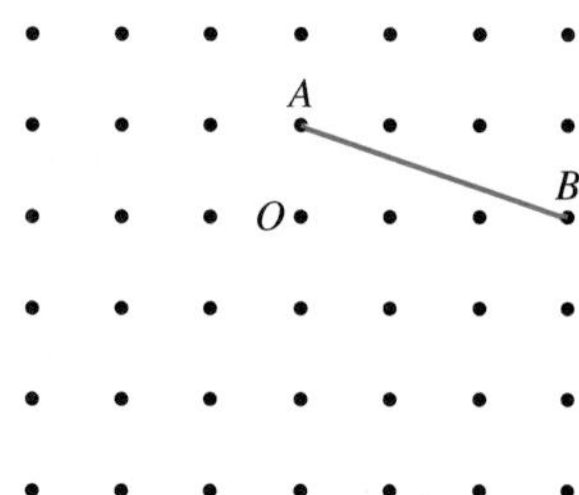

16. On the square lattice shown, draw the image of $\triangle ABC$ under a size transformation with center O and scale factor 2.

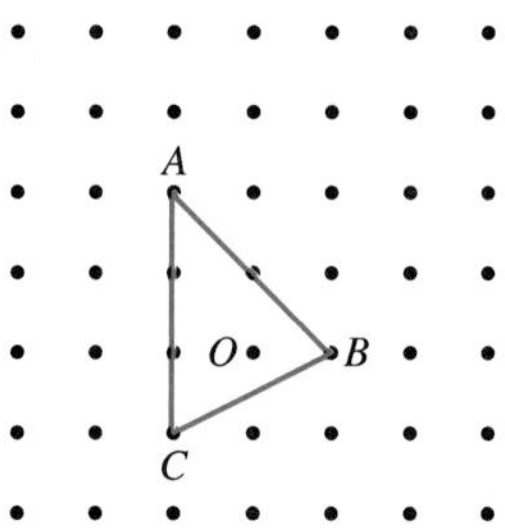

17. A triangle has vertices $A(1, 2)$, $B(4, 4)$, and $C(6, 3)$.

a. Sketch $\triangle A''B''C''$, the image of $\triangle ABC$ under the size transformation centered at the origin with scale factor 3 followed by a reflection with respect to the y-axis.

b. What are the coordinates of A'', B'', and C''?

c. If a point has coordinates (a, b), what are the coordinates of its image under this similitude?

18. Determine each type of isometry that takes the shape on the left (or top) to the shape on the right (or bottom).

a.

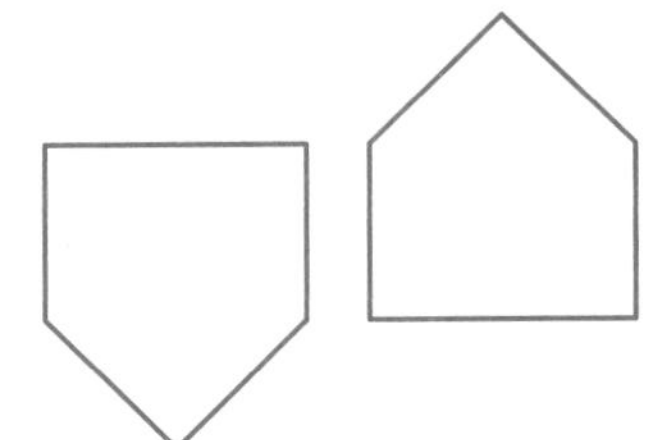

b.

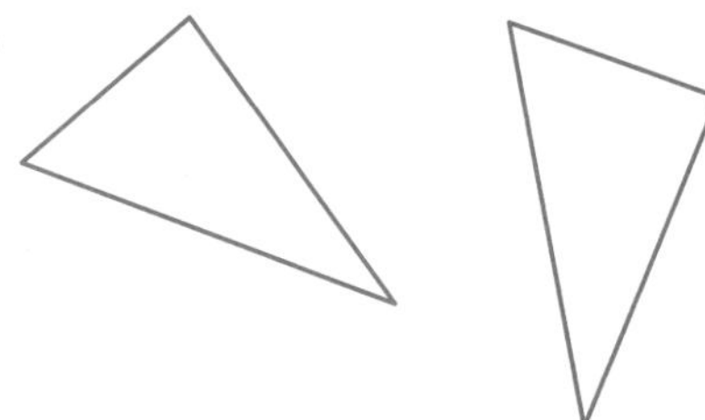

c.

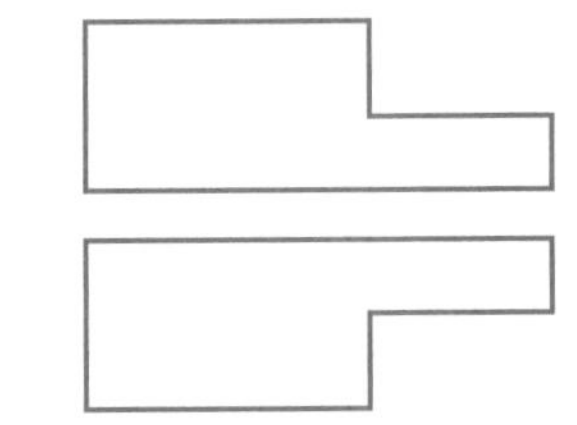

19. Describe the similitude that takes trapezoid $ABCD$ to trapezoid $A'B'C'D'$.

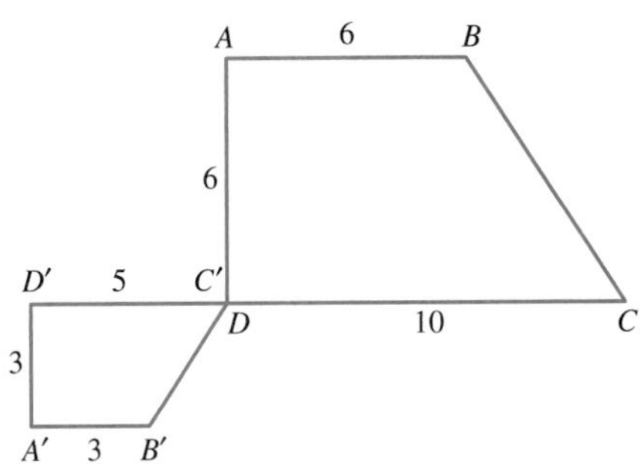

20. Suppose that a transformation is applied to all points in the plane and the image of $\overline{AB}$ is $\overline{AB}$. Under what kind of transformation might this occur?

PROOFS

21. Show that a glide reflection can be performed by first reflecting and then translating or by first translating and then reflecting. That is, show that the image of an arbitrary point is the same in each case. (NOTE: Recall that we assume that the translation is in a direction parallel to the line of reflection.)

22. Suppose that $ABCD$ is a rhombus. Use reflections with respect to lines $\overleftrightarrow{AC}$ and $\overleftrightarrow{BD}$ to show that $ABCD$ is a parallelogram.

APPLICATIONS

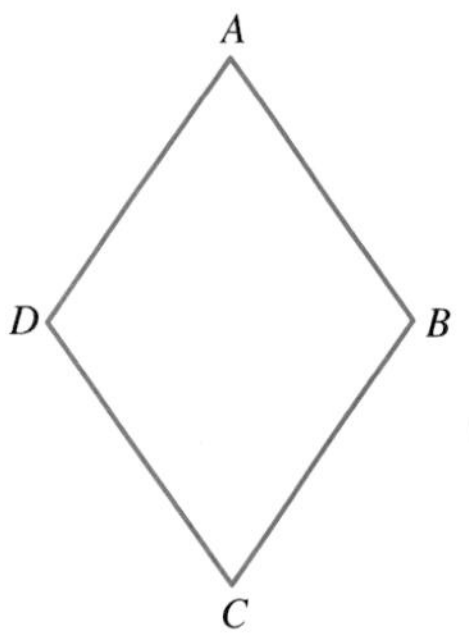

23. On the billiards table shown, ball A is to carom off four of the rails and then strike ball B.

Draw a path for a successful shot.

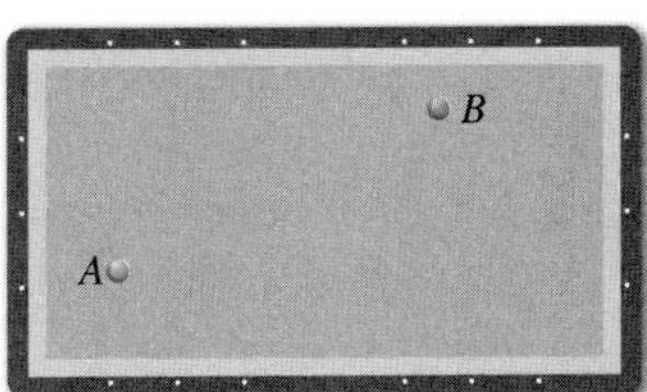

24. A graphic designer working on a logo needs to put the design shown on a coordinate system and then double its size. How can she accomplish this?

TOPIC 1

ELEMENTARY LOGIC

A knowledge of logic allows us to evaluate the validity of arguments, in and out of mathematics. The validity of an argument depends on its logical form and not on the particular meaning of the terms it contains. For example, the argument, "All *X*s are *Y*s; all *Y*s are *Z*s; therefore, all *X*s are *Z*s" is valid no matter what *X*, *Y*, and *Z* are. In this topic section, we will study how we can use logic to represent arguments symbolically and to analyze arguments using tables and diagrams.

STATEMENTS

Statements are the building blocks on which logic is built. A **statement** is a declarative sentence that is true or false but not both. Examples of statements are:

1. Alaska is the largest state of the United States, geographically. (True)
2. Based on population, Texas is the largest state of the United States. (False)
3. $2 + 3 = 5$ (True)
4. $3 < 0$ (False)

The following are not statements as defined in logic.

1. Oregon is the best state. (Subjective)
2. Help! (An exclamation)
3. Where were you? (A question)
4. The rain in Spain (Not a sentence)
5. This sentence is false. (Neither true nor false!)

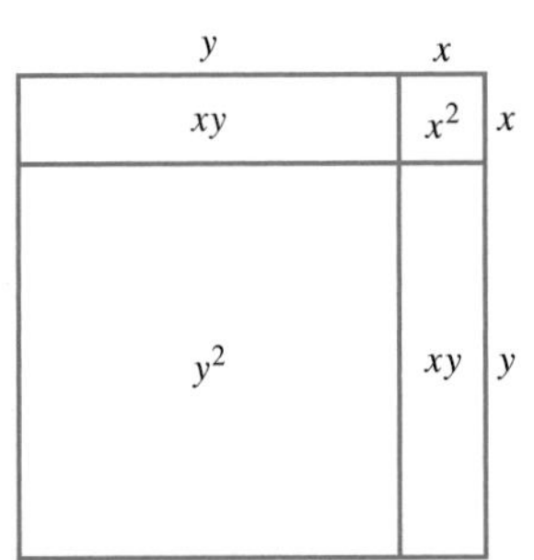

FIGURE T1.1

Often, we can make ideas in mathematics clearer by using variables and diagrams. For example, we can use the equation $2m + 2n = 2(m + n)$, where the variables m and n are whole numbers, to show that the sum of any two arbitrary even numbers, $2m$ and $2n$ here, is the even number $2(m + n)$. Figure T1.1 shows that $(x + y)^2 = x^2 + 2xy + y^2$, where each term in the expanded product is the designated area of the rectangular region. In particular, we can use symbols and diagrams to clarify the logic of verbal arguments. Statements are represented symbolically by lowercase letters (e.g., p, q, r, and s).

We can create new statements from existing statements in several ways. For example, if p represents the statement "The sun is shining," then the **negation** of p, written $\sim p$ and read "not p," is the statement "The sun is not shining." When a statement is true, its negation is false, and when a statement is false, its negation is true. That is, a statement and its negation have opposite truth values. We summarize this relationship between a statement and its negation using a **truth table**, shown next.

p	$\sim p$
T	F
F	T

This table shows that when the statement p is T, then $\sim p$ is F, and when p is F, then $\sim p$ is T.

LOGICAL CONNECTIVES

We can join, or connect, two or more statements to form **compound statements**. Next we study the four commonly used **logical connectives** "and," "or," "if-then," and "if and only if."

And

If p is the statement "It is raining" and q is the statement "The sun is shining," then the **conjunction** of p and q is the statement "It is raining *and* the sun is shining." The conjunction is represented symbolically as "$p \wedge q$." The conjunction of two statements p and q is true only when both p and q are true. That is, we say that the statement "It is raining and the sun is shining" is true only in the case where it is true that the sun is shining *and* it is raining. The next truth table displays the possible values of p and q with the corresponding values of $p \wedge q$.

p	q	$p \wedge q$
T	T	T
T	F	F
F	T	F
F	F	F

Notice that the two statements p and q each have two possible truth values, T and F. Hence, there are four possible combinations of T and F to consider.

Or

The **disjunction** of statements p and q is the statement "p or q," which is represented symbolically as "$p \vee q$". In practice, there are two common uses of "or": the exclusive "or" and the inclusive "or." The statement "I will go or I will not go" is an example of the use of the **exclusive "or"** because either "I will go" is true or "I will not go" is true, but both cannot be true simultaneously. The **inclusive "or"** (called "and/or" in everyday language) allows for the situation in which both parts are true. For example, the statement "It will rain or the sun will shine" uses the inclusive "or." It is true if (1) it rains, (2) the sun shines, or (3) it rains and the sun shines. That is, the inclusive "or" in $p \vee q$ allows for both p and q to be true. In mathematics, we agree to use the inclusive "or" whose truth values we summarize in the next truth table.

p	q	$p \vee q$
T	T	T
T	F	T
F	T	T
F	F	F

EXAMPLE T1.1 Decide if the following statements are true or false, where p represents the statement "Rain is wet" and q represents the statement "Black is white."

a. $\sim p$ **b.** $p \wedge q$ **c.** $(\sim p) \vee q$

d. $p \wedge (\sim q)$ **e.** $\sim(p \wedge q)$ **f.** $\sim[p \vee (\sim q)]$

SOLUTION

a. p is T, so $\sim p$ is F.

b. p is T and q is F, so $p \wedge q$ is F.

c. $\sim p$ is F and q is F, so $(\sim p) \vee q$ is F.

d. p is T and $\sim q$ is T, so $p \wedge (\sim q)$ is T.

e. p is T and q is F, so $p \wedge q$ is F and $\sim(p \wedge q)$ is T.

f. p is T and $\sim q$ is T, so $p \vee (\sim q)$ is T and $\sim[p \vee (\sim q)]$ is F. ■

If-Then

One of the most important compound statements is the implication. The statement "If p, then q," denoted by "$p \Rightarrow q$," is called an **implication** or **conditional** statement. The statement p is called the **hypothesis** and q is called the **conclusion**. To construct the truth table for $p \Rightarrow q$, consider the following conditional promise given to a math class: "If you average at least 90 percent on all tests, then you will earn an A." Let p represent "Your average is at least 90 percent on all tests" and q represent "You earn an A." Then there are four possibilities, as shown in the next table.

Average at Least 90%	Earn an A	Promise Kept
Yes	Yes	Yes
Yes	No	No
No	Yes	Yes
No	No	Yes

Notice that the only way the promise can be broken is in line 2. In lines 3 and 4, the promise is not broken since an average of at least 90 percent was not attained. (In these cases, a student may still earn an A—it does not affect the promise either way.) This example suggests the following truth table for the conditional $p \Rightarrow q$.

p	q	$p \Rightarrow q$
T	T	T
T	F	F
F	T	T
F	F	T

The truth values for $p \wedge q$ and $q \wedge p$ are always the same. Also, the truth tables for $p \vee q$ and $q \vee p$ are identical. (Why?) But the truth tables of $p \Rightarrow q$ and $q \Rightarrow p$ are not identical. Consider this example: Let p be "You live in New York City" and q be "You live in New York state." Then $p \Rightarrow q$ is true since if you live in New York City, then you must live in New York state. However, $q \Rightarrow p$ is not true because living in New York state does not necessarily imply that you live in New York City. For example, you may live in Albany. The conditional $q \Rightarrow p$ is called the converse of $p \Rightarrow q$. As the example shows, a conditional may be true and its converse false. On the other hand, a conditional and its converse may both be true. There are two other variants of a conditional, the contrapositive and the inverse, that are used in mathematics. A summary of these statements is given next.

Given conditional: $p \Rightarrow q$
The **converse** of $p \Rightarrow q$ is $q \Rightarrow p$.
The **inverse** of $p \Rightarrow q$ is $(\sim p) \Rightarrow (\sim q)$.
The **contrapositive** of $p \Rightarrow q$ is $(\sim q) \Rightarrow (\sim p)$.

The following truth table displays the various truth values for these four statements. Remember that the only case in which a conditional is false is when the hypothesis is true and the conclusion is false.

				Conditional	Contrapositive	Converse	Inverse
p	q	$\sim p$	$\sim q$	$p \Rightarrow q$	$\sim q \Rightarrow \sim p$	$q \Rightarrow p$	$\sim p \Rightarrow \sim q$
T	T	F	F	T	T	T	T
T	F	F	T	F	F	T	T
F	T	T	F	T	T	F	F
F	F	T	T	T	T	T	T

Notice that the columns of truth values under the conditional $p \Rightarrow q$ and its contrapositive are the same. When this is the case, we say that the two statements are logically equivalent. In general, two statements are **logically equivalent** when they have the same truth tables. In a similar manner, the converse of $p \Rightarrow q$ and the inverse of $p \Rightarrow q$ have the same truth table. Therefore, they, too, are logically equivalent. In mathematics, replacing a conditional with a logically equivalent conditional often facilitates the solution of a problem.

EXAMPLE T1.2 Prove that if x^2 is odd, then x is odd.

SOLUTION

Rather than trying to prove that the given conditional is true, consider its logically equivalent contrapositive: If x is not odd (i.e., x is even), then x^2 is not odd (i.e., x^2 is even). Even numbers are of the form $2m$, where m is a whole number. Thus, the square of $2m$, $(2m)^2 = 4m^2 = 2(2m^2)$, is also an even number because it is of the form $2n$. Thus, if x is even, then x^2 is even. Because this statement is true and it is logically equivalent to the original statement, the original statement is also true. ■

If and Only If

The connective "p if and only if q" is called the **biconditional** and is written $p \Leftrightarrow q$. It is the conjunction of $p \Rightarrow q$ and its converse $q \Rightarrow p$. That is, $p \Leftrightarrow q$ is logically equivalent to $(p \Rightarrow q) \wedge (q \Rightarrow p)$. The truth table of $p \Leftrightarrow q$ follows.

p	q	$p \Rightarrow q$	$q \Rightarrow p$	$(p \Rightarrow q) \wedge (q \Rightarrow p)$	$p \Leftrightarrow q$
T	T	T	T	T	T
T	F	F	T	F	F
F	T	T	F	F	F
F	F	T	T	T	T

Notice that the biconditional $p \Leftrightarrow q$ is true when p and q have the same truth values and false otherwise.

We can use the words "necessary" and "sufficient" to describe conditionals and biconditionals. For example, the statement "Water is *necessary* for the formation of ice" means "If there is ice, then there is water." On the other hand, the statement "Water is *sufficient* for the formation of ice" would be false, because it means that water guarantees the formation of ice. The statement "A rectangle with two adjacent sides the same length is a sufficient condition to define a square" means "If a rectangle has two adjacent sides the same length, then it is a square." We have the following, symbolically:

$p \Rightarrow q$ means q **is necessary for** p.
$p \Rightarrow q$ means p **is sufficient for** q.
$p \Leftrightarrow q$ means p **is necessary and sufficient for** q.

ARGUMENTS

Deductive or **direct reasoning** is a process of reaching a conclusion from one (or more) statements, called the hypothesis (or hypotheses). We can rephrase this somewhat informal definition using the language and symbolism in the preceding section. An **argument** is a set of statements where one of the statements is called the conclusion and the rest make up the hypothesis.

A **valid argument** is an argument in which the conclusion must be true whenever the hypothesis is true. When an argument is valid, we say that the conclusion follows from the hypothesis. For example, consider the following argument.

If it is snowing, then it is cold.
It is snowing.
Therefore, it is cold.

In this argument, when the two statements in the hypothesis, namely "If it is snowing, then it is cold" and "It is snowing" are both true, then one can conclude that "It is cold." That is, this argument is valid because the conclusion follows from the hypothesis.

An argument is an **invalid argument** if its conclusion can be false when its hypothesis is true. An example of an invalid argument follows.

If it is raining, then the streets are wet.
The streets are wet.
Therefore, it is raining.

For convenience, we will represent this argument as $[(p \Rightarrow q) \wedge q] \Rightarrow p$. This is an invalid argument because the streets could be wet from a variety of causes (e.g., a street cleaner, an open fire hydrant, etc.) without having had any rain. In this example, $p \Rightarrow q$ is true and q may be true, whereas p is false. The next truth table also shows that this argument is invalid, since it is possible to have the hypothesis $[(p \Rightarrow q) \wedge q]$ true with the conclusion p false.

p	q	$p \Rightarrow q$	$(p \Rightarrow q) \wedge q$	$[(p \Rightarrow q) \wedge q] \Rightarrow p$
T	T	T	T	T
T	F	F	F	T
F	T	T	T	**F**
F	F	T	F	T

The argument with hypothesis $[(p \Rightarrow q) \wedge \sim p]$ and conclusion $\sim q$ is another example of a common invalid argument form. Here, when p is F and q is T, $[(p \Rightarrow \sim q) \wedge p]$ is T and $\sim q$ is F.

There are three important valid argument forms that we use repeatedly in logic. The first two "laws" are discussed briefly in Chapter 4.

1. **Law of Detachment**: $[(p \Rightarrow q) \wedge p] \Rightarrow q$

In words, the law of detachment says that whenever a conditional statement and its hypothesis are true, the conclusion is also true. That is, the conclusion can be "detached" from the conditional. An example of the use of this law follows.

If a number ends in zero, then it is a multiple of 10.
Forty is a number that ends in zero.
Therefore, 40 is a multiple of 10.

The following truth table verifies that the law of detachment is a valid argument form. Notice that the final column consists of all T's.

p	q	$p \Rightarrow q$	$(p \Rightarrow q) \wedge p$	$[(p \Rightarrow q) \wedge p] \Rightarrow q$
T	T	T	T	T
T	F	F	F	T
F	T	T	F	T
F	F	T	F	T

In line 1 of the preceding truth table, when the hypothesis $(p \Rightarrow q) \wedge p$ is true, the conclusion, q, is also true. This law of detachment is used in everyday language and thought.

We can also use diagrams to decide the validity of arguments. Consider the following argument.

All mathematicians are logical.
Greta is a mathematician.
Therefore, Greta is logical.

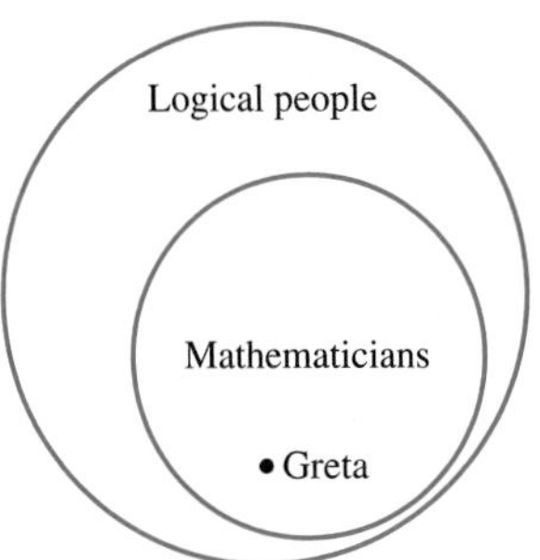

FIGURE T1.2

We can picture this argument using an **Euler diagram** (Figure T1.2). The "mathematician" circle within the "logical people" circle represents the statement "All mathematicians are logical." The point labeled "Greta" in the "mathematician" circle represents "Greta is a mathematician." Because the "Greta" point is within the "logical people" circle, we conclude that "Greta is logical."

The second common valid argument form is the law of syllogism (or hypothetical syllogism).

2. Law of Syllogism: $[(p \Rightarrow q) \wedge (q \Rightarrow r)] \Rightarrow (p \Rightarrow r)$

The following argument is an application of this law.

If a number is a multiple of 8, then it is a multiple of 4.
If a number is a multiple of 4, then it is a multiple of 2.
Therefore, if a number is a multiple of 8, then it is a multiple of 2.

We can verify this argument using an Euler diagram (Figure T1.3). The circle within the "multiples of 4" circle represents the "multiples of 8" circle. The "multiples of 4" circle is within the "multiples of 2" circle. Thus, from the diagram, it must follow that "If a number is a multiple of 8, then it is a multiple of 2" because all the multiples of 8 are within the "multiples of 2" circle.

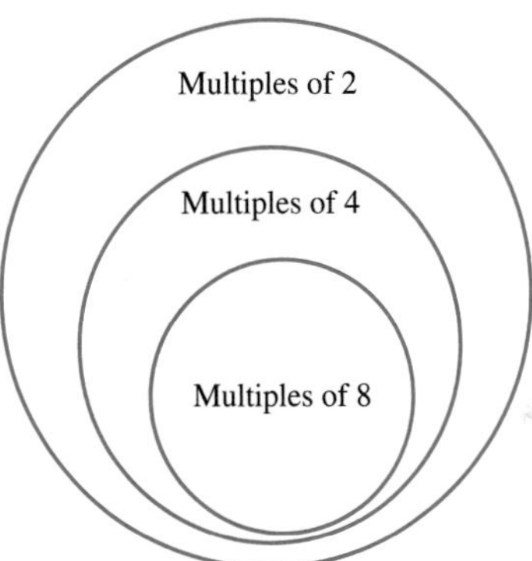

FIGURE T1.3

The following truth table also proves the validity of the law of syllogism.

p	q	r	$p \Rightarrow q$	$q \Rightarrow r$	$p \Rightarrow r$	$(p \Rightarrow q) \wedge (q \Rightarrow r)$	$[(p \Rightarrow q) \wedge (q \Rightarrow r)] \Rightarrow (p \Rightarrow r)$
T	T	T	T	T	T	T	T
T	T	F	T	F	F	F	T
T	F	T	F	T	T	F	T
T	F	F	F	T	F	F	T
F	T	T	T	T	T	T	T
F	T	F	T	F	T	F	T
F	F	T	T	T	T	T	T
F	F	F	T	T	T	T	T

Observe that in rows 1 and 5, the hypothesis $(p \Rightarrow q) \wedge (q \Rightarrow r)$ is true. In both cases, the conclusion, $p \Rightarrow r$, is also true. Thus, the argument is valid.

The final valid argument we study here is used often in mathematical reasoning.

3. Law of Contraposition: $[(p \Rightarrow q) \wedge (\sim q)] \Rightarrow \sim p$

Consider the following argument (Figure T1.4).

If a quadrilateral is a square, then it has four 90° angles.
ABCD does not have four 90° angles.
Therefore, $ABCD$ is not a square.

This argument is an application of the law of contraposition. Figure T1.4(b) illustrates this argument. All points outside the larger circle represent quadrilaterals that do *not* have four congruent angles such as the one in Figure T1.4(a). Any quadarilateral outside the larger circle clearly must be outside the smaller circle. Thus, because $ABCD$ would lie outside the circle labeled "Quadrilaterals with four 90° angles," it is not a square.

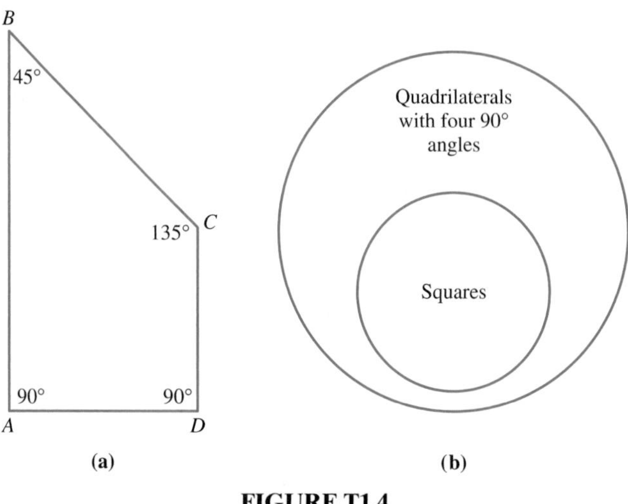

FIGURE T1.4

(NOTE: The sentence "a quadrilateral has four 90° angles" is called an "open" sentence, strictly speaking, because no particular quadrilateral is specified. Therefore, the sentence is neither true nor false as given. The sentence "Quadrilateral $ABCD$ in Figure T1.5 has four 90° angles" is a true statement. Because the use of open sentences is prevalent throughout mathematics, we will permit such "open" sentences in conditional statements without pursuing an in-depth study of such sentences.)

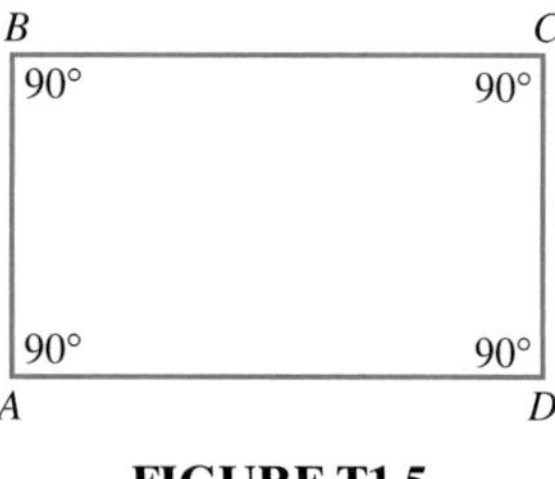

FIGURE T1.5

The next truth table provides a verification of the validity of this argument form.

p	q	$p \Rightarrow q$	$(p \Rightarrow q) \wedge \sim q$	$\sim p$	$[(p \Rightarrow q) \wedge (\sim q)] \Rightarrow \sim p$
T	T	T	F	F	T
T	F	F	F	F	T
F	T	T	F	T	T
F	F	T	T	T	T

Notice that row 4 is the only instance when the hypothesis $(p \Rightarrow q) \wedge \sim q$ is true. Here, the conclusion of $[(p \Rightarrow q) \wedge (\sim q)] \Rightarrow \sim p$, namely $\sim p$, is also true. Therefore, the argument is valid.

We can also apply the three argument forms we have been studying to statements that are modified by "quantifiers," that is, words such as "all," "some," "every," or their equivalents. Here, again, we can use Euler diagrams to decide the validity or invalidity of various arguments. Consider the following argument.

All logicians are mathematicians.
Some philosophers are not mathematicians.
Therefore, some philosophers are not logicians.

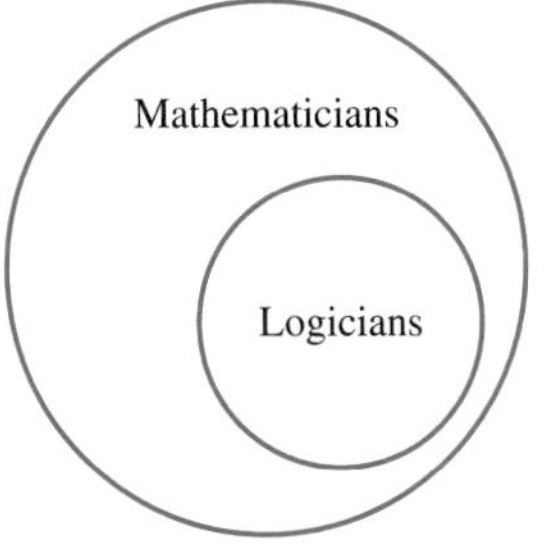

FIGURE T1.6

The Euler diagram in Figure T1.6 represents the first line of this argument. But, because the second line guarantees that there are "some" philosophers outside the "mathematician" circle, we use a dot to represent at least one philosopher who is not a mathematician (Figure T1.7). Observe that, due to the dot, there is always at least one philosopher who is not a logician. Therefore, the argument is valid. Note that the word "some" means "at least one."

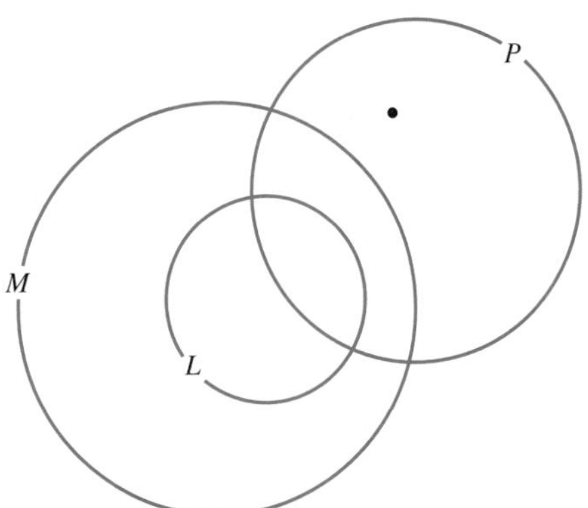

FIGURE T1.7

Next consider the following argument.

All rock stars have green hair.
No presidents of banks are rock stars.
Therefore, no presidents of banks have green hair.

An Euler diagram that represents this argument is shown in Figure T1.8, where G represents all people with green hair, R represents all rock stars, and P represents all bank presidents. Note that Figure T1.8 allows for presidents of banks to have green hair, because the circles G and P may have an element in common. Thus, the argument, as stated, is invalid because the hypothesis can be true while the conclusion is false. We can determine the validity or invalidity of the arguments given in the problem set in this way using Euler diagrams. Be sure to consider *all* possible relationships among the sets before drawing any conclusions.

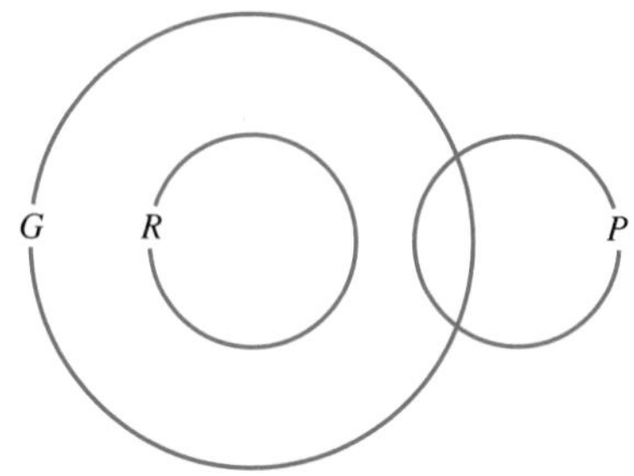

FIGURE T1.8

PROBLEM SET T1

EXERCISES/PROBLEMS

1. Determine which of the following are statements.

a. What's your name?

b. The rain in Spain falls mainly in the plain.

c. Happy New Year!

d. Five is an odd number.

2. Determine which of the following are statements.

a. Is $\triangle ABC$ isosceles?

b. Go away!

c. The sum of the angle measures

d. The area of a circle with radius r is $2\pi r$.

3. Let r, s, and t be the following statements,

r: Roses are red.

s: The sky is blue.

t: Turtles are green.

Translate the following statements into English.

a. $r \wedge s$ **b.** $r \wedge (s \vee t)$

c. $s \Rightarrow (r \wedge t)$ **d.** $(\sim t \wedge t) \Rightarrow \sim r$

4. Write the following in symbolic form using $p, q, r,$ $\sim, \wedge, \vee, \Rightarrow,$ and $\Leftrightarrow$, where $p, q,$ and r represent the following statements,

p: The sun is shining.

q: It is raining.

r: The grass is green.

a. If it is raining, then the sun is not shining.

b. It is raining and the grass is green.

c. The grass is green if and only if it is raining and the sun is shining.

d. Either the sun is shining or it is raining.

5. Fill in the headings of the following truth table using $p, q, \wedge, \vee, \sim,$ and $\Rightarrow$.

p	q		
T	T	T	F
T	F	F	T
F	T	T	T
F	F	T	T

6. Fill in the headings of the following truth table using $p, q, \wedge, \vee, \sim,$ and $\Rightarrow$.

p	q		
T	T	T	T
T	F	F	T
F	T	F	F
F	F	F	T

7. If p is T, q is F, and r is T, find the truth values for the following:

a. $p \wedge \sim q$ **b.** $\sim(p \vee q)$

c. $(\sim p) \Rightarrow r$ **d.** $(\sim p \wedge r) \Leftrightarrow q$

e. $(\sim q \wedge p) \vee r$ **f.** $p \vee (q \Leftrightarrow r)$

g. $(r \wedge \sim p) \vee (r \wedge \sim q)$

h. $(p \wedge q) \Rightarrow (q \vee \sim r)$

8. Suppose that $p \Rightarrow q$ is known to be false. Give the truth values for the following:

a. $p \vee q$ **b.** $p \wedge q$

c. $q \Rightarrow p$ **d.** $\sim q \Rightarrow p$

9. Construct one truth table that contains truth values for all of the following statements and determine which are logically equivalent.

a. $(\sim p) \vee (\sim q)$ **b.** $(\sim p) \vee q$
c. $(\sim p) \wedge (\sim q)$ **d.** $p \Rightarrow q$
e. $\sim(p \wedge q)$ **f.** $\sim(p \vee q)$

10. Use a truth table to determine which of the following are always true.

a. $(p \Rightarrow q) \Rightarrow (q \Rightarrow p)$
b. $p \Rightarrow p$
c. $[p \wedge (p \Rightarrow q)] \Rightarrow q$
d. $(p \vee q) \Rightarrow (p \wedge q)$
e. $(p \wedge q) \Rightarrow p$

11. State the hypothesis (or hypotheses) and conclusion for each of the following arguments.

a. All football players are introverts. Tony is a football player, so Tony is an introvert.
b. Bob is taller than Jim, and Jim is taller than Sue. So Bob is taller than Sue.

12. State the hypothesis (or hypotheses) and conclusion for each of the following arguments.

a. All penguins are elegant swimmers. No elegant swimmers fly, so penguins don't fly.
b. $x < y$ and $y < z$ implies that $x < z$.

13. Determine the validity of the following arguments.

a. All professors are handsome.
Some professors are tall.
Therefore, some handsome people are tall.

b. If I can't go to the movie, then I'll go to the park.
I can go to the movie.
Therefore, I will not go to the park.

c. If you score at least 90 percent, then you'll earn an A.
If you earn an A, then your parents will be proud.
You have proud parents.
Therefore, you scored at least 90 percent.

d. Some arps are bomps.
All bomps are cirts.
Therefore, some arps are cirts.

e. All equilateral triangles are equiangular.
All equiangular triangles are isosceles.
Therefore, all isosceles triangles are equilateral.

f. If you work hard, then you will succeed.
You do not work hard.
Therefore, you will not succeed.

g. Some girls are teachers.
All teachers are college graduates.
Therefore, all girls are college graduates.

h. If it doesn't rain, then the street won't be wet.
The street is wet.
Therefore, it rained.

14. Determine a valid conclusion that follows from each of the given statements and explain your reasoning.

a. If you study hard, then you will be popular. You will study hard.
b. If Scott is quick, then he is a basketball star. Scott is not a basketball star.
c. All friends are respectful. All respectful people are trustworthy.
d. Every square is a rectangle. Some parallelograms are rhombuses. Every rectangle is a parallelogram.

15. Using each pair of statements, determine whether
(i) p is necessary for q,
(ii) p is sufficient for q, or
(iii) p is necessary and sufficient for q.

a. p: Bob has some water.
q: Bob has some ice.
b. p: It is snowing.
q: It is cold.
c. p: It is December.
q: 31 days from today it is January.

16. Using each pair of statements, determine whether
(i) p is necessary for q,
(ii) p is sufficient for q, or
(iii) p is necessary and sufficient for q.

a. p: Janet is at least 16 years old.
q: Janet has her driver's license.
b. p: A square has a side that measures 5 inches.
q: A square has an area of 25 square inches.
c. p: Ted walked to work this morning.
q: Ted woke up this morning.

17. Which of the laws (detachment, syllogism, or contraposition) is being used in each of the following arguments?

a. If Joe is a professor, then he is learned. If you are learned, then you went to college. Joe is a professor, so he went to college.
b. If you have children, then you are an adult. Bob is not an adult, so he has no children.
c. If I am broke, I will ride the bus. When I ride the bus, I am always late. I'm broke, so I am going to be late.

18. Which of the laws (detachment, syllogism, or contraposition) is being used in each of the following arguments?

a. All women are smart. Helen of Troy was a woman. So Helen of Troy was smart.

b. If today is Tuesday, then tomorrow is Wednesday. Tomorrow is Saturday, so today is not Tuesday.

c. If I don't eat breakfast, I'll be hungry by 10:00. If I'm hungry before noon, I always snack before lunch. I skipped breakfast, so I will snack before lunch.

19. If possible, determine the truth value of each statement. Assume that a and b are true, p and q are false, and x and y have unknown truth values. If a value cannot be determined, write "unknown."

a. $p \Rightarrow (a \vee b)$ **b.** $b \Rightarrow (p \vee a)$

c. $x \Rightarrow p$ **d.** $a \vee p$

e. $b \Rightarrow q$ **f.** $b \Rightarrow x$

g. $a \wedge (b \vee x)$ **h.** $(y \vee x) \Rightarrow a$

i. $(y \wedge b) \Rightarrow p$ **j.** $(a \vee x) \Rightarrow (b \wedge q)$

k. $x \Rightarrow a$ **l.** $x \vee p$

m. $\sim x \Rightarrow x$ **n.** $x \vee (\sim x)$

o. $\sim[y \wedge (\sim y)]$

20. Decide the truth value of each statement.

a. Alexander Hamilton was once president of the United States.

b. The world is flat.

c. If dogs are cats, then the sky is blue.

d. If Tuesday follows Monday, then the sun is hot.

e. If Christmas Day is December 25, then Texas is the largest state in the United States.

21. Rewrite each argument in symbolic form and then check the validity of the argument.

a. If today is Wednesday (w), then yesterday was Tuesday (t). Yesterday was Tuesday, so today is Wednesday.

b. The plane is late (l) if it snows (s). It is not snowing. Therefore, the plane is not late.

22. Rewrite each argument in symbolic form and then check the validity of the argument.

a. If I do not study (s), then I will eat (e). I will not eat if I am worried (w). Hence, if I am worried, I will study.

b. Meg is married (m) and Sarah in single (s). If Bob has a job (j), then Meg is married. Hence, Bob has a job.

23. Draw an Euler diagram to represent the following argument and decide if it is valid.

All timid creatures (T) are bunnies (B).
All timid creatures are furry (F).
Some cows (C) are furry.
Therefore, all cows are timid creatures.

24. Use the following Euler diagram to determine which of the following statements are true. (Assume that there is at least one person in every region within the circles.)

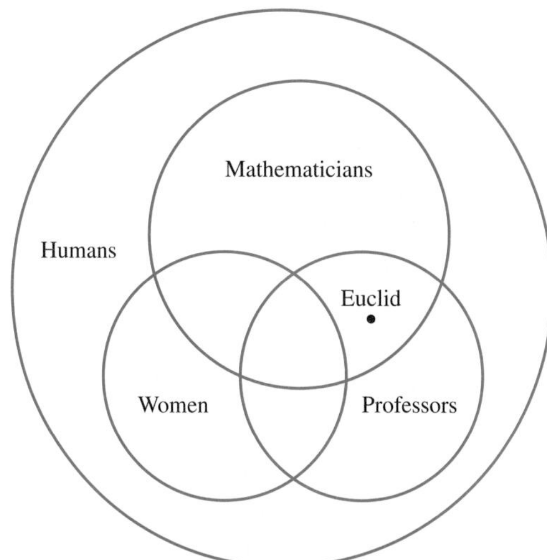

a. All women are mathematicians.

b. Euclid was a woman.

c. All mathematicians are men.

d. All professors are humans.

e. Some professors are mathematicians.

f. Euclid was a mathematician and human.

PROOF

25. Prove that the conditional $p \Rightarrow q$ is logically equivalent to $\sim p \vee q$.

TOPIC REVIEW

Next you will find a list of vocabulary and notation, a summary of key concepts, and practice test questions similar to the problems in this topic section. Review each item in the *Vocabulary/Notation* list mentally, and, if necessary, refer back to the indicated page and write a definition. Make certain that you can explain in your own words each of the *Key Concepts* and be able to provide examples. Then work the problems in the Topic Test to assess your understanding of the content. Answers to the Topic 1 Test can be found at the end of the book.

Vocabulary/Notation

Statement 527
Negation (~) 528
Truth table 528
Compound statements 528
Logical connective 528
Conjunction (and), $p \wedge q$ 528
Disjunction (or), $p \vee q$ 528
Exclusive "or" 528
Inclusive "or" 528
Implication/conditional (if . . . , then . . .), $p \Rightarrow q$ 529
Hypothesis 529
Conclusion 529
Converse 530
Inverse 530
Contrapositive 530
Logically equivalent statements 530
Biconditional (if and only if), $p \Leftrightarrow q$ 531
Is necessary for 531
Is sufficient for 531
Is necessary and sufficient for 531
Deductive/direct reasoning 531
Argument 531
Valid argument 531
Invalid argument 532
Law of Detachment 532
Euler diagram 533
Law of Syllogism 533
Law of Contraposition 534

Key Concepts

1. We use logic to determine if arguments are valid or invalid.
2. We can represent statements and arguments symbolically and pictorially.
3. We can identify valid arguments using truth tables and Euler diagrams.

TOPIC 1 TEST

TRUE-FALSE

Mark as true any statement that is always true. Mark as false any statement that is never true or that is not necessarily true. Be able to justify your answers.

1. The disjunction of p and q is true whenever p is true and q is false.
2. If $p \Rightarrow q$ is true, then $\sim p \Rightarrow \sim q$ is true.
3. "I am older than 20 or younger than 30" is an example of an exclusive "or."
4. The converse of $p \Rightarrow q$ is $q \Rightarrow p$.
5. $p \Rightarrow q$ means p is necessary for q.

EXERCISES/PROBLEMS

6. Find the converse, inverse, and contrapositive of each statement.
   a. $p \Rightarrow \sim q$
   b. $\sim p \Rightarrow q$
   c. $\sim q \Rightarrow \sim p$
7. Decide the truth value of each statement.
   a. $4 + 7 = 11$ and $1 + 5 = 6$.
   b. $2 + 5 = 7 \Leftrightarrow 4 + 2 = 8$.
   c. $3 \cdot 5 = 12$ or $2 \cdot 6 = 11$.
   d. If $2 + 3 = 5$, then $1 + 2 = 4$.
   e. If $3 + 4 = 6$, then $8 \cdot 4 = 31$.
   f. If 7 is even, then 8 is even.
8. Use Euler diagrams to check the validity of each argument.
   a. Some men are teachers.
   Sam Jones is a teacher.
   Therefore, Sam Jones is a man.
   b. Gold is heavy.
   Nothing but gold will satisfy Amy.
   Hence, nothing that is light will satisfy Amy.

c. No cats are dogs.
All poodles are dogs.
So some poodles are not cats.

d. Some cows eat hay.
All horses eat hay.
Only cows and horses eat hay.
Frank eats hay, so Frank is a horse.

e. All chimpanzees are monkeys.
All monkeys are animals.
Some animals have two legs.
So some chimpanzees have two legs.

9. Complete the following truth table.

p	q	$p \wedge q$	$p \vee q$	$p \Rightarrow q$	$\sim q \Leftrightarrow p$	$\sim q \Rightarrow \sim p$
T				T		
	T	F				
F			F			T
			T	F		

10. Using an Euler diagram, display an example of an invalid argument. Explain.

11. For the following pair of statements, determine whether p is necessary for q, p is sufficient for q, or p is both necessary and sufficient for q.

p: $\triangle ABC$ has exactly two congruent sides.
q: $\triangle ABC$ has exactly two congruent angles.

12. Determine a valid conclusion that follows from each pair of statements, if possible. Explain your reasoning.

a. All polyhedra have edges. A tetrahedron is a polyhedron.

b. If it is 5:00 P.M., then it is quitting time at work. It is 4:15 P.M.

13. Suppose that $\sim p \Rightarrow q$ is known to be false. Give truth values for each of the following.

a. $p \vee q$ **b.** $\sim p \Leftrightarrow q$ **c.** $\sim p \wedge \sim q$

14. Is it ever the case that the conjunction, disjunction, and implication of two statements are all true at the same time? All false? If so, what are the truth values of each statement? If not, explain why not.

TOPIC 2

INEQUALITIES IN ALGEBRA AND GEOMETRY

In Chapter 4, we studied triangle congruence and equations arising from corresponding parts of congruent triangles. In Chapter 7, we studied congruence relationships among central angles, arcs, and chords of circles. In this section, we study inequality relationships in triangles and circles that we prove using results from Chapters 4 and 7. We also review the skills we need to solve inequalities such as $5x - 8 > 37$.

INEQUALITIES

Sometimes we need to solve problems in geometry that involve inequalities. For example, we may have a situation in which the measure of one angle is larger than the measure of another, one line segment is shorter than another, and so on. We use the symbol "$>$" to indicate that one number is greater than another and "$<$" to indicate less than. These symbols are often combined with the equals sign to form the symbol "$\leq$," which means "is less than or equal to" and "$\geq$," which means "is greater than or equal to." For example, $n \leq 5$ means "n is less than 5 or n is equal to 5."

We can interpret the statement "$3 < 5$" in several ways. Using a number line, we see that $3 < 5$ and 3 is to the left of 5 (and 5 is to the right of 3) (Figure T2.1).

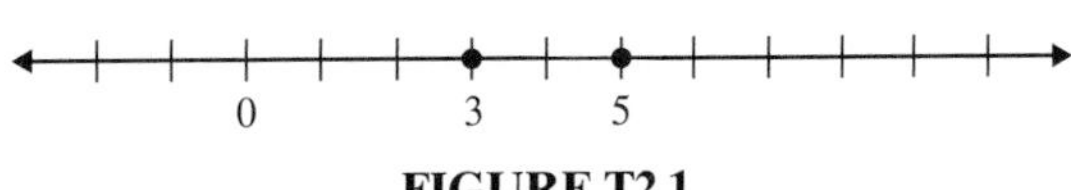

FIGURE T2.1

In general, for real numbers a and b, we say that a is less than b (and write $a < b$) if and only if a is to the left of b on the real number line.

We can also relate inequalities to equations. To say that $3 < 5$ means that there is a positive real number, 2 in this case, that we can add to 3 to obtain 5.

In other words, we say that $3 < 5$ because $3 + 2 = 5$. In general, we say that a **is less than** b if and only if there is a positive real number p such that $a + p = b$. We say that b **is greater than** a whenever a is less than b.

If a and b are real numbers represented on the real number line, either they share the same position (hence, $a = b$), or a is to the left of b ($a < b$), or b is to the left of a ($a > b$). This relationship is stated next.

TRICHOTOMY PROPERTY

For any real numbers a and b, exactly one of the following statements is true:

$$a = b \quad \text{or} \quad a > b \quad \text{or} \quad a < b$$

Often we compare more than two numbers. For example, if we know that $\angle A < \angle B$ and $\angle B < 90°$, we can conclude that $\angle A < 90°$. This is an example of the following theorem.

THEOREM T2.1 Transitive Property for Less Than

Let a, b, and c represent three real numbers. If $a < b$ and $b < c$, then $a < c$.

Although this property is stated for less than, similar results hold for "$>$," "$\leq$," and "$\geq$." That is, if $a > b$ and $b > c$, then $a > c$. We leave the proof of the transitive property for the problem set.

SOLVING INEQUALITIES

In geometry, we sometimes need to use algebra to solve an inequality such as $5x - 8 > 37$ for x. That is, we need to find all real number values for x so that the inequality is true. We can accomplish this by applying the following principles and using techniques similar to those we use when solving equations.

Inequality Principle 1: If a real number is added to or subtracted from both sides of an inequality, the direction of the inequality is unchanged.

If $a < b$, then $a + c < b + c$.

Inequality Principle 2: If both sides of an inequality are multiplied or divided by the same *positive* real number, the direction of the inequality is unchanged.

If $a < b$ and $c > 0$, then $ac < bc$.

Inequality Principle 3: If both sides of an inequality are multiplied or divided by the same *negative* real number, the direction of the inequality is *reversed.*

If $a < b$ and $c < 0$, then $ac > bc$.

In the case of Principles 1 and 2, we say that the inequality is preserved, and in Principle 3, we say that the inequality has been reversed.

EXAMPLE T2.1

Solve each of the following inequalities for x.

a. $5x - 8 > 37$ **b.** $5x + 4 \geq 9x - 52$

SOLUTION

a. To solve this inequality for x, we isolate x. First we simplify the inequality by using Inequality Principle 1 and adding 8 to both sides.

$$\begin{array}{rl} 5x - 8 > 37 & \\ +8 \quad +8 & \text{Inequality Principle 1} \\ \hline 5x \quad > 45 & \end{array}$$

Next we apply Inequality Principle 2 and divide both sides by 5.

$$\frac{5x}{5} > \frac{45}{5} \qquad \text{Inequality Principle 2}$$

$$x > 9$$

Notice that the direction of the inequality did not change in the second step since we divided by a *positive* number.

CHECK

There are infinitely many solutions to the inequality $x > 9$, all of which will satisfy the original inequality $5x - 8 > 37$. We can check such solutions by substituting some values into the original inequality. For $x = 12$, we have $5(12) - 8 = \mathbf{52}$ and $52 > 37$. Next we check one value that should *not* satisfy the original inequality. For $x = 7$, we have $5(7) - 8 = \mathbf{27}$ and $27 < 37$, rather than "greater than." Thus, 7 is *not* one of the solutions of $5x - 8 > 37$. Although it is impossible to check all values, checking one that works and one that does not often helps catch errors.

b. As we did in (a), we will isolate x by first subtracting and then dividing.

$$\begin{array}{rl} 5x + 4 \geq \quad 9x - 52 & \\ -9x \qquad\quad -9x \quad & \text{Inequality Principle 1} \\ \hline -4x + 4 \geq -52 \qquad & \\ -4 \qquad -4 \qquad & \text{Inequality Principle 1} \\ \hline -4x \geq -56 \qquad & \end{array}$$

$$\frac{-4x}{-4} \leq \frac{-56}{-4} \qquad \text{Inequality Principle 3 (Notice the inequality is reversed since we divided by a negative.)}$$

$$x \leq 14$$

CHECK

In the original inequality, try $x = 13$, which should be a solution, and $x = 15$, which should *not* be a solution.

$$5(13) + 4 = 69 \text{ and } 9(13) - 52 = \mathbf{65}, \text{ and } 69 \geq 65. \checkmark$$
$$5(15) + 4 = 79 \text{ and } 9(15) - 52 = \mathbf{83}, \text{ and } 79 < 83. \checkmark$$

TRIANGLES

Consider $\triangle ABC$ in Figure T2.2(a) where $AC > AB$. It appears that $\angle B > \angle C$. To prove this, find point D on $\overline{AC}$ where $AB = AD$ [Figure T2.2(b)].

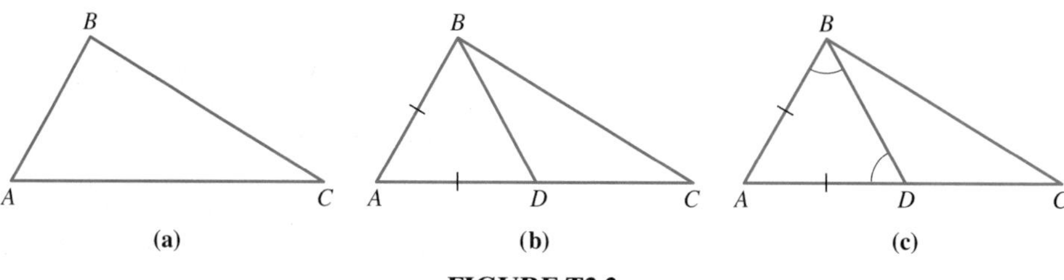

FIGURE T2.2

Because $AB = AD$, we have $\angle ABD = \angle ADB$ [Figure T2.2(c)]. We also know $\angle ABC = \angle ABD + \angle DBC$. Therefore, $\angle ABC > \angle ABD = \angle ADB > \angle C$. Justify this inequality. (HINT: Notice $\angle ADB$ is an exterior angle of $\triangle BDC$.) In other words, $\angle ABC > \angle C$. We summarize this result next.

THEOREM T2.2

If two sides of a triangle are not congruent, the measure of the angle opposite the longer side is greater than the measure of the angle opposite the shorter side.

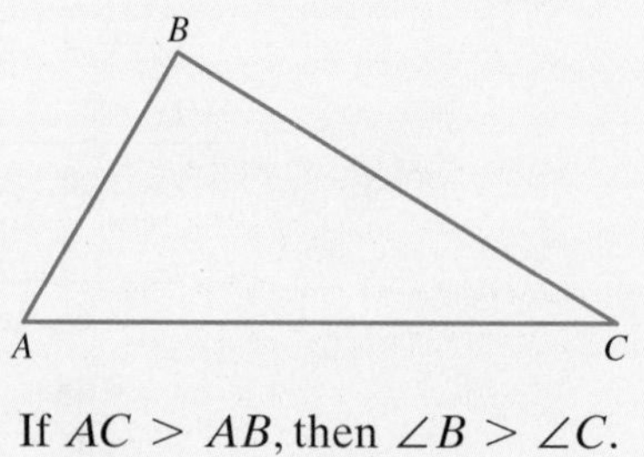

If $AC > AB$, then $\angle B > \angle C$.

The converse of Theorem T2.2 is also true. In fact, we can use Theorem T2.2 to help make the proof of its converse, which follows.

> **THEOREM T2.3**
>
> If two angles of a triangle are not congruent, then the side opposite the larger angle is longer than the side opposite the smaller angle.
>
> 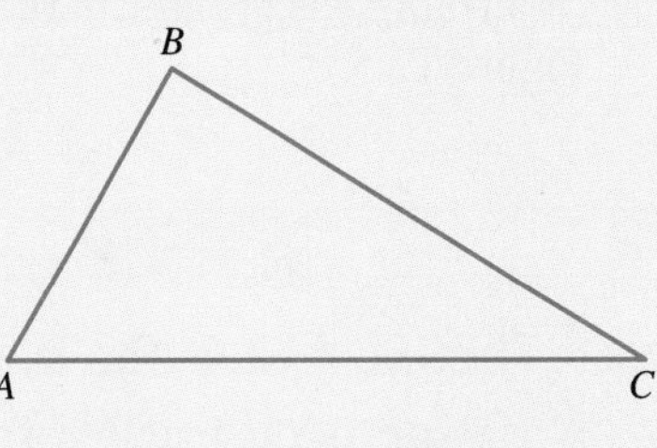
>
>
> If $\angle B > \angle C$, then $AC > AB$.

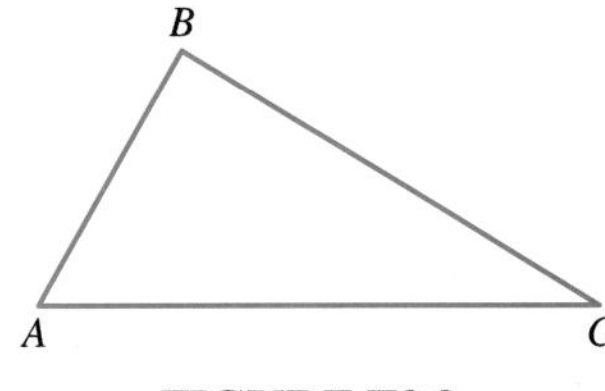

FIGURE T2.3

Given $\angle B > \angle C$ (Figure T2.3)

Prove $AC > AB$

Plan We will use indirect reasoning. First it is given that $\angle B > \angle C$. Assume that $AC = AB$ and show that this is impossible. Then assume that $AC < AB$ and show that this, too, is impossible. Thus, we will conclude that $AC > AB$ by the Trichotomy property.

Proof First assume that $AC = AB$. Then $\triangle ABC$ is isosceles and $\angle B = \angle C$, which contradicts our hypothesis $\angle B > \angle C$. Next assume that $AC < AB$. Then we could also write $AB > AC$. It follows from Theorem T2.2 that $\angle C > \angle B$, which contradicts our hypothesis. Therefore, because $AC = AB$ and $AC < AB$ are not possible when $\angle B > \angle C$, it follows that $AC > AB$. ■

EXAMPLE T2.2 Show that the shortest distance from a point to a line is the length of the perpendicular from the point to the line. That is, given line l and point P in Figure T2.4(a), prove that PA is the shortest distance from P to l.

SOLUTION

Let B be any other point on line l. Draw $\overline{PB}$. We will show that $PA < PB$, for any point B [Figure T2.4(b)]. Because $\triangle PAB$ is a right triangle, we have $\angle PAB > \angle PBA$. Thus, by Theorem 2.3, we know that $PB > PA$. So the shorter of the distances from P to l is PA, the perpendicular distance. ■

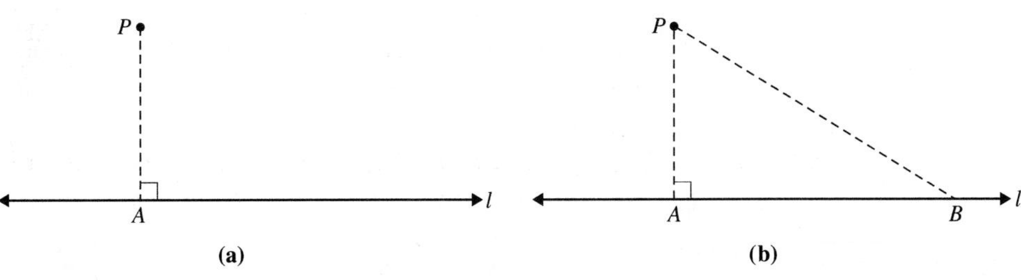

FIGURE T2.4

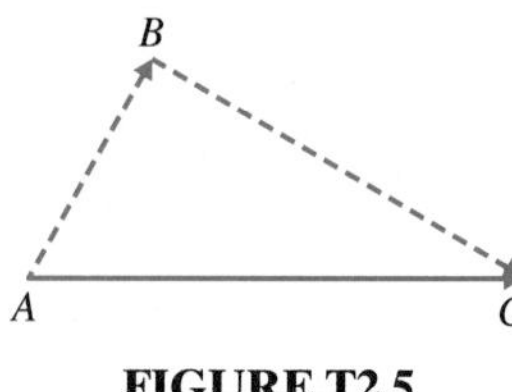

FIGURE T2.5

The next theorem, called the Triangle Inequality, shows that there is a relationship among the lengths of the sides of a triangle. For example, consider $\triangle ABC$ in Figure T2.5. If you were asked to walk from A to C, which would be the shorter route: (i) along $\overline{AC}$, or (ii) from A to B along $\overline{AB}$ and then to C along $\overline{BC}$? The next theorem answers this question. It states that the sum of the lengths of any two sides of a triangle is greater than the length of the third side, as appears to be the case in Figure T2.5.

THEOREM T2.4 The Triangle Inequality

The sum of the lengths of any two sides of a triangle is greater than the length of the third side.

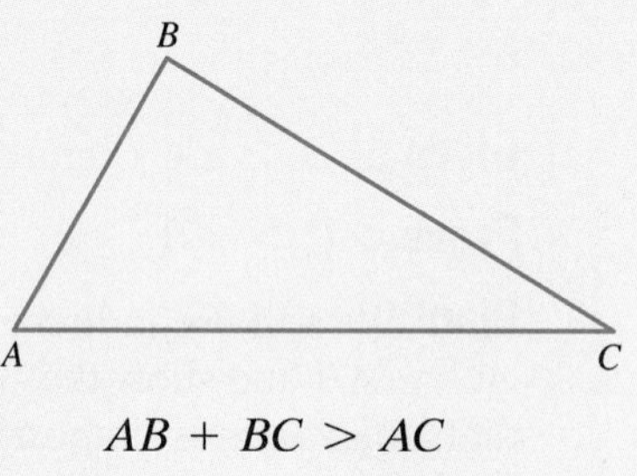

$AB + BC > AC$

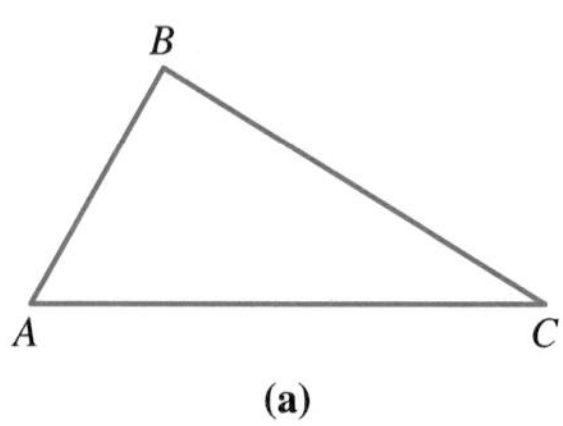

(a)

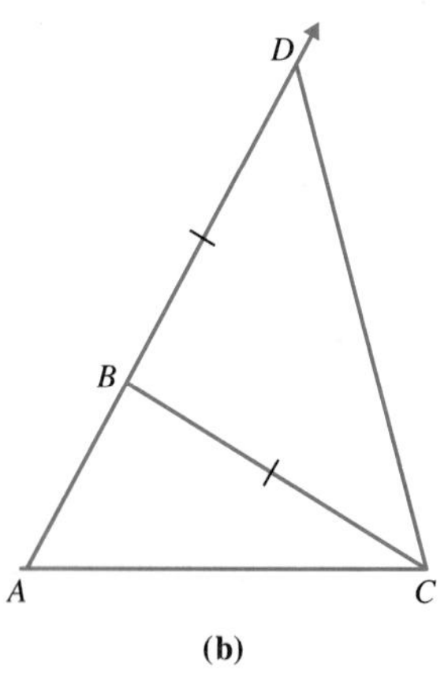

(b)

FIGURE T2.6

Given $\triangle ABC$ [Figure T2.6(a)]

Prove $AB + BC > AC$

Proof On $\overrightarrow{AB}$, find point D where B is between A and D, and $BD = BC$ (Figure T2.6(b)). Because $\triangle BCD$ is isosceles, $\angle D = \angle BCD$. Because $\angle ACD > \angle BCD$, we have $\angle ACD > \angle D$ by substitution. By the previous theorem, we conclude that $AD > AC$. But $AD = AB + BD$. So we have $AB + BD > AC$, and because $BD = BC$, we substitute BC for BD to have $AB + BC > AC$. ■

EXAMPLE T2.3 Given $\triangle ABC$ as shown in Figure T2.7, where $AB = 5$ m and $AC = 6$ m, what lengths are possible for side $\overline{BC}$?

SOLUTION

By the Triangle Inequality, we know that the sum of the lengths of any two sides must be greater than the length of the third side. So we can write

(i) $5 + 6 > BC$, **(ii)** $6 + BC > 5$, and **(iii)** $5 + BC > 6$

From inequality **(i)**, we know that $11 > BC$, or $BC < 11$.
From inequality **(ii)**, we know that $BC > -1$, but because BC must be nonnegative, this gives us no new information.
From inequality **(iii)**, we know that $BC > 1$.
Therefore, we have $BC > 1$ and $BC < 11$, which we write as $1 < BC < 11$.
That is, BC must be between 1 m and 11 m. ■

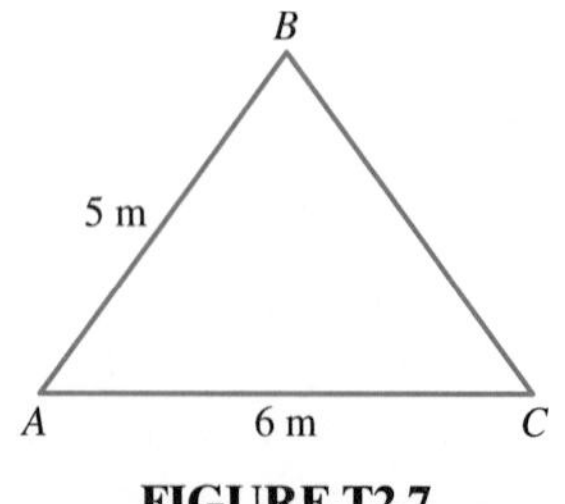

FIGURE T2.7

The next two theorems are consequences of the preceding ones. Due to the complexity of the proof of the first theorem, we simply state it without proof here.

THEOREM T2.5 SAS Inequality

If two sides of one triangle are congruent to two sides of another, but the measure of the included angle of the first triangle is less than the measure of the included angle of the other triangle, then the side opposite the included angle in the first triangle is shorter than the corresponding side of the second triangle.

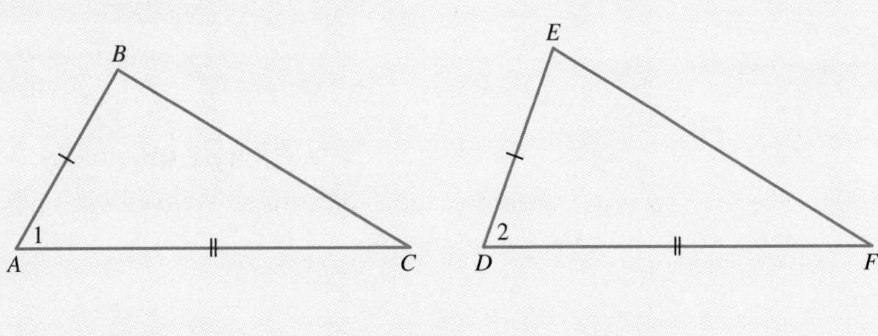

If $AB = DE$, $AC = DF$, and $\angle 1 < \angle 2$, then $BC < EF$.

We leave the proof of the next theorem, which uses the SAS Inequality Theorem, for the problem set.

THEOREM T2.6 SSS Inequality

If two sides of one triangle are congruent to two sides of another, but the third side of the first triangle is shorter than the third side of the other triangle, then the measure of the angle opposite the third side in the first triangle is less than the measure of the corresponding angle in the second triangle.

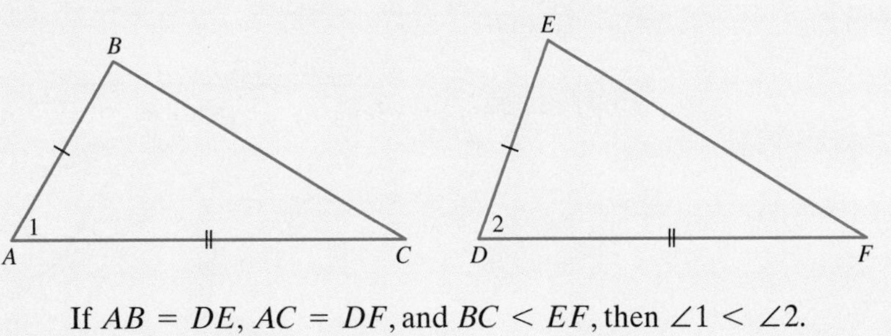

If $AB = DE$, $AC = DF$, and $BC < EF$, then $\angle 1 < \angle 2$.

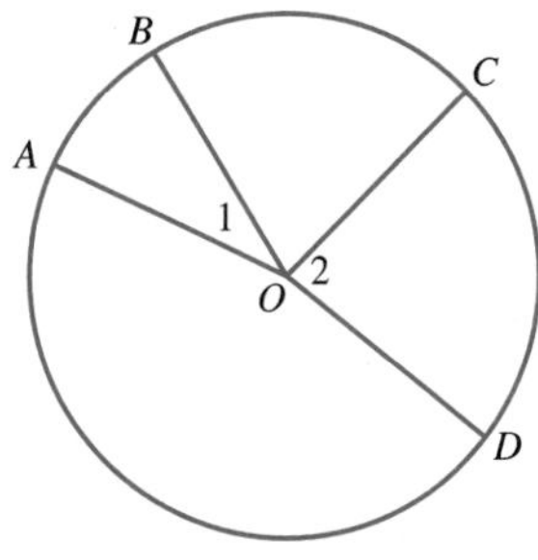

FIGURE T2.8

CIRCLES

Recall that congruent central angles intercept congruent arcs and chords. Similar inequality relationships also hold for circles. For example, $\angle 1 < \angle 2$ in circle O (Figure T2.8). Because $\angle 1 = \widehat{AB}$ and $\angle 2 = \widehat{CD}$ we have $\widehat{AB} < \widehat{CD}$ The converse of this argument is also true. We summarize these two results as follows.

THEOREM T2.7

In a circle or in circles with the same radius, the measure of one central angle is less than the measure of another if and only if the measure of the intercepted arc of the first angle is less than the measure of the intercepted arc of the second angle.

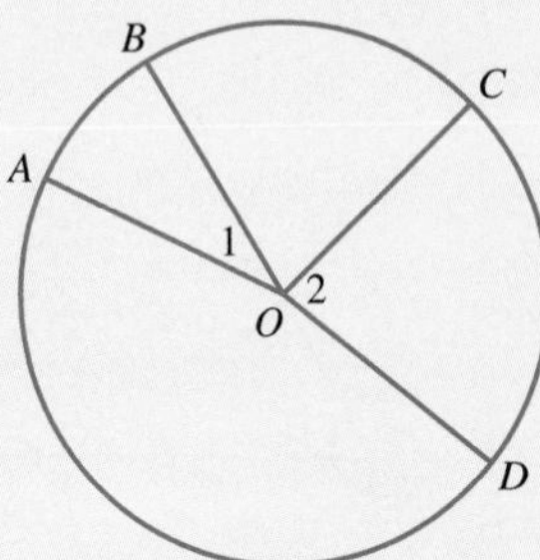

$\angle 1 < \angle 2$ if and only if $\widehat{AB} < \widehat{CD}$.

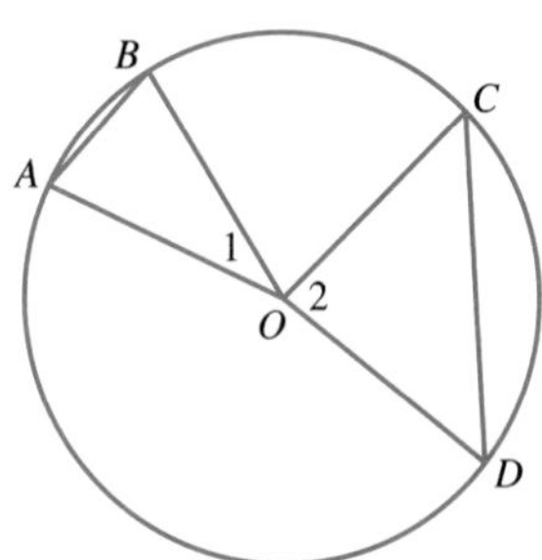

FIGURE T2.9

Next consider chords $\overline{AB}$ and $\overline{CD}$ in circle O of Figure T2.9. Suppose that $\angle AOB < \angle COD$. Because $OA = OB = OC = OD$ and $\angle 1 < \angle 2$, by the SAS Inequality Theorem, $AB < CD$. Conversely, if $AB < CD$, then by the SSS Inequality Theorem, $\angle 1 < \angle 2$. We summarize this result in the next theorem.

THEOREM T2.8

In a circle or in circles with the same radius, the measure of one central angle is less than the measure of another if and only if the length of the chord intercepted by the first angle is less than the length of the chord intercepted by the second.

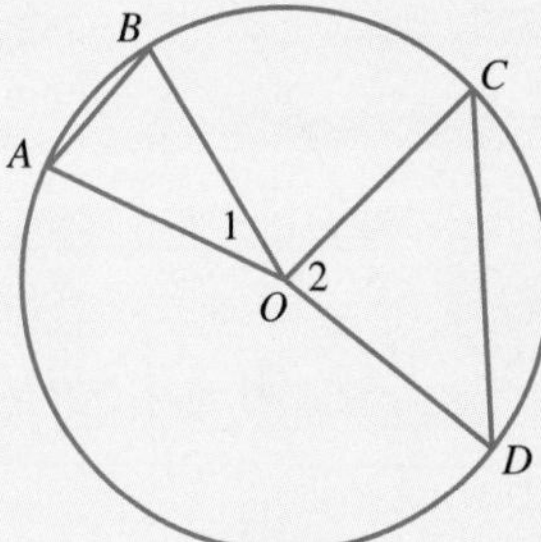

$\angle 1 < \angle 2$ if and only if $AB < CD$.

We can combine the two preceding theorems to form the next theorem.

THEOREM T2.9

In a circle or in circles with the same radius, the length of one chord is less than the length of a second chord if and only if the measure of the arc intercepted by the first chord is less than the measure of the arc intercepted by the second.

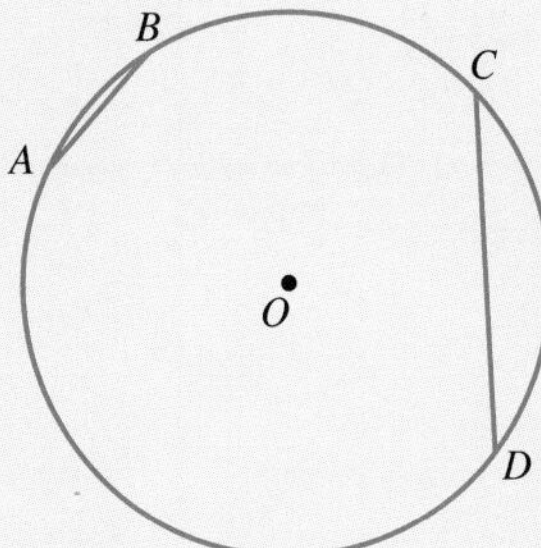

$AB < CD$ if and only if $\widehat{AB} < \widehat{CD}$.

In Figure T2.9 it appears that the chord nearer the center of the circle is longer than the one farther from the center. Next we prove this result.

Let $x < y$ in Figure T2.10(a), where $\overline{OE}$ and $\overline{OF}$ are perpendicular to $\overline{AB}$ and $\overline{CD}$, respectively.

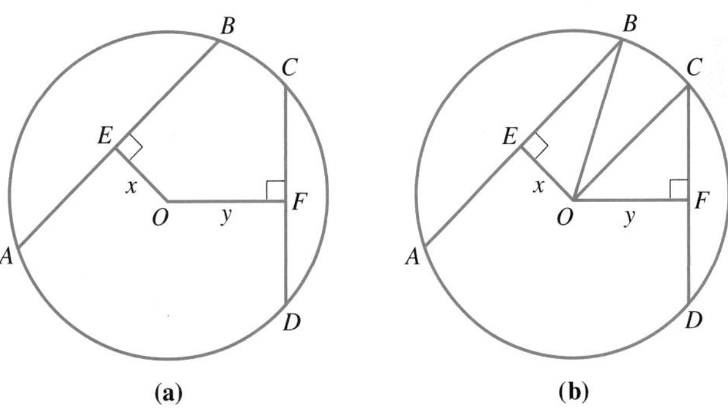

FIGURE T2.10

By Problem 38 of Section 7.2, E and F are midpoints of $\overline{AB}$ and $\overline{CD}$, respectively. By the Pythagorean theorem, $x^2 + (BE)^2 = (BO)^2$ and $y^2 + (CF)^2 = (CO)^2$ [Figure T2.10(b)]. Because $(BO)^2 = (CO)^2$, we have $x^2 + (BE)^2 = y^2 + (CF)^2$. Because $x < y$, we have $x^2 < y^2$ and $x^2 + (BE)^2 < y^2 + (BE)^2$. Finally, $y^2 + (CF)^2 < y^2 + (BE)^2$ by substitution, so $(CF)^2 < (BE)^2$, or $CF < BE$. This implies that $2CF < 2BE$, or that $CD < AB$, which was what we were trying to prove. We can reverse this argument, leading to the following result.

THEOREM T2.10

In a circle or in circles with the same radius, the length of one chord is greater than the length of a second chord if and only if the first chord is nearer to the center than the second.

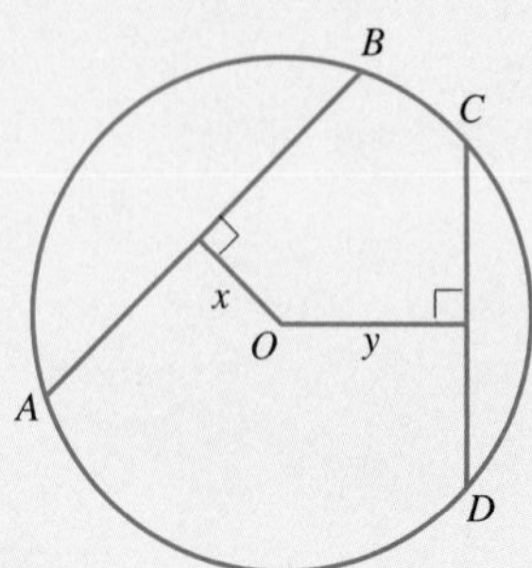

$AB > CD$ if and only if $x < y$.

Notice that an immediate consequence of this theorem is the fact that diameters are the longest chords in a circle.

PROBLEM SET T2

EXERCISES/PROBLEMS

1. Solve each inequality for x.

a. $4x - 1 < 23$

b. $9 + 6x \leq -6$

c. $3x - 8 > 7x + 12$

2. Solve each inequality for x.

a. $3x - 7 > 11$

b. $12 - x < 4$

c. $2x + 6 \leq 7x - 34$

3. Solve each inequality for x.

a. $2 - 3(x - 4) \leq 2(1 - x)$

b. $\frac{2x - 1}{7} \leq -3$

c. $x(x - 5) \leq x^2 - 3(x - 4)$

4. Solve each inequality for x.

a. $3 - (x - 1) > 3x + 4$

b. $\frac{6x - 5}{2} \geq -25$

c. $2x(x - 1) \geq x^2 + x(x + 2) + 8$

5. Given $\triangle ABC$ as shown, explain why $\angle B > \angle C$.

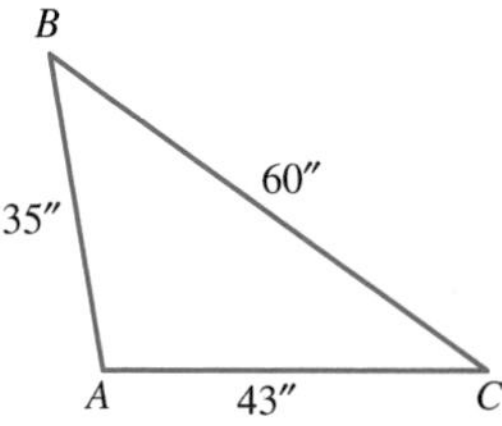

6. Given $\triangle ABC$ as shown, explain why $AB < BC$.

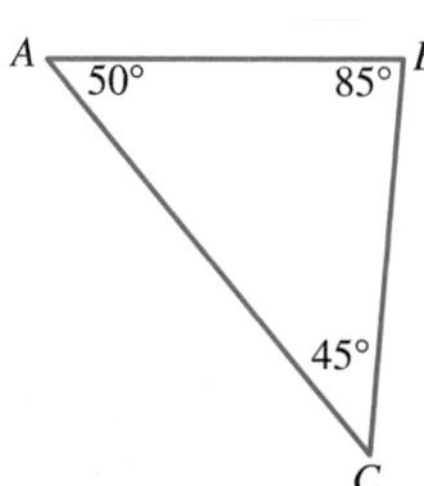

7. In $\triangle PQR$, $\angle P = 41°$ and $\angle Q = 80°$.

a. What is the measure of $\angle R$?

b. Which of the three sides is the longest?

c. Which of the three sides is the shortest?

8. In $\triangle VWX$, $\angle W = 26°$ and $\angle X = 105°$.

a. What is the measure of $\angle V$?

b. Which of the three sides is the longest?

c. Which of the three sides is the shortest?

9. In circle O, $\angle AOB = 95°$, $\angle BOC = 42°$, and $\angle COD = 60°$.

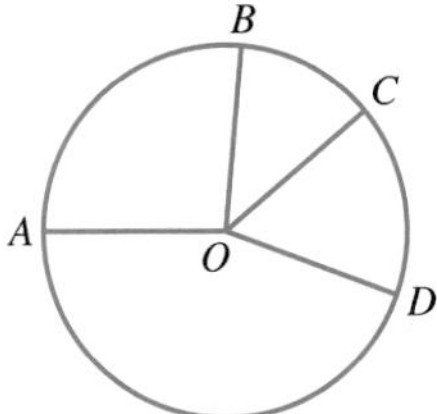

a. What is the measure of $\angle AOD$?

b. Which of $\widehat{AB}$ or $\widehat{BC}$ is larger? Why?

c. Which of $\widehat{AD}$ or $\widehat{AC}$ is larger? Why?

10. In circle O, $\widehat{AD}$ has measure 112°, $\widehat{BC}$ has measure 34°, and $\overline{BD}$ is a diameter.

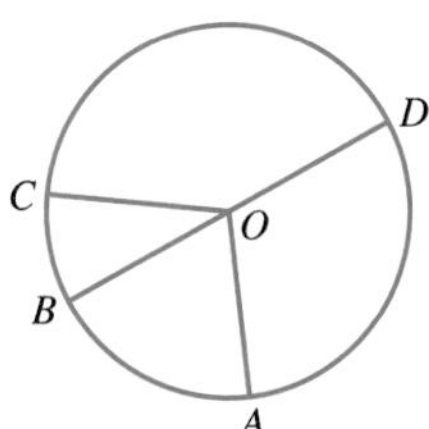

a. What is the measure of $\widehat{CD}$?

b. Which of $\angle AOC$ or $\angle AOD$ is larger? Why?

c. Which of $\angle DOA$ or $\angle DOC$ is larger? Why?

11. Answer the following questions about the circle O as shown.

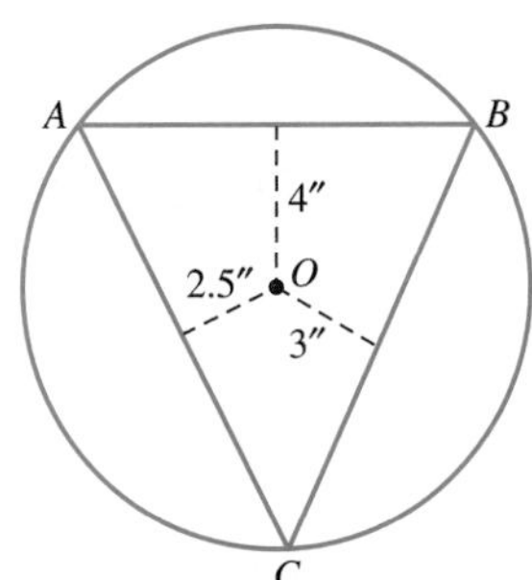

a. Which side of $\triangle ABC$ is longest? Explain.

b. Which angle of $\triangle ABC$ is the largest? Explain.

c. Which of arcs $\widehat{AB}$, $\widehat{BC}$ or $\widehat{AC}$ has the smallest measure? Explain.

12. Answer the following questions about circle O where $AB = 5''$, $BC = 2''$, and $DE = 9''$.

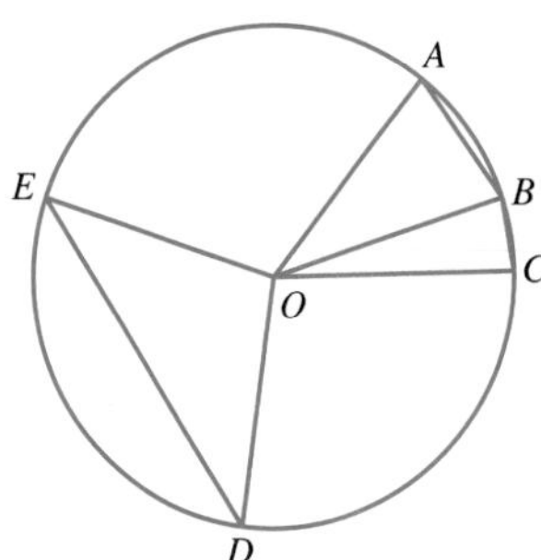

a. Which of $\angle AOB$, $\angle BOC$, or $\angle EOD$ is the largest? Explain.

b. Which of $\widehat{AB}$, $\widehat{BC}$, or $\widehat{DE}$ has the smallest measure? Explain.

c. Which of $\angle EOD$ or $\angle AOC$ is larger? Explain.

13. Assuming that the following figure is drawn precisely, write three valid inequality statements relating pairs of AB, BD, BE, or BC. For example, $AB < BC$.

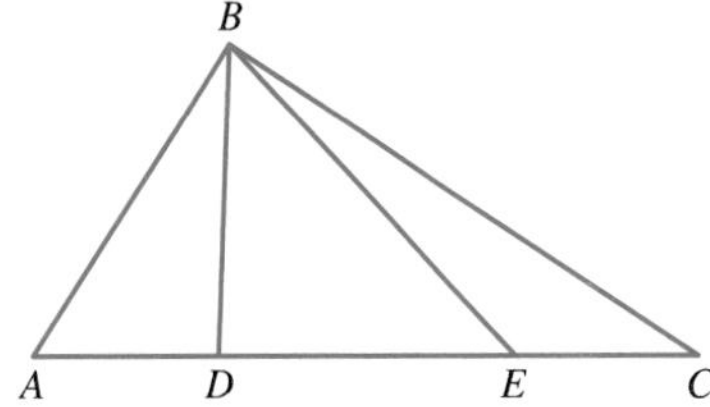

14. Using the figure in problem 13, write three valid inequality statements relating pairs of $\angle A$, $\angle ADB$, $\angle AEB$, or $\angle C$. For example, $\angle ADB > \angle AEB$.

15. Given parallelogram $ABCD$ with dimensions as shown, what lengths are possible for side $\overline{BC}$?

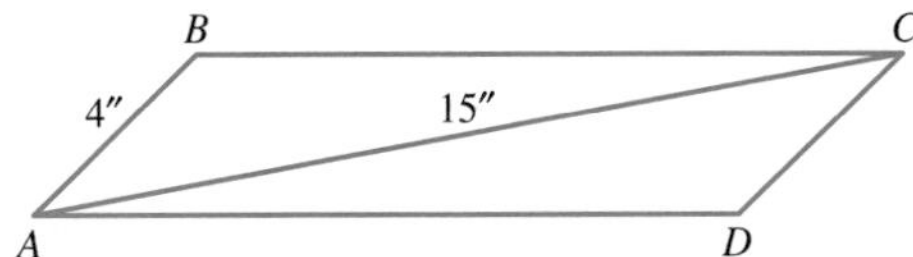

16. Given quadrilateral $PQRS$ with dimensions as shown, what lengths are possible for diagonal $\overline{QS}$?

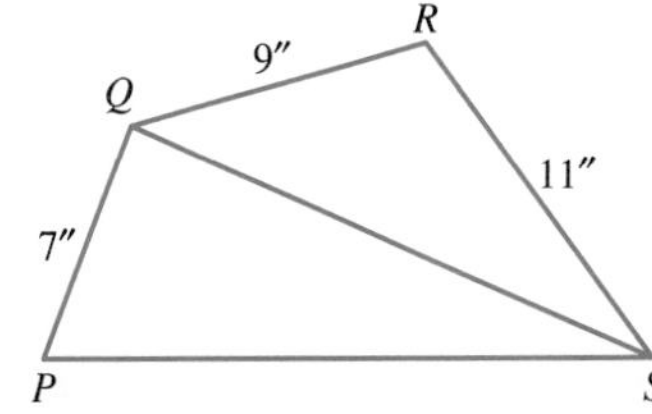

PROOFS

17. Prove that the hypotenuse is longer than either of the legs of a right triangle.

18. $\triangle ABC$ is a right triangle with D on $\overline{AC}$. Prove that $BD < BC$ using two different methods.

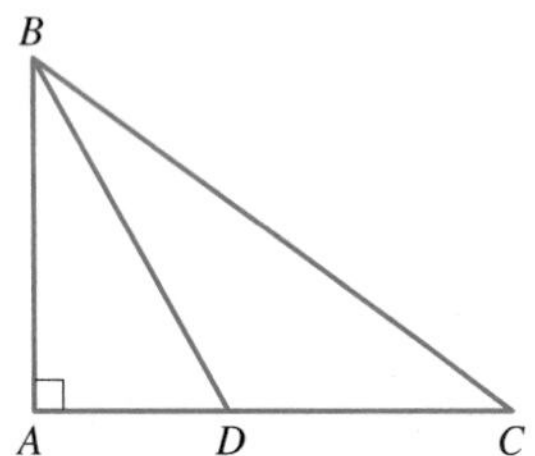

19. In $\triangle ABC$, $AB = BC$ and D is on $\overline{AC}$. Prove that $AB > BD$.

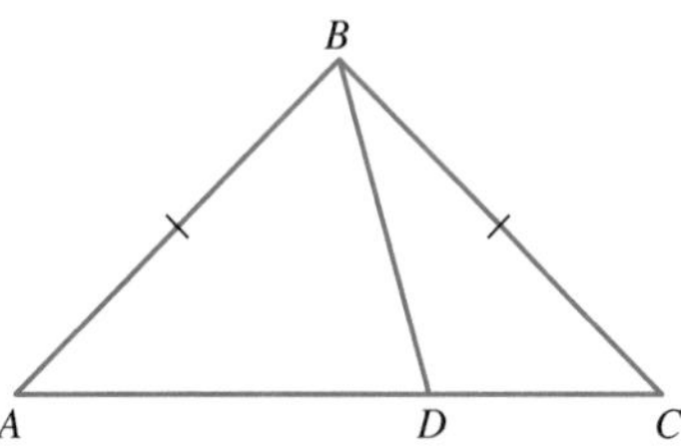

20. $ABCD$ is a kite with $AB = BC$, $AD = CD$, and $AB < AD$. Prove that $\angle D < \angle B$.

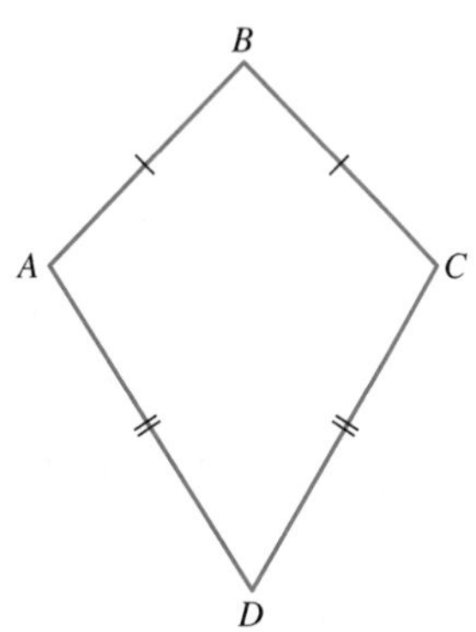

21. For rectangle $ABCD$, prove that the diagonal AC is longer than either of the sides using two different methods.

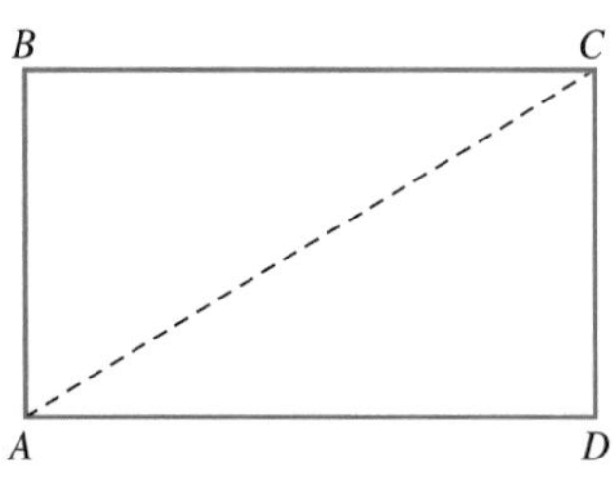

22. Let $\triangle ABC$ be any triangle with $\overline{BD}$ the median to $\overline{AC}$. Prove that $BD < \frac{1}{2}(AB + BC)$.

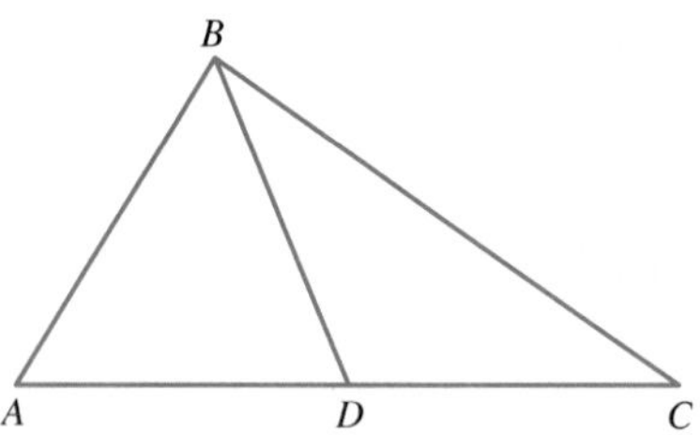

23. In isosceles trapezoid $ABCD$, $AB = BC = CD$ and $BC < AD$. Prove that $\angle A < \angle C$.

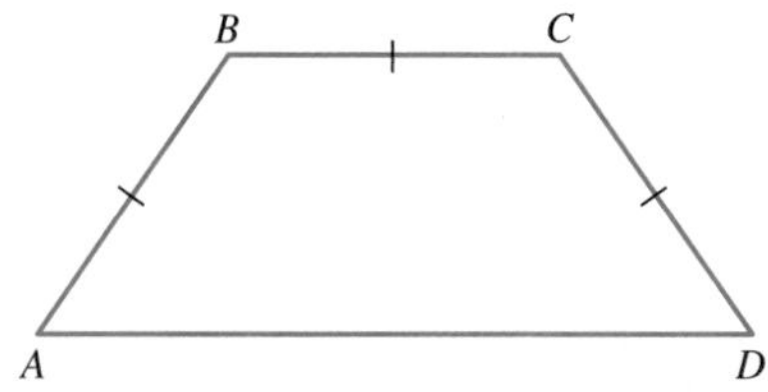

24. Let $ABCD$ be any quadrilateral. Prove that $AC + BD < AB + BC + CD + AD$. Restate this result in words.

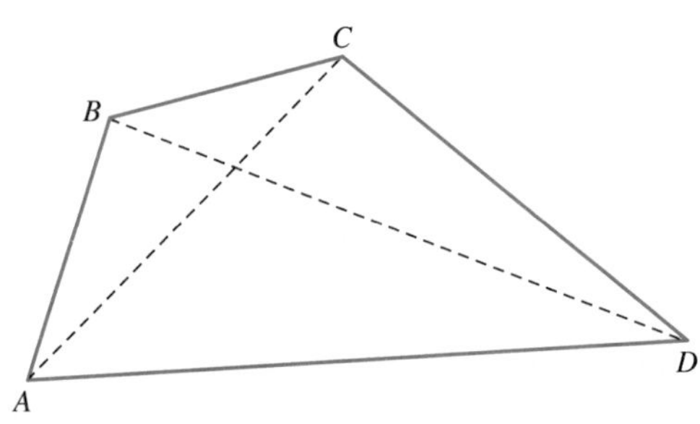

25. In the circle O as shown, $\overset{\frown}{AB} = 90°$ and $\overset{\frown}{CD} > 90°$, and D, O, and B are collinear. Prove that the area of $\triangle CDO$ is less than the area of $\triangle ABO$.

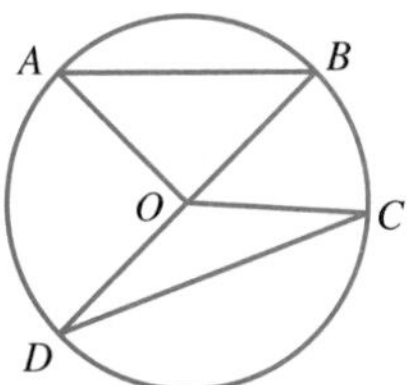

26. Let A be a point in the exterior of circle O. Prove that the shortest distance from A to the circle is AB, where B is the point of intersection of $\overline{OA}$ and circle O.

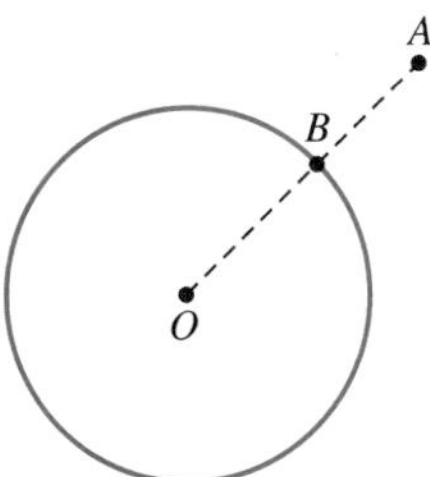

27. Let A be a point in the interior of circle O. Prove that the shortest distance from A to the circle is AB, where B is the point of intersection of ray $\overrightarrow{OA}$ and circle O.

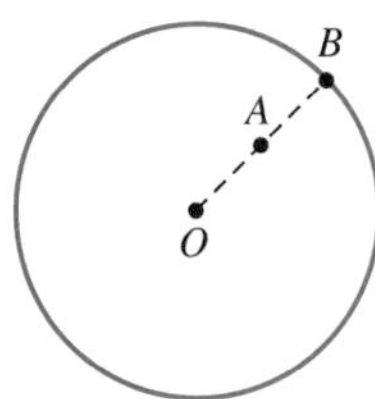

28. Let P be any point in the interior of circle O except O itself. Show that the shortest chord containing P must be the one where $\overline{OP}$ is perpendicular to the chord.

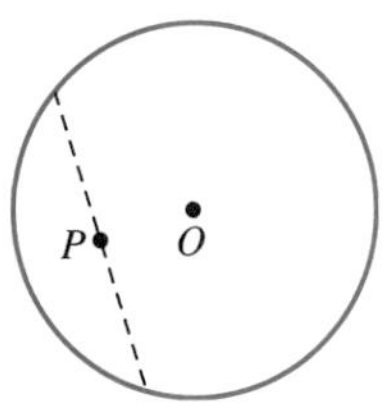

(HINT: Draw chord $\overline{AB}$ where P is on $\overline{AB}$ and $\overline{OP}$ is perpendicular to $\overline{AB}$. Then draw any other chord $\overline{CD}$ through point P. Then consider the perpendicular segment $\overline{OQ}$ from O to $\overline{CD}$.)

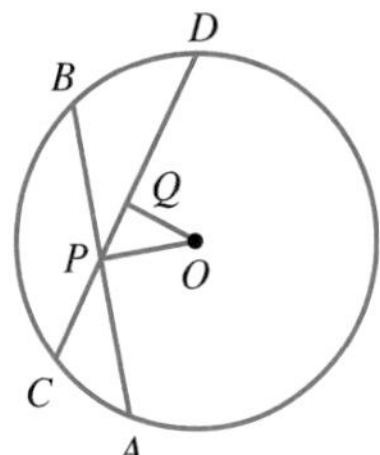

29. Prove the transitive property for less than. That is, prove that if $a < b$, and $b < c$, then $a < c$. (HINT: Use the definition of less than: $a < b$ if there is a positive p such that $a + p = b$.)
30. Prove Theorem T2.6: SSS Inequality—if two sides of one triangle are congruent to two sides of another, but the third side of the first triangle is shorter than the third side of the other triangle, then the measure of the angle opposite the third side in the first triangle is less than the measure of the corresponding angle in the second triangle.

TOPIC REVIEW

Next you will find a list of vocabulary and notation, a summary of key concepts, and practice test questions similar to the problems in this topic section. Review each item in the *Vocabulary/Notation* list mentally, and, if necessary, refer back to the indicated page and write a definition. Make certain that you can explain in your own words each of the *Key Concepts* and be able to provide examples. Then work the problems in the Topic Test to assess your understanding of the content. Answers to the Topic 2 Test can be found at the end of the book.

Vocabulary/Notation

Less then (<) 542
Greater than (>) 542
Trichotomy Property 542
Transitive Property 542
Triangle Inequality 546
SAS Inequality 547
SSS Inequality 547

Key Concepts

1. In triangles, there are inequality relationships between the lengths of the sides and the measures of angles opposite those sides.
2. In circles, there are inequality relationships among the measures of the central angles and the measures of their intercepted arcs and lengths of their chords.
3. The lengths of chords are related to their distances from the center of the circle.

TOPIC 2 TEST

TRUE-FALSE

Mark as true any statement that is always true. Mark as false any statement that is never true or that is not necessarily true. Be able to justify your answers.

1. If a, b, and c are real numbers such that $a < b$ and $b < c$, then $a < c$.
2. If a, b, c, and d are real numbers such that $a < b$ and $c < d$, then $ac < bd$.
3. If $\angle A \geq \angle B$ and $\angle B < 20°$, then $\angle A > 20°$.
4. If the three angles of a triangle have different measures, then the triangle is scalene.
5. A triangle cannot have sides of lengths 8 cm, 11 cm, and 20 cm.

EXERCISES/PROBLEMS

6. Solve for x: $17 - 3x > 20$
7. Solve for x: $\dfrac{x - 7}{2} < 3x + 4$
8. In $\triangle RST$, the sides have lengths as follows: $RS = 4$ cm, $ST = 12$ cm, $RT = 10$ cm.
   a. Which angle of $\triangle RST$ is the largest? Explain.
   b. Which angle of $\triangle RST$ is the smallest? Explain.
9. In $\triangle XYZ$, the angles have measures as follows: $\angle X = 32°$, $\angle Z = 84°$.
   a. Which side of $\triangle XYZ$ is the longest? Explain.
   b. Which side of $\triangle XYZ$ is the shortest? Explain.
10. In circle O, $\angle AOB = 115°$ and $\angle BOC = 80°$.

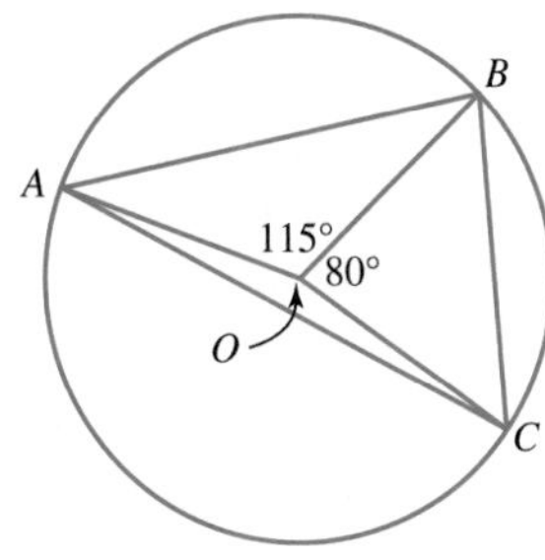

   a. Which has the greater measure: $\widehat{AB}$ or $\widehat{AC}$?
   b. Which is longer: $\overline{AO}$ or $\overline{BC}$?

11. Given $\triangle ABC$ with dimensions as shown, what lengths are possible for side $\overline{AC}$?

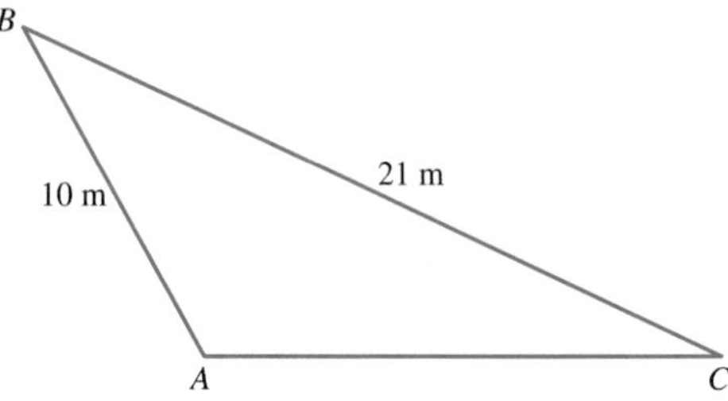

12. In circle O, $\overline{AC}$ is a diameter, $\overline{BO} \perp \overline{DO}$, $\angle AOD = 140°$, and $\overline{OE}$ bisects $\angle AOD$.

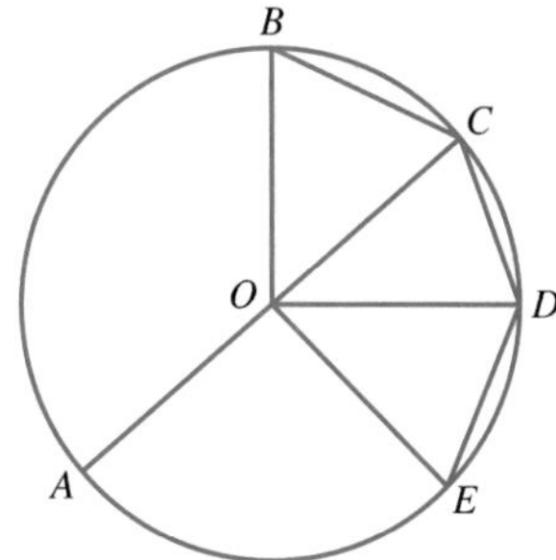

   a. Which chord is closest to O: $\overline{BC}$, $\overline{CD}$, or $\overline{DE}$? Explain.
   b. Which angle is the smallest: $\angle OBC$, $\angle OCD$, or $\angle OED$? Explain.

13. Let $\triangle ABC$ be any triangle with $\overline{BD}$ the angle bisector of $\angle B$. Prove that (i) $\angle ADB > \angle ABD$, and (ii) $\angle CDB > \angle CBD$.

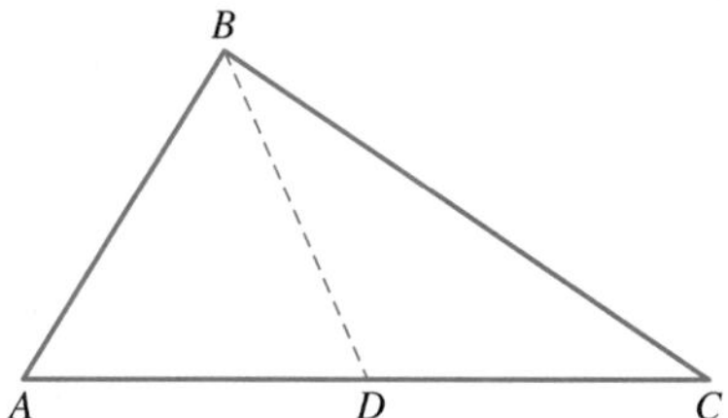

14. Prove that the sum of the lengths of the diagonals of a pentagon is less than twice its perimeter.

TOPIC 3

NON-EUCLIDEAN GEOMETRY

In his famous work, *The Elements*, Euclid stated many definitions, five postulates of a geometric nature, and five common notions that could be applied to both numbers and geometry. The five postulates of a geometric nature were as follows:

Postulate 1. A straight line can be drawn from any point to any point.

Postulate 2. It is possible to extend a finite straight line indefinitely.

Postulate 3. A circle can be drawn with any point as center and any distance as radius.

Postulate 4. All right angles are equal.

Postulate 5. If a straight line meets two straight lines, so as to make the two interior angles on the same side of it taken together less than two right angles, these straight lines, being continually produced, shall at length meet on the side on which are the angles that are less than two right angles.

Postulate 5, commonly called the **Parallel Postulate**, was the center of much debate. For over 2000(!) years, there were many attempts to prove the Parallel Postulate from the other postulates. However, it has been shown that it is impossible to prove it in this way.

Exploration of the Parallel Postulate stimulated interest in this area of geometry. It was demonstrated that, when assuming the first four postulates, some theorems that were proved using the Parallel Postulate could, in fact, be used to prove the Parallel Postulate. Thus, they were equivalent formulations of the Parallel Postulate. Four such results are listed next.

1. Playfair's Axiom: Given a line l and a point P not on l, there is only one line m containing P that is parallel to l.
2. Parallel lines are everywhere equidistant.
3. The sum of the measures of the angles in a triangle is 180°.
4. Proclus's Axiom: If a line intersects one of two parallel lines, then it must also intersect the other line.

In addition to motivating these equivalent formulations of the Parallel Postulate, the attempt to prove it led to the discovery of other geometries. In this section, we provide a look at several of these geometries.

ABSOLUTE GEOMETRY

In *The Elements*, Euclid proved many theorems before actually employing the Parallel Postulate. The body of knowledge that follows from the postulates that preceded the Parallel Postulate became known as **absolute geometry**. In the 17th century, Girolamo Saccheri (1676–1733), an Italian Jesuit, approached the dilemma of trying to prove the Parallel Postulate by first proving additional theorems that did not rely on the Parallel Postulate. In his study, Saccheri introduced some quadrilaterals, since named **Saccheri quadrilaterals**, that had a pair of opposite sides congruent and two right angles as shown in Figure T3.1. The other two angles were called **summit angles**. He was hoping that he could infer that these quadrilaterals were rectangles without using the Parallel Postulate. However, he could not. Later it was shown that Saccheri quadrilaterals were not rectangles unless the Parallel Postulate was assumed.

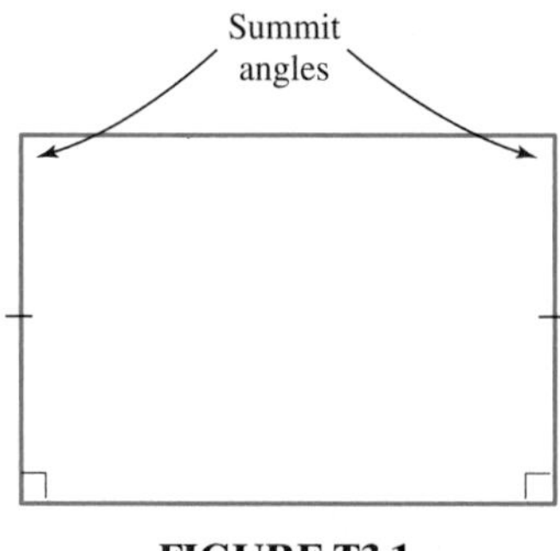

FIGURE T3.1

Although the Parallel Postulate seems "obviously" true, this is only because of our world of reference. That is, because we are generally limited to a small portion of the earth, it seems as if we are living on a plane. However, suppose we wanted to study the geometry of a sphere. The sphere itself would be our plane and points on the sphere our points. The shortest distance between two points on a sphere is an arc of a **great circle**, that is, a circle of the sphere that passes through both endpoints of a diameter of a sphere. The equator of the earth is one example of a great circle. In the geometry of a sphere, we define these great circles to be the lines. Figure T3.2(a) shows two examples of lines on a sphere.

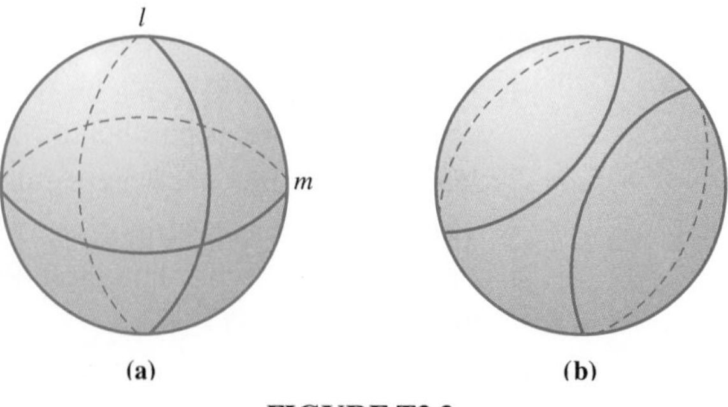

FIGURE T3.2

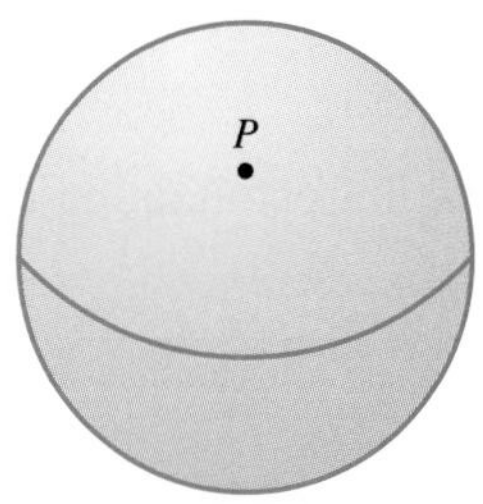

FIGURE T3.3

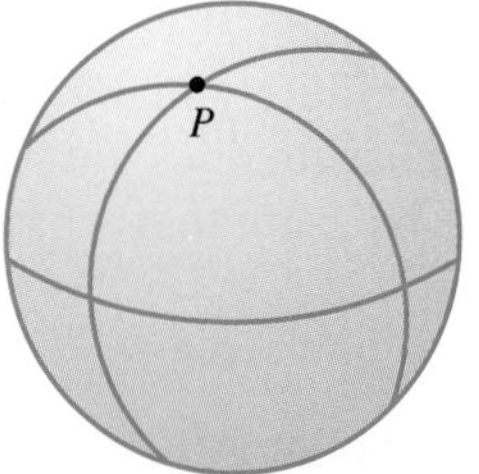

FIGURE T3.4

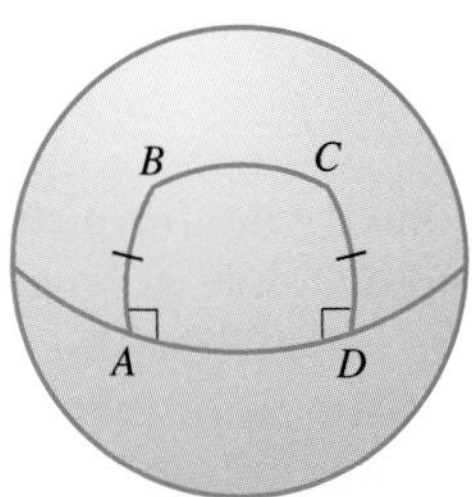

FIGURE T3.5

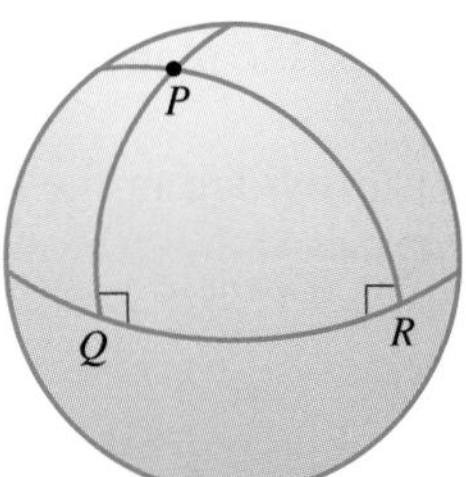

FIGURE T3.6

Figure T3.2(b) shows two examples of curves that would not be considered lines in this geometry.

Using the points of a sphere as our points, the great circles of a sphere as our lines, and the sphere as our plane, we can develop a geometry. It is interesting to note, however, that the Parallel Postulate does not hold in this geometry. Using Playfair's Axiom, consider any great circle (such as the equator) and any point P not on the equator (Figure T3.3). Then any other great circle containing P also will intersect the equator. That is, there is no "line" through P that is parallel to the equator. Figure T3.4 shows two such great circles. This model also presents a Saccheri quadrilateral that is not a rectangle because the angles at B and C are greater than 90° (Figure T3.5).

ELLIPTIC GEOMETRY

In 1854, George Riemann (1826–1866) formalized this idea of a spherical geometry by replacing the Parallel Postulate with the following.

ELLIPTIC PARALLEL POSTULATE

Given a line l and a point P not on l, there is no line containing P that is parallel to l.

We can show in absolute geometry that parallel lines do *exist* (only the uniqueness is guaranteed by the Parallel Postulate). Thus, we must modify Euclid's Postulate 1 as follows to be consistent with the Elliptic Parallel Postulate: Given any two points, there is *at least* one line containing the points.

Some of the results that follow from the Elliptic Parallel Postulate are quite surprising in that they differ so much from or actually contradict results derived in Euclidean geometry. For example, consider Figure T3.6. The great circles through P and Q and through P and R are perpendicular to the equator. So $\triangle PQR$ has two right angles, a property not shared by triangles in Euclidean geometry. This means that the sum of the angle measures in $\triangle PQR$ exceeds 180°. Some results of the Elliptic Parallel Postulate are listed next.

1. There are no parallel lines.
2. The sum of the measures of the angles in any triangle exceeds 180°.
3. There are no rectangles.
4. If two triangles are similar, they must be congruent.
5. Two lines intersect in two points.

HYPERBOLIC GEOMETRY

Even before Riemann developed his alternative to Euclidean geometry, Joano Bolyai (1802–1860), a Hungarian army officer, and Nichollas Lobachevsky (1793–1856), a Russian mathematician, each independently developed a geometry based on absolute geometry using the following postulate in place of the Parallel Postulate.

> **HYPERBOLIC PARALLEL POSTULATE**
>
> Given a line l and a point P not on l, there exist at least two lines containing P that are parallel to l.

We can use the following model to picture a geometry that satisfies this parallel postulate. Consider a plane to be the interior of a circle, points to be the points of this interior, and lines to be chords of the circle (without the endpoints). Lines l, m, and n and point P are shown in Figure T3.7. Notice how m and n contain P, yet do not intersect l. Thus, m and n are parallel to l. In fact, there are infinitely many lines through P in this plane that are parallel to l. The usual angle measure does not hold in this geometry. Instead, there is a formula to determine angle measure that is based on the usual Euclidean angle measure.

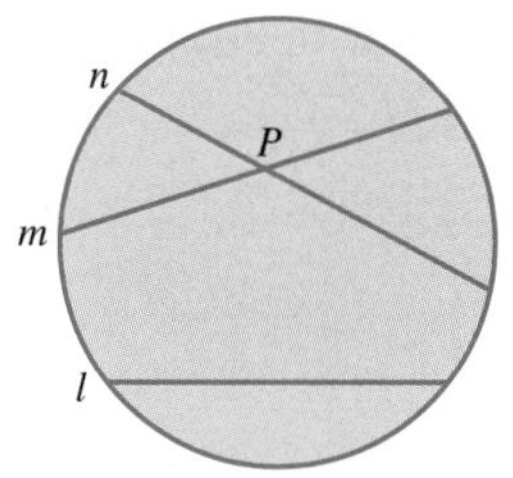

FIGURE T3.7

Some theorems that arise from the Hyperbolic Parallel Postulate follow.

1. The sum of the measures of the angles of any triangle is less than 180°.
2. Different triangles may have different angle sums.
3. There are no rectangles.
4. If two triangles are similar, they must be congruent.
5. Parallel lines are not everywhere equidistant.

Thus far, we have seen three ways in which Euclidean geometry has been studied based on the Parallel Postulate. First, Saccheri developed absolute geometry, that is, Euclidean geometry without the Parallel Postulate. Next, Bolyai and Lobachevsky developed an alternative geometry that allowed for more than one line to be parallel to a given line. Finally, elliptic geometry allowed for no parallel lines. The development of these latter two geometries had a profound impact on mathematics. It encouraged mathematicians to alter axioms and to study the resulting systems. These systems yielded important mathematical results even if they did not model any known physical situation as Euclid's system modeled the plane, which was considered to represent portions of the Earth.

A summary and comparison of three of these geometries is provided in the following table.

	Euclidean Geometry	**Elliptic Geometry**	**Hyperbolic Geometry**
Model	Plane	Sphere	Interior of a circle
Parallels to a line through a point	One	None	Infinitely many
Angle sum in a triangle	180°	>180°	<180°

The postulates of Euclidean geometry may be altered in other ways to produce other very different geometries. We will complete our treatment of non-Euclidean geometry by studying two other systems, finite geometries and taxicab geometry.

FINITE GEOMETRIES

Finite geometries are composed of a finite set of points. For example, consider points A, B, C, and D in Figure T3.8(a).

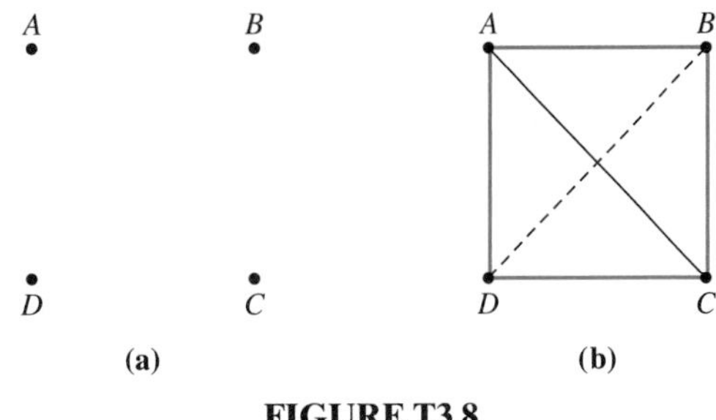

FIGURE T3.8

We define the set $\{A, B, C, D\}$ to be our only plane. The six possible pairs of points will be our lines. So the "line" AB in this geometry consists of only the points A and B. One way of modeling this geometry is to think of the points as vertices of a triangular pyramid with a pair of points forming a "line" if they are endpoints of the same side of the pyramid [Figure T3.8(b)]. Keep in mind that the pairs of points are the "lines" in this geometry, not the segments connecting them as shown in Figure T3.8(b).

Next we list three postulates that this example models.

Postulate 1. There are exactly four points.
Postulate 2. Every pair of points is on a line.
Postulate 3. Every line contains a pair of points.

We can deduce the next several theorems from these three postulates.

Theorem 1. If two lines have a point in common, they have exactly one point in common.
Theorem 2. There are exactly six lines.
Theorem 3. Each point contains exactly three lines.
Theorem 4. Given any line l and point P *not* on the line, there is one and only one line parallel to l that contains P.

We can prove these theorems by reasoning deductively as we do in Euclidean geometry. On the other hand, because this geometry contains a small finite number of points, we can also prove them by checking all possible cases. Although we may use the pyramid model to check cases, an alternative method is to assign letters to the points and to check the theorems using the letters. In the four-point geometry described by $\{A, B, C, D\}$, let the pairs AB, AC, AD, BC, BD, and CD be the lines. These points and "lines" satisfy Postulates 1, 2, and 3, so they represent a model for the four-point geometry. We can use this model to verify Theorems 1–4, using pairs of letters as lines. We prove Theorem 3 in the following example. We leave the proofs of the other three theorems for the problem set.

EXAMPLE T3.1 Prove that in the four-point geometry, each point contains exactly three lines.

SOLUTION

The four points in this geometry are A, B, C, and D. The lines in this geometry are the pairs AB, AC, AD, BC, BD, and CD. We must show that every point appears in exactly three of these pairs.

A appears in AB, AC, and AD. B appears in AB, BC, and BD.
C appears in AC, BC, and CD. D appears in AD, BD, and CD.

Thus, every point does appear in three pairs and so contains exactly three lines. ■

TAXICAB GEOMETRY

We studied Euclidean geometry from a coordinate point of view in Chapter 8. There we found the distance between two points (or the length of a segment) using the Pythagorean theorem (Figure T3.9). Given a coordinate system, there is another

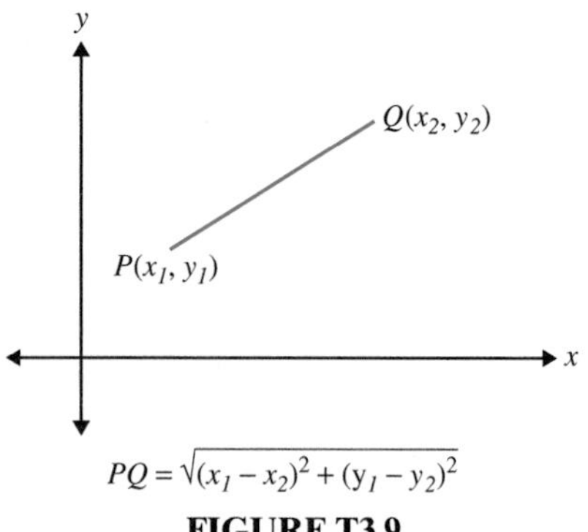

FIGURE T3.9

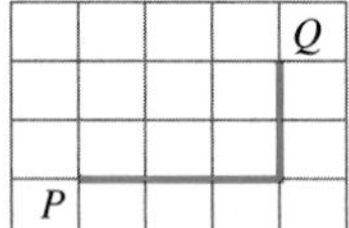

FIGURE T3.10

way to measure distance. Imagine a coordinate grid as intersecting streets, North/South and East/West (Figure T3.10). In taxicab geometry, the **taxicab distance** from P to Q is the shortest distance a taxi would travel along this grid of intersecting streets. In Figure T3.10, $PQ = 3 + 2 = 5$, as indicated by the blue line segments. If $P(a, b)$ and $Q(c, d)$ are two points in the plane, then, using absolute value, the distance from P to Q is given by $|a - c| + |b - d|$. In this geometry, the points, lines, and plane are the same as in Euclidean geometry. Just the distance formula has been changed. Moreover, taxicab geometry satisfies many of the same postulates as Euclidean geometry, even the Parallel Postulate.

In this book, we study a set of postulates that includes Euclid's postulates as well as triangle congruence postulates. Within this set of postulates, there is one that taxicab geometry does not satisfy. Consider the right triangles $\triangle ABC$ and $\triangle DEF$ in Figure T3.11. Notice that *using taxicab distance*, we find that $AB = 2 = AC$ and $DE = 2 = DF$. Because the triangles are right triangles, they should be congruent by the SAS Congruence Postulate. However, notice that they are not the same size (although they are the same *shape*); therefore, they are *not* congruent. Thus, the SAS Congruence Postulate does not hold in taxicab geometry.

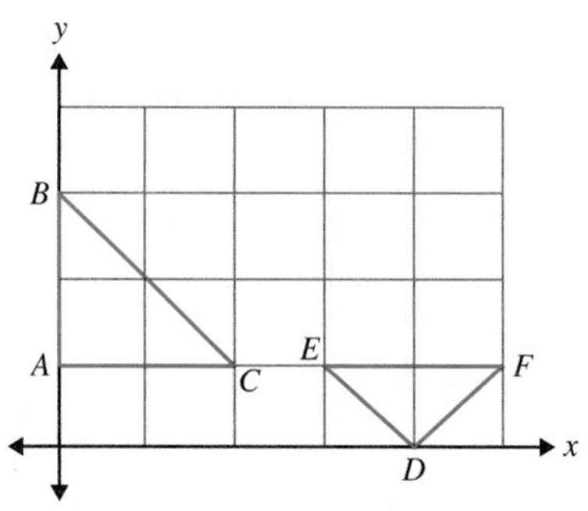

FIGURE T3.11

There are other surprising consequences of the altered distance formula in taxicab geometry. The next example illustrates one such result.

EXAMPLE T3.2

Find the "circle" with center $(0, 0)$ and radius 2 in taxicab geometry.

SOLUTION

In Figure T3.12(a), points $A(-2, 0)$, $B(0, 2)$, $C(2, 0)$, and $D(0, -2)$ are on the circle since their taxicab distance from the origin is 2.

Also, $E(-1, 1)$, $F(1, 1)$, $G(1, -1)$, and $H(-1, -1)$ are 2 taxicab units from the origin [Figure T3.12(b)]. In Figure T3.12(c), let (x, y) be a point on $\overline{BC}$. The equation of $\overline{BC}$ in the Euclidean sense is $y = -x + 2$. By substitution, the point $(x, -x + 2)$ represents an arbitrary point of $\overline{BC}$ for $0 \le x \le 2$. Thus, the taxicab distance from $(0,0)$ to $(x, -x + 2)$ is $x + (-x + 2) = 2$. This shows that any point on $\overline{BC}$ is on the "circle" of points at a distance of 2 from $(0, 0)$. Continuing in this manner, we can show that the "circle" with center $(0, 0)$ and radius 2 in taxicab geometry is actually the square $ABCD$! ■

(a)

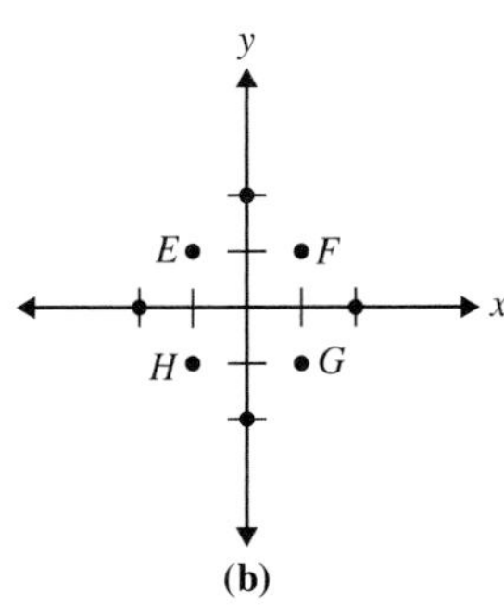

(b)

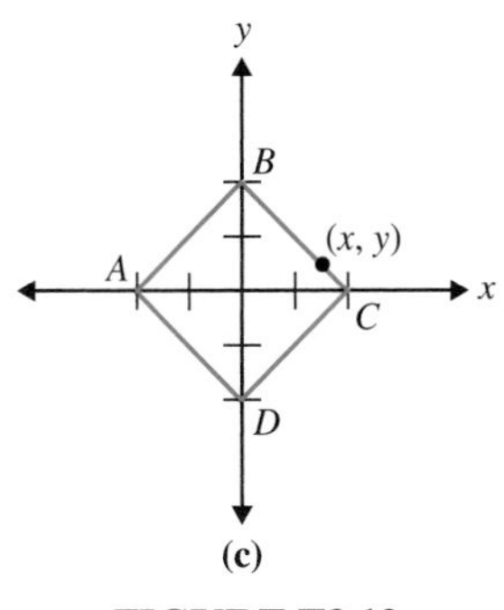

(c)

FIGURE T3.12

In Euclidean geometry, a point is equidistant from the endpoints of a segment if and only if it is on the perpendicular bisector of the segment. The next example explores this idea in taxicab geometry.

EXAMPLE T3.3

a. Find all points equidistant from $(-1, 0)$ and $(1, 0)$.

b. Find all points equidistant from $(0, 0)$ and $(1, 3)$.

SOLUTION

a. Let (x, y) be any point equidistant from $(-1, 0)$ and $(1, 0)$ (Figure T3.13). Then the distance from $(-1, 0)$ to (x, y) is $(x + 1) + y$, and the distance from $(1, 0)$ to (x, y) is $1 - x + y$. Because these distances are equal, we have $(x + 1) + y = 1 - x + y$, or $2x = 0$, or $x = 0$. Thus, any point equidistant from $(-1, 0)$ and $(1, 0)$ must be on the y-axis. That is, $(0, y)$ is equidistant from $(-1, 0)$ and $(1, 0)$, where the distance from either point to $(0, y)$ is $y + 1$.

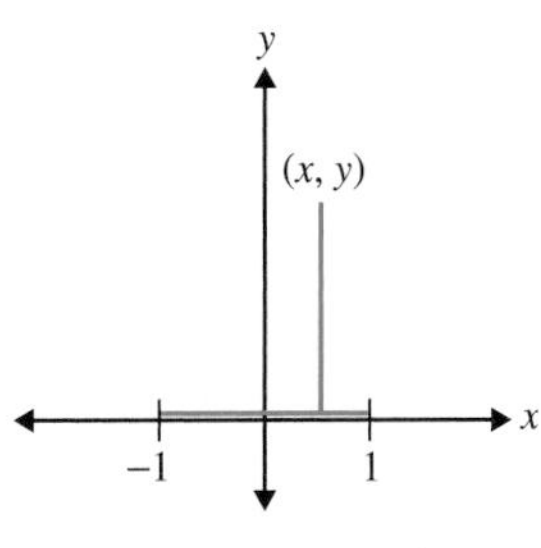

FIGURE T3.13

b. First, observe that $(1, 1)$ is 2 units from both $(0, 0)$ and $(1, 3)$ [Figure T3.14(a)].

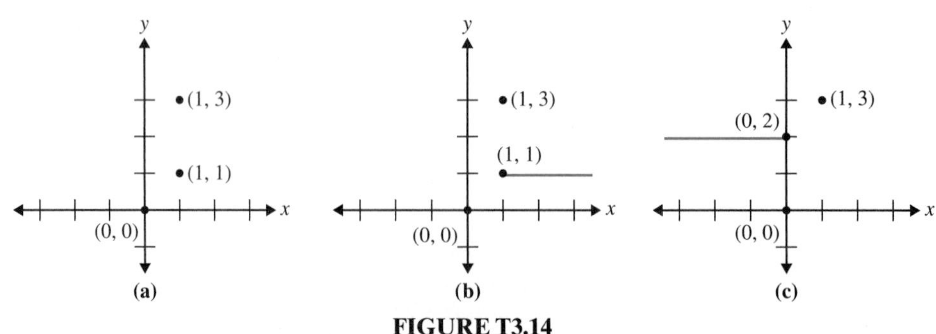

FIGURE T3.14

Also, in general, $(x, 1)$, for $x > 1$, is $x + 1$ units from both points [Figure T3.14(b)]. In a similar manner, the point $(x, 2)$, for $x < 0$, is equidistant from $(0, 0)$ and $(1, 3)$ [Figure T3.14(c)]. Last, the points on the segment whose endpoints are $(0, 2)$ and $(1, 1)$ are equidistant from $(0, 0)$ and $(1, 3)$ [Figure T3.14(d)]. We leave the verification of this result for the problem set.

Thus, the jagged line shown in Figure T3.14(e) is the set of points that are equidistant from $(0, 0)$ and $(1, 3)$.

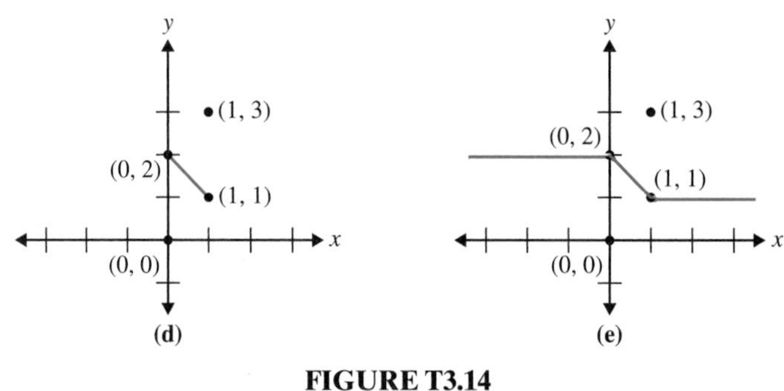

FIGURE T3.14

Other interesting properties of taxicab geometry are revealed in the problem set. A very nice treatment of this subject is found in the Dover Publications book *Taxicab Geometry* by Eugene F. Krause.

PROBLEM SET T3

EXERCISES/PROBLEMS

1. In Euclidean geometry, two distinct lines in a plane are either parallel or intersect in exactly one point. Discuss the intersection of two distinct lines in the spherical geometry of this section. Include a picture with your explanation.

2. In Euclidean geometry, given a line l and a point P not on the line, there is exactly one line through P that is perpendicular to l. Discuss the number of lines through P that are perpendicular to l in the geometry of the sphere. Include a picture with your explanation.

3. In the spherical geometry of this section, a "line" is a great circle. What will a "circle" look like in this model? (HINT: Think of a circle as the set of points equidistant from a fixed point.)

Problems 4–12 deal with taxicab geometry. Let $d_T(A, B)$ represent taxicab distance between A and B and $d_E(A, B)$ represent the usual Euclidean distance.

4. Find $d_T(A, B)$ and $d_E(A, B)$ for the following pairs.
   a. $A(0, 0)$, $B(0, 10)$
   b. $A(1, 1)$, $B(1, 8)$

c. $A(-2, -3), B(7, -3)$

d. $A(5, 4), B(5, -2)$

e. Using parts (a)–(d), describe, in general, pairs of points where the taxicab distance equals the Euclidean distance.

5. Find $d_T(A, B)$ and $d_E(A, B)$ for the following pairs.

a. $A(2, 3), B(-1, 6)$

b. $A(-1, -2), B(3, 4)$

c. $A(3, 5), B(9, -2)$

d. It is claimed that $d_T(A, B) \geq d_E(A, B)$ for all pairs A, B. Determine if this claim is correct for the pairs in parts (a)–(c). If it is, do you think that it is true for any pair of points?

6. Is the Pythagorean theorem true in taxicab geometry? Explain.

7. a. The length of the side of a square is 4 and the diameter of a circle is 4. Compare their areas using the Euclidean distance.

b. Repeat part (a) using the taxicab distance and Example T3.1.

8. Determine if the following statement is true or false for the given points.

$$d_T(A, B) + d_T(B, C) \geq d_T(A, C)$$

a. $A(1, 1), B(2, 3), C(5, 9)$

b. $A(-1, 4), B(4, 1), C(0, -5)$

9. Recall that in Euclidean geometry, P is between A and B if $AP + PB = AB$. Plot each of the following triples of points to see if one appears to be between the other two. Then check to see if

$$d_T(A, P) + d_T(P, B) = d_T(A, B)$$

a. $A(2, 3), B(3, 6), C(4, 9)$

b. $A(-1, -3), B(2, 0), C(5, 3)$

10. Complete Example T3.3 by verifying that every point on the segment with endpoints $(0, 2)$ and $(1, 1)$ is equidistant from the two points $(0, 0)$ and $(1, 3)$.

11. Figure T3.11 shows two triangles that satisfy SAS using taxicab distance but are not congruent. Find two triangles that satisfy ASA using taxicab distance, yet are not congruent.

12. Figure T3.11 shows two triangles that satisfy SAS using taxicab distance but are not congruent. Find two triangles that satisfy SSS using taxicab distance yet are not congruent.

Problems 13–15 apply to the four-point finite geometry discussed in this section. Use the points and lines associated with {A, B, C, D} to prove the following theorems.

13. Prove Theorem 1: If two lines have a point in common, they have exactly one point in common.

14. Prove Theorem 2: There are exactly six lines.

15. Prove Theorem 4: Given any line l and point P *not* on the line, there is one and only one line parallel to l that contains P.

Let {A, B, C, D, E, F, G} be a set of points and the following triples be "lines": ABC, AGF, AED, BGD, CGE, CFD, and EBF. Determine if the statements in problems 16–20 are true or false in this seven-point geometry. If a statement is true, prove that it is true.

16. Two distinct lines intersect in exactly one point.

17. Each point is on exactly three lines.

18. Each point is on at least one line with every other point.

19. For every pair of points, there are exactly two lines containing neither point.

20. If any three lines are not concurrent (that is, do not all share the same point), there is exactly one point that is not on any of the three lines.

Problems 21–23 deal with Saccheri quadrilaterals. You may use results about triangle congruence, but you may not use Euclid's Parallel Postulate or any of its equivalents.

21. Prove that the diagonals of a Saccheri quadrilateral are congruent.

22. Prove that the summit angles of a Saccheri quadrilateral are congruent.

23. A **Lambert quadrilateral** is one that has three right angles. It can be shown that the fourth angle is not obtuse. In the Lambert quadrilateral $ABCD$ shown, prove that $CD \leq AB$. (HINT: Suppose that $AB < CD$. Let E be a point on $\overline{CD}$, consider $\angle BED$, and observe that $ABED$ is a Saccheri quadrilateral.)

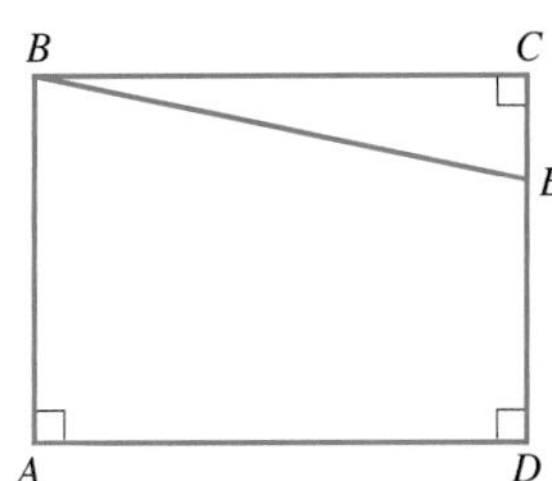

TOPIC REVIEW

Next you will find a list of vocabulary and notation, a summary of key concepts, and practice test questions similar to the problems in this topic section. Review each item in the *Vocabulary/Notation* list mentally, and, if necessary, refer back to the indicated page and write a definition. Make certain that you can explain in your own words each of the *Key Concepts* and be able to provide examples. Then work the problems in the Topic Test to assess your understanding of the content. Answers to the Topic 3 Test can be found at the end of the book.

Vocabulary/Notation

Parallel Postulate 555
Absolute geometry 556
Saccheri quadrilaterals 556
Summit angles 556
Great circle 556
Elliptic Parallel Postulate 557
Hyperbolic Parallel Postulate 558
Finite geometry 559
Taxicab geometry 560
Taxicab distance 560

Key Concepts

1. The Parallel Postulate was the focus of much attention in Euclidean geometry as well as the stimulus for the formation of many new geometries.
2. Saccheri developed absolute geometry in his attempt to prove the Parallel Postulate from the other postulates.
3. There are non-Euclidean geometries whose parallel postulates differ from the standard Euclidean parallel postulate in two ways,
   i. Hyperbolic geometry is an example of a non-Euclidean geometry where given a line l and a point P not on the line, there is more than one line through P parallel to l.
   ii. Elliptic geometry is a non-Euclidean geometry where given a line l and a point P not on the line, there are no lines through P parallel to l.
4. There are finite geometries that model many aspects of Euclidean geometry.
5. Taxicab geometry is a non-Euclidean geometry that satisfies all of Euclid's postulates. However, it does not satisfy the SAS Congruence Postulate.

TOPIC 3 TEST

TRUE-FALSE

Mark as true any statement that is always true. Mark as false any statement that is never true or that is not necessarily true. Be able to justify your answers.

1. Two points determine a unique line in the elliptic geometry on the sphere as described in this section.
2. Two points determine a unique line in the interior of a circle model of hyperbolic geometry as described in this section.
3. In the geometry on a sphere as described in this section, a triangle can have two right angles.
4. The taxicab distance from $A(-1,4)$ to $B(2,3)$ is greater than the Euclidean distance AB.
5. A 3-4-5 triangle in taxicab geometry is not necessarily a right triangle.

EXERCISES/PROBLEMS

6. In Euclidean geometry, the measure of an exterior angle of a triangle is equal to the sum of the measures of the two nonadjacent interior angles. Discuss whether this relationship holds in the geometry of the sphere. Include a picture with your explanation.
7. We can define a three-point geometry by letting the set of points $\{A, B, C\}$ represent the plane. Then the lines will be pairs of points. We will accept the following three postulates.

   Postulate 1: There are exactly three points.

   Postulate 2: Every pair of points is on a line.

   Postulate 3: Every line contains exactly two points.

Answer the following questions about this three-point geometry. Justify your answers.

a. How many lines are in this geometry?

b. If two lines intersect, how many points do they have in common?

c. Are there parallel lines in this geometry?

8. The length of a side of an equilateral triangle is 8.

a. Draw a sketch of the triangle and find the area of the triangle using Euclidean distance.

b. Draw a sketch of the triangle in taxicab geometry. (Be careful! More than one triangle is possible.)

c. Find the area of the triangle using taxicab distance.

9. Suppose that two triangles satisfy the SSS Similarity Theorem using taxicab distance. Must the two triangles be similar? Justify your answer or provide a counterexample.

10. Prove that if a quadrilateral is both a Saccheri and a Lambert quadrilateral, then it is a rectangle.

APPENDIX 1 Getting Started: Making Proofs

HOW TO USE THIS SUMMARY

Suppose that you need to show that two segments are congruent. Check the list of suggestions under "Congruent Segments." The third suggestion indicates that you may be able to show that the segments are congruent by showing that they are corresponding parts of congruent triangles. Another reminder is that opposite sides of a parallelogram are congruent. Can you show that the two segments are opposite sides of a parallelogram?

Be sure to note the page number after each result so that you do not use a result that has not yet been proven in your course.

Congruent Angles

Complements of the same or congruent angles 44
Supplements of the same or congruent angles 44
Vertical angles 190
Corresponding parts of congruent triangles 198
Base angles of an isosceles triangle 207
Alternate interior angles formed by a transversal and two parallel lines 245
Corresponding angles formed by a transversal and two parallel lines 246
Alternate exterior angles formed by a transversal and two parallel lines 249
Opposite angles of a parallelogram 258
Base angles of an isosceles trapezoid 273

Congruent Segments

Sides of an equilateral triangle 52
Adjacent sides of a rhombus 55
Corresponding parts of congruent triangles 198
Sides opposite congruent angles in a triangle 209
Opposite sides of a parallelogram 258
Diagonals of a rectangle 271
Sides opposite congruent angles in a trapezoid 274
Diagonals of an isosceles trapezoid 277

Congruent Triangles

SAS Congruence 197
LL Congruence 197
HL Congruence 198
ASA Congruence 199
SSS Congruence 201

AAS Congruence 252

HA Congruence 253

Formed by a diagonal of a parallelogram 258

Image under an isometry 448

Perpendicular Lines and Right Angles

Two lines that intersect to form congruent supplementary angles 44

Converse of the Pythagorean theorem: Show $a^2 + b^2 = c^2$ where a, b, and c are the lengths of the sides of a triangle 201

A bisector of the vertex angle of an isosceles triangle 210

A line containing two points: the midpoint of a segment and a point equidistant from the endpoints of a segment 211

Median from the vertex angle in an isosceles triangle 215

Diagonals of a rhombus 264

Angle inscribed in a semicircle 367

A radius and tangent line at the point of tangency 385

Lines whose slopes have a product of negative one 441

Parallel Lines

Congruent alternate interior angles 242

Lines perpendicular to the same line 243

Congruent corresponding angles 243

Supplementary interior angles on the same side of a transversal 243

Supplementary exterior angles on the same side of a transversal 249

Midsegment Theorem 322

Equal slopes or both slopes undefined 439

Parallelogram

A quadrilateral with opposite sides parallel 55

A quadrilateral with opposite sides congruent 259

A quadrilateral with opposite angles congruent 259

A quadrilateral with two sides congruent and parallel 260

A quadrilateral with diagonals that bisect each other 262

A quadrilateral with four congruent sides (a rhombus) 263

A quadrilateral with four congruent angles (a rectangle) 270

Midquad Theorem 323

Rectangle

A quadrilateral with four right angles 55

A parallelogram with one right angle 271

A parallelogram with congruent diagonals 271

Rhombus

A quadrilateral with four congruent sides 55

A parallelogram with diagonals that are perpendicular 264

A quadrilateral with diagonals that are perpendicular bisectors of each other 264

A parallelogram in which the diagonals bisect the opposite angles 265

Trapezoid

A quadrilateral with exactly one pair of parallel sides 55

Isosceles Trapezoid

A trapezoid with two congruent sides 55

A trapezoid with congruent base angles 274

A trapezoid with congruent diagonals 277

Complementary Angles

Two acute angles of a right triangle 44

Two adjacent angles whose nonadjacent sides form a right angle 44

Supplementary Angles

Two adjacent angles whose nonadjacent sides form a straight angle 44

Two interior angles on the same side formed by a transversal and two parallel lines 246

Proportional Segments

Corresponding parts of similar triangles 307

Side Splitting Theorem 321

Similar Triangles

AA Similarity 309

Right triangles whose acute angles are congruent 310

SAS Similarity 310

LL Similarity 311

SSS Similarity 312

Side Splitting Theorem 321

Image under a similitude 502

Congruent Central Angles

Intercept the same or congruent arcs 363

Intercept the same or congruent chords 380

Congruent Arcs

Congruent Inscribed Angles

Congruent Chords

Tangent Lines

APPENDIX 2 Summary of Postulates, Theorems, and Corollaries

Post 2.1 Every line contains at least two distinct points. 40

Post 2.2 Two points are contained in one and only one line. 40

Post 2.3 If two points are in a plane, then the line containing these points is also in the plane. 40

Post 2.4 Three noncollinear points are contained in one and only one plane, and every plane contains at least three noncollinear points. 40

Post 2.5 In space, there exist at least four points that are not all coplanar. 40

Post 2.6 **The Ruler Postulate**: Every line can be made into an exact copy of the real number line using a 1-to-1 correspondence. 41

Post 2.7 **The Protractor Postulate**: If we place one ray of an angle at 0° on a protractor and the vertex at the midpoint of the bottom edge, then there is a 1-to-1 correspondence between all other rays that can serve as the second side of the angle and the real numbers between 0° and 180° inclusive, as indicated by a protractor. 42

Thm 2.1 **Angle Measure in a Triangle (informal)**: The sum of the measures of the angles in a triangle is 180°. 66

Thm 2.2 **Angle Measure in a Polygon**: The sum of the measures of the vertex angles in a polygon with n sides is $(n - 2)180°$. 68

Thm 2.3 **Vertex Angle Measure in a Regular Polygon**: The measure of a vertex angle in a regular n-gon is $\frac{(n - 2)180°}{n}$. 69

Thm 3.1 **Perimeters of Common Quadrilaterals**

Figure	Perimeter
Square with sides of length s	4s
Rectangle with sides of lengths a and b	$2a + 2b$
Parallelogram with sides of lengths a and b	$2a + 2b$
Rhombus with sides of length s	$4s$
Kite with sides of lengths a and b	$2a + 2b$

107

Post 3.1 **Circumference of a Circle**: The circumference of a circle is the product of π and the diameter of the circle.

$$C = \pi d = 2\pi r$$ 108

Post 3.2 **Area of a Region Enclosed by a Simple Closed Curve**:

a. For every simple closed curve and unit square, there is a positive real number that gives the number of unit squares (and parts of unit squares) that exactly tessellate the region enclosed by the simple closed curve.

b. The area of a region enclosed by a simple closed curve is the sum of the areas of the smaller regions into which the region can be subdivided. 109

Post 3.3 Any two triangles whose corresponding angles and sides are congruent have the same area. 110

Thm 3.2 **Area of a Right Triangle**: The area of a right triangle is half the product of the lengths of its legs.

$$A = \frac{1}{2}ab$$ 111

Thm 3.3 **Area of a Triangle**: The area of a triangle is half the product of the length of one side and the height to that side.

$$A = \frac{1}{2}bh$$ 111

Thm 3.4 **Area of a Parallelogram**: The area of a parallelogram is equal to the product of the length of one of its sides and the height to that side.

$$A = bh$$ 122

Thm 3.5 **Area of a Trapezoid**: The area of a trapezoid is equal to half the product of the height and the sum of the lengths of its two bases.

$$A = \frac{1}{2}h(b_1 + b_2)$$ 123

Thm 3.6 **Area of a Regular Polygon**: The area of a regular polygon is equal to half the product of the perimeter of the polygon and the perpendicular distance from its center to one of its sides.

$$A = \frac{1}{2}Ph$$ 124

Thm 3.7 **Area of a Circle**: The area of a circle is equal to the product of π and the square of the radius.

$$A = \pi r^2$$ 125

Thm 3.8 **Pythagorean Theorem**: The sum of the squares of the lengths of the legs of a right triangle is equal to the square of the length of the hypotenuse.

$$a^2 + b^2 = c^2$$ 134

Thm 3.9 **30°-60° Right Triangle**: In a 30°-60° right triangle, the length of the leg opposite the 30° angle is half the length of the hypotenuse. The length of the leg opposite the 60° angle is $\frac{\sqrt{3}}{2}$ times the length of the hypotenuse. 138

Thm 3.10 **45°-45° Right Triangle**: In a 45°-45° right triangle, the length of the hypotenuse is $\sqrt{2}$ times the length of a leg. 139

Thm 3.11 **Surface Area of a Right Prism**: The surface area of a right prism is the sum of twice the area of its base plus the product of the perimeter of the base and the height of the prism.

$$SA = 2A + Ph$$ 148

Thm 3.12 **Surface Area of a Right Regular Pyramid**: The surface area of a right pyramid is the sum of the area of its base plus half the product of the perimeter of the base and the slant height.

$$SA = A + \frac{1}{2}Pl$$ 149

Thm 3.13 **Surface Area of a Right Circular Cylinder**: The surface area of a right circular cylinder is twice the area of its base plus the circumference of its base times its height.

$$SA = 2A + Ch = 2\pi r^2 + 2\pi rh$$ 149

Thm 3.14 **Surface Area of a Right Circular Cone**: The surface area of a right circular cone is the sum of the area of its base plus half the circumference of its base times its slant height.

$$SA = A + \frac{1}{2}Cl = \pi r(r + l)$$ 150

Thm 3.15 **Surface Area of a Sphere**: The surface area of a sphere is 4π times the square of its radius.

$$SA = 4\pi r^2$$ 151

Post 3.4 **Volume Postulate**

a. For every polyhedron and unit cube, there is a real number that gives the number of unit cubes (and parts of unit cubes) that exactly fill the region enclosed by the polyhedron.

b. The volume of the region enclosed by a polyhedron is the sum of the volumes of the smaller regions into which the region can be subdivided. 159

Thm 3.16 **Volume of a Right Rectangular Prism**: The volume of a right rectangular prism is the product of its length, width, and height.

$$V = lwh$$ 161

Cor 3.17 **Volume of a Cube**: The volume of a cube is the cube of the length of its side.

$$V = s^3$$ 161

Thm 3.18 **Volume of a Pyramid**: The volume of a pyramid is one-third the product of the area of its base and its height.

$$V = \frac{1}{3}Ah$$ 162

Thm 3.19 **Volume of a Right Circular Cylinder**: The volume of a right circular cylinder is the product of the area of its base and its height.

$$V = Ah = \pi r^2 h$$ 163

Thm 3.20 **Volume of a Right Circular Cone**: The volume of a right circular cone is one-third the product of the area of its base and its height.

$$V = \frac{1}{3}Ah = \frac{1}{3}\pi r^2 h$$ 163

Thm 3.21 **Volume of a Sphere**: The volume of a sphere is $\frac{4}{3}\pi$ times the cube of its radius.

$$V = \frac{4}{3}\pi r^3$$ 164

Thm 4.1 If x and y are the measures of a pair of vertical angles, then $x = y$. 190

Post 4.1 **SAS Congruence Postulate**: If two sides and the included angle of one triangle are congruent, respectively, to two sides and the included angle of another triangle, then the two triangles are congruent. 197

Thm 4.2 **LL Congruence Theorem**: If two legs of one right triangle are congruent, respectively, to two legs of another right triangle, then the two triangles are congruent. 197

Thm 4.3 **HL Congruence Theorem**: If the hypotenuse and a leg of one right triangle are congruent, respectively, to the hypotenuse and a leg of another right triangle, then the two triangles are congruent. 198

Post 4.2 **ASA Congruence Postulate**: If two angles and the included side of one triangle are congruent, respectively, to two angles and the included side of another triangle, then the two triangles are congruent. 199

Post 4.3 **SSS Congruence Postulate**: If three sides of one triangle are congruent, respectively, to three sides of another triangle, then the two triangles are congruent. 201

Thm 4.4 **Converse of the Pythagorean Theorem**: If the sum of the squares of the lengths of two sides of a triangle equals the square of the third side, then the triangle is a right triangle. 201

Thm 4.5 In an isosceles triangle, the angles opposite the congruent sides are congruent. 207

Cor 4.6 Every equilateral triangle is equiangular. 208

Thm 4.7 If two angles of a triangle are congruent, then the sides opposite those angles are congruent. 209

Cor 4.8 Every equiangular triangle is equilateral. 209

Thm 4.9 In an isosceles triangle, the ray that bisects the vertex angle bisects the base and is perpendicular to it. 210

Thm 4.10 **Perpendicular Bisector Theorem**: A point is on the perpendicular bisector of a line segment if and only if it is equidistant from the endpoints of the segment. 211

Thm 4.11 **Exterior Angle Theorem**: The measure of an exterior angle of a triangle is greater than the measure of either of the nonadjacent interior angles. 212

Thm 5.1 If two lines are cut by a transversal to form a pair of congruent alternate interior angles, then the lines are parallel. 242

Cor 5.2 If two lines are both perpendicular to a transversal, then the lines are parallel. 243

Cor 5.3 If two lines cut by a transversal form a pair of congruent corresponding angles with the transversal, then the lines are parallel. 243

Cor 5.4 If two lines cut by a transversal form a pair of supplementary interior angles on the same side of the transversal, then the lines are parallel. 243

Post 5.1 **The Parallel Postulate**: Given a line l and a point P not on l, there is only one line m containing P such that $l \parallel m$. 244

Thm 5.5 If a pair of parallel lines is cut by a transversal, then the alternate interior angles formed are congruent. 245

Cor 5.6 If two lines are parallel and a line is perpendicular to one of the two lines, then it is perpendicular to the other line. 246

Cor 5.7 If two parallel lines are cut by a transversal, then each pair of corresponding angles formed is congruent. 246

Cor 5.8 If two parallel lines are cut by a transversal, then both pairs of interior angles on the same side of the transversal are supplementary. 246

Thm 5.9 **Angle Sum in a Triangle Theorem**: The sum of the angle measures in a triangle is 180°. 250

Cor 5.10 **The Exterior Angle Theorem**: The measure of an exterior angle of a triangle is equal to the sum of the measures of the two nonadjacent interior angles. 251

Cor 5.11 **AAS Congruence Theorem**: If two angles and a side of one triangle are congruent, respectively, to two angles and a side of another triangle, then the two triangles are congruent. 252

Cor 5.12 **HA Congruence Theorem**: If the hypotenuse and an acute angle of one right triangle are congruent, respectively, to the hypotenuse and an acute angle of another right triangle, then the two triangles are congruent. 253

Thm 5.13 **Angle Bisector Theorem**: A point is on the bisector of an angle if and only if it is equidistant from the sides of the angle. 253

Thm 5.14 A diagonal of a parallelogram forms two congruent triangles. 258

Cor 5.15 In a parallelogram, the opposite sides are congruent and the opposite angles are congruent. 258

Cor 5.16 Parallel lines are everywhere equidistant. 258

Thm 5.17 If both pairs of opposite sides of a quadrilateral are congruent, then the quadrilateral is a parallelogram. 259

Thm 5.18 If both pairs of opposite angles of a quadrilateral are congruent, then the quadrilateral is a parallelogram. 259

Thm 5.19 If a quadrilateral has two sides that are parallel and congruent, then it is a parallelogram. 260

Thm 5.20 The diagonals of a parallelogram bisect each other. 261

Thm 5.21 If the diagonals of a quadrilateral bisect each other, then it is a parallelogram. 262

Thm 5.22 Every rhombus is a parallelogram. 263

Thm 5.23 The diagonals of a rhombus are perpendicular to each other. 264

Thm 5.24 If the diagonals of a parallelogram are perpendicular to each other, then the parallelogram is a rhombus. 264

Thm 5.25 A parallelogram is a rhombus if and only if its diagonals bisect the opposite angles. 265

Thm 5.26 Every rectangle is a parallelogram. 270

Thm 5.27 A parallelogram with one right angle is a rectangle. 271

Thm 5.28 The diagonals of a rectangle are congruent. 271

Thm 5.29 If a parallelogram has congruent diagonals, then it is a rectangle. 272

Thm 5.30 The base angles of an isosceles trapezoid are congruent. 273

Thm 5.31 If the base angles of a trapezoid are congruent, then it is isosceles. 274

Thm 5.32 If three parallel lines form congruent segments on one transversal, then they form congruent segments on any transversal. 282

Thm 6.1 **Cross Multiplication Theorem**: $\frac{a}{b} = \frac{c}{d}$ if and only if $ad = bc$. 300

Thm 6.2 $\frac{a}{b} = \frac{c}{d}$ if and only if $\frac{d}{b} = \frac{c}{a}$.

$\frac{a}{b} = \frac{c}{d}$ if and only if $\frac{a}{c} = \frac{b}{d}$.

$\frac{a}{b} = \frac{c}{d}$ if and only if $\frac{b}{a} = \frac{d}{c}$. 301

Thm 6.3 If $\frac{a}{b} = \frac{b}{c}$, then b is the geometric mean (mean proportional) of a and c. 302

Post 6.1 **AAA Similarity Postulate**: Two triangles are similar if and only if three angles of one triangle are congruent, respectively, to three angles of the other triangle. 309

Thm 6.4 **AA Similarity Theorem**: Two triangles are similar if two angles of one triangle are congruent, respectively, to two angles of the other triangle. 309

Cor 6.5 Two right triangles are similar if an acute angle of one triangle is congruent to an acute angle of the other triangle. 310

Thm 6.6 **SAS Similarity Theorem**: Two triangles are similar if two sides of one triangle are proportional, respectively, to two sides of another triangle and the angles included between the sides are congruent. 310

Cor 6.7 **LL Similarity**: Two right triangles are similar if the legs of one triangle are proportional, respectively, to the legs of the other triangle. 311

Thm 6.8 **SSS Similarity Theorem**: Two triangles are similar if three sides of one triangle are proportional to three sides of another triangle. 312

Thm 6.9 In a right triangle, the altitude to the hypotenuse is the mean proportional between the two segments formed by the altitude on the hypotenuse. 321

Thm 6.10 **Side Splitting Theorem**: A line parallel to one side of a triangle forms a triangle similar to the original triangle and divides the other two sides of the triangle into proportional corresponding segments. 321

Thm 6.11 **Midsegment Theorem**: A midsegment of a triangle is parallel to the third side and is half its length. 322

Cor 6.12 **Midquad Theorem**: If $ABCD$ is any quadrilateral and $E, F, G,$ and H are midpoints of the sides $\overline{AB}, \overline{BC}, \overline{CD}$, and $\overline{AD}$, respectively, then $EFGH$ is a parallelogram. 323

Thm 6.13 In a right triangle with acute angle A, $\frac{\sin A}{\cos A}$ is $\tan A$. 334

Thm 6.14 In a right triangle with acute angle A, $\sin^2 A + \cos^2 A = 1$. 334

Thm 6.15 **Law of Sines**: For $\triangle ABC$ with side lengths $a, b,$ and c,

$$\frac{\sin A}{a} = \frac{\sin B}{b} = \frac{\sin C}{c}$$ 342

Thm 6.16 **Law of Cosines**: For $\triangle ABC$ with side lengths $a, b,$ and c,

$$a^2 = b^2 + c^2 - 2bc\cos A$$
$$b^2 = a^2 + c^2 - 2ac\cos B$$
$$c^2 = a^2 + b^2 - 2ab\cos C$$ 344

Post 7.1 Length of an Arc: The ratio of the length, l, of an arc of a circle to the circumference, C, of the circle equals the ratio of the measure of the central angle, $x°$, of the arc to $360°$. $l = \frac{x}{360°}(2\pi r)$ 364

Post 7.2 **Area of a Sector**: The ratio of the area of a sector of a circle to the area of the circle is equal to the ratio of the measure of the central angle, $x°$, of the sector to $360°$.

$$A_{\text{sector}} = \frac{x}{360°}(\pi r^2)$$ 365

Thm 7.1 **Inscribed Angle Theorem**: The measure of an inscribed angle in a circle is equal to half the measure of its intercepted arc. 365

Cor 7.2 Inscribed angles that intercept the same arcs (or congruent arcs) are congruent. 367

Cor 7.3 An angle is inscribed in a semicircle if and only if it is a right angle. 367

Thm 7.4 The perpendicular bisector of a chord contains the center of the circle. 374

Cor 7.5 The intersection of the perpendicular bisector of any two nonparallel chords of a circle is the center of the circle. 374

Cor 7.6 If two circles, O and O', intersect in two points, A and B, then the line containing O and O' is the perpendicular bisector of $\overline{AB}$. 375

Thm 7.7 If two chords intersect, then the measure of any one of the vertical angles formed is equal to half the sum of the measures of the two arcs intercepted by the two vertical angles. 376

Thm 7.8 If two chords of a circle intersect, the product of the lengths of the two segments formed on one chord is equal to the product of the lengths of the two segments formed on the other chord. 377

Thm 7.9 If two secants intersect outside a circle, the measure of the acute angle formed is half the difference of the measures of the intercepted arcs. 383

Thm 7.10 If two secants intersect outside a circle, then the product of the lengths of the segments formed on one secant is equal to the product of the lengths of the corresponding segments on the other secant. 384

Thm 7.11 A radius or diameter of a circle is perpendicular to a tangent line at its point of tangency. 385

Thm 7.12 The measure of an angle formed when a chord intersects a tangent line at the point of tangency is half the measure of the arc intercepted by the chord and the tangent line. 386

Thm 7.13 If a secant and a tangent line intersect outside a circle, the measure of the angle formed is half the measure of the larger intercepted arc minus the measure of the smaller intercepted arc. 386

Thm 7.14 If two tangent lines intersect outside a circle, the measure of the angle formed is half the measure of the larger intercepted arc minus the measure of the smaller intercepted arc. 387

Cor 7.15 If two tangent lines are drawn to a circle from the same point in the exterior of the circle, the distances from the common point to the points of tangency are equal. 388

Thm 7.16 If we draw tangent and secant lines to a circle from the same point in the exterior of the circle, the length of the tangent segment is the mean proportional between the length of the external secant segment and the length of the secant. 389

Thm 7.17 **Circumcenter of a Triangle:** The perpendicular bisectors of the sides of a triangle intersect in a single point, the circumcenter. 398

Thm 7.18 **Orthocenter of a Triangle:** The altitudes of a triangle intersect in a single point, the orthocenter. 399

Thm 7.19 **Incenter of a Triangle:** The angle bisectors of a triangle intersect in a single point, the incenter. 400

Thm 7.20 **Centroid of a Triangle:** The medians of a triangle intersect in a single point, the centroid, which is two-thirds of the way from any vertex to the other endpoint of the median from that vertex. 401

Thm 8.1 **Distance Formula:** If $P(x_1, y_1)$ and $Q(x_2, y_2)$ are two points in the coordinate plane, then
$PQ = \sqrt{(x_2 - x_1)^2 + (y_2 - y_1)^2}$ 424

Thm 8.2 **Test for Collinearity:** The points P, Q, and R are collinear, with Q between P and R, if and only if $PQ + QR = PR$. 427

Thm 8.3 **Midpoint Formula:** If $P(x_1, y_1)$ and $Q(x_2, y_2)$ are endpoints of a line segment, then the midpoint of $\overline{PQ}$ is
$M\left(\frac{x_1 + x_2}{2}, \frac{y_1 + y_2}{2}\right)$. 428

Thm 8.4 **Slopes of Parallel Lines:** Two lines in a coordinate plane are parallel if and only if

a. their slopes are equal, or

b. their slopes are undefined. 439

Thm 8.5 **Slopes of Perpendicular Lines:** Two lines in a coordinate plane are perpendicular if and only if (a) one line is horizontal and the other is vertical, or (b) the product of their slopes equals -1. 441

Thm 8.6 **Forms of Equations of Lines:** Every nonvertical line in the plane can be expressed in either of the following two forms where m represents the slope of the line, b is the y-intercept of the line, and (x_1, y_1) is a point on the line.

Point-slope form: $y - y_1 = m(x - x_1)$

Slope-intercept form: $y = mx + b$ 448

Thm 8.7 **Solutions of a System of Two Linear Equations:** Two linear equations have 0, 1, or infinitely many simultaneous solutions if and only if the lines they represent are parallel, intersect in one point, or coincide, respectively. 451

Thm 8.8 **Equation of a Circle:** The circle with center (h, k) and radius r has the equation $(x - h)^2 + (y - k)^2 = r^2$. 453

Thm 9.1 **Properties of Translations**

1. Translations take lines to lines, rays to rays, and line segments to line segments.
2. Translations preserve distance.
3. Translations preserve angle measure.
4. Translations preserve perpendicularity.
5. Translations preserve parallelism.

480

Thm 9.2 **Properties of Rotations**

1. Rotations take lines to lines, rays to rays, and line segments to line segments.
2. Rotations preserve distance.

3. Rotations preserve angle measure.
4. Rotations preserve perpendicularity.
5. Rotations preserve parallelism. 484

Thm 9.3 **Properties of Reflections**

1. Reflections take lines to lines, rays to rays, and line segments to line segments.
2. Reflections preserve distance.
3. Reflections preserve angle measure.
4. Reflections preserve perpendicularity.
5. Reflections preserve parallelism. 485

Thm 9.4 **Properties of Glide Reflections**

1. Glide reflections take lines to lines, rays to rays, and line segments to line segments.
2. Glide reflections preserve distance.
3. Glide reflections preserve angle measure.
4. Glide reflections preserve perpendicularity.
5. Glide reflections preserve parallelism. 487

Thm 9.5 **Congruent Triangles and Isometries**: Two triangles are congruent if and only if there is an isometry that takes one triangle to the other. 488

Thm 9.6 **Congruent Polygons and Isometries**: Two polygons are congruent if and only if there is an isometry that takes one polygon to other. 488

Thm 9.7 **Properties of Size Transformations**

1. Size transformations take lines to lines, rays to rays, and line segments to line segments.
2. Size transformations preserve orientation.
3. Size transformations preserve ratios of distances.
4. Size transformations preserve angle measure.
5. Size transformations preserve perpendicularity.
6. Size transformations take lines to parallel lines. 501

Thm 9.8 **Similar Triangles and Similitudes**: $\triangle ABC \sim \triangle A'B'C'$ if and only if there is a similitude that takes $\triangle ABC$ to $\triangle A'B'C'$. 502

Thm 9.9 **Similar Polygons and Similitudes**: Two polygons are similar if and only if there is a similitude that takes one to the other. 504

Answers to Odd-Numbered Problems, Chapter Tests, and Topics Sections

SECTION 1.1

1. 13 **3.** 5

5. One solution is shown. Many solutions are possible.

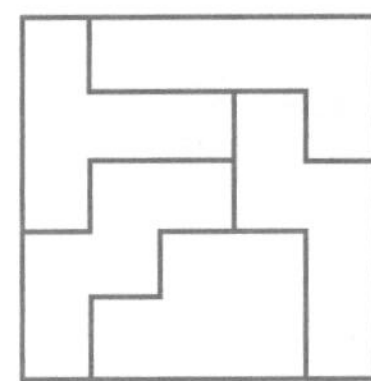

7. a. 4 **b.** 2 **c.** 5

9. One answer is shown. Other answers are possible.

Let R = red, BL = blue, G = green, W = white, and BK = black.

BL	G	R	W
R	W	BK	BL
BK	BL	G	R
G	R	W	BK

11. 9

13.

15. 40

17. a. 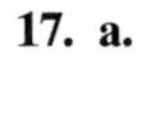 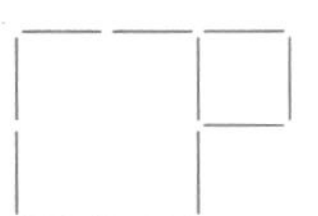 **b.** 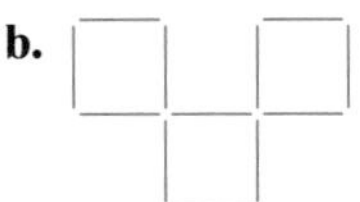

19. No; There are four points where three lines meet. For at least three of these points, your tracing will take you to the point, away from the point, and then back to the point. There is no way to leave the point without retracing a line.

21. One answer is shown. Other answers are possible. The numbers indicate the order in which the segments are traced.

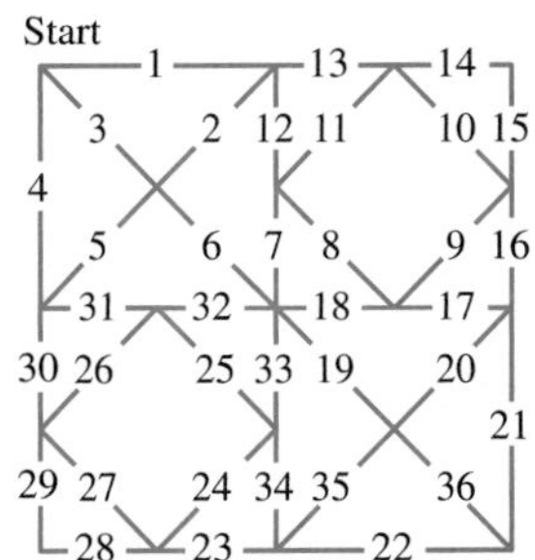

23. 27 inches by 9 inches **25.** 132 feet

27. It cannot be done unless he uses fewer than all of the blocks. Because the sum of all 10 heights is 55 cm, two equal towers would each be 27.5 cm tall, which is impossible.

29.

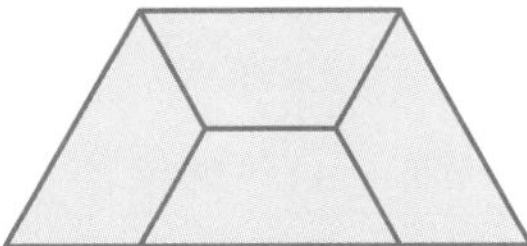

31. Make the following cut.

33. 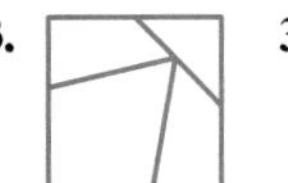 **35.**  

37. a. **b.**

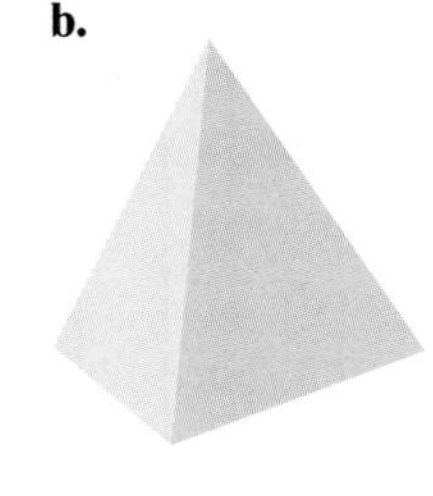

39. Many solutions are possible. She cannot begin and end at the same intersection. If she starts at B, then she must end at C, because there are an odd number of streets available at those intersections. If she starts at C, then she must end at B. If she starts at any other point, she must retrace a street. One route she can follow is $CBECDEFAB$, but other routes are possible.

41. a. No, there will always be at least one house not connected to one of the utilities.

b. Yes, one solution is shown next.

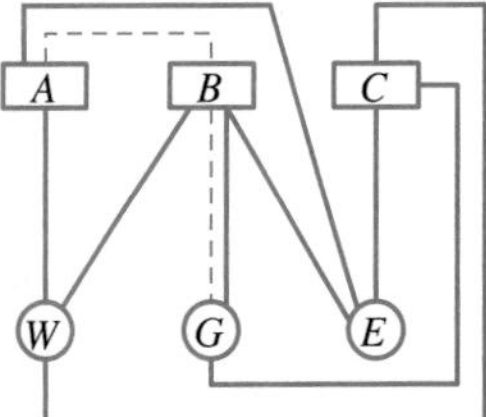

43. 15 inches

45. a. 4 **b.** 3 **c.** 2

SECTION 1.2

1. a.

X	X	X	X	X
X	X	X	X	X
X	X	X	X	X
X	X	X	X	X

b. $1 + 2 + 3 + 4 = \frac{1}{2}(4 \times 5)$

c. 1275, 2850 **d.** $\frac{n(n+1)}{2}$

3. $4, 6, 8, 2n + 2$ **5.** $6, 9, 12, 3n + 3$

7. $\frac{n(n+1)}{2}$ **9.** $2n^2 - n$

11. a. ii **b.** i **c.** i **d.** ii

13. a. **b.** **c.**

d. **e.**

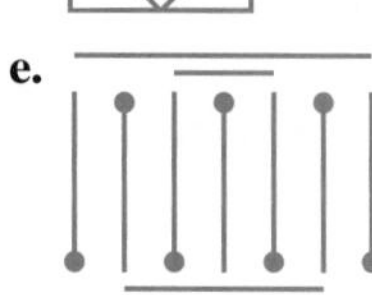

15. a. $36, 49, 144, n^2$ **b.** $-12, -14, -24, -2n$

c. $243, 729, 177147, 3^{n-1}$ **d.** $13, 15, 25, 2n + 1$

17. 40 **19.** 15 **21.** 55, 89, 144, 233, 377

23. 268 **25.** $28, 3n - 2$

27. 100 squares, $1 + 3 + 5 + \cdots + (2n - 1) = n^2$ squares

29. $46, 5n - 4$ **31.** 220

33. a. 27 **b.** "D" column **35.** $\frac{1}{3}$

37. a. Black: 1, 3, 9, and 27 White: 0, 1, 4, and 13 **b.** 81, 40 **c.** 3^{n-1} **39.** 7

41. a.

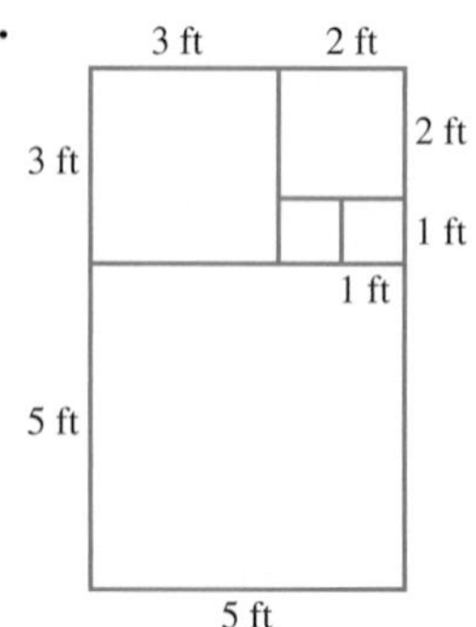

b. 2 by 3, 3 by 5, 5 by 8, 8 by 13, and 13 by 21

c. After the first two square slabs are poured, the side length of a new square is the sum of the side lengths of the previous two squares.

d. 10

Chapter Review Problems

SECTION 1.1

1.

2. 5 ways **3.** 8 inches

4.

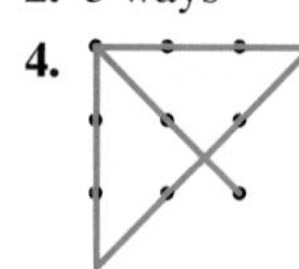

5.

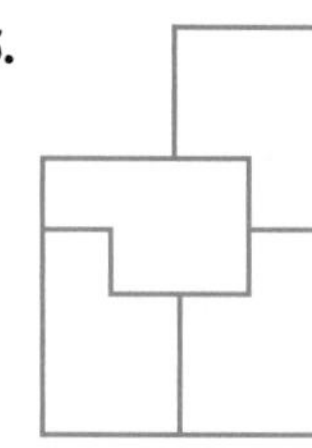

SECTION 1.2

1. a. 243; 6561; 1,594,323 **b.** 47, 58, 69

c.

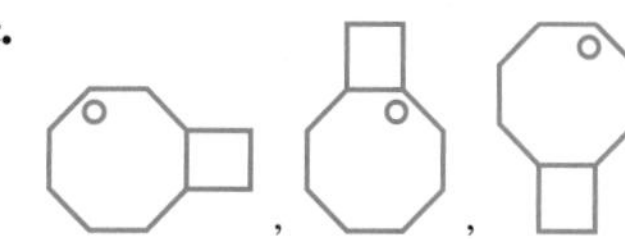

2. 24 ways **3.** 6

4. $6, 11, 16, 5n + 1$ **5.** $3, 6, 10, 36, \frac{n(n-1)}{2}$

CHAPTER 1 TEST

1. Draw a Picture, Guess and Test, Use a Variable, Look for a Pattern, Make a Table, Solve a Simpler Problem

2. Use a Variable

3.

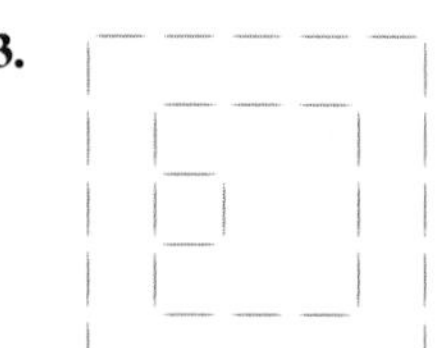

4.

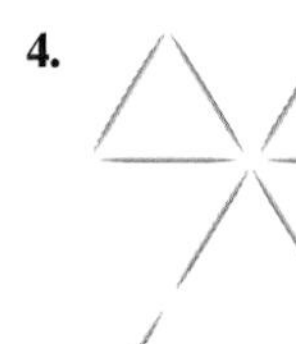

5. 37°, 37°, 106°

6.

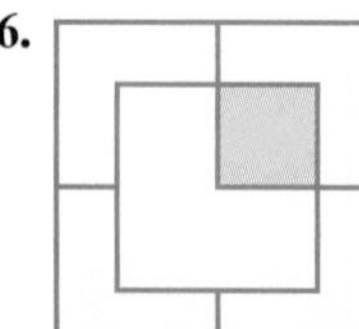

7. a. 7 pieces **b.** 11 pieces **c.** 16 pieces

8. 109 squares, $1 + 2(2) + 2(3) + \cdots + 2(n) = n^2 + n - 1$

9. a. 85, $a_n = 9n - 5$ **b.** 1025, $a_n = 2^n + 1$

10. a. 9 **b.** 36 **c.** 100 **d.** $1 + 8 + 27 + \cdots + n^3$

11.

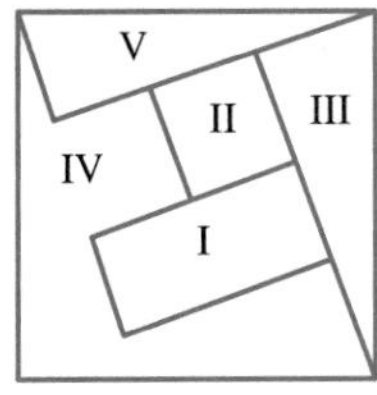

SECTION 2.1

1. 10 line segments: $\overline{AB}, \overline{AC}, \overline{AD}, \overline{AE}, \overline{BC}, \overline{BD}, \overline{BE}, \overline{CD}, \overline{CE}, \overline{DE}$

3. 8 rays: $\overrightarrow{AB}, \overrightarrow{BC}, \overrightarrow{CD}, \overrightarrow{DE}, \overrightarrow{ED}, \overrightarrow{DC}, \overrightarrow{CB}, \overrightarrow{BA}$

5. 9 ways: $\overleftrightarrow{AB}$ $\overleftrightarrow{AC}$ $\overleftrightarrow{AD}$ $\overleftrightarrow{BC}$ $\overleftrightarrow{BD}$ $\overleftrightarrow{BE}$ $\overleftrightarrow{CD}$ $\overleftrightarrow{CE}$ $\overleftrightarrow{DE}$

7. a. 4.34 **b.** 1.91 **c.** 6.65 **d.** 4.22

9. a. $\angle BAC$ or $\angle CAB$, $\angle BAD$ or $\angle DAB$, $\angle CAD$ or $\angle DAC$

b. $\angle BAC$ and $\angle CAD$

11. a. 10 **b.** 4 **c.** 6

13. a. 60° acute **b.** 114° obtuse **c.** 90° right

15. $\angle AFB = 60°$, $\angle CFD = 35°$

17. $\angle 1 = 54°$, $\angle 2 = 126°$ **19.** $\angle A = 72°$

21. 3 lines: 7 regions, 4 lines: 11 regions, 5 lines: 16 regions, 10 lines: 56 regions, n lines: $\dfrac{n(n+1)}{2} + 1$ regions

23. a. 19.05° **b.** 12.11° **c.** 247.933° **d.** 3.533°

25. a. 31°36′ **b.** 95°45′ **c.** 241°19′ **d.** 25°29′

27. a. 16° 30′ 36″ **b.** 0° 19′ 48″ **c.** 91° 59′ 35″ **d.** 58° 1′ 44″

29. a.

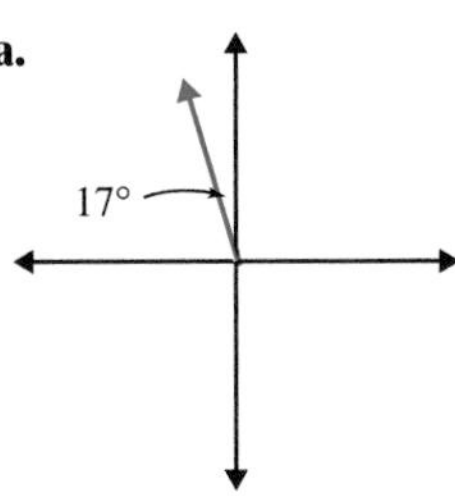

The bearing of $\overline{BA}$ is S17°E.

b.

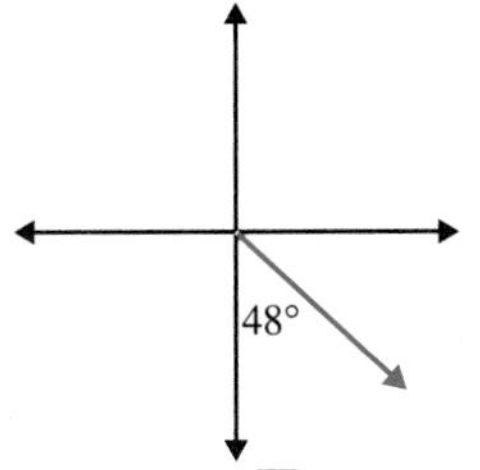

The bearing of $\overline{BA}$ is N48°W.

c.

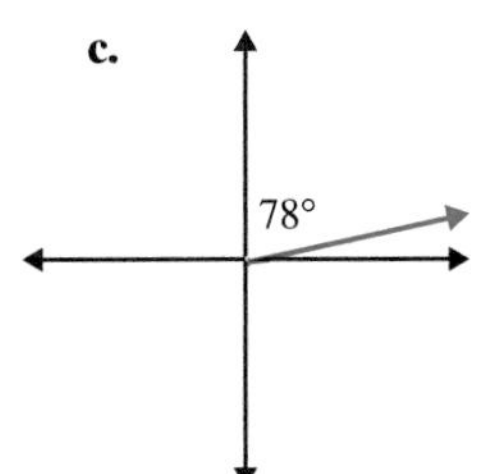

The bearing of $\overline{BA}$ is S78°W.

d.

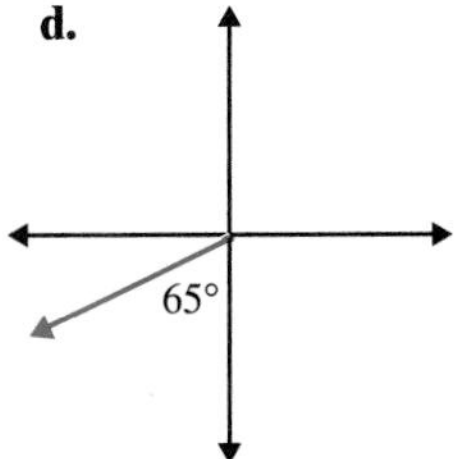

The bearing of $\overline{BA}$ is N65°E.

31. N 72° E

33.

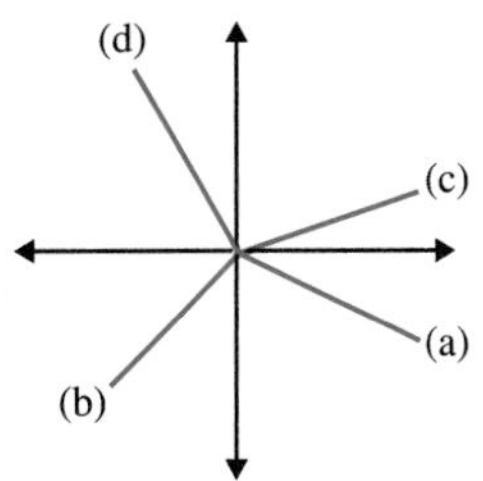

35. a. 40° **b.** 320° **c.** 277°43′ **d.** 155°25′

37. 109°6′

39. a. 30°, 360° **b.** 6°, 0.5° **c.** 145°, 158° **d.** 12:16 A.M.

41.

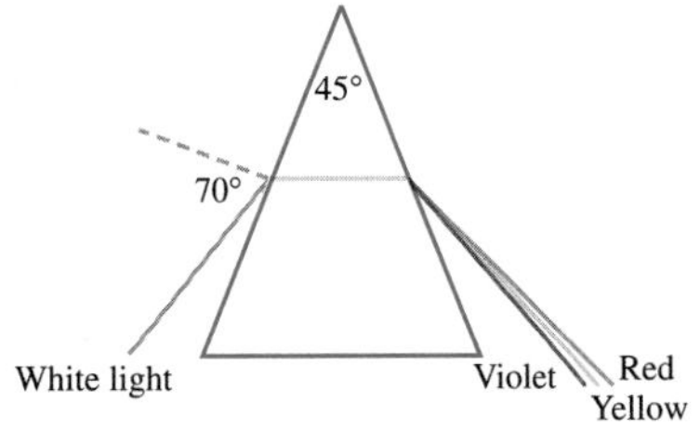

SECTION 2.2

1. a. $\triangle DEF, \triangle GHI, \triangle KJL$ **b.** $\triangle GHI$

c. $\triangle ABC, \triangle MNO$ **d.** $\triangle ABC, \triangle JKL$

e. $\triangle DEF$

3. a. One of many correct examples follows.

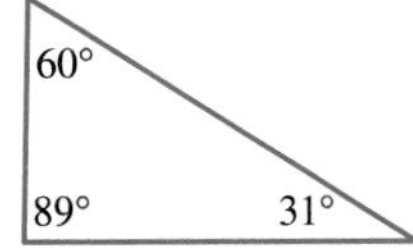

b. One of many correct examples follows.

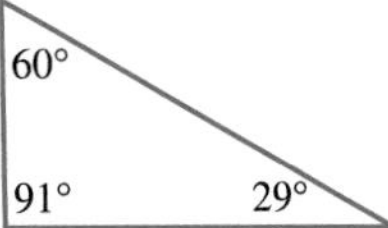

c. Not possible. An equilateral triangle has equal angles. If it has one right angle it must have three, which is impossible.

d. Not possible. If a triangle is equilateral, all sides would have the same length. Thus, it cannot be scalene.

5. a. Four triangles: 2-9-9, 4-8-8, 6-7-7, 8-6-6

b. Four triangles: 3-8-9, 4-7-9, 5-6-9, 5-7-8

c. None

7. a. 20 **b.** 6 **c.** 10 **d.** 2 **e.** 8

9. Many simple closed curves are possible

11. a.

b.

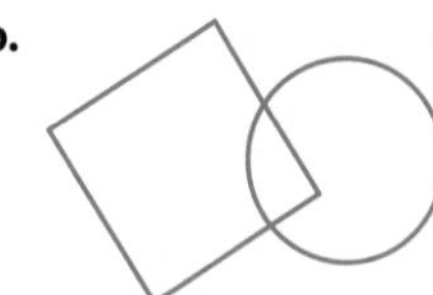

c.

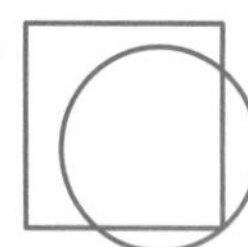

d.

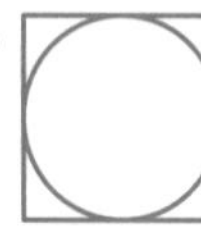

13. a. Yes **b.** No—not all line segments

c. No—part is retraced **d.** No—part is crossed

15. a.

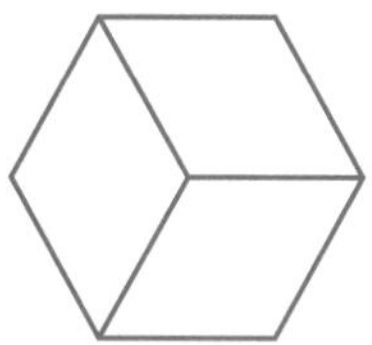

b.

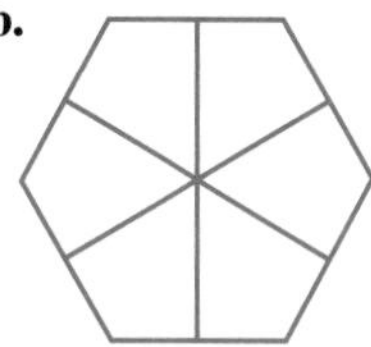

17. a. *OFHQ* **b.** *ADPK* or *AEGJ*

c. *MNOS*, *MOQK*, *KQHI*

d. $\triangle ACK$, $\triangle IJK$, $\triangle FGH$, $\triangle OCD$, $\triangle CEF$, ΔOPQ, $\triangle ABL$, $\triangle MCO$

e. $\triangle MNO$, $\triangle MSO$, $\triangle LBC$

f. *MNOS*, *KQHI* **g.** *COSM*

h. $\triangle ROQ$, $\triangle LKC$ **i.** $\triangle ROP$, $\triangle ACL$

j. *KCOQ*, *KQHJ*, *CDOM*, *FODE*, *KMOP*

k. *LBCK*

19. a. 11 **b.** 1 **c.** 3 **d.** 1

21. a. 10 **b.** 2 **c.** 0

23. a.

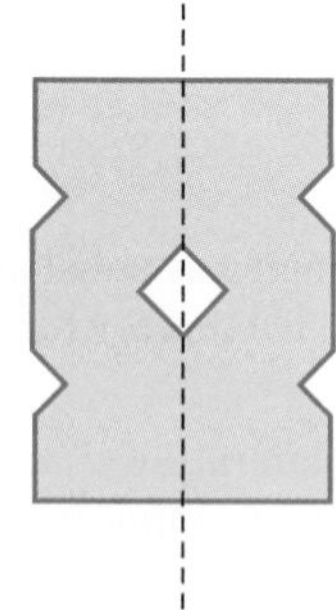

b.

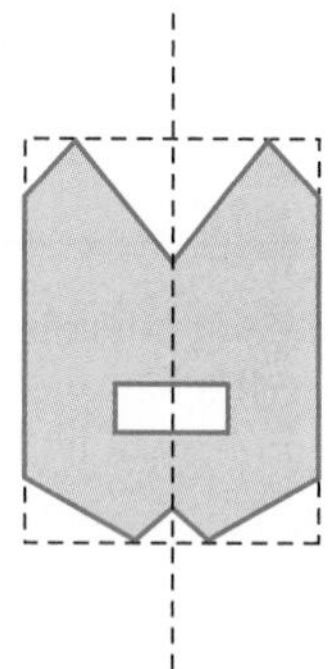

c.

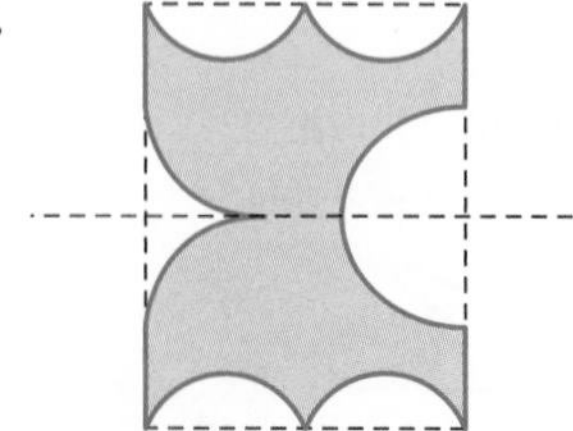

25. a.

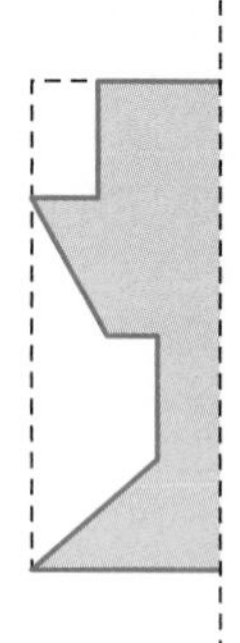

b.

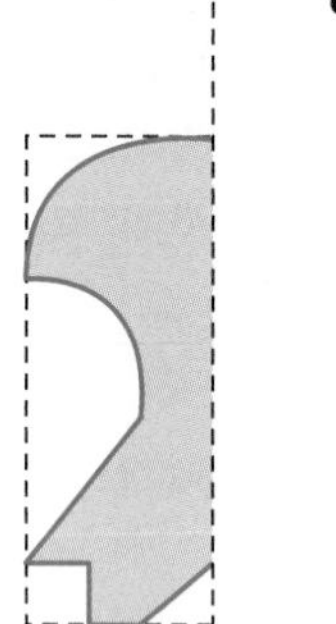

c.

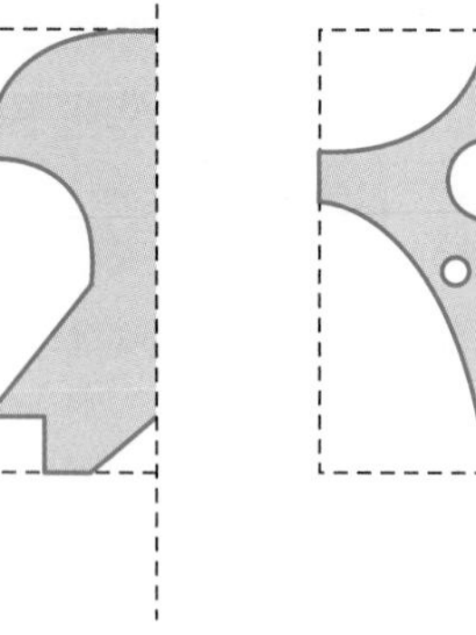

27. a. Yes, 2 **b.** Yes, 1

29. a. Yes, 1 **b.** No

31. a. Draw the lines of symmetry in the regular n-gons in (i)–(iii). How many does each have?

(i) Five

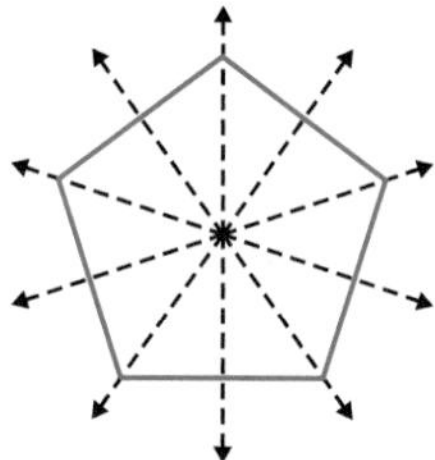

(ii) Six

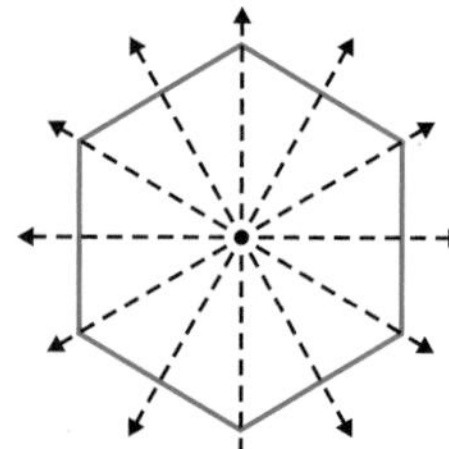

(iii) Eight

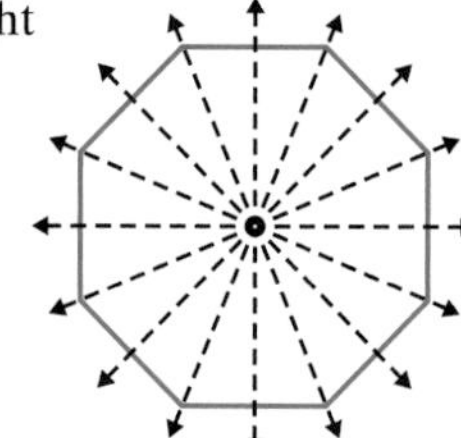

b. n

33. a. Reflection symmetry only

b. Both

c. Reflection symmetry only

d. Reflection symmetry only

e. Both

f. Reflection symmetry only

35. a. Four axes of reflection, three rotation symmetries (90°, 180°, and 270°)

b. Two axes of reflection, one rotation symmetry (180°)

c. Two axes of reflection, one rotation symmetry (180°)

d. Four axes of reflection, three rotation symmetries (90°, 180°, and 270°)

e. Four axes of reflection, three rotation symmetries (90°, 180°, and 270°)

f. Two axes of reflection, one rotation symmetry (180°)

37. a. 7 spokes, 6 rotation symmetries

b. 12 spokes, 12 axes of reflection symmetry

SECTION 2.3

1.

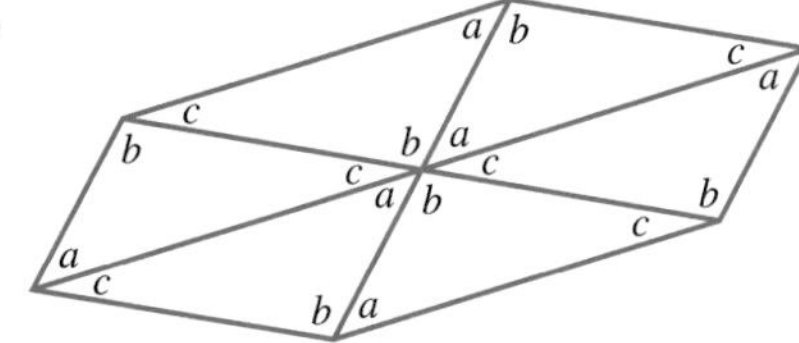

3. a.

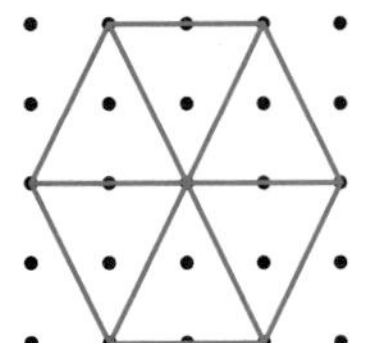

b.

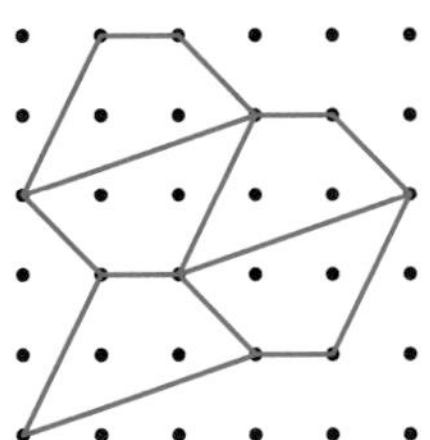

c.

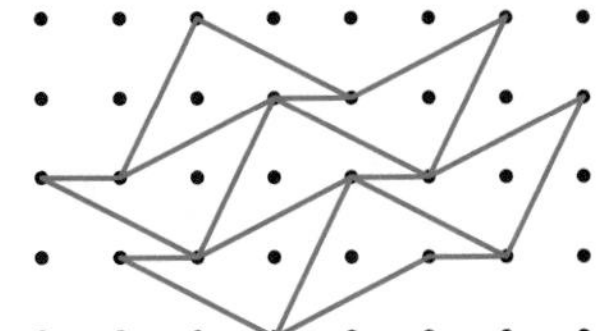

5. a. $\angle BAC = 54°$, $\angle ABC = 90°$, $\angle ACB = 36°$—the angle sum should be 180°

b. $\angle HDE = 100°$, $\angle DEF = 126°$, $\angle EFG = 89°$, $\angle FGH = 115°$, $\angle GHD = 110°$—the angle sum should be 540°

7. a. 30° **b.** 20° **c.** 62° **d.** 34°

9. a. 70° **b.** 130° **c.** 120° **d.** 20°

e. 20° **f.** 80° **g.** 60° **h.** 100°

11. a. 98° **b.** each 125°

c. 120° **d.** 95°, 100°, 80°, 85°

13. 11 sides **15.** 4860°

17. a. 150° **b.** 157.5° **c.** 144°

d. 162° **e.** 160° **f.** 170°

19. a. 48 **b.** 72 **c.** 40 **d.** 100

21. Equilateral triangle, square, regular hexagon, and regular 12-gon

23. a.

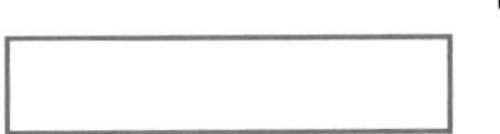

b.

25. The measure of one vertex angle in a regular heptagon is not a factor of 360°.

27.

Number of Sides and Angles in a Polygon	Number of Triangles into Which the Polygon Can Be Divided	Sum of Angle Measures in the Polygon	Measure of One Angle in the Polygon
3	1	180°	180°/3 = 60°
4	2	2(180°) = 360°	360°/4 = 90°
5	3	3(180°) = 540°	540°/5 = 108°
6	4	4(180°) = 720°	720°/6 = 120°
8	6	6(180°) = 1080°	1080°/8 = 135°
10	8	8(180°) = 1440°	1440°/10 = 144°
n	$n - 2$	$(n - 2)(180°)$	$\frac{(n-2)180°}{n}$

29. $\angle ACB = 69°24'10''$

31. a. 540°, no error **b.** 718°5′, error = 1°55′

33.

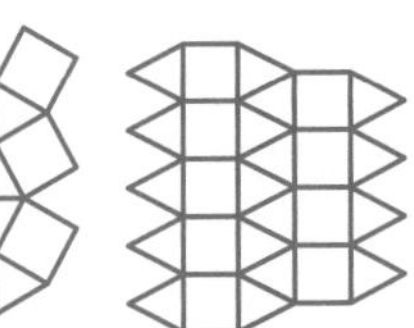

SECTION 2.4

1. a. A polyhedron **b.** A polyhedron

c. Not a polyhedron. The shape has circular regions.

3. a.

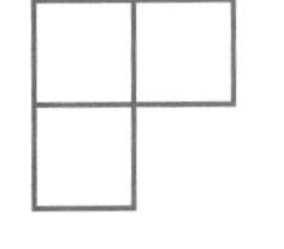

Top

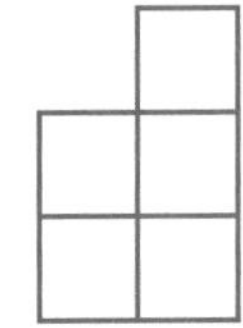

Front

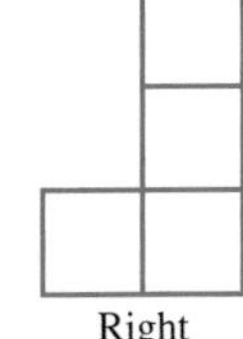

Right

b. 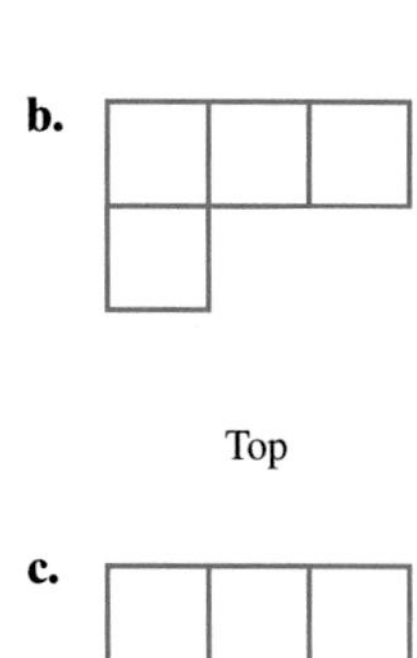 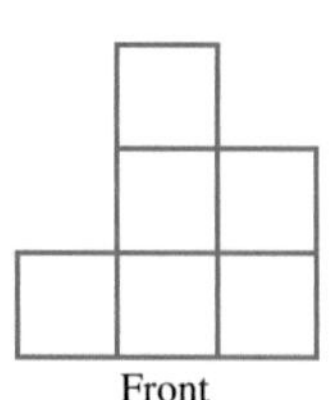 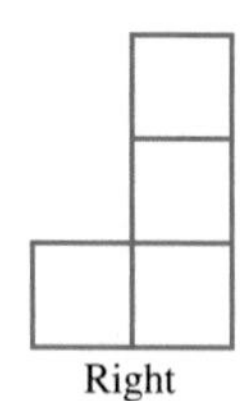

Top Front Right

c. 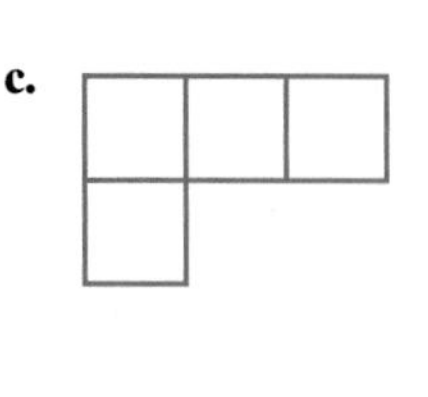 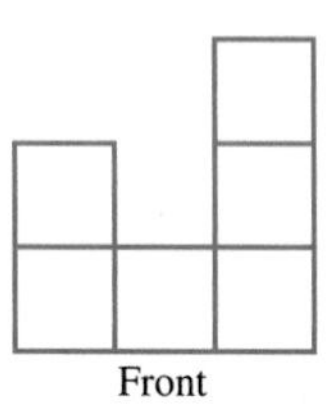

Top Front Right

d. 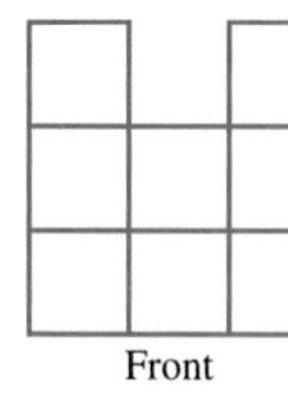 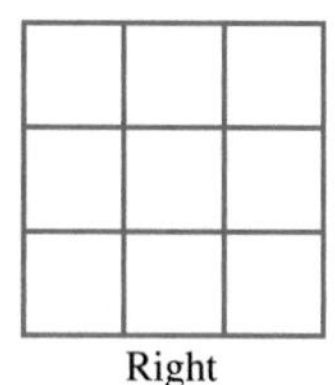

Top Front Right

5.

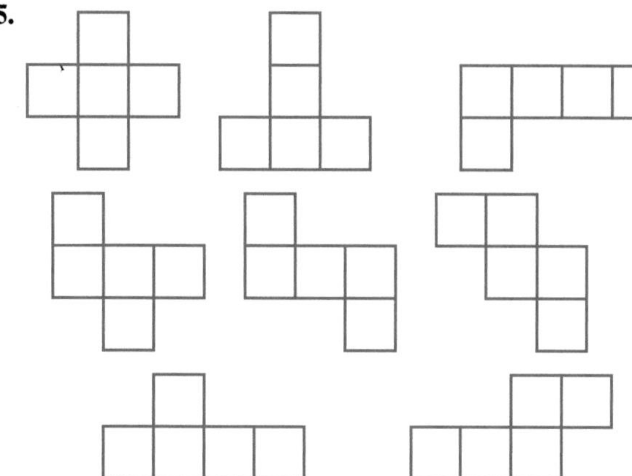

7. a. *ABCDE*, *FGHIJ*

b. *ABGF*, *BCHG*, *CDIH*, *EDIJ*, *AEJF*

c. *ABGF*, *BCHG*, *CDIH*, *FGHIJ*

d. Right pentagonal prism

9. a. Oblique rectangular prism **b.** Right octagonal prism

11. Right regular hexagonal prism

13. a. Oblique regular hexagonal pyramid

b. Right square pyramid

15. *a*, *b*

17. Right regular pentagonal pyramid

19.

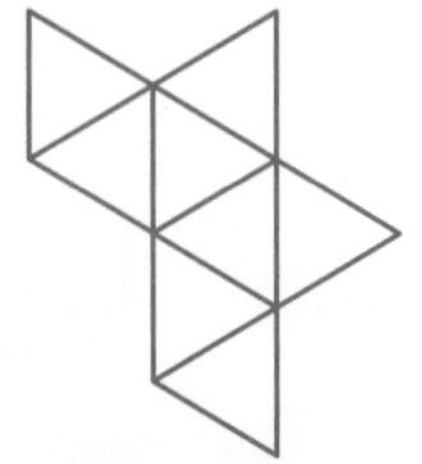

21. (iii)

23.

Base	F	V	F + V	E
Triangle	5	6	11	9
Quadrilateral	6	8	14	12
Hexagon	8	12	20	18
n-gon	$n + 2$	$2n$	$3n + 2$	$3n$

Yes, Euler's formula holds for these figures.

25.

Figure	F	V	F + V	E
(i)	7	10	17	15
(ii)	9	9	18	16
(iii)	8	12	20	18

Yes, Euler's formula holds for these figures.

27. 12 vertices—one possible answer is a right hexagonal prism

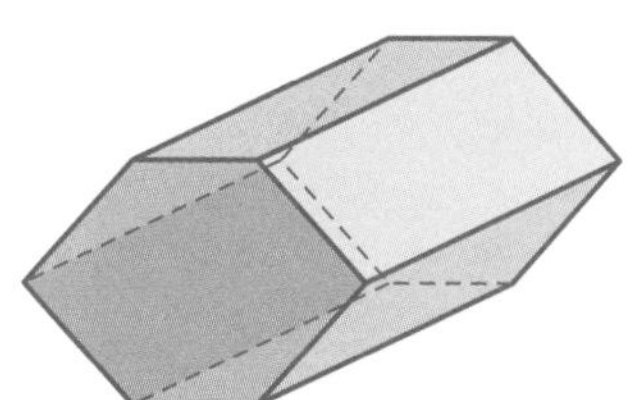

29. a. Oblique circular cone

b. Oblique circular cylinder

31. a. Circle **b.** Rectangle

33. a. 3 **b.** 6 **c.** 9

35. a. 6 **b.** 4 **c.** 6 **d.** Infinitely many

37. a. Equilateral triangle **b.** Square

39. a. Right circular cylinder

b. Right rectangular prism

c. Right square pyramid

d. Regular octahedron

41. Oblique dodecagonal (12-gon) pyramid. Euler's formula holds with $F = 13$, $V = 13$, and $E = 24$.

SECTION 2.5

1. a. $\frac{1 \text{ ft}}{12 \text{ in.}}$ **b.** $\frac{5280 \text{ ft}}{1 \text{ mi}}$ **c.** $\frac{16 \text{ oz}}{1 \text{ lb}}$ **d.** $\frac{1 \text{ gal}}{4 \text{ qt}}$

3. a. $\frac{60 \text{ min}}{1 \text{ hr}}$ **b.** $\frac{1 \text{ min}}{60 \text{ sec}}$ **c.** $\frac{1 \text{ yd}}{36 \text{ in.}}$ **d.** $\frac{1 \text{ work-day}}{24 \text{ work-hours}}$

5. $\frac{1 \text{ c}}{8 \text{ oz}}, \frac{1 \text{ qt}}{4 \text{ c}}, \frac{1 \text{ gal}}{4 \text{ qt}}, \frac{1 \text{ peck}}{2 \text{ gal}}$

7. a. 72 oz **b.** 62.4 degrees **c.** 1,584,000 in. **d.** 0.0105 kL

9. a. 15.24 cm **b.** 91.44 m **c.** 128.016 m **d.** 0.621 mi

11. a. 66 ft/sec **b.** $8\frac{1}{3}$ dollars/hour
c. 2.5 ft/century **d.** 16 m/mL

13. a. 30.48 m/sec **b.** 0.807 lb/yd

15. 105 km/hr **17.** 22.49 km/hr **19.** 2.5 in. per year

21. 3,178,807,947 L **23.** 1.17×10^{-8} mph

25. Light is 1,079,000 times faster than sound.

27. a. 5.88×10^{12} mi/yr **b.** 4.47×10^{14}

c. 0.716 hours

29. a. $\frac{1 \text{ smoot}}{67 \text{ in.}}$ **b.** $\frac{1.7018 \text{ m}}{1 \text{ smoot}}$ **c.** 620 m **d.** 64.5 smoots

Answers for the Additional Problems

1. 199 days **2.** 4950 angles

Chapter Review Problems

SECTION 2.1

1. Line

A B

Line segment

A B

Ray

A B

2.

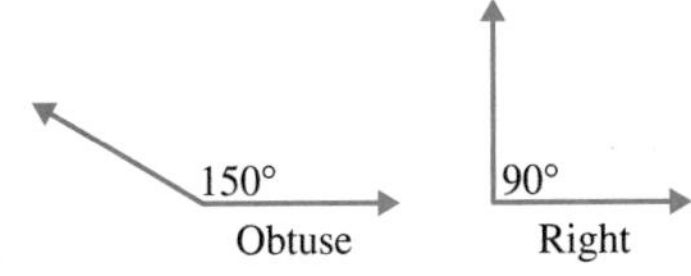

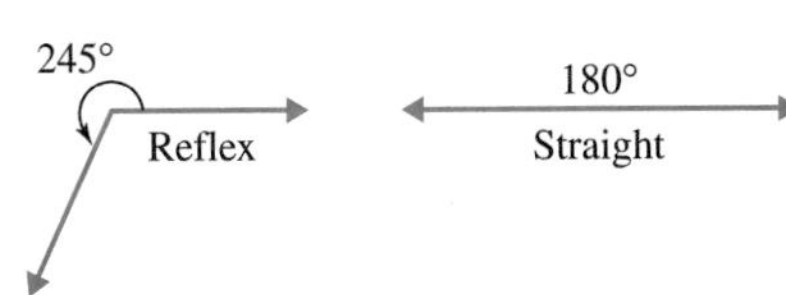

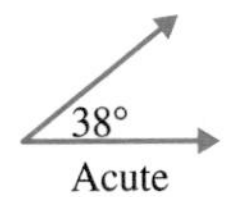

3. $A = 1.7, B = 0.4, C = -1.2;$
$AB = 1.3, CB = 1.6, AC = 2.9$

4. 45°, 45°

5. a. $\angle CFD$ **b.** $\angle CFD$ and $\angle DFE$ **c.** 28°

d. $\angle CFE$ **e.** $\angle BFD$ and $\angle AFE$ **f.** 34°

g. $\angle BFE$ and $\angle AFD$

SECTION 2.2

1. a. 5, 6 axes of symmetry **b.** 0, 1 axis of symmetry

c. 1, 2 axes of symmetry

2. a. 1 **b.** 3 **c.** 0 **d.** 2

3. a. 90°, 180°, 270° **b.** 180° **c.** None

4. a. Trapezoid and kite

b. Opposite sides of the quadrilateral must be parallel.

c. All four angles must be right angles.

d. All four sides must be congruent.

e. The four sides must be congruent and the four angles must be right angles.

5. a.

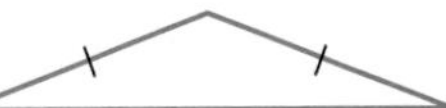

b. Impossible; An acute triangle has three angles with measures less than 90°.

c.

d. Impossible; No two sides are congruent in a scalene triangle.

6. Right isosceles triangles

SECTION 2.3

1. Many answers are possible.

a.

b.

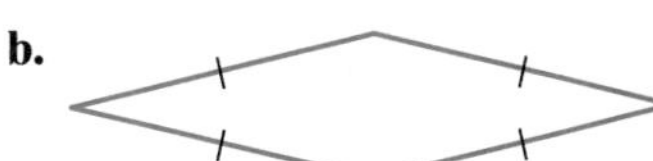

2. a. 16 **b.** 157.5°

3. $x = 30°, y = 15°$ **4.** 44.25°

5.

6. 180°

7. Since the sum of the measures of the vertex angles is 360°, we can arrange the four copies of the quadrilateral around a single vertex. Each vertex angle in the quadrilateral is represented and the figures do not overlap. The four copies form a pattern, which can be repeated.

SECTION 2.4

1. a. Neither **b.** Prism **c.** Pyramid **d.** Prism

2. a. Sphere **b.** Right rectangular prism

c. Right circular cylinder **d.** Regular hexahedron

3. Oblique square prism

4. The base is an equilateral triangle. Each lateral face is a parallelogram.

5. The base is a regular hexagon. Each lateral face is an isosceles triangle.

6. $V = 22, F = 11$

SECTION 2.5

1. 28.2 marlocks
2. 174.08 T
3. 692 km
4. 691 L/yr
5. 27.8 m/sec

CHAPTER 2 TEST

1. F—Three points may be collinear.
2. T
3. F—An isosceles triangle may also be equilateral.
4. F—The vertex angle of a *regular* hexagon measures 120°.
5. F—The diameter of a circle is twice its radius.
6. T
7. T
8. T
9. T
10. F—A polyhedron has faces that are polygonal regions.
11. Equiangular—all right angles
12. $a = 76°, b = 28°, c = 50°, d = 24°, e = 156°, f = 26°, g = 50°$.
13. 165°.
14. **a.** *IBMG, IABR, HIRG* **b.** *HIDF, GRDF, ABGH*
    c. *IGED, PQJO*
    d. $\triangle GHI, \triangle GRI, \triangle IAB, \triangle IRB, \triangle BRM, \triangle MRG, \triangle EFD$
    e. $\triangle ICG, \triangle QJO, \triangle JOP, \triangle OPQ, \triangle QPJ, \triangle BMG, \triangle MGI, \triangle GIB, \triangle IBM, \triangle LMN, \triangle CKL, \triangle BKJ$
    f. *PQJO* **g.** *GHIC, RNCJ*
    h. $\triangle ICD, \triangle NCD, \triangle GNM, \triangle GJC, \triangle GJI, \triangle IJQ, \triangle IJB, \triangle BCK, \triangle IKM, \triangle IJO$
    i. $\triangle LMG, \triangle IJR, \triangle QRP, \triangle PRO, \triangle JRO, \triangle JRQ, \triangle GRN, \triangle IBK$
    j. *IABM, IACD, HACE, HIDE, HIMG, GNDE, GMDF, GNDF, INGH, KCDM, LCDM, AINC*
    k. *IBCD, GMDE*
15. **a.** $\angle 1$ and $\angle 2$, $\angle 3$ and $\angle 4$, $\angle 1$ and $\angle 3$, $\angle 2$ and $\angle 4$, $\angle 6$ and $\angle 9$
    b. $\angle 7$ and $\angle 10$, $\angle 5$ and $\angle 8$, $\angle 5$ and $\angle 7$, $\angle 8$ and $\angle 10$
    c. $\angle 9$ or $\angle 6$
    d. $\angle 1$ and $\angle 2$, $\angle 2$ and $\angle 4$, $\angle 4$ and $\angle 3$, $\angle 3$ and $\angle 1$, $\angle 5$ and $\angle 6$, $\angle 6$ and $\angle 7$, $\angle 7$ and $\angle 10$, $\angle 10$ and $\angle 9$, $\angle 9$ and $\angle 8$, $\angle 8$ and $\angle 5$
    e. $\angle 1, \angle 4, \angle 7, \angle 10, \angle 8, \angle 5$ **f.** $\angle 2, \angle 3$
16. **a.** rotation 5, reflection 6
    b. rotation 1, reflection 2
    c. rotation and reflection—both infinitely many
17. 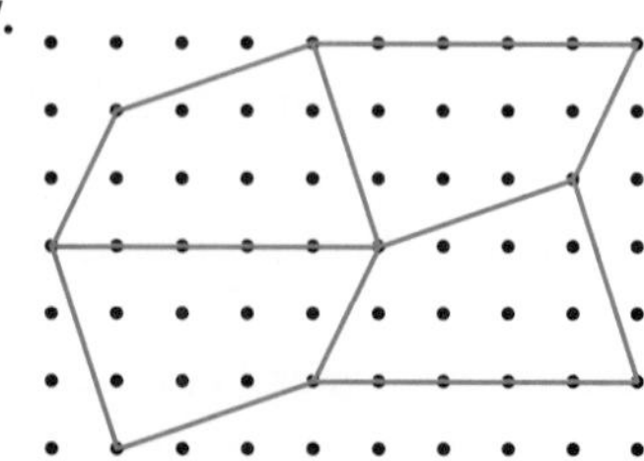
18. **a.** 24 **b.** 3 **c.** 15 **d.** 6
19. 11,700,000 milliseconds
20. **a.** One possible answer is shown.

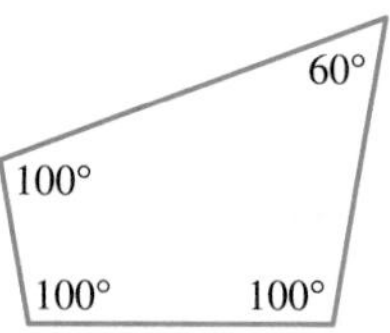

b. One possible answer is shown.

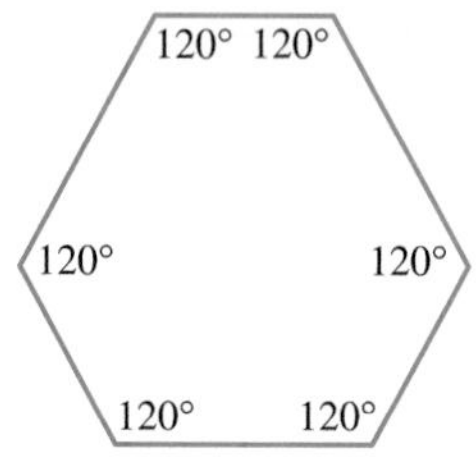

21. 6.02 g/L
22. 540°10′—the traverse does not close; error = 10′
23. 795 revolutions
24. Right trapezoidal prism

SECTION 3.1

1. **a.** $P = 22$ in. **b.** $P = 53.2$ m
3. **a.** $P = 20$ cm **b.** $P = 60$ in. **c.** $P = ns$
5. **a.** One tetromino with the maximum perimeter of 10 is shown. Others are possible.

b. The following tetromino has the minimum perimeter of 8.

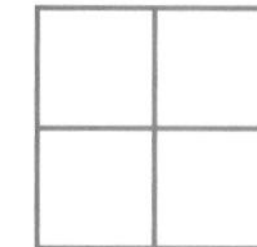

7. **a.** $n + 2$ **b.** $2n + 2$ **c.** $3n + 2$
   d. $4n + 2$ **e.** $6n + 2$ **f.** $(m - 2)n + 2$
9. **a.** $C \approx 25.13$ in. **b.** $C \approx 33.23$ m
11. $r \approx 1.59$ cm
13. **a.** 7 square units **b.** 11.5 square units
    c. 12 square units.
15. **a.** $P = 28$ in., $A = 40\text{ in}^2$
    b. $P = 22.8$ cm, $A = 32.49$ cm
17. Width = 14 in., length = 22 in.

19. The perimeter is 2 times the original perimeter, 3 times the original perimeter, 10 times the original perimeter, and n times the original perimeter, respectively.

21. No. Counterexample: If length is 15 m and width is 5 m, then the perimeter is 40 m and the area is 75 m^2. But if length is 10 m and width is 10 m, the perimeter is also 40 m, but the area is 100 m^2.

23. Many answers are possible.

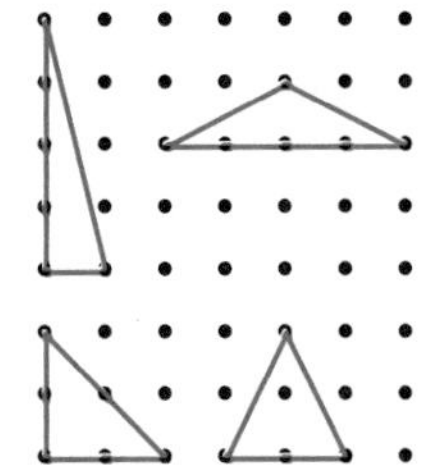

25. a. $P = 24$ m, $A = 24$ m^2 **b.** $P = 15.6$ cm, $A \approx 9.34$ cm^2
c. $P \approx 17.07$ in., $A = 12.5$ in^2

27. They are equal. The areas are equal because $\overline{ED}$ in $\triangle AED$ is twice the length of $\overline{CD}$ in rectangle $ABCD$ and because the two figures have the same base.

29.

Side	Height	Area
$a = 56$ mm	49 mm	1372 mm^2
$b = 51$ mm	54 mm	1377 mm^2
$c = 64$ mm	43 mm	1376 mm^2

31. a. 0.35 ft^2 **b.** 20,000,000 cm^2 **c.** 2.93 m^2

33. a. $A \approx 38.5$ ft^2 **b.** $A = 3.58$ m^2

35. The gap is 4.77 ft. You could crawl under it, but not walk under it unless you were less than 4.77 ft tall.

37. a. $A = 30$ cm^2 **b.** $A \approx 9.92$ m^2 **c.** $A \approx 8.18$ ft^2

39. a. $A = \frac{1}{2}b_1h$ **b.** $A = \frac{1}{2}b_2h$ **c.** $A = \frac{1}{2}b_1h + \frac{1}{2}b_2h$

d. $A = \frac{1}{2}(b_1 + b_2)h, b_1 + b_2 = b, A = \frac{1}{2}bh$

e. $A = \frac{1}{2}bh$; yes

41. No. Even if the basketballs were both only 29.5 in. in circumference, their diameters would be 9.39 in. each. So, side by side, the width would be 18.78 in. Thus, in order for the balls to fit through the ring side by side, the ring must have a diameter greater than 18.78 in.

43. a. 77 ft **b.** 42 yd^2

45. 0.5625 ft by 0.5625 ft, 1.6875 ft by 1.6875 ft

47. About 297 times **49.** 67.2 mph

51. 458. 4 revolutions per minute

SECTION 3.2

1. a. $A = 28$ cm^2 **b.** $A = 20$ in^2

3. a. $b \approx 6.67$ **b.** $A \approx 14.70$

5.

Side	Height	Area
$AD = 34$ mm	23 mm	782 mm^2
$AB = 27$ mm	29 mm	783 mm^2

7. $A = 215.35$ mm^2 **9.** $A = 148$ in^2

11. 6″ **13.** $A \approx 19.99$ mm^2

15. a. $A \approx 56.55$ cm^2 **b.** $A = 12.5\pi \approx 39.27$ in^2

17. a. $A \approx 412.44$ in^2 **b.** $A = 93.38$ cm^2

19. $A = 32\pi - 64$ in$^2 \approx 36.53$ in^2 **21.** Circle

23. The area of the circle is 4 times the original area, 25 times the original area, and n^2 times the original area, respectively.

25. a. $A = \frac{1}{2}bh$ **b.** Congruent, $A = \frac{1}{2}bh$

c. $A = \frac{1}{2}bh + \frac{1}{2}bh$ **d.** $A = bh$, yes

27. Area is between 1650 mm^2 and 1660 mm^2

29. Large. The large pizza costs \$0.08 per in^2 while the small costs \$0.14 per in^2 and the medium costs \$0.11 per in^2.

31. 16,327 ft^2

33. a. 125.66 acres **b.** 34.34 acres

35. 4.4 lbs **37.** 875,520 ft^2

39. a. 19.6 percent **b.** 16.0 ft

41. Probability = 0.497

43. a. $A \approx 201.06$ in^2 **b.** $A \approx 47.12$ in^2
c. $A \approx 21.21$ in^2 **d.** Probability ≈ 0.23
e. Probability ≈ 0.11

SECTION 3.3

1. a. 30 in. **b.** 8.5 cm **c.** $\sqrt{63} \approx 7.94$ m

3. a. $x = 11.0$ cm, $y = 10.2$ cm
b. $x \approx 23.3$ in., $y \approx 33$ in., $z \approx 22.6$ in.

5. a. $b \approx 9.22, c \approx 9.50$
b. $a = 10, c \approx 11.71, P \approx 27.81$
c. $c \approx 3.74, P \approx 9.03, A = 3.5$
d. $a \approx 1.73, b = 1, c = 2, P = 4.73$

7. Length = 8 in., width = 3 in.

9. a. Yes **b.** No **c.** Yes

11. a. $a^2 + b^2 = c^2$ **b.** $a^2 + b^2 > c^2$ **c.** $a^2 + b^2 < c^2$

13. a. Leg = $(7\sqrt{3})/3 \approx 4.04$ cm
Hyp. = $(14\sqrt{3})/3 \approx 8.08$ cm
b. Legs = $6\sqrt{2} \approx 8.49$ in. each

15. $h = 8.7\sqrt{3}$ cm, $A \approx 489.7$ cm^2 **17.** $A = 120$ cm^2

19. $A = 44.18\sqrt{2} \approx 62.48$ m^2 **21.** $A \approx 1039.2$ cm^2

23. $A = \frac{h^2}{\sqrt{3}} = \frac{h^2\sqrt{3}}{3}$ **25.** $d = x + x\sqrt{2}$

27. 10 cm

29. Area of large square is c^2.

Area of the four triangles is $4\left(\frac{1}{2}ab\right) = 2ab$.

Area of the small square is $(b - a)^2 = b^2 - 2ab + a^2$.

Area of the large square is equal to the area of the small square plus the area of the four triangles. So, $c^2 = b^2 - 2ab + a^2 + 2ab$, or $c^2 = a^2 + b^2$.

31. 4.6 in. **33.** 15.5 ft **35.** 0.87 mph **37.** 550 ft

39. 82 ft **41.** 189 ft **43.** 2.2 cm **45.** 120 ft

47. Yes. If the umbrella is placed from one corner to the farthest (diagonally opposite) corner, the length of the diagonal is 33.54 in.

49. 0.60 in. **51.** 28.54 in^2

SECTION 3.4

1. a. $SA = 800\ in^2$ **b.** $SA = 240\ cm^2$

3. a. $SA = 295\ ft^2$ **b.** $SA \approx 436\ cm^2$

5. 12 in. × 8 in. × 6 in.

7. a. $SA = 360\ ft^2$ **b.** $SA = 896\ cm^2$

9. $SA \approx 71\ cm^2$ **11.** $SA \approx 1002\ in^2$

13. a. $SA \approx 1141\ cm^2$ **b.** $SA \approx 276\ cm^2$

15. $d = 8$ units

17. a. $SA \approx 1206\ in^2$ **b.** $SA \approx 1612\ m^2$

19. a. $SA \approx 452\ in^2$ **b.** $SA \approx 1810\ m^2$ **c.** $SA \approx 154\ cm^2$

21. $d \approx 83$ mm

23. The radius should be increased by 1.66 ft.

25. $SA \approx 43\ cm^2$

27. a. $SA = 168\ cm^2$ **b.** $SA \approx 366\ in^2$

29. $SA \approx 17.44\ cm^2$

31. a. 6 × 6 × 1 **b.** 2 × 9 × 2

c. 3 × 3 × 4, $SA = 66$ square units

d. 1 × 1 × 36, $SA = 146$ square units

33. 43.4 ft^2 **35.** $SA \approx 4222.3\ ft^2$ **37.** 4.7 liters

39. $SA \approx 1.8\ in^2$ **41.** $32.76

43. One-third, or about 2094 cm^2

SECTION 3.5

1. a. $V = 14, SA = 46$ **b.** $V = 7, SA = 30$

c. $V = 21, SA = 54$

3. a. $V = 1500\ in^3$ **b.** $V = 200\ cm^3$

5. a. $V \approx 277\ ft^3$ **b.** $V = 576\ cm^3$

7. a. $V = 15\ ft^3$ **b.** $V = 114\ in^3$

9. $V = 1450.185\ cm^3$

11. a. $V \approx 2958\ cm^3$ **b.** $V \approx 342\ cm^3$

13. a. $V \approx 2827\ m^3$ **b.** $V \approx 2413\ in^3$

15. a. $V \approx 905\ in^3$ **b.** $V \approx 7238\ m^3$ **c.** $V \approx 3054\ cm^3$

17. a. 0.23 ft^3 **b.** 1,200,000 cm^3 **c.** 11,326,738.64 mm^3

19. $V \approx 478\ cm^3$, $SA \approx 404\ cm^2$

21. The new volume is 8 times the original volume.

23. a. The area of the larger square is 9 times the area of the smaller square.

b. The volume of the larger cube is 27 times the volume of the smaller cube.

c. The new volume is 8 times the original volume.

25. 2 × 9 × 18

27. The cylinder with the short side of the rectangle as its height has the greater volume. The cylinder with the short side of the rectangle as its height has volume 81.85 in^3, and the cylinder with the long side of the rectangle as its height has volume 63.24 in^3.

29. 230,992 m^3 **31.** $50 **33.** 318 cm^3

35. a. 6,272,640 in^3, 3630 ft^3 **b.** 112.53 tons **c.** 27,116 gal

37. a. 90 in^3 **b.** No

39. Approximately 36 kg **41.** 78 cm^3 **43.** 646 gallons

45. a. 120,000 cm^3 **b.** 31,000 cm^2

47. 5250 in^3 **49.** 369 ft^3 **51.** 42.65 hr

53. a. $h = 40$ ft **b.** Approximately 848,000 gal

c. Sphere.

Answers for the Additional Problems

1. $11^2 + 10^2 + \cdots + 1^2 = 506$

2. a. Cut through two opposite edges.

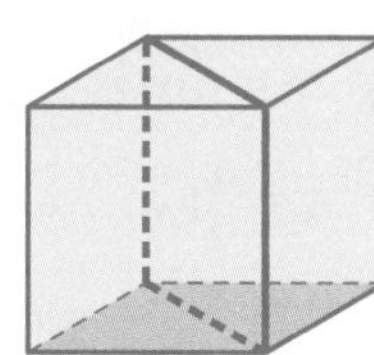

b. Cut with a plane through one vertex and at equal distances from an adjacent vertex. Other solutions are possible.

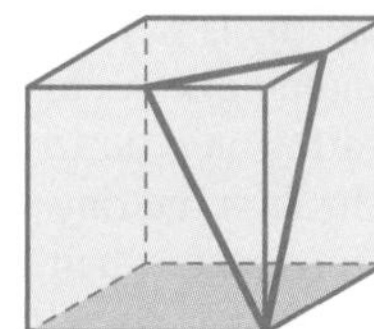

c. Cut with a plane through two opposite vertices and through the midpoints of parallel edges as shown.

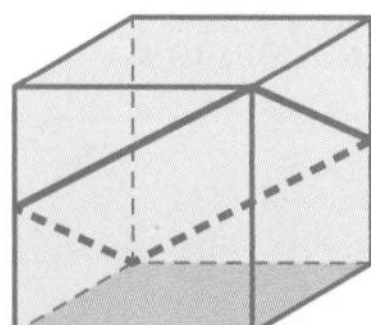

Chapter Review Problems

SECTION 3.1

1. a. 12.8 mm b. 23.2 cm c. 64 ft d. 18.8 cm e. 16.24 ft
2. a. 346.7 yd b. 1.32 acres
3. a. 843 in. b. 6.25 ft^2
4. 81 in^2
5. $r = \frac{5}{\pi}$ in.
6. The circumference is doubled.
7. a. $P = 30$ cm, $A = 30\ cm^2$ b. $P \approx 21.4$ mm, $A = 9\ mm^2$

SECTION 3.2

1. a. $A = 34.8\ mm^2$ b. $A = 9.28\ in^2$
2. a. $A = 38.055\ m^2$ b. $A = 79.005\ cm^2$
3. $A \approx 9.8\ cm^2$ 4. 29.49 in^2 5. 1431 in^2 6. 20.668 mm^2

SECTION 3.3

1. $12^2 + 35^2 = 1369 = 37^2$ 2. No, obtuse
3. a. 16.61 in. b. 4 cm
4. $x = 20$
5. a. Angles: 30°, 60° Legs $= 2.5\sqrt{3}$ in. b. Angles: 45°, 45° Hypotenuse $= 7\sqrt{2}$
6. $A \approx 93.5\ m^2$
7. $PQ \approx 8.5$ in. $PR \approx 16.4$ in.

SECTION 3.4

1. 134.26 m^2 2. $840 + 300\sqrt{3} \approx 1359.6\ in^2$
3. $d \approx 5.77$ cm 4. $A \approx 4166.12\ m^2$
5. 68 m^2 6. 28.7 in^2
7. a. 0.0385 m^2 b. 2,970,722.304 in^2 c. 11.9599 yd^2

SECTION 3.5

1. a. 101 cm^3 b. 1263 in^3
2. $h \approx 4.37$ cm
3. a. 15,000,000 m^3 b. 0.044 yd^3 c. 0.110 in^3
4. 0.17 m^3 5. 6.4 in^3
6. a. 31,665,152 ft^3 b. 457,769 ft^2
7. 9.5 cm

CHAPTER 3 TEST

1. F—Perimeter is $2l + 2w$
2. F—Only when the triangle is a right triangle
3. T
4. F—For example, rectangles 5 cm by 10 cm and 7 cm by 8 cm
5. F—The shape could be a pyramid.
6. F—The surface area of the cylinder is larger.
7. F—The altitude depends on which side is used as a base.
8. T
9. T
10. F—The area will be four times as large as the original area.
11. a. $\frac{\pi d^2}{4}$ b. $\frac{C^2}{4\pi}$
12. 8 cm
13. a. 162 in^3 b. Volume is multiplied by 4 c. Volume is multiplied by 9 d. 216 in^3
14. $A = \frac{3\sqrt{3}}{2}x^2$
15.

D	*R*	*SA*	*C*	*V*
a. 56	28	3136π	56π	$\frac{87{,}808\pi}{3}$
b. 10	5	100π	10π	$\frac{500\pi}{3}$
c. 18	9	324π	18π	972π
d. 6	3	36π	6π	36π

16. 105.86 in^2, 49.42 in. 17. 1364 kg, 13.1 ft, 1902 g/cm^2
18. 13 square units 19. 432 in^2, 545 in^3
20. Approximately 238 ft^2
21. Approximately 575 revolutions 22. 24,086 ft^3
23. 147,000,000 yd^3 24. Approximately 53 ft

SECTION 4.1

1. The vertices lie in one and only one plane.
3. *ABCDEFGH* has an edge length of 3 and/or a surface area of 54 in^2.
5. Lines *l* and *m* intersect.
7. The lines are perpendicular if and only if they form right angles.
9. $\triangle ABC$ has no two sides that are congruent.
11. No deduction possible
13. $\triangle QRS$ is acute.
15. *ABCD* is a quadrilateral with four congruent sides and/or *ABCD* is a rhombus.
17. T. Converse: If the length of a side of a square is 5 cm, then its area is 25 cm^2. T. Biconditional: A square has area 25 cm^2 if and only if the length of a side is 5 cm.
19. T. Converse: If two angles share a common side, they are adjacent. T. Biconditional: Two angles are adjacent if and only if they share a common side.
21. T. Converse: If the diagonal of a square measures $10\sqrt{2}$ in., then the side of the square measures 10 in. T. Biconditional: The diagonal of a square measures $10\sqrt{2}$ if and only if the sides measure 10 in.

23. T. Converse: If $a^2 + b^2 = c^2$, $\triangle ABC$ has a right angle at $\angle C$. T. Biconditional: $\triangle ABC$ has a right angle at $\angle C$ if and only if $a^2 + b^2 = c^2$.

25. T. Converse: If a triangle is a right triangle, then it has sides lengths 3, 4, and 5. F.

27. T. Converse: If a triangle has rotation symmetry, it is equilateral. T. Biconditional: A triangle has rotation symmetry if and only if it is equilateral.

29. T. Converse: If the triangle is equiangular, then it has three equal sides. T. Biconditional: A triangle is equiangular if and only if it has three equal sides.

31. T. Converse: If a pyramid has $n + 1$ vertices, then it has a base with n sides. T. Biconditional: A pyramid has a base with n sides if and only if it has $n + 1$ vertices.

33. s

35. Any one of $r \Rightarrow t$, $r \Rightarrow p$, $s \Rightarrow t$

37. Any one of $p \Rightarrow t$, $p \Rightarrow$ not r, not $q \Rightarrow t$

39. If $ABCD$ is a parallelogram, then $ABCD$ is a quadrilateral.

41. If a prism has a base with n sides, then it has $2n$ vertices.

43. If a polyhedron is regular, then its faces are regular polygons.

45. 1. Given
    2. Angle Sum in a Triangle Theorem
    3. Definition of a Right Angle
    4. Substitution [(3) into (2)]
    5. Subtraction
    6. Definition

47. The farmer knows she will have enough compost if and only if the volume of compost needed is less than or equal to 1300 ft^3. Since $(100)(150)\left(\frac{1}{12}\right) = 1250 \leq 1300$, she has enough.

49. a. "If you believe in peanut butter, then you gotta believe in Peter Pan." The statement is false. You can believe in peanut butter and not believe in Peter Pan. The converse is "If you believe in Peter Pan, then you believe in peanut butter." The converse is false. You might believe in Peter Pan and not believe in peanut butter.

    b. "If you've got questions, then we've got answers." The statement is not necessarily true. They might not have an answer to one of your questions. The converse is "If we've got answers, then you've got questions." The converse is not necessarily true. You may not have any questions.

    c. "If it takes a licking, then it keeps on ticking." The statement is not necessarily true. The watch could break after taking a licking. The converse is "If it keeps on ticking, then it takes a licking." The converse is not necessarily true. In order for a watch to continue functioning, it does not have to take a licking.

SECTION 4.2

1. $\overline{PQ} \cong \overline{XY}$, $\overline{QR} \cong \overline{YZ}$, $\overline{PR} \cong \overline{XZ}$, $\angle P \cong \angle X$, $\angle Q \cong \angle Y$, $\angle R \cong \angle Z$

3. $\triangle ABC \cong \triangle VWU$

5. $\triangle ABC \cong \triangle DEF$, SAS

7. Not necessarily congruent

9. $\triangle QRS \cong \triangle VTU$, SAS or $\triangle QRS \cong \triangle UTV$, SAS

11. $\triangle ABE \cong \triangle DBC$, ASA

13. $\triangle WXZ \cong \triangle YXZ$, LL

15. $\triangle RYV \cong \triangle TWX$, LL

17. $\triangle CBD \cong \triangle ABE$, ASA

19. $\triangle QSR \cong \triangle QTP$, SSS;
    $\triangle STP \cong \triangle TSR$, SSS

21. $\overline{DE}$ 23. $\angle B$ 25. $\overline{FE}$

27. $\angle D$, $\angle BCD$, $\angle DEF$, $\angle EFA$, $\angle FAB$

29. $\overline{EC}$, $\triangle ACB \cong \triangle ECD$ by SAS and corresponding parts of congruent triangles are congruent

31. $\angle FAC$

33. **Given** $\overline{UW} \cong \overline{YW}$ and $\overline{VW} \cong \overline{XW}$

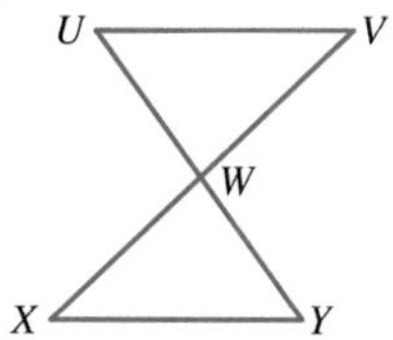

Prove $\angle U \cong \angle Y$

Proof

Statement	Reason
1. $\overline{UW} \cong \overline{YW}$	1. Given
2. $\overline{VW} \cong \overline{XW}$	2. Given
3. $\angle UWV \cong \angle YWX$	3. Vertical angles
4. $\triangle UWV \cong \triangle YWX$	4. SAS Congruence
5. $\angle U \cong \angle Y$	5. C.P.

35. Consider rhombus $ABCD$ as shown.

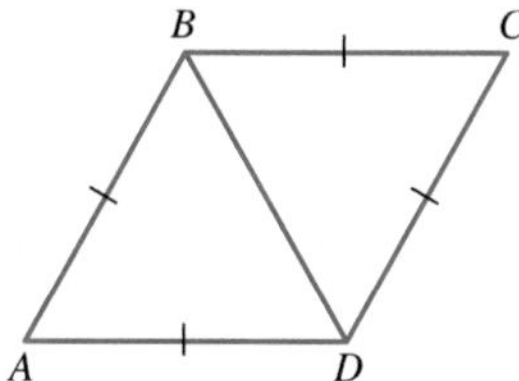

The four sides are congruent and the diagonal is congruent to itself. Thus, $\triangle ABD \cong \triangle CDB$ by SSS Congruence.

37. Outline: Use congruent segments formed by intersecting diagonals to show $\triangle BEC \cong \triangle DEA$. Then $\angle DBC \cong \angle BDA$ by C.P.

39. We are given $\overline{BE} \cong \overline{DE}$ and $\overline{AE} \cong \overline{CE}$. We know $\angle BEC \cong \angle DEA$ since they are vertical angles. Thus, $\triangle BEC \cong \triangle DEA$ by SAS. Then $\angle EBC \cong \angle EDA$ by C.P. We also know $\angle BEA \cong \angle DEC$ since they are vertical angles. Thus, $\triangle BEA \cong \triangle DEC$ by SAS. Then $\angle ABE \cong \angle CDE$ by C.P. Since $\overline{BD} \cong \overline{BD}$, we have $\triangle DBC \cong \triangle BDA$ by ASA.

41. Outline: Show that $\triangle ACF \cong \triangle AEB$ by HL and then use C.P.

43. Use the Pythagorean theorem to get the other pair of legs congruent. Then use LL to show that the triangles are congruent.

45. Because of the way beam DB is attached to beam AC, we know $\overline{AB} \cong \overline{CB}$ and $\angle DBA = \angle DBC = 90°$. Since $\overline{DB} \cong \overline{DB}$, we conclude $\triangle ABD \cong \triangle CBD$ by SAS or LL. For each truss, $\overline{DA} \cong \overline{DC}$ by C.P. so the beams can be cut the same length.

SECTION 4.3

1. $\angle Z = 70°$ because $\triangle XPY \cong \triangle ZPY$ by LL.

3. $\angle BCA = 51°$, $\angle ABE = 8°$

5. **a.** 76° **b.** 62° **c.** 14° **d.** 56° **e.** 34° **f.** 42°

7. **a.** 90° **b.** Theorem 4.9 **c.** Theorem 4.10

9. Outline: Use Theorem 4.7 to show $\triangle WVY$ and $\triangle YWX$ are isosceles. Then show that they are congruent. The result follows by C.P.

11. We have $\overline{PX} \cong \overline{RY}$. Because $\overline{PY}$ and $\overline{RX}$ are altitudes, we have $\angle PXR = 90° = \angle RYP$. Also, $\overline{PR} \cong \overline{PR}$. So $\triangle PXR \cong \triangle RYP$ by HL.

13. Outline: Since $\overline{BD}$ is an altitude, it is perpendicular to $\overline{AC}$. $\triangle ABD \cong \triangle CBD$ by ASA. $\overline{AD} \cong \overline{CD}$ by C.P. Thus, $\overline{BD}$ bisects $\overline{AC}$.

15. Outline: Use the fact that the median bisects the base to show that it divides the triangle into two congruent triangles. Then two congruent, adjacent, supplementary angles are formed.

17. Outline: If $\triangle ABC$ is equilateral, show that $\triangle ABC \cong \triangle CBA$. Then the angles are congruent by C.P.

19. Outline: If $\triangle ABC$ is equiangular show that $\triangle ABC \cong \triangle CBA$.

21. Each of the pictures must be centered on the perpendicular bisector of the base (or top) of the wall. That will be the case if and only if the distances from the hanger for a picture to points A and D (or B and C) are the same.

SECTION 4.4

1–23. Constructions

25. The perpendicular bisector of $\overline{AC}$ also bisects $\angle B$ in $\triangle ABC$.

27. The perpendicular bisectors of the three sides of a triangle intersect in a single point.

29. The three altitudes of a triangle, if extended, intersect in a single point.

31. Since all sides of the triangle constructed are the same length, then triangle is equilateral.

33. Since two sides and their included angle are used to construct the triangle, then any triangles constructed in this way must be congruent by SAS.

35. Outline: The altitude from B is perpendicular to its base $\overline{AC}$. Use construction 5 and its justification.

37. $\triangle ABC$ is a right isosceles triangle. If $\overline{CD}$ is perpendicular to $\overline{AB}$, then $\angle DCB = 45°$. This means $\overline{CD}$ bisects $\angle C$. By Theorem 4.9, the bisector of the vertex angle in an isosceles triangle bisects the base, so $AD = DB$.

39. **a.** The well should be located somewhere along the perpendicular bisector of $\overline{AB}$.

b. The well should be located somewhere along the perpendicular bisector of $\overline{AC}$.

c. If the well is located at the point of intersection of the perpendicular bisectors of $\overline{AB}$ and $\overline{AC}$, then it will be equidistant from A, B, and C.

Answers for the Additional Problems

1. $EC = \sqrt{52}$ in.

2. Approximately 0.025 in.

Chapter Review Problems

SECTION 4.1

1. **a.** A triangle is equilateral. **b.** The triangle is isosceles. **c.** The triangle is isosceles. **d.** No conclusion is possible.

2. The triangle has an area of $9\sqrt{3}$ cm^2.

3. **a.** $w, t \Rightarrow w$ **b.** $t, s \Rightarrow t$ **c.** $s \Leftrightarrow t$

4. **a.** If a polygon has congruent angles and congruent sides, then it is a regular polygon.

b. If a polygon is regular, then it has congruent angles and congruent sides. Yes, it is true.

5. If two line segments have the same length, then they are congruent. If two line segments are congruent, then they have the same length.

6. **1.** Given

2. Definition of congruent angles

3. Given

4. Definition of supplementary angles

5. Substitution [(2) into (4)]

6. Addition

7. Division

8. Substitution [(2) into (8)]

9. Definition of a right angle

SECTION 4.2

1. a. $\angle U$ b. $\overline{SU}$ c. $\overline{RQ}$
2. a. $\triangle EDF \cong \triangle TRS, ASA$ b. $\triangle BDC \cong \triangle HGJ$, HL
3. a. Congruent by SAS b. Not necessarily congruent.

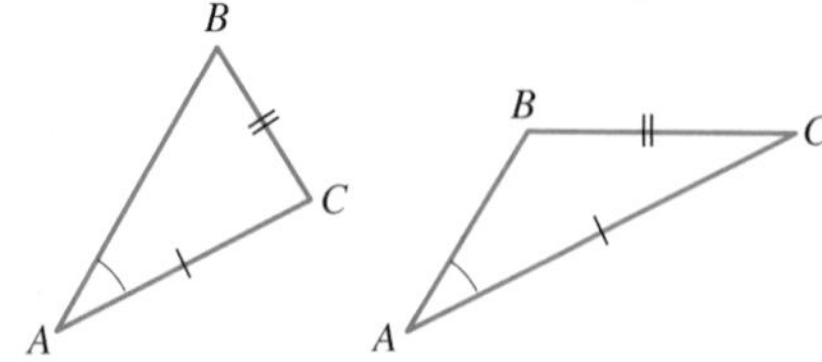

c. Congruent by LL
4. $\triangle ABE \cong \triangle DCE$ by SAS
$\triangle BAD \cong \triangle CDA$ by LL or ASA
5. **Proof**

Statement	Reason
1. $\angle D \cong \angle E$	1. Given
2. $\overline{DB} \cong \overline{EB}$	2. Given
3. $\angle DBC \cong \angle EBA$	3. Vertical angles
4. $\triangle DBC \cong \triangle EBA$	4. ASA Congruence
5. $\angle C \cong \angle A$	5. C.P.

SECTION 4.3

1. $\angle A = 50°$, $\angle B = 80°$
2. $SQ = 7.5$ cm, $\angle RSQ = 90°$
3. $\overline{EG}$ is the perpendicular bisector of $\overline{DF}$ by Theorem 4.10.
4. a. 54° b. 63° c. 63° d. 24° e. 15°
5. $\overline{XZ} \cong \overline{XZ}$ by the reflexive property. $\angle W \cong \angle Y$ because both are right angles. $\angle XZW \cong \angle XZY$ by the definition of an angle bisector. $\angle WXZ \cong \angle YXZ$ because they are both complementary to congruent angles. Thus, $\triangle XWZ \cong \triangle XYZ$ by ASA.

SECTION 4.4

1–6. Constructions

CHAPTER 4 TEST

1. T 2. T
3. F—We can write a biconditional when a conditional and its *converse* are both true.
4. T 5. T
6. F—It is called the hypothesis.
7. F—There is no AAA Congruence. At least one pair of corresponding sides must be congruent.
8. T 9. T
10. F—The congruent angles must be the *included* angles.
11. a. If a number is an integer, then it is a real number.
b. If a number is a real number, then it is an integer.
c. No, since the converse is not true.
12. 21.9 cm
13. She gains weight. Other valid conclusions include
She carries her work shoes,
She doesn't carry a lunch, etc.
14. $\triangle VWZ \cong \triangle YWX$ by SAS.
15. $\triangle ABC \cong \triangle ADC$ by HL.
16. $\triangle PQR \cong \triangle PTS$ by SAS. Also, $\triangle PTR \cong \triangle PQS$ by SAS.
17. Construct a pair of perpendicular lines to form a 90° angle. Bisect one of the 90° angles to form two 45° angles. Then bisect one of the 45° angles to form two 22.5° angles.

18–20. Constructions

21. $\angle PQT \cong \angle RQS$ because they are vertical angles. Therefore, $\triangle PQT \cong \triangle RQS$ by SAS. By C.P., $\overline{PT} \cong \overline{RS}$. Thus, $\triangle PST \cong \triangle RTS$ by SSS.
22. a. Because $ABCD$ is isosceles, $\angle BAD \cong \angle CDA$. Because $\overline{QA}$ and $\overline{PD}$ bisect $\angle BAD$ and $\angle CDA$, respectively, $\angle QAD \cong \angle PDA$. Therefore, $\triangle ARD$ is isosceles because $\overline{AR}$ and $\overline{DR}$ are opposite congruent angles.
b. Since $\angle C \cong \angle B$, $\overline{CD} \cong \overline{AB}$, and $\angle CDP \cong \angle BAQ$, $\triangle PCD \cong \triangle QBA$ by ASA.
c. From (a), $AR = DR$, and from (b), $AQ = DP$. By subtraction, $RQ = RP$, so $\triangle PQR$ is isosceles.
23. Approximately 1138 ft^2
24. He should travel along the path that is the perpendicular bisector of the line segment joining his home and his friend's house.

SECTION 5.1

1. $\angle 3$ and $\angle 5$, $\angle 9$ and $\angle 11$
3. $\angle 3$ and $\angle 8$, $\angle 11$ and $\angle 6$, $\angle 1$ and $\angle 5$, $\angle 9$ and $\angle 13$
5. 180°
7. $\angle 3 = 65°$, $\angle 8 = 65°$
9. $\angle 1 = 60°$, $\angle 2 = 120°$, $\angle 3 = 60°$, $\angle 4 = 120°$, $\angle 5 = 60°$, $\angle 6 = 45°$, $\angle 7 = 75°$, $\angle 8 = 60°$, $\angle 9 = 45°$, $\angle 10 = 75°$, $\angle 11 = 45°$, $\angle 12 = 135°$, $\angle 13 = 45°$, $\angle 14 = 135°$
11. $\angle 11$
13. $\angle 17$
15. $\angle 13 = 140°$, $\angle 5 = 40°$
17. $\angle 1 = 80°$, $\angle 2 = 100°$, $\angle 3 = 135°$, $\angle 4 = 125°$, $\angle 5 = 55°$, $\angle 6 = 125°$, $\angle 7 = 55°$, $\angle 8 = 45°$, $\angle 9 = 135°$, $\angle 10 = 45°$, $\angle 11 = 80°$, $\angle 12 = 100°$, $\angle 13 = 125°$, $\angle 14 = 55°$, $\angle 15 = 100°$, $\angle 16 = 80°$, $\angle 17 = 100°$, $\angle 18 = 80°$, $\angle 19 = 125°$, $\angle 20 = 55°$
19. F
21. T—by Corollary 5.3
23. $\angle B = 80°$—by Corollary 5.7

25. $\angle ADF = 80°$—by Corollary 5.7

27. $\angle DFE = 80°$—by Theorem 5.5

29. $\angle 1 = 71°$, $\angle 2 = 67°$, $\angle 3 = 71°$, $\angle 4 = 42°$

31. $\angle 1 = 45°$, $\angle 2 = 135°$, $\angle 3 = 55°$, $\angle 4 = 100°$

33. **Given** Lines l and m each perpendicular to line t

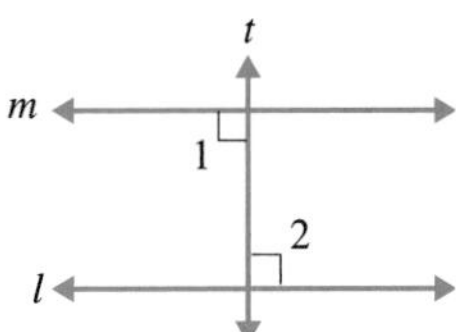

Prove $l \parallel m$

Proof

Statement	Reason
1. $l \perp t$ and $m \perp t$	1. Given
2. $\angle 1 = 90°$ and $\angle 2 = 90°$	2. Perpendicular lines form 90° angles.
3. $\angle 1 = \angle 2$	3. Substitution
4. $l \parallel m$	4. Theorem 5.1

35. Let l and m be two lines forming supplementary interior angles with transversal t as shown.

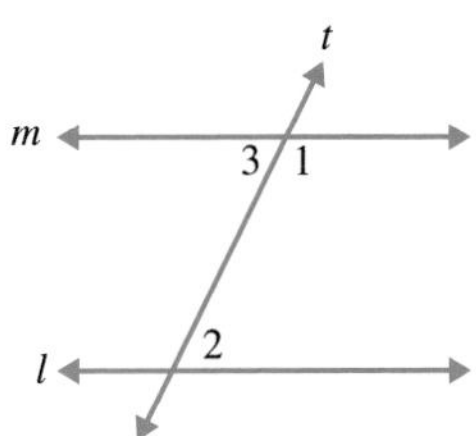

So $\angle 1 + \angle 2 = 180°$. Also, $\angle 1 + \angle 3 = 180°$ since they form a straight angle. Then, by substitution, we have $\angle 1 + \angle 2 = \angle 1 + \angle 3$, so $\angle 2 = \angle 3$. Since these are alternate interior angles, lines l and m are parallel by Theorem 5.1.

37. Let l and m be parallel and $t \perp l$.

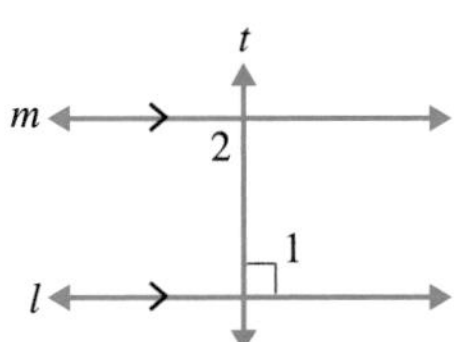

We know $\angle 1 = 90°$. By Theorem 5.5, $\angle 1 = \angle 2$. So $\angle 2 = 90°$. Thus, $m \perp t$.

39. Outline: Use Theorem 5.5 to show that a pair of alternate interior angles is congruent. Then use adjacent supplementary angles to show interior angles are supplementary.

41. **a.** 64.4° **b.** 40°

SECTION 5.2

1. T—Corollary 5.10

3. T—$\angle 2 + \angle 3 = \angle 5$ by Corollary 5.10 and $\angle 10 + \angle 11 = \angle 5$ by Corollary 5.10, so by substitution, $\angle 2 + \angle 3 = \angle 10 + \angle 11$.

5. T—Sum of angles in a triangle is 180°

7. T—Corollary 5.10

9. T—$\angle 3 = \angle 6$ by Theorem 5.5 and the sum of supplementary angles is 180°.

11. F **13.** F

15. T—$\angle 15 = \angle 8 + \angle 9 + \angle 11$ by Corollary 5.10.

17. T—$\angle 12 = \angle 8 + \angle 14$ by Corollary 5.10, and $\angle 14 = \angle 8$ because base angles of isosceles triangles are congruent, so, by substitution, $\angle 12 = 2\angle 8$.

19. T—$\angle 13 + \angle 12 = 180°$ because the sum of supplementary angles is 180°; $\angle 11 = \angle 12$ because base angles of isosceles triangles are congruent, so, by substitution, $\angle 13 + \angle 11 = 180°$.

21. $\angle 1 = 130°$, $\angle 2 = 35°$, $\angle 3 = 65°$, $\angle 4 = 115°$

23. Area $= 69 \text{ in}^2$

25. **Given** Congruent angles as marked in the following figure.

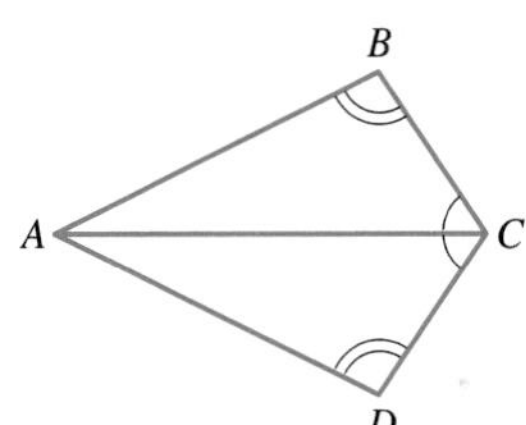

Prove $\overline{CD} \cong \overline{CB}$

Proof

Statement	Reason
1. $\angle ACD \cong \angle ACB$ and $\angle ADC \cong \angle ABC$	1. Given
2. $\overline{AC} \cong \overline{AC}$	2. Identity
3. $\triangle ACB \cong \triangle ACD$	3. AAS
4. $\overline{CD} \cong \overline{CB}$	4. C.P.

27. Given $\triangle ABC$ and $\triangle DEF$ as shown.

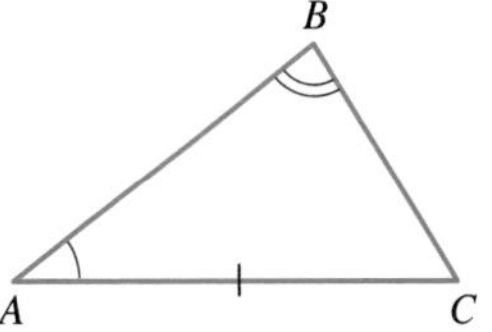

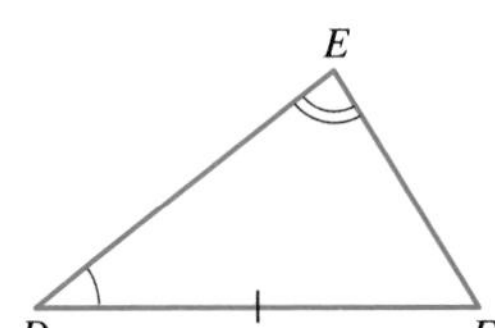

It is given that $\angle A \cong \angle D$, $\angle B \cong \angle E$, and $\overline{AC} \cong \overline{DF}$. Since $\angle A + \angle B + \angle C = 180°$ and $\angle D + \angle E + \angle F = 180°$ we have $\angle A + \angle B + \angle C = \angle D + \angle E + \angle F$. Subtracting equal angle measures from both sides, we get $\angle C = \angle F$. Thus, $\triangle ABC \cong \triangle DEF$ by ASA.

29. Outline: Use a picture similar to the one used to prove subgoal 1 and let $BP = PC$ and $\angle B = \angle C = 90°$. Use HL to show the triangles congruent and, thus, $\angle A$ is bisected.

31. Construct the angle bisector from each angle and put the marker at the intersection of these bisectors.

SECTION 5.3

1. F **3.** T—Corollary 5.15 **5.** F

7. F **9.** F **11.** T—Theorem 5.14

13. $\angle X = 115°$, $\angle Y = 65°$, $\angle W = 65°$, $\angle Z = 115°$

15. $\angle 1 = 35°$, $\angle 2 = 80°$, $\angle 3 = 125°$

17. 72°, 72°, 108°, 108° **19.** T—Theorem 5.22, Theorem 5.20

21. T—$AB = AD$, $BC = DC$, $CA = CA$, so $\triangle ADC \cong \triangle ABC$ by SSS Congruence.

23. T—Theorem 5.23, Theorem 5.9 (Angle Sum in a Triangle Theorem)

25. T—Opposite angles of a parallelogram are congruent.

27. $\angle 1 = 30°$, $\angle 2 = 60°$, $\angle 3 = 60°$

29. Side $= 4\sqrt{3} \approx 6.93$

31. $\angle 1 = 140°$, $\angle 2 = 30°$, $\angle 3 = 110°$, $\angle 4 = 40°$

33. Height = 8 cm, area = 160 cm^2 **35.** Diagonal = 16 cm

37. Given $ABCD$ as shown

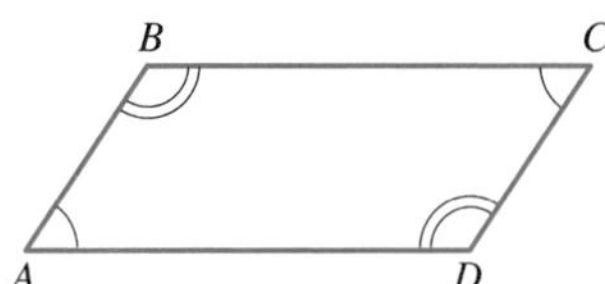

Prove $ABCD$ is a parallelogram.

Proof

Statement	Reason
1. $\angle A = \angle C$ and $\angle B = \angle D$	1. Given
2. $\angle A + \angle B + \angle C + \angle D = 360°$	2. Angle sum in a rectangle is 360°
3. $2(\angle A) + 2(\angle B) = 360°$	3. Substitution
4. $\angle A + \angle B = 180°$	4. Divide both sides by 2.
5. $\overline{BC} \parallel \overline{AD}$	5. Corollary 5.4
6. $\angle A + \angle D = 180°$	6. $\angle B = \angle D$
7. $\overline{AB} \parallel \overline{CD}$	7. Corollary 5.4
8. $ABCD$ is a parallelogram.	8. Definition

39. Given is rhombus $ABCD$ with diagonals $\overline{AC}$ and $\overline{BD}$.

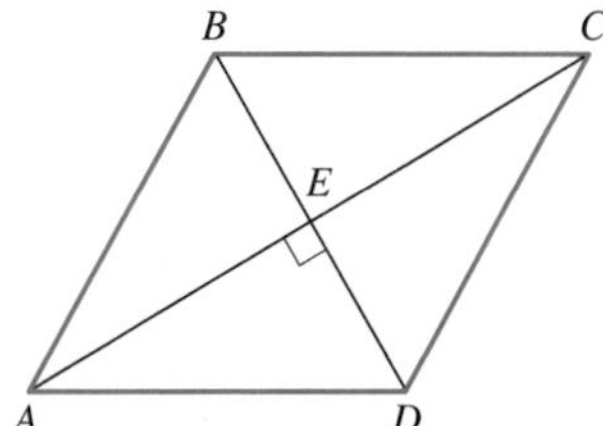

By Theorem 5.23, we know that $\overline{AC} \perp \overline{BD}$, so $\angle AED = 90°$. Since the sum of the angles in $\triangle AED$ is 180°, we know $\angle CAD + \angle BDA + 90° = 180°$, or $\angle CAD + \angle BDA = 90°$. Thus, $\angle CAD$ and $\angle BDA$ are complementary.

41. Outline: Draw in bisectors of two consecutive angles. Use the fact that any two consecutive angles of a parallelogram are supplementary and that the bisectors divide each angle into two congruent angles.

43. Since $\overline{AE}$ bisects $\angle A$, we have $\angle BAE = \angle EAF = \frac{1}{2}(\angle A)$. In a similar manner, $\angle ECF = \angle DCF = \frac{1}{2}(\angle C)$. Since $ABCD$ is a parallelogram, $\angle A = \angle C$, so $\frac{1}{2}(\angle A) = \frac{1}{2}(\angle C)$. Thus, $\angle EAF = \angle ECF$. Also, $\angle BEA = \angle EAF$ and $\angle ECF = \angle DFC$ since each is a pair of alternate interior angles formed by parallel lines. Thus, by substitution, $\angle BEA = \angle DFC$. Since these angles are equal, so are $\angle CEA$ and $\angle AFC$. Therefore, by Theorem 5.18, $AECF$ is a parallelogram since two pairs of opposite angles are congruent.

45. By Theorem 5.21, if the diagonals bisect each other, the quadrilateral is a parallelogram. Because the legs act as diagonals of a quadrilateral, the board will always be parallel to the ground.

47. 11 times

SECTION 5.4

1. F **3.** T—Theorem 5.26, Theorem 5.20

5. T—Theorem 5.28 **7.** F

9. T—$AB = BA$, $AC = BD$ by Theorem 5.28 and $BC = AD$, so $\triangle ABD \cong \triangle BAC$ by SSS Congruence.

11. Area = 108 cm^2

13. T—definition of rhombus, Theorem 5.23

15. T—Theorem 5.26, Theorem 5.20 **17.** T—Theorem 5.5

19. T—$AB = BC$, $BE = CE$, $EA = EB$ by Theorem 5.28, Theorem 5.20, so $\triangle ABE \cong \triangle BCE$ by SSS.

21. T—Theorem 5.25, definition of rectangle

23. Side $= 8\sqrt{2} \approx 11.31$ mm **25.** T—Theorem 5.30

27. T—$AB = CD$, $AD = AD$, $\angle BAD \cong \angle CDA$ by Theorem 5.30. So $\triangle BAD \cong \triangle CDA$ by SAS.

29. T—Problem 27; corresponding parts of congruent triangles are congruent.

31. T—$\angle ABD = \angle DCA$ by Problem 27, corresponding parts of congruent triangles are congruent; $\angle ABD = \angle DCA$ and $\angle ABC = \angle DCB$ by Theorem 5.30. So $\angle DBC = \angle ACB$, which implies that $\triangle BEC$ is isosceles. So $BE = EC$.

33. Legs $= \sqrt{20} \approx 4.47$ in. **35.** $EH = 45$ in.

37. $WZ = \sqrt{468}$ in. ≈ 21.63 in.

39. Let $ABCD$ be a rhombus with one right angle. Since a rhombus has four congruent sides, $ABCD$ must be a square.

41. Let $ABCD$ be a rhombus with congruent diagonals. By Theorems 5.22 and 5.29, $ABCD$ is a rectangle. A rhombus that is a rectangle is a square.

43. Given $ABCD$ is a parallelogram with $\angle A = 90°$.

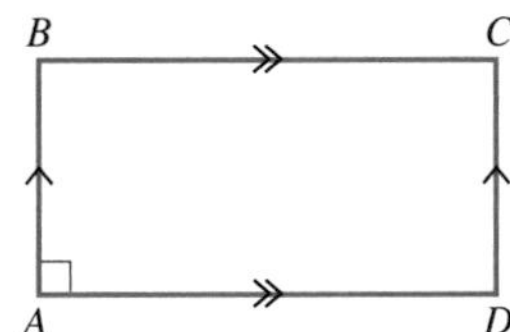

Prove $ABCD$ is a rectangle.

Proof

Statement	Reason
1. $ABCD$ is a parallelogram with $\angle A = 90°$.	1. Given
2. $\angle C = 90°$	2. Opposite angles of a parallelogram are equal.
3. $\angle A + \angle B = 180°$ and $\angle A + \angle D = 180°$	3. Interior angles on the same side of the transversal
4. $90° + \angle B = 180°$ and $90° + \angle D = 180°$	4. Substitution
5. $\angle B = 90°$ and $\angle D = 90°$	5. Subtracting equals from both sides
7. $ABCD$ is a rectangle.	6. Definition

45. Let $ABCD$ be a trapezoid with congruent diagonals as shown.

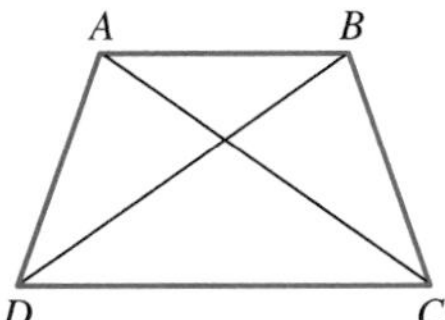

Extend $\overline{AB}$ and $\overline{DC}$ as shown to form trapezoids $EADH$ and $BFGC$ congruent to $ABCD$.

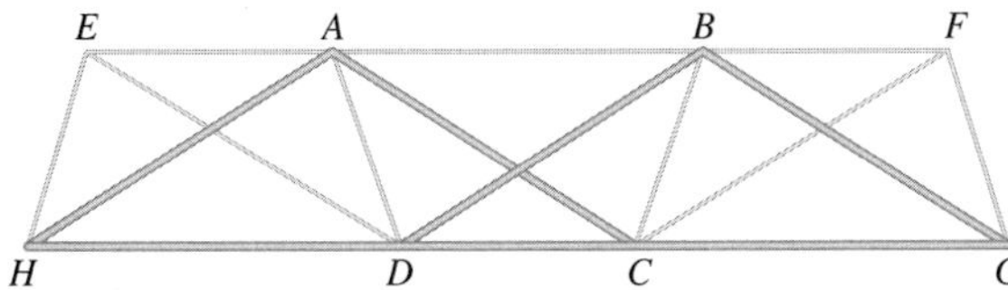

Then $\triangle HAC \cong \triangle GBD$ by SSS. Therefore, $\angle ACH \cong \angle BDG$ by C. P. So $\triangle ACD \cong \triangle BCD$ by SAS. Thus, $\overline{AD} \cong \overline{BC}$ by C.P.

47. a. The door frame could be a rectangle, but it also could be a parallelogram that is not a rectangle.

b. The door frame could be a rectangle, but it also could be an isosceles trapezoid.

c. The door frame could be a rectangle, but it also could be a trapezoid or a general quadrilateral with one right angle.

SECTION 5.5

1–19. Constructions

21. $A = 42.92\text{ cm}^2$

23. $M_1M_3 = M_2M_4$ and $\overline{M_1M_3} \perp \overline{M_2M_4}$

25. Let $ABCD$ represent the quadrilateral constructed. By construction, one angle should be congruent to $\angle A$ and the opposite sides should be parallel.

27. The same construction used in problem 6 could be used with $\angle A = 90°$. The resulting parallelogram will have four right angles, so it will be a rectangle.

29. One construction could be to first construct the 60° angle by constructing an equilateral triangle. Then bisect the angle making the length of the bisector d. Next construct a perpendicular bisector l of d. The points where l intersects the sides of the 60° angle will be two vertices of the rhombus. The fourth vertex will be at the end of d.

31. Copy one of the diagonals, say b. Then find the midpoint of b. Bisect the other diagonal of length c. Using SSS, form a triangle using lengths half of b, half of c, and a. Extending the diagonal sides of this triangle to lengths b and c will give the other two vertices of the parallelogram. Theorem 5.21 can be used to prove that the resulting figure is a parallelogram.

33. $CU = UQ$

35. Consider quadrilateral $CEFD$. Label the intersection of $\overline{CF}$ and $\overline{DE}$ as point G.

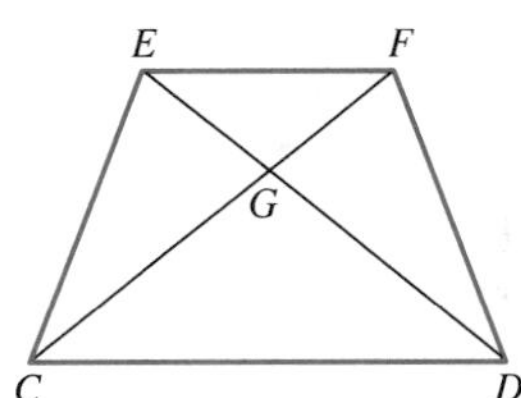

By construction, we know $\overline{EC} \cong \overline{FD}$, $\overline{CF} \cong \overline{DE}$, and $\overline{CD} \cong \overline{CD}$. Therefore, $\triangle ECD \cong \triangle FDC$ by SSS and $\angle EDC \cong \angle FCD$ by C.P. By Theorem 4.7, $\overline{CG} \cong \overline{DG}$ and $\triangle CDG$ is isosceles. Also by construction, we know $\overline{EC} \cong \overline{FD}$, $\overline{CF} \cong \overline{DE}$, and $\overline{EF} \cong \overline{EF}$. Therefore, $\triangle ECF \cong \triangle FDE$ by SSS and $\angle CFE \cong \angle DEF$ by C.P. By Theorem 4.7, $\overline{EG} \cong \overline{FG}$ and $\triangle EFG$ is isosceles. Because $\angle EGF$ and $\angle CGD$ are congruent vertical angles and because $\triangle EGF$ and $\triangle CGD$ are isosceles triangles, we know that $\angle FCD \cong \angle EDC \cong \angle CFE \cong \angle DEF$. Since alternate interior angles are congruent, $\overline{CD} \parallel \overline{EF}$.

Answers for the Additional Problems

1. Suppose c^2 is a square and a and b are both odd. Then $a = 2m + 1$ and $b = 2n + 1$, for some whole numbers m and n. So

$$\begin{aligned} a^2 + b^2 &= (2m + 1)^2 + (2n + 1)^2 \\ &= 4m^2 + 4m + 1 + 4n^2 + 4n + 1 \\ &= 4(m^2 + m + n^2 + n) + 2 \end{aligned}$$

This is not a perfect square because it has a factor of 2 but not a factor of 4. Therefore, it must be true that a and b cannot *both* be odd.

2. Suppose that $\overline{BD}$ and $\overline{BE}$ are altitudes to side $\overline{AC}$ as shown.

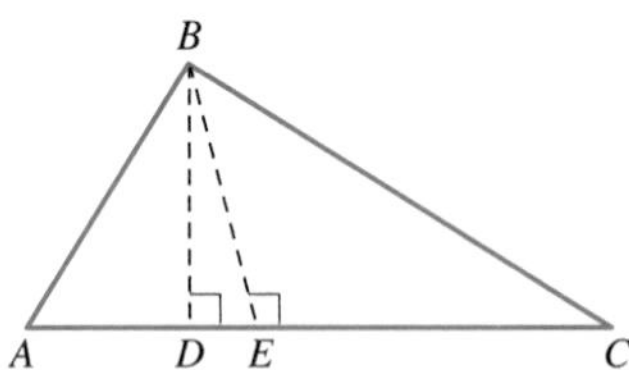

Then $\triangle BDE$ is a triangle with two right angles, which is impossible. Thus, $\overline{BD}$ is the only altitude to $\overline{AC}$.

Chapter Review Problems

SECTION 5.1

1. a. F

   b. T—By Theorem 5.5, $\angle 8 \cong \angle 15$. We know $\angle 8 \cong \angle 13$ and $\angle 17 \cong \angle 15$ by vertical angles. By substitution, $\angle 13 \cong \angle 17$.

   c. F

   d. T— $\angle 2$ and $\angle 8$ are supplementary by Corollary 5.8 and $\angle 8 \cong \angle 13$ by vertical angles.

2. a. $\angle 8$ and $\angle 15$, $\angle 7$ and $\angle 2$

   b. $\angle 13$ and $\angle 17$, $\angle 14$ and $\angle 16$

   c. $\angle 2$ and $\angle 14$, $\angle 8$ and $\angle 17$, $\angle 13$ and $\angle 15$, $\angle 7$ and $\angle 16$

3. $\angle 1 = 90°$, $\angle 2 = 50°$, $\angle 3 = 40°$, $\angle 4 = 125°$, $\angle 5 = 55°$, $\angle 6 = 140°$, $\angle 7 = 50°$, $\angle 8 = 130°$, $\angle 9 = 75°$, $\angle 10 = 105°$, $\angle 11 = 75°$, $\angle 12 = 105°$, $\angle 13 = 130°$, $\angle 14 = 50°$, $\angle 15 = 130°$, $\angle 16 = 50°$, $\angle 17 = 130°$, $\angle 18 = 90°$, $\angle 19 = 90°$, $\angle 20 = 90°$

4. Outline: Use Theorem 5.5 to show that a pair of alternate interior angles is congruent. Then use vertical angles to show a pair of alternate exterior angles is congruent. Last, use a pair of adjacent, supplementary exterior angles and substitution to show that a pair of exterior angles on the same side of the transversal is supplementary.

SECTION 5.2

1. $\angle CBD$, $\angle BDG$, and $\angle ADF$ **2.** $x = 13.5°$

3. a. T—The triangles are congruent by the HA congruence theorem (Corollary 5.12).

   b. F

   c. T—By Corollary 5.10, $\angle ABE + \angle EBD = \angle BCD + \angle CDB$ and $\angle CDB \cong \angle BAE$.

4. By Corollary 5.10, we know $\angle CDE = \angle CBD + \angle BCD$ and $\angle CBA = \angle CDB + \angle BCD$. We know $\angle CBD \cong \angle CDB$ by Theorem 4.5. By substitution, $\angle CDE = \angle CDB + \angle BCD = \angle CBA$. Thus, $\angle CBA \cong \angle CDE$.

SECTION 5.3

1. $\overline{AB} \cong \overline{DC}$, $\overline{AD} \cong \overline{BC}$, $\overline{BE} \cong \overline{DE}$, $\overline{AE} \cong \overline{CE}$

2. 146°, 34°, 146°, 34° **3.** 225 in^2

4. $\angle WXY = 126°$, $\angle XYZ = 54°$

5. $PQ = QR = RS = SP$ and $\angle Q \cong \angle S$ by the definition of a rhombus. $AQ = \frac{1}{2}PQ$ and $DS = \frac{1}{2}PS$. Since $PQ = PS$, we know $AQ = DS$. In a similar manner, $BQ = CS$. Therefore, $\triangle AQB \cong \triangle DSC$ by SAS. By corresponding parts, $AB = DC$. In a similar fashion, we can show that $AD = BC$. Because both pairs of opposite sides of $ABCD$ are congruent, $ABCD$ is a parallelogram.

SECTION 5.4

1. $ABCD$ can be a rectangle, a square, or an isosceles trapezoid.

2. Rectangles and squares

3. $CF = 6$ cm, $CB = 6$ cm, $CD = 6\sqrt{3}$ cm

4. 97.78 cm^2 **5.** 22 cm

SECTION 5.5

1–4. Constructions

5. Answers may vary depending on your construction. One solution is to construct an angle congruent to A, with sides of length a. Label the two endpoints of these sides B and D. Then locate point C by making arcs of radius a from points B and D.

Justification

Statement	Reason
1. $\angle BAD \cong \angle A$	1. Construction 2
2. $AD = AB = BC = CD = a$	2. Construction 1
3. $ABCD$ is a rhombus with angle congruent to $\angle A$.	3. Definition of a rhombus

6. Construction

CHAPTER 5 TEST

1. T
2. F—Corresponding angles must be congruent.
3. F—Angle sum is 360°
4. F—Bases of the trapezoid are not congruent
5. F—Any parallelogram has opposite angles congruent.
6. T **7.** T
8. F—The diagonals of an isosceles trapezoid are congruent.
9. T
10. F—Any point on the angle bisector is equidistant from the sides of the angle, but not from just any two points on the sides.

11. 20°, 160°, 160°

12. $\angle CBD = 30°$, $BD = 2$, $AC = 2$, $AD = \sqrt{3}$

13. $ABCD$ is a rhombus.

14. 24 yd

15. $\angle 6$ and $\angle 11$, $\angle 7$ and $\angle 10$, $\angle 8$ and $\angle 13$, $\angle 9$ and $\angle 12$

16. $\angle 6$ and $\angle 14$, $\angle 7$ and $\angle 15$, $\angle 8$ and $\angle 16$, $\angle 4$ and $\angle 12$, $\angle 5$ and $\angle 13$, $\angle 9$ and $\angle 17$

17. $\angle 1 = 30°$, $\angle 5 = 40°$, $\angle 10 = 75°$

18–20. Constructions

21. The diagonals of a square are congruent and bisect each other. Thus, $\triangle ADE \cong \triangle ABE$ by SSS.

22. Because E and F are midpoints of $\overline{AB}$ and $\overline{CD}$, respectively, and $ABCD$ is a parallelogram, $AE = BE = CF = DF$. Thus, $\triangle EBC \cong \triangle FDA$ by SAS. (Also $\angle B \cong \angle D$, and $\overline{BC} \cong \overline{AD}$.) By C.P., $\overline{CE} \cong \overline{AF}$. Because $\overline{CE} \cong \overline{AF}$ and $\overline{AE} \cong \overline{CF}$, $AECF$ is a parallelogram.

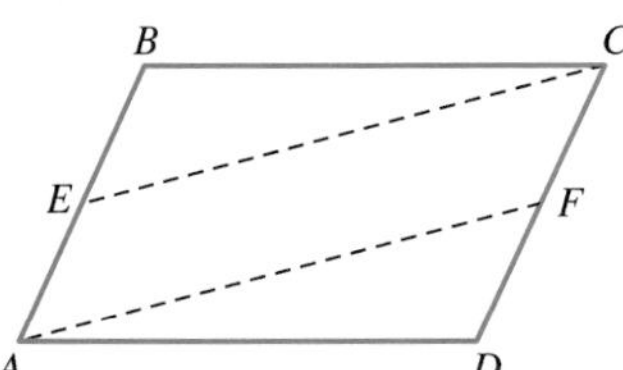

23. Approximately 34 ft^3

24. The two crossbars are the same length, as are the distances between them on the vertical supports. Thus, these four segments form a parallelogram. When one support bar descends, the other rises, but they remain parallel and vertical because of the parallel crossbars. The pans are perpendicular to the supports, so they remain horizontal.

SECTION 6.1

1. $\frac{1}{4}$ **3.** $\frac{5}{1}$

5. a. $\frac{11}{24} = \frac{11 \times 6}{24 \times 6} = \frac{66}{144}$ **b.** $\frac{14x}{91x} = \frac{4 \times 3.5x}{26 \times 3.5x} = \frac{4}{26}$

c. $\frac{3.8}{9.5} = \frac{3.8 \times 7}{9.5 \times 7} = \frac{26.6}{66.5}$

7. Yes, $18(8.5) = 153 = 51(3)$ **9.** $n = 30$

11. $n = 20$ **13.** $s = 30$

15. $d = \frac{5}{2}$ **17.** $x = 320$

19. $x = 15$ **21.** 6

23. $\sqrt{10\pi} \approx 5.6$

25. For a cylinder with a height, h, and a base with radius r, its volume is $\pi r^2 h$. The new cylinder has volume $\pi(2r)^2(3h) = 12\pi r^2 h$. Thus, the ratio of the volume of the original cylinder to the volume of the new cylinder is $\pi r^2 h{:}12\pi r^2 h$, or 1:12.

27. If $\frac{a}{b} = \frac{c}{d}$, then $ad = bc$ (by cross multiplication) and $da = bc$. Thus, $\frac{d}{b} = \frac{c}{a}$ (by cross multiplication).

29. If $\frac{a}{b} = \frac{c}{d}$, then $ad = bc$ (by cross multiplication) and $bc = ad$. Thus $\frac{b}{a} = \frac{d}{c}$ (by cross multiplication).

31. Given $\frac{a}{b} = \frac{c}{d}$

Prove $\frac{a+b}{b} = \frac{c+d}{d}$

Proof

Statement	Reason
1. $\frac{a}{b} = \frac{c}{d}$	**1.** Given
2. $ad = bc$	**2.** Cross multiplication
3. $ad + bd = bc + bd$	**3.** Addition
4. $(a + b)d = b(c + d)$	**4.** Distributive property
5. $\frac{a+b}{b} = \frac{c+d}{d}$	**5.** Cross multiplication

33. If $\frac{a}{b} = \frac{c}{b}$, then $ab = bc$ by cross multiplication. Dividing both sides by b yields $a = c$.

35. Outline: To show that $\frac{a}{b} = \frac{a+c}{b+d}$, cross multiply and simplify. Then work backward starting with $\frac{a}{b} = \frac{c}{d}$.

37. Outline: Show that if $\frac{a_1}{b_1} = \frac{a_2}{b_2} = \cdots = \frac{a_n}{b_n}$, then $\frac{a_1 + a_2 + \cdots + a_n}{b_1 + b_2 + \cdots + b_n} = \frac{a}{b}$. To prove, extend the proof for problem 32.

39. a. 105 mi, 350 mi, $35n$ mi **b.** 28.6 in.

41. 24,530 mi

43. a. 221 returns **b.** 27 more returns

45. 499.2 lb/ft^2 **47.** 29 in.

49. 17 in. **51.** 206 mph

SECTION 6.2

1. $\triangle ABC \sim \triangle DEC$ by AA Similarity

3. $\triangle LMN \sim \triangle OPQ$ by LL Similarity

5. $DE = 10$, $DF = 14$ **7.** $XY = \frac{27}{8}$ cm, $VW = \frac{128}{9}$ cm

9. $EG = 4$, $HI = 3\sqrt{3}$, $FH = 2\sqrt{3}$

11. $AC = \frac{56}{9}$ in., $BC = \frac{65}{4}$ in.

13. a. Bases: 1 : 2, Heights: 1 : 2, Areas: 1 : 4

b. Bases: 2 : 3, Heights: 2 : 3 Areas: 4 : 9

c. The ratio of the areas is equal to the square of the ratio of the linear dimensions.

15. Outline: Use SSS Similarity.

17. Outline: Since base angles in isosceles triangles are congruent, all *four* base angles are congruent. Use AA Similarity.

19. Outline: In trapezoid $ABCD$, show that transversals $\overline{AC}$ and $\overline{BD}$ form congruent alternate interior angles. Use AA Similarity.

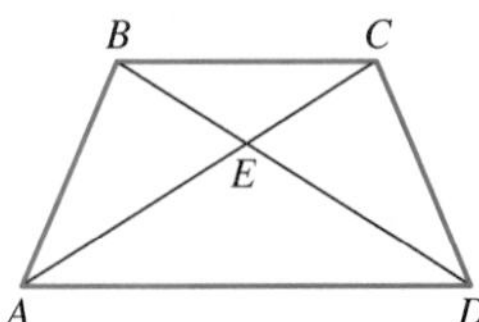

21. Given $\triangle ABC \sim \triangle A'B'C'$ with medians $\overline{AM}$ and $\overline{A'M'}$

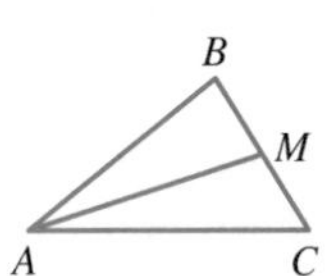

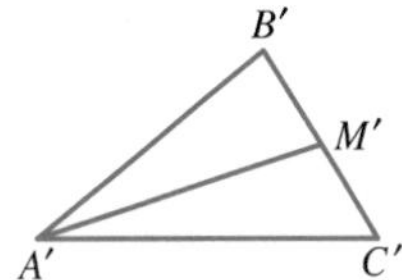

Prove $\dfrac{AM}{A'M'} = \dfrac{AB}{A'B'}$

Proof

Statement	Reason
1. $\triangle ABC \sim \triangle A'B'C'$	**1.** Given
2. $\angle B \cong \angle B'$	**2.** Definition of similar triangles
3. $\dfrac{BC}{AB} = \dfrac{B'C'}{A'B'}$	**3.** Definition of similar triangles
4. $\dfrac{BC}{2AB} = \dfrac{B'C'}{2A'B'}$	**4.** Dividing both sides by 2
5. $\dfrac{BM}{AB} = \dfrac{B'M'}{A'B'}$	**5.** Substitution
6. $\triangle ABM \sim \triangle A'B'M'$	**6.** SAS Similarity
7. $\dfrac{AM}{A'M'} = \dfrac{AB}{A'B'}$	**7.** Definition of similar triangles

23. Outline: Let $\triangle ABC$ be similar to $\triangle A'B'C'$ where $\overline{BM}$ and $\overline{B'M'}$ are altitudes.

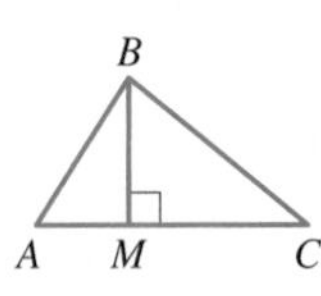

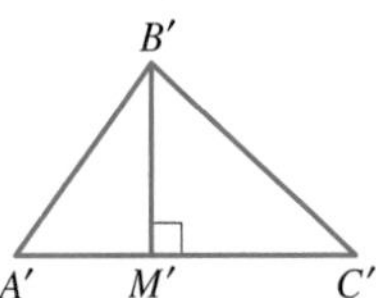

Use the result proven in problem 22.

25. Outline: The volumes of the cones will be $\frac{1}{3}\pi(AC)^2(BC)$ and $\frac{1}{3}\pi(A'C')^2(B'C')$. Using the fact that $\dfrac{AC}{BC} = \dfrac{A'C'}{B'C'}$, show that $\dfrac{\frac{1}{3}\pi(AC)^2(BC)}{\frac{1}{3}\pi(A'C')^2(B'C')} = \left(\dfrac{AC}{A'C'}\right)^3$.

27. Outline: Show that $\angle A \cong \angle A'$, and $\angle B \cong \angle B'$, and use AA Similarity.

29. Given: $\triangle ABC$ and $\triangle A'B'C'$ with parallel sides as shown. Extend $\overline{B'C'}$ to intersect $\overline{AB}$ at D.

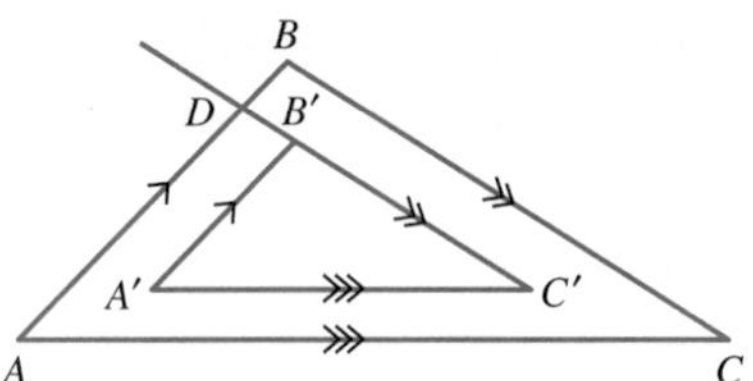

$\angle A'B'C' \cong \angle ADB'$ and $\angle ADB' \cong \angle ABC$ since they are pairs of corresponding angles. Therefore, $\angle A'B'C' \cong \angle ABC$. In a similar manner, $\angle B'C'A' \cong \angle BCA$. So $\triangle ABC \sim \triangle A'B'C'$ by AA Similarity.

31. 22 m

33. $\triangle AEB \sim \triangle CED$ by SAS Similarity, therefore $\angle DCE = \angle EAB$. Also, $\triangle ABE$ is isosceles, so $\angle EAB = \angle EBA$. This implies that $\overline{AB} \parallel \overline{DC}$ because alternate interior angles are congruent. So the board will always be parallel to the ground.

35. 6.53 ft

37. a. 60 cm^3 **b.** $d \approx 9.5$ cm

39. Carol's shadow $= 6\frac{9}{11}$ ft $\approx$ 6 ft, 10 in.

Tom's shadow $= 8\frac{2}{11}$ ft $\approx$ 8 ft, 2 in.

41. a. 8.75 ft **b.** 1.5 in.

SECTION 6.3

1. $6\sqrt{6}$ **3.** $\dfrac{8}{3}$ **5.** 2

7. $AC = \dfrac{15}{4}$ **9.** $AD = \dfrac{15}{4}$ **11.** $AB = \dfrac{28}{5}$

13. $EF = \dfrac{9}{4}, BC = \dfrac{8}{3}$ **15.** $AB = 6, AG = 4$

17. $MN = 6$ cm **19.** 105 cm^2

21. Given $\angle A \cong \angle C$, we know $\overline{AB} \cong \overline{CB}$ by Theorem 4.7. $\overline{AM} \cong \overline{CN}$ since M and N are midpoints of $\overline{AB}$ and $\overline{BC}$. Given $MNPQ$ is a rectangle, $\angle MQA \cong \angle NPC$ since they are both supplementary to right angles. Thus, $\triangle MQA \cong \triangle NPC$ by AAS. Thus, the area of $MNPQ$ is 118.79 m^2.

23. Observe that x is the mean proportion between a and 1. Thus, $\dfrac{a}{x} = \dfrac{x}{1}$, or $a = x^2$.

25. Given $\overline{DE} \parallel \overline{AC}$, we know corresponding angles are congruent, so show that $\triangle ABC \sim \triangle DBE$ by AA Similarity. Corresponding sides are proportional in similar triangles so $\dfrac{AB}{DB} = \dfrac{CB}{EB}$. Then use the result that if $\dfrac{a}{b} = \dfrac{c}{d}$, then $\dfrac{a-b}{b} = \dfrac{c-d}{d}$ to show $\dfrac{AB - DB}{DB} = \dfrac{CB - EB}{EB}$.

Simplify to show $\dfrac{AD}{DB} = \dfrac{CE}{EB}$.

27. Given M is the midpoint of $\overline{AB}$ in the right $\triangle ABC$.

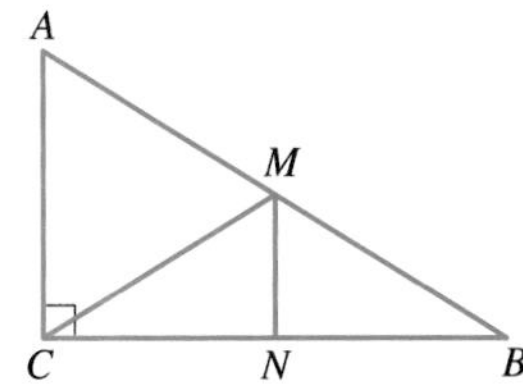

Prove $MA = MC = MB$

Proof

Statement	Reason
1. $MA = MB$	1. Definition of midpoint
2. Construct $\overline{MN}$ perpendicular to $\overline{CB}$.	2. Construction
3. $CN = NB$	3. Theorem 6.10
4. $\triangle MCN \cong \triangle MBN$	4. LL
5. $MC = MB$	5. C.P.

29. Given $\overline{BE}$ bisects $\angle ABC$ and $\dfrac{AB}{BE} = \dfrac{BD}{BC}$.

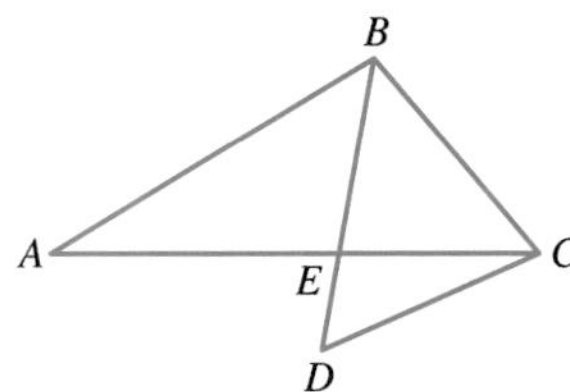

Prove $\angle A = \angle D$

Proof

Statement	Reason
1. $\angle ABE \cong \angle DBC$	1. Definition of angle bisector
2. $\dfrac{AB}{BE} = \dfrac{BD}{BC}$	2. Given
3. $\triangle ABE \sim \triangle DBC$	3. SAS Similarity
4. $\angle A = \angle D$	4. Definition of similar triangles

31. Given $\triangle ABC$ as shown

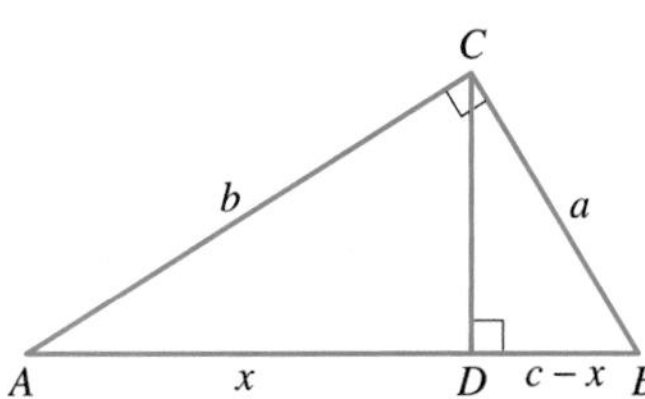

$\dfrac{c}{a} = \dfrac{a}{c - x}$ and $\dfrac{c}{b} = \dfrac{b}{x}$. Therefore, $c(c - x) = a^2$ and $cx = b^2$. Thus, $c^2 - cx = a^2$ and $cx = b^2$ leads to $c^2 - b^2 = a^2$ and $a^2 + b^2 = c^2$.

33. Given Equilateral $\triangle ABC$ with midsegment $\overline{MN}$.

Prove $\triangle MBN$ is equilateral.

Proof

Statement	Reason
1. $AB = BC = AC$ and $\overline{MN}$ a midsegment	1. Given
2. $MN = \frac{1}{2}AC$	2. Midsegment Theorem
3. $MB = \frac{1}{2}AB$ and $NB = \frac{1}{2}BC$	3. Definition of midpoint
4. $MB = NB = MN$	4. Substitution
5. $\triangle MBN$ is equilateral.	5. Definition of equilateral triangle

35. Outline: Draw in the diagonals of the rhombus. Then use the Midsegment Theorem and the fact that the diagonals of a rhombus are perpendicular.

37. Outline: Draw in the diagonals of the isosceles trapezoid. Then use the Midsegment Theorem to show opposite sides of the midquad are congruent. Show adjacent sides of the midquad are congruent by using Theorem 5.30 and congruent triangles.

39. Outline: Use the Midsegment Theorem and SSS Similarity.

41. 8.14 in. by 8.60 in.

SECTION 6.4

1. $\sin A = \dfrac{5}{13}$

3. $\tan A = \dfrac{5}{12}$

5. $\cos B = \dfrac{5}{13}$

7. $\tan A = \dfrac{7}{5} = 1.4$

9. $\cos A = \dfrac{7}{11} \approx 0.64$

11. $\cos B = \dfrac{\sqrt{2}}{2} \approx 0.71$

13. $a = 5.03$

15. $c \approx 9.72$

17. $\angle A \approx 35.5°$

19. $\angle A \approx 46.7°$

21. $\cos B = 0.5$

23. $c \approx 0.85$

25. $\angle A \approx 19.5°, \angle B \approx 70.5°, CA = 8\sqrt{2} \approx 11.3$

27. $AB \approx 13.7$ m, $BC \approx 7.8$ m, $\angle A = 35°$

29. Theorem 6.13: $\dfrac{\sin 45°}{\cos 45°} = \dfrac{0.7071}{0.7071} = 1 = \tan 45°$

Theorem 6.14: $\sin^2 45° + \cos^2 45° = (0.7071)^2 + (0.7071)^2 = 1$

31. Theorem 6.13: $\dfrac{\sin 30°}{\cos 30°} = \dfrac{0.5}{0.866} = 0.5774 = \tan 30°$

Theorem 6.14: $\sin^2 30° + \cos^2 30° = (0.5)^2 + (0.866)^2 = 1$

33. $a = 5, b = 5\sqrt{3} \approx 8.7, \angle B = 60°$

35. $c = 26, \angle A \approx 67.4°, \angle B \approx 22.6°$

37. $\angle B = 67°, a = 1.7, c \approx 4.3$

39. $\angle A = 78°, a \approx 10.1, b \approx 2.1$

41. Area $\approx 390.1\text{ ft}^2$ **43.** Area $\approx 66.5\text{ in}^2$

45. Area $\approx 337.4\text{ cm}^2$

47. $(\cos A)(\tan A) = \frac{b}{c} \cdot \frac{a}{b} = \frac{a}{c} = \sin A$

49. Let $\triangle ABC$ be an isosceles right triangle.

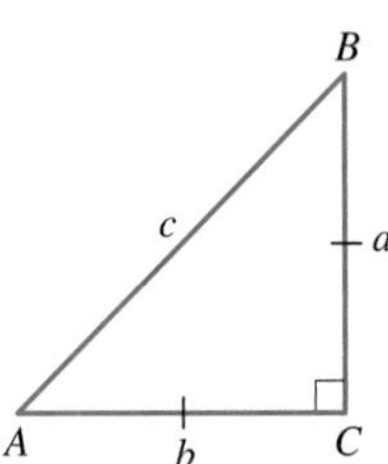

$\sin A = \frac{a}{c} = \frac{b}{c} = \cos A$

51. Let $\triangle ABC$ be as shown.

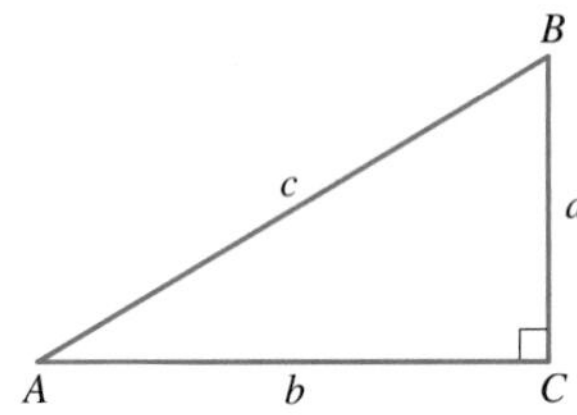

$$\sin A = \frac{a}{c} = \cos B = \cos(180° - 90° - A)$$
$$= \cos(90° - A)$$

53. Let $\triangle ABC$ be as shown.

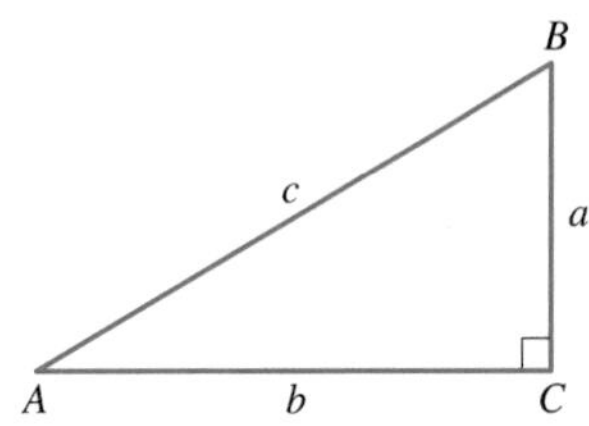

$$\frac{1}{\cos A} = \frac{1}{b/c} = \frac{c}{b} = \sec A$$

55. Let $\triangle ABC$ be as shown.

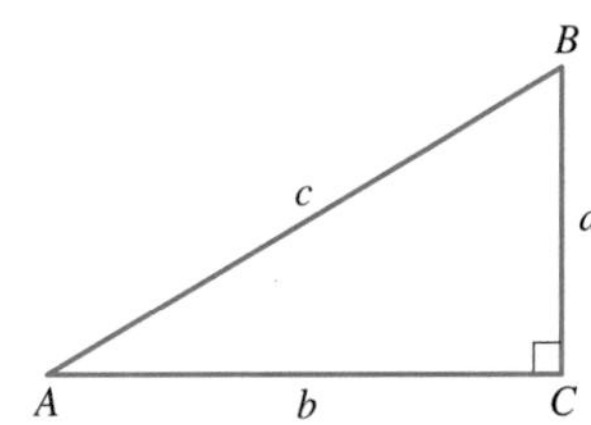

$$\frac{\cos A}{\sin A} = \frac{\frac{b}{c}}{\frac{a}{c}} = \frac{b}{a} = \cot A$$

57. Let $\triangle ABC$ be as shown.

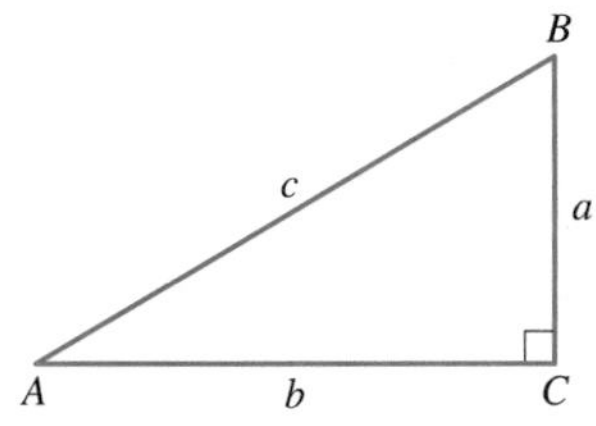

$$\csc^2 A - \cot^2 A = \left(\frac{c}{a}\right)^2 - \left(\frac{b}{a}\right)^2 = \frac{c^2 - b^2}{a^2} = \frac{a^2}{a^2} = 1$$

59. 92 ft **61.** 42.89 ft **63.** 80.30 ft

65. a. 50 ft **b.** 46.9 ft

67. 42.3 ft **69.** 126 m

SECTION 6.5

1. $c \approx 20.7$ **3.** $b \approx 3.7$ **5.** $a \approx 45.5$

7. $\angle B \approx 19.3°$ **9.** $\angle A \approx 58.6°$

11. $\angle B \approx 53.5°, b \approx 9.7, c \approx 7.3$

13. $\angle A \approx 69.0°$, $\angle C \approx 43.0°, a \approx 91.7$

15. $\angle A \approx 61.3°$, $\angle B \approx 45.7°, a \approx 6.6$

17. $a \approx 62.8$ **19.** $b \approx 40.5$ **21.** $c \approx 20.2$

23. $\angle B \approx 34.8°$ **25.** $\angle C \approx 68.9°$

27. $\angle B \approx 35.0°$, $\angle C \approx 60.0°, a \approx 62.8$

29. $\angle A \approx 54.9°$, $\angle B \approx 85.3°$, $\angle C \approx 39.8°$

31. $\angle A \approx 53.4°$, $\angle B \approx 78.6°$, $\angle C \approx 47.9°$

33. a. Law of Sines **b.** $b \approx 33.8$ cm

35. a. Law of Cosines **b.** $\angle A \approx 36.7°$

37. 363.39 in^2

39. By the Law of Cosines we know

$$b^2 = a^2 + a^2 - 2a^2 \cos B$$
$$= 2a^2 - 2a^2 \cos B$$
$$= 2a^2(1 - \cos B)$$

41. Area of a triangle is $A = \frac{1}{2}bh$. Using $\triangle ABC$, we can express h in terms of $\angle A$ and side c as follows:

$$h = c \sin A$$

Substituting this value of h into the area formula, we have

$$A = \frac{1}{2}bc \sin A$$

Next, use $\sin^2 A + \cos^2 A = 1$ to express $\sin A$ in terms of $\cos A$.

$$\sin A = \sqrt{1 - \cos^2 A}$$

Substitute this into the previous area equation.

$$A = \frac{1}{2}bc\sqrt{1 - \cos^2 \text{A}}$$

Now use the Law of Cosines to express cos A in terms of the sides.

$$\cos A = \frac{-a^2 + b^2 + c^2}{2bc}$$

Substitute the value of cos A into the previous equation.

$$A = \frac{1}{2}bc\sqrt{1 - \left(\frac{-a^2 + b^2 + c^2}{2bc}\right)^2}$$

43. 84° **45.** 337 mi **47.** 48 yards

49. Speed ≈ 450.14 mph; Direction: N 10.20°E

Answers for the Additional Problems

1. 16 cm, 30 cm, 34 cm **2.** $AC = 7.66$ in.

Chapter Review Problems

SECTION 6.1

1. 4 to 5, $\frac{4}{5}$, and 4/5 **2.** $\frac{2}{15}$

3. Yes, $(18)(48) = 864 = (32)(27)$

4. 4.4 cm **5.** 12.4

6. If $\frac{a}{b} = \frac{c}{d}$, then multiply both sides by bd.

$$\frac{a}{b} \times \frac{bd}{1} = \frac{c}{d} \times \frac{bd}{1}$$
$$\frac{abd}{b} = \frac{cbd}{d}$$
$$ad = cb$$

SECTION 6.2

1. $\angle A \cong \angle P, \angle B \cong \angle Q, \angle C \cong \angle R$

$$\frac{BC}{QR} = \frac{AC}{PR}$$

2. a. MNO **b.** MNO

3. a. $\triangle DFE \sim \triangle PQR$ by SSS Similarity

b. $\triangle MNO \sim \triangle TUS$ by AAA or SSS Similarity (Other similarity statements are possible.)

c. Not similar

d. $\triangle ABC \sim \triangle EDC$ by SAS Similarity

4. 7.5 cm **5.** 138.125 yd

SECTION 6.3

1. a. $BD = \sqrt{99} \approx 9.95, BC = \sqrt{220} \approx 14.83$

b. $DC = 6.4, AB = \sqrt{22.25} \approx 4.72$

2. a. $DE = 11.2$ **b.** $BC = 7$

3. $x = 3, AG = 57$

4. $P = 10\sqrt{5} \approx 22.36$ cm, $A = 25$ cm^2

5. Given $\angle BDE \cong \angle BED, \overline{DB} \cong \overline{EB}$ by Theorem 4.7. We know $\overline{AD} \cong \overline{DB}$ and $\overline{CE} \cong \overline{EB}$ by the definition of a midsegment. Thus, $\overline{AD} \cong \overline{CE}$ by substitution. $\overline{DE} \parallel \overline{AC}$ by Theorem 6.11. Therefore, $ADEC$ is an isosceles trapezoid by definition.

6. Given $ABCD$ is a square with midpoints M, N, O, and P.

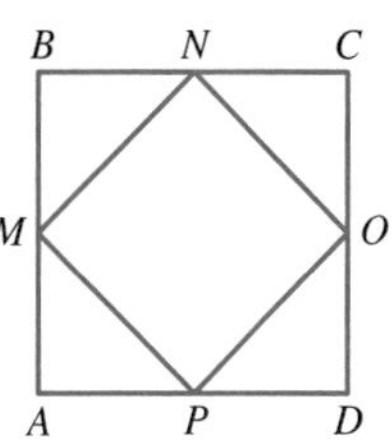

Prove $MNOP$ is a square.

Proof $\triangle AMP \cong \triangle BNM \cong \triangle CON \cong \triangle DPO$ by SAS. Thus, $\overline{MP} \cong \overline{MN} \cong \overline{ON} \cong \overline{OP}$ by C.P. Since $ABCD$ is a square, it has four right angles. Thus, $\triangle AMP \cong \triangle BNM \cong \triangle CON \cong \triangle DPO$ are isosceles right triangles so each acute angle measures 45° and $\angle NMP \cong \angle MPO \cong \angle PON \cong \angle ONM = 90°$. So $MNOP$ is a square.

SECTION 6.4

1. $\angle A \approx 51°, \angle B \approx 39°, c \approx 27$

2. a. $\tan A = \frac{\sqrt{21}}{2}$ **b.** $c \approx 43.51$

3. a. $\frac{\sqrt{2}}{2}$ **b.** $\frac{\sqrt{3}}{2}$ **c.** $\sqrt{3}$

4. 22.32 ft^2 **5.** 269.2 ft

SECTION 6.5

1. $\angle B = 88°, b \approx 29.22$ cm, $c \approx 23.65$ cm

2. $\angle B \approx 48.65°, \angle C \approx 58.15°, a \approx 17.27''$

3. $\angle GDC \approx 75.8°, DC \approx 7.0'$

4. 74.7°, 47.2°; and 58.1° **5.** 301.4 mph. S 12.4° W

CHAPTER 6 TEST

1. F—An equivalent proportion is $\frac{a}{c} = \frac{b}{d}$.

2. T

3. F—The included angles must also be congruent.

4. F—A midsegment is parallel to one side of the triangle.

5. T **6.** T

7. F—For example, some rectangles are not similar to a square.

8. T **9.** T

10. F—The cosecant ratio is the reciprocal of the sine ratio.

11. F—The Law of Cosines must be used in that situation.

12. T **13.** $x = 7.5$

14. $\sqrt{91} \approx 9.54$ **15.** $EF = 63$

16. a. $5^2 + 12^2 = 25 + 144 = 169 = 13^2$. So $a^2 + b^2 = c^2$.

b. For one angle, the sine is $\frac{5}{13}$ and the cosine is $\frac{12}{13}$. For the other angle the sine is $\frac{12}{13}$ and the cosine is $\frac{5}{13}$.

c. $\sin^2 A + \cos^2 A = \left(\frac{5}{13}\right)^2 + \left(\frac{12}{13}\right)^2 = \frac{25}{169} + \frac{144}{169} = \frac{169}{169} = 1.$

17. Perimeter = 40 cm

18. Hypotenuse ≈ 13.5

19. $x = 6$

20. $\triangle BAC \sim \triangle BDE$ by AA Similarity, because $\angle A \cong \angle BDE$ and $\angle B \cong \angle B$. $\triangle BAC \sim \triangle BFD$ by AA Similarity, because $\angle BFD \cong \angle BAC$ and $\angle DBF \cong \angle ABC$.

21. Approximately 1023 m^2

22. 8.4 blocks

23. $\angle B \approx 58.4°$,
$\angle C \approx 49.6°$,
$c \approx 9.9$ mm

24. $a = 48$, $\angle A = 41.1°$, $\angle B = 48.9°$

25. The volume of the original cone is $\frac{1}{3}\pi r^2 h$. The volume of the new cone is $\frac{1}{3}\pi\left(\frac{r}{2}\right)^2(2h) = \frac{1}{2}(\frac{1}{3}\pi r^2 h)$. Therefore, the ratio of the volume of the new cone to the volume of the original cone is 1 : 2.

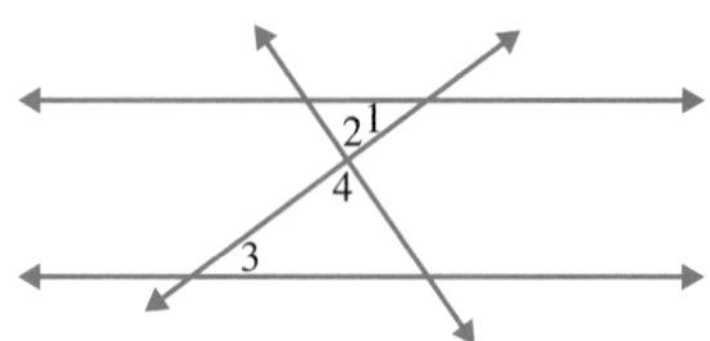

26. $\angle 1 = \angle 3$ since they are alternate interior angles and $\angle 2 = \angle 4$ since they are vertical angles. Therefore, the triangles are similar by AA Similarity.

27. Approximately 241 ft

28. Length ≈ 35.4 ft

29. 9.02 cm

SECTION 7.1

1. $\widehat{AB}, \widehat{AD}, \widehat{BC}, \widehat{BD}, \widehat{CD}$

3. $\widehat{AB}$ and $\widehat{BC}$

5. Central angle

7. $\angle COD$

9. 137°

11. 65°

13. 245°

15. $\widehat{AC} \approx 25.13$ cm and $\widehat{DC} \approx 3.49$ cm

17. 36.30 cm^2

19. $\angle BCD$ and $\angle BAD$

21. $\angle CEB$ is neither inscribed nor central.

23. $\angle C = 40°$

25. $\angle B = 35°$

27. $\angle CEB = 105°$

29. $\widehat{AD} = 110°$

31. $\angle A = 30°$, $\angle B = 30°$, $\angle C = 30°$

33. $\angle AEB = 120°$

35. We cannot necessarily conclude that $\widehat{AB} = 120°$ because we do not know that E is the center of the circle.

37. $\angle BOC = 80°$

39. Radius = 19.5″

41. $\angle B = 85°$

43. $PS = \sqrt{251} \approx 15.84$ in.

45. $\angle ABC = 111°$, $\angle BCD = 120°$, $\angle CDE = 129°$, $\angle DEA = 90°$, $\angle EAB = 90°$

47. Given Rectangle ABCD inscribed in a circle

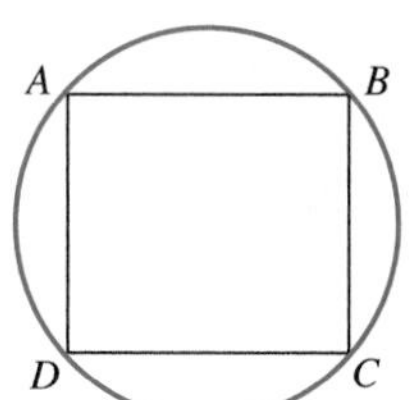

Prove $\overline{AC}$ and $\overline{BD}$ are diameters.

Proof

Statement	Reason
1. $\angle ADC$ and $\angle ABC$ are inscribed in the circle.	1. Given
2. $\angle ADC = \angle ABC = 90°$	2. $ABCD$ is a rectangle
3. $\widehat{ABC} = \widehat{ADC} = 180°$	3. Inscribed Angle Theorem
4. $\overline{AC}$ and $\overline{BD}$ are diameters.	4. Diameters have endpoints on the endpoints of semicircles.

49. $\overline{BD}$ is a diameter.

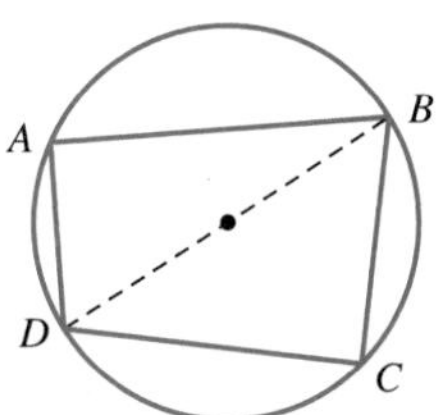

Thus, $\angle BAD = \angle BCD = 90°$. Since their sum is 180°, the sum of the measures of the remaining two angles is $360° - 180° = 180°$. Hence, $\angle ADC$ and $\angle ABC$ are supplementary.

51. Outline: Use the fact that $\widehat{AC} = \angle AOC$ and $\triangle AOB$ is isosceles to show that $\angle ABC = \frac{1}{2}\angle AOC$.

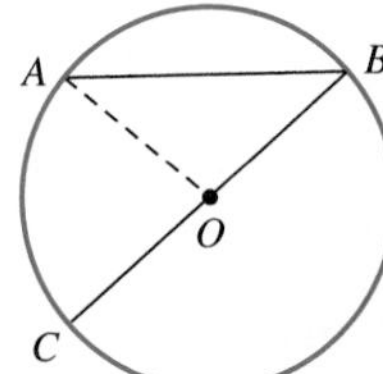

53. We are given that $\overline{BD}$ is the diameter of the circle, so $\angle DCB$ is a right angle by Corollary 7.3. By the definition of the sine of an acute angle, we know $\sin D = \frac{a}{d}$. By Corollary 7.2, $\angle A \cong \angle D$. Therefore, $\sin A = \frac{a}{d}$ by substitution and $d = \frac{a}{\sin A}$. By the Law of Sines, we know $\frac{a}{\sin A} = \frac{b}{\sin B} = \frac{c}{\sin C}$. Thus, $d = \frac{a}{\sin A} = \frac{b}{\sin B} = \frac{c}{\sin C}$.

55. Arc length $= \frac{16\pi}{9} \approx 5.59$ in.

57. Volume ≈ 59.13 in^3

59. To find the center of a circle, place the vertex of the right angle of the triangle on the circle. The two points where the circle intersects the triangle are endpoints of a diameter by Corollary 7.3. Draw the diameter. Repeat this process and draw another diameter. The intersection of the two diameters is the center of the circle.

SECTION 7.2

1. $GH = FE$, $\widehat{FGE} = 180°$

3. $\angle BEC = 149°$

5. $\angle CED = 70°$

7. $AF = 8.4$ in.

9. $BF = 4$ cm

11. $DE = 1.6$ cm

13. $CE = 5$ in.

15. $\angle AFE = 104.5°$

17. $\widehat{ED} = 95°$, $\widehat{BC} = 122°$, and $\widehat{CD} = 41°$

19. $AE = 12$ cm

21. $AD = 16$ ft

23. Use Corollary 7.5.

25. Use Corollary 7.5.

27. Given Chords $\overline{AB}$ and $\overline{CD}$ are congruent.

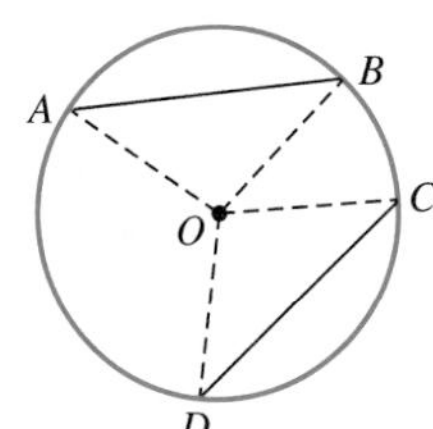

Prove $\widehat{AB} = \widehat{CD}$

Proof

Statement	Reason
1. $AB = CD$	1. Given
2. $AO = BO = CO = DO$	2. Radii of the same circle
3. $\triangle ABO \cong \triangle DCO$	3. SSS Congruence
4. $\angle AOB \cong \angle DOC$	4. C.P.
5. $\widehat{AB} = \widehat{CD}$	5. Definition of the measure of an arc

29. Outline: Use the fact that $\angle AOB = \angle COD$ to show $\triangle AOB \cong \triangle COD$ by SAS. Thus, $\overline{AB} \cong \overline{CD}$.

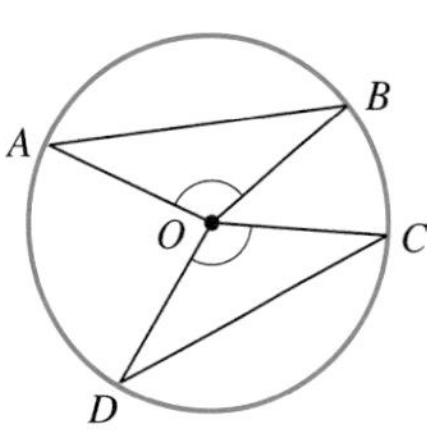

31. Let O and P be congruent circles. Then $AO = BO = CP = DP$. Since $\angle AOB = \angle CPD$, the measures of their arcs are the same.

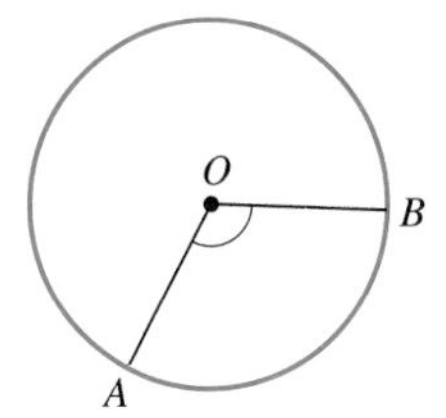

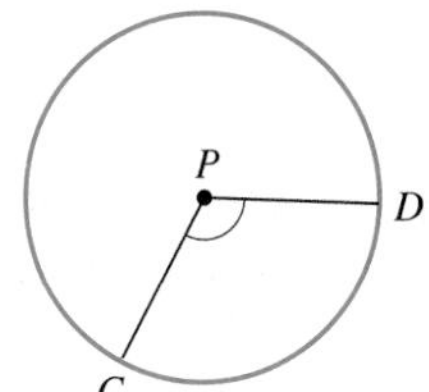

33. Given Circle O with $\overline{AB} \cong \overline{CD}$

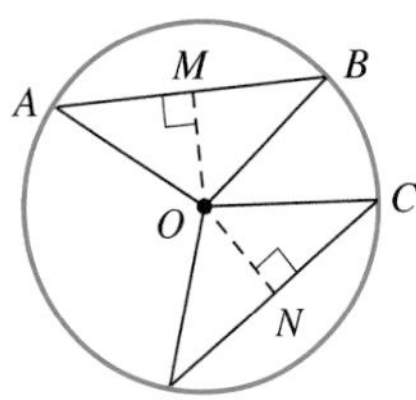

Prove $OM = ON$

Proof

Statement	Reason
1. $\overline{AB} \cong \overline{CD}$	1. Given
2. $\triangle AOB \cong \triangle COD$	2. SSS Congruence
3. $\angle B \cong \angle D$	3. C.P.
4. $\triangle BOM \cong \triangle DON$	4. H.A.
5. $OM = ON$	5. C.P.

35. $ABCD$ is a kite inscribed in a cricle.

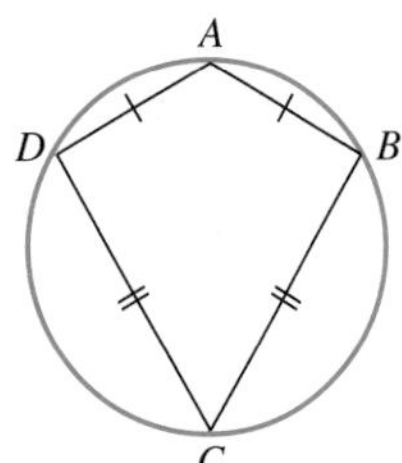

Since $\widehat{ABC} + \widehat{ADC} = 360°$, $\angle B + \angle D = 180°$. Since $\angle B = \angle D$, they must be right angles.

37. Outline: Let M be the midpoint of $\overline{AB}$.

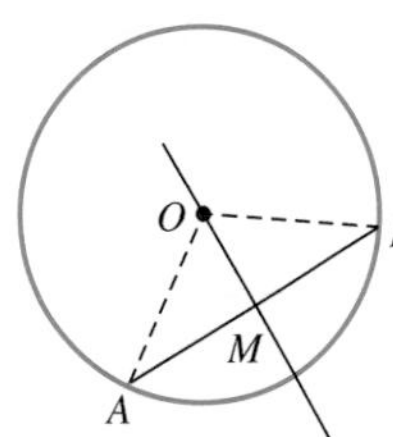

Show that $\triangle AMO \cong \triangle BMO$ by SSS Congruence. Then $\angle AMO \cong \angle BMO$ and $\angle AMO + \angle BMO = 180°$, so $\angle AMO = \angle BMO = 90°$.

39. Let l be the perpendicular bisector of $\overline{AB}$.

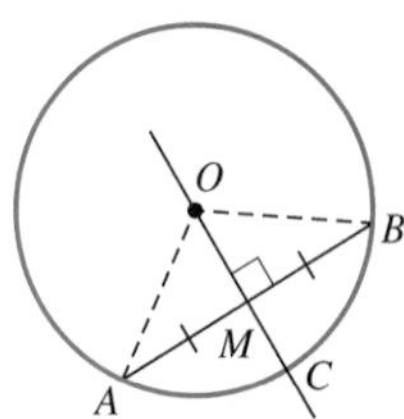

Then $\triangle AMO \cong \triangle BMO$ by LL Congruence. Therefore, $\angle AOM \cong \angle BOM$ and $\widehat{AC} \cong \widehat{BC}$. Thus, l bisects $\widehat{ACB}$.

41. The wastepaper basket should be placed at the point of intersection of the perpendicular bisectors of any two of the segments $\overline{JW}$, $\overline{WS}$, or $\overline{JS}$. This would be the center of the circle containing J, S, and W so each would be an equal distance from the wastepaper basket.

43. Radius $\cong$ 5.67 cm

SECTION 7.3

1. $\angle C = 26°$

3. $EC = \dfrac{\sqrt{249} + 3}{2}$ cm $\cong 9.39$ cm

5. $AB \approx 2.46$ in.

7. $\widehat{BD} = 45°$ and $\angle C = 22.5°$. By Theorem 7.9, we know $\angle C = \dfrac{1}{2}(\widehat{AE} - \widehat{BD})$.

9. $DE = \dfrac{47}{7} \approx 6.71$

11. $\angle SPQ = 65°$

13. $\widehat{PQ} = 216°$

15. $\widehat{PQR} = 88°$

17. $\angle CBO = 90°$

19. $AC = BC$ by Corollary 7.15 and $OC = OC$, so $\triangle ACO \cong \triangle BCO$ by HL.

21. $\widehat{AE} = 150°$

23. $AB = 8\sqrt{2} \approx 11.31$ cm

25. $AP = 3.5$ ft, $BQ = 8.5$ ft

27. $\widehat{EF} = 28°$, $\widehat{GH} = 35°$, $\widehat{DI} = 27°$

29. $A = 2 - \dfrac{\pi}{2}$ square units

31. Let $\overline{AB}$ and $\overline{AC}$ be tangent to a circle as shown.

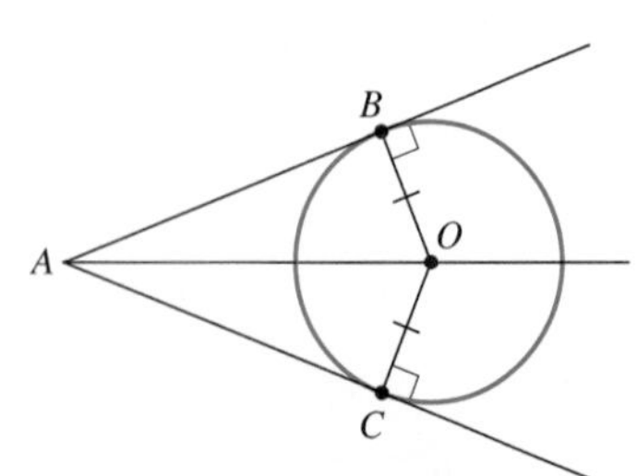

Since $BO = CO$ and $AO = AO$, we have $\triangle AOB \cong \triangle AOC$ by HL. Therefore, $\overline{AB} \cong \overline{AC}$ by C.P., so $AB = AC$.

33. Let A be on circle O and $\overleftrightarrow{AB} \perp \overline{AO}$.

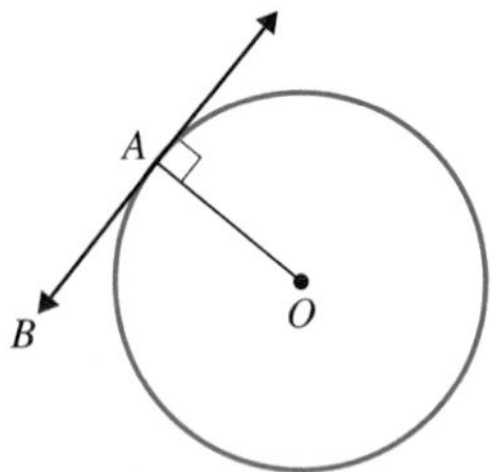

We will use an indirect proof to show that $\overleftrightarrow{AB}$ is tangent to circle O. Suppose $\overleftrightarrow{AB}$ is not tangent to circle O. Then it must intersect circle O in a second point C.

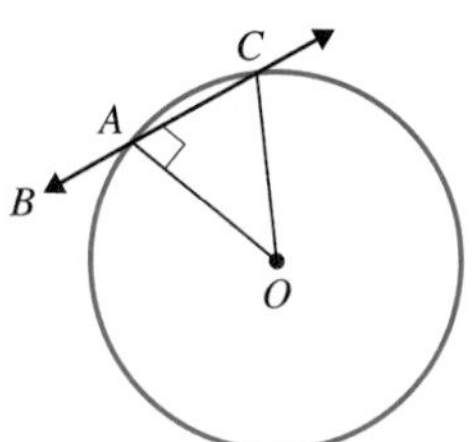

But $\triangle ACO$ is isosceles since $AO = CO$, and $\angle OAC = \angle OCA$. Thus $\triangle ACO$ has two right angles, which is impossible. Therefore, our assumption is false and $\overleftrightarrow{AB}$ is tangent to circle O at A.

35. Let $\overleftrightarrow{AD}$ and $\overleftrightarrow{BC}$ be two parallel secants of circle O. Draw secant $\overleftrightarrow{BD}$.

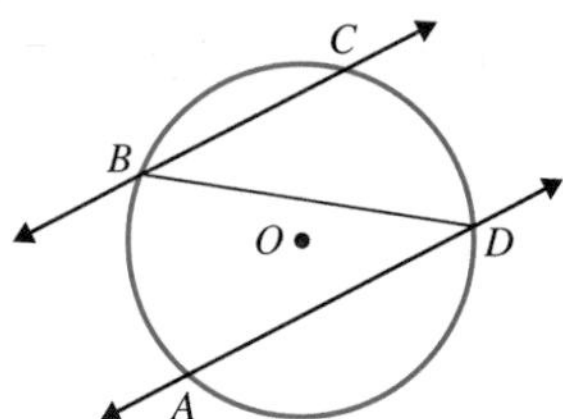

Then $\angle BDA \cong \angle CBD$ by Theorem 5.5. By Theorem 7.1, the measure of $\angle BDA$ is $\frac{1}{2}$ of the measure of $\widehat{AB}$ and the measure of $\angle CBD$ is $\frac{1}{2}$ of the measure of $\widehat{DC}$. Thus, the measure of $\widehat{AB} = 2\angle BDA = 2\angle CBD =$ the measure of $\widehat{DC}$, so $\widehat{AB} \cong \widehat{DC}$.

37. Outline: Draw tangent line $\overleftrightarrow{AB}$ with $\overleftrightarrow{AC} \perp \overleftrightarrow{AB}$.

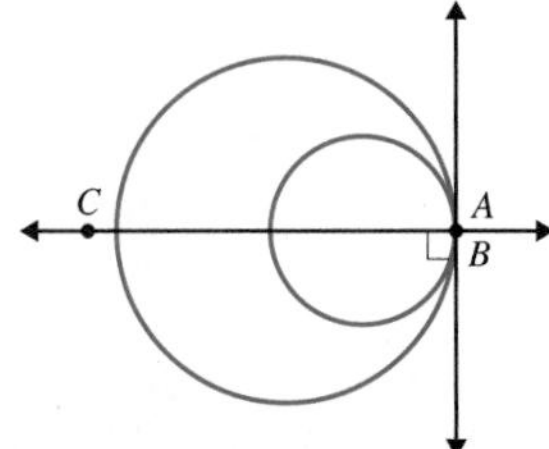

Use the fact that $\overleftrightarrow{AC} \perp \overleftrightarrow{AB}$, to show that the centers of both circles and A are collinear.

39. Diameter $= 2$ ft

41. 22,000 miles; there cannot be a total solar eclipse. Point P is about $22{,}000 - 3960 = 18{,}040$ miles above the surface of the earth, so no part of the umbra would cover the earth.

SECTION 7.4

1–5. Constructions

7. Point A is the orthocenter of $\triangle DBC$.

9–13. Constructions

15. Construct point E on $\overrightarrow{AP}$ so that $AP = \frac{2}{3}AE$. Use the fact that $BE = \frac{1}{2}BC$ to locate point C.

17–23. Construction

25. 90°, 90°, 120°, kite

27. Construction

29. Let $\overline{AE}$ and $\overline{CD}$ be angle bisectors.

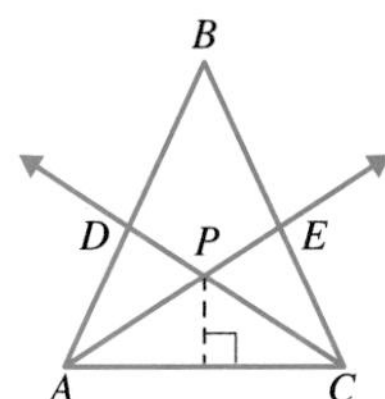

Each point on $\overline{AE}$ is equidistant from $\overline{AB}$ and $\overline{AC}$ and each point on $\overline{CD}$ is equidistant from $\overline{AC}$ and $\overline{BC}$. Thus, point P, the intersection of $\overline{AE}$ and $\overline{CD}$, must be equidistant from all three sides. Since it is equidistant from $\overline{BA}$ and $\overline{BC}$, it must also be on the bisector of $\angle ABC$. Thus, all three angle bisectors intersect in P.

31. All the points on an angle bisector are equidistant from the sides of the angle. Thus, the point of intersection of the three angle bisectors of a triangle is the same distance from all three sides. This distance is the radius of the inscribed circle.

33. Outline: Show that the two tangent lines form right angles with the diameter and that these are congruent alternate interior angles.

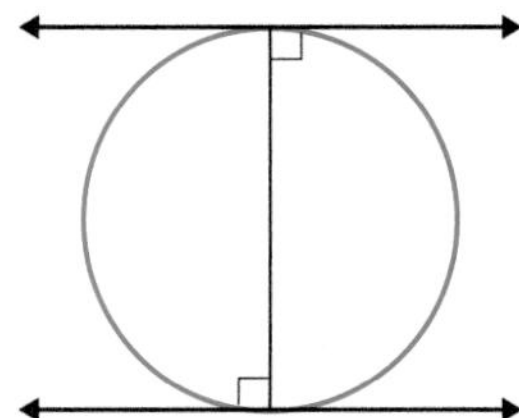

35. a. $(CE)^2 = (CB)^2 + (BE)^2 - 2(CB)(BE)\cos(B + 60°)$
$(CE)^2 = (CA)^2 + (AE)^2 - 2(CA)(AE)\cos(A + 60°)$

b. We know $BE = AE = AB$, so we substitute into the expression in (a).

$(CE)^2 = (CB)^2 + (AB)^2 - 2(CB)(AB)\cos(B + 60°)$

$(CE)^2 = (CA)^2 + (AB)^2 - 2(CA)(AB)\cos(A + 60°)$

c. $(HI)^2 = (BH)^2 + (BI)^2 - 2(BH)(BI)\cos(B + 60°)$

d.

$$BH = \frac{2}{3}\left(\frac{(AB)\sqrt{3}}{2}\right) \text{ and } BI = \frac{2}{3}\left(\frac{(CB)\sqrt{3}}{2}\right)$$

$$\begin{aligned}(HI)^2 &= (BH)^2 + (BI)^2 - 2(BH)(BI)\cos(B + 60°)\\ &= \left(\frac{2}{3}\cdot\frac{(AB)\sqrt{3}}{2}\right)^2 + \left(\frac{2}{3}\cdot\frac{(CB)\sqrt{3}}{2}\right)^2\\ &\quad - 2\left(\frac{(AB)(CB)}{3}\right)\cos(B + 60°)\\ &= \frac{(AB)^2}{3} + \frac{(CB)^2}{3} - 2\left(\frac{(AB)(CB)}{3}\right)\cos(B + 60°)\end{aligned}$$

$$3(HI)^2 = (AB)^2 + (CB)^2 - 2(AB)(CB)\cos(B + 60°)$$

e.

$$(GH)^2 = (AG)^2 + (AH)^2 - 2(AG)(AH)\cos(A + 60°)$$

f.

$$AG = \frac{2}{3}\left(\frac{(CA)\sqrt{3}}{2}\right) \text{ and } AH = \frac{2}{3}\left(\frac{(AB)\sqrt{3}}{2}\right)$$

$$\begin{aligned}(GH)^2 &= (AG)^2 + (AH)^2 - 2(AG)(AH)\cos(A + 60°)\\ &= \left(\frac{2}{3}\cdot\frac{(CA)\sqrt{3}}{2}\right)^2 + \left(\frac{2}{3}\cdot\frac{(AB)\sqrt{3}}{2}\right)^2\\ &\quad -2\left(\frac{(CA)(AB)}{3}\right)\cos(A + 60°)\\ &= \frac{(CA)^2}{3} + \frac{(AB)^2}{3} - 2\left(\frac{(CA)(AB)}{3}\right)\cos(A + 60°)\end{aligned}$$

$$3(GH)^2 = (CA)^2 + (AB)^2 - 2(CA)(AB)\cos(A + 60°)$$

g.

$$\begin{aligned}(CE)^2 &= (CB)^2 + (AB)^2 - 2(CB)(AB)\cos(B + 60°)\\ &= 3(HI)^2\end{aligned}$$

$$\begin{aligned}(CE)^2 &= (CA)^2 + (AB)^2 - 2(CA)(AB)\cos(B + 60°)\\ &= 3(GH)^2\end{aligned}$$

Thus, $3(HI)^2 = 3(GH)^2$, so $HI = GH$.

37. 64 cm

Answers for the Additional Problems

1. Consider the following three cases: (i) acute triangles, (ii) right triangles, and (iii) obtuse triangles. In each case the centroid lies inside the triangle because all three medians lie inside the triangle.

2. Consider the following two cases: (i) the first integer is odd and (ii) the first integer is even. Given that the three integers are consecutive integers, in case (i) we must have odd + even + odd and in case (ii) even + odd + even. Because the sum of the integers must be 180°, an even integer, only case (i) satisfies the conditions.

Chapter Review Problems

SECTION 7.1

1. a. Inscribed angle **b.** Central angle
c. $\angle AOB = 2\angle ACB$ **d.** 180° **e.** 90°

2. $\angle DOC = 139°$, $\widehat{AB} = 123°$

3. $\angle A = 100°$, $\angle C = 80°$, $\angle D = 65°$, $\widehat{AB} = 75°$, and $\widehat{AD} = 85°$

4. a. 810 ft^2 **b.** 36.7 ft

SECTION 7.2

1. 11.45 cm **2.** 109.5° **3.** $\frac{12}{7}$

4. By Theorem 7.1, we know $\angle BCA = \frac{1}{2}\widehat{AB}$ and $\angle BAC = \frac{1}{2}\widehat{BC}$. Thus, by substitution, $\angle BCA = \angle BAC$ and $\triangle ABC$ is isosceles.

5. Suppose that circles O and O' contain the same three noncollinear points A, B, and C. By Corollary 7.5, the perpendicular bisectors of the chords contain the centers of the circles and these perpendicular bisectors intersect in one point, so we must have $O = O'$. Hence, the circles are the same.

SECTION 7.3

1. $\widehat{ED} = 149°$, $\widehat{FE} = 6°$, $\widehat{BC} = 63°$

2. $DE \approx 3$ cm.

3. $\angle ACE = 85°$, $\widehat{CD} = 105°$, $\angle BCD = 52.5°$

4. a. $\widehat{DC} = 123.69°$, $\widehat{AE} = 56.31°$

b. $BE \approx 4.01$ cm

5. A circle would be formed, since all points will be equidistant from the center by the result proved in problem 25 in Section 7.2.

SECTION 7.4

1–2(a). Constructions **b.** C, B, A

3–6. Constructions

CHAPTER 7 TEST

1. T

2. F—The incenter of a triangle always lies inside the triangle.

3. T

4. F—The diameter (4 inches) is the longest chord.

5. F—Its measure is half the measure of the intercepted arc.

6. T **7.** T **8.** T **9.** T

10. F—$\angle AOB$ is larger than $\angle POQ$.

11. a. $\overleftrightarrow{AC}$ **b.** $\angle ECD, \angle BCE$ **c.** $\widehat{ECD}, \widehat{DEB}, \widehat{BDE}$

d. $\angle OEA$ **e.** $\overline{BC}, \overline{CD}, \overline{CE}$

12. 56° **13.** 47° **14.** 105°

15. a. Approximately 63.7 percent

b. Approximately 78.5 percent **c.** Circular plug

16. a. $\widehat{AB} = 30°$ **b.** $\angle EDF = 34°$

17. $TS = 15$ ft and $QT = 4$ ft

18. $\frac{40}{3}$ **19–20.** Constructions

21. Because $\widehat{AB} = \widehat{CB}$, we have $\angle AOB \cong \angle COB$. Therefore, $\triangle AOD \cong \triangle COD$ by SAS $(\overline{OA} \cong \overline{OC}, \angle AOB \cong \angle COB, \overline{OD} \cong \overline{OD})$. By C.P., $\angle ADO \cong \angle CDO$ and $AD = CD$. Because they form a straight angle, they are 90° angles. Thus, $\overline{OB}$ is the perpendicular bisector of $\overline{AC}$.

22. Because $\triangle ABO, \triangle BCO$, and $\triangle CDO$ are isosceles triangles, they have pairs of congruent angles as shown. We are given that $\angle 2 = 2(\angle 3)$.

a. Prove $\angle AOB = 4(\angle 3)$: Because $\angle AOB$ is an exterior angle of $\triangle BCO$, $\angle AOB = \angle 2 + \angle 2$, or $2(\angle 2)$. Because $\angle 2 = 2(\angle 3)$, we have by substitution $\angle AOB = 2[2(\angle 3)] = 4(\angle 3)$.

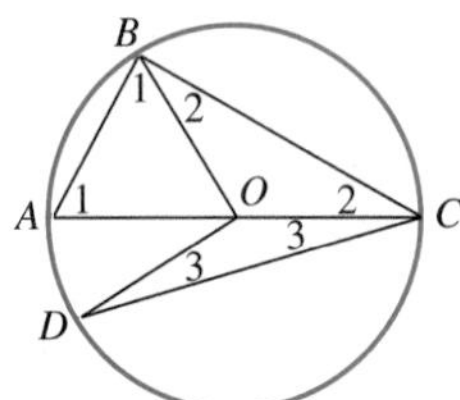

b. Prove $\angle AOD + \angle ABO = 90°$: $\angle AOD + \angle COD = 180°$. $\angle COD = 180° - 2(\angle 3) = 180° - \angle 2$. Therefore, $\angle AOD + 180° - \angle 2 = 180°$, so $\angle AOD = \angle 2$. Because $\angle ABC$ is inscribed in a semicircle, $\angle ABC = 90°$. Thus, $\angle 1 + \angle 2 = 90°$, so $\angle ABO = 90° - \angle 2$. Therefore, $\angle AOD + \angle ABO = \angle 2 + (90° - \angle 2) = 90°$.

23. $r \approx 278'$ **24.** 146,200 ft^2

SECTION 8.1

1. a. (−3, 2), (3, 2), (−3, −2), (3, −2)

b. (−2, 4), (0, 4), (4, 0), (3, −5)

3. **a.** 5 **b.** 5 **c.** 4 **d.** 9

5. $AB = CD = \sqrt{18}$ or $3\sqrt{2}$, $BC = AD = 5$. Because opposite sides are congruent, $ABCD$ is a parallelogram.

7. **a.** Yes **b.** No **c.** Yes

9. $n = 3$ or $n = \frac{9}{5}$

11. **a.** No **b.** No **c.** Yes **d.** Yes

13. **a.** $(-\frac{3}{2}, 2)$ **b.** $(-1, 2)$ **c.** $(-\frac{5}{2}, \frac{9}{2})$ **d.** $(3, 1)$ **e.** $(2, 7)$ **f.** $(\frac{3}{2}, \frac{3}{2})$

15. **a.** Right isosceles **b.** Right scalene **c.** Obtuse isosceles **d.** Acute scalene

17. **a.** Right triangle **b.** Right triangle **c.** Not a right triangle

19. Not a rectangle

21. $C(5\sqrt{3}, 5)$ or $C(-5\sqrt{3}, 5)$

23. Sides of $A'B'C'D'$ are one-third as long as sides of $ABCD$.

25. **a.** The sides of $A'B'C'$ are twice as long as the sides of ABC.
b. The area of $A'B'C'$ is four times the area of ABC.

27. **a.** $ABCD$ is a square. **b.** 90 square units
c. The area of $A'B'C'D'$ is one-ninth the area of $ABCD$.

29. Answers may vary. Possible coordinates include $(13, 0)$, $(0, 13)$, $(-13, 0)$, $(0, -13)$, $(5, 12)$, $(-5, 12)$, $(-5, -12)$, and $(5, -12)$.

31. **a.** **b.** **c.**

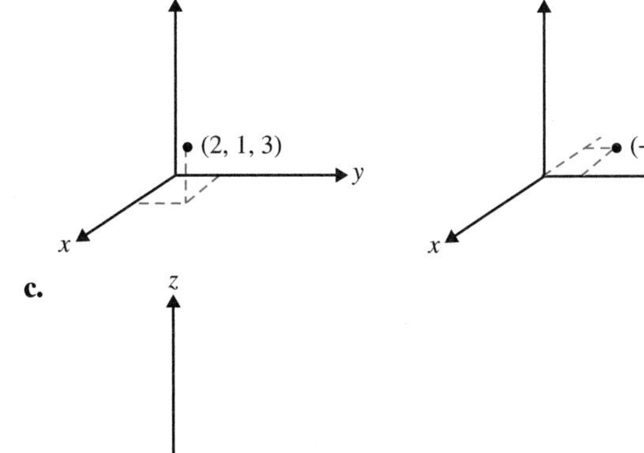

33. Octant 2: $(-, +, +)$ Octant 3: $(-, -, +)$
Octant 4: $(+, -, +)$ Octant 5: $(+, +, -)$
Octant 6: $(-, +, -)$ Octant 7: $(-, -, -)$
Octant 8: $(+, -, -)$

35. **a.** $\sqrt{6}$ **b.** $\sqrt{89}$

37. **a.** $\left(\frac{\sqrt{3}}{2}, \frac{1}{2}\right)$ **b.** $(0, 1)$
c. $\left(-\frac{\sqrt{2}}{2}, \frac{\sqrt{2}}{2}\right)$ **d.** $(1, 0)$

39. **a.** $\sin 30° = \frac{1}{2}$, $\cos 30° = \frac{\sqrt{3}}{2}$
b. $\sin 90° = 1$, $\cos 90° = 0$
c. $\sin 135° = \frac{\sqrt{2}}{2}$, $\cos 135° = -\frac{\sqrt{2}}{2}$
d. $\sin 0° = 0$, $\cos 0° = 1$

41. **a.** $\frac{\sqrt{3}}{2}, \frac{\sqrt{3}}{2}$ **b.** $\frac{\sqrt{2}}{2}, \frac{\sqrt{2}}{2}$
c. $\sin A = \sin(180° - A)$

43. **a.** 18.7 **b.** 43.3

45. **a.** The area of $\triangle ABC$ = area of $DABE$ + area of $EBCF$ − area of $DACF$

$$= \frac{1}{2}(x_2 - x_1)(y_1 + y_2) + \frac{1}{2}(x_3 - x_2)(y_2 + y_3) - \frac{1}{2}(x_3 - x_1)(y_1 + y_3)$$

$$= \frac{1}{2}(x_2y_1 - x_2y_2 - x_1y_1 - x_1y_2) + \frac{1}{2}(x_3y_2 + x_3y_3 - x_2y_2 - x_2y_3) - \frac{1}{2}(x_3y_1 + x_3y_3 - x_1y_1 - x_1y_3)$$

$$= \frac{1}{2}\left[\begin{array}{l}(x_2y_1 + x_2y_2 - x_1y_1 - x_1y_2 + x_3y_2 + x_3y_3 \\ - x_2y_2 - x_2y_3) - (x_3y_1 + x_3y_3 - x_1y_1 - x_1y_3)\end{array}\right]$$

$$= \frac{1}{2}(x_2y_1 - x_1y_2 + x_3y_2 - x_2y_3 - x_3y_1 + x_1y_3)$$

$$= \frac{1}{2}[x_1(y_3 - y_2) + x_2(y_1 - y_3) + x_3(y_2 - y_1)]$$

b. Area of $\triangle ABC$ = 5.5 square units. The areas are the same.

c. Area of $\triangle ABC$ = 58.5 square units. Add absolute values around the formula to make sure the area always ends up positive.
Area of $\triangle ABC$ =

$$\frac{1}{2}\left|x_1(y_3 - y_2) + x_2(y_1 - y_3) + x_3(y_2 - y_1)\right|$$

47. Outline: Use coordinates and the distance formula to show that $PM + MQ = PQ$. Then P, M, and Q must be collinear.

49. Let $P(x, y)$ be a point in the plane. Choose two values of x such that $r < s < x$. Then $A(r, mr + b)$ and $B(s, ms + b)$ are points on line l.
$AB = \sqrt{(s - r)^2 + [(ms + b) - (mr + b)]^2} = \sqrt{(s - r)^2 + m^2(s - r)^2} = \sqrt{(1 + m^2)(s - r)^2} = (s - r)\sqrt{1 + m^2}$. In a similar way, $BP = (x - s)\sqrt{1 + m^2}$ and $AP = (x - r)\sqrt{1 + m^2}$
So $AB + BP = (s - r + x - s)\sqrt{1 + m^2} = (x - r)\sqrt{1 + m^2} = AP$. Thus, A, B, and P are collinear.

51. $(3.5, 3.75)$, $(7, 2.5)$, $(10.5, 1.25)$

53. 4.8 km

SECTION 8.2

1. $-\frac{4}{5}, -\frac{4}{5}$: They are the same.

3. a. $\frac{1}{2}$ **b.** $\frac{4}{3}$ **c.** 0 **d.** Undefined

5. a. Positive **b.** 0

7. a. Falls **b.** Horizontal
c. Rises **d.** Vertical

9. Collinear

11. A, D, and C are collinear.

13. a. Yes **b.** Yes

15. a. $-\frac{9}{2}$ **b.** 4

17. a. Yes **b.** No

19. a. $(6, 6), (10, -4)$, or $(-14, 6)$
b. $(9, 4), (5, 0)$, or $(-9, 2)$

21. Slope of $\overline{AB}$ = slope of $\overline{CD} = 1$
Slope of $\overline{BC}$ = slope of $\overline{AD} = -\frac{3}{4}$
So $ABCD$ is a parallelogram.

23. a. 1 **b.** Undefined

25. a. $\overline{AB} \perp \overline{BC}$ **b.** $\overline{AB} \perp \overline{BC}$

27. $a = 10$ **29.** 11

31. a. No **b.** No **c.** Yes

33. a. No **b.** Yes **c.** No **d.** No

35. The midpoints of $\overline{AB}$ and $\overline{BC}$ are $M_1\left(-\frac{3}{2}, 2\right)$ and $M_2(3, 3)$, respectively. Using the distance formula, we have
$M_1M_2 = \sqrt{\left(3 + \frac{3}{2}\right)^2 + (3 - 2)^2} = \frac{\sqrt{85}}{2}$ and
$AC = \sqrt{85}$. Thus, $M_1M_2 = \frac{1}{2}AC$. Also, the slope of
$M_1M_2 = \frac{3 - 2}{3 + \frac{3}{2}} = \frac{2}{9}$ and the slope of $\overline{AC}$ is $\frac{2}{9}$.
So $\overline{M_1M_2} \parallel \overline{AC}$.

37. Outline: Determine the slopes of the diagonals of the quadrilateral. Use the fact that a rhombus is a parallelogram that has perpendicular diagonals.

39. a. $\frac{PO}{OQ} = \frac{RS}{PS}$

b. Since $\angle POQ = \angle RSP = 90°$ and $\frac{PO}{OQ} = \frac{RS}{PS}$.
$\triangle QOP \sim \triangle PSR$ by SAS Similarity.

c. $\frac{PO}{RS} = \frac{QO}{PS} = \frac{QP}{PR}$

d. $\frac{QO + PS}{PS} = \frac{PO + RS}{RS}$ by problems 31 in Section 6.1

e. $QO + PS = QO + OT = QT$ and $PO + RS = ST + RS = RT$. So from (d), we have $\frac{QT}{RS} = \frac{RT}{RS}$ and $\triangle PSR \sim \triangle QTR$ by SAS Similarity.

f. $\frac{QT}{PS} = \frac{RT}{RS} = \frac{QR}{PR}$

g. Since $\frac{QT}{PS} = \frac{QR}{PR}$, we have $QR = PR \cdot \frac{QT}{PS}$.

h. $\frac{QP}{RS} = \frac{QO}{PS} \Rightarrow \frac{QP + PR}{PR} + \frac{QO + PS}{PS}$
$\Rightarrow QP + PR = \frac{PR}{PS}(QO + PS)$
$\Rightarrow QP + PR = PR\left(\frac{PO + RS}{RS}\right)$ from (c)
$\Rightarrow QP + PR = \frac{PR \cdot RT}{RS}$
$\Rightarrow QP + PR = \frac{PR \cdot QT}{PS}$ from (f)

i. P, Q, and R are collinear since $QP + PR = QR$.

41. a. 1.56 in. **b.** 10 feet

43. 0.08, approximately 4.6°

45. 0.067 or 6.667 percent

SECTION 8.3

1. a. $5 = 7(1) - 2$ **b.** $-2(7.5) = 6(-3) + 3$

3. Answers will vary.

5. a. $y = -3x - 2$; Slope $= -3$, y-intercept $= -2$
b. $y = 2x - 5$; Slope $= 2$, y-intercept $= -5$
c. $y = \frac{5}{2}x + 3$; Slope $= \frac{5}{2}$, y-intercept $= 3$
d. $y = \frac{2}{7}x - \frac{8}{7}$; Slope $= \frac{2}{7}$, y-intercept $= -\frac{8}{7}$

7. a. $y = 3x + 7$ **b.** $y = -x - 3$

9. a. $y = \frac{1}{6}x + 2$ **b.** $y = -2x$

11. a. Rises to the right **b.** Falls to the right
c. Horizontal

13. a. $x = 0$ **b.** $y = 5$ **c.** $x = -9$ **d.** $y = 7$

15. a. $y = \frac{4}{3}x + 1$ **b.** $b = 1$ **d.** $m = \frac{4}{3}$

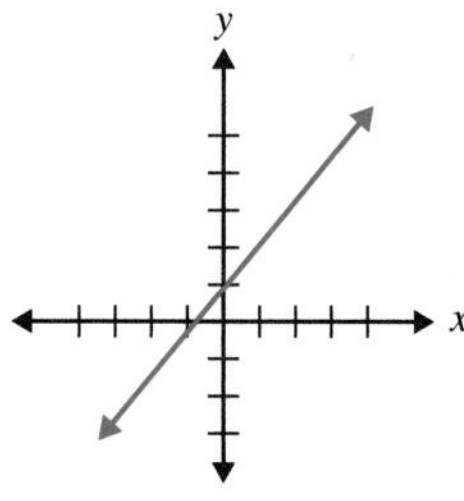

17. a.

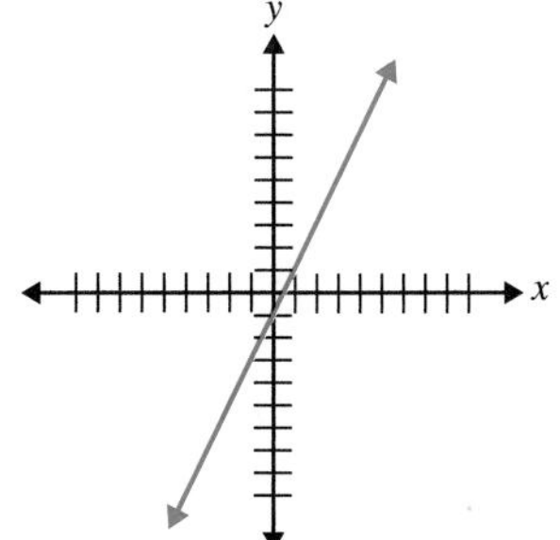

b.

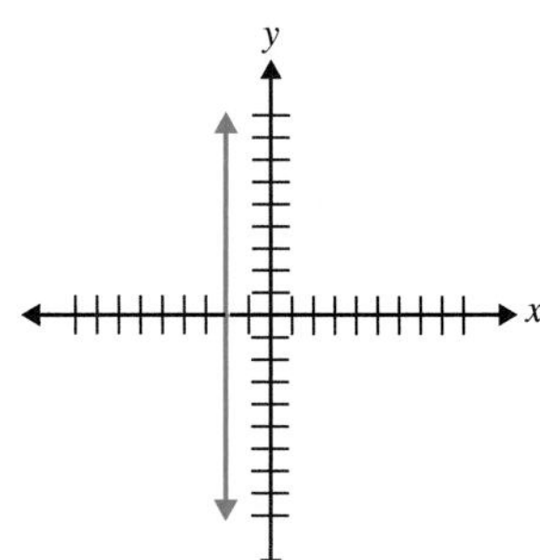

c.

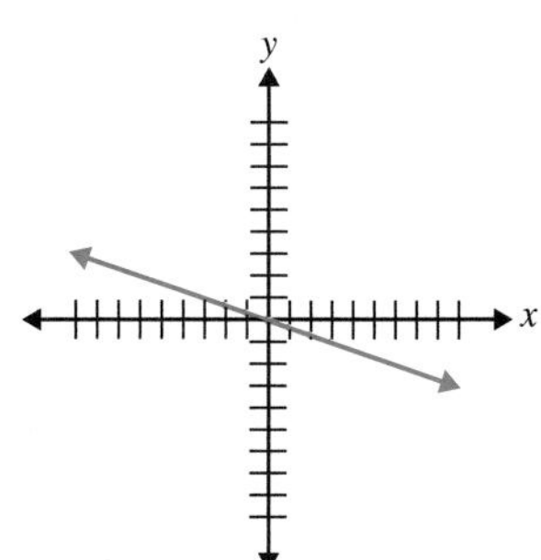

19. a.

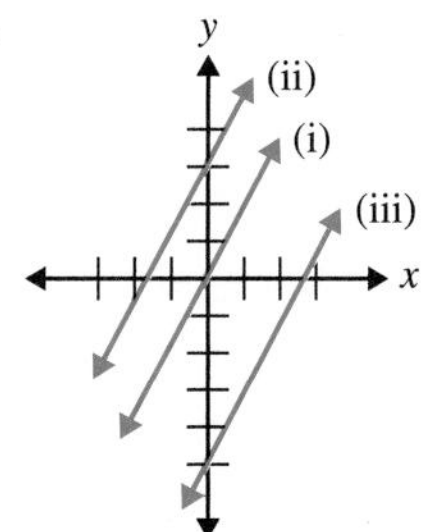

b. The lines are parallel with slope 2.

c. The lines all have the same slope c and so are parallel.

21. a. $y = 2x - 11$ **b.** $y = -x - 1$

c. $y = -\frac{3}{5}x - \frac{19}{5}$ **d.** $y = -\frac{3}{2}x - \frac{3}{2}$

23. $y = -3x + 5$ **25.** $y = -\frac{4}{3}x + 11$

27. a. $(3, 2)$ **b.** $(3, -4)$

c. No solution—parallel lines **d.** $\left(\frac{13}{2}, \frac{3}{2}\right)$

29. a. No solution—parallel lines

b. No solution—parallel lines

c. $(-3, 1)$

d. All points on the line $-2x + 3y = 9$

31. a. $5x = 10$ **b.** $x = 2$ **c.** $y = 3$

33. a. $y = \frac{1}{2}x$ and $y = -\frac{1}{2}x + 3$ **b.** $\left(3, \frac{3}{2}\right)$

35. a. $3^2 + 4^2 = 9 + 16 = 25$

b. $(-3)^2 + 5^2 = 9 + 25 = 34$

37. a. Center: $(-7, 1)$ Radius $= 11$ **b.** Center: $(5, 8)$ Radius $= \sqrt{27} = 3\sqrt{3}$

39. a. $(x + 1)^2 + (y + 2)^2 = 5$

b. $(x - 2)^2 + (y + 4)^2 = 41$

c. $(x - 1)^2 + (y - 2)^2 = 20$

41. $(1, -1)$

43. $\left(\frac{4}{9}, -\frac{17}{9}\right)$

45. a. 2 solutions **b.** $(0, 1)$ and $\left(-\frac{4}{5}, \frac{3}{5}\right)$

47. 3 square units

49. Outline: Use the distance formula and the fact that the distance from any point (x, y) on the circle to the center (h, k) of the circle is r.

51. By Theorem 7.20, we know the centroid of a triangle is located two-thirds of the way from the vertex to the side of the triangle. The coordinates of the centroids are $(-\sqrt{3}, 3)$, $(3, -\sqrt{3})$, and $(3 + \sqrt{3}, 3 + \sqrt{3})$.
Use the distance formula to verify the distance between each pair of centroids is the same.

Distance between $(-\sqrt{3}, 3)$ and $(3, -\sqrt{3})$:

$$d^2 = (3 + \sqrt{3})^2 + (3 + \sqrt{3})^2$$
$$d = \sqrt{24 + 12\sqrt{3}}$$

Distance between $(-\sqrt{3}, 3)$ and $(3 + \sqrt{3}, 3 + \sqrt{3})$:

$$d^2 = (3 + 2\sqrt{3})^2 + (3 + \sqrt{3} - 3)^2$$
$$d = \sqrt{24 + 12\sqrt{3}}$$

Distance between $(3, -\sqrt{3})$ and $(3 + \sqrt{3}, 3 + \sqrt{3})$:

$$d^2 = (3 + \sqrt{3} - 3)^2 + (3 + 2\sqrt{3})^2$$
$$d = \sqrt{24 + 12\sqrt{3}}$$

53. a. \$5.60, \$13.10 **b.** $y = 0.5x + 0.6$

55. a. $y = 435 + 14.5x$

b. 14.5—For each foot added to the length of the side fence, the cost increases by \$14.50.

c. 435—A fixed cost, meaning that the owner will pay this cost even if the side fence has length zero feet.

d. Approximately 87 feet

57. a. 100 **b.** 125 **c.** 225

SECTION 8.4

1. Because slope of $\overline{AD}$ = slope of $\overline{BC} = 0$, we have $\overline{AB} \parallel \overline{BC}$. Thus, C and D may be labeled as $C(c, b)$ and $D(d, 0)$. Also, $AB = \sqrt{a^2 + b^2}$ and $CD = \sqrt{(d - c)^2 + b^2}$. Because, in general, these two lengths are not the same, $ABCD$ is not isosceles.

3. $B\left(\frac{a}{2}, b\right)$

$$AB = \sqrt{\left(\frac{a}{2} - 0\right)^2 + (b - 0)^2} = \sqrt{\frac{a^2}{4} + b^2}$$
$$BC = \sqrt{\left(a - \frac{a}{2}\right)^2 + (0 - b)^2} = \sqrt{\frac{a^2}{4} + b^2}$$

Because $AB = BC$, $\triangle ABC$ is isosceles. However, because $\sqrt{\frac{a^2}{4} + b^2}$ is not necessarily equal to a, $\triangle ABC$ is not necessarily equilateral.

5. a. $C(0, c)$ and $D(a, b)$

$$AB = AD = \sqrt{a^2 + b^2}$$
$$BC = CD = \sqrt{(0 - a)^2 + (c - b)^2} = \sqrt{a^2 + (c - b)^2}$$

So $ABCD$ is a kite since two pairs of adjacent nonoverlapping sides are congruent.

b. No. In that case, $ABCD$ is s rhombus, a special case of a kite, since $AB = BC = CD = AD = \sqrt{a^2 + a^2} = a\sqrt{2}$. In fact, because $\overline{AB} \perp \overline{BC}$, $ABCD$ is a square.

7. Slope of $\overline{WZ} = \dfrac{\frac{b}{2} - 0}{\frac{a}{2} - \frac{e}{2}} = \dfrac{b}{a - e}$

Slope of $\overline{YX} = \dfrac{\frac{b + d}{2} - \frac{d}{2}}{\frac{a + c}{2} - \frac{c + e}{2}} =$

$$\frac{b + d - d}{a + c - c - e} = \frac{b}{a - e}$$

Because the slope of $\overline{WZ}$ equals the slope of $\overline{YX}$, we know that $\overline{WZ} \parallel \overline{YX}$

9. Outline: Given square $ABCD$

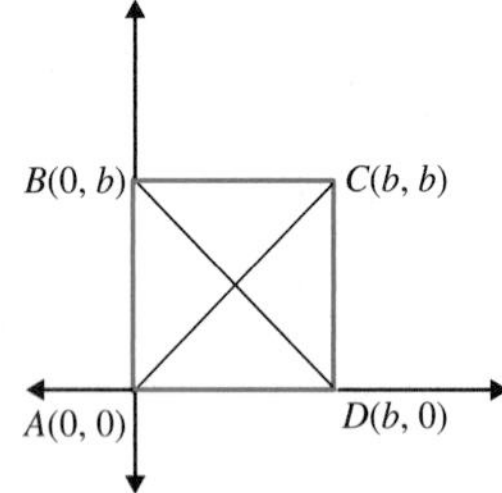

Use coordinates to show that the product of the slopes of $\overline{AC}$ and $\overline{BD}$ is -1.

11. Given parallelogram $ABCD$

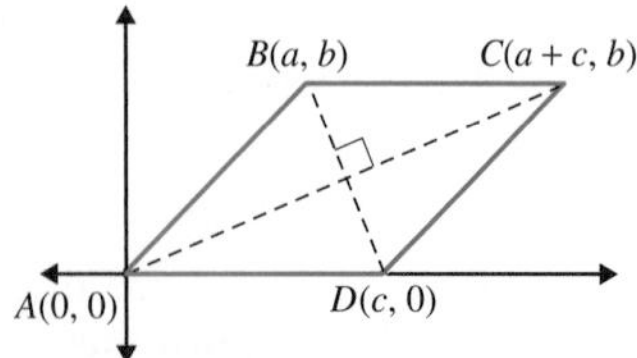

With coordinates shown, $ABCD$ is a parallelogram, because slope $\overline{AD}$ = slope $\overline{BC} = 0$ and slope $\overline{AB}$ = slope $\overline{DC} = \frac{b}{a}$.

$$AD = c - 0 = c$$
$$BC = (a + c) - a = c$$
$$CD = \sqrt{(a + c - c)^2 + (b - 0)^2} = \sqrt{a^2 + b^2}$$
$$AB = \sqrt{(a - 0)^2 + (b - 0)^2} = \sqrt{a^2 + b^2}$$

Thus, $\overline{AD} \cong \overline{BC}$ and $\overline{AB} \cong \overline{DC}$.

13. Given that M is the midpoint of $\overline{BC}$

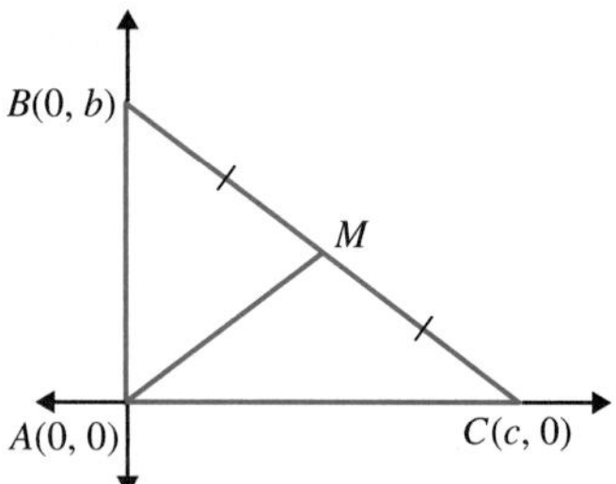

M, the midpoint of $\overline{BC}$, has coordinates $\left(\frac{c}{2}, \frac{b}{2}\right)$. By definition, $BM = CM$.

$$CM = \sqrt{\left(\frac{c}{2} - c\right)^2 + \left(\frac{b}{2} - 0\right)^2} = \sqrt{\left(\frac{c}{2}\right)^2 + \left(\frac{b}{2}\right)^2}$$
$$AM = \sqrt{\left(\frac{c}{2} - c\right)^2 + \left(\frac{b}{2} - 0\right)^2} = \sqrt{\left(\frac{c}{2}\right)^2 + \left(\frac{b}{2}\right)^2}$$

Therefore, $AM = CM = BM$.

15. Outline: Given isosceles trapezoid $ABCD$

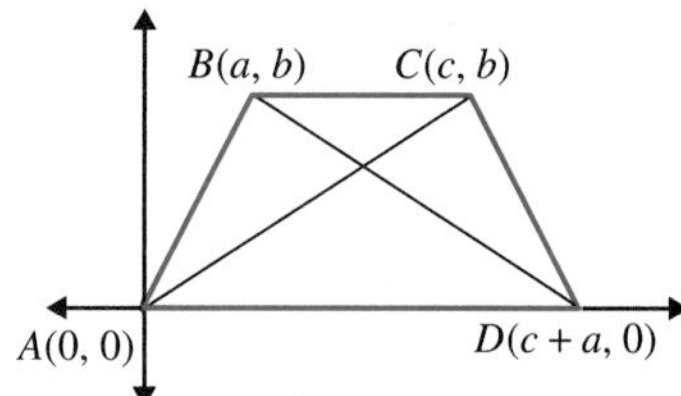

Use coordinates and the distance formula to show that $AC = BD$.

17. Outline: Given rectangle $ABCD$ with midpoints M, N, O, and P.

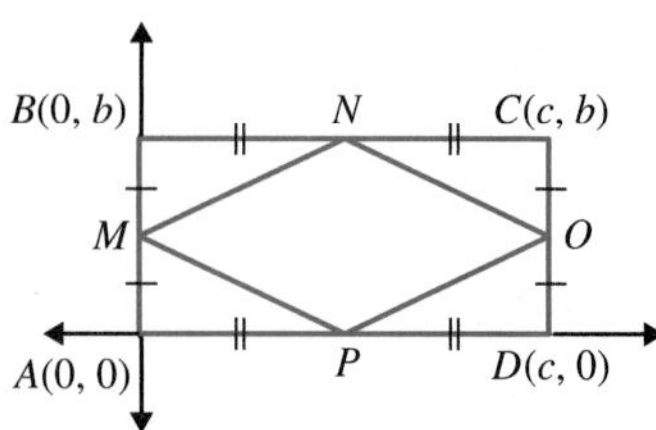

Use coordinates and the midpoint formula to find the coordinates of M, N, O, and P. Then use the distance formula to show $MN = NO = OP = PM$.

19. **Given** M, N, O, P midpoints of sides of rhombus $ABCD$

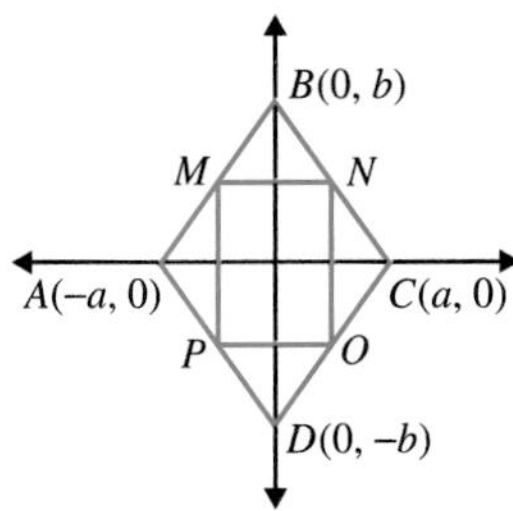

Prove $MNOP$ is a rectangle.

Proof

Statement	**Reason**
1. $AB = BC = CD = DA = \sqrt{a^2 + b^2}$	**1.** Distance formula
2. $ABCD$ is a rhombus	**2.** Definition
3. Midpoints are $M\left(-\frac{a}{2}, \frac{b}{2}\right), N\left(\frac{a}{2}, \frac{b}{2}\right), O\left(\frac{a}{2}, -\frac{b}{2}\right), P\left(-\frac{a}{2}, -\frac{b}{2}\right)$	**3.** Midpoint formula
4. Slope of $\overline{MN}$ = slope of $\overline{OP} = 0$	**4.** Slope formula
5. $\overline{MP}$ and $\overline{NO}$ are both vertical	**5.** Slope definition
6. $\overline{MN} \perp \overline{NO}, \overline{NO} \perp \overline{OP}, \overline{OP} \perp \overline{PM}$, and $\overline{PM} \perp \overline{MN}$	**6.** Theorem 8.5
7. $MNOP$ is a rectangle.	**7.** Definition

21. Outline: Given isosceles $\triangle ABC$ with M, N midpoints of $\overline{AB}, \overline{CB}$, respectively

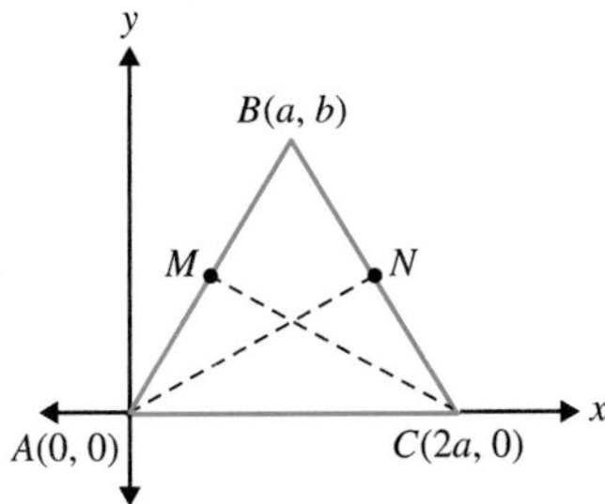

Use coordinates and the midpoint formula to find coordinates of M and N. Then show $AN = CM$ using the distance formula.

23. Given parallelogram $ABCD$ with $AC = BD$

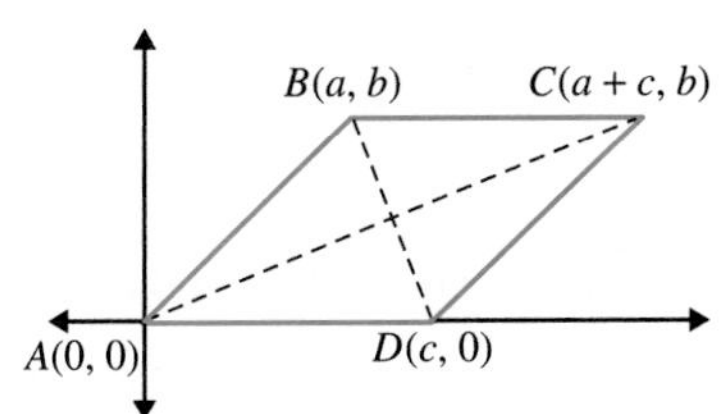

$BD = \sqrt{(a - c)^2 + b^2}$ and $AC = \sqrt{(a + c)^2 + b^2}$. Since $AC = BD$, $(a - c)^2 = (a + c)^2$. Therefore, $a^2 - 2ac + c^2 = a^2 + 2ac + c^2$, so $-2ac = 2ac$, or $4ac = 0$. Thus, either $a = 0$ or $c = 0$. If $c = 0$, then $A = D$ and $ABCD$ is not a parallelogram. Hence, $a = 0$. Thus, B and C have coordinates $B(0, b)$ and $C(c, b)$. So $\angle A = 90°$ and $ABCD$ is a rectangle.

25. Given that l_1, l_2, l_3 are the perpendicular bisectors of the sides of $\triangle ABC$

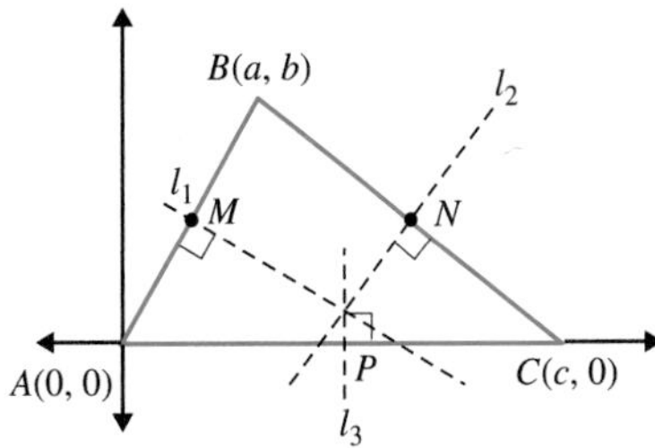

M, N, P have coordinates: $M\left(\frac{a}{2}, \frac{b}{2}\right), N\left(\frac{a + c}{2}, \frac{b}{2}\right), P\left(\frac{c}{2}, 0\right)$. Equation of l_3 is $x = \frac{c}{2}$. Slope of l_1 is $-\frac{a}{b}$, so its equation is $y - \frac{b}{2} = -\frac{a}{b}\left(x - \frac{a}{2}\right)$, or $y = -\frac{a}{b}x + \frac{a^2 + b^2}{2b}$. Slope of l_2 is $\frac{c - a}{b}$, so its equation is $y - \frac{b}{2} = -\left(\frac{c - a}{b}\right)\left(x - \frac{a + c}{2}\right)$, or $y = \frac{c - a}{b}x + \frac{a^2 + b^2 - c^2}{2b}$. l_1 and l_3 intersect where $x = \frac{c}{2}$ and $y = -\frac{a}{b}\left(\frac{c}{2}\right) + \frac{a^2 + b^2}{2b} = \frac{a^2 + b^2 - ac}{2b}$. l_2 and l_3 intersect where $x = \frac{c}{2}$ and $y = \left(\frac{c - a}{b}\right)\left(\frac{c}{2}\right) + \frac{a^2 + b^2 - c^2}{2b} = \frac{a^2 + b^2 - ac}{2b}$. Thus l_1, l_2, l_3 are concurrrent.

27. Outline: Given M, N, P midpoints of $\overline{AB}, \overline{BC}, \overline{AC}$

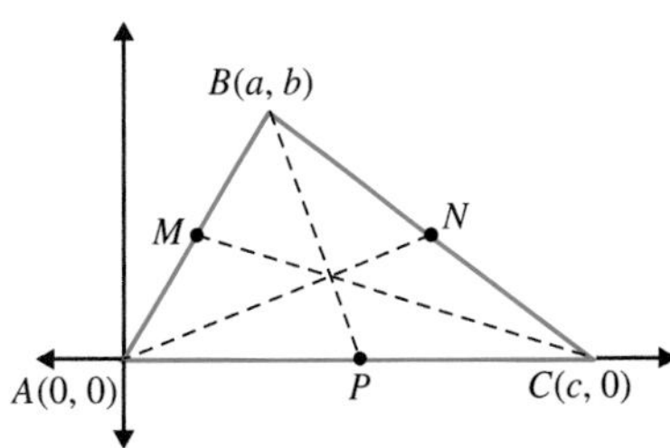

Using coordinates, find each midpoint. Using the coordinates of the vertices and the midpoints, write the equation of each median. Show that each pair of medians intersect in the same point.

Answers for the Additional Problems

1. $\sqrt{13} \approx 3.61$ km.
2. Place the figure on a coordinate system with $B = (0, 2a)$ and $C = (2a, 0)$. Then $D = (-a, a)$and $E = (a, -a)$. Using these coordinates, the slope of $\overline{BC}$ is -1 and the slope of $\overline{DE}$ is -1. Therefore, $\overline{BC} \parallel \overline{DE}$.

Chapter Review Problems

SECTION 8.1

1. **a.** Quadrant III **b.** Quadrant IV **c.** y-axis **d.** Quadrant II **e.** x-axis **f.** Quadrant I
2. $(13, -1.5)$
3. **a.** $ABCD$ is a rhombus. **b.** $(-1, -5)$
4. They are collinear. $AB + BC = AC$ since $AB = 2\sqrt{5}$, $BC = 3\sqrt{5}$, and $AC = 5\sqrt{5}$.
5. Let the vertices of $\triangle ABC$ have coordinates $A(a, b)$, $B(c, d)$, and $C(e, f)$. Given D is the midpoint of $\overline{AB}$ and E is the midpoint of $\overline{BC}$, we know $D = \left(\frac{a+c}{2}, \frac{b+d}{2}\right)$ and $E = \left(\frac{c+e}{2}, \frac{d+f}{2}\right)$. By the distance formula we have $AC = \sqrt{(a-e)^2 + (b-f)^2}$ and

$$\begin{aligned} DE &= \sqrt{\left(\frac{a+c}{2} - \frac{c+e}{2}\right)^2 + \left(\frac{b+d}{2} - \frac{d+f}{2}\right)^2} \\ &= \sqrt{\frac{(a+c-c-e)^2}{4} + \frac{(b+d-d-f)^2}{4}} \\ &= \frac{1}{2}\sqrt{(a-e)^2 + (b-f)^2} \\ &= \frac{1}{2}AC \end{aligned}$$

SECTION 8.2

1. $-\frac{37}{4}$
2. **a.** $9, -\frac{1}{9}$ **b.** 0, Undefined **c.** $\frac{2}{7}, -\frac{7}{2}$ **d.** $-\frac{5}{3}, \frac{3}{5}$ **e.** Undefined, 0
3. $ABCD$ is a rectangle.
4. **a.** $B(4, 4)$ **b.** $B(4, -4)$
5. Give M is the midpoint of $\overline{AB}$, we know the coordinates of M are $\left(\frac{a+c}{2}, \frac{b+d}{2}\right)$ by the definition of the midpoint. The slope of $\overline{AM}$ is $\dfrac{\frac{b+d}{2} - b}{\frac{a+c}{2} - a} = \dfrac{d-b}{c-a}$ and the slope of $\overline{BM}$ is

$$\frac{\frac{b+d}{2} - d}{\frac{a+c}{2} - c} = \frac{b-d}{a-c} = \frac{-(d-b)}{-(c-a)} = \frac{d-b}{c-a}$$

Since the slope of $\overline{AM}$ is the same as the slope of $\overline{BM}$, A, B, and M are collinear.

SECTION 8.3

1. **a.** Slope $= -\frac{3}{4}$, y-intercept $= 6$

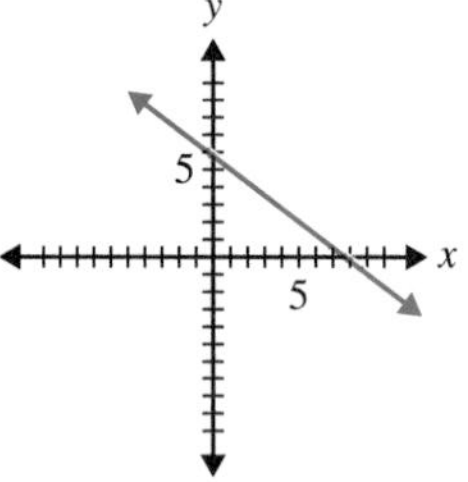

b. Slope $= \frac{1}{4}$, y-intercept $= 3$

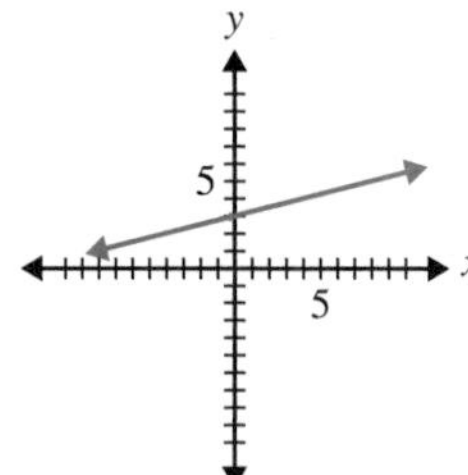

c. Slope $= -5$, y-intercept $= 10$

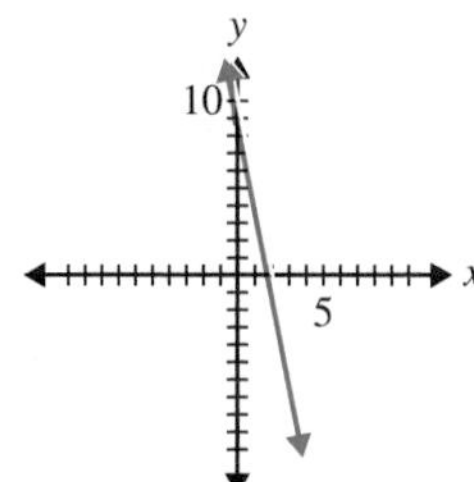

d. Slope $= 1$, y-intercept $= 7$

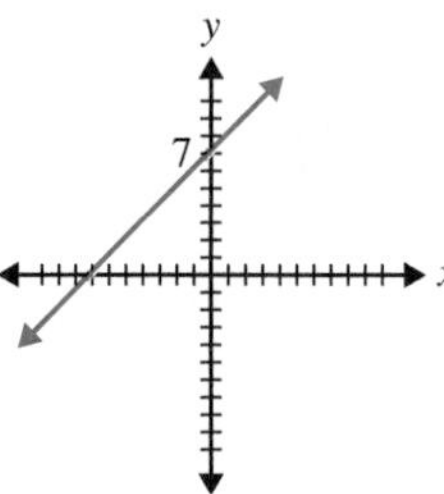

2. **a.** $y = 5$ **b.** $x = 4$ **c.** $y = -\frac{3}{2}x - 1$
3. **a.** $(-4, 1)$ **b.** No Solution—Parallel lines
4. $y = -\frac{2}{5}x + \frac{29}{10}$
5. $(x-2)^2 + (y-4)^2 = 50$

SECTION 8.4

1. $B(0, b), C(a, 0)$, and $D(0, -b)$
2. By the distance formula, $AB = BC = CD = DA = \sqrt{a^2 + b^2}$. Therefore, $ABCD$ is a rhombus.
3. Let the coordinates of A, B, and C be $A(0, 0)$, $B(a, b)$, and $C(2a, 0)$. Given $\overline{BM}$ is the median to $\overline{AC}$, the coordinates of M are $(a, 0)$.

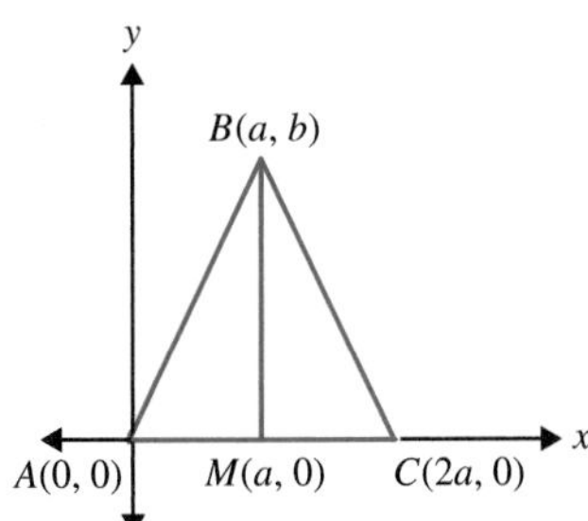

Since the x-coordinate of B and M is the same, $\overline{BM}$ is vertical. The slope of $\overline{AC}$ is $\frac{0 - 0}{2a - 0} = 0$. Thus, $\overline{AC}$ is horizontal. Therefore $\overline{BM} \perp \overline{AC}$.

4. Given: M, N, O, and P are midpoints of sides of isosceles trapezoid $ABCD$.

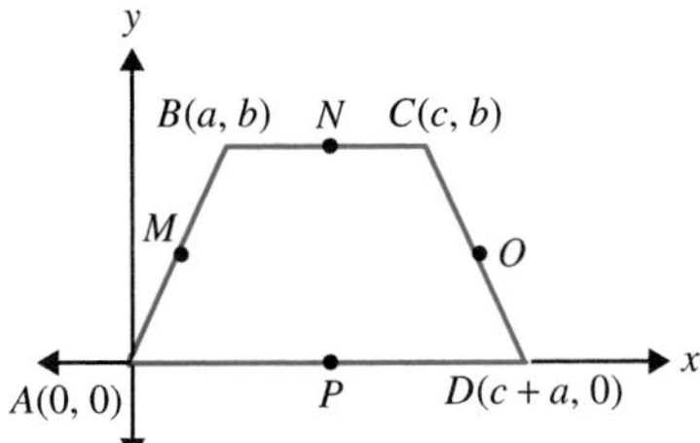

The coordinates of M, N, O, and P are $M\left(\frac{a}{2}, \frac{b}{2}\right), N\left(\frac{a + c}{2}, b\right), O\left(\frac{2c + a}{2}, \frac{b}{2}\right)$, and $P\left(\frac{c + a}{2}, 0\right)$. Using the distance formula, we have

$$MN = NO = OP = PM = \sqrt{\left(\frac{c}{2}\right)^2 + \left(\frac{b}{2}\right)^2}$$

Therefore, $MNOP$ is a rhombus.

CHAPTER 8 TEST

1. F—The slope of a perpendicular line is $-\frac{1}{2}$.
2. F—only true if B is between A and C
3. T
4. T
5. T
6. F—It might have zero or infinitely many solutions.
7. T
8. F—The slope of the line is $-\frac{a}{b}$.
9. F—The area of the new triangle will be four times the area of $\triangle ABC$.
10. T
11. **a.** $(6, 1)$ **b.** $\sqrt{72}$ or $6\sqrt{2}$ **c.** 1 **d.** $y = x - 5$
12. Yes, they are collinear.
    Slope of $\overline{AB}$ = slope of $\overline{AC}$ = slope of $\overline{BC}$. Also, $AB + BC = AC$.
13. **a.**

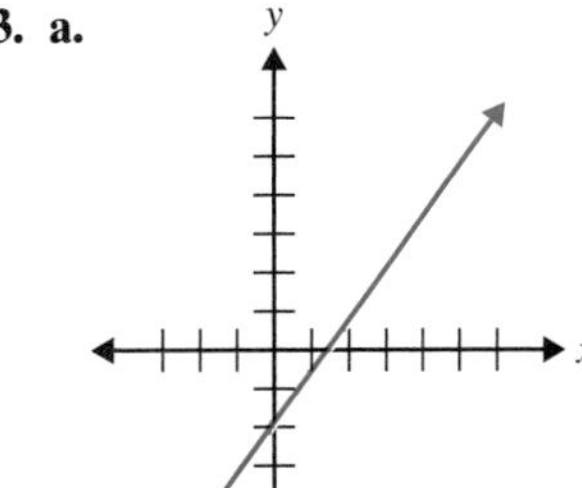

b.

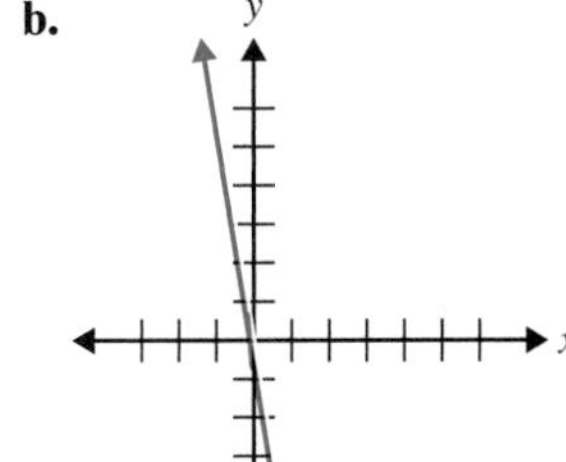

c.

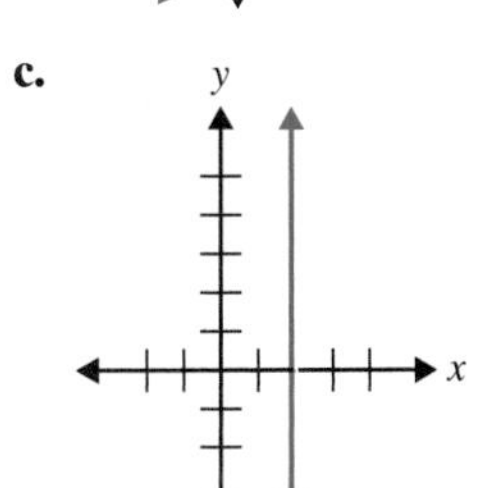

d.

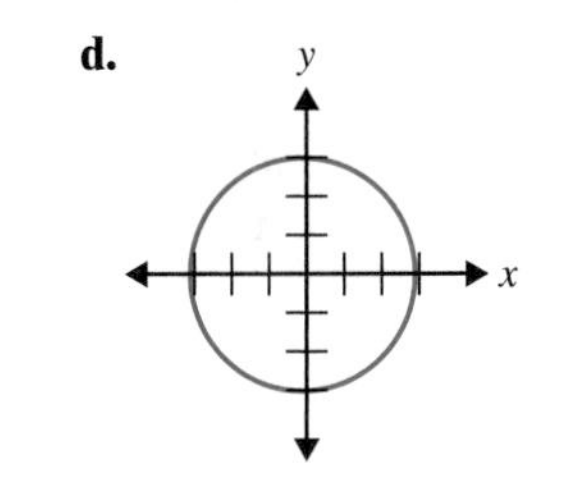

e.

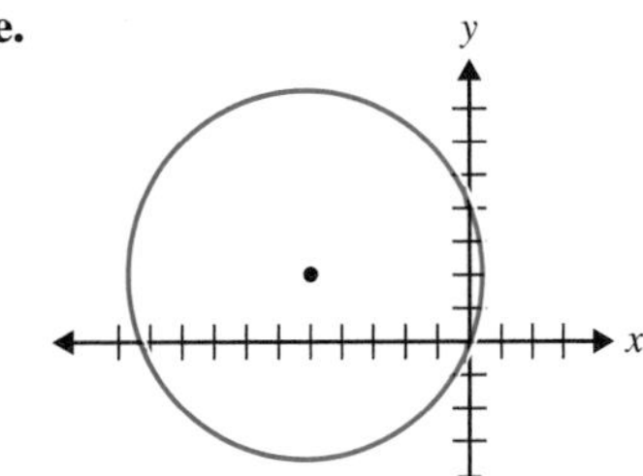

14. **a.** $(4, 1)$ **b.** No solution—parallel lines
    c. No solution—Line does not intersect the circle
15. $y = -\frac{1}{2}x - \frac{1}{2}$ or $x + 2y = -1$
16. $(c - b, a)$
17. $AC = \sqrt{(3 + 1)^2 + (7 - 1)^2} = \sqrt{16 + 36} = \sqrt{52}$ or $2\sqrt{13}$
    $BD = \sqrt{(4 + 2)^2 + (2 - 6)^2} = \sqrt{36 + 16} = \sqrt{52}$ or $2\sqrt{13}$
18. $(x + 1)^2 + (y - 4)^2 = 52$
19. $ABCD$ is a parallelogram because $\overline{AB}$ and $\overline{CD}$ have the same slope and $\overline{AD}$ and $\overline{BC}$ have the same slope. Thus, opposite sides are parallel.
20. $(-2, -1), (0, 11), (12, -1)$
21. M_1, M_2, and M_3 are the midpoints of the sides of equilateral $\triangle ABC$.

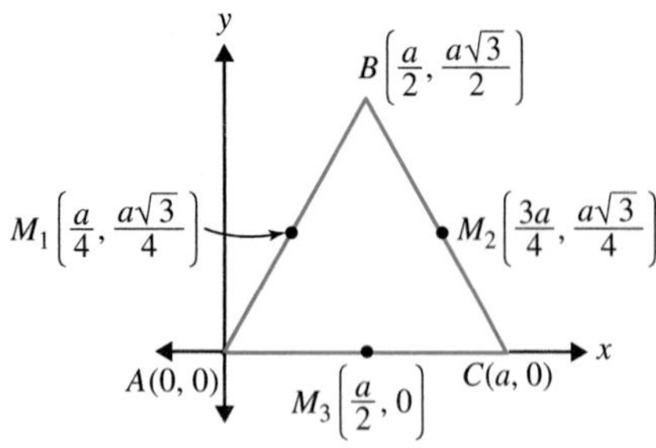

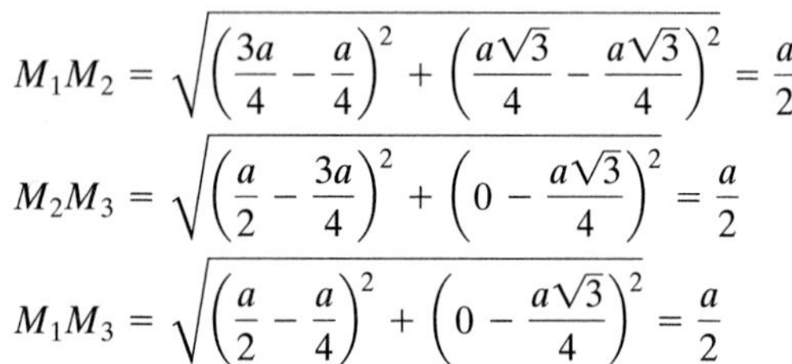

$$M_1M_2 = \sqrt{\left(\frac{3a}{4} - \frac{a}{4}\right)^2 + \left(\frac{a\sqrt{3}}{4} - \frac{a\sqrt{3}}{4}\right)^2} = \frac{a}{2}$$
$$M_2M_3 = \sqrt{\left(\frac{a}{2} - \frac{3a}{4}\right)^2 + \left(0 - \frac{a\sqrt{3}}{4}\right)^2} = \frac{a}{2}$$
$$M_1M_3 = \sqrt{\left(\frac{a}{2} - \frac{a}{4}\right)^2 + \left(0 - \frac{a\sqrt{3}}{4}\right)^2} = \frac{a}{2}$$

Therefore, $\triangle M_1M_2M_3$ is equilateral.

22. If the diagonals are perpendicular, then the product of their slopes is -1.

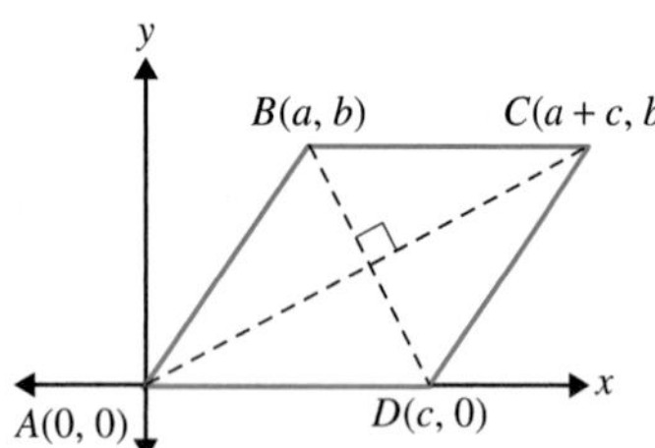

$$\left(\frac{b}{a-c}\right) \cdot \left(\frac{b}{a+c}\right) = -1$$
$$\frac{b^2}{a^2 - c^2} = -1$$
$$b^2 = c^2 - a^2$$
$$a^2 + b^2 = c^2$$

This shows that $AB = AD$. Because opposite sides of a parallelogram are congruent and $AB = AD$, all sides must be congruent. Therefore, $ABCD$ is a rhombus.

23. Slope is $\frac{5}{96}$; 5.2 percent grade

24. a. $y = (0.01)\pi x$ or $y \approx 0.0314x$

b. $(0.01)\pi$—For each cm added to the length of the wire, the volume increases by $(0.01)\pi$ cm^3.

c. $y = (0.0894)\pi x$

d. $(0.0894)\pi$—For each cm added to the length of the wire, the mass increases by $(0.0894)\pi$ g.

SECTION 9.1

1. Yes. Yes. Translations take lines to parallel lines.

3. a. Answers will vary.

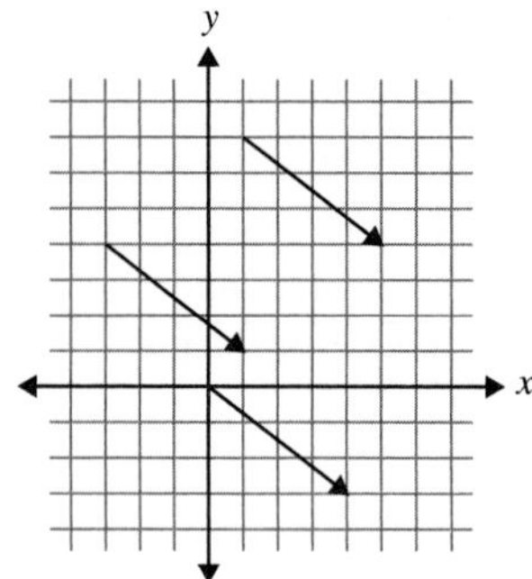

b. Many vectors are possible.

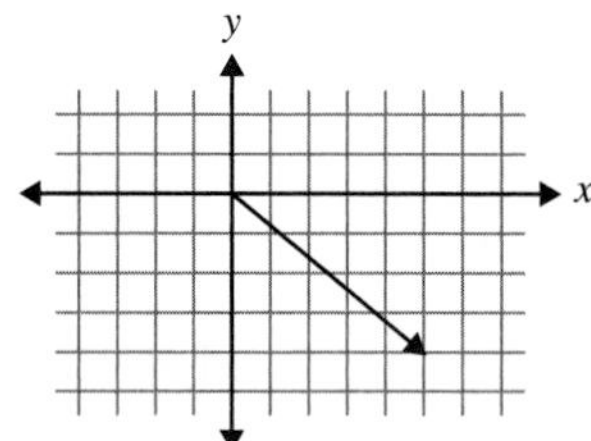

5. a. $A'(-2, -1), B'(2, -6)$ **b.** $A'(-3, 4), B'(1, -1)$

c. $A'(1, 6), B'(5, 1)$ **d.** $A'(2, 0), B'(6, -5)$

7. Yes. Yes. Each translation takes a triangle to a congruent triangle.

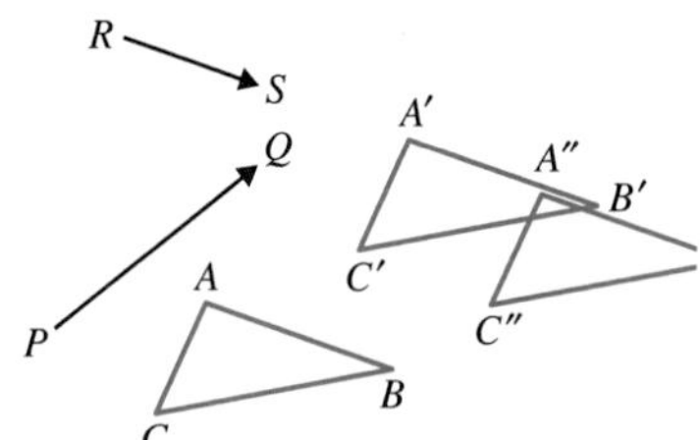

9.

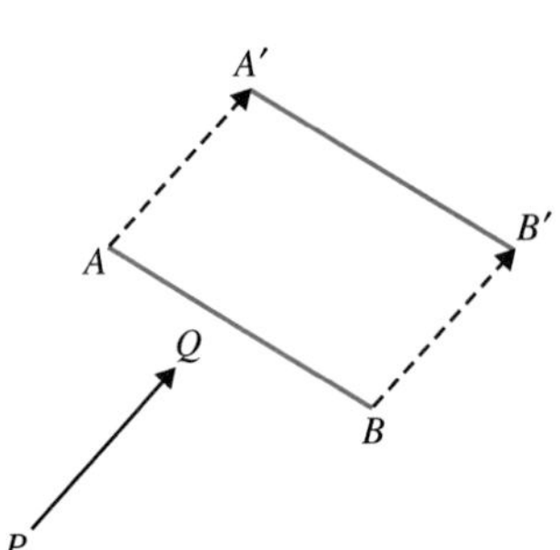

11. a. $A'(0, 4), B'(4, 6), C'(5, 1)$

b. $A'(3, 0), B'(7, 2), C'(8, -3)$

c. $A'(-2, 5), B'(2, 7), C'(3, 2)$

13. a P P' O **b.** P O P' **c.** P' P O

15. **a.** $-60°$ **b.** $110°$ **c.** $180°$ (or $-180°$) **d.** $-20°$

17.

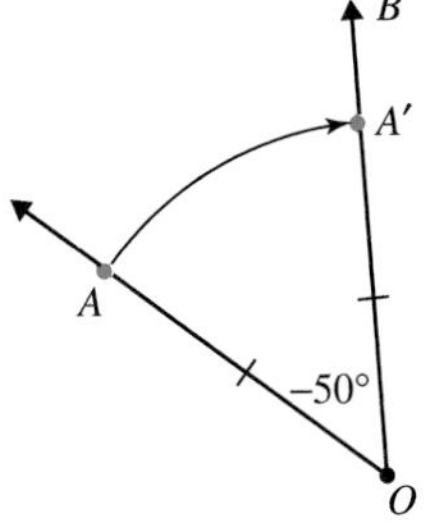

19. No. Rotations do not necessarily take lines to parallel lines.

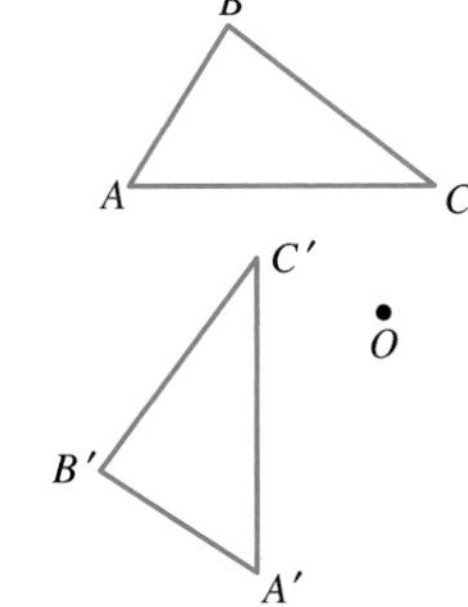

21.

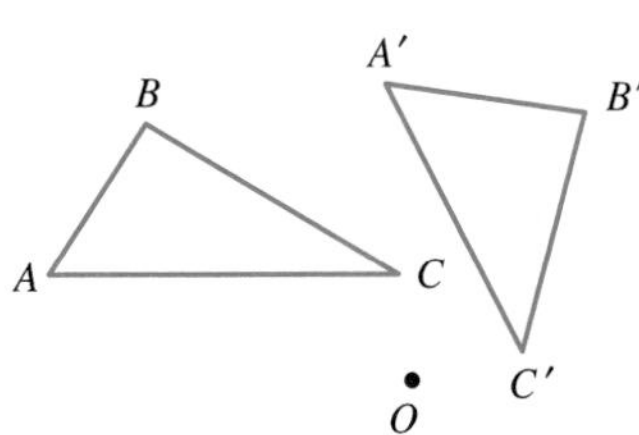

23. After the rotation, $\triangle A'B'C'$ would appear as follows.

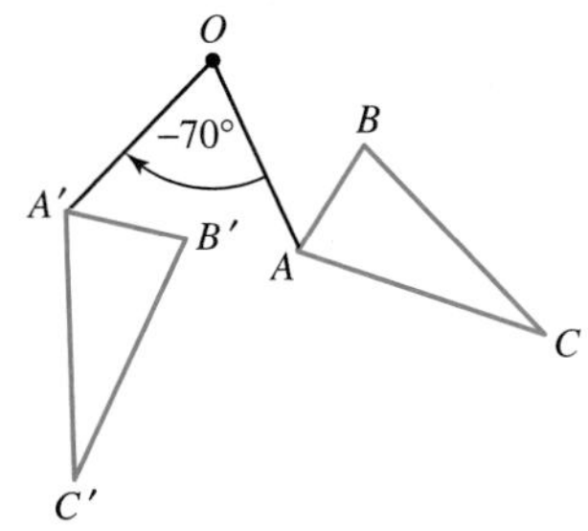

After the translation, $\triangle A''B''C''$ would appear as follows.

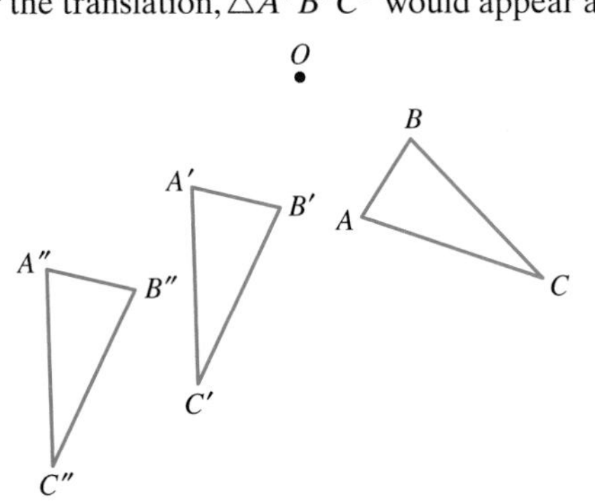

25. **a.** $(5, -1)$ **b.** $(3, 1)$ **c.** $(4, 2)$ **d.** $(-1, 3)$ **e.** $(-2, -5)$ **f.** $(y, -x)$

27.

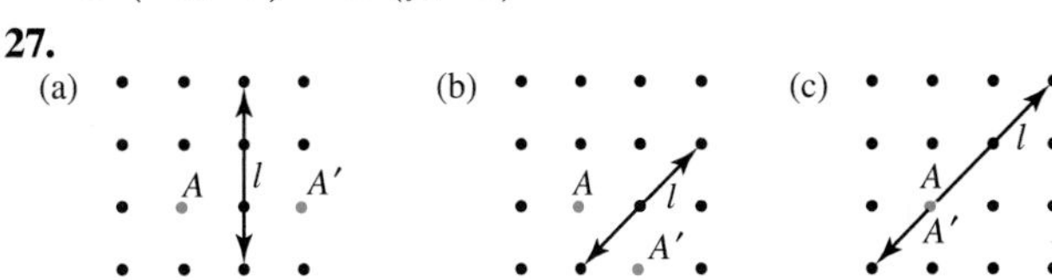

29. **a.** Isosceles because $QP = QP'$

b. A right triangle because $\overline{PP'} \perp l$

31. $ABCB'$ is a kite because $CB = CB'$ and $AB = AB'$.

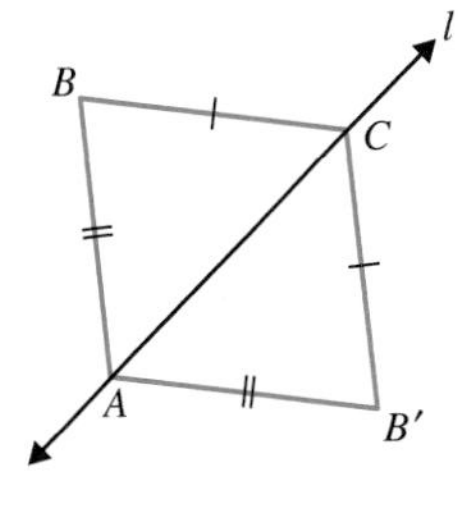

33.

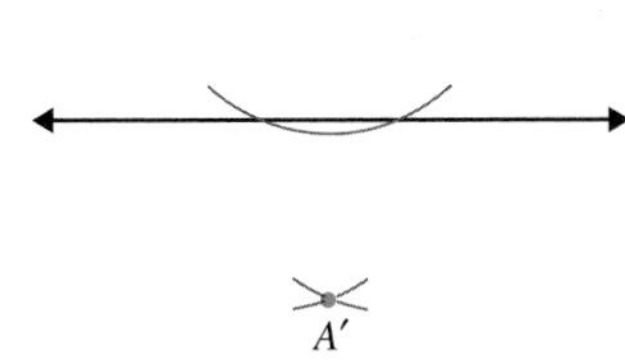

35. **a.**

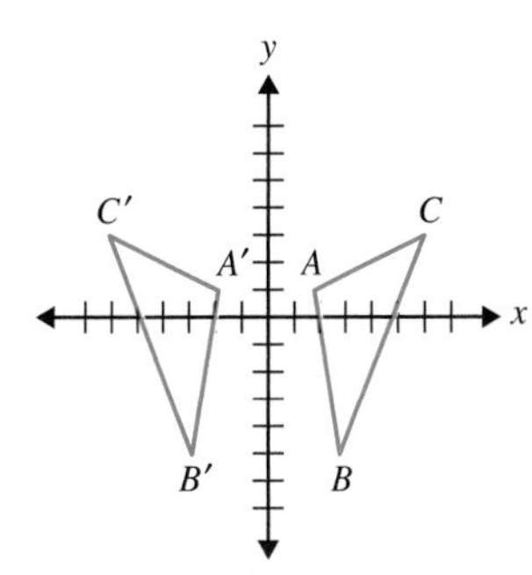

b. $A'(-2, 1), B'(-3, -5), C'(-6, 3)$

c. $(-a, b)$

37.

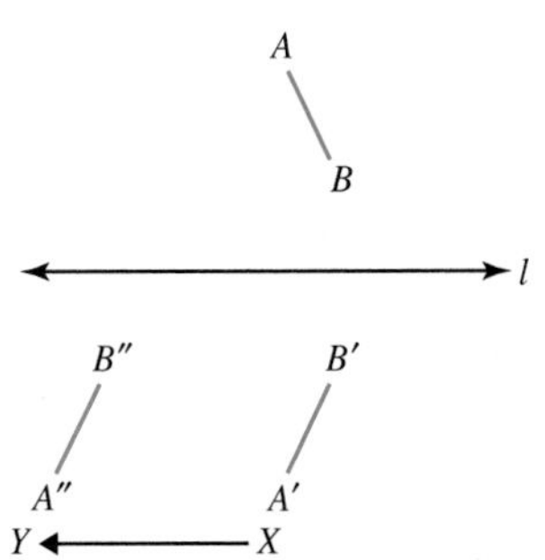

39. a.

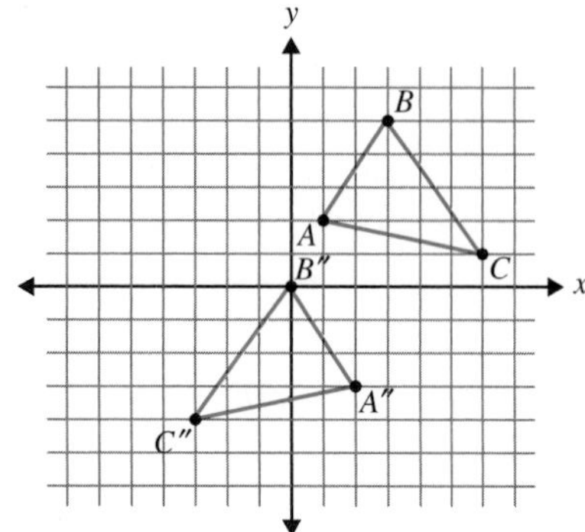

b. $A''(2, -3)$, $B''(0, 0)$, and $C''(-3, -4)$

c. $(3 - a, b - 5)$

41. a. Translation **b.** Reflection

43. Line l is the perpendicular bisector of $\overline{BB'}$, $\overline{AA'}$, and $\overline{CC'}$.

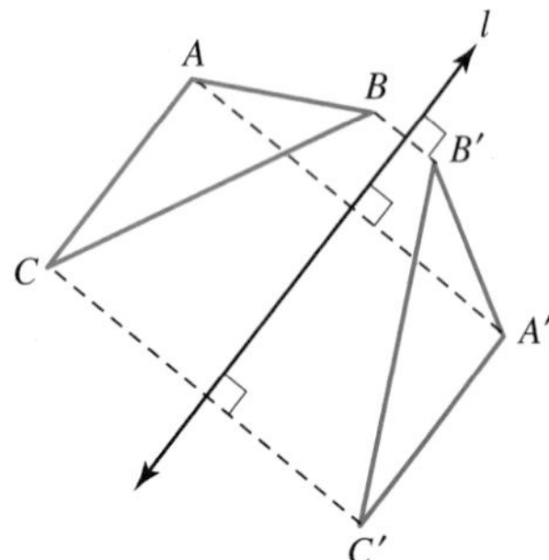

45. a. Glide reflection **b.** Rotation **c.** Translation

47. a.

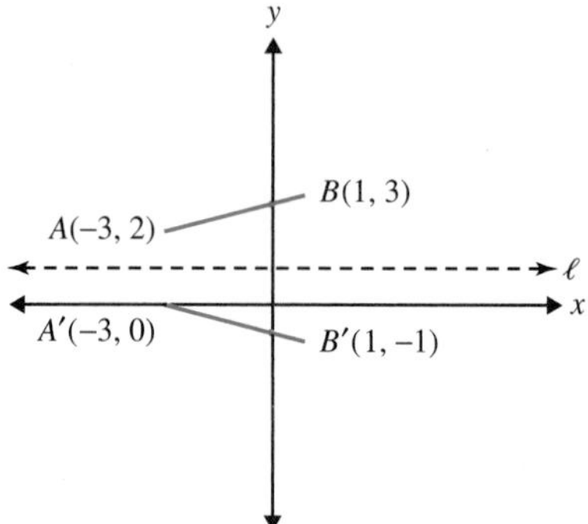

We could draw other figures to determine the line of reflection, which is the line $y = 1$.

b.

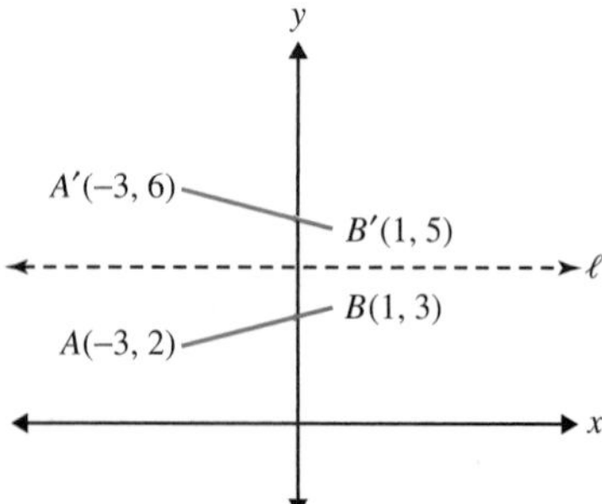

We could draw other figures to determine the line of reflection, which is the line $y = 4$.

c.

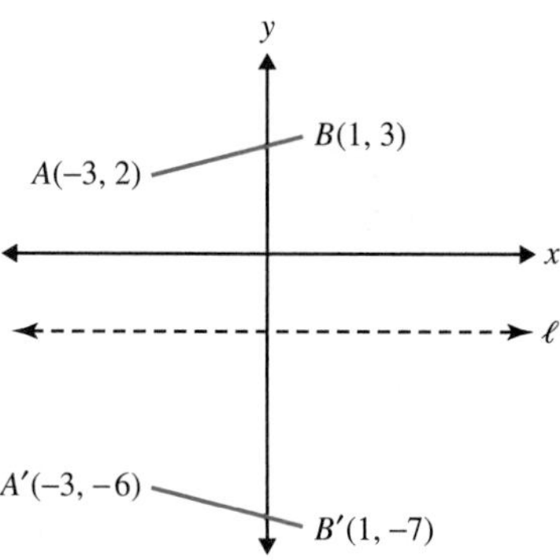

We could draw other figures to determine the line of reflection, which is the line $y = -2$.

49. Translations are used to position the tiles to form a tessellation.

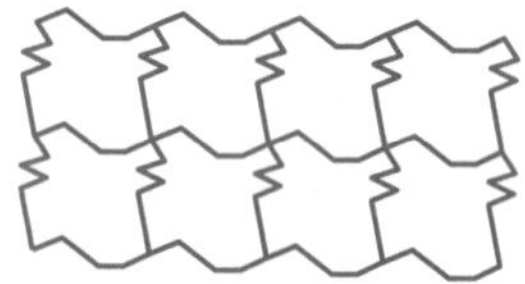

51. Let $\overline{A'B'}$ be the image of $\overline{AB}$ under the rotation through θ about O.

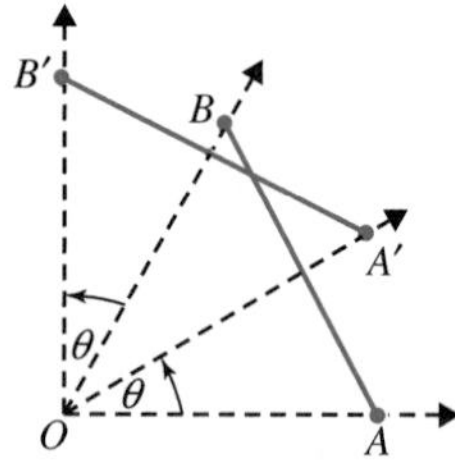

By definition of a rotation, $OA = OA'$ and $OB = OB'$. $\angle A'OB' = \angle AOB + \theta - \theta = \angle AOB$. Thus, $\triangle OA'B' \cong \triangle OAB$ (SAS) and $AB = A'B'$.

53. Outline: Suppose $\overline{AB} \perp \overline{CD}$ and they intersect at P. Let $\overline{A'B'}$ and $\overline{C'D'}$ be their images under the rotation about O. Then apply property 3 of Theorem 9.2.

55. Show that reflections preserve distance.

Given $\overline{A'B'}$ is the reflection of $\overline{AB}$ with respect to line l, $\overline{AA'}$ intersects l at P, and $\overline{BB'}$ intersects l at Q.

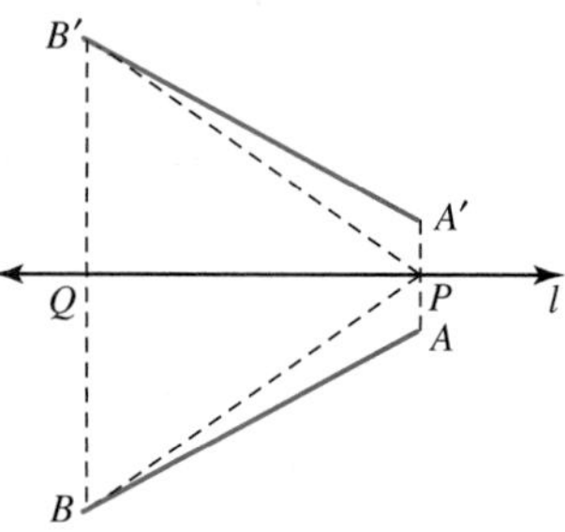

Prove $\overline{AB} \cong \overline{A'B'}$

Proof

Statement	Reason
1. $\overline{B'Q'} \cong \overline{BQ}$ and $\angle BQP \cong \angle B'QP$	1. Definition of reflection
2. $\triangle BQP \cong \triangle B'QP$	2. SAS Congruence
3. $BP = B'P$ and $\angle BPQ \cong \angle B'PQ$	3. C.P.
4. $\overline{AP} \cong \overline{A'P}$ and $\angle APQ \cong \angle A'PQ = 90°$	4. Definition of reflection
5. $\angle APB = \angle A'PB'$	5. Complements of congruent angles
6. $\angle APB = \angle A'PB'$	6. SAS Congruence
7. $\overline{AB} \cong \overline{A'B'}$	7. C.P.

57. Outline: Use the fact that angle measure is preserved.

59. Outline: Let $\overline{A''B''}$ be the image of $\overline{AB}$ under the glide reflection consisting of reflection with respect to line l and translation $\overrightarrow{PQ}$.

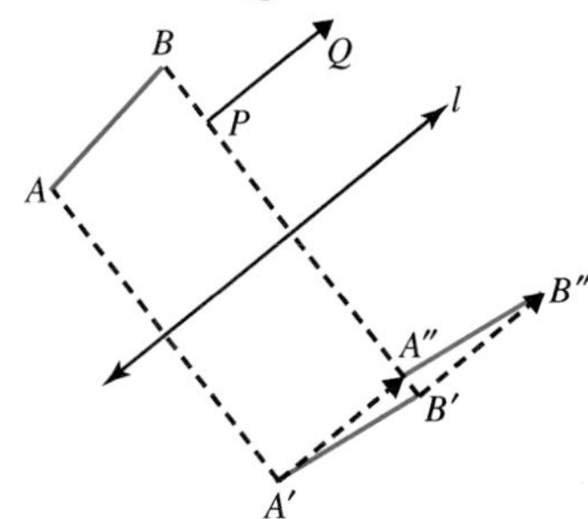

Use the fact that both reflections and translations preserve distance.

61. a. Translation and rotation
    b. Translation and rotation
    c. Translation
    d. Translation, rotation, reflection, glide reflection

63.

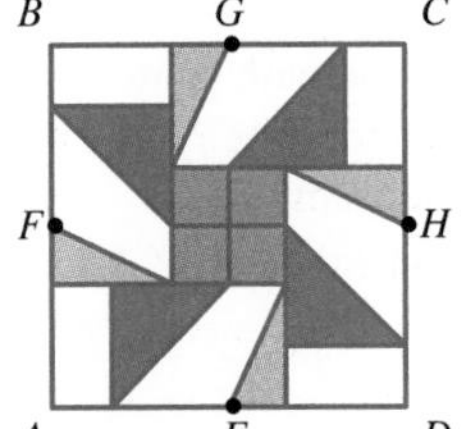

65. The three possible moves are shown. Each of these moves is equivalent to a glide reflection because it involves either a horizontal or vertical translation followed by a reflection.

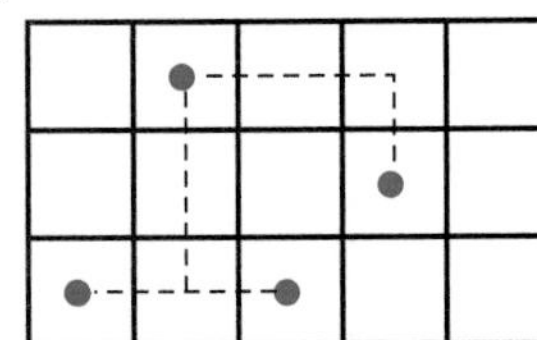

67. a. There are three unique moves for player O.
    The following four moves are equivalent.

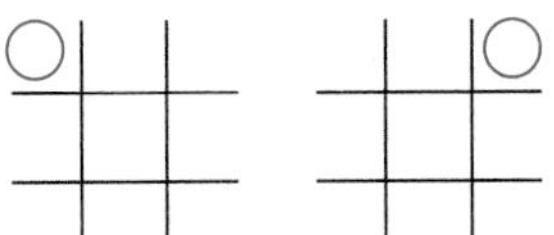

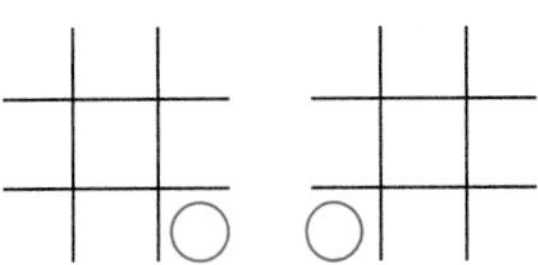

The following four moves are equivalent.

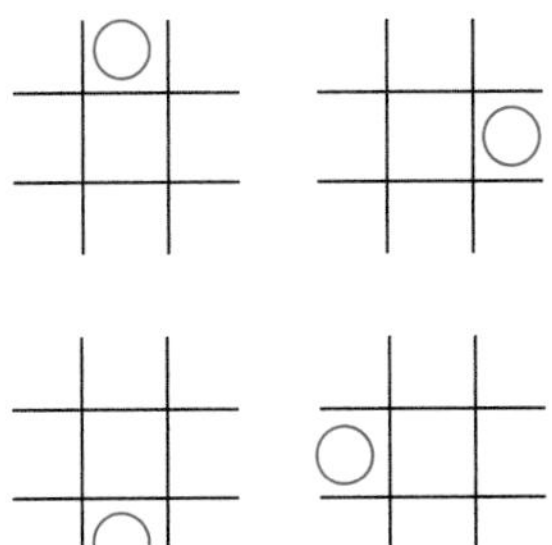

The following move has no other moves equivalent to it.

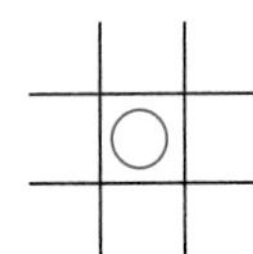

b. If player O places an O in the center, then there are two unique moves for player X. Otherwise, there are five unique moves.

SECTION 9.2

1. a. $\dfrac{OA'}{OA} = 4$ b. $\dfrac{OA'}{OA} = 3.75$

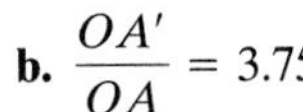

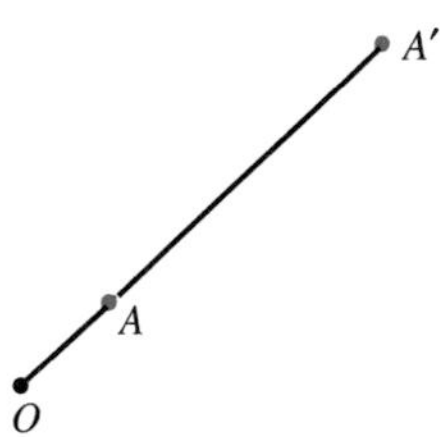

c. $\dfrac{OA'}{OA} = 2\dfrac{5}{8} = 2.625$ d. $\dfrac{OA'}{OA} = \dfrac{2}{3}$

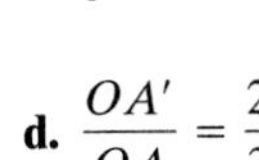

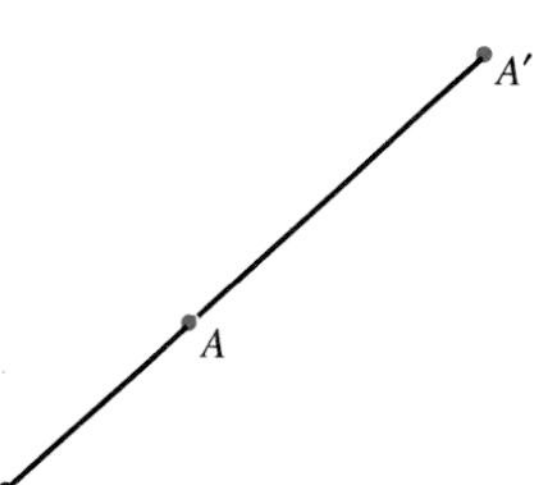

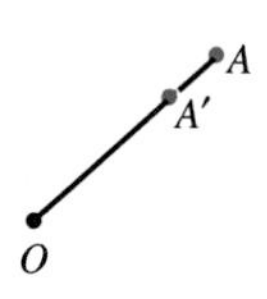

3. a. O P P′

b.

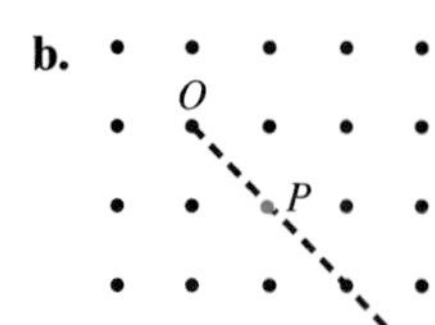

c.

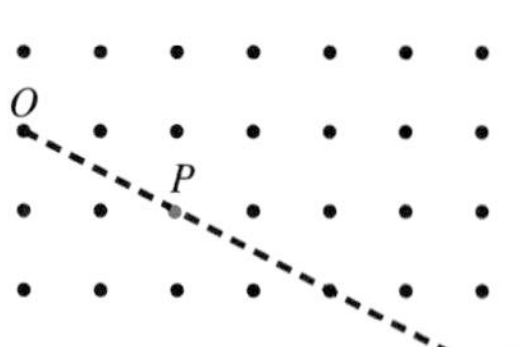

5. a.

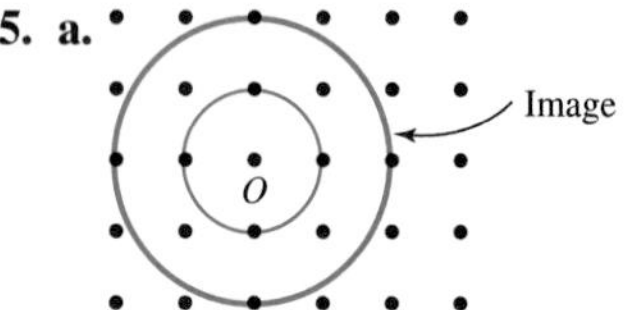

b.

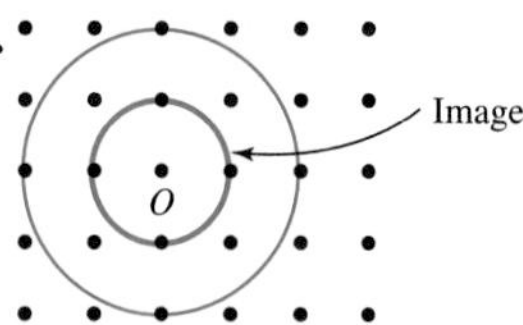

7. a.

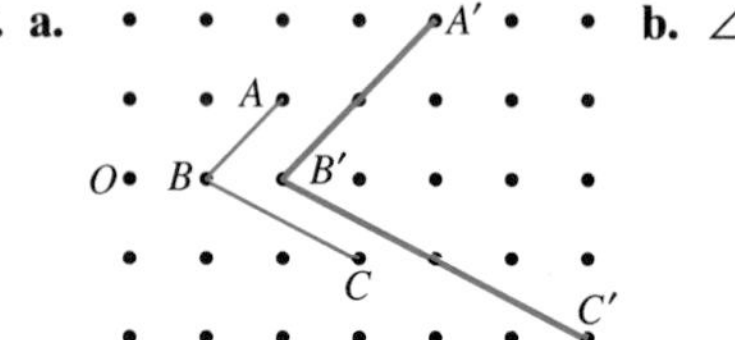

b. $\angle ABC \cong \angle A'B'C'$

9. a.

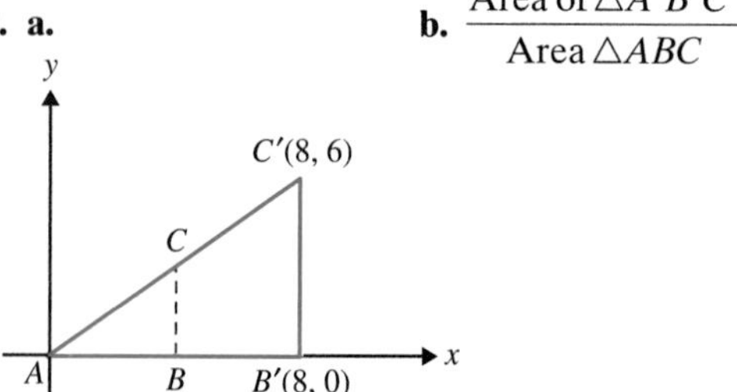

b. $\dfrac{\text{Area of } \triangle A'B'C'}{\text{Area } \triangle ABC} = \dfrac{24}{6} = 4$

11. The circumference of the image circle is three times the circumference of the original circle and the area of the image circle is nine times the area of the original circle.

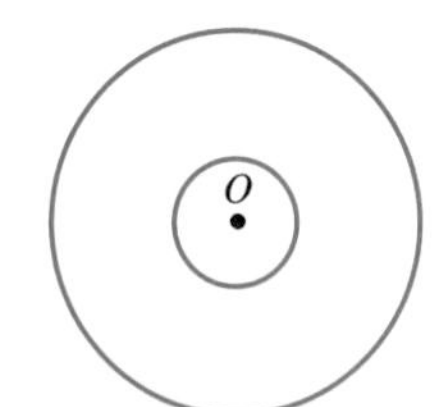

13. a.

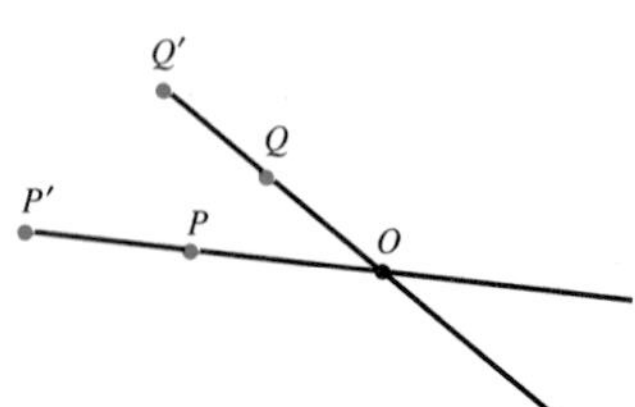

b. The scale factor is approximately 2.

15.

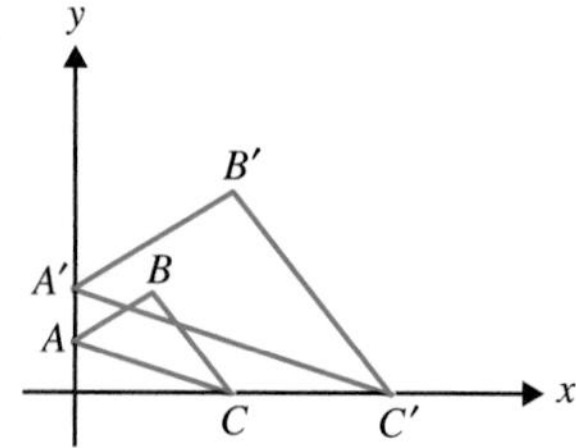

Yes, $\triangle A'B'C'$ is a size transformation image of $\triangle ABC$ with center $(0, 0)$ and scale factor 2.

17. a.

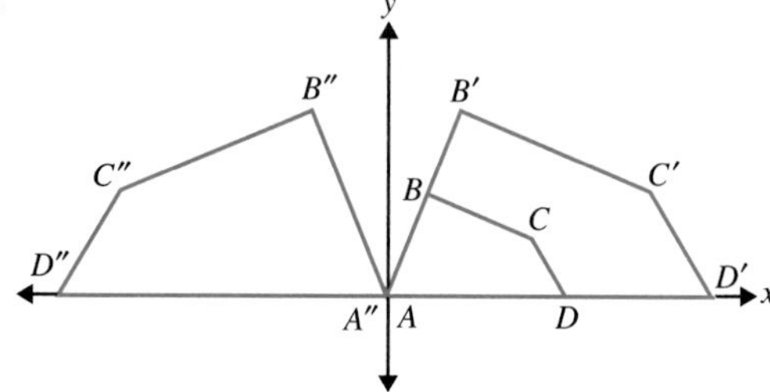

b. $A''(0, 0), B''(-2, 6), C''(-8, 4), D''(-10, 0)$

c. $(-2a, 2b)$

19. a.

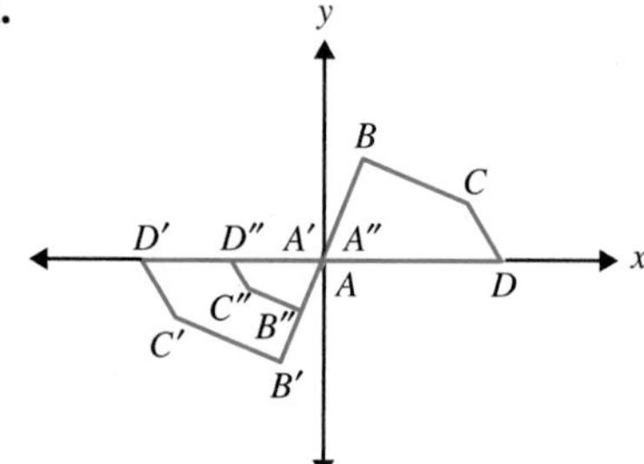

b. $A''(0, 0), B''(-\frac{1}{2}, -\frac{3}{2}), C''(-2, -1), D''(-\frac{5}{2}, 0)$

c. $\left(-\dfrac{a}{2}, -\dfrac{b}{2}\right)$

21. First perform the size transformation with center A and scale factor $\dfrac{9}{2}$. Then translate 3 units to the left.

23. First apply the size transformation with center A and scale factor $\dfrac{1}{3}$ to $\triangle ABC$. Call the image $\triangle A'B'C'$. Then reflect $\triangle A'B'C'$ across the line that is the perpendicular bisector of $\overline{C''C'}$.

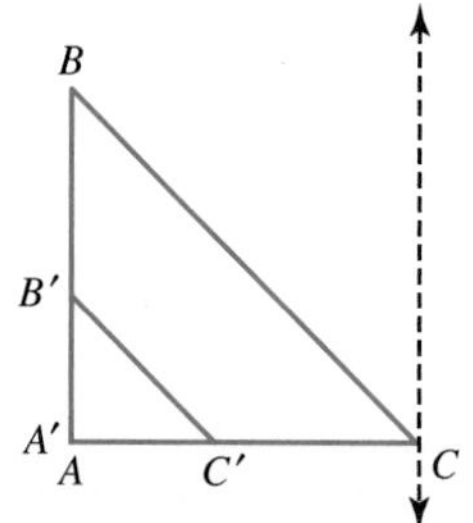

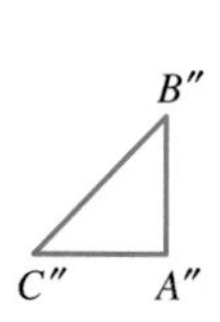

25. There is a size transformation with scale factor of 2 (in this case) that takes $\triangle ABC$ to $\triangle A'B'C'$. Since $\triangle A'B'C' \cong \triangle DEF$, there is an isometry that takes $\triangle A'B'C'$ to $\triangle DEF$. This size transformation and isometry constitute a similitude, so $\triangle ABC \sim \triangle DEF$.

27. Consider a size transformation with center O that takes A to A', B to B', C to C'. Then $\frac{OA'}{OA} = \frac{OB'}{OB} = \frac{OC'}{OC}$. We have $\angle AOB = \angle A'OB'$, $\angle BOC = \angle B'OC'$ since A and A', B and B', C and C' are collinear. Thus, $\triangle OAB \sim \triangle OA'B'$ and $\triangle OBC \sim \triangle OB'C'$ by SAS Similarity. Then $\frac{OA'}{OA} = \frac{A'B'}{AB}$ and $\frac{OB'}{OB} = \frac{B'C'}{BC}$. So $\frac{A'B'}{AB} = \frac{B'C'}{BC}$. Thus, $BC \cdot A'B' = AB \cdot B'C'$ and $\frac{A'B'}{B'C'} = \frac{AB}{BC}$.

29. Outline: Consider the size transformation with center O that takes A to A', B to B', C to C', with $\overline{AB} \perp \overline{BC}$. Use the fact that size transformations preserve angle measure.

31. Scale factor ≈ 2.12; The center is the oil drum.

33. Scale factor $= \frac{4}{3}$

35. a.

b.

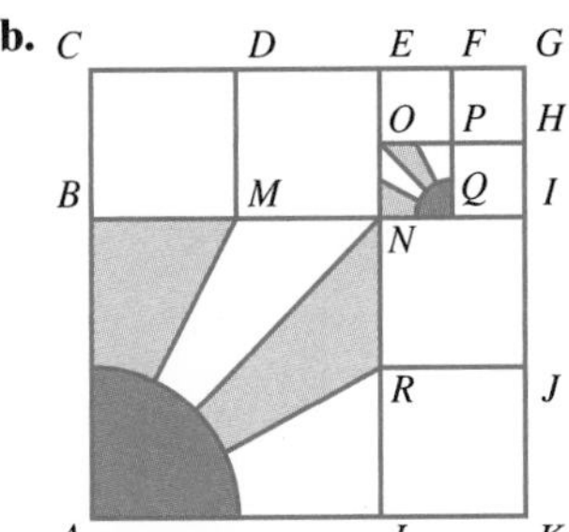

SECTION 9.3

1. Reflections with respect to the lines containing the following segments: $\overline{AC}$, $\overline{BD}$, $\overline{EG}$, $\overline{FH}$

3. Rotations of 180° and 360° around P

5. a. The reflections with respect to lines $\overleftrightarrow{AC}$ and $\overleftrightarrow{BD}$ and 180°and 360° rotations around the point of intersection of the diagonals

b. Because a rhombus is both a kite and a parallelogram, the rhombus must have at least the same isometries as the kite and parallelogram.

7. a. Yes. Reflections take triangles to congruent triangles.

b. Yes **c.** No

d. Yes. Both reflections preserve distance; thus, one reflection followed by another will preserve distance.

e. A translation in the direction perpendicular to the parallel lines and twice the distance between the lines

9. All. Parts (a) and (d) because they rotate one diagonal to the other.

11. Reflections with respect to $\overleftrightarrow{AI}$, $\overleftrightarrow{BJ}$, $\overleftrightarrow{CF}$, $\overleftrightarrow{DG}$, and $\overleftrightarrow{EH}$. Rotations of 72°, 144°, 216°, and 288° with respect to the center of the pentagon. Any of the rotations listed will rotate a diagonal to another diagonal. For example, for the rotation of 72°, $\overline{AC}$ rotates to $\overline{EB}$. Thus, they are congruent. In a similar way, $\overline{EB}$ rotates to $\overline{DA}$, and so on.

13. The reflection with respect to $\overleftrightarrow{AC}$ takes B to D. Therefore, $AB = AD$ and $BC = DC$. The reflection with respect to $\overleftrightarrow{BD}$ takes A to C. Therefore, $AB = CB$ and $AD = CD$. Combining these four equations, we have $AB = AD = DC = BC$. Thus, $ABCD$ is a rhombus.

15. Let M be the midpoint of $\overline{AC}$. Then the image of A under the 180° rotation with center M is C, and of C is A. If we apply this same rotation to B, the image is D since $ABCD$ is a parallelogram. Thus, this rotation takes $\triangle ABC$ to $\triangle CDA$, so they are congruent.

17. Outline: Use the fact that $\overline{AB}$ goes to $\overline{BC}$ under a reflection with respect to $\overleftrightarrow{BP}$. Then show $\triangle ABP \cong \triangle CBP$.

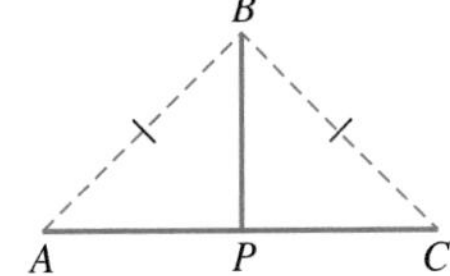

19. Given Point B is on the perpendicular bisector l of AC

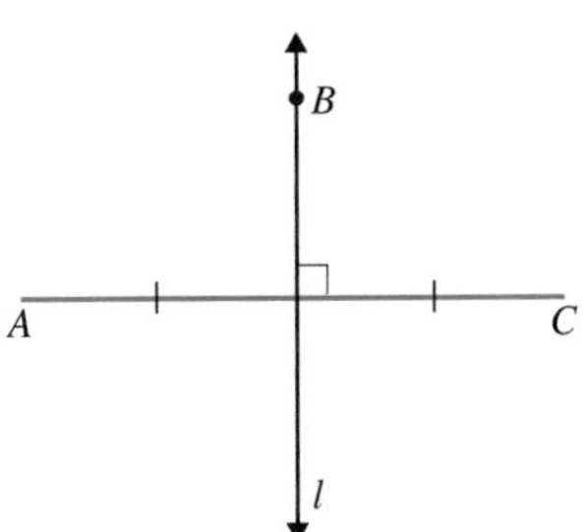

Prove $AB = CB$

Proof

Statement	Reason
1. l is the perpendicular bisector of AC.	**1.** Given
2. Reflection image of A with respect to l is C and reflection image of B with respect to l is B	**2.** Definition of reflection

3. Image of $\overline{AB}$ is $\overline{CB}$	**3.** Reflections take lines to lines.
4. $\overline{AB} \cong \overline{CB}$	**4.** Reflections preserve congruence.
5. $AB = CB$	**5.** Definition of congruence

21. Outline: Use the fact that A goes to C under the rotation and rotations take parallel lines to parallel lines to show that B must be on $\overleftrightarrow{DC}$ after the rotation. Show B must also be on $\overleftrightarrow{AD}$ after the rotation. Thus, B must go to D, and $\overline{BP}$ goes to $\overline{DP}$.

23. Given Points P, Q, and R are the midpoints of the sides of $\triangle ABC$.

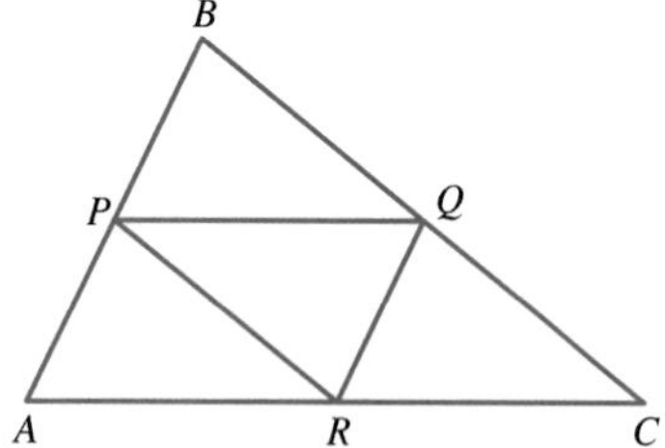

a. Since P and R are midpoints of $\overline{AB}$ and $\overline{AC}$, respectively, the size transformation with center A and scale factor 2 takes P to B and R to C. Therefore, $\triangle APR$ goes to $\triangle ABC$ under this size transformation.

b. $\triangle BAC$

c. The size transformation with center C and scale factor 2

25. In each move, the shaded region maintains the same area.

27. One path is shown. However, there are other correct solutions. First reflect B to B_1, then B_1 to B_2, then B_2 to B_3. Aim A for B_3.

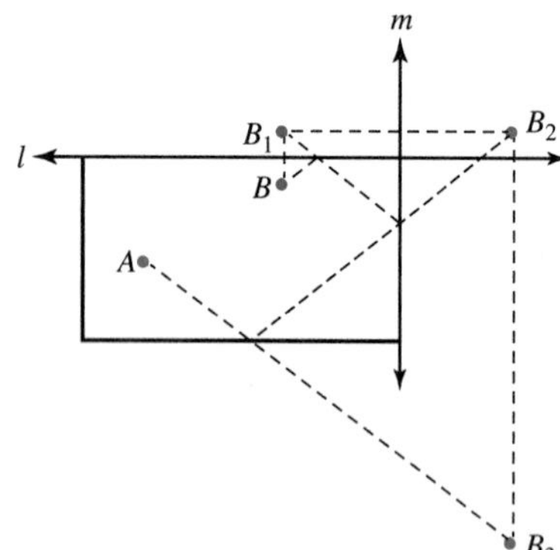

Answers for the Additional Problems

1. First draw a line of reflection (a vertical line, a horizontal line, or one of the two diagonal lines). Then shade pairs of squares and their images.

2. $R(-c, d), S(-a, b)$ and $R(c, -d), S(a, -b)$. Other answers are possible.

Chapter Review Problems

SECTION 9.1

1. a.

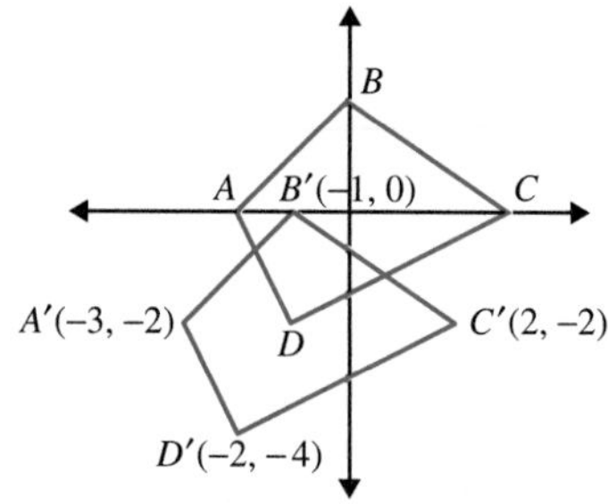

b. $(x - 1, y - 2)$

2. a. $(-3, 2)$ **b.** $(-3, -1)$ **c.** $(-4, -1)$
d. $(2, -4)$ **e.** $(4, 2)$ **f.** $(-y, x)$

3. a. $A'(2,0), B'(0,-2), C'(-3,0), D'(1,2)$
b. $A'(-2,0), B'(0,-2), C'(3,0), D'(-1,2)$

4.

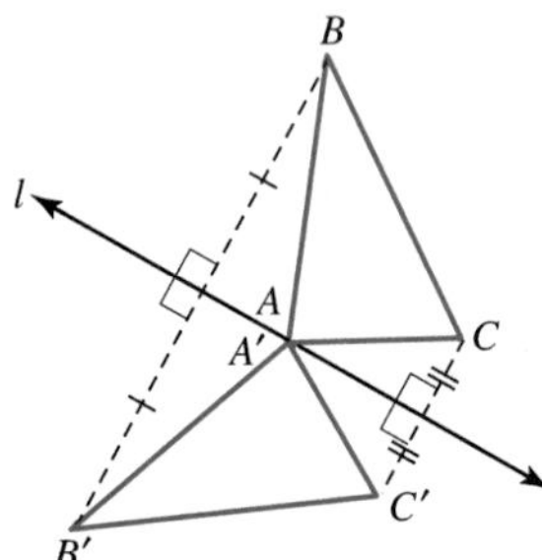

5. a. $(-1, 12)$ **b.** $(10, 3)$ **c.** $(9, -6)$ **d.** $(5 - x, y + 2)$

SECTION 9.2

1. a.

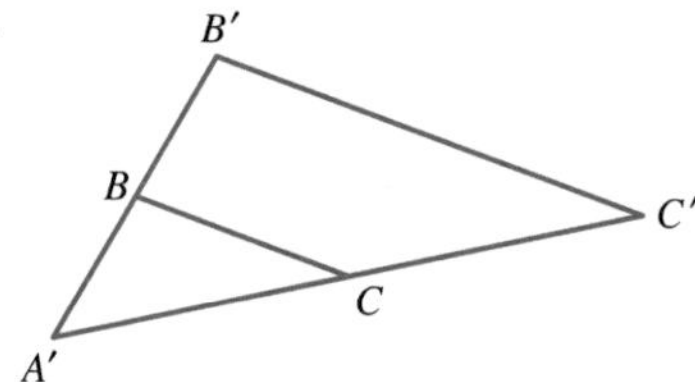

b.

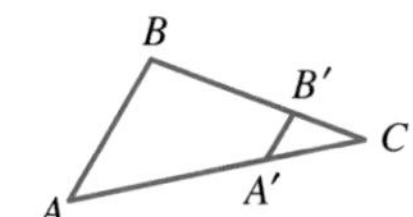

2. a. The center is O as shown in the following figure. The scale factor is 3.

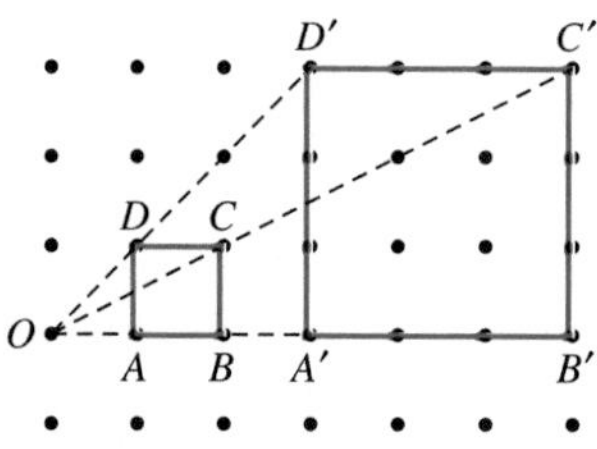

b. Each ratio is 3. **c.** 9

3. $A'(3, -2)$, $B'(11, 0)$, and $C'(7, 8)$.
   $A''(-3, -2)$, $B''(-11, 0)$, and $C''(-7, 8)$.
   $\triangle A''B''C''$ is similar to $\triangle ABC$ because there is a similitude that takes $\triangle ABC$ to $\triangle A''B''C''$.
4. Descriptions may vary. First perform a size transformation with the center at the center of the square and scale factor $\frac{\sqrt{2}}{2}$. Then rotate the image $-45°$ using the center of the square as the center of rotation.
5. First apply a size transformation with scale factor $\frac{GH}{AB}$. Then use an isometry, such as a translation, to take the image of $ABCDEF$ to $GHIJKL$. A size transformation together with an isometry is a similitude. Therefore, any two regular hexagons are similar.

SECTION 9.3

1. Reflections with respect to $\overleftrightarrow{EG}$ and $\overleftrightarrow{HF}$, and rotations of 180° and 360° around point O will map $ABCD$ onto itself.
2. Let E and F be midpoints of $\overline{BC}$ and $\overline{AD}$, respectively. Reflect trapezoid $ABCD$ across the line $\overleftrightarrow{EF}$. The image of point A will be point D, the image of point B will be point C, and the image of $\overline{AB}$ will be $\overline{DC}$. The reflection image of $\angle BAE$ with respect to $\overleftrightarrow{EF}$ is $\angle CDE$. Reflections preserve angle measure, so $\angle A \cong \angle D$.
3. It is possible. One path is shown. Let line m correspond to the right-hand side of the enclosure around the hole and line l correspond to the top side as shown. First reflect point B with respect to line m to point B_1, then reflect point B_1 with respect to line l to point B_2. When hitting the ball from point A, aim for B_2.

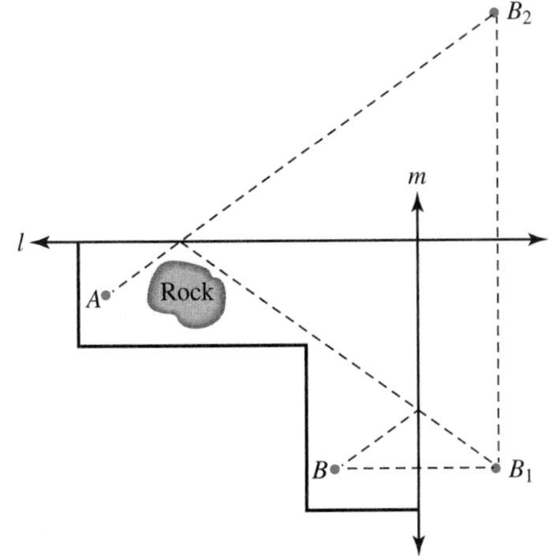

CHAPTER 9 TEST

1. T
2. T
3. T
4. F—The reflection portion of the glide reflection does not necessarily take a line segment to a parallel line segment.
5. T
6. F—A reflection reverses the orientation of a triangle.
7. T
8. T
9. F—The isometry could be a rotation.
10. F—There is a similitude, but not necessarily a size transformation.
11. **a.** $A'(0, 1), B'(3, 4), C'(7, 3)$
    b. $(a + 3, b + 2)$
12.

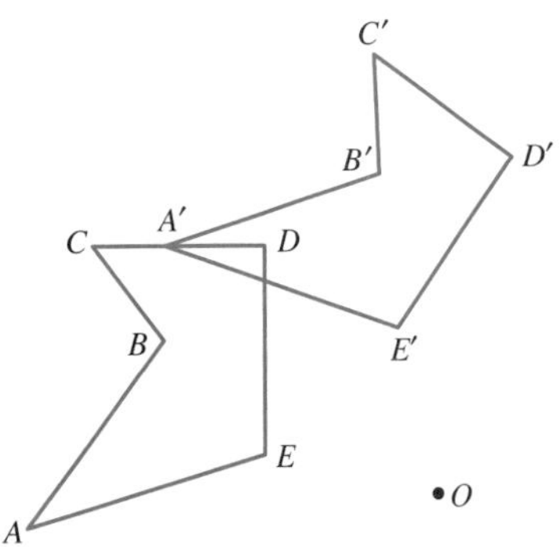

13.

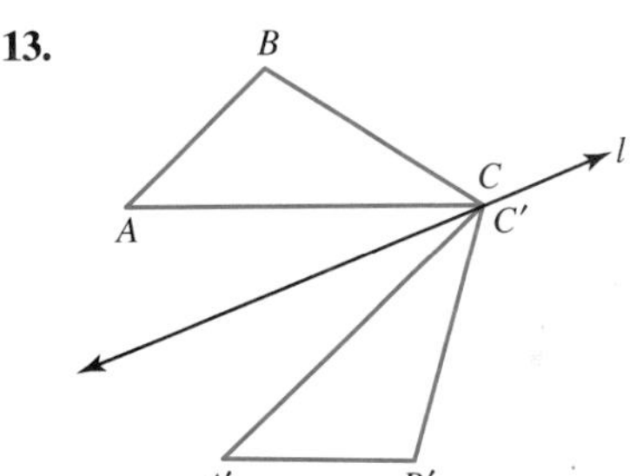

14.

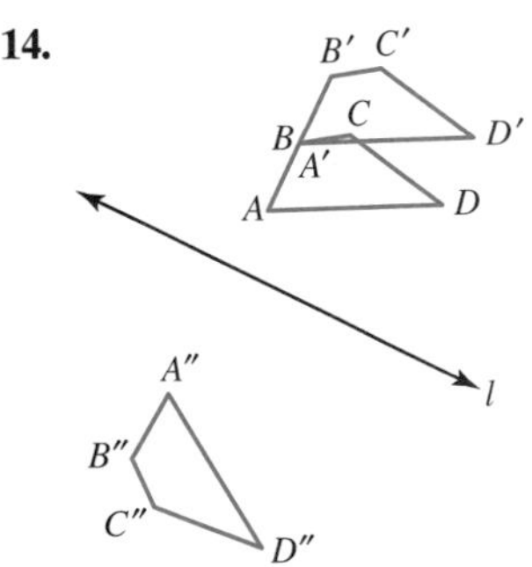

15.

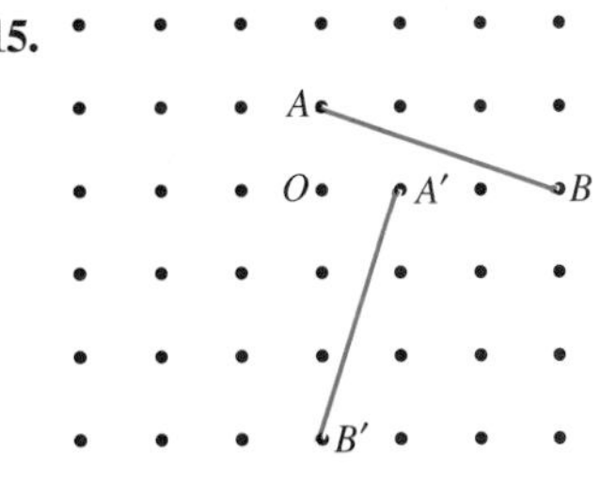

16.

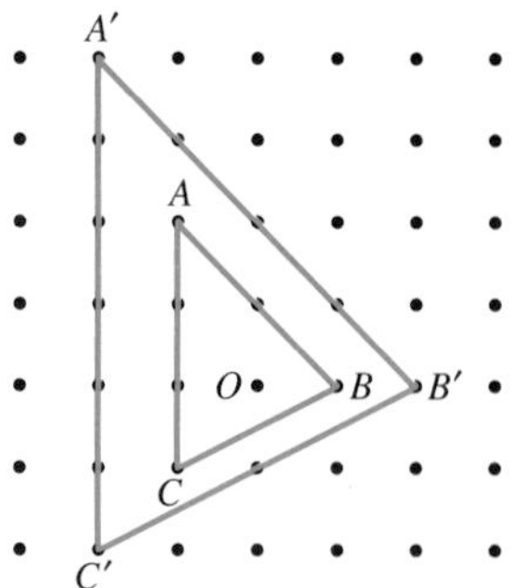

17. a.

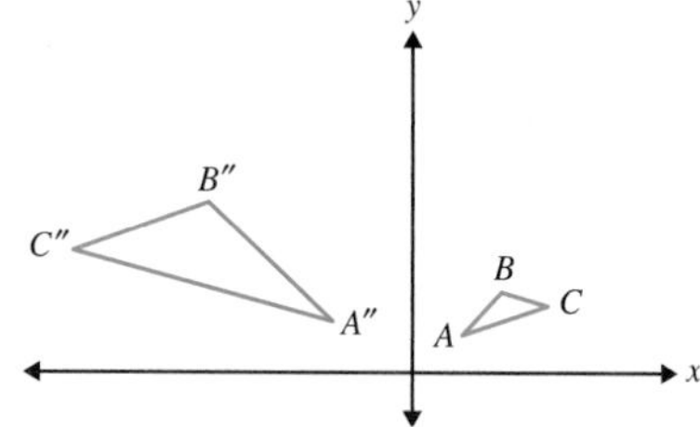

b. $A'(-3, 6), B'(-12, 12), C'(-18, 9)$

c. $(-3a, 3b)$

18. a. Glide reflection or rotation

b. Rotation

c. Reflection

19. First perform a size transformation with center D and scale factor $\frac{1}{2}$. Then reflect the resulting figure across $\overleftrightarrow{DC}$ and translate left 5 units. There are infinitely many other correct ways to describe this similitude as the composition of a size transformation followed by an isometry, depending on where the center of the size transformation is located.

20. The transformation could be the identity transformation (for example, a rotation of 0°). The transformation could be a reflection with respect to a line l that contains $\overline{AB}$.

21. We can accomplish this by interchanging the translation and the reflection in the definition of the glide reflection.

22. Because $ABCD$ is a rhombus, the reflection with respect to $\overleftrightarrow{AC}$ takes $ABCD$ to itself, as does the reflection with respect to $\overleftrightarrow{DB}$. Because the diagonals of the rhombus are perpendicular, the combination of these two reflections will be a 180° rotation around E. Because such a rotation takes lines to parallel lines, $\overline{AB} \parallel \overline{CD}$ and $\overline{AD} \parallel \overline{BC}$. Thus, $ABCD$ is a parallelogram.

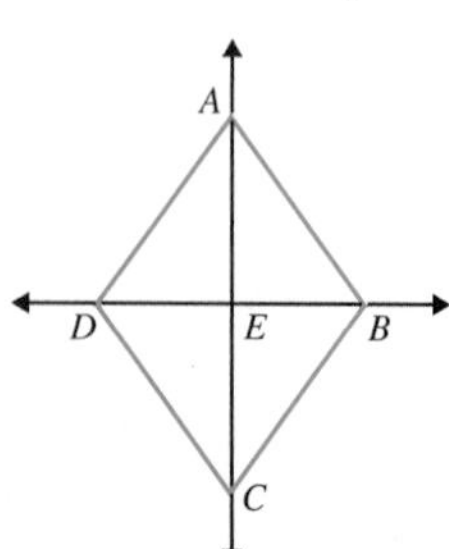

23. One successful path is shown.

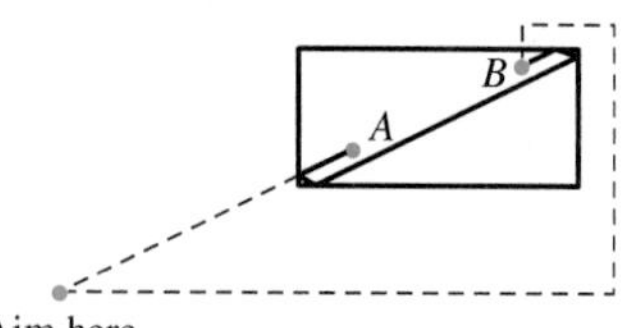

24. Place the logo at the origin and map a few key points using a size transformation of scale $\sqrt{2}$. This will have the effect of doubling the area of the logo.

TOPIC 1 PROBLEM SET

1. b. and **d.**

3. a. Roses are red, and the sky is blue.

b. Roses are red, and the sky is blue or turtles are green.

c. If the sky is blue, then both roses are red and turtles are green.

d. If it is not the case that turtles are not green and turtles are green, then roses are not red.

5. $p \Rightarrow q$ and $\sim(p \wedge q)$

7. a. T **b.** F **c.** T **d.** T

e. T **f.** T **g.** T **h.** T

9.

p	q	$\sim p$	$\sim q$	$(\sim p) \vee (\sim q)$	$(\sim p) \vee q$	$(\sim p) \wedge (\sim q)$
T	T	F	F	F	T	F
T	F	F	T	T	F	F
F	T	T	F	T	T	F
F	F	T	T	T	T	T

p	q	$\sim p$	$\sim q$	$p \Rightarrow q$	$\sim(p \wedge q)$	$\sim(p \vee q)$
T	T	F	F	T	F	F
T	F	F	T	F	T	F
F	T	T	F	T	T	F
F	F	T	T	T	T	T

11. a. Hypotheses: All football players are introverts, and Tony is a football player.

Conclusion: Tony is an introvert.

b. Hypotheses: Bob is taller than Jim, and Jim is taller than Sue.

Conclusion: Bob is taller than Sue.

13. a. Valid **b.** Invalid **c.** Invalid **d.** Valid

e. Invalid **f.** Invalid **g.** Invalid **h.** Valid

15. a. (i) only **b.** (ii) only **c.** (ii) only

17. a. Syllogism **b.** Contraposition **c.** Syllogism

19. a. T **b.** T **c.** Unknown

d. T **e.** F **f.** Unknown

g. T **h.** T **i.** Unknown

j. F	k. T	l. Unknown
m. Unknown	n. T	o. T

21. a. $w \Rightarrow t$ and t, therefore w. Invalid.
    b. $s \Rightarrow l$ and $\sim s$, therefore $\sim l$. Invalid.

23. Invalid.

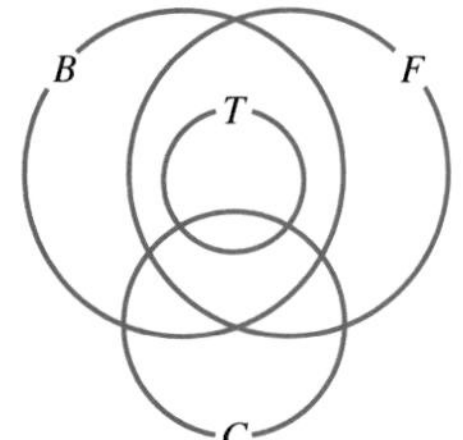

25.

p	q	$\sim p$	$\sim p \vee q$	$p \Rightarrow q$
T	T	F	T	T
T	F	F	F	F
F	T	T	T	T
F	F	T	T	T

The last two columns are the same.
Therefore, $\sim p \vee q$ is logically equivalent to $p \Rightarrow q$.

TOPIC 1 TEST

1. T
2. F—$\sim q \Rightarrow \sim p$ is true.
3. F—You may be older than 20 and younger than 30.
4. T
5. F—It means q is necessary for p.
6. a. $\sim q \Rightarrow p, \sim p \Rightarrow q, q \Rightarrow \sim p$
   b. $q \Rightarrow \sim p, p \Rightarrow \sim q, \sim q \Rightarrow p$
   c. $\sim p \Rightarrow \sim q, q \Rightarrow p, p \Rightarrow q$
7. a. T b. F c. F
   d. F e. T f. T
8. a. Invalid b. Valid c. Valid
   d. Invalid e. Invalid
9. Row 1: T T T T T F T
   Row 2: F T F T T T T
   Row 3: F F F F T F T
   Row 4: T F F T F T F
10. One example is: Some Bs and A. All Cs are Bs. Therefore all Bs are Cs.

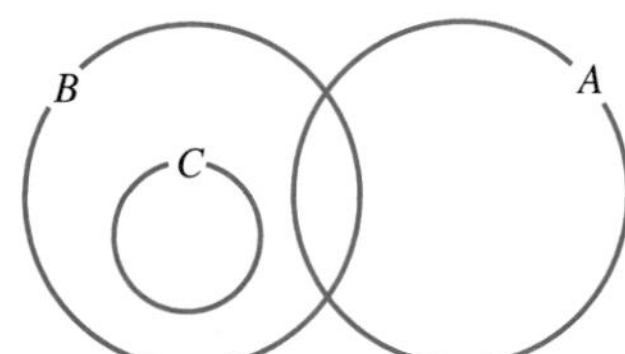

The Euler diagram represents the two hypotheses, but the conclusion is false.

11. All three
12. a. A tetrahedron has edges.
    b. No valid conclusion can be drawn.
13. a. F b. F c. T
14. All three are true where the two statements are both true. The only time that the implication is false is when the hypothesis is true and the conclusion is false. In this case, the disjunction is true. Thus, the conjunction, disjunction, and implication are never all false at the same time.

TOPIC 2 PROBLEM SET

1. a. $x < 6$ b. $x \leq -2.5$ c. $x < -5$
3. a. $x \geq 12$ b. $x \leq -10$ c. $x \geq -6$
5. Because the side opposite $\angle B$ is shorter than the side opposite $\angle C$
7. a. 59° b. $\overline{PR}$ c. $\overline{QR}$
9. a. 163°
   b. $\overset{\frown}{AB}$ because $\angle AOB$ is larger than $\angle BOC$
   c. $\overset{\frown}{AD}$ because $\angle AOD$ is larger than $\angle AOC$
11. a. $\overline{AC}$ because it is closest to the center
    b. $\angle B$ because $\overline{AC}$ is the longest chord
    c. $\overset{\frown}{AB}$ because $\overline{AB}$ is the shortest chord
13. $BD < BC$, $BE < BC$, and $AB < BE$. Other correct answers are possible.
15. $11 < BC < 15$. That is, BC must be between 11″ and 15″.
17. Because the hypotenuse is opposite the 90° angle and the 90° angle is the largest angle in a right triangle, the hypotenuse must be longer than the other two legs by Theorem T3.3.
19. First, $\angle BDC > \angle A$ because $\angle BDC$ is an exterior angle of $\triangle ABD$. Because $\triangle ABC$ is isosceles, $\angle A = \angle C$. Therefore, $\angle BDC > \angle C$. By Theorem T3.3, $BC > BD$. Because $AB = BC$, we have $AB > BD$.

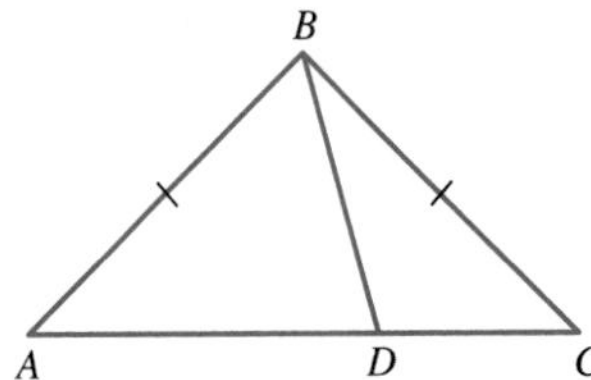

21. Method 1: Use the Pythagorean theorem: Because $\triangle ACD$ is a right triangle, $(AD)^2 + (CD)^2 = (AC)^2$. Therefore, $(AD)^2 < (AC)^2$ and $(CD)^2 < (AC)^2$, or $AD < AC$ and $CD < AC$.

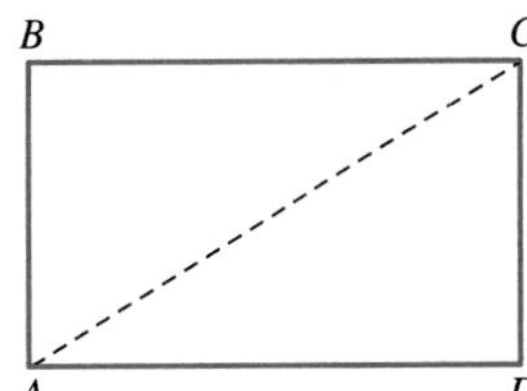

Method 2: Use Theorem T3.3: Because $\angle DCA < \angle D$ and $\angle DAC < \angle D$, by Theorem T3.2, we have that $AD < AC$ and $CD < AC$.

23. In $\triangle ACD$, (i) $\angle CAD < \angle ACD$ because $CD < AD$. In $\triangle ABC$, (ii) $\angle BAC = \angle BCA$ because they are opposite congruent sides. Adding the left sides and the right sides of (i) and (ii), we have $\angle BAC + \angle CAD < \angle BCA + \angle ACD$. Thus, $\angle BAD < \angle BCD$, or, equivalently, $\angle A < \angle C$.

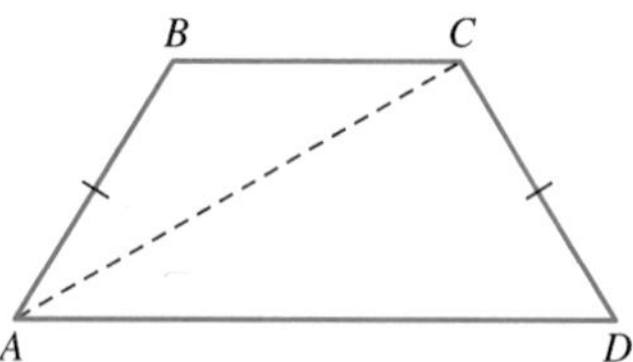

25. The area of $\triangle AOB$ is $\frac{1}{2}(OB)(OA)$. Extend $\overline{AO}$ through O to point E on $\overline{DC}$. Because $\overline{DC}$ is in the interior of the circle, $OE < OA$. Therefore, the area of $\triangle ODE$ is $\frac{1}{2}(OD)(OE) < \frac{1}{2}(OB)(OA)$, which is the area of $\triangle ABO$.

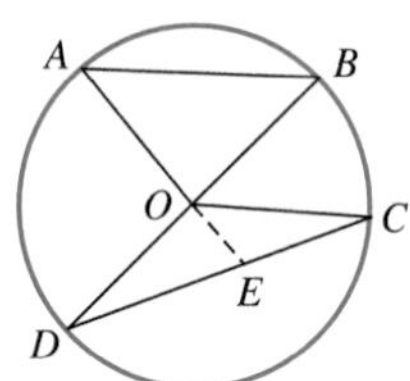

27. Let B be the point where ray $\overrightarrow{OA}$ intersects the circle and P be another point on the circle. By the Triangle Inequality, $OP < OA + AP$. However, $OB = OA + AB$. OB and OP are equal because they are radii. So $OA + AB < OA + AP$. Thus, $AB < AP$.

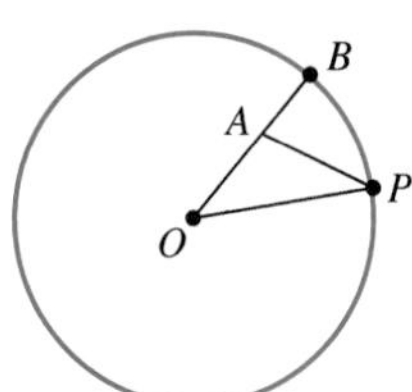

29. If $a < b$ and $b < c$, then there are positive real numbers p and q such that $a + p = b$ and $b + q = c$. Substituting $a + p$ in the second equation for b yields $a + p + q = c$ for the positive real number $p + q$. Thus, $a < c$.

TOPIC 2 TEST

1. T

2. F—since c and d may be negative

3. F—since both $\angle A$ and $\angle B$ may be less than 20°

4. T

5. T

6. $x < -1$

7. $x > -3$

8. **a.** $\angle R$ because $\overline{ST}$ is the longest side

b. $\angle T$ because $\overline{RS}$ is the shortest side

9. **a.** $\overline{XY}$ because Z is the largest angle

b. $\overline{YZ}$ because X is the smallest angle

10. **a.** $\widehat{AC}$ **b.** $\overline{BC}$

11. $11 < AC < 31$, or AC must be between 11″ and 31″.

12. **a.** $\overline{DE}$ because $\angle EOD = 70°$ is larger than $\angle BOC = 50°$ and $\angle COD = 40°$

b. $\angle OED$, $\triangle BOC$, $\triangle COD$, and $\triangle DOE$ are isosceles; hence, their base angles are congruent. Because $\triangle DOE$ contains the largest central angle, its base angles are the smallest.

13. **(i)** Because $\angle ADB$ is an exterior angle of $\triangle BCD$, $\angle ADB > \angle CBD$. However, $\angle CBD = \angle ABD$, so $\angle ADB > \angle ABD$. Case (ii) is proven similarly.

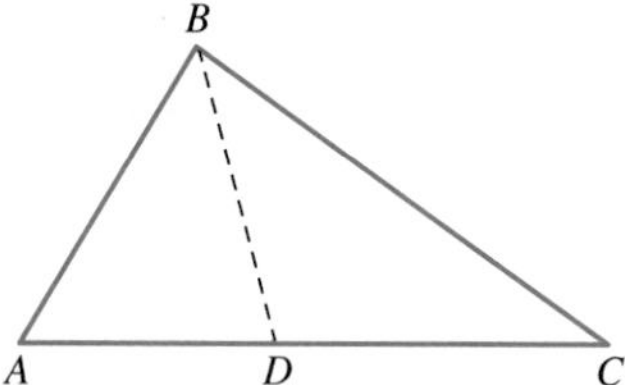

14. In pentagon $ABCDE$,

$AC < AB + BC$, $BD < BC + CD$,

$CE < CD + DE$, $AD < DE + AE$,

$BE < AE + AB$

Adding left sides and adding right sides, we have $AC + BD + CE + AD + BE$ is less than twice the perimeter.

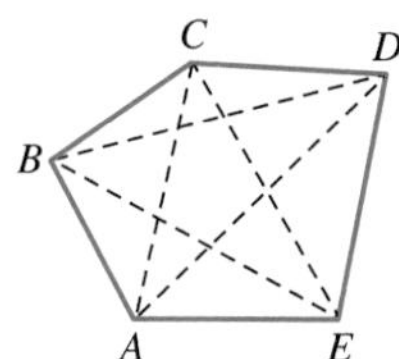

TOPIC 3 PROBLEM SET

1. Lines are great circles and no two great circles are parallel. In fact, any two great circles intersect in two points that are endpoints of a diameter of the sphere.

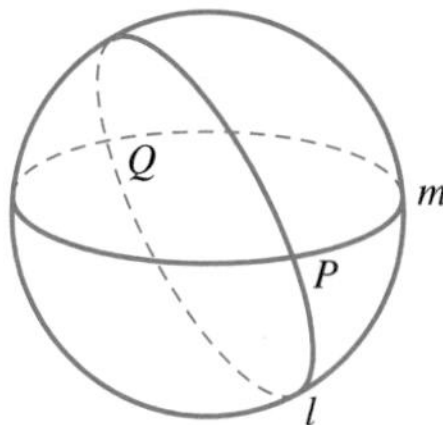

3. A "circle" on the sphere will look like a circle, but it will not necessarily be a great circle. For example, a circle of points equidistant from point P is shown.

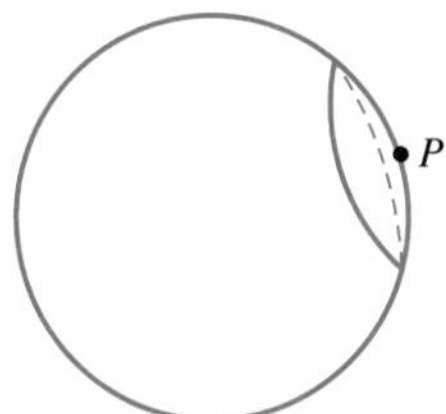

5. a. $d_T(A, B) = 6, d_E(A, B) = \sqrt{18}$ or $3\sqrt{2}$

b. $d_T(A, B) = 10, d_E(A, B) = \sqrt{52}$ or $2\sqrt{13}$

c. $d_T(A, B) = 13, d_E(A, B) = \sqrt{85}$

d. Yes, it is true for (a)–(c). Yes, true for all pairs A and B.

7. a. Area of the square is 16 square units. Area of the circle is 4π square units.

b. Area of the square varies depending on its placement on the axes. For example,

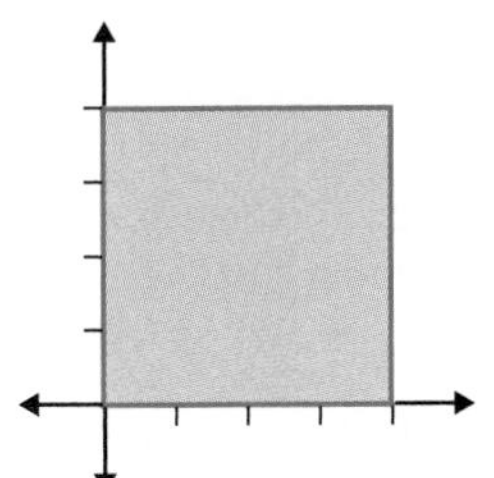

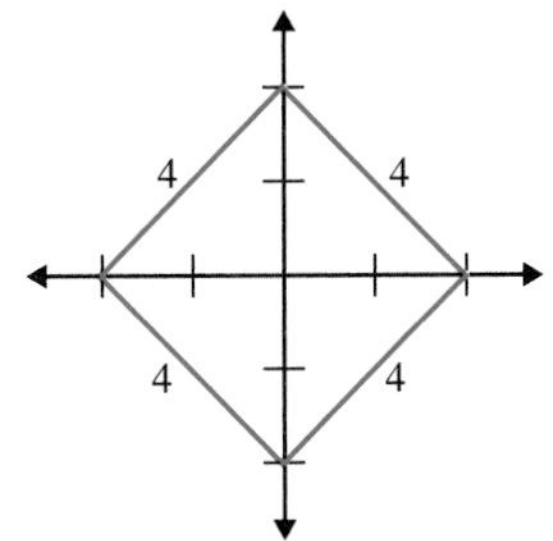

9. a. B lies between A and C.
$d_T(A, B) + d_T(B, C) = d_T(A, C)$

b. B lies between A and C.
$d_T(A, B) + d_T(B, C) = d_T(A, C)$

11. Both of the following triangles have two angles of 45° and an included side of 2 units. Yet they are not congruent. Other examples are possible.

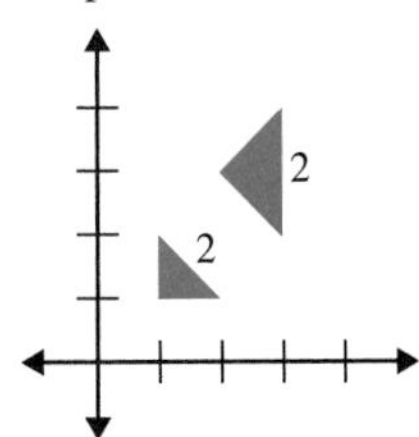

13. Lines in this geometry are the pairs of points AB, AC, AD, BC, BD, and CD. The lines that have a point in common are the pairs that share a letter. For example, AB and AC share point A. In every case, lines that have a point in common have exactly one point in common.

15. To verify this theorem, we list for each combination of a line and a point not on it, the one line in the set of six possible lines that does not intersect the given line.

Line AB and point C—line CD.
Line AB and point D—line CD.
Line BC and point A—line AD.
Line BC and point D—line AD.
Line CD and point A—line AB, and so on. In every case there is exactly one line parallel to a given line.

17. T—Each point occurs in exactly three lines. For example, A is on lines ABC, AED, and AGF.

19. T—For any given pair of points, there are exactly two lines that contain neither point. For example, for points B and G, the two lines that contain neither point are AED and CFD.

21. $\triangle ACD \cong \triangle DBA$ by LL, so $AC \cong BD$ by C.P.

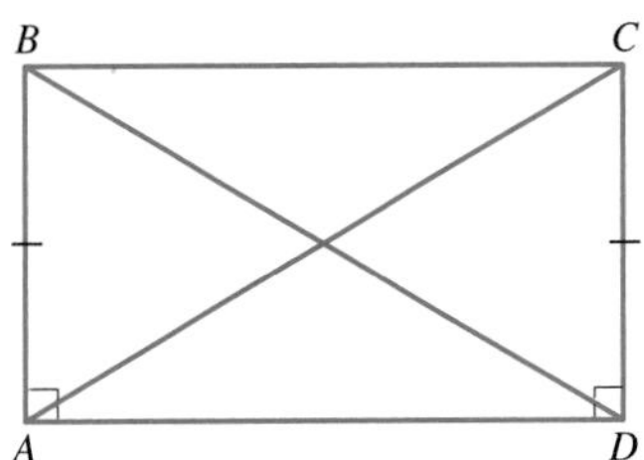

23. Assume $CD > AB$, and choose point E on $\overline{CD}$ such that $AB = DE$. Then $ABED$ is a Saccheri quadrilateral. Because $\angle BED$ is an exterior angle to $\triangle BCE$, $\angle BED > \angle C = 90°$. Thus, $\angle BED$ is obtuse. By problem 22, we know that the summit angles are congruent, so $\angle ABE$ is also obtuse. If $\angle ABE$ is obtuse, then $\angle ABC$ must also be obtuse. This results in a contradiction because the fourth angle in a Lambert quadrilateral is not obtuse. Therefore, our assumption that $CD > AB$ must be false, and we have that $CD \leq AB$.

TOPIC 3 TEST

1. F—There are infinitely many line (great circles) through each pair of points.

2. T **3.** T **4.** T **5.** T

6. No, this result does not hold in the geometry of the sphere. For example, in the figure shown, the exterior angle at C measures 90°, but the two nonadjacent interior angles at A and B have a sum of 180°.

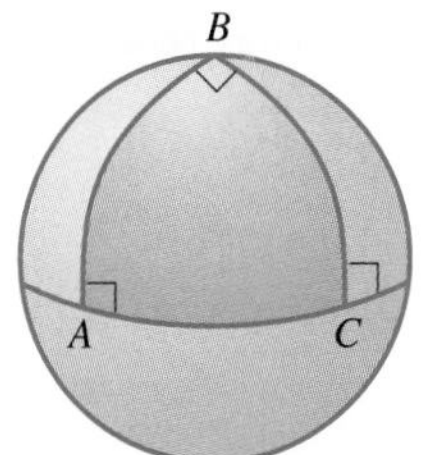

7. a. Three lines: AB, AC, BC

b. One point. For example, AB and AC intersect in A.

c. No, because a line must contain two points. Given line AB, only point C remains.

8. a.

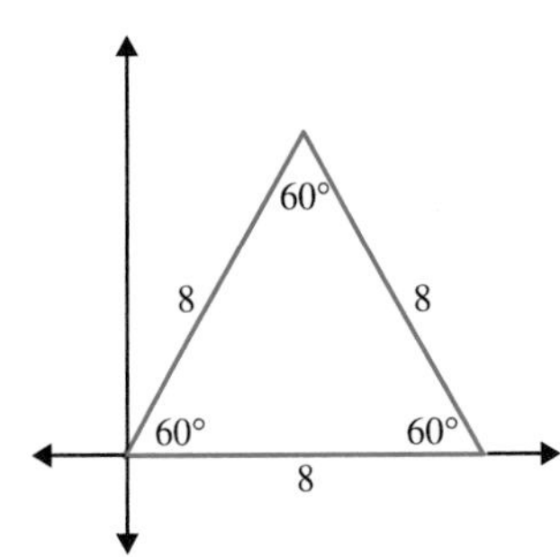

$$A = \frac{s^2\sqrt{3}}{4} = \frac{8^2\sqrt{3}}{4} = 16\sqrt{3} \text{ square units}$$

b. Three possibilities are

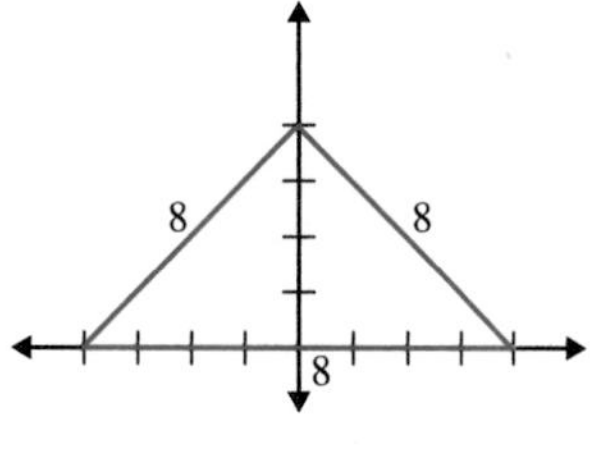

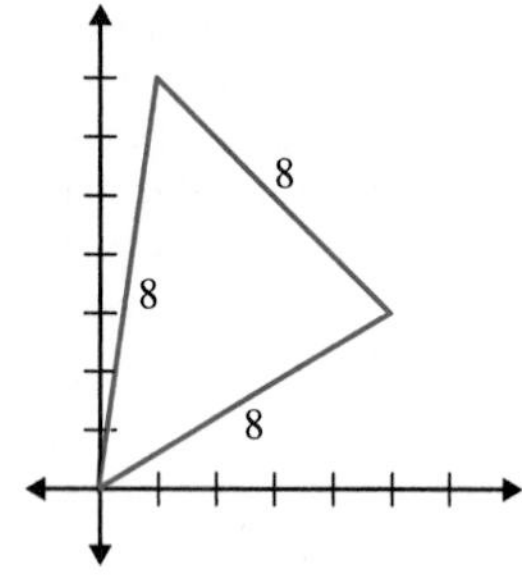

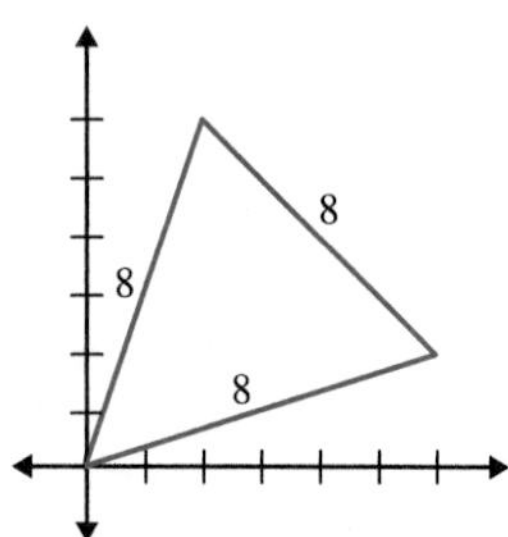

c. In each case, the area is 16 square units.

9. The triangles are not necessarily similar. One counterexample is shown.

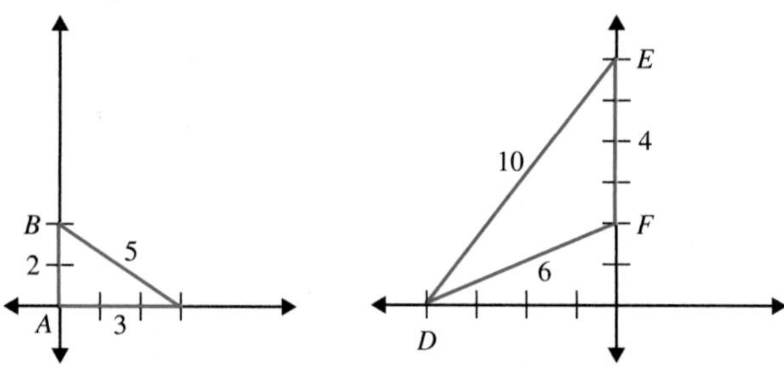

Each side of $\triangle FED$ is twice as long as a corresponding side of $\triangle ABC$, but $\triangle FED$ is not similar to $\triangle ABC$ because $\triangle ABC$ is a right triangle.

10. We have quadrilateral $ABCD$ in which $\angle A = \angle C = \angle D = 90°$ because $ABCD$ is a Lambert quadrilateral.

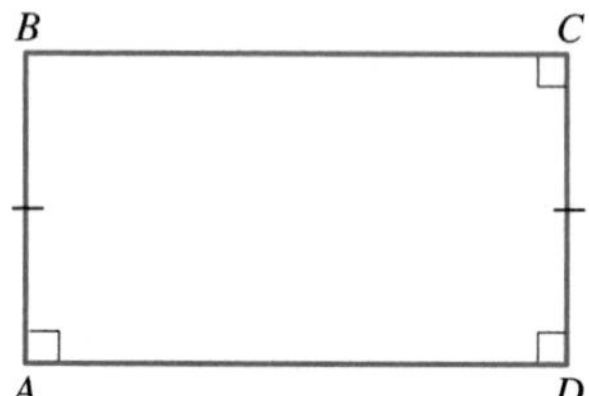

Also, $AB = CD$ because $ABCD$ is a Saccheri quadrilateral. By problem 19 of this section, the summit angles of a Saccheri quadrilateral are congruent Thus, $\angle B = 90°$, and so all four angles of $ABCD$ are right angles. Therefore, $ABCD$ is a rectangle.

INDEX

A

B

D

H

I

N

O

P

Q

R

S

T

U

V

W

X

Y

Z

Geometry Formulas and Relationships

Angle Sum in a Triangle

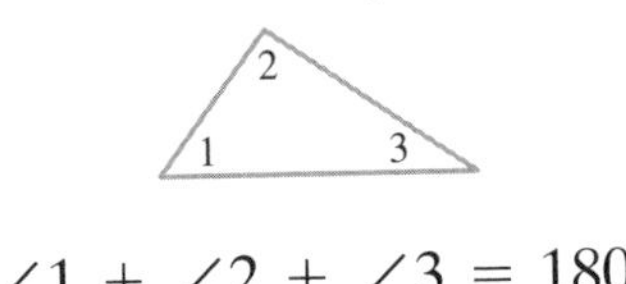

$$\angle 1 + \angle 2 + \angle 3 = 180°$$

Angle Sum in a Polygon

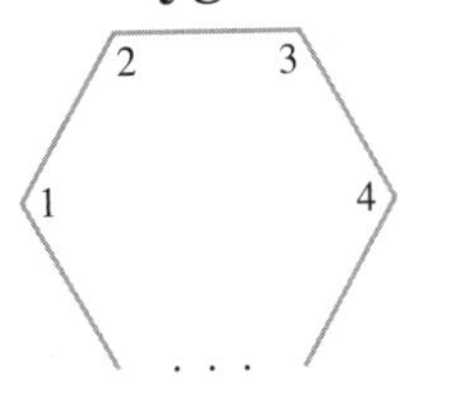

$$\angle 1 + \angle 2 + \angle 3 + \cdots \angle n = (n - 2) \cdot 180°$$

Pythagorean Theorem:

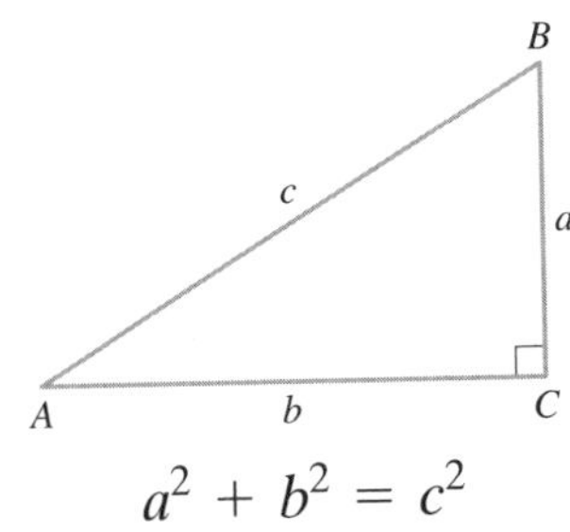

$$a^2 + b^2 = c^2$$

Special Right Triangles:

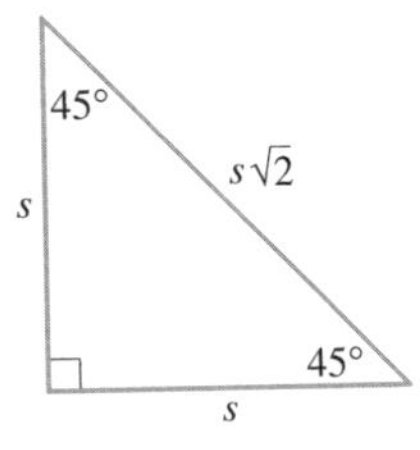

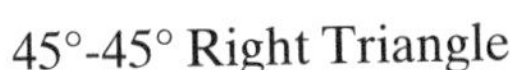

45°-45° Right Triangle

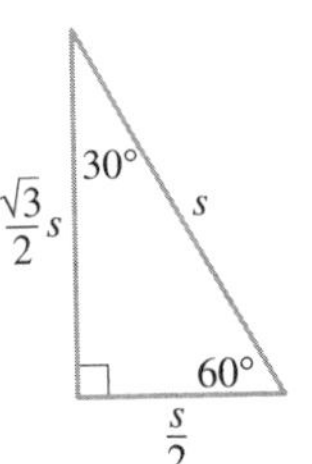

30°-60° Right Triangle

Heron's Formula: If a triangle has sides of length a, b, and c, then its area is given by

$$A = \sqrt{s(s - a)(s - b)(s - c)}, \text{ where } s = \frac{a + b + c}{2}$$

Euler's Formula: For any polyhedron with F faces, V vertices, and E edges, the following formula holds:

$$F + V = E + 2$$

Trigonometric Ratios:

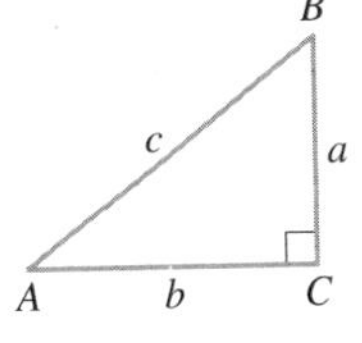

$$\tan A = \frac{a}{b} = \frac{\text{opposite side}}{\text{adjacent side}}$$

$$\sin A = \frac{a}{c} = \frac{\text{opposite side}}{\text{hypotenuse}}$$

$$\cos A = \frac{b}{c} = \frac{\text{adjacent side}}{\text{hypotenuse}}$$

Laws of Trigonometry

Law of Sines: $\dfrac{\sin A}{a} = \dfrac{\sin B}{b} = \dfrac{\sin C}{c}$

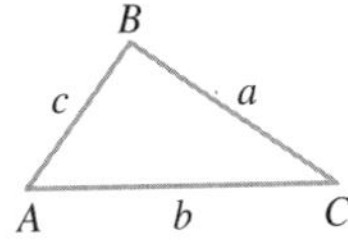

Law of Cosines: $c^2 = a^2 + b^2 - 2ab\cos C$

Coordinate Geometry:

Slope of the line through $P(x_1, y_1)$ and $Q(x_2, y_2)$: $m = \dfrac{y_2 - y_1}{x_2 - x_1}$

Distance from $P(x_1, y_1)$ to $Q(x_2, y_2)$: $d = \sqrt{(x_2 - x_1)^2 + (y_2 - y_1)^2}$

Midpoint of segment $\overline{PQ}$: $M = \left(\dfrac{x_1 + x_2}{2}, \dfrac{y_1 + y_2}{2}\right)$

Slope-Intercept Form of a Line: $y = mx + b$

Point-Slope Form of a Line: $y - y_1 = m(x - x_1)$

Units of Measurement

U. S. Customary System

Length

1 foot (ft) = 12 inches (in.)
1 yard (yd) = 3 ft
1 mile (mi) = 5280 ft = 1760 yd

Area

1 acre = 43,560 ft^2

Weight

1 pound (lb) = 16 ounce (oz)
1 ton = 2000 lb

Capacity

1 tablespoon (T) = 3 teaspoons (t)
1 cup (c) = 16 T = 8 ounces (oz)
1 pint (pt) = 2 cups (c)
1 quart (qt) = 2 pints (pt)
1 gallon (gal) = 4 quarts (qt)
1 $ft^3 \approx$ 7.48 gal

Metric System

Common units

Length—meter (m)
Area—are (a)
Capacity—liter (L)
Mass—gram (g)
Temperature—degree Celsius (°C)

Common prefixes

milli- (1/1000 of)
centi- (1/100 of)
deci- (1/10 of)
deca- (10 times)
hecto- (100 times)
kilo- (1000 times)

Conversions between Systems

1 in. = 2.54 cm (exactly)
1 m $\approx$ 39.37 in.
1 km $\approx$ 0.6214 mi
1 kg $\approx$ 2.2 lb
1 L $\approx$ 1.057 qt

List of Symbols, Abbreviations, and Notation

Symbol	Meaning	Page
$\overleftrightarrow{AB}$	line AB	40
$\overline{AB}$	line segment AB	41
AB	the length of $\overline{AB}$	41
$\overrightarrow{AB}$	ray AB	41
$\angle ABC$	angle ABC	42
$^\circ$	degree	42
⦜	right angle	43
$l \perp m$	l is perpendicular to m	43
$\triangle ABC$	triangle ABC	52
$l \parallel m$	l is parallel to m	55
$\approx$	approximately equal to	88
π	pi (3.14159...)	108
$p \Rightarrow q$	if p, then q	186
$\therefore$	therefore	187
$p \Leftrightarrow q$	p if and only if q	189
$\leftrightarrow$	correspondence	196
$\triangle ABC \cong \triangle DEF$	$\triangle ABC$ is congruent to $\triangle DEF$	196
SAS	Side-Angle-Side Congruence postulate	197
LL	Leg-Leg Congruence Theorem	197
HL	Hypotenuse-Leg Congruence Theorem	198
C.P.	**C**orresponding **P**arts of Congruent Triangles Are Congruent	198
ASA	Angle-Side-Angle Congruence Postulate	199
SSS	Side-Side-Side Congruence Postulate	201
$\rightrightarrows$	parallel lines	242
$a : b$	ratio $\frac{a}{b}$	298
$\triangle ABC \sim \triangle DEF$	$\triangle ABC$ is similar to $\triangle DEF$	307

Symbol	Meaning	Page
AAA	Angle-Angle-Angle Similarity Postulate	309
AA	Angle-Angle Similarity Theorem	309
SAS	Side-Angle-Side Similarity Theorem	310
LL	Leg-Leg Similarity Theorem	311
SSS	Side-Side-Side Similarity Theorem	312
tan	tangent	329
sin	sine	331
cos	cosine	331
$\tan^{-1}$	inverse tangent	333
$\sin^{-1}$	inverse sine	333
$\cos^{-1}$	inverse cosine	333
$\overparen{AB}$	arc AB	362
(x, y)	ordered pair	422
$\overrightarrow{AB}$	vector AB	479
$\sim p$	not p	528
$\wedge$	and	528
$\vee$	or	528
$>$	is greater than	541
$<$	is less than	541
$\leq$	is less than or equal to	541
$\geq$	is greater than or equal to	541
d_T	taxicab distance	562
d_E	Euclidean distance	562